C000186457

ISBN 978-1-5279-2152-8
PIBN 10902639

FB &c Ltd, Dalton House, 60 Windsor Avenue, London, SW19 2RR.
Company number 08720141. Registered in England and Wales.

For support please visit www.forgottenbooks.com

University of Kansas Publications
MUSEUM OF NATURAL HISTORY

VOLUME 14 · 1960-1965

Museum of Natural History
UNIVERSITY OF KANSAS
LAWRENCE
1965

Museum of Natural History
UNIVERSITY OF KANSAS
LAWRENCE

PRINTED BY
ROBERT R. (BOB) SANDERS, STATE PRINTER
TOPEKA, KANSAS
1965
30-7489

CONTENTS OF VOLUME 14

1. Neotropical bats from western México. By Sydney Anderson. Pp. 1-8. October 24, 1960.
2. Geographic variation in the harvest mouse, Reithrodontomys megalotis, on the Central Great Plains and in adjacent regions. By J. Knox Jones, Jr. and B. Mursaloğlu. Pp. 9-27, 1 fig. July 24, 1961.
3. Mammals of Mesa Verde National Park, Colorado. By Sydney Anderson. Pp. 29-67, pls. 1 and 2, 3 figs. July 24, 1961.
4. A new subspecies of the black myotis (bat) from eastern Mexico. By E. Raymond Hall and Ticul Alvarez. Pp. 69-72, 1 fig. December 29, 1961.
5. North American yellow bats, "Dasypterus," and a list of the named kinds of the genus Lasiurus Gray. By E. Raymond Hall and J. Knox Jones, Jr. Pp. 73-98, 4 figs. December 29, 1961.
6. Natural history of the brush mouse (Peromyscus boylii) in Kansas with description of a new subspecies. By Charles A. Long. Pp. 99-110, 1 fig. December 29, 1961.
7. Taxonomic status of some mice of the Peromyscus boylii group in eastern Mexico, with description of a new subspecies. By Ticul Alvarez. Pp. 111-120, 1 fig. December 29, 1961.
8. A new subspecies of ground squirrel (Spermophilus spilosoma) from Tamaulipas, Mexico. By Ticul Alvarez. Pp. 121-124. March 7, 1962.
9. Taxonomic status of the free-tailed bat, Tadarida yucatanica Miller. By J. Knox Jones, Jr., and Ticul Alvarez. Pp. 125-133, 1 fig. March 7, 1962.
10. A new doglike carnivore, genus Cynarctus, from the Clarendonian, Pliocene, of Texas. By E. Raymond Hall and Walter W. Dalquest. Pp. 135-138, 2 figs. April 30, 1962.
11. A new subspecies of wood rat (Neotoma) from northeastern Mexico. By Ticul Alvarez. Pp. 139-143. April 30, 1962.
12. Noteworthy mammals from Sinaloa, Mexico. By J. Knox Jones, Jr., Ticul Alvarez, and M. Raymond Lee. Pp. 145-159, 1 fig. May 18, 1962.
13. A new bat (Myotis) from Mexico. By E. Raymond Hall. Pp. 161-164, 1 fig. May 21, 1962.
14. The mammals of Veracruz. By E. Raymond Hall and Walter W. Dalquest. Pp. 165-362, 2 figs. May 20, 1963.
15. The Recent mammals of Tamaulipas, México. By Ticul Alvarez. Pp. 363-473, 5 figs. May 20, 1963.
16. A new subspecies of the fruit-eating bat, Sturnira ludovici, from western Mexico. By J. Knox Jones, Jr., and Gary L. Phillips. Pp. 475-481, 1 fig. March 2, 1964.
17. Records of the fossil mammal Sinclairella, family Apatemyidae, from the Chadronian and Orellan. By William A. Clemens, Jr. Pp. 483-491, 2 figs. March 2, 1964.
18. The mammals of Wyoming. By Charles A. Long. Pp. 493-758, 82 figs. July 6, 1965.

Index, pp. 759-784.

UNIVERSITY OF KANSAS PUBLICATIONS
MUSEUM OF NATURAL HISTORY

Volume 14, No. 1, pp. 1-8
October 24, 1960

Neotropical Bats from Western México

BY

SYDNEY ANDERSON

UNIVERSITY OF KANSAS
LAWRENCE
1960

University of Kansas Publications, Museum of Natural History

Volume 14, No. 1, pp. 1-8
Published October 24, 1960

University of Kansas
Lawrence, Kansas

PRINTED IN
THE STATE PRINTING PLANT
TOPEKA, KANSAS
1960

28-4805

Neotropical Bats from Western México

BY

SYDNEY ANDERSON

Tropical fruit-eating bats of the genus *Artibeus* reach their northern limits on the lowlands of the eastern and western coasts of México. Recent students have placed the species of Mexican *Artibeus* in two groups; one includes bats of small size and one includes bats of large size (Dalquest, 1953:61; Lukens and Davis, 1957:6; and Davis, 1958:163). Three of the small species (*A. cinereus phaeotis, A. aztecus,* and *A. turpis nanus*) and three of the large species (*A. hirsutus, A. jamaicensis jamaicensis,* and *A. lituratus palmarum*) have been reported as far north as Jalisco along the west coast. *A. cinereus phaeotis* and *A. turpis nanus* are known from as far north as southern Sinaloa, and *A. hirsutus* is known from as far north as southern Sonora (Hall and Kelson, 1959:140, 141). Additional specimens of *A. hirsutus* from Sonora, Sinaloa, and Chihuahua, and specimens of *A. lituratus* and *A. jamaicensis* from Sinaloa that extend the known ranges of these two species northward are reported here; data on variation, distribution, and reproduction concerning these three species are included. Also, specimens of *Sturnira lilium* and of the genus *Chiroderma* from Chihuahua that extend their known ranges northwestward are reported.

Support for field work that yielded the specimens reported came from the National Science Foundation, the American Heart Association, Inc., and the Kansas University Endowment Association. Catalogue numbers of The University of Kansas Museum of Natural History are cited. The latitude (N) and longitude (W) are recorded to the nearest minute for each locality mentioned.

Artibeus lituratus palmarum J. A. Allen and Chapman.—Specimens from Eldorado (24°19′, 107°20′), Sinaloa, extend the known range of the species approximately 265 miles northwestward from Huajimic (21°37′, 104°21′), Nayarit. Skins and skulls of 11 specimens (75211-75221, 7 males and 4 females) taken on November 13, 1957, 1 mi. S Eldorado, were prepared by William L. Cutter. Skeletons of 12 specimens (75222-75233, 3 males and 9 females) from Eldorado were obtained by Cutter on the same day. None of the 13 females was pregnant. One specimen (75211, female) is immature; it has open phalangeal ephiphyseal sutures (as do four other larger individuals); this specimen measured 83 mm. in total length, weighed 45 grams, and has a skull 26.6 mm. in greatest length, 22.4 mm. in condylocanine length, 13.4 mm. in lambdoidal breadth, and has unusually small second (last) upper molar teeth, each having about one half the occlusal area of the M2 of the average adult in the series. None of the 23 specimens has a third upper molar. All except one have both third lower molars; one (75233) lacks the third lower molar on both sides of the jaw. Facial stripes vary from conspicuous to inconspicuous, but are evident in each of the 11 skins. The two skins having the darkest pelage are both of males and are the only two skins having open epiphyseal sutures. Five adult males and three adult females are represented by skins. Three of the male skins are slightly darker and less reddish than those of the three females, and the contrast between paler neck and shoulders and other parts is slightly less marked. The other two males are paler and more rufous than the three females; the palest and most rufous of these two males is an old individual having well-worn teeth. Dichromatism is not correlated with age or with sex in this series, which, therefore, differs from specimens reported by Lukens and Davis (1957:9) who observed that dichromatism was correlated with sex. In size, as shown in measurements below, in darkness of ventral pelage, and in cranial features the specimens from Sinaloa agree with those from Guerrero, and differ from specimens of *Artibeus jamaicensis*, in the ways described by Lukens and Davis (*loc. cit.*).

Average measurements of males and females do not differ significantly. The following are average and extreme measurements in millimeters of 17 adults (lacking epiphyseal sutures): total length, 93.4 (90-99); length of hind foot, 19.8 (19-21); length of ear, 23.8 (23-26); length of forearm, 65.2 (60.1-70.6); greatest length of skull, 29.24 (28.0-30.2); condylocanine length, 25.25 (24.2-26.4); lambdoidal breadth, 15.92 (15.3-16.6); postorbital constriction, 6.29 (5.8-6.9); and weight (in grams), 63.2 (51-69).

Artibeus jamaicensis jamaicensis Leach.—A female (61088) obtained on June 18, 1954, by Albert A. Alcorn, from 32 mi. SSE Culiacan (24°26′, 107°07′), Sinaloa, extends the known range of the species approximately 415 miles northwestward from 2 mi. N. Ciudad Guzmán (19°43′, 103°28′), Jalisco. Two other females (61089-61090) from central Sinaloa, collected on June 19, 1954, by A. A. Alcorn, are from ½ mi. E Piaxtla (23°51′, 106°38′). Each of the three specimens contained a single embryo. The embryos (in the order the specimens are listed above) measured 28, 26, and 25 millimeters. The Sinaloan specimens are both paler and browner than specimens from Jalisco and from eastern México, and the facial stripes are more distinct, being as distinct in one specimen (61088) as in any of the *Artibeus lituratus* reported here. Four additional specimens from Jalisco are: 34232-34235, 3 males and 1 non-pregnant female taken by J. R. Alcorn at Hacienda San Martín (20°18′, 103°30′), 18 mi. W Chapala, 5000 feet, on July 12, 1949. Each specimen of *A. jamaicensis* listed above lacks epiphyseal sutures and both an upper and a lower third molar on each side. In size, coloration of ventral pelage, and configuration of skull, the specimens agree with the description of specimens from Guerrero and differ from other species as reported by Lukens and Davis (1957:7, 9).

Minimum and maximum measurements in millimeters for the three *A. jamaicensis* from Sinaloa, followed by corresponding figures for the four from Jalisco, are: total length, 80-82, 82-84; length of hind foot, 15-16, 16-17; length of ear, all 20, 20-21; length of forearm, 54.7-55.9, 54.7-58.5; greatest length of skull, 26.6-27.3, 26.5-28.2; condylocanine length, 22.6-23.2, 22.5-23.9; lambdoidal breadth, 13.9-14.5, 13.7-15.1; and postorbital constriction, 6.2-6.6; 6.3-6.7.

Artibeus hirsutus Andersen.—One specimen (25053, in preservative) of a series from ¼ mi. W Aduana (27°02′, 109°03′), 1600 feet, Sonora, was cited by Hall and Kelson (1959:136) and reproductive data from two skins (24841-24842) were mentioned earlier by Cockrum (1955:490). In addition to these three specimens the series includes 20 specimens in preservative (25052, 25054-25072). All were collected on May 16, 1948, by J. R. Alcorn. Number 25070, on deposit in the Institute of Biology in Mexico City, and two others (25053-25054) are not on hand as I write this, and have not been examined by me. *Artibeus hirsutus* has recently been found in northern Sinaloa and in southwestern Chihuahua. Three males (75208-75210) from El Fuerte (26°24′, 108°41′), Sinaloa, were obtained on December 10, 1957, by William L. Cutter. Four specimens (79441-79444, 2 males and 2 females) were captured in mist nets on the north bank of the Río Septentrión, 1½ mi. SW Tocuina (27°07′, 108°22′), 1500 feet, Chihuahua, on July 18, 20, and 21, 1958, by Kenneth E. Shain and me. I captured another (79445, a male) in a hand net in an abandoned, horizontal mine shaft on the north side of the Río Batopilas, at about 3500 feet elevation, across the canyon from the village of La Bufa (27°09′, 107°33′), Chihuahua, on July 10, 1958. Eight specimens (12406-12413) in the Museum of Natural History at the University of Illinois were collected on July 22 and 23, 1956, in Santo Domingo Mine (26°55′, 109°05′), 7 mi. SW Alamos, Sonora, by W. Z. Lidicker, W. H. Davis, and J. R. Winkelman. Eight specimens (9981-9988) in the Los Angeles County Museum were collected on July 26, 1953, by Kenneth E. Stager, 5 mi. W Alamos in an old mine tunnel at Aduana. One (36581) of six specimens (36581-36586, 4 males and 2 pregnant females) from 2 mi. ENE Tala (20°39′, 103°40′), 4500 feet, Jalisco, was reported by Hall and Kelson (1959:136); the locality being erroneously cited as 8 mi. ENE Tala. These six specimens were collected by J. R. Alcorn on February 28, 1950.

The 59 specimens from Guerrero are distributed by localities as follows: 8 mi. N, 1 mi. W Teloloapan (Teloloapan is at 18°18′, 99°54′), 3600 feet, 15 specimens (66432-66446, all males, including one skeleton and two in preservative) obtained by Robert W. Dickerman on February 7, 1955; Alpixafia, 4 kms. NW Teloloapan, 1540 M., 16 specimens (35219-35234, 5 males and 11 females) obtained by Bernardo Villa R. on May 22 and 23, 1949; 1 mi. N Teloloapan, 7 specimens (66447-66453, all males, including one in preservative) obtained by Dickerman on February 8, 1955; 4 kms. SE Teloloapan (Cerro Piedras Largas), 1760 M., 15 specimens (35310-35324, 12 males and 3 females) obtained by Villa R. on October 20 and 21, 1948; Puente de Dios, 1700 M., Yerbabuena (=8 mi. N, 1 mi. W Teloloapan), six specimens (28408-28413, 4 males and 2 females) obtained by Villa R. on July 25 and 29, 1948. Six of the 16 Guerreran specimens taken in May are young as shown by the open epiphyseal sutures; all other Guerreran specimens lacked these sutures. Of 13 adult females from Guerrero, only two, taken in May, contained embryos (one embryo in each).

Average and extreme measurements in millimeters of 28 adult Guerreran *A. hirsutus* of both sexes (the sexes are not significantly different) are as follows: total length, 79.5 (69-90); length of hind foot, 15.1 (12-17); length of ear, 21.1 (19-24); length of forearm, 55.8 (53.1-57.8); greatest length of skull, 27.11 (26.3-28.0); condylocanine length, 22.92 (22.1-23.5); lambdoidal breadth, 14.23 (13.7-14.6); postorbital constriction, 6.51 (6.2-6.8); and weight (in grams), 37.4 (34.0-42.6).

The presence or absence of the third molar tooth was recorded for 88 specimens (28 from Sonora, Sinaloa, and Chihuahua, and 60 from Guerrero and Jalisco). The third molar tooth is present on both sides of the lower jaw in all specimens except one (12413 Univ. Illinois) from Sonora which lacks both upper and lower third molars. The upper third molar is usually present on both sides. The exceptions are as follows: the above mentioned Sonoran specimen and one other Sonoran specimen, one specimen from La Bufa, Chihuahua, two from Jalisco, and five from Guerrero lack the tooth on both sides; two specimens from Guerrero and one from Sonora lack the tooth on only one side. Facial stripes are absent or present but inconspicuous in all specimens recorded here. The generally grayish hue, hairiness of interfemoral membrane, and configuration of skull described by Lukens and Davis (1957:7) for *A. hirsutus* are evident in all the specimens reported here. Skins of three adults from Sonora and Chihuahua are slightly browner and somewhat paler than skins of adults from Jalisco and Guerrero.

Reproductive data from Sonora and Chihuahua are as follows: of the five Chihuahuan specimens, two are immature (open epiphyseal sutures); the one adult female (79443) contained a single embryo 28 mm. in crown-rump length. Eight of 20 Sonoran specimens taken in May are females, each of which lacks epiphyseal sutures, and each contained one embryo. One embryo measured 8 mm. in length of uterine enlargement; all others are longer than 20 mm. from crown to rump, but vary in stage of development, some having no pigmentation in the membranes and others having pigmentation. The forearm is only 42 mm. long in one young male from Sonora. Three of 8 Sonoran specimens taken in July had open epiphyseal sutures but were of adult size. In summary of the reproductive data by states, *Artibeus hirsutus* is known to bear embryos in the following months: May in Sonora, July in Chihuahua, February in Jalisco, and May in Guerrero. These data, along with the presence of embryos and young of various ages among specimens taken at the same place and time, indicate that the species does not have a restricted breeding season.

A geographic overlap of the ranges of *A. hirsutus* and *A. jamaicensis* from Guerrero to central Sinaloa is now known. But the two species have not been taken at the same place within this region of overlap.

Other species.—At the locality on the Río Septentrión, 1500 feet, 1½ mi. SW Tocuina, Chihuahua, from which specimens of *A. hirsutus* were obtained as mentioned previously, several other species of tropical bats were captured, including *Desmodus rotundus murinus* Wagner, *Glossophaga soricina leachii* (Gray), *Chilonycteris parnellii mexicana* Miller, *Sturnira lilium parvidens* Goldman, and *Chiroderma* (specimens not yet certainly identified to species). The canyon of the Río Septentrión is steep and rocky, the tropical vegetation occurs only in the bottom of the canyon, and unless construction of a railroad had been in progress the area could have been reached only after several days by means of a pack train. From a distributional standpoint the occurrence of *Sturnira* and *Chiroderma* 1½ mi. SW of Tocuina is of unusual interest.

The published record of *Sturnira lilium* nearest to Tocuina is from 2 mi. N Ciudad Guzmán (19°43′, 103°28′), Jalisco, and the nearest published record of the genus *Chiroderma* is of *Chiroderma isthmicum* from Presidio (18°37′, 96°47′), Veracruz (Hall and Kelson, 1959:126, 134). The Chihuahuan specimens extend the known range of *Sturnira lilium* approximately 585 miles northwestward and that of the genus *Chiroderma* approximately 920 miles northwestward from the localities noted above. Five specimens (79434-79438) of *Sturnira lilium,* two adults and three immature individuals, were taken from July 18 to July 22, 1958, by the author and Kenneth E. Shain, as also were the two (79439-79440) *Chiroderma.*

To the list given by Koopman and Martin (1959:9) of neotropical genera known to range farther north on the west coast of North America than on the east coast there can now be added *Artibeus, Sturnira* and *Chiroderma* (as noted above), *Anoura, Choeronycteris* and *Leptonycteris* (Hall and Kelson, 1959:119, 120, 122; Hoffmeister, 1959:18), and *Liomys* (Hall and Kelson, 1959:536).

In view of these additional genera, and others that almost certainly remain to be discovered farther north on the west coast, the suggestion by Koopman and Martin (1959:11) that species inhabiting humid tropical habitats, in general extend farther north on the east coast of Mexico than on the west coast may need to be reconsidered. On the west coast, areas of more humid tropical vegetation and climate are more distant from the coastline as one proceeds northwestward from Nayarit to Sonora. The broad band of humid tropical vegetation along the coast is progressively reduced in width, and crowded back against the mountains, and still farther north consists of only small scattered remnants that are difficult to visit, in the bottoms of deep canyons.

LITERATURE CITED

COCKRUM, E. L.
1955. Reproduction in North American Bats. Trans. Kansas Acad. Sci., 58:487-511.

DALQUEST, W. W.
1953. Mexican bats of the genus *Artibeus.* Proc. Biol. Soc. Washington, 66:61-66.

DAVIS, W. B.
1958. Review of Mexican bats of the Artibeus "cinereus" complex. Proc. Biol. Soc. Washington, 71:163-166, 1 fig. in text.

HALL, E. R., and KELSON, K. R.
1959. The mammals of North America. The Ronald Press, N. Y., Vol. I, xxx + 1-546 + 1-79 pp., 312 figs. and 320 maps in text, unnumbered figures in text.

HOFFMEISTER, D. F.
1959. Distributional records of certain mammals from southern Arizona. Southwest. Nat., 4:14-19, 1 fig. in text.

KOOPMAN, K. F., and MARTIN, P. S.
1959. Subfossil mammals from the Gómez Farías region and the tropical gradient of eastern Mexico. Jour. Mamm., 40:1-12, 1 fig. and 2 tables in text.

LUKENS, P. W., JR., and DAVIS, W. B.
1957. Bats of the Mexican state of Guerrero. Jour. Mamm., 38:1-14.

Transmitted August 18, 1960.

University of Kansas Publications
Museum of Natural History

Volume 14, No. 2, pp. 9-27, 1 fig. in text
July 24, 1961

Geographic Variation in the Harvest Mouse, Reithrodontomys megalotis, On the Central Great Plains And in Adjacent Regions

BY

J. KNOX JONES, JR. AND B. MURSALOĞLU

University of Kansas
Lawrence
1961

University of Kansas Publications, Museum of Natural History

Editors: E. Raymond Hall, Chairman, Henry S. Fitch,
Robert W. Wilson

Volume 14, No. 2, pp. 9-27, 1 fig. in text
Published July 24, 1961

University of Kansas
Lawrence, Kansas

PRINTED IN
THE STATE PRINTING PLANT
TOPEKA, KANSAS
1961

28-7578

Geographic Variation in the Harvest Mouse, Reithrodontomys megalotis, On the Central Great Plains And in Adjacent Regions

BY

J. KNOX JONES, JR. AND B. MURSALOĞLU

The western harvest mouse, *Reithrodontomys megalotis,* inhabits most parts of the central Great Plains and adjacent regions of tall grass prairie to the eastward, shows a marked predilection for grassy habitats, is common in many areas, and is notably less variable geographically than most other cricetids found in the same region. R. *megalotis* occurs (see Hall and Kelson, 1959:586, map 342) from Minnesota, southwestern Wisconsin, northwestern Illinois, Iowa and Missouri westward to, but apparently not across, the Rocky Mountains from southeastern Alberta to Colorado; it is known in Oklahoma only from the Panhandle, thence southward through the Panhandle and Trans-Pecos areas of Texas to southern México, westward across the mountains in New Mexico to the Pacific Coast, and northward to the west of the Rockies to southern British Columbia.

Hoffmeister and Warnock (1955) studied western harvest mice from Illinois, Iowa, northeastern Kansas, Minnesota and Wisconsin, concluded that one subspecific name (*Reithrodontomys megalotis dychei* J. A. Allen, 1895, with type locality at Lawrence, Douglas Co., Kansas) applied to all, and relegated *Reithrodontomys megalotis pectoralis* Hanson, 1944 (type locality at Westpoint, Columbia Co., Wisconsin) to synonymy under *dychei.* Our study, based upon an examination of 1350 specimens, concerns the area west of the Missouri River from Kansas and Nebraska westward to Montana, Wyoming, Colorado and northern New Mexico. Our objectives were to study variation in *R. megalotis* in the region indicated and to decide what subspecific names properly apply to populations of the species that occur there.

Aside from the name R. *m. dychei,* currently applied to western harvest mice from a large part of the region here under study, three other subspecific names need consideration:

"*Reithrodontomys aztecus*" J. A. Allen, 1893 (type locality, La Plata, San Juan Co., New Mexico), currently applied to specimens from northern

New Mexico and southern Colorado (and adjacent parts of Arizona and Utah) east to southwestern Kansas and the Oklahoma Panhandle;

"*Reithrodontomys megalotis caryi*" A. H. Howell, 1935 (type locality, Medano Ranch, 15 mi. NE Mosca, Alamosa Co., Colorado), proposed for, and currently applied to, harvest mice from the San Luis Valley, Colorado, but possibly a synonym of *aztecus* according to Hooper (1952:218); and

"*Reithrodontomys dychei nebrascensis*" J. A. Allen, 1895 (type locality, Kennedy, Cherry Co., Nebraska), proposed for harvest mice from western Nebraska and adjacent areas, but regarded as a synonym of *dychei* by A. H. Howell (1914:30-31).

Our comments concerning the taxonomic status of these several names appear beyond.

We are grateful to Dr. W. Frank Blair, University of Texas, for the loan of a specimen from the Texas Panhandle (TU), and to Dr. Richard H. Manville, U. S. Fish and Wildlife Service, for the loan of specimens of *R. m. caryi* from the Biological Surveys Collection (USNM). We are grateful also to persons in charge of the following collections for allowing one of us (Jones) to examine Nebraskan specimens of *R. megalotis* in their care: University of Michigan Museum of Zoology (UMMZ); University of Nebraska State Museum (NSM); and U. S. National Museum (USNM). A research grant from the Society of the Sigma Xi facilitated travel to the institutions mentioned. Specimens not identified as to collection are in the Museum of Natural History of The University of Kansas. All measurements are in millimeters, and are of adults (as defined by Hooper, 1952:12) unless otherwise noted.

Secondary Sexual Variation

Hooper (1952) did not accord separate treatment to males and females in taxonomic accounts of Latin American harvest mice because (p. 11): "In no species . . . does sexual dimorphism in the measurements, if present at all, appear to be sufficient to warrant separating the sexes in the analysis." Hooper did not statistically test the validity of treating the sexes together in R. *megalotis*. He did test a series of R. *sumichrasti* from El Salvador, in which he found no basis for separate treatment of males and females.

Some authors (Verts, 1960:6, for instance) have recorded females of *R. megalotis* as larger than males in external measurements, whereas others (Dalquest, 1948:325, for instance) have recorded males as the larger. In order to learn something of secondary sexual variation, and to decide whether or not to separate the sexes in our study, we compared adult males and females from the southern part of the Panhandle of Nebraska (Cheyenne, Keith, Kimball, Morrill and Scotts Bluff counties) in four external and

TABLE 1. ANALYSIS OF SECONDARY SEXUAL VARIATION IN ADULT REITHRODONTOMYS MEGALOTIS FROM THE SOUTHERN PART OF THE NEBRASKA PANHANDLE. FOR EACH MEASUREMENT, THE NUMBER OF SPECIMENS USED, THE AVERAGE, THE EXTREMES, AND ONE STANDARD DEVIATION ARE GIVEN.

CHARACTER	Males				Females			
Total length	27	135.0	(121–149)	± 614	32	141.0	(127–149)	± 5.36
Length of tail-vertebrae	27	63.9	(56–74)	± 4.63	32	65.2	(58–73)	± 4.06
Length of hind foot	27	17.0	(16–18)	± 0.60	32	17.3	(15–19)	± 0.81
[illegible]gth of ear from notch	27	12.9	(12–14)	± 0.55	32	13.0	(12–14)	± 0.61
Greatest length of skull	27	21.0	(20.2–21.8)	± 0.43	28	21.3	(20.4–22.2)	± 0.48
Zygomatic breadth	25	10.7	(10.3–11.0)	± 0.21	28	10.9	(10.4–11.3)	± 0.25
Breadth of [illegible]e	27	10.0	(9.6–10.5)	± 0.22	28	10.1	(9.8–10.7)	± 0.18
Depth of cranium	26	7.9	(7.4–8.4)	± 0.20	28	7.9	(7.7–8.3)	± 0.15
Length of rostrum	27	7.3	(6.8–7.6)	± 0.21	28	7.4	(6.9–8.0)	± 0.27
Breadth of rostrum	27	3.8	(3.6–4.1)	± 0.11	28	3.8	(3.5–4.0)	± 0.12
Length of incisive foramen	27	4.4	(4.1–4.6)	± 0.10	28	4.5	(4.1–4.9)	± 0.19
Length of palate	26	3.5	(3.1–3.8)	± 0.18	28	3.5	(3.2–4.0)	± 0.15
[illegible]ar length of maxillary tooth-row	27	3.4	(3.2–3.7)	± 0.14	28	3.4	(3.2–3.7)	± 0.13
[illegible]al breadth	27	3.1	(2.9–3.3)	± 0.12	28	3.1	(2.8–3.3)	± 0.11
Breadth of [illegible]ic plate	27	1.9	(1.8–2.1)	± 0.10	28	2.0	(1.9–2.3)	± 0.12
Breadth of mesopterygoid fossa	26	0.9	(0.6–1.1)	± 0.12	28	0.9	(0.8–1.2)	± 0.12

twelve cranial measurements (see Table 1). The external measurements are those customarily taken by collectors and were read from the labels of the specimens; cranial measurements were taken to the nearest tenth of a millimeter by means of dial calipers, and are those described by Hooper (1952:9-11). Females from our sample averaged larger than males in all external and several cranial measurements, but individual variation greatly exceeded secondary sexual variation in each of these measurements and in no case was the greater size of females statistically significant. Therefore, and because we found no qualitative external or cranial differences between the sexes, males and females have been considered together in each population studied.

Pelage and Molt

Western harvest mice that attain adulthood acquire at least three distinct types of pelage in sequence in the course of their development. The first of these, the juvenal pelage, is short, relatively sparse, and characteristically grayish brown. The molt (post-juvenal molt) from juvenal pelage to subadult pelage seemingly occurs at an early age, perhaps frequently before the young leave the nest, as individuals in juvenal pelage are few among specimens studied by us. Judging from study skins alone, the progress of post-juvenal molt in R. *megalotis* is similar to that described for R. *humulis* by Layne (1959:69-71). The subadult pelage is thicker, longer and brighter than juvenal pelage and closely resembles the pelage of adults; it differs from adult pelage dorsally in being somewhat duller and in having less contrast between back and sides.

The pelage of adults varies depending on season. In summer the individual hairs are relatively short (5-6 mm. at the middle of the back) and sparse. The over-all color of the dorsum, sides and flanks is brownish to dark brownish, and the venter is grayish. In winter the pelage is dense, long (8-9 mm. at the middle of the back) and lax. The over-all color dorsally in fresh winter pelage in most specimens is paler (more buffy) than summer pelage, the sides are markedly buffy, and the venter is whitish; even the tail is more pilose and more sharply bicolored than in summer. Adults molt, usually completely but occasionally only partially, at least twice a year—once in spring (in May and June in Nebraskan specimens) from winter to summer pelage, and once in autumn (in October and November in Nebraskan specimens) from summer to winter pelage. Of the two molts, the one in spring is most easily discernible because the contrast in color between worn winter pelage and fresh summer pelage is considerably greater than that between worn summer pelage and fresh winter pelage, and because the progress of spring molt is seemingly more regular than that of autumn molt. In spring, molt proceeds posteriorly in a more or less regular line on both dorsum and venter; in most specimens it is completed first on the venter. In autumn, molt is irregular, or at best is coincident over large parts of the body, and frequently is seen only by searching through the pelage with a fine probe or dissecting needle. In both spring and autumn, molt seemingly is delayed in females that are pregnant or lactating.

In both winter pelage and summer pelage, the upper parts have blackish or grayish guard hairs and shorter, more numerous cover hairs. All the cover hairs are gray basally; some have a buffy band terminally and others have a buffy subterminal band with a terminal black tip. The generally darker over-all color of upper parts in summer pelage results (as seen in Nebraskan specimens) from a narrower band of buff on the cover hairs (only approximately one half the width of the band on hairs in winter pelage), a darker buffy band (ochraceous buff rather than pale ochraceous or straw color), and a relative sparseness of the pelage, which allows the gray basal portion of some hairs to show on the surface. The more grayish venter of summer-taken specimens results from much more of the grayish basal portion of the white-tipped hairs showing through than in the longer, denser pelage of winter.

Wear on the pelage seems in general to produce a paler over-all color of upper parts, evidently due mostly to abrasion of the terminal black tip of the cover hairs, but possibly actual fading of the pelage is involved also. Worn winter pelage is especially notable for its paleness; the buffy tones are accentuated and the upper parts, especially posteriorly, may even appear fulvous. The difference in color of upper parts between specimens in worn winter pelage and fresh summer pelage (or for that matter specimens in fresh *versus* worn winter pelage) from the same locality is greater in our material than the difference between some specimens in comparable pelages from localities more than 500 miles apart.

We have seen no specimens taken in winter in which we could discern that the autumn molt had been incomplete, but three old adult males in summer pelage indicate that spring molt is not always completed. KU 50154, obtained on August 14, 1952, 5 mi. N and 2 mi. W Parks, Dundy Co., Nebraska, has the entire posterior back and sides still in old winter pelage and does not appear to have been actively molting; the entire venter is in summer pelage. KU 50146, obtained on August 22, 1952, 3 mi. E Chadron, Dawes Co., Nebraska, has small patches or tufts of winter pelage remaining on the rump and likewise does not appear to have been actively molting. KU 72085, obtained on October 13, 1956, 4 mi. E Barada, Richardson Co., Nebraska, is in the process of molting from summer to winter pelage, but has tufts of old winter pelage on the rump.

Geographic Variation

Geographic variation, both in color of pelage and in external and cranial dimensions, is less in *R. megalotis* in the region studied than in most other cricetine species that occur there. Nevertheless, meaningful variation is present. The assumption that variation in *R. megalotis* paralleled in degree that of other species, *Peromyscus maniculatus* for example, led to untenable taxonomic conclusions by some previous workers.

Color of Pelage

Color of pelage is remarkably uniform, considering the geographic extent of the area involved, over most of the northern part of the central grasslands. Perhaps this uniformity results partly from the predilection of the western harvest mouse for grassy habitats, for in most areas on the Great Plains the species is restricted to riparian communities, principally along river systems, where soils, cover, and other conditions approximate those of corresponding habitats farther to the east to a much greater degree than do conditions in upland habitats. Differential selective pressure, therefore, theoretically would be less between eastern and western populations of *R. megalotis* than in an upland-inhabiting species. In any event, specimens from western Nebraska, Wyoming, northern Colorado, and adjacent areas average only slightly paler dorsally than specimens in corresponding pelages from the eastern parts of Nebraska and Kansas, and many individuals from the two areas can be matched almost exactly.

To the southwest, on the other hand, a trend toward paler (pale brownish, less blackish) upper parts is apparent. Specimens from southwestern Kansas and adjacent parts of Colorado and Oklahoma average slightly paler in comparable pelages than specimens from northeastern Kansas and eastern Nebraska, but most specimens from farther southwest, in northern New Mexico and southwestern

Colorado, are discernibly, although not markedly, paler than mice from northern and eastern populations.

A "pectoral spot," fairly common in some populations of *R. megalotis* east of the Missouri River (see Hoffmeister and Warnock, 1955:162-163), is present in only a small percentage of the specimens we have studied, and when present is usually only faintly developed.

External and Cranial Size

As seen in Figure 1, the tail and especially the ear are longer in mice from New Mexico and adjacent areas than in specimens

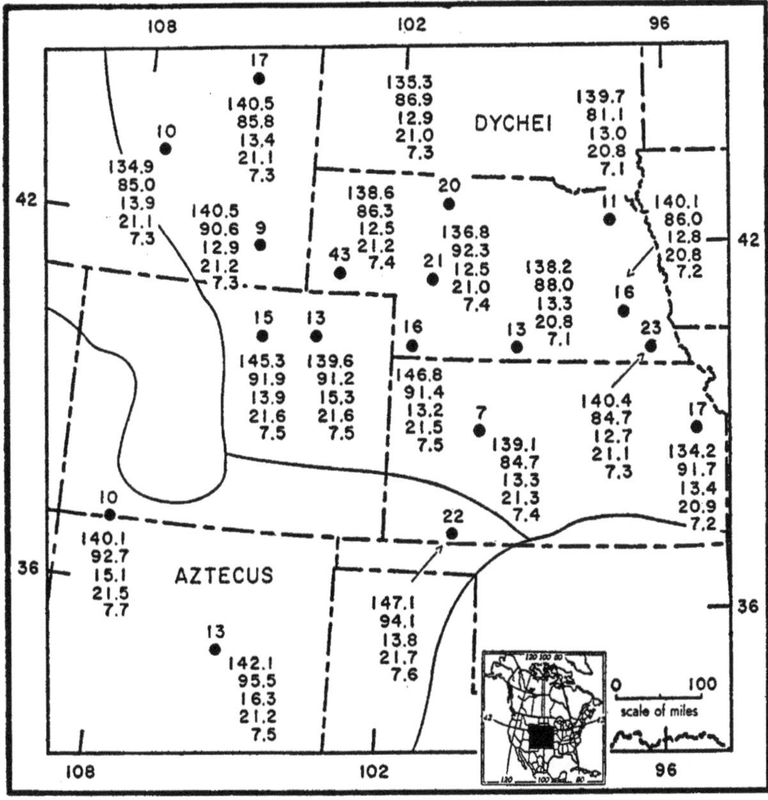

FIG. 1. Geographic variation in five measurements of *Reithrodontomys megalotis* on the central Great Plains. The size of each sample is given, along with total length, length of tail expressed as a percentage of the head and body, length of ear, greatest length of skull, and length of rostrum. The approximate distribution of the species in the region shown and the approximate boundary between the subspecies *R. m. aztecus* and *R. m. dychei* also are indicated.

from northern localities. The ear, only slightly variable in size in the northern part of the region, is markedly longer in the southwest, averaging more than 2 mm. longer in specimens from New Mexico and adjacent southwestern Colorado than in specimens from Nebraska and eastern Kansas; specimens in a zone from central Colorado through southwestern Kansas and adjacent Oklahoma generally have ears of a size between the two extremes. As concerns the tail we note a slight trend toward increasing length (best expressed as percentage of length of body) from north to south throughout the central plains, but in general the trend is more pronounced southwestwardly. Variation in length of tail and length of ear, therefore, appear to be in accord with Allen's Rule. Length of body and length of hind foot seem not to vary significantly in specimens we have studied.

The skulls of specimens examined differed only slightly, except that the rostrum is significantly longer and relatively, if not actually, narrower in specimens from the south and southwest than in mice from the rest of the region under study. The rostrum is longest (average 7.7 mm.) in specimens from the vicinity of the type locality of R. *m. aztecus*, but is relatively long (7.5-7.6 mm.) in populations from as far north as northeastern Colorado and southwestern Nebraska. An average greater occipitonasal length (greatest length of skull) in specimens from the south and southwest results mostly from the longer rostrum.

Recognition of two subspecies of R. *megalotis* on the central Great Plains seems justified on the basis of the geographic variation discussed above. One subspecies, for which the name R. *m. aztecus* is applicable, occurs in the southwest and is characterized by the culmination of trends in the region studied to paler upper parts, longer tail, longer ear, and longer, relatively narrower rostrum—characters that appear at least partly independent of each other as concerns gradation toward the smaller, darker-colored populations to the northward. The latter, while exhibiting some differences in color (slightly paler westwardly) and length of tail (shorter northwardly), stand more or less as a unit in contrast to the mice from the southwest, and represent, in our judgment, a single subspecies, R. *m. dychei.* The area of intergradation between the two subspecies is relatively broad, considering all the characters mentioned, and assignment of some intergrades is admittedly difficult.

Reithrodontomys megalotis aztecus J. A. Allen

Reithrodontomys aztecus J. A. Allen, Bull. Amer. Mus. Nat. Hist., 5:79, April 28, 1893 (type locality, La Plata, San Juan Co., New Mexico).

Reithrodontomys megalotis aztecus, A. H. Howell, N. Amer. Fauna, 36:30, June 5, 1914.

Reithrodontomys megalotis caryi A. H. Howell, Jour. Mamm., 16:143, May 15, 1935 (type locality, Medano Ranch, 15 mi. NE Mosca, Alamosa Co., Colorado).

Distribution.—Western and southern Colorado, southeastern Utah, northeastern Arizona and northern New Mexico, east to the panhandles of Texas and Oklahoma and to southwestern Kansas.

External measurements.—Average and extremes of 10 adults (5 males, 5 females) from San Juan County, New Mexico, and adjacent Montezuma County, Colorado, are: total length, 140.1 (126-150); length of tail-vertebrae, 67.4 (56-71); length of hind foot, 17.3 (16-18); length of ear from notch, 15.1 (13-17); tail averaging 92.7 per cent of length of body. Corresponding measurements of 13 adults (7 males, 6 females) from Bernalillo and Guadalupe counties, New Mexico, are: 142.1 (129-156); 69.4 (60-75); 17.9 (17-19); 16.3 (15-18); tail averaging 95.4 per cent of length of body. Corresponding measurements of 22 adults (17 males, 5 females) from Meade County, southwestern Kansas, are: 147.1 (139-162); 71.3 (65-77); 17.6 (17-19); 13.8 (13-15); tail averaging 94.1 per cent of length of body. For cranial measurements see Table 2.

Remarks.—For comparisons with *Reithrodontomys megalotis dychei*, geographically adjacent to the northeast, see account of that subspecies.

When Howell (1935:143) named *Reithrodontomys megalotis caryi* from the San Luis Valley of Colorado he compared it directly only with R. *m. megalotis* from southern New Mexico and northern Chihuahua. Few adults were available to Howell from the San Luis Valley, accounting for the fact, we think, that the published measurements of *caryi* average less than those given for R. *m. aztecus* by Howell (*op. cit.*:144) and herein. We have examined 16 of the 23 specimens from Medano Ranch and the single specimen from Del Norte that Howell listed. Unfortunately, none is fully adult. The specimens from Medano Ranch, collected in late October and early November, are mostly in fresh winter pelage or molting from subadult pelage, and closely resemble topotypes of *aztecus* in comparable pelages. Comparison of skulls of the specimens from Medano Ranch with skulls of topotypes and other individuals of *aztecus* of approximately equal age indicates that the Coloradan specimens may average slightly smaller and have somewhat shorter rostra. Externally, topotypes of *caryi* have the relatively long tail of *aztecus* and approach it in length of ear (measured on dry specimens). To us, they appear to be intergrades between *aztecus* and *dychei*, but to bear closer resemblance to the former, and we tentatively regard *caryi* as a synonym of *aztecus*. Benson (1935:140) noted that two adult topotypes of *caryi* were "similar to adult topotypes of *aztecus*." Specimens from southern Colorado east of the San Luis Valley, assigned to *aztecus*, are intergrades between it and *dychei*, as are two specimens from El Paso County, to the north, which resemble *aztecus* in color but resemble *dychei* in other characters and are tentatively assigned to the latter.

Specimens from southwestern Kansas and adjacent Oklahoma, herein referred to *aztecus*, also are intergrades with *dychei*. Individuals from Meade County, for example, are intermediate on the average between typical specimens of the two subspecies in color of upper parts (if anything, nearer *dychei*), resemble *dychei* in length of ear, but resemble *aztecus* in length of tail and rostral proportions (consequently also in length of skull). Although a case could be made for assignment of the specimens from Meade County (and elsewhere in southwestern Kansas) to *dychei*, they are, everything considered, nearer *aztecus*, to which subspecies they have been assigned consistently since first reported from the area by Hill and Hibbard (1943:24).

Of two specimens examined from 10 mi. S and 1 mi. W Gruver, Hansford Co., in the Panhandle of Texas, the one adult is clearly assignable to *aztecus* as is the specimen from 9 mi. E Stinnett, Hutchinson Co., Texas, that was referred to *dychei* by Blair (1954:249).

Reithrodontomys megalotis aztecus has had a rather unstable taxonomic history. Allen, who originally named the subspecies (1893:79), regarded it two years later (1895:125) as a synonym of *R. m. megalotis*, the subspecies with geographic range to the south and west of that occupied by *aztecus*. Howell (1914:30) recognized *aztecus* as valid, but he, too, questioned its distinctness from *megalotis* in a later paper (1935:144). Hooper (1952:218), the most recent reviewer, supported the validity of *aztecus* because specimens available to him averaged "distinctly larger in skull length and size of brain case" than specimens of *megalotis*. Our comparisons of typical specimens of *aztecus* with specimens of *megalotis* from southern New Mexico and southwestern Texas confirm Hooper's observations and indicate also that *aztecus* has a longer rostrum and slightly longer ear.

Specimens examined.—205, as follows:

COLORADO. *Alamosa County:* Medano Ranch, 15 mi. NE Mosca, 16 (USNM). *La Plata County:* 1 mi. NW Florida, 6700 ft., 1; Florida, 6800 ft., 1. *Las Animas County:* 1 mi. S, 7 mi. E Trinidad, 2. *Montezuma County:* 1 mi. W Mancos, 5; north end, Mesa Verde Nat'l Park, 7000 ft., 3; Far View Ruins, Mesa Verde Nat'l Park, 7700 ft., 3; Park Point, Mesa Verde Nat'l Park, 8525 ft., 2; within 3 mi. Rock Springs, Mesa Verde Nat'l Park, 7500-8200 ft., 6. *Prowers County:* Lamar, 2. *Rio Grande County:* Del Norte, 1 (USNM).

KANSAS. *Finney County:* 1 mi. S, 2 mi. E Garden City, 4. *Ford County:* ½ mi. NW Bellefont, 10; 6¼ mi. N Fowler, 2. *Grant County:* 2 mi. S, 9 mi. W Santanta, 1. *Kearney County:* 3½ mi. N, 4 mi. E Lakin, 4. *Meade County:* within 2½ mi. Fowler, 10; Meade County State Park, 14 mi. SW Meade, 48; 17 mi. SW Meade, 5. *Morton County:* 7½ mi. S Richfield, 4; 8 mi. N Elkhart, 1; 7½ mi. N, 1½ mi. W Elkhart, 2. *Seward County:* 3 mi. NE Liberal, 1. *Stanton County:* 1 mi. N, 6-7½ mi. W Manter, 2; dam of Lake Stanton, 1.

NEW MEXICO. *Bernalillo County:* 6½ mi. E Alameda, 11; 5 mi. W Albuquerque, 3. *Catron County:* 1 mi. NE Apache Creek, 4; Apache Creek, 2. *Guadalupe County:* 4 mi. SW Santa Rosa, 4700 ft., 10. *McKinley County:* Upper Nutria, 7200 ft., 2. *Rio Arriba County:* 4 mi. N El Rito, 1; 1 mi. SE El Rito, 1. *Sandoval County:* 3 mi. N La Cueva Rec. Area, 1. *San Juan County:* 2 mi. N La Plata, 15. *Santa Fe County:* 1 mi. W Santa Fe Municipal Airport, 1; La Bajada Grade, 20 mi. W Santa Fe, 1. *Socorro County:* 2 mi. S San Antonio, 4.

OKLAHOMA. *Beaver County:* 7 mi. S Turpin, 1. *Texas County:* 3½ mi. SW Optima, 8.

TEXAS. *Hansford County:* 10 mi. S, 1 mi. W Gruver, 2. *Hutchinson County:* 9 mi. E Stinnett, 1 (TU).

Reithrodontomys megalotis dychei J. A. Allen

Reithrodontomys dychei J. A. Allen, Bull. Amer. Mus. Nat. Hist., 7:120, May 21, 1895 (type locality, Lawrence, Douglas Co., Kansas).

Reithrodontomys megalotis dychei, A. H. Howell, N. Amer. Fauna, 36:30, June 5, 1914.

Reithrodontomys dychei nebrascensis J. A. Allen, Bull. Amer. Mus. Nat. Hist., 7:122, May 21, 1895 (type locality, Kennedy, Cherry Co., Nebraska).

Distribution.—Southwestern Wisconsin, southern Minnesota, northwestern Illinois, Iowa, Missouri and northwestern Arkansas, west through Kansas (except southwestern part), Nebraska and the Dakotas to the foothills of the Rocky Mountains from central Colorado to southeastern Alberta.

External measurements.—Average and extremes of 17 adults (11 males, 6 females) from Douglas County, Kansas, are: total length, 134.2 (115-151); length of tail-vertebrae, 64.2 (59-72); length of hind foot, 16.7 (15-18); length of ear from notch, 13.4 (12-15); tail averaging 91.7 per cent of length of body. Corresponding measurements of 20 adults (14 males, 6 females) from Cherry County, Nebraska, are: 135.3 (122-155); 62.9 (56-72); 17.5 (17-18); 13.0 (12-14); tail averaging 86.9 per cent of length of body. For cranial measurements see Tables 1 and 2.

Remarks.—From *Reithrodontomys megalotis aztecus*, geographically adjacent to the southwest, *R. m. dychei* differs as follows: upper parts averaging darker (especially in summer pelage), owing principally to more suffusion of blackish middorsally; tail slightly shorter; ears markedly shorter, rostrum shorter and relatively broader; occipitonasal length shorter owing to shorter rostrum.

"*Reithrodontomys dychei nebrascensis*," named by Allen (1895: 122) from Kennedy, Nebraska, was distinguished in the original description from *dychei* by "slightly larger size, relatively longer ears, and more strongly fulvous coloration." Allen applied the name *nebrascensis* to harvest mice from Montana south to central Colorado and western Nebraska. Howell (1914:30-31) placed *nebrascensis* in synonymy under *dychei* because he found specimens from Kennedy to be "indistinguishable from specimens of typical *dychei* in comparable pelage." We concur with Howell. Topotypes of *nebrascensis* that we have examined average only slightly paler than topotypes of *dychei* in the same pelage (some specimens from each series can be matched almost exactly), and do not differ significantly in any external or cranial measurements. The "fulvous" upper parts of the series from Kennedy (all taken in late April) that was available to Allen resulted from worn winter pelage. We think that Allen was led astray also by his erroneous assumption that geographic variation in color of R. *megalotis* on the Great Plains paralleled that found in *Peromyscus maniculatus*. Actually, R. *megalotis* varies in color much less geographically in the region concerned than does P. *maniculatus*.

Specimens from the northwestern part of the range of *dychei* (Wyoming, Montana and western South Dakota), like those from western Nebraska, average slightly paler dorsally than topotypes and other specimens from eastern Kansas and Nebraska (a few approach *aztecus* in this regard), but do not otherwise differ. Most specimens from northern Colorado, southwestern Nebraska (Hitchcock and Dundy counties) and western Kansas average slightly paler than typical specimens and have longer rostra, approaching *aztecus* in these particulars, but have the shorter ears and shorter tail of *dychei*. In general, these intergrades resemble *dychei* to a greater degree than *aztecus* and are accordingly assigned to the former. One exception is a series from Muir Springs, 2 mi. N and 2½ mi. W Ft. Morgan, Colorado. Specimens in this series approach typical *dychei* in color, but resemble *aztecus* in having long ears and long rostra (average 15.3 and 7.5, respectively, in 13 adults). The specimens from Muir Springs resemble *aztecus* to a greater degree than *dychei*, but are assigned to the latter because specimens from farther west and farther south in Colorado are assignable to *dychei*. Howell (1914:31) earlier noted that specimens from northern and central Colorado were intergrades between the two subspecies.

The geographic range occupied by R. *m. dychei* (from east of the Mississippi River in Illinois and Wisconsin to the foothills of the Rockies) is large (although not so large as that currently ascribed to R. *m. megalotis,* which ranges from southern British Columbia to central México). Most other small rodents that occur in the same geographic area occupied by *dychei* are represented there by at least two subspecies, a dark one in the east and a pale one in the west. Eastern populations of *dychei* have, it is true, somewhat darker upper parts than mice from western localities, but the differences are slight; also, judging from the literature, the "pectoral spot" is more common in eastern mice.

It should be noted that R. *m. dychei* probably has extended its range both eastward and westward in the last century as a result of agricultural practices—clearing of land in the east and irrigation in the west.

Specimens examined.—1145, as follows:

COLORADO. *Adams County:* South Platte River, 5 mi. N Denver, 1; 3 mi. S, 1 mi. W Simpson, 1. *El Paso County:* 5 mi. E Payton, 1; 4 mi. S maingate of Camp Carson, 1. *Larimer County:* 3 mi. N Loveland, 1; 9¼ mi. W, ½ mi. N Loveland, 5600 ft., 1; 16 mi. W Loveland, 6840 ft., 1; 3½-4½ mi. W Loveland, 5030 ft., 7; 6 mi. W, ½ mi. S Loveland, 5200 ft., 14; 7 mi. W, 2½ mi. S Loveland, 5370 ft., 1. *Morgan County:* Muir Springs, 2 mi. N, 2½ mi. W Ft. Morgan, 21. *Washington County:* Cope, 6. *Yuma County:* 1 mi. W to 1 mi. E Laird, 6.

KANSAS. *Atchison County:* 1½ mi. S Muscotah, 10; 4½ mi. S Muscotah, 2. *Barton County:* 3 mi. N, 2 mi. W Hoisington, 3. *Brown County:* 1 mi. E Reserve, 2; 5 mi. S Hiawatha, 4. *Cheyenne County:* 23 mi. NW St. Francis, 1; 1 mi. W St. Francis, 12; 8 mi. S, 1½ mi. W St. Francis, 1. *Decatur County:* 1 mi. N, 2 mi. E Oberlin, 4; 5 mi. S, 8 mi. W Oberlin, 1. *Doniphan County:* Geary, 2. *Douglas County:* 5 mi. N, ½ mi. E Lawrence, 1; 1 mi. NW Midland, 1; 4½ mi. N Lawrence, 2; 4 mi. N, 1¾ mi. E Lawrence (sec. 8, T. 12 S, R. 20 E), 10; ½ mi. NW Lecompton, 1; 2½ mi. N, 1 mi. W Lawrence, 2; 2 mi. N Lawrence, 2; U. P. Railroad tracks, N of Lawrence, 1; 9⅕ mi. W Lawrence, 1; 5 mi. W Lawrence, 1; 2 mi. W Lawrence, 4; 1 mi. W Lawrence, 4; Fort Lake, Lawrence, 1; Lawrence, 24; 1 mi. SW Lawrence, 2; 1 mi. S, 1½ mi. W Lawrence, 2; 1¾ mi. S, 3½ mi. E Lawrence, 1; 2 mi. SW Lawrence, 2; 7-7½ mi. SW Lawrence, 4; Rock Creek, 850 ft., 10 mi. SW Lawrence, 8; N end Lone Star Lake, 9 mi. S, 7 mi. W Lawrence, 1; no specific locality, 6. *Ellis County:* ½ mi. S, 3½-4 mi. W Hays, 2250 ft., 12. *Franklin County:* 4 mi. N Ottawa, 2; ½ mi. S, 1¾ mi. E Ottawa, 4. *Gove County:* Castle Rock, 4; no specific locality, 1. *Jackson County:* ½ mi. N, 3 mi. W Holton, 4. *Leavenworth County:* Ft. Leavenworth, 2; no specific locality, 3. *Logan County:* no specific locality, 2. *Marshall County:* 2 mi. N, 4 mi. E Oketo, 1; ½ mi. N, 1½ mi. E Waterville, 1; 1 mi. E Waterville, 5; ½ mi. SW Waterville, 4. *Mitchell County:* ½ mi. S, 3½ mi. W Beloit, 1500 ft., 4. *Nemaha County:* Nebraska-Kansas line, 7 mi. N Sabetha, 1; 10½ mi. N Seneca, 1; 2½ mi. S Sabetha, 6. *Norton County:* 1½ mi. N, ¼ mi. E Norton, 1; ½ mi. N, 4 mi. E Norton, 5; 1 mi. SW Norton, 10; 4 mi. W, 1 mi. S Logan, 3. *Osage County:* 3 mi. N Lyndon, 1. *Osborne County:* ½ mi. W Downs, 5. *Phillips County:*

2¼ mi. SE Long Island, 1. *Pottawatomie County:* 1 mi. NW Fostoria, 1. *Rawlins County:* 2 mi. NE Ludell, 17; 2 mi. S Ludell, 2; Atwood, 3; Atwood Lake, 2. *Republic County:* 1½ mi. S, 1 mi. E Belleville, 1; Rydal, 8. *Scott County:* State Park, 2. *Shawnee County:* 1 mi. S Silver Lake, 857 ft., 2. *Sherman County:* ½ mi. S, 1½ mi. E Edson, 1. *Smith County:* 2 mi. E Smith Center, 9. *Stafford County:* 16 mi. N, 4 mi. E Stafford, 1. *Thomas County:* 10 mi. N, 6 mi. E Colby, 5. *Trego County:* 16 mi. S, 4½ mi. E Wakeeney, 1. *Wichita County:* 15 mi. W Scott City, 5.

MONTANA. *Big Horn County:* Big Horn River, 14 mi. S Custer, 2750 ft., 4. *Dawson County:* 1 mi. W Glendive, 2070 ft., 3. *Phillips County:* 1 mi. N, 1 mi. W Malta, 2248 ft., 1. *Powder River County:* Powderville, 2900 ft., 1.

NEBRASKA. *Antelope County:* Neligh, 16 (6 NSM, 9 USNM). *Boyd County:* 5 mi. WSW Spencer, 1; 5 mi. S, 2 mi. E Spencer, 2; 6 mi. SSE Spencer, 1. *Box Butte County:* Alliance, 2 (USNM). *Buffalo County:* Kearney, 2 (USNM). *Burt County:* 1 mi. E Tekamah, 3. *Butler County:* 2 mi. N, 2 mi. W Bellwood, 2 (NSM); 4-5 mi. E Rising City, 11; 4 mi. E, 1 mi. S Rising City, 5. *Chase County:* 2 mi. SE Enders, 1. *Cherry County:* W of Crookston, 1 (NSM); Valentine, 2 (USNM); Ft. Niobrara Nat'l Wildlife Refuge, 4 mi. E Valentine, 5 (3 NSM); 3 mi. SSE Valentine, 4; 3 mi. S Valentine, 12; 8 mi. S Nenzel, 2; Niobrara River, 10 mi. S Cody, 2 (1 USNM); 11 mi. S, 2 mi. W Nenzel, 1; 18 mi. NW Kennedy, 8 (2 NSM, 6 USNM); Two Mile Lake, 6 (4 NSM, 2 USNM); Watt's Lake, Valentine Nat'l Wildlife Refuge, 3; Hackberry Lake, 12 (UMMZ); 2 mi. W to 4 mi. E Kennedy, 25 (4 UMMZ, 12 USNM); no specific locality, 1 (USNM). *Cheyenne County:* 15 mi. S Dalton, 4300 ft., 1; 3 mi. N Sidney, 6; 4 mi. E Sidney, 42. *Cuming County:* Beemer, 1 (USNM). *Custer County:* 7 mi. NW Anselmo, 1 (UMMZ); within 1 mi. Victoria Spring, 9 (UMMZ); 2 mi. E Lillian, 1 (UMMZ); Comstock, 1 (NSM); Callaway, 3 (USNM); 6 mi. SE Mason City, 1 (UMMZ). *Dawes County:* Wayside, 1; 3 mi. E Chadron, 2; 6 mi. S Chadron, 1 (NSM); 8 mi. S Chadron, 1 (NSM); 10 mi. S Chadron, 1 (UMMZ); 1 mi. W Crawford, 2 (NSM); Crawford, 2 (UMMZ). *Dawson County:* ½ mi. S Gothenburg, 5; 3 mi. SSE Gothenburg, 4. *Deuel County:* 1 mi. N, 2 mi. W Chappell, 3. *Dixon County:* 3 mi. NE Ponca, 4. *Dundy County:* Rock Creek Fish Hatchery, 5 mi. N, 2 mi. W Parks, 42; 2 mi. N, 2 mi. W Haigler, 1; Arikaree River, Parks, 2; 2 mi. SW Benkleman, 7; Haigler, 3 (1 NSM, 2 USNM). *Franklin County:* 1½-2 mi. S Franklin, 10. *Gage County:* 1 mi. SE DeWitt, 3; ¼ mi. W Homestead Nat'l Mon., 1; 1 mi. S, 1 mi. W Barnston, 1; 1½ mi. S, 2 mi. E Barnston, 18. *Garden County:* Crescent Lake Nat'l Wildlife Refuge, 1; ½ mi. S Oshkosh, 1. *Hall County:* 6 mi. S Grand Island, 5. *Harlan County:* 1 mi. W Alma, 17. *Hitchcock County:* Republican River, Trenton, 3. *Hooker County:* Kelso, 3 (UMMZ). *Holt County:* 6 mi. N Midway, 4; 1 mi. S Atkinson, 4 (2 NSM); Ewing, 1 (USNM). *Jefferson County:* 7 mi. S, 2 mi. W Fairbury, 6; 3 mi. S, 1 mi. W Endicott, 1. *Johnson County:* 1 mi. S, 1½ mi. E Burr, 1. *Kearney County:* 1¾-3¾ mi. S Kearney, 6. *Keith County:* 4 mi. WNW Keystone, 69. *Keya Paha County:* 12 mi. N Springview, 8; 12 mi. NNW Springview, 5. *Kimball County:* 3 mi. E Kimball, 1; Smeed, 40. *Knox County:* 3 mi. W Niobrara, 2; 1 mi. SE Niobrara, 5; 2 mi. S Niobrara, 2; Verdigre, 2 (USNM). *Lancaster County:* within 5 mi. Lincoln, 21 (8 NSM). *Lincoln County:* 2 mi. N North Platte, 1; Conroy Canyon, SW corner sec. 4, T. 11 N, R. 27 W (5 mi. S, 2½ mi. W Brady), 2 (NSM). *Logan County:* 1-2 mi. NE Stapleton, 11. *Madison County:* Norfolk, 1 (USNM). *Morrill County:* 1 mi. N Bridgeport, 4. *Nemaha County:* 2 mi. SW Peru, 6; 3 mi. S, 1½ mi. E Peru, 2. *Nuckolls County:* 2 mi. WSW Superior, 5; 1 mi. SSW Hardy, 9. *Otoe County:* 1 mi. SE Nebraska City, 3; 3 mi. S, 2 mi. E Nebraska City, 3. *Pawnee County:* Turkey Creek, 4 mi. NW Pawnee City, 2 (NSM); 4 mi. S, 8 mi. W Pawnee City, 7; 1 mi. S Du Bois, 4. *Platte County:* Columbus, 3 (USNM). *Polk County:* 15 mi. W Osceola, 2. *Red Willow County:* 5 mi. S, 2½ mi. E McCook, 2; 8 mi. S, 3 mi. E McCook, 2. *Richardson County:* 5 mi. N, 2 mi. W Humboldt, 2 (1 NSM); 4 mi. E Barada, 16; 3½ mi. S, 1 mi. W Dawson, 6; 2 mi. N Falls City, 2; 4-6 mi. W Falls City, 4; ½ mi. S,

1½ mi. W Rulo, 1. *Saline County:* 2 mi. NE Crete, 1; ½ mi. W DeWitt, 1. *Sarpy County:* 1 mi. W Meadow, 1. *Saunders County:* 2 mi. NW Ashland, 3. *Scotts Bluff County:* 8 mi. NNW Scottsbluff, 1; Mitchell, 1 (NSM); ½-1 mi. S Mitchell, 13; 5 mi. S Gering, 10; 7 mi. S Gering, 1; 11-12 mi. S Scottsbluff, 4600-4800 ft., 8; 12 mi. SSW Scottsbluff, 4700 ft., 5. *Sioux County:* 1 mi. S, 4 mi. W Orella, 1 (NSM); 8 mi. N Harrison, 2 (UMMZ); 6½-7 mi. W Crawford, 3 (1 NSM); 3½ mi. N, 1 mi. E Glen, 1 (NSM); 3 mi. NE Glen, 1 (NSM); Glen, 3 (NSM); Agate, 4600 ft., 1. *Stanton County:* 1½ mi. S Pilger, 3; 6 mi. SE Norfolk, 1. *Thomas County:* 1 mi. W Halsey, 2; Halsey, 1 (NSM). *Thurston County:* 1 mi. S Winnebago, 8. *Valley County:* 2 mi. W Ord, 1; 2 mi. S, 4 mi. E Ord, 6. *Washington County:* 1 mi. E Blair, 6; 3 mi. SE Blair, 2; 6 mi. SE Blair, 7; 3 mi. S, 2 mi. E Ft. Calhoun, 1 (NSM). *Wayne County:* ½ mi. W-2½ mi. E Wayne, 3. *Webster County:* 3 mi. S Red Cloud, 2.

SOUTH DAKOTA. *Buffalo County:* 2 mi. S, 3 mi. E Ft. Thompson, 1370 ft., 4. *Clay County:* 2½ mi. N, ½ mi. W Vermillion, 1. *Pennington County:* 2 mi. S, 3 mi. W Scenic, 1. *Stanley County:* 1.2 mi. S, 4 mi. W Ft. Pierre, 1484 ft., 1.

WYOMING. *Albany County:* 27 mi. N, 8 mi. E Laramie, 6420 ft., 2. *Big Horn County:* 7½ mi. E Graybull, 4050 ft., 1; 7 mi. S, ½ mi. E Basin, 3900 ft., 1. *Campbell County:* 4 mi. N, 3 mi. E Rockypoint, 3800 ft., 3; 1⅗ mi. N, ¾ mi. E Rockypoint, 2; Rockypoint, 5; 5 mi. S, 4 mi. W Rockypoint, 1; Ivy Creek, 5 mi. N, 8 mi. W Spotted Horse, 2. *Crook County:* 1½ mi. NW Sundance, 5000 ft., 3. *Fremont County:* 2 mi. N, 3 mi. W Shoshoni, 4650 ft., 1; 3/10 mi. NW Milford, 5357 ft., 1; Milford, 5400 ft., 1.

TABLE 2. CRANIAL MEASUREMENTS OF TWO SUBSPECIES OF REITHRODONTOMYS MEGALOTIS.

NUMBER AVERAGED AND SEX	Greatest length of skull	Zygomatic breadth	Breadth of braincase	Interorbital breadth	Depth of cranium	Length of rostrum	Breadth of rostrum	Length of incisive foramen	Length of palate	Alveolar length of maxillary tooth-row
R. m. dychei, Douglas County, Kansas										
Av. 17 (11♂, 6♀)	20.9	10.5	10.1	3.1	7.9	7.2	3.8	4.3	3.5	3.3
Minimum........	20.4	10.0	9.8	3.0	7.7	6.8	3.6	4.0	3.2	3.1
Maximum........	21.9	10.9	10.3	3.3	8.2	7.9	4.0	4.5	3.9	3.4
Cherry County, Nebraska										
Av. 20 (14♂, 6♀)	21.0	10.9	10.3	3.1	7.9	7.3	3.8	4.4	3.6	3.5
Minimum........	20.4	10.0	9.8	2.9	7.5	6.8	3.5	4.3	3.4	3.2
Maximum........	22.1	11.3	10.7	3.3	8.4	7.8	4.1	4.7	3.9	3.7
R. m. aztecus, San Juan County, New Mexico, and Montezuma County, Colorado										
Av. 10 (6♂, 4♀)	21.5	10.8	10.2	3.1	8.1	7.7	3.7	4.5	3.4	3.5
Minimum........	20.5	10.4	9.9	2.9	7.9	7.2	3.5	3.9	3.1	3.2
Maximum........	22.7	11.1	10.6	3.3	8.4	8.2	3.9	4.8	3.7	3.7

Hot Springs County: 3 mi. N, 10 mi. W Thermopolis, 4900-4950 ft., 7. *Johnson County:* 1 mi. W, 8/10 mi. S Buffalo, 4800 ft., 5; 6½ mi. W, 2 mi. S Buffalo, 5620 ft., 4; 1 mi. WSW Kaycee, 4700 ft., 8. *Laramie County:* Horse Creek, 5000 ft., 3 mi. W Meriden, 1; 1 mi. N, ½ mi. W Pine Bluffs, 5040 ft., 4; 1 mi. S Pine Bluffs, 5100 ft., 1; 2 mi. S Pine Bluffs, 5200 ft., 2. *Natrona County:* 1 mi. NE Casper, 5150 ft., 1; 2¼ mi. W Casper, 5250 ft., 1; 7 mi. S, 2 mi. W Casper, 6370 ft., 1. *Niobrara County:* 2 mi. S, ½ mi. E Lusk, 5000 ft., 1. *Park County:* 4 mi. N Garland, 2; 13 mi. N, 1 mi. E Cody, 5200 ft., 2; 6/10 mi. S, 3 2/10 mi. E Cody, 5020 ft., 1. *Platte County:* 2½ mi. S Chugwater, 5300 ft., 4. *Sheridan County:* 3 mi. WNW Monarch, 3800 ft., 4; 5 mi. NE Clearmont, 3900 ft., 6. *Washakie County:* 1 mi. N, 3 mi. E Tensleep, 4350 ft., 5.

LITERATURE CITED

ALLEN, J. A.
1893. List of mammals collected by Mr. Charles P. Rowley in the San Juan region of Colorado, New Mexico and Utah, with descriptions of new species. Bull. Amer. Mus. Nat. Hist., 5:69-84, April 28.
1895. On the species of the genus Reithrodontomys. Bull. Amer. Mus. Nat. Hist., 7:107-143, May 21.

BENSON, S. B.
1935. The status of Reithrodontomys montanus (Baird). Jour. Mamm., 16:139-142, 1 fig., May 15.

BLAIR, W. F.
1954. Mammals of the Mesquite Plains Biotic District in Texas and Oklahoma, and speciation in the central grasslands. Texas Jour. Sci., 6:235-264, 1 fig., September.

DALQUEST, W. W.
1948. Mammals of Washington. Univ. Kansas Publ., Mus. Nat. Hist., 2:1-444, 140 figs., April 9.

HALL, E. R., and K. R. KELSON
1959. The mammals of North America. Ronald Press, New York, vols. 1:xxx + 1-546 + 79 and 2:viii + 547-1083 + 79, 553 figs., 500 maps, 178 unnumbered text figs., March 31.

HILL, J. E., and C. W. HIBBARD
1943. Ecological differences between two harvest mice (*Reithrodontomys*) in western Kansas. Jour. Mamm., 24:22-25, February 20.

HOFFMEISTER, D. F., and J. E. WARNOCK
1955. The harvest mouse (Reithrodontomys megalotis) in Illinois and its taxonomic status. Trans. Illinois Acad. Sci., 47:161-164, 1 fig.

HOOPER, E. T.
1952. A systematic review of the harvest mice (genus Reithrodontomys) of Latin America. Misc. Publ. Mus. Zool., Univ. Michigan, 77: 1-255, 9 pls., 24 figs., 12 maps, January 16.

HOWELL, A. H.
1914. Revision of the American harvest mice (genus Reithrodontomys). N. Amer. Fauna, 36:1-97, 7 pls., 6 figs., June 5.
1935. The harvest mice of the San Luis Valley, Colorado. Jour. Mamm., 16:143-144, May 15.

LAYNE, J. N.
1959. Growth and development of the eastern harvest mouse, Reithrodontomys humulis. Bull. Florida State Mus., 4:61-82, 5 figs., April 27.

VERTS, B. J.
1960. Ecological notes on *Reithrodontomys megalotis* in Illinois. Nat. Hist. Misc., Chicago Acad. Sci., 174:1-7, 1 fig., July 25.

Transmitted March 30, 1961.

□

28-7578

University of Kansas Publications
Museum of Natural History

Volume 14, No. 3, pp. 29-67, pls. 1 and 2, 3 figs. in text
July 24, 1961

Mammals of Mesa Verde National Park, Colorado

BY

SYDNEY ANDERSON

University of Kansas
Lawrence
1961

UNIVERSITY OF KANSAS PUBLICATIONS
MUSEUM OF NATURAL HISTORY

Institutional libraries interested in publications exchange may obtain this series by addressing the Exchange Librarian, University of Kansas Library, Lawrence, Kansas. Copies for individuals, persons working in a particular field of study, may be obtained by addressing instead the Museum of Natural History, University of Kansas, Lawrence, Kansas. There is no provision for sale of this series by the University Library, which meets institutional requests, or by the Museum of Natural History, which meets the requests of individuals. However, when individuals request copies from the Museum, 25 cents should be included, for each separate number that is 100 pages or more in length, for the purpose of defraying the costs of wrapping and mailing.

* An asterisk designates those numbers of which the Museum's supply (not the Library's supply) is exhausted. Numbers published to date, in this series, are as follows:

Vol. 1. Nos. 1-26 and index. Pp. 1-638, 1946-1950.

*Vol. 2. (Complete) Mammals of Washington. By Walter W. Dalquest. Pp. 1-444, 140 figures in text. April 9, 1948.

Vol. 3. *1. The avifauna of Micronesia, its origin, evolution, and distribution. By Rollin H. Baker. Pp. 1-359, 16 figures in text. June 12, 1951.

*2. A quantitative study of the nocturnal migration of birds. By George H. Lowery, Jr. Pp. 361-472, 47 figures in text. June 29, 1951.

3. Phylogeny of the waxwings and allied birds. By M. Dale Arvey. Pp. 473-530, 49 figures in text, 13 tables. October 10, 1951.

4. Birds from the state of Veracruz, Mexico. By George H. Lowery, Jr., and Walter W. Dalquest. Pp. 531-649, 7 figures in text, 2 tables. October 10, 1951.

Index. Pp. 651-681.

*Vol. 4. (Complete) American weasels. By E. Raymond Hall. Pp. 1-466, 41 plates, 31 figures in text. December 27, 1951.

Vol. 5. Nos. 1-37 and index. Pp. 1-676, 1951-1953.

*Vol. 6. (Complete) Mammals of Utah, *taxonomy and distribution*. By Stephen D. Durrant. Pp. 1-549, 91 figures in text, 30 tables. August 10, 1952.

Vol. 7. *1. Mammals of Kansas. By E. Lendell Cockrum. Pp. 1-303, 73 figures in text, 37 tables. August 25, 1952.

2. Ecology of the opossum on a natural area in northeastern Kansas. By Henry S. Fitch and Lewis L. Sandidge. Pp. 305-338, 5 figures in text. August 24, 1953.

3. The silky pocket mice (Perognathus flavus) of Mexico. By Rollin H. Baker. Pp. 339-347, 1 figure in text. February 15, 1954.

4. North American jumping mice (Genus Zapus). By Philip H. Krutzsch. Pp. 349-472, 47 figures in text, 4 tables. April 21, 1954.

5. Mammals from Southeastern Alaska. By Rollin H. Baker and James S. Findley. Pp. 473-477. April 21, 1954.

6. Distribution of Some Nebraskan Mammals. By J. Knox Jones, Jr. Pp. 479-487. April 21, 1954.

7. Subspeciation in the montane meadow mouse, Microtus montanus, in Wyoming and Colorado. By Sydney Anderson. Pp. 489-506, 2 figures in text. July 23, 1954.

8. A new subspecies of bat (Myotis velifer) from southeastern California and Arizona. By Terry A. Vaughan. Pp. 507-512. July 23, 1954.

9. Mammals of the San Gabriel mountains of California. By Terry A. Vaughan. Pp. 513-582, 1 figure in text, 12 tables. November 15, 1954.

10. A new bat (Genus Pipistrellus) from northeastern Mexico. By Rollin H. Baker. Pp. 583-586. November 15, 1954.

11. A new subspecies of pocket mouse from Kansas. By E. Raymond Hall. Pp. 587-590. November 15, 1954.

12. Geographic variation in the pocket gopher, Cratogeomys castanops, in Coahuila, Mexico. By Robert J. Russell and Rollin H. Baker. Pp. 591-608. March 15, 1955.

13. A new cottontail (Sylvilagus floridanus) from northeastern Mexico. By Rollin H. Baker. Pp. 609-612. April 8, 1955.

14. Taxonomy and distribution of some American shrews. By James S. Findley. Pp. 613-618. June 10, 1955.

15. The pigmy woodrat, Neotoma goldmani, its distribution and systematic position. By Dennis G. Rainey and Rollin H. Baker. Pp. 619-624, 2 figures in text. June 10, 1955.

Index. Pp. 625-651.

(Continued on inside of back cover)

University of Kansas Publications
Museum of Natural History

Volume 14, No. 3, pp. 29-67, pls. 1 and 2, 3 figs. in text
July 24, 1961

Mammals of Mesa Verde National Park, Colorado

BY

SYDNEY ANDERSON

University of Kansas
Lawrence
1961

University of Kansas Publications, Museum of Natural History

Editors: E. Raymond Hall, Chairman, Henry S. Fitch, Robert W. Wilson

Volume 14, No. 3, pp. 29-67, pls. 1 and 2, 3 figs. in text
Published July 24, 1961

University of Kansas
Lawrence, Kansas

PRINTED IN
THE STATE PRINTING PLANT
TOPEKA, KANSAS
1961

28-7577

Mammals of Mesa Verde National Park, Colorado

BY

SYDNEY ANDERSON

INTRODUCTION

A person standing on the North Rim of the Mesa Verde in southwestern Colorado sees a vast green plain sloping away to the south. The plain drops 2000 feet in ten miles. On a clear evening, before the sun reaches the horizon, the rays of the sun are reflected from great sandstone cliffs forming the walls of deep canyons that appear as crooked yellow lines in the distance. Canyon after canyon has cut into the sloping green plain. These canyons are roughly parallel and all open into the canyon of the Mancos River, which forms the southern boundary of the Mesa Verde. If the observer turns to the north he sees the arid Montezuma Valley 2000 feet below. A few green streaks and patches in the brown and barren low country denote streams and irrigated areas. To the northeast beyond the low country the towering peaks of the San Miguel and La Plata mountains rise more than 4000 feet above the vantage point on the North Rim at 8000 feet. To the northwest, in the hazy distance 90 miles away in Utah, lie the isolated heights of the La Sal Mountains, and 70 miles away, the Abajo Mountains (see Fig. 1).

In the thirteenth century, harassed by nomadic tribes and beset by years of drouth, village dwelling Indians left their great cliff dwellings in the myriad canyons of the Mesa Verde, and thus ended a period of 1300 years of occupancy. The story of those 1300 years, unfolded through excavation and study of the dwellings along the cliffs and earlier dwellings on the top of the Mesa, is one of the most fascinating in ancient America. To stop destructive commercial exploitation of the ruins, to preserve them for future generations to study and enjoy, and to make them accessible to the public, more than 51,000 acres, including approximately half of the Mesa, have been set aside as Mesa Verde National Park, established in 1906. The policies of the National Park Service provide protection, not only for the features of major interest in each park, but for other features as well. Thus the policy in Mesa Verde National Park is not only to preserve the many ruins, but also the wildlife and plants.

Five considerations prompted me to undertake a study of the

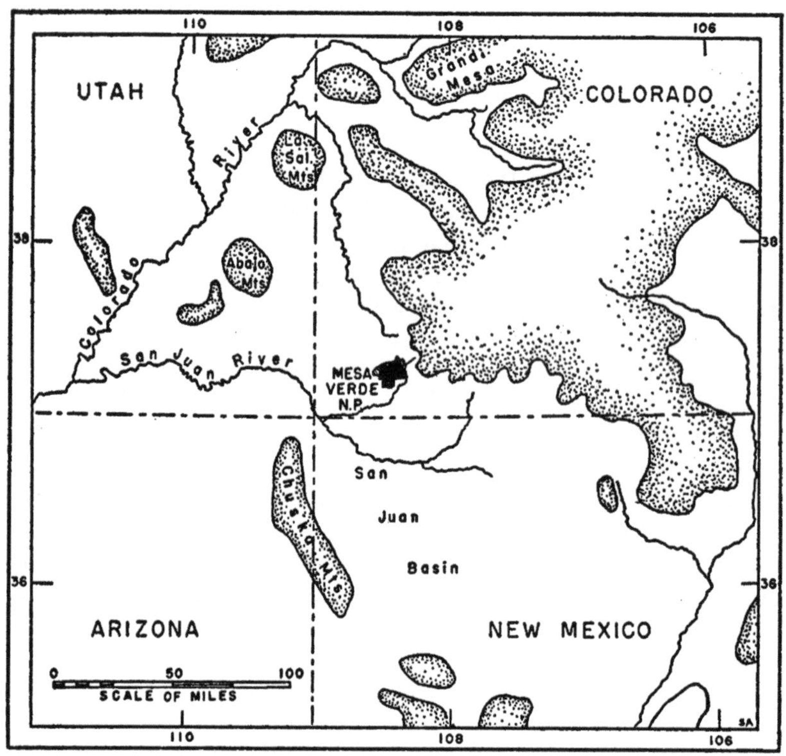

FIG. 1. Map of the "four corners" region showing the position of Mesa Verde National Park (in black) relative to the mass of the Southern Rocky Mountains above 8000 feet elevation (indicated by stippled border) to the northeast in Colorado, and the positions of other isolated mountains in the region.

mammals of Mesa Verde National Park: First, the relative lack of disturbance; second, the interesting position, zoogeographically, of the Mesa that extends as a spur of higher land from the mountains of southwestern Colorado and that is almost surrounded by arid country typical of much of the Southwest; third, the discovery in the Park of *Microtus mexicanus*, a species of the Southwest until then not known from Colorado; fourth, the co-operative spirit of the personnel at the Park when I visited there in 1955; and finally, the possibility of making a contribution not only to our knowledge of mammals, but to the interpretive program of the Park Service.

A Faculty Research Grant from The University of Kansas provided some secretarial help and field expenses for August and early September, 1956, when my wife, Justine, and I spent our vacation enjoyably collecting and studying animals in the Park. The co-operation of Dr. E. Raymond Hall is greatly appreciated; a grant to him from the American Heart Association provided field expenses for work by Mr. J. R. Alcorn, collector for The University of Kansas Museum of Natural History, in 1957.

Mr. Harold R. Shepherd of Mancos, Colorado (Senior Game Biologist for the State of Colorado, Department of Game and Fish), provided advice in the field, helped in identifying plants, and saved specimens of rodents (in 1958 and 1959) taken in his studies of the effect of rodents on browse utilized by deer. Mr. J. D. Hart, Assistant Director of the Department of Game and Fish, issued a letter of authority to collect in Colorado; and Superintendent O. W. Carlson approved my appointment as a collaborator. Mr. "Don" Watson, then Park Archeologist, and Mrs. Jean M. Pinkley, now Park Archeologist, assisted us in 1956, and since then have provided advice and assistance, and have reviewed the manuscript of this report.

Geologically, the Mesa Verde is the northern edge of a Cretaceous, coal-bearing, sandstone deposit called the Mesaverde group, which dips beneath the San Juan Basin of New Mexico. An abrupt retreating escarpment commonly forms on arid plateaus underlain by horizontal rocks of unequal strength, and characterizes the borders of mesas. Such an escarpment forms the North Rim of the Mesa Verde. However, the dip of the rocks has channelled drainage southward and erosion has cut numerous, deep, parallel-sided canyons rather than a simple, retreating escarpment. The Mesa Verde therefore is, technically speaking, a cuesta rather than a mesa. The remnants of the plateau left between the canyons are also (and again incorrectly in the technical sense) called mesas; Chapin Mesa and Wetherill Mesa are examples.

Climatically, the Mesa Verde is arid; precipitation averaged 18.41 inches per year for a period of 37 years. Precipitation may be scattered through the year, and more important, may be erratic from month to month and from year to year. In addition to low precipitation and periods of drouth, a great amount of sunshine, and thin, well-drained soils on all but the more sheltered parts of the Mesa favor vegetation that requires neither great amounts of, nor a continuous supply of, water.

The vegetation of the Mesa is illustrated in Plates 1 and 2, and consists predominantly of pinyon pine, *Pinus edulis* Engelm., and Utah juniper, *Juniperus osteosperma* (Torr.) Little. More sheltered areas along the North Rim and in most of the canyons support scattered small stands of Douglas fir, *Pseudotsuga menziesii*

(Merb.) Franco. These are the "spruce trees" of Spruce Tree Canyon. An occasional ponderosa pine, *Pinus ponderosa* Laws., represents a vestige of more montane species of plants and animals in the Park. The dusky grouse, *Dendragapus obscurus* (Say), occurs along the North Rim in oak-chaparral, and is one of the few montane species of birds; several montane mammals are discussed later. The vegetation of the Mesa Verde has not changed appreciably in the last thousand years. The tree rings of 13 centuries show that Douglas fir has grown essentially as it does now, varying with precipitation from year to year, and periodically suffering from drouth (Schulman, 1946:18). Surface ruins yield mostly pinyon and juniper; cave ruins yield more Douglas fir than surface ruins; and "only rarely does yellow pine [*Pinus ponderosa*] occur in the ruins, indicating that then, as now, this tree grew only in the northern and higher parts of the Mesa Verde, remote from most of the ruins" (Getty, 1935:21).

Not all areas within the Park are undisturbed. The rights of way of roads are kept clear, as are campgrounds and other facilities in the area of headquarters. Part of the Mancos Valley within the Park is privately owned and is still in agricultural use. Cattle from land belonging to the Ute Indians wander into the Park from the Mancos Canyon along the floor of the canyon above the mouth of Weber Canyon. In addition to the pasture near headquarters, Prater Canyon below a fence across the canyon above Middle Well is used to pasture horses used by visitors to the Park and belonging to the pack and saddle concessioner. In 1956, the floor of Long Canyon was grazed by stock belonging to Utes, and horses ranged freely onto Wetherill Mesa as far as the North Rim. Occasionally livestock enter the floor of other canyons, for example Navajo, Soda, Prater, Morfield, and Waters canyons, owing to inadequate fencing, or no fencing.

The first mammals from the Mesa to be preserved for scientific study were seven specimens in the United States National Museum (designated USNM in lists of specimens examined) obtained by Merritt Cary in 1907, and mentioned in his "Biological Survey of Colorado" (Cary, 1911). In 1931 and 1932, R. L. Landberg obtained a few specimens that are in the Denver Museum of Natural History. In 1935, C. W. Quaintance, Lloyd White, Harold P. Pratt, and A. E. Borell prepared specimens, some of which remain in the museum at the Park (all specimens in the museum at the Park are designated by "MV" for Mesa Verde and by their catalogue numbers), and some are in the Museum of Vertebrate Zoology at the University of California at Berkeley (designated "MVZ" in the following accounts). Specimens in The University of Kansas Museum of Natural History are referred to by catalogue numbers only. Specimens prepared by D. Watson bear dates from 1936

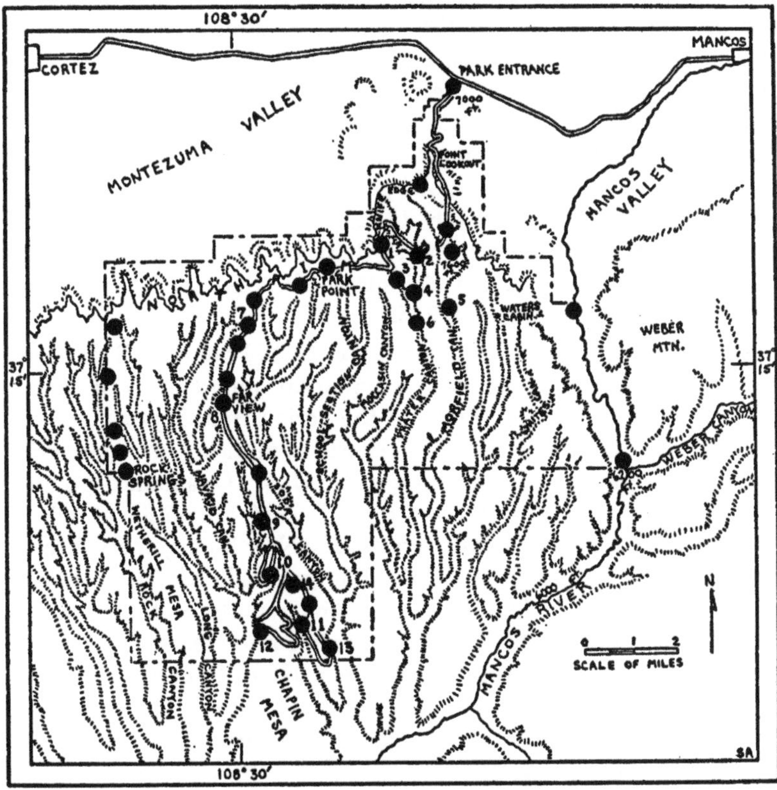

FIG. 2. Map of Mesa Verde National Park and vicinity. The map and this legend provide the names of places mentioned in the following accounts of mammals. Localities from which specimens have been preserved are indicated by dots. Localities within ¼ mile of each other are not indicated by separate dots. Unnumbered dots designate some of the places from which specimens were obtained. The numbered dots are: (1) Prater Grade; (2) Upper Well, Prater Canyon, 7575 ft.; (3) Chickaree Draw, 8200 ft.; (4) ¼ mi. N Middle Well, 7500 ft., Prater Canyon; (5) east side of Morfield Canyon about one mile below the well; (6) Lower Well, Prater Canyon; (7) Sect. 27, head of east fork Navajo Canyon; (8) Far View, designated on various specimens as Far View Ruins, Far View Point, and Far View House, 7700 ft.; (9) localities designated Utility Area, and Well, "Park Well," or "Old Park Well"; (10) Headquarters, including the designations 25 mi. [by road] SW Mancos, Museum, Hospital, head of Spruce Tree Canyon, Spruce Tree House, and Spruce Tree Lodge; (11) Cliff Palace, across the canyon about ¼ mile southwest are Sun Temple and Oak Tree Ruin; (12) Square Tower House; (13) Balcony House; (14) Indian Cornfield, "Cornfield," or "Garden."

until 1955. In 1938, Raymond F. Harlow prepared some specimens; his Student Technician's Report of 7 typescript pages, for July 8 to September 9, 1938, is on file at Mesa Verde National Park. In 1944 and 1945, Dr. D. A. Sutton, then a student at the University of Colorado, collected chipmunks for his own study, and also some other specimens that are in the University of Colorado Museum and the Park Museum. In 1949, Dr. R. B. Finley, then a student at The University of Kansas, collected in and near the Park and obtained a few specimens preserved in The University of Kansas Museum of Natural History. Rodents preserved by Harold R. Shepherd have been mentioned. I have examined 244 specimens that were collected by the above persons. Between August 8 and September 4, 1956, and on July 17, 1960, I collected 216 mammals from Mesa Verde National Park. Between November 3, and 12, 1957, J. R. Alcorn collected 275 mammals from the Mesa. The total of specimens examined is 735.

Written reports by C. W. Quaintance, H. P. Pratt, and R. Harlow have been of considerable use. A typescript report of 13 pages by Wildlife Technician H. P. Pratt for the period from September 9 to October 15, 1935, and monthly reports comprising 40 typescript pages and 4 pages with photographs by C. W. Quaintance for the period from February 18 through July 17, 1935, are on file at offices of Region Four, National Park Service, 180 New Montgomery Street, San Francisco 5, California. Chief Ranger Wade has kindly made available the files in his office, including reports of the Superintendent and reports of the Chief Ranger in earlier years, and Annual or Biennial Animal Census Reports since 1930. Special reports on prairie dogs, porcupines, and deer are in the files. These reports, and random reports that were regarded as reliable, are recorded on card files in both the Chief Ranger's office and Park Archeologist's office. Most of the information reported here on the larger mammals was gleaned from the above sources. A study of population fluctuations in porcupines by Donald A. Spencer and perhaps a study of movements of porcupines by Spencer, Wade and Fitch are to be published elsewhere. Other studies still in progress are mentioned in the following accounts.

ACCOUNTS OF SPECIES

Sorex merriami leucogenys Osgood
Merriam's Shrew

Specimen: MV 7898/507, head of Navajo Canyon (locality No. 7 in Fig. 2), October 21, 1954.

This was the third reported specimen of the rare Merriam's shrew from Colorado (Rodeck and Anderson, 1956:436).

Sorex vagrans obscurus Merriam
Wandering Shrew

Specimens examined.—Total, 8: Morfield Canyon, 7600 ft., 75972, 75973; Upper Well, Prater Canyon, 7575 ft., 69235-69238; ¼ mi. N Middle Well, Prater Canyon, 7500 ft., 69239-69240.

The specimens from Prater Canyon were trapped in the grasses and sedges of the meadow comprising the floor of the canyon. The ground and vegetation were dry at the time of capture, September 2, 3, and 4, 1956. *Microtus montanus* was the only other species taken in the mouse traps in the sedge and grass. Five of the six specimens from Prater Canyon are young, having slightly worn teeth; the sixth is an old adult male the teeth of which are so much worn that only a few traces of the reddish-brown pigment remain. His testes were 5 mm. long. These specimens are from an area of intergradation between *S. v. obscurus* and *S. v. monticola.* The length of the maxillary tooth-row in these six specimens averaged 6.23 (6.1-6.4) millimeters. Comparison with average measurements of 6.6 and 6.8 in samples of *S. v. obscurus,* and of 5.9 in a sample of *S. v. monticola* (Findley, 1955:64, 65) reveals the intermediate size of the specimens from the Mesa Verde. The gap between habitat suitable for *Sorex vagrans* on the Mesa Verde and the nearest record-station for *S. v. monticola* to the south and west in the Chuska Mountains is wider than the gap between the Mesa Verde and the nearest record-station for *S. v. obscurus* to the north and east, one mile west of Mancos, 75971, 7000 feet, or at Silverton. On geographic grounds the specimens from the Mesa Verde are referred to *S. v. obscurus.* The two specimens from Morfield Canyon were trapped on November 4, 1957, and are grayish above and silvery below. Their pelage contrasts markedly with the dorsally brownish and ventrally buffy pelage of the September-taken specimens from Prater Canyon.

Myotis californicus stephensi Dalquest
California Myotis

Specimens examined.—Total, 3: Rock Springs, 7400 ft., 69243, 69246, August 21 and 22, 1956; 4505 Denver Museum, within the Park (exact locality not recorded), R. L. Landberg, July 27, 1931.

The specimens from Rock Springs were an adult male and a non-pregnant adult female. Both were shot over the road in pinyon and juniper. The specimens are referred to *M. c. stephensi* on account of their paleness, *stephensi* being paler than *M. c. californicus* from east of Mesa Verde in Colorado.

Myotis evotis evotis (H. Allen)
Long-eared Myotis

Specimens examined.—Total, 4: Chickaree Draw, Prater Canyon, 8200 ft., MV 7841/507, probably in the summer of 1935; Rock Springs, 7400 ft., 69241, August 23, 1956, and 69249, August 18, 1956; Museum, Headquarters, 6950 ft., 69251, August 24, 1956.

An adult male (69241) was taken in a Japanese mist net stretched fifteen feet across a dirt road where it entered the stand of pinyon and juniper at the south edge of the burn on Wetherill Mesa between 7:20 and 8:30 p. m.; at the same place and time I captured five other bats of four species: *Myotis thysanodes, Myotis subulatus, Eptesicus fuscus,* and *Plecotus townsendii.* A piece of mist net attached to an aluminum hoop-net two and one half feet in diameter was used to good advantage in capturing bats rebounding from the larger mist net, and in frightening bats into the larger net when they approached closely. An adult male (69249) was shot at 7:20 p. m. while flying six to eight feet from the ground between pinyon trees up to 20 feet high; the air temperature was 70° F. A female (69251) was found seemingly exhausted on the floor in the museum at Park Headquarters in the daytime, and was immature as indicated by small size, open basicranial sutures, unworn teeth, weakly ossified zygoma, and open epiphyseal sutures of phalanges.

Myotis subulatus melanorhinus (Merriam)
Small-footed Myotis

Specimens examined.—Total, 8: Rock Springs, 7400 ft., 69242, 69244, 69245, 69247, 69248, August 21 to 23, 1956; Hospital, Park Headquarters, MV 7886/507, ♂, July 12, 1939; Headquarters, MV 7877/507, ♀, August 30, 1938; 4504 Denver Museum, within the Park (exact locality not recorded), R. L. Landberg, July 27, 1931.

The specimens from Rock Springs are two adult males that were shot, and one adult male, one adult female, and one young male that were netted at the place described in the account of *Myotis evotis*. The three adult males are near the average color of *M. s. melanorhinus*, and distinctly darker than the *Myotis californicus* from the Mesa Verde. In the female the pelage is paler and brighter, and the ears and membranes are darker, than in *M. californicus*.

Myotis thysanodes thysanodes Miller
Fringed Myotis

Specimen: Rock Springs, 7400 ft., 69250, ad. ♀, August 23, 1956; taken in net as noted in account of *Myotis evotis*.

Myotis volans interior Miller
Long-legged Myotis

Specimen: Rock Springs, 7400 ft., 69252, ad. ♀, August 21, 1956; shot over road.

Eptesicus fuscus pallidus Young
Big Brown Bat

Specimen: Rock Springs, 7400 ft., 69253, ad. ♀, August 23, 1956; taken in net as noted in account of *Myotis evotis*.

Plecotus townsendii pallescens (Miller)
Townsend's Big-eared Bat

Specimens examined.—Total, 5: Rock Springs, 7400 ft., 69254, ad. ♀, nonpregnant, August 23, 1956; Square Tower House, 6700 ft., 69255-69258, March, 1955.

The specimen from Rock Springs was taken in a net as noted in the account of *Myotis evotis*. The specimens from Square Tower House were obtained by D. Watson in a dimly lighted chamber formed by fracture in the rocks at the bottom of the canyon wall, above the talus slope. The bats were suspended from the wall of the chamber, which was at least six feet wide and fifteen feet long.

Tadarida brasiliensis mexicana (Saussure)
Brazilian Free-tailed Bat

Specimens examined.—Total, 2: Cliff Palace, 6800 ft., MV 7862/507 and 7863/507, males, both collected by A. E. Borell, on August 23, 1936.

Lepus californicus texianus Waterhouse
Black-tailed Jackrabbit

The black-tailed jackrabbit inhabits the Montezuma Valley to the north of the Mesa Verde and the Mancos Valley to the northeast, and has been seen occasionally on the top of the Mesa according to reports with date and locality noted in the files at the Park for the years 1941, 1942, 1947, 1948, 1950, and 1951. In 1942 four observations were made, in 1950 and 1951 two observations were recorded each year, and in other years only one observation was recorded each year. Nine observations are for Chapin Mesa south of Far View; only two observations are for higher elevations on the North Rim.

Sylvilagus audubonii warreni Nelson
Desert Cottontail

Specimens examined.—Total, 2: Head of Prater Canyon, MV 7850/507; Far View Ruins, 75974, ad. ♀, nonpregnant, November 8, 1957.

One specimen was shot, while it was sitting near a pile of logs, by J. R. Alcorn by means of a bow and arrow. Although S. *audubonii* occurs on the Mesa along with S. *nuttallii*, S. *audubonii* is the species of the lowlands throughout the western United States at the latitude of Mesa Verde National Park. For example, S. *a. warreni* (69260) but not S. *n. pinetis* was obtained along the east side of the Mancos River at 6200 feet elevation (less than 50 yards outside the Park) and the same was true at the same elevation at a place 4¼ mi. N of the Park (No. 69259 from 2 mi. E Cortez).

Sylvilagus nuttallii pinetis (J. A. Allen)
Nuttall's Cottontail

Specimens examined.—Total, 3: ad. ♂, 69263, skull only, dead on road, 1¼ mi. N Park Headquarters, 7275 feet, August 9, 1956; ad. ♀, 69261, no embryos, dead on road, ¾ mi. S and 1¼ mi. W Park Point, 8000 ft., August 8, 1956; ad. ♂, 69262, shot in brushy area on the burn on Wetherill Mesa 2 mi. NNW Rock Springs, 7900 ft., August 24, 1956.

Nuttall's cottontail in Colorado is in general the cottontail of the highlands, and the three localities just mentioned are on the top of the Mesa Verde.

Sciurus aberti mimus Merriam
Abert's Squirrel

Specimens examined.—Total, 2: ♂, MV 7872/507, prepared by D. Watson, killed by a car "near" the Park Well on September 24, 1937; ♀ (an unnumbered cased skin only), found dead "near" the Park Well on June 21, 1937.

Since 1934 these squirrels have been observed and recorded each year except in 1938, 1943, 1947, 1953, 1957, and 1958. The 77 reported observations can be grouped as follows: 11 from within a mile of the entrance to the Park, 14 from the North Rim or higher parts of canyons adjacent to it, 38 from Chapin Mesa south of Far View, and 14 not classifiable. The large number of observations on Chapin Mesa, chiefly in the vicinity of Park Headquarters, indicates the presence of more observers rather than more squirrels in this area.

Tamiasciurus hudsonicus fremonti (Audubon and Bachman)
Red Squirrel

Specimens examined.—Total, 2: MV 7843/507, Chickaree Draw, Prater Canyon, 1935, C. W. Quaintance and Lloyd White; ♀, 69264, no embryos, ¼ mi. NNW Middle Well, Prater Canyon, 7600 ft., August 31, 1956.

Red squirrels, or chickarees as they are called in Colorado, are known from only one place on the Mesa Verde, a side canyon on the west side of Prater Canyon above Middle Well. This side canyon has been named Chickaree Draw by C. W. Quaintance, who, with Lloyd White, studied the chickaree there in 1935. Quaintance reported the small colony at 7800 feet elevation in Douglas fir beneath which were found piles of cones from which the seeds had been eaten by the chickarees. On May 29, 1935, White observed a chickaree eating green oak leaves. On June 3, 1935, a nest was found in an old hollow snag up under the rim rock; there were four young squirrels in the nest. At least one nest was in a juniper and was composed mostly of oak leaves and grass. One nest twenty-five feet from the ground in a Douglas fir was composed of oak leaves and finely shredded cedar bark. In August, 1956, I found these squirrels in the same area and I shot one specimen. Other chickarees were seen and heard and the characteristic piles of parts of Douglas fir cones still attest to their presence. On September 1, 1953, D. Watson observed a pair of chickarees in Prater Canyon. The only other specific record in the files at the Park is of two seen in a branch of Soda Canyon in late 1956. Jean Pinkley tells me that chickarees have been observed in 1958 and 1959 at several other localities from Prater Canyon to the hill at the head of Navajo Canyon. The extent to which increased observations indicate an increase in number of chickarees is uncertain, since the amounts of time spent in the field and the percentage of observations recorded are not known.

Marmota flaviventris luteola A. H. Howell
Yellow-bellied Marmot

Records are available of observations at 14 different places in the Park and in 19 different years between 1930 and 1960. Approximately two-thirds of the observations have been on Prater Grade or in upper Prater Canyon or in upper Morfield Canyon. On the morning of August 24, 1956, Harold Shepherd and I heard the whistle of an animal that he was certain was a marmot, 2 mi. NNW of Rock Springs at the west rim of Wetherill Mesa. Mr. Shepherd has worked in areas occupied by marmots for years in southwestern

Colorado. Wetherill Mesa is the locality farthest west in the Park where marmots are known to occur. They occur as far south as Cliff Palace.

Cynomys gunnisoni zuniensis Hollister
Gunnison's Prairie Dog

Specimens examined.—Total, 3: MV 7836/507, Prater Canyon, 7600 ft., C. W. Quaintance and L. White, May 24, 1935; ♀, MV 7847/507, head of Prater Canyon, June 13, 1935, C. W. Quaintance (the skin is on display); MV 7887/507, Prater Canyon, September 1, 1939.

C. W. Quaintance in reports on the results of his work in 1935 included the following information:

On February 20 in Prater Canyon Ranger Markley noticed that prairie dogs were active although about three feet of snow lay on the ground. Between April 15 and May 15 approximately 500 prairie dogs were in Prater Canyon above Lower Well; through field glasses 350 were counted. Young were first noted in Prater Canyon on July 12. Quaintance and Lloyd White had under observation two bulky nests of the red-tailed hawk in the tops of tall Douglas firs in side draws of Prater Canyon. Quaintance found near the rimrock a quarter of a mile from the prairie-dog town the skeletons of two prairie dogs between a sliver of a dead pinyon branch and the branch itself. Another skeleton lay on a dead limb fifteen feet from the ground. A red-tailed hawk once was observed to swoop down, seize a prairie dog and fly down the canyon. The four colonies found in the Park were in Prater Canyon, in Morfield Canyon, in the east fork of School Section Canyon, and in Whites Canyon. The last two were smaller colonies than the first two.

Prairie dogs were observed away from these colonies. On June 20 a young prairie dog ran into a culvert on the Knife Edge Section of the road. Others were observed on the north side of the road, at the head of the east prong of School Section Canyon, on the road west of Park Point, and on the road at the head of Long Canyon five miles from the nearest known colony in the Park. Possibly this last individual came from the Montezuma Valley north of the Park. Mr. Prater, after whom Prater Canyon is named, homesteaded on the Mesa Verde in 1899. He informed Quaintance that prairie dogs were present in Morfield Canyon prior to 1900 but were not in Prater Canyon in 1899. Prater said he drowned out a few that came into Prater Canyon before 1914. In 1942, Chief Ranger Faha wrote in his Annual Animal Census Report that he had interviewed an old time resident (name not noted) who stated that prairie dogs were not present on the Mesa Verde until about 1905 or 1906 and that Helen Morfield, the daughter of Judge Morfield who homesteaded in Morfield Canyon, brought the first prairie dogs on the Mesa Verde. Estimates of the prairie-dog population in the Annual Animal Census Reports for 1935 through 1941 were: 1935—800, 1936—650, 1937—650, 1938—650, 1939—no report, 1940—1500 and increasing, 1941—slight decrease. After 1942 more adequate records were kept by Chief Ranger Wade and other Park Service personnel.

On August 9, 1943, occupied burrows of prairie dogs were found to be thinly scattered down Prater Canyon from the head of the canyon at the Maintenance Camp to a point about one hundred feet below the lower well. The largest concentration was in the vicinity of the upper well near Prater's Cabin. Little new digging that would indicate a spreading population was noticed. Seemingly desirable, but unoccupied, habitat extended at least two miles south of the inhabited area. In Morfield Canyon, burrows were found from a point one hundred yards north of the fence at the south boundary of Section 17, south for a mile and one-half to a point one-third of a mile into Section 29. The greatest concentration was in the vicinity of Morfield Well. South of this point the burrows were found only along the narrow dry sides of the canyon and in sage-covered areas at slightly higher elevations than the rest of the floor of the canyon. Seemingly desirable habitat extended at least three miles to the south and one mile to the north of the occupied area. The report of the study in 1943 concluded with the statement that artificial control by poisoning would be unwise and unnecessary. Requests were being made at that time to exterminate prairie dogs in the Park on the basis of the unproved assumption that prairie dogs move from the Park to surrounding range land where extermination was then being attempted by poisoning.

On August 10, 1944, no occupied burrows were found in Whites Canyon or the east fork of School Section Canyon. A heavy rain on August 9 made accurate count of occupied burrows possible. In Prater Canyon the occupied area extended 200 feet south of the area occupied in 1943. In Morfield Canyon no change had occurred. North of the fence in Morfield Canyon 130 occupied burrows were counted. More than one hole, if judged to be part of the same burrow system, were counted as one. The vegetation within the colony had continued to improve in spite of the large population of prairie dogs.

On August 8 and 14, 1945, although a careful search was made, the only prairie dogs found in Prater Canyon were living in one burrow fifty yards from the Maintenance Camp. In Morfield Canyon the colony had decreased. Occupied burrows were found on the west side of the canyon near the fence and above the well (17 burrows), and below the well on the west side (estimated 30 burrows). The total population in both canyons was estimated to be 100, compared with 800 in the preceding year. The ground-water table was thought to be rising, and vegetation was increasing.

On August 12, 1946, two prairie dogs were observed in Prater Canyon, one near the Maintenance Camp, and the other a mile to the south. In Morfield Canyon 18 occupied burrows were found north of the fence and 36 below the well, in the same two areas occupied in 1945.

On August 12, 1947, two animals were seen at one of the localities occupied a year earlier in Prater Canyon, and three burrows were occupied. In Morfield Canyon 119 occupied burrows were counted. At least 12 dens occupied by badgers were present in 1946, and four in 1947.

On August 9, 1948, no evidence of living prairie dogs was found in Prater Canyon. In Morfield Canyon 45 burrows were counted north of the fence. The grass had been increasing in abundance for several years.

On August 18, 1949, no evidence of living prairie dogs was found in either canyon. In 1951 five prairie dogs were said to have been seen in Prater Canyon in June and July. No other observations have been recorded.

On June 22, 1956, 13 pups and 7 adult prairie dogs were released in an enclosure in Morfield Canyon. Periodic inspections in the summer revealed that the colony was surviving and healthy. By the following spring no prairie dogs remained. Another reintroduction is planned this year (1960).

Both the history of the prairie dogs and the history of the viewpoint of people toward them are interesting. Individual views have ranged from a desire to exterminate all the prairie dogs to a desire to leave them undisturbed by man.

In review: The early history of prairie dogs on the Mesa Verde is not well documented but reports are available of the absence of prairie dogs before settlement by white men, and of introductions of prairie dogs. Other reports indicate that prairie dogs have been observed far from established colonies; therefore natural invasion may account for the establishment of prairie dogs on the Mesa.

Grazing of moderate to heavy intensity by livestock continued in Morfield Canyon until 1941. Cessation of grazing and above average precipitation were accompanied by increased growth of vegetation in the colonies of prairie dogs. Mr. Wade has suggested that flooding of burrows by ground water drove prairie dogs from some lower parts of the floors of the canyons, and that increased vegetation favored predators, primarily badgers and coyotes, which further reduced the population. The abruptness of the decline, especially in Prater Canyon, is consistent with the theory that some epidemic disease occurred. This possibility was considered at the time of the decline, and a Mobile Laboratory of the United States Public Health Service spent from June 5 to June 25, 1947, in the Park collecting rodents and their fleas for study. The primary concern was plague, which had been detected in neighboring states. No evidence of plague or of tularemia was reported after study of 494 small rodents obtained from 13 localities in the Park. Only six prairie dogs (all from Morfield Canyon) were studied. The negative report does not prove that tularemia or some other disease was not a factor in the decimation of the colony in Prater Canyon the year before.

If prairie dogs were able to survive primarily because of overgrazing by domestic animals, future introductions may fail. If disease was the major factor in their disappearance, reintroductions may succeed.

Spermophilus lateralis lateralis (Say)
Golden-mantled Ground Squirrel

Specimens examined.—Total, 10: highway at School Section Canyon, MV 7894/507; Sect. 27, head of east fork of Navajo Canyon, 7900 ft., 69265; and Prater Canyon, 7600 to 7800 ft,. MV 7835/507, 7837/507, 7846/507, 7874/507, 7875/507, MVZ 74411-74413.

In 1956, I observed S. *lateralis* ½ mi. W of Park Point, ¾ mi. WSW Park Point, in the public campground at Park Headquarters, at the lower well in Prater Canyon, and at two other places on the North Rim. Other observations on file were made at Prater Grade, Park Point, "D" cut (on North Rim 1 mi. WSW Park Point), and Morfield Canyon. A juvenile was noted at Park Point on June 28, 1952, by Jean Pinkley, and five young were seen together at "D" cut on July 3, 1935. The earliest observation, also recorded by Jean Pinkley, was on February 1, 1947. All of the localities with the exception of Park Headquarters are above 7500 feet, and most of the localities are in vegetation that is predominantly oak-brush.

Spermophilus variegatus grammurus (Say)
Rock Squirrel

Specimens examined.—Total, 6: Head of Prater Canyon, MV 7876/507; Chickaree Draw, Prater Canyon, MV 7843/507, 7844/507; Headquarters Area, MV 7888/507; Ruins Road ½ mi. NE of Cliff Palace, MV 7893/507; and Spruce Tree House, 4334 in Denver Museum.

Specimen number 7893/507 had 360 Purshia seeds in its cheek-pouches according to a note on the label. On July 18, 1960, I found a young male rock squirrel dead on the road a mile north of headquarters that had 234 pinyon seeds in its cheek-pouches. Young, recorded as "half-grown," have been observed in May and July. The first appearance may be as early as January. In 1950, D. Watson thought that they did not hibernate, except for a few days when the weather was stormy. I observed a rock squirrel in August in the public campground at Park Headquarters sitting on its haunches on a branch of a juniper some twelve feet from the ground and eating an object held in its forefeet. The rock squirrel ranges throughout the Park in all habitats.

Eutamias minimus operarius Merriam
Least Chipmunk

Specimens examined.—Total, 17: North Rim above Morfield Canyon, MV 7856/507; Morfield Canyon, 7600 ft. (obtained on Nov. 4, 1957), 75976; Middle Well in Prater Canyon, 7500 ft., MV 7855/507; Prater Canyon, 7600 ft., MVZ 74414; Park Point, 8525 ft., 69267-69270; ¼ mi. S, ¾ mi. W Park Point, 8300 ft., 69271-69272; Sect. 27, head of east fork of Navajo Canyon, 7900 ft., 69273; Far View Ruins, 7700 ft., 69274-69275, and two uncatalogued specimens in preservative; 3 mi. N Rock Springs, 8200 ft., 69276-69277.

Five of the fourteen specimens of known sex are females, all of which were taken in August and September, and none of which is recorded as having contained embryos. The skulls of the eight August-taken specimens also suggest that young are born in late spring or early summer: the largest skull had well-worn teeth that might indicate an age of more than one year; four others had complete adult dentitions that were barely worn; and three had not yet acquired complete adult dentitions.

The records of *E. minimus*, like those of *Spermophilus lateralis*, indicate greatest abundance in the higher parts of the Mesa Verde and in areas of predominantly brushy vegetation.

Eutamias quadrivittatus hopiensis Merriam
Colorado Chipmunk

Specimens examined.—Total, 13: Prater Canyon, 7600 ft., MV 7838/507; Lower Well, Prater Canyon, 69278; Park Headquarters, MV 7889/507; near the old Park Well, 7300 ft., 5468 in Univ. of Colorado collection; Utility Area,

5469 and 5470 in Univ. of Colorado collection; Spruce Tree House, 4352-4355 in Denver Museum; Mesa Verde, 25 mi. [by road] SW Mancos, 149080-149081 USNM; Square Tower House, 7000 ft., 5467 in Univ. of Colorado collection.

Although both species occur in some of the same areas, *E. q. hopiensis* is more abundant than is *E. minimus* in stands of pinyon and juniper, along cliffs, and at low elevations. (A specimen of *hopiensis*, MV 7849/507, from 3 mi. S of the Park boundary where the 6000 foot countour line cross the Mancos River is indicative of the occurrence at low elevations.)

Thomomys bottae aureus J. A. Allen
Botta's Pocket Gopher

Specimens examined.—Total, 35: Prater Canyon, 7600 ft., 74408-74410 MVZ; Upper Well, Prater Canyon, 7575 ft., 69279; ¼ mi. N Middle Well, Prater Canyon, 7500 ft., 69280; Middle Well, Prater Canyon, 7500 ft., 69281-69285, 75977; Morfield Canyon, 7600 ft., 75978; ¾ mi. S, 1¾ mi. W Park Point, 8000 ft., 69286-69288; 1¼ mi. S, 1¾ mi. W Park Point, 8000 ft., 69289; 1½ mi. S, 2 mi. W Park Point, 8075 ft., 69290; Sect. 27, head of east fork Navajo Canyon, 7900 ft., 69291-69292; ½ mi. N Far View Ruins, 7825 ft., 69293; Far View Ruins, 7700 ft., 69294, MV 7852/507, 7853/507; 3 mi. N Rock Springs, 8200 ft., 69295-69298; 2½ mi. N, ½ mi. W Rock Springs, 8100 ft., 69299-69301; 2 mi. N, ¼ mi. W Rock Springs, 69302-69303; 1 mi. NNW Rock Springs, 69304; ½ mi. NNW Rock Springs, 69305; Mesa Verde, northern end, 8100 ft., 149087 USNM.

The pocket gophers of the Mesa Verde and vicinity are of one species, *Thomomys bottae*. The distribution and variation of this species in Colorado have been studied recently by Youngman (1958) who referred all specimens from the Mesa Verde to *T. b. aureus*. He noted that some specimens have dark diffuse dorsal stripes that are wide in specimens from the Mancos River Valley. The generally darker color of the specimens from the Mancos Valley as compared with that of specimens from on the Mesa was noticed in the field, and is another example of the local variability of pocket gophers. The nine specimens listed by Youngman (1958:372) as from "Mesa Verde National Park," Mancos River, 6200 ft., are not here listed among "specimens examined" because possibly some, or all, of the nine were trapped on the east side of the River and therefore outside the Park. None was, however, farther than 30 yards east of the Park.

In the Park, pocket gophers occur both on mesa tops and in canyons. Most of the localities listed above and others at which mounds were seen are areas of disturbance such as the old burn on Wetherill Mesa, the rights of way for roads, the river valley, and the grazed floor of Prater Canyon. Little evidence of pocket gophers was found on unusually rocky slopes, steep slopes, or in stands of pinyon and juniper or in relatively pure stands of oak-brush. In addition to workability of the soil, the presence of herbaceous plants, many of them weedy annuals, is probably the most important factor governing the success of pocket gophers in a local area. No female was recorded to have contained embryos, but two had enlarged uteri or placental scars. This fact and the capture of nine half-grown individuals indicate breeding prior to late August when most specimens were trapped.

Dipodomys ordii longipes (Merriam)
Ord's Kangaroo Rat

Kangaroo rats have been seen crossing the highway in the Park less than one mile from the Park entrance by Jean Pinkley.

Castor canadensis concisor Warren and Hall
Beaver

In 1935 Quaintance and White spent June 16 to June 20 in the Mancos River Bottoms at the mouth of Weber Canyon, looking for sign of fresh beaver work. They found none. Annual Animal Census Reports include the following information based on patrols along the Mancos River at the east boundary of the Park: 1937—estimate 4 beaver present, 1938—8, 1941—numerous bank burrows, 1942—uncommon, 1944—uncommon, 1945—most concentrated at southeast corner, 1946—runs and two small dams seen (flood had washed out larger dams), 1947—only in 1½ miles north of boundary with Ute Reservation, 1949—two separate colonies (each with dams and one with a large house), 1950—none, owing to drouth and diversion of water upstream completely drying the river at times, 1951—none, 1953—present, 1955—present. On the Mancos River, 6200 ft., in late August, 1956, sign of beaver was abundant, numerous trees had been cut but none within a week, and a bank den was found on the west side of the river extending back 50 feet from the stream and caved in at three places. In 1959 dens were still present.

Reithrodontomys megalotis aztecus J. A. Allen
Western Harvest Mouse

Specimens examined.—Total, 38: North end Mesa Verde National Park, 7000 ft., 75984-75986; Park Point, 8525 ft., 69316-69317; Far View Ruins, 7700 ft., 69318-69319, 79220, MV 7897/507, and 23 uncatalogued specimens in preservative; 3 mi. N Rock Springs, 8200 ft., 69320-69321; 2 mi. NNW Rock Springs, 7900 ft., 69322-69323; 1 mi. NNW Rock Springs, 7600 ft., 69324; ½ mi. NNW Rock Springs, 7500 ft., 69325.

The specimen listed last (69325) was an adult male recovered from the stomach of a small (snout-vent length 334 mm., wt. 26.0 gms.) *Crotalus viridus* that was trapped in a Museum Special mouse-trap on a rocky slope mostly barren of vegetation. The availability of samples taken in August (by Anderson in 1956), in September (by Shepherd in 1958), and in November (by Alcorn in 1957) makes the following comparison of age and reproductive condition possible. The sample from November includes some specimens from outside the Park as follows: 1 mi. W Mancos, Colorado, 75979-75983, and 2 mi. N La Plata [not shown on Fig. 2], San Juan County, New Mexico, some 18 miles southeast of the Park, 75987-76000. The data shown in Figure 3 indicate that females are pregnant at least from in August into November. A smaller percentage of females was pregnant in November than in August or September. The fact that all females more than 130 mm. long were pregnant in September suggests an autumnal peak in breeding activity. A change in the ratio of small individuals (less than 130 mm. in length) to large individuals (130 mm. or more in length) is indicative of a sustained breeding period throughout the time shown. In August the ratio was 1 to 2.3, in September the ratio was 1 to 1.2, and the ratio was 1 to 0.7 in November. The western harvest mouse is found usually in grassy areas.

Peromyscus boylii rowleyi (J. A. Allen)
Brush Mouse

Specimens examined.—Total, 14: North end Mesa Verde National Park, 7000 ft., 76002-76003; Far View House, 7700 ft., MV 7851/507, 7854/507; Far View Point, 5 uncatalogued specimens in preservative; ½ mi. N Spruce Tree Lodge, 34742; 25 mi. [by road] SW Mancos, 149094 and 149096 USNM; Oak Tree Ruin, 6700 ft., MV 7870/507; and Cliff Palace, 6800 ft., MV 7864/507.

The specimens were taken in August, September, and November. One adult female trapped on September 10, 1958, had six embryos.

Peromyscus crinitus auripectus (J. A. Allen)
Canyon Mouse

Specimens examined.—Total, 3: Mesa Verde [Spruce Tree Cliff Ruins], 149095 USNM; Balcony House, MV 7865/507, 7866/507.

Peromyscus maniculatus rufinus (Merriam)
Deer Mouse

Specimens examined.—Total, 396: North end Mesa Verde National Park, 7000 ft., 76004-76100; Prater Canyon, 7600 ft., 76101-76144, MV 7839/507, 7840/507; Upper Well, Prater Canyon, 7575 ft., 69328-69329; Morfield Canyon, 7600 ft., 76145-76184; Park Point, 8525 ft., 69330-69342, 69344-69360; 1½ mi. E Waters Cabin, 6400 ft. (labels on some specimens read "West Bank

Mancos River, Northeast side Mesa Verde National Park"), 69361-69376, 76185-76204; Sect. 27, head of east fork Navajo Canyon, 7900 ft., 69377-69380, 69422-69426; 3 mi. N Rock Springs, 8200 ft., 69403-69410; 2 mi. NNW Rock Springs, 7900 ft., 69411-69412; 1 mi. NNW Rock Springs, 7600 ft., 69413-69418; ⅛ mi. NNW Rock Springs, 7500 ft., 69419-69421; Far View Ruins, 7700 ft., 69386-69402; Far View Point, 76530-76531, 79221 and 90 uncatalogued specimens in preservative; Mancos River, 6200 ft., 69382-69385; back of Park Museum, 6930 ft., MV 7857/507; Mesa Verde, 25 mi. [by road] SW Mancos, 149093 USNM; Cornfield, MV 7878/507.

The most abundant mammal is the ubiquitous deer mouse. Series of specimens taken in August (by Anderson in 1956), in September (by Shepherd in 1958 and 1959), and in November (by Alcorn in 1957) make possible the following comparisons of age, reproductive conditions, and molts.

The specimens obtained in August and November were placed in five categories according to age (as judged by wear on the teeth). These categories correspond in general to those used by Hoffmeister (1951:1) in studies of *Peromyscus truei*. From his descriptions I judge that wear in *Peromyscus maniculatus* differs from wear in *Peromyscus truei* in that the last upper molar is not worn smooth before appreciable wear appears on the first two molars, and the lingual and labial cusps wear more nearly concurrently. The five categories differ as follows: category 1, last upper molar in process of erupting, showing no wear; category 2, some wear apparent on all teeth, but most cusps little worn; category 3, greater wear on all teeth, lingual cusps becoming rounded or flattened; category 4, lingual cusps worn smooth, labial cusps show considerable wear; category 5, all cusps worn smooth. The condition of the pelage was noted for each prepared skin. Hoffmeister (*op. cit.*: 4) summarized changes in pelage that he observed in *Peromyscus truei*, and he summarized earlier work by Collins with *Peromyscus maniculatus*. In *P. maniculatus* a grayish juvenal pelage is replaced by a postjuvenal pelage in which the hairs are longer and have longer, pale, terminal or subterminal bands giving a paler and more buffy or ochraceous hue to the dorsal pelage. The postjuvenal pelage is replaced by an adult pelage that is either brighter or, in some cases, is not distinguishable with certainty from the postjuvenal pelage. Not only is the juvenal pelage distinguishable from the postjuvenal pelage, but the sequence of ingrowth of postjuvenal pelage follows a regular pattern that is usually different from that of subsequent molts. The loss of juvenal hair is less readily observed than the ingrowth of new postjuvenal hair on account of the greater time required for the growth of any individual hair than for the sudden loss of a hair.

Molt was observed in some individuals no longer having juvenal pelage; some new pelage was observed on the skins of seven mice collected in August. Each of these was in category 4 or 5 and probably had been born in the previous calendar year. These seven molting individuals make up nearly 17 per cent of 42 individuals that had completed the juvenal to postjuvenal molt. In November, 80 per cent of individuals (92 of 115) that had previously obtained their postjuvenal or adult pelage were molting. These mice were in age-categories 3, 4, and 5. Some of the individuals in category 3 were developing new hair beneath a relatively unworn bright pelage that I judge to be an adult pelage rather than a postjuvenal pelage. If this judgment be correct and if the relatively unworn dentition (category 3) means that these animals are young of the year, we must conclude that individuals born in early summer may molt from juvenal to postjuvenal, then to adult pelage, and finally in the autumn into another adult pelage. Other individuals, six in number and of categories 2 and 3, are simultaneously completing the juvenal to postjuvenal molt and beginning the postjuvenal to adult molt. The juvenal to postjuvenal molt begins, as has been described by various authors, along the lateral line and proceeds dorsally and ventrally and anteriorly and posteriorly, and the last patch to lose the gray juvenal color is the top of head and nape, or less frequently the rump. In some individuals a gray patch on the nape remained but emerging hair was not apparent; perhaps the molt had been halted just prior to completion. The progressing band of emerging hair is narrow in most specimens but in some up to one-fifth of the circumference of the body has hair at the same degree of emergence. Subsequent molts, both from postjuvenal to adult pelage and between adult pelages, are less regular in point, or points, of origin, width of progressing molt, and amount of surface molting at one time. Half or more of the dorsum is oftentimes involved in the same stage of molt at once. In some specimens the molt begins along the lateral line, and in others in several centers on the sides. In some skins distinct lines of molt are visible without parting the hair, and in some others the molt is patchy in appearance. Growth of new hair is apparent at various times of the year as a result of injury such as that caused by bot fly larvae, cuts, scratches, or bites of other mice. Abrasion, wear, irritation by ectoparasites, and other kinds of injury to the skin may play a part in the development of a patchy molt. Both breeding and molting are sources of considerable stress, and the delay of the peak of molting activity until November when breeding activity has decreased seems of benefit to

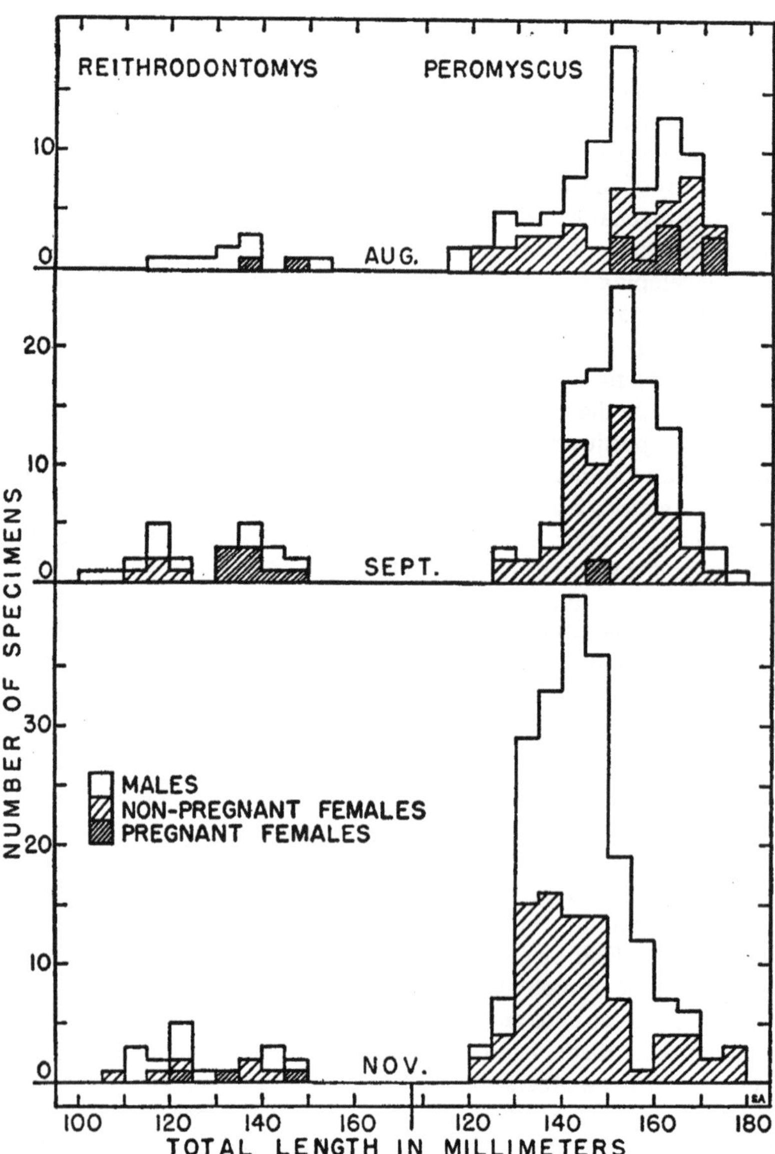

FIG. 3. Frequency distributions, according to size, of *Reithrodontomys megalotis* and *Peromyscus maniculatus* in three samples taken in August, September, and November. Sexes and pregnancy or nonpregnancy of females are indicated. See discussion in text.

the mice. A change in the ratio of young mice (categories 1, 2, and 3) to old mice (categories 4 and 5) between August and November was noted. In August, 29 per cent of the population is composed of old mice, and in November only 6 per cent. This change results from birth of young as well as death of old mice, but may indicate that a mouse in November has less than one chance in ten of being alive the following November. Some females born early in the reproductive season breed in their first summer or autumn. For example, a female of category 2, taken on August 12, and probably in postjuvenal pelage, had placental scars. Undoubtedly the young of the year contribute to the breeding population, especially late in the season.

In Figure 3 the proportion of females bearing embryos in August, September, and November is shown. Of the females trapped in August, 11 of 32 that were more than 144 mm. in total length contained embryos; an additional 14 females were lactating or possessed placental scars or enlarged uteri. Therefore, approximately 80 per cent of the larger females were reproducing in August. In September two females were pregnant and an additional sixteen of the 44 females examined showed other evidence of reproduction; these eighteen females make up 41 per cent of those more than 144 mm. in total length. The only reproductive data available for November pertain to the presence or absence of embryos. No female was pregnant although 35 females more than 144 mm. in total length were examined. Some of the skins show prominent mammae indicative of recent nursing, and juveniles less than a month old were taken. The reproductive activity of deer mice on the Mesa Verde seems to be greatly reduced in autumn.

Peromyscus difficilis nasutus (J. A. Allen)
Rock Mouse

Specimen: 1 mi. NNW Rock Springs, 7600 ft., 69413, a young individual completing the molt from juvenal to postjuvenal pelage.

Peromyscus truei truei (Shufeldt)
Pinyon Mouse

Specimens examined.—Total, 42: North end Mesa Verde National Park, 7000 ft., 76220-76232; Far View Ruins, 7700 ft., 69326-69327, 79222, and 8 uncatalogued specimens in preservative; Far View Point, 76532-76535; Far View House, 7700 ft., 74416 MVZ; ½ mi. NNW Rock Springs, 7500 ft., 69429-69430; Rock Springs, 7400 ft., 69431-69435; Park Well, 7450 ft., 69428; Headquarters, MV 7882/507; back of Museum, MV 7879/507, 7880/507, 7881/507; Square Tower House, 6700 ft., 69438.

In August three females were pregnant or lactating, or both. None of seven adult females taken in November was pregnant.

Neotoma cinerea arizonae Merriam
Bushy-tailed Wood Rat

Specimen: Head of Prater Canyon, MV 7873/507. Another, in the Denver Museum, from Spruce Tree House, was reported by Finley (1958: 270).

Neotoma cinerea prefers vertical crevices in high cliffs but occupies other areas.

Neotoma mexicana inopinata Goldman
Mexican Wood Rat

Specimens examined.—Total, 10: Headquarters, MV 7890/507 and probably 7861/507, 74421 MVZ; Spruce Tree Lodge, 6950 ft., 34802-34803; Spruce Tree House, 74419-74420 MVZ; Square Tower House, MV 7869/507; Cliff Palace, 74422 MVZ; Balcony House, MV 7868/507.

The Mexican wood rat is the most common species of wood rat on the Mesa Verde. The two specimens from Spruce Tree Lodge obtained by R. B. Finley on September 2, 1949, are young individuals.

Another species of the genus, the white-throated wood rat, *Neotoma albigula,* may occur within the Park, since three specimens (34757-34759) from the Mesa Verde were trapped on September 15, 1949, by R. B. Finley, approximately 4½ miles south of the Park [6 mi. E, 17 mi. S Cortez, 5600 ft.—south of the area shown in Figure 2]. Finley (1958:450) stated that at that locality he trapped *Neotoma mexicana* [No. 34801], that *N. albigula* was perhaps more common there than *N. mexicana,* that dens of *N. albigula* were more common than those of *N. mexicana* under large rocks in the talus on the south slope of the Mesa, and that dens of *N. mexicana* seemed to be more numerous in crevices of ledges in the bedrock and cliffs.

Ondatra zibethicus osoyoosensis (Lord)
Muskrat

D. Watson (in letter of January 16, 1957) reported that he has seen muskrat tracks many times along the Mancos River. He also relates a report received from Chief Ranger Wade and D. A. Spencer who saw a muskrat, no doubt a wanderer, on the Knife Edge Road on a cold winter night. These men, both reliable observers, stopped and saw the muskrat at a distance of two feet, where it took shelter under a power shovel parked beside the road. Reports of dens seen along the Mancos River are available for 1944, 1945, 1946, and 1947.

Microtus longicaudus mordax (Merriam)
Long-tailed Vole

Specimens examined.—Total, 36: North end Mesa Verde National Park, 7000 ft., 76233-76237; entrance to Mesa Verde National Park, 5123-5126 in Denver Museum; Prater Canyon, 7600 ft., 76238-76244; Upper Well, Prater Canyon, 7575 ft., 69441; Morfield Canyon, 7600 ft., 76245-76259, 76261-76263; west bank Mancos River, northeast side Mesa Verde National Park, 76260.

The vegetation at the above-named localities is a combination of brush and grasses that are both more luxuriant than in areas dominated by pinyon and juniper on the more southern and altitudinally lower part of the top of the Mesa where no *M. longicaudus* was taken.

Microtus mexicanus mogollonensis (Mearns)
Mexican Vole

Specimens examined.—Total, 22: Prater Canyon, 7600 ft., 76283-76287; Sect. 27, head of east fork of Navajo Canyon, 7900 ft., 69442; Far View Ruins, 7700 ft., 69443, 79223-79224; 2 mi. NNW Rock Springs, 7900 ft., 69444-69446; Park Well, 7450 ft., 69447-69453; rock ledge at head of Spruce Tree Canyon, unnumbered specimen in Denver Museum; Headquarters, MV 7895/507, 7896/507.

The first specimen of the Mexican vole from Colorado was obtained on the Mesa Verde and has been reported by Rodeck and Anderson (1956:436). Specimens have now been taken at seven localities on the Mesa. Prater Canyon is the only one of these localities at which any other species of vole was taken. There *Microtus longicaudus* and *Microtus montanus* were also obtained. Judging from the vegetation at the above localities, *M. mexicanus* is to be expected in drier areas with less cover than *M. montanus* inhabits, and in areas having less cover than those inhabited by *M. longicaudus.*

Microtus montanus fusus Hall
Montane Vole

Specimens examined.—Total, 16: Upper Well, 7575 ft., 69454-69465; ¼ mi. N Middle Well, 7500 ft., 69466-69469.

The voles were trapped in the dry but dense meadow of grass and sedge covering the floor of the canyon (see Plate 1). *Sorex vagrans* was trapped in the same places. Four of the females of *M. montanus* trapped on September 3, 1956, were pregnant.

Erethizon dorsatum couesi Mearns
Porcupine

Specimens examined.—Total, 2: 69470, old ♀, and 69471, her young male offspring, both obtained on August 28, 1956, in the canyon of the Mancos River, 6200 feet, along the western side of the River.

I saw no other porcupine in the Park.

In 1935, C. W. Quaintance took special notice of porcupines because of the possibility, then being considered, of their being detrimental to habitat conditions thought to be favorable to wild turkeys. Porcupines were suspected of killing ponderosa pine, which occurred in only a few places, and which was thought to be necessary for wild turkeys. Porcupines were recorded as follows: one found dead on the road at the North Rim on March 16; one killed in oak brush along the North Rim; one killed between April 15 and May 15; oak brush damaged by porcupines in Soda Canyon below the well; one seen on July 4 on the Poole Canyon Trail; one seen at the foot of the Mesa on June 26; one seen by Lloyd White in Moccasin Canyon on June 27; and one seen by Mrs. Sharon Spencer on July 1 in Prater Canyon. After four months on the Mesa Verde, Quaintance concluded that there were not so many porcupines as had been expected and that there were more ponderosa pines than had been expected.

In 1946, Donald A. Spencer began a study of porcupines on the Mesa Verde and in 1958 deposited, in the University of Colorado Library, his results in manuscript form as a dissertation in partial fulfillment of the requirements for a higher degree ("Porcupine population fluctuations in past centuries revealed by dendrochronology," 108 numbered and 13 unnumbered pages, 39 figures, and 13 tables). Dendrochronology, or the dating of trees by studying their rings, is a technique widely used in the southwest by archeologists, climatologists, and others. Spencer found that porcupines damage trees in a characteristic manner, and that damage to a pinyon pine was evident as long as the tree lived. By dating approximately 2000 scars and plotting the year for each scar, Spencer observed three peaks since 1865; these were in about 1885, 1905, and 1935. The increase and decrease each time were at about the same rate. The study did not yield precise population estimates. Some porcupines were destroyed but Spencer is of the opinion that the decline that came in following years was independent of the control measures. Spencer thinks that activities of porcupines on the Mesa Verde are a major factor in maintaining a forest cover of relatively young trees, and also in preventing invasion of trees into areas of brush.

The general policy in regard to porcupines from 1930 to 1946 was to kill them because they eat parts of trees. In at least the following years porcupines were killed: 1930, 1933, 1935, 1940, 1943, 1944, and 1946. The largest number reported killed in one year is 71 in 1933 when a crew of men was employed for this purpose. The amount of effort devoted to killing porcupines varied from year to year. The most frequently voiced alarm was that the scenic value of the areas along the entrance highway and near certain ruins was being impaired. The direst prediction was that all pine trees on the Mesa Verde were doomed to extinction in the near future. The last prediction has not come to pass, nor has this extinction occurred in the past thousand years and more during which pine trees and porcupines have existed together on the Mesa Verde.

In 1946 the studies of Spencer, Wade, and Fitch began. Much effort was expended in obtaining and dating scars for analysis, and the interesting results mentioned above were the reward. Also many porcupines were captured alive and marked with ear-tags so that they could be recognized later. For example, in the winter of 1946 and 1947, 117 were marked in Soda Canyon. A decline in numbers in recent years reduced the impetus for continuation of the study by reducing the results obtained for each day spent searching for porcupines. Information obtained on movements of porcupines relative to season and weather conditions in these studies may be summarized and published later. Data regarding ratio of young to adult animals from year to year are also of interest.

The effect of a porcupine on a single tree is often easy to assess. The effect of a fluctuating population of porcupines on a mixed forest is not so easy to assess, but is of more intrinsic interest. It is desirable that studies designed to evaluate the latter effect continue while the population remains low and also when the next cyclic increase begins. Publication of Spencer's results would be a major step forward.

Cahalane (1948:253) mentions the difficulty that has been experienced in protecting aesthetically desirable trees around cliff dwellings. Perhaps in a local area removal of porcupines is sometimes warranted, but control of the porcupine seems undesirable to me, as a general policy, because one purpose of a National Park is to preserve natural conditions and that implies naturally occurring changes.

What is needed is continued careful study of the ecological relationships of animals and of plants. National parks provide, to the extent that they are not disturbed or "controlled," especially favorable places for studies of this sort.

Mus musculus subsp.
House Mouse

Specimens examined.—Total, 7: North end Mesa Verde National Park, 7000 ft., 76290; west bank Mancos River northeast side Mesa Verde National Park, 76291-76296.

Canis latrans mearnsi Merriam
Coyote

Specimens examined.—Total, 3: 69472, skull only of a young individual, found dead at the top of the bank of the Mancos River, 1½ mi. E Waters Cabin, 6400 ft., August 29, 1956, probably killed by man; ad. ♂, 76298, taken by J. R. Alcorn, November 10, 1957, on the top of the Mesa at Square Tower House; and skin and skull, MV 7858/507, without data.

Tracks or scats of the coyote were seen in all parts of the Park visited. Coyotes range throughout the area. On September 3, 1956, 35 coyote scats were found on the dirt roads in Prater and Morfield canyons above 7300 feet elevation and on the road crossing the divide between these canyons. Probably none of these scats was more than a month old. Coyote tracks were seen at some of the fresher scats. Scats associated with fox tracks and scats of small size were not picked up. Nevertheless, a few of the scats studied may have been those of foxes. Judging from the contents of scats that were certainly from foxes, the effect of inadvertent inclusion of fox scats would be to elevate the percentage of scats containing berries (but not more than five percentage points). Each scat was broken up and the percentage of scats containing each of the following items was noted (figures are to the nearest per cent). Remains of deer occurred in 48 per cent of scats, gooseberries (*Ribes*) in 34 per cent, porcupines in 29 per cent, insects in 11 per cent, birds in 11 per cent, unidentified hair in 9 per cent, and unidentified material in 6 per cent. One scat (3 per cent) contained an appreciable amount of plant debris, one contained *Microtus* along with other items, and one contained only *Sylvilagus;* 14 scats had material of more than one category. The percentage in each category of the volume of each scat was estimated. Data on volume warrant no conclusion other than one that can be drawn from the percentages of occurrence, namely that the major food sources used in August, 1956, by coyotes in these canyons were deer, berries, and porcupines and that other sources, though used, were relatively unimportant. Deer were common in the area. It is fortunate that coyotes remain to help regulate the deer population. Wolves, *Canis lupus,* which at one time occurred in the Park, are now gone. The coyote and mountain lion are the only sizeable predators that remain.

Vulpes vulpes macroura Baird
Red Fox

D. Watson (in letter of January 16, 1957) reported that red foxes have been seen on the Mesa by several employees of the Park. These persons know the gray fox, which often is seen in winter feeding at their back doors, and Mr. Watson considers the reports reliable. In the early morning of October 24, 1943, a reddish-yellow fox having a white-tipped tail was observed by three men, one of whom was Chief Ranger Wade, at Park Point. In 1948, 1950, and 1953 black foxes have been reported.

Urocyon cinereoargenteus scottii Mearns
Gray Fox

Specimens examined.—Total, 3: ♂, MV 7867/507, 2 mi. N of Headquarters, 7400 ft., September 24, 1935, H. P. Pratt; ♂, 76299, November 9, and ♀, 76300, trapped on November 12, 1957, by J. R. Alcorn at Square Tower House.

The gray fox is common on the Mesa.

Ursus americanus amblyceps Baird
Black Bear

From 1929 through 1959 at least 151 observations of bears were recorded. Observations were unrecorded in only five years—1952, 1953, 1954, 1956, and 1958. Most observations were in the 1940's and the peak was in 1944 (18 observations) and 1945 (21 observations). Cubs have been recorded in 10 different years. If dated reports are tabulated by months the following figures are obtained for the 12 months beginning with January: 0, 0, 0, 4, 15, 19, 19, 9, 10, 9, 3, 0. The peak in the summer months and the absence of observations in the winter months are significant. Individual bears probably enter and leave the Park in the course of their normal wanderings; however bears probably hibernate, breed, and bear young within the Park and should not be regarded as merely occasionally wandering into the Park.

Procyon lotor pallidus Merriam
Raccoon

In December, 1959, three raccoons were seen on Prater Grade and later three were seen in Morfield Canyon near the tunnel. I saw a dead raccoon at the side of the highway 3 mi. WSW of Mancos, 6700 feet, on August 8, 1956. This locality is outside of the Park and not on the Mesa, but is mentioned because it indicates that the raccoon probably occurs along the Mancos River, which forms the eastern boundary of the Park. The raccoon is rare in the area. Some local persons were surprised to hear of its presence; other persons told me that raccoons were present, but rare.

Bassariscus astutus flavus Rhoads
Ringtail

Specimens examined.—Total, 4: MV 7884/507 and 7885/507, trapped in Balcony House and prepared by D. Watson in 1939; MV 7901/507 and 7902/507, without data.

The cliff dwellings are favored by ringtails and in some years they are common near occupied dwellings in the area of headquarters. Ringtails have been seen in each major habitat within the Park.

Mustela frenata nevadensis Hall
Long-tailed Weasel

Specimens examined.—Total, 5: MV 7891/507, ♂, from the "Garden" [= Indian Cornfield]; ♀, MV 7892/507, also from the "Cornfield"; MV 7859/507, "Killed by car on Prater Grade"; ♂, MV 7871/507, in winter pelage, from the North Rim; and ♂, 83464, killed on the road ½ mi. NE of the tunnel, Morfield Canyon.

C. W. Quaintance in 1935 reported that on January 11, he and Mr. Nelson saw a weasel attack a cottontail, and on March 9, while on the snow plow, Mr. Nelson witnessed another cottontail being killed by a weasel. Weasels in white winter pelage have been recorded in December and January. The brown pelage has been seen as late as November.

Mustela vison energumenos (Bangs)
Mink

D. Watson (in letter of January 16, 1957) wrote: "When Jack Wade, now Chief Ranger, was doing patrol work in the Mancos Canyon back in the 1930's, he saw mink along the river at the east side of the Park. Several years ago, the people who lived on the ranch where Weber Canyon joins the Mancos trapped a mink." Tracks have been reported along the Mancos River in several years.

Spilogale putorius gracilis Merriam
Spotted Skunk

Specimen: Immature ♂, MV 7860/507, Cliff Palace, August 22, 1936, prepared by A. E. Borrell.

In some years these little skunks have become so numerous in the area of headquarters that they were a nuisance, and were captured in garbage cans and released in other parts of the Park.

Mephitis mephitis estor Merriam
Striped Skunk

D. Watson advises me that striped skunks are fairly common around the entrance to the Park, along the foot of the Mesa, and along the Mancos River. Striped skunks have been reported in 1951

PLATE 1

Upper: View of the North Rim of Mesa Verde, looking west from Park Point, the highest place on the North Rim. The south-facing slope on the left is covered with brushy vegetation, mostly oak. Sheltered parts of the north-facing slope support stands of Douglas fir, and at a few places some ponderosa pines. Photo taken in August, 1956, by S. Anderson.

Lower Left: View of Rock Canyon from Wetherill Mesa, looking southwest from a point 2 mi. NNW Rock Springs. The area in the foreground on Wetherill Mesa was burned in 1934. Photo taken in August, 1956, by S. Anderson.

Lower Right: Prater Canyon, at Upper Well, 7575 feet. In the matted grasses and sedges on the floor of the canyon *Microtus montanus* and *Sorex vagrans* were captured. *Tamiasciurus hudsonicus* was found in a side canyon, Chickaree Draw, one half mile southwest of the place shown. Chickaree Draw is more sheltered than the slope in the background and has a denser stand of Douglas fir than occurs here. Photo taken in August, 1956, by S. Anderson.

PLATE 2

UPPER: Relatively undisturbed stand of pinyon pine and Utah juniper ¼ mi. N Rock Springs, at 7400 feet elevation on Wetherill Mesa along a service road. The vegetation shown is characteristic of the lower more exposed parts of the top of the Mesa Verde. Photo taken in August, 1956, by S. Anderson.

LOWER: Wetherill Mesa, ½ mi. NNW Rock Springs, 7500 feet elevation. This area burned in 1934. It contained no pine or juniper in 1956 despite attempted reforestation in the thirties and the presence of a stand of pinyon and juniper (shown above) only one quarter of a mile away. Possibly fire in the last three or four hundred years on the higher parts of the Mesa has been a factor in producing chaparral there, rather than pinyon and juniper. Photo taken in August, 1956, by S. Anderson.

in Morfield Canyon, in 1952 on the Knife Edge, in 1953 at Windy Point (¼ mi. N of Point Lookout), and in 1959 at the head of Morfield Canyon.

Taxidea taxus berlandieri Baird
Badger

Several reports, but no specimens, of the badger have been obtained. In 1935, C. W. Quaintance wrote that in School Section Canyon tracks of cougar, bobcat, coyote, and deer were found, and that pocket gophers, badgers, and cottontail rabbits were present. Later in 1935, H. P. Pratt wrote that he had found evidence of badgers "at the lower well in Prater Canyon, where on September 23, there were extensive badger diggings and fresh tracks in the vicinity of the prairie dog colony there." Badgers are common in the lowlands around the Mesa and they are common enough on the Mesa to be regarded as nuisances by archeologists on account of badgers digging in ruins. Badgers have been seen from three to six times each year from 1950 to this date, most of them in the vicinity of the North Rim.

Felis concolor hippolestes Merriam
Mountain Lion

Mountain lions range throughout the Park. There are reliable sight records of lions and lion tracks, but no specimen has been preserved. Early records of observations include the report of tracks seen in Navajo Canyon by Cary (1911:165), and a lion seen in 1917. Since 1930 the more adequate records include reports of from one to eight observations each year for 26 of the 30 years. Young animals (recorded as "half-grown") or cubs have been reported in four of these years. The tabulation of dated reports by month beginning with January is: 2, 0, 3, 2, 8, 4, 6, 7, 4, 9, 5, 7. Mountain lions range more widely than bears in their daily and seasonal activities, but like bears probably breed, bear young, and feed in the Park. Although at any one time lions may or may not be within the Park, it is part of their normal range and the species should be regarded as resident and is not uncommon.

Lynx rufus baileyi Merriam
Bobcat

Specimens examined.—Total, 2: A specimen (now mounted in Park Museum) from the Knife Edge Road; and ad. ♀, 76302, Prater Canyon, 7500 ft., November 12, 1957, obtained by J. R. Alcorn.

Bobcats are present throughout the Park. Approximately 80 observations of bobcats are on file, from all parts of the Park and in all months. Probably the bobcat and the gray fox are the most abundant carnivores in the Park. In addition to known predation by mountain lions and coyotes on porcupines, the bobcat kills porcupines. A dead porcupine and a dead bobcat with its face, mouth, and one foot full of quills were found together on January 31, 1952, under a boulder in front of Cliff Palace. On August 20, 1956, I saw a bobcat hunting in sage in a draw near a large clump of oakbrush, into which it fled, at the head of the east fork of Navajo Canyon, Sect. 21, near the North Rim, 8100 feet.

Odocoileus hemionus hemionus (Rafinesque)
Mule Deer

Specimens examined.—Total, 2: Young ♂, 76303, November 8, 1957, Far View Ruins; ♀, 76304, November 12, 1957, Spruce Tree House Ruin, both obtained by J. R. Alcorn.

In all parts of the Park, mule deer are common. Five projects concerning deer are in progress or have been concluded recently on the Mesa. One is a study of the responses of different species of plants to browsing and was begun in 1949 by Harold R. Shepherd for the Colorado Department of Game and Fish. A number of individual plants and in some instances groups of plants were fenced to exclude deer. Systematic clips of 20, 40, 60, 80, or 100 per cent of the annual growth are made each year. The results of the first ten years of this study are being prepared for publication by Shepherd.

A study of browsing pressure was initiated in 1952 by Regional Biologist C. M. Aldous, on eight transects in the Park. Each transect consists of 15 plots at intervals of 200 feet. The amount of use of each plant species was recorded from time to time. The study was terminated in 1955. I have seen no summary of results of this study.

A trapping program was begun in 1953 with the co-operation of the Colorado Department of Game and Fish. Deer are trapped, marked, and released. Some are released in areas other than where trapped. In this way the excessive size of the herd near headquarters has been reduced. Recoveries of marked deer outside the Park by hunters and retrapping results in the Park should provide information about movements of deer and about life expectancy.

The "Deer Trend Study" was initiated in 1954. From November to May, twice a day, at the same time, a count is made along the entrance road from the Park Entrance to Headquarters. Ten drainage areas traversed are tabulated separately. The results of four years of this study indicate that the greatest number of deer are present in November, December, and January, and that only about one-fourth as many are present in February and March. Depending on severity of weather, the yearly pattern varies, the deer arriving earlier, or leaving earlier. This change in numbers, the recovery outside of the Park of animals marked in the Park, and direct observations of movement indicate that the Mesa Verde is an intermediate range rather than a summer-range or winter-range. In summer deer tend to move northward and eastward out of the Park, and in winter they move back through the Park toward lower and more protected areas in canyons both in the Park and south of the Park on the Ute Reservation. Some deer remain in the Park the entire year. Close co-operation between personnel of the Park Service and of the Colorado Department of Game and Fish has regulated hunting outside the Park in such a way as to provide satisfactory control of the deer within the Park.

A study of the effect of rodents on plants used by deer was initiated in 1956 by Harold R. Shepherd. Three acres were fenced in a fashion designed to exclude rodents but not deer. An adjacent three acres were fenced as a control, but not so as to exclude rodents or deer. Eight trap lines nearby provide an index of rodent fluctuations from year to year. These studies will need to be continued for a period of ten years or more, and should provide much information concerning not only deer but also rodents and their effect on vegetation.

Cervus canadensis nelsoni V. Bailey
Wapiti

Wapiti are seen periodically; probably they wander in from the higher mountains to the northeast and do not remain for long. The following note was included in the 1921 report of Mr. Jesse L. Nusbaum, then Superintendent of the Park: "The first elk ever seen in the Park made his appearance near the head of Navajo Canyon, August 15 of this year, and travelled for two miles in front of a Ford car down the main road before another car, travelling in the opposite direction, scared him into the timber." Additional observations have been recorded as follows: School Section Canyon ("fall" 1935), Knife Edge Road (July, 1940), West Soda Canyon and Windy Point (December, 1949), Long Canyon (July, 1959), and Park Entrance (December, 1959). Three of the six observations are in July and August; therefore movement by wapiti into the Park can not be attributed entirely to disturbance during the hunting season.

Ovis canadensis canadensis Shaw
Bighorn

Some early records of the bighorn were mentioned by C. W. Quaintance (1935): In a letter of January 20, 1935, John Wetherill said that a "Mountain Sheep Canyon" (now Rock Canyon) was named for a bunch of sheep that wintered near their camp; and Sam Ahkeah, a Navajo, says the Indians occasionally find remnants of sheep on the Mesa, which they take back to their hogans. Cahalane (1948:257) reported that hunting presumably had eliminated bighorns from the Mesa by 1896; however Jean Pinkley reports that a large ram was killed on Point Lookout in 1906.

On January 30, 1946, 14 sheep (3 rams, 7 ewes, and 4 lambs) from the herd at Tarryall, Colorado, were obtained through the Colorado Department of Game and Fish and were released at 8:30 a. m. at the edge of the canyon south of Spruce Tree Lodge. The sheep, instead of entering the canyon as expected, turned north, passed behind the museum, and eventually disappeared northward on Chapin Mesa. The sheep evidently divided into at least two bands. On April 24, 1946, three sheep were seen 2½ mi. N of Rock Springs, and on June 19, 1947, tracks were seen in Mancos Canyon. In 1947, 1948, and 1949 farmers in Weber Canyon reported seeing sheep many times on Weber Mountain, and watering at the Mancos River. In May, 1949, an estimate of 27 sheep on Weber Mountain was made after several days study by men from the state game department. The herds continued to increase. In 1956 I saw two bighorns. On August 18, at 6:20 a. m., my wife and I briefly observed a bighorn on the rocks below Square Tower Ruins. On August 24, I was digging with a small shovel in rocky soil behind the cabin at Rock Springs, when hoof beats were heard approaching in the rocky head of the canyon to the east. An adult ewe came up to the fence around the cabin area and looked at me, seemingly curious about the noise my shovel had been producing. I remained motionless and called to my wife, Justine, to come from the cabin and see the sheep. The ewe seemed not to be disturbed by my voice, but took flight, returning in the direction from which she had come, the moment Justine appeared from behind the cabin. Sheep can now be seen on occasion in any of the deep canyons across the southern half of the Park. The sheep have caused slight damage in some of the ruins by bedding down there, and by climbing on walls. As the sheep increase in numbers this activity may be regarded as a problem. In 1959 an estimated 75 to 100 sheep were in the Park and adjacent areas.

DISCUSSION

The distributions of animals are influenced by geographic, vegetational, and altitudinal factors. The Mesa Verde is intermediate in geographic position and altitude between the high Southern Rocky Mountains and the low southwestern desert. For this reason, we find on the Mesa Verde (1) a preponderance of species having wide distributions in this part of the country, and having relatively wide ranges of tolerance for different habitats, (2) a lesser number of exclusively montane or boreal species than occur in the higher mountains to the northeast of the Mesa and that may reach the limits of their ranges here, and (3) a small number of species of southern or Sonoran affinities. Fifty-four species are recorded above.

Forty-one of these species are represented by specimens from the Park. Thirteen additional species in the list have been seen in the Park.

On the Grand Mesa, which is more elevated than, and some 110 miles north of, the Mesa Verde (see Figure 1), 55 per cent of the species of mammals have boreal affinities and the other 45 per cent are wide-spread species (Anderson, 1959:414). Boreal species from the Mesa Verde are *Sorex vagrans, Sylvilagus nuttallii, Spermophilus lateralis, Marmota flaviventris, Tamiasciurus hudsonicus, Microtus montanus,* and *Microtus longicaudus.* These seven species comprise only thirteen per cent of the mammalian fauna of the Mesa Verde. Other boreal species that occur in the mountains of Colorado on the Grand Mesa or elsewhere (Findley and Anderson, 1956:80) and that do not occur on the Mesa Verde are *Sorex cinereus, Sorex palustris, Ochotona princeps, Lepus americana, Clethrionomys gapperi, Phenacomys intermedius, Zapus princeps, Martes americana, Mustela erminea,* and *Lynx canadensis.* The 47 species from the Mesa Verde that are not exclusively boreal make up 87 per cent of the mammalian fauna. Most of these are wide-spread species and are more abundant in the deserts or other lowlands than in the coniferous forests of the highlands, for example the eight species of bats, and *Sylvilagus audubonii, Thomomys bottae, Taxidea taxus, Bassariscus astutus, Canis latrans, Cynomys gunnisoni, Reithrodontomys megalotis,* and *Lepus californicus.* A few of the wide-spread species are more common in the highlands than in the lowlands, for example *Ursus americanus, Felis concolor, Castor canadensis, Erethizon dorsatum,* and *Cervus canadensis,* and the ranges of three of these, the bear, mountain lion and wapiti, are more restricted today than formerly. A few species find their favorite habitat and reach their greatest abundance in altitudinally and vegetationally intermediate areas such as upon the Mesa Verde, or in special habitats, such as the rock ledges, and crevices that are so abundant on the Mesa. Examples of this group of species are *Spermophilus variegatus*, *Peromyscus crinitus*, *Peromyscus truei*, *Neotoma cinerea*, and *Neotoma mexicana*. One species, *Dipodomys ordii*, is restricted to the desert. Species that are restricted to the desert and that occur in Montezuma County, Colorado, but that are not known from the Mesa Verde are *Ammospermophilus leucurus*, *Perognathus flavus*, and *Onychomys leucogaster*.

Species known to have changed in numbers in the past 50 years are the mule deer that has increased, and the prairie dog that has decreased. Possibly beaver have increased along the Mancos River. The muskrat, mink, beaver, and raccoon usually occur only along the Mancos River, as there is no other permanent surface water in the Park.

Species such as the bighorn and the marmot that are rare within the Park, or those such as the chickaree, the prairie dog, the wandering shrew, the montane vole, and the long-tailed vole that occupy only small areas of suitable habitat within the Park are the species most likely to be eliminated by natural changes, or through the activities of man. For example parasites introduced through domestic sheep that wander into the range of bighorns within the Park might endanger the bighorn population. An increase in grazing activity, road building, and camping in Prater and Morfield canyons might eliminate the small areas of habitat occupied by the montane vole and the wandering shrew. Fire in Chickaree Draw could destroy all the Douglas fir there, and consequently much of the habitat occupied by the chickaree.

Probably some species inhabit the Mesa that have not yet been found, but they are probably few, and their discovery will not alter the faunal pattern in which the few boreal species occupy restricted habitats in the higher parts of the Mesa, and a preponderance of geographically wide-spread species occupy all or most of the Mesa, and surrounding areas. Additional bats are the species most likely to be added to the list.

LITERATURE CITED

ANDERSON, S.
1959. Mammals of the Grand Mesa, Colorado. Univ. Kansas Publ., Mus. Nat. Hist., 9(16):405-414, 1 fig. in text.

CAHALANE, V. H.
1948. The status of mammals in the U. S. National Park System, 1947. Jour. Mamm., 29(3):247-259.

CARY, M.
1911. A biological survey of Colorado. N. Amer. Fauna, 33:1-256, 39 figs., frontispiece (map).

FINDLEY, J. S.
1955. Speciation of the Wandering Shrew. Univ. Kansas Publ., Mus. Nat. Hist., 9(1):1-68, figs. 1-18.

FINDLEY, J. S. and ANDERSON, S.
1956. Zoogeography of the montane mammals of Colorado. Jour. Mamm., 37(1):80-82, 1 fig. in text.

FINLEY, R. B.
1958. The wood rats of Colorado, distribution and ecology. Univ. Kansas Publ., Mus. Nat. Hist., 10(6):213-552, 34 plates, 8 figs., 35 tables in text.

GETTY, H. T.
1935. New dates from Mesa Verde. Tree-ring Bulletin, 1(3):21-23.

HOFFMEISTER, D. F.
1951. A taxonomic and evolutionary study of the piñon mouse, Peromyscus truei. Illinois Biol. Monogr., vol. XXI(4), pp. ix + 104, 24 figs., 4 tables and 5 plates in text.

RODECK, H. G. and ANDERSON, S.
1956. *Sorex merriami* and *Microtus mexicanus* in Colorado. Jour. Mamm., 37(3):436.

SCHULMAN, E.
1946. Dendrochronology at Mesa Verde National Pork. Tree-ring Bulletin, 12(3):18-24, 2 figs., 1 table in text.

YOUNGMAN, P. M.
1958. Geographic variation in the pocket gopher, Thomomys bottae, in Colorado. Univ. Kansas Publ., Mus. Nat. Hist., 9(12):363-387, 7 figs. in text.

Transmitted April 11, 1961.

□
28-7577

(Continued from inside of front cover)

8. Nos. 1-10 and index. Pp. 1-675, 1954-1956.

9. 1. Speciation of the wandering shrew. By James S. Findley. Pp. 1-68, 18 figures in text. December 10, 1955.
2. Additional records and extensions of ranges of mammals from Utah. By Stephen D. Durrant, M. Raymond Lee, and Richard M. Hansen. Pp. 69-80. December 10, 1955.
3. A new long-eared myotis (Myotis evotis) from northeastern Mexico. By Rollin H. Baker and Howard J. Stains. Pp. 81-84. December 10, 1955.
4. Subspeciation in the meadow mouse, Microtus pennsylvanicus, in Wyoming. By Sydney Anderson. Pp. 85-104, 2 figures in text. May 10, 1956.
5. The condylarth genus Ellipsodon. By Robert W. Wilson. Pp. 105-116, 6 figures in text. May 19, 1956.
6. Additional remains of the multituberculate genus Eucosmodon. By Robert W. Wilson. Pp. 117-123, 10 figures in text. May 19, 1956.
7. Mammals of Coahuila, Mexico. By Rollin H. Baker. Pp. 125-335, 75 figures in text. June 15, 1956.
8. Comments on the taxonomic status of Apodemus peninsulae, with description of a new subspecies from North China. By J. Knox Jones, Jr. Pp. 337-346, 1 figure in text, 1 table. August 15, 1956.
9. Extensions of known ranges of Mexican bats. By Sydney Anderson. Pp. 347-351. August 15, 1956.
10. A new bat (Genus Leptonycteris) from Coahuila. By Howard J. Stains. Pp. 353-356. January 21, 1957.
11. A new species of pocket gopher (Genus Pappogeomys) from Jalisco, Mexico. By Robert J. Russell. Pp. 357-361. January 21, 1957.
12. Geographic variation in the pocket gopher, Thomomys bottae, in Colorado. By Phillip M. Youngman. Pp. 363-387, 7 figures in text. February 21, 1958.
13. New bog lemming (genus Synaptomys) from Nebraska. By J. Knox Jones, Jr. Pp. 385-388. May 12, 1958.
14. Pleistocene bats from San Josecito Cave, Nuevo León, México. By J. Knox Jones, Jr. Pp. 389-396. December 19, 1958.
15. New Subspecies of the rodent Baiomys from Central America. By Robert L. Packard. Pp. 397-404. December 19, 1958.
16. Mammals of the Grand Mesa, Colorado. By Sydney Anderson. Pp. 405-414, 1 figure in text. May 20, 1959.
17. Distribution, variation, and relationships of the montane vole, Microtus montanus. By Emil K. Urban. Pp. 415-511. 12 figures in text, 2 tables. August 1, 1959.
18. Conspecificity of two pocket mice, Perognathus goldmani and P. artus. By E. Raymond Hall and Marilyn Bailey Ogilvie. Pp. 513-518, 1 map. January 14, 1960.
19. Records of harvest mice, Reithrodontomys, from Central America, with description of a new subspecies from Nicaragua. By Sydney Anderson and J. Knox Jones, Jr. Pp. 519-529. January 14, 1960.
20. Small carnivores from San Josecito Cave (Pleistocene), Nuevo León, México. By E. Raymond Hall. Pp. 531-538, 1 figure in text. January 14, 1960.
21. Pleistocene pocket gophers from San Josecito Cave, Nuevo León, México. By Robert J. Russell. Pp. 539-548, 1 figure in text. January 14, 1960.
22. Review of the insectivores of Korea. By J. Knox Jones, Jr., and David H. Johnson. Pp. 549-578. February 23, 1960.
23. Speciation and evolution of the pygmy mice, genus Baiomys. By Robert L. Packard. Pp. 579-670, 4 plates, 12 figures in text. June 16, 1960.

Index Pp. 671-690.

10. 1. Studies of birds killed in nocturnal migration. By Harrison B. Tordoff and Robert M. Mengel. Pp. 1-44, 6 figures in text, 2 tables. September 12, 1956.
2. Comparative breeding behavior of Ammospiza caudacuta and A. maritima. By Glen E. Woolfenden. Pp. 45-75, 6 plates, 1 figure. December 20, 1956.
3. The forest habitat of the University of Kansas Natural History Reservation. By Henry S. Fitch and Ronald R. McGregor. Pp. 77-127, 2 plates, 7 figures in text, 4 tables. December 31, 1956.
4. Aspects of reproduction and development in the prairie vole (Microtus ochrogaster). By Henry S. Fitch. Pp. 129-161, 8 figures in text, 4 tables. December 19, 1957.
5. Birds found on the Arctic slope of northern Alaska. By James W. Bee. Pp. 163-211, pls. 9-10, 1 figure in text. March 12, 1958.
6. The wood rats of Colorado: distribution and ecology. By Robert B. Finley, Jr. Pp. 213-552, 34 plates, 8 figures in text, 35 tables. November 7, 1958.
7. Home ranges and movements of the eastern cottontail in Kansas. By Donald W. Janes. Pp. 553-572, 4 plates, 3 figures in text. May 4, 1959.

(Continued on outside of back cover)

(Continued from inside of back cover)

8. Natural history of the salamander, Aneides hardyi. By Richard F. Johnston and Schad Gerhard. Pp. 573-585. October 8, 1959.

9. A new subspecies of lizard, Cnemidophorus sacki, from Michoacán, México. By William E. Duellman. Pp. 587-598, 2 figures in text. May 2, 1960.

10. A taxonomic study of the Middle American Snake, Pituophis deppei. By William E. Duellman. Pp. 599-612, 1 plate, 1 figure in text. May 2, 1960.

Index Pp. 611-626.

11. 1. The systematic status of the colubrid snake, Leptodeira discolor Günther. By William E. Duellman. Pp. 1-9, 4 figs. July 14, 1958.

2. Natural history of the six-lined racerunner, Cnemidophorus sexlineatus. By Henry S. Fitch. Pp. 11-62, 9 figs., 9 tables. September 19, 1958.

3. Home ranges, territories, and seasonal movements of vertebrates of the Natural History Reservation. By Henry S. Fitch. Pp. 63-326, 6 plates, 24 figures in text, 3 tables. December 12, 1958.

4. A new snake of the genus Geophis from Chihuahua, Mexico. By John M. Legler. Pp. 327-334, 2 figures in text. January 28, 1959.

5. A new tortoise, genus Gopherus, from north-central Mexico. By John M. Legler. Pp. 335-343. April 24, 1959.

6. Fishes of Chautauqua, Cowley and Elk counties, Kansas. By Artie L. Metcalf. Pp. 345-400, 2 plates, 2 figures in text, 10 tables. May 6, 1959.

7. Fishes of the Big Blue River Basin, Kansas. By W. L. Minckley. Pp. 401-442, 2 plates, 4 figures in text, 5 tables. May 8, 1959.

8. Birds from Coahuila, México. By Emil K. Urban. Pp. 443-516. August 1, 1959.

9. Description of a new softshell turtle from the southeastern United States. By Robert G. Webb. Pp. 517-525, 2 pls., 1 figure in text, August 14, 1959.

10. Natural history of the ornate box turtle, Terrapene ornata ornata Agassiz. By John M. Legler. Pp. 527-669, 16 pls., 29 figures in text. March 7, 1960.

Index Pp. 671-703.

12. 1. Functional morphology of three bats: Eumops, Myotis, Macrotus. By Terry A. Vaughan. Pp. 1-153, 4 plates, 24 figures in text, July 8, 1959.

2. The ancestry of modern Amphibia: a review of the evidence. By Theodore H. Eaton, Jr. Pp. 155-180, 10 figures in text. July 10, 1959.

3. The baculum in microtine rodents. By Sydney Anderson. Pp. 181-216, 49 figures in text. February 19, 1960.

4. A new order of fishlike Amphibia from the Pennsylvanian of Kansas. By Theodore H. Eaton, Jr., and Peggy Lou Stewart. Pp. 217-240, 12 figures in text. May 2, 1960.

More numbers will appear in volume 12.

13. 1. Five natural hybrid combinations in minnows (Cyprinidae). By Frank B. Cross and W. L. Minckley. Pp. 1-18. June 1, 1960.

2. A distributional study of the amphibians of the isthmus of Tehuantepec, México. By William E. Duellman. Pp. 19-72, pls. 1-8, 3 figs. August 16, 1960.

3. A new subspecies of the slider turtle (Pseudemys scripta) from Coahuila, México. By John M. Legler. Pp. 73-84, pls. 9-12, 3 figures in text. August 16, 1960.

4. Autecology of the Copperhead. By Henry S. Fitch. Pp. 85-288, pls. 13-20, 26 figures in text. November 30, 1960.

5. Occurrence of the Garter Snake, Thamnophis sirtalis, in the Great Plains and Rocky Mountains. By Henry S. Fitch and T. Paul Maslin. Pp. 289-308, 4 figures in text. February 10, 1961.

6. Fishes of the Wakarusa River in Kansas. By James E. Deacon and Artie L. Metcalf. Pp. 309-322, 1 figure in text. February 10, 1961.

7. Geographic variation in the North American Cyprinid Fish, Hybopsis gracilis. By Leonard J. Olund and Frank B. Cross. Pp. 323-348, pls. 21-24, 2 figures in text. February 10, 1961.

8. Descriptions of two species of frogs, Genus Ptychohyla—studies of American hylid frogs, V. By William E. Duellman. Pp. 349-357, pl. 25, 2 figures in text. April 27, 1961.

More numbers will appear in volume 13.

14. 1. Neotropical bats from western México. By Sydney Anderson. Pp. 1-8. October 24, 1960.

2. Geographic variation in the harvest mouse Reithrodontomys megalotis on the central Great Plains and in adjacent regions. By J. Knox Jones, Jr. and B. Mursaloglu. Pp. 9-27, 1 figure in text. July 24, 1961.

3. Mammals of Mesa Verde National Park, Colorado. By Sydney Anderson. Pp. 29-67, pls. 1-2, 3 figures in text. July 24, 1961.

More numbers will appear in volume 14.

University of Kansas Publications
Museum of Natural History

Volume 14, No. 4, pp. 69-72, 1 fig.
December 29, 1961

A New Subspecies of the Black Myotis (Bat) From Eastern Mexico

BY

E. RAYMOND HALL AND TICUL ALVAREZ

University of Kansas
Lawrence
1961

University of Kansas Publications, Museum of Natural History

Volume 14, No. 4, pp. 69-72, 1 fig.
Published December 29, 1961

University of Kansas
Lawrence, Kansas

PRINTED BY
JEAN M. NEIBARGER, STATE PRINTER
TOPEKA, KANSAS
1961

28-8477

A New Subspecies of the Black Myotis (Bat) From Eastern Mexico

BY

E. RAYMOND HALL AND TICUL ALVAREZ

In 1928 when Miller and Allen (Bull. U. S. Nat. Mus., 144) published their revisionary account of American bats of the genus *Myotis,* the black myotis, *Myotis nigricans,* was known no farther north than Chiapas and Campeche. Collections of mammals made in recent years for the Museum of Natural History of The University of Kansas include specimens of *M. nigricans* from eastern Mexico as far north as Tamaulipas. Critical study of this newly acquired material reveals that it pertains to an hitherto unnamed subspecies that may be named and described as follows:

Myotis nigricans dalquesti new subspecies

Type.—Male, adult, skin and skull, No. 23839 Museum of Natural History, University of Kansas; from 3 km. E of San Andrés Tuxtla, 1000 ft., Veracruz; obtained on January 5, 1948, by Walter W. Dalquest, original No. 8444.

Range.—Tropical Life-zone of eastern México from southern Tamaulipas to central Chiapas.

Diagnosis.—Color black or dark brown, venter having brownish wash; size large (see measurements); M1 and M2 quadrangular; prominent protostyle on P4; P2 and P3 in straight line; sagittal crest absent.

Comparison.—Color almost as in *Myotis nigricans extremus,* the subspecies occurring adjacent to *dalquesti* in Chiapas and Tabasco. From *M. n. extremus, dalquesti* differs as follows: larger; hypocone in M1 and M2 broader making posterointernal part less rounded; protostyle of P4 prominent instead of absent; P3 in line with C and P2 instead of displaced lingually; sagittal crest absent instead of present posteriorly. *Myotis nigricans nigricans* and *M. n. dalquesti* are of approximately equal size; otherwise they differ in the same features as do *extremus* and *dalquesti.*

Measurements.—Average and extreme measurements of seven males from the type locality, followed by those of 19 females from 38 km. SE Jesús Carranza, and finally length of forearm and cranial measurements of eight female topotypes of *M. n. extremus,* are as follows: Total length, 80 (77-82), 76 (72-80); length of tail, 32.8 (30-35), 33.5 (31-35); hind foot, 7.9 (7-8), 8.0 (8-8); forearm, 34.2 (33.6-35.3), 35.1 (33.1-36.4), 33.1 (31.8-34.3); greatest length of skull (including incisors), 13.8 (13.3-14.1), 13.6 (13.2-14.1), 12.9 (12.6-13.1); zygomatic breadth, 8.1 (7.9-8.4), 8.1 (7.9-8.3), 8.0 (only one can be measured); width of rostrum above canines, 3.2 (3.1-3.3), 3.2 (3.0-3.4), 3.1 (3.0-3.2); interorbital constriction, 3.6 (3.5-3.7), 3.6 (3.5-3.8), 3.4 (3.3-3.4); occipital depth (excluding auditory bullae and sagittal crest), 4.6 (4.4-4.8), 4.6 (4.3-4.9), 4.3 (4.1-4.6); maxillary tooth-row (C-M3), 5.0 (4.8-5.1), 5.0 (4.8-5.2), 4.7 (4.6-4.8); maxillary breadth at M3, 5.2 (5.1-5.4), 5.3 (5.1-5.5), 5.1 (4.8-5.2).

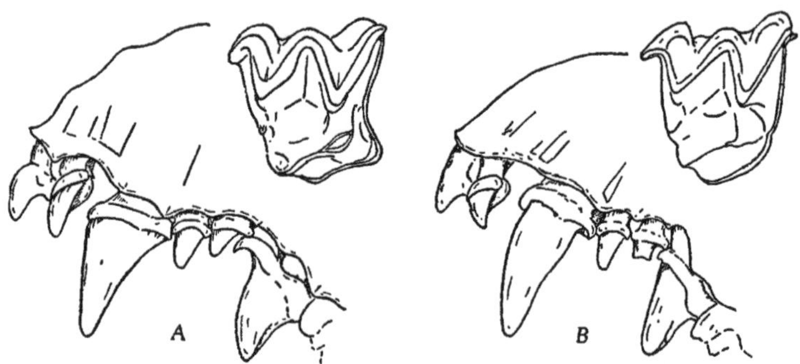

FIG. 1. Left side of skull, incisors, canine, and premolars × 11, and occlusal surface of left first upper molar × 20. A. *Myotis nigricans dalquesti*, holotype. B. *Myotis nigricans extremus* No. 77674 USNM, topotype.

Remarks.—The subspecific name *dalquesti* is given in recognition of Prof. Walter W. Dalquest who gathered the largest and most varied collection of mammals ever taken in the state of Veracruz.

Inspection of the measurements given above will reveal that there is no overlap between *extremus* and *dalquesti* in the interorbital constriction or occipital depth and only slight overlap in the length of the maxillary tooth-row and maxillary breadth.

In 10 adult females from Ocosingo, Chiapas, there is suggestion of intergradation between *dalquesti* and *extremus* in that one specimen (66515 KU) has the cranial characters of *extremus* except that it is large like *dalquesti;* in two other skulls P3 is slightly displaced lingually and two other skulls bear a slight sagittal crest. These are features characterizing *extremus*. Otherwise the specimens resemble *dalquesti*, to which subspecies they are here referred.

Three males from a place 8 km. W and 10 km. N El Encino, 400 ft., Tamaulipas, are the northernmost representatives of the species and differ from the other specimens of *dalquesti* in shorter forearm, shorter maxillary tooth-row and lesser maxillary breadth.

Study in the laboratory was supported by Grant No. 56 G 103 from the National Science Foundation. Field work was supported by a grant from the Kansas University Endowment Association. We thank Dr. David H. Johnson for lending eight topotypes of *M. n. extremus*. Other specimens of *extremus* available to us are as follows: 1 mi. E Teapa, Tabasco, 1 (7535 LSU—courtesy of Dr. George H. Lowery, Jr.); Cayo Dist. Augustine, British Honduras, 1 (9670 KU, in red phase); 12 km. NNW Chinajá, Guatemala, 4.

Specimens examined.—Total, 142, as follows: Tamaulipas: 8 km. W, 10 km. N El Encino, 400 ft., 5. Veracruz: 4 km. WNW Fortín, 3200 ft., 1; 2 km. N Motzorongo, 1500 ft., 1; 3 km. E San Andrés Tuxtla, 1000 ft., 7; 38 km. SE Jesús Carranza, 500 ft., 118. Chiapas: Ocosingo, 10.

Transmitted June 30, 1961.

□

University of Kansas Publications
Museum of Natural History

Volume 14, No. 5, pp. 73-98, 4 figs.
December 29, 1961

orth American Yellow Bats, "Dasypterus," And a List of the Named Kinds Of the Genus Lasiurus Gray

BY

E. RAYMOND HALL AND J. KNOX JONES, JR.

University of Kansas
Lawrence
1961

University of Kansas Publications, Museum of Natural History

Volume 14, No. 5, pp. 73-98, 4 figs.
Published December 29, 1961

University of Kansas
Lawrence, Kansas

PRINTED BY
JEAN M. NEIBARGER, STATE PRINTER
TOPEKA, KANSAS
1961

28-8516

North American Yellow Bats, "Dasypterus," And a List of the Named Kinds Of the Genus Lasiurus Gray

BY

E. RAYMOND HALL AND J. KNOX JONES, JR.

INTRODUCTION

Yellow bats occur only in the New World and by most recent authors have been referred to the genus *Dasypterus* Peters. The red bats and the hoary bat, all belonging to the genus *Lasiurus* Gray, also occur only in the New World except that the hoary bat has an endemic subspecies in the Hawaiian Islands.

The kind of yellow bat first to be given a distinctive name was the smaller of the two species that occur in North America. It was named *Nycticejus ega* in 1856 (p. 73) by Gervais on the basis of material from the state of Amazonas, Brazil, South America, but was early recognized as occurring also in North America (in the sense that México and Central America, including Panamá, are parts of North America). More than 40 years elapsed before subspecific names were proposed for the North American populations; Thomas named *Dasypterus ega xanthinus* in 1897 (p. 544) from Baja California, and *Dasypterus ega panamensis* in 1901 (p. 246) from Panamá.

The larger of the two North American species was named *Lasiurus intermedius* in 1862 (p. 246) by H. Allen on the basis of material from extreme northeastern México. Another alleged species, *Dasypterus floridanus*, was named in 1902 (p. 392) by Miller from Florida, but as set forth below it is only a subspecies of *L. intermedius*, a species that is seemingly limited to parts of the North American mainland and Cuba.

A third species, *Atalapha egregia*, allegedly allied to the small yellow bat, *L. ega*, was named in 1871 (p. 912) by Peters from Santa Catarina, Brazil, but Handley (1960:473) thinks that *L. egregius* is allied instead to the red bats. The species *L. egregius* has not been studied in connection with the observations reported below.

Bats of the genus concerned were given the generic name *Nycteris* by Borkhausen in 1797 (p. 66), and the name *Lasiurus* by Gray in 1831 (p. 38). For much of the latter part of the 19th century the generic name *Atalapha* proposed by Rafinesque in 1814 (p. 12) was used because it antedated the name *Lasiurus*. In this period Harrison Allen (1894:137) raised to generic rank the name *Dasypterus* that had been proposed by Peters in 1871 (p. 912) only as a subgenus for the yellow bats. Since 1894 the yellow bats ordinarily have borne the generic name *Dasypterus*. The red bats and the hoary bat continued to be referred to as of the genus *Atalapha* until early in the 20th century when it was decided that a European bat of another genus was technically the basis for the name *Atalapha*. Thereupon *Lasiurus* was again used in the belief that it was the earliest available name for the bats concerned. But in 1909 (p. 90) Miller showed that the name *Lasiurus* was preoccupied by *Nycteris* Borkhausen, 1797 (p. 66). From 1909 until 1914 in conformance with the Law of Priority *Nycteris* was used for the red bat and the hoary bat.

At this point it is desirable to digress and indicate why and how the Law of Priority came into being. In the 19th century different technical names were used for the same kind of animal depending on the opinions of individual authors. For example, one author used name A because it was most descriptive of the morphology of the animal, another author used name B because it had been used more often than any other, another author used name C because it was more euphonious, etc. In order to achieve uniformity and stability a set of rules was drawn up in 1901 at the International Zoological Congress in Berlin. Those rules were based principally on the rule, or law, of priority. In effect, the law stated that the technical name first given to a kind of animal (with starting date as of January 1, 1758, *Systema Naturae* of Linnaeus) would be the correct and official name. After the mentioned rules were adopted, some zoologists, mostly non-taxonomists, objected to the rules and in response to these objections a compromise was adopted in 1913 at the International Zoological Congress in Monaco and the International Committee on Zoological Nomenclature was authorized to set aside, at its discretion, the Law of Priority. In 1913 it was thought by everyone that the names conserved (*nomina conservanda*) by setting aside the rules would be few.

Returning now to the generic names applied to the bats concerned, it is to be noted that from 1803 until 1909 *Nycteris* had been used as the generic name of an African bat on the erroneous assumption that the name was first applied in a valid fashion to the African bat. With the aim of conserving the name *Nycteris* for the African bat, some zoologists petitioned the International Committee on Zoological Nomenclature to set aside the Law of Priority and petitioned also that the name *Lasiurus* be validated for use again as the generic name for New World bats. This petition was granted in 1914 in the first lot of names for which exception to the rules was made. As a result, since 1914 *Lasiurus* has been used with increasing frequency, and *Nycteris* with decreasing frequency, for New World bats.

The above explanation of the application of the generic names *Nycteris, Atalapha,* and *Lasiurus* is given for two reasons: First, study of more abundant material than was available to Harrison Allen in 1894 when he raised *Dasypterus* to generic rank reveals, as set forth beyond, that the yellow bats are not generically different from the red bats and hoary bat and so will bear the same generic name that is applied to the red bat and hoary bat; second, a choice of generic names has to be made. Actually, the International Commission on Zoological Nomenclature since 1913 has voted to make many, instead of only a few, exceptions to the rules. The number of names resulting from these exceptions is becoming so large that some zoologists fear that the chaotic condition of nomenclature in the previous century will return. Those who hold such fears maintain that adherence to the rules of 1901, or to the Law of Priority, or at least to some rules, clearly is desirable. Certainly there is much logic in that view. According to the rules, *Nycteris* is the correct name of the bats concerned. According to the Commission, it is well to use instead the name *Lasiurus*. Perhaps the time has come to follow the rules and use *Nycteris*. But, because of the possibility that the Commission will return to its policy of 1913 and recommend only a few instead of many exceptions to the rules, the generic name *Lasiurus* is tentatively used in the following accounts.

Genus Lasiurus Gray

Hairy-tailed Bats

1797. *Nycteris* B[orkhause]n, Der Zoologe (Compendiose Bibliothek gemeinnützigsten Kenntnisse für alle Stände, pt. 21), Heft 4-7, p. 66. Type, *Vespertilio borealis* Müller [= *Lasiurus borealis*]. *Nycteris* Borkhausen is a homonym of *Nycteris* G. Cuvier and É. Geoffroy St.-Hilaire, 1795, type *Vespertilio hispidus* Schreber, 1774 [= *Nycteris hispida*], from Senegal. Although *Nycteris* Cuvier and Geoffroy St.-Hilaire is a *nomen nudum*, Opinion 111 of the International Commission of Zoological Nomenclature establishes the name as available for a genus of Old World bats. On this basis, *Nycteris* Borkhausen is not available for the New World genus. *Nycteris* É. Geoffroy St.-Hilaire, 1803, is a synonym of *Nycteris* Cuvier and Geoffroy St.-Hilaire, 1795, as given status by the Commission.

1831. *Lasiurus* Gray, Zool. Misc., No. 1, p. 38. Type, *Vespertilio borealis* Müller.

1871. *Atalapha* Peters, Monatsber. K. Preuss. Akad. Wiss., Berlin, p. 907, and other authors [*nec Atalapha* Rafinesque, 1814].

Type species.—*Vespertilio borealis* Müller.

Diagnosis.—Interfemoral membrane large and most of its upper surface furred; mammae 4; third, fourth and fifth fingers progressively shortened; ear short and rounded; skull short and broad; nares and palatal emargination wide and shallow (width transversely exceeding length anteroposteriorly); sternum prominently keeled; i. $\frac{1}{3}$, c. $\frac{1}{1}$, p. $\frac{1}{2}$ or $\frac{2}{2}$, m. $\frac{3}{3}$; when two upper premolars present, anterior one minute, peglike, and displaced lingually; M3 much reduced, area of its crown less than a third that of M1.

Members of this genus are notable for having three and even four young (more than other bats). In North America at least *L. borealis* and *L. cinereus*, are migratory.

Provisional Key to the Recent Species of *Lasiurus*

1. Color reddish or grayish (not yellowish); normally two premolars on each side of upper jaw.
   2. Occurring on Antillean islands (color reddish).
      3. Length of upper tooth-row less than 4.5 mm. (occurring on Hispaniola and Bahamas) .... *L. minor.*
      3′. Length of upper tooth-row more than 4.5 mm. (not occurring on Hispaniola and Bahamas).
         4. Greatest length of skull less than 13.9 mm. (occurring on Cuba) .... *L. pfeifferi.*
         4′. Greatest length of skull more than 13.9 mm. (occurring on Jamaica) .... *L. degelidus.*
   2′. Occurring on mainland and coastal islands of North and South America; also on Galapagos and Hawaiian islands (color reddish or grayish).
      5. Total length more than 120 mm.; color grayish .... *L. cinereus.*
      5′. Total length less than 120 mm.; color reddish.

6. Upper parts brick red to rusty red, frequently washed with white; lacrimal ridge present.
   7. Not occurring on Galapagos Islands .... *L. borealis.*
   7′. Known only from Galapagos Islands (both ear of 7.6 mm. and thumb of 6.4 mm. allegedly shorter than in *L. borealis* of adjacent mainland; presence of lacrimal ridge not verified) .... *L. brachyotis.*
6′. Upper parts not brick red to rusty red; lacrimal ridge not developed.
   8. Forearm more than 46.5 mm. (48 in only known specimen, a male); dorsum bright rufous (absence of lacrimal ridge not verified) .... *L. egregius.*
   8′. Forearm less than 46.5 mm.; dorsum not bright rufous.
      9. Upper parts mahogany brown washed with white; forearm less than 43 mm. .... *L. seminolus.*
      9′. Upper parts deep chestnut; forearm more than 43 mm. (44.8 in only known specimen, a female) *L. castaneus.*

1′. Color yellowish; only one premolar on each side of upper jaw.
   10. Total length more than 119 mm.; length of upper tooth-row 6.0 mm. or more .... *L. intermedius.*
   10′. Total length less than 119 mm.; length of upper tooth-row less than 6.0 mm. .... *L. ega.*

Lasiurus intermedius

Northern Yellow Bat

Diagnosis.—Upper parts yellowish orange, or yellowish brown, or brownish gray faintly washed with black to pale yellowish gray; size large (forearm, 45.2-62.8; condylocanine length, 16.9-21.5).

Distribution and Geographic Variation

Lasiurus intermedius H. Allen, type from Matamoros, Tamaulipas, has been reported from the Rio Grande Valley of Texas southward to Honduras and in Cuba. *Lasiurus floridanus* (Miller), type from Lake Kissimmee, Florida, has been recorded from southeastern Texas, eastward along the Gulf of Mexico to Florida, and thence northward along the Atlantic Coast to extreme southeastern Virginia (see records of occurrence beyond and Fig. 2). Specimens of *intermedius* from the vicinity of the type locality and from other localities in México differ from specimens of *floridanus* (from Florida and southern Georgia) as follows: Larger, both externally (especially forearm) and cranially (see measurements); teeth larger and heavier; skull heavier and having more prominent sagittal and lambdoidal crests; braincase less rounded, more elongate; auditory bullae relatively smaller; upper parts averaging brighter (yellowish to yellowish-orange in general aspect, rather than yellowish brown to brownish gray).

The differences mentioned above are of the magnitude of those that ordinarily separate subspecies of a single species rather than two species. Miller (1902:392-393), in the original description of *floridanus,* noted that the differences between it and *intermedius* were slight and remarked (p. 393): "Indeed, it is probable that it intergrades with the Texas animal." Lowery (1936:17) also has suggested that intergradation might occur between *intermedius* and *floridanus* "in southwestern Louisiana or eastern Texas"; later (1943:223-224) he pointed out that specimens from Baton Rouge, Louisiana, averaged larger in cranial dimensions than typical *floridanus* and again mentioned the possibility of intergradation between the two kinds. Sanborn (1954:25-26) touched obliquely on the problem when he wrote: "In Florida, *Dasypterus intermedius* is referred to as a Florida yellow bat (*Dasypterus floridanus*)." Handley (1960:478) wrote that certain morphological similarities suggested "gene flow" between the two kinds.

Specimens examined from Louisiana resemble *floridanus* from Georgia and Florida to the eastward in external dimensions. Some of those specimens resemble *floridanus* in size of skull, but two skulls from Louisiana are inseparable from those of topotypes of *intermedius.* The upper parts of specimens from Louisiana are generally like those of animals to the east but average somewhat paler (less brownish). The specimens seen from Louisiana seem to be intergrades between *intermedius* and *floridanus* but clearly are assignable to the latter.

The picture is less clear as regards bats from southeastern Texas (one specimen each from Colorado and Travis counties, and four specimens from Harris County). Five of the specimens have skulls (the Travis County specimen is a skin only) and of these, four are clearly assignable, on the basis of size and shape of the skull, to *intermedius.* The fifth skull (specimen from Colorado County) is intermediate in size between *floridanus* and *intermedius* and on that basis alone could be assigned with equal propriety to either. All these specimens from Texas more closely resemble *floridanus* than *intermedius* in external size (forearms: 49.2, 49.6, 50.7, 49.9 (approximate), 49.6, 49.1). The pale yellowish-gray upper parts of the four adults, seemingly resulting from a dilution of the brown-

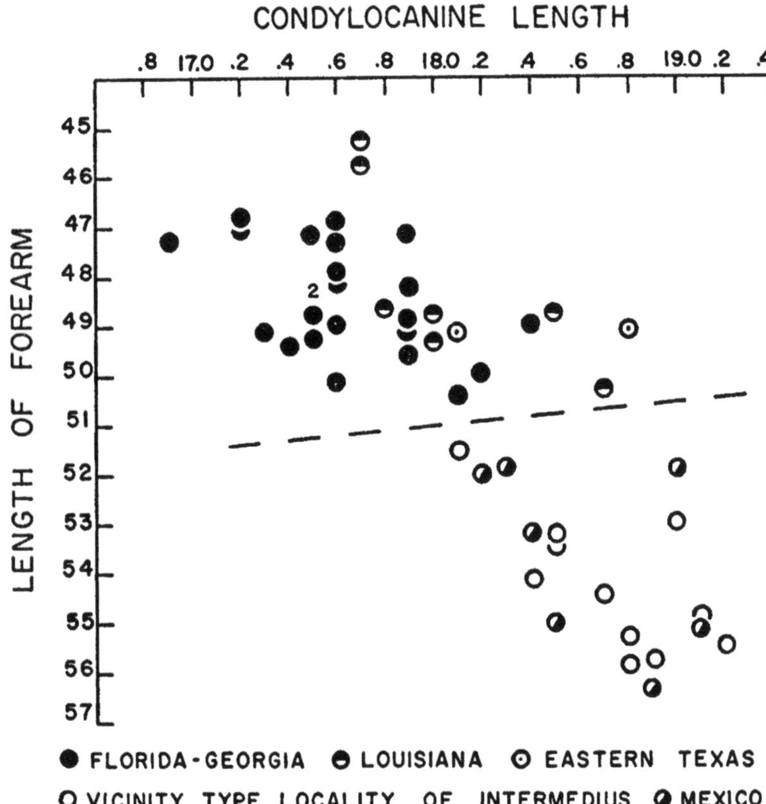

FIG. 1. Condylocanine length plotted against length of forearm for specimens of the species *Lasiurus intermedius.*

ish color found in *floridanus,* differ from the color of typical specimens of both *intermedius* and *floridanus,* but the average is nearer that of *floridanus* than that of *intermedius.* Color of pre-adult pelage in the one July-taken young of the year resembles the color of adults. An August-taken young of the year is in process of acquiring the adult pelage but the hairs have not reached their full growth; it is pale yellowish but not so grayish as the other specimens. All characters considered, the specimens from eastern Texas resemble *floridanus* more than they do *intermedius,* and so are provisionally assigned to *floridanus* (as was done by Taylor and Davis, 1947:19; Eads, *et al.,* 1956:440; and, Davis, 1960:59).

Additional material from southeastern Texas is needed. It will be remembered that the type locality of *intermedius* is in the Rio Grande Valley; all specimens seen, in the study here reported on, from the Texas side of the valley are unquestionably referable to that subspecies.

Intergradation, then, occurs between *L. intermedius* and *L. floridanus* in some degree in southern Louisiana and in more marked degree in southeastern Texas. Specimens from the area of intergradation vary more individually in many features than do specimens from other areas. In general the intergrades tend to resemble *floridanus* in small size externally and *intermedius* in large size of skull. The specimens from southeastern Texas differ from typical specimens of both subspecies in color, being pale yellowish gray (instead of yellowish to yellowish-orange as in *intermedius* or yellowish brown to brownish gray as in *floridanus*), and this difference is shared to some extent with animals from Louisiana, the latter being somewhat intermediate between bats from Texas and those from Florida and Georgia, although nearer those from Florida and Georgia.

An hypothesis to account for the variation noted is that in Wisconsin Time, and perhaps in earlier Pleistocene times, this yellow bat was (as it is now) a warmth-adapted animal as Blair (1959: 461) would term it. Some cool period forced the mainland populations of the two species into two refugia—peninsular Florida and eastern México—and the present area of intergradation is, therefore, of a secondary rather than a primary type. Possibly also the relatively treeless area of part of southern Texas has made for a sparse population there of *Lasiurus intermedius* and gene flow now may be, and long may have been, slight between the eastern and southern segments of the species.

It could be contended that the peculiar coloration of specimens from southeastern Texas, coupled with the tendency to have a large skull (as has *intermedius*) and small external dimensions (as has *floridanus*), justifies subspecific recognition for the animals that here are termed intergrades. But, judging by the specimens now available, such subspecific recognition would tend to obscure rather than clarify the geographic variation noted.

Life History

Probably bats of the species *Lasiurus intermedius* seek retreats primarily in trees (see Moore, 1949*a*:59-60) but Baker and Dickerman (1956:443) reported "approximately 45 yellow bats" concealed on July 22, 1955, "among dried corn stalks hanging from the sides of a large open tobacco shed" in the state of Veracruz. Young are born in late spring, three being the only number known except that Davis (1960:59) was told that in the vicinity of Mission, Texas, two was the usual number "born in May and June." Sherman (1945:194) reported a female with young (number not given) taken on June 7, 1918, at Seven Oaks, Florida, and another with three young taken on June 20, 1941, at Ocala, Florida. Lowery (1936:17) recorded a female, having three young, obtained on June 17, 1932, at Baton Rouge, Louisiana. A specimen taken on May 19, 1940, at Baton Rouge contained three embryos. Baker and Dickerman (*loc. cit.*) reported four adult females from Veracruz as lactating on July 22, 1955, but they were accompanied by flying young of the year and probably were near the end of the lactation period. Among specimens examined, juveniles are available by date as follows: 5 mi. N Baton Rouge, Louisiana (June 26, 1953); Palm Beach, Florida (July 6, 1950); and Izamal, Yucatán ("taken with mother" on July 28, 1910). Breeding probably takes place in autumn and winter; Sherman (*op. cit.*:196) reported males from Florida as sexually "mature" from the beginning of September to mid-February. Late winter segregation of sexes has been reported.

Subspecies

In the following accounts, localities of occurrence in each state are listed from north to south; if two lie in the same latitude, the westernmost is listed first. Localities that are italicized are not shown on the distribution map (Fig. 2), either because undue crowding of symbols would result or, in several cases, because we could not precisely place the localities. Length of forearm is the average of both forearms in individuals in which both forearms could be measured.

Lasiurus intermedius intermedius H. Allen

1862. *Lasiurus intermedius* H. Allen, Proc. Acad. Nat. Sci. Philadelphia, 14:246, "April" (between May 27 and August 1), type from Matamoros, Tamaulipas.

Geographic distribution.—Southern México (Yucatán, Chiapas and Oaxaca), northward along Gulf Coast to Rio Grande Valley of southern Texas (see Fig. 2).

Diagnosis.—Size medium (see measurements); sagittal crest present (height above braincase averaging 0.4 mm. in 12 from Brownsville, Texas); interorbital region relatively broad; M3 relatively broad (see comparisons in account of the Cuban subspecies beyond); mesostyle of M1 and M2 and 2nd commissure and cingulum of M3 large; pelage yellowish to yellowish-orange.

Comparisons.—See p. 79 and under accounts of *Lasiurus intermedius floridanus* and the Cuban subspecies.

External measurements.—Three adult males from the Sierra de Tamaulipas in Tamaulipas: Total length, 146, 136, 142; length of tail-vertebrae, 69, 67, 70; length of hind foot, 11, 11, 11; length of ear from notch, 17, 16, 17; length of forearm (dry), 53.2, 51.8, 51.9. Corresponding measurements for two adult females from 1 mi. SW Catemaco, Veracruz: 149, 155; 64, 69; 11, 12; 17, 17; 51.8, 55.2. Weight in grams of the Tamaulipan specimens, respectively: 24, 21, 24. For cranial measurements see Table 1.

Records of occurrence.—Specimens examined, 45, as follows: TEXAS: 5⅝ mi. N Mission, 2 (Texas A & M); *Santa Ana National Wildlife Refuge*, 1 (USNM); Brownsville, 13 (4 AMNH; 1 Texas A & M; 8 USNM). TAMAULIPAS: *Matamoros*, 2 (USNM); Sierra de Tamaulipas, 1200 ft., 10 mi. W, 2 mi. S Piedra, 1 (KU); *Sierra de Tamaulipas, 1400 ft., 16 mi. W, 3 mi. S Piedra*, 2 (KU). VERACRUZ: 16 mi. SW Catemaco, 15 (KU). OAXACA: Oaxaca, 1 (British Mus.). CHIAPAS: San Bartolomé, 1 (USNM). YUCATAN: Tekom, 1 (Chicago Mus.); Izamal, 5 (USNM). HONDURAS: Río Yeguare, between Tegucigalpa and Danli, 1 (MCZ).

Additional records: TEXAS: *Padre Island* (Miller, 1897:118); *Cameron County* (*ibid.*). OAXACA: Tehuantepec (Handley, 1960:478). YUCATAN: *Yaxcach* (not found, Gaumer, 1917:274).

Lasiurus intermedius floridanus (Miller)

1902. *Dasypterus floridanus* Miller, Proc. Acad. Nat. Sci. Philadelphia, 54:392, September 12, type from Lake Kissimmee, Oceola Co., Florida.

Geographic distribution.—Extreme southeastern Virginia, south along Atlantic Coast to and including peninsular Florida (except possibly extreme southern tip), thence westward to southern Louisiana and the southern part of eastern Texas (see Fig. 2).

Diagnosis.—Size small (see measurements); sagittal crest present but low; interorbital region relatively broad; teeth essentially as in *L. i. intermedius* except averaging smaller; pelage yellowish brown to grayish brown. For comparison with the Cuban subspecies, see account of that subspecies.

Comparisons.—From *Lasiurus intermedius intermedius, L. i. floridanus* differs as follows: averaging smaller (see measurements), especially in forearm and skull; teeth smaller; skull having less prominent sagittal and lambdoidal crests; braincase more nearly round; tympanic shields over petrosals approximately same size and therefore relatively larger; pelage of upper parts duller, yellowish brown to brownish gray instead of yellowish to yellowish orange.

External measurements.—Average (and extremes) of 14 February-taken males from along the Aucilla River, Jefferson Co., Florida: Total length, 126.8 (121-131.5); length of tail-vertebrae, 54.2 (51-60); length of hind foot, 9.8 (8-11); length of ear from notch (13 specimens), 16.3 (15-17); forearm (dry, 13 specimens), 48.1 (46.7-50.0). Corresponding measurements of the holotype, an adult female (after Miller, 1902:392): 129, 52, 9, 17, 49. Average (and extremes) weight in grams of the series of males: 17.7 (15.5-19.5). For cranial measurements see Table 1.

Records of occurrence.—Specimens examined, 65, as follows: TEXAS: Austin, 1 (Texas U.); *4 mi. N Huffman,* 1 (Texas A & M); Houston, 3 (1 KU; 2 MVZ); Eagle Lake, 1 (Texas A & M). LOUISIANA: 5 mi. N Baton Rouge, 1 (LSU); *1 mi. W LSU Campus, Baton Rouge,* 1 (LSU); *Baton Rouge,* 7 (1 AMNH; 5 LSU; 1 USNM); *½ mi. E Baton Rouge,* 1 (LSU); North Island, Grand Lake, 1 (LSU); Lafayette, 2 (USNM); Houma, 2 (USNM). GEORGIA: Beachton, 11 (6 Chicago Mus.; 5 USNM). FLORIDA: *2 mi. S Tallahassee,* 1 (AMNH); 5 mi. W Jacksonville, 1 (AMNH); Aucilla River, 15 mi. S Waukenna, 7 (Univ. Fla.); *Aucilla River, at U. S. Hgy. 98,* 8 (Univ. Fla.); *W of Gainesville,* 1 (Univ. Fla.); Gainesville, 3 (2 Univ. Fla.; 1 Univ. Mich.); *near Gainesville,* 1 (Univ. Fla.); *Alachua County,* 1 (Univ. Mich.); 2 mi. SW Deland, 2 (Univ. Fla.); head of Chassahowitzka River, 1 (USNM); Lakeland, 2 (Univ. Fla.); Seven Oaks [near present town of Safety Harbor], 2 (1 AMNH; 1 USNM); Lake Kissimmee, 1 (USNM); Palm Beach, 1 (Univ. Fla.); *Mullet Lake* (not found), 1 (USNM).

Additional records: VIRGINIA: Willoughby Beach (Rageot, 1955:456). SOUTH CAROLINA: 5 mi. NW Charleston (Coleman, 1940:90). LOUISIANA: New Orleans (Lowery, 1943:223). MISSISSIPPI: Hancock County (Hamilton, 1943:107). GEORGIA: W edge Camilla (Constantine, 1958:65). FLORIDA (Sherman, 1945:195, unless otherwise noted): *St. Marys River* [near Boulogue]; *vicinity Palm Valley* (Ivey, 1959:506); *6 mi. N Lake Geneva* (Sherman, 1937:108); Old Town; Welaka (Moore, 1949*a*:59); Bunnell; Ocala; *Davenport; Hillsborough River State Park;* 1 mi. NE Punta Gorda (Frye, 1948:182); Miami (Moore, 1949*b*:50).

Lasiurus intermedius insularis, new subspecies

Holotype.—Adult female, preserved in alcohol but having skull removed, formerly in the Poey Museum, University of Havana, now No. 81666, Museum of Natural History, University of Kansas, from Cienfuegos, Las Villas Province, Cuba; obtained on January 23, 1948, by D. Gonzáles Muñoz.

Geographic distribution.—Known only from the island of Cuba (see Fig. 2).

Diagnosis.—Large throughout (see measurements); sagittal crest enormously developed, especially posteriorly (height above braincase averaging 1.7 mm. in 4 specimens); interorbital region narrow; M3 narrow; mesostyle of M1 and M2 and 2nd commissure and cingulum of M3 small; pelage yellowish to reddish-brown.

Comparisons.—From *Lasiurus intermedius intermedius* of the adjacent mainland of México, *L. i. insularis* differs as follows: Larger, both externally and cranially; sagittal crest relatively higher, especially posteriorly; interorbital region relatively narrower; palate longer posterior to tooth-rows; teeth distinctly larger throughout except M3, which is relatively (frequently actually) narrower, averaging 66.1 (62.5-71.0) per cent width of M2 in *insularis* rather than 74.1 (66.6-79.3) per cent in 10 *intermedius* from Brownsville, Texas; mesostyle of M1 and M2 relatively smaller as are second commissure and cingulum of M3; coloration of No. 254714 USNM resembling that of *L. i. intermedius*, but coloration of three specimens, preserved in alcohol, averaging somewhat darker (more reddish-brown) than in *intermedius*.

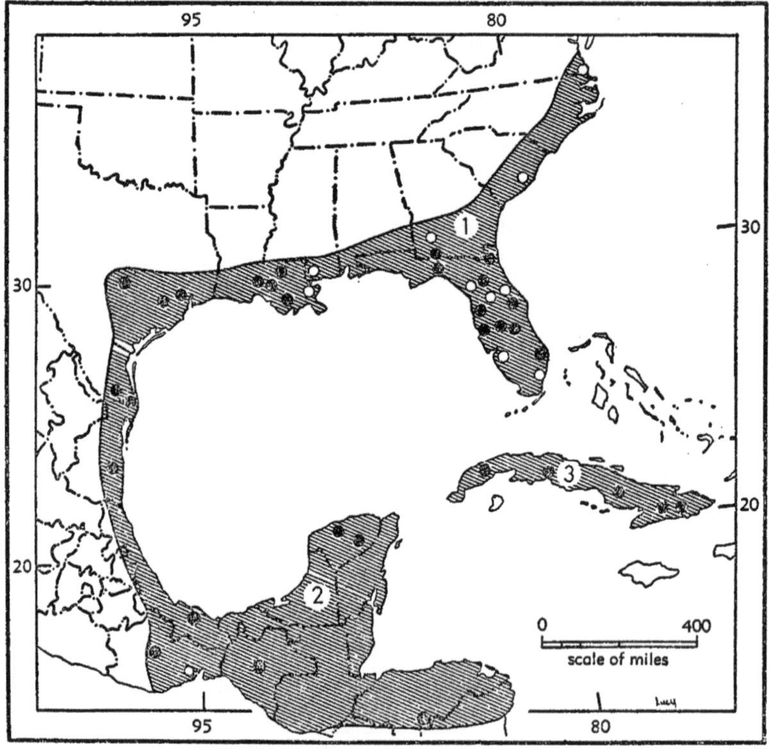

FIG. 2. Geographic distribution of the three subspecies of *Lasiurus intermedius*.
1. *L. i. floridanus* 2. *L. i. intermedius* 3. *L. i. insularis*

Black dots represent localities of capture of specimens examined. Hollow circles represent localities of capture of other specimens recorded in the literature but not examined by us (Hall and Jones).

From *Lasiurus intermedius floridanus* of the adjacent Floridan mainland, *L. i. insularis* differs in many of the same ways that it differs from *L. i. intermedius*, except that the differences are even more trenchant because *floridanus* is smaller than *intermedius*. Indeed, the difference in size between *floridanus* and *insularis* is approximately the same as between *Lasiurus borealis* and *Lasiurus cinereus*.

Measurements.—External measurements (all taken from specimens preserved in alcohol) of the holotype, followed by those of two other females, one from Laguna La Deseada, San Cristóbal, Pinar del Río Province, and the other from Bayate, Guantánamo, Oriente Province, are, respectively: Total length, 164, 161, 150; length of tail-vertebrae, 68, 76, 77; length of hind foot, 12, 12, 13; length of ear from notch, 20, 17, 19; length of forearm, 61.2, 62.6, 61.8. The length of forearm of a study skin from San Germán (that otherwise lacks external measurements) having wings spread is approximately 55.4. For cranial measurements see Table 1.

Remarks.—Four of the five specimens on which the name *L. i. insularis* is based differ to such a degree from mainland populations of the species *L. intermedius* that specific rather than subspecific recognition for the Cuban bat might seem warranted. It is because of the fifth specimen (USNM 254714) that we accord subspecific rank to *insularis*. It is smaller than the other Cuban specimens and except for longer condylocanine length, longer mandibular toothrows, narrower interorbital region, and heavier dentition is indistinguishable in measurements from the largest specimens of *L. i. intermedius* from the mainland. In addition, it appears not to have the enormously developed sagittal crest of the other specimens of *insularis* although posteriorly the dorsal part of the skull (where the crest is most prominent) is missing. USNM 254714 agrees with the other Cuban specimens in having the mesostyle of M1 and M2 somewhat reduced and in having a small M3 on which the cingulum and second commissure are poorly developed, and this specimen is regarded as representative of the lower size limits of the Cuban population.

The skull from San Blas was found in an owl pellet (see de Beaufort, 1934:316).

Records of occurrence.—Specimens examined, 5, all from Cuba, as follows: Pinar del Río Prov.: Laguna La Deseada, San Cristóbal, 1 (Poey Museum). Las Villas Prov.: Cienfuegos, 1 (KU, the holotype). Camaguey Prov.: San Blas, 1 (Amsterdam Zoological Museum). Oriente Prov.: San Germán, 1 (USNM); Bayate, Guantánamo, 1 (Ramsdem Museum, Univ. Oriente).

TABLE 1.—CRANIAL MEASUREMENTS (IN MILLIMETERS) OF THREE SUBSPECIES OF LASIURUS INTERMEDIUS

Catalogue number or number of specimens averaged	Museum	Sex	Locality	Condylocanine length	Zygomatic breadth	Interorbital breadth	Alveolar length C–M3	Breadth of rostrum (between anterior openings of infra-orbital canals)	Mastoid breadth	Length of mandibular tooth-row (i–m3)
			Lasiurus intermedius floridanus							
Ave.10	UF	♂♂	[1]Aucilla River, Florida	17.6	12.8	5.0	6.2	7.2	10.0[2]	8.0
Min.				17.0	12.6	4.7	6.0	6.9	9.6	7.8
Max.				18.2	13.0	5.3	6.4	7.5	10.2	8.2
1783	LSU	♀	Baton Rouge, La.	18.7	....	5.1	6.7	7.7	....	8.8
1820	LSU	♀	Baton Rouge, La.	18.5	....	....	6.7	7.2	10.1	8.7
1840	LSU	♂	Baton Rouge, La.	18.0	12.7	5.0	6.4	7.1	9.9	8.0
6790	LSU	♂	Baton Rouge, La.	18.0	12.8	4.9	6.5	7.2	9.9	8.2
3681	LSU	♂	7 mi. SE Baton Rouge, La.	17.7	12.6	5.0	6.4	7.0	9.8	8.2
6791	LSU	♀	[3]Grand Lake, La.	17.9	12.6	4.9	6.3	7.2	9.9	8.3
88	MVZ	♀	Houston, Texas	19.1	13.8	5.1	6.6	7.5	10.3	8.7
69	TAMC	♀	4 mi. N Huffman, Texas	18.8	13.4	5.0	6.7	7.7	....	8.7
85	TAMC	♂	Eagle Lake, Texas	18.1	12.9	4.8	6.6	7.2	9.8	8.5

Lasiurus intermedius intermedius

1437	USNM	?	Matamoros, Tamaulipas	18.9	13.6	5.1	6.6	7.5	10.7	8.9
1439	USNM	?	Matamoros, Tamaulipas	19.0	14.0	5.3	6.6	7.8	10.7	8.8
Ave. 12	USNM[4]	?[5]	Brownsville, Texas	18.7[6]	13.8[6]	5.2	6.6	7.7	10.4[6]	8.7
Min.				18.1	13.0	4.9	6.4	7.4	10.0	8.4
Max.				19.2	14.7	5.5	7.0	8.2	11.1	9.0
37	KU	♂	[7]Sierra de Tamaulipas	18.2	13.2	5.5	6.2	7.6	10.3	8.0
32	KU	♂	[8]Sierra de Tamaulipas	18.4	13.7	5.2	6.5	7.4	10.6	8.4
34	KU	♂	[8]Sierra de Tamaulipas	18.3	13.2	5.1	6.5	7.6	10.3	8.1
59	KU	♀	Catemaco, Veracruz	19.0	13.5	5.0	6.5	7.5	10.2	8.8
60	KU	♀	Catemaco, Veracruz	19.0	13.5	4.7	6.4	7.6	10.3	8.7

Lasiurus intermedius insularis (all from Cuba)

2395	AZM	?	Cave near San Blas	21.4	15.1	4.8	7.3	8.4	11.9	9.5+
254714	USNM	♂	San Germán, Oriente	19.5	14.1	4.8	6.9	7.8	11.0	9.3
81666	KU	♀	Cienfuegos, Las Villas	20.5	15.2	4.6	7.2	8.2	11.9	9.6
	Poey Mus.	♀	San Cristóbal, Pinar del Río	21.5	15.6	4.7	7.5	8.9	11.8	9.7
	Ramsdem Oriente Univ.	♀	Bayate, Guantánamo, Oriente	20.9	14.8	4.6	7.3	8.4	11.2	9.7

1. "Rt. 98" and "15 mi. S Waukenna" both in Jefferson Co.
2. Only nine specimens.
3. "N Island, Grand Lake, Iberville Parish."
4. Some in Amer. Mus. Nat. History.
5. Females, 8; males, 3; unsexed, 1.
6. Only 11 specimens.
7. 10 mi. W, 2 mi. S Piedra, Tamaulipas.
8. 16 mi. W, 3 mi. S Piedra, Tamaulipas.

Lasiurus ega

Southern Yellow Bat

Diagnosis.—Upper parts yellowish brown (much as in *Lasiurus intermedius floridanus* from Louisiana) having overlay of grayish or blackish anterior to shoulders; hair on basal half of interfemoral membrane more yellowish than elsewhere; size medium (forearm 42.7-52.2; condylocanine length 14.6-16.3).

This species occurs from the southwestern United States (Palm Springs, California, and Tucson, Arizona) southward into Uruguay and northeastern Argentina. Of the six currently (see Handley, 1960) recognized subspecies of *L. ega,* four occur only in South America, and two occur only in North America.

Cabrera (1958:115) regarded *Dasypterus ega fuscatus* Thomas (1901:246), based on three specimens from Río Cauquete, Río Cauca, Colombia, as a synonym of *Dasypterus ega panamensis* Thomas (*loc. cit.*) that was based on a specimen from Bogava, 250 meters elevation, Chiriqui, Panamá. The latter name has line priority over *fuscatus.* Cabrera (1958:116) remarked that: "Las diferencias que Thomas señaló entre el *Dasypterus* de Panamá y el de Colombia (*fuscatus*) nos parecen estar dentro de los límites de la variación individual, siendo además muy raro que una especie de quiróptero este representada en Colombia y en Panamá por razas diferentes."

On July 16, 1958, at the British Museum of Natural History, one of us (Hall) examined the holotypes of *panamensis* and *fuscatus,* as well as other materials used by Thomas, and readily perceived the differences that he pointed out. Thomas' description, although terse, is accurate. *L. e. fuscatus* is much more blackish than *panamensis.* We are inclined to retain the two names as applicable to two subspecies. Whether or not *fuscatus* is synonymized under *panamensis,* the holotype of *panamensis* is an intergrade between the almost black Colombian animal (*fuscatus*) and the paler individuals in Central America and territory north thereof. Even so, the holotype of *panamensis* more closely resembles the blackish Colombian population than the paler populations to the north and the name *panamensis,* therefore, is correctly applicable to the bat from Panamá, but not to bats of the species *Lasiurus ega* from farther north as most authors (see, for example, Hall and Kelson, 1959:194, map 143; and Handley, 1960:474) suggested was the case. For the populations north of Panamá the name *Lasiurus ega xanthinus* (Thomas) (1897:544) needs to be used.

Lasiurus ega xanthinus (Thomas)

1897. *Dasypterus ega xanthinus* Thomas, Ann. Mag. Nat. Hist., ser. 6, 20:544, December, type from Sierra Laguna, Baja California.

1953. *Lasiurus ega xanthinus,* Dalquest, Louisiana State Univ. Studies, Biol. Ser., 1:61, December 28.

Geographic distribution.—Southern California, southern Arizona, and northern Coahuila southward through México to southern Costa Rica.

Diagnosis.—Yellowish brown with an overlay of grayish anterior to the shoulders; forearm, 42.7-47.2.

Remarks.—Specimens from Baja California and the adjacent western part of the mainland of México average paler than specimens from Veracruz and some places in Central America but the differences are slight.

Records of occurrence.—Specimens examined, 21, as follows: BAJA CALIFORNIA.—Comondú, 1 (USNM); Sierra Laguna, 4 (1 USNM, 3 British Mus.). COAHUILA.—4 mi. W Hacienda La Mariposa, 2300 ft., 2 (KU). ZACATECAS.—Concepción del Oro, 7680 ft., 4 (KU). TAMAULIPAS.—Sierra de Tamaulipas, 1200 ft., 10 mi. W, 2 mi. S Piedra, 5 (KU); 16 mi. W, 3 mi. S Piedra, 1 (KU). SINALOA.—1 mi. S Pericos, 1 (KU). VERACRUZ.—Achotal, 1 (Chicago Mus.). YUCATAN.—Yaxcach, 1 (USNM). COSTA RICA.—Lajas, Villa Quesada, 1 (AMNH); San José, 1 (AMNH).

Additional records: CALIFORNIA: Palm Springs (Constantine, 1946:107). ARIZONA: Tucson (Cockrum, 1961:97). BAJA CALIFORNIA (Handley, 1960: 474): Santa Ana; Miraflores. SINALOA: Escuinapa (Handley, 1960:475). DURANGO: Aguajequiroz, 12 mi. SSW Mapimí, 5000 ft. (Greer, 1960:511). SAN LUIS POTOSI (Dalquest, 1953:62): 1½ mi. E Río Verde; 19 km. SW Ebano; 4 mi. SSW Ajinche. QUINTANA ROO: 7 mi. N, 37 mi. E Puerto de Morelos (Ingles, 1959:384). HONDURAS: Tegucigalpa (Handley, 1960:474).

Lasiurus ega panamensis (Thomas)

1901. *Dasypterus ega panamensis* Thomas, Ann. Mag. Nat. Hist., ser. 7, 8:246, September, type from Bogava [= Bugaba], Chiriquí, 250 meters, Panamá.

1960. *Lasiurus ega panamensis,* Handley, Proc. U. S. Nat. Mus., 112:474, October 6.

Geographic distribution.—Panamá; also recorded by Handley (1960:474) from Venezuela.

Diagnosis.—"General colour dark browish clay-color, something between Ridgway's 'raw-umber' and 'clay-color'. Fur black basally, then dull brownish buffy, the extreme tips black. Center of face similar to back, cheeks from eyes to lips contrasting black. Rump and hairy part of interfemoral verging toward brownish fulvous. Under surface similar to upper." (Thomas, 1901: 246.) Forearm of holotype, 46.5.

Remarks.—Notes taken down by one of us (Hall) on July 16, 1958, at the British Museum, Natural History, contain the following: "Color accurately described by Thomas. The blackish stands out. The difference between the types of D. *e. panamensis* and D. *e. xanthinus* is tremendous."

Record of occurrence.—Specimen examined, one, the type (British Mus.).

RELATIONS BETWEEN THE SPECIES OF LASIURUS

As suggested by Dalquest in 1953 (p. 62) and by Handley in 1959 (p. 119) and 1960 (p. 473), the yellow bats, *Lasiurus ega* (Gervais) and *Lasiurus intermedius* H. Allen, so closely resemble the hoary bat, *Lasiurus cinereus* (Palisot de Beauvois), and the red bats, *Lasiurus borealis* (Müller) and the seven related species listed below, that all are properly included in a single genus. Many of the common characteristics are enumerated above in the diagnosis of the genus (see also Handley, 1960:473).

A listing of the differences between the species is less impressive than a listing of the resemblances. The yellow bats differ less from the red bats than does the hoary bat, *L. cinereus,* which differs from all of the others as follows: talonid on m3 larger; p4 single-rooted instead of double-rooted; hypocone on M1 and M2 smaller; coronoid process lower; ossified part of tympanic ring, which shields

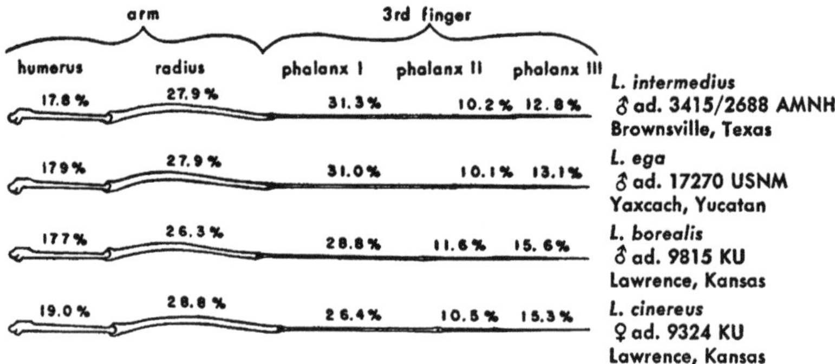

FIG. 3. Diagram of bones of right arm and third finger (middle digit) including cartilage on distal end of terminal (3rd) phalanx. Percentages are in terms of the over-all length of the arm and third finger.

the petrosal, larger; humerus relatively shorter; forearm relatively longer; first phalanx of middle finger relatively shorter; presternum including keel longer than wide instead of *vice versa.* The differences in the sternum and proportions of the forelimb reflect the more rapid flight of the hoary bat. The yellow bats differ from the red bats and hoary bat in long rostrum, pronounced sagittal crest, high coronoid process, absence of the first upper premolar, long first phalanx of the third digit and short terminal (3rd) phalanx of the same digit. Features in which the red bats are extreme in the genus are short rostrum, short forearm, and relatively longer second phalanx of the third finger. The red bats differ only slightly one from another.

Next to nothing is known of extinct Tertiary ancestors of species of the genus *Lasiurus*. Also relatively little is known about *Lasiurus* in the Pleistocene. Consequently, evolution of the living species has to be inferred almost entirely from what is known about their structure, habits, and geographic distribution. Figure 4 presents some ideas concerning relationships.

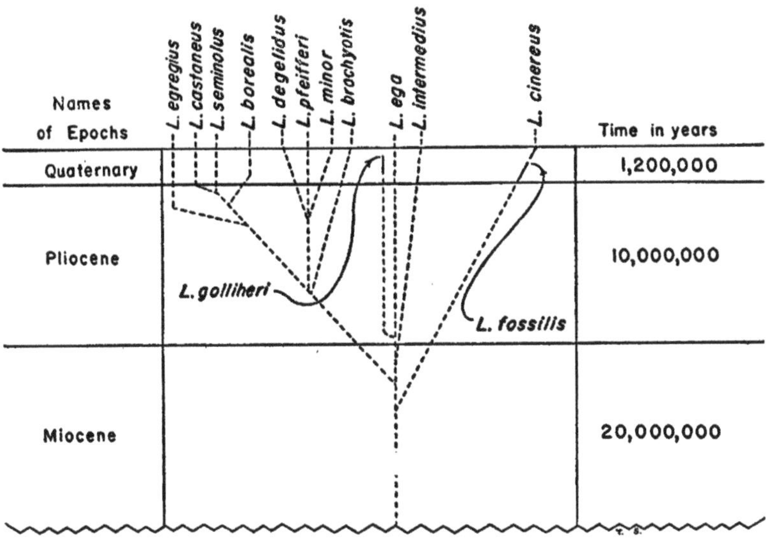

FIG. 4. Postulated relationships of species of the genus *Lasiurus*.

LIST OF NAMED KINDS OF THE GENUS LASIURUS

The words "type from" indicate that a specimen or specimens served as basis for the name. The words "type locality" signify lack of knowledge as to whether a specimen was preserved.

Red Bats

Lasiurus borealis borealis (Müller), 1776, type from New York.

[Vespertilio] noveboracensis Erxleben, 1777, based, in part, on "Der Neujorker" of Müller (*ante*).

Vespertilio lasiurus Schreber, 1781, type locality, North America.

Vespertilio rubellus Palisot de Beauvois, 1796, type locality unknown.

Vespertilio rubra Ord, 1815, based on the red bat of Wilson, Amer. Ornith., 6:60.

Vespertilio tesselatus Rafinesque, 1818, type locality unknown.

Vespertilio monachus Rafinesque, 1818, type locality unknown.

Vespertilio rufus Warden, 1820, based on the red bat of Wilson, *ibid.*

Lasiurus funebris Fitzinger, 1870, type locality, Tennessee.

Myotis quebecensis Yourans, 1930, type from Anse-à-Wolfe, Quebec.

Lasiurus borealis frantzii (Peters), 1871, type from Costa Rica.

Lasiurus borealis teliotis (H. Allen), 1891, type probably from California.

Lasiurus borealis ornatus Hall, 1951, type from Penuela, Veracruz.

Lasiurus borealis varius (Poeppig), 1835, type from Antuco, Provincia de Bió-Bió, Chile.

Nycticeus poepingii Lesson, 1836, type from Chile.

Lasiurus borealis salinae Thomas, 1902, type from Cruz del Eje, Cordoba, Argentina.

Lasiurus borealis blossevillii Lesson and Garnot, 1826, type from Montevideo, Uruguay.

Vespertilio bonariensis Lesson, 1827, type from Buenos Aires, Argentina.

Lasiurus enslenii Lima, 1926, type from São Lourenço, Rio Grande do Sul, Brazil.

Lasiurus pfeifferi (Gundlach), 1861, type from Cuba.

Lasiurus degelidus Miller, 1931, type from Sutton's, District of Vere, Jamaica.

Lasiurus minor Miller, 1931, type from "Voute l'Eglise," 1350 ft., a cave near the Jacmel road a few kilometers N Trouin, Haiti.

Lasiurus seminolus (Rhoads), 1895, type from Tarpon Springs, Pinellas Co., Florida.

Lasiurus castaneus Handley, 1960, type from Tacarcuna Village, 3200 ft., Río Pucro, Darién, Panamá.

Lasiurus egregius (Peters), 1871, type from Santa Catarina, Brazil.

Lasiurus brachyotis (J. A. Allen), 1892, type from San Cristóbal Island, Galapagos Islands.

Yellow Bats

Lasiurus golliheri (Hibbard and Taylor), Contributions Mus. Paleo., Univ. Michigan, 16:162, fig. 10F, July 1, 1960 [an extinct species], type from [a stratum of Late Pleistocene Age] "Below the caliche bed in the Kingsdown formation; Cragin Quarry local fauna, locality 1 (Sangamon age); Big Springs Ranch, SW ¼ sec. 17, T. 32 S., R. 28 W. (Kansas University Locality 6), Meade County, Kansas."

Lasiurus ega xanthinus (Thomas), 1897, type from Sierra Laguna, Baja California.

Lasiurus ega panamensis (Thomas), 1901, type from Bugaba, Chiriquí, Panamá.

Lasiurus ega fuscatus (Thomas), 1901, type from Río Cauquete, Colombia.

Dasypterus ega punensis J. A. Allen, 1914, type from Isla de Puná, Ecuador.

Lasiurus ega ega (Gervais), 1856, type from Ega, Estado de Amazonas, Brazil.

Lasiurus caudatus Tomes, 1857, type from Pernambuco, Brazil.

Lasiurus ega argentinus (Thomas), 1901, type from Goya, Province of Corrientes, Argentina.

Lasiurus intermedius intermedius H. Allen, 1862, type from Matamoros, Tamaulipas, México.

Lasiurus intermedius floridanus (Miller), 1902, type from Lake Kissimmee, Osceola Co., Florida.

Lasiurus intermedius insularis Hall and Jones, 1961, type from Cienfuegos, Las Villas Province, Cuba.

Hoary Bats

Lasiurus fossilis Hibbard, Contributions Mus. Paleo., Univ. Michigan, 8(No.6): 134, fig. 5, June 20, 1950 [an extinct species], type from [an early Pleistocene or a late Pliocene deposit] "Rexroad formation, Rexroad fauna. Locality UM-K1-47, Fox Canyon, XI Ranch, Meade County, Kansas."

Lasiurus cinereus cinereus (Palisot de Beauvois), 1796, type from Philadelphia, Pennsylvania. Known from Late Pleistocene time as well as from Recent time (see Hibbard and Taylor, Contributions Mus. Paleo., Univ. Michigan, 16:159, fig. 10A, July 1, 1960, for occurrence in Cragin Quarry local fauna, Sangamon Age, Meade County, Kansas).

Vespertilio pruinosus Say, 1823, type from Engineer Cantonment, Washington Co., Nebraska.

A[talapha]. mexicana Saussure, 1861, type from an unknown locality, probably from Veracruz, Puebla, or Oaxaca.

Lasiurus cinereus villosissimus É. Geoffroy St.-Hilaire, 1806, type locality, Asunción, Paraguay.

Lasiurus grayi Tomes, 1857, type from Chile.

Atalapha pallescens Peters, 1871, type from Paramo de la Culata, Andes de Mérida, Venezuela.

Atalapha cinerea brasiliensis Pira, 1905, type from Ignape, São Paulo, Brazil.

Lasiurus cinereus semotus (H. Allen), 1890, type from Hawaii.

EXPLANATION AND ACKNOWLEDGMENTS

Hall and Jones are jointly responsible for the accounts of the two species of yellow bats, but Hall alone assumes responsibility for the other parts of the paper. Thanks are extended to the National Science Foundation for financial support (Grant No. 56 G 103) of the study here reported on. We are grateful also to the following persons for the loan of specimens in their care: S. B. Benson, Museum of Vertebrate Zoology, University of California (MVZ); W. F. Blair, Department of Zoology, University of Texas (Univ. Texas); W. B. Davis, Dept. Wildlife Management, Agricultural and Mechanical College of Texas (TAMC or Texas A & M); D. H. Johnson, C. O. Handley, Jr., and W. H. Setzer, U. S. National Museum (USNM); Barbara Lawrence, Museum of Comparative Zoology at Harvard College (MCZ); J. N. Layne, Department of Biology, University of Florida (UF); G. H. Lowery, Jr., Museum of Natural History, Louisiana State University (LSU); P. J. H. van Bree, Department of Mammals, Zoölogisch Museum, Amsterdam (AZM); and R. G. Van Gelder, American Museum of Natural History (AMNH). Thanks are extended also to E. T. Hooper and W. H. Burt, Mus. Zoology, University of Michigan (Univ. Mich.), to Philip Hershkovitz, Chicago Natural History Museum (Chicago Mus.), and to Peter Crowcroft, British Museum, Natural History, for permission to examine specimens there. Mr. Gilberto Silva Taboada arranged the loan of specimens from the Poey Museum, University of Havana and from the Ramsdem Museum, University of Oriente, both in Cuba. Mr. Silva Taboada and Dr. Carlos G. Aguayo of the Poey Museum graciously arranged an exchange of specimens whereby the holotype of *L. i. insularis* became the property of the Museum of Natural History, University of Kansas. Specimens in the last mentioned institution are identified with the symbol KU.

LITERATURE CITED

ALLEN, H.

1862. Descriptions of two new species of Vespertilionidae, and some remarks on the genus Antrozous. Proc. Acad. Nat. Sci. Philadelphia, pp. 246-248, "April" but between May 27 and August 1.

1894. A monograph of the bats of North America. Bull. U. S. Nat. Mus., 43:i-ix + 1-198, pls. 1-38, March 14.

BAKER, R. H., and DICKERMAN, R. W.

1956. Daytime roost of the yellow bat in Veracruz. Jour. Mamm., 37: 443, September 11.

BLAIR, W. F.

1959. Distributional patterns of vertebrates in the southern United States in relation to past and present environments. Pp. 443-468, *in* Hubbs, C. L. (ed.), Zoogeography, Amer. Assoc. Adv. Sci. Publ. 51:x + 509, January 16.

BORKHAUSEN, M. B.

1797. Der Zoologe (Compendiose Bibliothek gemeinnützigsten Kenntnisse für alle Stände, pt. xxi), Heft iv-vii [including page 66; original not seen].

CABRERA, A.

1958. Catalogo de los mamíferos de America del Sur. Rev. Mus. Argentino Cienc., Nat. Cienc., Zool., 4(1):1-307, March 27.

COCKRUM, E. L.

1960. Southern yellow bat from Arizona. Jour. Mamm., 42:97, February 20.

COLEMAN, R. H.

1940. Dasypterus floridanus in South Carolina. Jour. Mamm., 21:90, February 15.

CONSTANTINE, D. G.

1946. A record of Dasypterus ega xanthinus from Palm Springs, California. Bull. Southern California Acad. Sci., Los Angeles, 45:107, September 20.

1958. Ecological observations on lasiurine bats in Georgia. Jour. Mamm., 39:64-70, 1 fig., February 20.

DALQUEST, W. W.

1953. Mammals of the Mexican state of San Luis Potosí. Louisiana State Univ. Studies, Biol. Ser., 1:1-229, 1 fig., December 28.

DAVIS, W. B.

1960. The mammals of Texas. Game and Fish Comm., Bull. 41:1-252, 73 figs., 64 maps.

DE BEAUFORT, L. F.

1934. Dasypterus intermedius H. Allen in Cuba. Jour. Mamm., 15:316, November 15.

EADS, R. B., MENZIES, G. C., and WISEMAN, J. S.

1956. New locality records for Texas bats. Jour. Mamm., 37:440, September 11.

FRYE, O. E., JR.

1948. Extension of range of two species of bats in Florida. Jour. Mamm., 29:182, May 14.

GAUMER, G. F.

1917. Monografía de los mamíferos de Yucatán. Dept. Talleres Gráficos Secretaría Fomento, México, xli + 331 pp., 57 pls., 2 photographs, 1 map.

GERVAIS, P.
1856. *In* Castelnau, F. L. de Laporte. Expédition dans les parties centrales de l'Amérique du Sud . . . pendant . . . 1843 a 1847 . . ., vol. for 1855 [part], pp. 25-88, pls. 7-15.

GRAY, J. E.
1831. Descriptions of some new genera and species of bats. Zoological Miscellany, No. 1, pp. 37-38.

GREER, J. K.
1960. Southern yellow bat from Durango, Mexico. Jour. Mamm., 41: 511, November 11.

HALL, E. R., and KELSON, K. R.
1959. The mammals of North America. The Ronald Press Co., New York, 1280 pp., 1231 illustrations, March 31.

HAMILTON, W. J., JR.
1943. The mammals of eastern United States. Comstock Publ. Co., Ithaca, New York, 432 pp., illustrated.

HANDLEY, C. O., JR.
1959. A revision of American bats of the genera Euderma and Plecotus. Proc. U. S. Nat. Mus., 110:95-246, 27 figs., September 3.
1960. Descriptions of new bats from Panama. Proc. U. S. Nat. Mus., 112:459-479, October 6.

INGLES, L. G.
1959. Notas acerca de los mamíferos Mexicanos. An. Inst. Biol., 29: 379-408, March 31.

IVEY, R. D.
1959. The mammals of Palm Valley, Florida. Jour. Mamm., 40:585-591, November 20.

LOWERY, G. H., JR.
1936. A preliminary report on the distribution of the mammals of Louisiana. Proc. Louisiana Acad. Sci., 3:11-39, 2 text figs., 4 pls., March.
1943. Check-list of the mammals of Louisiana and adjacent waters. Occas. Papers Mus. Zool., Louisiana State Univ., 13:213-257, 5 figs., November 22.

MILLER, G. S., JR.
1897. Revision of the North American bats of the family Vespertilionidae. N. Amer. Fauna, 13:1-140, 3 pls., 40 figs., October 16.
1902. Twenty new American bats. Proc. Acad. Nat. Sci. Philadelphia, 54:389-412, September 12.
1909. The generic name *Nycteris*. Proc. Biol. Soc. Washington, 22:90, April 17.

MOORE, J. C.
1949*a*. Putnam County and other Florida mammal notes. Jour. Mamm., 30:57-66, February 14.
1949*b*. Range extensions of two bats in Florida. Quart. Jour. Florida Acad. Sci., 11:50, March 22.

PETERS, W.
1871. "22 December Gesammtsitzung der Akademie. . . . eine monographische Übersicht der Chiropterengattungen *Nycteris* und *Atalapha*." Monatsberichte d. Konig. Preuss. Akad. d. Wiss. zu Berlin (for 1870), pp. 900-914, 1 pl.

RAFINESQUE, C. S.
1814. Précis des decouvertes et travaux somiologiques. Palerme, pp. 1-55 + 3.

RAGEOT, R. H.
1955. A new northeasternmost record of the yellow bat, *Dasypterus floridanus.* Jour. Mamm., 36:456, August 30.

SANBORN, C. C.
1954. Bats of the United States. Public Health Reports, 69(1):17-28, illustrated, January.

SHERMAN, H. B.
1937. A list of the Recent land mammals of Florida. Proc. Florida Acad. Sci. (for 1936), 1:102-128.
1945. The Florida yellow bat, *Dasypterus floridanus.* Proc. Florida Acad. Sci. (for 1944), 7:193-197, January 20.

TAYLOR, W. P., and DAVIS, W. B.
1947. The mammals of Texas. Bull. Texas Game, Fish and Oyster Comm., 27:1-79, illustrated, August.

THOMAS, O.
1897. *Descriptions of new bats and rodents from America.* Ann. Mag. Nat. Hist., ser. 6, 20:544-553, December.
1901. *New Neotropical mammals, with a note on the species of* Reithrodon. Ann. Mag. Nat. Hist., ser. 7, 8:246-254, September.

Transmitted June 30, 1961.

□

28-8516

UNIVERSITY OF KANSAS PUBLICATIONS
MUSEUM OF NATURAL HISTORY

Volume 14, No. 6, pp. 99-110, 1 fig.
December 29, 1961

Natural History of the Brush Mouse (Peromyscus boylii) in Kansas With Description of a New Subspecies

BY

CHARLES A. LONG

UNIVERSITY OF KANSAS
LAWRENCE
1961

University of Kansas Publications, Museum of Natural History

Editors: E. Raymond Hall, Chairman, Henry S. Fitch,
Theodore H. Eaton, Jr.

Volume 14, No. 6, pp. 99-110, 1 fig.
Published December 29, 1961

University of Kansas
Lawrence, Kansas

PRINTED BY
JEAN M. NEIBARGER, STATE PRINTER
TOPEKA, KANSAS
1961

28-8518

Natural History of the Brush Mouse (Peromyscus boylii) in Kansas With Description of a New Subspecies

BY

CHARLES A. LONG

In order to determine the geographic distribution of the brush mouse in the state, 15 localities, chosen on the basis of suitable habitat, were investigated by means of snap-trapping in the winter and spring of 1959, spring of 1960, and winter and spring of 1961. Variation in specimens obtained by me and in other specimens in the Museum of Natural History, The University of Kansas, was analyzed. Captive mice from Cherokee County, Kansas, were observed almost daily from March 27, 1960, to June 1, 1961. Captive mice from Chautauqua and Cowley counties were studied briefly. Contents of 38 stomachs of brush mice were analyzed, and diet-preferences of the captive mice were studied. Data from live-trapping and from snap-trapping are combined and provide some knowledge of size and fluctuation of populations in the species.

Examination of the accumulated specimens and the captive mice reveals the occurrence in southern Kansas of an unnamed subspecies, which may be named and described as follows:

Peromyscus boylii cansensis new subspecies

Type.—Male, adult, skin and skull; No. 81830, K. U.; from 4 mi. E Sedan, Chautauqua County, Kansas; obtained on December 30, 1959, by C. A. Long, original No. 456.

Range.—Known from 3 mi. W Cedar Vale, *in* Cowley County, Kansas, and from the type locality.

Diagnosis.—Size medium (see Table 1 beyond); underparts white; upper parts Ochraceous-Tawny laterally, becoming intermixed with black and approaching Mummy Brown dorsally (capitalized color terms after Ridgway, 1912); eye nonprotuberant; tail short but well-haired distally and usually less than half total length; nasals long; cranium large.

Comparisons.—From P. *b. attwateri*, the subspecies geographically nearest *cansensis*, the latter can be easily distinguished by the less protuberant eyes and relatively shorter tail (91 per cent of length of head and body; in topotypes of P. *b. attwateri* from Kerr County, Texas, 104 per cent; in specimens of P. *b. attwateri* from Cherokee County, Kansas, 103 per cent). P. *b. cansensis* is darker than P. *b. attwateri* and darker than P. *b. rowleyi*, the palest subspecies of brush mouse, which occurs to the westward. The skull and nasals (see Table 1) in adults of P. *b. attwateri* from Cherokee County average shorter than in *cansensis*.

Specimens examined.—Total, 26. *Cowley Co.:* 3 mi. W Cedar Vale, 16. *Chautauqua Co.:* type locality, 10.

TABLE 1. AVERAGE AND EXTREME MEASUREMENTS OF SPECIMENS OF P. B. CANSENSIS, OF P. B. ATTWATERI FROM CHEROKEE COUNTY, KANSAS, AND OF TOPOTYPES OF P. B. ATTWATERI LISTED BY OSGOOD, 1909.

	P. b. cansensis			*P. b. attwateri*	
	Three miles west of Cedar Vale	Type locality	Both localities	Two miles south of Galena	Type locality*
No. specimens....	11	7	18	20	10
Total length.....	180.5 170–199	176.7 166–188	179.1	186.2 170–210	196.0 ....
Tail-vertebrae....	85.5 72–101	85.0 75–93	85.3	94.5 83–104	100.0 ....
Hind foot.......	23.1 22–24	23.6 22–25	23.3	23.8 22–25	21.0 ....
Ear from notch...	18.2 17–19	19.1 18–21	18.5	18.4 14–21	....
Greatest length of skull.......	27.9 26.8–29.0	28.3 27.9–28.9	28.1	27.8 26.6–29.1	....
Length of nasals..	10.4 9.9–10.8	10.2 9.5–10.7	10.3	9.9 9.1–10.4	....
Zygomatic breadth......	14.3 13.9–15.0	13.5 13.0–13.9	13.9	13.8 13.3–14.4	....

* From Turtle Creek, Kerr County, Texas, after Osgood (1909:148).

Distribution of Peromyscus boylii *in Kansas*

The subspecies *Peromyscus boylii attwateri* is known in the state only from Cherokee County, the southeasternmost county in the state. Probably the only locality where the brush mouse occurs in that county is on the systems of cliffs along Shoal Creek, southward from Galena, to the eastward of Baxter Springs. This is the extent of the known range, and in my opinion the probable range, of P. *b. attwateri* in the state (see Fig. 1). Cockrum (1952:fig. 49) by mistake mapped the species from west of Baxter Springs in Cherokee County.

Osgood (1909:149) recorded the subspecies P. *b. attwateri* from Cedar Vale, Chautauqua County, Kansas, but the specimen from there must now be assigned to *cansensis* on geographic grounds. Probably the specimen was not obtained from Cedar Vale itself for the habitat is not suitable there. Numerous specimens are known from 3 mi. W Cedar Vale, *in* Cowley County, Kansas, all of which are assigned to *cansensis.* Osgood's recorded locality is situated between this locality and the type locality of *cansensis,* which is 4 mi. E Sedan, Chautauqua County, Kansas. The distribution of *cansensis* also is shown in Fig. 1.

The probable geographic range of P. *boylii* is based on trapping data (see Fig. 1). The brush mouse is confined to systems of

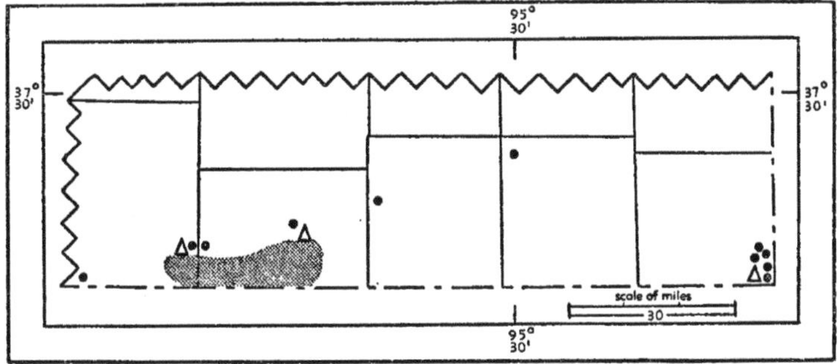

FIG. 1. Distribution of the brush mouse in Kansas. The southernmost row of counties includes from left to right Cowley, Chautauqua, Montgomery, Labette, and Cherokee. Black dots represent trapping localities from which brush mice were not obtained. Triangles represent localities from which brush mice were obtained. The stippled area contains suitable habitat for the brush mouse, but was not investigated. The easternmost triangle represents a place 2 mi. S Galena, Cherokee Co., Kansas, from which P. *b. attwateri* is known. The westernmost triangle represents a place 3 mi. W Cedar Vale, *in* Cowley Co., Kansas, from which P. *b. cansensis* is known. The triangle of intermediate position represents the type locality of P. *b. cansensis,* a place 4 mi. E Sedan, Chautauqua Co., Kansas. Many of the trapping localities have been investigated more than once.

wooded cliffs in Kansas. The two subspecies seem to be separated by more than 80 miles of grasslands. Blair (1959) has postulated that in the northeastern part of its range P. *b. attwateri* is represented by disjunct, relict populations formed by diminishing montane or cool, moist environmental conditions. He has implied that the critical climatic change occurred during post-Wisconsin times, and that the isolation of these populations occurred so recently that no morphological differentiation has resulted in them. Inasmuch as the species is widely distributed in México, the southwestern

United States, and in California, and has been recorded from the Pleistocene of California (Hay, 1927:323), it is reasonable to suppose that the species immigrated into Kansas from the southwest and that the immigration was in a generally northward or eastward direction. If long tail and large eyes are specializations for a scansorial mode of life (discussed below), then P. *b. cansensis* must be considered more primitive than P. *b. attwateri* for the eyes are less protuberant and the tail is shorter in P. *b. cansensis* than in the latter. I suggest that P. *b. cansensis* occurred in what is now known as Kansas before P. *b. attwateri* entered this area by way of the Ozark Mountains. The occurrence of a mouse of "the *truei* or *boylei* group" (Hibbard, 1955:213) in southwestern Kansas in the Jinglebob interglacial fauna of the Pleistocene adds little to support the thesis outlined above, but is not inconsistent with the thesis. Incidentally, the geographic distribution of P. *boylii* may differ somewhat from that shown by Blair (1959:fig. 5); whereas he has mapped the distribution of P. *boylii* to show disjunctivity in P. *b. attwateri* and homogeneity in the distribution of other subspecies of the brush mouse to the westward and southward, disjunctivity actually occurs frequently also in the western and southern subspecies.

Ecology

In Kansas the brush mouse is confined to systems of cliffs, the faces of which range in height to at least 40 feet. The highest cliffs—some approximately 100 feet—on which brush mice are known to occur in Kansas are along Shoal Creek, Cherokee County. The brush mouse is found on low bluffs that are parts of higher systems, but in Cherokee County the mouse was not obtained from low bluffs separated by even a few miles from the cliff-system along Shoal Creek. As implied above the brush mouse is adapted for a scansorial mode of life; but other mice and rats inhabit the rocky crevices of low bluffs. Whereas the brush mouse is well adapted for living on high cliffs it seems that the other rodents are better adapted for life on low cliffs. *Sigmodon hispidus* was obtained from the low, limestone cliffs mentioned previously. From most low bluffs in southeastern Kansas (and on some high bluffs outside the geographic range of *cansensis*) *Peromyscus leucopus* was obtained. In Cowley County the brush mouse was abundant when P. *leucopus* was not and *vice versa* during this study. *Sigmodon hispidus* did not associate with the brush mouse in any area, although S. *hispidus* was often trapped in grassy areas adjacent to cliffs and on the grassy crests of the hills. Except at the locality in Cherokee County, the pack rat, *Neotoma floridana,* was found in association with the brush mouse. *Microtus ochrogaster* was the must abundant rodent in adjacent southwestern Missouri (Jackson, 1907) before *Sigmodon* thoroughly infiltrated this area and southeastern Kansas. Activities of other rodents may have confined the brush mouse ecologically to cliffs. Although the grasslands are a barrier to further intrusion by the brush mouse into Kansas, one cannot assume that they alone confined the brush mouse to cliffs. Such an assumption would not explain its absence on systems of cliffs similar to and near other systems of cliffs on which it is found in the non-grassy Ozarkian habitats of Arkansas, as was noticed by Black (1937). Such an assumption would not indicate why the size of the cliff-systems is correlated with the absence or presence of the brush mouse on the northeastern margin of its geographic range.

Parasites found on P. *b. attwateri* include three individuals of the laelapid mite, *Haemolaelaps glasgowi.* Two of these mites were removed from a live mouse. Two larval Ixodid ticks, *Ixodes* possibly *cookei,* were removed from the pinnae of the ears of a specimen of *cansensis* from the type locality, 4 mi. E Sedan, Chautauqua County. Four larval Ixodid ticks, *Dermacentor* possibly *variabilis,* were removed from the pinnae of the ears of a live specimen of *cansensis* from 3 mi. W Cedar Vale, in Cowley County.

Black (1937:195) and Cockrum (1952:180-181) reported stomach

TABLE 2. STOMACH CONTENTS OF 38 BRUSH MICE FROM SOUTHEASTERN KANSAS IN WINTER AND SPRING.

Localities and number of stomachs	Month	Empty	Acorn pulp	Seeds
2 mi. S Galena				
10	May, 1959	2	6	2
11	December, 1959	1	10	0
3	March, 1960	1	2	0
4 mi. E Sedan				
3	December, 1959	3	0	0
2	April, 1961	1	1	0
3 mi. W Cedar Vale				
6	December, 1959	1	3	2
3	December, 1960	0	3*	0

* Judged to be acorn pulp or hickory nut pulp.

contents of P. *b. attwateri* from Cherokee County containing acorn pulp, seeds, and insects. Analysis of 38 stomachs of the brush mouse (Table 2) show acorns to be the most commonly used food in winter and spring. Seed coats were only rarely found, and insects were absent. Two captive females preferred acorns. Live beetles and grasshoppers of numerous kinds were decapitated and their inner parts eaten. Seeds (wheat, corn, and oats) were also eaten. Inasmuch as acorns appear to be the chief food, it is not surprising that the brush mouse is usually found on cliffs that support stands of blackjack oak (*Quercus marilandica*). Other oaks are present, but I have no evidence that the brush mouse eats their acorns. A. Metcalf told me that he observed in December, 1960, a released brush mouse interrupt its movement toward a hole in a cliff-face along Cedar Creek, Cowley County, in order to pick up an acorn (judged to be from the blackjack oak) in daylight. The mouse carried the acorn into the hole in the cliff. I have observed that captive brush mice eat acorns of the blackjack oak but not some other kinds of acorns.

Behavior

The chief differences observed between the brush mouse and other species of the genus *Peromyscus* in Kansas can be summarized as follows: the brush mouse is a superior and more cautious climber; seldom jumps from high places when under stress; is capable of finding its way better in partial darkness; has a stronger preference for acorns; and sometimes buries or hides seeds or acorns. These are all behavioral adaptations that seem in harmony with its mode of life.

Buck, Tolman, and Tolman (1925) showed the balancing function of the tail in *Mus musculus*. Climbers (for example, squirrels) often possess long, well-haired tails. It is reasonable to suggest (as did Hall, 1955:134) that the long, tufted tail is an adaptation for a scansorial existence. Little observation is necessary to observe how such a tail is used in balancing. Furthermore, it is used as a prop when the mouse is climbing a vertical surface. Dalquest (1955:144) mentioned tree-climbing in P. *boylii* from San Luis Potosi, México. It may occur in P. *b. attwateri* or in P. *b. cansensis* also, but there is no evidence as yet to prove it.

The brush mouse can seldom be induced to jump from heights of two feet or more. Rather it tends to scamper downward or to remain in place. It often swings itself over an edge, holding to it by its hind feet, and sometimes to it lightly with its tail, and reduces a short jump by almost the length of its body. Such caution seems to be an adaptation in a mouse that lives as a climber.

Many animals of cavernous habitats have small eyes (see Dobzhansky, 1951:284). Some nocturnal animals (for example, owls) have large eyes. The brush mouse has large, protuberant eyes; it lives in the deep crevices and fissures of the cliffs on which it is found, but it is not strictly a cave-dwelling animal. Perhaps large eyes aid the brush mouse in performing activities in the partial darkness of a deep crevice or hole in a cliff. Brush mice experimentally placed in what appeared to be total darkness fed, built houses of cotton, and ran and climbed in the usual manner.

On several occasions the captive brush mice hid surplus seeds and on other occasions hid acorns by burying them and sometimes by placing them in a small jar. The mice never carried the surplus food into their house.

Black (1937:195) has claimed that the brush mouse builds a nest similar to that of the nest of the pack rat, *Neotoma floridana*. Hall (1955:134) doubts this to be the case. Dalquest (1953:144) described a nest of P. *boylii* from San Luis Potosi as seven inches in diameter, made of leaves, and found in a hollow tree. Drake (1958:110) noted that P. *b. rowleyi* lives in holes and crevices in rocky bluffs in Durango, México. I have found this to be the case for P. *b. attwateri*, as did A. Metcalf (unpublished) for P. *b. cansensis*. Nests of sticks and leaves were taken apart by Metcalf, and all sign indicated only the presence of the pack rat. I have observed that there are no such houses on the cliffs along Shoal Creek, Cherokee County, and that no pack rats have been obtained from there (pack rats have not been reported from Cherokee County). Blair (1938) found two brush mice (P. *b. attwateri*) in the house of a pack rat in Oklahoma. Nests of the brush mice that occur in Kansas have not been found.

A lactating, pregnant female (KU 81833) of P. *b. attwateri*, containing three embryos, was obtained on December 24, 1959, and shows that this subspecies breeds in winter. Accumulated records for the subspecies indicates year-round breeding (see Cockrum, 1952:181). Another female obtained on March 27, 1960, was probably lactating.

Pregnant females of P. *b. cansensis* (KU 84892, 84895, and 84890) were obtained from the type locality on April 1-2, 1961, containing 3, 4, and 5 embryos respectively. This indicates, perhaps, increased breeding in spring; five was the highest number of embryos found in brush mice in Kansas.

Population Studies

In the period of my study the populations of brush mice became smaller, perhaps owing to the severe winter of 1959-1960. In Cowley County, P. *leucopus* is now abundant and P. *boylii* rare where in December of 1959, the opposite was true. It is also possible, of course, that trapping has depleted the populations.

Conclusions

1. A new subspecies of brush mouse is named and described from southern Kansas.
2. The new subspecies has smaller eyes and a shorter tail and may be more primitive than P. *b. attwateri.*
3. No significant sexual dimorphism was noted in P. *boylii.*
4. In Kansas, P. *b. attwateri* is known only from a single locality; P. *b. cansensis* is known from only two localities, both in Kansas.
5. The cliff-dwelling habit of P. *boylii* probably isolates populations from one another.
6. The grasslands constitute a barrier for the brush mouse.
7. In Kansas, P. *b. cansensis* probably is an older population than P. *b. attwateri.*
8. In Kansas the brush mouse is confined to systems of cliffs that are wooded and that are at least 40 feet in height.
9. The brush mouse may be confined to cliffs in part by activities of other rodents.
10. The brush mouse commonly associates with the pack rat.
11. Laelapid mites have been found on specimens of P. *b. attwateri.*
12. Larval ixodid ticks were found on specimens of P. *b. cansensis.*
13. Acorns seem to be the chief food of the brush mouse; insects and seeds are also commonly eaten.
14. The brush mouse is adapted for climbing and probably for a partly subterranean life.
15. P. *b. attwateri* breeds in winter, as well as in other parts of the year.

16. P. *b. cansensis* is known to breed in early April.
17. The highest number of embryos obtained from a brush mouse in Kansas is five.

Acknowledgments

I am indebted to Prof. E. Raymond Hall and to Mr. J. Knox Jones, Jr., for suggestions and editorial assistance. Prof. R. H. Camin identified the ticks and mites recorded herein. Mr. A. Metcalf, Mrs. C. F. Long, and Mr. D. L. Long helped with the field studies and in other ways.

Literature Cited

BLACK, J. D.
1937. Mammals of Kansas. 30th Biennial Report, Kansas State Board of Agri., 35:116-217.

BLAIR, W. F.
1938. Ecological relationships of the mammals of the Bird Creek Region, Northeastern Oklahoma. Amer. Midl. Nat., 20:473-526.

1959. Distributional patterns of vertebrates in the southern U. S. in relationship to past and present environment. Zoogeography, pp. 463-464 and Fig. 5, January 16.

BUCK, C. W., TOLMAN, N., and TOLMAN, W.
1925. The tail as a balancing organ in mice. J. Mamm., 6:267-271.

COCKRUM, E. L.
1952. Mammals of Kansas. Univ. Kansas Publ., Museum of Nat. Hist., 7:6, 180-181.

DALQUEST, W. W.
1953. Mammals of the Mexican state of San Luis Potosí. Louisiana State Univ. Studies, Biol. series No. 1, 232 pp.

DOBZHANSKY, T.
1951. Genetics and the origin of species, 3d ed. New York, Columbia Univ. Press, x + 364 pp.

DRAKE, J. D.
1958. The brush mouse, Peromyscus boylii, in southern Durango. Museum Publ., Michigan State Univ., 1:97-132.

HALL, E. R.
1955. Handbook of mammals of Kansas. Univ. Kansas Mus. Nat. Hist. Publ. No. 7, 303 pp.

HAY, O. P.
1927. The Pleistocene of the western region of N. America . . . . Carnegie Inst. Washington, 346 pp., 12 pls.

HIBBARD, C. W.
1955. The Jinglebob interglacial (Sangamon?) fauna from Kansas . . . . Museum of Paleo., Univ. Michigan, pp. 179-228, 2 pls.

JACKSON, H. H. T.
1907. Notes on some mammals of southwestern Missouri. Proc. Biol. Soc. Washington, 20:71-74.

OSGOOD, W. H.
1909. Revision of the mice of the American genus Peromyscus. N. Amer. Fauna, 28:1-285, April 17.

RIDGWAY, R.
1912. Color standards and color nomenclature. Washington, D. C., 43 pp., 53 pls.

Transmitted June 30, 1961.

□

28-8518

University of Kansas Publications
Museum of Natural History

Volume 14, No. 7, pp. 111-120, 1 fig.
December 29, 1961

Taxonomic Status of Some Mice of The Peromyscus boylii Group in Eastern Mexico, With Description of a New Subspecies

BY

TICUL ALVAREZ

University of Kansas
Lawrence
1961

University of Kansas Publications, Museum of Natural History

Volume 14, No. 7, pp. 111-120, 1 fig.
Published December 29, 1961

University of Kansas
Lawrence, Kansas

PRINTED BY
JEAN M. NEIBARGER, STATE PRINTER
TOPEKA, KANSAS
1961

29-393

Taxonomic Status of Some Mice of The Peromyscus boylii Group in Eastern Mexico, With Description of a New Subspecies

BY

TICUL ALVAREZ

Saussure (1860) described *Peromyscus aztecus* from southern México. Osgood (1909) by comparison of one of Saussure's specimens with some from Mirador, Veracruz, concluded that *aztecus* was a subspecies of *P. boylii*. Dalquest (1953) incorrectly reported specimens of P. *boylii* from San Luis Potosí as P. *b. aztecus*. Merriam (1898) named *Peromyscus levipes* from Mt. Malinche, Tlaxcala. Thomas (1903) described from Orizaba, Veracruz, P. *beatae*, which Osgood (1909) mistakenly thought was indistinguishable from P. *boylii levipes*. Therefore, Osgood in 1909 in his revision of the genus *Peromyscus* reported only two subspecies of P. *boylii* from eastern México: P. *b. levipes*, and P. *b. aztecus*. Study of Osgood's and Thomas' material, along with recently collected specimens from the states of eastern México, leads me to conclude that P. *aztecus* and P. *boylii* are different species; that P. *beatae* is a valid subspecies different from P. *b. levipes;* and finally that specimens of P. *boylii* from Nuevo León and northwestern Tamaulipas pertain to an hitherto unnamed subspecies.

Peromyscus aztecus Saussure

1860. *H[esperomys]. aztecus* Saussure, Revue et Mag. Zool., Paris, ser. 2, 12:105, type from southern México, probably from the vicinity of Mirador, Veracruz, according to Osgood (N. Amer. Fauna, 28:156-157, April 17, 1909).

1909. *Peromyscus boylei aztecus*, Osgood, N. Amer. Fauna, 28:156, April 17.

Geographic distribution.—Known only from Mirador and Jalapa in Veracruz, and Huachinango in Puebla.

Diagnosis.—Size medium for the genus (see measurements); tail about as long as head and body; dorsal coloration near Sayal Brown (capitalized color terms after Ridgway, 1912); sides reddish; underparts Light Buff; tail bicolored but not distinctly so; supraorbital border of skull angular, and bullae pointed anteriorly; anterior half of braincase nearly straight (not rounded) as viewed from above; upper molar series long (4.7-5.0); incisive foramina short in relation to length of skull.

Comparisons.—From *Peromyscus boylii*, P. *aztecus* differs as follows: Larger in most parts measured; maxillary tooth-row 4.7-5.0 instead of 4.0-4.6; color brighter on sides (reddish instead of ochraceous); supraorbital border angular instead of rounded; anterior border of zygomatic plate convex in upper half and almost straight in lower half as opposed to nearly straight throughout in *boylii;* pterygoid fossa broader; bullae more pointed anteriorly and less inflated; mesostyles of upper molars larger; surface between orbital region and nasals convex in lateral view instead of flat.

Remarks.—When Saussure (1860:105) described P. *aztecus* he did not designate a type or type locality. Osgood (1909:157) designated as lectotype the mounted specimen, in the Geneva Museum, which has the skull inside and of which Saussure figured the molar teeth. Osgood (*loc. cit.*) examined one of the three specimens (No. 3926 USNM) that Saussure used in describing P. *aztecus* and found that it agreed "in every respect with recently collected specimens from Mirador, Veracruz, which, in the lack of exact knowledge, may be assumed to be the type locality, as it is certain that some at least of Saussure's specimens were taken near there."

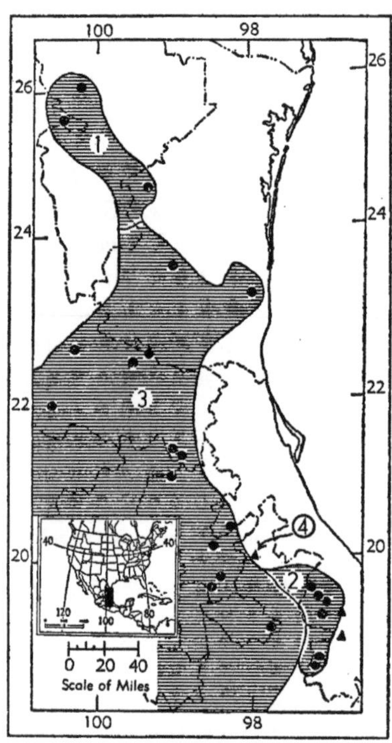

FIG. 1. Two species of *Peromyscus*.
1. P. *boylii ambiguus*
2. P. *boylii beatae*
3. P. *boylii levipes*
4. P. *aztecus* (triangles)

Osgood regarded P. *aztecus* as a subspecies of P. *boylii* because of the resemblance between *aztecus* and P. *b. evides,* but *evides* is far removed geographically (occurring only in western México) from *aztecus,* and is smaller. P. *aztecus* is larger than any known subspecies of P. *boylii,* and is not known to intergrade with P. *b. levipes* or P. *b. beatae* (with which *aztecus* occurs sympatrically at Jalapa, Veracruz), the two subspecies of *boylii* that are found nearest the geographic range of P. *aztecus.* Also, as mentioned previously, *aztecus* possesses distinctive characters that distinguish it from all subspecies of *boylii.* For these reasons I regard *aztecus* as a distinct species.

According to Hall and Kelson (1959:634), P. *aztecus* occurs in San Luis Potosi, Hidalgo, and west-central Veracruz, but their map 364 is based on the records of Osgood (1909:158) and Dalquest (1953:143). I have examined all the specimens reported by the two authors last named and find that those from San Luis Potosi are P. *boylii levipes.*

The diagnosis and comparisons here presented of *aztecus* were based on specimens from Mirador in comparison with all the specimens of P. *boylii* from eastern México listed beyond. The largest specimens of P. *boylii* that I have examined are from Las Vigas, Veracruz, and localities within a radius of five kilometers thereof. Some measurements of these large specimens of P. *boylii* overlap those of P. *aztecus* but the two kinds of mice differ greatly in characters of the skull, in color, and in length of tail.

The specimens (three adults and three juveniles) from Huachinango, Puebla, are slightly darker than specimens from Mirador but do not differ otherwise. Of two specimens reported from Jalapa, Veracruz, by Osgood (1909:158), one (108547 USNM) agrees with specimens from Mirador in color and cranial characteristics and is P. *aztecus,* whereas the other (108548 USNM) is P. *b. beatae.*

Specimens examined.—Total 16 (all USNM) as follows: PUEBLA: Huachinango, 6. VERACRUZ: Mirador, 9; Jalapa, 1.

Peromyscus boylii levipes Merriam

1898. *Peromyscus levipes* Merriam, Proc. Biol. Soc. Washington, 12:123, April 30, type from Mt. Malinche, 8400 ft., Tlaxcala.

1909. *Peromyscus boylei levipes,* Osgood, N. Amer. Fauna, 28:153, April 17.

Geographic distribution.—Southeastern Tamaulipas and eastern San Luis Potosí, south through the central states of México to Guatemala.

Diagnosis.—Size medium for the species; tail shorter or longer than head and body (83-112.3%); color variable according to locality but in general ochraceous, having some dusky on upper parts; supraorbital border not angular, almost rounded; auditory bullae large.

Comparisons.—For comparisons see accounts of the subspecies discussed beyond and Osgood (1909:145).

Remarks.—A precise diagnosis for P. *b. levipes* is difficult to prepare because some geographic variation in color and in the cranial characters is present within the range of the subspecies as here understood. For instance there is a gradual cline of decreasing size to the northward in nearly all measurements, but the ratio of length of tail to length of head and body does not present such a cline; mice from several localities in San Luis Potosí have a relatively shorter tail than do mice from farther north and from farther south. Also, specimens labeled in reference to Zacualpilla, Jacales, Jacala, Tulancingo, and San Miguel Regla average slightly darker dorsally than do typotypes. Some of these specimens are reddish on the cheek and lateral line. Specimens from San Luis Potosi resemble topotypes, but some specimens from northeastern localities in that state have cinnamon or brownish upper parts and are intermediate in coloration between populations of *levipes* to the south and populations of the same subspecies to the north from the Sierra Madre Oriental and the Sierra de Tamaulipas. Specimens from these two sierras have a cinnamon-reddish color that is more intense in specimens from the Sierra de Tamaulipas.

Osgood (1909:153) recorded P. *b. levipes* as occurring from central Nuevo León south through San Luis Potosí, Hidalgo, and Veracruz to southern Oaxaca. Actually specimens from Nuevo León and from most parts of Veracruz differ subspecifically from *levipes* and also from each other. In Veracruz, P. *b. levipes* is known only from the northwestern part.

Specimens examined.—Total 179 as follows: TAMAULIPAS: Sierra Madre Oriental, 5 mi. S, 3 mi. W Victoria, 1900 ft., 2; *8 mi. S, 6 mi. W Victoria, 4000 ft.*, 37; Sierra de Tamaulipas, 2000 ft., 8 mi. S, 11 mi. W Piedra, 13. SAN LUIS POTOSI: Villar, 11 (USNM); 10 km. E Platanito, 19 (LSU); *8 mi. E (by road) Santa Barbarita*, 12 (LSU); *Agua Zarca*, 3 (LSU); 6 km. NE Cd. Maíz, 13 (LSU); *Pendencia Region (Puerto Lobos)*, 1 (LSU); *Pendencia, 2½ mi. N Puerto Lobos*, 5 (LSU); 3 km. SW Sán Isidro, 15 (LSU); Cerro Coneja Region, Llano Coneja, 6100 ft., 2 (LSU); Xilitla, 4 (LSU). HIDALGO: 10 mi. NE Jacala, 5050 ft., 7; Regla (Sán Miguel), 2250 m., 4; Arroyo de las Tinajas, 2370 m., 9.5 km. SSW Tulancingo, 1; 10 mi. NW Apam, 7750 ft., 1. VERACRUZ: 3 km. N Zacualpan, 6000 ft., 1; *3 km. W Zacualpan, 6000 ft.*, 12; *2 km. N Los Jacales, 7500 ft.*, 8; *6 km. WSW Zacualpilla, 6500 ft.*, 5. TLAXCALA: Mt. Malinche, 3 (USNM).

Peromyscus boylii beatae Thomas

1903. *Peromyscus beatae* Thomas, Ann. Mag. Nat. Hist., ser. 7, 11:485, May, type from Xometla Camp, Mt. Orizaba, Veracruz.

Geographic distribution.—East side of the Sierra Madre Oriental in Veracruz, from Jalancingo south to Xuchil.

Diagnosis.—Size large for the species; tail no shorter than head and body (100-114.8%); dorsum dark (near Prout's Brown or Mummy Brown middorsally, Clay Color on sides); supraorbital border rounded; anterior palatine foramina long.

Comparisons.—*P. b. beatae* differs from other subspecies of P. *boylii* by the combination of large size, long tail, and dark color.

Remarks.—Thomas (1903:485) described P. *beatae* on the basis of five specimens from Xometla Camp (lat. 18° 59′ N, long. 97° 10′ W) and one juvenile from Santa Barbara Camp, both on the Volcán de Orizaba, Veracruz. Thomas thought that *beatae* was related to *aztecus*, but the differences relied on by him to distinguish the two are the same as those that distinguish *aztecus* from *boylii*. Osgood (1909:153) placed *beatae* in synonymy under P. *b. levipes* because Mount Orizaba (type locality of *beatae*) is "relatively very near" Mount Malinche (type locality of *levipes*), and Thomas had not compared *beatae* with *levipes*. Xometla, on the east side of the Volcán de Orizaba, is approximately 56 miles east of the Tlaxcalan part of Mount Malinche and is situated where the Tropical Life-zone begins, whereas Mount Malinche is in the Austral Life-zone on the Mexican Plateau; the difference in habitat between the two places is great. Topotypes of *levipes* differ from two topotypes of *beatae* in the same fashion as do other specimens of *levipes* (from San Luis Potosí) from other specimens of *beatae* (from Veracruz). Unfortunately, the topotypes of *beatae* lack external measurements and are subadults, but their coloration agrees with that of other specimens that are here referred to *beatae*.

Hall and Kelson (1959:634, map 364) incorrectly mapped the distribution of *levipes* in Veracruz. There are at least two places named Xuchil in the state of Veracruz and Hall and Kelson (*loc. cit.*) unfortunately plotted the one at lat. 20° 42′ N, long. 97° 42′ W whereas the specimens actually were collected at the Xuchil on the pleateau south of the Volcán de Orizaba (18° 53′ N, 97° 14′ W) in the west-central part of Veracruz. The specimens from Xuchil are P. *b. beatae*.

Intergradation in color between the two subspecies *levipes* and *beatae* is seen in specimens from Jalapa and Zacualpan (3 km. N, also others from 3 km. W), Veracruz. Intergradation between these two subspecies possibly will be found elsewhere along the Sierra Madre Oriental.

Specimens examined.—Total 60 as follows: VERACRUZ: 1 km. E Jalancingo, 6500 ft., 2; *2 km. S Jalancingo*, 2; 6 km. SSE Altotonga, 8000 ft., 8; *1 km. W Las Vigas, 8500 ft.*, 2; Las Vigas, 8500 ft., 13; *2 km. E Las Vigas, 8000 ft.*, 5; *3 km. E Las Vigas, 8000 ft.*, 8; *5 km. E Las Vigas*, 7 (TAM); *5 km. N Jalapa, 4500 ft.*, 2; *Jalapa*, 1 (USNM); 10 km. SE Perote, N slope Cofre de Perote, 10,500 ft., 1 (TAM); Xometla Camp, Mt. Orizaba, 8500 ft., 2 (BM); *Sta. Barbara, Mt. Orizaba, 12,000 ft.*, 1 (BM); Xuchil, 6 (CM).

Peromyscus boylii ambiguus new subspecies

Type.—Male, adult, skin and skull, No. 33092, United States National Museum, from Monterrey, Nuevo León; obtained on February 17, 1891, by Wm. Lloyd, original number 377.

Geographic distribution.—Eastern Coahuila, central Nuevo León, and the Sierra San Carlos, Tamaulipas.

Diagnosis.—Size small for the species; tail averaging longer than head and body (90-114%); dorsal coloration ochraceous, slightly darker middorsally; cheeks and lateral line Capucine Orange; skull small; supraorbital border rounded; anterior palatine foramina short.

Comparisons.—*P. b. ambiguus* differs from P. *b. levipes* in smaller size, longer tail relative to length of head and body, smaller incisive foramina, brighter and paler color, and relatively broader interorbital region. From *P. b. beatae*, P. *b. ambiguus* differs in being smaller in all parts measured and paler.

Remarks.—Osgood (1909:155) reported as P. *b. levipes* 37 specimens from Monterrey and 18 from Cerro de la Silla, Nuevo León, but noted that they were "aberrant." I have examined those same specimens and can hardly decide to which species, P. *boylii* or P. *pectoralis*, they belong. Everything considered I, as did Osgood, opine that the specimens are P. *boylii*. However, I do not rule out the possibility that in this area there is an unnamed species, because I find an unusually wide range of variation in such cranial characters as size of the bullae, width and form of the pterygoid fossa, and shape of the braincase. Extremes of these characters are not constantly associated except in one specimen (33124 USNM), which is the smallest of all the adults examined. It has small bullae, a short rostrum, widely spreading zygomatic arches anteriorly, and a narrow pterygoid fossa, but does not differ externally from the other specimens. Additional material from this area is needed in order to make out the systematic position of these mice.

Because of the wide range of variation in some of its characters, P. *b. ambiguus* is difficult to diagnose. Nevertheless, its small external and cranial size, short anterior palatine foramina, and bright color seem to separate it from other subspecies of P. *boylii* in the eastern part of the range of the species. These differences are most conspicuous when specimens from the northernmost part of the range of *levipes* are compared with specimens of *ambiguus*.

The specimens from the Sierra San Carlos, Tamaulipas, closely resemble *levipes* in color, but are referred to *ambiguus* on the basis of small size, as also are the two specimens from 12 km. E San Antonio de las Alazanas, Coahuila.

TABLE 1. MEASUREMENTS (IN MILLIMETERS) OF PEROMYSCUS

Number of specimens	Total length	Length of tail-vertebrae	Length of hind foot	Per cent length of tail to head and body	Greatest length of skull	Zygomatic breadth	Interorbital constriction	Length of nasals	Palatine slits	Maxillary tooth-row
P. aztecus Mirador, Veracruz										
7 mean	229	113	24.5	....	30.1	15.3	4.7	12.5	6.4	4.8
max.	238	121	26	....	30.9	15.8	5.0	13.5	6.8	5.0
min.	215	107	24	....	29.2	14.9	4.6	11.2	5.7	4.7
P. boylii beatae Las Vigas to 3 km. E thereof, Veracruz										
14 mean	219.3	116.7	23.8	113.7	28.9	14.4	4.5	11.5	6.3	4.5
max.	235	130	25	128.9	29.8	15.1	4.7	12.5	6.8	4.8
min.	204	107	22	100.0	27.9	13.8	4.2	10.7	5.9	4.4
6 km. SSE Altotonga, Veracruz										
5 mean	224.4	116.1	24.1	109.4	29.1	14.5	4.5	11.6	6.4	4.5
max.	241	126	25	114.8	30.1	15.2	4.6	12.0	6.7	4.7
min.	221	110	24	100.0	28.6	14.0	4.4	11.2	6.0	4.3
P. boylii levipes 3 km. SW San Isidro, San Luis Potosí										
11 mean	205.6	99.5	22.4	93.8	28.5	14.2	4.4	11.3	5.9	4.4
max.	219	114	23	108.6	30.5	14.5	4.6	12.2	6.4	4.6
min.	193	90	21	87.4	27.2	13.8	4.2	10.8	5.6	4.1
6 km. NE Cd. del Maíz, San Luis Potosí										
9 mean	198.7	96	22	93.4	28.1	14.0	4.4	11.2	6.0	4.5
max.	205	105	22	105.0	28.7	14.2	4.6	11.7	6.4	4.6
min.	187	90	22	85.7	27.3	13.4	4.3	10.6	5.7	4.3
11 mi. W, 8 mi. S Piedra, Tamaulipas										
5 mean	201.8	101.8	22.6	101.8	28.5	14.0	4.4	11.3	6.1	4.3
max.	214	110	23	109.3	29.0	14.1	4.6	11.5	6.2	4.7
min.	193	94	22	94.9	28.2	13.9	4.2	11.0	6.0	4.1
P. boylii ambiguus La Vegonia, Tamaulipas										
7 mean	199.1	101.6	21.3	104.3	26.9	13.4	4.3	10.5	5.6	4.3
max.	213	109	22.4	108.9	28.6	13.7	4.5	11.8	5.9	4.7
min.	188	97	20	98.0	26.4	13.2	4.2	9.5	5.3	4.1
Monterrey, Nuevo León										
16 mean	199.7	103.2	21.3	106.9	27.6	13.9	4.4	10.7	5.6	4.2
max.	216	114	22	125.6	28.2	14.9	4.6	11.5	5.8	4.5
min.	176	92	19	88.0	26.8	13.2	4.1	10.2	5.0	4.0

Specimens examined.—Total 64 as follows: NUEVO LEON (USNM): Monterrey, 37; *Cerro de la Silla*, 18. COAHUILA: 12 km. E San Antonio de las Alazanas, 9000 ft., 2. TAMAULIPAS: La Vegonia, Sierra San Carlos, 7 (UMMZ).

I am grateful to the following persons for the loan of specimens: G. B. Corbet, British Museum, Natural History (BM); David H. Johnson, United States National Museum (USNM); George H. Lowery, Jr., Louisiana State University (LSU); Philip Hershkovitz, Chicago Natural History Museum (CM); William B. Davis, Texas Agricultural and Mechanical College (TAM); W. H. Burt and Emmet T. Hooper, University of Michigan Museum of Zoology (UMMZ). Specimens lacking designation as to collection are housed in the Museum of Natural History of The University of Kansas. I am indebted to Professor E. Raymond Hall and Mr. J. Knox Jones, Jr. for the use of these specimens and for other assistance. It is appropriate to record also that the findings reported above are an outgrowth of related work done as a Research Assistant under Grant No. 56 G 103 from the National Science Foundation.

LITERATURE CITED

DALQUEST, W. W.
1953. Mammals of the Mexican state of San Luis Potosí. Louisiana State Univ. Biol. Sci. Ser., 1:1-233, December 28.

HALL, E. R., and KELSON, K. R.
1959. The mammals of North America. The Ronald Press, New York, vol. 2:ix + 547-1083 + 79, illustrated, March 31.

OSGOOD, W. H.
1909. Revision of the mice of the American genus Peromyscus. N. Amer. Fauna, 28:1-285, 8 pls., April 17.

RIDGWAY, R.
1912. Color standards and color nomenclature. Washington, D. C., iv + 43 pp., 53 pls.

SAUSSURE, M. H. de
1860. Note sur quelques mammifères du Mexique. Revue et Mag. Zool., Paris, ser. 2, 12:97-110, March.

THOMAS, O.
1903. *On three new forms of* Peromyscus *obtained by Dr. Hans Gadow, F. R. S., and Mrs. Gadow in Mexico.* Ann. Mag. Nat. Hist., ser. 7, 11:484-487, May.

Transmitted June 30, 1961.

□

29-393

UNIVERSITY OF KANSAS PUBLICATIONS
MUSEUM OF NATURAL HISTORY

Volume 14, No. 8, pp. 121-124
March 7, 1962

A New Subspecies of Ground Squirrel (Spermophilus spilosoma) from Tamaulipas, Mexico

BY

TICUL ALVAREZ

UNIVERSITY OF KANSAS
LAWRENCE
1962

UNIVERSITY OF KANSAS PUBLICATIONS, MUSEUM OF NATURAL HISTORY

Volume 14, No. 8, pp. 121-124
Published March 7, 1962

UNIVERSITY OF KANSAS
Lawrence, Kansas

PRINTED BY
JEAN M. NEIBARGER, STATE PRINTER
TOPEKA, KANSAS
1962

29-1505

A New Subspecies of Ground Squirrel (Spermophilus spilosoma) from Tamaulipas, Mexico

BY TICUL ALVAREZ

When A. H. Howell (N. Amer. Fauna, 56:1-256, 1938) revised the North American ground squirrels, he had no specimens of the spotted ground squirrel, *Spermophilus spilosoma*, from Tamaulipas. Thirteen years later, Hall (Univ. Kansas Publ., Mus. Nat. Hist., 5:38, 1951) listed the species for the first time from the state when he recorded as *S. s. annectens* 13 specimens from the barrier beach 88-89 miles south and 10 miles west of Matamoros.

In 1953, Mr. Gerd Heinrich collected 10 individuals of *S. spilosoma* from the coastal plain of eastern Tamaulipas that extend southward the known range of the species on the east coast of México, provide the first specimens from the mainland of Tamaulipas, and represent a new subspecies that is named and described below.

Spermophilus spilosoma oricolus new subspecies

Type.—Female, adult, skin and skull, No. 55497 Museum of Natural History, The University of Kansas; from one mile east of La Pesca, Tamaulipas, México; obtained on May 27, 1953, by Gerd Heinrich, original number 6933.

Diagnosis.—Size medium for species (see measurements); general color cinnamon buff, almost pure on dorsal surface of hind foot and dorsal and ventral midline of tail; spots well marked; postorbital constriction narrow; auditory bullae small; viewed from front, lower border of maxillary plate forming continuously concave line with jugal; zygomatic process of maxillary narrow, having anterior border concave.

Comparisons.—From *Spermophilus spilosoma annectens* (specimens from Padre and Mustang islands, Texas), *S. s. oricolus* differs as follows: larger in external dimensions but smaller in cranial dimensions; postorbital constriction narrower, 13.4 (13.2-13.8) instead of 14.0 (13.5-14.7); rostrum slightly broader; greatest distance between posterior border of maxillary plate and squamosal arm of zygoma longer, averaging 9.9 (9.7-10.2) instead of 9.6 (9.0-10.1); zygomatic process of maxillary, viewed dorsolaterally above lacrimal narrower, having anterior border concave instead of almost straight; when skull viewed from front, lower border of maxillary plate forming continuously concave line with jugal instead of almost a right angle where jugal and maxillary meet; ramus of lower jaw narrower; general color paler; upper surface of hind foot washed with cinnamon instead of yellowish; postauricular spot absent (usually present in *annectens*).

From *Spermophilus spilosoma pallescens* (specimens from southeastern Coahuila), *S. s. oricolus* differs as follows: color more cinnamon especially on upper surface of hind foot; dorsal spots more distinct; smaller externally, except hind foot, which measures the same; braincase and postorbital constriction narrower; nasals shorter; auditory bullae conspicuously smaller.

From *Spermophilus spilosoma cabrerai* (specimens from eastern San Luis Potosí), *S. s. oricolus* differs mainly in pale (cinnamon) rather than dark (blackish) upper parts, but differs also in being smaller externally (judging from measurements given in the original description by Dalquest, Proc. Biol. Soc. Washington, 64:107, 1951), and in having a greater zygomatic breadth, narrower braincase and postorbital constriction, and narrower maxillary process.

Measurements.—Average and extreme measurements of eight specimens from the type locality (four females, including type, and four males) are as follows: total length (only five specimens), 234 (212-245); length of tail vertebrae (five only), 67 (50-75); length of hind foot, 36.1 (35-37); length of ear from notch, 8.4 (7.5-10.0); length of head and body, 166 (155-171); greatest length of skull, 41.2 (40.6-42.7); zygomatic breadth, 24.4 (23.7-25.2); cranial breadth, 18.4 (17.8-18.9); interorbital constriction, 9.2 (8.5-9.8); postorbital constriction, 13.4 (13.2-13.8); length of nasals, 14.0 (13.6-14.8); length of maxillary tooth-row, 8.0 (7.7-8.4); greatest distance between posterior border of maxillary plate and squamosal arm of zygoma, 9.9 (9.7-10.2).

Remarks.—The type locality of *S. s. oricolus* is nearly at sea level on the coastal plain of eastern Tamaulipas. In so far as now known, the population of ground squirrel here named as new is isolated, the nearest records of occurrence being those reported by Hall (*loc. cit.*) from a place on the barrier beach approximately 80 miles north of La Pesca. The nearest record from the mainland (*S. s. annectens*) is from southern Texas.

A possible explanation for the presence of the species at La Pesca is that it dispersed southward along the barrier beach, and that an isolated or semi-isolated segment finally reached the mainland where the barrier beach rejoins it just northeast of La Pesca. This possibility is strengthened by study of the specimens already mentioned from 88-89 miles south and 10 miles west of Matamoros, because they combine many characters of *annectens* and *oricolus*. The insular specimens differ from both *annectens* and *oricolus* in having shorter nasals and a shorter skull, narrower zygomatic and interorbital regions, and a relatively broader interpterygoid space. Also, the zygomatic process of the maxillary, viewed dorsolaterally above the lacrimal, is even narrower that in *oricolus* and has the anterior border even more concave. In color, specimens from the barrier beach are pale as is *oricolus*, but the over-all color is reddish cinnamon rather than cinnamon buff. Concerning the juncture of the zygomatic plate with the jugal, 12 of the 13 specimens studied resemble *annectens* in this character and one resembles *oricolus*. The specimens from the barrier beach may themselves represent an unnamed subspecies; more material than now is available is needed, because most of the specimens are not fully adult. Because the 13 specimens from the barrier beach resemble *oricolus* slightly more than *annectens*, all characters considered, they are tentatively assigned to *oricolus*.

The author is grateful to Professor E. Raymond Hall and Mr. J. Knox Jones, Jr., for permission to examine the specimens here reported and for helpful suggestions. Field work that yielded the specimens was financed by the Kansas University Endowment Association. The laboratory phases of the study were made when the author was a half-time Research Assistant supported by a grant, No. 56 G 103, from the National Science Foundation.

Specimens examined.—A total of 23, all from Tamaulipas: 88 mi. S, 10 mi. W Matamoros, 12; 89 mi. S, 10 mi. W Matamoros, 1; 1 mi. E La Pesca, 10.

Transmitted November 8, 1961.

29-1505

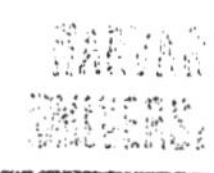

UNIVERSITY OF KANSAS PUBLICATIONS
MUSEUM OF NATURAL HISTORY

Volume 14, No. 9, pp. 125-133, 1 fig. in text
March 7, 1962

Taxonomic Status of the Free-tailed Bat, Tadarida yucatanica Miller

BY

J. KNOX JONES, JR., AND TICUL ALVAREZ

UNIVERSITY OF KANSAS
LAWRENCE
1962

University of Kansas Publications, Museum of Natural History

Volume 14, No. 9, pp. 125-133, 1 fig. in text
Published March 7, 1962

University of Kansas
Lawrence, Kansas

PRINTED BY
JEAN M. NEIBARGER, STATE PRINTER
TOPEKA, KANSAS
1962

29-1507

Taxonomic Status of the Free-tailed Bat, Tadarida yucatanica Miller

BY

J. KNOX JONES, JR., and TICUL ALVAREZ

The Yucatán free-tailed bat, *Tadarida yucatanica* Miller, apparently first was mentioned in the literature by Dobson (1876:731) under the name *Nyctinomus gracilis.* Later, Dobson (1878:437) and Alston (1879-1882:33) listed bats from Dueñas and Petén, Guatemala, under the same name, and Thomas (1888:129) employed it for a specimen from Cozumel Island, Quintana Roo. Twenty-six years after Dobson's first reference, Miller (1902:393) formally described *yucatanica* on the basis of specimens from Chichén-Itzá, Yucatán, and the name has stood unchallenged in the literature until now as a monotypic species. *T. yucatanica* seemingly is sedentary, not migratory. It was thought for many years to be restricted in distribution to the Yucatán region, but now is known from scattered localities in tropical and subtropical areas from Tamaulipas, México (Villa R., 1960:314) southeastward at least to central Panamá (Bloedel, 1955:235).

Our attention first was focused on the taxonomic status of the species when we attempted to identify specimens from southern Tamaulipas in the collections of the Museum of Natural History. Included in our material are the two specimens from a cave "10 kilometers north-northeast of the village of Antiguo Morelos" that were reported by Dalquest and Hall (1947:247) as *Tadarida femorosacca,* and a series of 34 individuals from the vicinity of Piedra in the Sierra de Tamaulipas that was collected by Mr. Gerd Heinrich in June, 1953.

In 1954, Goodwin (1954:2) named *Tadarida laticaudata ferruginea* based on 11 specimens from a cave "8 miles north of Antiguo Morelos," Tamaulipas. Goodwin claimed for *ferruginea* characters that distinguished it specifically from *yucatanica* and from *femorosacca,* and that allied it with *T. laticaudata,* a species previously not reported from North America. More recently, Villa R. (1960) reviewed the status of *ferruginea* and, on the basis of comparisons of specimens from "Cueva del Abra, 10 km. NNE Antiguo Morelos" with specimens from Yucatán, regarded *ferruginea* and *yucatanica* not only as conspecific, but not even subspecifically distinct. Villa R. apparently did not investigate the relationship of *yucatanica* to *laticaudata.* In view of the taxonomic history of the bats thus far reported from Tamaulipas, a review of the status of *T. yucatanica* seemed in order. Accordingly, pertinent specimens from Tamaulipas and the Yucatán region were borrowed to complement those already on hand, along with specimens of *T. laticaudata* and *T. femorosacca.*

Comparisons of the Tamaulipan specimens in our collection with bats from Goodwin's series of *ferruginea* reveal no differences whatsoever and clearly indicate that all are of the same species. Thus the names *Tadarida femorosacca* of Dalquest and Hall and *Tadarida laticaudata ferruginea* of Goodwin were applied to the same kind of bat. Comparisons of the combined specimens from Tamaulipas with topotypes and other specimens of *yucatanica* reveal only minor differences in size and color and we are inclined to agree with Villa R. that the two populations are conspecific. We do not agree, however, that "there are no significant differences between the samples from the Yucatán Peninsula and from Cueva del Abra, Tamaulipas" (Villa R., *op. cit.*:316).

Goodwin (*op. cit.*:3) noted that *ferruginea* was "similar" to *yucatanica* in external measurements, but that the color was a deeper red and the underparts were washed with Pinkish Buff (of Ridgway) rather than Wood Brown. The Tamaulipan specimens at hand average slightly larger externally (see measurements beyond) than those from Yucatán and Guatemala and also differ slightly in color. Perhaps the upper parts are a "deeper red" as Goodwin claimed, but to our eyes they simply average paler and slightly more reddish brown than the dark brown upper parts of *yucatanica.* We agree that the venter of *ferruginea* is washed with Pinkish Buff and distinctly, although not markedly, paler than that of *yucatanica;* the ears and membrances of *ferruginea* also are paler. The most significant differences between the two populations are in several cranial dimensions. Goodwin noted that *ferruginea* had a larger skull. The measurements published by Villa R., too, show Tamaulipan specimens to be the larger, but he did not recognize the magnitude of the difference for two reasons. First, the standard deviations given for some of his measurements seem to be in error, completely obscuring some important differences between the two kinds. Possibly the decimal point was misplaced and the standard deviation in each instance (greatest length, basal length, basilar length, and mandible in each sample, and zygomatic breadth in the Tamaulipan series) was recorded as 10 times greater than actually

TABLE 1. CRANIAL MEASUREMENTS OF THREE SUBSPECIES OF TADARIDA LATICAUDATA.

Number of specimens averaged or catalogue number, and sex	Greatest length of skull*	Zygomatic breadth	Least interorbital constriction	Maxillary breadth	Mastoid breadth	Alveolar length of maxillary tooth-row	Breadth across M3
Tadarida laticaudata ferruginea, vicinity type locality**							
Average 10 ♂♂	17.96	10.12	3.73	5.21	9.85	6.64	7.27
Minimum	17.8	9.9	3.7	5.0	9.7	6.4	6.9
Maximum	18.3	10.5	3.8	5.4	10.1	6.9	7.5
vicinity Piedra, Sierra de Tamaulipas, Tamaulipas**							
Average 30 ♀♀	17.69	10.01	3.71	5.12	9.78	6.43	7.14
Minimum	17.3	9.8	3.5	4.9	9.5	6.2	6.9
Maximum	18.2	10.4	3.9	5.3	10.1	6.7	7.8
Tadarida laticaudata laticaudata, Sapucay, Paraguay							
114953 USNM, ♂	17.9	10.1	3.7	5.3	9.6	6.5	7.4
114956 USNM, ♂	18.2	10.5	3.8	5.4	10.3	6.4	7.7
114955 USNM, ♀	18.0	10.6	3.6	5.2	10.4	6.7	7.9
114958 USNM, ♀	18.1	10.2	3.7	5.1	10.3	6.6	7.4
114964 USNM, ♀	17.6	10.1	3.7	5.3	9.8	6.6	7.8
Tadarida laticaudata yucatanica, topotypes							
108156 USNM, ♂	17.6	9.9	3.7	5.1	9.8	6.5	7.0
108157 USNM, ♂	17.4	10.0	3.6	5.0	9.6	6.5	7.1
108158 USNM, ♂	17.5	9.8	3.7	5.3	9.8	6.5	6.9
108159 USNM, ♀	16.7	9.4	3.6	4.9	9.3	6.1	6.8
108161 USNM, ♀	17.0	9.7	3.5	5.0	9.5	6.4	6.8
La Libertad and Flores, Petén, Guatemala							
Average 12 ♂♂	17.62	9.90	3.64	5.07	9.67	6.47	6.93
Minimum	17.3	9.5	3.4	4.8	9.5	6.3	6.8
Maximum	18.1	10.2	3.7	5.4	9.8	6.6	7.3
Average 8 ♀♀	17.16	9.62	3.55	4.85	9.46	6.37	6.91
Minimum	16.9	9.3	3.4	4.8	9.2	6.2	6.8
Maximum	17.5	9.8	3.7	5.0	9.6	6.6	7.0

* Exclusive of incisors.

** Same series for which external measurements are given.

was the case. Second, he did not separate the sexes in his samples. Males average larger than females in cranial measurements and, when the sexes are treated separately, Tamaulipan specimens average larger in all cranial measurements taken than specimens of *yucatanica*—significantly larger in most (see Figure 1 and Table 1). On the basis of the differences mentioned we regard *ferruginea* and *yucatanica* as distinct at the subspecific level.

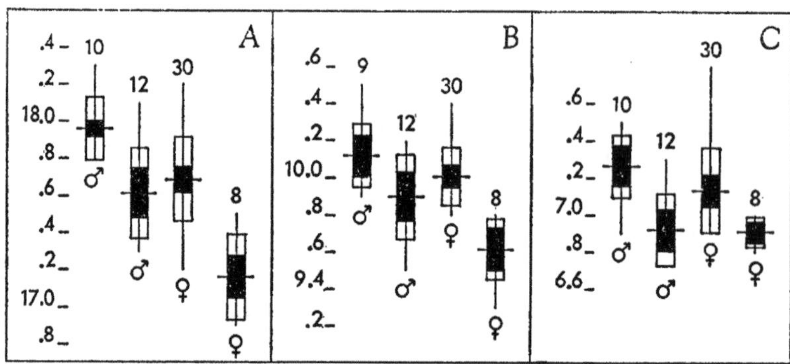

FIG. 1. Differences between specimens from Tamaulipas (same series for which external measurements are given) and specimens from Flores and La Libertad, Guatemala, in three cranial measurements: A, greatest length of skull exclusive of incisors; B, zygomatic breadth; C, breadth across M3. The sexes are separated in each measurement, with the larger Tamaulipan sample to the left in each instance. For each measurement, the horizontal line indicates the mean, the vertical line the observed range, the black rectangle two standard errors on either side of the mean, and the open rectangle one standard deviation on either side of the mean; size of sample is shown above each measurement.

As noted, *ferruginea* was originally described in the species *T. laticaudata*. Comparison of the North American specimens discussed above with specimens of *laticaudata* from Sapucay, Paraguay (USNM 114953, 114955-56, 114958-59, 114961-62, 114964), and Piracicaba, Brazil (USNM 123830), reveals that they are strikingly similar. The differences that separate *laticaudata* from *yucatanica*, and especially *ferruginea*, are no greater than those that separate the two North American kinds and are quantitative rather than qualitative in nature. On account of the morphologic similarity, and because the three kinds are allopatric and have geographic ranges that are more or less complementary, we regard *ferruginea* and *yucatanica* as subspecies of *laticaudata*. Thus we return essentially to the taxonomic arrangement of Dobson and other early workers, because the name *Nyctinomus gracilis* applied by them to North American bats was the name then current for the South American species now known as *laticaudata*. *T. laticaudata* is the oldest specific name proposed for a New World member of the genus.

Probably *T. laticaudata* is widespread in the South American tropics. The monotypic species *T. europs* and *T. gracilis* may well prove to be conspecific with it. In North America, we suspect that future zoological exploration will reveal *laticaudata* to be more or less continuously distributed from eastern México to South America, and possibly present on some of the Caribbean islands as well. It is to be remembered that as late as 15 years ago the species was known on this continent only from a restricted part of the Yucatán Peninsula.

T. laticaudata is related to another North American species, the larger *T. femorosacca*. Benson (1940:26-28) carefully compared Sonoran specimens of *femorosacca* with specimens of *laticaudata* from Paraguay and noted a number of differences. Some of these do not appear to be constant when North American examples of *laticaudata* are considered, but some clearly separate the two species. *T. femorosacca* is larger, both externally and cranially (although there is slight overlap), is distinctly more grayish (less reddish) throughout, has larger and more rugose ears, larger teeth, a longer, more cylindrical interorbital region, a tubelike extension of the anterior nares, and lacks the distinct emargination present in *laticaudata* at the dorsoposterior margin of the anterior nares. At present, *femorosacca* is known from the southwestern part of the United States (southern parts of California, Arizona, and New Mexico) and from the Mexican states of Baja California, Jalisco, and Sonora—only to the north and west of the Mexican Plateau.

Aside from *T. laticaudata*, *T. femorosacca*, and the large, distinctive *T. molossa*, one other species of the *T. laticaudata* group, *Tadarida aurispinosa*, has been reported from North America. Carter and Davis (1961) listed five specimens of *aurispinosa* from Cueva del Abra, where the species occurs along with *T. laticaudata*. Judging from the published measurements of *aurispinosa* it bears about the same relationship to *femorosacca* in size that the latter bears to *laticaudata*.

North American Subspecies of Tadarida laticaudata

We are grateful to authorities at The American Museum of Natural History (AMNH) and the United States National Museum (USNM) for the loan of specimens of *T. laticaudata*, and to authorities at the Museum of Vertebrate Zoology, University of California, and Department of Zoology, University of Arizona, for the loan of specimens of *T. femorosacca* for comparisons. Specimens in the Museum of Natural History (KU) also have been used. All measurements in the following accounts are in millimeters.

Tadarida laticaudata ferruginea Goodwin

Tadarida laticaudata ferruginea Goodwin, Amer. Mus. Novit., 1670:2, June 28, 1954 (type locality, 8 mi. N [= Cueva del Abra, 10 km. NNE] Antiguo Morelos, Tamaulipas).

Geographic distribution.—Presently known only from southern Tamaulipas.

External measurements.—Average and extreme measurements of 30 females from two localities labeled with reference to Piedra, in the Sierra de Tamaulipas, are as follows: total length, 103.4 (97-110); length of tail, 39.9 (35-45); length of hind foot, 10.0 (10); length of ear from notch, 19.1 (19-20); length of forearm (dry), 43.3 (41.6-45.0). Average and extremes of corresponding measurements of 12 males from the vicinity (within 20 kilometers) of the type locality are: 99.3 (95-112); 37.6 (34-42); 10.0 (9-11); 19.0 (17-21); 43.1 (41.8-44.4). See Table 1 for cranial measurements.

Comparisons.—From *Tadarida laticaudata yucatanica*, *T. l. ferruginea* differs as explained on p. 128. From *T. l. laticaudata*, *T. l. ferruginea* differs in being slightly but distinctly paler throughout (difference about the same as between *ferruginea* and *yucatanica*), and in having on the average a narrower skull and smaller teeth.

Records of occurrence.—Specimens examined, 65, as follows: TAMAULIPAS: Sierra de Tamaulipas, 2 mi. S, 10 mi. W Piedra, 1200 ft., 27 (KU); Sierra de Tamaulipas, 3 mi. S, 16 mi. W Piedra, 1400 ft., 7 (KU); 5 mi. S El Mante, 8 (AMNH); 11 mi. S El Mante, 13 (AMNH); 10 km. NNE Antiguo Morelos [= 8 mi. N Antiguo Morelos], 7 (5 AMNH, 2 KU); 20 mi. SW El Mante, 2 (AMNH).

Tadarida laticaudata yucatanica (Miller)

Nyctinomops yucatanicus Miller, Proc. Acad. Nat. Sci. Philadelphia, 54:393, September 12, 1902 (type locality, Chichén-Itzá, Yucatán).

Geographic distribution.—Presently known from the Yucatán Peninsula eastward to Panamá.

External measurements.—Available measurements of two males and a female from Chichén-Itzá, Yucatán, are, respectively: total length, 98, 101, 97; length of tail, 40, 40, 39; length of forearm (dry), 42.5, 42.5, 42.0. Average and extreme measurements of 10 males and six females (the latter all in alcohol) from La Libertad, Guatemala, are, respectively: total length, 101.0 (97-103), 94.7 (92-97); length of tail, 40.6 (37-45), 36.8 (36-39); length of hind foot, 10.7 (10-11), 10.1 (9.5-10.5); length of ear from notch, 20.4 (19-21), 17.6 (17-18.5); length of forearm (dry), 42.3 (41.0-43.6), 42.5 (40.8-44.3). See Table 1 for cranial measurements.

Comparisons.—From *Tadarida laticaudata ferruginea*, *T. l. yucatanica* differs as explained on p. 128. *T. l. yucatanica* resembles *T. l. laticaudata* in color of upper parts and membranes, but is paler ventrally, and *yucatanica* is smaller both externally and cranially.

Records of occurrence.—Specimens examined, 40, as follows: YUCATAN: Chichén-Itzá, 5 (USNM). GUATEMALA: Flores, 4 (AMNH); La Libertad, 31 (KU).

Additional records: YUCATAN (Villa R., 1960:315): Mérida; Uxmal. QUINTANA ROO: Cozumel Island (Thomas, 1888:129). CAMPECHE: San José Carpizo (Villa R., 1960:315). BRITISH HONDURAS: El Cayo (Murie, 1935:19). GUATEMALA: Uaxactún (Murie, 1935:19); Cobán (Goodwin, 1955:4); Dueñas (Dobson, 1878:437). EL SALVADOR: San Salvador (Felten, 1957:8). PANAMA: Pacora (Bloedel, 1955:235).

LITERATURE CITED

ALSTON, E. R.
1879-1882. Biologia Centrali-Americana. Mammalia. xx + 1-220 pp., 22 pls.

BENSON, S. B.
1940. Notes on the pocketed free-tailed bat. Jour. Mamm., 21:26-29, February 14.

BLOEDEL, P.
1955. Observations on the life histories of Panama bats. Jour. Mamm., 36:232-235, May 26.

CARTER, D. C., and DAVIS, W. B.
1961. Tadarida aurispinosa (Peale) (Chiroptera: Molossidae) in North America. Proc. Biol. Soc. Washington, 74:161-165, August 11.

DALQUEST, W. W., and HALL, E. R.
1947. Tadarida femorosacca (Merriam) in Tamaulipas, Mexico. Univ. Kansas Publ., Mus. Nat. Hist., 1:245-248, 1 fig., December 10.

DOBSON, G. E.
1876. A monograph of the group Molossi. Proc. Zool. Soc. London, pp. 701-734, 6 figs., November 7.
1878. Catalogue of the Chiroptera in the . . . British Museum. British Museum, xlii + 567 pp., 30 pls., after May 20.

FELTEN, H.
1957. Fledermäuse (Mammalia, Chiroptera) aus El Salvador. Senckenbergiana Biologica, 38:1-22, 2 pls., 3 figs., January 15.

GOODWIN, G. G.
1954. A new short-tailed shrew and a new free-tailed bat from Tamaulipas, Mexico. Amer. Mus. Novit., 1670:1-3, June 28.
1955. Mammals from Guatemala, with the description of a new little brown bat. Amer. Mus. Novit., 1744:1-5, August 12.

MILLER, G. S., JR.
1902. Twenty new American bats. Proc. Acad. Nat. Sci. Philadelphia, 54:389-412, September 12.

MURIE, A.
1935. Mammals from Guatemala and British Honduras. Misc. Publ., Univ. Michigan, 26:1-30, 1 pl., 1 fold-out map, July 15.

THOMAS, O.
1888. List of mammals obtained by Mr. G. F. Gaumer on Cozumel and Ruatan islands, Gulf of Honduras. Proc. Zool. Soc. London, p. 129, June.

VILLA R., B.
1960. Tadarida yucatanica in Tamaulipas. Jour. Mamm., 41:314-319, 1 fig., August 15.

Transmitted November 8, 1961.

□

29-1507

University of Kansas Publications
Museum of Natural History

Volume 14, No. 10, pp. 135-138, 2 figs.
April 30, 1962

A New Doglike Carnivore, Genus Cynarctus, From the Clarendonian, Pliocene, of Texas

BY

E. RAYMOND HALL and WALTER W. DALQUEST

University of Kansas
Lawrence
1962

University of Kansas Publications, Museum of Natural History
Editors: E. Raymond Hall, Chairman, Henry S. Fitch, Theodore H. Eaton, Jr.

Volume 14, No. 10, pp. 135-138, 2 figs.
Published April 30, 1962

University of Kansas
Lawrence, Kansas

PRINTED BY
JEAN M. NEIBARGER, STATE PRINTER
TOPEKA, KANSAS
1962

29-2890

A New Doglike Carnivore, Genus Cynarctus, From the Clarendonian, Pliocene, of Texas

BY

E. RAYMOND HALL and WALTER W. DALQUEST

A study of a right maxilla bearing P3-M1 and part of a right mandibular ramus bearing m2 (see figures) reveals the existence of an unnamed species of cynarctine carnivore. It may be known as:

Cynarctus fortidens new species

Holotype.—Right maxilla bearing P3, P4, and M1, No. 11353 KU; bluff on west side of Turkey Creek, approximately 75 feet above stream, Raymond Farr Ranch, Center NE, NE, S. 48 Blk. C-3, E. L. and R. R. Ry. Co., Donley County, Texas [approximately 6.5 miles north and 1 mile east of Clarendon], Clarendon fauna, Early Pliocene age. Obtained by W. W. Dalquest, on June 25, 1960.

Referred material.—Fragment of right lower mandible bearing m2, No. 11354 KU (see fig. 2), found about two feet horizontally distant from the holotype in the same stratum as the holotype and on the same date by the same collector (a staff member of the Department of Biology of Midwestern University, Wichita Falls, Texas).

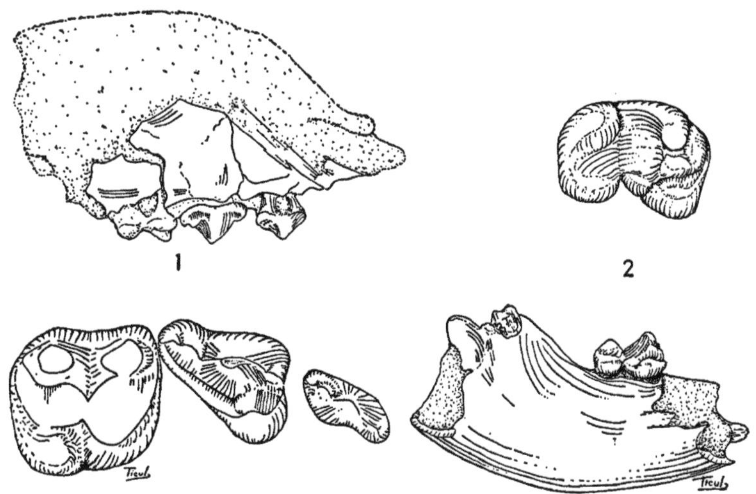

Fig. 1. *Cynarctus fortidens,* No. 11353 KU (Midwestern Univ. No. 2044). Lateral view of holotype × 1, and occlusal view of check-teeth × 2.

Fig. 2. *Cynarctus fortidens,* No. 11354 KU (Midwestern Univ. No. 2045). Lateral view of right lower mandible and m2 × 1, and oblique occlusal view of m2 × 2.

Diagnosis.—Size large (see measurements); no accessory cusp between protocone and paracone of fourth upper premolar; first upper molar longer than broad and lacking cingulum on part of tooth lingual to protocone.

Comparisons.—From *Cynarctus crucidens* Barbour and Cook (see page 225 of Two New Fossil Dogs of the Genus Cynarctus from Nebraska. Nebraska Geol. Surv., 4(pt. 15):223-227, 1914; also pages 330 and 338 of Dental Morphologie of the Procyonidae with a Description of Cynarctoides, Gen. Nov. Geol. Ser. Field Mus. Nat. Hist., 6:323-339, 10 figs., October 31, 1938) *C. fortidens* differs in lacking, instead of having, an accessory cusp between the protocone and paracone of the fourth upper premolar and in lacking, instead of having, a cingulum on the part of P4 that is internal (lingual) to the protocone.

Remarks.—The lower jaw and its second molar seem to be from an individual significantly larger than the holotype. Possibly the lower jaw and upper jaw are from two species but the lower jaw probably is from a male and the upper jaw from a female of the same species.

Reasons for regarding *Cynarctus* as belonging to the family Canidae instead of to the family Procyonidae have been stated recently in detail by E. C. Galbreath (Remarks on *Cynarctoides acridens* from the Miocene of Colorado. Trans. Kansas Acad. Sci., 59(3):373-378, 1 fig., October 31, 1956) and need not be repeated here. Although some uncertainty remains as to the familial position of *Cynarctus*, we favor Galbreath's view that the genus belongs in the family Canidae.

The holotype of *Cynarctus crucidens* is from Williams Canyon, Brown County, Nebraska. According to C. B. Schultz (*in litt.*, December 6, 1961), Williams Canyon is a tributary of Plumb Creek; the upper part of the Valentine formation and the younger lower part of the Ash Hollow formation are exposed in Williams Canyon; which one of these formations yielded the holotype of *C. crucidens* is unknown.

On the basis of the correlation chart (Pl. 1 in Nomenclature and Correlation of the North American Continental Tertiary. Bull. Geol. Soc. Amer., 52(pt. 1):1-48, 1941) by H. E. Wood 2nd *et al.*, *C. fortidens* and *C. crucidens* are equivalent in age or *C. fortidens* is the younger.

The rounded summits of the principal cusps of the teeth of *C. fortidens* suggests that it was mainly frugivorous instead of carnivorous—more frugivorous by far than the living gray fox, *Urocyon cinereoargenteus,* that is known to eat substantial amounts of fruits and berries. Indeed, no other canid that we know of has teeth so much adapted to a frugivorous diet as are those of *C. fortidens.* Its degree of adaptation to a frugivorous diet is more than in the procyonid genus *Nasua* but less than in the procyonid genus *Bassaricyon.*

Measurements (of crowns) of *C. fortidens.*—P3-M1, length, 25.8 (millimeters); P4-M1, 18.9; P3, length, 6.2; P3, breadth, 2.8; P4, length of outer border, 9.3; P4, breadth, 7.05; M1, length, 9.7; M1, breadth, 9.3; m2, length, 10.3; m2, breadth, 6.6; depth of mandible at posterior end of m2, 17; thickness of mandible, 7.1.

Transmitted February 21, 1962.

□

29-2890

UNIVERSITY OF KANSAS PUBLICATIONS
MUSEUM OF NATURAL HISTORY

Volume 14, No. 11, pp. 139-143
April 30, 1962

A New Subspecies of Wood Rat (Neotoma) from Northeastern Mexico

BY

TICUL ALVAREZ

UNIVERSITY OF KANSAS
LAWRENCE
1962

University of Kansas Publications, Museum of Natural History

Volume 14, No. 11, pp. 139-143
Published April 30, 1962

University of Kansas
Lawrence, Kansas

PRINTED BY
JEAN M. NEIBARGER, STATE PRINTER
TOPEKA, KANSAS
1962

29-2891

A New Subspecies of Wood Rat (Neotoma) from Northeastern Mexico

BY

TICUL ALVAREZ

The White-throated woodrat, *Neotoma albigula*, has been known previously from the Mexican state of Tamaulipas by only eight individuals reported by Goldman (N. Amer. Fauna, 31:37, October 19, 1910), which were assigned to *Neotoma albigula leucodon* (type locality, city of San Luis Potosí, México). Additional specimens from southwestern Tamaulipas, obtained in recent years by representatives of the Museum of Natural History, along with specimens from parts of Nuevo León and Coahuila, represent an unnamed subspecies, which is named and described as follows:

Neotoma albigula subsolana new subspecies

Type.—Male, adult, skin and skull, No. 56950, Museum of Natural History, The University of Kansas, from Miquihuana, 6400 ft., Tamaulipas; obtained on July 20, 1953, by Gerd H. Heinrich, original number 7553B.

Geographic distribution.—Sierra Madre Oriental from southeastern Coahuila to southwestern Tamaulipas.

Diagnosis.—Over-all size small for species (see measurements), but tail, maxillary tooth-row and incisive foramina relatively long; upper parts dark (individual hairs banded subterminally with cinnamon and tipped with grayish, yielding an over-all color of grayish brown); lips gray, especially anteriorly and medially; alveoli of incisors narrow (4.8-5.2); posterior branch of premaxilla extending only slightly behind nasals; rostrum short; braincase broad; mastoid breadth averaging 51.1 (47.8-52.7) per cent of basilar length.

Comparisons.—*Neotoma albigula subsolana*, differs from topotypes of *N. a. leucodon*, the subspecies geographically adjacent to the southwest, as follows: size smaller, especially length of palatal bridge (6.9-8.1 instead of 8.2-9.6), alveolar length of maxillary tooth-row (8.3-8.9 instead of 8.8-9.7), and greatest length of auditory bulla (7.3-7.9 instead of 8.2-8.9); mastoid breadth relatively greater, 51.1 (47.8-52.7) instead of 47.0 (45.5-49.1) per cent of basilar length; posterior process of premaxilla extending only slightly beyond posterior border of nasals; auditory bulla conspicuously smaller; upper parts darker, especially middorsally; over-all color grayish instead of ochraceous or yellowish; lips gray instead of nearly white.

Neotoma albigula subsolana differs from *N. a. albigula*, geographically adjacent to the northwest (specimens from Pima County, Arizona) as follows: size averaging slightly larger, except length of nasals; mastoid breadth averaging 18.8 (17.9-20.2) instead of 17.9 (17.7-18.2), its ratio to basilar length therefore greater, 51.1 (47.8-52.7) instead of 49.4 (47.9-50.0); zygomatic arches expanded posteriorly instead of nearly parallel as in *albigula;* interparietal longer and narrower; mesopterygoid fossa broader; auditory bulla slightly smaller; upper parts distinctly darker.

Remarks.—*N. a. subsolana* is characterized by the combination of small size, dark color, small auditory bulla and relatively broad braincase. Typical specimens have been collected only at higher elevations in the Sierra Madre Oriental where no other species of *Neotoma* is known to occur.

Intergradation between *N. a. subsolana* and *N. a. leucodon* occurs at lower elevations on the west side of the Sierra Madre Oriental as shown by specimens from nine miles southwest of Tula, Tamaulipas, and Sierra Guadalupe, Coahuila, from which places some specimens are paler than others, approaching *leucodon* in color, and are slightly larger than typical *subsolana.* Specimens assigned to *leucodon* from vicinity of Presa Guadalupe and from 1 to 6 kilometers south of Matehuala, San Luis Potosí, are typical of that subspecies in measurements but are darker than topotypes.

N. a. subsolana intergrades with *N. a. albigula* in southeastern Coahuila (specimens from 6 to 9 miles east of Hermanas and from Panuco) where some individuals average paler and smaller than topotypes of *subsolana* and some have skulls that combine characters of *subsolana* and *albigula.* These specimens, which were referred to *N. a. leucodon* by Baker (Univ. Kansas Publ., Mus. Nat. Hist., 9:281-282, June 15, 1956), are assigned to *subsolana* on the basis of relatively dark upperparts and broad mesopterygoid fossa (narrow in only one specimen).

On geographic grounds, specimens not studied by me from Municipio de Galeana, Nuevo León (Koestner, Great Basin Nat., 2:13, 1941), and those from Jaumave, Tamaulipas (Goldman, N. Amer. Fauna, 31:37, October 19, 1910), probably are referable to *N. a. subsolana.*

The subspecific name *subsolana* (Latin adjective for eastern) is proposed for this woodrat because of its eastern geographic occurrence.

Measurements.—Average and extreme measurements of nine topotypes (6 males and 3 females) are as follows: total length, 338 (315-370); length of tail-vertebrae, 157 (130-182); length of hind foot, 35 (33-37); length of ear from notch, 31 (29-34); basilar length, 36.5 (34.7-39.2); zygomatic breadth, 23.9 (22.5-25.0); interorbital constriction, 5.5 (5.7-6.2); length of nasals, 15.5 (15.2-16.5); length of incisive foramina, 9.4 (8.8-10.1); length of palatal bridge, 7.7 (6.9-8.1); alveolar length of maxillary tooth-row, 8.7 (8.3-8.9); length of auditory bulla, 7.6 (7.3-7.9); mastoid breadth, 18.7 (17.9-20.2).

Specimens examined.—A total of 124 (all from Mus. Nat. Hist., Univ. Kansas) from: COAHUILA: 6 mi. E Hermanas, 1; 9 mi. E Hermanas, 1; Panuco, 3000 ft., 4; 1 mi. S, 4 mi. W Bella Unión, 7000 ft., 3; 3 mi. S, 3 mi. E Bella Unión, 6750 ft., 1; 6 mi. E, 4 mi. S Saltillo, 7500 ft., 5; 7 mi. S, 4 mi. E Bella Unión, 7200 ft., 3; 14 mi. W, 1 mi. N San Antonio de las Alazanas, 6500 ft., 2; 12 mi. S, 2 mi. E Arteaga, 7500 ft., 5; north slope Sierra Guadalupe, 10 mi. S, 5 mi. W General Cepeda, 6500 ft., 26; 7 mi. S, 1 mi. E Gómez Farías, 6500 ft., 3; 8 mi. N La Ventura, 5500 ft., 1. NUEVO LEON: Iturbide, Sierra Madre Oriental, 5000 ft., 10; Laguna, 1; 9 mi. S Aramberri, 3900 ft., 3; 1 mi. W Doctor Arroyo, 5800 ft., 4. TAMAULIPAS: Miquihuana, 6400 ft., 22; Joya Verde, 35 km. SW Cd. Victoria (on Jaumave Road), 3800 ft., 2; Nicolás, 56 km. NW Tula, 5500 ft., 10; Tajada, 23 mi. NW Tula, 5200 ft., 2; 9 mi. SW Tula, 3900 ft., 15.

Comparative material.—*N. a. albigula,* 10 specimens (all KU) from: ARIZONA: 4 mi. S, 5 mi. E Continental, 4; 7 mi. E Tucson, 2500 ft., 1; 30 mi. S Tucson, 1; 14 mi. S, 3 mi. E Continental, 1; Sta. Catalina Mts., south slope Molino basin, 4200 ft., 2; Santa Rita Mts., northwest slope, near Sta. Rita Range, 4300 ft., 1.

N. a. leucodon, 46 specimens (in Mus. Nat. Hist., Univ. Kansas, unless otherwise noted) from: SAN LUIS POTOSI: 6 km. S Matehuala, 13 (LSU); 1 km. S Matehuala, 2 (LSU); 7 km. W Presa de Guadalupe, 5 (LSU); Presa de Guadalupe, 4 (LSU); 8 mi. SW Ramos, 6700 ft., 3; 10 mi. NE San Luis Potosí, 6000 ft., 2; San Luis Potosí, 9 (USNM); Hda. La Parada, 8 (USNM).

I am grateful to Prof. E. Raymond Hall and Mr. J. Knox Jones, Jr., for permission to examine critical specimens and for helpful suggestions. I am grateful also to Dr. George H. Lowery, Jr., of the Louisiana State University (LSU) and to Dr. David H. Johnson and Dr. Richard H. Manville of the United States National Museum (USNM) for the loan of specimens. Gerd H. Heinrich (in 1953) and Percy L. Clifton (in 1961) collected for the Museum of Natural History the Tamaulipan specimens herein reported. Fieldwork was supported by the Kansas University Endowment Association. Laboratory phases of the study were made when the author was a half-time Research Assistant supported by grant No. 56 G 103 from the National Science Foundation.

Transmitted February 21, 1962.

□

29-2891

UNIVERSITY OF KANSAS PUBLICATIONS
MUSEUM OF NATURAL HISTORY

Volume 14, No. 12, pp. 145-159, 1 fig. in text
May 18, 1962

Noteworthy Mammals from Sinaloa, Mexico

BY

J. KNOX JONES, JR., TICUL ALVAREZ, AND
M. RAYMOND LEE

UNIVERSITY OF KANSAS
LAWRENCE
1962

University of Kansas Publications, Museum of Natural History

Volume 14, No. 12, pp. 145-159, 1 fig. in text
Published May 18, 1962

University of Kansas
Lawrence, Kansas

PRINTED BY
JEAN M. NEIBARGER, STATE PRINTER
TOPEKA, KANSAS
1962

29-3000

Noteworthy Mammals from Sinaloa, Mexico

BY

J. KNOX JONES, JR., TICUL ALVAREZ, and M. RAYMOND LEE

In several of the past twelve years field parties from the Museum of Natural History have collected mammals in the Mexican state of Sinaloa. Most of the collections contained only a modest number of specimens because they were made by groups that stopped for short periods on their way to or from other areas, but several collections are extensive. Field work by representatives of this institution now is underway in Sinaloa with the aim of acquiring materials suitable for treating the entire mammalian fauna of that state.

Among the mammals thus far obtained are specimens of twenty species that represent significant extensions of known range, are of especial taxonomic or zoogeographic interest, or that complement published information, and it is these records that are reported herein.

The following persons obtained specimens mentioned beyond: J. R. Alcorn (1950); J. R. and A. A. Alcorn (1954 and 1955); R. H. Baker and a party of students (1955); W. L. Cutter (1957); S. Anderson and a party of students (1959); M. R. Lee (1960 and 1961); and J. K. Jones, Jr., accompanied by R. R. Patterson and R. G. Webb (1961). The Kansas University Endowment Association and the American Heart Association provided funds that helped to defray the cost of field operations.

In the accounts that follow, all measurements are in millimeters and all catalogue numbers refer to the mammal collection of the Museum of Natural History, The University of Kansas. Place-names associated with specimens examined are indicated on the accompanying map (Fig. 1).

Notiosorex crawfordi (Coues).—A non-pregnant female (75184) was obtained on November 29, 1957, at El Fuerte by W. L. Cutter. Comparison of this specimen with topotypes of *N. evotis* (see below) and with undoubted examples of *N. crawfordi* proves our specimen to be referable to the latter. We presume that the shrew reported as *evotis* on geographic grounds from El Carrizo by Hooper (1961:120) also is referable to *crawfordi*. External measurements of our female are: total length, 77; length of tail, 20 (tip missing); length of hind foot, 11; length of ear from notch, 8; weight in grams, 4. Cranial measurements of this individual are given in Table 1.

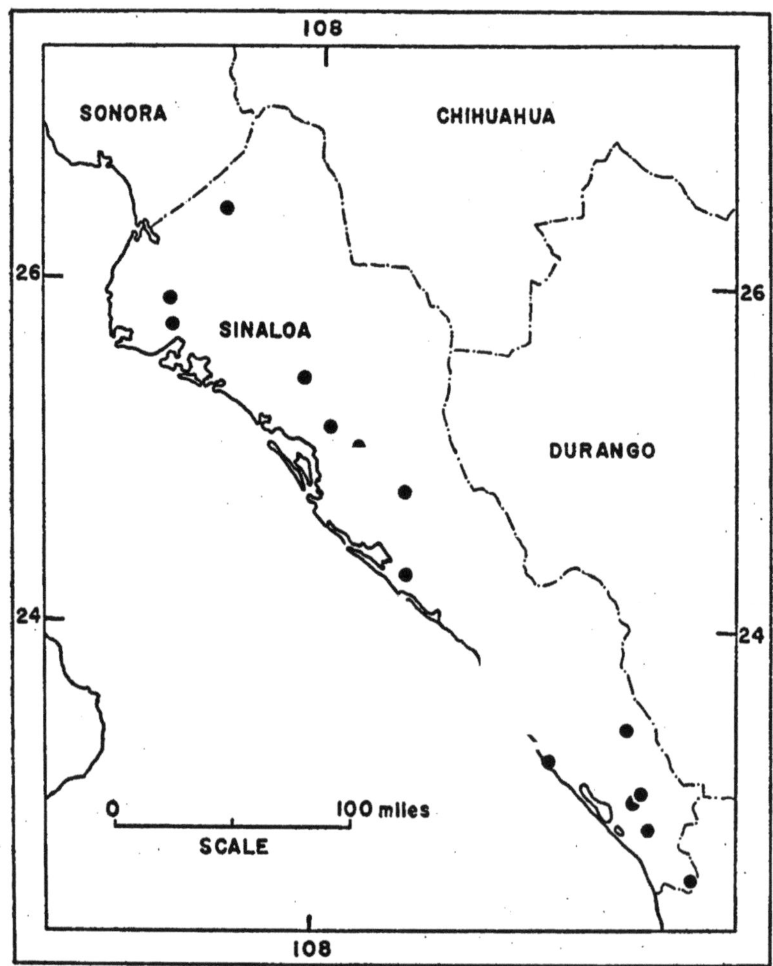

FIG. 1. Map of Sinaloa on which are plotted symbols representing place-names mentioned in text. From north to south, these are: El Fuerte; San Miguel; Los Mochis; Guamúchil; Terrero; Pericos; Culiacán; El Dorado; Piaxtla and Camino Reál (one symbol); Pánuco; Mazatlán; Matatán; Rosario; Escuinapa; Concepción.

Notiosorex evotis (Coues).—Four topotypes (85533-36), all males, were collected by Lee at Mazatlán. One was caught on December 17, 1960, in a museum special trap set "in low weeds near thorn bush" in a sandy field at the north edge of Mazatlán, less than a mile from the ocean. A few trees and some grasses grew in this area; *Mus musculus* and *Perognathus pernix* were taken in the same line of traps. Additional trapping at this locality failed to produce

TABLE 1. CRANIAL MEASUREMENTS OF TWO SPECIES OF NOTIOSOREX.

Catalogue number, or number of specimens averaged, and sex	Condylobasal length	Interorbital constriction	Maxillary breadth	Cranial breadth	Palatal length	Length of maxillary tooth-row
Notiosorex crawfordi, Huachuca Mts., Arizona*						
Average 6 (2♂, 4♀)	16.01	3.72	5.08	8.32	6.59	5.93
Minimum	15.7	3.6	4.9	7.8	6.3	5.8
Maximum	16.5	3.85	5.2	8.8	7.15	6.2
El Fuerte, Sinaloa						
75184 KU, ♀	16.5	3.7	5.0	8.4	6.9	6.1
SW Guadalajara, Jalisco						
33318 KU, ♂	....	3.6	4.9	....	7.1	5.7
42583 KU, ?	15.0+	3.5	4.6	....	6.6	5.4±
42584 KU, ?	....	3.6	4.9	....	7.1±	6.1±
2 mi. E La Palma, Michoacán						
42586 KU, ?	....	3.8	4.9	....	6.9	....
42587 KU, ?	....	3.8	4.8	....	6.9	6.0
42588 KU, ?	....	....	4.9	....	6.9	6.2
Notiosorex evotis, Mazatlán, Sinaloa						
Average 4 (♂)	17.68	4.05	5.37	8.68	7.60	6.58
Minimum	17.4	4.0	5.3	8.5	7.5	6.5
Maximum	17.9	4.1	5.4	8.8	7.7	6.7

* After Hoffmeister and Goodpaster, 1954:51.

more shrews. The other three specimens were captured alive on February 1 (one) and February 2 (two), 1961, in the wake of a bulldozer that was clearing land adjacent to the place where the first specimen was trapped. The ground cover being cleared away consisted mostly of dry, dense weeds and short, thorny scrub; the latter was sparse in some places and formed dense thickets in others. One individual that was kept alive for a short time in a can ate crickets and roaches readily and ate one spider, but refused isopods. On one occasion it ate six crickets in about three hours. Wet oatmeal and oatmeal mixed with peanut butter both were refused.

Average and extreme external measurements of the four males are as follows: total length, 93.2 (90-98); length of tail, 25.5 (23-27); length of hind foot, 11.9 (11-13); length of ear from notch, 7.7 (7-8); weight in grams, 5.4 (4.4-6.3). Cranial measurements are given in Table 1.

Notiosorex evotis was described by Coues (1877:652) on the basis of a single specimen, obtained at Mazatlán by Ferdinand Bischoff in 1868, that originally had at least the partial skull inside. Subsequently the skull was removed and evidently lost (Poole and Schantz, 1942:181). Coues named *evotis* as a species distinct from *crawfordi* (described by him in the same paper) on the basis of larger size, shorter tail, and alleged slight differences in color. He did not describe the skull, but did note that the dentition was "substantially the same as that of *N. crawfordi.*" Evidently, the only other correctly identified specimen of *evotis* on record is an individual from Mazatlán in the British Museum, the skull of which was figured by Dobson (1890:pl. 23, fig. 20).

Merriam (1895:34) characterized *evotis,* known to him by only the holotype, as: "Similar to *N. crawfordi,* but slightly larger and darker." He did not examine the skull, which by that time had been "lost or mislaid." Merriam reduced *evotis* to subspecific status under *crawfordi* with the following remarks: "In the absence of sufficient material of *N. evotis,* it is impossible to determine its exact relations to *crawfordi.* Dobson did not recognize it as distinct, but figured its teeth under the name *crawfordi* [*loc. cit.,* possibly a *lapsus*]. For the present it seems best to retain it as a subspecies."

Merriam's arrangement of *evotis* as a subspecies of *crawfordi* has been followed by subsequent workers, mostly, we suppose, because additional material of undoubted *evotis* has not until now been available. Comparisons of our four specimens with specimens (from Jalisco, Sinaloa and Tamaulipas) and published descriptions and measurements (see especially Hoffmeister and Goodpaster, 1954:46-47, 51) of *crawfordi* reveal that *evotis* has a longer body and hind foot than *crawfordi* but a relatively (sometimes actually) shorter tail and ear, and a distinctly larger, heavier skull (see Table 1). The upper parts of our specimens average pale brownish gray and are paler, not darker, than the upper parts of *crawfordi.* But, all of the latter were obtained in the warm months of the year except one November-taken individual from El Fuerte, Sinaloa, the dorsal pelage of which approaches in color that of the darkest of the *evotis.* The pelage of both kinds probably is paler in winter than in summer and may be indistinguishable in the same season. Ventrally, all four *evotis* are grayish white, faintly to moderately tinged with brownish buff.

Notiosorex evotis differs cranially from *Notiosorex crawfordi* as follows: larger (see measurements); mesopterygoid fossa squared rather than broadly U-shaped anteriorly; rounded process on maxillary at posterior border of infraorbital canal well developed (faint or lacking in *crawfordi*); occipital condyles smaller and, in lateral view, elevated above basal plane of skull; upper molars slightly more crowded in occlusal view. These differences, although admittedly slight, appear to be constant in the specimens we have seen, but ought to be used cautiously owing to the small samples studied.

Shrews of the genus *Notiosorex* have been reported twice previously from localities in west-central México, other than from Mazatlán, as follows: 21 mi. SW Guadalajara (remains from owl pellets) and 13 mi. S, 15 mi. W Guadalajara, Jalisco, by Twente and Baker (1951:120-121); and Cerrito Loco, 2 mi. E La Palma, Michoacán (remains from owl pellets), by Baker and Alcorn (1953:116). The remains were referred to *evotis* on geographic grounds in one instance and were so referred inferentially in the other. Examination of the specimens upon which these reports were based reveals that all are *crawfordi* on the basis of characters previously cited. As a result, *N. evotis* is known only from the type locality at Mazatlán, whereas *N. crawfordi* is widely distributed on the Mexican Plateau as far south as Jalisco and northern Michoacán, and occurs on the west side of the Sierra Occidental as far south as northern Sinaloa.

The two kinds obviously are closely related and intergradation eventually may be demonstrated between them. But, for the present, we adopt a conservative course and treat *evotis* as a full species owing to its distinctive features, restricted geographic distribution, and the lack of evidence of intergradation between it and *crawfordi*.

Balantiopteryx plicata pallida Burt.—Thirty-five specimens from two adjacent localities along the Río del Fuerte in northern Sinaloa, 3 mi. NE San Miguel, 300 ft. (84944-48) and 10 mi. NNE Los Mochis (60572-75, 60667-78, 60681-94), provide the first records of the subspecies from the state. Individuals from both localities were shot at dusk as they foraged among trees in the valley of the river. Fifteen of 18 females from 10 mi. NNE Los Mochis, collected on June 5, 6 and 7, 1955, were pregnant; each contained a single embryo, the embryos ranging from 7 to 15 mm. in crown-rump length. *B. p. pallida* previously has been reported from the southern parts of Baja California and Sonora.

Balantiopteryx plicata plicata Peters.—Specimens in the Museum of Natural History from the following localities, several of which are marginal, document better than previously has been done the distribution of this subspecies in Sinaloa: 32 mi. SSE Culiacán (60699); 10 mi. SE Escuinapa (68629); 17 mi. SSE Guamúchil (60576); 5 mi. NW Mazatlán (85537-61, 85901-04); 1 mi. SE Mazatlán, 10 ft. (39461-76); 1 mi. S Pericos (60697-98, 60700); ½ mi. E Piaxtla (60701); ½ mi. W Rosario, 100 ft. (39477-79); 5 mi. SSE Rosario (60702-03); 4 mi. N Terrero (60695-96).

Pregnant females, each with a single embryo, were recorded in 1954 from 4 mi. N Terrero, 2 (June 9), 1 mi. S Pericos, 2 (June 13), and 5 mi. SSE Rosario, 2 (June 20). None of 16 December-taken females from 5 mi. NW Mazatlán was pregnant.

The specimen from 17 mi. SSE Guamúchil, preserved in alcohol, is provisionally referred to *B. p. plicata* on geographic grounds inasmuch as specimens from the nearby localities of 1 mi. S Pericos and 4 mi. N Terrero, although more grayish on the average than specimens from southern Sinaloa, are somewhat darker and distinctly larger than specimens of *B. p. pallida* from along the Río del Fuerte in northern Sinaloa. Specimens from southern Sinaloa average only slightly paler than typical *plicata* examined from southern México and Nicaragua.

Pteronotus psilotis (Dobson).—A total of six specimens from two localities in southern Sinaloa provide the first records from the state and are the northernmost records in western México. The two localities are: ½ mi. S Concepción, 250 ft. (84987-90); 1 mi. W Matatán (84985-86). The two individuals from the last-mentioned place extend the known range of the species approximately 275 miles north-northwest from a locality 7 mi. W, ½ mi. S Santiago, Colima (Anderson, 1956:349), and place the limit of the known distribution of P. *psilotis* farther to the north in western México than in the eastern part of the country. We follow Burt and Stirton (1961:24-25) in use of the generic name *Pteronotus* for this species.

The two specimens from 1 mi. W Matatán were shot at late dusk as they foraged with other bats, presumably of the same species, low over water at the place where the Río San Antonio joins the larger Río Baluarte. The four individuals from ½ mi. S Concepción were captured in mist nets stretched across the Río de las Cañas at the Sinaloa-Nayarit border, and were taken shortly after dark at heights of three feet or less above the water. Our six specimens all are males. Five are in the reddish color phase and one is in the brownish phase.

Average and extreme measurements of the six males, which average slightly smaller than specimens examined from Colima and Guerrero, are as follows: total length, 66.8 (65-69); length of tail, 16.3 (15-18); length of hind foot, 11.8 (11-12); length of ear from notch, 16.9 (16.5-17.0); length of forearm (dry), 41.5 (40.6-42.4); weight in grams, 8.3 (6.9-9.8); greatest length of skull, 15.4 (15.2-15.5); zygomatic breadth, 8.3 (8.2-8.4); interorbital constriction, 3.4 (3.3-3.6); mastoid breadth, 8.7 (8.6-8.8); length of maxillary toothrow, 5.8 (5.8-5.9); breadth across M3, 5.4 (5.3-5.6).

Sturnira lilium parvidens Goldman.—The first specimens to be reported from Sinaloa are as follows: 32 mi. SSE Culiacán (61087); 1 mi. S El Dorado (75207); Pánuco, 22 km. NE Concordia (85648-50). The three bats from the last-mentioned locality were caught after midnight in a mist net stretched across a road adjacent to a nearly dry stream bed. The vegetation in the vicinity of the net consisted mostly of dry weeds and grass along with some low shrubs, but a tree-filled canyon was about one-fourth mile above the net. We lack details about the capture of the other two bats.

S. *l. parvidens* has been reported only once from farther north in western México than Sinaloa. Anderson (1960:7) recorded five specimens from along the Río Septentrión, 1½ mi. SW Tocuina, Chihuahua.

Artibeus lituratus palmarum Allen and Chapman.—This species has been reported once previously from Sinaloa (from 1 mi. S El Dorado by Anderson, 1960:3). Six specimens (85668-72, 85674), all males, collected on December 23 and 24, 1960, at Pánuco, 22 km. NE Concordia, provide the second known occurrence in the state.

Artibeus toltecus (Saussure).—A male (85666) from Pánuco, 22 km. NE Concordia, provides the first record of this species from Sinaloa and extends the known range northwestward approximately 182 miles from Ambas Aguas [= 6½ km. SW Amatlán de Jora], Nayarit (Andersen, 1908:300). Our specimen was taken on December 22, 1960, in a mist net placed across a road in an area where vegetation consisted mostly of weeds, grasses and shrubs. Two *Glossophaga soricina leachii* and two *Choeronycteris mexicana* were taken in the same net.

Davis' (1958:165-166) key is useful in separating the small Mexican members of the genus *Artibeus*, but we have found some adults of *toltecus* to be smaller than the key indicates. For example, in the 12 Mexican specimens (Oaxaca, 6, Tamaulipas, 3, Jalisco, 2, Sinaloa, 1) examined by us the total length of skull varies from 19.7 to 21.0 and the forearm from 36.3 to 42.6.

Dalquest (1953) and more recently Koopman (1961) regarded *A. toltecus* and the larger *A. aztecus*, which occurs in the same areas but at higher elevations than *toltecus*, as subspecies of the more southerly *A. cinereus*. Davis (*op. cit.*), on the other hand, recognized *toltecus*, *aztecus*, and *cinereus* as distinct species. More specimens of small and medium-sized *Artibeus* are needed from México before this baffling complex can be studied adequately, but on the basis of specimens examined we are inclined to agree with Davis as concerns the specific distinctness of *toltecus* and *aztecus.* In Tamaulipas (the mammalian fauna of which is currently under study by Alvarez) for example, *toltecus* is known from Rancho Pano Ayuctle at an elevation of 300 feet in tropical deciduous forest, whereas *aztecus* has been taken only four miles away at Rancho del Cielo, but at an elevation of 3000 feet in cloud forest. The altitudinal difference between ranges of the two kinds in Tamaulipas corresponds to that found in Sinaloa (see Koopman, *loc. cit.*) and is of approximately the same magnitude found by Davis at higher elevations in Guerrero. This relationship suggests that the two kinds are neither subspecies of a single species, nor individual variants of a widespread, monotypic species, but probably are two different species. We agree that one, most likely the smaller *toltecus,* may eventually prove to be a northern subspecies of *cinereus.*

Myotis occultus Hollister.—A single specimen of this species (67491) from 1 mi. N, ½ mi. E San Miguel provides the first certain record from Sinaloa, and is indistinguishable from specimens from Alamos, Sonora, that were referred to *occultus* by Hall and Dalquest (1950:587). Miller and Allen (1928:100) identified a skin alone from Escuinapa as *occultus,* but Hall and Dalquest (*loc. cit.*) later assigned this specimen provisionally to *M. fortidens* on geographic grounds and because it agreed in color with undoubted specimens of the latter from Guerrero. Specimens from south of San Miguel and north of the undoubted range of *fortidens* are needed in order to ascertain whether the two kinds are distinct species or instead only subspecies of a single species.

The Sinaloan bat was taken in a mist net stretched over a drainage ditch adjacent to the Río del Fuerte on the night of June 19-20, 1955, by R. H. Baker. Several other kinds of bats were obtained (shot or netted) at the same place, among which was one specimen of *Myotis velifer.* The specimens studied of *occultus* from Sinaloa and Sonora are clearly separable from specimens of *velifer* from the same region (Sonora and northern Sinaloa) in having paler (more reddish) pelage, shorter forearm, smaller skull, relatively broader rostrum, and four fewer teeth.

Myotis velifer velifer (J. A. Allen).—Three specimens from the following localities in northern Sinaloa provide the first records of the species from the state: El Fuerte (75234); Río del Fuerte, 1 mi. N, ½ mi. E San Miguel (67490); Río del Fuerte, 10 mi. NNW Los Mochis (61149). The subspecies *M. v. velifer* has been reported previously from the adjacent states of Chihuahua, Durango, and Sonora.

A female (61149) obtained on June 8, 1954, carried a single embryo that measured 3 mm. in crown-rump length.

Lasiurus borealis teliotis (H. Allen).—A female from 10 mi. NNW Los Mochis (61172), obtained on June 8, 1954, represents the first record of the species from Sinaloa, and is tentatively referred to this subspecies. It resembles cranially, but is paler than, Californian specimens seen of *teliotis*.

Molossus ater nigricans Miller.—This large free-tailed bat previously has been reported no farther north in western México than the type locality, Acaponeta, Nayarit. Nineteen specimens from four different localities in Sinaloa are as follows: 1 mi. SE Camino Reál, 400 ft. (85093-99); 32 mi. SSE Culiacán (61279-87); 1 mi. S Pericos (61277-78); ½ mi. E Piaxtla (61288). The specimens labeled with reference to Camino Reál and Piaxtla were obtained along the Río Piaxtla at approximately the same place. Those from 1 mi. S Pericos extend the known range of the species approximately 225 miles northwestward.

M. a. nigricans is characteristically an early flier. Along the Río Piaxtla, 1 mi. SE Camino Reál, where bats probably found daytime retreats in the rocky walls of the steep-sided valley of the river, individuals first appeared early in the evening when the sun was still on the western horizon, but were gone before other species of bats were seen. A female from 32 mi. SSE Culiacán, taken on June 18, 1954, contained one embryo that was 18 mm. in crown-rump length. Each of the color phases of the species, reddish (8) and black (11), are represented among our specimens. We follow Goodwin (1960) in the use of the specific name *ater* for this bat.

Dasypus novemcinctus mexicanus Peters.—Two armadillos (85402-03) from the valley of the Río del Fuerte, 3 mi. NE San Miguel, 300 ft., are the first of the species to be reported from northern Sinaloa. They extend the known range northwestward in the state approximately 285 miles from Escuinapa (Russell, 1953: 25) and signal the possible occurrence of *D. n. mexicanus* in southern Sonora. Sign of the armadillo was abundant at the place where our two specimens were collected. Because it was felt that the species possibly had been introduced along the Río del Fuerte, a number of local residents were questioned on the point, but all insisted that armadillos were native to the area.

External measurements of 85402 (female) and 85403 (male) are, respectively, as follows: total length, 725, 748; length of tail, 351, 357; length of hind foot, 87, 89; length of ear from notch, 39, 39.

Sylvilagus audubonii goldmani (Nelson).—This cottontail has been reported from Sinaloa only from Bacubirito, Culiacán (type locality), and Sinaloa (Nelson, 1909:226). Additional records are: 12 mi. N Culiacán (67561-62); 6 mi. N El Dorado (75263); 6 mi. N, 1½ mi. E El Dorado (75264-66); 7 mi. NE El Fuerte (81076-77); and 1 mi. S Pericos (61292-93). Specimens from the vicinity of El Dorado extend the known range some 30 miles southward from the type locality. A female from 1 mi. S Pericos that was taken on June 13, 1954, carried three embryos that measured 29 mm. in crown-rump length.

Sciurus truei Nelson.—Three specimens (61300-02) of this species collected by A. A. Alcorn on June 19, 1954, 32 mi. SSE Culiacán extend the known range approximately 210 miles south-southeast from Guirocoba, Sonora (Burt, 1938:38), and provide the first record from Sinaloa. Two of the specimens are females and each was pregnant, one with two embryos and the other with three.

Our specimens generally agree in color with *S. truei*, but are larger than typical individuals and in this respect approach *S. sinaloensis* of southern Sinaloa. Probably *truei* and *sinaloensis* both are only subspecies of the more southerly *S. colliaei*. The three nominal species currently constitute the *S. colliaei* group in which the presence or absence of P3 seems to vary geographically. The tooth frequently is absent in the northern *truei* and usually present (invariably in the specimens we have examined) in *colliaei*. Only one of our Sinaloan specimens is accompanied by a skull; in it P3 is present on the right side and absent on the left.

External measurements of the male and two females are, respectively: total length, 512, 508, 504; length of tail, 263, 263, 252; length of hind foot, 64, 63, 64; length of ear from notch, 28, 29, 28. Cranial measurements of 61300 (a female) are: greatest length of skull, 56.2; zygomatic breadth, 32.6; interorbital constriction, 17.9; postorbital constriction, 17.9; length of nasals, 17.3; alveolar length of maxillary tooth-row (on side lacking P3), 10.9.

Thomomys umbrinus atrovarius J. A. Allen.—Two specimens (85104-05) from the valley of the Río Piaxtla, 1 mi. SE Camino Reál, 400 ft., resemble the description of *atrovarius* and agree in size, color and most cranial details with a specimen (85744) from 5 mi. NW Mazatlán. The first-mentioned specimens extend the known range of the subspecies some 50 miles northward from Mazatlán (Bailey, 1915:96), and indicate the probable occurrence of the species at lower elevations in other parts of central Sinaloa.

Peromyscus merriami goldmani Osgood.—This subspecies has been reported previously only from the type locality, Alamos, Sonora. Eight specimens were collected in Sinaloa by W. L. Cutter in the autumn of 1957 as follows: 6 mi. N, 1½ mi. E El Dorado (75368-72); 2½ mi. N El Fuerte (75365-66); El Fuerte (75367). The first-mentional locality is approximately 200 miles south-southeast of the type locality. All specimens collected by Cutter were taken in lowland areas, supporting remarks by Commissaris (1960) concerning habitat preferences of P. *merriami* as compared with those of the closely related P. *eremicus.*

Two of three females from northeast of El Dorado were pregnant on November 18 and 19; one carried four embryos (8 mm. in crown-rump length) and the other three (11 mm.).

External and cranial measurements of P. *m. goldmani* previously were known only for the holotype (Osgood, 1909:252, 267). Measurements of five adults, a male (75370) and four females (75365, 75369, 75371-72) are, respectively, as follows: total length, 204, 225, 215, 214, 210; length of tail, 105, 120, 110, 108, 109; length of hind foot, 21, 23, 23, 22, 22; length of ear from notch, 21, 21, 21, 20, 21; weight in grams, 29, 19, 35 (pregnant), 33, 34 (pregnant); greatest length of skull, 26.6, 26.5, 26.9, 26.5, ——; zygomatic breadth, 13.8, 13.9, 14.1, 13.4, ——; interorbital constriction, 3.9, 3.8, 4.0, 4.0, ——; mastoid breadth, 11.8, 11.9, 11.8, 11.9, 11.5; length of nasals, 10.1, 9.4, 10.0, 10.0, ——; length of maxillary tooth-row, 4.5, 4.3, 4.1, 4.1, 4.1.

Onychomys torridus yakiensis Merriam.—Only one specimen of this grasshopper mouse has been reported previously from Sinaloa (from the town of Sinaloa by Hollister, 1914:471). Thirteen specimens in the Museum of Natural History better define the range of the species in the state as follows: 12 mi. N Culiacán (67981-82); 6 mi. N, 1½ mi. E El Dorado (75374-80); 2½ mi. N El Fuerte (75373); 1 mi. S Pericos (62118-20). The individuals from northeast of El Dorado extend the known range of the species some 115 miles south-southeast from Sinaloa.

A female taken on November 17, 1957, from 6 mi. N, 1½ mi. E El Dorado carried two embryos that measured 23 mm. in crown-rump length. A female obtained on November 18 at the same place carried four embryos that measured 10 mm.

Neotoma albigula melanura Merriam.—Four specimens from northern Sinaloa, two (85379-80) from 3 mi. N, 1 mi. E San Miguel, 350 ft., and two (75386-87) from 2½ mi. N El Fuerte, provide the first records of the species from the state. *N. a. melanura* has been known previously from adjacent parts of Sonora and Chihuahua (see Hall and Kelson, 1959:687-688). The specimens from northeast of San Miguel were trapped in runways under cholla cactus, in which nests also were found, on a slope above a rocky arroyo.

Spilogale pygmaea Thomas.—Two pygmy spotted skunks from 5 mi. NW Mazatlán (85898-99) are the fifth and sixth of the species to be reported (see Van Gelder, 1959:381) and the second and third taken in Sinaloa (the holotype of *pygmaea* was obtained at Rosario). One of our specimens, an adult male, was shot on the night of January 10, 1961, as it foraged near an old hollow tree in weedy-thorn bush habitat adjacent to the Pacific Ocean. The hollow tree contained the nest of a woodrat. The second, an adult female, was trapped nearby in a commercial rat trap baited with peanut butter and set near a burrow in a forested area having little undergrowth.

The two individuals here reported fit fairly well the description of color pattern given for the species by Van Gelder (*op. cit.*: 379), but are larger (considering sex), externally and cranially, than any of the four specimens reported previously. Measurements of the male and female are, respectively: total length, 291, 270; length of tail, 65, 58; length of hind foot, 38, 35; length of ear from notch, 25, 23; weight in grams, 247.0, 190.5; condylobasal length, 46.0, 42.9; occipitonasal length, 45.0, 41.4; zygomatic breadth, 29.0, 27.3; mastoid breadth, 23.9, 22.5; interorbital constriction, 14.3, 13.6; postorbital constriction, 14.8, 14.1; palatilar length, 15.6, 14.6; postpalatal length, 23.2; 22.4; cranial depth, 16.6, 15.2; length of maxillary tooth-row, 14.2, 13.4. Cranial measurements were taken in the manner described by Van Gelder (*op. cit.*:236-237).

LITERATURE CITED

ANDERSEN, K.
1908. A monograph of the Chiropteran genera *Uroderma, Enchistenes,* and *Artibeus.* Proc. Zool. Soc. London, pp. 204-319, illustrated, September.

ANDERSON, S.
1956. Extension of known ranges of Mexican bats. Univ. Kansas Publ., Mus. Nat. Hist., 9:347-351, August 15.
1960. Neotropical bats from western México. Univ. Kansas Publ., Mus. Nat. Hist., 14:1-8, October 24.

BAILEY, V.
1915. Revision of the pocket gophers of the genus Thomomys. N. Amer. Fauna, 39:1-126, 8 pls., 10 figs., November 15.

BAKER, R. H., and A. A. ALCORN
1953. Shrews from Michoacán, México, found in barn owl pellets. Jour. Mamm., 34:116, February 9.

BURT, W. H.
1938. Faunal relationships and geographic distribution of mammals in Sonora, Mexico. Misc. Publ. Mus. Zool., Univ. Michigan, 39:1-77, 26 maps, February 15.

BURT, W. H., and R. A. STIRTON
1961. The mammals of El Salvador. Misc. Publ. Mus. Zool., Univ. Michigan, 117:1-69, 1 fig., September 22.

COMMISSARIS, L. R.
1960. Morphological and ecological differentiation of *Peromyscus merriami* from southern Arizona. Journ. Mamm., 41:305-310, 2 figs., August 15.

COUES, E.
1877. Precursory notes on American insectivorous mammals, with descriptions of new species. Bull. U. S. Geol. Surv. Territories, 3:631-653, May 15.

DALQUEST, W. W.
1953. Mexican bats of the genus Artibeus. Proc. Biol. Soc. Washington, 66:61-65, August 10.

DAVIS, W. B.
1958. Review of the Mexican bats of the Artibeus "cinereus" complex. Proc. Biol. Soc. Washington, 71:163-166, 1 fig., December 31.

DOBSON, G. E.
1890. A monograph of the Insectivora, systematic and anatomical. Part III (includes only plates XXIII-XXVIII), Gurney and Jackson, London, May.

GOODWIN, G. G.
1960. The status of *Vespertilio auripendulus* Shaw, 1800, and *Molossus ater* Geoffroy, 1805. Amer. Mus. Novit., 1994:1-6, 1 fig., March 8.

HALL, E. R., and W. W. DALQUEST
1950. Pipistrellus cinnamomeus Miller 1902 referred to the genus Myotis. Univ. Kansas Publ., Mus. Nat. Hist., 1:581-590, 5 figs., January 20.

HALL, E. R., and K. R. KELSON
1959. The mammals of North America. Ronald Press, New York, 2:viii + 547-1083 + 79, illustrated, March 31.

HOFFMEISTER, D. H., and W. W. GOODPASTER
1954. The mammals of the Huachuca Mountains, southeastern Arizona. Illinois Biol. Monog., 24:v + 1-152, 27 figs., December 31.

HOLLISTER, N.
1914. A systematic account of the grasshopper mice. Proc. U. S. Nat. Mus., 47:427-489, pl. 15, 3 figs., October 29.

HOOPER, E. T.
1961. Notes on mammals from western and southern Mexico. Jour. Mamm., 42:120-122, February 20.

KOOPMAN, K. F.
1961. A collection of bats from Sinaloa, with remarks on the limits of the Neotropical Region in northwestern Mexico. Jour. Mamm., 42: 536-538, November 20.

MERRIAM, C. H.
1895. Revision of the shrews of the American genera Blarina and Notiosorex. N. Amer. Fauna, 10:5-34, pls. 1-3, 2 figs., December 31.

MILLER, G. S., JR., and G. M. ALLEN
1928. The American bats of the genera Myotis and Pixonyx. Bull. U. S. Nat. Mus., 144:viii + 1-218, 1 pl., 1 fig., 13 maps, May 25.

NELSON, E. W.
1909. The rabbits of North America. N. Amer. Fauna, 29:1-314, 13 pls., 19 figs., August 31.

OSGOOD, W. H.
1909. Revision of the mice of the American genus Peromyscus. N. Amer. Fauna, 28:1-285, 8 pls., 12 figs., April 17.

POOLE, A. J., and V. S. SCHANTZ
1942. Catalog of the type specimens of mammals in the United States National Museum, including the Biological Surveys collection. Bull. U. S. Nat. Mus., 178:xiii + 1-705, April 9.

RUSSELL, R. J.
1953. Description of a new armadillo (Dasypus novemcinctus) from Mexico with remarks on geographic variation of the species. Proc. Biol. Soc. Washington, 66:21-25, March 30.

TWENTE, J. W., and R. H. BAKER
1951. New records of mammals from Jalisco, Mexico, from barn owl pellets. Jour. Mamm., 32:120-121, February 15.

VAN GELDER, R. G.
1959. A taxonomic revision of the spotted skunks (genus *Spilogale*). Bull. Amer. Mus. Nat. Hist., 117:229-392, 47 figs., June 15.

Transmitted March 15, 1962.

29-3000

University of Kansas Publications
Museum of Natural History

Volume 14, No. 13, pp. 161-164, 1 fig.
May 21, 1962

A New Bat (Myotis) From Mexico

BY

E. RAYMOND HALL

University of Kansas
Lawrence
1962

University of Kansas Publications, Museum of Natural History
Editors: E. Raymond Hall, Chairman, Henry S. Fitch,
Theodore H. Eaton, Jr.

Volume 14, No. 13, pp. 161-164, 1 fig.
Published May 21, 1962

University of Kansas
Lawrence, Kansas

PRINTED BY
JEAN M. NEIBARGER, STATE PRINTER
TOPEKA, KANSAS
1962

29-3265

A New Bat (Myotis) from Mexico

BY

E. RAYMOND HALL

A single specimen of little brown bat from the northern part of the state of Veracruz seems to be of an heretofore unrecognized species. It is named and described below.

Myotis elegans new species

Holotype.—Female, adult, skin and skull, No. 88398 Museum of Natural History, The University of Kansas; 12½ mi. N. Tihuatlán, 300 ft. elevation, Veracruz, Mexico; obtained on September 24, 1961, by Percy L. Clifton, original No. 985.

Geographic distribution.—Known only from the type locality.

Diagnosis.—A small-footed species having a short tail and small skull. Pelage on upper parts near (16′ *l*) Prout's Brown (capitalized color terms after Ridgway, Color Standards and Color Nomenclature, Washington, D. C., 1912), and more golden on underparts; ears pale brownish and flight-membranes only slightly darker; thumb small (7.5 mm. including wrist); tragus slender but deeply notched. Longitudinal, dorsal profile of skull relatively straight but frontal region elevated from rostrum and lambdoidal region elevated from posterior part of parietal region; posterior margin of P4 (in occlusal view) notched.

Comparisons.—Among named kinds of *Myotis*, *M. elegans* shows most resemblance to the species *M. californicus* and *M. subulatus*. Differences from the latter include shorter tail and ear, more golden color on underparts, pale (not blackish) lips, ears and flight membranes, more slender tragus, shorter skull, posterior border of P4 (in occlusal view) more deeply notched, and longitudinal dorsal profile of skull higher in frontal and lambdoidal regions.

Differences from *M. californicus* include shorter tail, more golden color on underparts, deeper notch in tragus, shorter skull, notched instead of smooth posterior border of P4 (in occlusal view), longitudinal, dorsal profile of skull less abruptly elevated in frontal region and with (instead of without) prelambdoidal depression. From *M. c. mexicanus* that occurs to the north, west, and south of the type locality of *M. elegans* the latter further differs in darker color, paler ears, paler flight membranes, and lesser size, including skull.

Differences from *M. nigricans* of the same region include reddish instead of black pelage, smaller hind foot, smaller skull, rostrum smaller in relation to remainder of skull, narrower interorbital region, and absence of a sagittal crest.

Measurements.—Total length, 79; length of tail, 34; length of hind foot, 7.5; length of ear from notch, 12; length of tragus, 6.5; weight, 4 grams; length of forearm, 33.0; greatest length of skull, 12.4; condylobasal length, 11.9; interorbital constriction, 3.2; breadth of braincase, 6.1; occipital depth, 4.5;

length of mandible, 8.9; length of maxillary tooth-row, 4.6; maxillary breadth at M3, 4.9; length of mandibular tooth-row, 5.0. Degree of wear on teeth, stage 2 (in terminology of Miller and Allen, Bull. U. S. Nat. Mus., 144, May 25, 1928).

Remarks.—The longitudinal dorsal profile of the skull and the deeply notched posterior border of P4 seem to be distinctive of *elegans*. When the characters of *elegans* first were tabulated it was felt that it probably was only subspecifically different from some previously named species. But further study of the distinctive characters indicates that they are outside the range of variation of any near relative of *elegans* and it, therefore, is here accorded specific rank.

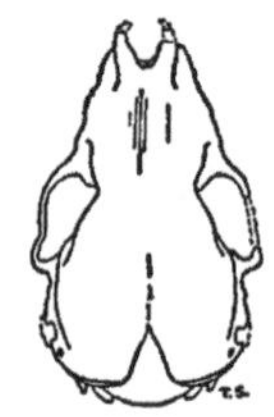

FIG. 1. Lateral view (left) and dorsal view (right) of the holotype of *Myotis elegans*, × 2.

Material examined.—Known only from the holotype.

Transmitted April 2, 1962.

□

29-3265

University of Kansas Publications
Museum of Natural History

Volume 14, No. 14, pp. 165-362, 2 figs.
May 20, 1963

The Mammals of Veracruz

BY

E. RAYMOND HALL AND WALTER W. DALQUEST

University of Kansas
Lawrence
1963

UNIVERSITY OF KANSAS PUBLICATIONS
MUSUEM OF NATURAL HISTORY

Institutional libraries interested in publications exchange may obtain this series by addressing the Exchange Librarian, University of Kansas Library, Lawrence, Kansas. Copies for individuals, persons working in a particular field of study, may be obtained by addressing instead the Museum of Natural History, University of Kansas, Lawrence, Kansas. There is no provision for sale of this series by the University Library, which meets institutional requests, or by the Museum of Natural History, which meets the requests of individuals. However, when individuals request copies from the Museum, 25 cents should be included, for each separate number that is 100 pages or more in length, for the purpose of defraying the costs of wrapping and mailing.

* An asterisk designates those numbers of which the Museum's supply (not the Library's supply) is exhausted. Numbers published to date, in this series, are as follows:

Vol. 1. Nos. 1-26 and index. Pp. 1-638, 1946-1950.

*Vol. 2. (Complete) Mammals of Washington. By Walter W. Dalquest. Pp. 1-444, 140 figures in text. April 9, 1948.

Vol. 3. *1. The avifauna of Micronesia, its origin, evolution, and distribution. By Rollin H. Baker. Pp. 1-359, 16 figures in text. June 12, 1951.

*2. A quantitative study of the nocturnal migration of birds. By George H. Lowery, Jr. Pp. 361-472, 47 figures in text. June 29, 1951.

3. Phylogeny of the waxwings and allied birds. By M. Dale Arvey. Pp. 473-530, 49 figures in text, 13 tables. October 10, 1951.

*4. Birds from the state of Veracruz, Mexico. By George H. Lowery, Jr., and Walter W. Dalquest. Pp. 531-649, 7 figures in text, 2 tables. October 10, 1951.

Index. Pp. 651-681.

*Vol. 4. (Complete) American weasels. By E. Raymond Hall. Pp. 1-466, 41 plates, 31 figures in text. December 27, 1951.

Vol. 5. Nos. 1-37 and index. Pp. 1-676, 1951-1953.

*Vol. 6. (Complete) Mammals of Utah, *taxonomy and distribution*. By Stephen D. Durrant. Pp. 1-549, 91 figures in text, 30 tables. August 10, 1952.

Vol. 7. Nos. 1-15 and index. Pp. 1-651, 1952-1955.

Vol. 8. Nos. 1-10 and index. Pp. 1-675, 1954-1956.

Vol. 9. *1. Speciation of the wandering shrew. By James S. Findley. Pp. 1-68, 18 figures in text, December 10, 1955.

2. Additional records and extension of ranges of mammals from Utah. By Stephen D. Durrant, M. Raymond Lee, and Richard M. Hansen. Pp. 69-80. December 10, 1955.

3. A new long-eared myotis (Myotis evotis) from northeastern Mexico. By Rollin H. Baker and Howard J. Stains. Pp. 81-84. December 10, 1955.

4. Subspeciation in the meadow mouse, Microtus pennsylvanicus, in Wyomir By Sydney Anderson. Pp. 85-104, 2 figures in text. May 10, 1956.

5. The condylarth genus Ellipsodon. By Robert W. Wilson. Pp. 105-116, figures in text. May 19, 1956.

6. Additional remains of the multituberculate genus Eucosmodon. By Robert W. Wilson. Pp. 117-123, 10 figures in text. May 19, 1956.

7. Mammals of Coahulia, Mexico. By Rollin H. Baker. Pp. 125-335, 75 figures in text. June 15, 1956.

8. Comments on the taxonomic status of Apodemus peninsulae, with description of a new subspecies from North China. By J. Knox Jones, Jr. Pp. 337-346, 1 figure in text, 1 table. August 15, 1956.

9. Extensions of known ranges of Mexican bats. By Sydney Anderson. Pp. 347-351. August 15, 1956.

10. A new bat (Genus Leptonycteris) from Coahulia. By Howard J. Stains. Pp. 353-356. January 21, 1957.

11. A new species of pocket gopher (Genus Pappogeomys) from Jalisco, Mexico. By Robert J. Russell. Pp. 357-361. January 21, 1957.

12. Geographic variation in the pocket gopher, Thomomys bottae, in Colorado. By Phillip M. Youngman. Pp. 363-387, 7 figures in text. February 21, 1958.

13. New bog lemming (genus Synaptomys) from Nebraska. By J. Knox Jones, Jr. Pp. 385-388. May 12, 1958.

14. Pleistocene bats from San Josecito Cave, Nuevo León, México. By J. Knox Jones, Jr. Pp. 389-396. December 19, 1958.

15. New subspecies of the rodent Baiomys from Central America. By Robert L. Packard. Pp. 397-404. December 19, 1958.

16. Mammals of the Grand Mesa, Colorado. By Sydney Anderson. Pp. 405-414, 1 figure in text, May 20, 1959.

17. Distribution, variation, and relationships of the montane vole, Microtus montanus. By Sydney Anderson. Pp. 415-511, 12 figures in text, 2 tables. August 1, 1959.

(Continued on inside of back cover)

University of Kansas Publications
Museum of Natural History

Volume 14, No. 14, pp. 165-362, 2 figs.
May 20, 1963

The Mammals of Veracruz

BY

E. RAYMOND HALL AND WALTER W. DALQUEST

University of Kansas
Lawrence
1963

University of Kansas Publications, Museum of Natural History

Editors: E. Raymond Hall, Chairman, Henry S. Fitch,
Theodore H. Eaton, Jr.

Volume 14, No. 14, pp. 165-362, 2 figs.
Published May 20, 1963

University of Kansas
Lawrence, Kansas

PRINTED BY
JEAN M. NEIBARGER, STATE PRINTER
TOPEKA, KANSAS
1963

29-4035

The Mammals of Veracruz

BY

E. RAYMOND HALL
(The University of Kansas)

AND

WALTER W. DALQUEST
(Midwestern University)

CONTENTS

	PAGE
INTRODUCTION	167
LIFE-ZONES	170
ITINERARY	174
GAZETTEER	182
CHECK LIST	186
ACCOUNTS OF SPECIES	192
LITERATURE CITED	356

INTRODUCTION

In the rich, tropical lowlands of Veracruz, mammals have long interested man. Several species had religious and ornamental significance. Statues and ornaments of clay and stone representing mammals are common. Four stone squirrel heads in the plaza at San Andrés Tuxtla weigh more than a hundred pounds each. The importance of hunting to the Indians before the Conquest is not known. They had few domestic animals and no draft animals. The Indians were a civilized, agricultural people; corn was a principal food crop.

The accounts of the Conquest, although well documented in many respects, provide relatively little information on the native wildlife. In one letter to Charles V, Cortés on July 10, 1519 (see pp. 20-21 of translation by J. B. Morris, 1928), mentioned that "All kinds of hunting is to be met with in this land and both birds and beasts similar to those we have in Spain, such as deer, both red and fallow, wolves, foxes, partridges, pigeons, turtle doves of several kinds, quails, hares and rabbits: so that in the matter of birds and beasts there is no great difference between this land and Spain, but there are in addition lions and tigers about five miles inland, of which more are to be found in some districts than in others." The situation is not greatly different today!

The colonial period was marked by the introduction of livestock (cattle, horses, burrows, sheep, swine, goats) and food plants (beans, sugar cane, rice). The impact of these introduced species on the native wildlife must have been great.

In the post-colonial period wildlife probably was affected by firearms becoming available to large numbers of persons. The Diaz period, in Veracruz, was marked by intensive exploitation of large parts of northern and southern Veracruz by foreign interests, especially British and American. The foreigners were far more effective in exterminating wildlife than was the native population.

In the modern period, since 1910, firearms and ammunition have been available to a large part of the population. New roads are constantly being constructed into previously inaccessible areas. Enforcement of game laws is difficult. Over large areas the larger mammals have been eliminated, and their extirpation at an accelerated rate in other areas seems inevitable.

Local names applied to mammals of Veracruz can be divided into three categories. The first category consists of Spanish names applied to mammals that closely resemble species in Spain, for example zorra (fox), ardilla (squirrel) and conejo (rabbit). The second category is made up of descriptive phrases in Spanish for animals that do not closely resemble any in Spain. Examples are cabeza de viejo (the tayra, a white-headed mustelid) and brazo fuerte (literally strong arm, referring to the tamandua—an anteater). The third category consists of names only slightly changed from the Indian (usually Nahuatl) language. Examples are coyote from the Nahuatl coyotl and mapache from the Nahuatl mapachtli.

In general, local names, distinctive of genera and often of species, are available for the mammals except kinds of bats, mice, and rats. All bats except the vampire are designated by the Spanish "murciélago." The vampire is "vampiro." Rat is "rata" and mouse is "ratón."

In 1947 and 1948 Walter W. Dalquest drafted an account, much more extensive than the present one, of the mammals of Veracruz. In the period 1959-1961 E. Raymond Hall modified the account into essentially its present form. In the latter period Bradford House and Ticul Alvarez re-examined the specimens, prepared new lists of "Specimens examined," and checked the specific and subspecific identifications. Hall and Alvarez studied the specimens from northern Veracruz obtained in 1960 by M. Raymond Lee and in 1961 by Percy L. Clifton and J. H. Bodley and incorporated the resulting information in the manuscript. In the itinerary and accounts of species the pronouns I, me, my, and mine refer to Dalquest. In the remarks on habits of the different species the quoted material that is not identified as from some previously published source is from Dalquest's field notebooks on file in the Museum of Natural History at the University of Kansas at Lawrence.

Mammals were collected in Veracruz for the Museum of Natural History principally to obtain topotypes and other specimens from there that would enable investigators at the Museum to have comparative materials for studying speciation and subspeciation in mammals from adjoining areas. Aims of the following account are to make known (1) the kinds of native mammals of Veracruz, and (2) the geographic distribution of each kind in Veracruz; and (3) to record natural history information as given beyond because such information is scarce or lacking for many of the species.

Presentation at this time of the results of study of the taxonomy and the geographic distribution of the mammals of Veracruz is timely because corresponding information is being organized by other authors for the mammals of Tamaulipas to the northward and for those of Tabasco to the southward. Also, the combined information from these three accounts will aid in investigations underway in Yucatán and adjoining areas.

For any technical (scientific) name of a species or subspecies beyond that differs from that in "The Mammals of North America" by Hall and Kelson (1959), an explanation or citation to an explanation immediately follows the listing of records of occurrence. Where no such citations are given, basis for a name can be found in the mentioned publication by Hall and Kelson.

For kinds of mammals known to occur in Veracruz, previously published information is cited in the accounts beyond (1) if the information relates to kinds found by other investigators but not by us, and (2) if specimens are listed from places additional to those from which the University of Kansas Museum of Natural History has specimens. Also, some previously unrecorded specimens in other museums are here recorded for the first time.

For the use of these specimens and for other assistance we thank David H. Johnson, Henry W. Setzer and Charles O. Handley, Jr., of the U. S. National Museum; Richard H. Manville, Viola S. Schantz, and John E. Paradiso, of the mammal collections of the U. S. Fish and Wildlife Service; Philip Hershkovitz and Karl F. Koopman of the Chicago Natural History Museum; and G. B. Corbet of the British Museum, Natural History. Some of the research in the laboratory was supported by Grant No. 56 G 103 from the National Science Foundation. Field work was supported by a grant from the American Heart Association, Inc., for a brief period by a grant from the Atomic Energy Commission, and principally by grants from the Kansas University Endowment Association.

LIFE-ZONES

Life-zones, from the Arctic-Alpine to the Tropical, are present in Veracruz. The physiography of the state is such, however, that the higher life-zones are all in a small area along the western edge of the central part of the state. There the eastern edge of the Mexican Plateau extends eastward into Veracruz, and the desert-inhabiting mammals of the Lower Sonoran Life-zone enter the state. At the eastern edge of the plateau in Veracruz, the land rises abruptly, and the top of the great volcanic cone of the Pico de Orizaba is 18,077 feet above sea level. Orizaba is the second highest mountain in North America, being exceeded only by Mount McKinley in Alaska. On the eastern side of Mt. Orizaba, and its neighboring mountain, Cofre de Perote, the land pitches down abruptly. The summit of Orizaba is only 70 miles from the Gulf of Mexico and the distance from its summit to the Tropical Life-zones at the city of Orizaba is only 18 miles.

The upper life-zones, in Veracruz, are present in only a small area. Much of this area is made up of the Lower Sonoran Life-zone, the desert of the Mexican Plateau, and even this zone is but faintly represented on the eastern slope of the Plateau. Above the Lower Sonoran Life-zone, it is difficult or impossible to delimit the individual life-zones customarily recognized farther northward; instead there is the area of pines, then the zacaton and firs, and finally the area above timberline which can be designated as the Arctic Alpine Life-zone.

The remainder, approximately 95 per cent of the state, lies in the Tropical Zone. For practical purposes we choose not to attempt to recognize an Upper Tropical Life-zone and a Lower Tropical Life-zone, but instead prefer to recognize three divisions of the Tropical Zone. The divisions are arid, lower humid, and upper humid.

The Arctic-Alpine Life-zone in Veracruz includes the peak of Mount Orizaba. Probably this life-zone normally has no distinctive mammalian fauna at this latitude.

The other boreal and temperate life-zones form a narrow belt round the upper parts of the slopes of the Cofre de Perote and the Pico de Orizaba, from near Tlapacoyan, on the north, to near Acultzingo, on the south. This belt is characterized by coniferous trees and a distinct mammalian fauna. Characteristic species, in Veracruz, are: *Sorex macrodon, Sorex saussurei, Cratogeomys perotensis, Reithrodontomys chrysopsis, Sciurus oculatus,* and *Neotomodon alstoni.* The parts of this belt at higher levels are more humid, and the species of conifers differ from those of the lower levels. The higher areas have numerous meadows of a bunchgrass (zacaton). Only a few kinds of mammals are restricted to the upper area; only the volcano mouse (*Neotomodon*) and the volcano harvest mouse (*Reithrodontomys chrysopsis*) seem to be confined to it.

The Mexican Plateau, intruding from the west brings the Sonoran life-zones into the area to the north and west of Cofre de Perote. This is an area of sandy desert, supporting mammals characteristic of the Sonoran life-zones: *Cratogeomys fulvescens, Dipodomys phillipsi, Perognathus flavus, Spermophilus perotensis,* and *Peromyscus difficilis.*

On the eastern slope, the Sonoran life-zones are almost pinched out, owing to the rapid drop of the terrain. There is an indication of the Sonoran zones near Las Vigas, and again in the valley below Acultzingo.

At the northern edge of the temperate area, near Tlapacoyan, there is an abrupt transition from the temperate area to the Tropical Zone. At Teziutlan in the state of Puebla there are extensive forests of conifers. Going eastward, and downward, a person passes through an area strongly reminiscent of the humid division of the Transition Life-zone of the northern Pacific Coast of the United States. Beneath broad-leaved trees of the alder type, a dense growth of mosses and sword ferns cover the ground. Continuing eastward, there is a gradual transition; the sword ferns give way to tree ferns, the alderlike trees give way to tropical trees, the stream courses have a distinct tropical flora, and finally the temperate forms of vegetation are found only on the higher hills, before disappearing completely in favor of a pure tropical forest.

At the southern edge of the temperate area, near Acultzingo, the conifers of the northern temperate area, the western desert flora, and the eastern tropical flora almost meet at an area of brushland. The entire area is somewhat humid, being cloud-covered much of the time as a result of the warm winds, which blow in from the gulf, rising and spilling over the lip of the plateau near Acultzingo. Nevertheless, the water runs off rapidly because the area is rocky and slopes steeply. The area of brushland has scattered coniferous trees in sheltered places. The trees support an extremely rich flora of bromeliads, mosses, ferns and orchids, although the ground beneath is relatively dry. The mammalian fauna is limited as to species and numbers of individuals, but is made up of species from the temperate zone, the desert, and the tropical jungles to the east. According to Dr. Edward H. Taylor of The University of Kansas, the area has a rich endemic fauna of reptiles and amphibians.

In Veracruz the arid division of the tropical zone is best seen on the coastal plain, which is relatively level, extending from sea level along the gulf to between 700 and 1700 feet, depending on the topography. Much of the plain is grassland that, by the later part of the dry season, is sparse, dead and brown. The earth is baked and cracked. Vegetation consists of scattered, bushlike trees with lacy foliage and thickets of flat-topped, thorny bushes. But along the water courses the jungle is thick and dense. There the trees are tall and abundantly hung with vines, parasites and epiphytes. The mammalian fauna of this division includes some species that are not found in other divisions of the Tropical Life-zone. Examples are *Reithrodontomys fulvescens, Sigmodon hispidus,* and *Liomys pictus.* Several species have subspecies in the arid division different from those in the upper humid division of the tropical zone.

The upper humid division of the tropical zone is the dense forest area, where the trees are tall and abundantly hung with many species of vines, mosses, ferns, cacti, bromeliads, orchids, and other parasites and epiphytes. This division has a distinct fauna. The mammals include several species that, in Veracruz, seem to be confined to it. These include: *Cryptotis mexicanus* and *Microtus quasiater.*

The line of contact between the upper humid and arid divisions of the tropical zone is sharp; in places, one can tell almost to the yard where the two meet. This arrangement has resulted from human activities. Originally, a band of oak forest separated the upper humid division and arid division of the tropical zone. This oak forest was transitional in nature, being neither strictly semiarid nor semihumid. Unfortunately the land most important for cultivation is at the junction of these two. There the land is level, and cultivation is practical; also this zone is the most humid part of the relatively arid coastal plain. As a result the land was cleared for cultivation and numerous towns and villages are situated there. A second factor that makes for rapid clearing is the oak itself. This wood makes a superior kind of charcoal. The trees grow on relatively level ground and therefore the charcoal made from them is easily transported to market. Consequently the oak forest has completely disappeared in most places. Only in the more inaccessible areas is there any extensive oak forest remaining. In so far as we were able to determine, the oak forest has no kinds of mammals endemic to it; those trapped were kinds which occurred also on the coastal plain and individuals were few.

The lower humid division of the tropical zone includes the dense, lowland jungle, and in most respects does not differ greatly from the upper humid division, but in detail it differs greatly. Many species of trees are confined to the lower humid division, as also are several species of mammals, for example *Alouatta villosa, Tapirus bairdii, Cyclopes didactylus* and *Tayassu pecari.*

The topography of Veracruz, except for the mountains and the edge of the Mexican Plateau already discussed, is relatively simple. The state is a low, level, tropical lowland, extending northward and westward between the Gulf of Mexico and the high, arid desert of the Mexican Plateau. The tropical part of the state has but two important topographic interruptions. One is the Tuxtla Mountains. They are a group of small volcanic peaks and basaltic intrusions. Although the peaks are not high, the highest, Volcán San Martín Tuxtla, being only 5000 feet above sea level, the area is much elevated above the surrounding plain. The area is an island, of the upper humid division of the tropical zone, completely surrounded by the lower humid division of the tropical zone less than 100 feet above sea level. The mammalian fauna of the Tuxtla Mountains has not been thoroughly studied, but the study which has been made reveals a certain amount of endemism among the mammals.

The second interruption of the coastal plain is the arm of high country which extends eastward along the twentieth parallel almost to the waters of the Gulf. There the lowlands average less than 10 kilometers wide for a distance of approximately 45 kilometers in a north to south direction along the coast. The high land that extends eastward to the coast seems to have served as a barrier to the dispersal of the spiny pocket mice since only *Liomys pictus* occurs to the southward and *Liomys irroratus* to the northward. For the most part, however, the barrier serves only to divide the ranges of subspecies.

ITINERARY

In the first season of field work in Veracruz by Dalquest he was accompanied by his wife, Peggy. On February 9, 1946, they traveled via Tlacotepec and Tehuacan in the state of Puebla to Potrero in the state of Veracruz. Potrero is the site of Ingenio El Potrero, a large sugar mill and sugar cane plantation. Mr. Dyfrig Forbes, the superintendent of Ingenio El Potrero, and his wife, Mrs. Leora Forbes, welcomed the Dalquests and invited them to make the beautiful Hacienda Potrero Viejo the base of their operations in Veracruz. Potrero Viejo is an old Spanish colonial village situated seven kilometers west of Potrero. Potrero Viejo, rather than the modern Potrero, is the "Potrero" of Sumichrast.

Several days were spent at Potrero Viejo. Trips were made to nearby caves where bats of several species were found. On February 15 camp was established on the north side of the Río Atoyac, approximately eight kilometers northwest of Potrero Viejo. There a local man named Valdo (killed in 1947) and two of Mr. Forbes' employees, Casildo Mazza and his nephew, Gerardo Mazza, assisted in the collecting of specimens. Work was continued in the vicinity of Potrero until March 24, except for a short trip on March 17 and 18, with Mr. Forbes, to the coastal plain near Piedras Negras.

On March 24 camp was moved to the Río Metlác, four kilometers west-northwest of the town of Fortín. Mr. Daniel Rabago, manager of the power plant of the Moctezuma Brewery, made available a splendid camp site, modern conveniences, and some of his employees as assistants. On April 8 the Dalquests returned to Potrero Viejo.

Further information, condensed from the field notes of W. W. Dalquest, is as follows: On April 10, we drove to Mexico City and then made a leisurely trip to the international border, collecting specimens at various localities along the way. On April 24 we shipped our specimens from Laredo, Texas, and on April 29 again started southward, reaching Potrero Viejo on May 5. There equipment and supplies were packed, and on May 9 we left for the coastal plain where we set up and maintained our camp on the Río Blanco 20 kilometers west-northwest of Piedras Negras. Gerardo Mazza again served as our assistant and camp was established on land belonging to his father. Except for one trip (May 20-23) to Cordoba and Potrero Viejo for supplies, we remained near Piedras Negras until May 31. The rainy season had begun by that time, and we had some difficulty in crossing the swollen Río Blanco and the muddy coastal plain. We remained at Potrero Viejo until June 6, then collected in central and northern México, and crossed the international boundary at Laredo on July 4.

The second season's work was begun on September 8, 1946, when I (W. W. Dalquest) and Allen Oleson of Boulder, Colorado, crossed the border at Laredo. Collections were made along the highway, and we did not arrive at Potrero Viejo until the afternoon of September 23. In the period September 26-28 we camped on the edge of an arroyo 15 kilometers east-southeast of San Juan de la Punta. The period September 28 to October 7 was spent on the Río Blanco, 20 kilometers west of Piedras Negras. Gerardo Mazza then was engaged and acted as assistant for the year. On October 7 we returned to Potrero Viejo, but left on October 9 and for one night camped 15 kilometers east-southeast of San Juan de la Punta, and then for two nights camped five kilometers southwest of Boca del Río, on the lower reaches of the Río Atoyac, near its mouth. On October 13 we drove through the city of Veracruz to Jalapa, and camped five kilometers north of that city. We remained until October 20. On October 20 we camped three kilometers west of Plan del Río and from the afternoon of October 21 until October 23 at Puente Nacional. On the 23rd we returned to Potrero Viejo and on October 24 we made a round trip to Cordoba. We left Potrero Viejo on October 21, camped on the beach at Mocambo on the night of the 28th, drove to Veracruz on the 29th, on to Puente Nacional and Jalapa to two kilometers west of Jico where we camped until October 31 (Oleson set traps on the evening of the 27th five kilometers north of Jalapa). On the 31st we drove back to Jalapa and eastward to where we camped three kilometers east of Las Vigas until November 5. From November 6 until November 9, we collected on the desert two kilometers east of Perote. From November 9 until 11 we camped and collected six kilometers south-southeast of Altotonga and on the 11th we drove *via* Tezuitlan, Puebla, to an overnight dry camp seven kilometers southeast of Jalacingo. On November 12 we drove high on the slopes of the Cofre de Perote, and camped at 10,550 feet elevation, 11 kilometers northwest of Pescados. Although it was bitterly cold, we remained in this camp on the Cofre de Perote until November 18 when we drove *via* a place designated as two kilometers south of Sierra de Agua, 8500 feet, to the desert, where we camped for one night three kilometers west of Limón, placing us one kilometer east of the border of Puebla. The next day we drove to the highway to Puebla, and then east, camping on the night of November 19 three kilometers west of Acultzingo, and reached Potrero next day, when Mr. Oleson left to return to the United States.

We (Gerardo Mazza and Dalquest) left Potrero Viejo on November 29, and managed to drive southward along the Tehuantepec railway and then south to where we camped three kilometers north of Presidio until December 4 when we moved camp to a point two kilometers north of Motzorongo, where we remained until December 11. On December 11 we returned to our earlier camp three kilometers north of Presidio, and worked there until returning to Potrero Viejo on December 13. On December 14 we made an overnight trip to a camp seven kilometers southeast of San Juan de la Punta. We returned on the 15th to Potrero Viejo and left there on December 17, drove *via* Cordoba to Orizaba, and camped three kilometers southeast of Orizaba until December 22 when we returned to Potrero Viejo.

On January 12, 1947, Gerardo Mazza left for the Río Blanco, to collect alone, and I left Potrero Viejo to go to the state of Tabasco. An illness contracted in Tabasco prevented field work for some days, and it was not until February 1 that Mazza and I together took the plane for Coatzacoalcos. We collected at places 14 kilometers southwest of Coatzacoalcos, at a place seven kilometers northwest of Paso Nuevo, and at a place 10 kilometers northwest of Minatitlan until February 12, when we took the plane for Cordoba. Unfortunately, weather conditions prevented the plane from landing in Veracruz, and we were taken to Mexico City. We were forced to live in Mexico City until money for our fare back to Potrero could be sent from there. We did not leave for the field again until February 26, and this time we went by train to Jimba where we remained until March 6 except for a side trip on March 4 and 5 to a place 15 kilometers to the southwest. We returned *via* Cordoba to Potrero Viejo leaving again for Cordoba on March 10, and from there on March 11, by train for Jesús Carranza (old name Santa Lucrecia) on the Isthmus of Tehuantepec. After several days spent near Jesús Carranza, we obtained passage on a dugout canoe, and lived with the Indians downstream from Jesús Carranza for more than a week. On March 23 we returned to Potrero Viejo. We made a short trip, March 31-April 2, to Cosamolapán, and then packed our materials and specimens. Mazza then returned to the Río Blanco while I left for the United States, crossing the international boundary on April 27.

In the third season, I (Dalquest) drove directly from Lawrence, Kansas, to Potrero Viejo, arriving on September 21, 1947. A local boy, having the given name of "Moises," there was engaged as assistant. Camp was established two kilometers north of Paraje Nuevo, a few kilometers from Potrero Viejo, on September 23, and collecting was done there until September 28. On September 28 we left for our old camp on Boca del Río, but stayed there only one night and collected specimens three kilometers west of Boca del Río before returning to Potrero Viejo. On October 6, we drove to the edge of the plateau and camped three kilometers west of Acultzingo where we remained until October 8, when we drove to Limón, on the Perote desert, and camped three kilometers west of the town until October 11. On that date we moved camp to two kilometers east of Perote and searched especially for ground squirrels. Leaving there on October 13, we went north, through Tezuitlan, Puebla, and then back east to the state of Veracruz, and camped four kilometers west of Tlapacoyan, October 13 to 17. On October 17 we drove farther east, and camped five kilometers east-northeast of El Jobo. There we found mammals so scarce that we left the area and returned to Potrero on October 20.

On October 27 we again went northward, through the city of Veracruz, and camped seven kilometers north-northwest of Cerro Gordo. We remained there until October 30, when we drove westward to the edge of the plateau and camped overnight four kilometers south of Jalacingo. On October 31 we drove on and camped three kilometers southwest of San Marcos, where we collected for several days. On November 5 we drove eastward to the gulf, and northward, stopping nine kilometers northwest of Nautla. Land crabs in great numbers here prevented successful collecting of mammals, and we drove, *via* San Marcos, north to three kilometers west of Gutierrez Zamora on November 6. We went still farther north on November 9, camping that night four kilometers east of Papantla and next day moved camp to 10 kilometers northwest of Papantla. On November 11 we moved camp to five kilometers south of Tehuantlan and remained there until November 15 when we started the return trip, trapping nine kilometers east of Papantla on the nights of November 15, 16 and 17. On November 18 we had heavy rains; on the 19th we reached Martínez de la Torre and remained there overnight. On November 20 we camped again four kilometers west of Tlapacoyan, remaining until November 25 when we drove *via* Tezuitlan, Jalacingo, Altotonga, Perote, and Jalapa to seven kilometers north-northwest of Cerro Gordo where we made collections until November 28 when we returned to Potrero Viejo.

On December 5, Angel Carrillo was employed as assistant, and he and I, on that date, left Potrero and drove over the newly constructed highway to Mirador. Hacienda Mirador still was inhabited by the Sartorius and Grohman families, as it was when Sumichrast obtained specimens there nearly 100 years before we did. E. W. Nelson collected there in 1894. The Grohman family kindly invited us to live at Hacienda Mirador while we were working in that area. We stayed at Mirador until December 10, and then moved east to a place 15 kilometers east-northeast of Tlacotepec, remaining there until December 14. We then started back to Potrero Viejo, trapping on December 14 to 16 four kilometers west of Paso de San Juan; on December 16 to 18 three kilometers west of Boca del Río; and on December 18 to 20 one kilometer east of Mecayucan. On December 20 we returned to Potrero Viejo. We left there on December 28, spent the night in Cordoba, and took the train southward the next day. We spent the following night at Rodriguez Clara, a division point, and the next morning took the train to San Andrés Tuxtla, arriving on the evening of December 30. The Tuxtla area was rich in mammals, and we remained there until January 21, 1948, when we returned *via* Cordoba to Potrero Viejo.

On February 2, Carrillo and I (Dalquest) left Potrero Viejo, took the train southward to the Isthmus, and descended the Río Jaltipec and Río Coatzacoalcos by dugout to a locality known as Zapotal, a collection of native houses on the riverbank. There we established a base, for operations in the Isthmus, living with the Indians whom I (Dalquest) had met on the visit to that area in the previous year. We reached Zapotal on February 5, and remained until May 3, except for the period of February 24 to March 19, when I (Dalquest) returned to the United States on business. The rainy season started early in May, 1948, and on May 20, 1948, I (Dalquest) reached Laredo on the return trip to the United States.

In May and June, 1948, Mr. Dyfrig Forbes obtained several valuable mammal specimens for us, in the immediate vicinity of Potrero Viejo. Actual collecting for the 1948-1949 season was started at Potrero Viejo on September 14, 1948, with Angel Carillo again as assistant. On September 19, we left for the desert near Perote, arriving that evening. From a base at the town of Perote we collected west and north of the town, and also two and three kilometers west of Limón. We remained at Perote until October 1. On that date we moved our base eastward to the pine forest and lava area about Las Vigas, and from a base at Las Vigas collected specimens from within a three kilometer radius. Mammals were abundant, in spite of the cold weather, and our collecting trunks were filled when we left there on October 21. We arrived at Potrero Viejo on October 22. We remained at Potrero Viejo until November 2, when we left and drove through Huatusco to Jalapa, Puebla, and Mexico City, northward through Pachuca, Real del Monte, and Atotonilco to collect in the extreme western part of the Chicontepec Rincón. We obtained specimens 10 kilometers southwest of Jacales on November 4 and 5, and at six kilometers west-southwest of Zacualpilla on November 6 to 10. On November 10, we drove eastward through Tulancingo, Hidalgo, and set traps seven kilometers west of El Brinco, Veracruz. Continuing southward from this locality, we trapped and hunted bats at Jalacingo on November 12 and 13. On November 14, we continued on to Potrero Viejo, obtaining specimens at Puente Nacional and Huatusco on the way. From November 17 to 27, we collected near Potrero Viejo, but on November 27, collected near Cautlapan, and spent the night at that town. On November 29, we drove to Coscomatepec, where we stayed until December 2, when we moved northward to Huatusco. We spent only one night at Huatusco, and returned to Cautlapan on December 3, and on to Potrero Viejo on December 4. On December 5 we made a trip to the Cumbres of Acultzingo with Mr. Forbes, collecting specimens, principally reptiles and amphibians, along the highway, and returned to Potrero Viejo that night. On December 8, we camped at Boca del Río, and the following day drove south along the Gulf to the port of Alvarado. There are no small mammals on the sands at Alvarado, because of the great numbers of land crabs. We spent several days trying to secure a manatee here, but failed. We did make arrangement with a fisherman to supply a specimen, and from him obtained a splendid skull a few weeks later. Leaving Alvarado on December 12, we drove through Veracruz and then westward to Puente Nacional, where we set traps that night. On December 13 we collected at Plan del Río, December 14 nine kilometers east of Totutla, and on December 15 at Cautlapan. We arrived at Potrero Viejo on December 16, and remained until December 29. On that date we drove to Teocelo where collecting was good, and we remained until January 10, 1949. On January 10 we drove to Las Vigas, where we remained until January 19, and returned to Potrero Viejo on January 20. On January 28 we started northward, stopping at Teocelo until February 1, then driving through Papantla to Tuxpan and west along the river to San Isidro, where we spent the night. The following day, February 4, we started northward along the road leading to Tampico, Tamaulipas. We collected at the following places: Potrero Llano, February 4; Cerro Azul, February 5; La Mar, February 6; El Cepillo, February 7; and Tampico Alto, February 8. On February 8, we drove back southward to La Mar, and then westward over a very bad road to Ozuluama, where we stayed until February 10. On February 10, we returned to La Mar, and on southward to Potrero Llano, where we remained until February 16. Along this entire route, we found small mammals to be rare, and only near Potrero Llano were specimens taken in fair numbers. We started back for Potrero Viejo on February 17, arriving on the 21st. From March 1 to 7, we were out of Veracruz, bringing the previous six months' catch to the border at Laredo. We returned to Potrero Viejo on March 11. We left Potrero on March 18, taking the railroad south to Jesús Carranza, on the Isthmus. We were fortunate in meeting an Indian friend there the following day, and he took us and our equipment to Zapotal, where we were greeted by our friends of the previous two years. We spent the time between March 20 and April 14 collecting on the Río Coatzacoalcos and Río Chalchijapa, east and south of Jesús Carranza. We returned to Potrero Viejo on April 2. On April 20, Carrillo was sent back to Zapotal, where I joined him on May 1. Between May 1 and May 16 we collected on the Río Coatzacoalcos and Río Solusuchil, southeast and east of Jesús Carranza. In 1949, collecting in Veracruz by Dalquest was terminated on June 5.

In the period April 22-26, 1949, W. W. Dalquest and E. R. Hall, for part of the time accompanied by Mrs. E. R. Hall, visited many of the places where mammals had been collected in the previous three years by Dalquest. Our itinerary follows:

April 22, from Mexico City east through Huauchinango, Puebla, Poza Rica (Veracruz), Papantla, Gutierrez Zamora, Tecolutla, to within two kilometers of Nautla, southwest through Tlapacoyan to Tezuitlan (Puebla) where we spent the night. April 23, to Jalacingo, Altotonga, Perote, Limón, to border of state of Puebla, back to Perote, Las Vigas, almost to city of Veracruz then southwest across Río Jamapa at Boca del Río, thence southwest through Peñuela, thence northeast seven kilometers to Potrero Viejo. April 24, caves (including Ojo de Agua) 13 miles west-northwest of Potrero Viejo, and another cave seven kilometers northwest of Potrero. April 25, Potrero Viejo, Cordoba, Orizaba, Acultzingo, Tehuacan, Limón, Perote, Las Vigas, Jalapa, Coatepec, Jico, Teocelo, back to Jalapa thence Puente Nacional, Huatusco, Coscomatepec, Fortín to Cordoba where we spent the night. April 26, to San Juan de la Punta, to within nine kilometers of Boca del Río, to Alvarado, back to place nine kilometers west of Boca del Río, to Boca del Río, then back west to Peñuela and to Potrero Viejo. April 27, *via* Cordoba and Orizaba to Mexico City.

In the period January 12 through January 24, 1951, W. W. Dalquest, Rollin H. Baker, Alford J. Robinson and George P. Young visited the southern part of the state of Veracruz and collected several mammals. With reference to the town of Jesús Carranza the localities of capture of mammals on this trip were as follows: 20 km. E (Boca del Río Chalchijapa); 37 km. E and 7 km. S (Arroyo Saoso); 20 km. ESE; 24 km. E and 7 km. S; 26 km. E and 8 km. S (Arroyo Azul).

On July 21, 1955, Rollin H. Baker, R. W. Dickerman, J. Keever Greer, DeLayne Hudspeth, John William Hardy, Charles M. Fugler, Robert L. Packard, Robert G. Webb, and South Van Hoose, drove northward, from Tollocita [= Tollosa], Oaxaca, taking the ferry over the Río Altapec near Jesús Carranza, *via* Acayucan, and camped five miles south of Catemaco. On July 22 the party drove to San Andrés Tuxtla, to five miles southeast of Lerdo de Tejada, to Alvarado, to the city of Veracruz. On July 24 the route was Veracruz, Jalapa, Perote, Tezuitlan (in Puebla), three miles northwest Nautla. On July 25 the party drove from three miles northwest Nautla to Tecolutla, Papantla, Poza Rica into the state of Puebla.

In 1960 M. Raymond Lee collected mammals for the University of Kansas Museum of Natural History in the northern part of the state of Veracruz—from February 27 to March 3 at places 10, 19 and 25 miles west of Tampico and a place five miles south of Tampico. From March 6 to April 10 he collected at Tuxpan, at places four and six kilometers north thereof, at places seven, nine, 12, 14, 15, 17, 25, 35 and 50 kilometers northwest of Tuxpan, at places four and five kilometers east of Tuxpan, and at places four and five kilometers northeast of Tuxpan. Again on April 18 he collected at the place four kilometers northeast. In this period, from March 29 to April 6, he collected at Hacienda Tamiahua, Cabo Rojo and on April 5 also on Isla Burros and the south end of Isla Juana Ramirez, both islands being in the Laguna de Tamiahua. From April 12 to 15 he collected at Zacualpan and three kilometers to the west; on April 16 two kilometers north of Los Jacales; on April 19 and 21 at Tlacolula; on April 20 at Ixcatepec; and on April 21 also at Piedras Clavadas and again at the place 35 kilometers northwest of Tuxpan.

In 1961 Percy L. Clifton and J. H. Bodley collected vertebrates for the University of Kansas Museum of Natural History in the northern part of the state of Veracruz—on September 10 and 11 at a place one mile east of Higo, 500 feet elevation; September 12-14, at Platón Sánchez, 800 feet elevation; and, September 18-25, twelve and one-half miles north of Tihuatlán, 300 feet elevation.

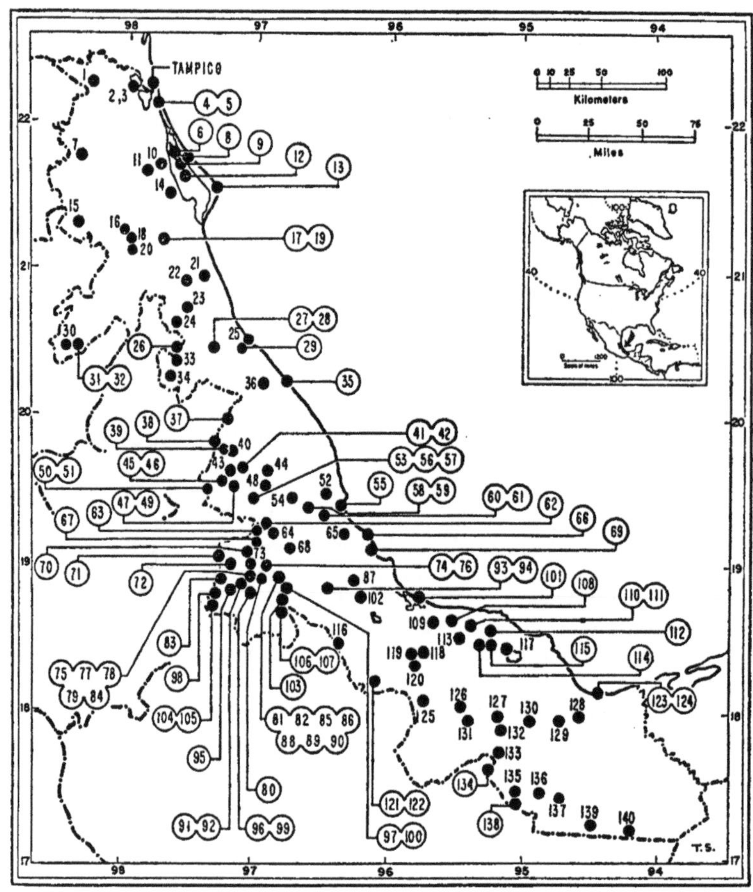

FIG. 1. Localities that are numbered in the Gazetteer (pp. 182-186). In the numerical sequence north takes precedence over south and west over east.

1. Chijol
2. Tamós
3. Hacienda El Caracol
4. Tampico Alto
5. Rivera
6. Isla Juana Ramírez
7. Higo
8. Hacienda Tamiahua
9. Isla Burros
10. El Cepillo
11. Ozuluama
12. Laguna Tamiahua
13. Cabo Rojo
14. La Mar
15. Platón Sánchez
16. Ixcatepec
17. Cerro Azul
18. Piedras Clavadas
19. Potrero Llano
20. Tlacolula
21. Tuxpan
22. San Isidro
23. Tihuatlán
24. Miahuapa

25. Tecolutla
26. El Brinco
27. El Tajín
28. Papantla
29. Gutiérrez Zamora
30. Jacales
31. Zacualpan
32. Zacualpilla
33. Tulapilla, La
34. Coyutla
35. Nautla
36. San Marcos
37. Tlapacoyan
38. Jalacingo
39. Altotonga
40. Acatlán
41. Las Vigas
42. Volcancillo (= Cerro de los Pajaros)
43. Sierra de Agua
44. Santa María
45. Perote
46. Guadalupe Victoria
47. Los Conejos
48. Jalapa
49. Pescados (= Los Pescados)
50. Limón
51. Cofre de Perote
52. Hacienda Tortugas
53. Jico (= Xico)
54. Cerro Gordo
55. San Carlos
56. Teocelo
57. Texolo
58. Plan del Río
59. Carrizal
60. Puente Nacional
61. Chichicaxtle
62. Mirador
63. Tотutla
64. Tlacotepec
65. Paso de San Juan
66. Veracruz
67. Huatusco
68. Río Jamapa
69. Boca del Río
70. Coscomatepec
71. Mt. Orizaba
72. Xometla Camp
73. Monte Blanco
74. Río Atoyac
75. Metlác
76. Ojo de Agua
77. Sumidero
78. Fortín
79. Río Metlác
80. Córdoba
81. Atoyac
82. Grutas Atoyac
83. Xuchil
84. Cuautlapan
85. Cueva de la Pesca, Potrero
86. Potrero
87. Mecayucan
88. Peñuela
89. Parajo Nuevo
90. Potrero Viejo
91. Orizaba
92. Cautlapan
93. Sala del Agua
94. El Maguey
95. Río Blanco
96. Tuxpango
97. San Juan de la Punta
98. Maltrata
99. Necostla (Necoxtla)
100. Dos Caminos
101. Alvarado
102. Piedras Negras
103. Omaelca
104. Río Blanco
105. Acultzingo
106. Presidio
107. Motzorongo
108. Lerdo de Tejada
109. Tlacotalpam
110. Uvero
111. Tula
112. Volcán de Tuxtla (Volcán San Martín Tuxtla)
113. San Juan de los Reyes
114. Santiago Tuxtla
115. San Andrés Tuxtla
116. Tierra Blanca
117. Catemaco
118. Río Tesechoacán
119. Cosamaloapan
120. Tilapa
121. Río Papaloapam
122. Otatitlán
123. Río Coatzacoalcos
124. Coatzacoalcos
125. Pérez
126. Paso Nuevo
127. Pasa Nueva
128. Minatitlán
129. Jaltipan
130. Acayucan
131. Jimba
132. San Juan Evangelista
133. Achotal
134. Buena Vista
135. Jesús Carranza
136. Río Chalchijapa
137. Arroyo Saoso
138. Arroyo Azul
139. Río Solosuchi (= Río Solosuchil)
140. Isthmus of Tehuantepec

GAZETTEER

The following names of places and geographical features are those to which reference is made in this paper. The spellings are based principally on the American Geographical Society's "Map of Hispanic America on the scale of 1:1,000,000 (Millionth Map)" and its accompanying Index (1944). Latitude north of the equator is followed by longitude west of Greenwich. Numbers in brackets refer to the position of the places on the accompanying map (Fig. 1).

Acatlán 19 43 N, 97 10 W [40]
Acayucan 17 57 N, 94 55 W [130]
Achotal 17 44 N, 95 08 W [133]
Acultzingo 18 42 N, 97 18 W [105]
Altotonga 19 46 N, 97 14 W [39]
Alvarado 18 47 N, 95 45 W [101]
Arroyo Azul 17 22 N, 95 01 W [138]
Arroyo Saoso (37 km. E and 7 km. S Jesús Carranza) 17 24 N, 94 41 W [137]
Atoyac 18 54 N, 96 47 W [81]
Boca del Río 19 06 N, 96 07 W [69]
Buena Vista 17 37 N, 94 14 W [134]
Cabo Rojo 21 34 N, 97 20 W [13]
Carrizal 19 22 N, 96 39 W [59]
Catemaco 18 25 N, 95 06 W [117]
Cautlapan (Ixtaczoquitlán) 18 51 N, 97 03 W [92]
Cerro Azul 21 12 N, 97 43 W [17]
Cerro de los Pajaros (see Volcancillo, authority of W. W. Dalquest)
Cerro Gordo 19 25 N, 96 42 W [54]
Chichicaxtle 19 20 N, 96 28 W [61]
Chijol (Chijal on some labels) 22 15 N, 98 16 W [1]
Coatzacoalcos (Puerto México) 18 08 N, 94 24 W [124]
Cofre de Perote 19 29 N, 97 21 W [51]
Córdoba 18 54 N, 96 56 W [80]
Cosamaloapan 18 22 N, 95 48 W [119]
Coscomatepec 19 04 N, 97 02 W [70]
Coyutla 20 15 N, 97 39 W [34]
Cuautlapan 18 53 N, 97 02 W [84]
Cueva de la Pesca, Potrero 18 53 N, 96 47 W [85]
Dos Caminos 18 47 N, 96 42 W [100]
El Brinco 20 27 N, 97 37 W [26]
El Cepillo 21 42 N, 97 45 W [10]
El Maguey (El Magay on some labels) 18 50 N, 96 43 W [94]
El Tajín 20 27 N, 97 23 W [27]
Fortín 18 54 N, 97 00 W [78]
Grutas Atoyac 18 54 N, 96 46 W [82]
Guadalupe Victoria (Agutepec) 19 32 N, 97 16 W [46]
Gutiérrez Zamora 20 27 N, 97 06 W [29]
Hacienda El Caracol, 5 km. SW Tamós Approximately 22 11 N, 98 01 W [3]

Hacienda Tamiahua 21 44 N, 97 33 W [8]

Hacienda Tortugas Possibly within 50 kilometers of 19 28 N, 96 28 W [52] (Hda. Tortugas is the type locality of *Eira barbara senex;* the type locality is at an elevation of approximately 600 feet in the District [= Municipio] of Jalapa.)

Higo 21 46 N, 98 22 W [7]

Huatusco 19 09 N, 96 57 W [67]

Isla Burros 21 43 N, 97 36 W [9]

Isla Juana Ramírez 21 47 N, 97 39 W [6]

Isthmus of Tehuantepec 17°-18° N, 94°-95° W [140]

Ixcatepec (Sta. María) 21 14 N, 98 01 W [16]

Jacales 20 26 N, 98 27 W [30]

Jalacingo 19 48 N, 97 18 W [38]

Jalapa 19 31 N, 96 55 W [48]

Jaltipan 17 58 N, 94 43 W [129]

Jesús Carranza (Santa Lucrecia) 17 26 N, 95 01 W [135]

Jico (Xico) 19 25 N, 97 00 W [53]

Jimba 17 56 N, 95 23 W [131]

Laguna Tamiahua 21 38 N, 97 35 W (central point of the lake) [12]

Lagunas (Not found; Osgood, 1909:201, lists *Peromyscus mexicanus mexicanus* from this locality.)

La Mar 21 31 N, 97 41 W [14]

Las Vigas 19 38 N, 97 05 W [41]

Lerdo de Tejada Approximately 18 36 N, 95 31 W [108] (Mentioned in R. H. Baker's itinerary on file in Museum of Natural History of The University of Kansas.)

Limón (= San Antonio Limón) 19 30 N, 97 21 W [50]

Los Conejos Approximately 19 31 N, 97 09 W (on north slope of Cofre de Perote) [47]

Los Pescados (see Pescados)

Maltrata 18 47 N, 97 16 W [98]

Mecayucan 18 53 N, 96 15 W [87]

Metlác Approximately 18 56 N, 97 00 W [75] (A power plant on the Río Metlác; approximately three kilometers northerly from Fortín according to field notes of W. W. Dalquest.)

Miahuapa 20 37 N, 97 37 W [24] (See also Tulapilla, La, in account of *Peromyscus leucopus mesomelas.*)

Minatitlán 17 59 N, 94 33 W [128]

Mirador 19 17 N, 96 54 W (W. W. Dalquest, field notes) [62]

Monte Blanco 18 59 N, 97 00 W [73]

Motzorongo 18 39 N, 96 45 W [107]

Mt. Orizaba 19 02 N, 97 16 W [71]

Nautla 20 13 N, 96 46 W [35]

Necostla (Necoxtla) 18 47 N, 97 01 W [99]

Necoxtla (see Necostla)

Ojo de Agua (a cave, and a spring which is the source of a river) 18 56 N, 96 54 W (W. W. Dalquest, field notes) [76]

Omaelca (Omaelco) 18 45 N, 96 46 W [103]

Orizaba 18 51 N, 97 05 W [91]

Otatitlán 18 11 N, 96 02 W [122]

Ozuluama (Ozulama on labels) 21 40 N, 97 51 W [11]

Papantla 20 27 N, 97 19 W [28]

Paraje Nuevo 18 52 N, 96 52 W (W. W. Dalquest, field notes) [89]

Paso de Ovejas 19 17 N, 96 26 W (Not shown on Fig. 1.)

Pasa Nueva Here (see map) is shown at 17 59 N, 95 11 W on the authority of Wetmore (1943:216, 217) who has told one of us (Hall) that he (Wetmore) had obtained verbal information from Colburn as to the location of the place concerned. Colburn collected the specimens that J. A. Allen (1904:29) recorded from "Pasa Nueva." Possibly Allen did not talk with Colburn about the exact position of the place and perhaps relied on some other source that caused him (Allen) to state that the location was "a short distance from Tlacotalpan, about 60 miles south of the city of Vera Cruz. . . ." Nava (Direc. Gen. Correos y Telegrafos de los Estados Unidos Mexicanos, 1892, p. 203) listed a "Paso Nuevo" as a rancho near Cosamaloapam. It was because Allen's statement as to location was plausible that Lowery and Dalquest (1951:541) tended to think of the locality as at approximately 18 23 N, 95 48 W. Possibly that is correct but because Wetmore had advice from the collector, Colburn, it is probable that the specimens were collected at 17 59 N, 95 11 W. [127]

Paso de San Juan 19 12 N, 96 19 W [65]

Paso Nuevo 18 01 N, 94 26 W [126]

Peñuela 18 52 N, 96 54 W [88]

Pérez 18 04 N, 95 43 W [125]

Perote 19 34 N, 97 14 W [45]

Pescados 19 30 N, 97 08 W (W. W. Dalquest, field notes) [49]

Piedras Clavadas "75 km. NW Tuxpan" 21 11 N, 97 59 W [18]

Piedras Negras 18 46 N, 96 11 W [102]

Plan del Río 19 23 N, 96 36 W (W. W. Dalquest, field notes) [58]

Platón Sánchez 21 17 N, 98 22 W [15]

Potrero 18 53 N, 96 47 W [86]

Potrero Llano (= Potrero del Llano) 21 10 N, 97 43 W (W. W. Dalquest, field notes) [19]

Potrero Viejo 18 52 N, 96 50 W [90]

Presidio 18 39 N, 96 46 W (W. W. Dalquest, field notes) [106]

Puente Nacional 19 20 N, 96 29 W [60]

Río Alvarado (see Río Papaloapam, W. W. Dalquest, field notes)

Río Atoyac 18 58 N, 96 54 W [74] SE to junction with Río Jamapa at 18 50 N, 96 40 W (W. W. Dalquest, field notes) Some maps give Río Atoyac also for the lower part of the river to which the name Río Jamapa is applied in our present account of the mammals of Veracruz.

Río Blanco (village) 18 49 N, 97 09 W [95]

Río Blanco 18 43 N, 97 18 W, east into Laguna Tlalixcoyan (a part of Laguna de Alvarado) at 18 45 N, 95 50 W [104]

Río Chalchijapa From across Oaxacan boundary at 94 46 N (where named Río Alegro) northerly to Boca Chalchijapa (17 26 N, 94 50 W) at confluence with Río Coatzacoalcos [136]

Río Coatzacoalcos From Oaxacan boundary at 17 20 N, 94 59 W, NE into sea at Puerto México (18 09 N, 94 25 W) [123]

Río Jamapa 19 06 N, 96 43 W, E into sea at Boca del Río (19 06 N, 96 07 W) [68]

Río Metlác A river, tributary to the Río Blanco, running from NW to SE at Fortín (W. W. Dalquest) [79]

Río Papaloapam From Oaxacan boundary at 18 11 N, 96 06 W, NE into Laguna de Alvarado (18 43 N, 95 45 W) [121]

Río Solosuchi (= Río Solosuchil) 17 14 N, 94 28 W, NW into Río Chalchijapa at 17 22 N, 94 47 W [139]

Río Tesechoacán This is a continuation northward of the Río Playa Vicente.

From Pérez the river continues northward to its junction with the Río Papaloapam at 18 24 N, 95 43 W [118]

Rivera 22 06 N, 97 46 W [5]

Sala del Agua 18 50 N, 96 43 W [93]

San Andrés Tuxtla 18 27 N, 95 13 W [115]

San Carlos 19 24 N, 96 21 W [55]

San Isidro 20 56 N, 97 33 W [22]

San Juan de la Punta 18 49 N, 96 44 W [97]

San Juan de los Reyes 18 31 N, 95 27 W [113]

San Juan Evangelista 17 53 N, 95 08 W [132]

San Marcos 20 12 N, 96 57 W [36]

Santa María Approximately 19 37 N, 96 54 W (see Goldman 1951:281) [44]

Santiago Tuxtla 18 28 N, 95 18 W [114]

Sierra de Agua 19 37 N, 97 11 W [43]

Sumidero 18 54 N, 97 01 W [77]

Tamós 22 13 N, 97 59 W [2]

Tampico (in state of Tamaulipas) 22 13 N, 97 51 W

Tampico Alto 22 06 N, 97 48 W [4]

Tecolutla 20 29 N, 97 01 W [25]

Teocelo 19 23 N, 96 58 W [57]

Texolo (possibly Teocelo; the Barranca of Texolo is one kilometer north of Teocelo but we do not know of any town or village bearing the name Texolo) 19 23 N, 96 58 W [57]

Tierra Blanca 18 26 N, 96 21 W [116]

Tihuatlán (Tehuatlan on specimen labels) 20 43 N, 97 33 W [23]

Tilapa 18 18 N, 95 47 W [120]

Tlacolula 21 06 N, 97 58 W [20]

Tlacotalpam (= Tlacotalpan) 18 37 N, 95 39 W [109]

Tlacotepec 19 12 N, 96 50 W [64]

Tlapacoyan 19 58 N, 97 13 W [37]

Totutla 19 13 N, 96 57 W [63]

Tula 18 36 N, 95 22 W [111]

Tulapilla, La 20 21 N, 97 37 W [33]

Tuxpan 20 57 N, 97 24 W [21]

Tuxpango 18 49 N, 97 01 [96]

Tuxtla (probably refers to Santiago Tuxtla)

Ubero (see Uvero)

Uvero (Ubero) Approximately 18 36 N, 95 25 W [110] Sumichrast (1882:228) identifies Uvero as a locality between Alvarado and Santiago Tuxtla. According to García y Cubas (Diccionario Geografico, Histórico y Biográfico de los Estados Unidos Mexicanos, México, 1881-1891), Uvero is 20 kilometers northwest of Santiago Tuxtla. Ubero, from which place *Oryzomys palustris couesi* has been recorded, may be another spelling for the same place.

Veracruz 19 12 N, 96 08 W [66]

Volcán de Tuxtla (see Volcán San Martín Tuxtla) 18 33 N, 95 13 W [112]

Volcán San Martín Tuxtla (see Volcán de Tuxtla)

Volcancillo (Cerro de los Parajos) 19 38 N, 97 04 W [42] A volcanic cone three kilometers east of Las Vigas (W. W. Dalquest).

Xico (see Jico)

Xometla Camp, Mt. Orizaba Probably is Xomitla at 18 59 N, 97 10 W [72]

Xuchil 18 53 N, 97 14 W [83] (Name appears on labels of specimens collected from June 14 to 18 inclusive by Edmund Heller and C. M. Barber.)
Zacualpan 20 26 N, 98 21 W [31]
Zacualpilla 20 25 N, 98 22 W [32]

CHECK LIST

The 198 kinds (subspecies and monotypic species) of 160 species which belong to 93 genera of 28 families of 11 orders are as follows:

Order MARSUPIALIA

Family Didelphidae

PAGE
Didelphis marsupialis—Opossum ... 192
Didelphis marsupialis californica Bennett ... 195
Didelphis marsupialis tabascensis J. A. Allen ... 195
Philander opossum pallidus (J. A. Allen)—Four-eyed Opossum ... 195
Marmosa mexicana mexicana Merriam—Mexican Mouse-opossum ... 199
Caluromys derbianus aztecus (Thomas)—Wooly Opossum ... 201

Order INSECTIVORA

Family Soricidae

Sorex vagrans orizabae Merriam—Vagrant Shrew ... 204
Sorex macrodon Merriam—Large-tooth Shrew ... 204
Sorex saussurei veraecrucis Jackson—Saussure's Shrew ... 205
Cryptotis mexicana mexicana (Coues)—Mexican Small-eared Shrew ... 205
Cryptotis nelsoni (Merriam)—Nelson's Small-eared Shrew ... 206
Cryptotis obscura (Merriam)—Dusky Small-eared Shrew ... 206
Cryptotis micrura (Tomes)—Guatemalan Small-eared Shrew ... 207

Order CHIROPTERA

Family Emballonuridae

Rhynchonycteris naso (Wied-Neuwied)—Brazilian Long-nosed Bat ... 208
Saccopteryx bilineata (Temminck)—Greater White-lined Bat ... 211
Peropteryx macrotis macrotis (Wagner)—Lesser Doglike Bat ... 212
Peropteryx kappleri kappleri Peters—Greater Doglike Bat ... 213
Centronycteris maximiliani centralis Thomas—Thomas' Bat ... 214
Balantiopteryx plicata plicata Peters—Peters' Bat ... 214
Balantiopteryx io Thomas—Thomas' Sac-winged Bat ... 215

Family Phyllostomidae

Pteronotus psilotis (Dobson)—Dobson's Mustached Bat ... 216
Pteronotus parnellii mexicana (Miller)—Parnell's Mustached Bat ... 217
Pteronotus davyi fulvus (Thomas)—Davy's Naked-backed Bat ... 218
Mormoops megalophylla megalophylla (Peters)—Peters' Leaf-chinned Bat, 219
Micronycteris megalotis mexicana Miller—Brazilian Small-eared Bat ... 221
Micronycteris sylvestris (Thomas)—Brown Small-eared Bat ... 222
Mimon cozumelae Goldman—Cozumel Spear-nosed Bat ... 223
Phyllostomus discolor verrucosus Elliot—Pale Spear-nosed Bat ... 224
Trachops cirrhosus coffini Goldman—Fringe-lipped Bat ... 224
Chrotopterus auritus auritus (Peters)—Peters' False Vampire Bat ... 225
Vampyrum spectrum nelsoni (Goldman)—Linnaeus' False Vampire Bat .. 226
Glossophaga soricina leachii (Gray)—Pallas' Long-tongued Bat ... 226

PAGE

Anoura geoffroyi lasiopyga (Peters)—Geoffroy's Tailless Bat 229
Hylonycteris underwoodi Thomas—Underwood's Long-tongued Bat 229
Leptonycteris nivalis nivalis (Saussure)—Long-nosed Bat 229
Carollia perspicillata azteca Saussure—Seba's Short-tailed Bat 230
Carollia castanea subrufa (Hahn)—Allen's Short-tailed Bat 233
Sturnira lilium parvidens Goldman—Yellow-shouldered Bat 234
Sturnira ludovici Anthony—Anthony's Bat 234
Chiroderma villosum jesupi J. A. Allen—Isthmian Bat 234
Artibeus jamaicensis jamaicensis Leach—Jamaican Fruit-eating Bat 234
Artibeus lituratus palmarum J. A. Allen and Chapman—Big Fruit-eating Bat 237
Artibeus cinereus phaeotis (Miller)—Gervais' Fruit-eating Bat 238
Artibeus toltecus (Saussure)—Toltec Fruit-eating Bat 238
Artibeus turpis turpis Andersen—Dwarf Fruit-eating Bat 238
Centurio senex Gray—Wrinkle-faced Bat 239

Family Desmodontidae

Desmodus rotundus murinus Wagner—Vampire Bat 239
Diphylla ecaudata Spix—Hairy-legged Vampire Bat 241

Family Natalidae

Natalus stramineus saturatus Dalquest and Hall—Mexican Funnel-eared Bat 242

Family Vespertilionidae

Myotis velifer velifer (J. A. Allen)—Cave Myotis 244
Myotis fortidens Miller and G. M. Allen—Cinnamon Myotis 246
Myotis keenii auriculus Baker and Stains—Keen's Myotis 247
Myotis thysanodes aztecus Miller and Allen—Fringed Myotis 247
Myotis volans amotus Miller—Long-legged Myotis 247
Myotis californicus mexicanus (Saussure)—California Myotis 247
Myotis elegans Hall—Graceful Myotis 248
Myotis nigricans dalquesti Hall and Alvarez—Black Myotis 248
Myotis argentatus Dalquest and Hall—Silver-haired Myotis 249
Pipistrellus subflavus veraecrucis (Ward)—Eastern Pipistrelle 249
Eptesicus fuscus miradorensis (H. Allen)—Big Brown Bat 250
Eptesicus brasiliensis propinquus (Peters)—Brazilian Brown Bat 250
Lasiurus borealis teliotis (H. Allen)—Red Bat 251
Lasiurus seminolus Rhoads—Seminole Bat 251
Lasiurus cinereus cinereus (Palisot de Beauvois)—Hoary Bat 251
Lasiurus intermedius intermedius H. Allen—Northern Yellow Bat 251
Lasiurus ega xanthinus (Miller)—Southern Yellow Bat 252
Nycticeius humeralis mexicanus Davis—Evening Bat 252
Rhogeëssa tumida tumida H. Allen—Little Yellow Bat 252
Plecotus mexicanus (G. M. Allen)—Mexican Big-eared Bat 252

Family Molossidae

Cynomops malagai Villa—Mexican Dog-faced Bat 254
Tadarida brasiliensis mexicana (Saussure)—Brazilian Free-tailed Bat ... 254
Eumops glaucinus (Wagner)—Wagner's Mastiff Bat 255
Molossus ater nigricans Miller—Red Mastiff Bat 255

Order PRIMATES

Family Cebidae

PAGE
Alouatta villosa mexicana Merriam—Howler Monkey 258
Ateles geoffroyi vellerosus Gray—Spider Monkey 260

Family Hominidae

Homo sapiens americanus Linnaeus—Man.......................... 262

Order EDENTATA

Family Myrmecophagidae

Tamandua tetradactyla mexicana (Saussure)—Tamandua 263
Cyclopes didactylus mexicanus Hollister—Two-toed Anteater 264

Family Dasypodidae

Dasypus novemcinctus mexicanus Peters—Nine-banded Armadillo 264

Order LAGOMORPHA

Family Leporidae

Sylvilagus brasiliensis truei (J. A. Allen)—Forest Rabbit 266
Sylvilagus floridanus—Eastern cottontail 267
Sylvilagus floridanus connectens (Nelson) 268
Sylvilagus floridanus orizabae (Merriam) 268
Sylvilagus floridanus russatus (J. A. Allen) 268
Sylvilagus audubonii parvulus (J. A. Allen)—Desert Cottontail 268
Sylvilagus cunicularius cunicularius (Waterhouse)—Mexican Cottontail .. 268

Order RODENTIA

Family Sciuridae

Spermophilus perotensis Merriam—Perote Ground Squirrel 269
Spermophilus variegatus variegatus (Erxleben)—Rock Squirrel 269
Sciurus deppei—Deppe's Squirrel 270
Sciurus deppei deppei Peters 271
Sciurus deppei negligens Nelson 272
Sciurus aureogaster—Red-bellied Squirrel 272
Sciurus aureogaster aureogaster Cuvier 274
Sciurus aureogaster frumentor Nelson 275
Sciurus oculatus oculatus Peters—Peters' Squirrel 275
Glaucomys volans herreranus Goldman—Southern Flying Squirrel 275

Family Geomyidae

Thomomys umbrinus—Southern Pocket Gopher 275
Thomomys umbrinus albigularis Nelson and Goldman 275
Thomomys umbrinus umbrinus (Richardson) 275
Heterogeomys hispidus—Hispid Pocket Gopher 275
Heterogeomys hispidus hispidus (Le Conte) 277
Heterogeomys hispidus isthmicus Nelson and Goldman 278
Heterogeomys hispidus latirostris Hall and Alvarez 278
Heterogeomys hispidus torridus Merriam 278

PAGE
Heterogeomys lanius Elliot—Big Pocket Gopher 278
Cratogeomys perotensis—Perote Pocket Gopher 278
Cratogeomys perotensis estor Merriam 280
Cratogeomys perotensis perotensis Merriam 280
Cratogeomys fulvescens subluteus Nelson and Goldman—Fulvous Pocket Gopher 280

Family Heteromyidae

Perognathus flavus mexicanus Merriam—Silky Pocket Mouse 281
Dipodomys phillipsii perotensis Merriam—Phillips' Kangaroo Rat 282
Liomys pictus—Painted Spiny Pocket Mouse 283
Liomys pictus obscurus Merriam 284
Liomys pictus veraecrucis Merriam 284
Liomys irroratus—Mexican Spiny Pocket Mouse 284
Liomys irroratus alleni (Coues) 286
Liomys irroratus pretiosus Goldman 286
Liomys irroratus torridus Merriam 286
Heteromys lepturus Merriam—Santo Domingo Spiny Pocket Mouse 286
Heteromys temporalis Goldman—Motzorongo Spiny Pocket Mouse 287

Family Cricetidae

Oryzomys palustris—Marsh Rice Rat 287
Oryzomys palustris couesi (Alston) 288
Oryzomys palustris peragrus Merriam 288
Oryzomys melanotis rostratus Merriam—Black-eared Rice Rat 289
Oryzomys alfaroi—Alfaro's Rice Rat 290
Oryzomys alfaroi chapmani Thomas 290
Oryzomys alfaroi palatinus Merriam 291
Oryzomys fulvescens fulvescens (Saussure)—Pygmy Rice Rat 291
Tylomys gymnurus Villa—Naked-tailed Climbing Rat 292
Nyctomys sumichrasti sumichrasti (Saussure)—Sumichrast's Vesper Rat. . 295
Reithrodontomys megalotis saturatus J. A. Allen and Chapman—Western Harvest Mouse 296
Reithrodontomys chrysopsis perotensis Merriam—Volcano Harvest Mouse, 297
Reithrodontomys sumichrasti sumichrasti (Saussure)—Sumichrast's Harvest Mouse 297
Reithrodontomys fulvescens—Fulvous Harvest Mouse 298
Reithrodontomys fulvescens difficilis Merriam 298
Reithrodontomys fulvescens tropicalis Davis 298
Reithrodontomys mexicanus mexicanus (Saussure)—Mexican Harvest Mouse 299
Peromyscus maniculatus fulvus Osgood—Deer Mouse 300
Peromyscus melanotis J. A. Allen and Chapman—Black-eared Mouse 302
Peromyscus leucopus—White-footed Mouse 303
Peromyscus leucopus affinis (J. A. Allen) 304
Peromyscus leucopus incensus Goldman 304
Peromyscus leucopus mesomelas Osgood 304
Peromyscus boylii—Brush Mouse 304
Peromyscus boylii beatae Thomas 305

PAGE

Peromyscus boylii levipes Merriam 305
Peromyscus aztecus (Saussure)—Aztec Mouse 305
Peromyscus bullatus Osgood—Perote Mouse 306
Peromyscus difficilis—Zacatecan Deer Mouse 307
Peromyscus difficilis amplus Osgood 307
Peromyscus difficilis saxicola Hoffmeister and de la Torre 308
Peromyscus simulatus Osgood—Jico Deer Mouse 308
Peromyscus furvus J. A. Allen and Chapman—Blackish Deer Mouse 308
Peromyscus angustirostris Hall and Alvarez—Narrow-nosed Mouse 309
Peromyscus mexicanus—Mexican Deer Mouse 309
Peromyscus mexicanus mexicanus (Saussure) 311
Peromyscus mexicanus teapensis Osgood 311
Peromyscus mexicanus totontepecus Merriam 311
Peromyscus nelsoni Merriam—Nelson's Deer Mouse 311
Baiomys taylori—Northern Pygmy Mouse 311
Baiomys taylori analogus (Osgood) 311
Baiomys taylori taylori Thomas 311
Baiomys musculus brunneus (J. A. Allen and Chapman)—Southern Pygmy Mouse 311
Sigmodon hispidus—Hispid Cotton Rat 312
Sigmodon hispidus saturatus Bailey 314
Sigmodon hispidus toltecus (Saussure) 314
Neotomodon alstoni perotensis Merriam—Volcano Mouse 315
Neotoma nelsoni Goldman—Nelson's Wood Rat 317
Neotoma mexicana—Mexican Wood Rat 317
Neotoma mexicana distincta Bangs 317
Neotoma mexicana torquata Ward 317
Microtus mexicanus mexicanus (Saussure)—Mexican Vole 318
Microtus quasiater (Coues)—Jalapan Pine Vole 320

Family Muridae

Rattus rattus—Black Rat 321
Rattus rattus alexandrinus (É. Geoffroy-Saint-Hilaire) 321
Rattus rattus rattus (Linnaeus) 321
Rattus norvegicus norvegicus (Berkenhout)—Norway Rat 321
Mus musculus and subspecies—House Mouse 322

Family Erethizontidae

Coendou mexicanus mexicanus (Kerr)—Mexican Porcupine 322

Family Dasyproctidae

Agouti paca nelsoni Goldman—Paca or Spotted Paca 324
Dasyprocta mexicana Saussure—Mexican Agouti 326

Order CARNIVORA

Family Canidae

Canis latrans cagottis (Hamilton-Smith)—Coyote 328
Urocyon cinereoargenteus—Gray Fox 329
Urocyon cinereoargenteus orinomus Goldman 330
Urocyon cinereoargenteus scottii Mearns 330

Family Procyonidae

PAGE
Bassariscus astutus astutus (Lichtenstein)—Ringtail 330
Bassariscus sumichrasti sumichrasti (Saussure)—Tropical Cacomixtle 331
Procyon lotor—Raccoon 332
Procyon lotor hernandezii Wagler 333
Procyon lotor shufeldti Nelson and Goldman 333
Nasua narica—Coati 333
Nasua narica molaris Merriam 335
Nasua narica narica (Linnaeus) 335
Potos flavus aztecus Thomas—Kinkajou 335

Family Mustelidae

Mustela frenata—Long-tailed Weasel 338
Mustela frenata macrophonius (Elliot) 338
Mustela frenata perda (Merriam) 338
Mustela frenata perotae Hall 338
Mustela frenata tropicalis (Merriam) 338
Eira barbara senex (Thomas)—Tayra 339
Galictis allamandi canaster Nelson—Grisón 340
Mephitis macroura—Hooded Skunk 340
Mephitis macroura eximius Hall and Dalquest 341
Mephitis macroura macroura Lichtenstein 341
Conepatus leuconotus leuconotus (Lichtenstein)—Eastern Hog-nosed Skunk 341
Conepatus semistriatus conepatl (Gmelin)—Striped Hog-nosed Skunk ... 342
Lutra annectens annectens Major—Southern River Otter 343

Family Felidae

Felis onca veraecrucis Nelson and Goldman—Jaguar 344
Felis concolor mayensis Nelson and Goldman—Mountain Lion 345
Felis pardalis pardalis Linnaeus—Ocelot 346
Felis wiedii oaxacensis Nelson and Goldman—Margay 346
Felis yagouaroundi—Jaguarundi 346
Felis yagouaroundi cacomitli Berlandier 347
Felis yagouraroundi fossata Mearns 347
Lynx rufus escuinapae J. A. Allen—Bobcat 347

Order SIRENIA

Family Trichechidae

Trichechus manatus latirostris (Harlan)—Manatee 348

Order PERISSODACTYLA

Family Tapiridae

Tapirus bairdii (Gill)—Baird's Tapir 348

Order ARTIODACTYLA

Family Tayassuidae

Tayassu tajacu crassus Merriam—Collared Peccary 350
Tayassu pecari ringens Merriam—White-lipped Peccary 352

Family Cervidae

	PAGE
Odocoileus virginianus—White-tailed Deer	353
Odocoileus virginianus thomasi Merriam	354
Odocoileus virginianus toltecus (Saussure)	354
Odocoileus virginianus veraecrucis Goldman and Kellogg	355
Mazama americana temama (Kerr)—Red Brocket	355

Order MARSUPIALIA

Family Didelphidae

Didelphis marsupialis

Opossum

The usual name in Veracruz is "tlacuache." In some places the name "zorro" is used, not to be confused with "zorra," or fox.

The opossum seems to be found in all life-zones save, probably, the Arctic-Alpine. The animal was reported to us from the high, coniferous forest, at 10,500 feet elevation on the Cofre de Perote, though we did not take it there. The opossum is common near Las Vigas, in the pine forest at 8000 feet elevation, and also on the sandy, arid desert near Perote, where we trapped one specimen along the cut-bank of a sandy arroyo. The opossum is most abundant in thickets and jungle near water, such as streams, rivers and lakes, and extensive cultivated fields of sugar cane or corn. At the upper edge of the upper humid division of the Tropical Life-zone we took specimens along cold, swift streams supporting rainbow trout. In the lower humid division of the Tropical Life-zone, opossums were obtained along shores of deep, slow rivers in steaming jungles.

Homes of several opossums were discovered in Veracruz. One was found, by a dog, in dense jungle, 20 kilometers west-northwest of Piedras Negras, on May 12, 1946. The dog began to bark at the bottom of a cut-bank of an arroyo where a log, about 15 inches in diameter and 35 feet long, lay parallel to the bank and about six inches away. Slumps of earth mixed with branches, roots and leaves formed a rough roof over the cavity between the bank and the log. Digging revealed a medium-sized opossum in the cavity, and a mass of dry, dead leaves.

On May 19, 1946, nests of the red-bellied squirrel (*Sciurus aureogaster*) were being examined in an arroyo near the Río Blanco, 20 kilometers west-northwest of Piedras Negras. A small opossum was frightened from one squirrel nest, a rounded ball of dry leaves approximately 15 inches in diameter. The opossum escaped by running along branches of the trees.

At Jesús Carranza the home of an opossum was in a hollow mahogany log that had been cut for lumber and left lying among many similar logs. The log was about three feet in diameter and 30 feet long. A cavity through the center was about eight inches in diameter. At the lower end, this cavity was partly filled with earth, seemingly tracked into the cavity by some animal. Investigation showed that the cavity extended the full length of the log, but light did not show through. When a long pole was pushed into the hollow, an opossum emerged from the other end. The cavity in the log had been partly obstructed by dry leaves, not, it seemed, in the form of a nest, but scattered along the entire length of the hollow.

The opossum seems to be completely nocturnal. Night hunting with a headlamp usually discloses one or two opossums each night in suitable habitat. At the Río Blanco, 20 kilometers west-northwest of Piedras Negras, eight were seen in a few hours in one night. One was in an open area of tall grass; one was at the base of a sandy cut-bank; six were in trees and vines in the jungle, five to 10 feet from the ground. Opossums wander from the jungle at times. At our camp, seven kilometers southeast of San Juan de la Punta, on December 15, 1946, they were common in the jungle along the Río Jamapa. One, however, was found in a flat-topped acacia tree isolated in an extensive grassland, fully 2500 feet from the jungle. The acacia is not a fruit tree, and it is not known why the animal was in the tree.

Opossums may be social to some extent. It is not unusual to catch two or more in traps set close together. Eight kilometers northwest of Potrero this was true. Four kilometers west-northwest of Fortín, at 3200 feet elevation, an opossum was obtained on March 28, 1946. Although traps were set daily, no other opossums were taken until April 2, when two males were found in traps set 50 feet apart.

Although a rather clumsy animal on the ground, the opossum is a swift, sure and agile climber. The brilliant red eyes seen by means of a reflected light at night are often at considerable heights above the ground. In the lowlands opossums rarely have any fat beneath the skin, but in the uplands characteristically have a deep layer of fat, especially in winter. Tails of younger animals are usually clear white and jet black; in older animals the white becomes duller and the black becomes grayish.

In jungle areas, the food of the opossum probably consists mainly of fruit; the jobo plum and several species of wild figs are favored food. To humans these fruits have a bitter or pitchlike taste. Some domestic fruits, such as banana, papaya and mango are eaten. No insects were found in stomachs examined. On a few occasions opossums followed trap lines, eating captured animals, and had to be caught before trapping for small species could be continued successfully. One opossum, obtained in the desert, contained remains of *Dipodomys phillipsii* and *Peromyscus maniculatus*. We found the opossum easy to trap by using carrion (bird, mammal or fish) as bait.

In some parts of Mexico and the United States the opossum is used as food. This is not true in Veracruz; when questioned about eating opossum, most residents were disgusted by the idea. The opossum does some damage to fruit and chickens, but is so easily captured, especially by dogs, that it is not a serious pest. The fur of the opossum in Veracruz is of no value, although the fur of the animals that live in the uplands possibly could be marketed.

We learned little concerning the breeding habits of the common opossum in Veracruz. Occasionally two specimens, a male and a female, were taken in adjacent traps. Three kilometers southeast of Orizaba, at 5500 feet elevation, a female was caught in a trap on December 9, 1946, about two hours after dark. Before she was removed from the trap, the beam of a flashlight was played on nearby trees, in order to reveal owls or other predators that might have been attracted by the struggles of the trapped animal. A large male opossum was found in the tree directly over the trapped female. An adult female with nine young in the pouch was obtained four kilometers west-northwest of Fortín, at 3200 feet elevation, on March 28, 1946. Fifteen kilometers east-southeast of San Juan de la Punta, on September 27, 1946, an adult female and a young male, about one-quarter grown, were shot from a strangler fig tree, where they were feeding at about 10:00 p. m.

Opossums obtained in the highlands of Veracruz were all in seemingly good health; they were fat and had long, soft fur. In the lowlands they were usually lean and had thin, coarse fur. Many specimens from the lowlands had scabby patches of bare skin; a nodulelike, crusted mass in the skin at the base of the fur, usually on the back, was common. Ectoparasites were not common on opossums in Veracruz. An animal taken two kilometers west of Jico, 4200 feet elevation, had several large ticks.

Several opossums, perhaps a dozen, were found dead along streams or in fields, but the cause of death was not determined.

Other animals of similar size, rabbits, for example, were almost never found dead.

Didelphis marsupialis californica Bennett

Specimens examined.—Total 14: Hacienda Tamiahua, Cabo Rojo, 3; 17 km. NW Tuxpan, 2; 9 km. NW Tuxpan, 1; Tuxpan, 1; 12½ mi. N Tihuatlán, 300 ft. 1; 2 km. E Perote, 8300 ft., 1; 5 km. N Jalapa, 4500 ft., 2; 2 km. W Jico, 4200 ft., 1; 3 km. SE Orizaba, 5500 ft., 3.

Additional records.—Under the name "*Didelphis marsupialis*," J. A. Allen (1901:168) recorded specimens from: Las Vegas [= Vigas]; Jico; and Maltrata. Ingles (1959:380) recorded the species from 25 mi. NW of the City of Veracruz. Ferrari-Pérez (1886:130) recorded the species from Jalapa.

This subspecies occurs in the state on the highlands along the western border and in the northern part of the state. Longer and "softer" fur, shorter tail, and longer nasals (relative to postnasal part of the skull) than in D. *m. tabascensis* seem to characterize D. *m. californica.* Our specimens, excepting the one from 2 km. E of Perote, are intergrades between the two mentioned subspecies. Measurements of our largest male, No. 19055, are 963; 488; 76; 57; basilar length, 104.8; length of nasals, 55.3; zygomatic breadth, 58.4.

Didelphis marsupialis tabascensis J. A. Allen

Specimens examined.—Total 57: 5 km. S Tehuatlán [= Tihuatlán], 700 ft., 6; 9 km. E Papantla, 300 ft., 1; 9 km. NW Nautla, 10 ft., 1; 3 km. SW San Marcos, 200 ft., 2; 4 km. W Tlapacoyan, 1700 ft., 2; Río Atoyac, 8 km. NW Potrero, 2; 4 km. WNW Fortín, 3200 ft., 3; Potrero Viejo, 1700 ft., 1; Río Blanco, 20 km. WNW Piedras Negras, 3; 15 km. ESE San Juan de la Punta, 2; 7 km. SE San Juan de la Punta, 2; Río Blanco, 20 km. W Piedras Negras, 3; 3 km. SE San Andrés Tuxtla, 1000 ft., 4; Coatzacoalcos, 1; Achotal, 17 (Chicago N. H. Mus.); 20 km. ENE Jesús Carranza, 200 ft., 1; 20 km. E Jesús Carranza, 300 ft., 2; 25 km. SE Jesús Carranza, 250 ft., 2; 34 km. SE Jesús Carranza, 400 ft., 1; 60 km. SE Jesús Carranza, 450 ft., 1.

Additional records.—Papantla (J. A. Allen, 1901:173); Mirador (*ibid.*); Boca del Río (Davis, 1944:374); Catemaco (J. A. Allen, 1901:173); Pasa Nueva (J. A. Allen, 1904:30); Minatitlán (J. A. Allen, 1901:168, as "*Didelphis marsupialis*").

The three largest skulls are Nos. 13776 and 13777, males, from Achotal and No. 23392, unsexed, from 5 km. S Tehuatlán. Meaurements, respectively, are: basilar length, 120, 115.8, approximately 115.4; length of nasals, 56.5, 56.3, 57.0. The three next largest specimens, all males, are No. 17684 from 4 km. WNW Fortín, No. 66269 from Coatzacoalcos, and No. 32050 from 25 km. S Jesús Carranza. Measurements are, respectively, as follows: 959,853,852; 473,386,400; 71,65,71; 59,54,54; basilar length, 111.3, 104.2, 111.7; length of nasals, 57.7, 54.3, 58.4; zygomatic breadth, 63.8, 65.8, 65.3.

Philander opossum pallidus (J. A. Allen)

Four-eyed Opossum

Specimens examined.—Total 50: 35 km. NW Tuxpan, 1; 25 km. NW Tuxpan, 1; Tuxpan, 2; 12½ mi. N Tihuatlán, 300 ft., 2; 3 km. SW San Marcos, 200 ft., 2; Teocelo, 4000 ft., 1; 3 km. W Boca del Río, 10 ft., 1; Boca del Río, 10 ft., 1; 5 km. SW Boca del Río, 2; Río Atoyac, 8 km. NW Potrero, 1700 ft., 8; 4 km. WNW Fortín, 3200 ft., 2; 7 km. W Potrero "1700 ft.," 1; Potrero Viejo, 1700 ft., 3; 3 km. SE Orizaba, 5500 ft., 5; Río Blanco, 20 km.

W Piedras Negras, 3; 2 km. N Motzorongo, 1500 ft., 2; 3 km. E San Andrés Tuxtla, 1000 ft., 6; Jimba, 350 ft., 1; 20 km. ENE Jesús Carranza, 200 ft., 2; 20 km. E Jesús Carranza, 300 ft., 2; 30 km. SSE Jesús Carranza, 300 ft., 2.

Additional records.—J. A. Allen (1901:216) lists specimens from Papantla; Chichicaxtle; Orizaba; Motzorongo; Catemaco. Córdoba and Huatusco (Sumichrast, 1882:32).

Our 24 males average larger than our 18 females. As would be expected, the sagittal crest increases with increasing age. Within each sex, the specimen having the highest sagittal crest has the greatest occipitonasal length. The two males (17672, 32057) having the highest sagittal crests yield measurements, respectively, as follows: 649, 530; 306, 280; 48, 40; 38, 34; occipitonasal length, 80.0, 76.7; nasal length, 37.1, 38.0. The two females having the highest sagittal crests are Nos. 19080 and 17676. They, respectively, yield corresponding measurements as follows: 612, 556; 300, 290; 44, 41; 37, 35; 78.0, 72.9; 37.7, 35.9.

Over most of Veracruz the four-eyed opossum is called "comadreja." In the uplands, near Jico and Jalapa, this name is applied to the weasel (*Mustela frenata*). Less commonly, the four-eyed opossum is called "ratón tlacuache," a name more usually used for the mouse-opossum, *Marmosa mexicana.*

The four-eyed opossum ranges throughout the tropics of Veracruz. At the extreme upper edge of the upper humid division of the Tropical Life-zone, it lives along cold, clear streams at the edge of the oak belt. Lower down, but still in the upper humid division, it was found along rivers and streams that flowed through dense jungle, where the tall, broad-leafed trees were thickly hung with orchids, vines, mosses and bromeliads. The four-eyed opossum was found living in thickets bordering the broad rivers of the coastal plain, in the arid division of the Tropical Life-zone, and along the marshy shores of rivers and streams of the lower humid division of the Tropical Life-zone, in the southern part of the state.

Most of our specimens were taken on the very shores of rivers or streams. Two examples show, however, that the species is not confined to such habitat. On May 21, 1946, at Potrero Viejo, seven kilometers west of Potrero, workers discovered a family of four young animals in a field of sugar cane, several kilometers from the nearest water at that time of the year. At Jimba, 350 feet elevation, in southern Veracruz, a four-eyed opossum was taken from a tree on a hillside fully three kilometers from the nearest water. These records are unusual, however. Against them are nearly 30 records from in and near water.

The four-eyed opossum seems to be entirely nocturnal. None was seen abroad by day, unless frightened from its daytime retreat, but one was seen shortly after dark.

At the Río Atoyac, eight kilometers northwest of Potrero, between February 19 and March 5, 1946, several four-eyed opossums were taken in a trail leading from a dense thicket of vines, bushes, thorny plants and creepers, to a stream ten feet away. Subsequently part of the thicket was cleared with machetes and a hole, about five inches in diameter that led downward beneath the roots of a tree, was discovered near the center of the thicket. This seemed to have been the home of at least one of the opossums. Several large basilisk lizards were living in the thicket.

Five kilometers southwest of Boca del Río, a four-eyed opossum was shot on the ground, near the base of a large "strangler fig" tree (*Ficus* sp.). As this tree was under observation for most of the day, it is assumed that the animal's home was in one of the numerous holes in the base of the tree, or in a hole in the ground beneath it.

Twenty kilometers east-northeast of Jesús Carranza, two of these opossums were found in nests that they had constructed in the palm thatch of the roofs of abandoned houses. These nests consisted of a handful of dry leaves, pushed in between the layers of palm fronds. From outside, a distinct spherical or oval lump in the thatch marked the site of the nest. Inside the house we could see no trace of the nests.

Twenty-five kilometers southeast of Jesús Carranza, the nest of a four-eyed opossum was found in a cavity in the side of a piece of tree trunk, 15 inches in diameter and three feet long, that was suspended in the air, by vines, seven feet over the surface of a dry wash. The nest was of dry leaves, about 11 inches deep and seven inches in diameter. When the vines supporting the section of tree trunk were slashed, the trunk fell to the ground and split open on the hard gravel. The opossum escaped.

Another four-eyed opossum was shot from the large hollow in the side of a giant "ligarón" tree. This tree was fully 12 feet in diameter at waist height, and contained a hollow about 60 feet high and five feet in diameter. The opossum was shot as it ran up the rough side of the hollow. The hollow was also the home of a colony of sac-winged bats of the species *Saccopteryx bilineata*.

On a few occasions, two four-eyed opossums were seen as close together as 50 feet, but otherwise they were solitary. Usually we saw one to three four-eyed opossums each night, while we hunted with a head lamp in suitable habitat. *Philander* seemed to be less common in most parts of Veracruz than *Didelphis*.

The actions of *Philander* differ from those of other marsupials that occur in Veracruz. *Philander* is quick and active. Trapped individuals are able to jump and twist about in surprising fashion. Care must be used to avoid being bitten when removing them from traps.

The four-eyed opossum is an agile climber and a skillful swimmer, but most of its hunting seems to be done on the ground, along the edges of streams. Along the Río Atoyac, eight kilometers northwest of Potrero, on the evening of February 18, 1946, a *Philander* was shot as it ran along a narrow trail halfway up a vertical bank about two meters in height. The trail was scarcely visible as it wound through moss and ferns. The animal fell into the stream below. A half hour later another *Philander* was shot only 50 to 60 feet distant from the first. The second one was on a horizontal branch or some dense vines about six feet above the surface of the stream, and also fell into the water.

Specimens are usually seen or trapped on the ground, but are sometimes seen in trees. Along the Río Atoyac several four-eyed opossums were taken in a trap set beneath the water level, at the base of a cut-bank. They could have reached the trap only by swimming. Seven kilometers southeast of San Juan de la Punta, on December 15, 1946, a four-eyed opossum was seen just before midnight, running swiftly over the large, rounded boulders (six to 18 inches in diameter) along the river bank. When frightened, the animal turned and made a smooth, clean dive into the swift water, and as it did not reappear, must have swum away under water.

In the lowlands these opossums seldom were fat but in the highlands, at 5000 feet elevation and higher, in winter, they had a deep layer of yellow fat immediately beneath the skin. *Didelphis* in the uplands likewise is fat in winter but in the lowlands usually is lean at all seasons.

The four-eyed opossum is omnivorous; it has been seen feeding on sweet-lemons, jobo plums, and the fruit of the Chico Zapote (*Sapote achras,* source of chewing gum). Five kilometers southwest of Boca del Río, at 10 feet elevation, on October 10, 1946, a four-eyed opossum was seen at the base of a large, hollow fig tree. The upper part of the hollow in the tree served as a retreat for a colony of the large fruit bat, *Artibeus jamaicensis.* Bats were bringing small, green figs into the hollow and feeding on them. Parts of the fruit, varying in size from almost whole figs to mere shreds, were dropped by the bats. The four-eyed opossum was feeding on these bits of figs.

On a number of occasions, four-eyed opossums followed our trap lines, eating mice and other small mammals that had been captured. These opossums are easily trapped by using flesh, preferably much decayed, for bait.

The four-eyed opossum probably breeds at all times of the year. A female having six young in her pouch was found at a place four kilometers west-northwest of Fortín, 3200 feet elevation, on March 28, 1946. Four young were found in a field at Potrero Viejo, 1700 feet elevation, on May 21, 1946. A female having four young in the pouch was obtained five kilometers southwest of Boca del Río, 10 feet elevation, on October 10, 1946. Males having enlarged testes were taken on December 7, 1946, December 20, 1946, and March 3, 1947.

The ear of one *Philander* was diseased and partly missing. Many individuals were examined for ectoparasites but no parasites were found. Like other opossums, *Philander* has a strong, unpleasant odor.

Marmosa mexicana mexicana Merriam

Mexican Mouse-opossum

Specimens examined.—Total 25: 4 km. W Tlapacoyan, 1700 ft., 5; 5 km. N Jalapa, 4500 ft., 8; 2 km. W Jalapa, 4200 ft., 1; 2 km. W Jico 4200 ft., 3; 4 km. WNW Fortín, 3200 ft., 1; 1 km. E Mecayucan, 200 ft., 1; Cautlapan, 1 (collection of E. H. Taylor); 3 km. SE Orizaba, 5500 ft., 1; 15 km. ESE San Juan de la Punta, 1; 25 km. ESE Jesús Carranza, 350 ft., 1; 25 km. SE Jesús Carranza, 250 ft., 1; 35 km. SE Jesús Carranza, 350 ft., 1.

Additional records (Tate, 1933:133, unless otherwise noted).—1½ mi. E Jalapa (J. A. Allen and Chapman, 1897:208); Texolo (not located; possibly = Teocelo); Veracruz; San André[s] Tuxtla; Pasa Nueva (J. A. Allen, 1904:29); Achotal.

The Mexican mouse-opossum occurs throughout the length of the state in the Tropical Life-zone.

The largest (17670, 4 km. W Fortín) of our 13 males and the largest female (32054, 35 km. SE Jesús Carranza), yield measurements, respectively, as follows: 273, 318; 149, 185; 20, 21; 22, 18; occipitonasal length, 33, 32.9; zygomatic breadth, 17.2, 17.6.

The Mexican mouse-opossum is not well known to the local residents of Veracruz. The only local name used seems to be "ratón tlacuache," literally, mouse-opossum.

The mouse-opossum reaches its maximum abundance in the densely forested areas at the upper edge of the upper humid division of the Tropical Life-zone. It is not a common species. Five kilometers north of Jalapa, at 5000 feet elevation, where it was found most abundantly it was in an approximate ratio of one mouse-opossum to eight mice. In most places it was far less common.

In the forested uplands this species was trapped in situations as follows: an area of giant elephant's ear in dense jungle, where the trees overhead were heavily draped with orchids, bromeliads, mosses and vines; in dense bushes, composed of many species, about five feet high and on a steep hillside; along a trail through dense forest on a hillside of 20 degree slope; under a stump in a dense forest on a hill; under logs, stumps and roots of trees in dense forest (four animals); on a slope of 15 degrees, covered with dense grass, coffee and low bushes; beneath tree-ferns at the foot of a cliff; under a log near a cliff; in a patch of dense vegetation drenched by the spray of a waterfall; in brush and weeds on the steep side of a valley.

Two specimens were taken on the arid coastal plain. Here this species was rare, and occupied a habitat unlike that of the upland forest. One individual was trapped in tall grass on a level plain, where there were scattered bushes about five feet tall, 15 kilometers east-southeast of San Juan de la Punta, at 400 feet elevation. This specimen was so extensively eaten by ants that only the skull was saved. Another was taken one kilometer east of Mecayucan, at 200 feet elevation, in an area where there were many low, thorny bushes and much open ground. This individual is not fully adult, but is much paler than any other specimen from Veracruz.

Three specimens were saved from the dense jungles of the southern part of Veracruz. One was trapped in tall, dense grass on a flood plain of the Río Solosuchil, 25 kilometers east-southeast of Jesús Carranza, at 350 feet elevation. Another was taken 25 kilometers southeast of Jesús Carranza on March 31, 1949. This animal was shot at night. The trees overhead were so dense that little sunlight reached the forest floor, and as a result there was no understory vegetation, save for a few low palms and bushes a foot or so in height. The ground was level, dry and covered with a thin layer of dry leaves. The small, red eyes of the mouse-opossum were first seen at a distance of about 20 feet. The animal was on a twig of a bush about 10 inches from the ground. It jumped to the ground and turned to look at the light.

Another mouse-opossum was obtained 35 kilometers southeast of Jesús Carranza on April 7, 1949. On this date, at about ten o'clock at night, we were hunting in deep forest, similar to that 25 kilometers east-southeast of Jesús Carranza. A pair of small, red eyes were seen at the mouth of a hole in a cut-bank. The bank was composed of sand, was almost vertical and about a foot in height. The hole was 30 millimeters in diameter and 50 millimeters from the base of the cut-bank. The burrow was opened and was found to contain a nest of dry leaves, about a handful in quantity, in a pocket about four inches in diameter and 16 inches from the entrance. An adult female mouse-opossum was captured in the nest. There were no young in the nest and the female was not lactating.

Mouse-opossums have an unpleasant, but not strong, musky odor. They are surprisingly tough; several that were caught just behind the necks in mousetraps were still alive the following morning, although the traps invariably killed mice larger than the mouse-opossums. One such animal, when removed from the trap, whipped its tail around a twig. When removed, its tail quickly fastened about the finger of its captor. It was much more dexterous with its tail than either *Didelphis* or *Philander*, but probably less so than *Caluromys.*

The mouse-opossum of Veracruz seems to be omnivorous. Animals were commonly taken in traps baited with banana, and one was taken in a meat-baited trap. Five kilometers north of Jalapa, the stomachs of two individuals contained only remains of insects, so finely chewed as to resemble the contents of the stomach of an insectivorous bat. Three stomachs held starchy, dull, grayish-white plant material, dark green in places. This may have been mostly acorns; acorns were abundant there at that time.

Two kilometers west of Jico, at 4200 feet elevation, the first three mice caught in a trap line had been eaten in the traps. The next trap held a *Marmosa.* Mice in the other traps, farther along the line, were untouched. I wondered if the mouse-opossum had eaten the three mice.

Mouse-opossums are rarely fat, but all of those taken seemed to be in good health. Only near Jalapa were ectoparasites found on this species. There one animal had a number of tiny yellow mites on the ears; one had a large, gorged tick clinging to the skin beneath one ear; one had a flea; two animals had three or four small mites each.

Caluromys derbianus aztecus (Thomas)

Wooly Opossum

Specimens examined.—Total 8: 3 km. E San Andrés Tuxtla, 1000 ft., 6; 20 km. ENE Jesús Carranza, 300 ft., 1; 20 km. E Jesús Carranza, 300 ft., 1.

Additional records.—Potrero (seen alive); San Juan de la Punta (Thomas, 1913:359).

Potrero is the northernmost place from which we have record of this species in Veracruz.

The largest adult male (23668) and largest adult female (23369), both from 3 km. San Andrés Tuxtla, yield measurements, respectively, as follows: 698, 742; 411, 442; 45, 45; 42, 41; occipitonasal length, 59.0, 60.5; zygomatic breadth, 35.1, 34.0.

The wooly opossum seems to have no distinctive vernacular name in Veracruz. Specimens taken in the San Andrés Tuxtla area were referred to by natives, but rather doubtfully, as comadreja. This is the local name of the four-eyed opossum. On the Río Coatzacoalcos the wooly opossum was called zorro or zorro colorado. There are no foxes (zorras) in this area, where the name zorro is usually used for the common opossum (*Didelphis*).

Our specimens were all taken in dense jungle where there were vines and tall trees. The species was associated with *Didelphis, Philander,* Potos *flavus, Tylomys gymnurus* and *Coendou mexicanus.* These are all nocturnal, fruit-eating mammals.

The wooly opossum seems to be completely nocturnal. With one exception, all of our specimens were taken at night. The one taken in the daytime probably was frightened from its daytime hiding place by dogs. Twice, individuals were found about 20 minutes after dark. Others were taken from one hour after dark until as late as midnight.

Caluromys seems to be entirely arboreal. One individual, however, was killed on the ground at night by dogs. A strong wind had been blowing, and many trees and large limbs had fallen earlier in the night. Perhaps the wooly opossum had fallen with a tree or branch or perhaps it had been snatched from a low branch by one of the dogs.

No homes of wooly opossums were found. Twenty kilometers east of Jesús Carranza, at 300 feet elevation, on February 6, 1948, a pack of dogs began to sniff and bark at the foot of a large mango tree. A few moments later a wooly opossum was seen in the tree approximately 20 feet from the ground. The mango tree had several hollows in the trunk immediately below the place where the animal was first seen and the opossum probably was sleeping in one of these when scented by the dogs.

This seems to be the rarest species of marsupial occurring in Veracruz. Hundreds of hours were spent in hunting with headlights in the Veracruz jungles, and several nights were spent at the type locality of *Caluromys d. aztecus* where we searched especially for wooly opossums. In spite of this, only a few specimens were collected in the four seasons of field work. It is significant also that the native hunters did not recognize the few specimens that we collected, and had no name for the species.

Caluromys, like other marsupials that occur in Veracruz, seems to be solitary. Never was more than one seen at a time, although on two occasions two individuals were shot from the same tree on the same night, but several hours apart.

In the jungle at night, the eyes of *Caluromys* glow a brilliant red, like the eyes of *Didelphis*. Animals were seen moving rather slowly in the tall jungle trees, or were motionless as they looked at the light. One animal was 35 feet from the ground on the limb of an "amate-capulín" tree, about 10 inches in diameter and another was taken in the same tree, about 50 feet from the ground. One that was shot and wounded, but not killed, moved swiftly. It was not vicious, as is *Philander*, nor dull and slow, like *Didelphis*. The general impression gained was that of an unusually long, slender, squirrellike animal having short legs.

The wooly opossum feeds on the berrylike fruit of the "amate-capulín" tree—a favorite of all fruit-eating birds and mammals. One specimen had the skins of two small, green berrylike fruits in the stomach. One old female was fat, but most of the specimens taken were lean.

At Potrero a wooly opossum was found by workmen in a pile of stored pipes. It was kept as a pet by Miss Marion Forbes of Potrero Viejo. The animal proved to be gentle and affectionate. Its ordinary movements were slow and deliberate, but when frightened it was able to run fairly fast. When running, the rear part of the body was held much higher than the shoulders; the nose almost touched the floor and the tail was held out almost straight. From the tip of the nose to the tip of the tail, the dorsal surface of the opossum formed a straight line at an angle of about 30 degrees with the floor.

This animal was a skillful climber, with a splendid sense of balance. It proved almost impossible to push the animal from the arm of a chair or a similar position; at least one of its feet or the tail would retain a secure grip. The tail of this opossum was one of its most remarkable features; it gripped objects almost as though the tail had an independent nervous system and eyes of its own. Even bas-relief carvings on furniture offered enough of a purchase to assist in supporting the animal. When suspended from the tip of its tail the animal twisted its body up, until it could grip its tail with its hands. It then climbed up, hand over hand, to the support on which the tip of the tail had a purchase.

When asleep the wooly opossum does not curl its body into a tight ball, but lies on its side in a loose semicircle. The tail is placed in one complete circle about the body, and the extreme tip is hooked over the tail about an inch from the base.

This individual was fed bananas, insects and mice. It refused live, adult mice, but ate those freshly killed. Young mice were eaten alive or dead. Live insects, such as large cockroaches, were eaten with every indication of excitement and pleasure. The insect was grasped in the hands and squeezed tightly. It was then usually transferred to the left hand and eaten, a bite at a time. Each bite was thoroughly chewed before another was taken.

The pouch of the wooly opossum is well developed in most females. One had a poorly developed pouch. Three young were found in the pouch of one female. No ectoparasites were found on any of our specimens of *Caluromys.*

Order INSECTIVORA

Family Soricidae

Sorex vagrans orizabae Merriam

Vagrant Shrew

Specimen examined.—Colfre de Perote, 9500 ft., 1 (U. S. N. M.).

Measurements of the adult female from Cofre de Perote, according to Jackson (1928:114), are: 98; 33.5; 13; condylobasal length, 16.5; cranial breadth, 7.8.

Sorex macrodon Merriam

Large-toothed Shrew

Specimens examined.—Total 4: Las Vigas, 8500 ft., 3; 3 km. W Acultzingo, 7000 ft., 1.

Additional records.—Jackson (1928:153) lists Orizaba and Xico.

So far this species is known only from the four places listed immediately above, all in west-central Veracruz at higher elevations.

The thickened borders of the premaxillae where they border the anterior nares distinguish each of the four skulls from those of the other *Sorex* that are here identified as *Sorex saussurei veraecrucis.*

A male with enlarged testes was captured in an oatmeal-baited trap set in dense, woody bushes along the mossy bank of a tiny stream three kilometers west of Acultzingo in the cloud brushland at the very edge of the Mexican Plateau. At Las Vigas three other males were trapped in runways of *Microtus mexicanus* in deep moss at the bottom of a valley on the northern edge of the town. Two other species of shrews, *Sorex saussurei* and *Cryptotis mexicanus,* were common in the same area.

Sorex saussurei veraecrucis Jackson

Saussure's Shrew

Specimens examined.—Total 10: 6 km. SSE Altotonga, 9000 ft., 1; 1 km. W Las Vigas, 8500 ft., 1; Las Vigas, 8500 ft., 7; 2 km. E Las Vigas, 8000 ft., 1.

Additional record.—Xico (= type locality).

On November 11, 1946, six kilometers south-southeast of Altotonga a large, adult female was trapped. She was not pregnant and was not nursing, but seemingly had raised a litter that year; the mammae were large. The trap was in a damp swale about 50 meters long by 20 meters wide. There were a few bushes two feet high, but the cover otherwise was grass about 3 inches high, low ferns, bracken, moss, and liverworts. The swale drained into a deep arroyo with mossy rocks and a cold, clear stream. Along the border of a cornfield only a few yards from where the shrew was caught, *Peromyscus melanotis* was trapped. In the arroyo we took only *Peromyscus boylii* and *Microtus mexicanus.* The surrounding area, for miles, is arid pine forest.

At Las Vigas seven specimens were obtained in runways of shrews and of *Microtus mexicanus* in deep, damp moss in a small valley in the pine forest. Three individuals of *Sorex macrodon,* numerous specimens of *Cryptotis mexicanus,* and *Microtus mexicanus* also were taken there. One kilometer west of Las Vigas a specimen was trapped in a small heap of boulders in a fence, or "cerca," of maguey plants. This individual was associated with *Peromyscus melanotis* and *Reithrodontomys megalotis.*

Two kilometers east of Las Vigas a shrew of this species was caught in a trap set in a damp place under a low cliff of lava in the "malpais," an extensive, recent flow of lava. *Peromyscus boylii* and *Neotoma mexicana torquata* were common there.

Cryptotis mexicana mexicana (Coues)

Mexican Small-eared Shrew

Specimens examined.—Total 54: 4 km. W Tlapacoyan, 1700 ft,. 3; 1 km. W Las Vigas, 8500 ft., 1; Las Vigas, 8500 ft., 44; Huatusco, 5000 ft., 3; Coscomatepec, 5000, ft., 3.

Additional records (Merriam, 1895:23).—Jalapa, type locality; Jico; Orizaba.

This shrew occupies a limited area in the state of Veracruz, but locally is abundant in favorable habitat.

On November 21, 1947, a dead individual was found in a trail four kilometers west of Tlapacoyan. Although much decayed, it was prepared as a specimen. The following night two others were taken in a line of mousetraps set in beds of six-inch high succulent vegetation along a small stream. The soil was soft and wet. There were numerous runways of small mammals, probably of *Microtus quasiater*. The traps along the streams took also *Microtus quasiater*, *Oryzomys palustris*, *Oryzomys alfaroi* and *Marmosa mexicana*.

At Las Vigas several individuals of *Cryptotis mexicana* were trapped along the long hedges of maguey plants, called "cercas," that separate the cornfields in this area. Here the shrew was associated with *Microtus mexicanus*, *Peromyscus melanotis* and *Reithrodontomys megalotis*. Many other individuals of *C. mexicana* were taken in the deep moss in a small, cold valley in the pine forest nearby. *Sorex macrodon*, *Sorex saussurei* and *Microtus mexicanus* were also found here, although the *Cryptotis* outnumbered the two species of *Sorex* together by ten to one.

At Huatusco, specimens were taken in a patch of wild bananas about 30 feet in length by 20 feet in width. At Coscomatepec, specimens were taken in dense, dry brush on an overgrown hillside.

We regularly trapped specimens of this species by using banana, peanut and walnut for bait. The stomachs of specimens examined in the field, however, contained remains of only insects and worms, as far as could be determined.

Several females taken at Las Vigas, between October 7 and 20, 1948, were nursing young. Other females were neither nursing nor pregnant. No females containing embryos were taken at that locality. One taken at Coscomatepec, 5000 feet elevation, contained three embryos, each five millimeters long, on December 2, 1948.

Cryptotis nelsoni (Merriam)

Nelson's Small-eared Shrew

Record.—The type locality, Volcán San Martín Tuxtla (Merriam, 1895:26). This shrew is known only from the type locality and may be only subspecifically separable from *Cryptotis mexicana*.

Cryptotis obscura (Merriam)

Dusky Small-cared Shrew

Specimen examined.—Zacualpan, 6000 ft., 1. The identification is tentative (specimens that might be useful in direct comparison are on loan to a prospective revisor of the genus, April 5, 1961). M. R. Lee, the collector, trapped the specimen near "a large deciduous tree among a mat of fallen leaves."

Cryptotis micrura (Tomes)

Guatemalan Small-eared Shrew

Specimens examined.—Total 11: 7 km. W El Brinco, 800 ft., 1; Las Vigas, 8500 ft., 2; 5 km. N Jalapa, 4500 ft., 1; 7 km. NNW Cerro Gordo, 3; Teocelo, 4500 ft., 3; 1 km. E Mecayucan, 200 ft., 1.

Additional records (Merriam, 1895:22, unless otherwise noted).—Jico; Boca del Río (Findley, 1955:615); Orizaba Valley; Catemaco.

Our specimens of *C. micrura* differ from those of *C. parva berlandieri* principally in darker pelage and larger skull, in which the top of the braincase, relative to the rostrum, is more elevated. Cranial measurements of eight of these specimens have been published by Findley (1955:616). When the species of the genus *Cryptotis* are revised we suppose that *C. micrura* will be arranged as a subspecies of *C. parva.*

This tiny shrew is the smallest mammal that occurs in the state of Veracruz. It is rare, and of rather erratic distribution through the Tropical Life-zone of the state. Little was learned of its habits.

A *Cryptotis micrura,* captured by hand five kilometers north of Jalapa, on October 29, 1946, was under a pile of rock, 30 feet long and 10 feet wide, in one of many trough-shaped runways that probably had been constructed by *Microtus quasiater.* These mice were abundant in the vicinity. The rocks were one to three feet in diameter and of irregular shape. The rock pile was in a grassy meadow that stretched away for 75 feet or more in every direction; cattle had grazed the grass here until it was only about three inches high.

Seven kilometers north-northwest of Cerro Gordo, at 1500 feet elevation, on October 28, 1947, 50 traps, set in low, thick, damp brush, took one shrew, along with *Reithrodontomys fulvescens* (5 individuals), *Peromyscus mexicanus* (1), *Sigmodon hispidus* (5), and *Baiomys musculus* (7). The shrew had no parasites. It was a female that had recently been nursing young. On October 29 another specimen was taken in the same area. On November 26, we returned to the same locality, and took a non-pregnant female in dense sawgrass. On the preceding night our traps had been set in the dense brush where two shrews were taken a month ago. The area had dried up and then had become wet again with the onset of the Norte Season. On the morning of the 26th our traps held only two individuals of *Baiomys* and a few individuals of *Sigmodon.* We immediately moved our traps to dense grass about three feet high (the flower stalks were six feet high) and set them in sheltered places to protect them from the rain. Sometimes we made a thatched roof over the trap site. We left the traps out all day, and took nine cotton rats. In the night the traps took *Cryptotis* (1), *Reithrodontomys* (3), and *Sigmodon* (3).

At Teocelo, along a low rock wall that was overgrown with brush and grasses and that served to separate two coffee groves, a shrew of this species was trapped on January 3, and another on January 7, 1949. *Reithrodontomys* and *Microtus quasiater* were common here.

On November 11, 1948, a *Cryptotis micrura* was obtained in a patch of elephant's ear and other succulent vegetation beside a small stream, seven kilometers west of El Brinco. It was associated with *Peromyscus leucopus* and *Peromyscus mexicanus.*

One kilometer east of Mecayucan, on December 19, 1947, a young *Cryptotis micrura* was taken on dry, hard ground beneath thorny bushes. The surface of the ground had some sparse grass and dry leaves but was mostly bare. This locality is on the arid coastal plain.

A skull of *Cryptotis micrura* was removed from the intestinal tract of a king snake (*Lampropeltis polyzona*) at Potrero Viejo, 1700 feet elevation, by Dr. E. H. Taylor in 1936.

Order CHIROPTERA

Family Emballonuridae

Rhynchonycteris naso (Wied-Neuwied)

Brazilian Long-nosed Bat

Specimens examined.—Total 59: 5 km. SW Boca del Río, 10 ft., 1; 1 km. E Mecayucan, 200 ft., 5; Río Blanco, 20 km. W Piedras Negras, 400 ft., 17; 15 km. W Piedras Negras, 300 ft., 6; 14 km. SW Coatzacoalcos, 100 ft., 16; 22 km. ESE Jesús Carranza, 300 ft., 4; 35 km. SE Jesús Carranza, 350 ft., 10.

Locally this species is abundant in the state of Veracruz, although its distribution is rather irregular. At times, none is found along miles of river in seemingly suitable habitat. Then a half dozen colonies may be found within the space of a mile. The following notes record typical finds of this bat.

"This afternoon Gerardo, Vicente, two other men, and I took a dugout canoe and went hunting for these small bats in the arroyo [an arroyo leading into the Río Blanco, 20 kilometers west of Piedras Negras, on October 4, 1946]. We paddled up the vine-hung stream, passing under many logs that lay across our path. Each log was carefully examined, and also the trees along the bank. We came upon a colony unexpectedly, and a swarm of mothlike flying bats, about 25 in all, went flitting upstream. At the next leaning tree about 15 had landed. They were spaced one to two inches apart in a vertical line on the under side of the tree. The lowest was about five inches from the water. My shot brought down four, and the group broke up, some flying downstream and others up. We followed them, and found them at odd places along the shore or on dead, leaning trees in the water. All were over water, and every one shot fell into the stream. After the first shot, I got no more chances at bunches. The bats were scattered singly or in groups of up to four. They usually allowed a fairly close approach. Many times we were within 10 feet of them before they flew. So well did the dull, grayish color with the four paler spots match lichens or mouldy spots on trees, that we often failed to see them." Of the 13 collected at this place, two are small and, as shown by the development of the skull, are young of the year. Of the seven adult females, none was pregnant but all had active mammary glands. The testes of the adult males were small.

A colony found on the north side of the Río Coatzacoalcos, 14 kilometers southwest of the city of Coatzacoalcos, on February 2, 1947, was described as follows: "The slow-moving rivers and streams of this area seem to be ideal for sharp-nosed bats, and we found them after a short search. We hunted them from a dugout canoe, looking carefully on the undersides of leaning tree trunks in the water and on the bank. One specimen, a male, was found singly. At another tree, on the bank but overhanging the stream, we saw a line of about 15. They were in a vertical line, along the lowest part of the circumference of the tree, which was about two and one half feet thick. The lowest bat was about two feet from the water—the highest about four and one half feet. I fired at the center of the line, getting eight specimens. The others flew upstream but we did not try to follow them. Five (of those taken) are non-pregnant females. The testes of the four males are small."

At the same locality, on February 7, 1947: "Yesterday afternoon we were poling a dugout along an arroyo of still water choked with water hyacinth, when a swarm of sharp-nosed bats came out from the underside of a dead tree lying across the arroyo. The underside of the tree was about two feet from the surface. There were about 50 bats in the group. Many lit on the underside of a similar tree, 75 feet on up the arroyo. We brought down 11 in one shot.

"The series includes six females and five males. The testes of the males vary from two by two to three by three millimeters in diameter. The testes are almost round. All the females were pregnant, with embryos (one each) from nine to ten millimeters in length."

The evening flight of this bat is mothlike, and the species can be recognized when the bats are in flight. They seem never to fly anywhere except over water. They fly rather slowly, in relatively straight lines, and usually from one to three feet above the surface of the water.

Like many other species of bats, the sharp-nosed bat retires to a convenient roost after the evening hunt. Surprisingly, the one night roost found was not over water. It was one kilometer east of Mecayucan. On September 20, 1947, it was described as follows: "Tonight I looked for bats under the bridge across the arroyo here. I saw nothing but what I took to be three big cockroaches, sticking flat to the concrete 30 feet over my head. Farther on, where the ground sloped up, was another. This I examined closely, and it was one of the large cockroaches. While passing under the others a few minutes later, I heard a bat flutter, and turned out my light. It seemed to alight over my head. When I turned my light on there were four cockroaches where, a moment before there had been three. They still looked like cockroaches, even having back legs, but with the light close to my eyes I could see four pairs of tiny red eyes. I felt almost foolish shooting at them, but with my shot a bat of this species fell. The others did not move (as I shot them all, one by one). By going along under the bridge I got another. One shot did drop a cockroach." The resemblance of these bats to cockroaches was remarkable. They stuck flat against the concrete. Their reversed wing tips lay back away from the body at a 30 degree angle, like the back legs of the big "queen" cockroaches. Their mottled bodies are similar in color.

Pregnant females of this species were taken on January 13 and February 7. When these bats are carrying young, they desert their roosts over water and retire to hollow trees. Thirty-five kilometers southeast of Jesús Carranza, on April 9, 1949, several were shot from colonies roosting on a cliff overhanging the Río Solosuchil. All were males. Later that same day, the hollow shell of a tree, lying on the ground in the shade of a deep arroyo, was broken open. A number of bats flew out. Eight were captured. Each was a female of *Rhynchonycteris,* having a single large young clinging to a nipple. Some of the young were so large that the females were unable to fly more than a few yards.

Saccopteryx bilineata (Temminck)

Greater White-lined Bat

Specimens examined.—Total 54: Boca del Río, 10 ft., 2; Río Blanco, 20 km. W Piedras Negras, 400 ft., 1; 2 km. N Motzorongo, 1500 ft., 2; 14 km. SW Coatzacoalcos, 100 ft., 6; 35 km. ENE Jesús Carranza, 150 ft., 1; 20 km. ENE Jesús Carranza, 200 ft., 8; 20 km. E Jesús Carranza, 300 ft., 10; 25 km. SE Jesús Carranza, 250 ft., 7; 20 km. S Jesús Carranza, 300 ft., 10; 34 km. SE Jesús Carranza, 400 ft., 1; 35 km. SE Jesús Carranza, 350-400 ft., 3; 38 km. SE Jesús Carranza, 500 ft., 2.

Additional records (Sanborn, 1937:330).—Veracruz; Achotal; Minatitlan.

The greater white-lined bat is abundant in the dense jungle of the southern part of Veracruz but is far less common farther north, in the central part of the state.

On the south side of the Río Blanco, 20 kilometers west of Piedras Negras, on October 1, 1946, a bat of this species was shot in flight. This area is part of the almost level coastal plain, forested only along water courses. The bat was flying about under the shade of a large mango tree along an arroyo. No other individuals were taken on the coastal plane, except at Boca del Río, 10 feet elevation, where two were found clinging to the side of a horizontal beam, 25 feet from the ground, in the ruins of an old building.

In the upper humid division of the Tropical Life-zone I found this species at only one locality. This was two kilometers north of Motzorongo. There two were shot as they hunted over the surface of a small pond in an arroyo. Nowhere else in Veracruz was this bat found at an elevation so high as 1500 feet.

Many daytime retreats of this species of bat were found in southern Veracruz. Almost all were in hollow trees; some were in shaded but well-lighted hollows and others in almost complete darkness. Some typical retreats were noted as follows:

On February 6, 1947, 14 kilometers southwest of Coatzacoalcos while searching for hollow trees in the dense jungle for topotypes of *Vampyrum spectrum*, we came across a colony of about 25 individuals of this bat (*Saccopteryx*) in a hollow tree, really just the shell of a tree, about three feet in diameter. The hollow was open on top and along one side. It was so light there that the bats were easily visible. They were not shy and I was able to catch four with my hands before the others took flight. They came back within five minutes, but this time were too shy to catch by hand. I shot three, but one was too smashed to save.

Twenty kilometers east of Jesús Carranza on February 7, 1948, we "Went hunting for bats in hollow trees. . . . A colony of seven of this species was found in a shell-like fig tree. The tree was about seven feet in diameter with a five-foot hollow that went up at least 40 feet. There were numerous large openings and the hollow was as light as day. The bats were scattered, clinging to the inside of the shell of the tree, 15 to 20 feet from the ground. All seven were taken—six females and one male. A single *Saccopteryx* was found in a small cavity no higher than my head, in a tall tree about 500 yards away. This cavity was rather dark."

At the same locality on February 7, 1948, "A few bats of this species, and about 20 *Desmodus*, were found in a large hollow tree . . . probably 20 feet in diameter at the base, and . . . isolated in a forest of palms that effectively shade it. The opening to the hollow is a narrow V, six feet high by three wide at its widest. I fired nine shots from my pistol into the hollow and one *Saccopteryx* and four vampires fell down. There were many other bats in the hole, but without my light I could do nothing."

Thirty-five kilometers southeast of Jesús Carranza three individuals of this bat were noted clinging to the back of a shaded aperture in the face of a limestone cliff. They were approximately 25 feet from the ground.

Most of the specimens of this species are females. Adult females with embryos were found only once, on February 18, 1948, when the two adults taken each had a single small embryo. The only parasites noted on *Saccopteryx bilineata* were small, red mites. Most of the specimens collected had a few such mites on the ears.

The flight of *Saccopteryx bilineata* closely resembles that of *Myotis* but is slower and more deliberate. Bats of this species usually fly only four to seven feet above the ground along water courses, trails or openings in the jungle, but once, specimens were shot while they were flying over the tops of small trees, fully 30 feet from the ground. They fly rather erratically, as they swerve and swoop in catching insects. They prefer to remain under the shade and cover of trees and branches while in flight. They emerge shortly after sunset, but while it is still almost daylight, and each individual seems to patrol a certain part of the jungle that I suppose is its home range or hunting ground.

Peropteryx macrotis macrotis (Wagner)

Lesser Doglike Bat

Specimens examined.—Total 15: 35 km. SE Jesús Carranza, 350 ft., 2; 38 km. SE Jesús Carranza, 500 ft., 13.

All of our specimens of this species were captured in caves in the limestone cliffs along the Río Chalchijapa, 35 to 38 kilometers southeast of Jesús Carranza.

The species was first found on April 27, 1948, when a colony was located in a shallow recess in a cliff. This recess was about four feet high, 50 feet long and 20 feet deep. The opening was flush with the face of the cliff and about 20 feet above the surface of the river. The bats were clinging to the roof of the recess, in shaded daylight. When frightened they took refuge in a small, roomlike cave about six feet high and 20 feet in diameter. The entrance to this cave was about 20 inches in diameter and the cave was dark. The bats were captured in a net hung over the entrance to the cave. One bat escaped and 10 were captured. On April 30 the cave was revisited and two more individuals were captured. In April, 1949, there were no bats of any kind in this cave.

On April 27, 1948, a specimen was taken in a long, tubelike cave about three feet in diameter and 100 feet long. An *Artibeus jamaicensis* was also found in the cave. The following day two others were taken in small caves in a limestone cliff. One bat was alone. The other was in a cave with several individuals of *Balantiopteryx io* and a *Chropoterus auritus.*

In April, 1949, the caves on the Chalchijapa were revisited. All the caves where *Peropteryx macrotis* had been found in 1948 were examined but only two bats of this species were taken.

Peropteryx kappleri kappleri Peters

Greater Doglike Bat

Specimens examined.—Four from 38 km. SE Jesús Carranza, 500 ft.

The four specimens were obtained on April 30, 1948, and were seen first as they flew from the underside of a large block of limestone that had fallen long before from a high cliff. The underside of this block was shaded but not dark. Two of the bats were captured when they entered a small cave formed by fallen boulders, The cave was rather dark, and approximately seven feet in diameter by 10 feet in height. The two remaining bats flew to a similar but smaller cave 30 feet away where they too were caught. All four were males having small testes. The flight of these bats was slow and deliberate. One was captured in full flight with a quick snatch of a hand.

Centronycteris maximiliani centralis Thomas

Thomas' Bat

Specimen examined.—One from 35 km. SE Jesús Carranza, 350 ft.

This rare bat is known from but seven specimens. Ours was shot in late afternoon while it was flying in dense jungle. The bat was about 10 feet from the ground and was following a regular course approximately 125 feet in length. At each end of its course, in approximately 10 trips up and down a small arroyo, the bat turned each time at almost exactly the same place. The flight was slow and fluttery.

Balantiopteryx plicata plicata Peters

Peters' Bat

Specimens examined.—Total 25: Puente Nacional, 500-1000 ft., 17; 9 km. E. Totutla, 2500 ft., 3; 7 km. SE San Juan de la Punta, 400 ft., 5.

These little bluish bats seem to prefer an arid habitat. On October 21 and 22, 1946, at Puente Nacional several colonies, numbering from two to 75 individuals, were in narrow, horizontal, slitlike caves, formed along the contact of a layer of soft conglomerate rock with a layer of hard sandstone. The bats were clinging to the ceiling of the cave with both feet and the thumbs, with bodies thrust out, away from the rock. The cave was light enough for a person to see the bats without the aid of an artificial light. When disturbed, they flew farther back into the cave. In the beam of a flashlight, their eyes gleamed like tiny rubies, far back in the low cave.

Near these caves, where the bats were resting by day, were several larger, more open caves. The floors of these were covered to a depth of six inches with guano. Seemingly the bats hung up at night in these larger caves, to rest after hunting.

Nine kilometers east of Totutla several bats of this species were found in a large, dry cave, along with several individuals of *Micronycteris sylvestris*. The three specimens taken were males. On December 15, 1946, seven kilometers southeast of San Juan de la Punta, a colony of about 200 bluish sac-winged bats and a few more than 50 common vampires, species *Desmodus rotundus*, were in a small, dark cave along the Río Blanco. The sac-winged bats were shy, and quick to take alarm.

Few females of *B. plicata* were found and none taken was pregnant. This species of bat is more often than not parasitized by a small orange or yellow mite. These mites live in clumps of from 15 to 30 or more on the ears and, less commonly, on the tail or antebrachial membranes.

Balantiopteryx io Thomas

Thomas' Sac-winged Bat

Specimens examined.—Total 107: Grutas Atoyac, 2 km. E Atoyac, 14; 5 km. S Potrero, 1700 ft., 11; 38 km. SE Jesús Carranza, 500 ft., 82.

We first found this species on March 6, 1946, in a cave known as Grutas Atoyac, two kilometers east of Atoyac, at about 1500 feet elevation. Approximately 25 individuals were in the innermost rooms of the cave, fully 300 meters from its mouth. The bats were hanging separately, most of them in small, pitlike depressions in the roof. A few hung from the walls. A few vampires (*Desmodus rotundus*) were in the same cave, but near its mouth. The 14 individuals of *Balantiopteryx io* were evenly divided as to sex. None of the females was pregnant. Each bat had from six to 10 small red mites attached to the membranes of the tail, wings and/or ears.

On April 25, 1948, in a large cave in a limestone cliff, 38 kilometers southeast of Jesús Carranza, along the Río Chalchijapa, this species was abundant. The mouth of the cave is about 30 feet high, 45 feet wide, 20 feet above the river, and the bottom of the entrance is seven feet below the base of the cliff. Inside, the cave varies from 20 to 50 feet in height and 30 to 60 feet in width. The main cave is about 500 feet in length, and there is a side passage about 75 feet long. In addition to *Balantiopteryx io, Pteronotus parnellii, Artibeus jamaicensis* and *Myotis nigricans* inhabited the cave.

Some individuals of *B. io* hung from openings in masses of wave-like stalactites 30 to 50 feet from the mouth of the cave, where there was daylight enough to permit a person to see the bats. They were about 20 feet from the floor. Approximately 75 feet from its mouth, the cave varies from 25 to 50 feet in height, and the bats were most abundant there. Most were hanging 25 to 35 feet from the floor, but some were as high as the cave extended. Only a few were seen as far as 100 feet from the mouth of the cave, and only one was noted more than 125 feet from the mouth. Without exception the bats hung singly, and usually more than nine inches from one another. They preferred to hang from the tops of pits and crevices but some hung from the open, flat ceiling. It was estimated that between 500 and 1000 individuals of *Balantiopteryx* were present in the cave. For so many bats, the quantity of guano in the cave was surprisingly small.

In addition to the large colony mentioned above, more than 1000 other individuals of *Balantiopteryx io* were seen at this locality. Most were in some of the thousands of small caves that pit the cliffs, but many were in deep, dark crevices and masses of stalactites that hung from the faces of the cliffs. A total of 89 specimens was saved from this locality, in 1948; only two were females. At the same locality in 1949, the bats were as abundant as the year before.

Two females taken on April 25, 1948, contained embryos.

Family Phyllostomidae

Pteronotus psilotis (Dobson)

Dobson's Mustached Bat

Specimens examined.—Seventy-five from 3 km. E San Andrés Tuxtla, 1000 ft.

Additional record.—Near Tuxpan (Malaga-Alba and Villa R., 1957:534).

For this species, long known as *Chilonycteris psilotis*, the older generic name *Pteronotus* is here used because Burt and Stirton (1961:24) regard all of the species heretofore referred to *Chilonycteris* and to *Pteronotus* as congeneric.

In early January, 1948, a large colony of this species was found along with other kinds of bats three kilometers east of San Andrés Tuxtla in a long, narrow, sinuous cave in a basaltic cliff. The entrance to the cave was triangular, about two meters on a side. The cave consisted of a tubelike lower portion about 200 feet long, abruptly blocked by a wall of rock, which probably resulted from a geologic fault. Ten feet above the lower level of the cave, a narrow opening allows entrance to the upper part of the cave, consisting of a series of low, shallow rooms, joined by passages.

The lower level was cool and relatively dry. There we caught *Pteronotus parnellii*, *Glossophaga soricina*, *Carollia perspicillata*, *Desmodus rotundus* and *Natalus stramineus*. The only odor noted in the lower level of the cave was the ammonia smell of the vampire bats.

The upper part of the cave, in strong contrast to the lower part, was hot and damp, with an almost overpowering stench. The temperature was estimated at 95 degrees Fahrenheit. The walls in many places were wet; drops of water hung from the roof before dropping to the floor. The floor was buried under many inches of semi-liquid bat guano, and the surface of the guano was so covered with the white larvae of a small species of fly that it appeared to have been whitewashed. The air of the cave contained millions of tiny black flies. In this upper level there were a few individuals of *Mormoops megalophylla* and thousands of individuals of *Pteronotus psilotis* and *Pteronotus davyi*.

Pteronotus psilotis was a weaker flier than *Pteronotus davyi*; a net that was hung in the cave took several specimens of P. *psilotis* but no P. *davyi*. P. *psilotis* was notably less agile in flight than P. *davyi*. At rest, *psilotis* preferred to lie on a horizontal or inclined surface; few hung from the roof of the cave. P. *psilotis* was notably swift on the ground, running and hopping with tightly folded wings, and resembled a large spider in motion.

This species is insectivorous. We noted individuals hunting in the evening in openings in the jungle, and among the branches of the tall jungle trees. On one rainy night, many small bats were seen hunting over the surface of a small lake, Laguna Encantada. Two of these were killed and proved to be P. *psilotis*. On later, nights, when it was not raining, no bats were seen over the lake.

Like many other species of bats, Dobson's mustached bat seems to fill its stomach quickly, and then to retire to some secluded place to rest. On one dark night, we spent some time in the cave mentioned above. A constant stream of bats was entering and another leaving the cave. With lights turned out, we stood by the side of the cave with small hand nets. When the bats began to pass, the net would be swung, and bats usually were captured. Then the lights would be turned on, and the captives were placed in a sack. The lights caused the bats to withdraw to the mouth of the cave, but shortly after the lights were turned out, the bats again began to stream past. In this fashion we captured *Pteronotus psilotis* (56), *Pteronotus davyi* (6), *Carollia* (1), and *Natalus* (1).

None of the females of this species taken in the first week of January was pregnant. There was an average of about four parasitic winged flies and three large black "stick-tight" fleas to each bat of this species that we examined.

Pteronotus parnellii mexicana (Miller)

Parnell's Mustached Bat

Specimens examined.—Total 7: 8 km. NW Potrero, 1700 ft., 1; 3 km. E San Andrés Tuxtla, 1000 ft., 2; 38 km. SE Jesús Carranza, 500 ft., 4.

Additional records.—Jalapa (Ward, 1904:639); Mirador (Rehn, 1904:204).

For this species, long known as *Chilonycteris parnellii*, the older generic name *Pteronotus* is here used because Burt and Stirton (1961:24) regard all of the species heretofore referred to *Chilonycteris* and to *Pteronotus* as congeneric.

This species seems to be rather rare in the State of Veracruz. Our specimens were all captured in caves surrounded by dense jungle. In a lava cave three kilometers east of San Andrés Tuxtla (see account of *Pteronotus psilotis*), two specimens of this bat were obtained by firing a pistol loaded with dust-shot into a wheeling mass of flying bats, that consisted principally of *Natalus stramineus* and *Glossophaga soricina.* This cave was searched on many occasions for other individuals of this species, but none was found.

In April, 1948, 38 kilometers southeast of Jesús Carranza, this species was found in a large limestone cave (see account of *Balantiopteryx io*). In this cave, from 75 to 150 feet from the entrance, there were six to eight colonies of *Myotis nigricans,* each colony consisting of a dense mass of from 50 to 150 pregnant females. Each of four colonies had a single *Pteronotus parnellii,* a nonpregnant female, packed tightly in with the mass of *Myotis.* The ground beneath each colony was examined, but nothing was found to support our suspicion that P. *parnellii* was preying on the *Myotis,* or in some fashion obtaining benefit from the association. On November 18, 1948, eight kilometers northwest of Potrero, an adult male P. *parnellii* was creeping about and over approximately 50 individuals of *Artibeus jamaicensis* that clung in a tight mass to the top of a small pit in the roof of a limestone cave.

Pteronotus davyi fulvus (Thomas)

Davy's Naked-backed Bat

Specimens examined.—Sixty-two from 3 km. E San Andrés Tuxtla, 1000 ft.

Additional records (Davis and Russell, 1952:235).—Mirador; San Andrés Tuxtla; South San Andrés Tuxtla; Achotal.

In Veracruz we found this species only in a cliff of basaltic rock (see account of *Pteronotus psilotis*) three kilometers east of San Andrés Tuxtla in the first week of January, 1948. On the ground, *Pteronotus davyi* acts much like *Pteronotus psilotis,* running and hopping with considerable agility, with wings tightly folded to the body, but in the air, P. *davyi* is far more skillful than P. *psilotis* and was far more difficult to capture in a butterfly net; also, P. *davyi* did not become entangled in the strands of a large hanging net. Like P. *psilotis,* P. *davyi* retires to a cave or similar retreat after its evening hunt. Swinging a net blindly at bats passing in the darkness yielded 56 individuals of P. *psilotis* as against six of P. *davyi.* Probably fewer specimens of P. *davyi* were taken because it could more often dodge the net; shooting blindly in places where both species were living brought down the two kinds in about equal numbers.

In the early evening, individuals of *Pteronotus davyi* emerged from the cave and began their evening hunt. Although they deviated much from a straight line of flight when hunting insects, their flight was rather slow and direct in comparison with that of other bats of about the same size. Most individuals of P. *davyi* flew only about seven feet from the ground, although some barely skimmed the ground and a few maintained an altitude of 25 feet or more. The line of flight of the bats from the mouth of the cave was consistently down the face of the cliff, through the jungle, across a small bay of the lake, over a meadow of tall sawgrass, and up a steep hillside, to the open, grassy hills. A few individuals drank water as they passed the lake, but most did not. Probably the flight broke up over the hills, but this could not be determined for it was usually too dark to watch the bats after they were followed to the hills. This flight was repeated night after night. The first bats reached the top of the hill, where we were camped, at dusk. This was about one kilometer from the cave. The flight continued for nearly an hour.

None of the females of *Pteronotus davyi* collected in the first week of January was pregnant. The testes of the numerous males collected were all small. Males outnumber females in our series by more than three to one. Our *Pteronotus davyi* had relatively few parasites. Most individuals had one small winged-fly or two such but many had none. No "stick-tight" fleas were noted, although these were common on the *Pteronotus psilotis* from the same cave.

Mormoops megalophylla megalophylla (Peters)

Peter's Leaf-chinned Bat

Specimens examined.—Total 6: 6 km. WSW Boca del Río, 10 ft., 1; 4 km. WNW Fortín, 3200 ft., 1; 3 km. E San Andrés Tuxtla, 1000 ft., 4.

Additional records.—Cañon of Actopan NE of Jalapa (Ward, 1904:634); Mirador (Rehn, 1902:170); Orizaba (Rehn, 1902:170).

Hall and Kelson (1959:96) list, under *M. m. megalophylla*, specimens recorded by Ward (1904:634) from the Cañon of Actopan and incorrectly plot the locality too far south. As clearly stated by Ward, the locality is "some miles northeast of Jalapa, Veracruz, in the bottom of the Cañon of Actopan." Therefore, the place is correctly to be plotted at the north margin of the black dot representing Mirador on Map 59, page 96 (Hall and Kelson, 1959).

Two subspecies of this bat, *Mormoops megalophylla megalophylla* (Peters) and *Mormoops megalophylla senicula* (Rehn), have been reported from Veracruz. They were originally described as being separable primarily on the basis of the structure of the second upper premolar (Rehn, 1902:169). In 1960, specimens, from northern and southern extremes of the range of the species, in the Museum of Natural History of the University of Kansas, were examined as to this dental character and other characters. Also, six specimens of *Mormoops megalophylla* from Coahuila, six from Guatemala, and one from Tabasco were examined. The structure of P2 did not serve to distinguish between these specimens by the methods employed. Villa and Jiménez by independent study later (1961:502) reported the same conclusion. But, in 1960, other differences in other parts of the specimens influenced us to recognize two subspecies. The pelage of the northern specimens was dull brown and they were more densely haired than the Guatemalan and Tabascan specimens. Mean length of the forearm on the dried skins was 55.1 mm. The color fitted the description given by Rehn (*op. cit.*) for *M. m. senicula*, including the "silvery suffusion." The Guatemalan and Tabascan specimens differed in color, from other specimens examined, primarily in that the hairs were ochraceous basally—as described by Rehn (*op. cit.*) for *M. m. megalophylla*. The hair was sparse to almost absent on the nape and crown allowing the color of the basal part of the hair to show at the edge of the "naked" area. The mean lengths of the forearms of the Guatemalan specimens and the Tabascan specimen were 54.4 and 54.7, respectively.

Subsequently, Davis and Carter studied a large number of specimens from north of Colombia. Although they noted (1962:65) the difference in color between specimens from Guatemala (chocolate brown) and specimens north thereof (wood brown), they agreed with Villa and Jiménez (1961:502) that *M. m. senicula* should be placed as a synonym of *M. m. megalophylla*. Because the study by Villa and Jiménez was based on a larger number of specimens than we had and because the study by Davis and Carter was of wider geographic scope than our study in 1960 we follow those four authors and apply the subspecific name *M. m. megalophylla* to all of our specimens from the state of Veracruz instead of recognizing the subspecific name *M. m. senicula* and applying it to specimens from Orizaba and others north thereof.

On March 25, 1946, four kilometers west-northwest of Fortín in the tropical-forest habitat, a dead, non-pregnant female was hanging head-down from a window screen. The cause of death could not be ascertained.

In a cave, described in the account of *Pteronotus psilotis*, three kilometers east of San Andrés Tuxtla in early January, 1948, four individuals were obtained in a low, shallow room in the upper level of the cave, where the temperature was estimated to be about 95 degrees Fahrenheit, and where the walls were not wet, and where the odor of the guano was almost overpowering. One was shot in flight among several thousand individuals of *Pteronotus psilotis* and *Pteronotus davyi*. One was asleep, hanging head-down from a rock wall, with its head bent under, almost to the chest, and the back somewhat bowed. The other two were awake, hanging from vertical walls of rock; they twisted about, watching the light, but did not attempt to fly. They squeaked when plucked from the wall.

In contrast to these occurrences, all in the upper humid division of the Tropical Life-zone, one was taken on the arid coastal plain at Boca del Río on April 23, 1949. The female at Boca del Río was hanging free, from the rough surface of an overhead joist in a fairly light room of an abandoned hacienda. In adjoinng rooms there were colonies of *Artibeus jamaicensis* and *Desmodus rotundus.* On other visits to this old building, a small colony of *Glossophaga soricina* was found hanging from the beam where the *Mormoops* was taken, but none was noted on the day that the *Mormoops* was found.

A female taken in early January was not pregnant, nor was a female taken in March. The one taken on April 23, 1949, contained an embryo 22 millimeters long. Two males taken in January had small testes. No parasites were ever noted on this species.

Micronycteris megalotis mexicana Miller

Brazilian Small-eared Bat

Specimens examined.—Total 16: Plan del Río, 1000 ft., 2; 4 km. W Paso de San Juan, 250 ft., 5; 14 km. SW Coatzacoalcos, 100 ft. 3; Achotal, 5 (Chicago N. H. Mus.); 35 km. SE Jesús Carranza, 400 ft., 1.

Additional record.—Cuesta de Don Lino, near Jalapa (Ward, 1904:653).

This species was found on the arid, open coastal plain of the central part of the state of Veracruz, and in the dense jungles of the southern part of the state. Suitable daytime retreats, rather than surrounding environments, seem to regulate the distribution of *Micronycteris megalotis.* As noted in Veracruz, and elsewhere in Mexico, bats of this species require narrow, dimly-lighted, horizontal, tubular hollows for their daytime retreats. In a natural state these are probably supplied by hollow logs and the burrows of large mammals.

On February 1, 1947, we looked into a large pipe, about two and one half feet in diameter, that passed beneath a railroad embankment, 14 kilometers southwest of Coatzacoalcos. Six bats were in the pipe, but only two, one *Micronycteris* and one *Carollia,* were obtained. On subsequent visits, three more individuals of *Micronycteris* and two more of *Carollia* were taken. Of these three individuals of *Micronycteris* one was found in the pipe after dark. As no bats were found in the pipe earlier that day, it is concluded that *M. megalotis,* like many other bats, retires to a suitable retreat to rest after feeding. The pipe where the bats were found was in dim light.

Four kilometers west of Paso de San Juan on the night of December 15, 1947, a *Micronycteris* was shot in an underpass under the highway from Jalapa to Veracruz. Another bat, probably of this species, was seen in the underpass.

Having become aware of the presence of *Micronycteris* at this locality, we made a systematic search of the pipelike drains that passed under the highway. Eight bats were found in one drain that was about 18 inches in diameter. Five of these were shot; four were of the species *Micronycteris megalotis* and one was a *Carollia perspicillata.*

Thirty-five kilometers southeast of Jesús Carranza, on February 15, 1948, our dogs had run an agouti to ground in a burrow about 10 inches in diameter. We had enlarged the mouth of the burrow and built a fire of dry palm fronds, forcing smoke into the burrow with a fan. Several small bats suddenly flew out into my face, probably blinded by the smoke. One must always be cautious in opening mammal burrows in the tropics because the dangerous fer-de-lance (*Bothrops atrox*) often lives in burrows of mammals. The sudden appearance of the bats caused me to jump back in alarm. A native assistant knocked one of the bats to the ground and it was a *Micronycteris megalotis.*

Stomachs of specimens taken in the daytime were empty, but stomachs of two shot at night were crammed with remains of insects.

None of the females taken on December 13, December 15, or February 7 was pregnant. The testes of two males taken on December 15 were of moderate size. The individuals from near Coatzacoalcos were parasitized by a few dark-colored, wingless flies.

Micronycteris sylvestris (Thomas)

Brown Small-eared Bat

Specimens examined.—Total 16: 15 km. ENE Tlacotepec, 1500 ft., 15; 9 km. E Totutla, 2500 ft., 1.

A colony of about 25 bats of this species was found in a cave in a sandstone and conglomerate cliff, 15 kilometers east-northeast of Tlacotepec on December 11, 1947. The cave was low and shallow, about six feet high by 30 feet wide and 30 feet deep, with a low, wide entrance. It was dimly lighted. Thirteen individuals were shot. None of the females taken was pregnant, and the testes of the males were small. All of the bats were parasitized by from three to five small, winged flies.

On December 14, 1948, another bat of this species was taken from a cave in a conglomerate cliff nine kilometers east of Totutla. This cliff was about 300 feet high, and the cave was approximately 50 feet from the top. The entrance was about 30 feet high by 40 feet wide. The cave was of about this dimension, and 75 feet deep. It was well lighted. The only bats present were one *Micronycteris sylvestris* and four individuals of *Balantiopteryx plicata.*

Mimon cozumelae Goldman

Cozumel Spear-nosed Bat

Specimens examined.—Total 10: 3 km. N Presidio, 1500 ft., 3; 35 km. SE Jesús Carranza, 350 ft., 4; 38 km. SE Jesús Carranza, 500 ft., 3.

On December 2, 1946, a deep, dark limestone cave, three kilometers north of Presidio, was the home of four of these large, handsome bats. This cave is situated in a small, jungle-covered hill, isolated in an extensive field of sugar cane. The bats were extremely shy, and were in flight when first seen, shortly after we entered the cave. Two escaped from us; the other two were captured.

At the same locality we watched this species feeding, and one was shot in flight. Immediately about our camp were several large trees of sweet oranges. Shortly after dusk the bats would appear from the direction of the hills. They arrived at elevations of from 20 to 35 feet from the ground. They were quite graceful, far more so than *Artibeus*. Their wing-beats were not rapid, and their flight was silent. The long interfemoral membrane could sometimes be seen against the clear sky. They approached the orange trees with a glide and a downward swoop. Numerous traps baited with banana and suspended in the orange trees, or tied to branches, were not disturbed by the bats. We were not able to see them in the trees, in the act of feeding. They ate only very ripe, sometimes spoiled, fruit. Possibly the bats were feeding on insects that were feeding on the fruit. Very ripe fruit was easily dislodged from the trees, and the sound of falling oranges attested to the activities of the bats until almost daybreak.

Two males taken at the above locality were lean, but a female was fat. The only ectoparasite noted was a small, winged fly.

On April 10, 1948, 38 kilometers southeast of Jesús Carranza five large bats were found in a small cave in a limestone bluff. The entrance was a hole a yard in diameter that descended vertically for a distance of 10 feet. The cave itself was about 10 feet high, four feet wide, 15 feet long and the bottom was covered by deep, dark water. A ledge along one wall of the cave allowed entrance. From the roof, three bats were taken. One other escaped out the entrance of the cave, and another was wounded but was lost when it entered a narrow crevice, a foot wide, that led off from one end of the cave. Two of the three bats taken were of the genus *Mimon* and the other a *Trachops cirrhosus*. One *Mimon* was a male and the other a female containing a large embryo. Another female, *Mimon*, with a full-term embryo, was one of two bats found in a small, room-like cave about 15 feet high and 12 feet in circumference. The bats were in a mass of stalactites in the highest part of the cave. The second bat escaped.

On April 4, 1949, 35 kilometers southeast of Jesús Carranza four individuals of *Mimon* and one *Trachops* were found and captured in a cave about 12 feet across and six feet high. All four were non-pregnant females.

Every cave in which *Mimon* lived was damp; the rock and stalactites were covered with a film of water. Only *Trachops* was found with *Mimon*. These two bats resemble each other closely, and cannot be distinguished in the caves. Both were completely silent. The droppings that littered the floor under their roosts were white, like the droppings of hawks and owls. Both genera are probably somewhat carnivorous.

Phyllostomus discolor verrucosus Elliot

Pale Spear-nosed Bat

Record.—Orizaba (Sanborn, 1936:97). The skull shown in figure 67 of Hall and Kelson (1959:108) is not P. *d. verrucosus* but instead is *Trachops cirrhosus coffini*.

Trachops cirrhosus coffini Goldman

Fringe-lipped Bat

Specimens examined.—Total 17: 38 km. SE Jesús Carranza, 500 ft., 2; 35 km. SE Jesús Carranza, 350 ft., 15.

One of three bats taken from a company of five in a limestone cave 38 kilometers southeast of Jesús Carranza on April 10, 1948, was this species (see account of *Mimon cozumelae*). It was an adult male.

Another was taken at the same locality on April 27 in a long, deep cave in a limestone cliff. About 50 bats, presumably of this species, were present. The cave was so dangerous, because of sink-holes covered with thin (½ inch thick) rock, that we were satisfied with one bat, a non-pregnant female. The bats in the cave were in a room entered by a hole descending for 10 feet at a 45 degree angle. The room was about 15 feet long by eight feet wide and seven feet high, but three holes in the floor descended 20 feet or more. A connected passage varied from a slit 10 feet high by six wide, to a hole scarcely large enough to allow a person passage by crawling. The floor was pitted with holes that went down to various depths. The floor was only a thin plate of rock. We went approximately 100 feet into the cave but the bats kept flying ahead of us. On later visits we found no bats in this cave.

On April 5, 1949, a colony of these bats was found 35 kilometers southeast of Jesús Carranza in a cave about 30 feet in length by 20 feet in width and height. The entrance was a small, narrow hole at the base of a limestone cliff. There were no bats in the cave proper but when a stone was dropped into a small hole, about a yard in diameter and four feet deep, we heard the flutter of a bat's wings. At the bottom of the hole, really a pit, a horizontal tube barely large enough to admit a man led off for a distance of about 10 feet. At the end of this tube we entered an almost circular chamber, 15 feet across, having a domed roof about 10 feet high. A female and 13 males of *Trachops* were in the chamber. We captured all 14. The floor was littered with the fecal droppings of this bat; they were white, and resembled the feces of birds more than those of bats. Possibly this species is carnivorous.

Chrotopterus auritus auritus (Peters)

Peters' False Vampire Bat

Specimens examined.—Three from 38 km. SE Jesús Carranza, 500 ft.

On April 27, 1948, at the mentioned locality, two individuals of this large bat were at the entrance of a short, narrow, high cave. The mouth of the cave is about five feet wide by 35 feet high. The bats were approximately three feet from the entrance, in a hole a foot in diameter and a foot deep, and were slow to fly. A net on the end of a long pole was pushed up to cover the hole, but the bats did not drop into the net until a shot was fired into the roof of the cave near the hole. The testes of the male were of moderate size and the female had a nearly full-term embryo. These large bats did not take flight as soon as did the several individuals of *Artibeus jamaicensis* and *Balantiopteryx io* that also were in the cave. At the same locality on the following day a male of *Chrotopterus*, whose testes were of moderate size, was found in the darkness at the base of a clump of yard-long stalactites at the edge of a cliff. The bat was about 25 feet from the ground. A net at the end of a long pole was placed over the bat, and it dropped into the net.

These bats were free of ectoparasites. White stains beneath the roosts of the bats resembled those left by the excreta of hawks and owls. The stomachs of the specimens were empty. The species is probably carnivorous.

Vampyrum spectrum nelsoni (Goldman)

Linnaeus' False Vampire Bat

Coatzacoalcos (Goldman, 1917:115), the type locality of the subspecies *V. s. nelsoni,* is the one record-station of occurrence in the state. This is the largest North American bat. Selected measurements, in millimeters, of the type are as follows: Total length, 130; length of tail, 0; length of hind foot, 31; height of ear from notch, 40; height of tragus, 12; height of noseleaf, 18; greatest length of skull, 51.0; zygomatic breadth, 23.6.

Glossophaga soricina leachii (Gray)

Pallas' Long-tongued Bat

Specimens examined.—Total 152: 35 km. NW Tuxpan, 1000 ft., 6; Potrero [del] Llano, 350 ft., 2; 17 km. NW Tuxpan, 7; Tuxpan, 3; 12½ mi. N Tihuatlan, 300 ft., 6; 10 km. NW Papantla, 750 ft., 1; 3 km. W Gutterez Zamora [= Gutierrez Zamora], 300 ft., 1; 3 mi. NW Nautla, 3; San Marcos, 200 ft., 3; Plan del Río, 1000 ft., 2; 4 km. W Paso de San Juan, 250 ft., 7; Boca del Río, 10 ft., 20; 8 km. NW Potrero, 1700 ft., 1; 7 km. NW Potrero, 1500 ft., 1; 4 km. WNW Fortín, 3200 ft., 4; Potrero Viejo, 7 km. W. Potrero, 10; 5 km. S Potrero, 1700 ft., 2; Alvarado, 10 ft., 1; 3 km. W Acultzingo, 7000 ft., 1; 3 km. E San Andrés Tuxtla, 1000 ft., 31; Cosamaloapán, 150 ft., 4; Cosamalapan [= Cosamaloapán], 150 ft., 8; 5 mi. S Catemaco, 2; 35 km. ENE Jesús Carranza, 150 ft., 1; Jesús Carranza, 250 ft., 1; 20 km E Jesús Carranza, 300 ft., 6; 38 km. SE Jesús Carranza, 500 ft., 6.

Additional records (Miller, 1913:419, unless otherwise noted).—Mirador; Córdoba (Sumichrast, 1882:202); Catemaco; Jaltipan; Achotal.

This long-tongued bat is fairly common in the lowlands of Veracruz. It makes its home in caves, old buildings, and similar retreats.

Four kilometers west-northwest of Fortín several specimens were taken between March 27 and April 3, 1946, from long, artificial tunnels constructed near the power plant of the Moctezuma Brewery. On the first visit to these tunnels, the bats were abundant; 13 individuals of *Carollia* and one *Glossophaga* were shot. When the tunnels were revisited the following day, the *Carollia* had left, but several individuals of *Glossophaga* remained.

The tunnel containing the most bats was about four feet wide, 25 feet high, and 600 feet long. It had been dug through a clay-colored, soft, crumbly rock of a claylike texture. Pallas' long-tongued bats were near one end of the tunnel, hiding, singly or in pairs, in small side niches, near the top of the tunnel, formed by the fall of loose concretions and earth. A colony of swallows nested at the other end of the tunnel. Bats of the genus *Carollia* were near the center of the tunnel, where it was darkest, and were hanging free from the roof.

The long-tongued bats could seldom be seen from the floor of the tunnel and were difficult to obtain. When disturbed in the little side-caves, as by a thrown pebble, the long-tongued bats would dart to another niche. They were rather noisy in flight, making a rustling whir.

At this same locality, on April 3, 1946, Mr. Rabago, manager of the power plant, told me of having seen bats fly out of small caves in the limestone cliff, immediately above the power plant. Although I had looked carefully for bats in these caves, I had found none. On this day, as I walked along the cliff, just above the plant, a bat flew out ahead of me. It took refuge in another cave, and when I searched for it there the bat flew out and returned to the cliff near where I had first seen it. Searching carefully, I found a hole, which I had overlooked before, about five feet from the ground. The hole, under an overhanging ledge, was about eight inches in diameter, coursed straight up for six inches, and then back into the cliff at a 45 degree angle for about seven feet. Turning my light into the opening, I saw two bats clinging to the roof near the back of the hole. Immediately beneath them was a pile of droppings about six inches in diameter and three inches high. One bat dropped when shot but the other flew out and escaped. The one killed lodged behind a pile of droppings, but was scraped out with a long stick.

At Potrero Viejo, seven kilometers west of Potrero, on September 24, 1946, small bats were reported in the old church and Mrs. Leora Forbes got permission, and the key, to enter. We opened the door to find eight of these small bats hanging in a row, each about three inches from the next one, to a rough overhead beam, about 12 feet from the floor. They were in dim light, but plainly visible. They were shy and took flight. Some went behind the altar. I found a small door in the altar and managed to squeeze through the intricately carved mass of 200-year-old wood. In back, between the altar and the end wall of the church, there was a dimly-lighted space about 15 inches wide. Some old wires hung down, and several bats clung to them. Two bracing boards along the back of the altar had other bats hanging head-down with wings wrapped around them, in the manner of *Balantiopteryx io*. Most of them hung by one foot. When my light struck them, they twisted about, peering at it. Most were shy and took flight. Some flew into a small room of the church, and there hung up in the dim light, 15 feet from the floor. One or two took refuge under a table. In all, about 18 to 20 were present. Some flew out, through a small passage, into the church tower, where they hung from the rough stone in a well-lighted place. In the church proper, we knocked down 11 bats with rubber bands. Two were males, each weighing ten grams. Of the females, three weighed ten grams; three weighed 11 grams; one 13 grams; one 14 grams; one 15 grams; and one 16 grams. The four heavier bats each had a single embryo 20 to 25 millimeters long. The testes of the two males were small. I saw no parasites on the bats.

At Jesús Carranza a colony of about 20 of these bats was found in a dimly-lighted house. The house was notably warm. *Molossus ater nigricans* also was present.

In the theater at Cosamaloapán on the night of April 1, 1947, we noted small bats flying through the beam of the projector. Returning next morning we saw six to eight of these bats hanging from the wires that support the screen and in the space between the elevated stage and the ground found about 700 more. In flight these bats made a dull roar, and a distinct wind. Three hundred more were in the dusky space above the low, partial ceiling of old signboards in the building. They hung head-down, with wings wrapped around the body, with a space of an inch or so separating each bat from its neighbors. Of the 14 saved, six were males and eight were females. One female had an embryo 10 mm. long.

This long-tongued bat seems to be a fruit eater as well as an eater of nectar (a flower bat). Some stomachs examined were full of fruit pulp. On the night of March 19, 1947, 35 kilometers east-northeast of Jesús Carranza, a rat trap baited with banana and set in a chico zapote (chicle) tree caught a *Glossophaga*.

Three kilometers east of San Andrés Tuxtla about 1000 individuals of *Glossophaga* were found in the attic of a partly ruined building of brick and stone, in which we were living. The evening flight of this colony was observed on January 9, 1948. All day the bats were completely silent but for a half hour before sunset there was a great deal of squeaking. The first bat emerged from the attic about 25 feet above the ground just as the sun disappeared behind a dark cloud on the horizon. The bat flew west, then made a complete semicircle about 100 feet in diameter, coming back past the house at the same level at which it left, but some 100 feet to the south. When it came back over the cliff south of the house it was about 75 feet from the tops of the jungle trees. It veered slightly southward and flew off to the lakeside jungle, still at the same level. The second bat emerged about three minutes later; it made a circle about 50 feet in diameter, and pitched down out of sight in a steep dive over the cliff south of the house. Five minutes later three or four bats came out, but all went back in again. Ten minutes later it was rather dark and bats were flying in and out. Ten minutes later still it was dark and I heard no more squeaking. Turning my light into the eaves revealed four bats that took flight. These were the last, for I looked into the attic and no others remained. The flight lasted about one-half hour in all.

Parasites noted on *Glossophaga* include only winged and wingless flies. Once three to eight small winged-flies to each bat were seen in a colony of seven bats. This was the heaviest infestation noted for this species.

Colonies of *Glossophaga* usually contain animals of both sexes. Breeding seems to continue over much of the year. Pregnant females (each with one embryo) were taken on March 12, 19, 29, April 1, 2 and September 24. On November 5, young were taken with the mothers and a young bat was found on November 18.

Anoura geoffroyi lasiopyga (Peters)

Geoffroy's Tailless Bat

Specimen examined.—State of Vera Cruz, 1 (U.S.N.M. 144470).
Additional record.—Texolo [= Teocelo] (Sanborn, 1933:27).

Hylonycteris underwoodi Thomas

Underwood's Long-tongued Bat

Specimen examined.—One from 15 km. ENE Tlacotepec, 1500 ft.
Additional record.—Metlác (G. M. Allen, 1942:97).

This is a rare species, known from but a few localities. Our single specimen was taken 15 kilometers east-northeast of Tlacotepec on December 12, 1947, when "Four or five of these little bats were found in a cave in the barranca here. The cave is only about four feet in diameter. . . . The entrance is about two by three feet. The bats were behind a ledge, clinging to the highest, darkest place. They were shy, and all but one escaped. Beneath their resting place was a pile of guano about three inches high by six in diameter. There were several pits of jobo plums on the pile, showing that some of this fruit is taken to the cave to be eaten. The cave is in a cliff of conglomerate rock, about 40 feet high. There is a stream of water below the cliff. The cave is about eight feet above the water. No parasites were found on the specimen taken. It is a male with small testes."

Leptonycteris nivalis nivalis (Saussure)

Long-nosed Bat

Specimens examined.—Twenty-nine from 3 km. W Boca del Río, 25 ft.

Additional records.—Pico de Orizaba, type locality (Saussure, 1860:492); "Veracruz" (Ward, 1904:653).

A colony of approximately 200 individuals of *Leptonycteris* was found in a dark, archlike tunnel in the basement of the old hacienda on the arid coastal plain, three kilometers west of Boca del Río on September 28, 1947. When disturbed, the bats retired to a dark room on the floor above. Also present in this room were about 50 individuals of *Artibeus jamaicensis* and a few vampires, *Desmodus rotundus*. The specimens collected were all parasitized, each bat having from three to 10 small winged-flies and several mites. Six of the females taken were pregnant, each having a single large embryo. The testes of the males were small.

Carollia perspicillata azteca Saussure

Seba's Short-tailed Bat

Specimens examined.—Total 79: 35 km. NW Tuxpan, 2; 25 km. NW Tuxpan, 3; 17 km. NW Tuxpan, 1; 3 mi. NW Nautla, 1; Mirador, 3500 ft., 6; 4 km. W Paso de San Juan, 250 ft., 1; Boca del Río, 10 ft., 1; Río Atoyac, 8 km. NW Potrero, 3; 4 km. WNW Fortín, 3200 ft., 28; Sumidero, 1150 m., 2; Alvarado, 10 ft., 12; Río Blanco, 20 km. W Piedras Negras, 400 ft., 3; 2 km. N Motzorongo, 1500 ft., 1; 3 km. E San Andrés Tuxtla, 1000 ft., 1; 5 km. S. Catemaco, 1; 14 km. SW Coatzacoalcos, 100 ft., 3; Achotal, 1; 35 km. ENE Jesús Carranza, 150 ft., 1; 20 km. E Jesús Carranza, 300 ft., 6; 35 km. SE Jesús Carranza, 400 ft., 1; 38 km. SE Jesús Carranza, 500 ft., 1.

Additional records (Hahn, 1907:112, unless otherwise noted).—Cuesta de Don Lino, near Jalapa (Ward, 1904:653); Jalapa (Ferrari-Pérez, 1886:128, probably this species); Río Tesechoacan, near Pérez, assumed to be the type locality; Jaltipan; Buena Vista.

This bat is one of the most abundant of the fruit bats that occur in Veracruz. Other abundant species are *Glossophaga soricina* and *Artibeus jamaicensis*. *Carollia* is primarily a cave-living species, although, on February 1, 1947, 14 kilometers southwest of Coatzacoalcos, and on December 15, 1947, 4 km. W Paso de San Juan, individuals were found in pipes beneath railroad embankments and a paved highway. On March 19, 1947, 20 kilometers east of Jesús Carranza, a colony was found in a hollow tree. In 1948, 38 kilometers southeast of Jesús Carranza, several small colonies were found in hollow trees in relatively open jungle.

Four kilometers west-northwest of Fortín, on March 27, 1946, Mr. Daniel Rabago, the engineer in charge of the power plant of the Moctezuma Brewery, guided me (Dalquest) to some artificial tunnels connecting the Río Metlac with the river to the west. The valley to the west is 120 meters higher and water is piped from a dam through an artificial tunnel and down to the power house. The main tunnel is 200 meters long, about seven feet high and seven wide, and is constructed through soft, shalelike rock. Midway, there is a side-tunnel, about 50 meters in length, similar to the main tunnel but containing no pipes or artificial lights. It opens on the side of the hill and the opening is barred with grill work. In this side-tunnel we found two bats. The one captured was a female of *Carollia.* About 200 meters farther downstream another tunnel penetrates the hillside. This is a slotlike tunnel about four feet wide and 25 feet high. At the west end of this tunnel, many swallows were nesting. From the center to the eastern end we found many individuals of *Carollia.* They were clinging either singly or in small groups. Perhaps 50 were present in the tunnel. It was so high that it was necessary to shoot the bats. I killed 10 of *Carollia* and one *Glossophaga* that were in a condition good enough to save. All of the specimens taken were females.

At the same locality, on April 1, 1946, one cave was reached only after considerable work with the machete. Only vampires were present. Entrance to the other cave was gained by sliding on our backs, down a 20 foot-long tunnel inclined at a 30 degree angle. The tunnel was about two feet high by six feet wide. The cave, about 50 feet wide and 70 to 80 feet long, is 12 to 20 feet in height. Falls of loose rocks and large boulders partition the space in the cave, and the floor is littered with angular pieces of limestone from a few inches to several feet in diameter, and is slippery because of bat guano, with many deep pits among the rocks. Great care is required in moving about in the cave. Probably about 100 individuals of *Carollia* were present of which I caught ten. All were males. The bats tended to hang together, though not in clusters, from low places in the ceiling. In pursuing them about the cave I discovered that their favorite perches were places where the roof was five to six feet high, usually near the side of the main room or under projecting ridges of large boulders. An interesting companionship between a *Carollia* and a *Natalus stramineus* was noted here, and is described in the account of *Natalus stramineus.*

Stomachs of *Carollia* taken at the Río Atoyac, eight kilometers northwest of Potrero, held a semi-liquid mass of yellow pulp, probably of the wild sweet-lemon or wild orange, both of which were ripe and common in the vicinity. One bat was taken in a banana-baited mouse trap set beside the trail in dense jungle at the above mentioned locality on February 25, 1946. Another was taken here in a trap set on the ground on March 25, 1946. Another was taken in a trap set on the ground on March 6, 1946, at the same locality, and under rather unusual circumstances. A line of traps baited with bananas had been set along the river for *Oryzomys.* One trap was set under an overhanging bush. The foliage of the bush formed a dense arch with a doughnut-shaped cavity a foot in diameter extending in a circle around the bush. The distance across the circle, from one side to the other, was about three feet. I broke an opening, about seven inches across, through the foliage and set a trap in the cavity. The following morning the trap held a *Carollia.* The bat must have smelled the banana, and entered through the small opening, landed on the ground beside the trap, and been caught while feeding on the bait. Seemingly this species has an excellent sense of smell.

On March 11, 1946, at the same place I noticed that in a bunch of sweet rotan bananas that we had hung in the shade of an orange tree a few meters from camp, a section about 20 millimeters in diameter had been eaten from the skin of a ripe banana and through this opening a section of banana 50 millimeters in length and 25 millimeters in diameter had been eaten out. The upper and lower ends of the cavity were smoothly concave, indicating that the center core of the banana had been eaten more extensively than the surrounding pulp, perhaps because it was more tasty or because it was more easily available to the bat. This banana was left on the stalk, and a mouse trap baited with banana was tied just above it. The next morning the trap and the banana previously eaten were undisturbed, but part of another ripe banana in the clump had been eaten. All bananas except this one were removed from the stalk, and the banana-baited trap was tied to it. Next morning a male *Carollia* was in the trap.

The following observations were made at a place two kilometers north of Motzorongo and entered in my field notebook on December 10, 1946. "For three nights now I have been watching fruit bats over a pool of water in the arroyo here. Three or more species are present, as well as *Saccopteryx bilineata* and *Myotis volans* [proved to be *Eptesicus propinquus*] and probably other bats. One small fruit bat is especially common, but only last night was I able to get one, and found it to be this species (*Carollia perspicillata*). . . . [Bats of this species] will hit a wire strung over the surface of the pool, and even fall into the water, but they are able to take off again. I caught one by setting banana-baited mouse traps tied to branches in their favorite resting tree. The specimen is a male with testes four by two millimeters. Two small winged-flies were the only parasites noted."

Bats of this species arrived at the pool shortly after it was completely dark, and some were present two hours later. They come in at a height of three to five feet making a swish like a bullet.

Possibly this is the end of a dive down to the pool from a higher level. They begin to flap their wings when within 10 feet of a bush or low-hanging branch of a tree on which they intend to land. The swish of their arrival, followed by a flapping, is easily heard and the air that is moved makes leaves and ferns rustle. It is my impression though that they never, or at least rarely, strike things with their wings. In this they are unlike some fruit bats. Arriving under a bush, they seem to drift up into it. This upward movement and alighting is in complete silence, in contrast with their noisy arrival. They usually rise two or three feet, until well into the network of vines and branches, before alighting.

When frightened, they leave their perch almost as silently as they arrive, but the wind from their wing beats causes the leaves to rustle. Once outside the bush, there is a great flapping; often they circle or spiral upwards, before flying off to a perch in another bush but sometimes leave with a swift, rushing sound.

Carollia is rarely parasitized. The only parasite noted was a species of small, winged fly. Four kilometers west-northwest of Fortín, on March 29, 1946, a bat that seemingly died from wounds received from another bat was clinging to the wall of a cave about six feet from the ground. It had not been dead for more than a few hours when found, for it was not at all decayed. The nose-leaf and skin of the muzzle had been torn off.

Males and females often are in separate colonies. Most of the females found between March 6 and March 29 were pregnant. Four taken on December 11 had swollen uteri but no recognizable embryos.

Carollia castanea subrufa (Hahn)

Allen's Short-tailed Bat

Records (Hahn, 1907:115).—Mirador; Otatitlán; Coatzacoalcos; Minatitlán; Achotal.

Hahn (1907) lists both *Carollia perspicillata azteca* and *Carollia subrufa* from localities in Veracruz. When our large series of *Carollia* from Veracruz is compared with his descriptions of *C. perspicillata azteca* and *C. subrufa,* and with a specimen from Veracruz identified by him as *C. subrufa,* we are unable to recognize two species. No specimen seen from Veracruz is so large as the largest *C. p. azteca* nor smaller than the smallest *C. subrufa* listed by Hahn from the general latitude of the localities of occurrence in Veracruz. Most of our specimens resemble the larger individuals of *C. subrufa* and the smaller individuals of *C. p. azteca* in size, both as to external measurements and cranial measurements. Light, heavy and intermediate dentition is found in our material. Consequently all of our material from Veracruz is referred to the earlier named *C. perspicillata azteca.*

Sturnira lilium parvidens Goldman

Yellow-shouldered Bat

Specimens examined.—Total 16: Hacienda Tamiahua, Cabo Rojo, 4; 12½ mi. N Tihuatlán, 300 ft., 11; 5 mi. S Catemaco, 1.

Additional record.—Mirador (Hershkovitz, 1949:442).

The sixteen specimens from Veracruz and the one specimen available from Tabasco (5 mi. SW Teapa) are dark as are seven of the 13 specimens available from Guatemala. The two specimens from the state of Puebla and many of the specimens from western México (Jalisco, Nayarit, and Chihuahua) are notably paler. But there is variation in each lot. Also there are two or more color phases. No consistent geographic variation in the skulls can be correlated with the dark color of the specimens from eastern Mexico. Therefore the specimens from the state of Veracruz are here referred to the subspecies *S. l. parvidens.*

Sturnira ludovici Anthony

Anthony's Bat

Record.—Mirador (Hershkovitz, 1949:442).

Chiroderma villosum jesupi J. A. Allen

Isthmian Bat

Records.—Presidio (G. M. Allen, 1927:158); Achotal (Sanborn, 1936:103).

Handley (1960:466) regards the name *Chiroderma isthmicum* Miller, 1912, previously used for this kind of bat, as a synonym of *Chiroderma jesupi* J. A. Allen, 1900, which is only subspecifically distinct from *Chiroderma villosum* Peters 1860.

Artibeus jamaicensis jamaicensis Leach

Jamaican Fruit-eating Bat

Specimens examined.—Total 154: 5 mi. S Tampico, 3; Hacienda Tamiahua, Cabo Rojo, 1; Tlacolula, 60 km. WNW Tuxpan, 6; Potrero Llano [= Potrero del Llano], 350 ft., 4; 12½ mi. N Tihuatlan, 300 ft., 2; 3 mi. NW Nautla, 1; Teocelo, 4000 ft., 1; 3 km. W Boca del Río, 10 ft., 34; Boca del Río, 10 ft. 30; 5 km. SW Boca del Río, 3; 8 km. NW Potrero, 1700 ft., 16; 5 km. N Potrero, 1500 ft., 17; 7 km. NW Potrero, 1500 ft., 3; Potrero Viejo, 5 km. W Potrero, 2; Cueva de la Pesca, Potrero, 650 m., 8; Cautlaupan [= Cuautlapan], 4000 ft., 1; Sala de Agua, 1500 ft., El Maguey, 8 km. S Potrero, 13; Río Blanco, 20 km. W Piedras Negras, 400 ft., 2; 5 mi. S Catemaco, 1; 38 km. SE Jesús Carranza, 500 ft., 6.

Additional records (Andersen, 1908:267).—Plan del Río; Mirador; Tuxtla [probably Santiago Tuxtla].

This is one of the most common fruit bats of Veracruz, and is probably the one most often seen. It is principally a cave-living species, but on the coastal plain, where caves are scarce, it lives in crevices in trees and even in wells.

This species was found in a limestone cave in a beautiful tropical setting at El Maguey, Sala de Agua, eight kilometers south of Potrero, on February 10, 1946. Approximately 200 feet in from the mouth, further progress is prevented by a swift river that, I was told, emerges from the mountain five kilometers away. The cave varies from 10 to 30 meters in width and six to 10 meters in height. The floor is damp earth with many boulders, but walking is not difficult. Bats were found only far back in the cave, on the very banks of the river. There many were heard and seen in flight. They were quite shy, and only once was a cluster seen. It was made up of perhaps a dozen individuals in a rough pit, two feet in diameter and one foot deep, in the roof of the cave. They flew before I could get a shot. Most of the other pits in the roof contained one, two or three bats each. They hung head downward, some against vertical rocks and others free. Sixteen were taken by shooting. Several of those killed continued to cling to the rock and had to be dislodged by throwing stones; others could not be reached at all.

For such heavy, stocky bats they fly very well. They make an audible swish as they fly, but they maneuver well and perch skillfully. No guano was noted on the floor, but numerous cores of the jobo plum were strewn about, obviously by the bats. The core is about the size and shape of a peanut without the constriction, and is not unlike a peanut shell in texture. The fresh fruit has a pulp layer one-quarter inch thick over the core.

About 75 bats were seen. Others may have been perched in the cave, back, over the river, or at least where I could not see them. Although they flew out over the river, none was actually seen perched there.

Five kilometers north of Potrero, in a beautiful tropical forest, on February 11, 1946, another large colony was found in a small cave on a limestone hillside. The mouth of the cave is an arch about 30 feet wide and 20 feet high. The floor is a jumble of fallen boulders, and the rejected cores of the jobo fruit almost cover some of the rocks. Several short, crevicelike tunnels lead from the entrance room, but no bats were found until I went through a small opening into a hemispherical room about 50 feet across. Bats by the thousands were present. Clusters of hundreds clung to the highest parts of the ceiling, and others clung to the walls, within reach of my hand. The wings of thousands in flight made a loud, flapping whir. In the center of the floor was a tarlike pool—the excrement of vampire bats. I estimated that 100 vampires and 2000 Jamaican fruit-eating bats were present. The two kinds remained separate, the vampires in the center of the ceiling and the fruit-eating bats elsewhere. The two kinds looked much alike but behaved differently. The vampires clung with thumbs as well as the feet, and moved about on the rock. The fruit-eating bats clung with the feet alone and did not shift their positions except to fly. The vampires crept into crevices to escape but the fruit-eating bats simply flew about; a few of the latter went into the entrance room, but most of them flew from perch to perch. All of the fruit-eating bats were adult, whereas the vampires were of several ages. The odor in the cave was that of ammonia from the vampire's pool. No droppings of fruit-eating bats were noted, although the cores of the jobo fruit were common on the floor. The cores were less common than in the entrance room, however.

These big bats are noisy feeders. On many occasions I watched bats that probably were this species, but that could not certainly be identified. The following notes, however, are accompanied by specimens, and the identity of the animals is known. On the Río Blanco, 20 kilometers west of Piedras Negras on October 6, 1946, where actual specimens were taken and saved, the bats fed on the jobo plums by flying into the clusters of fruit at the ends of the branches, with a great fluttering of wings and rustling of branches and leaves. Bats picked the fruits and flew away with them. The bats often alighted in the plum tree to eat the fruit; the pits dropped into the water below. Mostly, however, the bats flew to another tree or bush to eat. One of their favorite perches was in a bush not more than six feet high. There was a dense mass of vines and dead branches under the tree that formed a network, which, on first inspection, seemed to be too dense for such a large bat to penetrate. Yet they fluttered in, and clung to a branch, twig or vine. When a light was turned on they were seen hanging head down, each peering at me over the top of a fruit that was held in the mouth and by means of both wrists. When close to the bats I could see a dull red glow from their eyes. They were shy and usually flew as soon as the light struck them. Rarely would they wait until I started to put my gun to my shoulder. Only one was shot. Evidence showed that they alighted on the ground to feed on fallen fruit, although I was never able actually to see the bats on the ground. Their fluttering was noisy. One time, after watching them for about an hour a tropical downpour set in. Then the bats ceased to flutter, but were about as active [otherwise] as ever. They flew to the plum tree and returned, but were as silent as owls.

On October 10, 1946, five kilometers southwest of Boca del Río, a small colony was located among the roots of a tall strangler-fig. The roots were rather flat, and extended 25 feet up in the air. There were numerous bits of figs and some pits of jobo plums beneath the crevices. The bark at the tops of the crevices was scratched and showed the reddish inner layer. This was mainly a night resting place of fruit-eating bats, but about five were hidden in crevices. The one taken for a specimen was an adult with large testes.

On another night I put a chair under the most used crevices in the tree and noted that the bats eyes glowed red in the beam of a flashlight. Two males with testes of moderate size and a non-pregnant female were shot. Each dropped wild figs that were green in color and rather tough but seemingly ripe.

At Potrero Viejo, 1500 feet elevation, on the evening of February 14, 1946, Mrs. Leora Forbes heard that bats were occasionally seen in one of the rooms of the hospital of the Potrero Sugar Company. We visited the hospital about 10:00 p. m. and the intern told us that approximately 10 bats had been killed in one room a few nights earlier. The bats were killed because they annoyed the patients, who mistook the bats for vampires, and because the bats hung from crevices in walls near the ceiling to eat fruit. They left considerable debris. Only two bats were present when we examined the room—both of this species. They were collected and prepared. The walls were well marked with the pulp of a red fruit. The pulp contained a number of seeds. From the appearance of the wall, one could wrongly imagine that overripe raspberries had been thrown against it and left to dry. The bats visited a single dimly-lighted room and the corridor leading to it.

Embryos were found in this species on February 10 and 11. Young bats were seen in caves in March and late April. Small winged-flies are the common parasite on this species.

Artibeus lituratus palmarum J. A. Allen and Chapman

Big Fruit-eating Bat

Specimens examined.—Total 9: Tlacolula, 60 km. WNW Tuxpan, 2; 4 km. NE Tuxpan, 1; 12¼ mi. N Tihuatlán, 300 ft., 3; Mirador, 3500 ft., 1; 4 km. WNW Fortín, 3200 ft., 1; 20 km. E Jesús Carranza, 300 ft., 1.

Additional record.—Paso de Ovejas (Villa, R., and Jiménez, G., 1962:394).

This species was rare in the state of Veracruz, and our observations on it are scanty. Its habits seem to differ somewhat from those of *Artibeus jamaicensis*.

In early April, 1946, four kilometers west-northwest of Fortín, large bats were noted on several occasions along a limestone cliff. In spite of numerous attempts to stalk them, the bats were so shy that they flew before I could get within gunshot range. On April 5, however, one bat hesitated too long, and I shot it. It was hanging in the open, shaded but in full daylight, from a rough place on the ledge above a pile of "droppings." These, like the reddish-brown, flattened, ovoid "droppings" so often found beneath the roosts of *Artibeus* are actually balls of the skins or rinds of fruit. The bats chew the fruit while holding it between their wrists. Both pulp and rind are taken into the mouth. The pulp is swallowed and the rind is rejected in the form of a small pellet. In a few hours these pellets take on the characteristic reddish-brown color.

At Hacienda Mirador the skull of the big fruit-eating bat was obtained from a mummified animal that had been saved by Mr. Walter Grohman.

Twenty kilometers east of Jesús Carranza, while my companions and I were going along a jungle trail about midnight with a hunting light, a pair of tiny, bright-red eyes was seen in a small mango tree, about 15 feet from the ground. A shot brought down a specimen of *A. l. palmarum.*

A male taken on April 5, had enlarged testes, that measured six by eight millimeters. A female taken on February 7, was not pregnant.

Artibeus cinereus phaeotis (Miller)

Gervais' Fruit-eating Bat

Records (Davis, 1958:165, unless otherwise noted).—Minitatlán; Jesús Carranza; Río Solosuchil; Achotal (Sanborn, 1947:223—as *Dermanura fucundum*).

Artibeus toltecus (Saussure)

Toltec Fruit-eating Bat

Specimens examined.—Plan del Río, 1000 ft., 2 (catalogue Nos. 2914 and 2915 Texas Agric. and Mech. College; see Davis, 1944:378).

Additional record.—Mirador (Hershkovitz, 1949:449).

Artibeus toltecus is here accorded specific rank, instead of subspecific rank under *A. cinereus,* in accordance with the findings of Davis (1958:166).

Artibeus turpis turpis Andersen

Dwarf Fruit-eating Bat

Specimen examined.—Arroyo Azul, 1.

Additional records.—Plan del Río (Davis, 1958:163); Buena Vista (Andersen, 1908:310).

The trinomen instead of the binomen (*auct.*) is here used because Davis (1958:166) concluded that *Artibeus nanus* Andersen is a subspecies of *Artibeus turpis* Andersen that has one page of priority. Davis (1958:163) regards *A. t. nanus* as occurring only in western México and regards *A. t. turpis* as occupying eastern and southern México. Therefore, Andersen's (1908:310) record of a specimen from Buena Vista, Veracruz, may relate to *A. t. turpis* although recorded by Andersen (*loc. cit.*) under the name *Artibeus nanus.*

Centurio senex Gray

Wrinkle-faced Bat

Records.—Las Vigas (Ward, 1904:653); Mirador (Sanborn, 1949:198); Dos Caminos, km. 354, 4500 ft. (Sanborn, 1949:199); Orizaba (Sumichrast, 1882:203); Minatitlán (*ibid.*).

Family Desmodontidae

Desmodus rotundus murinus Wagner

Vampire Bat

Specimens examined.—Total 70: Tlacolula, 60 km. WNW Tuxpan, 1; 7 km. SW Tuxpan, 1; 10 km. SW Jacales, 6500 ft., 1; 1 km. E Jalacingo, 6500 ft., 3; 3 km. W Boca del Río, 10 ft., 9; 5 km. S Jalapa, 3 (Chicago N. H. Mus.); Boca del Río, 10 ft., 3; Ojo de Agua, 8 km. NW Paraje Nuevo, 9; 8 km. NW Potrero, 1700 ft., 1; 13 km. WNW Potrero, 2000 ft., 4; 5 km. N Potrero, 1500 ft., 12; 4 km. WNW Fortín, 3200 ft., 1; Grutas Atoyac, 2 km. E Atoyac, 1; Cueva de la Pesca, Potrero, 650 m., 1; 7 km. SE San Juan de la Punta, 400 ft., 2; 3 km. W Acultzingo, 7000 ft., 8; 3 km. E San Andrés Tuxtla, 1000 ft., 4; Achotal, 3 (Chicago N. H. Mus.); 20 km. E Jesús Carranza, 300 ft., 3.

Additional records.—5 km. N Jalapa (Davis, 1944:378); Portrero Viejo [=Potrero Viejo] (Hooper, 1947:43).

The vampire bat (*vampiro* in Spanish) in Veracruz is typically a cave-dweller but lives also in crevices in rocks, hollow trees, old wells and abandoned buildings. Probably these nontypical places are chosen when suitable caves are lacking. On the arid coastal plain, where caves are scarce, several colonies were found in old wells.

Five kilometers north of Potrero on February 11, 1946, approximately 100 vampires were found in a cave along with about 200 individuals of *Artibeus jamaicensis*. The vampires were in a cluster in almost the center of the ceiling. There was a pool of black, tar-like, digested blood on the floor beneath them. A strong odor of ammonia came from this pool. The vampires clung to the roof with both feet and the thumbs, and braced their bodies and raised their heads, peering at the intruder.

I fired twice into the cluster, and to my great disgust, the bats that fell plopped into the stinking pool, and were almost covered by the muck. The others flew and were lost to sight in the mass of wildly flying Jamaican fruit-eating bats (*Artibeus jamaicensis*). I then noted the vampire bats entering a narrow cleft into which they crept, far back out of my sight.

All of the fruit-eating bats were adult, but the vampires were of several sizes; some were almost full-grown and others must have been only a few days old. Some adult females were pregnant.

At the cave, "Ojo de Agua," eight kilometers northwest of Paraje Nuevo, on February 12, 1946, in one of the upper levels of the cave, where it is relatively dry, a companion and I found approximately 50 vampire bats hanging in a pit, a meter wide and a half a meter deep, in the limestone roof of the cave. On the floor four meters beneath, was a circular pool of excrement and blood. I fired three shots into the cavity and then quickly leaned forward over the pool. Falling bats bounced off my head and shoulders and only a few fell into the pool. The remainder of the bats flew farther into the cave and took refuge, singly or in small numbers, in pits and crevices, from which they flew with a loud swishing of wings at my approach. They were extremely agile and able to creep rapidly in crevices. One individual lay on a flat rock with its body on a horizontal plane. From this position it jumped, did not fly, horizontally to another rock eight inches away. It made this jump quite gracefully and landed with wings closed. It scuttled over the rock and into a crevice. We explored the cave rather thoroughly but found no other species of bats. The lower levels of the cave are cold and damp. About 75 meters from its mouth there is a rushing underground river and the air is full of spray. No bats were found in this part of the cave.

On the Río Blanco, 20 kilometers west of Piedras Negras, on September 30, 1946, we were told of a house with an old well, where bats were abundant. On peering into the well, which was perhaps 30 meters deep, about six vampires were seen clinging to the wall, and others flitted about beneath them. Attempt was made to catch some in a net, but they took refuge in small holes cut for footsteps. Then an old five-gallon can was filled with corncobs and red-hot coals; a handful of chili peppers was added. This was lowered on a long rope, and the well was filled with smoke. The bats could be heard fluttering about, but although my companion and I waited for more than an hour, none was driven out by the smoke.

It is well known that vampires feed on blood but it is not so well known that the animals preyed upon by the bats vary from place to place. In the state of San Luis Potosí, for example, vampires commonly attack chickens. In Veracruz, they seem never to do so, and people were unwilling to believe that they did so elsewhere. In Veracruz, the following notes were made at the Río Atoyac, eight kilometers northwest of Potrero, on March 21, 1946. "We have been in this area about six weeks now. I have examined many oxen, horses and burros for evidence of vampire attacks. No oxen bitten by vampires were seen by us, and relatively few horses with scars of vampire bats were noted. Almost all the burros seen had scars or wounds inflicted by these bats." Most of the bites were on the sides of the neck or shoulders. Some freshly bitten animals were noted. Numbers of flies buzzed about the wounds. Hornets were chewing on one fresh wound but the burro seemed to be unconscious of their bites or the presence of the flies. Masters of the burros seemed to be equally indifferent and I saw no evidence that they ever treated any of the bites. Seemingly the animals suffer no permanent ill-effects from a moderate number of vampire bites. Certainly, though, the wound is open to the attacks of parasitic insects and infection.

Resident people know of this bat from its attacks on animals, but most persons do not recognize the bat as a vampire when they see it. The common, large, Jamaican fruit-eating bat, *Artibeus*, was often confused with the vampire.

The vampire seems to have no regular breeding season. A few young bats, in various stages of development, as well as pregnant females and non-pregnant, non-lactating, females, were found in each large colony of vampires examined. The common parasite on the vampire bat is a small, winged fly, which, upon superficial inspection, seems to be the same as the fly found on *Artibeus jamaicensis*.

Diphylla ecaudata Spix
Hairy-legged Vampire Bat

Specimens examined.—Total 8: Ojo de Agua, 8 km. NW Paraje Nuevo, 1; 7 km. NW Potrero, 1700 ft., 3; 7 km. NW Potrero, 1500 ft., 2; 13 km. WNW Potrero, 1700 ft., 1; 5 km. S Potrero, 1700 ft., 1.

Additional record.—Orizaba (Málaga and Villa, 1957:530).

The name *Diphylla ecaudata centralis* Thomas, previously used for this bat, has been shown by Burt and Stirton (1961:37) to have been based on material that is inseparable from that on which the name *Diphylla ecaudata* Spix was based. *D. e. centralis*, therefore, is a synonym of *D. ecaudata*. The latter stands now as a monotypic species.

The hairy-legged vampire is a cave-living species. In Veracruz we found it associated with the vampire, *Desmodus*. The principal differences noted in the habits of the two kinds are that *Diphylla* is relatively solitary, prefers darker retreats, and leaves no pool of digested blood beneath its roosts; instead of a pool, we found only dry, brown stains beneath the perches of *Diphylla*.

Family Natalidae

Natalus stramineus saturatus Dalquest and Hall

Mexican Funnel-eared Bat

Specimens examined.—Total 110: 14 km. NW Tuxpan, 10; 9 km. NW Tuxpan, 7; 13 km. WNW Potrero, 2000 ft., 10; 4 km. WNW Fortín, 3200 ft., 1; 3 km. E San Andrés Tuxtla, 1000 ft., 82.

Additional records.—Jalapa (Ward, 1904:638); Mirador (Miller, 1902:400); Tilapa (Sumichrast, 1882:202); San Andrés Tuxtla (Miller, 1902:400).

The specific name *stramineus* instead of *mexicanus* is used here in conformance with the findings of Goodwin (1959).

The specimens from the northern part of Veracruz (14 km. NW Tuxpan, and 9 km. NW Tuxpan) differ notably in coloration from the specimens in the south. In coloration, specimens from the northern part of Veracruz (also specimens from Tamaulipas: 6 km. SW Rancho Santa Rosa; 16 mi. W, 3 mi. S Piedra; 14 mi. W, 3 mi. S Piedra; 20 mi. N, 3 km. W El Mantie; 8 km. NE Antiguo Morelos) are closer to *N. s. mexicanus* than to *N. s. saturatus* but resemble the latter in longer skull and longer maxillary tooth-row. Consequently all of the specimens from Veracruz are referred to *N. s. saturatus.*

This species was abundant in a lava cave three kilometers west of San Andrés Tuxtla (see account of *Pteronotus psilotis* for description of cave). About 200 feet from the entrance of this cave, a geologic fault separated the cave into upper and lower parts. A rock face or wall now interrupts the otherwise relatively level cave, and above this wall the cave continues for an undetermined distance. In the upper part of the cave we found thousands of individuals of *Pteronotus psilotis* and *Pteronotus davyi.* In the lower cave we found *Pteronotus parnellii, Glossophaga soricina, Carollia perspicillata, Desmodus rotundus* and *Natalus stramineus.* The numbers of individuals of *Natalus* in the cave varied greatly from day to day, reaching a low point of two bats on January 2, 1948, and a high of about 300 on January 10. On all days except January 10, all the funnel-cared bats (*Natalus*) in the cave were found in a single area, from five to 25 feet in front of the barrier wall. Here the cave was cool and damp. This was the part of the lower cave, fartherest from the mouth, where it was cool; above the rock wall in the upper cave the air was hot. The bats were from four to 15 feet from the floor. Some hung free from the ceiling and some were in cracks and crevices, but most of the bats clung to the vertical walls of the cave.

On January 10, when there were about 10 times as many funnel-eared bats in the cave as were ever noted before, approximately 50 of them were hanging from the roof in a wide part of the cave, only about 100 feet from the entrance. The bats had chosen a part of the room from which the lighted entrance of the cave was hidden by a bend.

The funnel-eared bats were not at all shy; individuals were usually plucked from the wall as they twisted about, watching the light, although on most occasions, a few took flight when an observer approached. It was noticed that the more bats there were in the cave, the more shy they were. When bats too high to be reached by hand were shot, the other bats hanging nearby rarely took flight at the sound of the gunshot.

All of the females taken in early January were non-pregnant and they were not lactating. The testes of the males were small. This species was commonly parasitized by a small winged-fly; three to seven flies were present on almost every bat taken.

On April 1, 1946, a *Natalus* was captured four kilometers west-northwest of Fortín in a cave occupied by a colony of *Carollia perspicillata* (see account of *Carollia* for description of the cave). The *Natalus* was in a most unusual association with one *Carollia*. "One *Carollia* hung from the center of the roof, usually in the center of a large open place. I climbed up on a boulder and shined my light directly on this bat. When I did so I noticed a tiny yellow bat clinging to the roof beside the *Carollia*. Unfortunately the two bats were too high to reach with my net, and I was too close to them to shoot. Time after time I would get within a few feet of them before they would fly. Always the two clung together and flew together. They did not associate with the other *Carollia* in the cave—by this I mean that they stayed near the ceiling in the open part of the cave. The presence of the little yellow bat enabled me to . . . [locate] the *Carollia*, and I could always find the little yellow bat by looking for a *Carollia* hanging from the open ceiling. Being unable to catch these bats with my net, I was forced to shoot them. The one shot brought down both. The little yellow bat is a *Natalus mexicanus* [= *N. stramineus*], a female. I examined the *Carollia* very carefully and found it to be a male, as were all the other *Carollia* in the cave, and to differ in no appreciable way from all the other *Carollia* present. The relationship between these two species is difficult to explain, but I had them under observation long enough to be sure that it existed."

No *Natalus* was observed feeding or hunting. The species is insectivorous, for one specimen taken in a cave at night, where it, like many other species of bats, had retired to rest after the evening hunt, had its stomach crammed with insect remains.

The male of *Natalus* possesses a glandular structure on the head not noted in any other kind of bat. It lies between the skin and the skull. It is attached to the skin between the eyes, and extends backwards from the point of attachment, becoming broader and thicker, and forming a cap over the top of the skull (see Dalquest, 1950:438-440). Nowhere is it attached to the skull. This glandular structure is absent in the female.

Family Vespertilionidae

Myotis velifer velifer (J. A. Allen)

Cave Myotis

Specimens examined.—Total 155: 4 km. E Las Vigas, 8500 ft., 143; 5 km. N Jalapa, 4500 ft., 1; 4 km. WNW Fortín, 3200 ft., 3; 3 km. SE Orizaba, 550 ft., 8.

Additional records.—Las Vegas [probably = Las Vigas] (Ward, 1904:647); 5 km. N Jalapa, 4500 ft. (Davis, 1944:378); Xuchil (Miller and Allen, 1928: 91); Orizaba (Miller and Allen, 1928:91).

This is a species of the upland, and it is common on the Mexican Plateau. It enters the upper humid division of the Tropical Life-zone in central Veracruz, and was taken as low as 3200 feet elevation.

In the daytime *Myotis velifer* retires, singly or in large colonies, to secluded retreats in buildings, crevices in cliffs, and caves.

Three kilometers southeast of Orizaba, on December 20, 1946, a colony of about 50 of these bats was living in crevices in the roof of a powerhouse where the machinery roared so loudly that persons had to shout to make one another hear. A few individuals of *Molossus ater* also lived there. We were told that the bats lived there the year around, emerging from the crevices in the evening. When they returned, they did not enter the crevices immediately, but hung from the ceiling in groups for an hour or so. The ceiling is about five meters high. By using a ladder, we caught five of the bats.

On March 28, 1946, along the face of a limestone cliff here, four kilometers west-northwest of Fortín, several hanging-up places of fruit bats were discovered but no bats were present. Near one of the resting places of fruit bats the droppings of an insectivorous species were noted. The cliff here is composed almost entirely of fossil trees of some palmlike species that had annual growth rings several inches thick. Many of these trees were hollow in life, the central cavity being two to four inches in diameter, circular in shape, and several feet in length. One of these cavities extended almost vertically from an overhanging wall. It was from this cavity that the droppings seemed to originate. With the aid of a ladder I got directly under the opening, and reflected a beam of light into it revealing two small bats about six feet from the opening and 12 feet from the ground. By means of a wire hook at the end of a bamboo pole one of the bats was obtained and proved to be *Myotis velifer velifer*.

This species is insectivorous. Like many bats, it retires to rest for an hour or so after hunting. About one hour after dark, near a waterfall, five kilometers north of Jalapa, at 4500 feet elevation, one of these bats was taken from a cavelike opening where no bats were found in the daytime. Two red-collared swifts were also caught here.

Myotis velifer emerges from its retreats, and begins its evening feeding rather late. The following notes were made on April 6, 1946, four kilometers west-northwest of Fortín. "Tonight I went hunting bats over a pool in the river. . . . In the past I have seen many *Myotis* here. They appear too late in the evening to shoot so we have twice strung wire over the pool. On each occasion, six to ten *Myotis* have struck the wire, often giving it a hard blow. . . . [but] none has been captured or, so far as I know, knocked into the water. In desperation tonight I focused the beam of my flashlight across the pool, and aiming down this lighted part, I pulled the trigger whenever a bat crossed the beam. After wasting several cartridges I finally . . . . killed another of these large *Myotis*, this time a male."

In the tropics this species probably does not hibernate. We were told that the colony near Orizaba is active the entire year. In the uplands, *Myotis velifer* hibernates in large colonies. Approximately four kilometers east of Las Vigas many colonies of this species lived in the lava caves on the small volcanic peak called Volcancillo, and in the lava flows below the peak. These caves were first visited on January 11, 1949. The temperature of the outside air was low but, save in early morning, above freezing. The caves were entered through holes. These are similar to those formed by the collapse of parts of the roofs of lava tubes. Inside these caves, the ceiling, walls, and floor are cool and wet. The floors are usually strewn with angular blocks of lava that have fallen from the ceiling, which make walking hazardous. The basaltic lava is fractured and soft; in places it is more scoria than basalt. The caves vary from simple tubes too small to permit a person to stand upright, to great halls where the beam of a flashlight scarcely reaches to the ceiling. *Myotis* was rarely less than 10 feet above the floor of the caves. Most individuals hung from damp places; many of the bats were "frosted" with droplets of water. Many hung singly; some were in pairs and small groups; most were in groups of 60 to 200. Nearly all of the bats were completely torpid; only a few were active in the cave and flying about. Some groups of bats were noted, day after day, in the same place in the same cave. Some of the bats, at least, remained in one position for more than a week. The torpid bats were dislodged with a pole, and fell to the floor. They were able to squeak, and to writhe about slowly; and after about 10 minutes in the collecting bag they became warm and completely active. *Plecotus mexicanus* and *Pipistrellus subflavus veraecrucis* were also present in these caves.

The breeding season of *Myotis velifer* in Veracruz may be irregular; pregnant females were taken on March 28, December 20, and December 21. The only parasites recorded were from bats taken in the tropical area of Veracruz. These were a flea, a winged fly, and what appeared to be a large louse.

Myotis fortidens Miller and G. M. Allen

Cinnamon Myotis

Specimens examined.—Total 7: Río Blanco, 20 km. WNW Piedras Negras, 3; 20 km. ENE Jesús Carranza, 200 ft., 4.

At our camp on the Río Blanco, 20 kilometers west-northwest of Piedras Negras, 200 feet elevation, we were told that small bats occasionally visited a nearby house in great numbers. We requested the persons who lived in the house to try and capture some of the bats. On the morning of May 13, 1946, two bats were brought to us and we were told that on the previous night the bats had been swarming at the house, alighting on the thatch of one end wall. The two bats brought into camp had been knocked down with a bamboo switch.

I visited the house and found it to be of the ordinary kind, with roof and end walls constructed of thatch of the fronds of the coyol palm. On the ground at the outside base of one end wall there was a considerable quantity of bat droppings, all fresh. We lifted the thatch, a layer at a time, but found no bats. Seemingly the thatch was used as a resting place only at night. The early part of the next night was spent at this house, but no bats were seen and we were told that the bats might not return for several nights. Although the inhabitants were asked to call us immediately when the bats did return, they failed to do so. However, when the bats next returned, on May 29, one of them was knocked down for us with a bamboo switch.

Twenty kilometers east-northeast of Jesús Carranza, on April 13, 1949, several small bats were noted flying about the palm-thatched roof of a house in the early evening. Two of these were shot; both proved to be of this species. On the evening of May 16, 1949, bats were noted about the thatch roof of the same house. They were not disturbed, but the following morning the thatch was lifted and seven bats emerged. Two of these were shot; each was a *M. fortidens.*

The evening flight of this species is rather slow. When hunting they fly from six to 12 feet above the ground, erratically, and with a fluttery wing beat. Those disturbed from their roosts in the thatch, however, came out of their retreats swiftly, and flew off into the jungle at high speed. Two of the seven found in the thatch were about one foot apart; the others were scattered, being 10 feet or more apart. The four specimens taken at this locality were all males.

No ectoparasites were noted on this species. A female from the Río Blanco had an embryo on May 13; one taken on May 29 was lactating.

Myotis keenii auriculus Baker and Stains

Keen's Myotis

Record.—Perote (Hoffmeister and Krutzch, 1955:3). For the nomenclatural history of this subspecies see Findley (1960:16-20).

Myotis thysanodes aztecus Miller and G. M. Allen

Fringed Myotis

Record.—3 mi. ESE Las Vigas (Davis and Carter, 1962:72).

Myotis volans amotus Miller

Long-legged Myotis

Record.—Cofre de Perote, 12,500 ft., the type locality (Miller, 1914:212).

Myotis californicus mexicanus (Saussure)

California Myotis

Record.—Mirador (Miller and G. M. Allen, 1928:160). Seemingly the only record of occurrence in the literature of this species in Veracruz is provided by the holotype of *Vespertilio agilis* H. Allen (Proc. Acad. Nat. Sci. Philadelphia, 18:282, 1866) from Mirador. Miller and G. M. Allen (1928:159) placed *V. agilis* in the synonymy of the earlier named *V[espertilio]. mexicanus* of Saussure (1860:282), the type locality of which is not precisely known. Since Miller and G. M. Allen (1928) seem not to have been able to examine the holotype of *mexicanus* and did not make it absolutely clear that they carefully examined the holotype of *V. agilis,* which they list (1928:160) as in the Academy of Natural Sciences at Philadelphia, some doubt remains as to both the specific and subspecific identity of *Vespertilio agilis* and, therefore, also, whether the species *Myotis californicus* actually was obtained at Mirador.

Myotis elegans Hall

Graceful Myotis

Specimen examined.—One from 12½ mi. N Tihuatlán, 300 ft.

This small bat, recently named (Hall, 1962:163), is known from a single individual caught on September 24, 1961, in a mist net, by Percy L. Clifton.

Myotis nigricans dalquesti Hall and Alvarez

Black Myotis

Specimens examined.—Total 159: 4 km. W Las Vigas, 33; 13 km. WNW Potrero, 2000 ft., 1; 4 km. WNW Fortín, 3200 ft., 1; 2 km. N Motzorongo, 1500 ft., 1; 3 km. E San Andrés Tuxtla, 1000 ft., 7; 38 km. SE Jesús Carranza, 500 ft., 116.

This subspecies was named by Hall and Alvarez (1961:71).

On January 5, 1948, about 12 of these little black *Myotis* were found in a narrow crevice between a brick wall and a wooden door frame at the top of a door in a partly ruined house three kilometers east of San Andrés Tuxtla. Eight were captured, and later that night the crevice was examined, and two bats were seen. The remainder of the colony did not return to the crevice in the following four days. All eight of the specimens taken were males, with small testes well out on the uropatagium. There were two to four small, wingless flies on each bat.

Thirty-eight kilometers southeast of Jesús Carranza numerous small colonies of from 10 to 50 bats were found in small, vertical crevices in limestone cliffs. Some of these crevices were over water. The bats were not far back in the crevices; usually their heads were about an inch from the opening. The bats were tightly packed in the crevices.

In a large limestone cave (for description see account of *Balantiopteryx io*) there were eight colonies of from 50 to 100 bats of this species. The bats were tightly packed together at the bases of stalactites, 10 to 15 feet from the floor of the cave, and were in a constant state of agitation. The bats were in the back part of the cave, 100 to 200 feet from the cave mouth, where it was very dark. All were on the sides of stalactites nearest the mouth of the cave. One or two colonies were not actually seen; they were over pools of deep water and were located by the constant squeaking.

All of the bats taken at this locality on April 10 were females with embryos. Those taken on April 27 had large embryos or were nursing young. No males were found among the 120 bats saved.

A bat of this species was shot in flight two kilometers north of Motzorongo on December 11, 1946. My notes record: "For two nights I have seen a small bat over the pool of water in the arroyo in the jungle here. It would arrive about 25 feet from the ground, make three or four circles about 30 to 50 feet in diameter, and vanish. . . . On the third night it came again, and I dropped it with a lucky shot. I did not find it until this morning. It was this species, *M. nigricans.* Ants had eaten the face somewhat, but the skull is perfect. It is a male with tiny testes—three by four millimeters. Other bats taken at the pool include: *Saccopteryx bilineata, Eptesicus* [*brasiliensis*] *propinquus* and *Carollia perspicillata.*"

Myotis argentatus Dalquest and Hall

Silver-haired Myotis

Specimens examined.—Two from 14 km. SW Coatzacoalcos, 100 ft.

This bat was found only once. On February 2, 1947, my companions and I were paddling a dugout canoe along one of the small, sluggish streams flowing into the Río Coatzacoalcos, 14 kilometers southwest of the city of Coatzacoalcos. I placed my hand on a dead stub, about eight inches in diameter and four feet high, to steady the canoe. The stub was riddled with termite- and other insect-holes, and in one such hole two bats were detected. One was an adult male; the other was a young female. The stub containing these bats was the limb of a tree that probably had fallen into the water. We searched in dozens of similar stubs, eroded by termites and sticking up from the water, but no other bats were found.

Pipistrellus subflavus veraecrucis (Ward)

Eastern Pipistrelle

Specimens examined.—Total 15: 4 km. E Las Vigas, 8500 ft., 14; 30 km. SSE Jesús Carranza, 300 ft., 1.

Additional records.—"Las Vegas [= Las Vigas?], Jalapa" (Miller, 1897:93).

On May 13, 1949, 30 kilometers south-southeast of Jesús Carranza, many bats, apparently all of the family Emballonuridae, as judged by the characteristic manner of flight, were noted here in the evening. About midnight, on this moonlight night, small bats of another kind were noted flying swiftly over the river and sandbar near our camp. After several attempts one was shot and it was an adult, non-pregnant female *Pipistrellus.*

Our largest series of this species was obtained in the caves four kilometers east of Las Vigas. This is the type locality of the subspecies (for description of these caves, see the account of *Myotis velifer*). Here we took bats of this species on January 15, 16 and 17. The bats were in hibernation at this time. They hung singly, well hidden, and usually one or two to a cave. A few were hanging from the walls of the caves, in the open, but the greater number were found under ledges or in holes or crevices in the lava walls of the caves. All were near the floor of the cave; only one was found as high as seven feet from the floor and several were only about three feet from the floor. All were cold and torpid. After being handled, and after remaining about 10 minutes in the collecting bag, the animals became completely active. Several specimens were "frosted" with droplets of water, and seemed to have been hanging in the same place in the cave for some time.

Eptesicus fuscus miradorensis (H. Allen)

Big Brown Bat

Specimens examined.—Total 3: 6 km. SSE Altotonga, 9000 ft., 1; 3 km. E Las Vigas, 8000 ft., 1; Potrero Viejo, 5 km. W Potrero, 1.

Additional records.—Las Vigas (Miller, 1897:100); 5 km. E Las Vigas, 8000 ft. (Davis, 1944:380); 5 km. N Jalapa, 4500 ft. (Davis, 1944:380); 3 mi. E Perote (Hooper, 1957:3); 4 mi. E Perote (Hooper, 1957:3); 1½ mi. S Perote (Hooper, 1957:3); Jico (Miller, 1897:100); Mirador, the type locality (H. Allen, 1866:287); Tuxpango (Miller, 1897:100); Tilapa (Sumichrast, 1882: 201); Cuautlapa (*ibid.*).

The big brown bat is fairly common in the pine forests at high elevations in Veracruz, but is rare in the lowlands. In the pine forest north of the Cofre de Perote, in early November, 1946, one bat was taken six kilometers south-southeast of Altotonga, at 9000 feet elevation, and another three kilometers east of Las Vigas, at 8000 feet elevation. These bats flew rather slowly, in straight lines or large circles, about 25 feet above the ground. The nights were cold, probably about 40 degrees Fahrenheit, and no insects were noted in the air. The bats, however, were remarkably fat.

The one specimen that I obtained from the tropical parts of Veracruz was taken in the village square of Potrero Viejo, 1700 feet elevation. Mr. Dyfrig Forbes obtained permission from the local authorities to shoot on the square. Such a crowd gathered, however, that the bat could not be found after it fell; several days later Miss Marion Forbes found it and it is preserved as a mummy.

Eptesicus brasiliensis propinquus (Peters)

Brazilian Brown Bat

Specimens examined.—Total 16: 4 km. NW Tuxpan, 5; Tuxpan, 2; 12½ mi. N Tihuatlan, 300 ft., 1; Potrero Viejo, 7 km. W Potrero, 1700 ft., 1; Potrero Viejo, 5 km. W. Potrero, 1700 ft., 2; Hacienda Potrero Viejo, 5 km. W. Potrero, 1700 ft., 1; Río Blanco, 20 km. W Piedras Negras, 400 ft., 2; 2 km. N Motzorongo, 1500 ft., 1; Achotal, 1 (14150 Chicago N. H. Mus.)

This small bat is not uncommon in the lowlands of central Veracruz, but the daytime retreat of the species was never found by us.

In the stable yard at Hacienda Potrero Viejo on March 24, 1946, the bats emerged at dusk and flew for some time about a large mango tree, across the wall from the observer and near buildings. Later they came near the stables and flew rather low and excessively fast. One was shot. Two additional specimens were secured here, under similar conditions, on April 4, 1946. On October 24, 1946, a bat of this species flew into the window of a house at Potrero Viejo, and was saved. Another was shot under the overhanging branches of a tall mango tree along the Río Blanco, 20 kilometers west of Piedras Negras, on October 4, 1946, where two days before a *Saccopteryx bilineata* was killed. Another *E. b. propinquus* was shot under the same tree on October 6, 1946. Two kilometers north of Motzorongo, on December 9, 1946, one of these bats was taken by means of a wire strung over the pool in the arroyo. The area was jungle, with dense vegetation all about. Over the pool were vines and tall trees. Two *Saccopteryx bilineata* were obtained at the pool, and fruit-eating bats of several kinds were present. The Brazilian brown bat hit the wire about one-half hour after dark, and fell into the water with a splash. It swam well and swiftly.

Lasiurus borealis teliotis (H. Allen)

Red Bat

Record.—Peñuela (Handley, 1960:472, regards *Lasiurus borealis ornatus* Hall, 1951, with type locality at Peñuela as indistinguishable from *Lasiurus borealis teliotis* H. Allen, 1891).

Lasiurus seminolus (Rhoads)

Seminole Bat

Record.—Vicinity of Tecolutla (Villa, 1955:238; Málaga Alba and Villa, 1957:552).

Villa (1955:238) reported one specimen from Tecolutla on the verbal authority of Málaga Alba as *Lasiurus borealis seminolus* but Villa stated that he personally did not check the identification. We have not been able to locate the specimen.

Lasiurus cinereus cinereus (Palisot de Beauvois)

Hoary Bat

Specimen examined.—One from 3 km. W Zacualpan, 6000 ft.

Additional record.—Jalapa (Ward, 1904:653).

Lasiurus intermedius intermedius H. Allen

Northern Yellow Bat

Specimens examined.—Fifteen from 1 mi. SW Catemaco.

For an explanation of the subspecific status of this bat, see Hall and Jones (1961:84).

The 15 specimens mentioned above were dislodged on July 22, 1955, from the south and east sides of two tobacco sheds, along with 35 other individuals. The outside of the walls was thatched with corn stalks. The 15 individuals collected were four females having worn teeth, six females having unworn teeth and five males having unworn teeth. The four females having worn teeth were lactating. The other 11 were smaller and seemed to be young of the year (Baker and Dickerman, 1956:443).

Lasiurus ega xanthinus (Miller)

Southern Yellow Bat

Specimen examined.—Achotal, 1 (F.M.N.H. 14151). For use of the subspecific name *xanthinus* instead of *panamensis* see Hall and Jones (1961:91).

Nycticeius humeralis mexicanus Davis

Evening Bat

Specimens examined.—Two from 4 km. NE Tuxpan. The bats were captured in a mist net at night.

Rhogeëssa tumida tumida H. Allen

Little Yellow Bat

Specimens examined.—Total 12: 25 km. W Tampico, 2; 12½ mi. N Tihuatlán, 300 ft., 8; Boca del Río, 10 ft., 1; Río Blanco, 20 km. W Piedras Negras, 400 ft., 1.

Additional record.—Mirador, the type locality (Goodwin, 1958:3).

At Río Blanco and Boca del Río specimens were shot in flight on the arid coastal plain. They were present in small numbers, at dusk, flew more rapidly than *Saccopteryx bilineata* or *Eptesicus brasiliensis propinquus*, and maintained a distance of five to 12 feet above the ground. Individuals observed were partial to the shade of trees and boughs; they left the shadow of tree limbs for short periods, seemingly to catch insects, but usually returned at once to the shadows. A similar restriction to shadow was noted in *Saccopteryx bilineata*, but to a lesser extent. The specimens from 12½ mi. N Tihuatlán were found by a native in a hollow tree. Those from 25 mi. W of Tampico were caught between 7:00 and 8:00 p. m. in a mist net.

Nothing was learned of the breeding habits of this species in Veracruz, and no ectoparasites were found on the specimens taken.

Plecotus mexicanus (G. M. Allen)

Mexican Big-eared Bat

Specimens examined.—Total 77: 6 km. WSW Zacualpilla, 6500 ft., 3; 4 km. E Las Vigas, 8500 ft., 74.

Additional records.—Las Vigas (Ward, 1904:653 as *Corynorhinus macrotis*—probably was *P. mexicanus*); Jico [5,500 ft.] (Handley, 1959:151).

Handley (1959:141) has arranged this named kind of bat as a monotypic species instead of as a subspecies of *Plecotus (Corynorhinus) townsendii.*

On November 6, 1948, small caves and mine tunnels in the hills six kilometers west-southwest of Zacualpilla were searched for bats. In one narrow, low cave, situated beside a stream at the base of a hillside, sparsely forested with pine trees, we found two Mexican big-eared bats. They were 35 to 50 feet from the entrance of the cave, and were clinging to the wall, head down at a place where the cave was only about one meter in diameter. Both were cold and torpid, but not hibernating. They clung by means of their feet and thumbs, with the backs bowed and the heads against their chests. The long ears of this species were closely coiled and the tail was folded under the body, covering the abdomen. Two days later another bat of this same kind was taken in the same cave, 40 feet from the entrance.

A large series of *Plecotus* was obtained four kilometers east of Las Vigas in mid-January, 1949. The bats were found in caves in the basaltic lava at a small volcanic cone called Volcancillo. The greater part of our material from that locality was taken in shallow lava tubes, from one meter to three meters in diameter, and from a few inches to a meter or so beneath the crust of the lava flow. These bats were not hibernating. They were found clinging to the walls of these relatively small, dry caves from 25 to 500 feet from the entrances. Certain caves seemed to be favored over others, and by returning to these caves each day, we secured specimens from places where, a day earlier, we had removed all the bats.

In the much larger, deeper, caves in the same area, we found this species in hibernation. These larger caves were colder and damper than the shallow caves. The fur of the big-eared bats taken in the deep caves was damp or wet. Some individuals were seen, day after day, in the same place in these caves. One individual did not change position in a week. The hibernating bats were distinctly fat; those from the shallow caves were less fat and most of them were lean. In the deep caves we found also *Myotis velifer* and *Pipistrellus subflavus.* Only *Plecotus* was found in the shallow caves.

We have no records of embryos from this bat in Veracruz. The testes of the specimens examined were all small. A species of winged fly is the common parasite on *Plecotus.* Almost all of the specimens taken in Veracruz had several of these parasites.

Family Molossidae

Cynomops malagai Villa

Mexican Dog-faced Bat

Records (Villa, 1955:1-2).—Tuxpan de Rodríguez Cano, 20° 57′, 97° 24′, four meters; Veracruz [city of].

Tadarida brasiliensis mexicana (Saussure)

Brazilian Free-tailed Bat

Specimens examined.—Total 26: 6 km. WSW Zacualpilla, 6500 ft., 9; Jalacingo, 6500 ft., 13; 4 km. WNW Fortín, 3200 ft., 1; Maltrata, 3 (U.S.N.M. 64340-64342 alcoholics).

Additional records.—5 km. E Las Vigas, 8000 ft. (Davis, 1944:380); 5 km. N Jalapa, 4500 ft. (Davis, 1944:380); Cofre de Perote, 13,000 ft. (Saussure, 1860:283); "Veracruz" (the state) (Shamel, 1931:5).

This is a common bat on the Mexican Plateau, where it seems to have taken the place of *Molossus ater nigricans.* We have only one specimen from the tropics, which may have strayed far from its normal range.

By searching for bat droppings on the ground beneath crevices, a small colony, six kilometers west-southwest of Zacualpilla, 6500 feet elevation, was found in a vertical crevice in a sandstone cliff in an arroyo. The base of the crevice was under an overhanging ledge, and was 15 feet from the ground. The bats were too far up in the crevice to see, and irregularities in the crevice made it impossible to shoot them. Accordingly the crevice was blasted open with dynamite, and nine bats were killed by the blast and subsequently recovered. One specimen did not fall to the ground until November 9 (the others were taken on November 6) and, as it was not decayed, we suppose that it had been wounded by the blast and died later. All of the individuals obtained were non-pregnant females.

On November 12, 1948, a large colony of Brazilian free-tailed bats was discovered in the Cathedral at Jalacingo. The bats were in crevices in the wooden roof of two narrow rooms, about 25 feet long and four feet wide, that extended from the floor of the cathedral up, parallel to the tower, to the ceiling of the cathedral, perhaps 75 feet above. Some of the bats were in shallow crevices, or in the angle between the roof and walls. The bats were semi-dormant, and by reaching from a small window in the tower wall, we knocked them loose from their perches with a pole. The bats opened their wings and being unable to fly sailed to the floor below, where they were placed in a collecting bag before they could become fully active. The specimens taken here included three males and ten non-pregnant females.

A male was obtained in the Tropical Life-zone, over a pool four kilometers west-northwest of Fortin on April 2, 1946, by placing a wire in the line-of-flight of bats that were drinking from the pool. Numerous specimens of *Molossus ater nigricans* were trapped in the wires over this pool.

Eumops glaucinus (Wagner)

Wagner's Mastiff Bat

Specimens examined.—Jesús Carranza, 250 ft., 3.

In the town of Jesús Carranza on March 12, 1947, a wooden building was occupied by a colony of about 25 individuals of *Glossophaga*. "The day was hot and the attic was very warm. I climbed to the roof to get a better swing at the active, shy *Glossophaga* and noted the head of a molossid emerging from between two boards that overlapped loosely forming a crevice about three-quarters of an inch wide. I pulled the bat out. It did not attempt to escape, or struggle much, or make a sound. I reached in and pulled out two more, all rather torpid in spite of the heat." Several individuals of *Molossus ater nigricans* were also found in this building.

Molossus ater nigricans Miller

Red Mastiff Bat

Specimens examined.—Total 117: Ozulama [= Ozuluama], 500 ft., 8; La Mar, 20 ft., 5; Tuxpan, 22; 12½ mi. N Tihuatlán, 300 ft., 6; Puente Nacional, 500 ft., 10; 4 km. W Paso de San Juan, 250 ft., 14 Río Atoyac, 8 km. NW Potrero, 16; 4 km. WNW Fortín, 3200 ft., 24; Potrero Viejo, 1700 ft., 3; 3 km. SE Orizaba, 3; Cosamaloapán, 150 ft., 3; Jesús Carranza, 250 ft., 3.

Additional records.—San Andrés Tuxtla (Miller, 1913:88); Tuxpango (Sumichrast, 1882:202); Catemaco (Miller, 1913:88).

Molossus ater instead of *Molossus rufus* is the correct specific name for this bat according to Goodwin (1960:6).

The red mastiff bat is one of the most common of the insectivorous bats in the Tropical Life-zone of Veracruz. Colonies were found in buildings, in hollow trees, and in crevices in cliffs.

Four kilometers west-northwest of Fortín on March 26, 1946, bats were reported to inhabit the hollow of a large tree. The tree was of the slim species with broad leaves and red bark, called "palo mulato." It was about two feet in diameter at the base. Approximately 25 feet from the ground there was a scar where the tree had been hit by lightning. Above this scar the tree was hollow for an unknown distance. Bat guano had formed a pile at the bottom of the lightning scar and had washed down the trunk of the tree. There was also a small pile at the base of the tree. Eleven specimens of *Molossus ater nigricans* in the black phase, were obtained from the colony of about 25 bats that lived there.

At Puente Nacional on October 21, 1946, bats of this species were seen emerging from crevices in a sandstone cliff and from small openings in a concrete bridge [the "National Bridge"]. On October 22, the bridge was investigated. The bridge is very old, and is of the series-of-arches type. There were originally drainage holes below the rail-wall, about every 25 feet, but those of the east side have become filled, on top, with rocks and soil. The lower, outer, ends of most of the holes are covered by plants. Most of the bats were found in holes that were still open on the bottom but plugged above. By digging away the earth and sod from above, we got into the tops of the holes. The bats were not clinging to the walls, but were lying flat on the ground, under stones. It appeared as if they had burrowed back under the rocks, but I think it more likely that they had dug the holes by continual use, loosening bits of the sandy soil that tumbled down the holes to the river below. Some of the burrows were all of six by five by one inches, in the shape of a pocket. Eight bats, all in the black color phase, were taken here.

At four places (Jesús Carranza; three kilometers southeast of Orizaba; Ozuluama; and at Cosamolapán) this species was found in hot, dry, and dusty crevices in the attics of buildings.

We watched the flight of this species of bat on several occasions, Usually the bats left their hiding places about one and one half hours before dark, usually before swallows and other birds had retired for the night. On dark days, they emerged earlier. Shortly after they emerged from their retreats they began the evening hunt. Usually they flew high, about 150 feet above the ground, in the early part of the evening. At this elevation, they were independent of forest, cliffs, clearings, and other ground relief. As dusk advanced they flew lower and at about the time it became completely dark the bats were hunting at approximately 35 feet above the ground. At this elevation, they must maneuver through clearings and other roads and trails, avoiding the forest. Because they hunt at a much higher speed than most insectivorous bats, they are less able to twist and turn in the pursuit of insects, and require large clearings or long stretches of narrow clearings, such as the space above a road, river, or field.

These bats do not drink until it becomes completely dark. Then they approach a pool at an elevation, above its surface, of about 25 feet. Usually they fly over and past the pool and, when just past its farther bank, they dive, turn, and sweep back, skimming the surface as they drink. They are easy to trap by stringing a wire an inch or so above the surface of the water. Approximately one half hour after complete darkness, the bats were seen returning to their retreats, and no specimens were taken in our wire traps more than one half hour after dark. Vespertilionid bats usually rest at night in some retreat other than their daytime roost. Red mastiff bats return to the daytime roost to rest at night.

These bats hunt also in the morning and often do not return until well after dawn. On several occasions we saw these bats returning to their roosts after the birds had been active for some time; perhaps one half hour after dawn. In the early evening and early morning, it is not unusual to see swallows and red mastiff bats hunting together.

Both males and females were found together in some colonies. Several times we found small groups of three to five males. Pregnant females were taken only in early spring: February 26, March 5, March 26, March 29, and April 5.

This bat has a strong, sweetish, musky odor, which seems to originate from the large gland in the chest. Commonly the fur of the chest is covered with the thin oil secreted by this gland. Several of these bats were fat, but most of the specimens were lean.

On November 24, 1946, one of these bats was flopping on the ground at the base of a tree, four kilometers west-northwest of Fortín, 3200 feet elevation. It was not wounded, but was unusually fat. The fur of the entire body was wet with the oil secretion from its chest gland which was soft and empty.

Beside the Río Atoyac, eight kilometers northwest of Potrero, at 1700 feet elevation, on March 1, 1946, a bat of this species was found hanging head-down from the weather-beaten post of an old, half-demolished dam across a small arroyo in the jungle. The bat was alive, but must have been ill, though no wounds were found on the body when the animal was skinned. At the same locality, on March 4, 1946, Mrs. Peggy Dalquest heard an animal squeaking in branches of a wild orange tree. When she turned the light from a flashlight into the tree, a bat dropped to the ground. The animal was a *Molossus* in the red color phase, and although apparently uninjured, it was not able to fly.

Parasites are rare on *Molossus ater nigricans*. Two species of mites were noted, a small red species and an even smaller yellow form. A small wingless fly was also noted.

Order PRIMATES

Family Cebidae

Alouatta villosa mexicana Merriam

Howler Monkey

Specimens examined.—Total 23: 10 km. NW Minititlán [= Minatitlán], 100 ft., 5; 35 km. ENE Jesús Carranza, 150 ft., 5; 20 km. E Jesús Carranza, 300 ft., 10; Arroyo Azul, 26 km. E and 8 km. S Jesús Carranza, 2; 35 km. SE Jesús Carranza, 400 ft., 1.

Additional records.—Pasa Nueva (Allen, J. A., 1904:40); Minatitlán (Merriam, 1902:67); Achotal (Lawrence, 1933:341).

In southern Veracruz, this species seems to live only at low elevations; our specimens were taken at elevations of between 100 and 400 feet. In the low, swampy jungles, near sea level, the howler was more common than the spider monkey. To the best of my knowledge, the spider monkey occurs wherever the howler does, but is found also in the higher areas, where the howler does not occur.

The howler is common in the low, dense jungles of broad-leafed trees and palms, as well as in the tall forest. We never saw monkeys on the ground, but on several occasions bands were found in small, isolated patches of jungle that they could have reached only by traveling over the ground.

Locally this species may be abundant. Normally it occurs in bands of from six to 50 individuals. In bands, females outnumber males by ten to one. Usually the spider monkeys and howlers remain separate, but in several bands of spider monkeys we noted a single howler. Such howlers that were shot were old males. One mixed band of about 100 spider and howler monkeys was seen.

Like the spider monkeys, the howler monkeys seem to feed principally in the early morning. They are commonly seen in the tops of tall trees throughout the day. Individuals sometimes remain motionless for hours. We have found what was probably the same band in the same general area on three consecutive days.

Howler monkeys are skillfull climbers but are far less active than the spider monkeys, and usually move through the trees rather slowly, although when frightened they are capable of swift movement. Their reactions differ from those of the spider monkey in many respects; howlers do not show the curiosity of the spider monkey. A man may walk along a trail beneath a band of howlers without becoming aware of their presence, unless they are moving about, breaking branches, or calling. When one member of a band is shot, the other monkeys usually hoot and roar, and move higher into the trees. Rarely do they swing off through the trees, as the spider monkeys almost always do.

Like the spider monkeys, howler monkeys throw things at intruders, and are more apt to do so than spider monkeys. Usually their missiles are too small to be dangerous, and their aim is poor. Old males have powerful jaws and large teeth, and doubtless could give a serious bite. On one occasion two howlers were shot, but, as is often the case with howlers, the dead animals failed to fall from the trees, for the bodies were supported by the animals' death grip with feet, hands, and tails. The remainder of the band, approximately 10 animals, were in the tops of nearby trees. When the collector climbed into the tree to dislodge the dead animals, "the other animals started roaring and, I think, were preparing to attack me. I went down the tree in a hurry. One old male followed me to within 30 feet of the ground." Although these animals did not actually come within 10 feet of me, they did make faces, gape their tremendous jaws, and make loud, booming roars that were disconcerting to me, high in a jungle tree.

Twice we found howler monkeys at night. On the first occasion, the grating of our dugout canoe on a sandbar caused one animal to make a low hoot. The monkeys, about 25 in all, were sleeping in crotches well out on the limbs of a tree overhead. We were able to see them in the beams of our flashlights by looking for the black tails tightly wrapped around branches. At close range, their eyes, in the beam of a light, had a dull red glow. When the band was disturbed by our shooting, the survivors made off along the palm fronds and branches. A few hid in masses of vines. On another occasion the whistle from a passing railroad train caused a monkey to hoot. We found the band sleeping in a group of tall trees, on slender limbs about 50 feet from the ground. A porcupine (*Coendou mexicanus*) was shot from the same tree.

The call of this monkey is a hooting roar or a barklike, coughing howl. The sound travels for long distances, and can easily be heard for more than a mile. In the mornings the calls from many bands of these monkeys may blend together to form a constant roar. Later in the day they call less often, but they may call at any hour of the day. On dull and rainy days they call less than on bright, clear days. At some times of the year, especially in late March, April, and the first of May, they seldom call. The call is given by the females as well as by males, but the loud roar, which carries to a great distance, is made by the old males.

This monkey is never eaten in Veracruz, and we were told that they never bother crops. Two females taken on February 4, 1947, had newly-born young, but young of various ages are seen in most bands of this species of monkey, and we suppose that they have no regular breeding season in Veracruz.

Ectoparasites are not common on howlers. One old male had a number of large nematode worms in the rectum. Under the skin, several howlers had fly larvae described, in the field notes, as "brown; heavily ridged; about ¾ inch long and ½ inch wide. They are only slightly flattened. They make a large lump on the animals, fully an inch in diameter. . . . [One animal had] some on the jaws, neck, chest, legs and feet."

Ateles geoffroyi vellerosus Gray
Geoffroy's Spider Monkey

Specimens examined.—Total 41: 10 km. NW Minatitlán, 100 ft., 4; San Juan Evangelista, 3 (U.S.N.M.); Achotal, 7 (Chicago N. H. Mus.); 35 km. ENE Jesús Carranza, 150 ft., 4; 20 km. ENE Jesús Carranza, 200 ft., 1; 20 km. E Jesús Carranza, 300 ft., 18; Arroyo Saoso, 37 km. E, 7 km. S Jesús Carranza, 1; 20 km. ESE Jesús Carranza, 1; 35 km. SE Jesús Carranza, 400 ft., 2.

Additional records (Kellogg and Goldman, 1944:35, unless otherwise noted). —Barranca de Boca, Canton de Jalapa; Mirador (Reinhardt, 1873:150); 15 mi. NE Huatasco, 2000 ft.; Volcán de Orizaba (Elliot, 1905:535); Pasa Nueva; Cuatotolapan.

According to the Rules of Zoological Nomenclature, *Sapojou* Lacépède is the correct generic name but the International Commission on Zoological Nomenclature (see Opinion 91) recommends the use of *Ateles*. The type locality of this subspecies is Mirador, Veracruz, by restriction. The spider monkey no longer occurs near Mirador, or elsewhere in central Veracruz. At Hacienda Mirador, 3500 feet elevation, on December 6, 1947, we were told by the elder members of the Grohman and Sartorius families that they had been told by their parents that monkeys once occurred there. There is still a pass to the northwest of Mirador that is called "Paso de los Monos" (Pass of the monkeys). The ancestors of the Sartorius and Grohman families were living in Hacienda Mirador when the collector, Sumichrast, obtained the specimens on which Reinhardt based the name *neglectus*.

This species once occurred in most or all of the forested parts of the Tropical Life-zone of Veracruz, but now is found only in the southern part of the state. I (Dalquest) saw individuals 15 kilometers southwest of Jimba in the Rincón area, and seven kilometers from Volcán San Martin Tuxtla.

In the lowlands of southern Veracruz, the spider and howler monkeys occur together in some localities. Where they occur together, the spider monkey is usually referred to as "mono" and the howler is usually referred to as "chango," but in some areas, this is reversed, the spider monkey being called chango. In higher areas, where only the spider monkey occurs, it is called either mono or chango.

The spider monkey is usually found in tall, extensive forest, but in some places we found spider monkeys to be abundant where the jungle was only about 50 feet in height. Usually the spider monkeys live in rather humid forest, but in the Rincón area, to the west of Jimba, they were in dry, though tall and dense, jungle. Near the mouth of the Río Coatzacoalcos they were in swampy forest, and they were abundant in the dense, humid forest and limestone cliffs to the southeast of Jesús Carranza.

The spider monkey seems to be entirely diurnal. We never saw or heard the species at night. Usually spider monkeys and howler monkeys remain in separate bands. On March 20, 1947, 35 km. east-northeast of Jesús Carranza, however, a band of about 100 monkeys was composed of about equal numbers of each of the two kinds. On several occasions we found bands of spider monkeys, with a single old male howler in the group. In general, the spider monkey was more common than the howler in southern Veracruz, except near the Gulf, where the howler was the more common.

The spider monkey usually lives in small bands of from 10 to 25 animals. The largest band noted contained approximately 50 animals, but this was unusual. On several occasions we found pairs and single animals. One solitary animal was an old female. One pair consisted of an old female and a young male. Several small groups of old males were seen, usually numbering three to five animals. The larger bands include animals of both sexes, with the females outnumbering the males.

Spider monkeys feed in the early morning. By two to three hours after sunrise they are usually rather inactive, sitting in the tops of the larger trees. Some bands remained in the same tree most of the day. Under such circumstances they are exceptionally inquisitive, and will descend to the lower parts of the trees to examine any strange animal that passes. Their alarm call is a loud, rather birdlike, double squeak, "EEEeek-eeaaAAK," repeated endlessly. Spider monkeys, calling together, in a large band make a great racket. We have heard them call continuously for more than an hour. The sound, to me, is irritating. Young spider monkeys make a long, drawnout squeal.

Young spider monkeys are commonly sold as pets. Those that I saw were gentle and affectionate. They are captured by shooting the mothers, and climbing after the babies until they are caught or the trees containing the young are chopped down. We were told by hunters that a baby would never leave the vicinity of its dead mother. Hunters told us also that on some occasions, when one monkey in a band was shot, the other monkeys would throw fruit and sticks and even large limbs at the hunters. We noted this habit only rarely, when monkeys were shot for specimens; the effort was rather half-hearted and the aim was poor. The howler monkey is more apt to throw things at an intruder than is the spider monkey. Almost invariably, when one or two spider monkeys of a band were shot, the others made off through the trees at high speed, leaving dead or injured animals to their fate. This is not true of the howler monkeys.

Except for the sale of young animals for pets, the spider monkey is of little economic importance in Veracruz. The flesh is never eaten. Dead animals have a strong and unpleasant smell. Even our dogs refused, at first, to eat the flesh. This odor was never detected from live animals kept as pets. These monkeys seem never to descend to the ground and they do not bother crops. The jungle fruits upon which they feed are rarely utilized by man.

The spider monkey has no regular breeding season. Young of various ages were seen in almost every large band examined. One man reported an albino spider monkey in the hills west of Jimba. Animals taken near Minatitlán in February had deep layers of yellow fat beneath the skin. Near Jesús Carranza, animals from some bands, young specimens as well as adults, were fat while animals from other bands in the same locality were lean. Parasites were rarely noted on spider monkeys, except that several animals had large fly-larvae in pockets under the skin. One old male had many nematode worms, about an inch long, in the rectum.

Family Hominidae

Homo sapiens americanus Linnaeus

Man

The native subspecies *H. s. americanus* has crossed extensively with the subspecies *H. s. sapiens* that invaded the area in the 16th century.

Order EDENTATA

Family Myrmecophagidae

Tamandua tetradactyla mexicana (Saussure)

Tamandua

Specimens examined.—Total 5: 20 km. ENE Jesús Carranza, 200 ft., 1; 20 km. E Jesús Carranza, 300 ft., 2; Arroyo Saoso, 37 km. E, 7 km. S Jesús Carranza, 1; 6 km. NW Paso Nuevo, 100 ft., 1.

Additional records.—"Passa Neuva" (Allen, J. A., 1904:394); Mirador (Hall and Kelson, 1959:239).

In Veracruz the four-toed anteater is usually called "brazo fuerte" but is referred to also as "oso hormiguero" and "chupa miel." The species occurs from one end of the state to the other, in the Tropical Life-zone at lower elevations. Though widespread, this anteater is not common, probably because it is killed by man at every opportunity.

This species is both aboreal and terrestrial. In trees it is said to prefer dense growths, such as the mango. Terrestrial habitat includes open ground under dense jungle, marshy ground, and open, arid grasslands.

On February 9, 1948, at a place 20 kilometers east of Jesús Carranza, a large animal was heard moving on the ground at midnight. It made considerable noise in the dense jungle. A few moments later an anteater, the source of the disturbance, climbed up on one of the aerial roots of a large tree, and began to climb the tree.

At the same locality on March 21, 1948, an anteater was found about noon, on an exceptionally hot day, in the top of a small coyol palm tree, on a drooping frond about eight feet from the ground.

Six kilometers northwest of Paso Nuevo, an anteater was followed by dogs to its burrow almost in the center of an extensive area of tall sawgrass. The burrow had an entrance about 10 inches in diameter, was approximately two meters long and at its end was approximately a half meter below the surface of the ground. No nest was present, but the end of the burrow was slightly enlarged.

There are a number of superstitions regarding the anteater. One is that it reaches its long tongue into a nostril of a dog, and so strangles it. Another relates to the destruction of sugar cane. On the lower reaches of the Río Coatzacoalcos, the people insist that the frequent destruction of small patches of sugar cane by some animal that tears and chews the stalks is the work of the anteater. We obtained a specimen and were able to prove to them that this was not so. The mouth of the anteater is too small to bite with, it has no teeth, and the lower jaw is soft and flexible. Chewing anything as hard as sugar cane would be impossible for an anteater.

Anteaters are killed whenever found. One excuse for this is that the animals are a danger to dogs. They are also killed for their skin, which is said to be relatively stronger than that of an ox or horse. Most are killed, however, by hunters who find an anteater in a burrow that they have worked diligently to open in the hopes of obtaining an armadillo.

We found no parasites on anteaters, not even ticks. The thick, tough hide may discourage ticks but the tapir in the same area is parasitized by a species of tick, in spite of its having an even thicker skin. Perhaps the strong, musky odor of the anteater is obnoxious to ticks.

None of the three anteaters taken was pregnant. What I took to be the salivary gland is worthy of mention because it is so greatly developed and enlarged as to extend back along the sides of the neck and then down, between the armpits, to the chest. When cut open it is seen to contain a thick, sticky, transparent liquid.

Cyclopes didactylus mexicanus Hollister

Two-toed Anteater

Record.—Minatitlán (Hall and Kelson, 1959:240).

In the area of the Río Coatzacoalcos, this little anteater is called "mico de noche." It seems to be rare, and we were unable to obtain a specimen. A few skins for sale were seen in shoemakers' shops, but the locality of origin of such skins was uncertain. Animals are occasionally taken alive when large trees are cut down. One was said to have been kept alive as a pet at Minatitlán.

Family Dasypodidae

Dasypus novemcinctus mexicanus Peters

Nine-banded Armadillo

Specimens examined.—Total 8: San Carlos, 1 (Chicago N. H. Mus.); Río Blanco, 20 km. W Piedras Negras, 400 ft., 1; 15 km. SW Jimba, 750 ft., 1; 20 km. ENE Jesús Carranza, 200 ft., 1; 20 km. E Jesús Carranza, 300 ft., 1; 25 km. SE Jesús Carranza, 250 ft., 1; 35 km. SE Jesús Carranza, 400 ft., 1; 30 km. SSE Jesús Carranza, 300 ft., 1.

Additional record.—Jalapa (Ferrari-Pérez, 1886:130).

Armadillos from Veracruz can be separated from those seen from Texas and elsewhere in México by examination of the lacrimal bone. In the armadillo, this bone makes up the anterior, dorsal part of the zygomatic arch. In skulls from Veracruz, the lower, anterior suture is on the lower edge or side of the zygomatic arch, and is visible from the lateral view. In other armadillo skulls examined, the lower, anterior suture is below the angle of the zygomatic arch, on the ventral side, and is not visible from the lateral aspect. This character serves to separate both adult and young animals. Age in armadillos can be determined by the presence or absence of certain sutures between bones of the skull. We consider as adult, animals having the sutures closed between the basiphenoid and the presphenoid, and between the basisphenoid and the basioccipital.

In addition, a skull from the Federal District is shorter and narrower than any from Veracruz and Texas. Two adults from Yucatán differ from other skulls examined, in that the presphenoid is much longer, extending forward beneath the palate as a long, pointed bone, and in that the greatest width of the braincase is at the squamosal, rather than at the suture between the squamosal and the frontal.

Specimens from the west coast of México have not been examined. Our material indicates but does not prove that specimens from Yucatán and Veracruz are subspecifically distinct from both *Dasypus novemcinctus mexicanus* and D. *n. texianus.*

In Veracruz, the armadillo is called both armadillo (pronounced armadeeo) and tochi. It is found throughout the Tropical Lifezone, and is everywhere hunted for food.

Favored habitat of the armadillo in Veracruz is dry jungle, either where the forest floor is open or covered with a tangle of vegetation. We found signs most abundant where the jungle was low but with a dense crown of branches, leaves and vines. As a result of the absence of sunlight in places of this kind, the ground beneath the trees was relatively open, and was littered with dry leaves. Dry ground seems to be preferred to damp ground, and level ground to rough ground.

Armadillos are not social. They are principally nocturnal; only one was seen abroad by day. Their eyes do not reflect the rays of a hunting light at night. As a result they are difficult to shoot at night. In the daytime they take refuge in burrows that they dig for themselves. A typical burrow was found 15 kilometers southwest of Jimba, after our dog began to bark a short distance from the trail. The brush and jungle were so dense that we had to chop our way by means of machetes in order to reach the dog. It was at the entrance to a burrow between two logs. The top log was so much decayed that we were able to cut it in two with our machetes; we moved the other log. The dog had enlarged the entrance of the burrow, and its original diameter could not be determined. When the dog was pulled away, we could hear the tochi scratching as it burrowed. We dug with our machetes, and within two feet found a mass of dry leaves—the nest. We could see the animal's tail, but it set its feet so securely in the earth that it took all of my strength to pull the animal from its burrow.

The armadillo is hunted by means of dogs especially trained for the work. In a good area, two hunters may catch as many as four armadillos in a day. The dog trails the animal to its hole, from which the armadillo is dug out. The flesh of the armadillo is held in great esteem. Correctly prepared, it is delicious, and resembles pork, but is sold for several times the price of pork or beef, in some areas. Only the agouti and paca are considered superior as food.

An armadillo taken on March 5, and another taken on April 24, each contained four large embryos. Large ticks are commonly found attached to the underside of the armadillo; each specimen examined had from 11 to 25 such ticks.

Order LAGOMORPHA

Family Leporidae

Sylvilagus brasiliensis truei (J. A. Allen)

Forest Rabbit

Specimens examined.—Total 5: 35 km. NW Tuxpan, 1; Potrero Llano [=Potrero del Llano], 350 ft., 1; 12½ mi. N Tihuatlán, 300 ft., 1; 2 km. N Motzorongo, 1500 ft., 1; 30 km. SSE Jesús Carranza, 300 ft., 1.

Additional records (Nelson, 1909:264).—Mirador; Motzorongo; Otatitlán; Buena Vista.

The forest rabbit is probably common in the dense jungle of Veracruz, but we found it extremely difficult to obtain. In field notes for December 6, 1946, it is recorded that at a place two kilometers north of Motzorongo "One of these small, dark rabbits was taken last night. It was in a beautiful setting. There is a grove of tall banana trees in a clearing, about 100 feet square, chopped out of the jungle. The area is near the stream in the arroyo, and is level. It was probably cleared last year, and has come up in elephant's ear. Some leaves are five feet long. I was hunting here about one hour after dark. Fruit-eating bats were making considerable noise, but I heard something else moving in the vegetation. My light picked out the dull, red glow of a single eye, and I had to shoot it at rather close range." The specimen was a male with testes measuring 30 by 12 millimeters.

An adult male taken along the Río Solosuchil, 30 kilometers south-southeast of Jesús Carranza on May 13, 1949, was shot approximately two hours after dark, when it came to drink at a sandbar at the edge of the river. Back of the bar the jungle was tall and dense except for patches of wild banana and occasional deep tangles of vines and brush, and the ground was open and level.

An adult female that was nursing young was shot in dry, low, dense and thorny jungle at Potrero Llano, 350 feet elevation, on

February 15, 1949. Eastern cottontails were found in a large cornfield here, and while hunting for them we walked along the border of the clearing. The light was shone into the jungle, wherever the vegetation would allow us to do so. In one place, the light reflected the glow of the eyes of a rabbit. It was shot, and found to be *S. brasiliensis*. Two individuals of *S. floridanus* were shot in the clearing 50 feet away.

Sylvilagus floridanus
Eastern Cottontail

The Eastern cottontail, known in Veracruz as "conejo," occurs in almost every part of the state. In the alpine meadows and brushlands of the high mountains, it occurs with the Mexican cottontail, *Sylvilagus cunicularius*, and in the deep jungle it occurs with the forest rabbit, *Sylvilagus brasiliensis*. It reaches its greatest abundance at lower elevations, in the extensive sugar cane fields and on the coastal plain. In these places, extensive grasslands and numerous thorny thickets offer ideal food and cover.

This Eastern cottontail is nocturnal and most of our specimens were shot while we were hunting at night with a headlamp. On dark and cloudy nights, the cottontails are not shy, and usually allow a hunter to approach within shotgun range. On clear, moonlit nights, the same animals are shy.

Locally, in ideal habitat, cottontails are abundant. While driving at night along roads through the sugar cane fields within a radius of seven kilometers of Potrero, as many as 10 cottontails to a mile were seen. On the coastal plain, near Piedras Negras, four to six were often seen on a night's hunt, but at most places they were far less common. In some places we obtained specimens only after many hours of hunting at night. A number of the specimens from the coastal plain were shot after they were flushed from thorny cover by dogs.

When abundant near habitations, cottontails commonly damage gardens, and may do great damage to sugar cane when the cane is small. Two kilometers north of Motzorongo we saw seven consecutive new shoots of cane, in one row, that had been cropped to the ground by a rabbit that we shot. Cornfields suffer heavily from the attacks of cottontails; stalks less than a foot high are commonly cut a few inches from the ground. In some areas we found whole fields in which almost every stalk had been cut by rabbits.

This cottontail is an important item of food and furnishes much sport in Veracruz. The cottontail and the chachalaca, in many parts of the state, are the only common wild species of game. The flesh of the cottontail is far less desirable than that of the paca and armadillo, but the cottontail is much more easily obtained than those species.

No pregnant females were obtained in Veracruz. Females that were nursing young were taken on May 27, February 12, and February 28. A male with greatly swollen testes was taken on December 5. A juvenile only 184 millimeters in total length was taken on December 17.

Most specimens had a few ticks, and some had many small ticks, but the infestations were small, considering the large numbers of ticks in the country that the cottontails inhabit. Several cottontails had a few fleas. One specimen had a fly larva under the skin on one side of the neck.

Sylvilagus floridanus connectens (Nelson)

Specimens examined.—Total 35: Hacienda Tamiahua, Cabo Rojo, 2; Ozulama [= Ozuluama], 500 ft., 3; Ixcatepec, 70 km. NW Tuxpan, 1; Tlacolula, 60 km. WNW Tuxpan, 2; Potrero Llano, 350 ft., 2; 50 km. NW Tuxpan, 2; 4 km. N Tuxpan, 1; 9 km. E Papantla, 300 ft., 1; Teocelo, 4000 ft., 1; 3 km. W Boca del Río, 10 ft., 1; 7 km. NW Potrero, 1700 ft., 2; 5 km. NW Potrero, 1500 ft., 1; Potrero Viejo, 1700 ft., 8; 15 km. W Piedras Negras, 300 ft., 2; Río Blanco, 20 km. WNW Piedras Negras, 5; 2 km. N Motzorongo, 1500 ft., 1.

Additional records (Nelson, 1909:186).—Jico; Chichicaxtle, type locality; Mirador; Orizaba.

Sylvilagus floridanus orizabae (Merriam)

Records (Nelson, 1909:185).—Las Vigas; Mount Orizaba.

Sylvilagus floridanus russatus (J. A. Allen)

Specimens examined.—Total 6: Jimba, 350 ft., 2; 7 km. NW Paso Nuevo, 100 ft,. 2; 14 km. SW Coatzacoalcos, 100 ft., 2.

Additional records (Nelson, 1909:187).—Catemaco; Coatzacoalcos; Minatitlán; Pasa Nueva, type locality.

Sylvilagus audubonii parvulus (J. A. Allen)

Desert Cottontail

Record.—Perote (Nelson, 1909:237).

Sylvilagus cunicularius cunicularius (Waterhouse)

Mexican Cottontail

Specimens examined.—Two from 3 km. W Acultzingo, 7000 ft.

Additional records (Nelson, 1909:241).—Las Vigas; Perote; Cofre de Perote; Orizaba.

On October 7, 1947, three kilometers west of Acultzingo, a small boy told us that he had seen a rabbit enter a hole. The burrow was excavated and a Mexican cottontail, a young, non-pregnant female, was captured. The burrow was about five inches in diameter, four feet long, and at its end two feet beneath the surface of the ground. There was no nest. At the same locality, on December 5, 1948, several Mexican cottontails were found with the aid of dogs, and one, an adult male, was obtained. No ectoparasites were found on it or on the young female.

In the winter of 1948-1949, at Las Vigas, Mexican cottontails were hunted on many nights. The animals were fairly common in the cool, damp meadows of zacatón and in thickets of brush on the mountainside nearby. The heavy fog, which occurs almost nightly at this elevation in the winter, greatly hampered us in our efforts to obtain specimens. No rabbits were seen in the early morning or daytime. Twice we found where some predator, probably a fox, had killed a rabbit the night before.

Order RODENTIA

Family Sciuridae

Spermophilus perotensis Merriam

Perote Ground Squirrel

Specimens examined.—Total 8: Perote, 7500 ft., 1; 2 km. E Perote, 7000 ft., 1; 2 km. W Limón, 7500 ft., 6.

Additional records (Davis, 1944:383).—Perote, 8300 ft.; Guadalupe Victoria.

This ground squirrel is called "moto" in the Perote area. In early October, 1947, we searched for ground squirrels near Perote, but most of them had gone into hibernation. One old female, that had been nursing young, was taken two kilometers east of Perote on October 13. It was fat, and had no ectoparasites.

In early October of 1948, a small series of specimens, principally young animals, was obtained in the same general area. All were fat. Two fleas were noted on the specimens taken.

Spermophilus variegatus variegatus (Erxleben)

Rock Squirrel

No specimens of this species were obtained in Veracruz, nor has the species been taken by other collectors. I saw two individuals in Veracruz. One was shot at and wounded on November 20, 1946, as it sat on the wall of an old ruin, seven kilometers west of Acultzingo, 7000 ft. elevation. The animal escaped and presumably died. A week was spent in the vicinity, but no other was seen. This locality is on the lip of the Mexican Plateau, about one kilometer from the Puebla boundary. Another specimen was seen on October 11, 1947, on the extensive lava flow, three kilometers west of Limón, 8000 feet elevation. This locality also is about one kilometer from the Puebla boundary.

Sciurus deppei

Deppe's Squirrel

In Veracruz this little squirrel is generally known as "ardilla montañera," or "ardilla chica." It lives in deep forest of broad-leafed trees, and prefers the dense jungle, where there is but little light. In tall forest it often occurs with the red-bellied squirrel, but seems never to enter the open palm-jungles, where the red-bellied squirrel is often common. Near the type locality, as at three kilometers west of Gutierrez Zamora, Deppe's squirrel was common in dense forest, the trees of which were seldom more than 30 feet in height or five inches in diameter at the base. Here the overhead vegetation was so dense that the ground beneath was only dimly lighted.

This species was never seen to enter holes in trees, but it probably does so. Leaf nests are common. Most nests that we were reasonably certain were of this species were about 12 inches in diameter and placed on larger limbs not less than 25 feet from the ground. Some were on slender main trunks, not far from the tops of trees.

Locally these little squirrels are abundant, but seem not to be social to any degree. They differ greatly from the red-bellied squirrel in their actions, in that they spend considerable time on the ground, are rather slow, and in trees move with little disturbance, rarely rustling leaves and branches. They usually are found on the tree trunks and larger branches. At the approach of a hunter, they slip around to the far side of a trunk or large limb, and remain motionless. Most attempts to "wait them out" failed. The most successful method of hunting them was that of moving steadily and slowly through the gloom of the deep woods, and watching carefully the main trunks and larger limbs 20 to 40 feet from the ground. There the squirrels could be seen in silhouette against the leaves overhead.

Nine kilometers east of Papantla we were told that on two hills, separated by only a few hundred yards of low, forested valley, the squirrels were different. Our own investigation confirmed this for on one hill only *deppei* was found whereas only *aureogaster* was found on the other hill. On other nearby hills, both species were found together. We can offer no explanation for this distribution.

On the slopes of Volcán San Martin Tuxtla, at 3000 feet elevation, a number of these squirrels were found on the ground in the early morning. A steady rain was falling, and we were hunting with the aid of dogs. Perhaps 10 little squirrels were flushed by the dogs. Some ran along the ground for several yards, not taking to the trees until closely pressed by the dogs. Because we were searching for larger animals, we shot only two Deppe's squirrels.

In the dense jungles at low elevation, to the east and south of Jesús Carranza, in extreme southern Veracruz, the little brown squirrel was abundant. It was found in the deepest part of the forest, where most of the trees were more than 100 feet high but most of the squirrels were seen on or near the ground; not one was noted more than 50 feet from the ground. They seem to prefer the trees that were heavily hung with vines, and trees with numerous cavities, such as the strangler fig.

These squirrels are of little economic value; they are too small to be worth hunting for food. The cost of the ammunition is greater than the value of their meat. They rarely come into contact with agriculture, but when they do so, are capable of doing great damage, especially to corn. This occurs where a milpa (cornfield) is situated in a clearing in dense forest. The squirrels climb the stalks, and without detaching the ears of corn, cut away the husks to reach the kernels. They begin to gnaw at the tip, and rarely eat more than half the kernels of one ear. It is a common sight, in such a milpa, to see ears of corn, half-eaten, with the bottom of the ear still swathed in its husks, but several inches of the bare cob, from which the corn has been eaten, projecting.

None of the females of this species that we examined was pregnant. Males having enlarged testes were taken in November, March and April. No parasites were recorded on this squirrel.

Sciurus deppei deppei Peters

Specimens examined.—Total 55: 12½ mi. N Tihuatlan, 300 ft., 2; 9 km. E Papantla, 300 ft., 8; Zacualpan, 6000 ft., 1; 3 km. W Guttierez [= Gutierrez] Zamora, 300 ft., 2; San Carlos, 4 (Chicago N. H. Mus.); 4 km. W Tlapacoyan, 1700 ft., 1; Dos Caminos, km. 354, alt. 4500 ft., 1 (Chicago N. H. Mus.); 7 km. SE Volcán San Martín Tuxtla, 3000 ft., 2; Achotal, 1 (Chicago N. H. Mus.); 20 km. ENE Jesús Carranza, 200 ft., 2; 20 km. E Jesús Carranza, 300 ft., 3; Arroyo Saoso, 37 km. E, 7 km. S Jesús Carranza, 2; 20 km. SE Jesús Carranza, 250 ft., 1; 25 km. SE Jesús Carranza, 250 ft., 3; 35 km. SE Jesús Carranza, 350 ft., 9; 30 km. SSE Jesús Carranza, 300 ft., 2; 60 km. SE Jesús Carranza, 450 ft., 1.

Additional records (Nelson, 1899:104).—Papantla, type locality; Las Vigas; Jalapa; Jico; Cordova [= Cordoba]; Montzorongo; Catemaco.

Sciurus deppei negligens Nelson

Specimens examined.—Platón Sánchez, 800 ft., 6.

Sciurus aureogaster

Red-bellied Squirrel

This squirrel is unusually variable in color. Over all of Vera cruz it is referred to as "ardilla," but the black phase is "ardilla negra," and the colored phase is "ardilla pinta." Most of the residents consider these color phases to be separate species.

The subspecies *frumentor* seems to be restricted to the pine forest of the mountains along the western border of the state. The habitat of the subspecies *aureogaster* is much more varied, and it is found in forests and jungles throughout the Tropical Life-zone of Veracruz. It probably reaches its maximum abundance in the tall, cool forests of the upper humid division of the Tropical Life-zone, at from 1500 to 4000 feet elevation. Where there are forests in the lower arid division of the Tropical Life-zone, as along streams and arroyos, this species is locally abundant. In the dense jungles of the extreme southern part of the state, the red-bellied squirrel is relatively scarce, as compared with farther north.

This diurnal species is far more active in the early morning than at other times of day. It probably lives in holes in trees, at least at times, but only one was seen to enter a hole. Leaf nests are commonly built. These are rather small nests for such a large squirrel. Most of the nests examined were from a foot to 18 inches in diameter; many were less than 35 feet from the ground. Opossums also utilize these nests. Nests in the lowlands were constructed principally of leaves of broad-leafed trees, with some twigs for strength. Nests of the subspecies *frumentor* were found in tall pine trees, were constructed of pine twigs and needles, were about 14 inches in diameter, and were placed 20 to 35 feet from the ground. No more than three individuals of this species were seen together, although several nests are commonly found within sight of each other in a small area.

The red-bellied squirrel is almost entirely arboreal. On a few occasions, we saw individuals descend to the ground to escape danger, but we never saw them feeding on the ground. They were often seen in low trees and bushes. Some hung head-down, by their back feet, from bushes as they fed on fruit at the ends of small twigs. Mostly, however, they live from 20 to 40 feet from the ground, but often higher. One, that was frightened at my approach, in a large tree, ran to the top of the tree and out along a branch, where it lay flat, parallel to the branch, fully 100 feet from the ground. It remained there as long as I watched. Had I not seen it hide, it would certainly have escaped notice. Although squirrels of this species are skillful climbers and are able to travel rapidly through the trees, they ordinarily do not make long leaps from one branch to another. One leap that spanned about five feet was the maximum observed.

Among the better known fruits on which this squirrel feeds are mango, wild green figs, jobo plums, tamarind pods, and chico zapote. Numerous other species were eaten, as described in our notes, but the plant species were not identified.

The subspecies *frumentor* normally eats seeds of the pine and other conifers as at Las Vigas where we saw numerous cores of pine cones from which the nuts had been eaten. In this area the red-bellied squirrel does much damage to corn. We were shown dozens of ears of corn, partly eaten and dragged back into the pines, by these squirrels. At least five per cent of the corn in one field had been taken by the squirrels. In the lowlands, the red-bellied squirrel may locally damage cornfields, but usually the squirrel is content with wild fruits.

In the highlands, the red-bellied squirrel is esteemed for food, and is hunted for that reason as much as to protect the cornfields. In the lowlands it is almost never eaten, and never, so far as known, is it hunted for sport. The standard of living of the people of the lowlands is generally higher than in the highlands, and more desirable game species are abundant in the jungle.

In the hills near Potrero, only the colored phase was seen. In extreme southern Veracruz, to the south and east of Jesús Carranza, all but about two per cent of the red-bellied squirrels seen were black. On the Río Blanco, west of Piedras Negras, the two color phases were found in about equal numbers.

A female containing two nearly full-term embryos was taken on March 3. A male, taken on January 9, had greatly swollen testes. Common parasites are large, leathery ticks. When present, these are almost always on the head and neck. Many of these squirrels are parasitized by fly warbles. In the lowlands, almost every squirrel obtained had warbles, or the scars of warbles, and as many as three were found on a specimen. Usually they were on the back, from between the shoulders to the rump. Less often they were on the abdomen; one male had been emasculated by a larva.

Two squirrels were shot that had large parts of their tails missing. The stubs had completely healed.

Remarks.—Kelson (1952:247) studied subspeciation in *Sciurus aureogaster* and recognized two subspecies. One, *S. a. aureogaster*, is characterized by a great amount of variation in color, color patterns and relative dimensions of the skull, and is widely distributed in Veracruz. Kelson states that the only evidence of geographic variation that he could detect was a slight increase southwardly in frequency and degree of melanism. He chose not to ascribe much taxonomic significance to this increase in melanism. The specimen (82965) from farthest north in Veracruz is from Hacienda Tamiahua and was not seen by Kelson. It has more white than any other specimen from Veracruz.

Kelson makes this statement about the other recognized subspecies: ". . . although the essential morphological characters of *S. a. frumentor* occur sporadically in other populations, the animals from the higher elevations above Jico and Las Vigas are notably homogeneous, differ collectively from surrounding populations, and occupy a logical geographic range. Therefore, *S. a. frumentor* is retained as a tenable subspecies, and [all other heretofore named variants] . . . are referred to *S. a. aureogaster.*"

All of our specimens with one exception are referred to *S. a. aureogaster* on geographical grounds. The one specimen referred to *S. a. frumentor* was obtained three kilometers east of Las Vigas. Davis (1944:384) reported a specimen from the Cofre de Perote and referred it to *S. a. aureogaster*. In deference to Kelson's (*op. cit.*) later taxonomic treatment of this species, this occurrence that was recorded by Davis is here tentatively listed under the subspecies *S. a. frumentor*.

Sciurus aureogaster aureogaster Cuvier

Specimens examined.—Total 75: 1 mi. E Higo, 500 ft., 1; Hacienda Tamiahua, Cabo Rojo, 1; 35 km. NW Tuxpan, 1; Platón Sánchez, 800 ft., 3; 17 km. NW Tuxpan, 2; 12½ mi. N Tihuatlán, 300 ft., 2; 5 km. S Tehuatlan [= Tihuatlán], 700 ft., 1; 4 km. E Papantla, 400 ft., 1; 9 km. E Papantla, 300 ft., 1; 3 km. SW San Marcos, 200 ft., 2; San Carlos, 2 (Chicago N. H. Mus.); Plan del Río, 1 (Chicago N. H. Mus.); Río Atoyac, 8 km. NW Potrero, 13; 7 km. NW Potrero, 1700 ft., 4; Río Blanco, 20 km. WNW Piedras Negras, 6; Río Blanco, 20 km. W Piedras Negras, 400 ft., 5; 15 km. W Piedras Negras, 300 ft., 2; 2 km. N Motzorongo, 1500 ft., 1; 3 km. E San Andrés Tuxtla, 1000 ft., 2; 5 mi. S Catemaco, 1; 14 km. SW Coatzacoalcos, 100 ft., 1; 10 km. NW Minititlán [= Minatitlán], 100 ft., 1; Jimba, 350 ft., 1; Achotal, 13 (Chicago N. H. Mus.); 35 km. ENE Jesús Carranza, 150 ft., 2; 20 km. ENE Jesús Carranza, 200 ft., 2; 20 km. E Jesús Carranza, 300 ft., 3.

Additional records (Nelson, 1899:42-44, unless otherwise noted.)—Papantla; 5 km. N Jalapa, 4500 ft. (Davis, 1944:384); Jalapa; Jico; Chichicaxtle; Puente Nacional, 500 ft. (Davis, 1944:384); Mirador; Orizaba; Motzorongo; San Andrés Tuxtla; Catemaco; Otatitlán; Coatzacoalcos; Minatitlán.

Sciurus aureogaster frumentor Nelson

Specimen examined.—One from 3 km. E Las Vigas, 8000 ft.

Additional records (Nelson, 1899:46, unless otherwise noted).—"Near" Las Vigas; N slope Cofre de Perote, 10,500 ft. (Davis, 1944:384—as S. *a. aureogaster*); "above" Jico.

Sciurus oculatus oculatus Peters

Peters' Squirrel

Records (Nelson, 1899:89).—Las Vigas; Cofre de Perote.

Glaucomys volans herreranus Goldman

Southern Flying Squirrel

Records.—Mountains of Veracruz, the type locality of the subspecies G. *v. herreranus* (Goldman, 1936:463); Los Pescados, Cofre de Perote (Hooper, 1952:110).

Family Geomyidae

Thomomys umbrinus

Southern Pocket Gopher

Thomomys umbrinus albigularis Nelson and Goldman

Specimen examined.—One from 2 km. N Los Jacales, 7500 ft.

Our adult female has no white on the throat or midline of the venter but neither does a female topotype of *T. u. albigularis.* Longer hind foot (29 mm.) and less of, and fainter, ochraceous on the sides of the neck on the specimen from Veracruz are the only features not duplicated in specimens from the type locality of *T. u. albigularis.*

Thomomys umbrinus umbrinus (Richardson)

Record.—Boca del Monte (Bailey, 1906:6).

Heterogeomys hispidus

Hispid Pocket Gopher

The hispid pocket gopher, like other species of pocket gopher, is called "tuza" in Veracruz. It seems to be confined to the Tropical Life-zone. Usually gophers of this kind are found in clearings, either natural or artificial, but are found also beneath trees in the forest and jungle. In especially favored habitat they become locally abundant, but usually are not common. We found them in greatest abundance in the sugar cane fields of the upper humid division of the Tropical Life-zone, and in cornfields on the arid coastal plain.

Burrows of this species are usually large—in sand or loose soil they are as much as four inches in diameter. In firm, packed soil they are smaller. Mounds are usually small, and contain less than a cubic foot of earth. Burrows are plugged at a distance of from one to two feet from the mound. The gophers are principally nocturnal. In places, however, we have seen and heard them feeding by day. None was taken in traps set in the daytime, however.

At times these gophers wander about on the surface of the ground. On December 6, 1946, two kilometers north of Motzorongo, a gopher was found shortly after midnight and my (Dalquest's) field notes of the next day read as follows: "Last night, returning from a hunt, along a trail, I heard a strange clicking, rattling noise at my feet. I looked down and saw a gopher, reared back on its hind legs, with front legs back between the hind legs, and head up, chattering at me. It looked as if it were actually looking for trouble. It could easily have escaped. It is a male of medium size (343 mm.). Its testes are large—15 by 9 mm., and probably in breeding condition. It seems to be healthy and normal." A young male of this species was taken under similar circumstances at about midnight on February 10, 1947, 14 kilometers southwest of Coatzacoalcos.

This species of gopher inhabits some of the most important agricultural land in México, and its depredations on crops are serious. On the south bank of the Río Blanco, 20 kilometers west-northwest of Piedras Negras, on May 14, 1946, we found gophers to be numerous in the cornfields. "So numerous are the mounds that it appears as if the ground had been ploughed." Gophers regularly take about one-fifth of the corn crop in that area. Near Potrero, this species of gopher is abundant in some sugar cane fields. Mr. Dyfrig Forbes, manager of a large plantation, states that about 50 per cent of the cane is destroyed, in some fields, each year. This is an average loss of some 22 tons of cane to a hectare, or (at 1947 prices and exchange) about 300 pesos (60 dollars) per hectare. Control by poison, gas, and traps, is only moderately effective. Two kilometers north of Motzorongo, banana trees were seen that had fallen to the ground because gophers had cut the roots and undermined the trees. About 15 trees had fallen in one grove, as a result of gophers' activities.

The ability of these gophers to live away from fields means that they can probably never be completely eradicated. Where we found workings of gophers common on the south bank of the Río Blanco, in May, 1946, mounds were conspicuous only in the cornfields. In the extensive grassy plain surrounding the fields, mounds were rarely seen. When the grass was burned away, hundreds of new mounds were formed on the plain. The gophers were not destroyed by the fire. The mounds of white sand were conspicuous on the blackened ground. In two days in a small area near our camp more than 100 new gopher mounds appeared in an area that we had not previously been aware was inhabited by gophers.

The hispid pocket gopher seems to breed throughout the year. We took a young animal on March 13, nursing females on March 31 and October 5, a female containing two large embryos on November 23, and a young male on December 5. No parasites were found on gophers of this species save at three kilometers east of San Andrés Tuxtla, 1000 feet elevation. There the gophers were heavily infested with small mites. The stomach of a spectacled owl shot at Motzorongo, 1500 feet elevation, on December 6, 1946, contained the remains of a hispid pocket gopher.

Heterogeomys hispidus hispidus (Le Conte)

Specimens examined.—Total 12: 4 km. W Tlapacoyan, 1700 ft., 2; 5 km. N Jalapa, 4500 ft., 1; 4 km. WNW Fortín, 3200 ft., 2; Potrero Viejo, 7 km. W Potrero, 1700 ft., 1; Potrero Viejo, 5 km. W Potrero, 1; 3 km. N Presidio, 1500 ft., 2; 2 km. N Motzorongo, 1500 ft., 1; Motzorongo, 1; 5 mi. SE Lerdo de Tejada, 1.

Additional records (Merriam, 1895:183).—Jico; Huatusco; Necostla [= Necoxtla].

The specimens, in the Museum of Natural History of the University of Kansas, of *Heterogeomys hispidus* from central and southern Veracruz were studied by W. W. Dalquest who wrote as follows concerning the subspecies *H. h. hispidus:* Specimens from Jalapa, Tlapacoyan, and Fortín are typical *hispidus.* Specimens from Potrero, Presidio, and Motzorongo are intermediate between *hispidus* and *torridus* and could almost as well be placed with one subspecies as with the other. These specimens are from the upper humid division of the Tropical Life-zone, although from the lower edge of it.

Approximately 10 years later Hugh B. House studied the same material and wrote as follows: On geographical grounds, the specimen from 5 km. N Jalapa is referable to *H. h. hispidus.* Indeed, it is from the area of the type locality for the subspecies, which is designated as "near Jalapa" (Merriam, 1895:181). On anatomical grounds, as set forth by Merriam (*op. cit.*) for what he considered to be a separate species, our specimen seems to be referable to *H. h. torridus.* However, juxtaposition of temporal impressions dorso-medially on the skull, a character used by Merriam to distinguish *H. h. torridus,* is found in specimens in our collection from many parts of the range of the species. Instead of being correlated with geography, the condition seems to be associated with large size and general angularity of the skull, and in some places with relatively great age, independent of sex. This seems to apply to many of the cranial distinctions made between *H. h. hispidus* and *H. h. torridus.* Furthermore, *H. h. hispidus* is supposedly characterized by the presence of a ridge or point on the auditory bulla extending toward the hamular process of the pterygoid. The only specimens found which lacked this formation are from the geographic range of this subspecies, whereas it was found on all specimens examined of the subspecies *H. h. torridus,* supposedly characterized by the absence of this spine. Disposition here of our series of specimens from Veracruz among the various subspecies reported follows that by Dalquest and as far as I can tell was made largely on a geographical basis.

Heterogeomys hispidus isthmicus Nelson and Goldman

Specimens examined.—Total 20: Tula, 7 (U. S. N. M.); 5 mi. SE Lerdo de Tejada, 7; 3 km. E San Andrés Tuxtla, 1000 ft., 4; 14 km. SW Coatzacoalcos, 100 ft., 1; Jesús Carranza, 250 ft., 1.

Additional records (Nelson and Goldman, 1929:150).—Catemaco; Jaltipan.

Notes made by W. W. Dalquest are as follows: "Our specimens from the Tuxtla Mountains are not typical of *isthmicus.* Except in slightly smaller size they resemble typical *hispidus,* but are isolated from that subspecies by the lowlands that, I suppose, are inhabited by the small and relatively hairless *H. h. torridus.* Nelson and Goldman (1929:150) referred a specimen from Catemaco, in the Tuxtla Mountains, to *H. h. isthmicus* and until more material is available from the Isthmus of Tehuantepec it seems best to apply that name to the gophers from the Tuxtla Mountains."

Hugh B. House approximately 10 years later studied the same material and in manuscript wrote as follows: "Nelson and Goldman (1929:149) named *H. h. isthmicus* and distinguished it from other subspecies (1929:150) by the 'abrupt median, crescent-shaped, forward deflection of the lambdoidal crest.' This characterization holds up only fairly well in our series and cannot be considered diagnostic."

The geographic location of a specimen of this species from Jesús Carranza is such that it might be an intergrade between the subspecies *H. h. tehuantepecus* and *H. h. isthmicus* but since it is immature it is of little taxonomic worth.

Heterogeomys hispidus latirostris Hall and Alvarez

Specimens examined.—Total 3: Hacienda El Caracol, Tamós, 2 (U. S. N. M. 159499, 159500); Hacienda Tamiahua, Cabo Rojo, 1.

For this subspecies that is not in Hall and Kelson (1959) see Hall and Alvarez (1961:121).

Heterogeomys hispidus torridus Merriam

Specimens examined.—Total 13: 4 km. N Tuxpan, 1; 12½ mi. N Tihuatlán, 300 ft., 8; 3 km. W Guttierez [=Gutierrez] Zamora, 300 ft., 1; Río Blanco, 20 km. WNW Piedras Negras, 2; 15 km. W Piedras Negras, 300 ft., 1.

Additional records.—Chichicaxtle (Merriam, 1895:184); 4 km. S Veracruz, 30 ft. (Davis, 1944:388); Boca del Río, 10 ft. (Davis, 1944:388).

Heterogeomys lanius Elliot

Big Pocket Gopher

This species is known from a single specimen taken at Xuchil.

Cratogeomys perotensis

Perote Pocket Gopher

This large pocket gopher is called "tuza" in Veracruz, as are all pocket gophers. In some areas it is called "tuza de tierra." It seems to be confined to the pine forest of the mountains. We found it in the cool, humid meadows of zacaton grass on the Cofre de Perote, and in cornfields in the arid, lava-flow area at Las Vigas. Although its range lies entirely within the pine forest area, this gopher is not found in the forest itself, but rather in openings, natural and artificial, in the forest. Near Las Vigas we found mounds of pocket gophers principally in cornfields, or in fields that had been planted to corn the year before. Residents told us that the gophers were concentrated in the cornfields. Corn has, it should be remembered, been grown in the area occupied by this gopher for hundreds of years.

We caught this species only at night, although traps were often left in the burrows during the day. Near Las Vigas, burrows were about three inches in diameter and constructed through firm, black soil, and soft, red earth. On the Cofre de Perote and the area to the south, burrows were smaller, usually about two inches in diameter. At Altotonga and Jalacingo, burrows were in hard, red earth. Mounds of this species are usually rather small, containing about one cubic foot of earth. Near Las Vigas, the mounds averaged bigger. Burrows were usually difficult to find, for they were usually about one foot beneath the surface and were solidly plugged near the mound.

Near Las Vigas, we found these gophers feeding extensively on the roots and lower stalks of corn. Sections of roots three to four inches long and a half inch in diameter were found strewn along their burrows. Near Altotonga they were also feeding on the roots of corn. We did not discover any food plants in burrows near Pescados, on the Cofre de Perote. Probably this gopher had been a serious pest to the cornfields for centuries before the discovery of the new world by Europeans.

The gopher is a serious pest throughout the state of Veracruz, and probably does more damage than any other single kind of animal. There is no adequate means of control available to the people owning small areas of land; they depend on the crops from their land for food for themselves and their families. Traps, poison, and gas are rarely available, are expensive, and are not very effective. When the depredations of the gophers become serious enough, men stand motionless, sometimes for hours, before an open gopher burrow, ready to impale the animal on a machete.

Seemingly this species of gopher does not breed in winter. We found no embryos in animals taken in November, and trapped no very young animals in the winter months.

Gophers are destroyed by man, and probably also by other animals. Near Las Vigas, we lost one gopher that was destroyed in the trap by a gray fox. This species of gopher is characteristically parasitized by a species of small, yellowish louse. From 30 to 50 lice were estimated to occur on most individuals. Fleas also occur on this gopher, but are rare.

Two of the three recognized subspecies of this species occur in Veracruz, *C. p. perotensis* and *C. p. estor*. "*C. p. peraltus* was described from Mount Orizaba 'Veracruz,' but the specimens actually came from Puebla, as shown [by] the labels and the collector's field catalogue" (W. W. Dalquest, MS.).

Among our specimens, those that are geographically referable to *C. p. perotensis* are, as a group, darker than *C. p. estor*, and more often have a sagittal crest. When the crest is present it is more fully developed anteriorly than in the specimens geographically referable to *C. p. estor*. Neither of these two characteristics, however, can be considered diagnostic either alone or in combination.

Cratogeomys perotensis estor Merriam

Specimens examined.—Total 27: 7 km. SE Jalacingo, 8000 ft., 1; 6 km. SSE Altotonga, 9000 ft., 2; 2 km. S Sierra de Agua, 8500 ft., 1; 2 km. W Las Vigas, 8000 ft., 1; Las Vigas, 8500 ft., 17; 2 km. E Las Vigas 8000 ft., 1; 3 km. E Las Vigas, 8000 ft., 4.

Additional records.—Las Vigas, 8000 ft., type locality (Merriam, 1895:155); 5 km. E Las Vigas, 8000 ft. (Davis, 1944:388).

Cratogeomys perotensis perotensis Merriam

Specimens examined.—Five from 1 km. NW Pescados, 10,500 ft., 5.

Additional records.—N slope Cofre de Perote, 10,500 ft. (Davis, 1944:387); Cofre de Perote, 9500 ft., type locality (Merriam, 1895:154).

Cratogeomys fulvescens subluteus Nelson and Goldman
Fulvous Pocket Gopher

Specimens examined.—Total 17: 2 km. N Perote, 8000 ft., 2; 2 km. E Perote, 8300 ft., 10; 3 km. W Limón, 7500 ft., 3; 2 km. W Limón, 7500 ft., 2.

Additional records.—Perote, type locality, 7800 ft. (Nelson and Goldman, 1934:152); Guadalupe Victoria, 8300 ft. (Davis, 1944:388).

The fulvous pocket gopher, like its neighbors, is termed "tuza" or "tuza de tierra," in Veracruz, and is abundant on the sandy Perote Desert, north of the Cofre de Perote. Their fresh mounds dotted the flats and, with the mounds of the kangaroo rats and ground squirrels, were observed along the highway through the Perote Desert from Sierra de Agua to the Puebla boundary, a distance of some 40 kilometers.

The fulvous pocket gopher seems to be diurnal as well as nocturnal. We took specimens in the daytime as well as at night. We had some difficulty in setting traps for this species because the soft, loose sand in which the burrows were constructed caved in at the slightest careless move. Mounds were large; well over a cubic foot of sand was included in each of most of them. The burrows were usually about three inches in diameter, and most of them were less than a foot under the ground. They were plugged near the mound.

Little was learned of the food habits of this species. In some places we found the gophers living in numbers where there seemed to be little food—only scattered grasses, weeds and cactus. In two places we found these gophers doing considerable damage to cornfields. In one field we estimated that one-fifth of the crop had been taken by gophers. The corn raised on the Perote Desert is small and poor. In several places, beans were destroyed by the mounds of gophers. Wheat also is destroyed, by mounds which cover the shoots when the plants are small, and is eaten by the gophers as well. In some places we found where the gophers were feeding on the maguey plants, a cultivated crop. As these gophers are numerous and large, they are a serious pest where cultivation is carried on, and cultivation is extensive in the desert areas they occupy.

The series of specimens trapped in November consists principally of males. None of the females taken was pregnant. This species, like the Perote gopher, probably does not breed in winter.

Of the specimens taken, 13 were examined for parasites. Only one had lice, and that one only a few. In contrast, all of the *perotensis* examined had numerous lice.

Family Heteromyidae

Perognathus flavus mexicanus Merriam

Silky Pocket Mouse

Specimens examined.—Total 2: 2 km. W Perote, 8000 ft., 1; 3 km. W Limón [= San Antonio Limón], 7800 ft., 1.

On October 10, 1947, three kilometers west of Limón, a female of this tiny mouse was trapped on an almost white sand dune. She was not pregnant or lactating. There were scattered clumps of bunchgrass about and prickly pear (yellow fruit). Cholla, and tree yuccas were not far away. Another female taken in sand along a row of maguey plants two kilometers west of Perote on September 28, 1948, was lactating. No ectoparasites were found on these two mice.

Dipodomys phillipsii perotensis Merriam

Phillips' Kangaroo Rat

Specimens examined.—Total 16: 2 km. N Perote, 8000 ft., 1; 2 km. W Perote, 8000 ft., 1; 2 km. E Perote, 8300 ft., 7; 3 km. W Limón, 7500 ft., 3; 2 km. W Limón, 7500 ft., 4.

Additional records.—Perote, type locality (Merriam, 1894:111); Guadalupe Victoria (Davis, 1944:391).

In Spanish, the kangaroo rat of the Perote Desert is called "ratón de cola grande." We failed to record the local Indian name. This species was found only on the open, sandy desert, where there were some shrubs, weeds, cactus and maguey plants. Grasses were scattered and sparse. Much of the desert is planted in corn, beans and wheat. In the same areas where we took kangaroo rats, we found the fulvous pocket gopher (*Cratogeomys fulvescens*) and the deer mouse (*Peromyscus maniculatus*) to be abundant.

The kangaroo rat is almost entirely nocturnal. We saw one dash across the road on a bright afternoon, but this individual had probably been frightened from its burrow by a snake or other predator. The rats were difficult to obtain, although present in numbers. They refused most trap-baits, including rolled oats, banana, walnuts, peanuts and peanut butter. We laboriously thrashed out the seeds of some native weeds, and so trapped two or three specimens. Most of our series, however, was taken by concealing traps at the entrances of their burrows or by shooting the animals at night, when their eyes reflected a dull red glow in the beam of a hunting light.

In 1947 we noted that burrows of this species were not plugged. They were about three inches in diameter at the mouth. There were two to five separate entrances a few feet apart. The tunnels from these entrances joined within a meter, where the burrow was about two inches in diameter. Scratchings were seen in the sand, but few tail marks were noted. A burrow excavated in September, 1948, was a foot below the surface. The main entrance was marked by a small mound of fresh sand. The main burrow was a slightly curved tube, four feet in length, and ending in a swollen chamber. There was no nest. About midway in the burrow several leaves of a kind of dandelion were within a few inches of each other. Approximately two feet from the end of the main burrow it branched, one part consisting of a two-foot long tube terminally vertical that opened on the surface where there was no pile of sand. The kangaroo rat escaped through this entrance when we opened its burrow.

Our specimens were taken from the last week of September to mid-November. No female taken at that time was pregnant. Several males were noted as having enlarged testes. Several of the rats taken were young. We suppose that litters are raised throughout spring and summer, probably regularly until August, and perhaps occasionally later. We found no ectoparasites on kangaroo rats in Veracruz.

Liomys pictus

Painted Spiny Pocket Mouse

Where common this species sometimes is found in grassy fields, but seemingly is only a visitor there. It is more commonly found in brushland, under the dense, thorny scrub of palms and bullhorn acacia on the coastal plain, and in brushy, weed-grown borders of cornfields. The mouse prefers relatively open ground, littered with twigs and dry leaves—a habitat rarely occupied by mice in the United States. We took mice of this species by walking along the dry beds of arroyos and setting traps in the brush along the banks, where the brush was too dense to enter easily.

This species is common in the arid lower division of the Tropical Life-zone, and was occasionally taken in the humid lower division. On a few occasions it was taken in the humid upper division, but only at the very lowest edge of the zone, and where extensive sugar cane fields, or pastures extended into the plain, and presented a path of favorable habitat to the mice.

Liomys pictus is principally nocturnal. None was trapped in the daytime, but on the south bank of the Río Blanco, 20 kilometers west-northwest of Piedras Negras, on May 18, 1946, a boy hunting for lizards saw one of these mice run under a log. The time was about midday. He pulled away the log and killed the mouse by striking it with the switch of bamboo, used to kill lizards. The mouse was an adult female.

This species was rarely found in abundance. In suitable habitat from one to three specimens could be taken in a line of 100 mousetraps each night. Often only one or two could be taken at a locality, in spite of much trapping. Three kilometers east of San Andrés Tuxtla the species was abundant and we took an average of about five specimens each day from a line of 60 traps. The traps were set along the borders of fields and under dense, low bushes on a hillside, and under low bushes beside a trail near a small lake.

Contents of the cheek pouches of captured animals consisted principally of seeds. On one occasion, two specimens had in their cheek pouches the seeds of seven or more species of plants. All of the seeds were from four to six millimeters in diameter. Seeds of about the size of a garden pea seemed to be preferred. Another mouse had been collecting small seeds, about the size of small peas, that were brilliant red on one side and jet black on the other. Soft, green leaves were often found in cheek pouches, and one animal was carrying the dry, branching joint of a dead weed in its pouch.

Breeding seems to occur throughout the year. The testes of males in breeding condition become greatly swollen, and the scrotum projects back under the tail. Some breeding records include:

Sept. 28—Breeding male;
Oct. 2—Young mouse;
Oct. 3—Breeding male;
Oct. 22—Nursing female;
Jan. 8—Breeding males;
Jan. 9—Breeding males;
Jan. 10—Four young mice;
March 2—Four embryos;
May 15—Breeding male;
May 18—Breeding male;
May 28—Breeding male;
May 29—Five embryos.

The only parasites noted on this species were small mites, pale brown or red above, and white below. These were about the size of a pinhead up to two millimeters in length. They were hard, and moved swiftly. Almost every spiny pocket mouse had a few of these mites, and some individuals had as many as 50.

Liomys pictus obscurus Merriam

Specimens examined.—Total 44: Santa Maria, 4 (U. S. N. M.); San Carlos, 1 (U. S. N. M.); Carrizal, 1 (U. S. N. M.); Puente Nacional, 500 ft., 5; 4 km. W. Paso de San Juan, 250 ft., 3; Boca del Río, 10 ft., 4; 2 km. N Paraje Nuevo, 1700 ft., 1; Orizaba, 3 (U. S. N. M.); 15 km. ESE San Juan de la Punta, 400 ft., 2; Río Blanco, 20 km. WNW Piedras Negras, 8; Río Blanco, 20 km. W Piedras Negras, 400 ft., 2; 3 km. N Presidio, 1500 ft., 2; Otatitlán, 8 (U. S. N. M.)

Additional records.—Plan del Río (Davis, 1944:390); Pasa Nueva (Allen, J. A., 1904:31).

Liomys pictus veraecrucis Merriam

Specimens examined.—Total 39: 3 km. E San Andrés Tuxtla, 1000 ft., 34; 5 mi. S Catemaco, 2; 14 km. SW Coatzacoalcos, 100 ft., 2; Jimba, 350 ft., 1.

Additional records.—Santiago Tuxtla (Goldman, 1911:43); Pasa Nueva (Allen and Chapman, 1904:31); San Andreas [= Andrés] Tuxtla (Merriam, 1902:47); Catemaco (Goldman, 1911:43).

Liomys irroratus

Mexican Spiny Pocket Mouse

A single specimen of the subspecies *torridus* was captured four kilometers west of Acultzingo, 7500 feet elevation, on June 9, 1946, in a locality almost alpine in aspect, although above to the west, and below to the east, the country was arid and desertlike. A low, dense stand of bushes was five feet high, and tall trees had the most luxurious growths of mosses, orchids and bromeliads seen in México.

Two specimens of the subspecies *alleni* were trapped 10 miles southwest of Jacales, 6500 feet elevation, on the "Chicontepec Rincón," of the Mexican Plateau. They were in low, thorny bushes on a dry hillside, among cactus plants and other desert flora.

Habitats for the subspecies *pretiosus* were listed as follows: "base of a wild orange tree on a hillside slope of 30°. Cover was dense grass and weeds about three feet high"; "dense cover of grass, weeds and succulent plants"; "open jungle of low trees and brush. Much of the ground was relatively open, with only dead leaves"; "in a cornfield"; "tall, dense grass and weeds"; "open, clear ground in a cornfield"; "a marshy place along a stream." This subspecies, in contrast to *alleni* and *torridus*, is tropical, inhabiting the lowlands. In the state of Veracruz at least, *alleni* and *torridus* are races of the highlands, where they live under almost desert conditions.

Mice of this species seem to be strictly nocturnal. Specimens were trapped only at night. Burrows noted in a cornfield near Gutierrez Zamora in November were open and about one inch in diameter. There were usually three to five within a square meter. One mouse trapped there had half its body still in its burrow. The soil was dark brown, crumbly, and claylike in texture. By setting traps only near burrows in cornfields we appreciably increased our catch of this species.

A mouse of this species seems to live 30 or more feet away from any other individual of its kind. The three to five holes mentioned above are almost certainly the work of a single mouse, for traps set near the groups of burrows (perhaps entrances to a single burrow) took only a single mouse each time.

In the aggregate these little mice must do considerable damage. They undoubtedly eat corn, though we have never found corn in their pouches. Seeds of native grasses and plants often were found in the cheek pouches of the mice. They seem to prefer seeds about the size of small peas.

We have few breeding records for this species. Some records for the subspecies *pretiosus* are as follows: Sept. 20—two females, each with 4 embryos; Oct. 16—male with enlarged testes; Nov. 7—male with enlarged testes; Nov. 8—two females, each with 3 embryos; Nov. 23—young mouse.

The common parasite on this species of mouse is a small, hard mite, described in our notes as "brown above and white beneath, rather slow-moving." One mouse had a large, white tick. Several individuals were taken that had lost parts of their tails.

Liomys irroratus alleni (Coues)

Specimens examined.—Two from 10 km. SW Jacales, 6500 ft.

Liomys irroratus pretiosus Goldman

Specimens examined.—Total 78: 5 mi. S Tampico, 4; Ozulama [= Ozuluama], 500 ft., 3; La Mar, 20 ft., 1; Piedras Clavadas, 75 km. NW Tuxpan, 2; Platón Sánchez, 8000 ft., 4; Ixcatepec, 70 km. NW Tuxpan, 3; 35 km. NW Tuxpan, 2; Potrero [del] Llano, 350 ft., 4; 25 km. NW Tuxpan, 2; 17 km. NW Tuxpan, 1; 15 km. NW Tuxpan, 1; 14 km. NW Tuxpan, 1; 6 km. N Tuxpan, 1; 5 km. NE Tuxpan, 1; 4 km. NE Tuxpan, 3; 12½ mi. N Tihuatlán, 300 ft., 8; 4 km. E Papantla, 400 ft., 1; 9 km. E Papantla, 300 ft., 6; 3 km. W Guttierez [= Gutierrez] Zamora, 300 ft., 7; Miahuapa, 8 km. N Coyutla, 8 m., 6; Miahuapa (La Tulapilla), 15 km. NNE Coyutla, 80 m., 3; 1 km. ENE Coyutla, 120 m., 5; 3 km. SW San Marcos, 200 ft., 2; 4 km. W Tlapacoyan, 1700 ft., 7.

Additional records.—Nautla "near sea level" (Hooper and Handley, 1948: 20); near El Tajín Ingles (1959:394).

Liomys irroratus torridus Merriam

Specimen examined.—One from 4 km. W Acultzingo, 7500 ft.

Additional record.—Acultzingo, 7000 ft. (Hooper and Handley, 1948:13).

Heteromys lepturus Merriam

Santo Domingo Spiny Pocket Mouse

Specimens examined.—Total 14: Achotal, 11 (Chicago N. H. Mus.); 25 km. SE Jesús Carranza, 250 ft., 3.

Additional record.—San Andrés Tuxtla (Goldman, 1911:26).

On March 29, 1949, at our camp 25 kilometers southeast of Jesús Carranza an adult female spiny mouse got into the mosquito net of one of the native hunters, who was sleeping on the ground. In his attempt to capture the animal, the man accidentally lifted the edge of the net and the mouse escaped. An hour later it, or another individual, was back, and inside the mosquito net of another man. This time the mouse was captured. The same night a young male was taken in a trap, baited with peanut and set on open ground in the deep jungle. Two nights later a spiny mouse was shot as it was hunting for food on the open, leaf-littered ground beneath the tall trees in the dense jungle. The spiny pocket mouse is at its northern limit of distribution in the state of Veracruz, and is rare at the few places where it has been found.

We suppose that study of adequate material will show that *Heteromys lepturus* intergrades with *H. temporalis* and with *H.*

desmarestianus, in which case *lepturus* and *temporalis* will be arranged as subspecies of *H. desmarestianus*.

Heteromys temporalis Goldman

Motzorongo Spiny Pocket Mouse

Specimens examined.—Three from 2 km. N Motzorongo, 500 ft.

Additional record.—Motzorongo, the type locality (Goldman, 1911:27).

Our three specimens were obtained on December 6, 8, and 9, 1946. The first was taken in the dense grass and weeds of an overgrown cornfield bordered on three sides by extensive fields of sugar cane and on the other by dense jungle. The only other mammal taken in the milpa on that date was a rice rat (*Oryzomys palustris*). Further trapping yielded no other *Heteromys*, in this milpa, although other small mammals were taken. The second specimen was trapped under coffee bushes where the ground was free of grasses and weeds. None of 40 other traps had been disturbed. Fifty traps set near this place in the abundant understory vegetation of a dense, poorly tended grove of coffee bushes, yielded the third *Heteromys*. The cheek pouches of one of these mice contained two coffee berries, 29 small round berries six millimeters in diameter, and a brown husk. The testes of one male measured 20 by 11 millimeters; the animal was probably in breeding condition. No parasites were found on any of the three specimens.

Family Cricetidae

Oryzomys palustris

Marsh Rice Rat

For use of the specific name *palustris* instead of *couesi* see Hall (1960:171).

This species prefers damp, marshy habitat, but occurs elsewhere, if there is sufficient cover. On the arid coastal plain we trapped a few of the rats beneath dense, dry, thorny bushes. Where the forest of the upper humid division of the Tropical Life-zone is dense, but open enough to support grasses, brush and annuals, rice rats are often abundant.

This species is principally, though not entirely, nocturnal. Of about 150 specimens trapped, only about five were taken in the daytime. All those taken in the daytime were young males.

At times marsh rice rats become relatively abundant. Catches of one rat to each three traps are not uncommon, but the specimens taken in these periods of abundance are almost all young, half-grown animals. Usually rice rats are less common; ordinarily about one specimen is taken in 50 traps. These rats usually do not leave distinct runways in dense vegetation. Often the species is trapped in runways of other small mammals, such as *Sigmodon*. The marsh rice rat eats principally vegetation; nothing else was in stomachs of specimens trapped. One old female was taken in a steel trap baited with fish, and there was evidence that she had eaten some of the fish before becoming caught.

Breeding probably continues throughout the year. Young rats are taken at all times of the year, and always outnumber adults by about 10 to one. Some breeding records are as follows:

March 6—3 embyros;
March 19—4 embryos;
Sept. 12—7 embryos;
Sept. 19—5 embryos;
Oct. 11—4 embryos;
Oct. 15—breeding male;
Oct. 21—breeding male;
Oct. 26—3 embryos;
Dec. 6—breeding male;
Dec. 9.—breeding male.

Parasites were uncommon on marsh rice rats, but the following were recorded: two small brown mites in ear; one flea; six to eight small brown mites; three medium-sized ticks; two small ticks; a small tick, brown above, white below, that moved rapidly; a translucent bot fly larva under the skin of the abdomen, measuring 10 by 4 by 3 millimeters.

Oryzomys palustris couesi (Alston)

Specimens examined.—Total 94: 5 km. N Jalapa, 4500 ft., 8; 7 km. NNW Cerro Gordo, 1500 ft., 4; 2 km. W Jico, 4200 ft,. 1; Teocelo, 4500 ft., 7; 3 km. W Plan del Río, 1000 ft., 1; Mirador, 3500 ft., 3; Boca del Río, 10 ft., 1; Coscomatopec [= Coscomatepec], 5000 ft., 5; 5 km. SW Boca del Río, 1; Monte Blanco, 1300 m., 1; Río Atoyac, 8 km. NW Potrero, 13; 4 km. WNW Fortín, 3200 ft., 2; 1 km. W Mecayucan, 200 ft., 2; Potrero Viejo, 7 km. W Potrero, 1; Potrero Viejo, 1700 ft., 4; Cautlapan, 4000 ft., 2; 3 km. SE Orizaba, 5500 ft., 1; 7 km. SE San Juan de la Punta, 400 ft., 1; 15 km. W Piedras Negras, 300 ft., 1; 3 km. N Presidio, 1500 ft., 4; 2 km. N Motzorongo, 1500 ft., 6; 3 km. E San Andrés Tuxtla, 1000 ft., 4; Coatzacoalcos, 3; 14 km. SW Coatzacoalcos, 100 ft., 2; 35 km. ENE Jesús Carranza, 150 ft., 1; Jesús Carranza, 250 ft., 2; 20 km. E Jesús Carranza, 300 ft., 6; 25 km. ESE Jesús Carranza, 350 ft., 2; 30 km. ESE Jesús Carranza, 300 ft., 1; 55 km. ESE Jesús Carranza, 450 ft., 1; 35 km. SE Jesús Carranza, 400 ft., 2; 30 km. SSE Jesús Carranza, 300 ft., 1.

Additional records (Goldman, 1918:31).—Jalapa, 4400 ft.; Jico; San Carlos; Orizaba; Motzorongo; Tlacotalpam; Pasa Nueva; Catemaco; Achotal; Buena Vista; 75 mi. S Rivera; Ubero.

For use of the specific name *palustris* instead of *couesi* see Hall (1960:171).

Oryzomys palustris peragrus Merriam

Specimens examined.—Total 64: 5 mi. S Tampico, 1; Hacienda Tamiahua, Cabo Rojo, 10; 1 mi. E Higo, 500 ft., 1; Platón Sánchez, 800 ft., 3; Cerro Azul, 350 ft., 1; Potrero [del] Llano, 350 ft., 4; 35 km. NW Tuxpan, 1000 ft., 2; 17 km. NW Tuxpan, 2; 5 km. NE Tuxpan, 1; 4 km. NE Tuxpan, 2; San Isidro, 100 ft., 2; 12½ mi. N Tihuatlán, 300 ft., 12; 5 km. S Tehuatlan [= Tihuatlán], 700 ft., 2; Miauapa (La Tulapilla), 15 km. NNE Coyutla, 80 m., 1; 3 km. SW San Marcos, 200 ft., 2; 4 km. W Tlapacoyan, 1700 ft., 18.

Additional record.—Papantla (Goldman, 1918:31).

For use of the specific name *palustris* in place of *couesi* see Hall (1960:171).

The specimens from Hda. Tamiahua differ some from the others from Veracruz, being ochraceous-reddish instead of ochraceous-yellowish. Their skulls resemble those of *O. p. aquaticus* (from Tamaulipas) more than they do those of *O. p. peragrus*. Both *aquaticus* and the specimens from Hda. Tamiahua have more widely spreading zygomata, and the ascending branches of the premaxillae tend to exceed the nasals posteriorly. The specimens from Hda. Tamiahua are slightly larger than those from the mainland between Tampico and Tuxpan. The specimens from Hda. Tamiahua are nevertheless referred to *peragrus* because of the resemblance of the two in size and in color.

One specimen, No. 24050, from 5 km. S Tihuatlan is larger than any other from Veracruz, including those from Hda. Tamiahua.

Oryzomys melanotis rostratus Merriam

Black-eared Rice Rat

Specimens examined.—Total 34: Cerro Azul, 350 ft., 1; Tlacolula, 60 km. WNW Tuxpan, 1; Potrero [del] Llano, 350 ft., 1; 35 km. NW Tuxpan, 1000 ft., 5; 17 km. NW Tuxpan, 1; 5 km. NE Tuxpan, 1; 4 km. NE Tuxpan, 2; Tuxpan, 2; 3 km. SW San Marcos, 200 ft., 2; Mirador, 3500 ft., 2; Cautlapan, 4000 ft., 1; 3 km. N Presidio, 1500 ft., 5; 2 km. N Motzorongo, 1500 ft., 2; 3 km. E San Andrés Tuxtla, 1000 ft., 5; 20 km. E Jesús Carranza, 300 ft., 2; 35 km. SE Jesús Carranza, 400 ft., 1.

Additional records (Goldman, 1918:54).—San Carlos; Motzorongo; Pasa Nueva; Achotal.

At several localities we took one black-cared rice rat, and at other places two rats, but concentrated and continued trapping at these places seldom succeeded in capturing more. The species seems to be rather generalized as to its habits. Some data on habitat are as follows: Three kilometers north of Presidio specimens were trapped at the edge of a dense hillside jungle. An extensive sugar cane field extended up to the scattered limestone boulders at the foot of the hills. Our traps took *Oryzomys palustris* and *Sigmodon* in addition to *O. melanotis*. Two kilometers north of Motzorongo, two were trapped under weeds and coffee bushes on a hillside, along with *Oryzomys palustris* and *Heteromys*. Three kilometers southwest of San Marcos, a specimen was taken in deep jungle, along with *Peromyscus mexicanus*. At Mirador, two were trapped in an oak forest, where there was considerable thin, open underbrush and a few fallen logs. Three kilometers east of San Andrés Tuxtla, three were taken in deep jungle and another in dense bushes about three feet high. A few breeding notes are as follows: November 5, female with three small embryos; December 9, male with enlarged testes; December 12, one female with three embryos and another with four embryos. We recorded no parasites from this species.

Oryzomys alfaroi

Alfaro's Rice Rat

Alfaro's rice rat, or the blackish rice mouse as we called it in the field, inhabits dense vegetation, especially in damp places. Four kilometers west of Tlapacoyan and also 1 kilometer east of Jalacingo, the species was obtained in dense growth of low, succulent plants along small streams. Usually, in the uplands, this species is found with *Crytotis mexicana, Microtus quasiater* and *Oryzomys palustris.* One individual was taken in a small, marshy swale. Near (two kilometers west) Jico, a small series was trapped in marshy places and in damp meadows of tall grass. At Mirador, specimens were trapped in succulent vegetation along the border of a tiny stream and in mossy, overgrown places in an oak forest at the edge of a coffee grove. At Teocelo, specimens were obtained in tall grass at the edge of a sugar cane field.

Oryzomys afaroi palatinus is a lowland, jungle-inhabiting subspecies. We took it in small areas of tall grass, on the flood plains of rivers, and in deep jungle. In the jungle, the rice rats were found on bare, dry, leaf-littered ground. Two were shot on the floor of the jungle at night. They were heard rustling the leaves, probably hunting for food. Another was found under a log in the jungle. The log was beside a trail, on level ground. A growth of bushes and vines made a dense wall along each side of the log for half of its length. About six feet of the other half was slightly elevated above the ground, and there was no dense growth of vegetation along that part of the log. When the log was rolled to one side, a nest was revealed. It consisted of a loose ball of soft, brown vegetation, probably shredded bark, about five inches in diameter and three inches high, and was situated where the elevated part of the log came into contact with the ground. The nest was open on top, the log having formed the roof. The half-grown, male Alfaro's rice rat obtained from under the log was presumed to have used the nest.

Few parasites were recorded from *Oryzomys alfaroi.* One had a medium-sized tick. Each of two specimens taken on April 8 had four embryos.

Oryzomys alfaroi chapmani Thomas

Specimens examined.—Total 37: 4 km. W Tlapacoyan, 1700 ft., 3; 1 km. E Jalacingo, 6500 ft., 6; 5 km. N Jalapa, 4500 ft., 1; 2 km. W Jico, 4200 ft., 13; Teocelo, 4500 ft., 6; Mirador, 3500 ft., 3; Huatusco, 5000 ft., 3; Coscomatopec [= Coscomatepec], 5000 ft., 2.

Additional records (Goldman, 1918:68).—Jalapa; Jico; Mirador.

On geographic grounds the specimens from the summit and northern side of the barrier range (1 km. E Jalacingo, and 4 km. W Tlapacoyan) would be expected to resemble *O. a. dilutior* but instead, in small and lightly-constructed skull, resemble *O. a. chapmani.*

Oryzomys alfaroi palatinus Merriam

Specimens examined.—Total 13: 20 km. ENE Jesús Carranza, 200 ft., 1; 22 km. ESE Jesús Carranza, 300 ft., 1; 25 km. ESE Jesús Carranza, 350 ft., 4; 25 km. SE Jesús Carranza, 250 ft., 4; 63 km. ESE Jesús Carranza, 500 ft., 1; 35 km. SE Jesús Carranza, 400 ft., 1; 35 km. SE Jesús Carranza, 350 ft., 1.

Our specimens from the jungles of southern Veracruz are indistinguishable from the holotype of *O. a. palatinus.*

Oryzomys fulvescens fulvescens (Saussure)

Pygmy Rice Rat

Specimens examined.—Total 54: Ozulama [=Ozuluama], 500 ft., 1; Platón Sánchez, 800 ft., 3; Tlacolula, 60 km. WNW Tuxpan, 1; Cerro Azul, 350 ft., 1; 12½ mi. N Tihuatlán, 300 ft., 2; 5 km. S Tehuatlan [=Tihuatlán], 700 ft., 1; 4 km. W Tlapacoyan, 1700 ft., 1; 5 km. N Jalapa, 4500 ft., 1; Jalapa, 1 (Chicago N. H. Mus.); 2 km. W Jico, 4200 ft., 1; Jico, 1 (Chicago N. H. Mus.); Teocelo, 4500 ft., 4; Teocelo, 4000 ft., 2; Mirador, 3500 ft., 1; 15 km. ENE Tlacotepec, 1500 ft., 1; 4 km. W Paso de San Juan, 250 ft., 1; 3 km. W Boca del Río, 10 ft., 2; Boca del Río, 10 ft., 1; 3 km. SW Boca del Río, 10 ft., 1; Coscomatopec [=Coscomatepec], 5000 ft., 2; Potrero Viejo, 1700 ft., 9; 3 km. SE Orizaba, 5500 ft., 2; Monte Blanco, 1300 m., 1; Cautlapan, 4000 ft., 3; 2 km. N Motzorongo, 1500 ft., 1; 3 km. E San Andrés Tuxtla, 1000 ft., 3; Achotal, 4 (Chicago N. H. Mus.); 20 km. E Jesús Carranza, 300 ft., 2; 25 km. ESE Jesús Carranza, 200 ft., 1.

Additional records (Goldman, 1918:91).—Mirador; Orizaba; Santiago Tuxtla; Pasa Nueva.

Specimens from northern Veracruz are smaller than specimens from central Veracruz, but are about equal in size to specimens from the southern parts of the state. We are not able to separate into subspecies Veracruzian specimens of this species on the basis of size or cranial characters, and specimens from the entire state are referred to the subspecies *O. f. fulvescens.*

The pygmy rice rat, termed pygmy rice mouse by us in the field, prefers a drier habitat than other members of its genus in Veracruz. It closely resembles the harvest mice (genus *Reithrodontomys*) in habits, as it does in appearance, and is usually taken in the trap lines that yield harvest mice. In the field, pygmy rice rats can be distinguished from *Reithrodontomys* by the plain, rather than grooved, anterior surfaces of the upper incisors.

We found the pygmy rice rat in dense vegetation, such as tall grass, brush, weeds, sugar cane, weedy stands of corn, meadows of sawgrass, patches of succulent plants, and thickets of tree fern. No distinct runways were identified with this species, but in its usual habitat, the cotton rat, *Sigmodon*, is common, and the pygmy rice rat, as well as harvest mice, *Reithrodontomys,* and pygmy mice, *Baiomys,* were commonly trapped in the runways of cotton rats.

Like other species of the genus *Oryzomys*, the young of this species are far more common than adults, although the disparity in numbers of the two ages was less than in *Oryzomys palustris*. Some breeding records for the pygmy rat are: November 15, three embryos; December 5, four embryos; December 17, four embryos; January 9, four embryos.

Parasites are usually present on this species. Some recorded are as follows: black larve of a large bot fly, under the skin of the abdomen, the larva being 14 mm. long, eight mm. wide, six mm. high, and having seven deep ridges or rings; two small ticks; mites.

Tylomys gymnurus Villa

Naked-tailed Climbing Rat

Specimens examined.—Total 6: Río Atoyac, 8 km. NW Potrero, 1; 3 km. N Presidio, 1500 ft., 1; 2 km. N Motzorongo, 1500 ft., 1; 3 km. E San Andrés Tuxtla, 1000 ft., 2; 25 km. SE Jesús Carranza, 250 ft., 1.

Additional record.—Presidio (Villa, 1941:763).

All of our specimens from Veracruz here are referred to *T. gymnurus*, the type locality of which is Presidio. This is done in spite of the fact that the two specimens from 3 km. E San Andrés Tuxtla differ considerably from material from Presidio and vicinity. Two other nominal species of *Tylomys* have been named by Merriam (1901). They are *T. tumbalensis* from Tumbala, Chiapas, and *T. bullaris* from Tuxtla Gutierrez, Chiapas, based on specimens so young as to permit little if any more than generic identification. Each of the two is known only from the holotype. When adults of *tumbalensis* and *bullaris* are obtained, the single specimen from 25 km. SE Jesús Carranza probably will prove to be referable to *tumbalensis* or *bullaris*, and the specimens from 3 km. E San Andrés Tuxtla may be found to represent an unnamed kind. We suppose that all of the kinds mentioned above are subspecies of a single species, but specimens have yet to be collected that will permit an objective decision.

Climbing rats of the genus *Tylomys* seem to be uncommon in Veracruz. They are difficult to trap, usually refusing to take bait of any kind. We found them only in the tall, dense forests, and in most cases, near rocks and cliffs. Several were seen in caves, and in a number of other caves piles of the droppings of a large rat, almost certainly of the climbing rat, were seen.

Since almost nothing has been recorded of the habits of this rat, the following notes taken at the time of capture of specimens are included:

At the Río Atoyac, eight kilometers northwest of Potrero, 1500 feet elevation, on March 12, 1946, I set a line of steel traps on the hillside in extremely dense, junglelike forest. Vines and creepers were so dense that it was almost impossible to penetrate the area without a machete. Quail doves were common here, and on a previous day I had shot an owl here in the daytime. A band of *Nasua* was seen and the traps were set for this animal. On the following morning a large rat was found in a trap set in a narrow pass between two limestone boulders. Under the overhang of one of the boulders each one of about a double handful of dry coffee berries had a large hole eaten in one side. Only a few rat droppings were noted in the vicinity. The specimen was a female containing two large embryos.

Three kilometers north of Presidio, on December 3, 1946, "at the type locality of this rat, we took a specimen after four days of trapping for it. Traps were set in dense jungle on the hillside. They were baited with prunes, raisins, and rolled oats without success. On the fourth day we tried our old standby—banana, though it is rather soft for rat trap bait. It was successful, however. Traps were set in trails, under and on limestone boulders, on root masses, and tied in trees. The rat was taken in a trap set on a ledge, about four inches wide, on a limestone outcrop or cliff about 10 feet high and 50 feet long. The entire area where the traps were set is densely grown with vegetation. Trees are tall, about 100 feet. The crowns of the trees are grown into great masses of vines and creepers, with much parasitic and epiphytic growth—orchids, bromeliads, ferns and mosses. Lower is dense brush, some coffee but mainly other plants. This reaches an average height of seven feet and it too is laced with vines. The ground is rather dark and free of grasses, etc.

"The rat is a male. Unfortunately the skull was broken by the trap. The tail is very hard-looking. The black portion has an enameled look, almost polished. The white is ivory-color except near the tip, where it is flesh tinted. No ectoparasites were found. The stomach contained finely-chewed, olive-green, odorless vegetation. The testes were large (20 by 11 mm.) and probably in breeding condition."

Two kilometers north of Motzorongo on December 11, 1946, I wrote that we had not found sign of this rat among the limestone boulders in the jungle that I considered typical habitat for the rat and had not caught even one rat in rat traps set there for several days. "Last night I set ten rat traps, tied to branches and vines, in low, dense trees along the arroyo here, and baited them with bananas in hope of catching a fruit bat. The area is fully 100 meters from limestone, though the jungle is all about. This morning one trap held a *Tylomys*. The trap was over the water of a pool, in dense vines and branches, about 15 feet from the water. Something had eaten the head completely off, but I saved the body skin It is a small male with testes only ten by five millimeters."

Three kilometers east of San Andrés Tuxtla on January 10, 1948. "Last night a large rat was seen in the jungle. I am almost sure it was this species. It ran across a horizontal vine about three inches in diameter and vanished in the branches of a tree. It moved jerkily, almost like a *Peromyscus*, but almost unbelievably fast. I could not get a shot at it.

"This afternoon we set six traps baited with walnuts, in the trees and vines where the rat was seen."

In the next three days the traps were visited every morning, and again in the evening, and various baits were tried. Climbing the trees was difficult, and was made unpleasant by the swarms of ants that viciously bit every exposed part of the body. Bananas and walnuts failed to take the rat, but on January 11, it, a fine adult female, was caught in a trap baited with a prune. Immediately anterior to the vagina a penislike structure was 20 mm. long, four wide, and three thick, a flattened structure without ossification. The mammae were all abdominal, close to the anus and vagina. The female contained two large embryos. The animal was not fat and no ectoparasites were noted on it.

Its stomach was full, not of fruit as I had expected, but of finely-chewed, brownish green, fibrous vegetation. It looked like lichen, had a rank odor, and may have been bark.

At the same locality on January 14 about one hour after dark, I saw a tree rat about 60 feet from the ground in a capulin tree in the jungle. I was able to recognize it as a *Tylomys* by the large amount of white, its long body, and small, bright red eyes. It ran along a horizontal branch about four inches in diameter to an old squirrel's nest. As I lost one rat by waiting for it to stop, I shot this one as it ran. It stuck in the tree. Next morning the rat still was in the tree. After we tried several means of dislodging it, and failed, we had to cut down the tree. This took all morning, working by turns with machetes. The rat was a fine large male with large testes.

Although he was in a capulin tree with ripe fruit, where birds and squirrels fed by day and marsupials by night, the rat had no fruit in its stomach, but only the yellowish and dark green pasty mass. No ectoparasites were noted, but the rat had been dead in the tree for 16 hours before we handled it.

A small male, seemingly in adult pelage, was obtained 25 kilometers southeast of Jesús Carranza, in extreme southern Veracruz, on August 31, 1949. He was caught in a live trap, baited with banana, tied in a tree about 10 feet from the ground.

Nyctomys sumichrasti sumichrasti (Saussure)

Sumichrast's Vesper Rat

Specimens examined.—Total 14: Coscomatepec, 5000 ft., 1; 20 km. E Jesús Carranza, 300 ft., 1; 25 km. SE Jesús Carranza, 250 ft., 9; 35 km. SE Jesús Carranza, 350 ft., 1; 38 km. SE Jesús Carranza, 500 ft., 2.

Additional records.—Eastern slope of the mountains in Veracruz (Saussure, 1860:107); Uvero (Sumichrast, 1882:325); Jalapa (Allen and Chapman, 1897:204).

The type locality of the subspecies *N. s. sumichrasti* was recorded in the original description as the eastern slope of mountains in Veracruz. Sumichrast's (1882:325) account suggests that Uvero is the type locality. Our only record of *Nyctomys* from the eastern slope of the principal mountain range is a tail, taken in a mouse trap at Coscomatepec, 5000 feet elevation. It is unlikely that our specimens from near sea level in southern Veracruz are of the same subspecies as that living in the upland forest of central Veracruz, but in the absence of specimens from the latter area we use the name *N. s. sumichrasti* for all of our specimens.

The tail referred to above was in a trap set beside a tiny stream, under low, dense, woody plants and low, twisted trees.

In southern Veracruz, the vesper rat is common but difficult to catch. Twenty kilometers east of Jesús Carranza, an adult female was shot at night in a palm jungle. She was on the branch of a bush, about five or six feet from the ground, and ran along a horizontal vine and palm frond. Traps set in the area took no additional specimens. Traps tied in trees and bushes 38 kilometers southeast of Jesús Carranza took the tail of one specimen and two young animals.

At places 25 and 35 kilometers southeast of Jesús Carranza, in the last week of March and the first weeks of April, 1949, we hunted vesper rats in hollow trees in the jungle. We took our first specimen while hunting for the two-toed anteater (*Cyclopes*). A fire was built in the hollow of a tree. The rat ran from a small opening in the trunk about 10 feet from the ground. Several young were clinging to its teats. We found that by investigating every hole in every forest tree, especially holes from one to three inches in diameter, leading to relatively small hollows, nests of this species could be found in fair numbers. The hollows were probed with sticks, or, if the hollow was sinuous, with sections of flexible vine. When necessary, fires were built and the hollows were smothered in smoke. The rats were forced to flee and often emerged from the hollows and climbed the tree trunk or attempted to escape by running along horizontal limbs. They ran at fair speed, not so slowly as the red tree mouse (*Phenacomys longicaudus*) of California, but more slowly than one would expect. Five specimens, all adult females with small young clinging to their teats, were shot while attempting to escape.

The nests of *Nyctomys*, as we found them, and including only brood nests as far as we know, were irregular masses of shredded bark that formed platforms at the bottoms of, or across, hollows in trees. The hollows varied from six to 13 inches in diameter. The nests were often superimposed on other, older nests. One hollow contained five nests, one on top of another. Only the upper nest was new and in use. Each of the nests contained a central cavity in the form of a flattened sphere, or oval, about four inches in diameter and three inches high. The nests, especially the old nests, were the homes of many insects, spiders and scorpions, and a lizard was in one. In spite of this we found no ectoparasites on any of the specimens examined.

Reithrodontomys megalotis saturatus J. A. Allen and Chapman

Western Harvest Mouse

Specimens examined.—Total 54: 2 km. N Los Jacales, 7500 ft., 1; 4 km. S Jalacingo, 1; 1 km. W Las Vigas, 8500 ft., 14; Las Vigas, 8500 ft., 22; 3 km. E Las Vigas, 8000 ft., 2; 1 km. NW Pescados, 10,500 ft., 4; 3 km. W Limón, 7500 ft., 4; 3 km. W Limón, 7800 ft., 2; 2 km. W Limón, 7500 ft., 3; 3 km. W Acultzingo, 7000 ft., 1.

Additional records (Hooper, 1952:60).—Jalacingo, 5500-6000 ft.; near Altotonca [= Altotonga], 6000 ft.; vicinity of Las Vigas, 7500-8000 ft.; Perote, 8000 ft.; Guadalupe Victoria, 8300 ft.; N slope Cofre de Perote, 10,000-10,200 ft.; Cofre de Perote; Volcán de Orizaba, timber line; Xuchil. (The specimens recorded by Howell, 1914, page 37, as from the State of Veracruz, at Huachinango and Mount Orizaba, seem actually to be from the State of Puebla.)

In Veracruz the western harvest mouse is confined to the higher western part of the state. It occurs in a number of habitats. Within a three-mile radius of Las Vigas we took specimens in a marshy place, along the bank of a tiny stream, in zacaton grass, in an old field overgrown with weeds, in brush along the border of a cornfield, and once in a pine forest. The same trap lines usually took *Cryptotis mexicanus* and *Microtus mexicanus*. On the Cofre de Perote, we took this mouse in meadows of tall zacaton grass, and two to three kilometers west of Limón on the desert we captured specimens among rocks, cactus, yuccas, and on the open, sand flats. The habitat range of the species is wide, but its altitudinal range is limited. In Veracruz we found it from 7000 to 10,500 feet.

At most places the western harvest mouse was uncommon, although in some habitats, for example in an old, weed-grown field three kilometers east of Las Vigas, it was abundant. Generally speaking, it was more abundant than other species of harvest mice in Veracruz.

Most of our specimens were taken in winter, and we, therefore, have few breeding records. On October 4, a female had four embryos. On November 3, a female had recently given birth to young.

Several nests of this mouse were found within a three-mile radius of Las Vigas. Two were in a field under old boards so rotten that they crumbled when overturned. The nests were rounded, and set in small cavities in the ground. Outside diameters were five and five and one half inches. The central cavities were small, only about two inches in diameter. Other nests were found in low, stone walls, when the stones were moved in our search for snakes and other animals. Well-worn trails, runways and burrows formed a network among the stones, and several harvest mice were captured by hand when they tried to escape along these runways. The nests in the rock walls resembled those under the boards, but were more irregular in shape. No young mice were found in these nests.

Reithrodontomys chrysopsis perotensis Merriam

Volcano Harvest Mouse

Specimen examined.—Cofre de Perote, 9500 ft., 1 (type in U. S. National Museum).

Additional record (Hooper, 1952:90).—N slope Cofre de Perote, Los Conejos, 10,600 ft.

Reithrodontomys sumichrasti sumichrasti (Saussure)

Sumichrast's Harvest Mouse

Specimens examined.—Total 8: 1 km. E Jalacingo, 6500 ft., 1; Las Vigas, 8500 ft., 3; 5 km. N Jalapa, 4500 ft., 1; 3 km. W Acultzingo, 7000 ft., 1; 4 mi. SW Acultzingo, 7000 ft., 2.

Additional records (Hooper, 1952:74).—Jalacingo, 5500-6000 ft.; Altotonca [= Altotonga], 6000 ft.; vicinity of Jalapa, 4500 ft.; vicinity of Jico, between 4500 and 6500 ft.; Orizaba; Maltrata, 6000 ft.; vicinity of Acultzingo.

This species of harvest mouse seems to be rather rare. We took only a few specimens. Five kilometers north of Jalapa, on October 20, 1946, one was taken in a swamp supporting purple-flowered plants having stiff stalks about five feet high; fuzzy, tall, stiff grass, about four feet high; thorny trees; vines and other dense vegetation. Beneath the dense mat of roots, two feet or more deep, the soil was cold and damp. *Oryzomys alfaroi*, *Oryzomys palustris* and *Microtus quasiater* were taken in the same trap line.

Three kilometers west of Acultzingo, one specimen of *R. s. sumichrasti* was obtained in brush as were two specimens of *Peromyscus difficilis.* The bushes were high and dense. Numerous large clumps of ferns, mosses, bromeliads, and orchids grew there. A tall (eight foot) bush was in fruit, and provided abundant small, yellowish, applelike fruit.

Reithrodontomys fulvescens

Fulvous Harvest Mouse

We took a specimen 10 kilometers southwest of Jacales and another six kilometers west-southwest of Zacualpilla in the dry desert of the Chicontepec Rincón amongst dry grass, maguey and cacti. In the upper humid division of the Tropical Life-zone, a few of our specimens were taken in sugar cane fields, but most were taken in the tangles of weeds and grasses along the borders of cornfields. Only at Teocelo, did we find the upland subspecies, *difficilis,* to be common.

The lowland subspecies, *tropicalis,* occupies dry, grassy areas, and is readily taken also in the weeds and grasses along the borders of fields. Generally, a few fulvous harvest mice are to be found in the extensive fields of sawgrass on the coastal plain.

Wherever we found the species in Veracruz, whether in arid desert, coastal plain or humid jungle, it was in weeds, grasses, vines, or other dense low-growing vegetation. Other rodents most commonly associated with this species were *Sigmodon hispidus, Oryzomys fulvescens* and *Baiomys.* Some breeding records for the subspecies *tropicalis* are:

Oct. 28—four embryos; also nursing female;
Oct. 30—four embryos; also three embryos;
Nov. 14—nursing female;
Nov. 27—two nursing females;
Dec. 12—three embryos;
Jan. 17—breeding male.

Parasites are not common on this species. Small mites were noted on some specimens and a tick was attached to the head of one mouse.

Reithrodontomys fulvescens difficilis Merriam

Specimens examined.—Total 32: 3 km. W Zacualpan, 6000 ft., 1; 10 km. SW Jacales, 6500 ft., 1; 6 km. WSW Zacualpilla, 6500 ft., 1; Teocelo, 4500 ft., 24; 15 km. ENE Tlacotepec, 1500 ft., 2; Potrero Viejo, 1700 ft., 3.

Additional records (Hooper, 1952:111).—1½ mi. N Jalapa, 4500 ft.; 4 mi. SE Jalapa, 4800 ft.; Jico, 4300 ft.; Mirador, 3500 ft.; Orizaba, 4000-4200 ft.; Río Blanco, 4200 ft.

Reithrodontomys fulvescens tropicalis Davis

Specimens examined.—Total 63: Chijal [= Chijol], 3; 19 mi. W Tampico, 7; 5 mi. S Tampico, 4; Tampico Alto, 50 ft., 2; 1 mi. E Higo, 500 ft., 1; Platón

Sánchez, 800 ft., 2; Potrero [del] Llano, 350 ft., 3; 35 km. NW Tuxpan, 1000 ft., 3; 15 km. NW Tuxpan, 6; 6 km. N Tuxpan, 1; 3 km. E Tuxpan, 1; San Isidro, 100 ft., 2; 12½ mi. N Tihuatlan, 300 ft., 3; 5 km. S Tehuatlan [=Tihuatlán], 700 ft., 1; 7 km. NNW Cerro Gordo, 20; Río Blanco, 20 km. WNW Piedras Negras, 3; 15 km. W Piedras Negras, 300 ft., 1.

Additional records (Hooper, 1952:109).—Nautla, 75 ft.; San Carlos, 25 ft.; vicinity of Plan del Río, 900-1000 ft.; Carrizal, 1000 ft.; Boca del Río, 10 ft.; Presidio, 1000 ft.; Catemaco, 1100 ft.

Reithrodontomys mexicanus mexicanus (Saussure)

Mexican Harvest Mouse

Specimens examined.—Total 17: 35 km. NW Tuxpan, 1000 ft., 1; 4 km. W Tlapacoyan, 1700 ft., 5; 1 km. E Jalacingo, 6500 ft., 2; 2 km. W Jico, 4200 ft., 2; Teocelo, 4500 ft., 1; Coscomatapec [= Coscomatepec], 5000 ft., 3; Río Atoyac, 8 km. NW Potrero, 1; Cuautlapan, 1; 3 km. SE Orizaba, 5500 ft., 1.

Additional records.—Altotonca [= Altotonga], 6000 ft. (Hooper, 1952:144); Jalapa, 4000 ft. (Merriam, 1901:552); Córdoba, 3000 ft. (Hooper, 1952:144).

This is a rare species of harvest mouse. A few specimens were taken, as often as not only one at a place. In each of several localities where a specimen was obtained intensive trapping usually failed to bring additional individuals to light.

Along the Río Atoyac, eight kilometers northwest of Potrero, 1700 feet elevation, one of these mice was taken in a clearing that had been a banana grove. Numerous limestone boulders lay among the thistles and other herbaceous plants. Two kilometers west of Jico one was taken among scattered boulders in a coffee grove at the foot of a cliff, and another was taken, along with several individuals of *Oryzomys alfaroi,* in a line of 125 traps set at the edge of a cleared coffee grove in bushes alongside a stone fence and a cliff. Three kilometers southeast of Orizaba one was taken in brush and grasses at the rim of the gorge of the Río Blanco. At Teocelo one was trapped in the runway of a pine mouse (*Microtus quasiater*) through the lush, herbaceous vegetation at the edge of a tiny stream on a steep hillside.

On December 1, 1948, at Coscomatepec, we were removing large bromeliads from trees, in search of reptiles and amphibians, when the nest of a Mexican harvest mouse was found. It was a ball of grasses and root fibers, about nine inches in diameter, rather loosely constructed, in plain view about eight feet from the ground on top of a large bromeliad growing from the trunk of a thorny tree. There were no twigs or limbs below the nest and none for several feet above it. When first sighted, the nest was mistaken for an old nest, of a bird, that had fallen from the tree above. Not until the bromeliad had been removed from the tree, and the nest partially pulled apart, did we find it to be a fresh nest. The three large young mice in the nest were two males and one female. All were captured and prepared. The young measured from 160 to 168 mm. in total length.

A female of this species, from eight kilometers northwest of Potrero, contained four small embryos on February 28, 1946.

The specimen (No. 83125) from 35 km. NW Tuxpan, 1000 ft., is *R. mexicanus,* but differs from other specimens of R. *mexicanus mexicanus* from Veracruz. The pelage is paler, lacks any of the darker color in the middle dorsal area, and the sides are more cinnamon. The skull is 0.8 mm. longer than the maximum re-

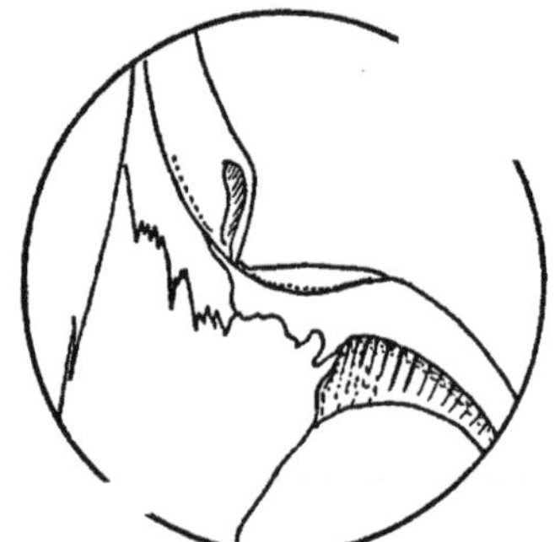

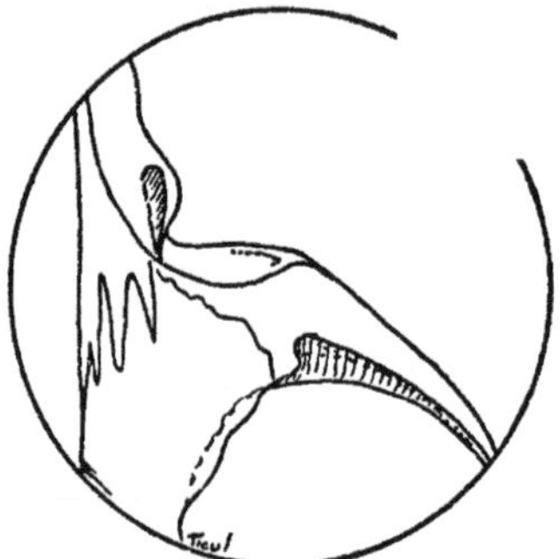

FIG. 2. Dorsal view of right root of zygomatic arch in two adult females of *Reithrodontomys mexicanus mexicanus* from Veracruz. × 6.
Left. No. 83125, from 35 km. NW Tuxpan, 1000 ft.
Right. No. 19396, from 27 km. W Jico, 4200 ft.
Note difference in depth and shape of zygomatic notch, and difference in shape of maxillary root of zygoma.

corded by Hooper (1952), and from other specimens of *R. m. mexicanus* differs as follows: deeper; rostrum broader and longer; maxillary tooth-row shorter than the minimum; incisive foramina smaller; keel of zygomatic plate scarcely visible when the skull is viewed from directly above; zygomatic notch different in form as shown in figure 2.

Peromyscus maniculatus fulvus Osgood

Deer Mouse

Specimens examined.—Total 184: 1 km. W Las Vigas, 8500 ft., 14; Las Vigas, 8500 ft., 11; 2 km. W Perote, 8000 ft., 55; 2 km. E Perote, 8300 ft., 56; 2 km. E Perote, 7000 ft., 8; 1 km. NW Pescados, 10,500 ft., 4; 3 km. W Limón, 7800 ft., 6; 3 km. W Limón, 7500 ft., 5; 2 km. W Limón, 7500 ft., 23; 4 mi. SW Acultzingo, 7000 ft., 2.

Additional records (Osgood, 1909:87, unless otherwise noted).—Perote, 1½ mi. S Perote, 8500 ft. (Hooper, 1957:5); Guadalupe Victoria (Davis, 1944:394); Cofre de Perote; Xuchil.

The deer mouse occupies open habitats at high elevations. We found it to be more abundant on the sandy soil of the desert north of the Cofre de Perote than elsewhere. There, on the open desert flats, mice, principally of this species, were more abundant per unit of area than anywhere else in Veracruz where we collected. More than 50 deer mice were taken in 100 traps in each of several nights.

The deer mouse is nocturnal. Of the several hundred trapped, we caught only one in the daytime. It was young, in the gray coat, beneath a maguey plant. This species begins to appear on the surface of the ground shortly after dark sets in, but does not emerge in abundance until about one half hour after dark. From then until midnight, deer mice are extremely active, and on a dark, moonless night may be seen by the hundreds. After midnight they are less active. Traps that were examined at several different times in the night caught by far more mice between one half hour after dark and midnight than later.

The homes of these mice near Perote, in early November, 1946, were burrows in the soft sand. More individuals were seen and trapped near maguey plants than on open sand, but the mice were present on the open sand, free of any but the most sparse cover, in great numbers. Mice were seen to enter burrows on a number of occasions, when watched by the aid of a flashlight. Attempts to excavate these burrows, both at night and in the following day, failed, because the soft sand caved in and the burrows were lost in less than a meter. One kilometer northwest of Pescados the burrows of these mice were open holes, leading down vertically into the ground, in a rather dry meadow. Here the soil was too hard and too rocky to excavate for any considerable distance.

In their ordinary movements at night, these mice are rather slow and deliberate. They climb well, and many times were seen on the big leaves of the maguey plants, several feet from the ground. When startled, the mice made a series of erratic leaps, with many sidewise hops, and were difficult to catch with the hands. When captured, they bit viciously, and their long, sharp incisors were capable of inflicting painful wounds.

At Las Vigas, these mice were fairly common in the long rows, or "cercas," of maguey plants. Here they were found with the Mexican vole (*Microtus mexicanus*), harvest mice (*Reithrodontomys*) and shrews (*Cryptotis* and *Sorex*). On the Perote Desert deer mice were caught in the same line of traps that caught the kangaroo rat (*Dipodomys phillipsii*) and the Perote mouse (*Peromyscus bullatus*). These are all species found on the loose sand. We took no deer mice among rocks nearby; the rocks were occupied by the big-eared rock mouse, *Peromyscus difficilis*.

Deer mice eat principally weed seeds, and probably some soft plant material. They eat some meat; several small mammals were eaten in the traps near Perote, and the tracks on the loose sand showed that they were eaten by this species. On the Perote Desert, we found many deer mice parasitized by the warble of a large fly. These warbles were chestnut brown, had heavy tranverse ridges, and were three-quarters of an inch long, one half inch wide, and one-quarter of an inch high. One large flea was noted. One mouse had a large swelling on the hind leg, and another had a large internal cyst. The remains of deer mice were found in the stomach of an opossum (*Didelphis*) from Perote.

Seemingly this species rarely breeds in winter. In November, 1946, northeast of Cofre de Perote (see list of specimens examined) only a few of the many mice trapped were young. Also, only one of more than 50 females examined was pregnant. She had three small embryos. At Las Vigas, in late October, 1948, only a few of the mice collected were young, and no pregnant females were taken.

Peromyscus melanotis J. A. Allen and Chapman

Black-eared Mouse

Specimens examined.—Total 27: 10 km. SW Jacales, 6500 ft., 1; 6 km. SSE Altotonga, 9000 ft., 2; 1 km. W Las Vigas, 8500 ft., 10; Las Vigas, 8500 ft., 3; 2 km. E Las Vigas, 8000 ft., 1; 4 km. SE Las Vigas, 9500 ft., 1; 1 km. NW Pescados, 10,500 ft., 9.

Additional records (Osgood, 1909:112, unless otherwise noted).—Perote; 1½ mi. S Perote, 8500 ft. (Hooper, 1957:5); N slope Cofre de Perote (Davis, 1944:395); Cofre de Perote; Santa Barbara Camp, Mount Orizaba.

We found this species to be relatively rare in Veracruz. Four kilometers southeast of Las Vigas, in 1946, we took only one specimen in 100 traps set in a deep canyon under logs, rocks and clumps of zacaton grass. A small series of topotypes was taken in October, 1948, principally in brush and under logs on the borders of cornfields one kilometer west of Las Vigas. This mouse did not occur in the lava rocks, where *Peromyscus boylii* was common, nor did we take it along the rows of maguey plants where we found many individuals of *Peromyscus maniculatus*. A few black-eared mice were taken in brush along the border of a cornfield six kilometers south-southeast of Altotonga.

On the Cofre de Perote, one kilometer northwest of Pescados, we found these mice in November. They did not emerge from their holes until after dark, and were so active that we were unable to catch them by hand. Their holes were less than one inch in diameter and descended vertically into the ground. The ground had been covered with a species of short grass, but this had been grazed by sheep and goats to one inch of the ground. We found 20 burrows in an area of an acre, and probably there were many others. Both *Peromyscus maniculatus* and *melanotis* were living in these burrows. Specimens were trapped in nearby zacaton grass, along with *Neotomodon alstoni.*

The black-eared mouse was breeding in November, one kilometer northwest of Pescados, when the ground was covered with frost almost every night. On November 4, the nest of a black-eared mouse was found 10 kilometers southwest of Jacales, at 6500 feet elevation, under a rock in an arid pine forest. The ground was covered with short, dry grass. Rocks were few. The rock under which the nest was found was about three feet long, two feet wide, and a foot thick. It was well embedded in the ground. The nest was at the end of a runway or burrow two feet long, that ran under the center of the rock, parallel with its long axis. The oval nest was open on top where an area of two and three-fourths square inches of it touched the rock. The nest itself was about three inches in length and two and one half inches in width and height, and formed of dry grass. The walls were only about one-half inch thick. The occupant, a male, was captured.

Peromyscus leucopus

White-footed Mouse

The white-footed mouse in Veracruz inhabits principally brush, rarely enters deep forest, and where it does so usually occurs only along the edges of fields or brushland, or where there are numerous boulders. This species is often common in brush, fields and weeds, where the vegetation is dense, and is often trapped also in sugar cane fields, and in newly cleared areas where there are many fallen logs. On the coastal plain, this mouse seems to be rare, and only a few specimens were taken in tall grass. In northern Veracruz, most of our specimens were taken in the brush and weeds along the edges of fields and in cornfields.

At Potrero Llano we found several nests of the white-footed mouse in a 20-acre cornfield that had been cleared about two years previously. The larger tree trunks were left lying on the ground, although all smaller trunks and limbs had been cut away and burned. The bark of the large trees had become loosened from the trunks in many places and when we opened these bark "blisters" by means of our machetes, we found numerous lizards, a few snakes, and about a dozen mouse nests. Seemingly all of the nests were of this species of mouse. The few mice captured in the nests were all P. *leucopus*. The nests were irregular masses of grass and inner bark of trees and varied from five to eight inches in diameter. They usually were flattened and fitted between the tree and the bark. The central cavities were large for such small mice, usually about four inches in diameter. No young mice were found in the nests.

Some breeding records for this species are as follows:

Sept. 26—Young mouse (gray coat);
Oct. 12—four embryos;
Oct. 25—male in breeding condition; four embryos; six embryos;
Dec. 2—six embryos;
Dec. 3—four embryos; male in breeding condition; two young mice.

Peromyscus leucopus affinis (J. A. Allen)

Record.—Pasa Nueva (Hall and Kelson, 1959:628).

Peromyscus leucopus incensus Goldman

Specimens examined.—Total 149: 19 mi. W Tampico, on Highway 110, 10; 16 mi. W Tampico, on Highway 110, 3; 5 mi. S Tampico, 5; 1 mi. E Higo, 500 ft., 3; El Cepillo, 20 ft., 2; S end Isla Juana Ramírez, Laguna Tamiahua, 3; Hacienda Tamiahua, Cabo Rojo, 19; La Mar, 20 ft., 3; Platón Sánchez, 800 ft., 8; Ixcatepec, 70 km. NW Tuxpan, 1; Tlacolula, 60 km. NW Tuxpan, 1; Cerro Azul, 350 ft., 1; 35 km. NW Tuxpan, 1000 ft., 8; Potrero [del] Llano, 350 ft., 48; 17 km. NW Tuxpan, 1; 15 km. NW Tuxpan, 2; 14 km. NW Tuxpan, 2; 6 km. N Tuxpan, 4; 4 km. NE Tuxpan, 2; Tuxpan, 3; 3 km. E Tuxpan, 1; 4 km. E Tuxpan, 3; 12½ km. N Tihuatlán, 300 ft., 8; 5 km. S Tehuatlan [= Tihuatlán], 700 ft., 1; 3 km. W Guttierez [= Gutierrez] Zamora, 300 ft., 1; 3 km. W Boca del Río, 10 ft., 5; 5 km. SW Boca del Río, 1.

Additional records (Goldman, 1942:158).—Otatitlan; San Andrés Tuxtla.

Peromyscus leucopus mesomelas Osgood

Specimens examined.—Total 65: 7 km. W El Brinco, 800 ft., 6; Miahuapa (La Tulapilla), 15 km. NNE Coyotla [probably = Coyutla], 80 m., 1; Miahuapa,, 8 km. N Coyutla, 80 m., 1; 4 km. W Tlapacoyan, 1700 ft., 5; Teocelo, 4500 ft., 6; Mirador, 3500 ft., 4; Río Atoyac, 8 km. NW Potrero, 13; 4 km. WNW Fortín, 3200 ft., 3; Potrero Viego [= Potrero Viejo], 7 km. W Potrero, 3; Potrero Viejo, 1700 ft., 10; 3 km. SE Orizaba, 5500 ft., 7; 3 km. N Presidio, 1500 ft., 6.

Additional records (Osgood, 1909:132, unless otherwise noted).—5 km. N Jalapa, 4500 ft. (Davis, 1944:396); Orizaba; Río Blanco.

Peromyscus boylii

Brush Mouse

The brush mouse occurs in brush, thickets, and especially rocky places. Five kilometers north of Jalapa we took specimens in thickets of a four-foot high, purple-flowered plant. Six kilometers south-southeast of Altotonga specimens were collected in brush along a small stream, and six kilometers southwest of Zacualpilla they were in the same habitat. Four kilometers south of Jalacingo they were found in piles of mossy boulders in a pine forest. In all of these habitats they were rather scarce. Only in the extensive lava beds within a radius of five kilometers of Las Vigas did we find this species to be common. Even there, in seemingly optimum habitat, it could not be termed abundant.

Strangely enough, this species seems rarely to occur with other species of *Peromyscus*. Six kilometers south-southeast of Altotonga, a few individuals of *Peromyscus melanotis* were taken with *boylii*. Except at that locality, *boylii* not only was the only species of *Peromyscus* taken in the particular areas where the traps were set, but was the only genus of mouse.

This species seems to be entirely nocturnal. We learned little of its habits. A male having enlarged testes was taken on October 16; a nursing female on October 15; a female containing three embryos on November 1. Within a radius of five kilometers of Las Vigas, several specimens were taken that had large cysts of tapeworms under the skin of the neck. The cysts were sacs filled with a colorless liquid and contained three or four tapeworms. Near Jalapa what seemed to be this same species of tapeworm was found in *Peromyscus boylii* and *Peromyscus furvus*.

Peromyscus boylii beatae Thomas

Specimens examined.—Total 45: 1 km. W Las Vigas, 8500 ft., 2; Las Vigas, 8500 ft., 13; 2 km. E Las Vigas, 8000 ft., 5; 3 km. E Las Vigas, 8000 ft., 8; 5 km. E Las Vigas, 7 (Texas Agric. and Mech. College); 10 km. SE Perote, N slope Cofre de Perote, 10,500 ft., 1 (Texas Agric. and Mech. College); Xometla Camp, Mount Orizaba, 8500 ft., 2 (British Mus.); Sta. Barbara, Mount Orizaba, 12,500 ft., 1 (British Mus.); Xuchil, 6 (Chicago N. H. Mus.).

Additional records (Osgood, 1909:155).—Perote; Orizaba; Maltrata.

For the characters of this subspecies, named in 1903 by Thomas but long regarded as inseparable from P. *b. levipes*, see Alvarez (1961:116-117).

Peromyscus boylii levipes Merriam

Specimens examined.—Total 39: 2 km. N Los Jacales, 7500 ft., 8; 3 km. W Zacualpan, 6000 ft., 13; 6 km. SW Zacualpilla, 6500 ft., 5; 1 km. E Jalacingo, 6500 ft., 3; 4 km. S Jalacingo, 2; 6 km. SSE Altotonga, 9000 ft., 8.

Peromyscus aztecus (Saussure)

Aztec Mouse

Specimens examined.—Total 10 (U. S. N. M.): Jalapa, 1; Mirador, 9.

The name *Hesperomys aztecus* Saussure, 1860, was applied by Osgood (1909:156) to two specimens from Jalapa, nine specimens from Mirador, six specimens from Huachinango in the state of Puebla, and one specimen from "Mexico." He tentatively placed *aztecus* as a subspecies of *Peromyscus boylii*. He had no intergrades and recorded no specimens of P. *b. levipes* [= P. *b. beatae*] and *P. b. aztecus* geographically nearer each other than Las Vigas (*levipes*) and Jalapa (*aztecus*). These two localities are 23 kilometers apart. We have specimens of P. *b. beatae* [= *levipes* of Osgood] from only five kilometers north of Jalapa. There is no evidence of intergradation in the specimens from Jalapa (*aztecus*) and five miles north thereof (*P. b. beatae*). Actually (see Alvarez, 1961:115) one of the two specimens that Osgood had from Jalapa was *Peromyscus boylii* (misidentified by Osgood as *P. aztecus*) and he therefore had the two species at one locality. Consequently we think that *P. aztecus* is specifically distinct from *Peromyscus boylii*. Larger size, more cinnamon coloration, less inflation anteriorly of tympanic bullae, sharply angled (*versus* rounded) supraorbital border of frontals, elongate (*versus* round) braincase and, more evenly convex dorsal outline of skull in lateral view (nasals and frontal tending to form straight line, lowest anteriorly, in *boylii*) are characters distinguishing *Peromyscus aztecus* from all specimens from Veracruz of *Peromyscus boylii*.

Peromyscus bullatus Osgood

Perote Mouse

Specimens examined.—Total 6: 3 km. W Limón, 7800 ft., 1; 3 km. W Limón, 7500 ft., 1; 2 km. W Limón, 7500 ft., 4.

Additional record.—Perote (Osgood, 1909:184).

Previous to 1947, this species was known only from the type locality. Nothing was known of its habits, but we supposed that it lived among rocks. Superficially it resembles the rock-loving species *Peromyscus difficilis*. In 1946 and 1947, we trapped in vain on rocky cliffs and outcrops for this species two to three kilometers west of Limón. Many specimens of *Peromyscus difficilis* were captured, but none of *bullatus*. On October 11, 1947, a small mouse resembling *difficilis* was taken in a trap set on a sandy desert flat, west of Limón. This specimen was seemingly in adult pelage, but was smaller than even the young *difficilis* in the gray coat, and differed in other respects.

When the skull was cleaned and studied, the mouse was identified as the rare *bullatus*. Returning in September, 1948, we trapped intensively on the same desert flat and obtained five additional specimens, including one adult male and an adult female.

The Perote mouse was rare; only one *bullatus* to about 50 *Peromyscus maniculatus* was taken in our traps. We took many *Peromyscus difficilis* among rocks a hundred feet from where we took *bullatus*, but it was not taken among the rocks, and not one *difficilis* was taken on the sandy flats. The specimens of *bullatus* were taken in mouse traps, baited with rolled oats and walnut, set near the sparse growth of grasses and desert weeds on the flats and dunes of fine sand. The only adult female taken was not pregnant. Three rather young mice were taken in the last week of September, indicating that the breeding season of this species is a little later than in *difficilis*.

In relation to size of skull, the auditory bullae are the largest among the species of *Peromyscus*. The greatly enlarged bullae may be a response to living on the open sandy desert. *Dipodomys*, among the heteromyids, has huge auditory bullae and lives in wind-drifted sand.

Peromyscus difficilis

Zacatecan Deer Mouse

For a review of the subspecies of *Peromyscus difficilis* see Hoffmeister and de la Torre (1961:1-13).

This species of deer mouse is restricted to rocky places, and prefers an arid habitat. Six kilometers west-southwest of Zacualpilla we took a series of specimens in a rocky canyon and along the cliffs of a gorge. In the pine forest of the hills a few hundred yards away, only *Peromyscus boylii* was found. Two to three kilometers west of Limón, on the Perote Desert, this species was abundant in the rocky cliffs and in a lava flow, but in the sandy flats a few yards away, only *Peromyscus maniculatus* and *P. bullatus* were found. Three kilometers west of Limón, 7500 feet elevation, the mice were taken among typical desert shrubs on a white, brecciated stone cliff. Plants included yuccas of two or three species, grasses, and at least five kinds of cacti. Most prominent was the prickly pear, of the yellow-fruited type.

Three and four kilometers west of Acultzingo, a small series of these mice was taken along old rock walls. This area is a brushland, but there is arid desert to the east and west.

Many young mice of this species were taken six kilometers west-southwest of Zacualpilla in November. Relatively few young mice were taken two to three kilometers west of Limón in November, but several males there were recorded as having enlarged testes, and one female had three embryos on November 19.

Peromyscus difficilis amplus Osgood

Specimens examined.—Total 90: 3 km. W Limón, 7500 ft., 34; 2 km. W Limón, 7500 ft., 30; Perote, 10 (U. S. N. M.); Maltrata, 3 (U. S. N. M.); 4 km. W Acultzingo, 7500 ft., 1; 3 km. W Acultzingo, 7000 ft., 12.

Additional record.—1½ mi. S Perote, 8500 ft. (Hooper, 1957:7).

Specimens from three kilometers west and from two kilometers west of Limón are notable for long external ears and large tympanic bullae. In 30 adults (16 males and 14 females) the ear averages fully as long as the hind foot. Actual measurements are as follows: ♂, hind foot, 25.8 (25-28); ear from notch, 26.0 (25-28); ♀, hind foot, 25.6 (25-27); ear from notch, 25.5 (23-27). Measured dry, the length of the ear is in males 21.5 (19.5-23.9), and in females 21.9 (20.0-24.2). Ten of the specimens from nearby Perote

(Osgood, 1909:182) have been examined and have similarly long external ears and large tympanic bullae. Hooper (1957:7) characterized seven specimens from 1½ miles south of Perote as having larger tympanic bullae and external ears than any other specimens seen by him of the species P. *difficilis*.

Nine adults (four males and five females) from three and four kilometers west of Acultzingo have shorter ears as is characteristic of P. *d. amplus* in several parts of its geographic range. Actual measurements are: ♂, 22.5 (21-24); ♀, 22.0 (21-23). Measured dry, the length of the ear is in males 19.4 (18.2-20.4), and in females 17.6 (16.5-18.8).

Although the specimens labeled with reference to Limón and Perote are notable for long external ears and large tympanic bullae, a few individuals from two kilometers west of Limón have shorter ears and smaller tympanic bullae, as small as certain specimens from three and four kilometers west of Acultzingo. Also, some of the 10 topotypes of P. *d. amplus* from Coixtlahuaca, Oaxaca, have long external ears and large tympanic bullae. It is concluded that the mice from Perote, 1½ mile south thereof, and from two and three miles west of Limón are not unique but are more nearly uniform in having long external ears and large tympanic bullae than are specimens from other areas.

Peromyscus difficilis saxicola Hoffmeister and de la Torre

Specimens examined.—Total 53: 6 km. WSW Zacualpilla, 6500 ft., 51; 10 km. SW Jacales, 6500 ft., 2.

For an account of this subspecies see Hoffmeister and de la Torre (1961:10).

Peromyscus simulatus Osgood

Jico Deer Mouse

Specimens examined.—Teocelo, 4500 ft., 3.

Additional record.—Near Jico, 6000 ft. (Osgood, 1909:193).

Peromyscus furvus J. A. Allen and Chapman

Blackish Deer Mouse

Specimens examined.—Total 31: 5 km. N Jalapa, 4500 ft., 5; 2 km. W Jico, 4200 ft., 26.

Additional records (Osgood, 1909:197).—Jalapa; Jico.

This large mouse seems to occupy but a tiny range. Though common at Jico and Jalapa, it has not been taken elsewhere. There is no obvious barrier, geographical or ecological, to the south or north. Yet the blackish deer mouse does not occur where we trapped in the vicinity of Jico, Huatusco or Coscomatepec to the south, or at Tlapacoyan to the north. We took the species in the cool forests at the very upper edge of the upper humid division of the Tropical Life-zone. There it was trapped along rocky cliffs, in canyons, in the forest, and in coffee thickets and brushy places. It occurred with *Marmosa mexicana, Oryzomys palustris, Oryzomys alfaroi* and *Microtus quasiater.*

The blackish deer mouse seems to be entirely nocturnal; we captured no specimens in the daytime. Several were trapped in small caves in a cliff and at the entrances to holes under logs and roots of overturned trees. This species may breed all year around. In late October many young mice of various ages were trapped. Most of the larger males trapped had enlarged testes. Several females were nursing young. Each of two pregnant females contained two embryos.

Parasites seem to be rare in this species. One infestation of a tapeworm, in a cyst under the skin of the neck, was noted. Several mice had a few tiny, hard mites.

Peromyscus angustirostris Hall and Alvarez

Narrow-nosed Mouse

Specimens examined.—Total 31: 3 km. W Zacualpan, 6000 ft., 24; Zacualpan, 6000 ft., 7.

The distinguishing characteristics of this recently discovered species are enumerated in the original description (Hall and Alvarez, 1961:203). M. Raymond Lee obtained all of the known specimens from an area supporting long-needled pine and trapped most of the specimens around rocks and water seeps.

Peromyscus mexicanus

Mexican Deer Mouse

The Mexican deer mouse is a forest-living species. It is the commonest mammal taken in the deep jungle of the upper humid division of the Tropical Life-zone, and in many places is the only small mammal taken in the collectors' traps. In the limestone areas, where numerous cliffs and boulders lie scattered on the ground beneath tall trees, a mouse of this species may be taken in almost every trap set. On the open, bare forest floor, where there are few fallen logs and little underbrush, Mexican deer mice are much scarcer. On the coastal plain they are generally uncommon, occurring in thickets and jungles that fringe the rivers and watercourses.

When these mice are unusually abundant in the forest, a few individuals may wander into thickets and weeds at the forest edge, and even into clearings. Mexican deer mice found in such locations are usually young. Not uncommonly, the species is taken in cornfields in which the ground is clean and the cornstalks are tall and dense. There we took this species along with *Liomys pictus* and *Liomys irroratus*—species with which the Mexican deer mouse otherwise rarely comes in contact. Later in the year, when the milpas grow up in grass and weeds, the Mexican deer mice leave the fields, and grass-loving forms of small mammals enter the fields. In their forest habitat, the Mexican deer mice commonly are taken with *Oryzomys alfaroi* and *Oryzomys melanotis* and along the edges of the forests with *Peromyscus leucopus.*

This species is nocturnal. Twice we saw individuals in the day, but both times in the gloom of caves. One very young animal was taken in a trap at midday.

The Mexican deer mouse feeds on seeds and fruits. Coffee berries are eaten, and small caches of dry coffee husks, each husk having a hole eaten in the side, are often found in the forest. Usually these caches are beneath rocks. The inside of the pit of the jobo plum is eaten, as well as the pulp. Several specimens had their chins and chests stained a deep red or deep purple, seemingly from the juice of some fruit or berry. Several mice were taken that had lost parts of their tails and the stumps had healed completely.

Some breeding records for this species are as follows:

Jan. 11—four breeding males;
Feb. 2—two breeding males; baby mouse;
March 5—baby mouse;
May 29—two embryos; two embryos;
Sept. 28—two embryos; nursing female; five breeding males;
Sept. 30—two embryos; two embryos;
Oct. 22—nursing female; two breeding males;
Oct. 28—nursing female;
Nov. 2—two breeding males;
Nov. 3—three embryos;
Nov. 4—three embryos;
Nov. 16—three embryos;
Dec. 1—baby mouse;
Dec. 2—two embryos; nursing female;
Dec. 5—nursing female;
Dec. 12—two breeding males;
Dec. 13—three embryos.

One specimen "had a strange growth on the head, which came off and was lost. The skull beneath was much distorted." Several mice were taken that had growths, perhaps parasites or eggs of parasites, at the bases of the whiskers, in the flesh. These were white granules, one to two millimeters in diameter. No structure could be noted. These objects were free in the flesh, and readily slipped out when flesh of a mouse's nose was pinched. Another mouse had "two large tapeworms, about four inches long, white, thin, and with a scolex about two millimeters in diameter, under the skin of the throat. These were not in a cyst, but free and active." Other records of parasites include:

"number of small ticks";
"one flea";
"two small black fleas";
"a few small, fast-moving mites";
"one small tick";
"three tiny red mites in ear";
"several small, dark-brown, hard, slow-moving mites";
"eight or ten small red mites";
"two small, fast-moving red mites";
"about four small mites on each mouse".

Peromyscus mexicanus mexicanus (Saussure)

Specimens examined.—Total 200: Piedras Clavadas, 75 km. NW Tuxpan, 6; 35 km. NW Tuxpan, 1; 25 km. NW Tuxpan, 2; 14 km. NW Tuxpan, 1; 3 km. W Gutierrez Zamora, 300 ft., 3; 9 km. E Papantla, 300 ft., 4; 3 km. SE San Marcos, 200 ft., 23; 7 km. NNW Cerro Gordo, 1; Teocelo, 4500 ft., 21; Puente Nacional, 500 ft., 6; Mirador, 3500 ft., 7; 15 km. ENE Tlacotepec, 1500 ft., 11; Río Atoyac, 8 km. NW Potrero, 29; 7 km. NW Potrero, 1700 ft., 5; 4 km. WNW Fortín, 3200 ft., 11; 15 km. ESE San Juan de la Punta, 400 ft., 9; Río Blanco, 20 km. WNW Piedras Negras, 4; Río Blanco, 20 km. W Piedras Negras, 400 ft., 11; 3 km. N Presidio, 1500 ft., 12; 2 km. N Motzorongo, 1500 ft., 15; 3 km. E San Andrés Tuxtla, 1000 ft., 13; 5 mi. S Catemaco, 5.

Additional records (Osgood, 1909:201).—Otatitlán; Lagunas; Pasa Nueva; Achotal; Carrizal; Catemaco; between Papantla and El Tajin (Ingles, 1959:397).

Peromyscus mexicanus teapensis Osgood

Specimens examined.—Total 9: 15 km. SW Jimba, 750 ft., 1; 14 km. SW Coatzacoalcos, 100 ft., 6; Jesús Carranza, 250 ft., 1; 63 km. ESE Jesús Carranza, 500 ft., 1.

Peromyscus mexicanus totontepecus Merriam

Specimens examined.—Total 22: 7 km. W El Brinco, 800 ft., 8; Cautlapan [= Cuautlapan], 4000 ft., 2; 3 km. SE Orizaba, 5500 ft., 12.

Peromyscus nelsoni Merriam

Nelson's Deer Mouse

Specimens examined.—Jico, 2 (U. S. N. M.).

Baiomys taylori

Northern Pygmy Mouse

See Packard (1960:579-670) for details of the taxonomy and geographic distribution of *Baiomys.*

At Potrero [del] Llano we found two of these mice in the stomach of a boa constrictor and another in the stomach of a king snake (*Lampropeltis polyzona*).

Baiomys taylori analogus (Osgood)

Record.—Acultzingo (Packard, 1960:639; see his account for a treatment of this subspecies).

Baiomys taylori taylori (Thomas)

Specimens examined.—Total 24: 19 mi. W Tampico, 1; 5 mi. S Tampico, 5; Tampico Alto, 50 ft., 2; 1 mi. E Higo, 500 ft., 3; Ozuluama, 500 ft., 2; Platón Sánchez, 800 ft., 1; Cerro Azul, 350 ft., 4; Potrero [del] Llano, 350 ft., 6.

Baiomys musculus brunneus (J. A. Allen and Chapman)

Southern Pygmy Mouse

Specimens examined.—Total 91: 7 km. NNW Cerro Gordo, 19; Teocelo, 4500 ft., 1; Puente Nacional, 500 ft., 2; 3 km. W Boca del Río, 10 ft., 8; 4 km. WNW Fortín, 3200 ft., 6; Río Atoyac, 8 km. NW Potrero, 1; 2 km. N Paraje Nuevo, 1700 ft., 13; Potrero Viejo, 1700 ft., 15; Cautlapan [= Cuautlapan], 4000 ft., 16; 1 km. E Mecayucan, 200 ft., 1; 3 km. SE Orizaba, 5500 ft., 3; Río Blanco, 20 km. WNW Piedras Negras, 4; 3 km. N Presidio, 1500 ft., 2.

Additional records (Packard, 1960:614).—2 mi. NW Plan del Río, 1000 ft.; Plan del Río, 1000 ft.; Carrizal; Chichicaxtle; Sta. María, near Mirador, 1800 ft.; Boca del Río, 10 ft.; Cordoba; 29 km. SE Cordoba; Presidio. (Packard's, *op. cit.*, systematic treatment of *Baiomys* should be consulted for nomenclature and information on geographic distribution.)

Pygmy mice prefer low, dense vegetation such as grasses, sawgrass, weeds, succulent plants along the borders of small streams, and dense thickets. Occasionally they are found in relatively open brushland, especially if the brush be thorny species, but rarely are common there. At the edges of clearings or other places where the species may be locally abundant, they may stray into the jungle for short distances, but rarely for more than a few yards.

Perhaps the favored habitat of this species in Veracruz, and the place where the greater part of our specimens was taken, is in the narrow borders of brush, weeds and grasses at the edges of fields. In Veracruz, the distribution of this species is irregular; in many apparently suitable localities we failed to find it. Usually, however, if present at all, the mice were numerous.

Usually the pygmy mouse was found with other grass-living forms, such as harvest mice (*Reithrodontomys*), the pygmy rice rat (*Oryzomys fulvescens*) and the cotton rat (*Sigmodon hispidus*). When pygmy mice are abundant, their presence may be detected by their tiny runways in the grass and weeds. Piles of tiny droppings, characteristically green, occur at intervals along these runways. Often this species is trapped in the larger runways of *Sigmodon hispidus*. In a few places, where pygmy mice were rare, we found no evidence of their presence, such as runways or piles of droppings. In such cases we usually took only one or two specimens at a locality. The distribution by months of embryos (one to four per female) suggests that the species breeds throughout the year.

At Potrero Viejo this species was commonly killed by house cats.

We found few parasites on *Baiomys*. One individual was noted as having a few mites.

Sigmodon hispidus

Hispid Cotton Rat

The hispid cotton rat is a grass-loving species that occupies grasslands, overgrown clearings, weed-grown borders of fields and brushy areas. It prefers dense growths of sacate (sawgrass or bunchgrass). Rarely it strays a short distance, usually not more than 50 feet, into jungle or forest when these border the more normal habitat.

The cotton rat is both diurnal and nocturnal. More specimens can be trapped by day than by night. In one day, a "few" were trapped between 9:00 a. m. and 1:00 p. m.; seven between 1:00 p. m. and 5:00 p. m.; three in the following night; four the next morning.

These rats construct broad trails through dense grass and other thick cover. In bunchgrass, there is usually enough space between the clumps of grass and roots to serve as runways, and runways are not extensively constructed there.

In some places the cotton rat does considerable damage to sugar cane, especially young plants. When the rats become extremely abundant, as they were in 1947 near Potrero, control measures sometimes are taken. The Potrero Sugar Company resorted to poison. Cotton rats also destroy a great deal of stored corn, and damage beans. The natives trap the rats by means of a clever deadfall. This consists of two sticks pushed into the ground about six inches apart. A strand of sawgrass is tied between these sticks, about six inches from the ground. In the center of the strand of grass, a single kernel of corn is tied in a half-hitch in the grass. A flat rock is leaned against the grass with the part above the grass about four inches from the ground. The cotton rat enters beneath the rock, cuts the strand of grass in order to obtain the corn, and thus releases the rock that falls and crushes the rat.

Cotton rats breed throughout the year. Young are more common than adults in every locality where we found this species. Some breeding records are as follows:

Jan. 10—breeding male;
Feb. 2—four embryos;
Feb. 9—breeding male;
Feb. 12—three embryos;
March 13—breeding male;
March 23—two embryos;
May 15—breeding male;
May 16—three breeding males;
May 30—nursing female;
Sept. 27—two embryos; five females with three embryos each;
Oct. 12—seven embryos;
Nov. 2—three breeding males;
Nov. 4—three embryos;
Nov. 14—three embryos; four embryos;
Nov. 15—three embryos;
Dec. 1—breeding male;
Dec. 11—two breeding males.

Three trapped cotton rats had lost parts of their tails in life, and the stumps had completely healed. One specimen had an infection of the shoulder, seemingly the result of a fly larva. Ectoparasites are not common on cotton rats. Listed were: "two or three small ticks"; "six to ten small mites, seemingly of two species"; "large bot fly larva in center of back"; "translucent bot fly larva, 10 millimeters long, four millimeters wide, three millimeters high, in center of back"; "a large tick in the fur of the back"; "two or three tiny brown and white mites."

At a place 15 kilometers east-northeast of Tlacotepec an assistant brought into camp about 25 small mammal skulls and fragments of skulls. These he had found in a cave, beneath the roost of an owl. The skulls had weathered out of owl pellets. All the skulls were of the cotton rat.

Sigmodon hispidus saturatus Bailey

Specimens examined.—Total 58: 3 km. E San Andrés Tuxtla, 1000 ft., 11; Catemaco, 1 (U. S. N. M.); Coatzacoalcos, 6 (U. S. N. M.); 14 km. SW Coatzacoalcos, 100 ft., 9; 10 km. NW Minititlan [= Minatitlán], 100 ft., 1; Achotal, 8 (Chicago N. H. Mus.); 35 km. ENE Jesús Carranza, 150 ft., 6; Jesús Carranza, 250 ft., 1; 20 km. E Jesús Carranza, 250 ft., 10; 55 km. ESE Jesús Carranza, 450 ft., 3; 25 km. SE Jesús Carranza, 250 ft., 2.

Additional record.—Pasa Nueva (Allen, 1904:31).

Sigmodon hispidus toltecus (Saussure)

Specimens examined.—Total 304: Tampico Alto, 50 ft., 1; 19 mi. W Tampico on Highway 110, 17; 16 mi. (by road) W Tampico on Highway 110, 5; 5 mi. S Tampico, 11; S end Isla Juana Ramírez, in Laguna Tamiahua, 7; Hacienda Tamiahua, Cabo Rojo, 10; Isla Burros, Laguna de Tamiahua, 5; La Mar, 20 ft., 1; Platón Sánchez, 800 ft., 7; Ixcatepec, 70 km. NW Tuxpan, 1; Potrero Llano [= Potrero del Llano], 350 ft., 17; 35 km. NW Tuxpan, 4; 17 km. NW Tuxpan, 2; 6 km. N Tuxpan, 6; 4 km. N Tuxpan, 2; 4 km. NE Tuxpan, 4; San Isidro, 100 ft., 3; 12½ mi. N. Tihuatlán, 300 ft., 20; 5 km. S Tehuatlan [= Tihuatlán], 700 ft., 10; 3 km. W Gutierrez Zamora, 300 ft., 1; 3 km. SW San Marcos, 200 ft., 13; Chichicaxtle, 4 (U. S. N. M.); 7 km. NNW Cerro Gordo, 32; Jico 6 (U. S. N. M.); Mirador, 3500 ft., 2 (1 in U. S. N. M.); 15 km. ENE Tlacotepec, 1500 ft., 15; 3 km. SW Boca del Río, 10 ft., 1; 5 km. SW Boca del Río, 1; Río Atoyac, 8 km. NW Potrero, 4; 7 km. NW Potrero, 1700 ft., 1; 2 km. N Paraje Nuevo, 1700 ft., 4; Potrero Viejo, 1700 ft., 3; Orizaba, 24 (U. S. N. M. 17, Chicago N. H. Mus. 7); 15 km. ESE San Juan de la Punta, 400 ft., 14; 7 km. SE San Juan de la Punta, 400 ft., 6; Río Blanco, 20 km. WNW Piedras Negras, 8; 3 km. N Presidio, 1500 ft., 9; 2 km. N Motzorongo, 1500 ft., 4; Motzorongo, 12 (U. S. N. M.); Tlacotalpan, 2 (U. S. N. M.); Otatitlán, 5 (U. S. N. M.).

Additional records.—El Tajín (Ingles, 1959:398); Jalapa (Allen and Chapman, 1897:207); Mt. Orizaba (Elliot, 1907:247).

Of the 14 skins from 19 miles west of Tampico, three are pale as is characteristic of *S. h. berlandieri;* the other 11 are darker as is characteristic of *S. h. toltecus.*

Three insular populations deserve comment. The ten individuals from the population on Cabo Rojo (Hacienda Tamiahua), the big off-shore barrier beach, have dark-colored sides, are larger than individuals of the other two insular populations, seem to have larger external ears than even the mainland animals, and in comparison with the latter further differ as follows: anterior part of glenoid surface of squamosal bone more angled; squamosal arm of zygomatic arch deeper (vertically); auditory bullae larger; bulla in contact with paroccipital process (rarely so in animals of same size from mainland); supraorbital crest higher.

. Seven specimens from the south end of Isla Juana Ramírez in the Laguna de Tamiahua, have paler sides than animals from either of the other two insular populations or than animals from the mainland.

Five specimens from Isla Burros, in the Laguna de Tamiahua, are the smallest, and have more ochraceous underparts (on the average) than other specimens.

The differences mentioned above exist between specimens of the same sex having about the same degree of wear on their molar teeth, and obtained and prepared at the same time of year (March 31-April 5).

The variations may deserve subspecific recognition.

Neotomodon alstoni perotensis Merriam

Volcano Mouse

Specimens examined.—Total 68: Las Vigas, 1 (U. S. N. M.); Perote, 1 (U. S. N. M.); Cofre de Perote, 5 (U. S. N. M.); 1 km. NW Pescados, 10,500 ft., 61.

Additional record.—Cofre de Perote, 10,500 ft. (Davis, 1944:398).

The volcano mouse is most abundant in the meadows of zacaton grass at high elevations on the mountains. It is less common about rocky outcrops and in marshes, but does occur in such places, and even was found on some occasions in the open pine forest. At the high altitudes where this species lives, the nights are cool or cold; frost occurs almost nightly in winter and snow is not uncommon. On the Cofre de Perote we found the volcano mouse associated with the black-eared deer mouse (*Peromyscus melanotis*) and the western harvest mouse (*Reithrodontomys megalotis*). The volcano mouse was more common than the other species, outnumbering the two combined by more than five to one in our traps. In mid-November we took an average of one volcano mouse to each five traps set.

On the Cofre de Perote, one kilometer northwest of Pescados, we took six volcano mice in a small area of zacaton grass, isolated from the main meadow. Because we had taken no other species in this area, we searched there for the nest of the volcano mouse that afternoon. We picked up the trail of the mouse where it had been trapped, between two clumps of zacaton beside a pine log. We first cut away all the grass of the clump nearest the pine log. A runway completely circled the clump at its base, inside the circle of down-drooping dead grass. An entrance to this circle was on the side away from the pine log, but no others were found. We cut the head of grass off, level with the ground, and bisected it. Much soil was mixed with the grass, and the punky center would have made a good place for the nest, as would the punky root mass. We scraped away the earth, and found, beneath an inch of packed loam, packed sand. There was no underground burrow. We then repeated the process with the adjoining clump of zacaton. This one was larger. Three holes entered the clump, and joined with the ring runway. This clump also had two chambers outside the ring, about six inches long by three wide, with a dry, soft earth floor. These were on the surface of the ground, just inside the down-drooping circle of grass blades. This clump also was excavated and bisected, but no nest or underground burrow was found. The third clump was double, with a narrow neck so riddled with burrows as almost to form two separate clumps. The ring runway about the smaller was not complete. It ended to the outside at one end, and in several places it joined the runways through the neck. The other section had a complete ring, joined to the runway through the neck, two chambers, and at least three runways from the ring to the outside. It was excavated but no burrow or nest was found. Between the clump and a rock two feet long by one foot long, we found a burrow, only partially concealed by grass. It measured one and seven-eighths inches at the mouth, and about one and one half inches throughout its length of about six feet. It descended to a depth of about four inches, passed around the rock, and branched. One branch ended eight inches away. The other continued for a foot, then abruptly went down between two rocks. At a depth of about 14 inches, it branched again. One branch rose gradually, in a distance of two feet, and stopped under a rock about one foot by six inches by six inches. The other branch passed beneath a rock about two feet by 18 inches, after which the branch ascended on the other side, and at a depth of four inches, entered the nest, which was a hollow ball of dry grass, five inches in diameter, having a central cavity two and one half inches in diameter. The nest was notably soft, and could not be lifted out in one piece. It was in a cavity, half in the earth and half in the roots of a clump of zacaton. By cutting away the opposite side of the clump of zacaton, we exposed the nest for a photograph.

In November, few evidences of breeding were noticed in the mice trapped. One female had two embryos. One young mouse was taken. The testes of all of the males were small and abdominal.

Few parasites were noted on volcano mice. There was an average of one small mite to each three mice. One rattlesnake (*Crotalus triseriatus anahuacus*) was found in the zacaton meadow. No other signs of predators were noted. We saw no hawks or owls in the course of our stay at the Cofre de Perote.

Neotoma nelsoni Goldman

Nelson's Wood Rat

Specimens examined.—Perote, 11 (U. S. N. M., including type).

Nelson and Goldman found this rat in the cactus at Perote. The cactus is now nearly gone, as a result of clearing the desert for wheat, corn and maguey. We failed to take specimens.

Neotoma mexicana

Mexican Wood Rat

On the desert three kilometers west of Limón we found signs of wood rats in small caves in the conglomerate rock of the small canyons and in an extensive lava flow. The rats refused all baits, even dry prunes. One young animal was shot at night.

Two and three kilometers east of Las Vigas, wood rats were fairly common in the recent lava flow. Their distribution was rather spotty. We found numerous signs about some large piles of boulders and in caves in the low cliffs. In other places, seemingly identical in all aspects, and nearby, there were no signs of occupancy by the rats. The rats were difficult to trap. They refused all baits save dry prunes, and even prunes were often refused. We took a small series after considerable effort.

In the lava within three kilometers of Las Vigas we found a few small, old signs of wood rats in deep caves. A few small stick piles, doubtless made by wood rats, were seen in crevices in rocks. These seemed not to be nests, but piles made only in response to the nest-building habit or instinct, so highly developed in some species of wood rats farther north.

Our specimens labeled with respect to Las Vigas include four adult females. None was pregnant on October 5 or 13. One old male had enlarged testes on November 3. A young rat was taken on October 12. No ectoparasites were noted on wood rats from Veracruz.

Neotoma mexicana distincta Bangs

Specimen examined.—Texolo [= Teocelo], 1 (U. S. N. M.).

Neotoma mexicana torquata Ward

Specimens examined.—Total 12: 3 km. W Limón, 7800 ft., 1; Cofre de Perote, 9500 ft., 2 (U. S. N. M.); Las Vigas, 8500 ft., 6; 2 km. E Las Vigas, 8000 ft., 2; 3 km. E Las Vigas, 8000 ft., 1.

Microtus mexicanus mexicanus (Saussure)

Mexican Vole

Specimens examined.—Total 128: 6 km. SSE Altotonga, 9000 ft., 2; Las Vigas, 8500 ft., 67; 1 km. W Las Vigas, 8500 ft., 10; Las Vigas, 7400-8000 ft., 11 (U. S. N. M.); 3 km. E Las Vigas, 8000 ft., 3; 2 km. N Perote, 8000 ft., 2; Perote, 3 (U. S. N. M.); 2 km. W Perote, 8000 ft., 3; Cofre de Perote, 26 (U. S. N. M.); 4 mi. SW Acultzingo, 7000 ft., 1.

In Veracruz this vole lives in cool, damp places along the rim of the plateau and occupies a zonally higher area than does the pine vole. The latter lives at the upper edge of the upper humid division of the Tropical Life-zone, whereas *M. mexicanus* lives in the next higher zone. *M. quasiater* lives in the oak forest association of tropical vegetation. Commonly a few pine trees extend down to or nearly to *quasiater* habitat. *M. mexicanus* occupies the pine forest habitat. The two species closely resemble each other in both habits and general appearance.

Our largest series of *mexicanus* was taken in the northern part of Las Vigas. There a few were trapped in a zacaton meadow, a number along the long rows of maguey plants and the greatest number in a deep, cool canyon. This canyon, at the northern edge of the town of Las Vigas, is in a beautiful pine forest. The canyon is deep, with steep sides about 200 feet high. These canyon sides are covered, beneath the tall pines, with a deep mat of thin-bladed grasses and moss. Near the stream, the ground is covered with a mat of almost pure moss about a foot deep. In the grass and moss we found literally thousands of tiny runways. Traps set in these runways took large numbers of shrews (*Cryptotis* and *Sorex*) and voles. Other than these mammals, the only species taken was the western harvest mouse (*Reithrodontomys megalotis*), of which we took two.

Our attempts to map the runways of *Microtus* were given up; there were simply too many intersections and runways in the area. In addition to the runways through the moss, there were many burrows in the ground. These probably led to nests, but we did not find any; the rocks in this area were too heavy and deeply sunken to move.

In the pine forest above, however, voles were less common. There we were able to follow the runways in the short grass that was grazed by goats to within about two inches of the ground. Most of the runways in the grass were short, three to four feet in length, with several branches not more than two feet in length. We judged that there were about four branches to the average runway. One runway was nearly 20 feet long. Most runways were sinuous, but rarely had sharp turns, and usually each runway kept its general direction throughout its length. Each runway and subsidiary runway terminated in a shallow trough, showing that the vole there emerged to move about on, rather than through, the grass. The runways were trough-shaped, about one and one-quarter inches wide. Usually three or four radiated from a common center, the burrow.

The burrows were about one inch in diameter, widening in places to as much as three inches. We did not excavate those that led into the ground. Many led under logs that we rolled aside. The runways beneath the logs resembled those in the grass except that there were more branches beneath the logs. Many of the short branches, or pockets, and the runways themselves were wider than the runways through the grass. In general, there was one long burrow parallel with the long axis of the log, with two or three short branches leading to the edge of the log. From each of these, runways radiated off through the grass.

Some of these runways had no nests, or if present, the nests were underground. Often there were burrows leading down into the ground beneath the logs. Many runways did have nests. In some of these we were able to catch voles. The nests were all of a common pattern: a relatively deep pocket in the ground, near the center of the log. One side of this pocket would be dug in sideways, and roofed over with a thin layer of earth between the nest and the log. This side pocket was filled with a large ball of soft, dry grass. Nests were about six to eight inches in diameter transversely and four to five inches in depth. There was a single entrance, and an interior cavity about three and one half inches in diameter.

Two kilometers west of Perote, we found voles living in maguey cercas, on the open desert, but the voles were rather uncommon in such places.

Some breeding records for this species are:

Sept. 28—three embryos

Oct. 8—nursing female; three embryos;

Oct. 14—three embryos;

Oct. 15—nursing female having one embryo; female with two embryos;

Oct. 19—nursing female;

Oct. 20—three embryos; nursing female;

Nov. 3—nursing female having three embryos;

Nov. 11—breeding male.

We took one or two young voles almost every day near Las Vigas, but it was obvious that breeding was not common in winter.

Microtus quasiater (Coues)

Jalapan Pine Vole

Specimens examined.—Total 74: 4 km. W Tlapacoyan, 1700 ft., 11; 5 km. N Jalapa, 4500 ft., 32; Teocelo, 4500 ft., 28; Huatusco, 5000 ft, 1; 4 km. WNW Fortín, 3200 ft., 1; 3 km. SE Orizaba, 5500 ft., 1.

Additional records.—Jalapa (Coues, 1874:192, type locality); Jico (Bailey, 1900:67); Orizaba (Hall and Cockrum, 1953:452); Tuxpango (Coues, 1874:192).

The normal habitat of the Jalapan pine vole in Veracruz is marshy meadows and grassy swales, from 3000 to 5500 feet in elevation. Four miles west-northwest of Fortin, one was taken in a marsh. Three kilometers southeast of Orizaba, one was trapped in damp grass beside a sugar cane field. Four kilometers west of Tlapacoyan, a small series of specimens was taken in damp grasses and succulent plants beside a tiny stream in a cool canyon. At Teocelo, specimens were taken in a meadow, beside a small, cool stream, and in grasses and weeds along the borders of fields. Ordinarily, the pine vole is found in the same places as rice rats (*Oryzomys*), harvest mice (*Reithrodontomys*) and shrews (*Cryptotis*). Five kilometers north of Jalapa, in October, 1946, the pine voles were extraordinarily abundant. From their more normal habitat they had spread out to thickets, dry hillsides, rock piles and rock walls, and even to the forest.

This species is principally nocturnal. About 75 per cent of our specimens were taken in the night. The habits of the pine vole in Veracruz are much like those of some meadow voles (subgenus *Microtus*) in the United States. They make distinct, groovelike runways in dense vegetation, and dig burrows under rocks and logs. Their runways are surprisingly narrow, being only about an inch in width. A nest found in a pine vole runway under a log was old and abandoned, although the burrow leading to it seemed to have been in use recently. The nest was a ball of dead grasses, about three inches in diameter.

The food of the pine vole consists principally of the roots and bases of the stalks of annuals. Some grass must be eaten, for we found cut blades lying in runways on numerous occasions. All stomachs that were examined had only granular, starchy vegetation, usually dull white but often stained dark brown or blackish in places. No green food was noted.

In mid-October we took a number of pregnant and nursing female pine voles. Twelve pregnant females, taken from October 14 to October 20, had from one to four embryos, with an average (mean) of 2.1.

Ectoparasites are not common on pine voles. On several occasions a few small brown mites were noticed. Tiny, orange-colored mites were seen in clusters on the ears of three or four voles. One pine vole had a tick in the ear and another had a translucent cyst under the skin of the foreleg. When opened, the cyst was found to contain seven tapeworms with large scolices, five millimeters in diameter, and short bodies, ten millimeters in length. One vole had a scar on the shoulder, perhaps from a fight with another pine vole. Two drowned pine voles were found in a tiny, steep-walled pool of water.

Family Muridae

Rattus rattus

Black Rat

In Veracruz, as far as our records go, this species occupies a rather unusual habitat. We never found it in the larger towns and cities. Neither did we find it, or any other house rat, living apart from man's dwellings. All of the specimens of black rats collected and observed were in small villages and isolated houses in the country. In the village of Potrero Viejo, 1700 feet elevation, both the black rat (R. *r. rattus*) and the roof rat (R. *r. alexandrinus*) are found. The black rats were found in smaller houses and buildings at the edge of the town, and the roof rats in the town itself. Neither was taken in the cane fields near the village.

Rattus rattus alexandrinus (É. Geoffroy Saint-Hilaire)

Specimens examined.—Total 8: Las Vigas, 8500 ft., 1; Potrero Viejo, 1700 ft., 7.

Rattus rattus rattus (Linnaeus)

Specimens examined.—Total 4: Río Atoyac, 8 km. NW Potrero, 1700 ft., 1; Jesús Carranza, 250 ft., 2; 20 km. E Jesús Carranza, 300 ft., 2.

Rattus norvegicus norvegicus (Berkenhout)

Norway Rat

We obtained no specimens of the Norway rat in the state of Veracruz, and doubt that it occurs in the tropics there. The species is included here on the basis of a dead specimen, whose head had been crushed by a car, seen and examined but not saved, on the street at Perote.

Mus musculus and subspecies

House Mouse

Specimens examined.—Total 24: El Cepillo, 20 ft., 1; Hacienda Tamiahua, Cabo Rojo, 1; 10 km. SW Jacales, 6500 ft., 2; 3 km. SW San Marcos, 200 ft., 1; Las Vigas, 8500 ft., 2; 1 km. W Las Vigas, 8500 ft., 1; Perote, 8000 ft., 1; 5 km. N Jalapa, 4500 ft., 4; Teocelo, 4000 ft., 1; Mirador, 3500 ft., 2; 3 km. SE Orizaba, 5500 ft., 1; Río Atoyac, 8 km. NW Potrero, 1; Potrero Viejo, 1700 ft., 2; 3 km. W Acultzingo, 7000 ft., 2; 3 km. E San Andrés Tuxtla, 1000 ft., 2.

House mice are found in the wild almost throughout the state of Veracruz. Only in the jungles of the extreme south did we fail to find them.

Five kilometers north of Jalapa we made special efforts to secure house mice from the fields, and four specimens were taken. These are topotypes of the subspecies *M. m. jalapae* Allen and Chapman. But they do not answer the description of *jalapae,* lacking the broad band of black on the back. Instead they are like *M. m. brevirostris.* Specimens from elsewhere in the state of Veracruz do have the color of *jalapae,* but are smaller than that subspecies as it was described. Further, at Potrero Viejo, specimens having the color pattern ascribed to *jalapae* and that ascribed to *brevirostris* both were taken in the same house.

Probably the house mice of Veracruz have been present for more than four hundred years. Much shifting of the populations probably has taken place. However that may have been, our small series of house mice from Veracruz shows wide variation in color.

Family Erethizontidae

Coendou mexicanus mexicanus (Kerr)

Mexican Porcupine

Specimens examined.—Total 14: Río Blanco, 20 km. WNW Piedras Negras, 300 ft., 4; 15 km. W Piedras Negras, 300 ft., 1; 3 km. E San Andrés Tuxtla, 1000 ft., 2; Catemaco, 2 (U. S. N. M.); 10 km. NW Minititlan [= Minatitlán], 100 ft., 1; Minititlan [= Minatitlán], 1 (U. S. N. M.); Achotal, 1 (Chicago N. H. Mus.); 20 km. E Jesús Carranza, 300 ft., 1; 25 km. SE Jesús Carranza, 250 ft., 1.

Additional record.—Jalapa (Ferrari-Pérez, 1886:130).

The abundant, short, yellow or white spines give the Mexican porcupine the appearance of being a large black animal with a white head. This is especially so at night or when the animal is some distance away from the observer.

The only name we heard applied to the porcupine in Veracruz was "puercoespin." We found the animal to be uncommon, or at least inconspicuous. Several were obtained in the dense jungles that flank the arroyos and water courses on the coastal plain. A few were taken in the dense jungles of the lower humid division of the Tropical Life-zone in southern Veracruz. Two were taken in the upper humid division of the Tropical Life-zone of the Tuxtla Mountains.

All but one of our porcupines were taken in tall trees, 60 to 100 feet from the ground. All were in trees densely hung with vines. One female was in a tree only about 40 feet high, and the porcupine was only about 20 feet from the ground. One man showed us a tree not over 10 feet high, in which he had seen a porcupine.

On the south bank of the Río Blanco, 20 kilometers west-northwest of Piedras Negras, on March 17, 1946, "I was hunting chachalacas and other specimens with some Mexican friends. We were going along the bank of a small arroyo where thorny trees were quite dense and rather high. One of the men pointed into a tree and began to talk rather excitedly. I thought it must be an iguana in the tree. They said no, that it was another animal and tried to point it out to me. I was unable to see it, and told them to shoot it with a .22 rifle. They fired and a large porcupine came into sight from the top of a large limb, where it had been lying vertically, hidden from view unless one moved about a good deal. It slowly began to slide down the tree, its prehensile tail tip curling. It was very hard to skin because there are more quills on the belly of this genus than there are on the North American porcupine. The tail skin was extremely thick and tough. I had to cut it out to the tip, and even then it would skin only with great difficulty, and the tip actually was broken off. I broke two needles in trying to sew up the tail. I had to use the pliers to drive the needle through the tail skin."

Another specimen was taken at the same locality a month later. "It was in an extremly dense tree. There was no evidence from the ground that there was an animal in the tree. I looked and found no droppings or scraps of bark. It was such a good looking animal tree that Vicente volunteered to climb it. I was not particularly surprised when he shouted that there was a porcupine. His instinct as to which trees contain animals is surprising. He forced the animal out on a limb by jerking on vines and giving a menacing growl. The animal came out on Gerardo's side of the tree, and he fired, bringing it down with one shot. It is a large non-pregnant female." The following day, May 19, a very young porcupine and a half-grown animal were obtained.

At a place ten kilometers northwest of Minatitlán, on February 6, 1947, "While night-hunting, we heard howler monkeys in the jungle and found one in a tall tree. I shot it. A few moments later Gerardo shouted that he saw another one in the same tree, and that it had a white head. I thought at once of capuchins, and told him to shoot it. He did not want to, saying that it was small. I told him to go ahead, that it might be a different kind. He fired, but nothing happened. A few minutes later I saw a pair of bright red eyes, high in the tree. The eyes were far too bright for a howler, so I fired, and brought down a porcupine. Two more monkeys were taken from the same tree."

Three kilometers east of San Andrés Tuxtla on January 13, 1948, "A large porcupine was shot last night while it was feeding in a capulin tree, between showers of heavy rain. The animal's eyes had a dull red glow.

"It is a female with one nearly full-term embryo, which I saved in alcohol. The mammae are all pectoral. There is a penislike structure anterior to the vagina—a flattened cartilaginous structure similar to that of *Tylomys*. The animal was rather fat. The stomach was crammed with capulin fruit. No ectoparasites were noted on it."

At the same locality on January 18, "another female porcupine was shot from the same tree as the above. It was feeding on the fruit about one hour after dark. It was on a branch about four inches in diameter and 15 feet from the main trunk and 30 feet from the ground. Its eyes had a dull, red glow. The white head was conspicuous in the beam of the light.

"This specimen is not pregnant, and lacks the penislike structure that the pregnant female had. The animal is not fat. One large tick was the only ectoparasite noted. A number of white nematodes about an inch long were free in the body cavity."

A specimen taken 20 kilometers east of Jesús Carranza on April 6, 1948, yielded a large, tough-skinned, dark brown, bot fly larva and a large tick.

Family Dasyproctidae

Agouti paca nelsoni Goldman

Paca or Spotted Paca

Specimens examined.—Total 10: Río Blanco, 15 km. W Piedras Negras, 350 ft., 1; Catemaco, 3 (U. S. N. M.); Boca del Río Chalchijapa, 20 km. E Jesús Carranza, 1; 20 km. ENE Jesús Carranza, 200 ft., 2; 13 km. ESE Jesús Carranza, 350 ft., 1; 25 km. SE Jesús Carranza, 250 ft., 1; 30 km. SSE Jesús Carranza, 300 ft., 1.

Additional record.—Chichicaxtle (Goldman, 1913:10).

In central and southern Veracruz, the paca is called "tepezcuintle," but in northern Veracruz is called "tuza," "tuza real" or "cuautuza." In the same areas the pocket gopher (*Heterogeomys*) is called "tuza de tierra."

The paca inhabits the tropical parts of Veracruz, from sea level to at least 5000 feet elevation. It prefers low, dense jungle and cover near streams. It is rare almost everywhere, and especially so in inhabited areas, because it is much hunted for food. The price of the flesh of the paca far exceeds that of any other meat, domestic or wild. The few specimens obtained by us were taken only after a great amount of hunting, and by offering large rewards to native hunters.

About 3:00 p. m. on April 4, 1948, 13 kilometers east-southeast of Jesús Carranza, a large female containing a nearly full-term embryo was taken on a small island in the river (Río Coatzacoalcos). From the canoe we saw her walking along the sandy beach but before we could get within gunshot range she dived into the river and vanished. She crossed an arm of the river, reappeared, walked along the sandy beach, and waded to the bank where she was shot.

On the night of May 18, 1949, a small male paca was shot at a place 30 kilometers south-southeast of Jesús Carranza. The night was dark, and the full moon had not yet come over the horizon. We were hunting from a dugout, when the canoe passed the mouth of a small arroyo. The delta of the arroyo made a small, flat, marshy place at the edge of the river (Río Solosuchi). In the beam of the hunting light, a single large, dull red eye was seen. The color of the eye resembled that of a large rabbit, *Sylvilagus,* which also often shows only a single eye in a beam of light. The eye was far too dull for that of a pauraque, the common large goatsucker. Our shot brought down the paca. It had been feeding on the tender, succulent vegetation on the muddy bank of the arroyo.

On many occasions our dogs ran pacas, but usually lost the trails when the animals entered water. On April 1, 1949, 25 kilometers southeast of Jesús Carranza, our dogs startled a paca in a patch of thorny bushes about 10:00 a. m. The paca burst out of the thorns, dashed down the slope, and entered a small almost vertical burrow. Its entrance was about 30 feet from the bank of a small arroyo in which there was a pool of semi-stagnant water about 20 feet long and six feet wide. The bank beside the pool was about four feet high, and of dark, solid earth. Just above the water a horizontal burrow in the bank was in line with the vertical hole about 30 feet away on the floor of the jungle above. We built a fire at the lower entrance, forcing smoke into the burrow, and a few moments later a small paca emerged from the lower entrance and was shot.

Several days later another paca was driven to ground in a burrow beside this same arroyo. This paca dashed through a pool of water and entered a burrow in the vertical side of a bank about six feet high. About 40 feet back from the arroyo we found a smaller, vertical burrow, but were unable to force smoke from one entrance to the other; perhaps the two were not connected. We excavated the burrow near the water. The burrow coursed upward at a 30 degree angle from the water for a distance of two feet and then leveled off for a distance of some 20 feet. At this point the burrow pitched down again at a 30 degree angle and we ceased digging. Indeed, we never did succeed in completely excavating any burrow of a paca. Many hours were spent working with machetes, but little was accomplished except the removal of many cubic yards of earth. All the burrows excavated were simple tubes, but their principal extent lay at a distance of six feet or more beneath the surface of the ground. One other paca was caused to flee its burrow because of smoke of dry palm fronds, liberally mixed with powdered sulfur. At least ten other burrows were filled with smoke without success.

All of our pacas were in good condition. No parasites of any kind were noted on them. Most of them were taken in an area where the agouti, *Dasyprocta,* was numerous. Almost every agouti that we obtained had some parasitic fly larvae under the skin. These were never seen on pacas.

Dasyprocta mexicana Saussure

Mexican Agouti

Specimens examined.—Total 46: 3 km. N Paraje Nuevo, 1700 ft., 3; 2 km. N Paraje Nuevo, 1700 ft., 1; 2 km. S Paraje Nuevo, 1; Buena Vista, 6 (U. S. N. M.); Catemaco, 3 (U. S. N. M.); Achotal, 8 (Chicago N. H. Mus.); 20 km. ENE Jesús Carranza, 200 ft., 7; 20 km. E Jesús Carranza, 300 ft., 2; 25 km. SE Jesús Carranza, 400 ft., 4; 35 km. SE Jesús Carranza, 400 ft., 4; 38 km. SE Jesús Carranza, 400 ft., 1; 30 km. SSE Jesús Carranza, 300 ft., 4; 60 km. SE Jesús Carranza, 450 ft., 2.

We are not able to find any difference between the agoutis of the upper humid division of the Tropical Life-zone of central Veracruz and the agoutis of the deep jungles of southern Veracruz.

Agoutis from the Isthmus area of Veracruz are the black *mexicana.* There is no evidence of intergradation between them and the brown agoutis of the *punctata* type, to the east in Campeche, or to the southwest in Chiapas. Because the ranges of the two kinds (*mexicana* and *punctata*) seem to be mutually exclusive, one would expect the two to intergrade. If they do intergrade, the area of intergradation must be much smaller than one would expect.

The agouti is called "Cerreti" over most of Veracruz, though locally the name "cuacechi" is used. Although much hunted, this species has not been so greatly reduced in numbers as the paca.

In some places agoutis hold their own, even near villages. They become very shy, will run for long distances when chased by dogs, and will take refuge in holes only as a last resort.

Near our base on the Río Coatzacoalcos, 20 kilometers east of Jesús Carranza, our pack of dogs chased agoutis almost daily, but only in a few instances were we able to capture the animals. Often the chase would continue for more than an hour. On the other hand, in the uninhabited country a short distance to the south, agoutis were abundant and the dogs usually brought them to bay after a relatively short chase, usually lasting about 15 minutes.

Near Potrero, where the agoutis are much hunted for food, they are crepuscular, and perhaps even partially nocturnal. They live in the cool, damp forest among the limestone boulders and make narrow trails to their feeding places. Our specimens from central Veracruz were all taken by a skillful Indian hunter, who set out small bunches of ripe bananas and visited them in early morning and late evening. In southern Veracruz, the agouti is entirely diurnal. We found it most active in the late morning and afternoon, less active in early morning and late evening, and never found any evidence of activity at night. Usually an agouti, at the approach of danger, stands motionless until the danger is past, and then dashes rapidly for the nearest shelter in a hollow log or burrow in the ground. When an agouti runs, a fan of long hair stands out around the animal's rump, somewhat reminiscent of the fan of the northern porcupine (*Erethizon*). The dash of an agouti for safety is rapid and erratic. It zigs and zags, without swerving greatly from its line of escape. Rarely do the feet come down twice in the same line. The animal makes a noise that is loud, out of all proportion to the agouti's size, as it rushes through the dead leaves and brush.

Some accounts of typical agouti hunts on the Río Chalchijapa and the Río Solosuchi, southeast of Jesús Carranza, are as follows:

March 27, 1948: Today the dogs ran several animals—probably all agoutis. Only one was caught. It took refuge in its burrow under a log and parallel to it. The burrow was a straight tube about 15 feet long, five inches in diameter, and a foot beneath the surface. Both ends opened out from under the log. The country was generally open, beneath tall, jungle trees, with some brush and many dead fallen leaves and twigs.

March 29: Today after a short chase, the dogs cornered an agouti in its burrow that was about 10 feet long, nine inches in diameter, and 30 inches beneath the surface. It had a single opening. The animal is a large male.

April 24: Today, after a short chase of about 15 minutes, the dogs brought a small agouti to bay in a hole in the ground at the base of a tall tree. The hole was large enough for one dog to enter, and it killed the agouti at the end of the burrow, about five feet from the mouth. Twice, now, we have taken very small specimens in very large holes. Larger animals seem to have smaller burrows than younger animals.

April 30: We found abundant agouti sign here a week ago, and so returned in order to obtain a series of specimens. The dogs worked well, running about 12 animals. Of these we got five; one was a skunk and four were agoutis. Most of the other animals were probably agoutis—three or four certainly were. The four obtained all took refuge in hollow logs. These hollows, in every case, were just large enough to admit the agoutis. If we were able to, we opened the logs with a machete. Some logs of very hard wood could not be opened. On these logs we used smoke of burning sulfur. Even this did not budge some animals, and we had to leave them. Of the four taken, one is a male with large testes; two are nursing females; and one is a female containing two full-term embryos. All have two to six fly warbles, usually on the backs or shoulders.

May 1: A large female was taken today after a long chase that lasted nearly an hour. As usual, the animal entered a hollow log and was chopped out.

About 4:00 p. m. a large male agouti ran along the edge of the river at the base of a sand bluff. It must have been hiding behind a clump of earth, for it was not seen until it dashed away with the erected "fan" of hairs. It did not try to enter the water. I shot it with the rifle. It was heavily infested with fly warbles. One area three inches in diameter on its shoulders was a solid mass of warbles and had to be cut out of the skin. Two or three black fleas were also present.

Twice we found remains of agoutis in stomachs of ocelots.

Order CARNIVORA

Family Canidae

Canis latrans cagottis (Hamilton-Smith)

Coyote

Specimens examined.—Total 2: 15 km. W Piedras Negras, 300 ft., 1; Río Blanco, 20 km. W Piedras Negras, 400 ft., 1.

This animal is known throughout Veracruz by its Spanish name, coyote, derived from the Nahuatl "coyotl". The coyote lives on the arid coastal plain and is said to occur as far west, along the coast, as the sugar cane fields near Potrero. In northern Veracruz it occurs principally along the sea beaches, moving inland wherever there are clearings. The coyote occurs also on the Perote Desert and in the pine forests of the mountains of the west-central part of the state. It avoids the dense jungle completely.

We saw no coyotes in the daytime, but often heard them howling at night on the coastal plain, and on the desert near Limón. Tracks were seen on the dust of trails. Coyotes avoided our traps, however, as did most carnivores in Veracruz. One of our specimens was shot at night when it attempted to steal a chicken. When it was skinned, an old pellet of buckshot was found lodged in one front foot. The second specimen was shot in an arroyo at night, with the aid of a hunting light.

The habits of the coyote in Veracruz seem to be much like those of the species in the United States. The coyote is universally accused of stealing chickens, although we never heard of one killing larger domestic animals on the coastal plain. Such animals as horses, burros, cattle, pigs, sheep and goats are raised in numbers on the coastal plain. Coyotes are said to feed on carrion whenever the opportunity offers.

Our specimens were in good pelage but had a large number of ticks—small to large, reddish, and present over the entire body, but especially on the ears.

Urocyon cinereoargenteus

Gray Fox

The gray fox is called zorra in Veracruz. In southern Veracruz, where there are no foxes, the opossum (*Didelphis*) is called zorro.

In Veracruz, foxes live in open country, including clearings in the jungle, from sea level at least to 8000 feet elevation, and they were reported from the Cofre de Perote at 10,500 feet elevation. They are probably most common in the lower arid division of the Tropical Life-zone of the coastal plain, but are also fairly common in the upper humid division forest and in the pine forests of high elevation. Only in the dense jungles of the lower humid division of the Tropical Life-zone do they seem to be absent.

Probably these foxes are principally nocturnal. We shot two at night, in the beams of hunting lights, and trapped another at night. To a certain extent, however, they are diurnal, for we saw several out at midday. One was hunting on an exceptionally hot, bright, sunlit day.

Foxes were accused of stealing chickens, and probably do so. We examined fox droppings near Las Vigas, 8000 feet elevation, and found them to contain only vegetable matter, except for one that held the remains of a mouse, *Peromyscus melanotis*. Most common, in droppings collected in early November, was a small, purple berry that grows in spikes, only a few inches high. One dropping was composed entirely of corn. Another was principally of the purple berry, but held also a single large blackish bean. At that locality, a pocket gopher, *Cratogeomys perotensis*, was eaten in the trap by a fox.

The only parasites noted on gray foxes were on specimens from the tropics. These all had the common large tick, locally called garrapata. Foxes are shot whenever opportunity arises, in Veracruz, and when killed are sometimes eaten by man. They are rarely hunted, however, and the steel trap is almost unknown to the natives. Consequently the gray fox exists throughout its range in fair numbers.

Urocyon cinereoargenteus orinomus Goldman

Specimens examined.—Total 11: Las Vigas, 1 (U. S. N. M.); 3 km. E Las Vigas, 8000 ft., 1; Jalapa, 1 (U. S. N. M.); 7 km. NW Potrero, 1700 ft., 1; Potrero, 1700 ft., 1; Orizaba, 3 (U. S. N. M.); 15 km. W Piedras Negras, 300 ft., 1; Río Blanco, 20 km. WNW Piedras Negras, 1; 24 km. E, 7 km. S Jesús Carranza, 1.

Urocyon cinereoargenteus scottii Mearns

Specimen examined.—One from 3 km. W Zacualpan, 6000 ft.

Family Procyonidae

Bassariscus astutus astutus (Lichtenstein)

Ringtail

Specimens examined.—Total 8: Acatlán, 4100 ft., 1; Las Vigas, 8500 ft., 1; 1 km. W Las Vigas, 8500 ft., 1; 2 km. E Las Vigas, 8000 ft., 1; 2 km. W Jico, 4200 ft., 1; Maltrata, 1 (U. S. N. M.); Orizaba, 2 (U. S. N. M.).

Additional record.—Xico (Sanborn, 1947:270).

In Veracruz the ringtail occurs in the uplands, principally on the Mexican Plateau. Where conditions are suitable it descends to the upper edge of the upper humid division of the Tropical Life-zone. Our lowest record altitudinally is 4200 feet, at Jico. Zonally the species ranges from the upper edge of the Tropical Life-zone, through the pine forests near Las Vigas, and onto the desert near Perote. The ringtail is rather restricted as to habitat. It is specialized for life in rocky places, such as cliffs, outcrops and lava flows. This seems to be true, to a greater or lesser degree, over the entire range of the species.

Two ringtails were seen on top of a rocky cliff two kilometers west of Jico on October 30, 1946. This area is tropical forest, with the pines of the next higher life-zone beginning to appear as outliers. The animals were found about 10 p. m,. along a rocky ledge near the cliff, when the beam of my headlight picked up two pairs of eyes. They had an unusually bright shine—one pure yellow and the other yellow tinged with orange. I fired a charge of buckshot between them. When I climbed up to the ledge, I found nothing but heard a loud chirping, rather birdlike. I saw an animal back in the brush, at a range of about four feet, but it slipped over the cliff. I assumed that it was wounded. We called the dog from camp, and he found one ringtail. This one was well shot—not the one I saw first. We searched in vain for an hour for the other and got well scratched with malamujer. The specimen found was a male. It had a strong skunklike odor. Its stomach was full of mice—seemingly of the species *Peromyscus furvus* that was abundant there. Three of the large ticks, locally known as garrapatas, were on the animal. No other parasites were seen. Its testes were rather small.

On the night of October 5, 1948, along the edge of the fresh lava flow two kilometers east of Las Vigas we were searching for wood rats wherever holes opened in the dense, drifting fog. A pair of greenish yellow eyes were seen; the animal was running swiftly over the rough lava. A whistle did not cause the animal to stop. A shot brought down an old female ringtail that was nursing young. Her stomach was filled with the fruit of the prickly pear cactus (locally called tuna de nopal). We saw no plants of this cactus in our wanderings near Las Vigas. The specimen had no ectoparasites.

On October 14, 1948, a full grown young ringtail was captured in a house in Las Vigas. The animal had entered in the night, and created a disturbance in trying to find a way out. The ringtail was kept alive for a short time. It had a harsh, menacing growl and a shrill, chattering squeal. It was an agile climber on the walls of the house.

Bassariscus sumichrasti sumichrasti (Saussure)

Tropical Cacomixtle

Specimens examined (U. S. N. M.).—Total 3: Jalapa, 2 (1 a skull only); Mirador, 1.

This species is the jungle equivalent of the ringtail, specialized for an arboreal life, rather than life in rocky places. We failed to take specimens in Veracruz.

Procyon lotor

Raccoon

In central and extreme southern Veracruz the raccoon is called "mapache." In the Tuxtla Mountains area it is called "mapachina." Both of these words are corruptions, probably Totonac and Zapotec, of the Nahuatlan word, "mapachtli," which is still used occasionally in that part of Veracruz on the edge of the Mexican Plateau.

Zonally the raccoon ranges throughout the Tropical Life-zone, from the upper edge of the upper humid division to the lowest part of the lower humid and lower arid division. Altitudinally we have records from 4200 feet to sea level. In Veracruz the species is less restricted to the vicinity of water than it is in the United States. Many of our specimens were captured several miles from water. As elsewhere in Mexico and North America, tracks are seen in the early morning, in the mud and sand along the edges of streams, ponds, and rivers. In the day, raccoons find refuge in hollows of trees, and later, when the tropical sun makes their hollows too warm, they emerge and rest on the limbs of tall, vine-laden trees, where the raccoons are invisible from the ground, but able to catch any cooling breeze. Local people are aware of this habit and when the sun is hottest visit the arroyos having vine-laden trees, which they patiently climb in search of the animals. Two of our specimens were obtained in this way.

One was taken on the south side of the Río Blanco, 20 kilometers west-northwest of Piedras Negras, on May 14, 1946. This area is on the arid coastal plain, and we went hunting for raccoons in the extreme heat of midday. We had gone some distance up the arroyo, examining each vine-hung tree carefully, without seeing even a squirrel. When we reached an especially tall, densely vine-hung tree, we looked for some time. Finally Vicente said that it was a perfect animal tree, and he was going to climb it. After some hard work he got up about 50 feet, and called that he saw a porcupine. Then he said no, it was a raccoon. We forgot our heat exhaustion, and even the dogs caught some of our excitement, for they gathered around the foot of the tree expectantly. Vicente climbed higher, growling menacingly and shaking vines. After several minutes the raccoon came plunging down through the vines and branches. Two weeks later at the same locality another raccoon was taken under similar circumstances.

Four kilometers west of Tlapacoyan on November 24, 1947, while hunting with several of the local people, we for a long time found nothing, until the dogs barked at the base of a large, hollow tree.

We heard animals in the tree, and built a fire and smoked out an old female raccoon and three young. The tree was in dense jungle near a small arroyo. I think it was an oak.

Twenty kilometers east of Jesús Carranza on February 8, 1948, the dogs treed something in a tree in the low, palm jungle. When we came up we found six raccoons in the tree, a low, bushy species about 50 feet high. We got all six animals. Two were large females, not pregnant. The others, one female and three males, were smaller, not fully adult animals. No ectoparasites were noted.

The raccoon is much hunted for food in Veracruz, although it is not considered as first-class meat. Even old animals, when properly prepared, are tender and well flavored. The chili and other spices used in cooking seem to kill any strong taste that might be present.

Procyon lotor hernandezii Wagler

Specimens examined.—Total 15: 4 km. W Tlapacoyan, 1700 ft., 2; Río Blanco, 20 km. WNW Piedras Negras, 300 ft., 2; Jico, 3 (U. S. N. M.); 20 km. ENE Jesús Carranza, 200 ft., 1; 20 km. E Jesús Carranza, 300 ft., 7.

Additional record.—Mirador (Goldman, 1950:65).

Taxonomic appraisal of specimens was summarized in 1948 in manuscript by Dalquest as follows: "Comparison of the skulls of raccoons from Veracruz with skulls from Campeche and from the Mexican Plateau, shows that our material most closely resembles those from the plateau. The skulls of our specimens are smaller, and narrower in the interorbital region, as compared with *shufeldti* from Campeche, and thus resemble *hernandezii.* The skins are more or less intermediate between those of the two subspecies, but more nearly like those of *hernandezii.* The fur is longer than in *shufeldti* but slightly shorter and thinner than in *hernandezii.* In color, all but three of our skins resemble *hernandezii.* Veracruzian raccoons are distinctly intermediate between the two subspecies, and show no characters worthy of separate recognition. In Veracruz, material from the high, cool tropics and the low, hot jungles, is essentially the same."

Procyon lotor shufeldti Nelson and Goldman

Record.—Minatitlán (Goldman, 1950:67).

Nasua narica

Coati

The coatimundi is termed "tejón" throughout Veracruz. The geographic range includes the Tropical Life-zone of the State, although the species is rare at high elevations and absent from the arid coastal plain. The coatimundi, or coati, seems to be entirely diurnal. We found all of our animals in the daytime, and never found the species at night.

The coati was most commonly found in low, brushy jungle and palm jungle, and rarely in deep forest. The jungle of coyol palm was especially favored; this palm furnishes shelter, shade, and a fruit of which the animal is especially fond. The coati likes also boulder-strewn areas, cliffs, and the dense tangles of low vegetation that fringe the water courses.

In Veracruz, as elsewhere in México, the natives maintain that there are two species of the coati. In Veracruz "tejón" and "tejón solo" are the names used. The "tejónes" travel in bands, whereas the "tejónes solos" are solitary. We took five "tejónes solos." Three of these were females and two were males. They were all adult animals, and rather old. But, our oldest and largest coati was with a band of other animals and was not a "tejón solo." In general the coati is a social animal. Bands vary greatly in size. Some bands include only three to 10 animals. Rarely a band numbers 50 or more. Bands of 100 were reported.

The coati spends much of its time on the ground. The bands usually travel rather slowly, and spread out when hunting, but the individuals proceed in single file when traveling. They keep the long tail upright, with the extreme tip bent backwards, almost horizontally. When hunting food on the forest floor they make considerable noise, scratching, rustling leaves and branches. They do a great deal of piglike rooting in the leaves and ground. When a band of coatis has passed, the characteristic signs are numerous scratched and rooted places in the leaves.

The food of the coati consists principally of vegetable matter. One favored food is the fruit of the coyol palm. This is a nut about two inches in length and an inch wide. It has a tough rind, and a thin layer of pulp over a rock-hard seed. The nuts grow in great, pendant clumps, weighing many pounds. All trees do not come into fruit at the same time; the season lasts from early March into June. The seeds beneath a coyol palm often show the tooth marks of the coati, where the oily pulp has been chewed away, leaving the tough fibers attached to the seed.

Corn is also eaten, and coatis occasionally do a great deal of damage to milpas. Bananas are eaten, even before they are ripe.

The coati is much hunted for food in Veracruz. Dogs are trained to tree the animal, which local people consider more of a squirrel than a carnivore. The flesh of the older animals is sometimes tough and rank, but that of younger animals is tender and of good flavor. One female taken on March 15, contained four embryos. One old animal had lost the terminal part of its tail. Parasites are uncommon on coatis. Most of the specimens examined by us had a few ticks and one old male had a heavy infestation of fly larvae. There were more than 50 larvae, some even on the feet, tail and jaws.

Nasua narica molaris Merriam

Specimen examined.—Hacienda Tamiahua, Cabo Rojo, 1.

Nasua narica narica (Linnaeus)

Specimens examined.—Total 35: Jico, 1 (U. S. N. M.); Carrizal, 1 (U. S. N. M.); 7 km. NW Potrero, 1; 3 km. N Paraje Nuevo, 1700 ft., 1; 2 km. N Paraje Nuevo, 1700 ft., 1; 2 km. E Paraje Nuevo, 1700 ft., 3; Potrero Viejo, 1700 ft., 5; Orizaba, 1 (U. S. N. M.); 14 km. SW Coatzacoalcos, 100 ft., 1; 35 km. ENE Jesús Carranza, 150 ft., 7; 20 km. ENE Jesús Carranza, 300 ft., 1; 20 km. E Jesús Carranza, 300 ft., 11; 35 km. SE Jesús Carranza, 400 ft., 1.

Additional records.—Pasa Nueva (Allen, J. A., 1904:39); Jalapa (Allen, J. A., 1879:165).

Potos flavus aztecus Thomas

Kinkajou

Specimens examined.—Total 31: 12½ mi. N Tihuatlán, 300 ft., 2; 7 km. SE San Juan de la Punta, 400 ft., 1; 3 km. N Presidio, 1500 ft., 1; Río Blanco, 20 km. WNW Piedras Negras, 1; Coatzacoalcos (Region), 4; 3 km. E San Andrés Tuxtla, 1000 ft., 1; 20 km. E Jesús Carranza, 300 ft., 7; 25 km. SE Jesús Carranza, 250 ft., 7; 35 km. SE Jesús Carranza, 1; 38 km. SE Jesús Carranza, 500 ft., 1; Arroyo Saoso, 37 km. E, 7 km. S Jesús Carranza, 3; 30 km. SSE Jesús Carranza, 300 ft., 1; 55 km. ESE Jesús Carranza, 450 ft., 1.

Additional record.—Jalapa (Ferrari-Pérez, 1886:128).

A single adult female kinkajou is available from the Tuxtla Mountains of Veracruz and differs from all other kinkajous examined from México, especially in large size and large skull. This single specimen is here regarded as an extreme variant of P. *f. aztecus*, but if additional material from the same area resembles our specimen, the kinkajou from the Tuxtla Mountains will require a new name.

The kinkajou is generally known as "marta" in Veracruz. Locally the name "martucha" may be used, and in northern Veracruz we heard it called "mico de noche." In southern Veracruz, "mico de noche" is the native name for the two-toed anteater.

The kinkajou is arboreal. Probably it descends to the ground. It prefers the tall trees of the deep jungle, but visits lower growth and riverside forest in search of fruits, and may live permanently in some place of that kind. We found kinkajous most common where trees were more than 150 feet in height. Rarely did we find them in palm jungle.

Kinkajous are nocturnal, but, like some other nocturnal and arboreal animals, emerge from their retreats at midday to take advantage of cooling breezes. Probably the hollows in which they live become unbearably hot in the strong tropical sun. Because daytime encounters with this species are unusual they are recounted here.

On the south side of the Río Blanco, 20 kilometers west-northwest Piedras Negras, 300 feet elevation, on May 15, 1946, while hunting with two companions about midday in a forest-fringed arroyo, we found several partly eaten fruits under one large tree. I picked up one and found the unmistakable tooth marks of a large fruit bat. Vicente walked ahead and picked up several which he looked at and remarked "this is not bat." He called to Gerardo, that there was an animal in the tree. Gerardo peered about a moment and then shouted "Yes, a martucha." The animal was out of our sight about 40 feet up in the vine-hung tree. Gerardo fired twice and it descended and ran along a horizontal limb about 40 feet from me and 20 feet from the opposite bank of the arroyo. It was running smoothly but rather slowly. I fired and hit it in the chest with a load of number five shot. It started to fall, then caught its tail around a branch and hung for a moment. It released its hold and crashed into the brush at the very lip of the arroyo.

Twenty kilometers west of Piedras Negras, on October 2, 1946, Mr. Allen Oleson found two kinkajous clinging together in a ball, nearly in the top of a tree, about 50 feet from the ground. He fired at them with his pistol and secured the larger, an adult female. The smaller escaped, making a mewing cry. The smaller was presumably the young of the larger, but the mammae of the larger animal were not active.

Twenty kilometers east of Jesús Carranza, early in the morning of March 21, 1948, a male kinkajou was found in a tree in low palm jungle. The day was very hot. The animal was probably driven from its hollow by the heat. It was in a tree of moderate size—about two feet in diameter at the base. The kinkajou lay on one limb immediately above a fork about 20 feet from the ground. Its head was up as it alertly watched us.

Most of our specimens of kinkajou were obtained at night. In the beam of a headlight, the kinkajou's eyes reflect a bright yellow glow. The kinkajou also makes a call that is distinctive and that, once heard, is not forgotten. Kinkajous call often, and a calling animal can usually be located. The loud, plaintive, somewhat quavering scream is made by both males and females.

On one or two occasions we thought that more than one kinkajou was in a tree, but never did we actually see more than one. In most instances the animals were almost certainly solitary. Even when locally abundant, they seem to remain 100 yards or more apart. They make considerable noise in their movements aloft, and there is often a shower of leaves, twigs and fruit falling beneath a tree where a kinkajou is feeding.

Kinkajous feed where fruit is found. We took one specimen only 20 feet from the ground, in a mass of vines and creepers. Others were fully 100 feet from the ground. Probably about 60 feet is the elevation at which they usually live. They prefer large limbs, and may carry fruit there to eat it. We have seen them well out on slender limbs, however. Twice, individuals were seen hanging by their tails and swinging in wide arcs.

When kinkajous find a tree having an abundance of fruit, they return to the same tree, night after night. We hunted one kinkajou in a patch of jungle for several successive nights, even shooting at it on some occasions. It remained at a height of about 100 feet. Although it became more shy, it returned each night to the fruit tree, until it was shot.

The food of the kinkajou seems to be entirely fruit. In addition to the mango, it eats: a tough-skinned green fruit; a green berry resembling the blue elderberry in size, shape and general appearance; a nutlike fruit; the fruit of the amata fig; and a round vine fruit, about three inches in diameter, having a hard rind and a pale pink pulp that looked like a melon and had a delicious odor.

Some kinkajous were shot over water. In every case the animal fell into the water and sank like a stone, not floating for even a moment. The body of the kinkajou is astonishingly heavy for its volume.

No parasites of any kind were found by us on the kinkajou. The animal has a rather strong, musky odor. The flesh is lean and rather rank. The kinkajou is not eaten in most parts of Veracruz, but one man requested the meat of a kinkajou that we shot. His companions made fun of him, but he maintained that it was choice food. One dead kinkajou was found on the bank of a river. It had no wounds of any kind, but was too decayed to dissect. The skull of this animal was saved.

A female taken on December 1 had a small embryo, five millimeters long, in the right horn of the uterus. Another had two nearly full-term embryos on January 20. In this species there are two mammae, in an area free of fur, well forward on the inguinal region.

Family Mustelidae

Mustela frenata

Long-tailed Weasel

The long-tailed weasel is called "oncilla" in Veracruz. It occurs from the Tropical Life-zone to the Canadian Life-zone, from 300 to 12,500 feet in elevation. It is rare, occurring locally where mice are abundant—about cliffs, in old rock walls, in brushy places, and in sugar cane fields. It is principally, if not entirely, diurnal. Several were seen in the daytime in Veracruz, and one at night.

Two specimens taken two kilometers west of Jico, were killed along a rocky, boulder-strewn coffee grove at the foot of a cliff. Their stomachs were crammed with the remains of deer mice (*Peromyscus furvus*).

Testes were moderately large, but probably not in breeding condition, on October 30. Specimens from seven kilometers west of Potrero were shot in sugar cane fields by local hunters and saved for us. The specimen from Las Vigas was taken in a trap set for a ringtail beside a lava flow.

Several weasels had ticks in the ears, but no other parasites were noted.

Mustela frenata macrophonius (Elliot)

Specimens examined.—Total 3: Achotal, 1 (Chicago N. H. Mus., type); Pérez, 2 (U. S. N. M.).

Mustela frenata perda (Merriam)

Specimens examined.—Total 2: Catemaco, 1 (U. S. N. M.); 35 km. SE Jesús Carranza, 400 ft., 1.

Mustela frenata perotae Hall

Specimen examined.—Cofre de Perote, 1 (U. S. N. M.).

Additional record.—Perote (Hall, 1951:355).

Mustela frenata tropicalis (Merriam)

Specimens examined.—Total 6: Hacienda Tamiahua, Cabo Rojo, 1; 2 km. W Jico, 4200 ft., 2; Las Vigas, 8500 ft., 1; 2 km. W Potrero, 1700 ft., 1; 3 km. N Paraje Nuevo, 1700 ft., 1.

Additional records.—Jico (Merriam, 1896:30); 5 km. N Jalapa, 4500 ft. (Davis, 1944:381); Jalapa (Hall, 1951:366); Orizaba (Hall, 1951:366).

The specimen from Las Vigas has some characteristics of *M. f. frenata*, which suggest intergradation between *frenata* and *tropicalis* in the highlands of Veracruz.

Eira barbara senex (Thomas)

Tayra

Specimens examined.—Total 8: Mirador, 1 (U. S. N. M.); Catemaco, 1 (U. S. N. M.); Pérez, 1 (U. S. N. M.); 20 km. ENE Jesús Carranza, 200 ft., 3; 20 km. E Jesús Carranza, 300 ft., 1; 38 km. SE Jesús Carranza, 500 ft., 1.

Additional records.—Hacienda Tortugas (Thomas, 1900:146); Pasa Nueva (Allen 1904:36).

The tayra is known throughout Veracruz as "cabeza de viejo," literally, head of old in reference to the white or gray head. It is well known to the natives, but we found it difficult to obtain specimens.

The tayra seems to be normally a forest animal. We heard no reports of its occurrence on the coastal plain, unless the type locality is there. As we observed it, it is diurnal. Several times, however, we heard a cry at night that we were told by natives was made by a tayra. This cry may or may not have been made by a tayra. Tayras are not social to any extent. We were told that four or five sometimes go together to raid sugar cane patches and cornfields. Perhaps these are family groups. On April 9, 1948, we saw two animals together in a fig tree overhanging a small arroyo 38 kilometers southeast of Jesús Carranza, when we were poling a canoe upstream. By the time we had got the canoe ashore, they had vanished. We spread out; I went up the arroyo. A few minutes later Chico shot. Then a tayra came running along a log that bridges the arroyo. It was wounded. I shot, but it kept on going. A minute later the dogs came along its trail and crossed the log. About 100 feet away they brought it to bay at the foot of a large tree. Though it was much wounded it was very vicious. It was very quick. When I tried to step on its chest, it bit my foot. When Casculo tried, it tore a long strip from his shoe, ruining it. The dogs and I finally killed it. It was an adult but non-pregnant female. It had very little of the musk odor of the weasel, but had a faint, unpleasant smell.

The natives accuse the tayra of destroying corn. One of our specimens was shot while tearing down cornstalks in a field. It is also accused of destroying sugar cane. We saw sugar cane patches that had been almost completely ruined by some animal. The stalks had been ripped, torn and chewed, and it seemed to be the work of a flesh-eating mammal, to judge from the tooth marks. Sugar cane seems to be an unusual diet for a mustelid. These stories of destruction of cane by the tayra are so widespread that there seems to be no reasonable doubt that the tayra is responsible. Probably fruit, birds and small mammals are the more usual diet of the tayra. One of our native helpers said he had seen tayras chasing squirrels.

Galictis allamandi canaster Nelson

Grisón

Specimens examined.—Total 2: Potrero Viejo, 1700 ft., 1; 20 km. ENE Jesús Carranza, 200 ft., 1.

Additional record.—Orizaba (Goodwin, 1953:435).

The grisón is one of the rarest carnivores that occurs in Veracruz. One of our specimens was shot by a native hunter in a sugar cane field at Potrero Viejo. He recognized the animal as unusual, and brought it to Mr. Dyfrig Forbes, who had it stuffed for us. The other specimen is a skull-only that was among several dry skulls saved for us by people on the Río Coatzacoalcos, east of Jesús Carranza. This particular lot included skulls of tamandua, agouti, armadillo, opossum, tayra, and coati. The hunter knew that we were anxious to obtain skulls of the tayra, and sold the grisón skull to us as a tayra skull. Not until the skull was cleaned in the laboratory was it identified as from a grisón.

Mephitis macroura

Hooded Skunk

"Zorrillo" is the vernacular name used for this species and for the hog-nosed skunk throughout Veracruz. The hooded skunk inhabits open areas. We have no evidence of its presence in the jungle. The species is found in the arid pine forest and open desert in the west-central part of the state, and on the arid coastal plain to the east. Howell (1901:42) records a specimen from Orizaba, but the specimen may have come from the highlands to the west, or even from the mountains of that name, rather than from the city.

Seemingly hooded skunks are entirely nocturnal. We caught none in traps. On one occasion a skunk was seen at night in a clump of thorny plants, and the area was surrounded with steel traps, baited with meat. The skunk did not visit the traps. Other meat-baited traps, set where skunks were common, took no skunks. We collected the needed specimens by hunting them at night with headlights. Their eyes have a greenish glow, as seen in the reflected light.

Hooded skunks from the coastal plain had relatively little skunk odor, far less than a striped skunk of equal size from the United

States. Skunks from the highlands, near Las Vigas, however, had the usual strong skunk odor.

In the pine forest area, we found hooded skunks along rock walls, along cercas of maguey, and along the edge of an extensive lava flow. Usually they were in relatively open country, rather than in the pine forest. On the coastal plain they were found in brush, in tall sawgrass meadows, and along fence lines of thorny, yucca-like plants. Near Las Vigas, the skunks were eating grasshoppers and a small purple berry. On the coastal plain, one was found feeding on mangos that had fallen to the ground, and two were feeding on grasshoppers in a recently plowed field.

One old male hooded skunk had enlarged testes on January 13, but none of our other specimens showed evidence of breeding. At Las Vigas two young, three-quarters grown, specimens were taken in mid-October.

Mephitis macroura eximius Hall and Dalquest

Specimens examined.—Total 3: Río Blanco, 20 km. WNW Piedras Negras, 2; 15 km. W Piedras Negras, 1.

Mephitis macroura macroura Lichtenstein

Specimens examined.—Total 11: Perote, 1 (U. S. N. M.); Las Vigas, 8500 ft., 6 (2 in U. S. N. M.); Jico, 4 (U. S. N. M.).

Additional record.—Orizaba (Hall and Dalquest, 1950:578).

Conepatus leuconotus leuconotus (Lichtenstein)

Eastern Hog-nosed Skunk

Specimens examined.—Total 3: Potrero Llano [= Potrero del Llano], 350 ft., 1; Río Blanco, 20 km. WNW Piedras Negras, 300 ft., 2.

Additional records.—Near Martinez de la Torre (N of Perote) (Ingles, 1959:405); Río Alvarado (now known as Río Papaloapan), type locality.

The specimen from northern Veracruz has the color pattern of *leuconotus* but is larger than *leuconotus,* with a larger skull, thus showing intergradation with *texensis.* Probably the range of this species is continuous along the gulf coast from Veracruz to Brownsville, Texas, and intergradation probably takes place gradually between the two subspecies.

The hog-nosed skunk is called "zorrillo" in Veracruz. Local residents recognize it as being a species different from the hooded skunk but used the same name for both so far as we could determine. The hog-nosed skunk and hooded skunk occur together on the coastal plain.

We found this species only at night. It seems to be rather rare in Veracruz. Our three specimens were all found in cornfields.

One of the fields was surrounded by low brushland and the other two by extensive grassy plains. On the plains, scattered thickets of bull-horn acacia and other thorny plants offered cover. The stomachs of all three specimens were filled with remains of insects. A female taken on May 24 was nursing young.

Individuals of this species emitted only a little musk when shot. Their musk glands were large and full, as full and large as the glands of a large *Mephitis* from the United States.

Conepatus semistriatus conepatl (Gmelin)

Striped Hog-nosed Skunk

Specimens examined.—Total 7: Motzorongo, 1 (U. S. N. M.); Catemaco, 2 (U. S. N. M.); Pérez, 1 (U. S. N. M.); 14 km. SW Coatzacoalcos, 100 ft., 1; 20 km. ENE Jesús Carranza, 200 ft., 1; 35 km. SE Jesús Carranza, 400 ft., 1.

Additional record.—Pasa Nueva (Allen, J. A., 1904:36).

This skunk, like the other skunks, is called "zorrillo," sometimes modified to "zorrillo que apesta mucho" or "zorrillo pijón." It is a jungle-inhabiting species and although widespread is rare. We hunted in vain for it at Motzorongo, the type locality. In southern Veracruz, we found it to be more common. One was shot there at night, 14 kilometers southwest of Coatzacoalcos, on February 12, 1947, at the base of an isolated tree where the shade of the tree and a few bushes beneath it allowed enough open ground in an extensive area of sawgrass for our lights to pick out the animal's eyes. They shine bright green.

The animal was an adult female, notably fat, and not pregnant. No parasites were noted on it. It had a strong odor, stronger than the odor of any other skunk obtained in Veracruz.

Thirty-five kilometers southeast of Jesús Carranza on April 30, 1948, shortly after dawn, we came upon a small opening in the forest where a year or so before a large tree had fallen smashing down others in its fall. The small opening had allowed the wind a longer reach, and other old, rotten trunks had fallen, all inwards. As a result there was a dense tangle of trunks, vines, bushes and grasses about 100 feet in diameter. This type of opening is fairly common in the jungle, and agoutis commonly live in such places. We were not surprised, therefore, when the dogs found an animal's trail. But, the trail, unlike that of an agouti, wound back and forth within the clearing before the dogs bayed at a hole that was under the branches of a windfall at almost the center of the clearing. The dogs had ruined the original entrance and we chopped our way in to the burrow. It was about six inches in diameter and five feet long. The skunk was found at the end, about a foot beneath the surface of the ground. The animal had the usual skunk odor, but not very strong. It was a male with large testes. Several large black ticks were noted on it. It had no fly larvae, although almost every other mammal taken at this locality had several.

Twenty kilometers east-northeast of Jesús Carranza on April 12, 1949, an old male was brought to bay in a burrow in a patch of wild bananas, locally called platano tuno. This variety of banana grows in extensive clumps, but between the clumps the ground is open. The burrow was a simple tube about four feet long and 10 inches beneath the surface, with an entrance about six inches in diameter. There was no nest.

Lutra annectens annectens Major

Southern River Otter

Specimens examined.—Total 4: 20 km. W Piedras Negras, 300 ft., 1; 20 km. ENE Jesús Carranza, 200 ft., 1; 35 km. SE Jesús Carranza, 400 ft., 1; 38 km. SE Jesús Carranza, 400 ft., 1.

Additional records.—Orizaba (Pohle, 1920:95); Papaloapan (Ingles, 1959: 406). Río Jamapa, near Huatusco (Sumichrast, 1882:213); Río Blanco, near Omealca (*ibid.*).

The otter is called "perro de agua" in Veracruz. It is not common, and occurs in the lower, more sluggish parts of the larger rivers. Otters were reported from Puente Nacional, in the Río Antiguo, at 500 feet elevation. They rarely leave the rivers, although on one occasion, 14 kilometers southwest of Coatzacoalcos, 100 feet elevation, we were told that one had strayed into a small stream, no more than six feet across and fully one kilometer from the river. We hurried to the place, but the otter had left. The local people told us that the occurrence of an otter in that stream was not unheard of, but was unusual.

Otters in Veracruz are diurnal. They are most active in the gray light of the very early morning, but may be seen at almost any hour of the day. In literally hundreds of hours of night hunting from a canoe, we saw not a single pair of eyes that we thought were of an otter.

On the Río Blanco, near Piedras Negras, we found otters to be extremely shy, and we got only a few glimpses of the animals. Gerardo Mazza and his father came upon one large and one smaller otter in the Río Blanco on June 3, 1946. They shot the larger animal, which sank from sight. The water of the river was greatly discolored from a recent rainstorm in the mountains and, although both men rushed into the river, not waiting to remove their clothing, they were unable to recover the animal. Our only specimen from that locality was shot by a hunter and lacks the skull.

On April 8, 1948, 35 kilometers southeast of Jesús Carranza I was up on a limestone bluff, hunting bats, when a large male otter appeared in a deep hole in the river and dived before Chico could shoot. He saw it go into a cave, almost completely under the surface. The mouth of the cave was about 18 inches in diameter. Peering into a large crevice I saw the otter about 15 feet below and as it turned to go over a projecting rock I shot the otter in the chest with a load of buckshot. Chico recovered it by diving.

We examined many otter scats there. All consisted only of fish bones and scales. One head of a needle gar was seen; it may have been discarded by the otter or may have been part of a scat. The scats and piles of scales were mainly of *bobo, zoro* and *mojarra.* One large pile of the suckerlike *puespuerco* was seen. Two or three seemed to be *robalo.* No remains of other animals or small fish were seen.

The only parasite found on this otter was a large, black tick, unlike any tick that I ever saw before.

Field notes for April 30 mention that in the same general area we had seen several otter in the last few days, had lost one that was dead or wounded, and that on April 30 as we were coming through a wide pool of shallow water with a shallow rapid at each end, we obtained another. It was a male, much smaller than the other one, but with fine fur.

On April 10, 1949, 20 kilometers east-northeast of Jesús Carranza a native hunter shot an adult female containing four large embryos and saved for us the skin and perfect skull.

Family Felidae

Felis onca veraecrucis Nelson and Goldman

Jaguar

Specimens examined.—Total 8: San Juan de los Reyes, 1 (U. S. N. M.); San Andrés Tuxtla, 1 (U. S. N. M., type); Pérez, 1 (U. S. N. M.); 20 km. ENE Jesús Carranza, 200 ft., 2; 20 km. E Jesús Carranza, 300 ft., 3.

Additional records.—"Norte de Veracruz" (Ingles, 1959:406); Orizaba (Nelson and Goldman, 1933:237); Achotal (Elliot, 1907:38).

The jaguar is known as "tigre" or "tigre real," in Veracruz. The term tigre is used in a rather general sense, however, and may refer to any of the cats.

The jaguar is much hunted in Veracruz. It has been almost extirpated in the heavily populated central part of the state. In the tick-infested, thorny brushlands of northern Veracruz, especially near the gulf, we were told that the jaguar was fairly common. We saw no signs of jaguars there, however, nor were skins offered for sale. In the south it is common, and in the uninhabited area south of the isthmus, it is about as abundant as the game supply will allow.

The jaguar is largely nocturnal in the areas that are most inhabited by man. At our base on the Río Coatzacoalcos, 20 kilometers east of Jesús Carranza, jaguars were entirely nocturnal. None was seen by day, but tracks from the night before were often seen. It is noteworthy that the Virginia deer, *Odocoileus virginianus*, was common there and probably was the principal source of food, although we found no deer killed by jaguars. This deer is nocturnal.

In the uninhabited jungles to the south, there are no deer. Instead the brocket and peccary are common, and these are diurnal species. In this area we twice saw jaguars in the daytime. The stomach of one jaguar taken in the daytime was empty. Jaguar scats examined along the rivers of southern Veracruz were made up of the hair of collared peccary and one was made up of agouti. One fresh jaguar-kill was found 35 kilometers southeast of Jesús Carranza. It was an adult male brocket, and we judged that it had been killed in the late afternoon of the preceding day. Judging from the sign, the brocket had come to the river on a narrow trail, through an area of sawgrass about 50 yards wide. The jaguar had caught the brocket about 15 feet from the river. Only the abdomen and one flank were torn. The viscera and part of the brisket were eaten. There were no marks on the head, neck or shoulders. The jaguar must have hit the brocket from behind and on one side, killing it with a bite in the chest or abdomen. The skull of the brocket was saved.

It is said that hunters have been mauled, even killed, by jaguars. Jaguars occasionally do considerable damage to livestock, especially cattle. Such damage is usually traceable to an individual jaguar. While we were at our base on the Río Coatzacoalcos, jaguar tracks were seen almost every day in the jungle a mile south of the house. There were more than 100 cattle in the area between the house and the jungle. Yet no livestock was damaged by jaguars there in the several months of our stay that was divided between three different years. Jaguar skins are sold for about 50 pesos, and find a ready market.

One jaguar was examined for ectoparasites, but none was found.

Felis concolor mayensis Nelson and Goldman

Mountain Lion

Specimens examined.—Total 4: Catemaco, 2 (U. S. N. M.); 20 km. ENE Jesús Carranza, 200 ft., 2.

Felis pardalis pardalis Linnaeus

Ocelot

Specimens examined.—Total 12: Catemaco, 3 (U. S. N. M.); Pérez, 5 (U. S. N. M.); Tierra Blanca, 1; 15 km. SW Jimba, 750 ft., 1; 14 mi. S Acayucan, 1; 20 km. E Jesús Carranza, 300 ft., 1.

Additional records.—Mirador (Goldman, 1943:377); Jalapa (Ferrari-Pérez, 1886:128 may be *F. pardalis* or *F. wiedii*).

The ocelot is often termed "tigre" in Veracruz but more correctly is called "tigrillo."

Ocelots are fairly common in the forest and jungles of the tropical parts of the state, but are seldom in evidence. In southern Veracruz they seem to be diurnal and solitary. Never did we shine the eyes of an ocelot when hunting at night. It is suggestive that the only stomach of an ocelot examined contained the remains of an agouti. The agouti, in southern Veracruz, is diurnal. The ocelot mentioned was an adult female. It was not fat, and had a single embryo 37.5 millimeters in length on April 22. It had a few of the common field ticks, called conchuelas.

Our dogs refused to trail ocelots, or other cats. The few specimens obtained were received from native hunters. Ocelot skins had a ready sale, and usually brought about eight pesos.

Felis wiedii oaxacensis Nelson and Goldman

Margay

Specimen examined.—Cordoba, 1 (A. M. N. H.).

Additional record.—Alvarado, type locality of *Felix mexicana* Saussure (1860:4) that currently is regarded as indistinguishable from *Felis wiedii oaxacensis*. Saussure's name is unavailable for use, being preoccupied by *Felis mexicana* Desmarest 1814.

Felis yagouaroundi

Jaguarundi

This cat is called "onca" throughout Veracruz. On a few occasions we heard it called "tigre" or "tigrillo," but these names are usually restricted to the larger, spotted cats.

We saw but three individuals. At Jimba, 350 feet elevation, one leaped from the cover of dense brush to the center of a railroad embankment, where it paused for a moment, and then leaped to the dense cover on the other side of the tracks. We were unable to shoot it, because a small boy was in the line of fire. On March 16, 1947, 35 kilometers east-northeast of Jesús Carranza a jaguarundi cat made a raid on the chickens at the house in which we were living. Dogs were put on the animal's trail, and although we twice got long-range shots at the animal, when it crossed a meadow, we failed to get it. The dogs lost it, after trailing it for an hour. On the Río Chalchijapa, southeast of Jesús Carranza, a jaguarundi cat was seen drinking from the river, at the edge of a sand bar. The animal was crouched, and to us, closely resembled a large weasel.

The jaguarundi is a highly specialized cursorial cat. We have discovered from natives that birds form its principal diet. All of our specimens were shot by natives, while the cats were raiding chickens. Usually, after catching a chicken in a quick dash, the cat stops to kill and eat it. There are many ground-living species of birds in the jungles where this cat is found, such as meadowlarks, tinamous of several kinds, quail of two species, currasows, guans, ant thrushes, and numerous others. Skins of jaguarundi cats have little value, and are often offered for sale for a few pesos in shoe stores.

Felis yagouaroundi cacomitli Berlandier

Specimens examined.—Mirador, 2 (U. S. N. M.).

Felis yagouaroundi fossata Mearns

Specimens examined.—Total 6: Río Blanco, 20 km. WNW Piedras Negras, 300 ft., 1; 15 km. SW Jimba, 750 ft., 1; Pérez, 1 (U. S. N. M.); 20 km. ENE Jesús Carranza, 200 ft., 1; 20 km. E Jesús Carranza, 300 ft., 2.

Lynx rufus escuinapae J. A. Allen
Bobcat

Specimen examined.—One from 3 km. W Limón, 7500 ft.

The bobcat is called "gato montez" in Veracruz. It is well known on the Perote Desert and in the pine forest area near Las Vigas. Our only specimen was taken on a sandy desert flat, a kilometer in diameter, three kilometers west of Límon, 7500 feet elevation, on September 29, 1948, about two hours after dark. The land is covered by fine, wind-drifted sand that supports some maguey and cactus, and a few scattered juniper trees. There is one large cornfield and several bean fields on the sand flat. The bobcat was probably hunting for mice at the edge of one of the bean fields. The specimen was shot when its eyes were seen in the reflected beam of a hunting light. It is an adult female that was still nursing young. No ectoparasites were noted on it. The specimen was skinned at the town of Perote, where several persons requested the meat.

At Las Vigas we were told that bobcats were fairly common near town. One man brought in an old skin, stuffed with ashes, for sale. The animal was said to have been killed in the lava flow east of Las Vigas the year before.

Order SIRENIA

Family Trichechidae

Trichechus manatus latirostris (Harlan)

Manatee

Specimen examined.—Alvarado, 1.

Additional records (Ingles, 1959:408, on testimony of natives).—Nautla; Coatzacoalcos; Río Papaloapan.

Manatees are reported at the mouths of the larger rivers in Veracruz. They are not common, and we never saw one alive. Fishermen at Alvarado reported that the manatees were once fairly common in the Bay of Alvarado, but were now scarce. Only about ten were taken in a year at the Port of Alvarado, although no specific hunting for them is carried on. Specimens, especially small individuals, are taken occasionally in fish nets. One fisherman was commissioned to obtain a skull for us, and after some two months, did so. The manatee is known as both "manati" and "malachin" in Veracruz.

Order PERISSODACTYLA

Family Tapiridae

Tapirus bairdii (Gill)

Baird's Tapir

Specimens examined.—Total 5: Buena Vista, 1 (U. S. N. M.); 20 km. ENE Jesús Carranza, 200 ft., 1; Arroyo Azul, 20 km. E, 8 km. S Jesús Carranza, 1; 32 km. ESE Jesús Carranza, 350 ft., 1; 60 km. SE Jesús Carranza, 450 ft., 1.

The tapir is called "anteburro" over most of Veracruz, though the names "tapir" (Spanish pronunciation) and "elefante" are also, though rarely, used.

Tapirs are still common in southern Veracruz. They are most abundant in the dense jungles on rolling hills and relatively level land, at fairly low elevations. However, they were found in the rough hills, among the limestone cliffs, and even where the jungle was of almost alpine aspect, though still at low elevations. Near human habitations, tapirs become extremely shy. However, they are able to persist even near towns. In one area, where we hunted almost every day and saw tracks of tapirs daily, the local people had rarely seen the animal. One of the best hunters had seen but two tapirs in the area. On one occasion, we found the tracks of a tapir within two kilometers of Jesús Carranza, and from the locality where the tracks were found we could plainly hear the railroad trains and switch engines.

The tapir seems to be both nocturnal and diurnal. Near human habitations it probably feeds at night. On one occasion, while we were hunting from a canoe at night, a large animal was heard moving about in an area of sawgrass near the river. No reflection of eyes could be seen in the beam of our hunting light. A few moments later a rather mournful scream was heard. My native companion imitated the call, though he had never heard it before. A few moments later a large tapir came to the river bank. At close range its eyes had a dull white glow. My companion said that it was calling for its calf, which had probably been eaten by a jaguar. We heard this call repeated through most of the night. This was the only occasion on which we saw a tapir at night.

Although usually found near water, the tapir is found far from water if other conditions are favorable. In the Rincón area, west of Jimba, we saw numerous tapir trails several kilometers from water. Probably these localities have numerous water holes over most of the year, but in the last two months of the dry season, this is not true.

In and along the rivers of the uninhabited areas of extreme southern Veracruz, we saw tapirs almost daily. Most were in the water of the rivers. They were skillful swimmers. Usually they lay with only their heads exposed above the surface, and probably remained motionless in the water for long periods in the heat of the day. When approached, they swam off, with only the head showing above the surface. They swam swiftly and smoothly. If frightened, they dive from sight beneath the surface, as skillfully as a muskrat, and remain beneath the surface for a considerable time. On land, the tapir looks clumsy. The legs seem too close together. The long neck is held upright, and the large head with the trunk all combine to give the animal a surprisingly giraffelike appearance.

The tapir is a powerful animal, as may be determined from its trails. It pushes through sawgrass, brush, vines, and other vegetation like an army tank, giving way only to trees. On one occasion, a tapir was frightened in a small arroyo and fled through the thorny bamboo, called jimba in Veracruz. This plant grows in dense patches, sometimes acres in extent. It is about 12 feet high, and each stalk is about three inches in diameter at the base, and as strong as most bamboo. At each joint there are three or four long, strong, hooked thorns. This plant is almost impenetrable for man, and all other mammals save the tapir. The animal mentioned crashed off through the jimba, tearing a trail through the tough growth as easily as a man might through tall grass. Natives told us that, occasionally, a tapir in its mad flight from danger, overruns a man and kills him. We were told that a female defending its young will try to trample an enemy.

About noon on May 25, 1948, at a place 32 kilometers east-southeast of Jesús Carranza, we were traveling up the river (Solosuchil) in a dugout when a motionless tapir was seen. It was in the water, about four feet deep, and 10 feet from shore. Only the head and neck were above water. When the canoe was about 20 feet distant, the tapir lunged to its feet. Angel fired a load of buckshot into the middle of its back. I hit it in the shoulder with a .22 slug. It dived completely under water and vanished. About two minutes later it appeared near shore about 50 feet upstream. Chico shot it with a load of buckshot, but it lunged to shore. I shot it in the neck with the .22, and it slumped. A total of 14 buckshot had hit it in the neck and shoulders and back. All had glanced off, after gouging small holes. My first shot had also glanced from the shoulder but entered the mandible, and lodged there. My second shot had entered the spinal column between the third and fourth vertebrae, killing the animal. It might have weighed 500 pounds. We rolled and pried it to shore and high enough for a picture. The animal, a non-pregnant female, had a few conchuelas (ticks) on the eyelids and ears, but about 100 large garrapatas (ticks) on the breast, abdomen, and medial side of the upper part of each foreleg. The skin of the back, neck and shoulders was more than one half inch thick.

Probably only the jaguar is an enemy of the tapir. We doubt that a jaguar ordinarily could kill a full-grown tapir. One of our hunters told of seeing a large tapir standing in the river, with blood pouring from many deep scratches in its face and neck. This animal was too sick to move, and allowed him to approach within a few feet in his canoe, without moving. When he returned an hour later, the animal was gone. The tapir has little economic value. We never heard of a tapir disturbing crops. The hide is so thick as to be worthless. The flesh is rank and red, resembling horsemeat. It is often eaten, but is considered poor food. Tapirs are often killed by natives, seemingly for no particular reason.

Order ARTIODACTYLA

Family Tayassuidae

Tayassu tajacu crassus Merriam

Collared Peccary

Specimens examined.—Total 9: Potrero Llano [= Potrero del Llano], 350 ft., 2; 22 km. ESE Jesús Carranza, 300 ft., 2; 32 km. ESE Jesús Carranza, 350 ft., 1; 20 km. SE Jesús Carranza, 200 ft., 1; 25 km. SE Jesús Carranza, 250 ft., 1; 30 km. SSE Jesús Carranza, 300 ft., 2.

Additional record—Pasa Nueva (Allen, 1904:30).

The collared peccary, called "javalin" or "javalina" in Veracruz, usually travels in bands of from five to ten individuals. It is piglike in its actions, rooting about in the leaves, and constantly hunting for food. Fallen fruit from the jungle trees probably furnishes most of its food. Corn from the natives' fields is also eagerly eaten and a band of peccaries can completely ruin a milpa in a few days.

The presence of a band of peccaries can often be detected by the chomping and clicking of the animals' tusks. When frightened they move off at a sharp trot, and their small hoofs make a distinct clicking noise, even on the soft floor of the jungle. When much frightened they make a dash for deep cover with surprising speed. At times they begin a dash with a few leaps of several yards. They are swift runners, and in dense cover can outdistance a pack of dogs in a short time. In one particular coyol palm jungle, where the undercover was thick and thorny, our dogs chased a pack of javalinas almost daily, but we never came up with the animals.

Ordinarily a hunter without dogs comes upon a band of these peccaries unexpectedly. Sometimes he is forewarned by hearing the clicking of tusks, or, if the wind is in the right direction, by the unmistakable strong odor of peccary. If not so warned, and he comes close to the animals, there is a loud snort, and there is a trotting noise as some of the animals flee. Usually, however, one animal is left, standing motionless, and looking like a small black ball on legs, with each hair standing upright. We believe that the females ordinarily run and leave the males on guard, for every one of our specimens of peccaries is a male.

Two of our specimens were taken in brush and vines at the edge of an isolated corn patch. The stomachs of these animals were filled with chunks of corn cobs. No attempt had been made to chew the corn from the cobs. One peccary was taken in a dense clump of wild bananas. The others were taken in deep jungle.

We were surprised that our native companions considered this species of peccary to be a somewhat fossorial animal. They spoke of running peccaries into their holes, and occasionally looked into holes in the ground for peccaries. In the spring of 1949, however, an old boar peccary did run into a large hole in the ground, where it was shot. A few days later one of our assistants, returning from a hunt, called that he had found a peccary's burrow, and to bring machetes. We found him at the edge of a small cut-bank in the forest, making a series of growling and barking noises. There was a burrow in the bank, about two feet in diameter. The man had been returning to camp, when he had walked over the burrow. The loud clicks from a peccary in the hole below him had given the first evidence of the presence of the animal. Because of sloppy work by the collectors, this animal was allowed to escape from its hole, and it severely wounded a dog before being killed. It was proven, however, that the collared peccary in Veracruz does not only take refuge in a burrow in the ground when in danger, but actually spends some time in the burrow, perhaps sleeping there.

Our specimens from southern Veracruz were almost covered with ticks. On some individuals there were hundreds, large, medium and small. The fewest ticks recorded from any specimen numbered 50. The collared peccary has a strong musky odor, emanating from a large gland along the mid-dorsal line of the rump, immediately under the skin. The musk gland of one individual measured 75 millimeters long by 35 millimeters wide. The flesh of some animals is so strongly saturated with this odor as to be scarcely edible, but if care is used in skinning the animal little or no musky odor can be detected on the flesh. We found the flesh to be porklike, and never fat. The flesh of the peccary is in demand for food in the native villages and sells for about three pesos per kilogram. On the hides the bristlemarks are distinct and the leather makes a good grade of pigskin. Large skins bring the native hunters about three pesos each.

Tayassu pecari ringens Merriam
White-lipped Peccary

Specimens examined.—Total 7: 20 km. ENE Jesús Carranza, 3; 20 km. E Jesús Carranza, 300 ft., 1; Arroyo Saoso, 37 km. E, 3 km. S Jesús Carranza, 3.

The white-lipped peccary is called "marina" in Veracruz. It is found only in the extreme southern parts of the state: The Tuxtla Mountains, the Rincón area, and the Isthmus of Tehuantepec.

Our specimens of this species were all obtained from native hunters, after we offered suitable rewards for specimens. On several occasions we followed small bands of this species in the jungle, but failed to come up with them.

The white-lipped peccary is more social than the collared peccary. Bands of 50 or more animals were reported to us, but the largest band followed by us, to judge from the tracks, numbered about 25. The smallest band followed numbered about 12. We were told that the odor of this species is like the odor of the collared species, and that white-lipped peccaries chomp and click their tusks in the same way and call like pigs, though what this call might be, we do not know. Hunters often bag a number of animals from a band; on one occasion about eight were obtained. These animals seem to be rather local in their distribution, abundant in some relatively small areas, and absent over great areas of seemingly similar habitat. They are dangerous when wounded, and in the Río Coatzacoalcos area, hunters have been killed by them.

Family Cervidae

Odocoileus virginianus

White-tailed Deer

According to the Rules of Zoological Nomenclature, *Dama* Zimmermann 1780 is the correct generic name but the International Commission on Zoological Nomenclature recommends (Opinion 581 of September 16, 1960) the use of *Odocoileus.*

The whitetail is called "venado" throughout Veracruz. It is much hunted, but is still fairly common in some places. In areas where the human population is dense, it has been completely exterminated, for example on the coastal plain near Veracruz.

The habits of this species are much like those of the whitetail of the United States, but modified somewhat to the tropical habitat. It is not found in the jungle or deep forest, but lives in the brush and thickets of the coastal plain, and near the flood plains of rivers in the southern part of the state. We studied and collected deer on the Río Coatzacoalcos, east of Jesús Carranza. Except at one locality, the deer were confined to the large areas of grassland, called sacatales, on the flood plains of the big river. Farther up the river, and on tributary rivers, where the jungle came down to the banks, there were no deer. We were told, however, that deer did occur at the mouth of one arroyo on the Río Solosuchil, called Arroyo de Zouza. Many persons had seen their tracks, but none of the jungle people had ever seen a deer there. We stopped at the mouth of this arroyo, which has a broad, marshy delta and a deep, thicket-filled valley, and found the unmistakable tracks of a large whitetail buck. Seemingly there is a small colony of deer at this locality, isolated by a large extent of jungle.

Along the Río Coatzacoalcos, the whitetail is nocturnal. A few were seen in the early mornings, or late afternoons, on dark, rainy days. Ordinarily they spend the day in dense thickets of wild bananas or in the dense coyol palm jungle at the edge of the grass fields. They are usually solitary, although on occasions we saw two, three or four together.

Over most of Veracruz, this deer is hunted with dogs in the daytime or with hunting lights at night. The whitetail is said to be preyed upon by the jaguar. Specimens taken by us were heavily parasitized by ticks. The antlers of a specimen taken in March, and another taken in April, were in the velvet stage.

In 1948 in manuscript Dalquest summarized results of his study of available specimens as follows: "Goldman and Kellogg (1940:89) described a subspecies of deer from northern Veracruz as new, under the name [*O. v.*] *veraecrucis*. They stated that this subspecies intergraded with the subspecies [*O. v.*] *thomasi* to the south. In the same paper they mention that the deer of the Isthmus of Tehuantepec are [*O. v.*] *thomasi*. Although no ranges or lists of specimens examined are included in the paper, it would seem that the authors would restrict the name *toltecus* to [deer of] the high parts of central Veracruz, and perhaps to the area farther westward.

"The type locality of [*O. v.*] *toltecus* is in the tropics, at Orizaba. Specimens, presumably of *toltecus*, have been examined from Mirador, Veracruz. I can see no essential differences between them and the specimens from Chijol, the type locality of *veraecrucis*. On geographic grounds I would not expect any difference between the two.

"Specimens from extreme southern Veracruz are smaller and more reddish thus agreeing with *thomasi*. They agree with specimens from the coastal plain of Veracruz. In this area I examined numerous skins of deer taken by hunters from the coastal plain as well as from the Isthmus area. Variation in color is great, although the average is rather bright red. Deer from both areas are small.

"In the Tuxtla Mountains I saw whitetails that were much larger than any deer observed on the coastal plain or the Isthmus.

"Possibly five minor geographic variants ought to be recognized in Veracruz, although specimens supporting such an arrangement are not available. The variants would be from northern Veracruz (*veraecrucis*), the uplands of central Veracruz (*toltecus*), the coastal plains (unnamed variant), the Tuxtla Mountains (unnamed variant) and the Isthmus of Tehuantepec (*thomasi*).

"The recognition of any of them, on the basis of the scanty material available, and considering the individual variation to be found in this species, seems hazardous."

Odocoileus virginianus thomasi Merriam

Record.—Catemaco (Miller and Kellogg, 1955:807).

Odocoileus virginianus toltecus (Saussure)

Specimens examined.—Total 13: Mirador, 4 (U. S. N. M.); Buena Vista, 4 (U. S. N. M.); 35 km. ENE Jesús Carranza, 150 ft., 1; 20 km. ENE Jesús Carranza, 200 ft., 3; 20 km. E Jesús Carranza, 300 ft., 1.

Additional records.—Near Orizaba (Saussure, 1860:247); Pasa Nueva (J. A. Allen, 1904:30).

Odocoileus virginianus veraecrucis Goldman and Kellogg

Specimens examined.—Total 3: Chijol, 2 (U. S. N. M.); Potrero Llano [= Potrero del Llano], 350 ft., 1.

Additional record.—Near port of Veracruz (Miller and Kellogg, 1955:806).

Mazama americana temama (Kerr)
Red Brocket

Specimens examined.—Total 13: Potrero Llano [= Potrero del Llano], 350 ft., 1; 3 km. SW San Marcos, 200 ft., 1; Mirador, 5 (U. S. N. M.); Catemaco, 2 (U. S. N. M.); Arroyo Saoso, 37 km. E, 7 km. S Jesús Carranza, 2; 35 km. SE Jesús Carranza, 400 ft., 1; 30 km. SSE Jesús Carranza, 300 ft., 1.

The brocket is called "temazate" in Veracruz. Seemingly it is never confused with the whitetail, or "venado." The brocket lives in the deep forest and jungle. It is shy, and quick to sneak away at the least alarm. Often it moves off a short distance and stands motionless, invisible in the gloom and reddish shades of the jungle floor. When chased by dogs it runs to dense thickets in the arroyos, where it winds about until the dogs are exhausted and give up the chase.

Brockets seem to have small individual ranges. Those chased by our dogs usually circled about in an area of a kilometer or so in diameter. Brockets are principally diurnal. One was seen at night, but the others seen by us were active in the daytime. Several were seen as they came to drink at the edges of rivers. One of these was at a level sand bar, but three or four others had chosen steeply sloping banks.

In the hills of central Veracruz, the brocket is still to be found, although it is scarce because of excessive hunting. At Mirador, the type locality of *Mazama sartorii* (now arranged as a synonym of *M. a. temama*), the brocket seems to have been extirpated. Mr. Walter Sartorius showed us a collection of antlers of deer and brocket from the area around Mirador but said that no brocket had been seen there for many years. It was the grandfather of Walter Sartorius after whom the species *Mazama sartorii* was named in 1860 by Saussure. In northern Veracruz, the brocket was said to be common in deep jungles, but we did not investigate. One skin was sold to us by a native hunter. In extreme southern Veracruz, the brocket is abundant.

Brockets are solitary animals. We never saw more than one at a time, nor did we see tracks of more than one animal in a place. One of the strangest things about this species is the scarcity of females. All of our specimens are males and we saw no females in Veracruz. Of the seven specimens in the United States National Museum from Veracruz, one is a female, and it is very young. Native hunters told us that females were rare.

Concerning a specimen taken three kilometers southwest of San Marcos in central Veracruz, on November 1, 1947, I wrote in my field notes as follows: "While walking along the bed of an arroyo today, in deep jungle, I heard a rock roll on the hillside above me. The slope was about 30°, and the hill was surprisingly open. The ground had seemingly recently slid, and since become covered with a dense mat of morning-glorylike vine. A male brocket was feeding on the hill. . . . The specimen was fat and in perfect health. The coat is fine, and deep, rich red in color. The horns are small, but sharp and gnarled. Its testes measured 38 mm. in length. It had no ectoparasites that we could find. The hoof-glands were of a dark olive-brown color. . . . I estimated its weight at 35 pounds."

An adult male shot in a dense tangle of wild banana and thorny bamboo in southern Veracruz was older than the specimen taken near San Marcos and had numerous small and medium-sized ticks, as well as hundreds of deer lice.

Flesh of the brocket tastes much like that of the whitetail and perhaps is even superior. It brings a good price in the native villages. The skins are too small and thin to provide good leather, and so have no commercial value, although a few are sold as trophies. Because brockets live in the deep forest and jungles, they do not come into contact with the crops of the natives. One brocket that had been killed by a jaguar was found (see account of the jaguar).

LITERATURE CITED

ALLEN, G. M.

1916. A third species of *Chilonycteris* from Cuba. Proc. New England Zool. Club, 6:1-7, 1 pl., February 8.

1927. The range of Chiroderma isthmicum Miller. Jour. Mamm., 8:158, May 11.

1942. *Hylonycteris underwoodi* in Mexico. Jour. Mamm., 23:97, February 16.

ALLEN, H.

1866. Notes on the Vespertilionidae of tropical America. Proc. Acad. Nat. Sci. Philadelphia, 18:279-288.

ALLEN, J. A.

1879. On the coatis (genus Nasua, Storr). Bull. U. S. Geol. & Geog. Survey Territories, 5:153-174, September 6.

1901. A preliminary study of the North American opossums of the genus Didelphis. Bull. Amer. Mus. Nat. Hist., 14:149-188, pls. 22-25, June 15.

1901. Descriptions of two new opossums of the genus Metachirus. Bull. Amer. Mus. Nat. Hist., 14:213-218, July 3.

1904. Mammals from southern Mexico and Central and South America. Bull. Amer. Mus. Nat. Hist., 20:29-70, 18 figs., February 29.

1904. The tamandua anteaters. Bull. Amer. Mus. Nat. Hist., 20:385-398, October 29.

ALLEN, J. A., and CHAPMAN, F. M.

1897. On a collection of mammals from Jalapa and Las Vigas, state of Vera Cruz, Mexico. Bull. Amer. Mus. Nat. Hist., 9:197-208, June 16.

1904. Mammals from southern Mexico and Central and South America. Bull. Amer. Mus. Nat. Hist., 20:29-80, 18 figs., February 29.

ALVAREZ, T.

1961. Taxonomic status of some mice of the Peromyscus boylii group in eastern Mexico, with description of a new subspecies. Univ. Kansas Publ., Mus. Nat. Hist., 14:111-120, 1 fig., December 29.

ANDERSEN, K.

1908. A monograph of the chiropteran genera *Uroderma, Enchisthenes,* and *Artibeus.* Proc. Zool. Soc. London, pp. 204-319, 59 figs., September 7.

BAILEY, V.

1900. Revision of American voles of the genus Microtus. N. Amer. Fauna, 17:3-88, 5 pls., 17 figs., June 6.

1906. Identity of *Thomomys umbrinus* (Richardson). Proc. Biol. Soc. Washington, 19:3-6, January 29.

BAKER, R. H., and DICKERMAN, R. W.

1956. Daytime roost of the yellow bat in Veracruz. Jour. Mamm., 37:443, September 11.

BURT, W. H., and STIRTON, R. A.

1961. The mammals of El Salvador. Miscl. Publ. Mus. Zool., Univ. Michigan, 117:1-69, 2 figs., September 22.

COUES, E.

1874. Synopsis of the Muridae of North America. Proc. Acad. Nat. Sci. Philadelphia, 26:173-196, December 15.

DALQUEST, W. W.

1950. The genera of the chiropteran family Natalidae. Jour. Mamm., 31:436-443, November 21.

DAVIS, W. B.

1944. Notes on Mexican mammals. Jour. Mamm., 25:370-403, December 12.

1958. Review of Mexican bats of the Artibeus "cinereus" complex. Proc. Biol. Soc. Washington, 71:163-166, December 31.

DAVIS, W. B., and CARTER, D. C.

1962. Notes on Central American bats with description of a new subspecies of Mormoops. Southwestern Nat., 7:64-74, 1 fig., June 1.

DAVIS, W. B., and RUSSELL, R. J., JR.

1944. Bats of the Mexican state of Morelos. Jour. Mamm., 33:234-239, May 14.

ELLIOT, D. G.

1903. Descriptions of apparently new species of mammals of the genera Heteromys and Ursus from Washington and Mexico. Field Columbian Mus. Publ., 80, Zool. Ser., 3:233-237, August 27.

1905. A checklist of mammals of the North American Continent, the West Indies and the neighboring seas. Field Columbian Mus. Publ. 105, Zool. Ser., 6:iv + 761, frontispiece.

1907. A catalogue of the collection of mammals in the Field Columbian Museum. Field Columbian Mus. Publ. 115, Zool. Ser., 8:viii + 694, 92 figs., February 9.

FELTEN, H.
1956. Fledermäuse (Mammalia, Chiroptera) aus El Salvador. Teil 3. Senckenbergiana Biologica, 37:179-212, pls. 24-27, 7 figs., April 15.

FERRARI-PÉREZ, F.
1886. Catalogue of animals collected by the Geographical and Exploring Commission of the Republic of Mexico. Proc. U. S. Nat. Mus., 9:125-199 [pp. 125-128, September 13; pp. 129-144, September 15; pp. 145-160, September 17; pp. 161-199, September 28].

FINDLEY, J. S.
1955. Taxonomy and distribution of some American shrews. Univ. Kansas Publ., Mus. Nat. Hist., 7:613-618, June 10.
1960. Identity of the long-eared myotis of the Southwest and Mexico. Jour. Mamm., 41:16-20, February 20.

GOLDMAN, E. A.
1911. Revision of the spiny pocket mice (genera Heteromys and Liomys). N. Amer. Fauna, 34:1-70, 3 pls., 6 figs., September 7.
1913. Descriptions of new mammals from Panama and Mexico. Smithsonian Miscl. Coll., 60(No. 22):1-20, February 28.
1917. New mammals from North and Middle America. Proc. Biol. Soc. Washington, 30:107-116, May 23.
1918. The rice rats of North America (genus Oryzomys). N. Amer. Fauna, 43:1-100, 6 pls., 11 figs., September 23.
1936. Two new flying squirrels from Mexico. Jour. Washington Acad. Sci., 26:462-464, November 15.
1942. A new white-footed mouse from Mexico. Proc. Biol. Soc. Washington, 55:157-158, October 17.
1943. The races of the ocelot and margay in Middle America. Jour. Mamm., 24:372-385, August 18.
1950. Raccoons of North and Middle America. N. Amer. Fauna, 60:vi + 153, 22 pls., 2 figs., November 7.

GOLDMAN, E. A., and KELLOGG, R.
1940. Ten new white-tailed deer from North and Middle America. Proc. Biol. Soc. Washington, 53:81-89, June 28.

GOODWIN, G. G.
1946. Mammals of Costa Rica. Bull. Amer. Mus. Nat. Hist., 87:271-473, pl. 17, figs. 1-50, 1 map, December 31.
1958. Bats of the genus *Rhogeëssa*. Amer. Mus. Novitates, 1923:1-17, December 31.
1959. Bats of the subgenus *Natalus*. Amer. Mus. Novitates, 1977:1-22, 2 figs., December 22.
1960. The status of *Vespertilio auripendulus* Shaw, 1800, and *Molossus ater* Geoffroy, 1805. Amer. Mus. Novitates, 1994:1-6, 1 fig., March 8.

HAHN, W. L.
1907. A review of the bats of the genus *Hemiderma*. Proc. U. S. Nat. Mus., 62:103-118, February 8.

HALL, E. R.
1951. A new name for the Mexican red bat. Univ. Kansas Publ., Mus. Nat. Hist., 5:223-226, December 15.
1951. American weasels. Univ. Kansas Publ., Mus. Nat. Hist., 4:1-466, 41 pls., 31 figs., December 27.
1960. Oryzomys couesi only subspecifically different from the marsh rice rat, Oryzomys palustris. Southwestern Nat., 5:171-173, November 1.
1962. A new bat (Myotis) from Mexico. Univ. Kansas Publ., Mus. Nat. Hist., 14:161-164, May 21.

HALL, E. R., and ALVAREZ, T.
1961. A new species of mouse (Peromyscus) from northwestern Veracruz, Mexico. Proc. Biol. Soc. Washington, 74:203-206, August 11.
1961. A new subspecies of pocket gopher (*Heterogeomys*) from northern Veracruz. Anal. Escuela Nac. Ciencias Biol., 10:121-122, December 20.
1961. A new subspecies of the black myotis (bat) from eastern Mexico. Univ. Kansas Publ., Mus. Nat. Hist., 14:69-72, 1 fig., December 29.

HALL, E. R., and COCKRUM, E. L.
1953. A synopsis of the North American microtine rodents. Univ. Kansas Publ., Mus. Nat. Hist., 5:373-498, 149 figs., January 15.

HALL, E. R., and DALQUEST, W. W.
1950. Geographic range of the hooded skunk, Mephitis macroura, with description of a new subspecies from Mexico. Univ. Kansas Publ., Mus. Nat. Hist., 1:575-580, 1 fig., January 20.

HALL, E. R., and JONES, J. K., JR.
1961. North American yellow bats, "Dasypterus," and a list of named kinds of the genus Lasiurus Gray. Univ. Kansas Publ., Mus. Nat. Hist., 14:73-98, 4 figs., December 29.

HALL, E. R., and KELSON, K. R.
1959. The mammals of North America. The Ronald Press, New York: 2 vols., xxx + 1083 + 79, illustrated, March 31.

HANDLEY, C. O., JR.
1959. A revision of American bats of the genera Euderma and Plecotus. Proc. U. S. Nat. Mus., No. 3417, 110:95-246, 27 figs., September 3.
1960. Descriptions of new bats from Panama. Proc. U. S. Nat. Mus., 112:459-479, October 6.

HERSHKOVITZ, P.
1949. Mammals of northern Columbia, preliminary report No. 5: Bats (Chiroptera). Proc. U. S. Nat. Mus., 99:429-454, fig. 38, May 10.

HOFFMEISTER, D. F., and KRUTZSCH, P. H.
1955. A new subspecies of *Myotis evotis* (H. Allen) from southeastern Arizona and Mexico. Chicago Acad. Sci., Nat. Hist. Miscellanea, 151:1-4, December 28.

HOFFMEISTER, D. F., and DE LA TORRE, L.
1961. Geographic variation in the mouse Peromyscus difficilis. Jour. Mamm., 42:1-13, 2 figs., February 20.

HOOPER, E. T.
1947. Notes on Mexican mammals. Jour. Mamm., 28:40-57, February 17.
1952. Records of the flying squirrel (*Glaucomys volans*) in México. Jour. Mamm., 33:109-110, February 18.
1952. A systematic review of the harvest mice (genus *Reithrodontomys*) of Latin America. Miscl. Publ. Mus. Zool., Univ. Michigan, 77: 1-255, 9 pls., 24 figs., 12 maps, 7 tables, January 16.
1957. Records of Mexican mammals. Occas. Papers Mus. Zool., Univ. Michigan, 586:1-9, April 30.

HOOPER, E. T., and HANDLEY, C. O., JR.
1948. Character gradients in the spiny pocket mouse, *Liomys irroratus*. Occas. Papers Mus. Zool., Univ. Michigan, 514:1-34, 1 fig., 1 map, 2 tables, October 29.

HOWELL, A. H.
1901. Revision of the skunks of the genus Chincha. N. Amer. Fauna, 20:1-62, 8 pls., 1 table, August 31.
1914. Revision of the American harvest mice (genus *Reithrodontomys*). N. Amer. Fauna, 36:1-97, 7 pls., 6 figs., 1 table, June 5.

INGLES, L. G.
1959. Notas acerca de los mamíferos Mexicanos. Anal. Inst. Biol., México, 29:379-408, March 31.

JACKSON, H. H. T.
1928. A taxonomic review of the American long-tailed shrews (genera Sorex and Microsorex). N. Amer. Fauna, 51:vi + 238, 13 pls., 24 figs., 15 tables, July 24.

KELLOGG, R., and GOLDMAN, E. A.
1944. Review of the spider monkeys. Proc. U. S. Nat. Mus., 96:1-45, 2 figs., November 2.

KELSON, K. R.
1952. The subspecies of the Mexican red-bellied squirrel, *Sciurus aureogaster*. Univ. Kansas Publ., Mus. Nat. Hist., 5:243-250, 1 fig., April 10.

LAWRENCE, B.
1933. Howler monkeys of the palliata group. Bull. Mus. Comp. Zool., 75:313-354, numerous tables, November.

LOWERY, G. H., JR., and DALQUEST, W. W.
1951. Birds from the state of Veracruz, Mexico. Univ. Kansas Publ., Mus. Nat. Hist., 3:531-649, 7 figs., October 10.

MÁLAGA ALBA, A., and VILLA R., B.
1957. Algunas notas acerca de la distribución de los murciélagos de America del Norte relacionados con el problema de la rabia. Anal. Inst. Biol., México, 27:528-569, 8 figs., 10 maps, September 30.

MERRIAM, C. H.
1894. Preliminary descriptions of eleven new kangaroo rats of the genera Dipodomys and Perodipus. Proc. Biol. Soc. Washington, 9:109-116, June 21.

1895. Monographic revision of the pocket gophers, family Geomyidae, exclusive of the species of Thomomys. N. Amer. Fauna, 8:1-258, frontispiece, 19 pls., 81 figs., 4 maps, January 31.

1895. Revision of the shrews of the American genera Blarina and Notiosorex. N. Amer. Fauna, 10:5-34, 3 pls., December 31.

1896. Synopsis of the weasels of North America. N. Amer. Fauna, 11: 1-44, frontispiece, 5 pls., 16 figs., June 30.

1901. Descriptions of 23 new harvest mice (genus Reithrodontomys). Proc. Washington Acad. Sci., 3:547-558, November 29.

1901. Seven new mammals from Mexico, including a new genus of rodents. Proc. Washington Acad. Sci., 3:559-563, November 29.

1902. Twenty new pocket mice (*Heteromys* and *Liomys*) from Mexico. Proc. Biol. Soc. Washington, 15:41-50, March 5.

1902. Five new mammals from Mexico. Proc. Biol. Soc. Washington, 15: 67-69, March 22.

MILLER, G. S., JR.
1897. Revision of the North American bats of the family Vespertilionidae. N. Amer. Fauna, 13:1-141, 3 pls., 40 figs., October 16.

1902. Twenty new American bats. Proc. Acad. Nat. Sci. Philadelphia, 54:389-412, September 12.

1913. Notes on the bats of the genus Molossus. Proc. U. S. Nat. Mus., 46:85-92, August 23.

1913. Revision of the bats of the genus Glossophaga. Proc. U. S. Nat. Mus., 46:413-429, 1 fig., 1 table, December 31.

1914. Two new North American bats. Proc. Biol. Soc. Washington, 27: 211-212, October 31.

MILLER, G. S., JR., and ALLEN, G. M.
1928. The American bats of the genera Myotis and Pizonyx. Bull. U. S. Nat. Mus., 144:viii + 218, 1 pl., 1 fig., 13 maps, numerous tables, May 25.

MILLER, G. S., JR., and KELLOGG, R.
1955. List of North American Recent mammals. Bull. U. S. Nat. Mus., 205:xii + 954, March 3.

MORRIS, J. B.
1928. Hernando Cortés. Live letters 1519-1526. Translated by J. Bayard Morris with an introduction. George Routledge & Sons, Ltd., London. Pp. xlvii + 388, frontispiece, 7 illustrations.

NELSON, E. W.
1899. Revision of the squirrels of Mexico and Central America. Proc. Washington Acad. Sci., 1:15-110, 2 pls., May 9.
1909. The rabbits of North America. N. Amer. Fauna, 29:1-314, 13 pls., 19 figs., numerous tables, August 31.

NELSON, E. W., and GOLDMAN, E. A.
1929. Four new pocket gophers of the genus Heterogeomys from Mexico. Proc. Biol. Soc. Washington, 42:147-152, March 30.
1933. Revision of the jaguars. Jour. Mamm., 14:221-240, August 17.
1934. Revision of the pocket gophers of the genus Cratogeomys. Proc. Biol. Soc. Washington, 47:135-153, 1 table, June 13.

OSGOOD, W. H.
1909. Revision of the mice of the American genus Peromyscus. N. Amer. Fauna, 28:1-285, 8 pls., 12 figs., several tables, April 17.

PACKARD, R. L.
1960. Speciation and evolution of the pygmy mice, genus Baiomys. Univ. Kansas Publ., Mus. Nat. Hist., 9:579-670, 4 pls., 12 figs., June 16.

POHLE, H.
1920. Die Unterfamilie der Lutrinae. . . . Archiv für Naturgeschichte, 85. Jahrgang 1919, Abt. A, 9 Heft, pp. 1-247, pls. 1-9, figs. 1-19, November.

REHN, J. A. G.
1902. A revision of the genus Mormoops. Proc. Acad. Nat. Sci. Philadelphia, 54:160-172, June 11.
1904. A study of the mammalian genus Chilonycteris. Proc. Acad. Nat. Sci. Philadelphia, 56:181-207, March 26.

REINHARDT, J.
1873. Et Bidrag til Kundskab om Aberne i Mexiko og Centralamerika. Vid. Medd. Nat. Foren. Kjobenhavn, 4 (ser. 3):150-158.

SANBORN, C. C.
1933. Bats of the genera Anoura and Lonchoglossus. Field Mus. Nat. Hist., Zool. Ser., 20:23-28, December 11.
1936. Records and measurements of Neotropical bats. Field Mus. Nat. Hist., Zool. Ser., 20:93-106, August 15.
1937. American bats of the subfamily Emballonurinae. Field Mus. Nat. Hist., Zool. Ser., 20:321-354, figs. 37-48, December 28.
1947. Catalogue of type specimens of mammals in Chicago Natural History Museum. Fieldiana: Zoology, 32(4):209-293, August 28.
1949. Mexican records of the bat, *Centurio senex*. Jour. Mamm., 30:198-199, May 23.

SAUSSURE, H. DE
1860. Note sur quelques mammifères du Mexique. Revue et Mag. Zool., ser. 2, 12:first article, pp. 3-11, January; third article, pp. 97-110, March; fourth article, pp. 241-254, June; fifth article, pp. 281-293, July; seventh article, pp. 479-494, November.

SHAMEL, H. H.
1931. Notes on the American bats of the genus Tadarida. Proc. U. S. Nat. Mus., 78:1-27, May 6.

SUMICHRAST, F.
1882. Enumeración de las especies de mamíferos, aves, reptiles y batracios observados en la parte central y meridional de la República Mexicana. La Naturaleza, 5:199-213 and 322-328.

TATE, G. H. H.
1933. A systematic revision of the marsupial genus *Marmosa*. . . . Bull. Amer. Mus. Nat. Hist., 66:1-250, 26 pls., 29 figs., 9 tables, August 10.

THOMAS, O.
1900. *The geographical races of the tayra* (Galictis barbara), *with notes on abnormally coloured individuals.* Ann. Mag. Nat. Hist., ser. 7, 5:145-148, January.

1913. *The geographical races of the woolly opossum* (Philander laniger). Ann. Mag. Nat. Hist., ser. 8, 12:358-361, October.

VILLA R., B.
1941. Una nueva rata de campo (*Tylomys gymnurus* sp. nov.). Anal. Inst. Biol., México, 12:763-766, 4 figs., November 18.

1955. *Cynomops malagai* sp. nov. y genero nuevo para la fauna de murciélagos de México. Acta Zoologica Mexicana, 1(4):1-6, 2 figs., September 15.

1955. El murciélago colorado de Seminola, *Lasiurus borealis seminolus* (Rhoads), en México. Anal. Inst. Biol., México, 26: 237-238, September 26.

VILLA R., B., and JIMÉNEZ G., A.
1961. Acerca de la posición taxonomica de Mormoops megalophylla senicula Rehn, y la presencia de virus rabico en estos murciélagos insectivoros. Anal. Inst. Biol., México, 31:501-509, 1 fig., April 17.

1962. Tres casos mas de rabia en los murciélagos de México. Anal. Inst. Biol., México, 32:391-395, March 30.

WARD, H. L.
1891. Descriptions of three new species of Mexican bat. Amer. Nat., 25:743-753, April 20.

1904. A study in the variations of proportions in bats. Trans. Wisconsin Acad. Sci., Arts and Letters, 14:630-654, pls. 50-55, 3 tables.

WETMORE, A.
1943. The birds of southern Veracruz. Proc. U. S. Nat. Mus., 93:215-340, 3 pls., 1 fig. (map), May 25.

Transmitted June 21, 1962.

29-4035

(Continued from inside of front cover)

18. Conspecificity of two pocket mice, Perognathus goldmani and P. artus. By E. Raymond Hall and Marilyn Bailey Ogilvie. Pp. 513-518, 1 map. January 14, 1960.
19. Records of harvest mice, Reithrodontomys, from Central America, with description of a new subspecies from Nicaragua. By Sydney Anderson and J. Knox Jones, Jr. Pp. 519-529. January 14, 1960.
20. Small carnivores from San Josecito Cave (Pleistocene), Nuevo León, México. By E. Raymond Hall. Pp. 531-538, 1 figure in text. January 14, 1960.
21. Pleistocene pocket gophers from San Josecito Cave, Nuevo León, México. By Robert J. Russell. Pp. 539-548, 1 figure in text. January 14, 1960.
22. Review of the insectivores of Korea. By J. Knox Jones, Jr., and David H. Johnson. Pp. 549-578. February 23, 1960.
23. Speciation and evolution of the pygmy mice, genus Baimoys. By Robert L. Packard. Pp. 579-670, 4 plates, 12 figures in text. June 16, 1960.

Index. Pp. 671-690

l. 10. 1. Studies of birds killed in nocturnal migration. By Harrison B. Tordoff and Robert M. Mengel. Pp. 1-44, 6 figures in text, 2 tables. September 12, 1956.
2. Comparative breeding behavior of Ammospiza caudacuta and A. maritima. By Glen E. Woolfenden. Pp. 45-75, 6 plates, 1 figure. December 20, 1956.
3. The forest habitat of the University of Kansas Natural History Reservation. By Henry S. Fitch and Ronald R. McGregor. Pp. 77-127, 2 plates, 7 figures in text, 4 tables. December 31, 1956.
4. Aspects of reproduction and development in the prairie vole (Microtus ochrogaster). By Henry S. Fitch. Pp. 129-161, 8 figures in text, 4 tables. December 19, 1957.
5. Birds found on the Arctic slope of northern Alaska. By James W. Bee. Pp. 163-211, plates 9-10, 1 figure in text. March 12, 1958.
*6. The wood rats of Colorado: distribution and ecology. By Robert B. Finley, Jr. Pp. 213-552, 34 plates, 8 figures in text, 35 tables. November 7, 1958.
7. Home ranges and movements of the eastern cottontail in Kansas. By Donald W. Janes. Pp. 553-572, 4 plates, 3 figures in text. May 4, 1959.
8. Natural history of the salamander, Aneides hardyi. By Richard F. Johnston and Gerhard A. Schad. Pp. 573-585. October 8, 1959.
9. A new subspecies of lizard, Cnemidophorus sacki, from Michoacán, México. By William E. Duellman. Pp. 587-598, 2 figures in text. May 2, 1960.
10. A taxonomic study of the middle American snake, Pituophis deppei. By William E. Duellman. Pp. 599-610, 1 plate, 1 figure in text. May 2, 1960.

Index. Pp. 611-626.

l. 11. Nos. 1-10 and index. Pp. 1-703, 1958-1960.

l. 12. 1. Functional morphology of three bats: Sumops, Myotis, Macrotus. By Terry A. Vaughan. Pp. 1-153, 4 plates, 24 figures in text. July 8, 1959.
*2. The ancestry of modern Amphibia: a review of the evidence. By Theodore H. Eaton, Jr. Pp. 155-180, 10 figures in text. July 10, 1959.
3. The baculum in microtine rodents. By Sydney Anderson. Pp. 181-216, 49 figures in text. February 19, 1960.
4. A new order of fishlike Amphibia from the Pennsylvanian of Kansas. By Theodore H. Eaton, Jr., and Peggy Lou Stewart. Pp. 217-240, 12 figures in text. May 2, 1960.
5. Natural history of the bell vireo. By Jon C. Barlow. Pp. 241-296, 6 figures in text. March 7, 1962.
6. Two new pelycosaurs from the lower Permian of Oklahoma. By Richard C. Fox. Pp. 297-307, 6 figures in text. May 21, 1962.
7. Vertebrates from the barrier island of Tamaulipas, México. By Robert K. Selander, Richard F. Johnston, B. J. Wilks, and Gerald G. Raun. Pp. 309-345, pls. 5-8. June 18, 1962.
8. Teeth of Edestid sharks. By Theodore H. Eaton, Jr. Pp. 347-362, 10 figures in text. October 1, 1962.

More numbers will appear in volume 12.

l. 13. 1. Five natural hybrid combinations in minnows (Cyprinidae). By Frank B. Cross and W. L. Minckley. Pp. 1-18. June 1, 1960.
2. A distributional study of the amphibians of the Isthmus of Tehuantepec, México. By William E. Duellman. Pp. 19-72, pls. 1-8, 3 figures in text. August 16, 1960.
3. A new subspecies of the slider turtle (Pseudemys scripta) from Coahulia, México. By John M. Legler. Pp. 73-84, pls. 9-12, 3 figures in text. August 16, 1960.
4. Autecology of the copperhead. By Henry S. Fitch. Pp. 85-288, pls. 13-20, 26 figures in text. November 30, 1960.
5. Occurrence of the garter snake, Thamnophis sirtalis, in the Great Plains and Rocky Mountains. By Henry S. Fitch and T. Paul Maslin. Pp. 289-308, 4 figures in text. February 10, 1961.
6. Fishes of the Wakarusa river in Kansas. By James E. Deacon and Artie L. Metcalf. Pp. 309-322, 1 figure in text. February 10, 1961.
7. Geographic variation in the North American cyprinid fish, Hybopsis gracilis. By Leonard J. Olund and Frank B. Cross. Pp. 323-348, pls. 21-24, 2 figures in text. February 10, 1961.

(Continued on outside of back cover)

(Continued from inside of back cover)

8. Decriptions of two species of frogs, genus Ptychohyla; studies of American hylid frogs, V. By William E. Duellman. Pp. 349-357, pl. 25, 2 figures in text. April 27, 1961.
9. Fish populations, following a drought, in the Neosho and Marais des Cygnes rivers of Kansas. By James Everett Deacon. Pp. 359-427, pls. 26-30, 3 figs. August 11, 1961.
10. Recent soft-shelled turtles of North America (family Trionychidae). By Robert G. Webb. Pp. 429-611, pls. 31-54, 24 figures in text. February 16, 1962.

Index. Pp. 613-624.

14. 1. Neotropical bats from western México. By Sydney Anderson. Pp. 1-8. October 24, 1960.
2. Geographic variation in the harvest mouse. Reithrodontomys megalotis, on the central Great Plains and in adjacent regions. By J. Knox Jones, Jr., and B. Mursaloglu. Pp. 9-27, 1 figure in text. July 24, 1961.
3. Mammals of Mesa Verde National Park, Colorado. By Sydney Anderson. Pp. 29-67, pls. 1 and 2, 3 figures in text. July 24, 1961.
4. A new subspecies of the black myotis (bat) from eastern Mexico. By E. Raymond Hall anad Ticul Alvarez. Pp. 69-72, 1 figure in text. December 29, 1961.
5. North American yellow bats, "Dasypterus," and a list of the named kinds of the genus Lasiurus Gray. By E. Raymond Hall and J. Knox Jones, Jr. Pp. 73-98, 4 figures in text. December 29, 1961.
6. Natural history of the brush mouse (Peromyscus boylii) in Kansas with description of a new subspecies. By Charles A. Long. Pp. 99-111, 1 figure in text. December 29, 1961.
7. Taxonomic status of some mice of the Peromyscus boylii group in eastern Mexico, with description of a new subspecies. By Ticul Alvarez. Pp. 113-120, 1 figure in text. December 29, 1961.
8. A new subspecies of ground squirrel (Spermophilus spilosoma) from Tamaulipas, Mexico. By Ticul Alvarez. Pp. 121-124. March 7, 1962.
9. Taxonomic status of the free-tailed bat, Tadarida yucatanica Miller. By J. Knox Jones, Jr., and Ticul Alvarez. Pp. 125-133, 1 figure in text. March 7, 1962.
10. A new doglike carnivore, genus Cynaretus, from the Clarendonian Pliocene, of Texas. By E. Raymond Hall and Walter W. Dalquest. Pp. 135-138, 2 figures in text. April 30, 1962.
11. A new subspecies of wood rat (Neotoma) from northeastern Mexico. By Ticul Alvarez. Pp. 139-143. April 30, 1962.
12. Noteworthy mammals from Sinaloa, Mexico. By J. Knox Jones, Jr., Ticul Alvarez, and M. Raymond Lee. Pp. 145-159, 1 figure in text. May 18, 1962.
13. A new bat (Myotis) from Mexico. By E. Raymond Hall. Pp. 161-164, 1 figure in text. May 21, 1962.
14. The mammals of Veracruz. By E. Raymond Hall anad Walter W. Dalquest. Pp. 165-362, 2 figures. May 20, 1963.

More numbers will appear in volume 14.

15. 1. The amphibians and reptiles of Michoacán, México. By William E. Duellman. Pp. 1-148, pls. 1-6, 11 figures in text. December 20, 1961.
2. Some reptiles and amphibians from Korea. By Robert G. Webb, J. Knox Jones, Jr., and George W. Byers. Pp. 149-173. January 31, 1962.
3. A new species of frog (Genus Tomodactylus) from western México. By Robert G. Webb. Pp. 175-181, 1 figure in text. March 7, 1962.
4. Type specimens of amphibians and reptiles in the Museum of Natural History, the University of Kansas. By William E. Duellman and Barbara Berg. Pp. 183-204. October 26, 1962.

More numbers will appear in volume 15.

UNIVERSITY OF KANSAS PUBLICATIONS
MUSEUM OF NATURAL HISTORY

Volume 14, No. 15, pp. 363-473, 5 figs.
May 20, 1963

The Recent Mammals of Tamaulipas, México

BY

TICUL ALVAREZ

UNIVERSITY OF KANSAS
LAWRENCE
1963

UNIVERSITY OF KANSAS PUBLICATIONS
MUSEUM OF NATURAL HISTORY

Institutional libraries interested in publications exchange may obtain this series by addressing the Exchange Librarian, University of Kansas Library, Lawrence, Kansas. Copies for individuals, persons working in a particular field of study, may be obtained by addressing instead the Museum of Natural History, University of Kansas, Lawrence, Kansas. There is no provision for sale of this series by the University Library, which meets institutional requests, or by the Museum of Natural History, which meets the requests of individuals. However, when individuals request copies from the Museum, 25 cents should be included, for each separate number that is 100 pages or more in length, for the purpose of defraying the costs of wrapping and mailing.

* An asterisk designates those numbers of which the Museum's supply (not the Library's supply) is exhausted. Numbers published to date, in this series, are as follows:

Vol. 1. Nos. 1-26 and index. Pp. 1-638, 1946-1950.

*Vol. 2. (Complete) Mammals of Washington. By Walter W. Dalquest. Pp. 1-444, 140 figures in text. April 9, 1948.

Vol. 3. *1. The avifauna of Micronesia, its origin, evolution, and distribution. By Rollin H. Baker. Pp. 1-359, 16 figures in text. June 12, 1951.

*2. A quantitative study of the nocturnal migration of birds. By George H. Lowery, Jr. Pp. 361-472, 47 figures in text. June 29, 1951.

3. Phylogeny of the waxwings and allied birds. By M. Dale Arvey. Pp. 473-530, 49 figures in text, 13 tables. October 10, 1951.

4. Birds from the state of Veracruz, Mexico. By George H. Lowery, Jr., and Walter W. Dalquest. Pp. 531-649, 7 figures in text, 2 tables. October 10, 1951.

Index. Pp. 651-681.

*Vol. 4. (Complete) American weasels. By E. Raymond Hall. Pp. 1-466, 41 plates, 31 figures in text. December 27, 1951.

Vol. 5. Nos. 1-37 and index. Pp. 1-676, 1951-1953.

*Vol. 6. (Complete) Mammals of Utah, *taxonomy and distribution*. By Stephen D. Durrant. Pp. 1-549, 91 figures in text, 30 tables. August 10, 1952.

Vol. 7. Nos. 1-15 and index. Pp. 1-651, 1952-1955.

Vol. 8. Nos. 1-10 and index. Pp. 1-675, 1954-1956.

Vol. 9. 1. Speciation of the wandering shrew. By James S. Findley. Pp. 1-68, 18 figures in text. December 10, 1955.

2. Additional records and extension of ranges of mammals from Utah. By Stephen D. Durrant, M. Raymond Lee, and Richard M. Hansen. Pp. 69-80. December 10, 1955.

3. A new long-eared myotis (Myotis evotis) from northeastern Mexico. By Rollin H. Baker and Howard J. Stains. Pp. 81-84. December 10, 1955.

4. Subspeciation in the meadow mouse, Microtus pennsylvanicus, in Wyoming. By Sydney Anderson. Pp. 85-104, 2 figures in text. May 10, 1956.

5. The condylarth genus Ellipsodon. By Robert W. Wilson. Pp. 105-116, 6 figures in text. May 19, 1956.

6. Additional remains of the multituberculate genus Eucosmodon. By Robert W. Wilson. Pp. 117-123, 10 figures in text. May 19, 1956.

7. Mammals of Coahuila, Mexico. By Rollin H. Baker. Pp. 125-335, 75 figures in text. June 15, 1956.

8. Comments on the taxonomic status of Apodemus peninsulae, with description of a new subspecies from North China. By J. Knox Jones, Jr. Pp. 337-346, 1 figure in text, 1 table. August 15, 1956.

9. Extension of known ranges of Mexican bats. By Sydney Anderson. Pp. 347-351. August 15, 1956.

10. A new bat (Genus Leptonycteris) from Coahuila. By Howard J. Stains. Pp. 353-356. January 21, 1957.

11. A new species of pocket gopher (Genus Pappogeomys) from Jalisco, Mexico. By Robert J. Russell. Pp. 357-361. January 21, 1957.

12. Geographic variation in the pocket gopher, Thomomys bottae, in Colorado. By Phillip M. Youngman. Pp. 363-387, 7 figures in text. February 21, 1958.

13. New bog lemming (genus Synaptomys) from Nebraska. By J. Knox Jones, Jr. Pp. 385-388. May 12, 1958.

14. Pleistocene bats from San Josecito Cave, Nuevo León, México. By J. Knox Jones, Jr. Pp. 389-396. December 19, 1958.

15. New subspecies of the rodent Baiomys from Central America. By Robert L. Packard. Pp. 397-404. December 19, 1958.

16. Mammals of the Grand Mesa, Colorado. By Sydney Anderson. Pp. 405-414, 1 figure in text, May 20, 1959.

17. Distribution, variation, and relationships of the montane vole, Microtus montanus. By Sydney Anderson. Pp. 415-511, 12 figures in text, 2 tables. August 1, 1959.

18. Conspecificity of two pocket mice, Perognathus goldmani and P. artus. By E. Raymond Hall and Marilyn Bailey Ogilvie. Pp. 513-518, 1 map. January 14, 1960.

(Continued on inside of back cover)

University of Kansas Publications
Museum of Natural History

Volume 14, No. 15, pp. 363-473, 5 figs.
May 20, 1963

The Recent Mammals of Tamaulipas, México

BY

TICUL ALVAREZ

University of Kansas
Lawrence
1963

University of Kansas Publications, Museum of Natural History
Editors: E. Raymond Hall, Chairman, Henry S. Fitch,
Theodore H. Eaton, Jr.

Volume 14, No. 15, pp. 363-473, 5 figs.
Published May 20, 1963

University of Kansas
Lawrence, Kansas

PRINTED BY
JEAN M. NEIBARGER, STATE PRINTER
TOPEKA, KANSAS
1963

29-4228

The Recent Mammals of Tamaulipas, México

BY

TICUL ALVAREZ

CONTENTS

	PAGE
Introduction	365
Physiography	366
Climate	368
Affinities of Tamaulipan Mammals	370
Plant-Mammal Relationships	371
Barriers and Routes of Movement	376
History of Mammalogy	379
Conservation	381
Methods and Acknowledgments	384
Gazetteer	386
Check-list	388
Accounts of Species and Subspecies	393
Literature Cited	467

INTRODUCTION

From Tamaulipas, the northeasternmost state in the Mexican Republic, 146 kinds of mammals, belonging to 72 genera, are here reported. Mammals that are strictly marine in habit are not included. The state is crossed in its middle by the Tropic of Cancer. Elevations vary from sea level on the Golfo de México to more than 2700 meters in the Sierra Madre Oriental; most of the state is below 300 meters in elevation. Its area is 79,602 square kilometers (30,732 square miles).

Tamaulipas, meaning "lugar en que hay montes altos" (place of high mountains), was explored in 1516 by the Spaniard Francisco Fernández de Córdoba, but it was not until the 18th century that José de Escandón established several villages in the new province of Nueva Santender from which, in the time of Iturbide's Empire, Tamaulipas was separated as a distinct political entity, with about the same boundaries that it now has.

My first contact with the state of Tamaulipas, as a mammalogist, was in 1957, when in company with Dr. Bernardo Villa R. I visited the Cueva del Abra in the southern part of the state. On several occasions since then I have been in the state, especially when employed by the Dirección General de Caza of the Mexican Government. In 1960-1962 I had the opportunity of studying the mammalian fauna of Tamaulipas at the Museum of Natural History of the University of Kansas. The approximately 2000 specimens there represent many critical localities, but are not sufficient to make this report as complete as could be desired. Consequently the following account should be considered as a contribution to the knowledge of the mammals of México and is offered in the hope that it will stimulate future studies of the Mexican fauna, especially that of the eastern region.

PHYSIOGRAPHY

Tamaulipas can be divided into three physiographic regions, which from east to west are Gulf Coastal Plain, Sierra Madre Oriental, and Central Plateau or Mexican Plateau (Fig. 1).

Gulf Coastal Plain

This physiographic region covers most of the state and extends northward into Texas and a short distance southward into Veracruz.

According to Tamayo (1949) and Vivo (1953), the Gulf Coastal Plain is formed by sedimentary rocks from Mesozoic to Pleistocene in age. The most common type of soil is Rendzin, especially in the coastal area. Elevations range from sea level to 300 meters. The area is in general a flat plain inclined to the sea but this plain is broken by several small sierras. The more important of these are the Sierra de Tamaulipas, which rises to more than 1000 meters, and the Sierra San Carlos, which has a maximum elevation of approximately 1670 meters. The Sierra de San José de las Rucias is smaller.

Sierra Madre Oriental

This physiographic region is represented in Tamaulipas by a small part of the long Sierra Madre Oriental that extends from the Big Bend area in Texas southward to the Trans-volcanic Belt of central México. The Sierra Madre Oriental is in the southwestern part of Tamaulipas. The Sierra was formed by folding of the Middle and Upper Cretaceous and Cenozoic deposits that now are 400 to 2700 meters in elevation. In general, the soils are Chernozems.

This physiographic region is situated between the other two physiographic regions in Tamaulipas and represents a barrier to the

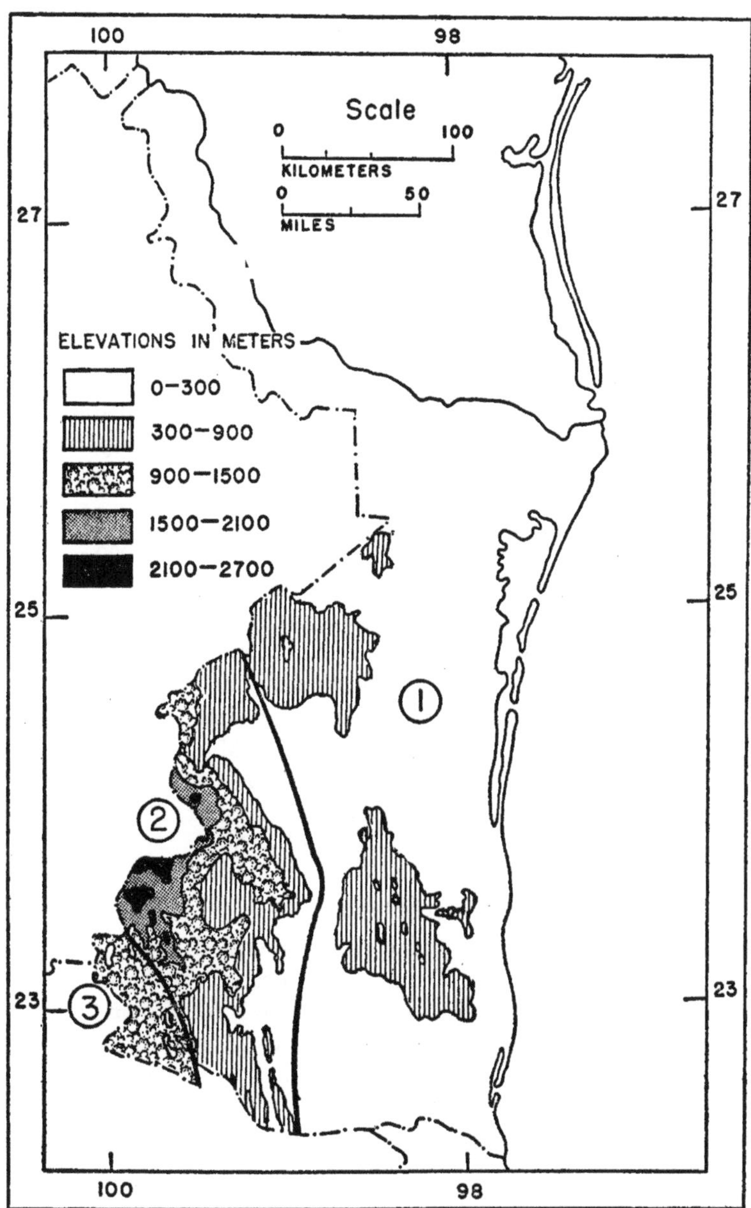

FIG. 1. Three physiographic regions: 1 Coastal Plain; 2 Sierra Madre Oriental; 3 Central Plateau.

distribution of some tropical mammals on the one hand and to those from the Mexican Plateau on the other.

Central Plateau

This physiographic region, commonly termed the Mexican Plateau, occupies only a small area of Tamaulipas in its southwesternmost part. The plateau is approximately 900 meters above sea level. In general, the Mexican Plateau was formed by Cretaceous sediments. The most common type of soil is Chestnut.

CLIMATE

Owing to the differences in elevations and varying distances from the sea, the climate of Tamaulipas is varied. Tamayo (1949), following the Koeppen System, assigned to Tamaulipas 10 different climate types that result principally from differences in temperature, precipitation, and humidity.

Temperature

The annual mean temperature for the lands less than 1000 meters in elevation, which make up most of the state, is between 20° and 25° C.; and the difference in monthly means is 5° C.

In the areas above 1000 meters, the annual mean is between 15° and 20° C., and the difference in the monthly means is 15° C.

The maximum temperature recorded in the state is 45° C. in the region of Ciudad Victoria, between the Sierra Madre Oriental, the Sierra San Carlos, and the Sierra de Tamaulipas. Minima recorded are between 0° and 5° C. on the southeastern coast, 0° to —5° C. between 98° 20′ long. and 99° 00′ long., and —5° to —10° C. in the Sierra Madre Oriental.

Precipitation

Rainfall varies seasonally and can be described as follows: In January it amounts to 25 to 50 mm. in the coastal region and 10 to 25 mm. in the rest of the state. In April there is more than 25 mm. to the north of about 23° north latitude, 10 to 25 mm. in the Sierra de Tamaulipas and Sierra Madre Oriental, and less than 10 mm. in the extreme southwestern part of the state.

In July rainfall amounts to less than 25 mm. in Nuevo Laredo and San Fernando, is from 25 to 50 mm. in the northeastern and central parts of the state, 50 to 100 mm. in the Sierra San Carlos and Sierra Madre Oriental, and 100 to 200 mm. in the area south of Soto la Marina and east of the Sierra Madre Oriental. In October rainfall is less than 50 mm. in the northern half of the state, including the Sierra de Tamaulipas, and 50 to 100 mm. in the rest of the state, except on the east side of the Sierra Madre Oriental and in the area near Tampico, which receive between 100 and 200 mm.

The number of rainy days per year varies from 60 to 90 at Sierra San Carlos, Sierra Madre Oriental, and in the lowlands south of 23° north latitude; the rest of the state has about 60 rainy days, excepting the Mexican Plateau, which has fewer than 60.

Although Tamayo (1949) followed the Koeppen System in classifying types of climate and thereby recognized 10 different kinds of climate in Tamaulipas, these can be grouped into three major categories as follows:

Steppe Dry Climate (Clima Seco de Estepa)

This kind of climate can be divided into two categories based on the average annual temperature.

Warm

The average annual temperature exceeds 18° C. but the mean of the coolest month is less than 18° C. This sub-climate is characterized by a short rainy season in summer and occurs on the west side of the southern part of the Sierra Madre Oriental and on the Mexican Plateau; it occurs also in the area northwest of Reynosa and on the east side of the Sierra Madre Oriental but in these areas the rainfall is irregularly distributed in the year.

Cool

The average annual temperature is less than 18° C. but the mean of the warmest month exceeds 18° C. This sub-climate occurs only on the west side of the northern part of the Sierra Madre Oriental.

Moderate Rainy Temperature Climate (Clima Templado Moderato Lluvioso)

This type of climate is characterized by the coolest month having a temperature of between —3° and 18° C. In the northeastern and central parts of Tamaulipas, including the Sierra de Tamaulipas, Ciudad Victoria, Gómez Farías, Rancho Pano Ayuctle, and Llera, the average temperature of the warmest month is less than 22° C.; the winters are dry and not rigorous, and the wettest month has ten times as much rain as the driest. In the Sierra San Carlos the average temperature of the warmest month is less than 22° C., and the rainy season is in the autumn.

Tropical Rainy Climate (Clima Tropical Lluvioso)

This climate is characterized by the average temperature of all months being above 18° C. and the mean-annual rainfall being above 75 cm. According to the distribution of precipitation this type of climate can be divided into: (1) areas having periodic rain and wet winters (southeastern Tamaulipas, south of 22° north latitude and east of 99° west longitude), and (2) areas having an irregular rainy season and dry winters (area around Ciudad Mante, between 99° 30′ and 98° 30′ west longitude and south of 22° 30′ north latitude).

AFFINITIES OF TAMAULIPAN MAMMALS

Owing to the differences in climate from one region to another, the flora and fauna also differ, especially in the southern part of the state as compared with the northern part.

For expressing the taxonomic resemblance of mammalian faunas having nearly equal numbers of taxa, Burt (1959:139) recommended the following formula: $C \times 100/(N_1 + N_2 - C)$ (where C is the number of taxa common to the two faunas, N_1 is the number of taxa in the smaller fauna, and N_2 is the number of taxa in the larger fauna). For non-flying mammals the resemblance of the Tamaulipan fauna to that of Texas, adjacent to the north, and Veracruz, adjacent to the south, is as follows:

Genera.—Texas 65 per cent, Veracruz 60 per cent.

Species.—Texas 45 per cent, Veracruz 39 per cent.

For bats the resemblance of the Tamaulipan fauna to those of Texas and Veracruz is as follows:

Genera.—Texas 40 per cent, Veracruz 51 per cent.
Species.—Texas 24, Veracruz 39.

TABLE 1.—NUMBER OF GENERA AND SPECIES OF NON-INTRODUCED LAND MAMMALS IN THREE STATES.

	Number of taxa				Number of taxa in common			
	genera		species		genera		species	
States	non-bats	bats	non-bats	bats	non-bats	bats	non-bats	bats
Texas	51	12	103	25	39	10	58	12
Tamaulipas	48	23	83	36	......	......	......	......
Veracruz	53	36	94	60	38	20	50	27

For all of the land mammals of Tamaulipas, the resemblance is as follows:
Genera.—Texas 58, Veracruz 57.
Species.—Texas 40, Veracruz 39.

On the whole, the fauna of Tamaulipas resembles faunas of both the Brazilian Subregion and the North American part of the Nearctic Subregion (see Hershkovitz, 1958:611). Considering the 48 genera of non-flying land mammals of Tamaulipas, 24 genera occur in habitats from the North American part through habitats of northern México into the Brazilian Subregion. Of the remaining 24 genera, 16 occur in the North American part of the Nearctic Subregion or in it and the part of northern México north of the Brazilian boundary, whereas eight occur in the Brazilian Subregion or in it and the northern part of México. None occurs only in Tamaulipas or only in northern México.

The non-flying fauna of the coastal plain east of the Sierra Madre Oriental and south of the Sierra de Tamaulipas and Soto la Marina is mainly tropical in affinities; only 27 per cent of that fauna (at the subspecific level) resembles the fauna north of Soto la Marina, which is Nearctic in its affinities. The fauna of the Sierra de Tamaulipas has a greater taxonomic resemblance (20.4 per cent at subspecific level) to that of the Sierra Madre Oriental, than does the fauna of the Sierra San Carlos (17.6 per cent). Taxonomic resemblance between the faunas from the Sierra San Carlos and the Sierra de Tamaulipas amounts to only 16.1 per cent. Therefore, the faunas of these two Sierras (both are included in the same zoogeographic unit) resemble each other less than either resembles the fauna of the Sierra Madre Oriental (in another zoogeographic unit). Of the three sierran faunas, those of the Sierra Madre Oriental and the Sierra de Tamaulipas have most in common. Migration from one to the other in relative recent time may account for the resemblance. The Sierra San Carlos may have been isolated for a long time and interchange between its fauna and those of the other two sierras, therefore, may have been slight.

Study of the taxonomic resemblance shows that the dividing line, in eastern México, between Nearctic and Neotropical faunas is along the eastern base of the Sierra Madre Oriental, the southern base of the Sierra de Tamaulipas and thence to the coast at or near Soto la Marina.

PLANT-MAMMAL RELATIONSHIPS

Merriam (1898) assigned to Tamaulipas four Life-zones. There were: Transitional on the highest elevations of the Sierra Madre;

Upper Austral at lower elevations on the Sierra Madre; Lower Austral over most of the state; and Tropical in the coastal areas.

Dice (1943) outlined Biotic Provinces on a map of North America and in the northern part of Tamaulipas showed two Biotic Provinces, Tamaulipan and Potosian. He did not show the southeastern limits of the Chihuahuan Biotic Province nor any of the limits of the Veracruzian Biotic Province and in text mentioned nothing about the limits of these two provinces with reference to Tamaulipas. Later, Goldman and Moore (1946) divided Tamaulipas in three Biotic Provinces: Tamaulipas, Sierra Madre, and Veracruz. Still later (1949), Smith published a map of Mexican Biotic Provinces based on the herpetofauna of the Republic. He divided Tamaulipas among four Provinces. Two were Nearctic (Austro-oriental and Tamaulipan) and the other two were Neotropical (Veracruzian and Cordoban).

Leopold (1950 and 1959) recognized five principal vegetational types in Tamaulipas as follows: Mesquite-grassland; Pine-oak Forest; Thorn Forest; Tropical Deciduous Forest; and Desert.

For dealing with the mammals of Tamaulipas in the following accounts the four Biotic Provinces (Tamaulipan, Potosian, Veracruzian, and Chihuahuan) of Dice are the most useful. For dealing with types of vegetation in the accounts that follow, Leopold's (1950) system is employed although reference is made to other associations and formations that have been reported in Tamaulipas.

Tamaulipan Biotic Province

This Province is recognized by most authors who have written about the zoogeography of México. It is the most extensive in the state and includes the northern part of the Coastal Plain (see Fig. 2).

The vegetation of the Tamaulipan Biotic Province is in general Mesquite-grassland but in the Sierra San Carlos and Sierra de Tamaulipas other types of vegetation are found.

Two formations occur in the Mesquite-grassland. The first is the Mesquite Scrub, in which the dominant plant is the mesquite (*Prosopis juliflora*), associated with *Cordia boissieri*, several species of *Acacia*, and in some areas with *Opuntia* and *Yucca treculeana*. The dominant grasses are of the genera *Bouteloua* and *Andropogon*. The second formation is the Gulf Bluestem Prairie, where species of *Andropogon* are the dominants on the well-drained sites.

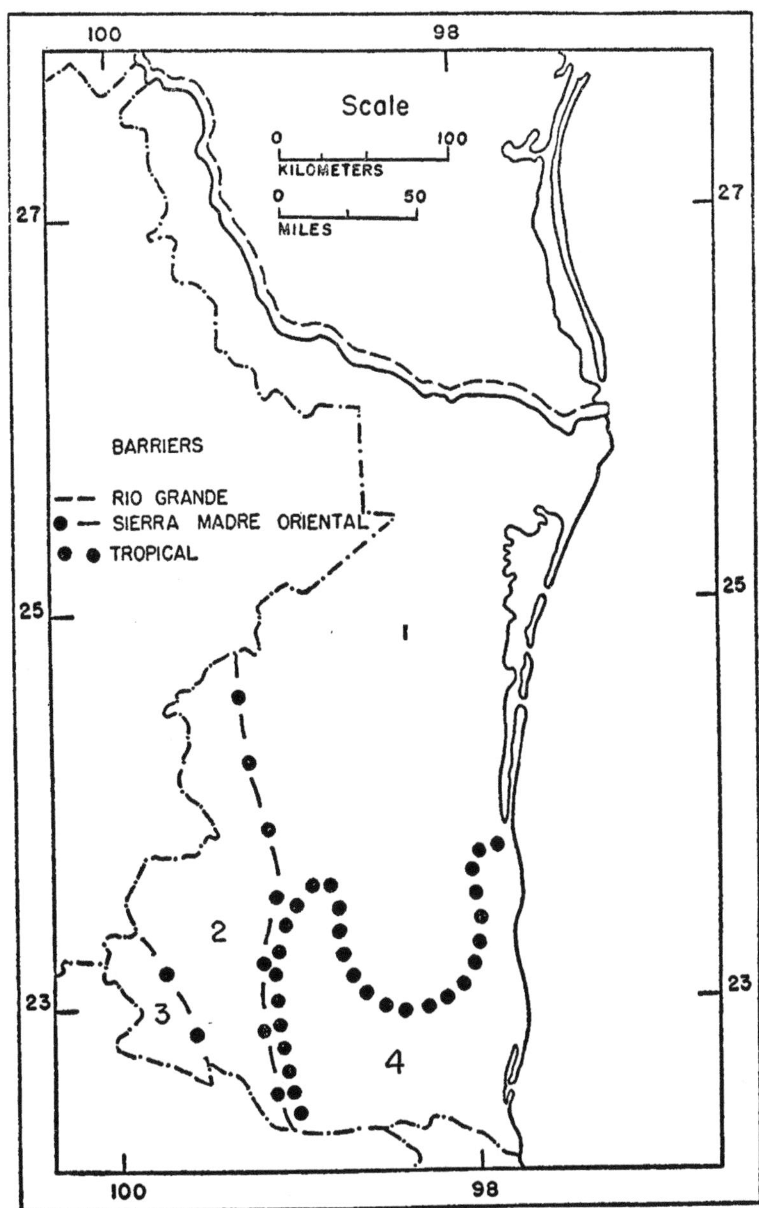

FIG. 2. Four biotic provinces: 1 Tamaulipan; 2 Potosian; 3 Chihuahuan; 4 Veracruzian.

Sloughs and depressions are occupied by cordgrass, *Spartina spartinae.* Many areas have been invaded by mesquite and other shrubs.

Around the Sierra de Tamaulipas and in the area between it and the Sierra San Carlos the vegetation is Thorn Forest (Tropical Thorn Forest of Martin *et al.*, 1954), in which the dominant plants are *Acacia, Ichthyomethia, Ipomea, Prosopis*, and *Cassia.* Another type of vegetation in the Sierra de Tamaulipas is the Tropical Deciduous Forest at 300 to 700 meters elevation, the trees of which are 20 meters high with a canopy averaging eight meters high (Martin *et al., op. cit.*). The common species of trees belong to the genera *Tabebuia, Ipomea, Bombax*, and *Conzattia.* Species of *Bursera, Acacia*, and *Cassia* are less abundant. In the low canyons *Bursera, Ceiba*, and *Psidium*, draped with lianas and various epiphytes, can be found.

The Pine-oak Formation grows above an elevation of 800 meters in the Sierra de Tamaulipas and is characterized by *Pinus cembroides, P. nelsonii, P. teocote*, and *Quercus arizonica.* Martin *et al.* (*op. cit.*) recorded Montane Scrub from the dry areas, between elevations of 600 and 900 meters. That scrub is formed by huisaches (*Acacia farnesiana*) along with a few oaks and some trees of the Tropical Deciduous Forest.

The vegetation of the Sierra San Carlos was studied by Dice (1937) and divided into three life belts, each with several associations. For more information about the plants of each association and their related mammals see the publication of the mentioned author.

Endemic mammals of the Tamaulipan Biotic Province, in the part of it that is in Tamaulipas, are the following: *Scalopus inflatus; Lepus californicus curti; Spermophilus spilosoma oricolus; Cratogeomys castanops tamaulipensis; Dipodomys ordii parvabullatus;* and *Sigmodon hispidus solus.* Other characteristic mammals of this Province in the state of Tamaulipas are: *Sylvilagus floridanus connectens; S. audubonii parvulus; Lepus californicus merriami; Perognathus merriami merriami; Dipodomys ordii compactus; Orzomys melanotis carrorum; Reithrodontomys fulvescens intermedius; Peromyscus boylii ambiguus; Canis latrans texensis; C. l. microdon; C. lupus monstrabilis; Taxidea taxus berlandieri; Mephitis mephitis varians; Felis pardalis albescens; Trichechus manatus latirostris;* and *Odocoileus virginianus texanus.*

Many other kinds of mammals occur mainly in the Tamaulipan Province but are not listed above because they occur also in one or more of the other provinces.

The Sierra de Tamaulipas is placed in the Tamaulipan Biotic Province because the fauna, especially of non-flying mammals, is closely related to that of the rest of the Province. Nevertheless, many mammals found in this Sierra are tropical in relationship. This is especially true of the bats. Therefore, most of the tropical bats that occur in Tamaulipas occur in the Veracruzian Biotic Province and in the Sierra de Tamaulipas.

Potosian Biotic Province

This Province occupies all of the Sierra Madre Oriental and, therefore, the southwestern part of the state.

The vegetation in general is Pine-oak Forest, in which the most common trees are *Abies religiosa, Pinus flexilis, P. patula, P. mon-*

tezumae, P. *teocote*, *Populus tremuloides*, *Juniperus flaccida*, *Quercus arizonica*, *Q. clivicola* and *Q. polymorpha*.

In his study of plants of the Gómez Farías area, Martin (1958) recorded several different types of vegetation, which in part can be placed in the Potosian Biotic Province, especially those types that occur to the northwest of the Cloud Forest. In addition to the Cloud Forest, Martin recognized Humid Pine-oak Forest, Dry Oak-pine Forest, Chaparral, Thorn Forest and Scrub, and Thorn Desert.

The only mammal endemic to the Potosian Province in Tamaulipas is *Cryptotis pergracilis pueblensis*. Other mammals that occur mainly in this Province are: *Sorex saussurei; Notiosorex crawfordi; Glaucomys volans herreranus; Cratogeomys castanops planifrons; Perognathus nelsoni; Liomys irroratus alleni; Reithrodontomys fulvescens griseoflavus; Microtus mexicanus subsimus; Ursus americanus eremicus; Conepatus leuconotus texensis;* and *Odocoileus hemionus*.

The fauna of this Province is a mixture of elements with tropical affinities on the east side of the Sierra Madre and with those of the Mexican Plateau on the west side.

Chihuahuan Biotic Province

This Province occurs in Tamaulipas only in a small portion of the Central Plateau physiographic region and occupies the southwesternmost part of the state.

The vegetation is of two types: Desert or Mesquite-grassland. The last is like that described for the Tamaulipan Biotic Province. In the Desert type the dominant plants are the cactus, *Opuntia leptocaulis*, and yuccas, *Yucca filifera* and *Y. potosina*. Subdominants are mariola, guayule, *Agave lechugilla*, *A. stricta* or *Larrea divaricata*. Along stream banks mesquite, *Prosopis juliflora*, can be found.

No endemic mammals of the Chihuahuan Province are known in Tamaulipas. Mammals that occur principally in this Province are: *Dipodomys merriami atronasus*; *D. ordii durranti*; *Peromyscus melanophrys consobrinus*; *P. difficilis petricola*; *Onychomys torridus subrufus*; and *Neotoma albigula subsolana*.

Veracruzian Biotic Province

This Province includes the southern part of the Coastal Plain physiographic region, south of the Sierra de Tamaulipas and Soto la Marina. But the exact line between this Province and the Tamaulipan Province to the north is difficult to draw. The northern boundary of the Veracruzian Province is the line between the Nearctic and Neotropical regions in eastern México.

Vegetation of most of the Veracruzian Biotic Province is Tropical Deciduous Forest. This Forest is made up of *Tabebuia*, *Ipomea*, *Bombax*, and *Conzattia*, along with some *Ceiba*, *Bursera*, and *Psidium*.

The mammalia fauna of the Veracruzian Biotic Province is tropical in nature. This is especially true of the bats. Representatives of the tropical genera *Micronycteris, Sturnira, Artibeus, Enchistenes, Desmodus, Diphylla*, and *Molossus* have their northern distributional limits in this Province. The non-flying mammals characteristic of the Province in Tamaulipas are: *Philander opossum pallidus; Marmosa mexicana; Ateles geoffroyi velerosus; Geomys tropicalis; Oryzomys melanotis rostratus; O. alfaroi huastecae; O. fulvescens engracie* (endemic to this Province in Tamaulipas); *O. f. fulvescens; Reithrodontomys mexicanus; Peromyscus orchraventer* (endemic); *Neotoma micropus angustapalata; Eira barbara senex; Felis wiedii oaxacensis;* and *Mazama americana temama.*

BARRIERS AND ROUTES OF MOVEMENT

The distributional patterns and affinities of the mammalian fauna of Tamaulipas suggest possible routes of migration and barriers that limited or controlled movements of the mammals.

Mammals may have reached Tamaulipas by way of a Northern route, a Trans-plateau route, a Montane route, or a Tropical route (Fig 3).

The Northern route permitted species of mammals from the temperate region to the north to enter the Tamaulipan Biotic Province from or *via* Texas. Several came from the Great Plains, and a few came from the eastern part of the United States. Also, a few mammals that may have originated in the Tamaulipan Province moved northwards. Some of these, according to Dice (1937:267) were *Liomys irroratus texensis, Peromyscus leucopus texensis*, and *Lepus californicus merriami.* Other mammals thought to have moved north by this route are *Didelphis marsupialis, Dasypus novemcinctus, Oryzomys palustris, Nasua narica*, and *Tayassu tajacu.* Some mammals that passed through Tamaulipas into Texas have extended their geographic ranges far north of Texas.

Mammals that came *via* the Trans-plateau route (name proposed by Baker, 1956:146) came no farther into Tamaulipas than the Chihuahuan Biotic Province. They encountered the barrier formed by the Sierra Madre Oriental. These mammals were listed in the account of the Chihuahuan Biotic Province.

The route that Baker (1956:146) termed the "Southern Route" I here term the Montane route because I think it was used for movement southward as well as northward.

The Montane route was used by mammals of boreal affinities (*Microtus* and *Neotoma*), that moved into Tamaulipas from the north; also in this category are bats of the family Vespertilionidae. For movement from south to north, the route was used by several species native to México, for example, *Cratogeomys castanops*. The

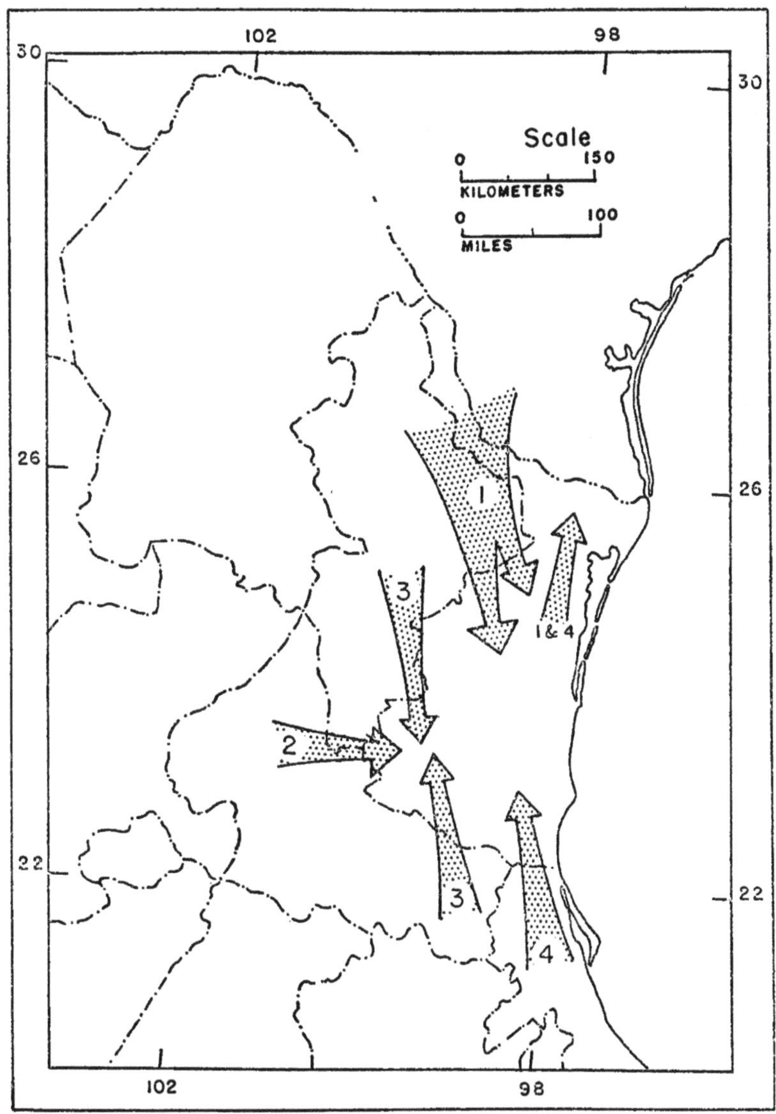

FIG. 3. Routes of movement: 1 Northern; 2 Trans-Plateau; 3 Montane; 4 Tropical.

seaward slope of the montane area has enabled some tropical mammals to move farther north than they have done at higher and lower elevations. *Philander opossum* seems to be an example.

The fourth route, the Tropical one, was used by mammals of tropical origin. Most moved into Tamaulipas only as far as the Veracruzian Biotic Province. The principal mammals that have used this route are the bats and marsupials, but *Sylvilagus brasiliensis, Ateles geoffroyi, Heterogeomys hispidus, Eira barbara,* and *Mazama americana* also can be included here. Some tropical mammals, as was pointed out previously, not only reached Tamaulipas but have moved through the state and far northward.

The major barriers to dispersal of mammals in Tamaulipas are three (see Fig. 2). Two of them, the Río Grande Barrier and the Sierra Madre Barrier, are physiographical, but the Tropical Barrier is maintained by a combination of environmental factors. The three barriers separate the four Biotic Provinces in Tamaulipas. The Sierra Madre Oriental, which forms the Potosian Biotic Province, lies between the Tamaulipan and Chihuahuan provinces. The Tropical barrier separates the Tamaulipan and Veracruzian biotic provinces.

The Río Grande, as was pointed out by R. H. Baker (1956:146), has low banks, is relatively shallow, and does not form an effective barrier for most mammals. For only two species, insofar as I know, has the Río Grande constituted a barrier. *Cratogeomys castanops* has not entered southeastern Texas from México, and *Spermophilus spilosoma* has not entered México from southeastern Texas except on the coastal barrier beach. Alvarez (1962:124) postulated that the beach was the route by which S. *spilosoma* arrived at La Pesca where the barrier beach meets the mainland.

The Sierra Madre Barrier is a good filter for some small mammals, especially for those that occur on the Mexican Plateau and those of tropical origin. The mammals that occur on each side of the Sierra are listed in accounts of the Chihuahuan (west side), Veracruzian and Tamaulipan (east side) biotic provinces.

The Tropical Barrier is formed mainly by a climatic complex (probably a change in temperature and rainfall) in the coastal region at or about the latitude of Soto la Marina, where no geographic barrier is found. In the western and central part of the Tropical Barrier, the climatic factor is supported by a geographic factor. The Sierra Madre Oriental is in the west and the Sierra de Tamaulipas is in the center. The several mammals that are affected by this barrier are listed in the accounts of the Veracruzian and Tamaulipan biotic provinces.

A peculiar pattern of distribution is that presented by *Scalopus inflatus* and *Geomys tropicalis*. Both are the only known species of their genera in northeastern México. Each is isolated from other species of its genus. The nearest known record of *Scalopus* is 45 miles northward and the nearest record of *Geomys* is approximately 165 miles northward. A possible explanation for the distribution of these two kinds is that each was widely distributed in one of the glacial periods and when the glacier receded to the north these animals remained in Tamaulipas, where they evolved and formed distinct species. The two species, *G. tropicalis* and *S. inflatus*, are fossorial and for this reason probably were able to resist inhospitable climates better than non-burrowing species.

HISTORY OF MAMMALOGY

In Tamaulipas the first exploration directed in substantial measure toward finding out about the mammalian fauna, at least as far as I know, was made by Dr. L. Berlandier, who traveled mainly in the northern half of the state. His collections provided specimens of several previously unknown mammals, which were described by Baird (1858). The original manuscript of Berlandier never has been published. About 1880 Dr. E. Palmer collected mammals in the southern part of Tamaulipas, in the area around Tampico. The results of his exploration were reported by J. A. Allen (1881). E. W. Nelson and E. A. Goldman twice collected in Tamaulipas (Goldman, 1951). In 1898 they visited and collected mammals in the southern part of the state, around Tampico, Altamira, Victoria, Forlón, and Miquihuana. In 1901-1902 they visited the area between Nuevo Laredo and Bagdad, then went south to Soto la Marina and Victoria. From their collections several species and subspecies have been described. Between 1910 and the early 1920's little was done in the way of scientific exploration because of the Mexican Revolution.

From 1930 on, several expeditions yielded new information about the native mammals. In that year L. B. Kellum visited the Sierra San Carlos. The results were reported by Dice (1937). Another important collection from Tamaulipas was made by Marian Martin in the area of Gómez Farías. Mammals collected by her were reported by Goodwin (1954). Hooper (1953) also reported specimens from Gómez Farías but included in his report records of mammals collected in other areas as well. In 1950 E. R. Hall and C. von

Wedel made a trip to the barrier beach in the northeastern part of the state and collected several kinds of mammals among which three were described as new by Hall (1951).

The report here presented is based upon specimens in the Museum of Natural History of The University of Kansas that were collected mainly by the persons named beyond. Gerd H. Heinrich and his wife Hilda collected in 1952 and 1953 in the areas around Miquihuana, Ciudad Victoria, Soto la Marina, Sierra de Tamaulipas, and Altamira. W. J. Schaldach collected in 1949 and 1950 in the Sierra Madre Oriental south of Ciudad Victoria; he returned to Tamaulipas in 1954 in company with V. Grissino and worked in the Sierra Madre Oriental south and north of Ciudad Victoria. In 1961 P. L. Clifton and J. H. Bodley collected in the northwestern part of the state and in the western part, around Tula, Nicolás, and Tajada. Some students and staff members of the Museum have occasionally collected in Tamaulipas.

As a result of all the mentioned expeditions and others, 32 species and subspecies have been described with type localities in Tamaulipas. They are:

Altamira
Lepus californicus altamirae Nelson
Sciurus aureogaster aureogaster (Cuvier) (by restriction)
Sciurus deppei negligens Nelson
Geomys tropicalis Goldman

Antiguo Morelos, 8 mi. N of
Tadarida laticaudata ferruginea Goodwin

Brownsville (Texas), 45 mi. from
Scalopus inflatus Jackson

Charco Escondido
Perognathus hispidus hispidus Baird
Neotoma micropus micropus Baird

El Carrizo
Peromyscus ochraventer Baker

Gómez Farías
Heterogeomys hispidus negatus Goodwin

Hacienda Santa Engracia
Oryzomys fulvescens engracia Osgood

Jaumave
Dipodomys ordii durranti Setzer

La Pesca, 1 mi. E of
Spermophilus spilosoma oricolus Alvarez

Matamoros
Cryptotis parva berlandieri (Baird)
Lasiurus intermedius intermedius (H. Allen)
Dasypus novemcinctus mexicanus Peters (by restriction)

Cratogeomys castanops tamaulipensis Nelson and Goldman
Felis yagouaroundi cacomitli Berlandier

Matamoros, 88 mi. S, 10 mi. W of
Lepus californicus curti Hall
Dipodomys ordii parvabullatus Hall
Sigmodon hispidus solus Hall

Mier
Canis latrans microdon Merriam

Miquihuana
Idionycteris mexicanus Anthony (*Plecotus phyllotis*)
Cratogeomys castanops planifrons Nelson and Goldman
Onychomys torridus subrufus Hollister
Neotoma albigula subsolana Alvarez
Odocoileus virginianus miquihuanensis Goldman and Kellogg

Rancho del Cielo, 5 mi. NW Gómez Farias
Cryptotis mexicana madrea Goodwin
Reithrodontomys megalotis hooperi Goodwin

Rancho Santa Ana, about 8 mi. SW Padilla
Oryzomys melanotis carrorum Lawrence

Sierra de Tamaulipas, 10 mi. W, 2 mi. S Piedra
Myotis keenii auriculus Baker and Stains

Sierra San Carlos, 12 mi. NW San Carlos
Peromyscus pectoralis collinus Hooper

CONSERVATION

A relatively large number of the species of Mexican big game occurs in Tamaulipas because its geographic position permits it to have species from the tropics and those from the northern plains and mountains. Eight of the 11 Mexican species that are considered as Big Game are recorded from the state. Until this century Tamaulipas was not densely populated by man either in the precolonial period or thereafter. Therefore many species of game are still relatively abundant.

Of the eight species that originally lived in Tamaulipas, the mule deer, brocket, and black bear never have been abundant there and now are in danger of extirpation. The pronghorn was also rare in the state and now has been extirpated as it has been in many other parts of México. The white-tailed deer, javalin, jaguar, and puma are still abundant in suitable habitats. The white-tailed deer is found almost everywhere in the state; in some areas it damages cornfields, and for this reason is killed by natives who eat the meat and sell the skins. The price of skins is low; in 1959 at Ciudad Mante tanners paid natives less than one dollar (10.00 Mexican pesos) per hide. Some idea of the abundance of deer in Tamaulipas is provided by our having found in one tanner's shop, in 1959 at

Ciudad Mante, about 500 deer skins. Besides these, we found about 65 skins of other species—jaguar, bear, ocelot, puma, margay, and raccoon. Additionally there was a large number of coati skins. Considering that México has no professional trappers and that commerce in skins of wild animals is illegal, it is felt that the number of skins found in the tanner's shop indicated a relative large population of game mammals.

The number of species of small game also is large. Some species are killed by natives for food, but most are killed in order to protect the cultivated crops, which are injured mainly by rabbits and squirrels.

Baker (1958) pointed out that the future of the game species in the northern part of México was not encouraging. He gave valid reasons for his view. In Tamaulipas, however, in some respects the outlook is more encouraging because there are many areas in which with a minimum of effort the authorities can save a good number of species.

As Baker (*op. cit.*) remarked, the fauna in México is declining mainly because many areas recently have been cultivated for the first time. Also, better roads have enabled hunters to reach areas that formerly were natural refuges for wild animals. Many times it has been said that the populations of wild animals were declining in México because the number of game wardens is too small to protect game in all parts of the country. In some ways this is true but it seems that the problem is really one of education. The people do not realize that the animals are part of nature and therefore have the same right to live that man has. Most people see only the bad side of the animals' activities and never consider the benefit that wild mammals provide for man. A typical case is that of the coyote, which is oftentimes killed only because it is a coyote. Sometimes individual coyotes do kill domestic animals, but the people seem never to understand that the coyote destroys a large number of mice, rabbits, and insects as has been shown by studies of the contents of coyote stomachs.

The Mexican Government at this time is making a concentrated effort to provide schools in all parts of the country and is formulating new programs of education. In this official program some lectures in conservation are needed with reference to the animal life. I know that some education now is given to people with respect to conservation of the water, soil, and forest, but gather that there is little that covers also conservation of animals.

I do not deny the necessity for some natives to kill wild animals.

People need to eat fresh meat and for some it is almost impossible to obtain meat in any other way than by killing wild animals. Some natives cannot afford to purchase meat in the markets or they live too far from any village or city to do so. Also, natives need to protect their cultivated areas; some of them have only four to six acres of land, on which corn is the only crop. When one deer in a night can destroy part of the corn, and in some areas not only one deer but several invade a field, and when one considers that besides deer there are rabbits, squirrels, raccoons, and coati, to name only some animals that feed on the corn, we find that the small cornfield at the end of the season may not contain any corn to harvest. It is understandable, therefore, that the natives kill the animals. In this way they protect their cultivated fields, obtain food and sometimes money for the skins. Many natives, however, destroy the wildlife only for pleasure or to obtain money for skins and meat, which sometimes is sold to restaurants.

Probably the best solution for the problem of conservation of wild animals is the establishment of wildlife refuges. In Tamaulipas, at least three refuges are needed in order to preserve the mammalian wildlife. These areas would serve also as a refuge for game birds and other vertebrates. A large area with suitable habitat for white-tailed deer, brocket, jaguar, puma, javalin, and fox could be established in the Sierra de Tamaulipas, which presents favorable habitat for all of the species named. A second area that does not need to be so large as the first could be established in the Sierra Madre Oriental, probably including some part of Nuevo León, where the black bear and the mule deer find suitable habitat. Probably the beaver can be introduced in the streams of the high mountains; beaver live in the same Sierra a little farther north in Nuevo León. The three species mentioned are in imminent danger of disappearing from Tamaulipas, if they have not already disappeared. The third refuge could be in some area of the northern part of the state near the Río Grande. This refuge should give protection to the beaver—a rare animal in México and in danger of extirpation over all the country. The pronghorn also would find suitable habitat in this area, but would have to be reintroduced there. With the establishment of these three refuges and with good management the fauna of Tamaulipas could be saved from extinction, would provide some recreation for sportsmen, and especially for the people in general who wish to study, photograph, or merely observe the native animal life.

The time is excellent for the establishment of the wildlife refuges in Tamaulipas because large areas are still in Federal ownership and because a considerable number of animals remain. Other favorable factors are that roads are not yet good in the areas proposed for refuges, the human population is low, and agriculture consequently is not practiced. But, with the rapid increase in population in México, these favorable conditions will change in a few years and it will be almost impossible to establish the refuges then.

METHODS AND ACKNOWLEDGMENTS

The families, genera, and species recorded in this report are arranged following Hall and Kelson (1959). Subspecies are in alphabetical order under the species. Remarks are given on natural history in each species account, if information is available. Discussion of subspecies known from the state is included. Under each subspecies, the citation to the original description is given with mention of type locality. Next is the citation to the first usage of the current name-combination. Then, synonyms are listed if there be such in the sense that original descriptions of the alleged species or subspecies had type localities in Tamaulipas.

Measurements, unless otherwise noted, are of adults and are given in millimeters. External measurements are in the following order: total length; length of tail vertebrae; length of hind foot; length of ear from notch. Capitalized color terms are those of Ridgway, Color Standards and Color Nomenclature, Washington, D. C., 1912. Capital letters designate teeth in the upper jaws and lower case letters designate teeth in the lower jaws; for example, M2 refers to the second upper molar and m2 refers to the second lower molar.

The localities of specimens examined and additional records are listed from north to south and their geographic positions can be found in the gazetteer and on the map (Fig. 4).

Most of the specimens examined are in the Museum of Natural History of the University of Kansas. Unless otherwise indicated, catalogue numbers relate to that collection. A few specimens from other collections were seen. Abbreviations identifying those collections are: UMMZ, the University of Michigan Museum of Zoology; AMNH, the American Museum of Natural History; and GMS, George M. Sutton collection (University of Oklahoma).

I am grateful to Prof. E. Raymond Hall and Dr. J. Knox Jones, Jr., for their advice and kind help that have enabled me to complete this work. I thank Dr. William E. Duellman for his advice concerning Zoogeography and Biologist Gastón Guzmán for help with the names of plants. For the loan of specimens I am grateful to Dr. George M. Sutton of the University of Oklahoma, to Dr. David H. Johnson and Dr. Richard H. Manville of the United States National Museum, to Drs. William H. Burt and Emmet T. Hooper of the University of Michigan Museum of Zoology, and to Dr. Richard Van Gelder of the American Museum of Natural History. I thank, also, Dr. William Z. Lidicker, Jr., for information about the locality called Lulú, and the collectors from the Museum of Natural History, especially Gerd H. Heinrich, William J. Schaldach, Percy L. Clifton, and John H. Bodley. I am grateful also to Charles A. Long and to several other persons, not named here, who helped me in some way to complete my study of the mammals of Tamaulipas.

Most of the field work was financed by the Kansas University Endowment Association. Some laboratory work was done when the author was half-time Research Assistant under Grant No. 56 G 103 from the National Science Foundation.

GAZETTEER

The specimens examined and additional records are listed with reference to the following place names. The geographic position of each was taken from the maps of the American Geographical Society of New York, scale 1:1,000,000, and the Atlas Geográfico de la República Mexicana, scale 1:500,000.

Acuña.—23°26′, 98°25′.
Agua Linda.—23°05′, 99°14′.
Aldama.—22°55′, 98°04′.
Alta Cima.—23°05′, 99°11′.
Altamira.—22°23′, 97°56′.
Antiguo Morelos.—22°33′, 99°05′.
Aserradero del Infernillo [Infiernillo]. —23°04′, 99°13′.
Aserradero del Pariso.—22°59′, 99°15′.
Bagdad.—25°57′, 97°09′.
Camargo.—26°20′, 98°50′.
Cerro del Tigre.—23°04′, 99°17′.
Chamal, 22°49′, 99°14′.
Charco Escondido.—25°46′, 98°22′.
Ciudad Victoria.—23°45′, 99°07′.
Cueva de Quintero.—22°39′, 99°02′.
Cueva La Esperanza.—23°55′, 99°17′.
Cueva La Mula.—see La Mula.
Cueva Los Troncones.—23°49′, 99°15′.
Cues.—22°58′, 98°13′.
Ejido Santa Isabel.—23°14′, 99°00′.
El Carrizo.—23°15′, 99°05′.
El Encino.—23°08′, 99°07′.
El Mante (Cd. Mante).—22°45′, 99°01′
El Mulato.—24°54′, 98°57′.
El Pachón.—22°36′, 99°03′.
Forlón.—23°14′, 98°49′.
Gómez Farías.—23°02′, 99°10′.
Guemes.—23°55′, 99° 00′.
Guerrero.—26°48′, 99°20′.
Hacienda Santa Engracia.—24°02′, 99°12′.
Hidalgo.—24°15′, 99°26′.
Jaumave.—23°24′, 99°23′.
Joya de Salas.—23°11′, 99°17′.
Joya Verde.—23°35′, 99°14′.
La Azteca (Ejido).—23°05′, 99°08′.
La Mula.—23°36′, 99°17′.
La Pesca.—23°47′, 97°48′.
La Purisima.—24°18′, 99°28′.
La Vegonia.—24°40′, 99°05′.
Limón.—22°49′, 99°00′.
Marmolejo.—24°38′, 99°00′.
Matamoros.—25°55′, 97°30′.
Mesa de Llera.—23°20′, 99°01′.
Mier.—26°27′, 99°09′.
Miquihuana.—23°27′, 99°46′.
Nicolás.—23°21′, 100°04′.
Nuevo Laredo.—27°30′, 99°30′.
Ocampo.—22°50′, 99°21′.
Ojo de Agua.—22°35′, 98°58′.
Padilla.—24°01′, 98°46′.
Palmillas.—23°18′, 99°33′.
Piedra.—23°30′, 98°06′.
Rancho del Cielo.—23°04′, 99°12′.
Rancho Pano Ayuctle.—23°07′, 99°13′.
Rancho Santa Rosa.—23°58′, 99°16′.
Rancho Tigre.—22°54′, 99°20′.
Rancho Viejo.—23°02′, 99°13′.
Reynosa.—26°06′, 98°15′.
Río Bravo (Town).—26°04′, 98°08′.
Río Corono [Corona].—23°50′, 98°50′.
San Antonio.—23°08′, 99°23′.
San Carlos.—24°35′, 98°57′.
San Fernando.—24°51′, 98°09′.
San José.—24°41′, 99°06′.
San Miguel.—24°45′, 99°05′.
Santa Maria.—23°31′, 98°41′.
Santa Teresa.—25°27′, 97°29′.
Savinito.—(?)23°43′, 98°51′.
Soto la Marina.—23°46′, 98°15′.
Tajada.—23°16′, 99°55′.
Tamaulipeca.—24°45′, 99°05′.
Tampico.—22°12′, 97°51′.
Tula.—23°00′, 99°42′.
Villagran.—24°29′, 99°29′.
Villa Mainero.—24°34′, 99°36′.
Washington Beach.—25°53′, 97°09′.
Xicotencatl.—23°00′, 98°57′.
Zamorina.—23°20′, 97°58′.

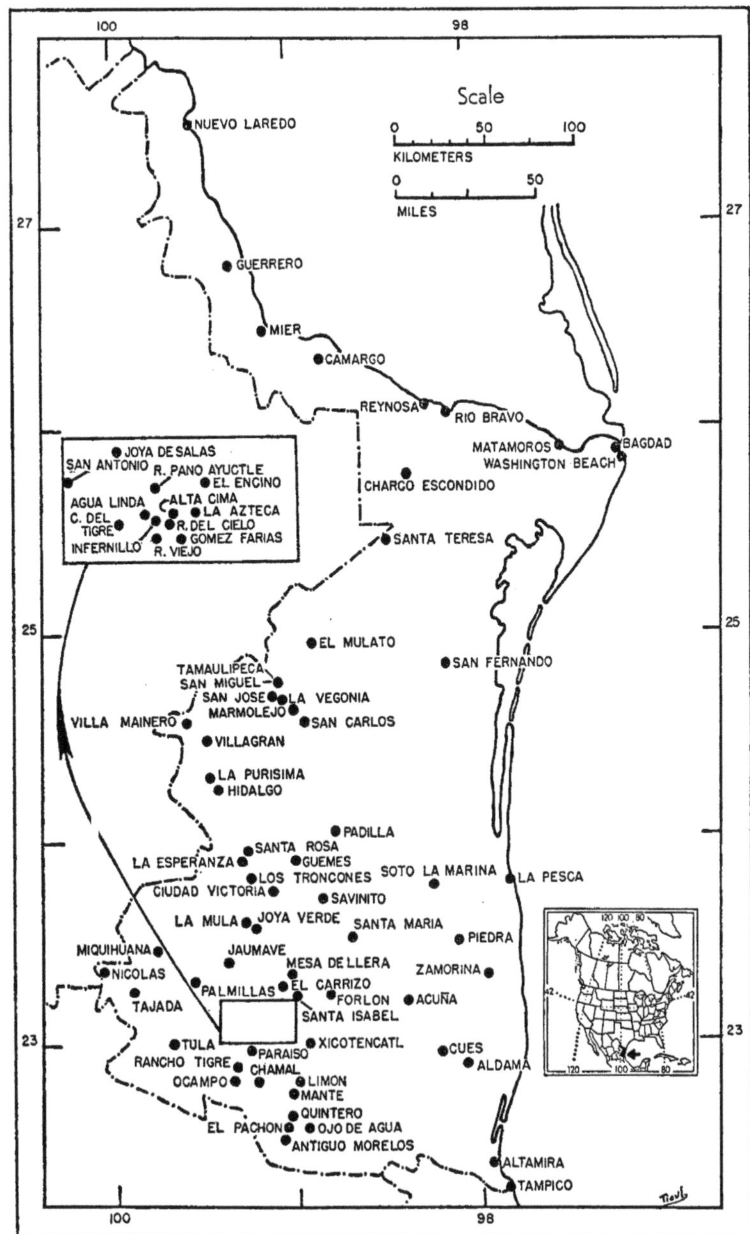

FIG. 4. Place names, in Tamaulipas, mentioned in text.

CHECK-LIST

The 146 kinds of native mammals of 120 species found in Tamaulipas belong to 72 genera of 25 families of 10 orders. Non-native mammals introduced by man are not included.

Class MAMMALIA

Order MARSUPIALIA

Family Didelphidae PAGE

Didelphis marsupialis californicus Bennett 393
Didelphis marsupialis texensis J. A. Allen 394
Philander opossum pallidus (J. A. Allen) 394
Marmosa mexicana mexicana Merriam 395

Order INSECTIVORA

Family Soricidae

Sorex saussurei saussurei Merriam 396
Cryptotis parva berlandieri (Baird) 396
Cryptotis pergracilis pueblensis Jackson 396
Cryptotis mexicana madrea Goodwin 396
Notiosorex crawfordi (Goues) 397

Family Talpidae

Scalopus inflatus Jackson 397

Order CHIROPTERA

Family Phyllostomatidae

Pteronotus rubiginosus mexicana (Miller) 398
Pteronotus davyi fulvus (Thomas) 398
Choeronycteris mexicana Tschudi 399
Mormoops megalophylla megalophylla (Peters) 399
Micronycteris megalotis mexicana Miller 400
Glossophaga sorocina leachii (Gray) 400
Leptonycteris nivalis nivalis (Saussure) 401
Sturnira lilium parvidens Goldman 401
Artibeus jamaicensis jamaicensis Leach 402
Artibeus lituratus palmarum Allen and Chapman 402
Artibeus toltecus (Saussure) 403
Artibeus aztecus Andersen 403
Enchistenes hartii (Thomas) 404
Centurio senex Gray 404

Family Desmodontidae

Desmodus rotundus murinus Wagner 405
Diphylla ecaudata Spix 406

Family Natalidae

Natalus stramineus saturatus Dalquest and Hall 407

Family Vespertilionidae PAGE
Myotis velifer incautus (J. A. Allen) 407
Myotis keenii auriculus Baker and Stains 408
Myotis californicus mexicanus (Saussure) 408
Myotis nigricans dalquesti Hall and Alvarez 409
Pipistrellus subflavus subflavus (F. Cuvier) 409
Pipistrellus hesperus potosinus Dalquest 410
Eptesicus fuscus miradorensis (H. Allen) 410
Lasiurus borealis borealis (Müller) 411
Lasiurus borealis teliotis (H. Allen) 412
Lasiurus cinereus cinereus (Palisot and Beauvois) 412
Lasiurus intermedius intermedius H. Allen 412
Lasiurus ega xanthinus (Thomas) 413
Nycticeus humeralis humeralis (Rafinesque) 413
Nycticeus humeralis mexicanus Davis 413
Rhogeëssa tumida tumida H. Allen 414
Plecotus phyllotis (G. M. Allen) 415
Antrozous pallidus pallidus (Le Conte) 415

Family Molossidae
Tadarida brasiliensis mexicana (Saussure) 415
Tadarida aurispinosa (Peale) 415
Tadarida laticaudata ferruginea Goodwin 416
Molossus ater nigricans Miller 417

Order PRIMATES

Family Cebidae
Ateles geoffroyi velerosus Gray 417

Order EDENTATA

Family Dasypodidae
Dasypus novemcinctus mexicanus Peters 418

Order LAGOMORPHA

Family Leporidae
Sylvilagus brasiliensis truei (J. A. Allen) 418
Sylvilagus audubonii parvulus (J. A. Allen) 418
Sylvilagus floridanus chapmani (J. A. Allen) 419
Sylvilagus floridanus connectens (Nelson) 419
Lepus californicus altamirae Nelson 420
Lepus californicus curti Hall 420
Lepus californicus merriami Mearns 421

Order RODENTIA

Family Sciuridae
Spermophilus mexicanus parvidens Mearns 421
Spermophilus spilosoma oricolus Alvarez 422
Spermophilus variegatus couchii Baird 422

PAGE

Sciurus aureogaster aureogaster Cuvier 423
Sciurus deppei negligens Nelson 424
Sciurus alleni Nelson 424
Glaucomys volans herreranus Goldman 425

Family Geomyidae

Geomys personatus personatus True 425
Geomys tropicalis Goldman 426
Heterogeomys hispidus negatus Goodwin 427
Cratogeomys castanops planifrons Nelson and Goldman 428
Cratogeomys castanops tamaulipensis Nelson and Goldman 428

Family Heteromyidae

Perognathus merriami merriami J. A. Allen 429
Perognathus hispidus hispidus Baird 429
Perognathus nelsoni nelsoni Merriam 430
Dipodomys ordii durranti Setzer 431
Dipodomys ordii parvabullatus Hall 431
Dipodomys ordii compactus True 431
Dipodomys merriami atronasus Merriam 432
Liomys irroratus alleni (Coues) 433
Liomys irroratus texensis Merriam 433

Family Castoridae

Castor canadensis mexicanus V. Bailey 434

Family Cricetidae

Oryzomys palustris aquaticus J. A. Allen 435
Oryzomys palustris peragrus Merriam 435
Oryzomys melanotis carrorum Lawrence 436
Oryzomys melanotis rostratus Merriam 437
Oryzomys alfaroi huastecae Dalquest 437
Oryzomys fulvescens fulvescens (Saussure) 438
Oryzomys fulvescens engracie Osgood 438
Reithrodontomys megalotis hooperi Goodwin 438
Reithrodontomys fulvescens griseoflavus Merriam 438
Reithrodontomys fulvescens intermadius J. A. Allen 439
Reithrodontomys fulvescens tropicalis Davis 439
Reithrodontomys mexicanus mexicanus (Saussure) 440
Peromyscus maniculatus blandus Osgood 440
Peromyscus melanotis J. A. Allen and Chapman 440
Peromyscus leucopus texanus (Woodhouse) 441
Peromyscus boylii ambiguus Alvarez 443
Peromyscus boylii levipes Merriam 443
Peromyscus pectoralis collinus Hooper 444
Peromyscus pectoralis eremicoides Osgood 445
Peromyscus melanophrys consobrinus Osgood 445
Peromyscus difficilis petricola Hoffmeister and de la Torre 446
Peromyscus ochraventer Baker 446
Baiomys taylori taylori (Thomas) 447

PAGE

Onychomys leucogaster longipes Merriam 447
Onychomys torridus subrufus Hollister 448
Sigmodon hispidus berlandieri Baird 449
Sigmodon hispidus solus Hall 450
Sigmodon hispidus toltecus (Saussure) 450
Neotoma albigula subsolana Alvarez 450
Neotoma angustapalata Baker 451
Neotoma micropus littoralis Goldman 453
Neotoma micropus micropus Baird 453
Microtus mexicanus subsimus Goldman 454

Order CARNIVORA

Family Canidae

Canis latrans microdon Merriam 454
Canis latrans texensis V. Bailey 455
Canis lupus monstrabilis Goldman 455
Urocyon cinereoargenteus scottii Mearns 455

Family Ursidae

Ursus americanus eremicus Merriam 456

Family Procyonidae

Bassariscus astutus flavus Rhoads 456
Procyon lotor fuscipes Mearns 457
Procyon lotor hernandezii Wagler 457
Nasua narica molaris Merriam 458
Potos flavus aztecus Thomas 458

Family Mustelidae

Mustela frenata frenata Lichtenstein 458
Mustela frenata tropicalis (Merriam) 459
Eira barbara senex (Thomas) 459
Taxidea taxus berlandieri Baird 460
Taxidea taxus littoralis Schantz 460
Spilogale putorius interrupta (Rafinesque) 461
Mephitis mephitis varians Gray 461
Mephitis macroura macroura Lichtenstein 461
Conepatus mesoleucus mearnsi Merriam 462
Conepatus leuconotus texensis Merriam 462

Family Felidae

Felis concolor stanleyana Goldman 462
Felis onca veraecrucis Nelson and Goldman 463
Felis pardalis albescens Pucheran 463
Felis wiedii oaxacensis Nelson and Goldman 464
Felis yagouaroundi cacomitli Berlandier 464
Lynx rufus texensis J. A. Allen 464

Order SIRENIA

Family Trichechidae

Trichechus manatus latirostris (Harlan) 465

Order ARTIODACTYLA

Family Tayassuidae

PAGE

Tayassu tajacu angulatus (Cope) 465

Family Cervidae

Odocoileus hemionus crooki (Mearns) 465

Odocoileus virginianus miquihuanensis Goldman and Kellogg 466

Odocoileus virginianus texanus (Mearns) 466

Odocoileus virginianus veraecrucis Goldman and Kellogg 466

Mazama americana temama (Kerr) 466

Family Antilocapridae

Antilocapra americana mexicana Merriam 467

ACCOUNTS OF SPECIES AND SUBSPECIES

Didelphis marsupialis

Opossum

The opossum occurs throughout Tamaulipas but is commonest in the south, especially in the areas of tropical forest and along water courses. Most of the specimens examined were caught in steel traps baited with remains of small animals (mostly mammals and birds, but one trap was baited with the head of a black bass). At Villa Mainero five individuals were caught in one night in five of seven traps scented with spilogale musk. These traps were set in runways along a thick thorn-brush fence, which separated a cornfield from thorn-brush desert. Along the Río Purificación 36 kilometers north and 10 kilometers west of Victoria an opossum was eaten in a trap by a small carnivore, probably a felid judging from tracks around the trap.

A female with 14 pouch young was taken in June in the Sierra de Tamaulipas and weighed 1350 grams; a March-taken female with nine small young in her pouch, from Soto la Marina, weighed 1800 grams. A male from the Sierra de Tamaulipas also weighed 1800 grams.

Didelphis marsupialis californica Bennett

1833. *Didelphis Californica* Bennett, Proc. Zool. Soc. London, p. 40, May 17, type locality restricted to Sonora by Hershkovitz (*infra*).

1951. *Didelphis marsupialis californica*, Hershkovitz Fieldiana-Zool., Chicago Nat. Hist. Mus., 31(47):548, July 10.

Distribution in Tamaulipas.—Southeastern part of state, north at least to Soto la Marina.

In studying Tamaulipan specimens, I was mindful that Hershkovitz (1951:550) regarded all opossums of this species in México as a single subspecies, even though J. A. Allen (1901) recognized two subspecies in the northeastern part of the Republic. According to Allen (p. 172), *D. m. texensis* (to which he ascribed a distribution in Texas and adjoining Tamaulipas) was described as: "Similar in coloration to *D. marsupialis* (*typica*) [*D. m. californica*], but with a relatively longer tail, longer nasals, usually terminating posteriorly in an acute angle, instead of being rounded or more or less abruptly truncated on the posterior border." The available material from Tamaulipas can be divided into two groups on the basis of shape and proportion of the nasals. In opossums from the southeast the nasals are truncate posteriorly and average 47.0 (45.1-48.4) per cent of the condylobasal length, whereas in specimens from elsewhere the nasals are acute posteriorly and average 50.7 (49.7-51.8) per cent of the condylobasal length. Tentatively, therefore, I follow Allen in recognizing two subspecies in northeastern México.

I note no especial difference in length of tail between *texensis* and *californica.* Hooper (1951:3) followed Hershkovitz in reporting as *californica* a specimen from Rancho del Cielo; to me, specimens from this area are referable to *texensis.*

One of the specimens from two miles south and 10 miles west of Piedra (54917) has a supernumerary tooth lingual and anterior to the last upper molar. The tooth is small (2.7 mm. long) and peglike.

Records of occurrence.—Specimens examined, 8: 3 mi. N Soto la Marina, 1; 2 mi. S, 10 mi. W Piedra, 12,000 ft., 7.

Additional records: Matamoros (Baird, 1858:234); Altamira (J. A. Allen, 1901:167).

Didelphis marsupialis texensis J. A. Allen

1901. *Didelphis marsupialis texensis* J. A. Allen, Bull. Amer. Mus. Hist., 14:172, June 15, type from Brownsville, Cameron County, Texas.

Distribution in Tamaulipas.—Northern, central and southwestern parts of state.

Records of occurrence.—Specimens examined, 7: San Fernando, 180 ft., 1; Villa Mainero, 1700 ft., 2; 36 km. N, 10 km. W Cd. Victoria (1 km. E El Barretal), on Río Purificación, 1; 12 km. N, 4 km. W Cd. Victoria, 1; Ejido Santa Isabel (12 km. S Llera), 2 km. W Pan-American Highway, 2000 ft., 1; 4 mi. N Jaumave, 2500 ft., 1.

Additional records: Matamoros (J. A. Allen, 1901:173); El Mulato, San Carlos Mts. (Dice, 1937:249); Rancho del Cielo (Hooper, 1953:3).

Philander opossum pallidus (J. A. Allen)
Four-eyed Opossum

1901. *Metachirus fuscogriseus pallidus* J. A. Allen, Bull. Amer. Mus. Nat. Hist., 14:215, July 3, type from Orizaba, Veracruz.

1955. *Philander opossum pallidus,* Miller and Kellogg, Bull. U. S. Nat. Mus., 205:8, March 3.

Distribution in Tamaulipas.—Known only from along eastern side of Sierra Madre Oriental, north to vicinity of La Purisima.

In Tamaulipas, the four-eyed opossum is seemingly common at relatively low elevations in the Tropical Deciduous Forest along the eastern side of the Sierra Madre Oriental, but the species is not restricted to this area as one specimen is available from a place seven kilometers southwest of La Purisima, in the drier forest of west-central Tamaulipas. The highest elevation at which individuals have been taken in the state is approximately 2500 feet.

Specimens obtained two kilometers west of El Carrizo were caught in steel traps that were baited with the bodies of small birds and mammals and that were set in trails leading through a fence of piled logs that separated a cornfield from adjacent forest. At Rancho Pano Ayuctle, some individuals were trapped in steel sets baited with scraps of meat; others were shot at night in the forest along the Río Sabinas. Schaldach reported in his notes that four-eyed opossums robbed trap lines set for small mammals at Rancho Pano Ayuctle. W. W. Dalquest trapped an individual seven kilometers southwest of La Purisima using the body of an armadillo as bait. The natives of southern Tamaulipas refer to this animal as "tlacuache cuatrojos."

Tamaulipan specimens of P. *o. pallidus* differ from topotypes and other specimens from the vicinity of the type locality in averaging somewhat paler dorsally and slightly smaller in cranial dimensions when specimens of equal age are compared. They differ also in having a longer terminal area of white on the tail, 53.1 per cent (43.3-62.8) of the length of the tail in 13 specimens from Tamaulipas, and 38.7 (30.9-48.2) per cent in 14 specimens from the vicinity of the type locality of *pallidus* in Veracruz; specimens from northern Veracruz are intermediate between the two mentioned populations in amount of white on the tail. Baker (1951:210) noted that the specimens from two kilometers west of El Carrizo had "proportionately longer tails than typical P. *o. pallidus* from central Veracruz," but I do not find this character to be consistent in the more abundant material now available.

Measurements.—External and cranial measurements of three adults, a male and female from Rancho Pano Ayuctle and a male from two kilometers west of El Carrizo, respectively, are as follows: 577, 580, 568; 294, 288, 290; 46, 43, 43; 40, 42, 37; condylobasal length, —, 70.1, 69.9; palatal length, 43.2, 42.3, 41.9; lambdoidal breadth, 23.6, 22.0, 22.7; alveolar length of maxillary toothrow, 29.5, 28.4, 29.0.

Records of occurrence.—Specimens examined, 15: 7 km. SW La Purisima, 1; Rancho Pano Ayuctle, 6 mi. N Gómez Farías, 300 ft., 1; Rancho Pano Ayuctle, 25 mi. N Mante and 3 km. W Pan-American Highway, 300 ft., 7; 10 km. N, 8 km. W El Encino, 400 ft., 3; 2 km. W El Carrizo, 2500 ft., 3 (one specimen deposited in Instituto de Biología, México).

Marmosa mexicana mexicana Merriam

Mexican Mouse-opossum

1897. *Marmosa murina mexicana* Merriam, Proc. Biol. Soc. Washington, 11:44, March 16, type from Juquila, 1500 m., Oaxaca.

1902. *Marmosa mexicana,* Bangs, Bull. Mus. Comp. Zool., 39:19, April.

Distribution in Tamaulipas.—Known only from Aserradero del Infernillo (Goodwin, 1954:3) in southwestern part of state.

Marmosa has been reported from Tamaulipas only by Goodwin (1954:3), who examined "15 rami, and one fragment of maxillary" that were found in a cave. Possibly they were remains from owl pellets.

Sorex saussurei saussurei Merriam

Saussure's Shrew

1892. *Sorex saussurei* Merriam, Proc. Biol. Soc. Washington, 7:173, September 29, type from N slope Sierra Nevada de Colima, approximately 8000 ft., Jalisco.

Distribution in Tamaulipas.—Known only from Miquihuana.

Jackson (1928:156) reported four specimens from Miquihuana, which he incorrectly located in Nuevo León.

Cryptotis parva berlandieri (Baird)

Least Shrew

1858. *Blarina berlandieri* Baird, Mammals, *in* Repts. Expl. Surv. . . ., 8(1):53, July 14, type from Matamoros, Tamaulipas.

1941. *Cryptotis parva berlandieri*, Davis, Jour. Mamm., 22:413, November 13.

Distribution in Tamaulipas.—Throughout state.

A female taken on July 5, one mile south of Altamira, carried three embryos 5 mm. in crown-rump length. A female from the same locality and another taken on June 6 in the Sierra de Tamaulipas were lactating. Weight of each of six males was 5.0 grams.

Records of occurrence.—Specimens examined, 9: Sierra de Tamaulipas, 10 mi. W, 2 mi. S Piedra, 1200 ft., 1; 1 mi. S Altamira, 8.

Additional records: Matamoros (Baird, 1858:53); 9 km. N Rancho Tigre (Goodwin, 1954:3).

Cryptotis pergracilis pueblensis Jackson

Slender Small-eared Shrew

1933. *Cryptotis pergracilis pueblensis* Jackson, Proc. Biol. Soc. Washington, 46:79, April 27, type from Huachinango, 5000 ft., Puebla.

Distribution in Tamaulipas.—Known only from Aserradero del Paraiso.

The only report from Tamaulipas of this small shrew is that of Goodwin (1954:3) who listed a cranium and mandible, possibly of the same individual, found on the floor of a cave. Goodwin referred the remains to *pueblensis* because of the "noticeably broader and heavier rostrum than in . . . *C. parva berlandieri* from Rancho Tigre."

Cryptotis mexicana madrea Goodwin

Mexican Small-eared Shrew

1954. *Cryptotis mexicana madrea* Goodwin, Amer. Mus. Novit., 1670:1, June 28, type from Rancho del Cielo, 5 mi. NW Gómez Farías, 3500 ft., Tamaulipas.

Distribution in Tamaulipas.—Known only from the type locality and vicinity thereof.

This subspecies is known only from two complete specimens, six crania and four rami collected in two different localities—the type locality and Aserradero del Infernillo, only seven kilometers from the type locality. All the specimens were examined and reported by Goodwin (1954:1; 1954:4). The type specimen "was taken in a low section of an overgrown ditch" and the other complete specimen was trapped in a stone wall that separated an orchard from a pasture. The six skulls were found in owl pellets.

Notiosorex crawfordi (Coues)

Crawford's Desert Shrew

1877. *Sorex (Notiosorex) crawfordi* Coues, Bull. U. S. Geol. and Geog. Surv. Territories, 3:651, May 15, type from near old Fort Bliss, approximately 2 mi. above El Paso, El Paso Co., Texas.

1895. *Notiosorex crawfordi*, Merriam, N. Amer. Fauna, 10:32, Dec. 31.

Distribution in Tamaulipas.—Known only from two localities in southwestern part of state.

The two specimens examined were collected in July, one in tropical forest and the other in pine-oak forest; each was a lactating female and each weighed 5 grams.

Judging from Merriam's (1895:32) description, the two females differ from the type and three specimens from San Diego, Texas, in having a unicolored tail and in being slightly larger externally. When more abundant material is available the *Notiosorex crawfordi* of northeastern México probably will be found to represent a new subspecies; for the present I follow Findley (1955:616) in referring Tamaulipan specimens to *N. crawfordi*.

Measurements.—External measurements of the specimens from Jaumave and Palmillas, respectively: 90, 90; 28, 31; 11, 11.5; 8, 8. For cranial measurements see Findley (1955:32).

Records of occurrence.—Specimens examined, 2: Jaumave, 2400 ft., 1; Palmillas, 4400 ft., 1.

Scalopus inflatus Jackson

Tamaulipan Mole

1914. *Scalopus inflatus* Jackson, Proc. Biol. Soc. Washington, 27:21, February 2, type from Tamaulipas, 45 miles from Brownsville, Texas.

Distribution in Tamaulipas.—Known only from the type locality.

Scalopus inflatus is known only from the type specimen, which is imperfect and lacks complete data according to Jackson (1914:21). The type locality is in Tamaulipas, 45 miles from Brownsville, Texas, but the exact direction from Brownsville is unknown; probably the locality was on the road between that town and San Fernando, Tamaulipas, which is south-southwest of Brownsville.

Pteronotus rubiginosus mexicanus (Miller)

Mustached Bat

1902. *Chilonycteris mexicana* Miller, Proc. Acad. Nat. Sci. Philadelphia, 54:401, September 12, type from San Blas, Nayarit.

Distribution in Tamaulipas.—Southern part of state in areas of tropical forest.

Most individuals of this species were taken in mist nets. Northwest of El Encino for example, bats were collected from a net placed in "a strategic position across a narrow opening" (Schaldach, fieldnotes) in a cave near the headwaters of the Río Sabinas; along the same river at Rancho Pano Ayuctle some were taken in a net stretched across a little creek (arroyo). In the cave near El Encino the collector (Schaldach) estimated the population of P. *rubiginosus* at between two and three hundred; at Ojo de Agua this bat was found in the deepest part of a cave in association with *Myotis nigricans.*

Two June-taken females from the Sierra de Tamaulipas were lactating, and weighed 17 and 18 grams.

The generic name *Pteronotus* is employed instead of *Chilonycteris* following Burt and Stirton (1961:24-25). The specific name *rubiginosus* is used in accordance with de la Torre (1955:696). Tamaulipan specimens are assigned to P. *r. mexicana* because they do not differ from specimens of that subspecies from Nayarit, except that the coloration of Tamaulipan specimens averages slightly darker in both color phases.

Specimens of this subspecies from the Sierra de Tamulipas, previously recorded by Anderson (1956:349), are the northernmost reported in eastern México.

Records of occurrence.—Specimens examined, 31: Sierra de Tamaulipas, 2 mi. S, 10 mi. W Piedra, 1200 ft., 1; Sierra de Tamaulipas, 3 mi. S, 10 mi. W Piedra, 1400 ft., 3; Rancho Pano Ayuctle, 25 mi. N El Mante, 3 mi. W Pan-American Highway, 300 ft., 3; Ojo de Agua, 20 mi. N El Mante, and 3 km. W Pan-American Highway, 300 ft., 2; 10 km. N, 8 km. W El Encino, 400 ft., 22.

Additional records (Goodwin, 1954:4): Aserradero del Paraiso; El Pachón.

Pteronotus davyi fulvus (Thomas)

Davy's Naked-backed Bat

1892. *Chilonycteris davyi fulvus* Thomas, Ann. Mag. Nat. Hist., ser. 6, 10:410, November, type from Las Peñas, Jalisco.

1912. *Pteronotus davyi fulvus*, Miller, Bull. U. S. Nat. Mus., 79:33, December 31.

Distribution in Tamaulipas.—Known only from the two localities reported in this paper.

According to field-notes of Schaldach *et al.*, individuals of P. *d. fulvus* appear when it is almost dark (about 6:30 p. m. in December and January), ordinarily fly about 25 feet above the ground, but occasionally are seen at heights of between 60 and 70 feet (near tops of the largest cypress trees). Most bats flew in a straight line for 10 to 20 yards, then zig-zagged, and repeated the same movements. All specimens examined are in the brown color phase.

Records of occurrence.—Specimens examined, 11: Rancho Santa Rosa, 25 km. N, 13 km. W Cd. Victoria, 260 m., 10; Rancho Pano Ayuctle, 6 mi. N Gómez Farías, 300 ft., 1.

Choeronycteris mexicana Tschudi

Mexican Long-tongued Bat

1844. *Choeronycteris mexicana* Tschudi, Untersuchungen über die fauna Peruana . . ., p. 72, type from México.

Distribution in Tamaulipas.—East side of Sierra Madre in southwestern part of state.

Specimens from La Mula were obtained in a small cave, which was inhabited also by *Desmodus rotundus* and *Tadarida brasiliensis*. The specimens from Miquihuana were captured in a mine by a native. Those from four kilometers north of Joya Verde also were taken from a mine. Females obtained in August at La Mula were lactating.

Specimens examined are indistinguishable from *C. mexicana* from Oaxaca and Jalisco. Baker (1956:172) found no differences between Coahuilan and Tamaulipan specimens. Most Tamaulipan specimens are dark grayish, but some are brownish and some are intermediate between the two colors mentioned. Fourteen adults weighed an average of 16.0 (12-18) grams.

Records of occurrence.—Specimens examined, 19: 4 km. N Joya Verde, 4000 ft., 3; La Mula, 13 mi. N Jaumave, 4; Cueva La Mula, 10 km. W Joya Verde, 2400 ft., 2; Miquihuana, 6500 ft., 10.

Mormoops megalophylla megalophylla (Peters)

Peters' Leaf-chinned Bat

1864. *Mormops megalophylla* Peters, Monatsb. preuss. Akad. Wiss., Berlin, p. 381, type from southern México.

Distribution in Tamaulipas.—Throughout state, except possibly west of the Sierra Madre Oriental.

Specimens from the Sierra de Tamaulipas were taken in mist nets in which *Pteronotus rubiginosus, Lasiurus borealis,* or *Centurio senex* also were captured. The specimen from Rancho Santa Rosa was shot as it flew at a height of six feet.

Tamaulipan specimens of *Mormoops megalophylla* are here assigned to *M. m. megalophylla* instead of to *M. m. senicula* following Villa and Jimenez (1961:503), who regarded *senicula* as indistinguishable from *megalophylla.*

Weight of four specimens from the Sierra de Tamaulipas averaged 16.2 (15-18) grams.

Records of occurrence.—Specimens examined, 5: Sierra de Tamaulipas, 3 mi. S, 16 mi. W Piedra, 1300 ft., 2; Sierra de Tamaulipas, 3 mi. S, 14 mi. W Piedra, 1400 ft., 1; Sierra de Tamaulipas, 3 mi. S, 10 mi. W Piedra, 1400 ft., 1; Rancho Santa Rosa, 25 km. N, 13 km. W Cd. Victoria, 260 m., 1.

Additional records: Cueva de Los Troncones, 7.5 km. NNW, 3.5 km. S Cd. Victoria (Villa and Jimenez, 1961:503); Cueva de Quintero, 15 km. SSW Cd. Mante (*ibid*); Tampico (Davis and Carter, 1962:67).

Micronycteris megalotis mexicana Miller

Brazilian Small-eared Bat

1898. *Micronycteris megalotis mexicana* Miller, Proc. Acad. Nat. Sci. Philadelphia, 50:329, August 2, type from Platanar, Jalisco.

Distribution in Tamaulipas.—Known only from Rancho Pano Ayuctle (Goodwin, 1954:4). The single specimen of this species presently known from Tamaulipas was shot while it was roosting in a ranch house.

Glossophaga soricina leachii (Gray)

Pallas' Long-tongued Bat

1844. *Monophyllus leachii* Gray, *in* The zoology of the voyage of H. M. S. Sulphur . . ., 1 (1, Mamm.): 18, April, type from Realego, Chinandega, Nicaragua.

1913. *Glossophaga soricina leachii*, Miller, Proc. U. S. Nat. Mus., 46:419, December 31.

Distribution in Tamaulipas.—Tropical region of southern part of state.

Specimens from the Sierra de Tamaulipas were taken in a cave along with *Desmodus rotundus* and *Tadarida laticaudata.* Specimens from 20 miles north of El Mante were collected from a cave about 50 yards deep. Weights of two females from the Sierra de Tamaulipas were 9 and 12 grams. Tamaulipan specimens examined do not differ from specimens from Nicaragua that were used in comparison.

Records of occurrence.—Specimens examined, 6: Sierra de Tamaulipas, 3 mi. S, 16 mi. W Piedra, 1400 ft., 2; 10 km. N, 8 km. W El Encino, 400 ft., 1; Ojo de Agua, 20 mi. N El Mante, and 3 km. W Highway, 300 ft., 2; 8 km. NE Antiguo Morelos, 500 ft., 1.

Additional records: 5 mi. NE Antiguo Morelos, near El Pachón (de la Torre, 1954:114); Altamira (Miller, 1913:420).

Leptonycteris nivalis nivalis (Saussure)

Long-nosed Bat

1860. *M.* [=*Ischnoglossa*] *nivalis* Saussure, Revue et Mag. Zool., Paris, ser. 2, 12:492, November, type from near snow line of Mt. Orizaba, Veracruz.

1900. *Leptonycteris nivalis*, Miller, Proc. Biol. Soc. Washington, 13:126, April 6.

Distribution in Tamaulipas.—Probably throughout southern part of state, but presently known only from one locality.

The specimens herein reported were taken in a cave. They provide the first record of the species from Tamaulipas and are assigned to the subspecies *nivalis* on the basis of their brownish color and small size in comparison with specimens of *L. n. longala* from Coahuila (see also description and measurements of *longala* given by Stains, 1957:356). None of the specimens suggests intergradation in color between *nivalis* and *longala,* but some are slightly larger than specimens of the former from Veracruz.

Twelve females taken on August 27, 1961, were pregnant. Each carried a single embryo, the embryos averaging 15.7 (12-20) mm. in crown-rump length. The average weight of the 12 females was 26.9 (24.5-30.0) grams; 10 males weighed an average of 24.6 (21-28) grams.

Measurements.—Average and extremes of ten specimens (5 males and 5 females) are as follows: 78.2 (76-80); 0.0; 16.4 (15-17); 16.7 (16-19); length of forearm, 48.4 (45.2-54.3); length of third finger, 100.8 (99.2-103.7); greatest length of skull, 26.8 (25.9-27.6); zygomatic breadth (6 only), 10.9 (10.7-11.1); least interorbital constriction, 4.6 (4.5-4.9); mastoid breadth, 10.8 (10.5-11.2); length of maxillary tooth-row, 8.7 (8.4-9.0).

Records of occurrence.—Specimens examined, 28: all from 6.5 mi. N, 13 mi. W Jimenez, 1250 ft.

Sturnira lilium parvidens Goldman

Yellow-shouldered Bat

1917. *Sturnira lilium parvidens* Goldman, Proc. Biol. Soc. Washington, 30:116, May 23, type from Papayo, about 25 mi. NW Acapulco, Guerrero.

Distribution in Tamaulipas.—Known presently only from Rancho Pano Ayuctle.

The two specimens from Tamaulipas were reported by de la Torre (1954:114) and in eastern México are the northernmost yet reported of the genus.

Artibeus jamaicensis jamaicensis Leach

Jamaican Fruit-eating Bat

1821. *Artibeus Jamaicensis* Leach, Trans. Linn. Soc. London, 13:75, type from Jamaica.

Distribution in Tamaulipas.—Tropical region of southern part of state.

The specimens from northwest of El Encino were shot deep (250 yards) in a cave; specimens of *Myotis nigricans* were obtained in the same cave. A female taken on May 24 carried a single embryo that was 43 mm. in crown-rump length. Six March-taken females reported by de la Torre (1954:114) had one embryo each that varied from 20 to 38 mm. in length.

Artibeus jamaicensis and *A. lituratus* are the largest bats known from Tamaulipas. In addition to the differences between the two species pointed out by Lukens and Davis (1957:9), I note, in Tamaulipas at least, that the postorbital constriction is narrower in relation to the condylobasal length in *lituratus*, 24.6 (23.7-26.0) per cent as compared to 27.9 (26.7-29.9) per cent in *jamaicensis*.

Records of occurrence.—Specimens examined, 19: 10 km. N, 8 km. W El Encino, 400 ft., 10; Aserradero del Paraiso, 19 km. N Chamal (by road), 8 (AMNH); Cueva El Pachón, 5 mi. N Antiguo Morelos, 1 (AMNH).

Additional records: Rancho Pano Ayuctle (de la Torre, 1954:114); 4 mi. N Antiguo Morelos, near El Pachón (*ibid.*).

Artibeus lituratus palmarum J. A. Allen and Chapman

Big Fruit-eating Bat

1897. *Artibeus palmarum* J. A. Allen and Chapman, Bull. Amer. Mus. Nat. Hist., 9:16, February 26, type from Botanical Gardens at Port of Spain, Trinidad.

1949. *A[rtibeus]. l[ituratus]. palmarum*, Hershkovitz, Proc. U. S. Nat. Mus., 99:447, May 10.

Distribution in Tamaulipas.—Tropical region in southern part of state.

Two specimens from the Río Sabinas were taken in a mist net placed across the small, crevicelike entrance to a cave. Ten pregnant females taken in late May each contained a single embryo; average crown-rump length of the 10 embyos was 43 (35-55) mm.

Tamaulipan specimens of *lituratus* do not differ appreciably in color from topotypes except that the facial stripes are narrow and, in three individuals, poorly marked. Lukens and Davis (1957:9) reported that females from Guerrero were paler than the males, but the male examined in this study does not differ in color from the females seen.

Records of occurrence.—Specimens examined, 15: Rancho Pano Ayuctle, 6 mi. N Gómez Farías, 300 ft., 13; cave at headwaters of Río Sabinas, 10 km. N, 8 km. W El Encino, 400 ft., 2.

Artibeus toltecus (Saussure)

Toltec Fruit-eating Bat

1860. *Stenoderma toltecus* Saussure, Revue et Mag. Zool., Paris, ser. 2, 12:427, October, type from México. Type locality restricted to Mirador, Veracruz, by Hershkovitz, Proc. U. S. Nat. Mus., 99:449, May 10, 1949.

1908. *Artibeus toltecus*, Andersen, Proc. Zool. Soc. London, p. 296, April 7.

Distribution in Tamaulipas.—Probably lowlands of southern part of state; known presently only from Rancho Pano Ayuctle.

Artibeus toltecus is closely related to another species, *A. aztecus*, that occurs also in Tamaulipas. Externally, *toltecus* differs from *aztecus* in being smaller and darker; cranially, *toltecus* also is the smaller and the P2 and M2 are more angular lingually than in *aztecus*, in which the teeth are rounded. One of the most important differences between these two species is that they occur at different altitudes. Davis (1958:165) reported that *toltecus* occurred at elevations below 5000 feet at more southerly localities in México, whereas *aztecus* occurred above 5000 feet. In Tamaulipas the two species probably have parallel distributions from south to north but *A. toltecus* is known from Rancho Pano Ayuctle at an elevation of 300 feet in rain forest, whereas *A. aztecus* is known from Rancho del Cielo at an elevation of 3300 feet in cloud forest. The two localities are only four miles apart.

One of the specimens examined (GMS 10640) is smaller, cranially and externally (see beyond), than any recorded by Davis (1958: 165).

Measurements.—Some external and cranial measurements of two females and a male (GMS 10668, 10646 and 10640) are, respectively, as follows: length of hind foot, 12.5, 12.0, 11.0; length of ear from notch, 15, 17, 15; length of forearm, 40.5, 40.0, 36.5; greatest length of skull, 20.9, 20.7, 19.7; zygomatic breadth, 12.3, 12.3, 11.7; least interorbital constriction, 5.2, 5.0, 5.0; length of maxillary tooth-row, 6.8, 6.8, 6.5; breadth of braincase, 9.3, 9.2, 9.1.

Records of occurrence.—Specimens examined, 3 from Río Sabinas, near Gómez Farías (Rancho Pano Ayuctle) (GMS).

Artibeus aztecus Andersen

Aztec Fruit-eating Bat

1906. *Artibeus aztecus* Andersen, Ann. Mag. Nat. Hist., ser. 7, 18:422, December, type from Tetela del Volcán, Morelos.

Distribution in Tamaulipas.—Probably higher areas of southern part of state; known presently only from Rancho del Cielo.

I follow Davis (1958:165) in treating *A. aztecus* and *A. toltecus* as distinct species. Differences between the two are discussed in the preceding account of *toltecus*.

One specimen examined (AMNH 146980) is distinctly larger than the others here assigned to *A. aztecus*, but does not exceed the maximal measurements given by Davis (*loc. cit.*) for the species. This specimen also has a narrower M2, and relatively and actually narrower braincase than other specimens (see measurements).

Specimens from Rancho del Cielo were collected in a limestone cave in the cloud forest. A female taken on July 2 carried a small embryo and another obtained on August 14 had an embryo that appeared to be nearly ready for birth.

Measurements.—Respective external and cranial measurements of three males (AMNH, uncatalogued) and a female (AMNH 146980) are as follows: total length, 58, 65, 66, 73; length of hind foot, 13, 12, 12, 13; length of forearm, —, 43, 40, 41; greatest length of skull, 21.6, 22.4, 21.5, 23.0; zygomatic breadth, 13.0, 12.8, 13.0, 12.4; least interorbital constriction, 5.2, 5.7, 5.5, 6.0; length of maxillary tooth-row, 7.0, 7.1, 6.9, 7.1; breadth of braincase, 10.0, 9.8, 10.0, 9.5.

Records of occurrence.—Specimens examined, 7, all from Rancho del Cielo, 3300 ft., (AMNH).

Enchistenes hartii (Thomas)

Little Fruit-eating Bat

1892. *Artibeus hartii* Thomas, Ann. Mag. Nat. Hist., ser. 6, 10:409, November, type from Trinidad, Lesser Antilles.

1908. *Enchistenes hartii*, Andersen, Proc. Zool. Soc. London, 2:224, September 7.

Distribution in Tamaulipas.—Known only from Aserradero del Infernillo.

Enchistenes hartii is known from Tamaulipas only by the cranium reported by Goodwin (1954:5), and this is the northernmost known occurrence. The bat has not been reported from any other Mexican state bordering on the Gulf of Mexico.

Centurio senex Gray

Wrinkle-faced Bat

1842. *Centurio senex* Gray, Ann. Mag. Nat. Hist., ser. 10, 10:259, December, type locality erroneously given as Amboyna, East Indies; subsequently restricted to Realejo, Chinandega, Nicaragua, by Goodwin (Bull. Amer. Mus. Nat. Hist., 87:327, December 31, 1946).

Distribution in Tamaulipas.—Tropical areas of southern part of state.

The single specimen examined, a female weighing 23 grams that carried an embryo (17 mm. crown-rump length), was taken on June 14 in a mist net stretched between oak trees in the Sierra de Tamaulipas. One other female and one cranium have been reported from Tamaulipas.

The specimen examined differs from two seen from southern México (5 mi. SW Teapa, Tabasco, and 2 mi. S Tollosa, Oaxaca) in being brownish instead of grayish, but resembles in color two specimens from Cozumel Island, Quintana Roo.

Measurements.—A female from the Sierra de Tamaulipas affords the following measurements: Total length, 67; length of hind foot, 13; length of ear from notch, 15; length of forearm, 43.1; condylobasal length, 15.0; zygomatic breadth, 5.1; palatal length, 4.1; least interorbital constriction, 5.3; length of maxillary tooth-row, 5.1.

Records of occurrence.—Specimen examined, one from the Sierra de Tamaulipas, 3 mi. S, 14 mi. W Piedra, 1300 ft.

Additional records: Rancho Pano Ayuctle (de la Torre, 1954:114); Aserradero del Infernillo (Goodwin, 1954:5).

Desmodus rotundus murinus Wagner

Vampire

1840. *D[esmodus]. murinus* Wagner, *in* Schreber, Die Säugthiere . . ., Suppl., 1:337, type from México.

1912. *Desmodus rotundus murinus*, Osgood, Field Mus. Nat. Hist., Publ. 155, Zool. Ser., 10:63, January.

Distribution in Tamaulipas.—Southern part of state, north at least to vicinity of Jiménez.

Hall and Kelson (1959:151) listed a place 12 kilometers west and 8 kilometers north of Ciudad Victoria as the northernmost locality of record for *Desmodus*, but three specimens from Cueva La Esperanza, 6 kilometers southwest of Rancho Santa Rosa, are from a site slightly to the northwestward (12 mi.) of the locality first mentioned and a specimen from 13 miles west and six and a half miles north of Jiménez represents the northeasternmost known occurrence of *Desmodus* in eastern México.

Most of the vampires examined in this study were taken in caves; those from four miles southwest of Padilla were obtained from a hollow tree. Nine specimens were collected in a small cave 70 kilometers south of Ciudad Victoria on January 18, when water on the floor of the cave was frozen; the bats were congregated on the ceiling at a height of 20 feet. In a cave in the Sierra de Tamaulipas, 16 miles west and three miles south of Piedra, females and young were found some 50 yards from the entrance; *Natalus stramineus* and *Glossophaga soricina* were obtained from the same cave. In another cave only half a kilometer distant, 12 males were collected. In Cueva La Mula, *Desmodus* was found near the mouth, whereas *Choeronycteris mexicana* and two *Tadarida brasiliensis* were collected in the deepest part. At Cueva La Esperanza, 300 feet deep and on the east side of the Sierra Madre Oriental, four different congregations of vampires were found along with about 400 *Natalus*. A male *Desmodus* obtained in a cave 13 miles west and six and a half miles north of Jiménez also was associated with *Natalus*.

Females with embryos or in lactation were collected as follows:

Rancho Pano Ayuctle, March 10, one pregnant female (embryo 40 mm. in crown-rump length); Río Sabinas, May 23, two pregnant females (embryos 36 and 43 mm.); Sierra de Tamaulipas, June 13, five lactating females and one female taken alive that gave birth on June 16 to one young; Cueva La Mula, August, nine lactating females. A male from the Sierra Madre that was obtained on January 5 had testes 8 mm. long.

The average weight of 21 adults from four miles southwest of Padilla was 39.1 (32.0-44.5) grams.

Records of occurrence.—Specimens examined, 107: 3 mi. W, 6.5 mi. N Jiménez, 1250 ft., 1; Río Soto la Marina, 4 mi. SW Padilla, 800 ft., 23; Cueva La Esperanza, 6 km. SW Rancho Santa Rosa, 360 m., 3; Cueva Los Troncones, 8 km. N, 12 km. W Cd. Victoria, Sierra Madre Oriental, 2500 ft., 2; Cd. Victoria, 1; Sierra Madre Oriental, 1900 ft., 5 mi. S, 3 mi. W Cd. Victoria, 3; La Mula, 13 mi. N Jaumave, 19; Cueva La Mula, 10 km. W Joya Verde, 2400 ft., 16; Joya Verde, 35 km. SW [Cd.] Victoria, 3800 ft., 6; Sierra de Tamaulipas, 1400 ft., 3 mi. S, 16 mi. W Piedra, 10; 70 km. S Cd. Victoria (*via* Highway), 6 km. W of Highway, 5; Rancho Pano Ayuctle, 6 mi. N Gómez Farías, 300 ft., 7; cave near headwaters Río Sabinas, 10 km. N, 8 km. W El Encino, 400 ft., 11.

Additional records (Malaga and Villa, 1957:539): Cueva La Sepultura, 7.5 km. NNW and hence 7 km. SSW (*via* highway) Cd. Victoria; El Ojo de Agua, at km. 10 on Valles-Tampico highway; Cueva del Abra, 2 km. SSW Cd. Mante.

Diphylla ecaudata Spix

Hairy-legged Vampire

1823. *Diphylla ecaudata* Spix, Simiarum et vespertilionum Brasiliensium . . ., p. 68, type locality, Brazil, restricted to Rio San Francisco, Baía, by Cabrera (Rev. Mus. Argentino Cien. Nat., 4:94, March 27, 1958).

Distribution in Tamaulipas.—Southern and central parts of state.

The hairy-legged vampire was first reported from Tamaulipas by de la Torre (1954:114), who recorded a male from five miles northeast of Antiguo Morelos, near El Pachón. Later in the same year Martin and Martin (1954:585) listed another male from El Pachón. Subsequently, Malaga and Villa (1957:543) reported specimens from two additional localities in the state, one of which (Cueva de la Sepultura) provides the northernmost place from which the species has been recorded. Malaga and Villa remarked that the species was abundant at Cueva de la Sepultura, being found in small groups clinging to the roof of the cave. Two females taken there on November 11 carried one embryo each; a lactating female was taken on November 14. The vampire, *Desmodus rotundus*, also was taken at Cueva de la Sepultura.

I follow Burt and Stirton (1961:37) in treating *Diphylla ecaudata* as a monotypic species.

Records: Cueva de la Sepultura, 7.5 km. NNW and hence 7 km. SSW (*via* highway) Cd. Victoria (Malaga and Villa, 1957:543); 5 mi. NE Antiguo

Morelos, near El Pachón (de la Torre, 1954:114); El Pachón (Martin and Martin, 1954:585); Cueva de Quintero, 4 km. SSW Quintero (Malaga and Villa, 1957:543).

Natalus stramineus saturatus Dalquest and Hall

Mexican Funnel-eared Bat

1949. *Natalus mexicanus saturatus* Dalquest and Hall, Proc. Biol. Soc. Washington, 62:153, August 23, type from 3 km. E San Andrés Tuxtla, 1000 ft., Veracruz.

1959. *Natalus stramineus saturatus*, Goodwin, Amer. Mus. Novit., 1977:7, December 22.

Distribution in Tamaulipas.—Central and southwestern parts of state.

All specimens examined were obtained from caves. At Cueva la Esperanza, approximately 400 individuals were found along with individuals of *Desmodus rotundus; Natalus* and *Desmodus* also were collected together in a cave approximately 30 yards deep three miles south and 14 miles west of Piedra, and in a cave six and a half miles north and 13 miles west of Jiménez, the northernmost locality from which *N. stramineus* is presently known.

Tamaulipan specimens do not differ significantly in external or cranial measurements in comparison with the specimens from Veracruz reported by Dalquest and Hall (1949:154), but do differ in color. Most are in the gray phase and are Avellaneus (grayish with yellowish hairs mixed) instead of Clay Color as are specimens from Veracruz; those few in the red phase are between Clay Color and Tawny-Olive instead of between Burnt Sienna and Chestnut. By consequence, bats from Tamaulipas resemble in color the smaller *N. s. mexicanus* of western México to a greater degree than they resemble *N. s. saturatus,* but I follow Goodwin (1959:7).

Dalquest and Hall (1949:154) reported the specimen from eight kilometers northeast of Antiguo Morelos as from San Luis Potosí, from which state the collector (Dalquest) evidently thought it had originated. Actually the place eight kilometers northeast of Antiguo Morelos is in Tamaulipas.

Records of occurrence.—Specimens examined, 64: 6.5 mi. N, 13 mi. W Jiménez, 1250 ft., 14; Cueva de la Esperanza, 6 km. SW Rancho Santa Rosa, 360 m., 20; Sierra de Tamaulipas, 3 mi. S, 16 mi. W Piedra, 1400 ft., 7; 3 mi. S, 14 mi. W Piedra, 2; Ejido Ojo de Agua, 20 mi. N, 3 km. W El Mante, 300 ft., 20; 8 km. NE Antiguo Morelos, 500 ft., 1.

Additional records (Goodwin, 1959:8): Antiguo Morelos; El Pachón.

Myotis velifer incautus (J. A. Allen)

Cave Myotis

1896. *Vespertilio incautus* J. A. Allen, Bull. Amer. Mus. Nat. Hist., 8:239, November 21, type from San Antonio, Bexar Co., Texas.

1928. *Myotis velifer incautus,* Miller and Allen, Bull. U. S. Nat. Mus., 144:92, May 25.

Distribution in Tamaulipas.—Probably most of northern part of state; presently known only from three localities.

The two specimens examined from the Sierra de Tamaulipas were taken in a mist net in which *Eptesicus fuscus*, *Myotis keenii*, and *Tadarida brasiliensis* also were captured. Both are females, one of which was lactating (June 20). Specimens from San Fernando probably were taken in houses by natives, who brought the bats to the collectors (Clifton and Bodley). The maxillary tooth-row and tibia are shorter, breadth across M3 narrower, and ear slightly longer in Tamaulipan specimens than in those for which measurements were given by Miller and Allen (1928:95), but the Tamaulipan specimens do not differ otherwise. The color in general is slightly more brownish than in Texan *incautus*, but about as in Oklahoman specimens examined. Three from San Fernando, Tamaulipas, are darker than others from that state.

The average weight of 12 non-pregnant females from San Fernando was 11.0 (9.5-13) grams. The only male obtained at the same locality weighed 12 grams.

Measurements.—Six females from San Fernando afford the following measurements: 100.0 (95-107); 42.5 (38-46); 10.3 (10-11); 15.3 (14.5-16); length of tibia, 17.4 (16.5-18.9); length of forearm, 44.8 (43.4-45.7); greatest length of skull, 16.5 (16.1-16.9); condylobasal length, 15.6 (15.3-15.8); least interorbital constriction, 4.0 (3.9-4.1); mastoid breadth, 8.3 (8.1-8.6); length of maxillary tooth-row, 6.5 (6.3-6.7); breadth across M3, 6.5 (6.0-6.9).

Records of occurrence.—Specimens examined, 15: San Fernando, 180 ft., 13; Sierra de Tamaulipas, 10 mi. W, 2 mi. S Piedra, 1200 ft., 2.

Additional record: Soto la Marina (Miller and Allen, 1928:93).

Myotis keenii auriculus Baker and Stains
Keen's Myotis

1955. *Myotis evotis auriculus* Baker and Stains, Univ. Kansas Publ., Mus. Nat. Hist., 9:83, December 10, type from 10 m. W, 2 mi. S Piedra, 1200 ft., Sierra de Tamaulipas, Tamaulipas.

1960. *Myotis keenii auriculus*, Findley, Jour. Mamm., 41:18, February.

Distribution in Tamaulipas.—Known only from type locality (2 specimens), but probably widely distributed in western part of state.

The two specimens known from Tamaulipas were caught in a mist net stretched across a narrow, brush-bordered arroyo in the Sierra de Tamaulipas. I tentatively follow Findley (1960) in arranging *auriculus* as a subspecies of *M. keenii.*

Records of occurrence.—Specimens examined, the holotype and one topotype.

Myotis californicus mexicanus (Saussure)
California Myotis

1890. *V[espertilio]. mexicanus* Saussure, Revue et Mag. Zool., Paris, ser. 2, 12:282, July, type from an unknown locality, but Dalquest (Louisiana State Univ. Studies, Biol. Ser., 1:49, December 28, 1953) restricted the type locality to the "desert (warmer part) of the state of México, México."

1897. *Myotis californicus mexicanus*, Miller, N. Amer. Fauna, 13:73, October 16.

Distribution in Tamaulipas.—Western mountains of state in pine-oak forest.

Only ten specimens of this species, five from Nicolás, two from Miquihuana and the other three, each from a different locality, have been reported from Tamaulipas. The specimen examined from 14 miles north and six miles west of Palmillas, a young female that still has deciduous incisors, was obtained on July 24. Of the five specimens from Nicolás, which represent the largest series of *M. californicus* ever reported from eastern México, some were caught in mist nets and others were shot over a water-hole.

Measurements.—Five skins and four skulls from Nicolás afford the following measurements: 86.0 (80-94); 39.0 (36-41); 7.4 (7-8.5); 13.7 (13.5-14.0); length of forearm, 33.0 (31.8-34.2); weight, 3.6 (3-4) grams; greatest length of skull, 13.9 (13.8-14.1); least interorbital constriction, 3.2 (3.1-3.3); breadth of braincase, 6.5 (6.4-6.5); length of maxillary tooth-row, 5.2 (5.1-5.3); breadth across M3, 5.1 (5.0-5.3).

Records of occurrence.—Specimens examined, 6: Nicolás, 56 km. NW Tula, 5500 ft., 5; 14 mi. N, 6 mi. W Palmillas, 5500 ft., 1.

Additional records: San José (Dice, 1937:249); Miquihuana (Miller and Allen, 1928:160); La Joya de Salas (Goodwin, 1954:5).

Myotis nigricans dalquesti Hall and Alvarez

Black Myotis

1961. *Myotis nigricans dalquesti* Hall and Alvarez, Univ. Kansas Publ., Mus. Nat. Hist., 14:71, December 29, type from 3 km. E of San Andrés Tuxtla, 1000 ft., Veracruz.

Distribution in Tamaulipas.—Tropical part of state, presently known only from two localities.

For taxonomic remarks concerning this bat see Hall and Alvarez (1961:72).

Records of occurrence.—Specimens examined, 5, from 8 km. W, 10 km. N El Encino, 400 ft.

Additional record: Cave in canyon of Río Boquillas, 8 km. SW Chamal (Goodwin, 1954:6).

Pipistrellus subflavus subflavus (F. Cuvier)

Eastern Pipistrelle

1832. *V[espertilio]. subflavus* F. Cuvier, Nouv. Ann. Mus. Hist. Nat. Paris, 1:17, type locality restricted to 3 mi. SW Riceboro, Liberty Co., Georgia, by W. H. Davis, Jour. Mamm., 40:522, November 20, 1959.

1897. *Pipistrellus subflavus*, Miller, N. Amer. Fauna, 13:90, October 16.

Distribution in Tamaulipas.—Presently known only from three localities, but probably occurs in most of eastern part of state.

Specimens examined are intermediate in color and measurements between *Pipistrellus subflavus subflavus* and *P. s. veraecrusis*, but the color resembles that of individuals of *subflavus* from Kansas more than that of specimens of *veraecrusis* from Las Vigas, Veracruz.

The two males from eight kilometers west and 10 kilometers north of El Encino represent the southernmost record of the subspecies.

Measurements.—External measurements of two males (58849, 58848) from 8 km. west and 10 km. north of El Encino and a male (60296) from Rancho Pano Ayuctle are, respectively, as follows: 78, 81, 83; 36, 38, 36; 10, 10, 9; 11, 11, 11; length of forearm, 33.1, 32.0, —; length of tibia, 14.6, 13.4, 13.0. Some cranial measurements of the two specimens from northwest of El Encino are: greatest length of skull, 12.8, 12.9; breadth of braincase, 6.5, 6.5; length of maxillary tooth-row, 4.0, 4.1.

Records of occurrence.—Specimens examined, 3: 8 km. W, 10 km. N El Encino, 400 ft., 2; Rancho Pano Ayuctle, 6 mi. N Gómez Farías, 300 ft., 1.

Additional record: Matamoros (H. Allen, 1894:128).

Pipistrellus hesperus potosinus Dalquest

Western Pipistrelle

1951. *Pipistrellus hesperus potosinus* Dalquest, Proc. Biol. Soc. Washington, 64:105, August 24, type from Presa de Guadalupe, San Luis Potosí.

Distribution in Tamaulipas.—Probably occurs throughout southwest part, but presently known only from Joya Verde.

The specimens reported herein were shot in July in a canyon that contained some standing water. According to the field notes of the collector (Schaldach), individuals of this bat in Tamaulipas flew later, in his experience, than bats of the same species in Sonora, Arizona and Coahuila, not emerging until it was almost fully dark.

Pipistrellus hesperus from Tamaulipas is identified as *P. h. potosinus* owing to the dark color, but the averages of some measurements differ slightly from those given by Dalquest (1951:106) for *potosinus* as follows: tail and ear shorter; foot larger; condylobasal length and cranial breadth less.

Measurements.—Average and extreme external and cranial measurements of five males from Joya Verde are: 73.2 (70-75); 27 (26-28); 7 (7); 12.4 (12-13); length of forearm, 31.0 (29.5-31.5); greatest length of skull, 12.4 (12.2-12.8); condylobasal length, 11.8 (11.4-12.3); breadth of braincase, 6.3 (6.0-6.5). Corresponding measurements of three females (60204, 60209, 60210) from the same locality are: 72, 78, 76; 27, 33, 35; 7, 7, 7; 12, 12, 12; 31, 31, 32; 12.3, 12.9, 13.5; 11.7, 12.2, —; 6.0, 6.6, 6.1.

Records of occurrence.—Specimens examined, 8, from Joya Verde, 35 km. SW Cd. Victoria, 3800 ft.

Eptesicus fuscus miradorensis (H. Allen)

Big Brown Bat

1866. *S[cotophilus]. miradorensis* H. Allen, Proc. Acad. Nat. Sci. Philadelphia, 18:287, type from Mirador, Veracruz.

1812. *Eptesicus fuscus miradorensis,* Miller, Bull. U. S. Nat. Mus., 79:62, December 31.

Distribution in Tamaulipas.—Southern part of state, north at least to Miquihuana.

Specimens from Miquihuana, Palmillas, and Nicolás were shot in flight at dusk; those from the Sierra de Tamaulipas were collected in a mist net. Five females, all taken in June, were lactating.

Judging from Hall and Kelson's (1959:185) distribution map for the species, two subspecies, *E. f. fuscus* and *E. f. miradorensis*, possibly occur in Tamaulipas, the former in the north and the latter in the south. Comparison of specimens presently available from the state (all from the southern part) with typical individuals of the two subspecies mentioned reveal that they resemble *miradorensis* to a greater degree than *fuscus* and they accordingly are assigned to the former. In measurements, the Tamaulipan specimens agree closely with *miradorensis;* in color, some resemble *miradorensis* but others approach *fuscus,* possibly indicating intergradation between the two subspecies in the material at hand. Probably *E. f. fuscus* will be found in the northern part of the state.

Measurements.—Average and extreme measurements of nine females from the Sierra de Tamaulipas and three males, two from Miquihuana (55137, 55138) and one from Palmillas (55139), are respectively: 121.3 (111-127), 115, 107, 115; 51.9 (50-56), 50, 45, 52; 10.9 (9.5-11.0), 10, 10, 11; 17.8 (17-18), 18, 18, 18; length of forearm, 49.6 (48-52.6), 48.9, 49.1, 49.1; length of tibia, 18.8 (18.2-19.3), 20.5, 17.3, 18.0; condylobasal length, 18.9 (18.5-19.3), 19.3, —, 18.8; zygomatic breadth, 13.1 (12.7-13.5), —, 13.0, 13.3; interorbital constriction, 4.2 (3.7-4.4), 4.0, 4.3, 4.1; length of maxillary tooth-row, 7.3 (7.1-7.5), —, 7.2, 7.2. Five lactating females weighed 20 (17-23) grams, and three males 17.5 (17-8) grams.

Records of occurrence.—Specimens examined, 17: Miquihuana, 6200 ft., 2; 14 mi. N, 6 mi. W Palmillas, 5500 ft., 1; Nicolás, 56 km. NW Tula, 5500 ft., 1; Sierra de Tamaulipas, 2 mi. S, 10 mi. W Piedra, 1200 ft., 12; Joya Verde, 35 km. SW [Cd.] Victoria, 3800 ft., 1.

Additional record: Aserradero del Paraiso (Goodwin, 1954:186).

Lasiurus borealis

Red Bat

Two subspecies of *Lasiurus borealis* have been reported from Tamaulipas. One, *L. b. borealis,* is known only from Matamoros, whereas the other, *L. b. teliotis,* is widely distributed in the central and southern parts.

A young animal from Ciudad Victoria was captured inside a house. All specimens taken in the Sierra de Tamaulipas were caught in mist nets, in which *Centurio senex, Pteronotus parnelli,* and *Mormoops megalophyla* also were taken.

Lasiurus borealis borealis (Müller)

1776. *Vespertilio borealis* Müller, Des Ritters Carl von Linné . . . vollständiges Natursystem . . ., Suppl., p. 20, type from New York.

1897. *Lasiurus borealis,* Miller, N. Amer. Fauna, 13:105, October 16.

Distribution in Tamaulipas.—Known only by two specimens from Matamoros (Miller, 1897:108).

Lasiurus borealis teliotis (H. Allen)

1891. *Atalapha teliotis* H. Allen, Proc. Amer. Philos. Soc., 29:5, April 10, type from an unknown locality, probably some part of California.

1897. *Lasiurus borealis teliotis*, Miller, N. Amer. Fauna, 13:110, October 16.

Distribution in Tamaulipas.—Generally distributed in higher parts of state.

Eight June-taken females, all lactating, from the Sierra de Tamaulipas averaged 10.0 (8-12) grams; five males from there weighed 9.2 (8-10) grams. According to Hall and Kelson (1959:188), males of this species usually are more brightly colored than females but this phenomenon is not evident in the Tamaulipan specimens. Males do, however, average slightly smaller than females.

The name *Lasiurus borealis teliotis* is employed following Handley (1960:472); formerly *L. b. ornatus* Hall was applied (Hall and Kelson, 1959:190) to bats here referred to as *teliotis*.

Records of occurrence.—Specimens examined, 7: Cd. Victoria, 1800 ft., 1; Sierra de Tamaulipas, 2 mi. S, 10 mi. W Piedra, 1200 ft., 1; Sierra de Tamaulipas, 3 mi. S, 14 mi. W Piedra, 1200 ft., 1; Sierra de Tamaulipas, 3 mi. S, 16 mi. W Piedra, 1400 ft., 4.

Lasiurus cinereus cinereus (Palisot de Beauvois)

Hoary Bat

1776. *Vespertilio cinereus* (misspelled *linereus*) Palisot de Beauvois, Catalogue raisonné du muséum de Mr. C. W. Peale, Philadelphia, p. 18, type from Philadelphia, Pennsylvania.

1864. *Lasiurus cinereus* H. Allen, Smiths. Misc. Coll., 7 (publ. 165): 21, June.

Distribution in Tamaulipas.—Probably state-wide but so far reported only from Matamoros (Miller, 1897:114), and Aserradero del Infernillo (Goodwin, 1954:6—cranium only).

Lasiurus intermedius intermedius H. Allen

Northern Yellow Bat

1862. *Lasiurus intermedius* H. Allen, Proc. Acad. Nat. Sci. Philadelphia, 14:246, "April" (between May 27 and August 1), type from Matamoros, Tamaulipas.

Distribution in Tamaulipas.—Eastern half of state, known only from three localities.

The three specimens examined were taken in mist nets along with *Lasiurus ega, Pteronotus rubiginosus* and *Mormoops megalophylla*.

The generic name *Lasiurus* is used instead of *Dasypterus* following Hall and Jones (1961).

Records of occurrence.—Specimens examined, 3: Sierra de Tamaulipas, 2 mi. S, 10 mi. W Piedra, 1200 ft., 1; Sierra de Tamaulipas, 3 mi. S, 16 mi. W Piedra, 1400 ft., 2.

Additional record: Matamoros (H. Allen, 1862:246).

Lasiurus ega xanthinus (Thomas)

Southern Yellow Bat

1897. *Dasypterus ega xanthinus* Thomas, Ann. Mag. Nat. Hist., ser. 6, 20:544, December, type from Sierra Laguna, Baja California.

1953. *Lasiurus ega xanthinus*, Dalquest, Louisiana State Univ. Studies, Biol. Ser., 1:61, December 28.

Distribution in Tamaulipas.—Probably occurs in southern and western parts of state; certainly known only from the Sierra de Tamaulipas.

Three June-taken females, all captured in mist nets, were lactating. Hall and Jones (1961:91) assigned all Mexican specimens of the southern yellow bat to *Lasiurus ega xanthinus*, but remarked that specimens from western México were paler than those from the east. Of the six specimens examined from Tamaulipas, four are dark, resembling in color specimens from Veracruz, Yucatán and Costa Rica, and the other two are somewhat paler, approaching specimens from Baja California, Zacatecas and Coahuila. In measurements, Tamaulipan specimens of *Lasiurus ega* generally resemble specimens from the west, but differ from any other *L. ega* seen in having a longer tail, longer ear, and shorter maxillary tooth-row.

Records of occurrence.—Specimens examined, 6: Sierra de Tamaulipas, 10 mi. W, 2 mi. S Piedra, 1200 ft., 4; 10 mi. W, 3 mi. S. Piedra, 1200 ft., 1; 16 mi. W, 3 mi. S. Piedra, 1400 ft., 1.

Nycticeius humeralis

Evening Bat

Nycticeius humeralis has the same distributional pattern in Tamaulipas as has *Lasiurus borealis* in that both are represented there by two subspecies, one known only from Matamoros and the other occurring in the rest of the state. Bats of this species (*N. h. mexicanus*) from Ciudad Victoria and some from the Sierra de Tamaulipas were shot in flight in evening; others from the last-mentioned locality were taken in mist nets. Lactating females (22 specimens) were collected in June and July.

Nycticeius humeralis humeralis (Rafinesque)

1818. *Vespertilio humeralis* Rafinesque, Amer. Monthly Mag., 3(6):445, October, type from Kentucky.

1819. *N[ycticeius]. humeralis* Rafinesque, Jour. Phys. Chim. Hist. Nat. et Arts, Paris, 88:417, June.

Distribution in Tamaulipas.—Matamoros (Miller, 1897:120), one specimen.

Nycticeius humeralis mexicanus Davis

1944. *Nycticeius humeralis mexicanus* Davis, Jour. Mamm., 25:380, December 12, type from Río Ramos, 1000 ft., 20 km. NW Montemorelos, Nuevo León.

Distribution in Tamaulipas.—Known certainly only from central part, but probably occurs at suitable places in all but extreme northern Tamaulipas.

Twenty-seven of 37 adults of *N. humeralis* examined from Tamaulipas are pale as is *N. h. mexicanus*, but 10 are darker and approach *N. h. humeralis* in this respect. Twenty-two females averaged 10.3 (9-13) grams and eight males averaged 9.5 (8-11) grams in weight.

Records of occurrence.—Specimens examined, 45: Cd. Victoria, 10; Sierra de Tamaulipas, 2-3 mi. S, 10 mi. W Piedra, 1200 ft., 31; 3 mi. S, 16 mi. W Piedra, 1400 ft., 4.

Rhogeëssa tumida tumida H. Allen

Little Yellow Bat

1866. *R[hogeëssa]. tumida* H. Allen, Proc. Acad. Nat. Sci. Philadelphia, 18:286, type from Mirador, Veracruz.

Distribution in Tamaulipas.—Southeastern part of state.

Specimens obtained from the vicinity of La Pesca were shot as were some from the Sierra de Tamaulipas. Others from the Sierra de Tamaulipas were taken in mist nets that were stretched across a small pool in an arroyo; *Eptesicus fuscus, Myotis velifer, M. keenii* and *Nycticeus humeralis* were captured in the same nets.

Females evidently bear young in Tamaulipas in April and May. Fourteen of 15 females collected at La Pesca in May were lactating, as were five of 31 taken in the Sierra de Tamaulipas in June. The weight of 46 females averaged 5.5 (4-7) grams, and that of nine males, 4.5 (4-5) grams.

Comparison of specimens from Tamaulipas with individuals from Veracruz reveals little difference in general color between the two samples. Most Tamaulipan specimens examined are dull yellowish brown, but some are darker. Goodwin (1954:6) reported a specimen from Santa María as being dark brown. Measurements of 10 females (see below) from the Sierra de Tamaulipas average a little larger than those reported by Miller (1897:123-124), Hall (1952: 232), and Goodwin (1958:10-12). I follow the last author in using the specific name R. *tumida* for this bat.

Measurements.—Average and extreme measurements of 10 females from the Sierra de Tamaulipas are as follows: 80.1 (78-83); 35.5 (33-37); 7.9 (7.5-8.0); 13.1 (13-14); length of forearm, 31.9 (30.6-33.0); greatest length of skull, 13.4 (13.1-13.8); zygomatic breadth, 8.6 (8.2-8.8); mastoid breadth, 5.6 (5.3-5.8); breadth across M3, 5.7 (5.5-6.0); length of maxillary tooth-row, 4.8 (4.7-4.9).

Records of occurrence.—Specimens examined, 59: 4 mi. N La Pesca, 1; 3 mi. N La Pesca, 3; 2 mi. N La Pesca, 11; 1 mi. N La Pesca, 4; La Pesca, 1; Sierra de Tamaulipas, 2 mi. S, 10 mi. W Piedra, 1200 ft., 39.

Additional record: Santa María (Goodwin, 1958:3).

Plecotus phyllotis (G. M. Allen)

Allen's Big-eared Bat

1916. *Corynorhynus phyllotis* G. M. Allen, Bull. Mus. Comp. Zool., 60:352, April, type from San Luis Potosí, probably near city of same name.

1959. *Plecotus phyllotis*, Handley, Proc. U. S. Nat. Mus., 110:130, Sept. 3.

1923. *Idionycteris mexicanus* Anthony, Amer. Mus. Novit., 54:1, January 17, type from Miquihuana, Tamaulipas.

Distribution in Tamaulipas.—Known only from Miquihuana.

The only specimen of this bat known from Tamaulipas was reported by Anthony (1923:1), and formed the basis of his description of *Idionycteris mexicanus*, a synonym of *Plecotus phyllotis* according to Handley (1956:53 and 1959:130).

Antrozous pallidus pallidus (Le Conte)

Pallid Bat

1856. *V[espertilio]. pallidus* Le Conte, Proc. Acad. Nat. Sci. Philadelphia, 7:437, type from El Paso, El Paso Co., Texas.

1864. *Antrozous pallidus*, H. Allen, Smiths. Misc. Coll., 7 (Publ. 165): 68, June.

Distribution in Tamaulipas.—Known only from a single ramus from Aserradero del Infernillo (Goodwin, 1954:6).

Tadarida brasiliensis mexicana (Saussure)

Brazilian Free-tailed Bat

1860. *Molossus mexicanus* Saussure, Revue et Mag. Zool., Paris, ser. 2, 12:283, July, type from Cofre de Perote, 13,000 ft., Veracruz.

1955. *Tadarida brasiliensis mexicana*, Schwartz, Jour. Mamm., 36:108, February 28.

Distribution in Tamaulipas.—Probably state-wide, but presently known from only five localities.

A female taken on June 21 in a mist net on the Sierra de Tamaulipas carried an embryo that was 29 mm. in crown-rump length. Two specimens were shot in flight in the deepest part of Cueva La Mula.

Records of occurrence.—Specimens examined, 4: 8 km. S Cd. Victoria, 1; Sierra de Tamaulipas, 10 mi. W, 2 mi. S Piedra, 1200 ft., 1; Cueva La Mula, 10 km. W Joya Verde, 2400 ft., 2.

Additional records: Río Bravo (town) (Villa, 1956:8); Rancho "La Isla," 3 km. N El Limón (Malaga and Villa, 1957:560); Cueva del Abra (*ibid.*); no specific locality (Shamel, 1931:6).

Tadarida aurispinosa (Peale)

Peale's Free-tailed Bat

1848. *Dysopes aurispinosus* Peale, U. S. Expl. Exp., 8:21, type taken on board the U. S. S. Peacock at sea, approximately 100 mi. S Cape San Roque, Brazil.

1931. *Tadarida aurispinosa*, Shamel, Proc. U. S. Nat. Mus., 78:11, May 6.

Distribution in Tamaulipas.—Known only from Cueva del Abra, six miles north-northeast of Antiguo Morelos.

Carter and Davis (1961) recorded for the first time this species from North America, on the basis of five specimens collected at Cueva del Abra. From the same locality P. L. Clifton collected several owl pellets which provide, besides many skulls of *Tadarida laticaudata*, four crania of *T. aurispinosa.* Available measurements of three, of the four *T. aurispinosa,* resemble those given by Carter and Davis (*op. cit.*) for their specimens. Measurements of the fourth cranium are smaller (greatest length of skull, 19.4; zygomatic breadth, 11.1; interorbital constriction, 3.7; cranial breadth, 9.1; mastoid breadth, 10.7; basal length, 16.3; length of maxillary toothrow, 7.4; breadth across M3, 7.9), but not outside the expected range of individual variation if we can judge by the range recorded by Jones and Alvarez (1962) for the related *Tadarida laticaudata.*

Records of occurrence.—Specimens examined, 4, from [Cueva del Abra], 6 mi. (by road) NNE Antiguo Morelos.

Tadarida laticaudata ferruginea Goodwin

Geoffroy's Free-tailed Bat

1954. *Tadarida laticaudata ferruginea* Goodwin, Amer. Mus. Novit., 1670:2, June 28, type from 8 mi. N Antiguo Morelos, Tamaulipas.

Distribution in Tamaulipas.—Known only from southeastern part of state.

Specimens from three miles south and 16 miles west of Piedra were found in a crevice inside a cave. Two days previously *Desmodus rotundus* and *Natalus stramineus* were obtained from the same cave. All other specimens from the Sierra de Tamaulipas were caught in mist nets. *Nycticeus humeralis, Myotis velifer, Eptesicus fuscus, Lasiurus borealis* and *L. intermedius* were taken in nets that also captured *T. laticaudata.*

All specimens taken (June 19-23) in the Sierra de Tamaulipas were females, except one. Of 33 females taken, 27 carried a single embryo each, the embryos averaging 27.0 (25-28) mm. in crown-rump length; the other five were lactating. Weight of the pregnant females averaged 16.0 (13-18) grams and that of the five lactating individuals averaged 13.0 (12-14) grams. A male weighed 22 grams.

For the taxonomic status of this species in North America see Jones and Alvarez (1962).

Records of occurrence.—Specimens examined, 65: Sierra de Tamaulipas, 2 mi. S, 10 mi. W Piedra, 1200 ft., 27; Sierra de Tamaulipas, 3 mi. S, 16 mi. W Piedra, 1400 ft., 7; 5 mi. S El Mante, 8 (AMNH); 11 mi. S El Mante, 13 (AMNH); 10 km. NNE Antiguo Morelos, 1; 8 mi. N Antiguo Morelos, 7 (5 AMNH, 2 KU); 20 mi. SW El Mante, 2 (AMNH).

Molossus ater nigricans Miller

Red Mastiff Bat

1902. *Molossus nigricans* Miller, Proc. Acad. Nat. Sci. Philadelphia, 54:395, September 12, type from Acaponeta, Nayarit.

Distribution in Tamaulipas.—Southern part of state, north at least to Guemes.

At Rancho Pano Ayuctle, according to the field notes of the collector (Schaldach), the red mastiff bat was common, and found daytime retreats in hollows in cypress trees. Schaldach twice found groups of bats in such hollows. *M. a. nigricans* is an early forager and most individuals seen were in flight before sunset, usually flying in a more or less straight line at heights of 25 to 60 feet above the ground. The odor of the chest gland was described by Schaldach as "strong" and "geranium-like." A female obtained three miles northeast of Guemes on August 19 carried a single embryo that was 33 mm. in crown-rump length.

Specimens examined average slightly smaller than the type specimen, especially in total length, length of hind foot, length of skull and length of maxillary tooth-row. Davis (1951:219) also noted some of these same differences in a specimen examined by him from two miles south of Ciudad Victoria. The variation in color is great among Tamaulipan specimens. Of the 15 examined, two are Dark Mummy Brown, six are Mummy Brown, six are Sudan Brown, and one is paler than Sudan Brown.

I follow Goodwin (1960:6) in using the specific name *ater*.

Records of occurrence.—Specimens examined, 15: 3 mi. NE Guemes, 2; Rancho Santa Rosa, 25 km. N, 13 km. W Cd. Victoria, 260 m., 2; Rancho Pano Ayuctle, 6 mi. N Gómez Farías, 300 ft., 1; Rancho Pano Ayuctle, 25 mi. N El Mante and 3 km. W Pan-American Hwy., 2200 ft., 8; 8 km. W, 10 km. N El Encino, 400 ft., 2.

Additional records (Davis, 1951:219): 2 mi. S Cd. Victoria; Altamira.

Ateles geoffroyi velerosus Gray

Spider Monkeys

1866. *Ateles vellerosus* Gray, Proc. Zool. Soc. London, p. 773 (for 1865), April, type locality "Brasil?"; restricted to Mirador, 2000 ft., about 15 mi. NE Huatusco, Veracruz, by Kellogg and Goldman, Proc. U. S. Nat. Mus., 96:33, November 2, 1944.

1944. *Ateles geoffroyi vellerosus*, Kellogg and Goldman, Proc. U. S. Nat. Mus., 96:32, November 2.

Distribution in Tamaulipas.—Probably extreme southern part.

No specimens of this monkey have been taken in Tamaulipas although Kellogg and Goldman (1944:34) pointed out that it probably occurred in the tropical forest of the southern part of the state. Later, Villa (1958:347) reported that A. Malaga Alba saw monkeys in 1954 at Barranca de Caballeros, approximately 25 kilometers north-northwest of Ciudad Victoria. No other report of their occurrence in the state has been forthcoming.

Dasypus novemcinctus mexicanus Peters

Nine-banded Armadillo

1864. *Dasypus novemcinctus* var. *mexicanus* Peters, Montsb. preuss Akad. Wiss., Berlin, p. 180, type from Matamoros, Tamaulipas (see Hollister, Jour. Mamm., 6:60, February 9, 1925).

1920. *D[asypus]. novemcinctus mexicanus*, Goldman, Smiths. Misc. Coll., 69 (5):66, April 24.

Distribution in Tamaulipas.—Probably state-wide except on Mexican Plateau; presently known only from five localities.

A 13-pound female from four kilometers west-southwest of La Purisima was captured after it was forced by the collector (Dalquest) and his dog out of the burrow that was under a log. A young specimen examined from seven kilometers southwest of La Purisima was captured by a dog. A partial skeleton including the skull was picked up on the barrier beach at a place 33 miles south of Washington Beach.

Records of occurrence.—Specimens examined, 3 (see text immediately above).

Additional records: Matamoros (Hollister, 1925:60); Rancho del Cielo (Hooper, 1953:11).

Sylvilagus brasiliensis truei (J. A. Allen)

Forest Rabbit

1890. *Lepus truei* J. A. Allen, Bull. Amer. Mus. Nat. Hist., 3:192, December 10, type from Mirador, Veracruz.

1950. *Sylvilagus brasiliensis truei*, Hershkovitz, Proc. U. S. Nat. Mus., 100: 351, May 26.

Distribution in Tamaulipas.—Southern part of state; known only from Rancho del Cielo (Goodwin, 1954:7).

Sylvilagus audubonii parvulus (J. A. Allen)

Desert Cottontail

1904. *Lepus (Sylvilagus) parvulus* J. A. Allen, Bull. Amer. Mus. Nat. Hist., 20:34, February 29, type from Apam, Hidalgo.

1909. *Sylvilagus audubonii parvulus*, Nelson, N. Amer. Fauna, 29:236, August 31.

Distribution in Tamaulipas.—Western part of state.

The specimen examined, a male that weighed 646 grams, was shot at night.

This species occurs only in western Tamaulipas. Hall and Kelson (1959:267, map 187) mistakenly plotted El Mulato, as being in the eastern part of the state; actually this locality is in the San Carlos

Mountains of the west, near the boundary between Tamaulipas and Nuevo León.

Records of occurrence.—One specimen examined from 4 mi. SW Nuevo Laredo, 900 ft.

Additional records (Nelson, 1909:237, unless otherwise noted): Nuevo Laredo; Guerrero; Mier; Camargo; El Mulato (Dice, 1937:256); Miquihuana.

Sylvilagus floridanus

Eastern Cottontail

This species occurs throughout Tamaulipas. A female from Soto la Marina, obtained on May 17, was lactating; another from 12 miles northwest of San Carlos, on August 23, carried two embryos that were 15 mm. in crown-rump length.

Sylvilagus floridanus chapmani (J. A. Allen)

1899. *Lepus floridanus chapmani* J. A. Allen, Bull. Amer. Mus. Nat. Hist., 12:12, March 4, type from Corpus Christi, Nueces Co., Texas.

1904. *Sylvilagus (Sylvilagus) floridanus chapmani,* Lyon, Smith. Misc. Coll., 45:336, June 15.

Distribution in Tamaulipas.—Northern two-thirds of state.

A male and pregnant female from 12 miles northwest of San Carlos weighed, respectively, 650 and 690 grams.

Records of occurrence.—Specimens examined, 17: San Fernando, 180 ft., 3; 12 mi. NW San Carlos, 1300 ft., 3; La Pesca, 3; Soto la Marina, 500 ft., 6; Ejido Eslabones, 2 mi. S, 10 mi. W Piedra, 1200 ft., 2.

Additional record: Jaumave (Nelson, 1909:178).

Sylvilagus floridanus connectens (Nelson)

1904. *Lepus floridanus connectens* Nelson, Proc. Biol. Soc. Washington, 17: 105, May 18, type from Chichicaxtle, Veracruz.

1909. *Sylvilagus floridanus connectens,* Lyon and Osgood, Bull. U. S. Nat. Mus., 62:32, January 28.

Distribution in Tamaulipas.—Southern part of state.

This subspecies has been reported previously from Tamaulipas only from Altamira. Specimens from 10 kilometers north and eight kilometers west of El Encino and 70 kilometers south of Ciudad Victoria, judging by their large size, dark color, and ochraceous brown (rather than pale ochraceous as in *S. f. chapmani*) upper sides of the hind feet are assignable to *connectens.*

Goodwin (1954:7) reported specimens from Chamal, Joya de Salas, Gómez Farías, and Pano Ayuctle as *S. f. chapmani,* remarking that they were intergrades between *chapmani* and *connectens.* Specimens reported by Goodwin are here assigned to *S. f. connectens* because the measurements of the specimen from eight kilometers west of El Encino are typical of that subspecies.

Records of occurrence.—Specimens examined, 4: 10 km. N, 8 km. W El Encino, 400 ft., 1; 2 km. W El Carrizo, 2; 9 mi. SW Tula, 5200 ft., 1.

Additional records (Goodwin, 1954:7, unless otherwise noted): Chamal; La Joya de Salas; Gómez Farías; Rancho Pano Ayuctle; Altamira (Nelson, 1909: 186).

Lepus californicus

Black-tailed Jack Rabbit

The black-tailed jack rabbit is the only species of *Lepus* known from Tamaulipas and is represented there by three subspecies, *L. c. merriami* of the northern part of the state, *L. c. altamirae* of the southeastern coastal plains, and *L. c. curti* of the barrier beach south of Matamoros. The known ranges of the three subspecies are not presently known to meet in Tamaulipas.

Lepus californicus altamirae Nelson

1904. *Lepus merriami altamirae* Nelson, Proc. Biol. Soc. Washington, 17: 109, May 18, type from Altamira, Tamaulipas.

1951. *Lepus californicus altamirae*, Hall, Univ. Kansas Publ., Mus. Nat. Hist., 5:45, October 1.

Distribution in Tamaulipas.—Southern coastal plain north certainly to vicinity of Soto la Marina.

The two specimens examined in this study (see below) are intermediate between *L. c. altamirae* and *L. c. curti*, but show greater resemblance to the former. In measurements they resemble *altamirae* rather than the smaller *curti.* They approach the latter in length of hind foot and are intermediate between the two subspecies in basilar length; in one specimen, the dimensions of the rostrum are as in *curti* and the other has the black patch on the posterior surface of the ear well developed, as in *altamirae*, but in the other the black is reduced. *L. c. altamirae* has been known previously only from Altamira.

Measurements.—Two male adults (55415, 55416) from north of Soto la Marina, afford the following external measurements: 610, 590; 100, 100; 124, 125; 124, 122 (length of ear from notch, dry, 114, 110). Cranial measurements are: basilar length, 75.1, 74.4; length of nasals, 46.1, 41.9; width of rostrum at PM, 25.1, 28.7; height of rostrum in front of PM, 25.2, 21.5; diameter of auditory bulla, 14.1, 13.0.

Records of occurrence.—Specimens examined, 2: 3 mi. N Soto la Marina, 1; 2 mi. NW Soto la Marina, 1.

Additional record: Altamira (Nelson, 1904:109).

Lepus californicus curti Hall

1951. *Lepus californicus curti* Hall, Univ. Kansas Publ., Mus. Nat. Hist., 5:42, October 1, type from barrier beach 88 mi. S, 10 mi. W Matamoros, Tamaulipas.

Distribution in Tamaulipas.—Known only by the three specimens mentioned in the original description from two barrier islands in northeastern part of state.

Records of occurrence.—Specimens examined, 3: 88 mi. S, 10 mi. W Matamoros, 2; 90 mi. S, 10 mi. W Matamoros, 1.

Lepus californicus merriami Mearns

1896. *Lepus merriami* Mearns, Preliminary diagnoses of new mammals from the Mexican border of the United States, p. 2, March 25, type from Fort Clark, Kinney Co., Texas.

1909. *Lepus californicus merriami*, Nelson, N. Amer. Fauna, 29:148, August 31.

Distribution in Tamaulipas.—Northern and western parts of state.

The two specimens examined, an adult female and a young male, from the barrier beach 33 miles south of Washington Beach are intergrades between *L. c. merriami*, reported from the mainland from as near as Matamoros, and *L. c. curti*, which occurs farther to the south on the same series of barrier beaches. Of seven characters that seem to differentiate the two subspecies, the adult female from 33 miles south of Washington beach resembles *merriami* in four as follows: tips of ears black (white in *curti*); nasals long; hind foot long; and supraoccipital process broad. The specimen resembles *curti* in shortness of tail and in having small auditory bullae. Breadth of rostrum above premolars, the seventh character, is less than in typical specimens of either of the two subspecies. More material is needed from the barrier beach in order to establish with certainty the relationships between jack rabbits occurring there.

Records of occurrence.—Specimens examined, 4: 33 mi. S Washington Beach, 2; 12 mi. NW San Carlos, 1300 ft., 2.

Additional records: Nuevo Laredo (Nelson, 1909:150); Mier (*ibid.*); Camargo (*ibid.*); Matamoros (Hall, 1951:185); Tamaulipeca, San Carlos Mts. (*ibid.*).

Spermophilus mexicanus parvidens Mearns

Mexican Ground Squirrel

1896. *Spermophilus mexicanus parvidens* Mearns, Preliminary diagnoses of new mammals from the Mexican border of the United States, p. 1, March 25, type from Fort Clark, Kinney Co., Texas.

Distribution in Tamaulipas.—Northern part of state, south at least to Xicotencatl.

Most of the specimens examined from Tamaulipas are in the brown phase (Howell, 1938:121) and differ from *S. m. parvidens* from Texas, Coahuila, and Nuevo León in being darker dorsally. Nevertheless, some individuals are as pale as those examined from the mentioned states. Measurements of Tamaulipan specimens average smaller than those given by Howell (1938:121) and Baker (1956:205) for *parvidens*.

Specimens from San Fernando differ slightly from those from Soto la Marina in having a relatively long tail (average 69.2 instead of 62.1 per cent of length of head and body) and in having the upper parts of the hind feet ochraceous instead of nearly white.

Two May-taken females from Soto la Marina carried 5 and 7 embryos that were 10 mm. in crown-rump length; another taken there was lactating. Weight of six non-pregnant females from San Fernando averaged 160.6 (129-197) grams. Two males from the same locality weighed 164 and 145 grams.

Measurements.—Average and extreme measurements of four males and three females from Soto la Marina are, as follows: 312.6 (296-330); 119.8 (110-130); 41.6 (38-43). Average cranial measurements of five specimens (two males, three females) from same locality are: greatest length of skull, 44.7 (43.7-47.4); zygomatic breadth, 26.9 (25.3-28.6); breadth of braincase, 19.4 (19.2-19.5); interorbital constriction, 13.3 (12.5-14.1); length of nasals, 15.9 (14.6-17.5); length of maxillary tooth-row, 8.3 (8.0-8.5).

Records of occurrence.—Specimens examined, 20: San Fernando, 180 ft., 12; Soto la Marina, 500 ft., 8.

Additional records (Howell, 1938:121 unless otherwise noted): Nuevo Laredo; Mier; Camargo; Reynosa; Bagdad; Victoria; Xecotencatl [=Xicotencatl] (J. A. Allen, 1891:223).

Spermophilus spilosoma oricolus Alvarez

Spotted Ground Squirrel

1962. *Spermophilus spilosoma oricolus* Alvarez, Univ. Kansas Publ., Mus. Nat. Hist., 14:123, March 7, type from 1 mi. E La Pesca, Tamaulipas.

Distribution in Tamaulipas.—Known only from the type locality and from parts of the barrier beach, but possibly occurs at other places in northeastern parts of state.

The 10 specimens from the type locality were trapped or shot on the beach, which was covered by thick, low, scattered bushes and grass. Of the many holes found there, some probably were used by ground squirrels and others by crabs. A female, taken on July 7 with two young at a place 33 miles south of Washington Beach, weighed 133 grams and had six placental scars. This specimen (reported as *Spermophilus spilosoma annectens* by Selander *et al.*, 1962:335) resembles others examined from the barrier beach (see Alvarez, 1962:124) and is therefore assigned to *S. s. oricolus*.

Records of occurrence.—Specimens examined, 24: 33 mi. S Washington Beach, 1; 88 mi. S, 10 mi. W Matamoros, 12; 89 mi. S, 10 mi. W Matamoros, 1; 1 mi. E La Pesca, 10.

Spermophilus variegatus couchii Baird

Rock Squirrel

1855. *Spermophilus couchii* Baird, Proc. Acad. Nat. Sci. Philadelphia, 1:332, April, type from Santa Catarina, a few miles west of Monterrey, Nuevo León.

1955. *Spermophilus variegatus couchii*, Baker, Univ. Kansas Publ., Mus. Nat. Hist., 9:207, June 15.

Distribution in Tamaulipas.—Possibly in southwestern part; reported only from Ciudad Victoria (Howell, 1938:141).

Since Baird (1855:332) described S. *v. couchii* and mentioned a specimen from Ciudad Victoria that was obtained by Berlandier, no other record from Tamaulipas has come to light. Probably the species obtained by Berlandier was introduced at Ciudad Victoria by man.

Sciurus aureogaster aureogaster Cuvier

Red-bellied Squirrel

1829. [*Sciurus*] *aureogaster* Cuvier, *in* Geoffroy St.-Hilaire, and F. Cuvier, Hist. Nat. Mamm., 6, livr. 59 pl. with text, September (binomen published only at end of work, table générale et méthodique, 7:4, 1842), type locality "California"; restricted to Altamira, Tamaulipas, by Nelson (Proc. Washington Acad. Sci., 1:38, May 9, 1899).

Distribution in Tamaulipas.—Tropical forest of southern part; north at least to Rancho Santa Rosa.

According to one collector (Schaldach), natives referred to *Sciurus aureogaster* as "ardilla pinta" or "ardilla colorada." He recorded in his field notes that S. *aureogaster* was most active between 7:00 and 9:00 a. m. and again from 3:00 to 5:00 p. m., that the nest was constructed of green oak leaves, and that the nest resembles somewhat in size and form that of S. *carolinensis*.

Of 53 specimens examined, 17 are black and one from 70 kilometers south of Ciudad Victoria is clearly more whitish than the others. Specimens from the northeastern part of the range of the species (= southeastern Tamaulipas) average darker than those from the south and west. In individuals that are not black, the ventral reddish color covers the shoulders and in some it extends between the shoulders to the median dorsal area.

Among females collected from December through May, only one, taken 43 kilometers south of Ciudad Victoria on March 17, was pregnant (one embryo).

The weight of seven adult males from Soto la Marina and the Sierra de Tamaulipas averaged 492.5 (400-575) grams.

Specimens herein reported from San Fernando provide the northernmost record of the species.

Records of occurrence.—Specimens examined, 53: San Fernando, 180 ft., 5; 9½ mi. SW Padilla, 800 ft., 3; Rancho Santa Rosa, 25 km. N, 13 km. W Cd. Victoria, 260 m., 8; 3 mi. NE Guemes, 5; Soto la Marina (3 mi. N), 500 ft., 6; Sierra de Tamaulipas, 10 mi. W, 8 mi. S Piedra, 1200 ft., 6; 43 km. S Cd. Victoria, 1; Ejido Santa Isabel, 2 km. W Pan-American Highway, 2000 ft., 5; 70 km. (by highway) S Cd. Victoria, 6 mi. W of Pan-American Highway, 3; 2 mi. W El Carrizo, 7; Rancho Pano Ayuctle, 6 mi. N Gómez Farías, 300 ft., 2; Rancho Pano Ayuctle, 25 mi. N, 3 km. W El Mante, 300 ft., 1; 8 km. W, 10 km. N El Encino, 400 ft., 1.

Additional records: Río Corono (= Corona) (J. A. Allen, 1891:222); Victoria (Kelson, 1952:249); Santa María (Goodwin, 1954:8); 3 mi. NW Acuña,

3500 ft. (Hooper, 1953:4); Forlón (Nelson, 1899:42); NE Zamorina (Hooper, 1953:4); Gómez Farías (Goodwin, 1954:8); Altamira (Nelson, 1899:42); Tampico (J. A. Allen, 1891:222).

Sciurus deppei negligens Nelson

Deppe's Squirrel

1898. *Sciurus negligens* Nelson, Proc. Biol. Soc. Washington, 12:147, June 3, type from Altamira, Tamaulipas.

1953. *Sciurus deppei negligens*, Hooper, Occas. Papers Mus. Zool., Univ. Michigan, 544:4, March 25.

Distribution in Tamaulipas.—Tropical forest in southern part of state, north to Rancho Santa Rosa and Padilla.

In Tamaulipas this squirrel is called "ardilla chica" or "ardilla barcina," and is abundant in areas where tall trees and dense brush prevail. This species evidently does not have restricted periods of activity, as does S. *aureogaster*, but is active throughout the day. At El Carrizo a nest, nine to 10 inches in diameter and constructed of leaves and small sticks, was in a thick tangle of branches 25 feet above the ground. A male having testes 11 mm. long was in the nest. Among 16 females collected in the months of February, May and June, only two, taken in February, were lactating. A female from 70 kilometers south of Ciudad Victoria, had four placental scars, three on the right side and one on the left, along with a resorbed embryo on the right side; according to the collector "the scars appeared quite recent, as evidenced by the fact that not all of the blood had been resorbed yet."

The northernmost localities from which S. *d. negligens* has been reported are nine and a half miles southwest of Padilla in the east, and Rancho Santa Rosa in the west.

Three males from the vicinity of Padilla weighed 309, 276, and 261 grams.

Records of occurrence.—Specimens examined, 92: 9½ mi. SW Padilla, 800 ft., 3; Rancho Santa Rosa, 25 km. N, 13 km. W Cd. Victoria, 260 m., 8; 3 mi. NE Guemes, 1; Sierra de Tamaulipas, 10 mi. W, 2 mi. S Piedra, 1200 ft., 3; Ejido Santa Isabel, 2 km. W Pan-American Highway, 2000 ft., 20; 70 km. (by highway) S Cd. Victoria and 6 mi. W Pan-American Highway, 43; 2 km. W El Carrizo, 12; 8 km. W, 10 km. N El Encino, 400 ft., 2.

Additional records: Victoria (Nelson, 1898:147); Santa María (Goodwin, 1954:8); Rancho Viejo (*ibid.*); Rancho del Cielo (*ibid.*); 3 mi. NW Acuña (Hooper, 1953:4); Pano Ayuctle (*ibid.*); Gómez Farías (Goodwin, 1954:8); Mesa de Llera, 10 mi. NE Zamorina (Hooper, 1953:4); Altamira (Nelson, 1898:147).

Sciurus alleni Nelson

Allen's Squirrel

1898. *Sciurus alleni* Nelson, Proc. Biol. Soc. Washington, 12:147, June 3, type from Monterrey, Nuevo León.

Distribution in Tamaulipas.—Along Sierra Madre Oriental in southwestern part of state.

This squirrel occurs in stands of oak and "nogalillos" (hickory) trees that grow along streams and arroyos. Individuals are active from sunrise to about 10:00 a. m. and again late in the afternoon. They give a soft "chirring" call.

Nelson (1899:92) noted that specimens from Miquihuana were smaller than those from the type locality. Among specimens I have examined, some are as large as topotypes and two females are larger (total length, 486 and 490) than measurements given for the species by Nelson (*op. cit.*).

Record of occurrence.—Specimens examined, 11, from Joya Verde, 35 km. SW Cd. Victoria, 3800 ft.

Additional records: Near Victoria (Nelson, 1899:92); Miquihuana (*ibid.*); Joya de Salas (Goodwin, 1954:8).

Glaucomys volans herreranus Goldman

Southern Flying Squirrel

1936. *Glaucomys volans herreranus* Goldman, Jour. Washington Acad. Sci., 26:463, November 15, type from Mts. of Veracruz.

Distribution in Tamaulipas.—Known only from Aserradero del Infernillo (Goodwin, 1954:9 and 1961:9).

Geomys personatus personatus True

Texas Pocket Gopher

1889. *Geomys personatus* True, Proc. U. S. Nat. Mus., 11:159 for 1888, January 5, type from Padre Island, Cameron County, Texas.

Distribution in Tamaulipas.—Known only from the barrier beach in northeastern part of state.

The specimens examined are referred, tentatively, to *Geomys personatus personatus* on geographic grounds. They average smaller in all measurements than *personatus* (but are larger than *G. p. megapotamus*), do not have the sagittal crest that usually is present in *personatus*, and the shape of the pterygoid bones is distinctive. In *personatus* and *megapotamus* the ventral border of the pterygoids (in lateral view) is convex instead of nearly straight as in specimens from the barrier beach. The specimens recorded here are all that are known of G. *personatus* (see account of G. *tropicalis*) from México.

Measurements.—Average and extreme external measurements of five females from 73 miles south of Washington Beach are as follows: 266.8 (263-271); 94.8 (91-98); 34 (33-35). Cranial measurements of two males (89038, 89032) and average and extremes of five females are respectively: basal length, 49.1, 46.6, 45.9 (44.2-46.8); basilar length, 42.9, 40.0, 39.8 (38.0-40.8); zygomatic breadth, 29.6, 28.3, 28.0 (25.7-29.9); squamosal breadth, 27.8, 25.9, 26.2 (23.8-25.4); interorbital constriction, 7.4, 6.9, 7.3 (6.7-7.8); alveolar length of maxillary tooth-row, 10.3, 9.2, 9.4 (9.1-9.7).

Records of occurrence.—Specimens examined, 17: 35 mi. SSE Matamoros, 8; 33 mi. S Washington Beach, 1; 73 mi. S Washington Beach, 8.

Additional record: 4 mi. S Washington Beach (Selander *et al.*, 1962:335 —possibly fragmentary skeletal remains never catalogued in any research collection).

Geomys tropicalis Goldman

Tropical Pocket Gopher

1915. *Geomys personatus tropicalis* Goldman, Proc. Biol. Soc. Washinton, 28:134, June 29, type from Altamira, Tamaulipas.

Distribution in Tamaulipas.—Known only from vicinity of type locality, in southeastern part of state.

Geomys tropicalis was named as a subspecies of G. *personatus* in 1915 by E. A. Goldman. To my knowledge, no one other than Goldman has critically studied specimens of this pocket gopher, nor have specimens other than those listed in the original description been reported up to now. In 1953, Gerd H. Heinrich collected a series of 19 individuals one mile south of Altamira. These specimens were compared (by E. R. Hall in March, 1962) with the holotype and paratypes of G. *p. tropicalis* and were found to be indistinguishable.

Careful comparisons of the specimens from one mile south of Altamira with topotypes of G. *personatus personatus* (and specimens of other subspecies) indicate that *tropicalis* differs from *personatus* in a number of important characters, some of which *tropicalis* shares with *Geomys arenarius* of the Rio Grande Valley and adjacent areas in Texas, New Mexico, and Chihuahua (see Table 2).

As can be seen in the accompanying table *tropicalis* resembles *arenarius* in half of the eight characters considered, especially in the presence of a knob on the zygomatic process of the squamosal (the diagnostic character of *arenarius* according to Merriam, 1895:140) and in the shape of the mesopterygoid fossa. G. *tropicalis* differs from *arenarius* principally in having a low sagittal crest in adult males (lacking in *arenarius*) and in the shape of the interparietal

TABLE 2.—DIFFERENCES BETWEEN THREE SPECIES OF GEOMYS.

	G. arenarius	*G. personatus*	*G. tropicalis*
Zygomatic arches	parallel	narrower posteriorly	narrower posteriorly
Sagittal crest	absent	present	small
Squamosal knob	present	absent	present
Interparietal	subquadrant	triangular	triangular
Mesopterygoid fossa	V-shaped	U-shaped	V-shaped
Ratio, zygomatic breadth to basal length	63.7–66.6	66.3–67.2	60.8–66.2
Ratio, mastoid breadth to basal length	58.0–60.4	59.8–63.1	58.0–59.6
Border of premaxilla at incisive foramina	wedge-shaped	subquadrate	subquadrate

bone, which in *tropicalis* is small (in some skulls difficult to see) and triangular instead of being relatively large and subquadrate as in *arenarius.*

G. *tropicalis* resembles *personatus* in half of the characters considered, notably in shape of the interparietal bone, outline of zygomatic arches, and constriction of the premaxillae where they border the incisive foramina.

Considering the distinctive combination of characters possessed by *tropicalis,* and its isolated, restricted geographic range (the nearest known record of *Geomys* is approximately 165 miles to the north), *tropicalis* is here regarded as a full species. A skull alone examined from 10 miles northwest of Tampico does not differ from those of other specimens studied.

The average weight of five non-pregnant July-taken females was 189.4 (180-200) grams. Weights of three males were 280, 270, and 255 grams. Females are in all measurements smaller than males.

Measurements.—Average and extreme measurements of five females and three males from one mile south of Altamira are, respectively, as follows: 243.5 (235-250), 260, 260, 265; 82.0 (78-85), 87, 93, 89; 32.2 (31-33), 35, 35, 33; ear from notch in both sexes, 5; condylobasal length, 42.3 (41.3-43.1), 46.0, 48.0, 46.2; zygomatic breadth, 26.6 (25.1-27.7), 30.4, 31.2, 30.5; interorbital constriction, 6.2 (6.1-6.3), 6.0, 6.2, 6.3; length of nasals, 14.6 (14.0-15.3), 17.0, 16.8, 15.9; alveolar length of maxillary tooth-row, 9.0 (8.6-9.3), 9.9, 10.0, 9.4.

Records of occurrence.—Specimens examined, 19: 1 mi. S Altamira, 18; 10 mi. NW Tampico, 1.

Additional record: Altamira (Goldman, 1915:134).

Heterogeomys hispidus negatus Goodwin

Hispid Pocket Gopher

1953. *Heterogeomys hispidus negatus* Goodwin, Amer. Mus. Novit., 1620:1, May 4, type from Gómez Feras [Farías], 1300 ft., Tamaulipas.

Distribution in Tamaulipas.—Known only from the vicinity of the type locality.

Specimens of this pocket gopher were taken in large Macabee traps, at night with the aid of a dog, and by natives using slingshots. Mounds of *H. hispidus* were common two miles west of El Carrizo near banana trees; the mouths of burrows were four to five inches in diameter. Two females collected at this locality on April 16 and 17 were lactating.

Specimens examined of *H. hispidus* from Tamaulipas resemble the description of *H. h. negatus* more than that of *H. h. concavus,* and are referred, therefore, to *negatus.* I assume, on geographic grounds, that the individuals reported by Hooper (1953:5) as *concavus* are *negatus;* they are here referred to as *negatus.* If this referral is correct, the subspecies *concavus* probably does not occur in Tamaulipas.

Records of occurrence.—Specimens examined, 6: Ejido Santa Isabel, 2 km. W Pan-American Highway, 2000 ft., 1; 2 km. W El Carrizo, 1; 5 km. W El Carrizo, 4.

Additional records: Rancho Pano Ayuctle (Hooper, 1953:5); Gómez Farías (Goodwin, 1953:1).

Cratogeomys castanops

Yellow-faced Pocket Gopher

Two subspecies of *Cratogeomys castanops* occur in Tamaulipas, *C. c. planifrons* in the higher elevations of the Sierra Madre Oriental in the western part of the state, and *C. c. tamaulipensis* on the plains of the Río Grande.

Specimens from Miquihuana were trapped in tunnels at 6400 feet elevation. At Palmillas, individuals were trapped in an area of mesquite, other bushes and "lechuguilla." Three specimens from southeast of Reynosa were collected in traps set along the dikes of irrigation ditches. Most specimens from Nicolás were brought by natives to the collector, but some were caught in traps set in tunnels among the desert bushes.

Cratogeomys castanops planifrons Nelson and Goldman

1943. *Cratogeomys castanops planifrons* Nelson and Goldman, Proc. Biol. Soc. Washington, 47:146, June 13, type from Miquihuana, 5000 ft., Tamaulipas.

Distribution in Tamaulipas.—Higher elevations in southwestern part of state.

Specimens from four miles north of Jaumave do not differ from specimens from Miquihuana. The weights of nine females averaged 146.4 (110-210) grams; three males weighed 178, 203, and 215 grams.

Records of occurrence.—Specimens examined, 29: Miquihuana, 6400 ft., 9; 4 mi. N Jaumave, 2500 ft., 5; Nicolás, 56 km. NW Tula, 5500 ft., 15.

Cratogeomys castanops tamaulipensis Nelson and Goldman

1934. *Cratogeomys castanops tamaulipensis* Nelson and Goldman, Proc. Biol. Soc. Washington, 47:141, June 13, type from Matamoros, Tamaulipas.

Distribution in Tamaulipas.—Known only from two localities in extreme northern part of state, but probably occurs throughout northeastern part of state.

Three specimens from three miles southeast of Reynosa are referred to *C. c. tamaulipensis* on geographic grounds. They are tawny brown dorsally instead of cinnamon brown or pinkish cinnamon as Nelson and Goldman (1943:141) described *tamaulipensis,* and the basioccipital bone (in one male) is parallel-sided instead of wedge-shaped. Possibly this difference is owing to sex; Nelson and

Goldman studied only one adult, a female (the type), and the only adult seen by me was a male.

Measurements.—An adult male (58118) from three miles southeast of Reynosa, measured as follows: 301; 81; 40; 7; condylobasal length, 57.0; zygomatic breadth, 41.2; palatal length, 36.1; breadth of rostrum, 11.8; length of nasals, 22.0; squamosal breadth, 34.0; alveolar length of maxillary tooth-row, 10.8.

Records of occurrence.—Specimens examined, 3, from 3 mi. SE Reynosa. Additional record: Matamoros (Nelson and Goldman, 1934:140).

Perognathus merriami merriami J. A. Allen

Merriam's Pocket Mouse

1892. *Perognathus merriami* J. A. Allen, Bull. Amer. Mus. Nat. Hist., 4:45, March 25, type from Brownsville, Cameron Co., Texas.

Distribution in Tamaulipas.—State-wide except southwestern part.

Most of the available specimens of P. *m. merriami* were collected in the semi-arid areas of mesquite and grasses. At Soto la Marina P. *m. merriami* was abundant in open fields surrounded by brush. One female, collected on July 4, one mile south of Altamira was lactating. Weights of 16 adults from Soto la Marina and that of nine adults from the vicinity of San Fernando are, respectively: 8.2 (7-10) and 8.1 (7-9) grams.

Specimens from Tamaulipas are darker than those examined from Coahuila and southern Texas. A skull picked up on the barrier beach, 73 miles south of Washington Beach, differs from all other skulls examined in having the rostrum (3.6 mm.) and M1 (4.3) wider, auditory bullae relatively smaller, and glenoid fossa larger (2.6 instead of less than 2.3 in specimens from Soto la Marina).

Records of occurrence.—Specimens examined, 46: 4—4.5 mi. S Nuevo Laredo, 900 ft., 4; 10 mi. S, 11 mi. E Nuevo Laredo, 600 ft., 2; 1 mi. S Santa Teresa, 1; San Fernando, 180 ft., 1; 2 mi. W San Fernando, 180 ft., 14; 73 mi. S Washington Beach, 1; 12 mi. NW San Carlos, 1300 ft., 1; Soto la Marina, 19; Ciudad Victoria, 1; 17 mi. SW Tula, 3900 ft., 1; 1 mi. S Altamira, 1.

Additional records (Osgood, 1900:22, unless otherwise noted): Mier; Reynosa; Matamoros; 40 mi. S Matamoros (Hooper, 1953:5); Hidalgo; Altamira.

Perognathus hispidus hispidus Baird

Hispid Pocket Mouse

1858. *Perognathus hispidus* Baird, Mammals, in Repts. Expl. Surv. . . ., 8(1):421, July 14, type from Charco Escondido, Tamaulipas.

Distribution in Tamaulipas.—Central and northern parts of state.

Two specimens examined from the vicinity of Nuevo Laredo were trapped in weeds and tall grass along an irrigation ditch that ran between desert and a cornfield. One was a lactating female (November 15) and weighed 31 grams; the other, an immature male, weighed 23 grams. A May-taken specimen from Soto la Marina possesses a broader and more ochraceous lateral line than the other three individuals examined from Tamaulipas and the Texan specimens seen.

Records of occurrence.—Specimens examined, 4: 10 mi. S, 11 mi. E Nuevo Laredo, 600 ft., 2; Soto la Marina, 500 ft., 1; 9½ mi. SW Padilla, 800 ft., 1.

Additional records (Osgood, 1900:44, unless otherwise noted): Mier; Matamoros; Charco Escondido (Baird, 1858:422); 3 mi. W Soto la Marina (Hooper, 1953:5).

Perognathus nelsoni nelsoni Merriam

Nelson's Pocket Mouse

1894. *Perognathus (Chaetodipus) nelsoni* Merriam, Proc. Acad. Nat. Sci. Philadelphia, 46:266, September 27, type from Hacienda La Parada, about 25 mi. NW Cd. San Luis Potosí, San Luis Potosi.

Distribution in Tamaulipas.—Known only from the west side of the Sierra Madre Oriental in southwestern part of state.

Most of the specimens examined were taken in semi-arid habitats where the dominant plants were cactus, weeds and bushes.

In Tamaulipas, specimens from the southern localities (places labeled with reference to Tula) are darker than those from the two northernmost localities (Miquihuana and four miles north of Jaumave). Most measurements are about equal in the southern and northern specimens, but in some measurements southern specimens average slightly smaller than those from the north. Greatest length of skull is a case in point. The difference in size is reflected in the weights. Average weights of nine males and nine females from southern localities are, respectively, 14.7 (12-16.5) and 13.8 (12-15.5) instead of 18.5 (17-20) and 17.0 (15-18) grams for four males and six females from the northern localities. In general, Tamaulipan specimens average somewhat smaller than those from other localities in eastern México (see measurements given by Baker, 1956:238, Dalquest, 1953:107, and Osgood, 1900:53).

Measurements.—Average and extreme measurements of six specimens (2 males and 4 females) from Miquihuana, three males from four miles north of Jaumave, and five (3 males and 2 females) from nine miles southwest of Tula are, respectively, as follows: 176.2 (163-185), —, 170, 173, (4 specimens only) 179.0 (165-186); 99.8 (97-105), ——, 90, 93, (4 specimens only) 96.7 (88-104); 22.5 (21-23), 23, 23, 24, 22.6 (22-23); 8 (8), 8, 8, 8, 8.8 (8-9); greatest length of skull, 26.1 (25.6-26.6), 25.8, 26.5, 26.9, 25.2 (24.9-25.7); mastoid breadth, 13.3 (12.9-13.6), 13.2, 13.8, 13.6, 13.1 (12.9-13.4); interorbital constriction, 6.4 (6.1-6.6), 5.9, 6.3, 6.3, 6.3 (6.1-6.8); interparietal breadth, 7.4 (6.8-7.9), 7.7, 7.2, 7.2, 7.6 (7.3-7.9); alveolar length of maxillary tooth-row, 3.7 (3.5-4.0); 3.6, 3.5, 3.6, 3.6 (3.5-3.8).

Records of occurrence.—Specimens examined, 42: Miquihuana, 6300 ft., 7; 4 mi. N Jaumave, 2500 ft., 5; Nicolás, 56 km. NW Tula, 5500 ft., 10; Tajada,

23 mi. NW Tula, 5200 ft., 6; 8 mi. N Tula, 4500 ft., 1; 9 mi. SW Tula, 3900 ft., 13.

Additional record: Jaumave (Miller, 1924:284).

Dipodomys ordii

Ord's Kangaroo Rat

This species has a restricted geographic distribution in Tamaulipas, although three subspecies occur in the state; two of them occur in the extreme northeast and the other in the far west.

Dipodomys ordii durranti Setzer

1949. *Dipodomys ordii fuscus* Setzer, Univ. Kansas Publ., Mus. Nat. Hist., 1:555, December 27, type from Jaumave, Tamaulipas.

1952. *Dipodomys ordii durranti* Setzer, Jour. Washington Acad. Sci., 42: 391, December 17, a renaming of D. *o. fuscus* Setzer, 1949.

Distribution in Tamaulipas.—Semi-desert areas in western part of state.

The specimen examined from four miles north of Jaumave was trapped in a xeric area in which the vegetation consisted of mesquite, high palmlike yuccas, and "lechugilla." Specimens from the vicinity of Tula were trapped along bushy fence rows and adjacent to clumps of bushes and cactus, or shot at night in an area in which the soil was a sandy loam having relatively large amounts of gravel. The average weight of seven specimens from Nicolás was 50.3 (42-60) grams.

According to Lidicker (1960:178 and in *litt.*), the place called Lulú that was ascribed to Tamaulipas by Setzer (1949:550), and from which *D. o. durranti* was reported, actually is in Zacatecas.

Records of occurrence.—Specimens examined, 19: Miquihuana, 6200 ft., 2; 4 mi. N Jaumave, 2500 ft., 3; Nicolás, 56 km. NW Tula, 12; 8 km. N Tula, 4500 ft., 2.

Additional records (Setzer, 1949:556): Nuevo Laredo; Jaumave.

Dipodomys ordii parvabullatus Hall

1951. *Dipodomys ordii parvabullatus* Hall, Univ. Kansas Publ., Mus. Nat. Hist., 5:38, October 1, type from 88 mi. S and 10 mi. W Matamoros, Tamaulipas.

Distribution in Tamaulipas.—Known only from two islands off the barrier beach.

Weight of four adults averaged 49.2 (44-60) grams.

Records of occurrence.—Specimens examined, 17: 33 mi. S Washington Beach, 4; 88 mi. S, 10 mi. W Matamoros, 7; 90 mi. S, 10 mi. W Matamoros, 6.

Dipodomys ordii compactus True

1889. *Dipodomys compactus* True, Proc. U. S. Nat. Mus., 11:160, January 5, type from Padre Island, Cameron Co., Texas.

1942. *Dipodomys ordii compactus*, Davis, Jour. Mamm., 23:332, August 13.

Distribution in Tamaulipas.—Reported only from Bagdad (Hall, 1951:41).

Dipodomys merriami atronasus Merriam

Merriam's Kangaroo Rat

1894. *Dipodomys merriami atronasus* Merriam, Proc. Biol. Soc. Washington, 9:113, June 21, type from Hacienda La Parada, about 25 mi. NW San Luis Potosí, San Luis Potosí.

Distribution in Tamaulipas.—Mexican Plateau in western part of state.

Specimens examined are tentatively assigned to *Dipodomys merriami atronasus.* They differ from typical *atronasus* as pointed out by Lidicker (1960:177). He noted that individuals from the eastern edge of the range of *D. m. atronasus* were slightly paler than typical specimens, but I found Tamaulipan material to be much darker, especially behind the nose and ears (blackish instead of brownish), than specimens from Aguascalientes, San Luis Potosí and Zacatecas.

Specimens examined were collected under the same conditions and in the same areas as D. *ordii durranti.* The average weight of 20 adults (11 females and nine males) was 46.6 (38-50) grams.

Records of occurrences.—Specimens examined, 27: Nicolás, 56 km. NW Tula, 5500 ft., 16; Tajada, 23 mi. NW Tula, 5200 ft., 4; 15 mi. N Tula, 1; 8 mi. N Tula, 4500 ft., 3; 9 mi. SW Tula, 3900 ft., 3.

Additional record: Tula (Lidicker, 1960:178).

Liomys irroratus

Mexican Spiny Pocket Mouse

This species is probably the most common rodent in Tamaulipas. It was taken at almost every locality sampled and was associated with many other kinds of rodents. Its distribution is state-wide with the exception of the extreme northwestern part. Two subspecies are represented in Tamaulipas, *L. i. alleni,* which occurs in the western side of the Sierra Madre Oriental in the southwest part of the state, and *L. i. texensis,* which occupies the rest of the range of the species in the state.

At Soto la Marina specimens were taken in dense brush, around the cultivated fields; no burrows were seen and all specimens were trapped before 10:00 p.m. On the Sierra de Tamaulipas, *Liomys* was collected in practically all microhabitats. In the vicinity of San Fernando, individuals were trapped in a dry area in which vegetation consisted of mesquite, cactus and chollas; the ground there was covered with dry leaves and small sticks, and burrows were found near the base of the mesquite bushes. One specimen was taken near the house of a woodrat. Two kilometers west of El Carrizo, where *Liomys irroratus* is called "ratón tuza," specimens were collected on rocks inclined at an angle of about twenty-five degrees that were covered with zacatón grass and some bushes. Some individuals were taken in a sugar cane field that was surrounded by bushes and tall grass; *Baiomys taylori, Sigmodon hispidus,* and *Peromyscus leucopus* were taken in the line of traps. One specimen was caught in a trap baited with banana.

Some dates concerning reproduction of *Liomys irroratus* in Tamaulipas are as follows: La Pesca, May 25, one female lactating and one female pregnant with 4 embryos that measured 8 mm.; Jaumave, July 26-29, three females lactating and three pregnant females that carried 6 embryos (6 mm.), 6 embryos (15 mm.), and 5 embryos (15 mm.); Palmillas, July 23, a female with 1 embryo measuring 6 mm.; Nicolás, October 19, a female carrying 4 embryos measuring 3 mm.

Liomys irroratus alleni (Goues)

1881. *Heteromys alleni* Coues, Bull. Mus. Comp. Zool., 8:187, March, type from Río Verde, San Luis Potosí.

1911. *Liomys irroratus alleni,* Goldman, N. Amer. Fauna, 34:56, September 7.

Distribution in Tamaulipas.—Extreme southwestern part of state.

This subspecies is easily distinguished from *L. i. texensis* by the following features: hind foot larger, 31.5 (30-33.5) instead of 27.8 (27-29); skull longer, 34.2 (32.4-36.4) instead of 31.5 (30.0-32.5); maxillary tooth-row longer, 5.4 (5.0-5.8) instead of 5.0 (4.8-5.1); interorbital constriction relatively narrower in *alleni.* Intergradation between *L. i. alleni* and *L. i. texensis* takes place at Rancho Santa Rosa (where, of the two specimens, one is conspicuously larger than the other), eight kilometers northeast of Antiguo Morelos, El Encino, and Ejido Santa Isabel. All specimens from the localities mentioned are here assigned to *texensis.*

Weight of three pregnant females averaged 68.9 (64-78) grams, that of non-pregnant females, 65.6 (64-68), and that of six males 73.0 (65-80).

Records of occurrence.—Specimens examined, 34: Villa Mainero, 1700 ft., 2; Nicolás, 56 km. NW Tula, 5500 ft., 6; Jaumave, 2400 ft., 23; 16 mi. N, 6 mi. W Palmillas, 5500 ft., 1; 14 mi. N, 6 mi. W Palmillas, 5500 ft., 2.

Additional records: Miquihuana (Goldman, 1911:56); Tula (Hooper and Handley, 1958:18).

Liomys irroratus texensis Merriam

1902. *Liomys texensis* Merriam, Proc. Biol. Soc. Washington, 15:44, March 5, type from Brownsville, Cameron Co., Texas.

1911. *Liomys irroratus texensis,* Goldman, N. Amer. Fauna, 34:59, September 7.

Distribution in Tamaulipas.—State-wide except extreme southwestern and northwestern parts.

Intergradation occurs between *L. i. texensis* and *L. i. pretiosus* in southeastern Tamaulipas as noted previously by Hooper (1953:5). Individuals from Altamira and one mile south thereof are small and dark as in *pretiosus*, but cranial measurements are as in *texensis* to which they are here assigned. Specimens from the vicinity of Tampico are typical *texensis*.

Average weight of the specimens from three different localities are as follows: Soto la Marina, seven males, 42.7, 14 females, 36.9; Sierra de Tamaulipas, 12 males, 47.3, 20 females, 40.7; Sierra Madre Oriental, eight males, 45.5, nine females, 37.0 grams.

The specimens reported by Ingles (1959:394) from two miles south of El Mante as *L. irroratus* are here referred to *texensis* on geographic grounds.

Records of occurrence.—Specimens examined, 121: 7 km. S, 2 km. W San Fernando, 7; 7 km. SW La Purisima, 1; Rancho Santa Rosa, 25 km. N, 13 km. W Cd. Victoria, 260 m., 2; 36 km. N, 10 km. W Cd. Victoria, 1; 15 mi. N Cd. Victoria, 2; 4 mi. N La Pesca, 5; Soto la Marina, 25; Sierra Madre Oriental, 5 mi. S, 3 mi. W Cd. Victoria, 1900 ft., 18; Sierra de Tamaulipas, 2 mi. S, 10 mi. W Piedra, 1200 ft., 36; Sierra de Tamaulipas, 3 mi. S, 10 mi. W Piedra, 1200 ft., 1; Ejido Santa Isabel, 2 km. W Pan-American Highway, 2000 ft., 3; Rancho Pano Ayuctle, 25 mi. N, 3 km. W El Mante, 300 ft., 1; Rancho Pano Ayuctle, 6 mi. N Gómez Farías, 300 ft., 8; 10 km. N, 8 km. W El Encino, 400 ft., 1; 2 km. W El Carrizo, 6; 53 km. N El Limón, 4; 8 km. NE Antiguo Morelos, 2; Altamira, 1; 1 mi. S Altamira, 3; 10 mi. NW Tampico, 1; 7 km. N Tampico, 2.

Additional records: Hidalgo (Goldman, 1911:59); Matamoros (*ibid.*); Bagdad (*ibid.*); Sierra de San Carlos (Hooper and Handley, 1948:20); 3 mi. W Soto la Marina (Hooper, 1953:5); [Cd.] Victoria (Goldman, 1911: 59); Acuña (Hooper and Handley, 1948:20); Mesa de Llera (Hooper, 1953:5); Gómez Farías (Goodwin, 1954:9); 2 mi. S Cd. Mante (Ingles, 1959:394); Antiguo Morelos (Hooper and Handley, 1948:20).

Castor canadensis mexicanus V. Bailey

Beaver

1913. *Castor canadensis mexicanus* V. Bailey, Proc. Biol. Soc. Washington, 26:191, October 23, type from Ruidoso Creek, 6 mi. below Ruidoso, Lincoln Co,. New Mexico.

Distribution in Tamaulipas.—Probably in the Río Grande drainage.

The beaver has been reported in Tamaulipas only from Matamoros (Baird, 1858:355—three specimens) and from 12 miles below, south of, Matamoros (V. Bailey, 1905:124). In Tamaulipas the beaver may occur only in the Río Grande drainage.

Oryzomys palustris

Marsh Rice Rat

Previous to this report only one subspecies of *Oryzomys palustris* had been recorded from Tamaulipas. Careful examination of the available material from the state shows that *O. p. aquaticus* occurs in the east and *O. p. peragrus* lives in the southwestern part of the state.

In general, specimens examined were trapped in dense brush alongside waterholes as at Altamira, or around cornfields as at the place 36 kilometers north and 10 kilometers west of Ciudad Victoria, where the bushes were mesquite and other kinds of Acacias. There the ground was covered by cat claw, and no grass was seen near the traps in which *O. palustris* was caught. In the Sierra de Tamaulipas a specimen was caught among rocks and bushes. Ingles (1959:395) reported that his specimens were trapped alive in dense brush and "tules."

A female taken at Jaumave on July 25 had 5 embryos, each 20 mm. in crown-rump length.

Oryzomys palustris aquaticus J. A. Allen

1891. *Oryzomys aquaticus* J. A. Allen, Bull. Amer. Mus. Nat. Hist., 3:289, June 30, type from Brownsville, Cameron Co., Texas.

1918. *Oryzomys couesi aquaticus*, Goldman, N. Amer. Fauna, 43:39, September 23.

1960. *Oryzomys palustris aquaticus*, Hall, The Southwestern Nat., 5:173, November 1.

Distribution in Tamaulipas.—North part of state, and coastal area south to Tampico.

Weights of two males were 80 and 82, and of a female 66 grams.

Oryzomys palustris aquaticus differs from *O. p. peragrus* in having a rich cinnamon, reddish color and the interorbital region constricted to less than 14.7 per cent of the greatest length of the skull. *O. p. peragrus* is ochraceous and grayish. The least width of its interorbital region is more than 14.5 per cent of the greatest length of the skull. Individuals studied from the Sierra de Tamaulipas are typical *aquaticus*. Of those from Altamira, one has the color as in *aquaticus*, but the color of the other two resembles that of *peragrus;* nevertheless, all of the mentioned specimens are here assigned to *aquaticus*.

Records of occurrence.—Specimens examined, 4: Sierra de Tamaulipas, 10 mi. W, 2 mi. S Piedra, 1200 ft., 1; 6 mi. N, 6 mi. W Altamira, 2; 5 mi. N, 5 mi. W Altamira, 1.

Additional records: Camargo (Goldman, 1918:40); Matamoros (*ibid.*); near Cd. Tampico (Ingles, 1958:395).

Oryzomys palustris peragrus Merriam

1901. *Oryzomys mexicanus peragrus* Merriam, Proc. Washington Acad. Sci., 3:283, July 26, type from Río Verde, San Luis Potosí.

1918. *Oryzomys couesi perargrus*, Goldman, N. Amer. Fauna, 43:39, September 23.

1960. *Oryzomys palustris peragrus*, Hall, The Southwestern Nat., 5:173, November 1.

Distribution in Tamaulipas.—Western part of state, along Sierra Madre Oriental.

Two males from Jaumave weighed 62 and 65 and one pregnant female weighed 67 grams.

Most records of *O. p. peragrus* are from places along the Sierra Madre Oriental, but Lawrence (1947:103) recorded a specimen from the Río Corona, which is east of, but not far from the mentioned Sierra. Baker (1951:215) reported two specimens from two different localities labeled with reference to Ciudad Victoria (same specimens reported here) as *O. p. aquaticus,* but pointed out that they tended "toward the darker *O. c. peragrus.*" Examination of more material and taking into consideration the relation between the interorbital constriction and the greatest length of skull, cause me here to refer those specimens to *peragrus.*

Hooper (1953:8) reported three young specimens from Rancho Pano Ayuctle as of the subspecies *aquaticus,* but study of two adults from the same locality reveals that this locality should be included within the geographic range of *peragrus.*

Records of occurrence.—Specimens examined, 9: 36 km. N, 10 km. W Cd. Victoria, 1; Jaumave, 2400 ft., 5; Rancho Pano Ayuctle, 25 mi. N, 3 km. W El Mante, 2; 70 km. S Cd. Victoria (by highway) and 6 km. W of Highway, 1.

Additional records: Río Corana (Lawrence, 1947:103); Pano Ayuctle (Hooper, 1953:8).

Oryzomys melanotis

Black-cared Rice Rat

Oryzomys melanotis occurs in Tamaulipas from Soto la Marina southward. Two subspecies are recorded: *O. m. carrorum* in the north and *O. m. rostratus* in the tropical area from Rancho Pano Ayuctle to Altamira.

Specimens from the Sierra de Tamaulipas were trapped along a stream, edged with trees, bushes and rocks; at Rancho Pano Ayuctle the animals were in grass between banana groves. The specimen from 70 kilometers south of Ciudad Victoria was taken in tall grass near a field of sugar cane in a line of traps that yielded also *Peromyscus leucopus, Sigmodon hispidus, Liomys irroratus,* and *Oryzomys fulvescens.* Hooper (1953:8) and Ingles (1959:395) reported *O. melanotis* as caught at the edges of cane fields.

Oryzomys melanotis carrorum Lawrence

1947. *Oryzomys rostratus carrorum* Lawrence, Proc. New England Zool. Club, 24:101, May 29, type from Rancho Santa Ana, about 8 mi. SW Padilla, Río Soto la Marina, Tamaulipas.

1959. *Oryzomys melanotis carrorum,* Hall and Kelson, The Mammals of North America, 2:560, March 21.

Distribution in Tamaulipas.—Southeast part of state; known only from the type locality and the Sierra de Tamaulipas.

The original description of this subspecies was based on three specimens collected at Rancho Santa Ana. Specimens examined from the Sierra de Tamaulipas extended the known range 45 miles southeast of the type locality, and also extend the previously known altitudinal range of 300-350 feet elevation to 1200 feet.

Specimens examined correspond in color and measurements to those recorded by Lawrence (1947:102-103). Of 12 specimens studied, the tympanic bullae of six touch the surface of the table when the skull rests on the tips of the incisors and the occipital condyles. In the other six the bullae are 0.3 to 1.3 mm. above the table top. The mesopterygoid space in the specimens examined are broad and U-shaped and not V-shaped as in the three specimens examined by Lawrence (*op. cit.*). Weight of six males was 52.5 (48-63) and of four females 44.7 (40-49) grams.

Measurements.—Average and extreme measurements of six males are as follows: 255.3 (240-269); 135.7 (120-147); 135.7 (120-147); 30.4 (30-31); 21 (20-22); greatest length of skull, 31.6 (30.9-32.5); zygomatic breadth, 15.3 (14.7-16.1); interorbital constriction, 4.8 (4.5-5.1); breadth of skull, 31.6 (30.9-32.5); length of nasals, 12.9 (12.4-13.4); length of anterior palatine foramina, 5.5 (5.2-5.7); length of palatal bridge, 6.1 (5.8-6.4); length of maxillary tooth-row, 4.0 (3.9-4.1). The females average slightly smaller.

Records of occurrence.—Specimens examined, 12 from Sierra de Tamaulipas, 10 mi. W, 2 mi. S Piedra, 1200 ft.

Additional record: Type locality (Lawrence, 1947:102).

Oryzomys melanotis rostratus Merriam

1901. *Oryzomys rostratus* Merriam, Proc. Washington Acad. Sci., 3:293. July 26, type from Metlatoyuca, Puebla.

1953. *Oryzomys melanotis rostratus,* Hooper, Occ. Papers Mus. Zool., Univ. Michigan, 544:8, March 25.

Distribution in Tamaulipas.—Extreme southeastern part of state, in tropical area.

Ingles (1959:395) reported one specimen from two miles north of Ciudad Mante as *O. melanotis;* here it is referred to *O. m. rostratus* on geographic grounds.

Records of occurrence.—Specimens examined, 2: 2 km. W El Carrizo, 1; Rancho Pano Ayuctle, 25 mi. N El Mante and 3 km. W Highway, 1.

Additional records: 2 mi. N Cd. Mante (Ingles, 1959:395); Altamira (Goldman, 1918:54).

Oryzomys alfaroi huastecae Dalquest

1951. *Oryzomys alfaroi huastecae* Dalquest, Jour. Washington Acad. Sci., 41:363, November 14, type from 10 km. E Platanito, San Luis Potosi.

Distribution in Tamaulipas.—Known only from Rancho del Cielo (Hooper, 1953:8).

Oryzomys fulvescens

Pygmy Rice Rat

The pygmy rice rat in Tamaulipas was collected in grass. Two kilometers west of El Carrizo in grass around a sugar cane field, traps, baited with scraps of deer meat, caught *Oryzomys fulvescens,*

Sigmodon hispidus, Peromyscus leucopus and *Liomys irroratus.* Seven kilometers north of Tampico, *O. fulvescens* was taken along with *Peromyscus leucopus, Sigmodon hispidus* and *Baiomys taylori.*

A female obtained on March 2, at Rancho Pano Ayuctle, had 4 embryos 16 mm. in crown-rump length.

Oryzomys fulvescens fulvescens (Saussure)

1860. *H[esperomys]. fulvescens* Saussure, Revue et Mag. Zool., Paris, ser. 2, 12:102, March, type from Veracruz; fixed by Merriam (Proc. Washington Acad. Sci., 3:295, July 26, 1901) at Orizaba.

1897. *Oryzomys fulvescens,* J. A. Allen and Chapman, Bull. Amer. Mus. Nat. Hist., 9:204, June 16.

Distribution in Tamaulipas.—Reported only from Rancho del Cielo (Goodwin, 1954:10).

Oryzomys fulvescens engracie Osgood

1945. *Oryzomys fulvescens engracie* Osgood, Jour. Mamm., 26:300, November 14, type from Hacienda Santa Engracia (32 km. N), NW of Cd. Victoria, Tamaulipas.

Distribution in Tamaulipas.—Central and southeast parts of state.

Records of occurrence.—Specimens examined, 13: 2 km. W El Carrizo, 5; Rancho Pano Ayuctle, 25 mi. N, 3 km. W El Mante, 6; 10 km. N, 8 km. W El Encino, 1; 7 km. N Tampico, 1.

Additional record: Altamira (Osgood, 1945:300).

Reithrodontomys megalotis hooperi Goodwin

Western Harvest Mouse

1954. *Reithrodontomys megalotis hooperi* Goodwin, Amer. Mus. Novit., 1660:1, May 25, type from Rancho del Cielo, 5 mi. NW Gómez Farías, 3500 ft., Tamaulipas.

Distribution in Tamaulipas.—Known only from type locality.

Reithrodontomys fulvescens

Fulvous Harvest Mouse

This is the most common species of *Reithrodontomys* in Tamaulipas; it occurs in almost all parts of the state, from sea level to high up in the mountains and from the tropical forest to the desert plain.

The three subspecies in the state are *R. f. intermedius* in the northern half, *R. f. griseoflavus* in the high parts of the Sierra Madre Oriental, and *R. f. tropicalis* in the southeast. The lines between these subspecies are difficult to establish because the zones of intergradation are broad. Characters for separating the three subspecies in Tamaulipas are listed by Hooper (1952).

Reithrodontomys fulvescens griseoflavus Merriam

1901. *Reithrodontomys griseoflavus* Merriam, Proc. Washington Acad. Sci., 3:553, November 29, type from Ameca, 4000 ft., Jalisco.

1952. *Reithrodontomys fulvescens griseoflavus*, Hooper, Miscl. Publ. Mus. Zool., Univ. Michigan, 77:98, January 16.

Distribution in Tamaulipas.—Known only from Jaumave.

Only specimens from Jaumave are clearly R. *f. griseoflavus;* all others east of this locality are intergrades between *griseoflavus* and *tropicalis,* under which latter subspecies they are included. In *griseoflavus* the tail is longer in relation to the head and body, 141.2 (135-153) per cent, than in the other two subspecies that occur in Tamaulipas. The average weight of 14 males was 14 (12-16) grams.

Record of occurrence.—Specimens examined, 15, from Jaumave, 2400 ft.

Reithrodontomys fulvescens intermedius J. A. Allen

1895. *Reithrodontomys mexicanus intermedius* J. A. Allen, Bull. Amer. Mus. Nat. Hist., 7:136, May 21, type from Brownsville, Cameron Co., Texas.

1914. *Reithrodontomys fulvescens intermedius,* A. H. Howell, N. Amer. Fauna, 36:47, June 5.

Distribution in Tamaulipas.—Northern half of state.

No specimen of this subspecies has been examined. Jones and Anderson (1958:447) reported specimens from Rancho Pano Ayuctle as *R. f. intermedius,* but here those same specimens are assigned to *R. f. tropicalis.* J. A. Allen (1891:223) recorded specimens from Santa Teresa as *Ochetodon mexicanus.* According to Hooper (1952: 142) that name was used by Allen for R. *fulvescens.* Allen's specimens from Santa Teresa are here referred to R. *f. intermedius* on geographic grounds.

Records (Hooper, 1952:108): Camargo, 200 ft.; 20 mi. S Reynosa, Charco Escondido; Matamoros, 30 ft.; 7.5 mi. S Matamoros; 29 mi. S Cd. Victoria, 800 ft.; Hacienda Santa Engracia, 800 ft.; Santa Teresa (50 mi. SW Matamoros); Sierra San Carlos (El Mulato, Tamaulipeca, 1500 ft.).

Reithrodontomys fulvescens tropicalis Davis

1944. *Reithrodontomys fulvescens tropicalis* Davis, Jour. Mamm., 25:393, December 12, type from Boca del Río, 8 km. S city of Veracruz, Veracruz.

Distribution in Tamaulipas.—Tropical area in southeastern part of state.

Most of the specimens examined of R. *fulvescens* are included in this subspecies, principally because of their reddish coloration that is characteristic of *R. f. tropicalis.* According to the original descripton by Davis (1944:393) this subspecies is smaller than *griseoflavus* and the posterior border of the incisive foramina terminate anterior to the plane of the molars. But, these characteristics are not found in any specimen examined from Tamaulipas and the average of external measurements is more than those given by Hooper (1952: 109) for *tropicalis.* Of all specimens from Tamaulipas, those from the vicinity of Altamira and Tampico are most nearly typical of *tropicalis.* Weights of seven males and five females, from the Sierra de Tamaulipas, were, respectively, 13 (11-15), and 11 (9-14) grams.

Records of occurrence.—Specimens examined, 51: Rancho Santa Rosa, 25 km. N, 13 km. W Cd. Victoria, 1; Cd. Victoria, 3; Sierra de Tamaulipas, 10 mi. W, 2 mi. S Piedra, 1200 ft., 12; 2 km. W El Carrizo, 1; Ejido Santa Isabel, 2 km. W Pan-American Highway, 2000 ft., 14; Rancho Pano Ayuctle, 25 mi. N, 3 km. W El Mante, 300 ft., 4; Rancho Pano Ayuctle, 6 mi. N Gómez Farías, 300 ft., 4; 6 mi. N, 6 mi. W Altamira, 2; 1 mi. S Altamira, 3; 16 km. N Tampico, 3; 7 km. N Tampico, 4.

Additional records: Hidalgo (Hooper, 1952:110); 5 mi. NE Gómez Farías, 1100 ft. (*ibid.*); La Azteca, 5 km. NNE Gómez Farías (Goodwin, 1954:11); Gómez Farías (*ibid.*); Antiguo Morelos (Hooper, 1952:110); 2 mi. W Tampico (Ingles, 1959:396).

Reithrodontomys mexicanus mexicanus (Saussure)

Mexican Harvest Mouse

1860. *R [eithrodon]. mexicanus* Saussure, Revue et Mag. Zool., Paris, ser. 2, 12:109, type from mountains of Veracruz; restricted to Mirador, Veracruz, by Hooper, Miscl. Publ. Mus. Zool., Univ. Michigan, 77:140, January 16.

1914. *Reithrodontomys mexicanus mexicanus,* A. H. Howell, N. Amer. Fauna, 36:70, June 5. Not *Reithrodontomys mexicanus* (Saussure), being instead of J. A. Allen, 1895:135, which in part equalled *Reithrodontomys fulvescens difficilis.*

Distribution in Tamaulipas.—Known from two localities, but probably occurs in all tropical areas in south part of state.

As noted before, J. A. Allen (1891:223) reported specimens from Rancho Santa Rosa as *Ochetodon mexicanus,* but he used this name for the species now known as R. *fulvescens.*

The specimen examined, previously reported by Jones and Anderson (1958:447), represents the northernmost occurrence of the species.

Records of occurrence.—One specimen examined from Rancho Pano Ayuctle, 6 mi. N Gómez Farías, 300 ft.

Additional record: Rancho del Cielo, 3500 ft. (Hooper, 1952:144).

Peromyscus maniculatus blandus Osgood

Deer Mouse

1904. *Peromyscus sonoriensis blandus* Osgood, Proc. Biol. Soc. Washington, 17:56, March 21, type from Escalón, Chihuahua.

1909. *Peromyscus maniculatus blandus* Osgood, N. Amer. Fauna, 28:84, April 17.

Distribution in Tamaulipas.—Reported only from Miquihuana (Osgood, 1909:86).

Peromyscus melanotis J. A. Allen and Chapman

Black-eared Mouse

1897. *Peromyscus melanotis* J. A. Allen and Chapman, Bull. Amer. Mus. Nat. Hist., 9:203, June 16, type from Las Vigas, Veracruz.

Distribution in Tamaulipas.—Known only from Miquihuana (Osgood, 1909: 112).

Peromyscus leucopus texanus (Woodhouse)
White-footed Mouse

1853. *Hesperomys texana* Woodhouse, Pioc. Acad. Nat. Sci. Philadelphia, 6:242, type probably from vicinity of Mason, Mason Co., Texas.

1909. *Peromyscus leucopus texanus*, Osgood, N. Amer. Fauna, 28:127, April 17.

Distribution in Tamaulipas.—Over all of state.

This is the most common species of the genus *Peromyscus* in Tamaulipas. It and *Liomys irroratus* are the two rodents most easily trapped throughout the state. In general P. *l. texanus* occurs in forested and brushy areas especially under 1200 feet in elevation, as was noted in the Sierra de Tamaulipas, where P. *l. texanus* was taken commonly at elevations of up to 1200 feet. Above this elevation the species was rare and P. *pectoralis* and P. *boylii* were more abundant than at lower elevations. The three specimens of P. *l. texanus* from 12 kilometers north and four kilometers west of Ciudad Victoria were trapped in a line of 110 traps set near tree stumps. Small burrows in the ground were noted here. The forest at this locality was composed of mesquite, ebony, acacias, a few yuccas and "nopales" (= cactuses); the ground was covered by cat claw.

Of the many young taken, 15 specimens were saved from Ejido Santa Isabel where *P. leucopus* was abundant in an area of chaparral consisting of wild "tomate," "zapote," "huizache" and "salvadora." Most of the specimens caught at this locality were taken between 7:30 and 9:30 p. m. in traps baited with a mixture of rolled oats, peanut butter and banana. Specimens from 53 kilometers north of El Limón were taken along with *Liomys irroratus;* the specimen from two kilometers west of El Carrizo was trapped near a dead mesquite log. *Reitrodontomys fulvescens* was taken in the same area. Four specimens of P. *leucopus* were taken at Rancho Pano Ayuctle, around a big pile of old firewood in an abandoned sugar mill. At the locality six miles north and six miles west of Altamira, P. *leucopus* was found in cultivated fields and along the grassy roadsides; in the vicinity of Tampico specimens were taken in an area of forested cactus-thorn. The specimen from seven kilometers south and two kilometers west of San Fernando was found in a trap set at the base of "nopal" cactus, which was surrounded by bushes and small trees (10-12 feet high).

Breeding records are as follows: Rancho Pano Ayuctle, on February 15, one female carried 2 embryos of 23 mm. in crown-rump length; Jaumave, July 26 to 29, five females, averaging 4.6 (3-6) embryos of 7 (3-15) mm., two females lactating, one on May 25 and the other on July 26; Ejido Santa Isabel, on January 20 to 25, three females lactating; Soto la Marina, on May 16, one female lactating.

Average weights were as follows: from Jaumave four pregnant females, 28.0 (25-33), eight males, 23.4 (21-27); from the Sierra de Tamaulipas, eight females non-pregnant, 21.2 (18-26), 14 males, 22.0 (19-27); from 6 mi. N, 6 mi. W Altamira, six males, 23.5 (21-27).

All specimens examined from Tamaulipas are assigned to P. *l. texanus* because their coloration is pale. Even so the color varies some according to locality; specimens from Rancho Pano Ayuctle and the Sierra de Tamaulipas have much of the cinnamon color that is characteristic of P. *l. incensus* from farther south, but even so specimens from the two localities last mentioned are paler than those from Veracruz that are typical *incensus.*

Goldman (1942:158) reported specimens from Altamira as P. *l. incensus,* in which subspecies Ingles (1959:397) included specimens from two miles west of Tampico, but specimens examined from the same area do not differ from individuals from far north thereof; for this reason I identify specimens from these localities as *texanus.* Osgood (1909:131) and Hooper (1953:7) also referred specimens from the southern part of Tamaulipas to *texanus.* These two authors examined 156 specimens and did not find any intergradation between *texanus* and *incensus,* but to me, the cinnamon tones of specimens from Rancho Pano Ayuctle and the Sierra de Tamaulipas, suggest intergradation between the two subspecies.

Osgood's (1909:265) measurements of P. *l. texanus,* from Brownsville, Texas, and those of 40 specimens from different localities in Tamaulipas are about the same except that the anterior palatine foramina average longer in Tamaulipas. Baker's (1956:262) specimens from Coahuila, averaged larger even than Tamaulipan specimens. Another difference between Osgood's measurements and Baker's was the shorter 3.4 (3.0-3.7) maxillary tooth-row in Tamaulipan specimens.

Hooper (1953:7) recorded specimens from General Terán, as in Tamaulipas; actually this locality is in Nuevo León.

Records of occurrence.—Specimens examined, 149: 4.5 mi. S Nuevo Laredo, 1; 3 mi. SE Reynosa, 2; 7 km. S, 2 km. W San Fernando, 1; Villa Mainero, 1700 ft., 1; Rancho Santa Rosa, 25 km. N, 13 km. W Cd. Victoria, 260 m., 2; 9.5 mi. SW Padilla, 800 ft., 2; 15 mi. N Cd. Victoria, 2; 4 mi. N La Pesca, 1; Soto la Marina, 11; La Pesca, 1; 12 km. N, 4 km. W Cd. Victoria, 3; 7 km. NE Cd. Victoria, 1; Sierra de Tamaulipas, 10 mi. W, and 2 mi. S Piedra, 1200 ft., 31; Ejido Eslabones, 10 mi. W, 2 mi. S Piedra, 1200 ft., 6; Jaumave, 20; Ejido Santa Isabel, 2 km. W Pan-American Highway, 2000 ft., 15; 53 km. N El Limón, 12 km. S Río Guayalejo, 5; Rancho Pano Ayuctle, 25 mi. N El Mante, 3 km. W Highway, 300 ft., 16; Rancho Pano Ayuctle, 6 mi. N Gómez Farías, 300 ft., 7; 8 km. W, 10 km. N El Encino, 400 ft., 3; 8 mi. N Tula, 4500 ft., 2; 2 km. W El

Carrizo, 3; 6 mi. N, 6 mi. W Altamira, 9; 16 km. N Tampico, 1; 7 km. N Tampico, 3.

Additional records (Osgood, 1909:131, unless otherwise noted): Nuevo Laredo; Mier; Camargo; near Bagdad; Sierra San Carlos (Hooper, 1953:7); Matamoros-Victoria Highway (*ibid.*); Charco Escondido (Baird, 1858:464); Hidalgo; Cd. Victoria; 10 mi. NE Zamorina (Hooper, 1953:7); Gómez Farías (Goodwin, 1954:12); Chamal (*ibid.*); Tula (Hooper, 1953:7); Antiguo Morelos (*ibid.*); Altamira (Goldman, 1942:158); 2 mi. W Tampico (Ingles, 1959:397); Tampico.

Peromyscus boylii

Brush Mouse

Specimens examined were obtained at higher elevations in the oak-tree zone of the Sierras in traps set among rocks, trees and in grassy areas. *Peromyscus boylii* was trapped in the same area as was P. *pectoralis* and no habitat distinction between the two was noted. Some behavioral differences, however, are pointed out in the account of P. *pectoralis.* Morphological differences between these two species in Tamaulipas were reported by Hooper (1952: 372).

A female taken on August 5 in the Sierra Madre Oriental carried two embryos 15 mm. in crown-rump length.

For the taxonomic status of P. *boylii* in Tamaulipas see Alvarez (1961).

Peromyscus boylii ambiguus Alvarez

1961. *Peromyscus boylii ambiguus* Alvarez, Univ. Kansas Publ. Mus. Nat. Hist., 14:118, December 29, type from Monterrey, Nuevo León.

Distribution in Tamaulipas.—Known only from the Sierra San Carlos.

Record of occurrence.—Specimens examined, 7 (UMMZ), all from La Vegonia, Sierra San Carlos.

Peromyscus boylii levipes Merriam

1898. *Peromyscus levipes* Merriam, Proc. Biol. Soc. Washington, 12:123, April 30, type from Mt. Malinche, 8400 ft., Tlaxcala.

1909. *Peromyscus boylii levipes,* Osgood, N. Amer. Fauna, 28:153, April 17.

Distribution in Tamaulipas.—Central and southern parts of state.

Weights of 19 males and 18 females from the Sierra Madre Oriental are, respectively, 25.2 (22-30) and 23.6 (20-29); weights of eight males and five females from the Sierra de Tamaulipas are 24.9 (22-32) and 29.6 (24-31).

Records of occurrence.—Specimens examined, 54: Sierra Madre Oriental, 8 mi. S, 6 mi. W Victoria, 4000 ft., 37; 5 mi. S, 3 mi. W Victoria, 1900 ft., 2; Ejido Eslabones, 10 mi. W, 2 mi. S Piedra, 1200 ft., 1; Sierra de Tamaulipas, 11 mi. W, 8 mi. S Piedra, 2000 ft., 13; 2 km. W El Carrizo, 1.

Additional records: Rancho del Cielo (Hooper, 1953:7); 3 mi. NW Acuña (*ibid.*); Rancho Viejo (Goodwin, 1954:12); Santa María (*ibid.*); Joya de Salas (*ibid.*).

Peromyscus pectoralis

White-ankled Mouse

Peromyscus pectoralis and P. *boylii* are closely related morphologically and seem to occupy the same habitat. In the Sierra Madre Oriental, according to the field notes of the collector (Heinrich, June 6 to August 5, 1953), individuals of P. *pectoralis* had a pinkish coloration on the mouth and forefeet produced by the juice of the "nopal" cactus fruit, on which obviously the mice feed, whereas only a few specimens of *boylii* were thus discolored. It was noted that *boylii* was feeding on acorns. Furthermore, the two species may differ in time of breeding; in August, males of *pectoralis* had the testes well developed when those organs were small in *boylii* collected at the same locality.

A specimen from 53 kilometers north of El Limón, was shot at a height of 10 feet on a concrete underpass. Other specimens were taken in a trap line that yielded *Peromyscus boylii*, P. *leucopus* and *Liomys irroratus*.

Two subspecies of P. *pectoralis* occur in Tamaulipas: P. *p. collinus* is widely distributed in the central and western parts of the state and P. *p. eremicoides* occurs only in the western "corner" of the state.

Peromyscus pectoralis collinus Hooper

1952. *Peromyscus pectoralis collinus* Hooper, Jour. Mamm., 33:372, August 19, type from San José, 2000 ft., Sierra San Carlos, 12 mi. NW San Carlos, Tamaulipas.

Distribution in Tamaulipas.—Along the central and western mountains.

A female obtained on January 21 at a place 53 kilometers north of El Limón, contained three embryos. A lactating female was taken on August 2 in the Sierra Madre Oriental. Males, as previously noted, had well-developed testes in August. The weights of 17 males and 20 females from the Sierra de Tamaulipas were, respectively, 26.6 (24-33), and 25.6 (21-31) grams.

Measurements of specimens from different localities in Tamaulipas averaged about the same, except that those of specimens from Palmillas, averaged smaller. The small size suggests intergradation between the subspecies *collinus* and *eremicoides*. The latter occurs to the west and differs from *collinus* in smaller size, more grayish coloration, completely white tarsal joint and relatively longer tail. Hooper (1952:374) reported specimens from Jaumave as intergrades between the two subspecies before mentioned and Osgood (1909: 164) identified two specimens from there as *eremicoides*. In the present account, individuals from Palmillas and Jaumave are referred to *collinus*.

Records of occurrence.—Specimens examined, 101: 7 km. SW La Purisima, 1; Sierra Madre Oriental, 5 mi. S, 3 mi. W Victoria, 1900 ft., 12; Sierra Madre Oriental, 8 mi. S, 6 mi. W Victoria, 4000 ft., 16; Sierra de Tamaulipas, 2 mi. S, 10 mi. W Piedra, 1200 ft., 36; Sierra de Tamaulipas, 3 mi. S, 14 mi. W Piedra, 1200 ft., 14; 14 mi. N, 6 mi. W Palmillas, 5500 ft., 1; Palmillas, 4400 ft., 3; 53 km. N El Limón, 12 km. S Río Guayalejo, 5; Joya Verde, 35 km. SW Victoria, 3800 ft., 9; 10 km. N, 8 km. El Encino, 400 ft., 1; 8 km. NE Antiguo Morelos, 500 ft., 3.

Additional records (Hooper, 1952:374, unless otherwise noted): Sierra San Carlos (Marmolejo, 1700 ft., San José, 2000 ft., Tamaulipeca, 1500 ft., La Vegonia, 2900 ft.); Villagran, 1300 ft.; Cd. Victoria; near Jaumave, 2400 ft.; Sierra de Tamaulipas, near Acuña, 1600 ft.; La Joya de Salas (Goodwin, 1954:12).

Peromyscus pectoralis eremicoides Osgood

1904. *Peromyscus attwateri eremicoides* Osgood, Proc. Biol. Soc. Washington, 17:60, March 21, type from Mapimi, Durango.

1909. *Peromyscus pectoralis eremicoides*, Lyon and Osgood, Bull. U. S. Nat. Mus., 62:128, January 28.

Distribution in Tamaulipas.—Known only from Miquihuana and vicinity of Tula.

The two specimens from Miquihuana are typical P. *pectoralis eremicoides* in external and cranial measurements. Specimens from nine miles southwest of Tula are characteristic of *eremicoides* in cranial measurements but the tail is shorter than usual for this subspecies, in this respect approaching P. *p. lacianus.*

Measurements.—Average and extreme measurements of 10 specimens from nine miles southwest of Tula and measurements of two males (56169, 56415) from Miquihuana are, respectively, as follows: 181.5 (173-197), 180, 197; 96.2; (87-110), 103, 113; 20.2 (19.0-21.5), 21, 21; 18.1 (16.5-19.0), 18, —; greatest length of skull, 24.8 (24.1-25.6), 25.5, 25.6; length of nasals, 9.0 (8.6-9.3), 9.3, 9.3; zygomatic breadth, 12.2 (11.7-12.8), 12.3, 12.9; interorbital constriction, 3.8 (3.7-4.0), 3.7, 3.9; length of maxillary tooth-row, 3.6 (3.5-3.7), 3.6, 3.8. Weights of the 10 specimens from nine miles southwest of Tula average 17.9 (16-24) grams.

Records of occurrence.—Specimens examined, 28: Miquihuana, 6200 ft., 2; Nicolás, 56 km. NW Tula, 5500 ft., 1; Tajada, 23 mi. NW Tula, 5200 ft., 1; 8 mi. N Tula, 4500 ft., 2; 9 mi. SW Tula, 3900 ft., 19; 17 mi. SW Tula, 3900 ft., 3.

Peromyscus melanophrys consobrinus Osgood

Plateau Mouse

1904. *Peromyscus melanophrys consobrinus* Osgood, Proc. Biol. Soc. Washington, 17:66, March 21, type from Berriozabal, Zacatecas.

Distribution in Tamaulipas.—Mexican Plateau part of state.

A lactating female caught on July 20 and four males from Miquihuana weighed, respectively, 51, and 50.2 (47-54) grams. A female, taken on July 24, 14 miles north and six miles west of Palmillas in a valley covered by mesquite and other bushes, had 3 embryos 10 mm.

in crown-rump length, and weighed 60 grams. One specimen from nine miles southwest of Tula was caught in an outcrop of rocks and two others were taken among bushes on the desert. A female on October 10 carried 4 embryos 2 mm. in crown-rump length.

Specimens of P. *melanophrys* here listed are the first to be reported from Tamaulipas. They are assigned to the subspecies *consobrinus* on the basis of dark color and because their size closely corresponds to that of the holotype. The specimen from the vicinity of Palmillas and one from Miquihuana (56408) are larger than the others and grayish.

A specimen (56413) from Miquihuana lacks all the molariform teeth. Its alveoli in one maxilla are closed and those in the opposite maxilla are more open than is normal.

Measurements.—Average and extreme measurements of four males, two females (56413, 56408) from Miquihuana, and a female (56414) from 14 miles north and 6 miles west of Palmillas, are, respectively, as follows: total length (two males only), 249, 245, 265, 247, 280; length of tail vertebrae (two males only), 137, 134, 141, 131, 157; length of hind foot, 26.7 (26-27), 27, 27, 27; ear from notch, 23.7 (23-24), 25, 24, 25; greatest length of skull, 30.3 (29.5-31.0), 31.2, 31.8, 32.2; interorbital constriction, 4.8 (4.7-4.9), 4.9, 4.8, 5.0; length of palatine slits, 6.6 (6.2-6.8), 6.9, 6.9, 6.8; length of diastema, 8.1 (8.0-8.3), —, 8.5, 8.5; alveolar length of maxillary tooth-row, 4.5 (4.3-4.7), —, 4.3, 4.6.

Records of occurrence.—Specimens examined, 16: Miquihuana, 6200 ft., 6; 14 mi. N, 6 mi. W Palmillas, 5500 ft., 1; Nicolás, 56 km. NW Tula, 5500 ft., 6; 9 mi. SW Tula, 3900 ft., 3.

Peromyscus difficilis petricola Hoffmeister and de la Torre

Zacatecan Deer Mouse

1959. *Peromyscus difficilis petricola* Hoffmeister and de la Torre, Proc. Biol. Soc. Washington, 72:167, November 4, type from 12 mi. E San Antonio de las Alazanas, 9000 ft., Coahuila.

Distribution in Tamaulipas.—Westernmost part of state.

The three specimens from Miquihuana were collected among rocks and stumps, in an oak forest. The specimens from 20 miles north of Tula were collected after midnight on a hillside covered mainly with juniper brush. A female (October 11) carried 3 embryos 26 mm. in crown-rump length.

Records of occurrence.—Specimens examined, 6: Miquihuana, 8500 ft., 3; 20 mi. N Tula, 5800 ft., 3.

Peromyscus ochraventer Baker

El Carrizo Deer Mouse

1951. *Peromyscus ochraventer* Baker, Univ. Kansas Publ., Mus. Nat. Hist., 5:213, December 15, type from 70 km. (by highway) S Ciudad Victoria, 6 km. W Pan-American Highway at El Carrizo, Tamaulipas.

Distribution in Tamaulipas.—Vicinity of the type locality.

The series of specimens examined was the same used by the original describer of the species. He (1951:214-215) pointed out that the mice were taken in junglelike forest among rocks and adjacent to logs. Burrows extended beneath large blocks of limestone, and each burrow where a mouse was caught was marked by a pile of excavated earth resembling a tiny mound left by a pocket gopher. These burrows were at an elevation of approximately 2800 feet above sea level on the steep sides of a small hill in an area where the vegetation was intermediate between that of the arid and humid subdivisions of the tropical region. Each of two females, captured on January 13, carried five placental scars; one of the females was lactating.

Records of occurrence.—Specimens examined, 24, from the type locality.

Additional records (Goodwin, 1954:12): Gómez Farías; Rancho del Cielo; Joya de Salas.

Baiomys taylori taylori (Thomas)

Northern Pygmy Mouse

1887. *Hesperomys (Vesperimus) taylori* Thomas, Ann. Mag. Nat. Hist., ser. 5, 19:66, January, type from San Diego, Duval Co., Texas.

1907. *Baiomys taylori* Mearns, U. S. Nat. Mus., Bull. 56:381, April 13.

Distribution in Tamaulipas.—All of state, except southwestern desert part.

The species of this genus have been revised recently by Packard (1960) and the specimens from Tamaulipas are arranged according to his systematic findings. The weight of 35 specimens labeled with reference to Altamira are 7.6 (6.0-9.0) grams; 15 from Jaumave weigh 6.9 (6.0-9.0) grams. Pregnant females were collected as follows: February 22, Ejido Santa Isabel, 3 (embryos x 4 mm. in crown-rump length); March 2, Rancho Pano Ayuctle, 6 x 16; July 9, six miles north and six miles west of Altamira, 1 x 4; July 28 and 29, Jaumave, 2 x 8 and 3 x 9. The average number of embryos was 2.8 (1-5).

Records of occurrence.—Specimens examined, 83: 4 mi. N La Pesca, 1; Cd. Victoria, 3; Jaumave, 2400 ft., 17; Ejido Santa Isabel, 2 km. W Pan-American Highway, 2000 ft., 7; Rancho Pano Ayuctle, 25 mi. N, 3 km. W El Mante, 300 ft., 4; Rancho Pano Ayuctle, 6 mi. N Gómez Farías, 300 ft., 1; Río Sabinas, 8 km. N El Encino, 400 ft., 1; 2 km. W El Carrizo, 2; 6 mi. N, 6 mi. W Altamira, 33; 5 mi. N, 5 mi. W Altamira, 4; 1 mi. S Altamira, 3; 16 km. N Tampico, 4; 10 mi. NW Tampico, 1; 7 mi. S Altamira, 1; 1 km. N Tampico, 1.

Additional records (Packard, 1960:654): Camargo; Charco Escondido, 20 mi. S Reynosa; Matamoras (= Matamoros); Hidalgo; 29 mi. N Cd. Victoria; Antiguo Morelos.

Onychomys leucogaster longipes Merriam

Northern Grasshopper Mouse

1889. *Onychomys longipes* Merriam, N. Amer. Fauna, 2:1, October 30, type from Concho County, Texas.

1913. *Onychomys leucogaster longipes,* Hollister, Proc. Biol. Soc. Washington, 26:216, December 20.

Distribution in Tamaulipas.—From Ciudad Victoria northward.

Only a young female was examined; she weighed 22 grams and extends the known range 59 miles eastward from Ciudad Victoria.

Record of occurrence.—One specimen examined from Soto la Marina, 500 ft. Additional records (Hollister, 1914:253): Camargo; Reynosa; [Cd.] Victoria.

Onychomys torridus subrufus Hollister

Southern Grasshopper Mouse

1914. *Onychomys torridus subrufus* Hollister, Proc. U. S. Nat. Mus., 47:472, October 29, type from Miquihuana, Tamaulipas.

Distribution in Tamaulipas.—West of Sierra Madre Oriental.

The six specimens examined were collected in the desert area west of the Sierra Madre Oriental. At Nicolás a trap set in front of a hole held one specimen, and another was trapped beneath a brush fence that inclosed a cornfield. *Dipodomys merriami* and *Perognathus penicillatus* also were trapped beneath the fence.

A subadult from Nicolás is slightly larger (see measurements) than either of two subadults from four miles north of Jaumave and an old specimen from eight miles north of Tula, except in the interorbital constriction, which is narrower. Nevertheless measurements of Tamaulipan *Onychomys torridus* resemble those given by Hollister (1914:483) for *O. t. subrufus*. A specimen from Nicolás is also darker than other individuals examined.

A female taken on July 15, four miles north of Jaumave, was lactating.

Measurements.—Measurements of a female from Nicolás, a male from eight miles north of Tula, and a female and a male from four miles north of Jaumave are as follows: 158, 147, 145, 144; 59, 58, 55, 55; 22, 21, 22, 22; 21, 20.5, 18, 18; condylobasal length, 24.4, 23.1, 23.9, 23.7; interorbital constriction, 4.1, 4.4, 4.3, 4.5; length of nasals, 10.6, 10.5, 10.5, 10.1; length of maxillary toothrow, 3.8, 3.6, 3.7, 3.7; breadth of braincase, 11.8, 11.3, 11.3, 11.0; weight in grams, 32.5, 26.0, 25.0, 25.0.

Records of occurrence.—Specimens examined, 6: 4 mi. N Jaumave, 2; Nicolás, 56 km. NW Tula, 5500 ft., 2; Tajada, 23 mi. NW Tula, 5200 ft., 1; 8 mi. N Tula, 4500 ft., 1.

Additional records (Hollister, 1914:475): Miquihuana; Jaumave.

Sigmodon hispidus

Hispid Cotton Rat

This species, as is known, is active by day and by night. It occurs mainly in grassy areas and most of the specimens examined were trapped there. But, one mile east of La Pesca, specimens were taken on a beach having sparse grass. *Neotoma micropus* and *Spermophilus spilosoma*, but no smaller rodents, were taken there. Also, many crabs were found in the traps. Possibly only the relatively large rodents are able to compete successfully with the crabs. The specimen from one kilometer east of El Barretal was caught in a rat-trap set in front of small hole in a fence of dead brush that surrounded a cornfield. The area outside the fence supported mesquite and ebony trees (10-12 feet high) and the ground was covered with cat claw. Six miles north and six miles west of Altamira, the two young specimens were taken on a small grassy island surrounded by mud.

According to natives, *Sigmodon* injures corn and sugar cane. Probably other species of rodents are responsible for some or all of such damage since other kinds of rodents were taken in the same areas.

Dice (1937:245) reported females from the Sierra San Carlos that carried 8 embryos of 18 mm., 5 x 33, 7 embryos very small, and 8 x 20. Females were collected on July 22, 29, and 30.

Sigmodon hispidus berlandieri Baird

1855. *Sigmodon berlandieri* Baird, Proc. Acad. Nat. Sci. Philadelphia, 7:333, type from Río Nazas, Coahuila.

1902. *Sigmodon hispidus berlandieri,* V. Bailey, Proc. Biol. Soc. Washington, 15:106, June 2.

Distribution in Tamaulipas.—From Jaumave and Llera to north.

This subspecies is distinguished from *S. h. toltecus* by larger size and paler, grayish coloration.

Baker (1951:216) reported a specimen from 35 kilometers north and 10 kilometers west of Ciudad Victoria (= 1 km. E El Barretal) as *S. h. toltecus.* Comparison of its skull with those from the vicinity of Altamira (*S. h. toltecus*) and those from Jaumave (*S. h. berlandieri*) shows that the skull from El Barretal closely resembles those

TABLE 3.—DATA ON REPRODUCTION.

LOCALITY	Date	Embryos	Size in mm.
4 mi. N La Pesca	May 26	4	30
Sierra de Tamaulipas	June 10	3	10
Sierra de Tamaulipas	June 11	4	10
Sierra de Tamaulipas	June 20	2	20
Ciudad Victoria	July 12	5	5
Jaumave	July 28	4	14
Jaumave	July 29	6	25
San Fernando	August 30	7	20
San Fernando	August 31	8	11
Vicinity of Nuevo Laredo	November 15	3	5
Vicinity of Nuevo Laredo	November 16	5	2

from Jaumave, in having the zygomatic arches more nearly parallel and the braincase more rounded than in skulls from Altamira. Therefore the specimen from the vicinity of El Barretal is here assigned to *S. h. berlandieri.*

Records of occurrence.—Specimens examined, 64: 4½ mi. S Nuevo Laredo, 600 ft., 1; 10 mi. S, 11 mi. E Nuevo Laredo, 8; San Fernando, 180 ft., 8; 4 mi. N La Pesca, 10; 3 mi. N La Pesca, 1; 1 mi. E La Pesca, 3; Soto la Marina, 500 ft., 1; 36 km. N, 10 km. W Cd. Victoria, 1 km. E El Barretal, Río Purificación, 1; Cd. Victoria, 1; 2 km. W Pan-American Highway (12 km. S Llera), Ejido Santa Isabel, 2000 ft., 1; Jaumave, 2400 ft., 29.

Additional records: Matamoros (Baird, 1858:506); Sierra San Carlos (El Mulato, Tamaulipeca, San Miguel) (Dice, 1937:254); Mesa de Llera (Hooper, 1953:9); Tamaulipas [state?] (Baird, 1858:506).

Sigmodon hispidus solus Hall

1951. *Sigmodon hispidus solus* Hall, Univ. Kansas Publ., Mus. Nat. Hist., 5:42, October 1, type from island 88 mi. S, 10 mi. W Matamoros, Tamaulipas.

Distribution in Tamaulipas.—Known only from two specimens from the type locality.

Sigmodon hispidus toltecus (Saussure)

1860. [*Hesperomys*] *toltecus* Saussure, Revue et Mag. Zool., Paris, ser. 2, 12:98, type from mountains of Veracruz [probably near Mirador, Dalquest, Louisiana State Univ. Studies, Biol. Sci. Series, 1:163, December 28, 1953].

1902. *Sigmodon hispidus toltecus*, V. Bailey, Proc. Biol. Soc. Washington, 15:110, June 2.

Distribution in Tamaulipas.—Tropical region in southern part of state. The specimen reported by Baker (1951:216) from one mile east of El Barretal is here referred to *S. h. berlandieri.*

Records of occurrence.—Specimens examined, 69: Sierra de Tamaulipas, 10 mi. W, 2 mi. S Piedra, 1200 ft., 24; Sierra de Tamaulipas, 11 mi. W, 8 mi. S Piedra, 2000 ft., 1; Rancho Pano Ayuctle, 25 mi. N El Mante, 3 km. W highway, 300 ft., 3; Rancho Pano Ayuctle, 6 mi. N Gómez Farías, 300 ft., 3; 8 km. W, 10 km. N El Encino, 400 ft., 2; 2 km. W El Carrizo, 2100 ft., 20; 6 mi. N, 6 mi. W Altamira, 8; 6 mi. N, 4 mi. W Altamira, 1; 5 mi. N, 5 mi. W Altamira, 3; 1 mi. S Altamira, 1; 16 km. N Tampico, 3.

Additional records: Rancho del Cielo, 15 to 20 mi. S Mesa de Llera (Hooper, 1953:9); Cd. Mante (Ingles, 1959:398); Tampico (Booth, 1957:15).

Neotoma albigula subsolana Alvarez

White-throated Woodrat

1962. *Neotoma albigula subsolana* Alvarez, Univ. Kansas Publ. Mus. Nat. Hist., 14:141, April 30, type from Miquihuana, 6400 ft., Tamaulipas.

Distribution in Tamaulipas.—Western side of Sierra Madre Oriental.

At Nicolás specimens were taken in traps set along a thorn fence and at Tajada two specimens were trapped along a rock wall. At other places some specimens were brought in by natives who captured the rats by tearing apart their houses.

Five females taken on October 18 at Nicolás carried embryos (one to two per female), which averaged 22.2 (11-45) mm. in crown-rump length. Another female, taken nine miles southwest of Tula on October 13, carried 2 embryos that were 35 mm. in crown-rump length. The average weight of the five pregnant females was 196.7 (183-207) grams. The average weights of nine adult males and six non-pregnant females from Miquihuana were, respectively, 215.6 (175-250) and 162.5 (155-175) grams.

Records of occurrence.—Specimens examined, 51: Miquihuana, 6400 ft., 22; Joya Verde, 35 km. SW Cd. Victoria (on Jaumave Road) 3800 ft., 2; Nicolás, 56 km. NW Tula, 5500 ft., 10; Tajada, 23 mi. NW Tula, 5200 ft., 2; 9 mi. SW **Tula, 3900 ft., 15.**

Additional record: Jaumave (Goldman, 1910:37).

Neotoma angustapalata Baker

Tamaulipas Wood Rat

1951. *Neotoma angustapalata* Baker, Univ. Kansas Publ., Mus. Nat. Hist., 5:217, December 15, type from 70 km. by highway S Ciudad Victoria, and 6 km. W Pan-American highway at El Carrizo, Tamaulipas.

Distribution in Tamaulipas.—Southern part of state; presently known from two localities.

Baker (1951:218) reported that specimens from the type locality were taken in crevices among rocks on a small hillside that supported a sparse cover of vegetation growing from a deep layer of humus. The specimen from eight kilometers west and 10 kilometers north of El Encino was shot about 40 yards from the entrance to a large cave, but no sign of wood rats were found there. Hooper (1953:9) reported that *N. angustapalata* occupied caves at Rancho del Cielo, where a female with two nursing young was taken.

When Baker (*op. cit.*) described *Neotoma angustapalata* on the basis of two specimens from El Carrizo, he assigned the species to the *N. mexicana* group because of the deep anterointernal re-entrant angle of M1. The deep angle found in *N. mexicana* differs markedly from the typical condition in either *N. micropus* or *N. albigula.* Study of the cranial characters and bacula of specimens of *N. micropus* and *N. angustapalata* tends to corroborate the statement of Hooper (1953:10), who commented on the taxonomic relationships of *N. angustapalata* as follows: "It should be pointed out that all characters considered . . . the specimens [*angustapalata*] appear to be large, deeply pigmented examples of the species *N. micropus* notwithstanding the deep anterior fold in M1. The presence of that deep fold is far from an absolute character in the *mexicanus* [*sic*] group."

My study of 48 crania of *N. micropus* from Tamaulipas reveals that the depth of the re-entrant angle of M1 is extremely variable, from almost absent in some individuals to deep (as in *angustapalata*) in others. Four specimens, one (56958) from the Sierra de Tamaulipas and three (56960, 56965, 56966) from the vicinity of Altamira, have the re-entrant angle as deep as in the holotype and topotype of *angustapalata.*

Comparison of the bacula of the holotype and one topotype of *angustapalata* with 15 bacula of *N. micropus* reveal that on the average the baculum of *angustapalata* differs from that of *micropus* in being longer, and narrower at the base (greatest length, 7.1, width at base, 3.4 mm., in the topotype). One specimen of *N. micropus littoralis* from the vicinity of Altamira, however, has a baculum of the same shape as in *angustapalata* (this same specimen is one of the three from there in which the re-entrant angle of the M1 is deep). The shape of the baculum among specimens of *micropus* is highly variable and bacula of specimens from different localities frequently are slightly different (see Fig. 5).

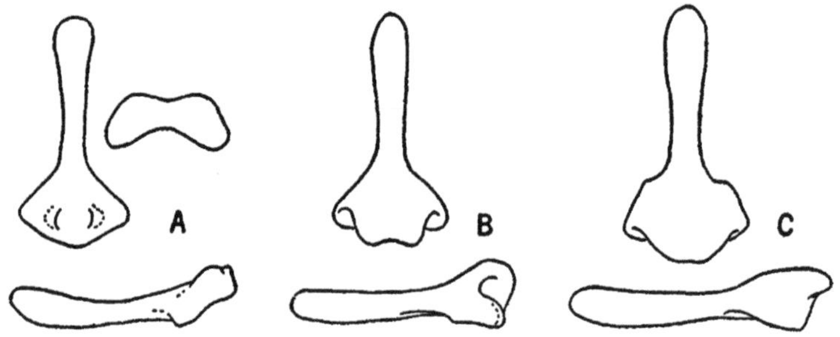

FIG. 5. Bacula of *Neotoma.* All × 4.
A, *Neotoma angustipalata* (topotype, 37062).
B, *Neotoma micropus micropus* (4 mi. SW Nuevo Laredo, 89147).
C, *Neotoma micropus littoralis* (Sierra de Tamaulipas, 2 mi. S, 10 mi. W Piedra, 56957).

The known distributions of *N. micropus* and *N. angustapalata* do not overlap (neither does the distribution of *N. albigula* overlap with either in Tamaulipas). The four specimens of *N. micropus* having the deep re-entrant angle in M1 are from localities near where the ranges of *angustapalata* and *micropus* probably meet. This could be interpreted in two ways: (1) these four specimens can be regarded as intergrades between *angustapalata* and *micropus,* in which case the former species should be placed as a subspecies of the latter. Or the four specimens, which were collected along with other specimens that lack deep re-entrant angles in the M1, can be assigned, on the basis of the deep angle, to *angustapalata,* in which case the species *micropus* and *angustapalata* would be in part sympatric. Until more material from critical areas is available for study, I continue to recognize *angustapalata* as a monotypic species. I agree with Hooper that it is closely related to *N. micropus.*

Measurements.—A female (58865) from 8 km. west and 10 km. north of El Encino, measured as follows: 404; 198; 41; 32; greatest length of skull, 49.7; basilar length, 40.8; zygomatic breadth, 25.9; length of nasals, 18.8; length of incisive foramina, 10.8; length of maxillary tooth-row, 9.9; greatest breadth of interpterygoid space, 4.0.

Records of occurrence.—Specimens examined, 3: 8 km. W, 10 km. N El Encino, 400 ft., 1; type locality, 2.

Neotoma micropus

Southern Plains Wood Rat

Most of the specimens examined were trapped in brushy areas. On the Sierra de Tamaulipas, wood rats were caught in steel traps set near or between rocks. In the vicinity of La Pesca, specimens were trapped on the beach where *Spermophilus spilosoma* and *Sigmodon hispidus* were taken also.

Two females, obtained on May 19 and June 10 at Soto la Marina and on the Sierra de Tamaulipas, respectively, each carried 2 embryos that were 40 mm. in crown-rump length. Dice (1937:254) reported that two females collected on July 24 and August 16 on the Sierra San Carlos each carried 2 embryos that ranged from 34 to 36 mm. in crown-rump length.

Neotoma micropus occurs throughout the Tamaulipan Biotic Province and is represented in Tamaulipas by two subspecies, each of which has its type locality in the state. Intergradation between the two takes place at Soto la Marina.

Neotoma micropus littoralis Goldman

1905. *Neotoma micropus littoralis* Goldman, Proc. Biol. Soc. Washington, 18:31, February 2, type from Altamira, 100 ft., Tamaulipas.

Distribution in Tamaulipas.—From the Sierra de Tamaulipas southward.

Weight of two males and three non-pregnant females was 248, 254, 185, 210, 240 grams, respectively.

Records of occurrence.—Specimens examined, 14: Sierra de Tamaulipas, 2 mi. S, 10 mi. W Piedra, 1200 ft., 6; 6 mi. N, 6 mi. W Altamira, 8.

Additional record: Altamira (Goldman, 1910:29).

Neotoma micropus micropus Baird

1855. *Neotoma micropus* Baird, Proc. Acad. Nat. Sci. Philadelphia, 7:333, April, type from Charco Escondido, Tamaulipas.

Distribution in Tamaulipas.—From Soto la Marina northward.

The weight of five males and four females from Soto la Marina averaged, respectively, 256.4 (210-317) and 233.0 (195-274) grams.

A specimen (56924) from La Pesca differs from all other specimens of *N. micropus* examined in being smaller, having a conspicuously shorter rostrum, broader intraorbital canal, and lower broader braincase. External measurements of this specimen are as follows: 347; 155; 39; —. Its cranial measurements are: greatest length, 44.8; basilar length, 34.3; zygomatic breadth, 23.6; interorbital constriction, 6.2; incisive foramina, 6.5; length of maxillary tooth-row, 8.7; width of mesopterygoid fossa, 4.1.

Records of occurrence.—Specimens examined, 58: 4 mi. SW Nuevo Laredo, 900 ft., 14; 4½ mi. S Nuevo Laredo, 1; 3 mi. SE Reynosa, 1; 3 mi. S Matamoros, 2; 33 mi. S Washington Beach, 1; San Fernando, 180 ft., 1; 7 km. S, 2 km. W San Fernando, 2; 12 mi. NW San Carlos, 1300 ft., 4; 9½ mi. SW Padilla, 800 ft., 3; 3 mi. N Soto la Marina, 3; Soto la Marina, 500 ft., 12; 4 mi. N La Pesca, 3; 1 mi. E La Pesca, 1; La Pesca, 2; 3 mi. NE Guemes, 1; 7 mi. NE Cd. Victoria, 1; Cd. Victoria, 6.

Additional records (Goldman, 1910:28, unless otherwise noted): Nuevo Laredo; 10 mi. S Nuevo Laredo (Booth, 1957:15); Camargo; Matamoros; Bagdad; 40 mi. S Matamoros (Hooper, 1953:9); Sierra San Carlos (El Mulato, Tamaulipeca) (Dice, 1937:254); San Fernando (J. A. Allen, 1891:224); Forlón.

Microtus mexicanus subsimus Goldman

Mexican Vole

1938. *Microtus mexicanus subsimus* Goldman, Jour. Mamm., 19:494, November 14, type from Sierra Gaudalupe, southeastern Coahuila.

Distribution in Tamaulipas.—Reported only from mountains near Miquihuana (Goldman, 1938:495).

Canis latrans

Coyote

In Tamaulipas two and possibly three subspecies of *Canis latrans* occur. *C. l. texensis* is known only from the northwesternmost part of the state, and *N. l. microdon* occurs from Camargo south to Nicolás. Hall and Kelson (1959:845) guessed that *C. l. cagottis* would be found in the southern third of the state; as yet specimens from there have not been obtained and the subspecific identity of the coyotes there, if any are present, remains in doubt.

Canis latrans microdon Merriam

1897. *Canis microdon* Merriam, Proc. Biol. Soc. Washington, 11:29, March 15, type from Mier, on Río Grande, Tamaulipas.

1932. *Canis latrans microdon*, Nelson, Proc. Biol. Soc. Washington, 45:224, November 26.

Distribution in Tamaulipas.—Probably state-wide, reported only from the northern half of the state.

Three specimens were examined. One is a pup from the vicinity of Padilla which is assigned to this subspecies on geographic grounds. The other two are skins, collected at Nicolás by natives, who deceived the collector by providing dog skulls with the coyote skins. These two specimens are referred to *C. l. microdon* on the basis of their dark color and dusky shading on the throat and chest. One has a rufous over-all color and the other is ochraceous yellowish. This difference in color suggests intergradation at this place between *C. l. microdon* that ranged to the northeast, *C. l. cagottis* to the south, and probably with *C. l. impavidus* distributed to the west.

Records of occurrence.—Specimens examined, 3: 9½ mi. SW Padilla, 800 ft., 1; Nicolás, 53 km. N Tula, 2.

Additional record: Camargo (Jackson, 1951:305); 20 mi. W Reynosa (Ingles, 1959:401); Matamoros (Jackson, 1951:305); Bagdad (*ibid.*); Sierra San Carlos (San Miguel, El Mulato) (Dice, 1937:251).

Canis latrans texensis V. Bailey

1905. *Canis nebrascensis texensis* V. Bailey, N. Amer. Fauna, 25:175, October 24, type from 45 mi. SW Corpus Christi at Santa Gertrudis, Kleberg Co., Texas.

1932. *Canis latrans texensis* V. Bailey, N. Amer. Fauna, 53:312, March 11.

Distribution in Tamaulipas.—Extreme northwest, known only from Nuevo Laredo (Jackson, 1951:279).

Canis lupus monstrabilis Goldman

Gray Wolf

1937. *Canis lupus monstrabilis* Goldman, Jour. Mamm., 18:42, February 11, type from 10 mi. S Rankin, Upton Co., Texas.

Distribution in Tamaulipas.—Probably extinct, recorded only from Matamoros (Goldman, 1944:468).

On the maps of distribution of *C. l. monstrabilis* published by Leopold (1959:400) and Baker and Villa (1960:370), Tamaulipas is included in the region in which the wolf is considered to be extinct.

Urocyon cineroargenteus scottii Mearns

Gray Fox

1891. *Urocyon virginianus scottii* Mearns, Bull. Amer. Mus. Nat. Hist., 3:236, June 5, type from Pinal Co., Arizona.

1895. *Urocyon cinereo-argenteus scottii*, J. A. Allen, Bull. Amer. Mus. Nat. Hist., 7:253, June.

Distribution in Tamaulipas.—All of state in suitable habitats.

The specimen from the Sierra Madre Oriental was obtained by a collector who used a rabbit call. Leopold (1959:408) reported that the highest elevation [about 2800 feet] at which he found gray fox in México was at Hacienda de Acuña, in the Sierra de Tamaulipas, where "dense, brushy draws and oak openings made ideal habitat." At this place Leopold saw, in early August, a family of foxes, four well-grown young and their parents. Dice (1937:250) reported *U. c. texensis* (a junior synonym of *U. c. scottii*), as abundant in the Sierra San Carlos.

The six specimens examined do not present any significant difference in size and shape of the skull from specimens of *scottii* from Arizona, except that one skull from the Sierra de Tamaulipas is smaller than the others, suggesting intergradation between the subspecies *scottii* and *tropicalis* from farther south.

Records of occurrence.—Specimens examined, 6: 2 mi. W San Fernando, 180 ft., 1; 15 km. W Rancho Santa Rosa, Sierra Madre Oriental, 4500 ft., 1; Ejido Santa Isabel, 2000 ft., 1; Sierra de Tamaulipas, 2 mi. S, 10 mi. W Piedra, 1200 ft., 2; Joya Verde, 35 km. SW Victoria, 3800 ft., 1.

Additional records: Near Marmolejo, San Carlos Mts. (Dice, 1937:250); Hacienda Acuña, Sierra de Tamaulipas (Leopold, 1959:408, only seen); La Joya de Salas (Goodwin, 1954:14).

Ursus americanus eremicus Merriam

Black Bear

1904. *Ursus americanus eremicus* Merriam, Proc. Biol. Soc. Washington, 17:154, October 6, type from Sierra Gaudalupe, Coahuila.

Distribution in Tamaulipas.—Probably in high and remote parts of the Sierra Madre Oriental; recorded only from Agua Linda (Goodwin, 1954:14).

Bassariscus astutus flavus Rhoads

Ringtail

1894. *Bassariscus astutus flavus* Rhoads, Proc. Acad. Nat. Sci. Philadelphia, 45:417, January 30, type from Texas, exact locality unknown.

Distribution in Tamaulipas.—Western half of state.

The two specimens examined provide the second record of this species in Tamaulipas; they were shot in the bottom of an arid canyon. One animal was about 30 feet up from the ground in an oak tree, and the other was along a small arroyo containing pools of water.

From Rhoads' paper (1893:416-417) on the genus *Bassariscus* it would seem that *B. astutus flavus* differs from *B. a. astutus* in smaller size, especially of the skull, shorter tail (shorter than head and body in *flavus* and longer than head and body in *astutus*) and the presence of fulvous color. Comparison of 10 specimens of *B. a. flavus* from Coahuila and Texas with two of *B. a. astutus* (Distrito Federal, 1; Las Vigas, Veracruz, 1) from central México reveals that the skulls do not differ qualitatively and that the skull of *flavus* tends to be smaller and relatively wider, but that there is overlap in size. In all *flavus* that I measured and in the two adults of *astutus* the tail is shorter than the head and body. The only real difference is the color; ringtails from Texas are deep fulvous instead of grayish as is *astutus* from the Distrito Federal and Veracruz. But the specimen from Veracruz has much fulvous and on the other hand specimens from Coahuila are more grayish than those from Texas.

The two specimens from Tamaulipas can be assigned to either subspecies *astutus* or *flavus* with almost equal propriety. Here they are referred to *B. a. flavus* on the basis of their relatively small skull, short tail, and presence of some fulvous color.

Measurements.—Measurements of female and male (60239, 60240), both adult, from Joya Verde, are, respectively: 745, 760; 370, 385; 70, 75; 47, 56; greatest length of skull (excluding incisors), 81.9, 83.1; zygomatic breadth, 46.1, 51.9; interorbital constriction, 16.3, 16.3; postorbital constriction, 19.5, 18.5; breadth of braincase, 33.7, 36.6; length of maxillary tooth-row, 31.5, 32.0; breadth across postorbital processes (tip to tip), 25.3, 26.8.

Records of occurrence.—Two specimens examined from Joya Verde, 35 km. SW Victoria, 3800 ft.

Additional record: Joya de Salas (Goodwin, 1954:14).

Procyon lotor

Racoon

Racoons occur all through the state. The one specimen examined was shot about 11:00 p. m. in a cypress tree. Its mouth contained fresh corn. The animal was notably fat and weighed 11 pounds. According to the natives the racoons do much damage in cornfields.

Procyon lotor fuscipes Mearns

1914. *Procyon lotor fuscipes* Mearns, Proc. Biol. Soc. Washington, 27:63, March 20, type from Las Moras Creek, 1011 ft., Fort Clark, Kinney Co., Texas.

Distribution in Tamaulipas.—Practically all of state, except western part.

Records (Goldman, 1950:51, unless otherwise noted): Camargo; Matamoros; Bagdad; Marmolego; Camp 2 (= 73 mi. S Washington Beach, Selander *et al.*, 1962:338, recorded only to species); Gómez Farías (Goodwin, 1954:14); Altamira.

Procyon lotor hernandezii Wagler

1831. *Pr [ocyon]. hernandezii* Wagler, Isis von Oken, 24:514, type from Tlalpan, Valley of Mexico.

1890. *Procyon lotor hernandezi*, J. A. Allen, Bull. Amer. Mus. Nat. Hist., 3:176, December 10.

Distribution in Tamaulipas.—Western part of state; known only from Rancho Santa Rosa.

The specimen examined is identified as P. *l. hernandezii* because the animal differs from specimens of P. *l. fuscipes* from southern Texas and Coahuila in the same way that Goldman (1950:50) noted that P. *l. hernandezii* differs from P. *l. fuscipes.* For example, in the specimen from Rancho Santa Rosa the interorbital region is lower, the braincase is less depressed near the fronto-parietal suture, the postorbital process is longer and more pointed, and the upper carnassial is longer. The color is the same as in specimens of *fuscipes* from Texas except that the postauricular spot is smaller, and the ground color is slightly more grayish. The median dorsal area is black, forming a longitudinal band about 3 cm. wide.

Record of occurrence.—One specimen examined from Rancho Santa Rosa, 25 km. N, 13 km. W Cd. Victoria.

Nasua narica molaris Merriam

Coati

1902. *Nasua narica molaris* Merriam, Proc. Biol. Soc. Washington, 15:68, March 22, type from Manzanillo, Colima.

Distribution in Tamaulipas.—Over all of state.

A male and female, both adults, from the same locality in the Sierra de Tamaulipas weighed, respectively, 3,150 grams and 4,836 grams. Three young from the same place weighed 2,250, 2,250, and 2,650 grams.

Records of occurrence.—Specimens examined, 7: Sierra de Tamaulipas, 10 mi. W, 2 mi. S Piedra, 1200 ft., 5; Rancho Pano Ayuctle, 25 mi. N El Mante, 3 km. W Pan-American Highway, 2200 ft., 1; 2 km. W El Carrizo, 1.

Additional records: Sierra San Carlos (San José, El Mulato) (Dice, 1937: 249); Soto la Marina (Goldman, 1942:81); Cd. Victoria (*ibid.*); 10 mi. NE Zamorina (Hooper, 1953:3); 3 mi. NW Acuña (*ibid.*); 19 km. SW Mante (Davis, 1944:381).

Potos flavus aztecus Thomas

Kinkajou

1902. *Potos flavus aztecus* Thomas, Ann. Mag. Nat. Hist., ser. 7, 9:268, April, type from Atoyac, Veracruz.

Distribution in Tamaulipas.—Uncertain; one specimen was seen by Leopold (1959:437) near Acuña.

Mustela frenata

Long-tailed Weasel

This species occurs in practically all of the state, but as in most other areas actual records are few; only two specimens, both males, have been examined. One was taken at Jaumave, in a steel-trap baited with fresh egg. It weighed 325 grams. The other was taken in the vicinity of Altamira and weighed 434 grams.

Two subspecies have been reported from Tamaulipas; *Mustela frenata frenata* that occurs in the central and northern parts of the state and *M. f. tropicalis* that occurs in the tropical area in the southern part of the state.

Mustela frenata frenata Lichtenstein

1831. *Mustela frenata* Lichtenstein, Darstellung neuer oder wenig bekannter Säugethiere . . ., pl. 42 and corresponding text, unpaged, type from Ciudad México, México.

1877. *Putorius mexicanus* Coues, Fur-bearing animals, U. S. Geol. Surv. Territories, Misc. Publ., 8:42, a *nomen nudum* [cited by Coues in synonmy as "*Putorius mexicanus*, Berlandier, MMS. ic. ined. 4 (Tamaulipas and Matamoras)"].

Distribution in Tamaulipas.—Central and northern parts of state.

The specimen from Jaumave is clearly *M. f. frenata,* but the other from northwest of Altamira has many characters of the subspecies *M. f. tropicalis* and is an intergrade between the two subspecies. In cranial features and in measurements the animal is like *frenata.* For example: least width of palate more than length of P4; distance between anterior border of auditory bulla and foramen ovale equal to the width of four (including I3) upper incisors; depth of tympanic bulla less than distance between it and foramen ovale; length of tail amounting to 82 per cent of length of head and body. The coloration is more nearly like that of *tropicalis.* For example, the region between the ears and the region behind the ears as far as the shoulders is almost black; hairs of the soles of the forefeet are of the same color as in *tropicalis.* But, width of the whitish underparts amounts to 53 per cent of the circumference of the body; in this respect the specimen is like *frenata.* I refer the specimen to *frenata* because, to me, it is slightly more nearly like it.

Measurements.—The male from 6 mi. N, 6 mi. W Altamira affords measurements as follows: 500; 226; 53; 23; basilar length (Hensel), 49.5; breadth of rostrum, 14.3; interorbital constriction, 11.9; orbitonasal length, 15.2; mastoid breadth, 27.2; zygomatic breadth, 32.4; tympanic bullae, length, 16.8; breadth, 7.5; length of m1, 5.7; P4, lateral length, 5.4, medial, 5.8; M1, breadth, 4.6, length, 2.4; depth of skull at anterior edge of basioccipital, 14.7.

Records of occurrence.—Specimens examined, 2: Jaumave, 2400 ft., 1; 6 mi. N, 6 mi. W Altamira, 1.

Additional records (Hall, 1951:347): Matamoros; Miquihuana.

Mustela frenata tropicalis (Merriam)

1896. *Putorius tropicalis* Merriam, N. Amer. Fauna, 11:30, June 30, type from Jico, Veracruz.

Distribution in Tamaulipas.—Tropical area in south part of state; reported only from 50 mi. south of Ciudad Victoria (Hall, 1951:366).

Eira barbara senex (Thomas)

Tayra

1900. *Galictis barbara senex* Thomas, Ann. Mag. Nat. Hist., ser. 7, 5:146, January, type from Hacienda Tortugas, approximately 600 ft., Jalapa, Veracruz.

1951. *Eira barbara senex,* Hershkovitz, Fieldiana-Zool., 31:561, July 10.

Distribution in Tamaulipas.—Known only from Pano Ayuctle (Hooper, 1953:4).

Taxidea taxus

Badger

The badger in Tamaulipas is poorly known because only a few specimens have been reported from the state. I have examined only two; one is the skull of a juvenile picked up in the sea along the barrier beach and the other is the skull of an adult male taken in a steel-trap baited with a bird body and rabbit meat. The trap was set in front of a hole in the semidesert area 12 miles south of San Carlos.

On their map 471 Hall and Kelson (1959:927) show a total of five subspecies of *Taxidea taxus*. They include the northern part of Tamaulipas in the geographic range of *T. t. berlandieri.* On page 926 Hall and Kelson (*op. cit.*) list ten additional subspecies described by Shantz. One of them *T. t. littoralis* (Shantz, 1949:301) was based on specimens from southeastern Texas and Matamoros, Tamaulipas. Of the two specimens examined by me the one from the barrier beach is here assigned to *T. l. littoralis* on geographic grounds, and the other one from the vicinity of San Carlos to *T. l. berlandieri.*

Taxidea taxus berlandieri Baird

1858. *Taxidea berlandieri* Baird, Mammals, *in* Repts. Expl. Surv. . . ., 8(1):205, July 14, type from Llano Estacado, Texas, near boundary of New Mexico.

1895. *Taxidea taxus berlandieri*, J. A. Allen, Bull. Amer. Mus. Nat. Hist., 7:256, June 29.

Distribution in Tamaulipas.—Reported from only one locality, in northwestern part of state.

The skull examined, of an adult male, differs from Coahuilan and New Mexican skulls in having a broad rostrum, better developed sagittal and lambdoidal crests, and smaller tympanic bullae. The measurements are greater than those given by Shantz (1949:302) for *T. l. littoralis* and it is for that reason that the skull examined is assigned to *T. l. berlandieri.*

Measurements.—The adult male measured as follows: 710; 115; 110; 55; condylobasal length, 123.1; zygomatic breadth, 81.1; mastoid breadth, 75.5; interorbital constriction, 29.3; least postorbital constriction, 27.6; length of maxillary tooth-row, 42.7; P4, length, 11.9, width, 10.7; M1, length, 11.7, width, 11.7; tympanic bulla, length, 23.3, depth (from basioccipital), 12.8.

Record of occurrence.—One specimen examined from 12 mi. S San Carlos, 1300 ft.

Taxidea taxus littoralis Schantz

1949. *Taxidea taxus littoralis* Schantz, Jour. Mamm., 30:301, August 17, type from Corpus Christi, Nueces Co., Texas.

Distribution in Tamaulipas.—Known only from two localities in northeastern part of state.

Records of occurrence.—One specimen examined from 33 mi. S Washington Beach.

Additional record: Matamoros (Schantz, 1949:302).

Spilogale putorius interrupta (Rafinesque)

Eastern Spotted Skunk

1820. *Mephitis interrupta* Rafinesque, Ann. Nat. . . ., 1:3. Type locality, Upper Missouri River?.

1952. *Spilogale putorious interrupta*, McCarley, Texas Jour. Sci., 4:108. March 30.

Distribution in Tamaulipas.—From Sierra de Tamaulipas northward.

The young male from La Pesca weighed 480 grams. In the Sierra de Tamaulipas a lactating female was taken (June 9) in a steel trap. A young male from there weighed 275 grams. The young male from three miles north of La Pesca weighed 520 grams.

Specimens from Tamaulipas are assigned to the subspecies *interrupta* following Van Gelder (1959:270-279). He regarded specimens from Tamaulipas as intergrades between *S. p. interrupta* and *S. p. leucoparia.*

Records of occurrence.—Specimens examined, 6: 9½ mi. SW Padilla, 1; 3 mi. N La Pesca, 1; La Pesca, 1; Rancho Santa Rosa, 2 km. N, 13 km. W Cd. Victoria, 260 m., 1; Sierra de Tamaulipas, 2 mi. S, 10 mi. W Piedra, 1200 ft., 2.

Additional records (Van Gelder, 1959:279): "Tamaulipas"; Cd. Victoria.

Mephitis mephitis varians Gray

Striped Skunk

1837. *Mephitis varians* Gray, Charlesworth's Mag. Nat. Hist., 1:581. Type locality, Texas.

1936. *Mephitis mephitis varians*, Hall, Carnegie Inst. Washington, Publ., 473:66, November 20.

Distribution in Tamaulipas.—North half of state.

Measurements.—An adult female from San Fernando measured as follows: 710; 360; 70; 30; basilar length, 56.2; condylobasal length 64.2; zygomatic breadth, 41.3; interorbital constriction, 19.0; length of maxillary tooth-row, 20.7.

Records of occurrence.—One specimen examined from San Fernando, 180 ft.

Additional records: Mier (A. H. Howell, 1901:32); Matamoros (*ibid.*); 2 mi. up stream from Marmolejo (Dice, 1937:250).

Mephitis macroura macroura Lichtenstein

Hooded Skunk

1832. *Mephitis macroura* Lichtenstein, Darstellung neuer oder weing bekannter Säugethiere . . ., pl. 46, type from mountains northwest of the city of México.

1877. *Mephitis edulis* Coues, Berlandier Mss., Fur-bearing Animals: . . ., U. S. Geol. Surv. Territories, Miscl. Publ., 8:236. Type locality, "Inhabits most of Mexico. I have found it around San Fernando de Bexar . . ."

Distribution in Tamaulipas.—West of Sierra Madre Oriental.

The two specimens from Jaumave are young; they were taken on different nights but in the same place. Weights of male and female, respectively, are 195 and 290 grams. The other three specimens, two young and an adult male, were brought to the collector (Bodley) by natives.

Records of occurrence.—Specimens examined, 5: San Fernando, 180 ft., 2; Jaumave, 2400 ft., 2; Nicolás, 56 km. NW Tula, 5500 ft., 1.

Conepatus mesoleucus mearnsi Merriam

Hog-nosed Skunk

1902. *Conepatus mesoleucus mearnsi* Merriam, Proc. Biol. Soc. Washington, 15:163, August 6, type from Mason, Mason Co., Texas.

Distribution in Tamaulipas.—Probably western part of state, but presently known only from Nicolás.

The specimens herein assigned to this species, represented by the skull only, differ conspicuously from those assigned to *C. leuconotus* only in breadth of M1.

Measurements.—Measurements of a skull (sex undetermined) from Nicolás are as follows: condylobasal length, 77.1; zygomatic breadth, 52.9; postorbital constriction, 21.1; mastoid breadth, 43.7; length of maxillary tooth-row, 23.4; breadth of M1, 7.1.

Records of occurrence.—Two specimens examined from Nicolás, 56 km. NW Tula, 5500 ft.

Conepatus leuconotus texensis Merriam

Eastern Hog-nosed Skunk

1902. *Conepatus leuconotus texensis* Merriam, Proc. Biol. Soc. Washington, 15:162, August 6, type from Brownsville, Cameron Co., Texas.

Distribution in Tamaulipas.—State-wide, except western part.

Three specimens are assigned to this species on the basis of the breadth of M1. In comparison with skulls from the type locality, those of Tamaulipan specimens are slightly smaller and narrower.

Measurements.—Some cranial measurements of a male adult (old) from ten miles west and two miles south of Piedra are: condylobasal length, 79.0; zygomatic breadth, 52.3; postorbital constriction, 22.0; mastoid breadth, 44.2; length of maxillary tooth-row, 24.4; breadth of M1, 9.3.

Records of occurrence.—Specimens examined, 2: La Pesca, 1; Ejido Eslabones, 10 mi. W, 2 mi. S Piedra, 1200 ft., 1.

Additional record: Near El Mulato (Dice, 1937:250).

Felis concolor stanleyana Goldman

Puma

1938. *Felis concolor stanleyana* Goldman, Proc. Biol. Soc. Washington, 51:63, March 18 (renaming of *F. c. youngi* Goldman, Proc. Biol. Soc. Washington, 49:137, August 22, type from Bruni Ranch, near Bruni, Webb Co., Texas).

Distribution in Tamaulipas.—Restricted to mountains of state.

The two specimens examined are skulls only, which were picked up in the field. In general the measurements are like those given by Goldman (1946:233) for the males of *Felis concolor stanleyana*. But the skull from Miquihuana yielded measurements that suggest intergradation between *F. c. stanleyana* and *F. c. azteca* of the western mountains of Tamaulipas.

Measurements.—Two skulls, one from Miquihuana and the second from 9½ mi. SW Padilla, yield measurements as follows: greatest length, 214.0, 213.0; condylobasal length, 195.0, 190.0; zygomatic breadth, 146.0, 140.1; height of skull (frontals to palate), 70.0, 72.4; interorbital constriction, 41.6, 41.4; breadth of nasals (at posterior union between premaxilla and maxilla), 20.1, 17.9; length of maxillary tooth-row, 62.7, 63.3; crown length of P3, 23.3, —; breadth of P3, 11.9, 12.2; anteroposterior diameter of upper canine, 15.1, 15.3.

Records of occurrence.—Specimens examined, 2: 9½ mi. SW Padilla, 800 ft., 1; Miquihuana, 6400 ft., 1.

Additional records: Matamoros (Goldman, 1946:234); Zamorina (Hooper, 1953:4).

Felis onca veraecrucis Nelson and Goldman

Jaguar

1933. *Felis onca veraecrucis* Nelson and Goldman, Jour. Mamm., 14:236, August 17, type from San Andrés Tuxtla, Veracruz.

Distribution in Tamaulipas.—Originally all of state; now restricted to sparsely populated areas.

Only one cranium, from the Sierra de Tamaulipas, was examined. It is in good condition but lacks all the teeth except P3 and P4 on the right side. The measurements are larger than those given by Goodwin (1954:15) for a skull from five miles north of Gómez Farías.

Measurements.—The cranium, sex undetermined, from the Sierra de Tamaulipas, affords measurements as follows: greatest length, 238.0; condylobasal length, 204.0; zygomatic breadth, 166.0; breadth of rostrum, 66.1; interorbital constriction, 48.2; mastoid breadth, 100.7; crown length of carnassial, 24.1.

Records of occurrence.—One specimen examined from Sierra de Tamaulipas, 2 mi. S, 10 mi. W Piedra.

Additional records: between Aldama and Soto la Marina (Nelson and Goldman, 1933:237); 5 km. N Gómez Farias (Goodwin, 1954:15).

Felis pardalis albescens Pucheran

Ocelot

1855. *Felis albescens* Pucheran, in I. Geoffroy Saint-Hilaire, Mammiferes, in Petit-Thoaurs, Voyage autor du monde sur . . . *la Venus* . . ., Zoologie, p. 149, type locality, Arkansas.

1906. *Felis pardalis albescens*, J. A. Allen, Bull. Amer. Mus. Nat. Hist., 22: 219, July 25.

Distribution in Tamaulipas.—All of state, except part west of Sierra Madre Oriental.

Hall and Kelson (1959:961) reported from Tamaulipas two subspecies of *Felis pardalis*. According to Goldman (1943:379) the more northern of the two, *F. p. albescens*, is smaller than the more southern one, *F. p. pardalis*. The skull examined, of a young female, from 10 miles north of Altamira, in southern Tamaulipas, is small, smaller even than skulls of *albescens* from Texas used in comparison. For this reason I here assign the specimen examined to *F. p. albescens* instead of *F. p. pardalis* as did Hall and Kelson (*op. cit.*). Hooper (1953:4) and Dice (1937:251) report as *F. p. pardalis* specimens from 10 miles northeast of Zamorina and others from the Sierra San Carlos. I assume that specimens from these two places should be referred to *albescens* since the specimen from 10 miles north of Altamira, the southernmost locality represented in Tamaulipas, is here referred to *albescens*.

Measurements.—Skull, from 10 mi. N of Altamira, measured as follows: condylobasal length, 97.3; zygomatic breadth, 77.6; squamosal constriction, 50.5; interorbital constriction, 22.2; postorbital constriction, 32.1; length of maxillary tooth-row, 34.7; length of upper carnassial crown (outer side), 13.6.

Records of occurrence.—One specimen examined, from 10 mi. N Altamira.

Additional records: Matamoros (Goldman, 1943:379); Sierra San Carlos (El Mulato and San José) (Dice, 1937:251); Soto la Marina (Goldman, 1943:379); 10 mi. NE Zamorina (Hooper, 1934:4).

Felis wiedii oaxacensis Nelson and Goldman

Margay

1931. *Felis glaucula oaxacensis* Nelson and Goldman, Jour. Mamm., 12: 303, August 24, type from Cerro San Felipe, 10,000 ft., near Oaxaca, Oaxaca.

1943. *Felis wiedii oaxacensis*, Goldman, Jour. Mamm., 24:383, August 17.

Distribution in Tamaulipas.—Probably along Sierra Madre Oriental; known only from Rancho del Cielo (Goodwin, 1954:15).

Felis yaguaroundi cacomitli Berlandier

Yaguaroundi

1895. *Felis cacomitli* Berlandier, *in* Baird, Mammals of the boundary, *in* Emory, Rept. U. S. and Mexican boundary survey 2(2):12, January, type from Matamoros, Tamaulipas.

1905. *Felis yaguaroundi cacomitli*, Elliot, Field Columb. Mus. Publ. 105, Zool. Ser., 6:370, December 6.

1901. *Felis apache* Mearns, Proc. Biol. Soc. Washington, 14:150, August 9, type from Matamoros, Tamaulipas.

Distribution in Tamaulipas.—Eastern and northern parts of Sierra Madre Oriental; known only from type locality and near Gómez Farías (Goodwin, 1954:15).

Lynx rufus texensis J. A. Allen

Bobcat

1895. *Lynx texensis* J. A. Allen, Bull. Amer. Mus. Nat. Hist., 7:188, June 20, based on the description of a bobcat by Audubon and Bachman, The viviparous quadrupeds of North America, 2:293, 1851, from "the vicinity of Castroville, on the headwaters of the Medina [River]," Medina Co., Texas.

1897. *Lynx rufus texensis*, Mearns, Preliminary diagnoses of new mammals . . . from the Mexican boundary line, p. 2, January 12 (preprint of Proc. U. S. Nat. Mus., 20:458, December 24).

Distribution in Tamaulipas.—Probably occurs in western half of state; known only from two localities.

The specimen examined was shot at night at about 3:00 a. m. in the beam of a headlight in typical scrub "monte." The native name for this bobcat in Tamaulipas is "gato rabón."

Measurements.—A male, from Rancho Santa Rosa, measured as follows: 885; 170; 172; 71; condylobasal length, 105.2; interorbital constriction, 22.5; postorbital constriction, 34.6; zygomatic breadth, 83.5; squamosal constriction, 51.7; length of maxillary tooth-row (C-P2), 38.2; length of upper carnassial (outer side), 14.5.

Record of occurrence.—One specimen examined from Rancho Santa Rosa, 360 m.

A itional records: Matamoros (Baird, 1858:96); El Mulato (Dice, 1937: 251)dd

Trichechus manatus latirostris (Harlan)

Manatee

1823. *Manatus latirostris* Harlan, Jour. Acad. Nat. Sci. Philadelphia, 3(1): 394. Type locality, near the capes of East Florida.

1934. *Trichechus manatus latirostris*, Hatt, Bull. Amer. Mus. Nat. Hist., 66:538, September 10.

Distribution in Tamaulipas.—Reported from mouth of Río Grande (Miller and Kellogg, 1955:791); probably extirpated in state.

Tayassu tajacu angulatus (Cope)

Collared Peccary

1889. *Dicotyles angulatus* Cope, Amer. Nat., 23:147, February, type from Guadalupe River, Texas.

1953. *Tayassu tajacu angulatus,* Dalquest, Louisiana State Univ. Studies, Biol. Sci. Ser., 1:207, December 28.

Distribution in Tamaulipas.—All of state, in suitable habitats.

Records: Near El Mulato (Dice, 1937:256); Alta Cima (Goodwin, 1954 15); Rancho del Cielo (*ibid.*); approx. 10 mi. N Cues (Leopold, 1947:443 map).

Odocoileus hemionus crooki (Mearns)

Mule Deer

1897. *Dorcelaphus crooki* Mearns, Preliminary diagnoses of new mammals of the genera *Mephitis, Dorcelaphus* and *Dicotyles,* from the Mexican border . . ., p. 2, February 11, type locality summit Dog Mtns., 6129 ft., Hidalgo Co., New Mexico.

1939. *Odocoileus hemionus crooki,* Goldman and Kellogg, Jour. Mamm., 20:507, November 14.

Distribution in Tamaulipas.—Reported only from Cerro del Tigre (Leopold, 1959:504), but probably throughout western part of state. Now rare in the state.

Odocoileus virginianus

White-tailed Deer

This species is relatively abundant in Tamaulipas from where three subspecies have been reported. Two specimens examined were shot at ni ht.

Odocoileus virginianus miquihuanensis Goldman and Kellogg

1940. *Odocoileus virginianus miquihuanensis* Goldman and Kellogg, Proc. Biol. Soc. Washington, 53:84, June 28, type from Sierra Madre Oriental, 6000 ft., near Miquihuana, Tamaulipas.

Distribution in Tamaulipas.—Throughout Sierra Madre Oriental.

An adult male, having two points on each antler, and a young male were examined and identified as this subspecies because of their small size and dark color.

Measurements.—A male from 15 km. W Rancho Santa Rosa affords measurements as follows: 1385; 245; 330; 154; condylobasal length, 234; length of maxillary tooth-row, 76.3; width across orbits at frontal-jugal suture, 100.9.

Records of occurrence—Specimens examined, 2: 15 km. W Rancho Santa Rosa, 4500 ft., 1; Ejido Santa Isabel, 2000 ft., 1.

Additional records (Goodwin, 1954:15): San Antonio, 11 km. SW Joya de Salas; Rancho Pano Ayuctle.

Odocoileus virginianus texanus (Mearns)

1898. *Dorcelaphus texanus* Mearns, Proc. Biol. Soc. Washington, 12:23, January 27, type from Fort Clark [north of Eagle Pass on Big Bend of Rio Grande], Kinney Co., Texas.

1902. *Dama v[irginiana]. texensis [sic]*, J. A. Allen, Bull. Amer. Mus. Nat. Hist., 16:20, February 1.

1901. *Odocoileus texensis* Miller and Rehn, Proc. Boston Soc. Nat. Hist., 30:17, December 27, an accidental renaming of *texanus*.

Distribution in Tamaulipas.—Probably all of northern part of state.

Two fragments of lower jaw from the barrier beach were examined and assigned to this subspecies on geographic grounds.

Records of occurrence.—Specimens examined, 2, fragments from 33 mi. S Washington Beach.

Additional records: Sierra San Carlos (El Mulato and Sardinia) (Dice, 1937:256).

Odocoileus virginianus veraecrucis Goldman and Kellogg

1940. *Odocoileus virginianus veraecrucis* Goldman and Kellogg, Proc. Biol. Soc. Washington, 53:89, June 28, type from Chijol, 200 ft., Veracruz.

Distribution in Tamaulipas.—Tropical area, reported only from Soto la Marina (Miller and Kellogg, 1955:806) and Savinito Tierre [= Tierra] Caliente (J. A. Allen, 1881:184) and Tampico (*ibid.*) as *Cariacus virginianus mexicanus.*

Mazama americana temama (Kerr)

Red Brocket

1782. *Cervus temama* Kerr, The Animal kingdom . . ., p. 303. Type locality, restricted to Mirador, Veracruz, by Hershkovitz (Fieldiana-Zool., Chicago Nat. Hist. Mus., 31:567, July 10, 1951).

1951. *Mazama americana temama*, Hershkovitz. Fieldiana-Zool., Chicago Nat. Hist. Mus., 31:567, July 10.

Distribution in Tamaulipas.—Southern part of state in tropical area.

The specimen examined is conspicuously darker than specimens from Veracruz and Chiapas, being especially more brownish and less reddish.

Records of occurrence.—One specimen examined from Rancho Pano Ayuctle (skin only).

Additional records: Alta Cima (Goodwin, 1954:15); Rancho del Cielo (Hooper, 1953:10).

Antilocapra americana mexicana Merriam

Pronghorn

1901. *Antilocapra americana mexicana* Merriam, Proc. Biol. Soc. Washington, 14:31, April 5, type from Sierra en Media, Chihuahua.

Distribution in Tamaulipas.—Originally in the northern part of state; now absent from Tamaulipas.

Antilocapra is here included on the basis of a skull recorded by Baird (1858:669) from Matamoros. J. A. Allen (1881:184) doubted the occurrence of this animal in Tamaulipas because Dr. Palmer found no indications of the presence of *Antilocapra* in any portion of the area that he traversed, which apparently was only southern Tamaulipas.

I am sure that the pronghorn is extinct in Tamaulipas, but its occurrence in the northern part of the state in relatively recent time (more than 100 years ago) seems possible because the habitat in northern Tamaulipas is suitable for the pronghorn.

LITERATURE CITED

ALLEN, H.

1862. Descriptions of two new species of Vespertilionidae, and some remarks on the genus Antrozous. Proc. Acad. Nat. Sci. Philadelphia, pp. 246-248, between May 27 and August 1.

1894. A monograph of the bats of North America. Bull. U. S. Nat. Mus., 43:ix + 198, 38 pls., March 14.

ALLEN, J. A.

1881. *List of mammals collected by Dr. Edward Palmer in north-eastern Mexico, with field-notes by the collector.* Bull. Mus. Comp. Zool., 8:183-189, March.

1891. *On a collection of mammals from southern Texas and northeastern Mexico.* Bull. Amer. Nat. Hist., 3:219-229, December.

1891. A preliminary study of the North American opossums of the genus Didelphis. *Ibid.*, 14:149-188, 4 pls., June 15.

ALVAREZ, T.

1961. Taxonomic status of some mice of the Peromyscus boylii group in eastern México, with description of a new subspecies. Univ. Kansas Publ., Mus. Nat. Hist., 14:111-120, 1 fig., December 29.

1962. A new subspecies of ground squirrel (Spermophilus spilosoma) from Tamaulipas, México. *Ibid.*, 14:121-124, March 7.

ANDERSON, S.

1956. Extensions of known ranges of Mexican bats. *Ibid.*, 9:347-351, August 15.

ANTHONY, H. E.

1923. Mammals from Mexico and South America. Amer. Mus. Novit., 54:1-10, 2 figs., January 17.

BAILEY, V.
1895. Biological survey of Texas. N. Amer. Fauna, 25:1-222, 23 figs., 8 pls., October 24.

BAIRD, S. T.
1855. *Characteristics of some new species of Mammalia, collected by the U. S. and Mexican Boundary Survey, Major W. H. Emory, U. S. A. Commissioner.* Proc. Acad. Nat. Sci. Philadelphia, 7:331-333, April.
1858. Mammals. *In* General report upon the Zoology of the Several Pacific railroad routes. U. S. P. R. R. Exp. and Surveys, pp. xlviii + 757, 60 pls., July 14.

BAKER, R. H.
1951. Mammals from Tamaulipas, México. Univ. Kansas Publ., Mus. Nat. Hist., 5:207-218, December 15.
1956. Mammals of Coahuila, México. *Ibid.*, 9:125-335, 75 figs., June 15.
1958. El futuro de la fauna silvestre en el norte de México. Anal. Inst. Biol., México, 28:349-357, June 14.

BAKER, R. H., and VILLA R., B.
1960. Distribución geographica y población actuales del lobo gris en México. *Ibid.*, 30:369-374, 1 map, March 31.

BOOTH, E. S.
1957. Mammals collected in Mexico from 1951 to 1956 by the Walla Walla College Museum of Natural History. Walla Walla College Publ., 20:1-19, 3 maps, July 10.

BURT, W. H.
1959. The history and affinities of the Recent land mammals of western North America. *In* Zoogeography. Amer. Assoc. Adv. Sci. Publ., 116, February 10.

BURT, W. H., and STIRTON, R. A.
1961. The mammals of El Salvador. Misc. Publ. Mus. Zool., Univ. Michigan, 117:1-69, 2 figs., September 22.

CARTER, D. C., and DAVIS, W. B.
1961. *Tadarida aurispinosa* (Peale) (Chiroptera: Molossidae) in North America. Proc. Biol. Soc. Washington, 74:161-165, August 11.

DALQUEST, W. W.
1951. Two new mammals from Central Mexico. *Ibid.*, 64:105-107, August 24.
1953. Mammals of the Mexican state of San Luis Potosí. Louisiana St. Univ. Press, pp. 1-133, 1 fig., December 28.

DALQUEST, W. W., and HALL, E. R.
1949. A new subspecies of funnel-eared bat (Natalus mexicanus) from eastern Mexico. Proc. Biol. Soc. Washington, 62:153-154, August 23.

DAVIS, W. B.
1944. Notes on Mexican mammals. Jour. Mamm., 25:270-403, December 12.
1951. Bat, *Molossus nigricans,* eaten by the rat snake, *Elaphe laeta. Ibid.*, 32:219, May 21.
1958. Review of Mexican bats of the Artibeus "cinereus" complex. Proc. Biol. Soc. Washington, 71:163-166, December 31.

DAVIS, W. B., and CARTER, D. C.
1962. Notes on Central American bats with description of a new subspecies of Mormoops. Southwestern Nat., 7:64-74, 1 fig., June 1.

DE LA TORRE, L.
1954. Bats from southern Tamaulipas, Mexico. Jour. Mamm., 35:113-116, February 10.
1955. Bats from Guerrero, Jalisco and Oaxaca, Mexico. Fieldiana-Zool., 37:695-701, 1 fig., 2 pls., June 19.

DICE, L. R.
1937. Mammals of the San Carlos Mountains and vicinity. Univ. Michigan Studies Sci. Ser., 12:245-268, 3 pls.
1943. The Biotic Provinces of North America. Univ. Michigan Press, pp. viii + 78, 1 map.

FINDLEY, J. S.
1955. Taxonomy and distribution of some American shrews. Univ. Kansas Publ., Mus. Nat. Hist., 7:613-618, June 10.
1960. Identity of the long-eared Myotis of the southwest and Mexico. Jour. Mamm., 41:16-20, 1 fig., 1 pl., February 20.

GOLDMAN, E. A.
1911. Revision of the spiny pocket mice (Genus Heteromys and Liomys). N. Amer. Fauna, 34:1-70, 6 figs., 3 pls., September 7.
1915. Five new mammals from Mexico and Arizona. Proc. Biol. Soc. Washington, 28:133-137, June 29.
1918. The rice rats of North America (Genus Oryzomys). N. Amer. Fauna, 43:1-100, 11 figs., 6 pls., September 23.
1938. Three new races of Microtus mexicanus. Jour. Mamm., 19:493-495, November 14.
1942. A new white-footed mouse from Mexico. Proc. Biol. Soc. Washington, 55:157-158, October 17.
1942. Notes on the coatis of the Mexican mainland. Proc. Biol. Soc. Washington, 55:79-82, June 25.
1943. The races of the ocelot and margay in Middle America. Jour. Mamm., 24:372-385, August 18.
1946. *Classification of the races of the puma*, pp. 175-302, pls. 46-93, fig. 6, tables 12-13, *in* Young, S. P., and Goldman, E. A., *The puma*, mysterious American cat. Amer. Wildlife Inst., xiv + 358 pp., 93 pls., 6 figs., 13 tables, November 16.
1950. Raccoons of North and Middle America. N. Amer. Fauna, 60: vi + 153, 2 figs., 22 pls., November 7.
1951. Biological investigations in Mexico. Smithsonian Misc. Coll., 115: xiii + 476, 71 pls., 1 map, July 31.

GOLDMAN, E. A., and MOORE, R. T.
1946. The Biotic Provinces of Mexico. Jour. Mamm., 26:347-360, 1 fig., February 12.

GOODWIN, G. G.
1954. Mammals from Mexico collected by Marian Martin for the American Museum of Natural History. Amer. Mus. Novit., 1689:1-16, November 12.
1958. Bats of the genus *Rhogeëssa*. *Ibid.*, 1923:1-17, December 31.
1959. Bats of the genus *Natalus*. *Ibid.*, 1977:1-22, 2 figs., December 22.
1960. The status of *Vespertilio auripendulus* Shaw, 1800, and *Molossus ater* Geoffroy, 1805. *Ibid.*, 1994:1-6, 1 fig., March 8.
1961. Flying squirrel (*Glaucomys volans*) of Middle America. *Ibid.*, 2059:1-22, 7 figs., November 29.

HALL, E. R.
1951. Mammals obtained by Dr. Curt von Wedel from the barrier beach of Tamaulipas, México. Univ. Kansas Publ., Mus. Nat. Hist., 5:33-47, 1 fig., October 1.
1951. A synopsis of the North American Lagomorpha. *Ibid.*, 5:119-202, 68 figs., December 15.
1951. American weasels. *Ibid.*, 4:1-466, 31 figs., 41 pls., December 27.
1952. Taxonomic notes on Mexican bats of the genus Rhogeëssa. *Ibid.*, 5:227-232, April 10.

HALL, E. R., and ALVAREZ, T.
1961. A new subspecies of the black Myotis (bat) from eastern México. *Ibid.*, 14:69-72, 1 fig., December 29.

HALL, E. R., and JONES, J. K., JR.
1961. North American yellow bats, "Dasypterus," and a list of the named kinds of the genus Lasiurus Gray. *Ibid.*, 14:73-98, 4 figs., December 29.

HALL, E. R., and KELSON, K. R.
1959. The mammals of North America. The Ronald Press Co., vol. 1: xxx + 546 + 1-79, vol. 2:viii + 547 + 1-79, 724 figs., 500 maps, March 31.

HANDLEY, C. O., JR.
1956. The taxonomic status of the *Corynorhinus phyllotis* G. M. Allen and *Idionycteris mexicanus* Anthony. Proc. Biol. Soc. Washington, 69:53-54, May 21.
1959. A revision of the American bats of the genera Euderma and Plecotus. Proc. U. S. Nat. Mus., 110:95-246, 47 figs., September 3.
1960. Descriptions of new bats from Panama. *Ibid.*, 112:459-479, October 6.

HERSHKOVITZ, P.
1951. Mammals from British Honduras, Mexico, Jamaica and Haiti. Fieldiana-Zool., 31:547-569, July 10.
1958. A geographic classification of Neotropical mammals. *Ibid.*, 36:583-620, 2 figs., July 11.

HOLLISTER, N.
1914. A systematic account of the grasshopper mice. Proc. U. S. Nat. Mus., 47:427-489, 1 pl., October 29.
1925. The systematic name of the Texas armadillo. Jour. Mamm., 16:60, February 9.

HOOPER, E. T.
1952. A systematic review of the harvest mice (Genus Reithrodontomys) of Latin America. Misc. Publ. Mus. Zool., Univ. Michigan, 77: 1-255, 23 figs., 9 pls., 12 maps, January 16.
1952. Notes on mice of the species *Peromyscus boylei* and *P. pectoralis*. Jour. Mamm., 33:371-378, 2 figs., August 19.
1953. Notes on mammals of Tamaulipas, Mexico. Occas. Papers Mus. Zool., Univ. Michigan, 544:1-12, March 25.

HOOPER, E. T., and HANDLEY, C. O., JR.
1948. Character gradients in the spiny pocket mouse, *Liomys irroratus*. *Ibid.*, 514:1-34, 1 map, October 29.

HOWELL, A. H.
1901. Revision of the skunks of the genus Chincha. N. Amer. Fauna, 20:1-62, 8 pls., August 31.
1938. Revision of the North American ground squirrels, with a classification of the North American Sciuridae. N. Amer. Fauna, 56:1-256, 20 figs., 32 pls., May 18.

JACKSON, H. H. T.
1914. New moles of the genus Scalopus. Proc. Biol. Soc. Washington, 27:19-21, February 2.
1928. A taxonomic review of the American long-tailed shrews (Genus Sorex and Microsorex). N. Amer. Fauna, 51:vi + 238, 24 figs., 13 pls., July 24.
1951. Classification of the races of the coyote, pt. 2, pp. 227-341, pls. 58-81, figs. 20-28, *in* Young, S. P., and Jackson, H. H. T., The clever coyote. Stackpole Co., Harrisburg, Pa., and Wildlife Manag. Inst., Washington, D. C., xv + 411 pp., 81 pls., 28 figs., 11 tables, November 29.

JONES, J. K., JR., and ALVAREZ, T.
1962. Taxonomic status of the free-tailed bat, Tadarida yucatanica Miller. Univ. Kansas Publ., Mus. Nat. Hist., 14:125-133, 1 fig., March 7.

JONES, J. K., JR., and ANDERSON, S.
1958. Noteworthy records of harvest mice in México. Jour. Mamm., 39:446-447, August 20.

KELLOGG, R., and GOLDMAN, E. A.
1944. Review of the spider monkeys. Proc. U. S. Nat. Mus., 96:1-45, November 2.

KELSON, K. R.
1952. The subspecies of the Mexican red-bellied squirrel, Sciurus aureogaster. Univ. Kansas Publ., Mus. Nat. Hist., 5:243-250, April 10.

LAWRENCE, B.
1947. A new race of Oryzomys from Tamaulipas. Proc. New England Zool. Club, 24:101-103, May 29.

LEOPOLD, A. S.
1947. Status of Mexican Big-game herds. Trans. 12th N. Amer. Wild. Conference, pp. 437-448.
1950. Vegetation zones of Mexico. Ecology, 31:507-518, 1 fig., October.
1959. Wildlife of Mexico. The Game birds and mammals. Univ. California Press, pp. xiii + 568, 193 figs.

LIDICKER, W. Z., JR.
1960. An analysis of intraspecific variation in the kangaroo rat Dipodomys merriami. Univ. California Publ. Zool., 67:125-218, 20 figs., 4 pls., August 4.

LUKENS, P. W., JR., and DAVIS, W. B.
1957. Bats of the Mexican state of Guerrero. Jour. Mamm., 38:1-14, February 25.

MALAGA A., A., and VILLA R., B.
1957. Algunas notas acerca de la distribución de los murciélagos de America del Norte relacionados con el problema de la rabia. Anal. Inst. Biol., México, 27:529-568, 8 figs., 10 maps, September 30.

MARTIN, M., and P. S.
1954. Notes on the capture of tropical bats at cuevo [sic] El Pachon, Tamaulipas, Mexico. Jour. Mamm., 35:584-585, November.

MARTIN, P. S.
1958. A biogeography of reptiles and amphibians in the Gomez Farias region, Tamaulipas, Mexico. Misc. Publ. Mus. Zool., Univ. Michigan, 101:1-102, 7 figs., 7 pls., 4 maps, April 15.

MARTIN, P. S., ROBINS, C. R., and HEED, W. B.
1954. Birds and biogeography of the Sierra de Tamaulipas, an isolated pine-oak habitat. Wilson Bull., 66:38-57, 2 figs., 1 map, March.

MERRIAM, C. H.
1895. Revision of the shrews of the American genera Blarina and Notiosorex. N. Amer. Fauna, 10:1-34, 2 figs., December 31.
1895. Monographic revision of the pocket gophers, family Geomydae (Exclusive of the species Thomomys). *Ibid.*, 8:1-258, 10 figs., 19 pls., 3 maps, January 31.
1898. Life Zones and Crop Zones of the United States. U. S. Dept. Agriculture, Bull., 10:1-79, 1 map, June.

MILLER, G. S., JR.
1897. Revision of the North American bats of the family Vespertilionidae. N. Amer. Fauna, 13:1-140, 40 figs., 3 pls., October 16.
1913. Revision of the bats of the genus Glossophaga. Proc. U. S. Nat. Mus., 46:413-429, 1 fig., December 31.

1924. List of North American Recent mammals, 1923. Bull. U. S. Nat. Mus., 128:xvi + 673, April 29.

MILLER, G. S., JR., and ALLEN, G. M.
1928. The American bats of the genera Myotis and Pizonyx. *Ibid.*, 144: vii + 217, 13 maps, May 25.

MILLER, G. S., JR.. and KELLOGG, R.
1955. List of North American mammals. *Ibid.*, 205:xii + 954, March 3.

NELSON, E. W.
1898. Description of the squirrels from Mexico and Central America. Proc. Biol. Soc. Washington, 12:145-156, June 3.
1899. Revision of the squirrels of Mexico and Central America. Proc. Washington Acad. Sci., 1:15-106, 2 pls., May 9.
1904. Descriptions of seven new rabbits from Mexico. Proc. Biol. Soc. Washington, 17:103-110, May 18.
1909. The rabbits of North America. N. Amer. Fauna, 29:1-314, 8 pls., August 31.

NELSON, E. W., and GOLDMAN, E. A.
1933. Revision of the jaguars. Jour. Mamm., 14:221-240, August 17.
1934. Revision of the pocket gophers of the genus Cratogeomys. Proc. Biol. Soc. Washington, 47:135-153, June 13.

OSGOOD, W. H.
1900. Revision of the pocket mice of the genus Perognathus. N. Amer. Fauna, 18:1-72, 15 figs., 4 pls., September 20.
1909. Revision of the mice of the American genus Peromyscus. *Ibid.*, 28: 1-285, 12 figs., 8 pls., April 17.
1945. Two new rodents from Mexico. Jour. Mamm., 26:299-301, November 14.

PACKARD, R. L.
1960. Speciation and evolution of the pygmy mice, genus Baiomys. Univ. Kansas Publ., Mus. Nat. Hist., 9:579-670, 12 figs., 4 pls., June 16.

RHOADS, S. N.
1893. Geographic variation in Bassariscus astutus, with description of a new subspecies. Proc. Acad. Nat. Sci. Philadelphia, 45:413-418, January 30.

SCHANTZ, V. S.
1949. Three new races of badgers (Taxidea) from southwestern United States. Jour. Mamm., 30:301-305, August 17.

SELANDER, R. K., JOHNSTON, R. F., WILKS, B. J., and RAUN, G. G.
1962. Vertebrates from the barrier islands of Tamaulipas, México. Univ. Kansas Publ., Mus. Nat. Hist., 12:309-345, 4 pls., June 18.

SETZER, H. S.
1949. Subspeciation in the kangaroo rat Dipodomys ordii. Univ. Kansas Publ., Mus. Nat. Hist., 1:473-573, 27 figs., December 27.

SHAMEL, H. H.
1931. Notes on the American bats of the genus Tadarida. Proc. U. S. Nat. Mus., 78:1-27, May 6.

SMITH, H. M.
1949. Herpetogeny in Mexico and Guatemala. Assn. Amer. Geographers, 39:219-238, 1 fig., September.

STAINS, H. J.
1957. A new bat (Genus Leptonycteris) from Coahuila. Univ. Kansas Publ., Mus. Nat. Hist., 9:353-356, January 21.

TAMAYO, J. L.
1949. Geografia general de México. Talleres Graficos de la Nación, México, vol. 1:vii + 628, vol. 2:1-583.

VAN GELDER, R. G.
1959. A taxonomic revision of the spotted skunks (Genus *Spilogale*). Bull. Amer. Mus. Nat. Hist., 117:233-392, 47 figs., June 15.

VILLA R., B.
1954. Distribución actual de los castores en México. Anal. Inst. Biol., México, 25:443-450, 2 pls., 1 map, November 9.
1956. Tadarida brasiliensis mexicana (Saussure), el murciélago guanero, es una subespecie migratoria. Acta Zool. Mex., 1:1-11, 2 figs., September 15.
1958. El mono araña (*Ateles geoffroyi*) encontrado en la costa de Jalisco y en la región central de Tamaulipas. Anal. Inst. Biol., México, 28:345-347, June 14.

VILLA R., B., and JIMENEZ G., A.
1961. Acerca de la posición taxonomica de *Mormoops megalophyla senicula* Rehn, y la presencia de virus rabico en estos murciélagos insectivoros. *Ibid.*, 31:501-509, 1 fig., April 17.

VIVO, J. A.
1953. Geografia de México. Fondo de Cultura Economica, México. 3er. Ed., pp. 1-338, 37 pls.

Transmitted June 28, 1962

29-422S

(Continued from inside of front cover)

19. Records of harvest mice, Reithrodontomys, from Central America, with description of a new subspecies from Nicaragua. By Sydney Anderson and J. Knox Jones, Jr. Pp. 519-529. January 14, 1960.
20. Small carnivores from San Josecito Cave (Pleistocene), Nuevo León, México. By E. Raymond Hall. Pp. 531-538, 1 figure in text. January 14, 1960.
21. Pleistocene pocket gophers from San Josecito Cave, Nuevo León, México. By Robert J. Russell. Pp. 539-548, 1 figure in text. January 14, 1960.
22. Review of the insectivores of Korea. By J. Knox Jones, Jr., and David H. Johnson. Pp. 549-578. February 23, 1960.
23. Speciation and evolution of the pygmy mice, genus Baiomys. By Robert L. Packard. Pp. 579-670, 4 plates, 12 figures in text. June 16, 1960.

Index. Pp. 671-690.

l. 10. 1. Studies of birds killed in nocturnal migration. By Harrison B. Tordoff and Robert M. Mengel. Pp. 1-44, 6 figures in text, 2 tables. September 12, 1956.
2. Comparative breeding behavior of Ammospiza caudacuta and A. maritima. By Glen E. Woolfenden. Pp. 45-75, 6 plates, 1 figure. December 20, 1956.
3. The forest habitat of the University of Kansas Natural History Reservation. By Henry S. Fitch and Ronald R. McGregor. Pp. 77-127, 2 plates, 7 figures in text, 4 tables. December 31, 1956.
4. Aspects of reproduction and development in the prairie vole (Microtus ochrogaster). By Henry S. Fitch. Pp. 129-161, 8 figures in text, 4 tables. December 19, 1957.
5. Birds found on the Arctic slope of northern Alaska. By James W. Bee. Pp. 163-211, plates 9-10, 1 figure in text. March 12, 1958.
6. The wood rats of Colorado: distribution and ecology. By Robert B. Finley, Jr. Pp. 213-552, 34 plates, 8 figures in text, 35 tables. November 7, 1958.
7. Home ranges and movements of the eastern cottontail in Kansas. By Donald W. Janes. Pp. 553-572, 4 plates, 3 figures in text. May 4, 1959.
8. Natural history of the salamander, Aneides hardyi. By Richard F. Johnston and Gerhard A. Schad. Pp. 573-585. October 8, 1959.
9. A new subspecies of lizard, Cnemidophorus sacki, from Michoacán, México. By William E. Duellman. Pp. 587-598, 2 figures in text. May 2, 1960.
10. A taxonomic study of the Middle American Snake, Pituophis deppei. By William E. Duellman. Pp. 599-610, 1 plate, 1 figure in text. May 2, 1960.

Index. Pp. 611-626.

l. 11. 1. The systematic status of the colubrid snake, Leptodeira discolor Günther. By William E. Duellman. Pp. 1-9, 4 figures. July 14, 1958.
2. Natural history of the six-lined racerunner, Cnemidophorus sexlineatus. By Henry S. Fitch. Pp. 11-62, 9 figures, 9 tables. September 19, 1958.
3. Home ranges, territories, and seasonal movements of vertebrates of the Natural History Reservation. By Henry S. Fitch. Pp. 63-326, 6 plates, 24 figures in text, 3 tables. December 12, 1958.
4. A new snake of the genus Geophis from Chihuahua, Mexico. By John M. Legler. Pp. 327-334, 2 figures in text. January 28, 1959.
5. A new tortoise, genus Gopherus, from north-central Mexico. By John M. Legler. Pp. 335-343. April 24, 1959.
6. Fishes of Chautauqua, Cowley and Elk counties, Kansas. By Artie L. Metcalf. Pp. 345-400, 2 plates, 2 figures in text, 10 tables. May 6, 1959.
7. Fishes of the Big Blue river basin, Kansas. By W. L. Minckley. Pp. 401-442, 2 plates, 4 figures in text, 5 tables. May 8, 1959.
8. Birds from Coahuila, México. By Emil K. Urban. Pp. 443-516. August 1, 1959.
9. Description of a new softshell turtle from the southeastern United States. By Robert G. Webb. Pp. 517-525, 2 plates, 1 figure in text. August 14, 1959.
10. Natural history of the ornate box turtle, Terrapene ornata ornata Agassiz. By John M. Legler. Pp. 527-669, 16 pls., 29 figures in text. March 7, 1960.

Index Pp. 671-703.

l. 12. 1. Functional morphology of three bats: Eumops, Myotis, Macrotus. By Terry A. Vaughan. Pp. 1-153, 4 plates, 24 figures in text. July 8, 1959.
2. The ancestry of modern Amphibia: a review of the evidence. By Theodore H. Eaton, Jr. Pp. 155-180, 10 figures in text. July 10, 1959.
3. The baculum in microtine rodents. By Sydney Anderson. Pp. 181-216, 49 figures in text, February 19, 1960.
4. A new order of fishlike Amphibia from the Pennsylvanian of Kansas. By Theodore H. Eaton, Jr., and Peggy Lou Stewart. Pp. 217-240, 12 figures in text. May 2, 1960.
5. Natural history of the bell vireo. By Jon C. Barlow. Pp. 241-296, 6 figures in text. March 7, 1962.
6. Two new pelycosaurs from the lower Permian of Oklahoma. By Richard C. Fox. Pp. 297-307, 6 figures in text. May 21, 1962.
7. Vertebrates from the barrier island of Tamaulipas, México. By Robert K. Selander, Richard F. Johnston, B. J. Wilks, and Gerald G. Raun. Pp. 309-345, pls. 5-8. June 18, 1962.
8. Teeth of Edestid sharks. By Theodore H. Eaton, Jr. Pp. 347-362, 10 figures in text. October 1, 1962.

More numbers will appear in volume 12.

(Continued on outside of back cover)

(Continued from inside of back cover)

l. 13. 1. Five natural hybrid combinations in minnows (Cyprinidae). By Frank B. Cross and W. L. Minckley. Pp. 1-18. June 1, 1960.

2. A distributional study of the amphibians of the Isthmus of Tehuantepec, México. By William E. Duellman. Pp. 19-72, pls. 1-8, 3 figures in text. August 16, 1960.

3. A new subspecies of the slider turtle (Pseudemys scripta) from Coahuila, México. By John M. Legler. Pp. 73-84, pls. 9-12, 3 figures in text. August 16, 1960.

4. Autecology of the copperhead. By Henry S. Fitch. Pp. 85-288, pls. 13-20, 26 figures in text. November 30, 1960.

5. Occurrence of the garter snake, Thamnophis sirtalis, in the great plains and Rocky mountains. By Henry S. Fitch and T. Paul Maslin. Pp. 289-308, 4 figures in text. February 10, 1961.

6. Fishes of the Wakarusa river in Kansas. By James E. Deacon and Artie L. Metcalf. Pp. 309-322, 1 figure in text. February 10, 1961.

7. Geographic variation in the North American Cyprinid fish, Hybopsis gracilis. By Leonard J. Olund and Frank B. Cross. Pp. 323-348, pls. 21-24, 2 figures in text. February 10, 1961.

8. Descriptions of two species of frogs, genus Ptychohyla; studies of American Hylid frogs, V. By William E. Duellman. Pp. 349-357, pl. 25, 2 figures in text. April 27, 1961.

9. Fish populations, following a drought, in the Neosho and Marais des Cygnes rivers of Kansas. By James Everett Deacon. Pp. 359-427, pls. 26-30, 3 figures in text. August 11, 1961.

10. North American recent soft-shelled turtles (family Trionychidae). By Robert G. Webb. Pp. 429-611, pls. 31-54, 24 figures in text. February 16, 1962.

Index. Pp. 613-624.

l. 14. 1. Neotropical bats from western México. By Sydney Anderson. Pp. 1-8. October 24, 1960.

2. Geographic variation in the harvest mouse, Reithrodontomys megalotis, on the central great plains and in adjacent regions. By J. Knox Jones, Jr., and B. Mursaloglu. Pp. 9-27, 1 figure in text. July 24, 1961.

3. Mammals of Mesa Verde national park, Colorado. By Sydney Anderson. Pp. 29-67, pls. 1 and 2, 3 figures in text. July 24, 1961.

4. A new subspecies of the black myotis (bat) from eastern México. By E. Raymond Hall and Ticul Alvarez. Pp. 69-72, 1 fig. in text. December 29, 1961.

5. North American yellow bats, "Dasypterus," and a list of the named kinds of the genus Lasiurus Gray. By E. Raymond Hall and J. Knox Jones, Jr. Pp. 73-98, 4 figs. in text. December 29, 1961.

6. Natural history of the brush mouse (Peromyscus boylii) in Kansas with description of a new subspecies. By Charles A. Long. Pp. 99-110, 1 fig. in text. December 29, 1961.

7. Taxonomic status of some mice of the Peromyscus boylii group in eastern México, with description of a new subspecies. By Ticul Alvarez. Pp. 111-120, 1 fig. in text. December 29, 1961.

8. A new subspecies of ground squirrel (Spermophilus spilosoma) from Tamaulipas, México. By Ticul Alvarez. Pp. 121-124. March 7, 1962.

9. Taxonomic status of the free-tailed bat, Tadarida yucatanica Miller. By J. Knox Jones, Jr., and Ticul Alvarez. Pp. 125-133, 1 figure in text. March 7, 1962.

10. A new doglike carnivore, genus Cynarctus, from the Clarendonian, Pliocene, of Texas. By E. Raymond Hall and Walter W. Dalquest. Pp. 135-138, 2 figures in text. April 30, 1962.

11. A new subspecies of wood rat (Neotoma) from northeastern Mexico. By Ticul Alvarez. Pp. 139-143. April 30, 1962.

12. Noteworthy mammals from Sinaloa, Mexico. By J. Knox Jones, Jr., Ticul Alvarez, and M. Raymond Lee. Pp. 145-149, 1 figure in text. May 18, 1962.

13. A new bat (Myotis) from Mexico. By E. Raymond Hall. Pp. 161-164, 1 figure in text. May 21, 1962.

14. The Mammals of Veracruz. By E. Raymond Hall and Walter W. Dalquest. Pp. 165-362, 2 figures in text. May 20, 1963.

15. The Recent mammals of Tamaulipas, Mexico. By Ticul Alvarez. Pp. 363-473, 5 figures in text. May 20, 1963.

More numbers will appear in volume 14.

l. 15. 1. The amphibians and reptiles of Michoacán, México. By William E. Duellman. Pp. 1-148, pls. 1-6, 11 figures in text. December 20, 1961.

2. Some reptiles and amphibians from Korea. By Robert G. Webb, J. Knox Jones, Jr., and George W. Byers. Pp. 149-173. January 31, 1962.

3. A new species of frog (Genus Tomodactylus) from western México. By Robert G. Webb. Pp. 175-181, 1 figure in text. March 7, 1962.

4. Type specimens of amphibians and reptiles in the Museum of Natural History, The University of Kansas. By William E. Duellman and Barbara Berg. Pp. 183-204, October 26, 1962.

More numbers will appear in volume 15.

University of Kansas Publications
Museum of Natural History

Volume 14, No. 16, pp. 475-481, 1 fig.
March 2, 1964

A New Subspecies of the Fruit-eating Bat, Sturnira ludovici, from Western Mexico

BY

J. KNOX JONES, JR., AND GARY L. PHILLIPS

University of Kansas
Lawrence
1964

University of Kansas Publications, Museum of Natural History

Volume 14, No. 16, pp. 475-481, 1 fig.
Published March 2, 1964

University of Kansas
Lawrence, Kansas

PRINTED BY
HARRY (BUD) TIMBERLAKE, STATE PRINTER
TOPEKA, KANSAS
1964

29-8529

A New Subspecies of the Fruit-eating Bat, Sturnira ludovici, from Western Mexico

BY

J. KNOX JONES, JR., AND GARY L. PHILLIPS

The fruit-eating bats of the genus *Sturnira* are represented on the North American mainland by two species, *S. lilium* and *S. ludovici*. The former, in most areas the smaller of the two, is widely distributed in México and Central America and is common in many places. On the other hand, *S. ludovici*, described by Anthony (1924:8) from near Gualea, Ecuador, generally has been regarded as rare; insofar as we can determine only 20 specimens of the species have been recorded previously from North America (Costa Rica, Honduras, and México).

In 1961 (M. Raymond Lee) and 1962 (Percy L. Clifton), field representatives of the Museum of Natural History collected mammals in western México. Among the bats obtained by them were 23 specimens of *S. ludovici*, which represent an heretofore undetected subspecies that is named and described below.

Sturnira ludovici occidentalis, new subspecies

Holotype.—Adult female, skin and skull, no. 92798 Museum of Natural History, The University of Kansas, from Plumosas, 2500 feet elevation, Sinaloa; obtained on August 31, 1962, by Percy L. Clifton (original no. 2939).

Distribution.—Western México; known certainly from southwestern Durango south to southern Jalisco (see Fig. 1).

Diagnosis.—Size small both externally and cranially (forearm in adults 40.4-44.1 mm., greatest length of skull 21.7-22.9); rostrum short and abruptly elevated; skull relatively broad; dorsal pelage drab brownish over-all, usually lacking epaulets (pale yellowish brown when present); ventral pelage brownish gray.

Comparisons.—From *Sturnira ludovici ludovici*, the only other subspecies of the species, *S. l. occidentalis* differs in averaging smaller in most external and cranial dimensions (in some measurements the upper size limits of *occidentalis* barely overlap the lower limits in specimens of *ludovici* examined), in having a relatively broader skull with a shorter, more abruptly elevated rostrum, and in being paler both dorsally and ventrally.

From *Sturnira lilium parvidens*, with which it is sympatric, *S. l. occidentalis* usually (but not always) differs in being brownish (rather than yellowish to yellowish orange) dorsally and in lacking epaulets, and differs in the following cranial features: first upper incisors simple (rather than weakly bifid in unworn condition), larger, and more nearly straight when viewed from the front; second upper incisors reduced; lower incisors bilobate rather than trilobate; lingual cusps on m1 and m2 greatly reduced; M2 usually turned inward from M1 at distinct angle. The two species have approximately the same external and cranial dimensions in western México.

Measurements (in millimeters).—External measurements of the holotype are as follows: total length, 58; length of hind foot, 15; length of ear, 18; forearm (average of both), 42.5. Corresponding average and extreme measurements of 11 adults from 4 km. N Durazno, Jalisco, followed by those of eight adults from 17 km. SE Talpa, Jalisco, are: 61.9 (59-65), 60.9 (57-68); 14.1 (12-15), 13.0 (13); 16.1 (15-18), 16.0 (15-17); 42.2 (40.4-43.8), 42.9 (41.6-44.1); weight in grams, 16.8 (15-19, six specimens only), 19.2 (16.3-22.5).

Cranial measurements of the holotype additional to those given in Table 1 are: condyloincisive length, 19.7; breadth across upper canines, 5.5; length of mandibular tooth-row (c-m3), 6.7.

Remarks.—The pattern of geographic variation in size in *Sturnira ludovici* resembles that in many other species of tropical bats in North America in that individuals from the northern parts of the range are smaller than those from the south. Mexican specimens herein assigned to *S. l. ludovici* average somewhat smaller than specimens from Central America and the northern part of South America (but are within the currently understood size limits of that subspecies) and average paler as well. Additional material is needed from central and eastern México before the limits of distribution of the two subspecies of *ludovici* can be determined accurately.

All specimens examined of the new subspecies were trapped in mist nets. The holotype was captured in a net stretched across an old road among large fruit trees situated along a small river (a tributary of the Río del Baluarte). Tropical deciduous vegetation grew in the narrow valley of the river but the adjacent hills supported oak. A specimen of *Artibeus jamaicensis jamaicensis* was netted along with the holotype and on the previous night, August 30, one individual each of *Glossophaga soricina leachii* and *Sturnira*

TABLE 1.—SOME MEASUREMENTS OF ADULTS OF TWO SUBSPECIES OF STURNIRA LUDOVICI.

Number of specimens averaged, or catalogue number, and sex	Length of forearm	Greatest length of skull	Zygomatic breadth	Mastoid breadth	Interorbital constriction	Length of maxillary tooth-row	Breadth across upper molars
Sturnira ludovici occidentalis, holotype							
92798 KU, ♀	42.5	22.0	12.5	11.4	5.3	6.1	7.5
½ mi. W Revolcaderos, Durango							
5698 MSU, ♀	43.7	22.6	13.1	11.9	6.0	6.3	7.8
5699 MSU, ♀	42.3	22.2	12.7	11.3	5.6	5.9	7.5
17 km. SE Talpa, Jalisco							
Average 8 (4♂, 4♀)	42.9	22.5	12.9	11.5	5.9	6.2	7.7
Minimum	41.6	21.7	12.6	10.9	5.7	6.0	7.5
Maximum	44.1	22.9	13.5	11.8	6.3	6.4	7.9
20 km. WNW Purificación, Jalisco							
92811 KU, ♂	42.0	22.6	13.2	12.0	6.0	6.2	7.7
4 km. N Durazno, Jalisco							
Average 11 (1♂, 10♀)	42.4	22.5	13.0	11.4	5.8	6.2	7.7
Minimum	40.4	21.8	12.6	10.8	5.3	5.8	7.5
Maximum	43.8	22.9	13.4	11.8	6.1	6.3	8.0
Sturnira ludovici ludovici, 10 mi. SW Villa Juárez, Puebla							
67399 KU, ♀	44.2	24.0	13.3	11.8	5.9	6.4	8.1
67400 KU, ♀	42.9	23.2	13.7	11.9	6.0	6.3	8.1
11 km. W Quiroga, Michoacán							
95703 UMMZ, ♂	......	23.5	13.5	11.6	5.9	6.2	7.6
95704 UMMZ, ♀	......	23.0	12.8	11.0	5.7	6.3	8.0
Vista Hermosa, Oaxaca							
91635 KU, ♀	45.1	23.9	13.4	11.8	6.0	6.5	8.0
91636 KU, ♀	46.0	23.6	13.1	11.9	5.7	6.7	8.0
La Cruz Grande, La Paz, Honduras							
126791 AMNH, ♀	44.0	23.6	13.5	11.8	6.1	6.3	8.1
126811 AMNH, ♀*	45.5	24.6	13.2	12.0	6.3	7.2	8.2
Sierra Negra, Sierra de Perijá, Colombia (after Hershkovitz, 1949)							
Minimum (2♂, 2♀)	44.2	22.9	13.2	......	6.2	6.5	.....
Maximum	46.0	24.2	13.8	......	6.7	7.0	.....
near Gualea, Ecuador							
67328 AMNH, ♂**	......	25.0	14.0	12.4	6.3	.....	8.4
67329 AMNH, ♂	45.3	24.9	13.9	12.2	6.1	7.0	8.4

* Holotype of *Sturnira hondurensis* (measurements after Goodwin, 1940:2).
** Holotype of *Sturnira ludovici ludovici* (measurements after Anthony, 1924:9).

lilium parvidens were taken in the same net. Baker and Greer (1962:69) also reported the two species of *Sturnira* as netted together 6 mi. S Pueblo Nuevo in adjacent Durango.

Other specimens of *S. l. occidentalis* were taken under the following circumstances: 17 km. SE Talpa, Jalisco (night of November 3-4, 1962)—nine individuals netted over the Río Mascota in "pine-oak zone" along with representatives of *S. l. parvidens, Artibeus toltecus, Chiroderma salvini, Eptesicus fuscus miradorensis, Lasiurus borealis teliotis,* and *Rhogeëssa gracilis;* 20 km. WNW Purificación, Jalisco (night of November 20-21, 1962)—two specimens captured in a mist net stretched beneath branches of a fig tree at the edge of the Río Jicote in which *Glossophaga commissarisi, S. l. parvidens, Artibeus turpis nanus,* and *Artibeus lituratus palmarum* also were taken; 4 km. N Durazno, Jalisco (nights of November 21-22 and 22-23, 1961)—11 specimens, of which 10 were females, netted in company with *G. s. leachii, S. l. parvidens, A. j. jamaicensis, A. toltecus, Centurio senex,* and *L. b. teliotis* over a stream in a small canyon that supported "fairly dense stands of very tall deciduous trees." Five of the 10 females from 4 km. N Durazno were pregnant; each contained a single embryo. Crown-rump length of the embryos averaged 26.8 (24-30) mm. No gross reproductive activity was evident in other females of *S. l. occidentalis* collected.

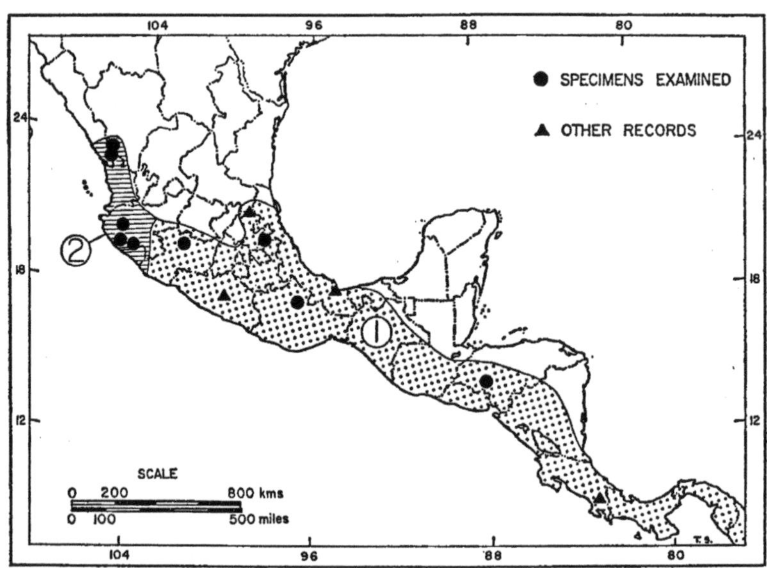

FIG. 1. Distribution of *Sturnira ludovici* in North America. 1. *S. l. ludovici.* 2. *S. l. occidentalis.*

Specimens examined.—A total of 26, arranged from north to south, as follows: DURANGO: ½ mi. W Revolcaderos, 6600 ft., 2 (MSU); 6 mi. S Pueblo Nuevo, 3000 ft., 1 (MSU). SINALOA: Plumosas, 2500 ft., 1 (the holotype). JALISCO: 17 km. SE Talpa, 5200 ft., 9; 20 km. WNW Purificación, 1400 ft., 2; 4 km. N. Durazno, 11.

Specimens of *S. l. ludovici* used in comparisons included a paratype (AMNH) from near Gualea, Ecuador, a specimen from Mindo, Ecuador, two specimens from La Cruz Grande, La Paz, Honduras (AMNH—paratypes of "*Sturnira hondurensis*"), and the following from México: 10 mi. SW Villa Juárez, 4850 ft., Puebla, 2; 11 km. W Quiroga, about 7000 ft., Michoacán, 2 (UMMZ); and Vista Hermosa, 1500 meters, Oaxaca, 5.

Acknowledgements.—For the loan of comparative materials we are grateful to R. H. Baker of The Museum, Michigan State University (MSU), W. H. Burt of the Museum of Zoology, University of Michigan (UMMZ), and R. G. Van Gelder of the American Museum of Natural History (AMNH). Specimens listed above that bear no designation as to collection are in the Museum of Natural History of The University of Kansas.

Literature Cited

ANTHONY, H. E.
1924. Preliminary report on Ecuadorean mammals. No. 6. Amer. Mus. Novit., 139:1-9, October 20.

BAKER, R. H., and J. K. GREER
1962. Mammals of the Mexican state of Durango. Publ. Mus., Michigan State Univ., Biol. Ser., 2:25-154, 4 pls., 6 figs., August 27.

GOODWIN, G. G.
1940. Three new bats from Honduras and the first record of *Enchisthenes harti* (Thomas) for North America. Amer. Mus. Novit., 1075:1-3, June 27.

HERSHKOVITZ, P.
1949. Mammals of northern Colombia. Preliminary report no. 5: Bats (Chiroptera). Proc. U. S. Nat. Mus., 99:429-454, fig. 38, May 10.

Transmitted June 24, 1963.

29-8529

University of Kansas Publications
Museum of Natural History

Volume 14, No. 17, pp. 483-491, 2 figs.
March 2, 1964

Records of the Fossil Mammal Sinclairella, Family Apatemyidae, From the Chadronian and Orellan

BY

WILLIAM A. CLEMENS, JR.

University of Kansas
Lawrence
1964

University of Kansas Publications, Museum of Natural History

Volume 14, No. 17, pp. 483-491, 2 figs.
Published March 2, 1964

University of Kansas
Lawrence, Kansas

PRINTED BY
HARRY (BUD) TIMBERLAKE, STATE PRINTER
TOPEKA, KANSAS
1964

29-8587

Records of the Fossil Mammal Sinclairella, Family Apatemyidae, From the Chadronian and Orellan

BY

WILLIAM A. CLEMENS, JR.

Introduction

The family Apatemyidae has a long geochronological range in North America, beginning in the Torrejonian land-mammal age, but is represented by a relatively small number of fossils found at a few localities. Two fossils of Orellan age, found in northeastern Colorado and described here, demonstrate that the geochronological range of the Apatemyidae extends into the Middle Oligocene. Isolated teeth of *Sinclairella dakotensis* Jepsen, part of a sample of a Chadronian local fauna collected by field parties from the Webb School of California, are also described.

I thank Mr. Raymond M. Alf, Webb School of California, Claremont, California, and Dr. Peter Robinson, University of Colorado Museum, Boulder, Colorado, for permitting me to describe the fossils they discovered. Also Dr. Robinson made available the draft of a short paper he had prepared on the tooth found in Weld County, Colorado; his work was facilitated by a grant from the University of Colorado Council on Research and Creative Work. I also gratefully acknowledge receipt of critical data and valuable comments from Drs. Edwin C. Galbreath, Glenn L. Jepsen, and Malcolm C. McKenna who is currently revising the Paleocene apatemyids and studying the phylogenetic relationships of the family. The prefixes of catalogue numbers used in the text identify fossils in the collections of the following institutions: KU, Museum of Natural History, The University of Kansas, Lawrence; Princeton, Princeton Museum, Princeton, New Jersey; RAM-UCR, Raymond Alf Museum, Webb School of California, Claremont, California (the permanent repository for these specimens will be the University of California, Riverside); and UCM, University of Colorado Museum, Boulder, Colorado. The system of notations for teeth prescribed for use here is as follows: teeth in the upper half of the dentition are designated by a capital letter and a number; thus M2 is the notation for the upper second molar; teeth in the lower half of the dentition are designated by a lower-case letter and a number; thus p2 is the notation for the lower second premolar.

Family APATEMYIDAE Matthew, 1909

Genus **Sinclairella** Jepsen, 1934

Sinclairella dakotensis Jepsen, 1934

The type of the species, Princeton no. 13585, was discovered in Chadronian strata of the upper part of the Chadron Formation cropping out in Big Corral Draw, approximately 13 miles south-southwest of Scenic, in southwestern South Dakota (Jepsen, 1934, p. 291). Detailed descriptions of the type specimen are given in papers by Jepsen (1934) and Scott and Jepsen (1936). Isolated teeth of Chadronian age referable to *Sinclairella dakotensis* have been discovered subsequently at a locality in Nebraska and fossils of Orellan age, also referable to *S. dakotensis*, have been collected at two localities in Colorado. The sample from each locality is described separately.

Sioux County, northwestern Nebraska

Material.—RAM-UCR nos. 381, left M1; 598, left m2; 1000, right m1; 1001, right m2; 1079, right m2; 1674, right M2; and 3013, left m2.

Locality and stratigraphy.—These Chadronian fossils were discovered by Raymond Alf and members of his field parties in several harvester ant mounds built in exposures of the Chadron Formation in Sec. 26, T 33 N, R 53 W, Sioux County, Nebraska (Alf, 1962, and Hough and Alf, 1958). This is UCR locality V5403. The collectors carefully considered the possibility that some of the fossils found in the ant mounds were collected from younger strata by the harvester ants and concluded this was unlikely (Alf, personal communication).

Description and comments.—The cusps of RAM-UCR no. 381, a left M1, are sharp and the wear-facets resulting from occlusion with the lower dentition are small. The paraconule is a low, ill-defined cusp on the anterior margin of the crown; a metaconule is not present. A smooth stylar shelf is present labial to the metacone. The crown was supported by three roots. There are no interradicular crests.

The crown of RAM-UCR no. 1674, a right M2, is heavily abraded and many morphological details of the cusps have been destroyed. Low interradicular crests linked the three roots of the tooth with a low, central prominence. As was the case with RAM-UCR no. 381, no significant differences could be found in comparisons with illustrations of the teeth preserved in Princeton no. 13585.

RAM-UCR nos. 598, 1001, 1079, and 3013 all appear to be m2's. The talonids of these teeth are not elongated, their trigonids have quadrilateral outlines, and the paraconids are small but prominent, bladelike cusps. The trigonid of RAM-UCR 1000 is elongated and the paraconid is a minute cusp; the tooth closely resembles the m1 of the type of *Sinclairella dakotensis*.

Logan County, northeastern Colorado

Material.—KU no. 11210 (fig. 1), a fragment of a left maxillary containing P4 and M1-2.

Locality and stratigraphy.—The fossil was found in the center of the W½, Sec. 21, T 11 N, R 53 W, Logan County, Colorado, ". . . in the bed below *Agnotocastor* bed, Cedar Creek Member . . ." (Ronald H. Pine, 1958, field notes on file at the University of Kansas). The bed so defined is part of unit 3 in the lower division of the Cedar Creek Member, as subdivided by Galbreath (1953:25) in stratigraphic section XII. The fauna obtained from unit 3 is of Orellan age.

Description and comments.—P4 of KU no. 11210 has a large posterolingual cusp separated from the main cusp by a distinct groove, which deepens pos-

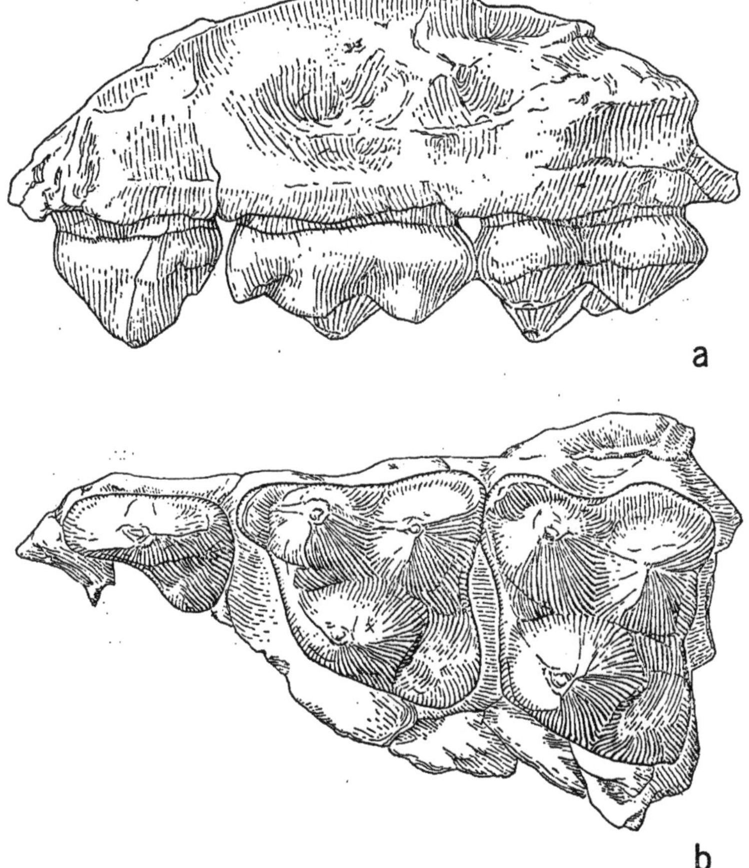

FIG. 1. *Sinclairella dakotensis* Jepsen, KU no. 11210, fragment of left maxillary with P4 and M1-2; Orellan, Logan County, Colorado; drawings by Mrs. Judith Hood: a, labial view; b, occlusal view; both approximately × 9.

teriorly. The posterolingual cusp is supported by the broad posterior root. P4 of the type specimen of *Sinclairella dakotensis* is described (Jepsen, 1934, p. 392) as having an oval outline at the base of the crown, and a small, posterolingual cusp. A chip of enamel is missing from the posterior slope of the main cusp of the P4 of KU no. 11210. The anterior slope of the main cusp is flattened, possibly the result of wear, and there is no evidence of a groove like that present on the P4 of the type specimen.

Only a few differences were found between the molars preserved in KU no. 11210 and their counterparts in the type specimen. A stylar shelf is present labial to the metacone of M1 of KU no. 11210, but, unlike the type, its surface is smooth and there is no evidence of cusps. Of the three small stylar cusps on the stylar shelf of M2 the smallest is in the position of a mesostyle. The M2 lacks a chip of enamel from the lingual surface of the hypocone. Unlike the M2 of Princeton no. 13585, in occlusal view the posterior margin of the M2 of KU no. 11210 is convex posterior to the metacone. The anterior edge of the base of the zygomatic arch of KU no. 11210 was dorsal to M2. The shallow oval depression in the maxillary dorsal to M1 might be the result of post-mortem distortion.

The molars preserved in KU no. 11210 and their counterparts in the type specimen do not appear to be significantly different in size (table 1) or morphology of the cusps. The only difference between the two specimens that might be of classificatory significance is the difference in size of the posterolingual cusp of P4. At present the range of intraspecific variation in the morphology of P4 has not been documented for any species of apatemyid. The evolutionary trend or trends of the apatemyids (McKenna, 1960, p. 48) for progressive reduction of function of p4 probably were paralleled by similar trends in the evolution of the P4. If so, the intraspecific variation in the morphology of P4 could be expected to be somewhat greater than that of the upper molars, for example. The morphological difference between the P4's of the type of *Sinclairella dakotensis* and KU no. 11210 is not extreme and does not exceed the range of intraspecific variation that could be expected for this element of the dentition. The close resemblances in size and morphology between the M1-2 of Princeton no. 13585 and KU no. 11210 also favor identification of the latter as part of a member of an Orellan population of *Sinclairella dakotensis*.

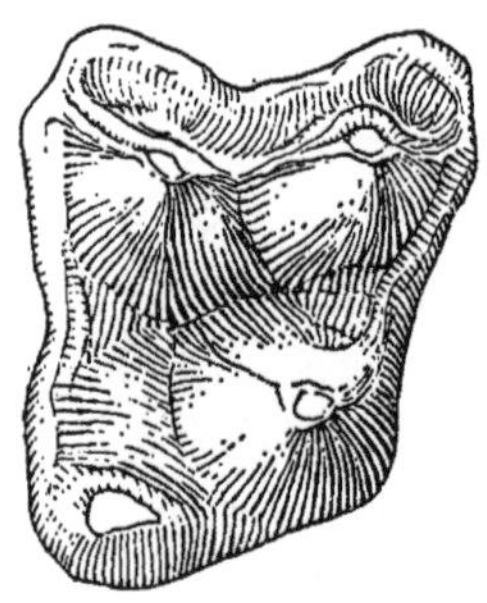

FIG. 2. *Sinclairella dakotensis* Jepsen, UCM no. 21073, right M2; Orellan, Weld County, Colorado; drawing by Mrs. Judith Hood: occlusal view, approximately × 9.

Weld County, northeastern Colorado

Material.—UCM no. 20173 (fig. 2), is a right M2.

Locality and stratigraphy.—The tooth was discovered at the Mellinger locality, Sec. 17, T 11 N, R 65 W, Weld County, Colorado. The Mellinger locality is in the Cedar Creek Member, White River Formation, and its fauna is considered to be of Orellan age (Patterson and McGrew, 1937, and Galbreath, 1953).

Description and comments.—UCM no. 21073, which is more heavily abraded than KU no. 11210, shows no evidence of a stylar cusp either anterolabial to the metacone or in the position of a mesostyle. A small stylar cusp is present anterolabial to the paracone. A notch that appears to have been cut through the enamel of the posterolabial corner of the crown could have received the parastylar apex of M3. A similar notch is not present on the M2 of KU no. 11210 nor indicated in the illustrations of the M2 of Princeton no. 13585. The coronal dimensions of UCM no. 21073 (table 1) do not appear to differ significantly from those of the M2's of KU no. 11210 and the type specimen of *Sinclairella dakotensis*.

Comments

With the discovery of Orellan apatemyids the geochronological range of the family in North America is shown to extend from the Torrejonian through the Orellan land-mammal ages. The discoveries reported here enlarge the Oligocene record of apatemyids to include not only the type specimen of *Sinclairella dakotensis*, a skull and associated mandible from South Dakota, but also seven isolated teeth, representing at least two individuals, from a Chadronian fossil locality in Nebraska and one specimen from each of two Orellan fossil localities in northeastern Colorado. Simpson (1944:73, and 1953:127) presented tabulations of the published records of American apatemyids and suggested the data indicated the populations of these mammals were of small size throughout the history of the family. The few pre-Oligocene occurrences of apatemyids described subsequently (note McKenna, 1960, figs. 3-10, and p. 48) and occurrences described here tend to reinforce Simpson's interpretation. This interpretation may have to be modified to some degree, however, when current studies of collections of pre-Oligocene apatemyids are completed (McKenna, personal communication).

Although information concerning the evolutionary trends of American apatemyids has been published, no data on the morphological variation in a population are available in the literature. An adequate basis for evaluating the significance of the morphological differences between the P4's of Princeton no. 13585 and KU no. 12110 coupled with the similarities of their M1-2's is lacking. In the evolution of American apatemyids the P4 underwent reduction in size and, apparently, curtailment of function. This history suggests the range of morphological variation of P4 in populations of *Sinclairella dakotensis* could be expected to be greater than that of the molars and encompass the morphological differences between the P4's of Princeton no. 13585 and KU no. 12110. The difference in age of the Chadronian and Orellan fossils does not constitute proof that they pertain to different species. Although the identification is admittedly provisional until more fossils including other parts of the skeleton are discovered, the Orellan fossils described here are referred to *Sinclairella dakotensis.*

TABLE 1.—MEASUREMENTS (IN MILLIMETERS) OF TEETH OF SINCLAIRELLA DAKOTENSIS JEPSEN.

	P4		M1		M2	
	length	width	length [1]	width [1]	length [1]	width [1]
Princeton no. 13585 [2]	2.1	1.1	4.0	3.7	3.4	4.7
RAM no. 381	....	....	4.1	3.5	....	....
RAM no. 1674	....	....	....	....	3.4	4.2
KU no. 11210	2.4	1.6	3.9	3.5	3.8	4.1+
UCM no. 21073	....	....	....	....	3.6	4.1
			m1		m2	
			length	width	length	width
Princeton no. 13585 [3]			3.5	2.4	3.7	2.8
RAM no. 1000			3.5	2.2	....	....
RAM no. 598			....	....	3.8	2.6
RAM no. 1001			....	....	3.6+	2.6
RAM no. 1079			....	....	4.0	2.8
RAM no. 3013			....	....	3.6	2.8

1. Length defined as maximum dimension of the labial half of the crown measured parallel to a line drawn through the apices of paracone and metacone. Width defined as maximum coronal dimension measured along line perpendicular to line defined by apices of paracone and metacone.

2. Dimensions provided by Dr. Glenn L. Jepsen.

3. Dimensions taken from Jepsen (1934:300).

Literature Cited

ALF, R.
1962. A new species of the rodent *Pipestoneomys* from the Oligocene of Nebraska. Breviora, Mus. Comp. Zool., no. 172, pp. 1-7, 3 figs.

GALBREATH, E. C.
1953. A contribution to the Tertiary geology and paleontology of northeastern Colorado. Univ. Kansas Paleont. Cont., Vertebrata, art. 4, pp. 1-120, 2 pls., 26 figs.

HOUGH, J., and ALF, R.
1958. A Chadron mammalian fauna from Nebraska. Journ. Paleon. 30: 132-140, 4 figs.

JEPSEN, G. L.
1934. A revision of the American Apatemyidae and the description of a new genus, *Sinclairella*, from the White River Oligocene of South Dakota. Proc. Amer. Philos. Soc., 74:287-305, 3 pls., 4 figs.

MCKENNA, M. C.
1960. Fossil Mammalia from the early Wasatchian Four Mile fauna, Eocene of northwest Colorado. Univ. California Publ. in Geol. Sci., 37:1-130, 64 figs.

MATTHEW, W. D.
1909. The Carnivora and Insectivora of the Bridger Basin, Middle Eocene. Mem. Amer. Mus. Nat. Hist., 9:289-567, pls. 42-52, 118 figs.

PATTERSON, B. and MCGREW, P. O.
1937. A soricid and two erinaceids from the White River Oligocene. Geol. Ser., Field Mus. Nat. Hist., 6:245-272, figs. 60-74.

SCOTT, W. B. and JEPSEN, G. L.
1936. The mammalian fauna of the White River Oligocene—Part I. Insectivora and Carnivora. Trans. Amer. Philos. Soc., n. s., 28:1-153, 22 pls., 7 figs.

SIMPSON, G. G.
1944. Tempo and mode in evolution. New York: Columbia Univ. Press, xviii + 237 pp., 36 figs.
1953. The major features of evolution. New York: Columbia Univ. Press, xx + 434 pp., 52 figs.

Transmitted June 24, 1963.

29-8587

University of Kansas Publications
Museum of Natural History

Volume 14, No. 18, pp. 493-758, 82 figs.
July 6, 1965

The Mammals of Wyoming

BY

CHARLES A. LONG

University of Kansas
Lawrence
1965

UNIVERSITY OF KANSAS PUBLICATIONS
MUSEUM OF NATURAL HISTORY

Institutional libraries interested in publications exchange may obtain this series by addressing the Exchange Librarian, University of Kansas Library, Lawrence, Kansas. Copies for individuals, persons working in a particular field of study, may be obtained by addressing instead the Museum of Natural History, University of Kansas, Lawrence, Kansas. When individuals request copies from the Museum, 25 cents should be included, for each 100 pages or part thereof, for the purpose of defraying the costs of wrapping and mailing. For certain longer papers an additional amount, indicated below, toward some of the costs of production, is to be included.

* An asterisk designates those numbers of which the Museum's supply is exhausted.

Vol. 1. Nos. 1-26 and index. Pp. 1-638, 1946-1950.

*Vol. 2. (Complete) Mammals of Washington. By Walter W. Dalquest. Pp. 1-444, 140 figures in text. April 9, 1948.

Vol. 3. *1. The avifauna of Micronesia, its origin, evolution, and distribution. By Rollin H. Baker. Pp. 1-359, 16 figures in text. June 12, 1951.

*2. A quantitative study of the nocturnal migration of birds. By George H. Lowery, Jr. Pp. 361-472, 47 figures in text. June 29, 1951.

3. Phylogeny of the waxwings and allied birds. By M. Dale Arvey. Pp. 473-530, 49 figures in text, 13 tables. October 10, 1951.

*4. Birds from the state of Veracruz, Mexico. By George H. Lowery, Jr., and Walter W. Dalquest. Pp. 531-649, 7 figures in text, 2 tables. October 10, 1951.

Index. Pp. 651-681.

*Vol. 4. (Complete) American weasels. By E. Raymond Hall. Pp. 1-466, 41 plates, 31 figures in text. December 27, 1951.

Vol. 5. Nos. 1-37 and index. Pp. 1-676, 1951-1953.

*Vol. 6. (Complete) Mammals of Utah, *taxonomy and distribution*. By Stephen D. Durrant. Pp. 1-549, 91 figures in text, 30 tables. August 10, 1952.

Vol. 7. Nos. 1-15 and index. Pp. 1-651, 1952-1955.

Vol. 8. Nos. 1-10 and index. Pp. 1-675, 1954-1956.

Vol. 9. Nos. 1-23 and index. Pp. 1-690, 1955-1960.

Vol. 10. Nos. 1-10 and index. Pp. 1-626, 1956-1960.

Vol. 11. Nos. 1-10 and index. Pp. 1-703, 1958-1960.

Vol. 12. *1. Functional morphology of three bats: Eumops, Myotis, Macrotus. By Terry A. Vaughan. Pp. 1-153, 4 plates, 24 figures in text. July 8, 1959.

*2. The ancestry of modern Amphibia: a review of the evidence. By Theodore H. Eaton, Jr. Pp. 155-180, 10 figures in text. July 10, 1959.

3. The baculum in microtine rodents. By Sydney Anderson. Pp. 181-216, 49 figures in text. February 19, 1960.

*4. A new order of fishlike Amphibia from the Pennsylvanian of Kansas. By Theodore H. Eaton, Jr., and Peggy Lou Stewart. Pp. 217-240, 12 figures in text. May 2, 1960.

5. Natural history of the bell vireo. By Jon C. Barlow. Pp. 241-296, 6 figures in text. March 7, 1962.

6. Two new pelycosaurs from the lower Permian of Oklahoma. By Richard C. Fox. Pp. 297-307, 6 figures in text. May 21, 1962.

7. Vertebrates from the barrier island of Tamaulipas, México. By Robert K. Selander, Richard F. Johnston, B. J. Wilks, and Gerald G. Raun. Pp. 309-345, pls. 5-8. June 18, 1962.

8. Teeth of Edestid sharks. By Theodore H. Eaton, Jr. Pp. 347-362, 10 figures in text. October 1, 1962.

9. Variation in the muscles and nerves of the leg in two genera of grouse (Tympanuchus and Pedioecetes). By E. Bruce Holmes. Pp. 363-474, 20 figures. October 25, 1963. $1.00.

10. A new genus of Pennsylvanian fish (Crossopterygii, Coelacanthiformes) from Kansas. By Joan Echols. Pp. 475-501, 7 figures in text. October 25, 1963.

11. Observations on the Mississippi kite in southwestern Kansas. By Henry S. Fitch. Pp. 503-519. October 25, 1963.

12. Jaw musculature of the mourning and white-winged doves. By Robert L. Merz. Pp. 521-551, 22 figures in text. October 25, 1963.

13. Thoracic and coracoid arteries in two families of birds, Columbidae and Hirundinidae. By Marion Anne Jenkinson. Pp. 553-573, 7 figures in text. March 2, 1964.

14. The breeding birds of Kansas. By Richard F. Johnston. Pp. 575-655, 10 figures in text. May 18, 1964. 75 cents.

15. The adductor muscles of the jaw in some primitive reptiles. By Richard C. Fox. Pp. 657-680, 11 figures in text. May 18, 1964.

Index. Pp. 681-694.

(Continued on inside of back cover)

University of Kansas Publications
Museum of Natural History

Volume 14, No. 18, pp. 493-758, 82 figs.
July 6, 1965

The Mammals of Wyoming

BY

CHARLES A. LONG

University of Kansas
Lawrence
1965

University of Kansas Publications, Museum of Natural History

Volume 14, No. 18, pp. 493-758, 82 figs.
Published July 6, 1965

University of Kansas
Lawrence, Kansas

PRINTED BY
HARRY (BUD) TIMBERLAKE, STATE PRINTER
TOPEKA, KANSAS
1965

30-2329

The Mammals of Wyoming

BY

CHARLES A. LONG

CONTENTS

PAGE

INTRODUCTION 495
ACKNOWLEDGMENTS 496
MATERIALS AND METHODS 497
PHYSIOGRAPHY AND CLIMATE 498
GLACIATION 500
CHECK LIST OF MAMMALS OF WYOMING 510
ACCOUNTS OF SPECIES AND SUBSPECIES 515
DISCUSSION 723
  Faunal Areas 726
  Speciation, Subspeciation, and Zoogeography 729
SOME EFFECTS OF MAN IN HISTORIC TIME 739
SUMMARY 740
SPECIES AND SUBSPECIES OF UNVERIFIED OCCURRENCE 741
TYPE LOCALITIES IN WYOMING 743
LITERATURE CITED 747

INTRODUCTION

Mention of Wyoming to many persons brings to mind the bears, the pronghorn, the moose, the elk, the porcupine, and other kinds of mammals that occur in this state. Forested, snow-capped mountain ranges, shallow rivers and cascading streams, vast rolling prairies, and arid, rocky deserts also are remembered, and these habitats are the homes of Wyoming's mammals. Yellowstone Park, in northwestern Wyoming, is world-famous for its natural wonders, and many persons have felt that the animals alone are worth a trip to Yellowstone.

The mammals of Wyoming have been studied to a greater or lesser extent by most of North America's noted mammalogists, and mammals from Wyoming have been collected and preserved in considerable numbers for many years by many investigators. Even so, no comprehensive account of them has been published. Indeed no list of the kinds of mammals that occur in the state has been published. Nevertheless, studies of mammals in nearby states and numerous studies of particular kinds of mammals in Wyoming have contributed to our knowledge of the fauna of the state.

The chief aims of my study were to list and describe the kinds of mammals found in the state, to analyze the mammalian fauna as a whole with regard to patterns of distribution and relationships of kinds, and to explain their distributions and relationships in the light of Quaternary history.

ACKNOWLEDGMENTS

For permission to study specimens and other materials in their care in Washington, D. C. I am indebted to Drs. David H. Johnson, Henry W. Setzer and Charles O. Handley, Jr. of the United States National Museum proper, and to Mr. John L. Paradiso, Dr. Richard H. Manville and Dr. Hartley H. T. Jackson of The Biological Surveys Collections. Also some specimens were lent to me by some of the persons mentioned above and by Dr. S. D. Durrant of the University of Utah, Dr. R. Fautin of the University of Wyoming, Dr. R. S. Hoffman of Montana State University, and Drs. R. G. Van Gelder and S. Anderson of the American Museum of Natural History.

To the many collectors and investigators, past and present, including such well-known persons as Theodore Roosevelt, Ernest Thompson Seton, and C. Hart Merriam I am grateful for having preserved the specimens used in preparing the following account of the mammals of Wyoming. Approximately two-thirds of these specimens are in the Museum of Natural History at The University of Kansas. This collection was made mainly by advanced students who were members of field parties from The University of Kansas when Wyoming was being used as an "out-door laboratory" for their training in vertebrate zoology. Professor E. Raymond Hall led the first parties there in 1945-1946, and the work was continued by other zoologists of The University of Kansas, especially by Professor E. Lendell Cockrum who led parties in 1947-1948 and Professor Rollin H. Baker who led parties in the years 1949-1951.

This collection, now in the care of Professor E. R. Hall and Dr. J. Knox Jones, Jr., was generously relinquished for my use in 1960. I am particularly grateful to Professor Emeritus Arthur B. Mickey of the University of Wyoming for making his personal collection of mammals available to me for study by placing them on permanent deposit in the Museum of Natural History at The University of Kansas. I am grateful to Messrs. W. Charles Kerfoot, Roger Beers, and David L. Long, who accompanied me in the summer of 1961 to Wyoming and neighboring states, helping me in my field studies.

Mr. Thomas Swearingen drew figures 1, 2, and 81, and assisted me in making the other figures. I am grateful also to my fellow student, Charles Douglas, for varied assistance.

I am grateful to my wife Claudine F. Long for help in Wyoming in August, 1961, for help in listing specimens throughout my study, and for assisting in many other ways essential to completion of the following account. I am grateful to Professors E. Raymond Hall, A. Byron Leonard and Charles W. Pitrat, who constituted a committee that guided me as a graduate student in my research. Dr. G. M. Richmond of the U. S. Geological Survey read the section on glaciation. Dr. J. Knox Jones, Jr., also gave much appreciated advice. Finally, it goes without saying that the one person who has done so many things to make such a study as this possible, who has shown continuous interest in this study, and who has gladly shared the benefits of his invaluable experience is my advisor Professor E. Raymond Hall.

Financial assistance, for which I am grateful, was received from The Society of the Sigma Xi, the Kansas Heart Foundation, the National Science Foundation (including NSF GB 100), and The University of Kansas Endowment Association. Also gratefully acknowledged is assitance (to the museum in obtaining specimens) from the Office of Naval Research, U. S. Navy in 1947, and the American Heart Association Inc. in the period 1957-1961.

MATERIALS AND METHODS

In this study 12,807 specimens of Wyoming mammals were examined. More than 8,000 of these are in the Museum of Natural History at the University of Kansas and a few more than 4,200 in the United States National Museum. Approximately 300 specimens listed herein are in the collection at the University of Wyoming. Three of the specimens examined are in the American Museum of Natural History, one is in the collection at the University of Utah, and 23 are in the collection at Montana State University. I visited Wyoming and neighboring states for two days in 1960 and for two and a half months in the summer of 1961. In the period September 1960 to January 1963 I examined the materials at the University of Kansas, and in nine weeks in the summer of 1962 I examined specimens at the United States National Museum.

The arrangement here of the families and higher categories of Recent mammals closely follows that of Simpson (1945). For characters of the several taxonomic categories, for arrangement of genera, and for distribution maps of each species and subspecies in all of North America the reader is referred to Hall and Kelson (1959). Keys to species herein are provided where it is judged that they are necessary or useful.

In the text beyond a generic heading, with comments, is included if so doing avoids repetition of information in the accounts of two or more species. A specific heading, with comments, is included in the text for each polytypic species.

Subspecies are listed alphabetically, except one of *Ochotona* newly named beyond, under each species. The scientific name judged to be correct according to the International Rules of Zoological Nomenclature, or instead (rarely) a replacement-name recommended by the International Commission on Zoological Nomenclature, heads the account of each subspecies, and is followed on the same line by the name of the author of that scientific name. Immediately below is the vernacular name (Hall, *et al.*, 1957) that applies to the species unless the species is polytypic and represented in Wyoming by more than one subspecies, in which case the vernacular name is listed below the scientific name of the species. A synonymy follows consisting of a citation to the original proposal of a name used, a citation to the first usage of the accepted or current name combination, and a citation to any junior synonym having its type locality in Wyoming. If judged pertinent, other name combination are listed.

Comments on the holotype or type locality, a description of the subspecies or monotypic species, and in some instances a section of comparisons follow in that order. Measurements are listed to aid the researcher in identification and to reinforce conclusions made in the text. The geographic distribution is stated and for most of the named kinds is shown on a map. Under the subheading of "*Remarks*" various data are presented. Finally, records of occurrence are listed consisting first of specimens examined (the localities of which are usually plotted as black dots on a distribution map) and second of additional records (represented by open symbols on the distribution map) found in the literature. Under county headings (see Fig. 2), arranged alphabetically, the localities in any given county are arranged from north to south. When localities lie on the same latitude, they are listed from west to east. Unplotted localities, in italicized type, are ordinarily within an eight mile radius of a plotted locality or the locality is of a general nature and nonmarginal (for example, *Platte River*). Several specimens labeled merely as Yellowstone National Park are listed beyond. I have assumed that each of these is from the part of the Park that lies in Wyoming. Specimens examined from collections other than the University of Kansas Museum of Natural History are identified as to collection by explanatory abbreviations. Capitalized color terms are those of Ridgway (1912). In dental formulae, capital letters refer to teeth in the upper jaw, whereas lower case letters refer to those in the lower jaw.

Most place-names used in the following accounts were found in atlases, on current road-maps, and especially on the map (frontispiece, Plate 1) in Merrit Cary's (1917) "Life zone investigations in Wyoming." That map excellently portrays the life-zones in Wyoming and shows many old locality names. A map (entitled "Wyoming Fishing Orders") showing streams in Wyoming that was published in 1961 by the Wyoming Game and Fish Commission also proved helpful.

A list of kinds of mammals that may occur in Wyoming, but of which satisfactory record is lacking, is to be found following the accounts and discussion. A list (with map) of type localities in Wyoming is also included.

PHYSIOGRAPHY AND CLIMATE

Much of the following information was gathered and published by Merrit Cary (1917), whose work was in all ways excellent. Wyoming is a rectangular state somewhat more than 97,000 square miles in area. The physiography and climate are varied. Surface features may be classified broadly as mountains, plains, and valleys or basins. The continental divide extends southeasterly from the midwestern margin of the Yellowstone plateau along the lofty crests of the Absaroka and Wind River ranges, across the Red Desert, and along the summits of the Sierra Madre Mountains, where the divide crosses the southern boundary of Wyoming.

The names of physiographic features mostly are those used by Fenneman (1931: 134). Some of the features are shown also on Figure 1, page 506.

The mountains of Wyoming are mostly in the northwest and occupy approximately a fourth of the total area of the state. Chief among these mountain ranges are the Absaroka, Wind River, Gros Ventre, and Teton ranges in the northwest; the Bighorn Mountains in the north-central part; and the Sierra Madre and Medicine Bow mountains in the south. Most of these are heavily forested ranges of high elevation, whose summits and crests rise far above timberline and often are covered by snow the year round. Glaciers are extensive and abundant in the Wind River Mountains. The "glacier group" probably is larger than the "glacier group" in Glacier National Park, Montana (Wentworth and Delo, 1931:608, 611); the existing glaciers of the Wind River Mountains have a total surface area of 15.14 square miles, extending along the continental divide for a distance of 15 miles, and number more than 50. The Absaroka Mountains and the Teton Mountains support small glaciers (Wentworth and Delo, 1931:610), as also do deep and sheltered canyons in the Bighorn Mountains (Salisbury and Blackwelder, 1903:217; Salisbury, 1906). All of these ranges belong to the Rocky Mountain system except the Bighorn Mountains, which are separated from it although a hook of low, but uplifted, hills connects them to the Rocky Mountain chain. This hook of elevated land separates the Bighorn Basin from the Wind River Basin to the south. Gannett Peak in the Wind River Mountains is the highest point in the state (13,785 feet) closely approached in elevation by Fremont Peak (13,730 feet, Wind River Mountains) and the Grand Teton (13,747 feet, Teton Mountains).

The low, isolated ranges, Ferris, Green, Seminole, Shirley, and Rattlesnake, are in the central part of the state and are timbered on their northern slopes (see Cary, 1917:10). The timbered foothills of the Uinta Mountains extend into the southwestern part of Wyoming. From them, arid, barren ridges and lowlands extend toward the mountains of northwestern Wyoming; these ridges give way to the timbered slopes of the Wyoming and Salt River ranges.

In southeastern Wyoming the low, timbered Laramie Mountains are a spur of the rugged Medicine Bow Mountains. The lofty peaks of the Medicine Bow Mountains, approximately six miles northwest of Centennial, are known as the "Snowy Range." The Medicine Bow Mountains are east of the rugged Sierra Madre Mountains mentioned previously.

In the northeast the timbered Black Hills and the Bear Lodge Mountains, closely associated and often considered as the "Black Hills" (Guthe, 1935), are surrounded by the Great Plains.

Wyoming, according to Cary (1917:10), although well supplied with mountains is better known for its vast open plains. These are either level or rolling, lying usually between 4,500 and 7,000 feet elevation, and are characterized by their types of vegetation: sage plains of the high arid plateaus, and grassy plains to the east and northeast of the continental divide. The grassy plains are part of the Great Plains. On the eastern border of Wyoming these grassy plains become sandy, constituting a small westward extension of the Sand Hills in Nebraska.

Rivers that have cut broad valleys in Wyoming are as follows: the North Platte and Laramie in southeastern Wyoming; the Belle Fourche and Powder in northeastern Wyoming; the Bighorn and Wind in northwestern Wyoming; and the Sweetwater and Green in southwestern and south-central Wyoming. The area where the Belle Fourche River crosses the eastern boundary of Wyoming has the lowest elevation (3,100 feet) in the state. The valleys of the Snake and Yellowstone rivers are small; the rivers that have cut them leave the state before attaining much size. The Snake River drains part of the Yellowstone Plateau and also the low, arid basin, Jackson Hole, situated between an upthrusted, magnificently dissected and weathered fault-block, the Teton Range, and several ranges overlooking the basin from the eastward.

The Red Desert in southwestern Wyoming is an extensive, barren, alkaline area mainly west of the continental divide. Higher ground (some of it timbered) is to be seen south and west in Wyoming and beyond. The Green River leaves the state southwest of the Red Desert and in doing so passes through the higher ground to the southward.

The climate of Wyoming is mainly arid; the precipitation ranges from less than 10 inches in the Red Desert and Bighorn Basin to about 15 inches in the other lowlands. In the Black Hills there are 15 to 20 inches of precipitation, and there is much more precipitation on the highest mountains of the northwestern part of the state. An elevated base level (averaging about 6,000 feet according to Cary, 1917:11) is associated with cool climate. Summers are short and cool (55° F. average) in high mountain valleys, but longer and warmer (65° F.) on the Great Plains. Much snow falls in the mountains, but winter is more severe on the plains (Cary, 1917:11-12) because of low temperatures and strong winds.

GLACIATION

Wyoming is noted for its beautiful and rugged mountains, although vast prairies are also present in the state. Much of the ruggedness of the mountains results from past glaciation. There is no evidence that continental glaciers entered the state.

Mountain glaciers indicate cool climate and, of course, sufficient precipitation to support their formation and regimen. Glaciation produces many physiographic features, and study of these features can reveal the nature of the climate in the past. Darwin (1859:290), Deevey (1949:1321), and others have presented evidence that glacial advances displaced animals and plants southward; during interglacial periods warmth-adapted animals and plants moved northward. Relict populations or isolated populations that may have undergone evolution, especially at the subspecific level, were isolated by glaciation.

Glaciation in Wyoming occurred in mountains in the Pleistocene and probably later. Another indication of Pleistocene climate in Wyoming is provided in two deposits of pollen. Sears mentioned to Ray (1940) and Mears (1953) that peat deposits are thin in the Medicine Bow Mountains, and that the thinness indicated lack of vegetation there from the time of deposition until 1500 years ago. Hansen (1951:115-116) studied peat deposits obtained from a bog in Bridger Basin and reported that in the "Postglacial" time the climate became warm, then somewhat warmer still and more moist, dryer again and even warmer, and finally cooler. He stated that 70 pollen profiles in the Pacific Northwest indicate a warm, dry "climatic maximum" in the "Postglacial."

Dorf (1959:195) suggested that when continental glaciers were advancing, the temperature in Wyoming was lower than the subarctic temperature of today. Glaciers occurred on high, mountainous areas, and tundra surrounded them.

Interglacial periods were comparatively warm, and the glacial ice receded (Dorf, 1959:196-197). Dorf indicated that in most interglacial ages the temperature probably was little different than it is at present, but that it was even higher in one "late interglacial age." Even during interglacial ages in Wyoming a diagonal zone of "subarctic" temperature was supposed to have remained along the Rocky Mountains resembling the zone existing there today.

The glacial history of the Wind River Mountains, which are in the northwestern part of the state (Fig 1), is the standard for comparative studies in western Wyoming and some adjacent areas. Glaciers in the Bighorn Mountains of north-central Wyoming greatly deepened some canyons. Some glaciation also occurred on the north flank of the Uinta Mountains in southwestern Wyoming and on the Medicine Bow Range in south-central Wyoming. I know of no mention of glaciation in the Black Hills of the northeastern part of the state.

Field studies by numerous geologists have yielded many data on mountain glaciation in Wyoming. Some important evidences of glaciation that have been studied are glacial moraines and rock striae. The relative ages of moraines have been determined by studying the effects on them of weathering, erosion, and vegetation and by comparing relative ages of glacier-affected drainages. Such studies indicate several past glaciations in Wyoming. Observations of glacial deposits and striae on rocks indicate the direction and extent of glacial movement. Other geologic techniques for studying glaciation (for example, radiocarbon dating of fossils or peat analysis in glacial deposits) have been less useful in Wyoming than in some other areas.

Studies of mountain glaciation in Wyoming may be divided into four phases: First, finding and describing evidences of glaciation; second, learning that glaciation in Wyoming occurred more than once; third, correlating stages and substages of one area with those of another; fourth, determining the climatic history of glaciated areas. Summations of studies in mountainous areas of Wyoming are discussed somewhat chronologically beyond. Strict adherence in the discussion of any one physiographic unit independently of others is not followed in this paper mainly because it is more convenient to combine some units that have been studied together.

Glaciation in the Wind River Mountains

The highest mountains in the Wind River Range were severely glaciated. Blackwelder (1915:307-340) named three stages of glaciation based on relative ages of glacial deposits. The oldest drift, referred to the Buffalo stage (1915-328), was extensive and perhaps resulted from recurrent glacial advances. Blackwelder referred fresh moraines to the youngest stage termed by him (1915: 324) Pinedale. An intermediate stage was termed the Bull Lake stage (1915:325). Alden (1926:73) attempted to correlate Blackwelder's stages with continental stages of glaciation as follows: Buffalo-Nebraskan; Bull Lake-Iowan?; and Pinedale-Wisconsinan. Richmond (1941:1929-1930) verified Blackwelder's findings. He also found that ice caps had covered the interior portion of the Wind River Range. During Bull Lake time, long ice tongues from an ice cap extended northeast to the Wind River and dammed it at three points. This ice cap was smaller in Pinedale time. Glaciers from the nearby Absaroka Range "failed to reach the main mass" of ice from the Wind River Mountains.

Hack (Howard and Hack, 1943:239-240) found a young, prominent moraine north of Temple Lake in these mountains. He postulated that the moraine represented a stage of glaciation later than

Pinedale, and used the term Temple Lake for the young moraine. In the nearby Du Noir Valley lying northeast of the Wind River Mountains, Miner and Delo (1943:131-137) recognized Bull Lake and Pinedale moraines; and more recently Miner and Apfel (1946: 1218-1219) and Keefer (1957:213-215) described glacial effects of Buffalo, Bull Lake, and Pinedale glaciers in this area. In these mountains Branson and Branson (1945:1148-1149) described small glaciers enclosed by small moraines, and Richmond (1948:1400-1401) again mentioned small moraines in the vicinity of existing glaciers. He stated that two advances of ice occurred in the Buffalo stage. Richmond suggested that Blackwelder's sequence of stages be modified to account for the described glaciations. Moss (1949: 1911) mentioned five glacial advances along Big Sandy Creek in the southern part of the Wind River Mountains. In order of decreasing age the advances were assigned to: Buffalo; "double maxima" of Bull Lake; Pinedale; Temple Lake; and "neoglaciation," the last including moraines of the "fifth advance" lying within 1,000 feet of small existing ice masses.

Moss (1951:865-883) formally assigned some recently deposited drift in the Wind River Mountains to the two youngest glacial stages. One stage was termed Temple Lake stage and the other Little Ice Age (= his neo-glaciation). Consequently, five stages of glaciation are recognized in these mountains. Holmes and Moss (1955) reviewed studies in this area describing deposits of all five stages; their important correlations appear in Table 1.

Glaciation on the Yellowstone Plateau and in the nearby Beartooth, Absaroka, and Owl Creek Mountains

W. H. Holmes (1881:204-205) mentioned a lack of moraines in Yellowstone Park, but he mentioned some evidence (numerous boulders) for past glaciation. Weed (1893) described glaciation of the Yellowstone Valley on the northern part of the Yellowstone Plateau (Fig. 1). The sources of the Yellowstone Glacier were "confluent ice sheets" on the northwestern part of the park, where all of the névé fields lay. Each mountain gorge in the park was the bed of a glacier. Darton (1906:25-26), Sinclair (1912:314-315), Dake (1919), and Parsons (1939) described effects of glaciation in the Owl Creek, Beartooth, and Absaroka Mountains to the eastward (boulders were transported eastward to a point 12 miles east of Cody). Schlundt and Moore (1909:34) provided a rare example of dating a glacial epoch in Wyoming by analyzing radioactive substances in Yellowstone Park. In fact, their work on radium in travertine was done more than 50 years ago, only seven years after radium was discovered. They stated: "The travertine of Terrace Mountain is overlain by glacial boulders. Since its activity is only 1 per cent of that of the recent deposits, its age is about 20,000 years." They also mentioned a shorter time period (14,000 years) based on a possible half-life for radium of 2,000

TABLE 1.—CLASSIFICATION OF GLACIAL DEPOSITS IN WYOMING AS A MEANS OF CORRELATING GLACIAL EPISODES. AFTER FIG. 5, HOLMES AND MOSS, 1955.

Area	Age or kind of deposits				
	Buffalo	Bull Lake	Pinedale	Temple Lake	Little Ice Age
Yellowstone Valley	X [1]	X	X		
Beartooth Mountains	Drift	Broad deposits	Canyon Moraines		
Absaroka Mountains		Rolling Moraines	Hum-mocky Moraines	Rock glaciers and small moraines, Little Ice Age?	
Bighorn Mountains	Isolated boulders	Lateral Moraines	Terminal Moraines		
Western Wyoming	X	X	X		
Teton Mountains	X triple [2]	X double?	X triple [2]	X?	Moraines in cirques
Gros Ventre Mountains	X	X double?	X		
N. Wind River Mountains	X	X	X		
S. Wind River Mountains	X	X	X	X	X
Medicine Bow Mountains [3]		Weath-ered moraines	X		

1. X refers to deposits identified (with regard to age) and known by the name under which they are listed.

2. Buffalo till consists of "three tills of different glaciation" and the Pinedale had "three advances" (letter from G. M. Richmond to E. R. Hall, August 2, 1963).

3. In the Medicine Bow Mountains, Hares (1948) found glacial till and identified it as Kansan or older. Ray (1940) reported five glacial stages, which he identified as Wisconsin I-V; Richmond (1953) identified these five stages as Wisconsin I-III and late Recent 1-2.

years. Inasmuch as the half-life of radium is now known to be less than 2,000 years, the time of 14,000 years is probably too long.

Alden (1928) described three glacial epochs on the Yellowstone Plateau. Fenneman (1931:153) also stated that glacial ice covered the Yellowstone Plateau as an ice sheet or piedmont glacier at least three times. Fenneman stated that "interglacial epochs were long and the larger valleys were greatly deepened between successive invasions of ice." He stated that the first epoch occurred in early Pleistocene, but that present glacial features resulted from ice in a late Pleistocene epoch. The ice (epoch not mentioned) at some places on the plateau attained a maximum thickness of approximately 2,000 feet. Three lobes of ice extended into three valleys, Yellowstone River Valley, Madison River Valley, and Snake River Valley, on the plateau. The Yellowstone Glacier, almost 3,000 feet thick, extended 36 miles north of the park.

Glaciation in the Bighorn Mountains

Early work in the Bighorn Mountains was done by Matthes (1900), Salisbury and Blackwelder (1903), and Salisbury (1906). The last-named person reported that an area of 360 square miles in these mountains was once covered by ice, and that associated snow fields may have covered considerable additional terrain. He found evidences of 19 glacial systems from approximately 100 sources. The elevations of termini of all glaciers ranged from 6,200 to 10,400 feet. Two distinct glacial stages are known and probably a third stage should be recognized.

Glaciation in the Medicine Bow Mountains

W. W. Atwood, Jr. (1937) first found evidence for the existence of complex glaciation in the Medicine Bow Mountains (Fig. 1). He identified deposits of a Wisconsin stage and older deposits that he assumed to be pre-Wisconsin in age. Five glacial tongues originated in a catchment area at the base of the Snowy Range, but none of them extended onto the surrounding plains. Ray (1940), who had worked out a complex of glacial stages in the nearby Cache la Poudre area of Colorado and found no evidence for pre-Wisconsin glaciation, visited the Medicine Bow Mountains and subsequently assigned Atwood's pre-Wisconsin deposits to an early Wisconsin substage (Wisconsin I). He recorded a glacial history for the Medicine Bow Mountains comparable to that made out by him in the Cache la Poudre area (Table 2). Glaciers were not

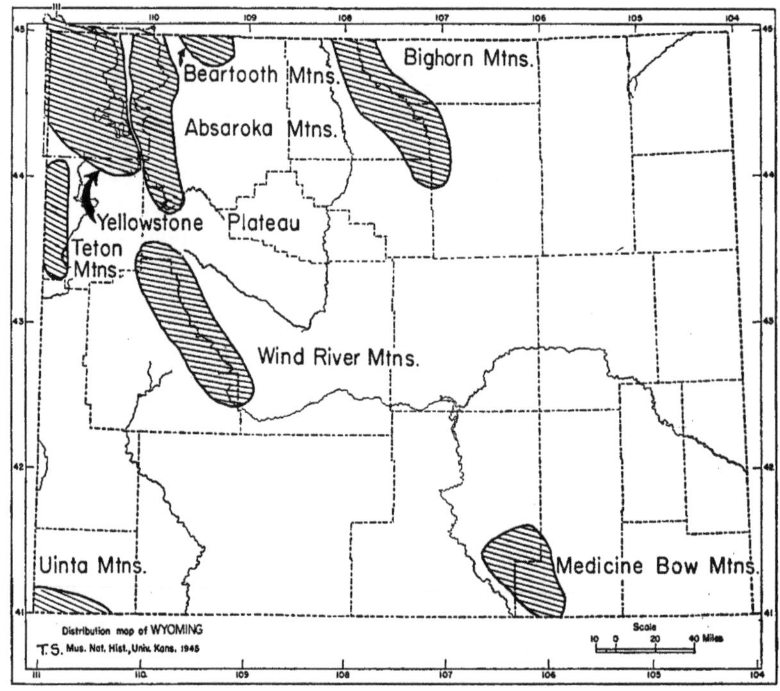

FIG. 1. Glaciated areas in Wyoming. After Holmes and Moss, 1955. The Yellowstone Valley extends northward from the Yellowstone Plateau. The southern extension of the Absaroka Mountains, south of the Shoshone River, frequently is termed the Shoshone Mountains. The low Owl Creek Mountains lie approximately 25 miles southeast of the Absaroka Mountains. Jackson Hole lies east of the Teton Mountains in the vicinity of Jackson Lake and southward. Snowy Range is a term applied to the high peaks of the northern part of the Medicine Bow Mountains.

formed in the fifth Wisconsin substage described by Ray but a big snow field at that time lay at the southeastern base of Medicine Bow Peak, as indicated by protalus ramparts there. Richmond (1953) assigned similar deposits in the nearby Rocky Mountain National Park of Colorado to late Recent time. If Ray is followed, five cold Wisconsin stages occurred in the Medicine Bow Mountains, of which the latest was nonglacial. Richmond thinks the last two substages described by Ray are in fact late Recent. Mears (1953:81) thinks there were at least four Wisconsin or later stages. He thinks that materials left by pre-Wisconsin glaciers probably were "largely obscured or destroyed by the extensive Wisconsin ice advances." The glacier in each Wisconsin substage was less extensive than its predecessor, as was determined by observations on relative weathering, soil development, and drainage patterns above them. A record by Hares (1948:1329) has seemingly been overlooked. He found a deposit of striated boulders thought by him to be "older" than "Kansan" age. Flint (1947:228-229) cites Atwood (1937) and Ray (1940) in listing the lowest elevation recorded for glaciers in these mountains as 7,500 feet, which is the lowest elevation recorded in the Cache la Poudre area; the lowest elevation recorded for a glacier in the Medicine Bow Mountains is 8,100 feet.

TABLE 2.—RAY'S CORRELATION OF DEPOSITS IN THE MEDICINE BOW MOUNTAINS WITH THOSE IN THE CACHE LA POUDRE AREA, COLORADO. AFTER RAY, 1940.

Elevation of moraine (ft.)	Atwood's assignment	Ray's assignments:		Elevation of moraine (ft.)
		Medicine Bow Range, Wyoming	Cache la Poudre, Colorado	
11,000*...		Wisc. V	Protalus	11,500
10,500....	Wisconsin undifferentiated	Wisc. IV	Long Draw	10,200
10,000....		Wisc. III	Corral Creek	9,100-10,100
8,500....		Wisc. II	Home	7,600
8,100....	pre-Wisconsin	Wisc. I	Twin Lakes	7,500

* Mears, 1953.

Glaciation in Other Areas

In the Jackson Hole-Teton Range area early work was accomplished by Bradley (1872), St. John (1877:444), Blackwelder (1915:325-328), and Alden (1928). Fryxell (1930) found evidences of Buffalo, Bull Lake, and Pinedale stages. In assigning deposits to these stages Fryxell, of course, did not know about younger stages subsequently discovered by Moss (1951). Buffalo ice was the most extensive. "Pinedale" glaciers were less extensive than preceding glaciers. Additional work was done by Fryxell, Horberg and Edmund (1941) and Montagne and Love (1957); in the latter paper ice covering 1300 square miles in Jackson Hole in Buffalo time is mentioned.

On the north flank of the Uinta Mountains, W. W. Atwood, Sr. (1909) suggested three "epochs" of glaciation, and Bradley found three stages represented (1936).

One mention is made by Bradley (1936:196) of an old glacial stage in the Wyoming Range correlated with Buffalo time. In the Gros Ventre Mountains, Swenson (1949:56-63) described glaciations assigned to Buffalo, Bull Lake, and Pinedale times.

Correlation

Now five glacial stages are recognized in the Wind River Mountains, a standard used in other Wyoming studies. Tentative assignments of stages elsewhere in older works need verification or revising. Tentative recent correlations are shown in Table 1.

It is important to correlate glaciation in Wyoming with continental glaciation; combining data on both kinds of glaciation probably more accurately reflects past world-wide climatic conditions than either alone would do. Alden (1926:73) attempted to do this by matching Buffalo with Nebraskan, Bull Lake with Iowan?, and Pinedale with Wisconsinan stages. Recent workers tentatively match Buffalo with pre-Wisconsin and the other stages with Wisconsin or Recent.

Holmes and Moss (1955:651) regarded the Bull Lake and Pinedale complexes of glaciation as Wisconsin in age. They thought that Temple Lake was comparable to either late Mankato or even early Little Ice Age. Only Buffalo was regarded by them as pre-Wisconsin. Opinions held by workers in the Medicine Bow Mountains would not be inconsistent with this point of view. Most deposits of the four glacial advances in the vicinity of the Snowy Range in these mountains are held to be Wisconsin or later by several workers (Ray, 1940; Richmond, 1953; and Mears, 1953). Only one record of a single deposit may indicate a pre-Wisconsin stage (Hares, 1948).

Many workers in western Wyoming, however, hold to other correlations like those of Alden (1926), in which the Bull Lake stage is considered to be a pre-Wisconsin stage. Dating of radioactive substances may solve the problem of correlation when such substances associated with glacial deposits (or interglacial deposits) have been accumlated. In summary, several things are important in analyzing animal distributions. Glaciation, at its height, was extensive enough to depress the tundra zone of the peaks of mountains of Wyoming. Glaciation occurred only on high mountain ranges (Fig. 1); the lowest glacial terminus was 6,200 feet elevation. Five glacial epochs occurred in the Wind River Mountains; named in order of decreasing age they are the Buffalo, Bull Lake (double), Pinedale, Temple Lake, and Little Ice Age stages. The Buffalo stage was most extensive and the moraines encompassing existing ice masses represent the least extensive stage. In other nearby mountain ranges in northwestern Wyoming data are comparable. In the Medicine Bow Mountains four glaciations have been regarded as Wisconsin, and by Richmond the last and a non-glacial snow-field have been considered to be late Recent. Subarctic temperatures prevalent during glacial stages have, in general, ameliorated since Pinedale time; interglacial temperatures resembled, somewhat, those of today. Displacement of animals northward and southward by interglacial and glacial ages, respectively, has been noted; east and west displacements are not so well known. Neither are the effects of the erratic amelioration of cold climate well known on animal distribution. A list of mammals and their distributions follow.

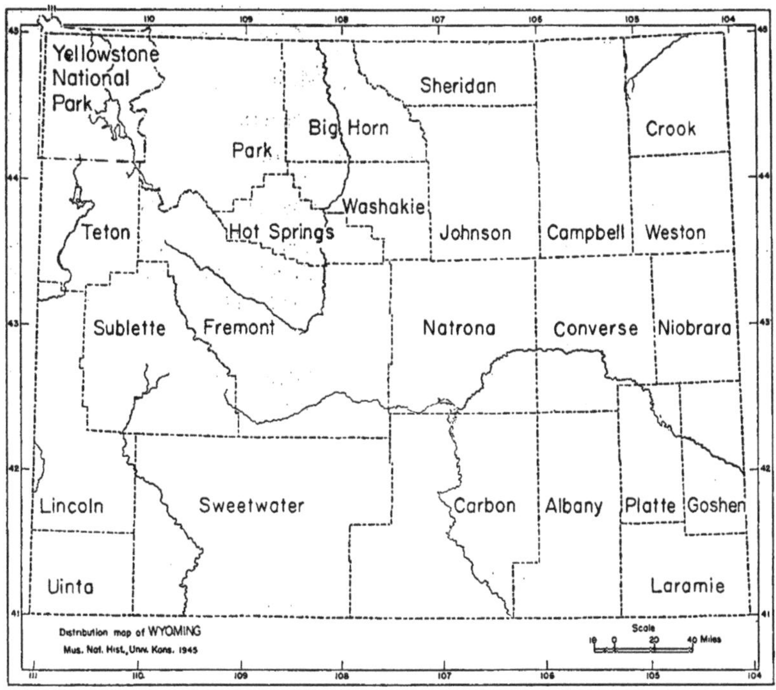

FIG. 2. Counties of Wyoming including Yellowstone National Park.

CHECK LIST OF MAMMALS OF WYOMING

The mammalian fauna of Wyoming consists of 172 kinds (subspecies and monotypic species) of which five were introduced directly or indirectly by man; these 172 kinds belong to 101 species, 58 genera, and 22 families of 7 orders.

Order MARSUPIALIA—Marsupials

Family DIDELPHIDAE—Opossums

		PAGE
Didelphis marsupialis virginiana Kerr	Opossum	515

Order INSECTIVORA—Insectivores

Family SORICIDAE—Shrews

		PAGE
Sorex cinereus cinereus Kerr *Sorex cinereus haydeni* Baird	Masked shrew	517
Sorex vagrans obscurus Merriam *Sorex vagrans vagrans* Baird	Vagrant shrew	520
Sorex nanus Merriam	Dwarf shrew	523
Sorex palustris navigator (Baird)	Water shrew	524
Sorex merriami leucogenys Osgood	Merriam's shrew	526

Family TALPIDAE—Moles

		PAGE
Scalopus aquaticus caryi Jackson	Eastern mole	528

Order CHIROPTERA—Bats

Family VESPERTILIONIDAE—Vespertilionid bats

		PAGE
Myotis lucifugus carissima Thomas	Little brown myotis	529
Myotis subulatus subulatus (Say)	Small-footed myotis	530
Myotis volans interior Miller	Long-legged myotis	531
Myotis evotis evotis (H. Allen)	Long-eared myotis	532
Myotis keenii septentrionalis (Trouessart)	Keen's myotis	532
Lasionycteris noctivagans (Le Conte)	Silver-haired bat	533
Lasiurus borealis borealis (Müller)	Red bat	533
Lasiurus cinereus cinereus (Palisot de Beauvois)	Hoary bat	534
Plecotus townsendii pallescens (Miller)	Townsend's bat	534
Euderma maculatum (J. A. Allen)	Spotted bat	535
Eptesicus fuscus pallidus Young	Big brown bat	536

Order LAGOMORPHA—Pikas, Rabbits, and Hares

Family OCHOTONIDAE—Pikas

		PAGE
Ochotona princeps new subspecies *Ochotona princeps figginsi* J. A. Allen *Ochotona princeps saxatilis* Bangs *Ochotona princeps ventorum* A. H. Howell	Pika	537

Family LEPORIDAE—Rabbits and Hares

Scientific name	Common name	Page
Sylvilagus floridanus similis Nelson	Eastern cottontail	542
Sylvilagus nuttallii grangeri (J. A. Allen)	Nuttall's cottontail	543
Sylvilagus audubonii baileyi (Merriam)	Desert cottontail	544
Lepus americanus bairdii Hayden	Snowshoe rabbit	546
Lepus americanus seclusus Baker and Hankins		
Lepus townsendii campanius Hollister	White-tailed jack rabbit	549
Lepus townsendii townsendii Bachman		
Lepus californicus melanotis Mearns	Black-tailed jack rabbit	552

Order RODENTIA—Rodents

Family SCIURIDAE—Squirrels

Scientific name	Common name	Page
Eutamias minimus confinis Howell	Least chipmunk	554
Eutamias minimus consobrinus (J. A. Allen)		
Eutamias minimus minimus (Bachman)		
Eutamias minimus operarius Merriam		
Eutamias minimus pallidus (J. A. Allen)		
Eutamias minimus silvaticus White		
Eutamias amoenus luteiventris (J. A. Allen)	Yellow-pine chipmunk	562
Eutamias dorsalis utahensis Merriam	Cliff chipmunk	564
Eutamias umbrinus fremonti White	Uinta chipmunk	565
Eutamias umbrinus montanus White		
Eutamias umbrinus umbrinus (J. A. Allen)		
Marmota flaviventris dacota (Merriam)	Yellow-bellied marmot	568
Marmota flaviventris luteola A. H. Howell		
Marmota flaviventris nosophora A. H. Howell		
Spermophilus richardsonii elegans Kennicott	Richardson's ground squirrel	572
Spermophilus armatus Kennicott	Uinta ground squirrel	575
Spermophilus spilosoma obsoletus Kennicott	Spotted ground squirrel	577
Spermophilus tridecemlineatus alleni Merriam	Thirteen-lined ground squirrel	577
Spermophilus tridecemlineatus olivaceus J. A. Allen		
Spermophilus tridecemlineatus pallidus J. A. Allen		
Spermophilus tridecemlineatus parvus J. A. Allen		
Spermophilus lateralis castanurus (Merriam)	Golden-mantled ground squirrel	583
Spermophilus lateralis cinerascens (Merriam)		
Spermophilus lateralis lateralis (Say)		
Spermophilus lateralis wortmani (J. A. Allen)		
Cynomys ludovicianus ludovicianus (Ord)	Black-tailed prairie dog	590
Cynomys leucurus Merriam	White-tailed prairie dog	591
Tamiasciurus hudsonicus baileyi (J. A. Allen)	Red squirrel	593
Tamiasciurus hudsonicus dakotensis (J. A. Allen)		
Tamiasciurus hudsonicus fremonti (Audubon and Bachman)		
Tamiasciurus hudsonicus ventorum (J. A. Allen)		

Family GEOMYIDAE—Pocket gophers

Scientific name	Common name	Page
Sciurus niger rufiventer É. Geoffroy St.-Hilaire	Fox squirrel	598
Glaucomys sabrinus bangsi (Rhoads)	Northern flying squirrel	599
Thomomys talpoides attenuatus Hall and Montague	Northern pocket gopher	600
Thomomys talpoides bridgeri Merriam		
Thomomys talpoides bullatus V. Bailey		
Thomomys talpoides caryi V. Bailey		
Thomomys talpoides cheyennensis Swenk		
Thomomys talpoides clusius Coues		
Thomomys talpoides meritus Hall		
Thomomys talpoides nebulosus V. Bailey		
Thomomys talpoides ocius Merriam		
Thomomys talpoides pygmaeus Merriam		
Thomomys talpoides rostralis Hall and Montague		
Thomomys talpoides tenellus Goldman		
Geomys bursarius lutescens Merriam	Plains pocket gopher	612

Family HETEROMYIDAE—Heteromyids

Scientific name	Common name	Page
Perognathus fasciatus callistus Osgood	Olive-backed pocket mouse	613
Perognathus fasciatus litus Cary		
Perognathus fasciatus olivaceogriseus Swenk		
Perognathus flavus piperi Goldman	Silky pocket mouse	616
Perognathus parvus clarus Goldman	Great Basin pocket mouse	617
Perognathus hispidus paradoxus Merriam	Hispid pocket mouse	617
Dipodomys ordii luteolus (Goldman)	Ord's kangaroo rat	618
Dipodomys ordii priscus Hoffmeister		
Dipodomys ordii terrosus Hoffmeister		

Family CASTORIDAE—Beavers

Scientific name	Common name	Page
Castor canadensis missouriensis Bailey	Beaver	621
Castor canadensis concisor Warren and Hall		

Family CRICETIDAE—Cricetids

Scientific name	Common name	Page
Reithrodontomys montanus albescens Cary	Plains harvest mouse	625
Reithrodontomys megalotis dychei J. A. Allen	Western harvest mouse	626
Peromyscus crinitus doutti Goin	Canyon mouse	627
Peromyscus maniculatus artemisiae (Rhoads)	Deer mouse	628
Peromyscus maniculatus nebrascensis (Coues)		
Peromyscus leucopus aridulus Osgood	White-footed mouse	634
Peromyscus truei truei (Shufeldt)	Piñon mouse	634
Onychomys leucogaster arcticeps Rhoads	Northern grasshopper mouse	635
Onychomys leucogaster brevicaudus Merriam		
Onychomys leucogaster missouriensis (Audubon and Bachman)		
Neotoma cinerea cinerea (Ord)	Bushy-tailed wood rat	638
Neotoma cinerea cinnamomea J. A. Allen		
Neotoma cinerea orolestes Merriam		
Neotoma cinerea rupicola J. A. Allen		

Clethrionomys gapperi brevicaudus (Merriam)
Clethrionomys gapperi galei (Merriam)
Clethrionomys gapperi idahoensis (Merriam) } Gapper's red-backed vole 642
Phenacomys intermedius intermedius Merriam ... Heather vole 646
Microtus pennsylvanicus insperatus (J. A. Allen)
Microtus pennsylvanicus pullatus S. Anderson } Meadow vole 647
Microtus montanus codiensis S. Anderson
Microtus montanus nanus (Merriam)
Microtus montanus zygomaticus S. Anderson } Montane vole 649
Microtus longicaudus longicaudus (Merriam) ... Long-tailed vole 653
Microtus richardsoni macropus (Merriam) ... Richardson's vole 656
Microtus ochrogaster haydenii (Baird) ... Prairie vole 659
Lagurus curtatus levidensis (Goldman) ... Sagebrush vole 660
Ondatra zibethicus cinnamominus (Hollister)
Ondatra zibethicus osoyoosensis (Lord) } Muskrat 661

Family MURIDAE—Murids

Rattus norvegicus norvegicus (Berkenhout) ... Norway rat 663
Mus musculus domesticus Rutty ... House mouse 663

Family ZAPODIDAE—Jumping mice

Zapus hudsonius campestris Preble
Zapus hudsonius preblei Krutzsch } Meadow jumping mouse 664
Zapus princeps idahoensis Davis
Zapus princeps princeps J. A. Allen
Zapus princeps utahensis Hall } Western jumping mouse 665

Family ERETHIZONTIDAE—Porcupines

Erethizon dorsatum bruneri Swenk
Erethizon dorsatum epixanthum Brandt } Porcupine 668

Order CARNIVORA—Carnivores

Family CANIDAE—Canids

Canis latrans latrans Say
Canis latrans lestes Merriam } Coyote 672
Canis lupus irremotus Goldman
Canis lupus nubilus Say
Canis lupus youngi Goldman } Gray wolf 675
Vulpes vulpes macroura Baird
Vulpes vulpes regalis Merriam } Red fox 678
Vulpes velox velox (Say) ... Swift fox 680
Urocyon cinereoargenteus ocythous Bangs ... Gray fox 682

Family URSIDAE—Bears

Ursus americanus cinnamomum Audubon and Bachman ... Black bear 682
Ursus arctos imperator Merriam ... Grizzly bear 685

Family PROCYONIDAE—Procyonids

Species	Common name	Page
Bassariscus astutus arizonensis Goldman	Ringtail	687
Procyon lotor hirtus Nelson and Goldman	Raccoon	688

Family MUSTELIDAE—Mustelids

Species	Common name	Page
Martes americana origenes (Rhoads) *Martes americana vulpina* (Rafinesque)	Marten	689
Martes pennanti columbiana Goldman	Fisher	691
Mustela erminea muricus (Bangs)	Ermine	692
Mustela frenata alleni (Merriam) *Mustela frenata longicauda* Bonaparte *Mustela frenata nevadensis* Hall *Mustela frenata oribasus* (Bangs)	Long-tailed weasel	693
Mustela vison energumenos (Bangs) *Mustela vison letifera* Hollister	Mink	695
Mustela nigripes (Audubon and Bachman)	Black-footed ferret	697
Gulo gulo luscus (Linnaeus)	Wolverine	697
Taxidea taxus taxus (Schreber)	Badger	699
Spilogale putorius gracilis Merriam *Spilogale putorius interrupta* (Rafinesque)	Spotted skunk	700
Mephitis mephitis hudsonica Richardson	Striped skunk	702
Lutra canadensis nexa Goldman	River otter	703

Family FELIDAE—Cats

Species	Common name	Page
Felis concolor hippolestes Merriam *Felis concolor missoulensis* Goldman	Mountain lion	705
Lynx canadensis canadensis Kerr	Lynx	707
Lynx rufus pallescens Merriam	Bobcat	708

Order ARTIODACTYLA—Artiodactyls

Family CERVIDAE—Deer

Species	Common name	Page
Cervus canadensis nelsoni V. Bailey	Wapiti or American elk	710
Odocoileus hemionus hemionus (Rafinesque)	Mule deer	712
Odocoileus virginianus dacotensis Goldman and Kellogg *Odocoileus virginianus ochrourus* V. Bailey	White-tailed deer	713
Alces alces shirasi Nelson	Moose	715

Family ANTILOCAPRIDAE—Pronghorn

Species	Common name	Page
Antilocapra americana americana (Ord)	Pronghorn	716

Family BOVIDAE—Bovids

Species	Common name	Page
Bison bison athabascae Rhoads *Bison bison bison* (Linnaeus)	Bison	717
Oreamnos americanus missoulae J. A. Allen	Mountain goat	720
Ovis canadensis audubonii Merriam *Ovis canadensis canadensis* Shaw	Mountain sheep	720

ACCOUNTS OF SPECIES AND SUBSPECIES

KEY TO ORDERS OF MAMMALS IN WYOMING

1. Incisor teeth $\frac{5}{4}$; marsupium present . . . . MARSUPIALIA, p. 515
1′. Incisor teeth not $\frac{5}{4}$; marsupium lacking 2
2. Flight-membrane present; fingers elongated and longer than forearm, CHIROPTERA, p. 528
2′. Flight-membrane lacking (membrane for volant gliding in *Glaucomys*); fingers not elongated and shorter than forearm 3
3. Upper incisors present; digits provided with claws 4
4. Diastema lacking; canines present 5
5. Middle incisors larger than canines . . . . INSECTIVORA, p. 516
5′. Middle incisors smaller than canines . . . . CARNIVORA, p. 671
4′. Diastema present; canines absent 6
6. Incisors 4 above . . . . LAGOMORPHA, p. 536
6′. Incisors 2 above . . . . RODENTIA, p. 552
3′. Upper incisors absent; digits provided with hooves . . . ARTIODACTYLA, p. 710

Order MARSUPIALIA

Marsupials

Family DIDELPHIDAE—Opossums

Didelphis marsupialis virginiana Kerr

Opossum

1792. *Didelphis virginiana* Kerr, The animal kingdom . . ., p. 193.
1952. *Didelphis marsupialis virginiana*, Hall and Kelson, Univ. Kansas Publs., Mus. Nat. Hist., 5:322, December 5.

Type.—From Virginia.

Description.—Maximum total length barely exceeding 1000 mm; upper parts blackish, grayish, or brownish strongly grizzled with white; tail haired at base but naked distally; hallux (on hind foot) clawless; sagittal crest frequently high (braincase low); female having well-developed marsupium containing, usually, 13 teats; 50 teeth; 5 incisors occurring on each side of upper jaw.

Measurements.—Measurements are listed by Hall and Kelson (1959:5) as follows: Total length, 645-1017; length of tail, 255-535; hind foot, 48-80; total length of skull, 80-139.

Distribution.—Known only from eastern Wyoming in northeastern Converse County. See *Remarks*. Not mapped.

Remarks.—A letter published by *Wyoming Wildlife* (February, 1963) mentioned that Mr. Hans W. Larsen captured a live opossum in northern Converse County. Correspondence with him provides additional information as follows: "This is in reply to your letter of Feb. 7 in regard to the opossum I captured alive last fall. I caged this opossum in Newcastle and it escaped by cutting a wire apparently with its teeth. I only had it a couple of weeks. . . . It was captured in the northeast corner of Converse County near the site of the old Dull Center post office. This is about 30 miles north and east of Bill, Wyo., which is only a post office. . . . I know this country very well and am positive it couldn't have escaped from anyone around there. It was believed to be a young 'possum by all who saw it. . . . The date was Aug. 31, 1962. It was evening, of course, when it crossed the country road in front of my pickup. When my coon dogs ran up to it, it played dead and I picked it up by the tail. . . . It liked corn on the cob, eggs whole in the shell, and other raw vegetables. . . ."

The opossum was captured in the watershed of the Cheyenne River, and may have been a young of the year that wandered into Wyoming from the eastward. Intensive collecting by investigators in Wyoming through the past 80 years has yielded no other opossum.

Order INSECTIVORA

Insectivores

Size of most kinds small; feet plantigrade and pentadactyl; digits clawed; snout slender and pointed.

KEY TO FAMILIES OF INSECTIVORES

1. Feet adapted for digging; eyes and ears not evident externally; tail less than ¾ as long as hind foot. . . . . TALPIDAE, p. 528

1′. Feet not adapted for digging; eyes and ears tiny but evident; tail more than ¾ as long as hind foot. . . . . SORICIDAE, p. 516

Family SORICIDAE—Shrews

Size small; not adapted for digging; ears and eyes evident; zygomata absent; usually predaceous.

Genus **Sorex** Linnaeus

Long-tailed Shrews

No significant secondary-sexual differences were found in Wyoming shrews, but they can be divided into two age groups on the basis of amount of tooth-wear and cranial differences (see Findley, 1955). Most specimens of shrews were collected in spring or summer and are either first or second year animals (see Findley, 1955, on *Sorex vagrans*; Pearson, 1945, on *Blarina*; Hamilton, 1940, on *Sorex fumeus*; and Conaway, 1952, on *Sorex palustris*). Younger shrews having no observable tooth-wear, of a size comparable with older shrews having observable tooth-wear, yield more nearly uniform cranial measurements because the relative proportions of cranial measurements in older shrews vary more. Thus, inasmuch as enough specimens of most species were available for this study, cranial measurements of younger shrews of adult size, designated as class 1, were chosen for comparative use. Older shrews are here designated as class 2; their skulls are broader, shorter, and have sagittal and lambdoidal ridges (see Findley, 1955, and Pruitt, 1954).

Measurements beyond were taken in the same fashion that Jackson (1928:12) and Findley (1955:5) took measurements.

KEY TO SPECIES AND SUBSPECIES OF SHREWS

1. Pelage flecked with silvery-white hairs on dark, grayish background; hind foot fringed with stiff hairs; tail scaly and scantily haired; total length more than 130 mm. . . . . **Sorex palustris navigator,** p. 524
1′. Pelage not flecked with silvery-white hairs; hind foot not fringed with stiff hairs; tail haired; total length less than 130. 2
2. Third unicuspid larger than fourth. 3
2′. Third unicuspid smaller than fourth. 5
3. Whitish underparts; crowded unicuspid row; postmandibular foramen well developed. . . . . **Sorex merriami leucogenys,** p. 526
3′. Brownish underparts; uncrowded unicuspid row; postmandibular foramen not well developed. 4
4. Tricolored pattern laterally; known from northeastern Wyoming, **Sorex cinereus haydeni,** p. 519
4′. Tricolored pattern obscure or lacking; not known from northeastern Wyoming. . . . . **Sorex cinereus cinereus,** p. 517
5. Tail less than 40; hind foot less than 11; braincase flattened, **Sorex nanus,** p. 523
5′. Tail more than 40 (ordinarily); hind foot more than 11; braincase convex viewed laterally. 6
6. Reddish-brown pigment evident dorsally; tail 41 or less; total length 108 or less. . . . . **Sorex vagrans vagrans,** p. 523
6′. Reddish-brown pigment lacking dorsally; tail frequently more than 41; total length often more than 108. . . . . **Sorex vagrans obscurus,** p. 521

Sorex cinereus Kerr

Masked Shrew

Small; brownish; resembling *Sorex vagrans;* fourth unicuspid tooth smaller than third in *S. cinereus,* but not in *S. vagrans;* nose and rostrum unusually elongated and pointed; upper parts dusky brownish; underparts grayish brown; total length 87-109 mm; tail almost half total length.

The masked shrew in Wyoming is most numerous in montane habitats. Three pregnant females were collected in July, at a place eight miles north and 16 miles east of Encampment, 8400 feet. They contained 7, 8, and 12 embryos. The female (KU 25220) having 12 was trapped on July 8, 1948, by W. M. Good.

Sorex cinereus cinereus Kerr

1792. *Sorex arcticus cinereus* Kerr, The Animal Kingdom . . ., p. 206.
1925. *Sorex cinereus cinereus,* Jackson, Jour. Mamm., 6:55-56, Feb. 9.

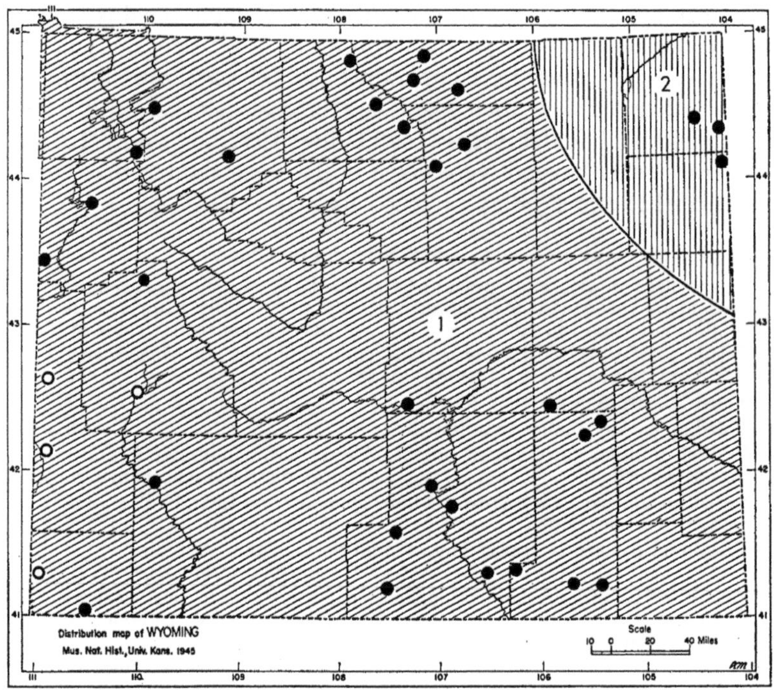

FIG. 3. Distribution of *Sorex cinereus*.
1. *S. c. cinereus* 2. *S. c. haydeni*

Type.—Name based on description by J. P. Forster, 1772, of shrew from Fort Severn, Canada (see Jackson, 1925:56).

Comparisons.—From *Sorex cinereus haydeni, S. c. cinereus* differs as follows: darker; larger (cranially and externally); relatively longer tail; less of tricolor pattern laterally; relatively longer palate; relatively narrower rostrum (see *Measurements*).

Measurements.—External and cranial measurements of seven adults from three miles ESE Browns Peak are as follows: Total length, 91.7 (88-94); tail vertebrae, 37.3 (35-39); hind foot, 10.8 (10-11); condylobasal length, 15.4 (14.9-15.8); maxillary tooth-row, 5.5 (5.4-5.8); palatal length, 7.7 (7.2-7.9); cranial breadth, 2.6 (2.4-2.8); least interorbital breadth, 2.6 (2.4-2.8); breadth maxillary processes, 4.0 (3.9-4.1).

Distribution.—Most of state; unrecorded in eastern fifth. See Fig. 3. In Wyoming *Sorex cinereus cinereus* has been found at elevations of 3800 to 9150 feet.

Remarks.—Jackson (1928:39) assigned shrews from western Wyoming to *Sorex cinereus cinereus* and those from eastern Wyoming to *S. c. haydeni*. He separated the two subspecies along an S-shaped curve, extending from north to south. I have assigned shrews from northeastern Wyoming to *S. c. haydeni* and those from elsewhere to *S. c. cinereus*.

Shrews from northeastern Wyoming are shorter in body, tail, condylobasal length, maxillary tooth-row, and palate than shrews to the south-southwestward. The shrews from northeastern Wyoming are narrow interorbitally; they are indistinguishable from *S. c. cinereus* on basis of breadth across the maxillary processes. In addition, shrews from northeastern Wyoming are paler, Buffy Brown to Olive Brown, whereas other masked shrews from Wyoming approach Blister Brown or Clove Brown. The southern shrews agree with *S. c. cinereus;* the northeastern shrews agree with *S. c. haydeni.* More specimens are needed from the vicinity of Laramie Peak for a more nearly accurate subspecific appraisal of specimens from there.

Records of occurrence.—Specimens examined, 227, as follows: **Albany Co.:** Springhill, 1 USNM; Laramie Peak, 8000 ft., 2 USNM; Libby Flats, Snowy Range, 1 Univ. Wyo.; *Libby Creek, 1; Nash Fork Creek, 1;* 3.6 mi. SW Laramie, 1; *Blair, Pole Mtn., 2; Pole Mtn., Ranger Sta., 8500 ft., 1;* 2 mi. SW Pole Mtn., 8300 ft., 4; *1 mi. SSE Pole Mtn., 8350 ft., 2; 6½ mi. S, 8¾ mi. E Laramie, 8200 ft., 1; 3 mi. S. Pole Mtn., 8100 ft., 1; Telephone Canyon, SE Laramie, 1.*

Big Horn Co.: Medicine Wheel Ranch, 28 mi. E Lovell, 9000 ft., 7; 2 mi. N, 12 mi. E Shell, 7900 ft., 1; Head Trappers Creek, 8500 ft., 20 USNM; *Bighorn Mtns., 4 USNM.*

Carbon Co.: 18 mi. NNE Sinclair, 6500 ft., 1; Ft. Steele, 1 USNM; Bridger Pass, 7500 ft., 3; 6 mi. S, 14 mi. E Saratoga, 8800 ft., 2; *10 mi. N, 14 mi. E Encampment, 8000 ft., 3; 9 mi. N, 4 mi. E Encampment, 6500 ft., 3; 8 mi. N, 14 mi. E Encampment, 8400 ft., 10; 8 mi. N, 19 mi. E Encampment, 9150 ft., 3;* Sierra Madre Mtns., 9038 ft., 1.

Converse Co.: 21½ mi. S, 24½ mi. W Douglas, 7600 ft., 1.

Johnson Co.: 9/10 mi. S, 1 mi. W Buffalo, 4800 ft., 1; Head N. Fork Powder River, 1.

Natrona Co.: 5 mi. W Independence Rock, 6000 ft., 1.

Park Co.: Grinnell Creek, Pahaska, 6300-7000 ft., 30 USNM; Valley, 7500 ft., 2 USNM.

Sheridan Co.: 3 mi. WNW Monarch, 3800 ft., 4; Eatons Ranch, Wolf, 1 USNM; 4 mi. NNE Banner, 4100 ft., 6.

Sublette Co.: 32 mi. N, 1 mi. W Pinedale, 8000 ft., 1; *31 mi. N Pinedale, 8025 ft., 4.*

Sweetwater Co.: Jct. Big Sandy and Green rivers, 6400 ft., 1.

Teton Co.: *1 mi. N Moran, 5 USNM; Lake Emma Matilda, Moran, 10 USNM; Jackson Hole Wildlife Park, 58 USNM;* 1 mi. S, 3¾ mi. E Moran, 6200 ft., 1; *Moose Creek, Teton Mtns., 6800 ft., 8 USNM;* Teton Pass, 7200 ft., 11 USNM.

Uinta Co.: 8 mi. S, 2½ mi. E Robertson, 8300 ft., 1; *9 mi. S Robertson, 8000 ft., 1.*

Yellowstone Nat'l Park: Mtn. Creek, 1 USNM.

Additional records (Jackson, 1928:50).—**Carbon Co.:** *S. base Bridger Peak, Sierra Madre Mtns., 8800 ft.* **Lincoln Co.:** 10 mi. SE Afton, 7500 ft.; Cokeville. **Sublette Co.:** Big Piney. **Teton Co.:** *Pacific Creek; Black Rock Creek.* **Uinta Co.:** Evanston.

Sorex cinereus haydeni Baird

1858. *Sorex haydeni* Baird, Mammals, *in* Repts. Expl. Survey . . ., 8(1):29, July 14.

1925. *Sorex cinereus haydeni,* Jackson, Jour. Mamm., 6:55-56, Feb. 9.

Type.—From "Fort Union, Neb." (Baird, 1858:30, 708) [=Mondak, Montana].

Comparisons.—*Sorex cinereus haydeni* differs from *S. c. cinereus* in the following characters: paler; smaller (cranially and externally); relatively shorter tail; tricolored laterally; relatively shorter palate; and broader rostrum.

Measurements.—Measurements of an adult male and adult female from 1½ miles east of Buckhorn are as follows: Total length, 93, 94; length of tail, 36, 35; hind foot, 11, 11; condylobasal length, 14.9, 14.6; maxillary tooth-row, 5.4, 5.3; palatal length, 5.8, 5.5; cranial breadth, 7.1, 7.7; least interorbital breadth, 2.5, 2.7; breadth maxillary processes, 4.1, 4.1.

Distribution.—Northeastern part of state. See Fig. 3 and account of *Sorex c. cinereus.*

Records of occurrence.—Specimens examined, 18, as follows: **Crook Co.:** *Warren Peak, 6000 ft., 4 USNM;* 3 mi. NW Sundance, 5900 ft., 9; *Sundance, 2 USNM;* Rattlesnake Creek, Black Hills, 6000 ft., 1 USNM. **Weston Co.:** 1½ mi. E Buckhorn, 6150 ft., 2.

Sorex vagrans Baird

Vagrant Shrew

Small; brownish; some individuals difficult to distinguish from *Sorex cinereus* (see account of that species).

Findley (1955) revised the species *Sorex vagrans*, and my findings agree with his concerning *S. vagrans* in Wyoming.

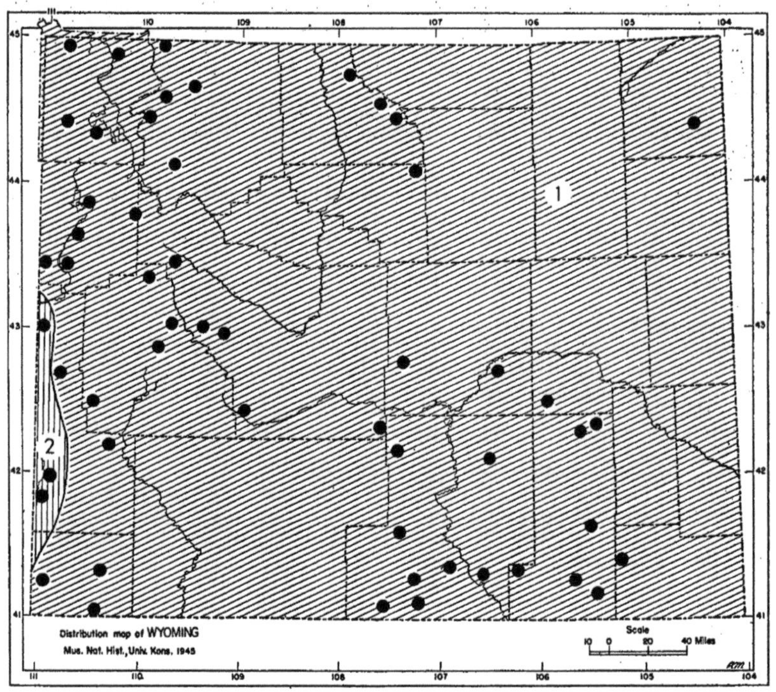

FIG. 4. Distribution of *Sorex vagrans*.
1. *S. v. obscurus* 2. *S. v. vagrans*

Sorex vagrans obscurus Merriam

1891. *Sorex vagrans similis* Merriam, N. Amer. Fauna, 5:34, July 30 (not *Sorex similis* Hensel 1855 from Sardinia).

1895. *Sorex obscurus* Merriam, N. Amer. Fauna, 10:72, December 31, a renaming of *Sorex similis* Merriam.

1955. *Sorex vagrans obscurus*, Findley, Univ. Kansas Publs., Mus. Nat. Hist., 9:43, December 10.

Type.—From Timber Creek, 8200 ft., Salmon Mountains [now Lemhi Mountains], 10 miles west of Junction, Lemhi County, Idaho.

Comparisons.—Larger than *S. v. vagrans* (see *Measurements*); dorsum (in fresh summer pelage) less reddish; said to be paler in winter (Findley, 1955).

Distribution.—Most of state; lacking in most arid areas and replaced by *S. v. vagrans* in western half of Lincoln County; known from elevations of 5400 to 10,500 feet. See Fig. 4.

TABLE 3.—AVERAGE AND EXTREME EXTERNAL AND CRANIAL MEASUREMENTS OF ADULTS OF *Sorex vagrans*

Locality	Number of specimens	Total length	Tail vertebrae	Hind foot	Condylobasal length	Maxillary tooth-row	Cranial breadth	Least interorbital breadth
				S. v. obscurus				
Sublette Co.	5	107 95–109	45.6 41–48	11.7 8–13	16.5 16.1–16.9	6.4 6.0–6.9	8.6 8.4–8.7	3.4 3.2–3.5
Ft. Bridger, Uinta Co.	4	110 102–119	44.8 42–47	13.3 13–14	16.8 16.6–17.0	6.6 6.5–6.7	8.4 8.1–8.8	3.4 3.3–3.4
Laramie Co.	2	107–118	41–43	12–12	16.2–16.4	6.2–6.3	8.7–8.9	3.5–3.7
				S. v. vagrans				
Lincoln Co.	15	102 95–108	39.1 36–41	11.8 8–13	15.9 15.6–16.1	5.9 5.3–6.2	8.2 7.7–8.7	3.1 2.9–3.3

Remarks.—*S. v. obscurus* maintains its distinctive characters, especially large size, throughout its geographic range in Wyoming and can be distinguished easily from the smaller *S. v. vagrans* of Lincoln County (see *Comparisons*).

A pregnant female (KU 32276) obtained by R. H. Baker on June 14, 1949, 22 miles south and 24.5 miles west of Douglas, 7600 feet, contained eight embryos.

Records of occurrence.—Specimens examined, 429, as follows: **Albany Co.**: Springhill, 6300 ft., 10 USNM; Laramie Peak, 8800 ft., 7 USNM; *3 mi. SW Eagle Peak, 6 USNM; 30 mi. N, 10 mi. E Laramie, 6760 ft., 3; 29 mi. N, 8¾ mi. E Laramie, 6420 ft., 4;* 27 mi. N, 5 mi. E Laramie, 6960 ft., 2; *26¾ mi. N, 6½ mi. E Laramie, 6700 ft., 1; 26 mi. N, 4½ mi. E Laramie, 6960 ft., 2;*

2¼ mi. ESE Browns Peak, 10,500 ft., 2; 3 mi. ESE Browns Peak, 10,000 ft., 7; Univ. Wyoming Sci. Camp, 1 Univ. Wyo.; Centennial, 8120 ft., 1; *Woods P. O., 1 USNM; Nelson Park, Medicine Bow Mtns., 1;* 2½ mi. NW Laramie, 1; *4 mi. SW Laramie, 1; 6 mi. SW Laramie, 7200 ft., 3; 6½ mi. S, 8¾ mi. E Laramie, 8200 ft., 4; 1 mi. ESE Pole Mtn., 8250 ft., 4; Blair, 1;* Wallis Camp Ground, Pole Mtn., 1; *1 mi. SSE Pole Mtn., 8350 ft., 3; 2 mi. SW Pole Mtn., 8350 ft., 3; 3 mi. S Pole Mtn., 8100 ft., 2; Vedauwoo, 1.*

Big Horn Co.: Medicine Wheel Ranch, 28 mi. E Lovell, 9000 ft., 14; *1 mi. N, 12 mi. E Shell, 7600 ft., 1;* 4½ mi. S, 17½ mi. E Shell, 8500 ft., 1; Head Trappers Creek, 8500 ft., 6 USNM.

Carbon Co.: Ferris Mtns., 8500 ft., 12 USNM; Shirley Mtns., 7600 ft., 7 USNM; Bridgers Pass, 18 mi. SW Rawlins, 7500 ft., 2, 2 USNM; 6 mi. S, 14 mi. E Saratoga, 8800 ft., 1; *1 mi. NW Silver Lake, 9620 ft., 1;* S Base Bridgers Peak, 3 USNM; *10 mi. N, 12 mi. E Encampment, 7200 ft., 2; 10 mi. N, 14 mi. E Encampment, 8000 ft., 6; 9 mi. N, 3 mi. E Encampment, 6500 ft., 1; 9 mi. N, 8 mi. E Encampment, 7000 ft., 1; 8 mi. N, 14½ mi. E Encampment, 8100 ft., 5;* 8 mi. N, 16 mi. E Encampment, 8400 ft., 4; 8 mi. N, 21½ mi. E Encampment, 9400 ft., 5; *8 mi. N, 19½ mi. E Savery, 8800 ft., 8; 7 mi. N, 17 mi. E Savery, 8300 ft., 1; 6½ mi. N, 16 mi. E Savery, 8300 ft., 1;* 6 mi. N, 14 mi. E Savery, 8000 ft., 2.

Converse Co.: 22 mi. S, 24½ mi. W Douglas, 7600 ft., 14.

Crook Co.: 3 mi. NW Sundance, 5900 ft., 1.

Fremont Co.: Jackeys Creek, 1 USNM; Moccasin Lake, 4 mi. N, 19 mi. W Lander, 10,000 ft., 1; *Milford, 5400 ft., 2;* 2½ mi. N Lander, 1; *17 mi. S, 6½ mi. W Lander, 8450 ft., 2;* 23½ mi. S, 5 mi. W Lander, 8600 ft., 1; 8 mi. E Rongis, Green Mtns., 8000 ft., 4 USNM.

Laramie Co.: 1 mi. N, 5 mi. W Horse Creek P. O., 7200 ft., 2.

Lincoln Co.: 10 mi. SE Afton, 7500 ft., 5 USNM; LaBarge Creek, 9000 ft., 1 USNM.

Natrona Co.: Rattlesnake Mtns., 7300-7500 ft., 18 USNM; 7 mi. S, 2 mi. W Casper, 6370 ft., 1; *7 mi. S, 1 mi. W Casper, 6370 ft., 1; 7 mi. S Casper, 6000 ft., 6 USNM.*

Park Co.: Beartooth Mtns., 6 USNM; Black Mtn., 2 USNM; Whirlwind Peak, 9000 ft., 1; Grinnell Creek, Pahaska, 37 USNM; 25 mi. S, 28 mi. W Cody, 6350 ft., 1; *Needle Mtn., 2 USNM; Valley, 7500 ft., 14 USNM.*

Sublette Co.: 31 mi. N Pinedale, 8025 ft., 3; 12 mi. NE Pinedale, 8000 ft., 2 USNM; *N Half Moon Lake, 7900 ft., 1;* 2¼ mi. NE Pinedale, 7500 ft., 2; 3 mi. W Stanley, 8000 ft., 3 USNM.

Teton Co.: *Pacific Creek, 3 USNM; 1 mi. N Moran, 3 USNM;* ½ mi. N Moran, 6750 ft., 1; *¼ mi. N, 2½ mi. E Moran, 6200 ft., 4; Moran, 6244-6742 ft., 1;* 1 Univ. Wyo.; 6 USNM; *½ mi. E Moran, 6742 ft., 1;* Jackson Hole Wildlife Park, 26 USNM; *3 mi. E Moran, 6210 ft., 9; 3¾ mi. E Moran, 6300 ft., 1; 1 mi. S, 3¾ mi. E Moran, 6200 ft., 9; Black Rock Creek, 2 USNM;* Togwotee Pass, 9500 ft., 1 Univ. Wyo.; *Timbered Island, 4 mi. N Moose, 6750 ft., 3;* 2½ mi. NE Moose, 6500 ft., 1; Teton Pass, 7200 ft., 15 USNM; *Teton Mtns., S Moose Creek, 6800 ft., 9 USNM;* Jackson, 3; locality not specified, 1 USNM.

Uinta Co.: 1 mi. N Ft. Bridger, 6650 ft., 1; *Ft. Bridger, 6650 ft., 3;* Evanston, 1 USNM; *9 mi. S Robertson, 8000 ft., 8; 10 mi. S, 1 mi. W Robertson, 8700 ft., 5;* 10½ mi. S, 2 mi. E Robertson, 8900 ft., 1; *13 mi. S, 1 mi. E Robertson, 9000 ft., 1; 13 mi. S, 2 mi. E Robertson, 9200 ft., 1;* 14 mi. S, 2 mi. E Robertson, 9000 ft., 1.

Washakie Co.: 5 mi. N, 9 mi. E Tensleep, 7400 ft., 2; *4 mi. N, 9 mi. E Tensleep, 7000 ft., 2.*

Yellowstone Nat'l Park: Mammoth Hot Springs, 1 USNM; Tower Falls, 1 USNM; *Willow Park, 2 USNM; Astringent Creek, 1 USNM;* Old Faithful, 3 USNM; Shoshone Lake, 2 USNM.

Additional records (Findley, 1955:44).—**Teton Co.:** *Two Ocean Lake; 7 mi. S Moran; Beaver Dick Lake; Whetstone Creek; Flat Creek-Gravel Creek divide; Flat Creek-Granite Creek divide.* **Yellowstone Nat'l Park:** *Flat Mtn.*

Sorex vagrans vagrans Baird

1858. *Sorex vagrans* Baird, Mammals, in Repts. Expl. Surv. . . . 8(1): 15, July 14.

1955. *Sorex vagrans vagrans*, Findley, Univ. Kansas Publs., Mus. Nat. Hist., 9:52-57, December 10.

Type.—From Shoalwater Bay [= Willapa Bay], Pacific County, Washington.

Comparisons.—Smaller than *Sorex vagrans obscurus*, more nearly red in summer, darker in winter.

Measurements.—See Table 3.

Distribution.—Western half of Lincoln County in western part of state. See Fig. 4.

Remarks.—One female (KU 37291) obtained by S. Anderson on July 16, 1950, seven miles north and one mile west of Afton, 6100 ft., contained five embryos.

Records of occurrence.—Specimens examined, 17, as follows: **Lincoln Co.:** 13 mi. N, 2 mi. W Afton, 6100 ft., 6; *10 mi. N Afton, 6200 ft.*, 2 *USNM; 7 mi. N, 1 mi. W Afton, 6100 ft.*, 5; 15 mi. N, 3 mi. E Sage, 6100 ft., 1; *12 mi. N, 2 mi. E Sage, 6100 ft.*, 2; 6 mi. N, 2 mi. E Sage, 6050 ft., 1.

Additional record (Findley, 1955:57).—**Lincoln Co.:** *Cokeville*.

Sorex nanus Merriam

Dwarf Shrew

1895. *Sorex tenellus nanus* Merriam, N. Amer. Fauna, 10:81, December 31.

1928. *Sorex nanus*, Jackson, N. Amer. Fauna, 51:174, July 24.

Type.—From Estes Park, Colorado.

Description.—One of the smallest American shrews; larger than *Microsorex;* resembles *S. vagrans* in having third unicuspid smaller than fourth, but braincase nearly flat, smaller, teeth smaller, and foramen magnum more ventral in *S. nanus*. Upper parts and top of tail sepia brown; tail whitish below and darkened distally; underparts grayish; snout less attenuate than in *S. cinereus*.

Measurements.—Mickey (1948) and Durrant and Lee (1955:560-561) recorded measurements of two specimens from Wyoming, respectively, as follows: Total length, 91, —; tail vertebrae, 39, —; hind foot, 10.5, —; ear from notch, 7, —; condylobasal length, 14.1, —; palatal length, 5.3, 5.5; cranial breadth, 7, —; least interorbital constriction, 3.0, 2.75; maxillary breadth, 4.0, 3.65; maxillary tooth-row, 5.1, 5.2; weight, 2.5 grams, —. Cranial measurements of 14 adults and adult-subadults (sexes rarely determined) from the Beartooth Plateau are as follows: greatest length of skull, 14.3 (13.6-14.9); breadth braincase, 6.4 (6.1-7.0); maxillary breadth (of processes), 3.7 (3.5-4.0); interorbital constriction, 2.6 (2.5-2.9); cranial depth, 2.9 (2.5-3.5) in 13 specimens.

Distribution.—Known from only three places, in Teton County (Durrant and Lee, 1955:560-561), Albany County (Mickey, 1948:294-295) and more recently from Park County (Hoffmann and Taber, 1960:232). See Fig. 5.

Remarks.—This rare shrew, known by 15 specimens before 1960, has been taken frequently by Hoffmann and Taber (1960:232) and their students on the Beartooth Plateau at high elevation (10,900 feet).

Records of occurrence.—Specimens examined, 25, as follows: **Albany Co.:** North Fork Camp Ground, T. 16 N, R. 78 W, Medicine Bow Nat'l Forest, 1. **Park Co.:** Approximately 30 mi. SW Red Lodge, Montana, Beartooth Plateau, "Polygon Can," 10,900 ft., 23 Montana State Univ. **Teton Co.:** South Cascade Canyon, Grand Teton Nat'l Park, 10,050 ft., 1 Univ. Utah.

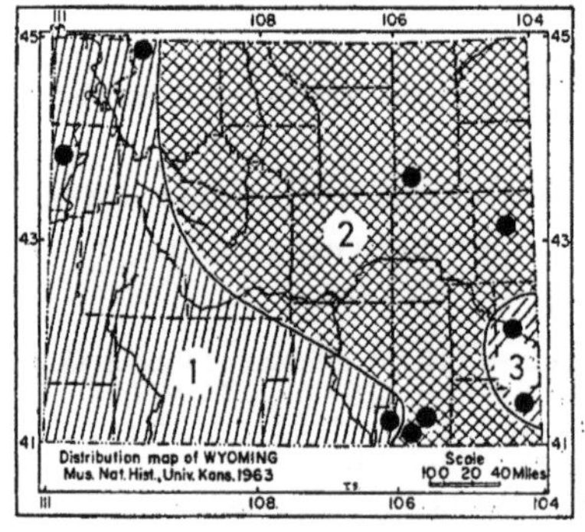

FIG. 5. Distribution of *Sorex nanus*, *Sorex merriami*, and *Scalopus*.
1. *Sorex nanus*
2. *Sorex merriami*
3. *Scalopus aquaticus*

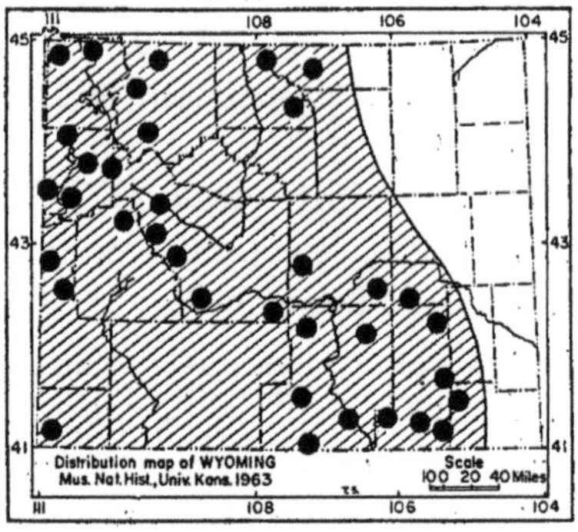

FIG. 6. Distribution of *Sorex palustris navigator*, the water shrew, in Wyoming.

Sorex palustris navigator (Baird)

Water Shrew

1858. *Neosorex navigator* Baird, Mammals, *in* Repts. Expl. Surv. . . ., 8(1):11, July 14.

1895. *Sorex* (*Neosorex*) *palustris navigator*, Merriam, N. Amer. Fauna, 10:92, December 31.

Type.—From near head Yakima River, Cascade Mtns., Washington.

Description.—Largest shrew in state (see *Measurements*); dorsum almost black but having a few scattered white hairs; underparts whitish or silvery-gray, rarely brownish; hind foot large, fringed with stiff hairs; third unicuspid smaller than fourth; post mandibular foramen not well developed and irregular in shape.

Measurements.—See Table 4.

Distribution.—Known throughout most of state except extreme southeastern areas and Green River watershed (see Fig. 6). Usually occurs near streams and lakes at elevations of 5400 to 10,600 feet.

Remarks.—Jackson (1928:186) contends that geographic variation in *Sorex palustris navigator* is of a microgeographic sort and that the subspecies "retains its characters" throughout its geographic range. Specimens from Park, Big Horn, and Lincoln counties aver-

TABLE 4.—AVERAGE AND EXTREME EXTERNAL AND CRANIAL MEASUREMENTS OF ADULTS OF *Sorex palustris navigator* FROM WYOMING.

County No. of Specimens	Total length	Length tail	Length hind foot	Palatal length	Cranial breadth	Least interorbital breadth	Maxillary breadth	Maxillary tooth-row	Condylobasal length
Albany 9	158, 150–164	76, 74–81	21.0, 19–21.5	8.6, 8.3–8.9	10.1, 10.0–10.3	4.0, 3.9–4.1	6.2, 5.8–6.5	7.0, 6.8–7.0	21.4, 20.8–21.9
Big Horn 5	149, 142–147	74, 72–78	20, 19–21	8.1, 7.8–8.4	9.7, 9.3–10.0	3.6, 3.4–3.8	5.8, 5.5–5.9	6.6, 6.2–6.8	20.4, 20.0–20.9
Carbon 5	150, 137–168	77, 71–84	20, 19–21	8.3, 8.2–8.4	10.0	3.8, 3.7–3.9	6.1, 6.0–6.2	6.6, 6.5–6.9	21.2, 20.0–21.5
Converse 3	156, 139–167	75, 72–79	20, 20–20	8.7, 8.5–8.9	10.5, 10.4–10.5	4.2, 4.1–4.2	6.3, 6.3–6.3	7.0, 6.9–7.0	21.7, 21.5–21.9
Fremont 1	163	73	19	8.4	9.8	3.6	5.9	7.0	20.8
Laramie 6	159, 150–161	80, 79–85	20.6, 19–23	8.5, 8.1–8.6	10.2, 9.8–10.5	3.8, 3.7–3.8	6.2, 6.0–6.4	6.6, 6.5–6.9	21.7, 21.1–22.0
Lincoln 5	147, 139–152	72, 69–73	20, 19–20	7.7, 7.5–8.0	9.8, 9.7–9.9	3.6, 3.5–3.7	5.7, 5.6–5.8	6.5, 6.5–6.6	20.9, 20.7–21.1
Natrona 1	155	74	21	8.5	9.9	3.9	6.2	7.0	21.4
Park 2	148, 146–149	72, 71–72	21, 20–21	8.0, 8.0–8.0	10.0, 10.0–10.0	3.9, 3.8–3.9	6.1, 6.0–6.2	6.7, 6.5–6.8	20.3, 20.2–20.3
Sublette 2	149, 146–152	74, 72–75	20, 20–20	8.4, 8.3–8.5	10.1	3.6, 3.5–3.6	5.9, 5.7–6.0	6.6, 6.5–6.7	21.3

age shorter in total length, length of tail, and condylobasal length than specimens from southeastern Wyoming, which are darker and often have the venter brownish. Those from central Wyoming (Fremont Co., 1; Natrona Co., 1) appear to be intermediate; they resemble specimens from Park, Big Horn, and Lincoln counties in total length, but have longer tails. I follow Jackson in assigning specimens from Wyoming to *S. p. navigator*. No significant cranial differences were observed.

A female (KU 16775) obtained by E. R. Hall on August 25, 1946, two miles southwest of Pole Mountain, 8300 ft., had eight embryos. A male (KU 37306) obtained by R. H. Baker on July 15, 1950, 13 miles north and two miles west of Afton, 6100 ft., had a tiny, anomalous right third upper unicuspid (Long, 1961).

Records of occurrence.—Specimens examined, 155 as follows: **Albany Co.:** Laramie Peak, 8000 ft., 1 USNM; 26¾ mi. N, 5½ mi. E Laramie, 1; 3 mi. NW Centennial, 8200 ft., 1 Univ. Wyo.; *Centennial, 8120 ft., 1; Hq. Park Medicine Bow Mtns., 1 USNM;* Laramie, 1 USNM; *10 mi. E Laramie, 8500 ft., 3 USNM; Wallis Camp Ground, near Pole Mtn., 8350 ft.,* 1; 1 mi. SW Pole Mtn., 8300 ft., 9; 1 mi. SSE Pole Mtn., 8350 ft., 1; *3 mi. S Pole Mtn., 8100 ft., 2; 3 mi. ESE Browns Peak, 10,000 ft., 1.*

Big Horn Co.: Medicine Wheel Ranch, 28 mi. E Lovell, 9000 ft., 5; Head Trappers Creek, 8500 ft., 3 USNM.

Carbon Co.: Ferris Mtns., 7800-8500 ft., 9 USNM; Shirley Mtns., 7600 ft., 1 USNM; 2 mi. S Bridgers Peak, 9300 ft., 2; 9½ mi. N, 11½ mi. E Encampment, 7200 ft., 1; *8 mi. N, 14½ mi. E Encampment, 8400 ft., 4; 8 mi. N, 16 mi. E Encampment, 8400 ft., 5;* 8 mi. N, 20 mi. E Savery, 8800 ft., 2.

Converse Co.: 21½ mi. S, 24½ mi. W Douglas, 7600 ft., 1.

Fremont Co.: 3 mi. S Dubois, 3 USNM; Lake Fork, 9600 ft., 1 USNM; Mosquito Park Ranger Sta., 2½ mi. N, 17½ mi. W Lander, 1; *Milford, 5400 ft., 1;* South Pass City, 1 USNM; 8 mi. E Rongis, 1 USNM.

Laramie Co.: 5 mi. W Horse Creek P. O., 7200 ft., 7.

Lincoln Co.: 13 mi. N, 2 mi. W Afton, 6100 ft., 4; *13 mi. N, 1 mi. W Afton, 6100 ft., 1; 10 mi. N Afton, 5 USNM; 6 mi. N, 1 mi. W Afton, 6100 ft., 1; Greys River, 7000 ft., 1;* 10 mi. SE Afton, 7500 ft., 2.

Natrona Co.: Rattlesnake Mtns., 7000 ft., 1 USNM; *7 mi. S Casper, 6000 ft., 3 USNM;* 10 mi. S Casper, 7750 ft., 1.

Park Co.: 16¼ mi. N, 17 mi. W Cody, 5625 ft., 2; *Black Mtn., 5 USNM;* Grinnell Creek, Pahaska, 6300 ft., 8 USNM; Valley, 7000 ft., 3 USNM.

Sheridan Co.: Eatons Ranch, Wolf, 5 USNM.

Sublette Co.: 31 mi. N Pinedale, 8025 ft., 2.

Teton Co.: 18 mi. N, 9 mi. W Moran, 1 Univ. Wyo.; Pacific Creek, 1 USNM; *1 mi. N Moran, 2 USNM; Moran, 5 USNM;* 1 mi. E Moran, 6742 ft., 1 Univ. Wyo.; *Jackson Hole Wildlife Park, 6 USNM;* Togwotee Pass, 9200 ft., 1 Univ. Wyo.; Teton Pass, 12 USNM; *S Moose Creek, 1.*

Uinta Co.: Evanston, 2 USNM.

Yellowstone Nat'l Park: NW Corner, 1 USNM; Mammoth Hot Springs, 4 USNM; *Glen Creek, 3 USNM.*

Sorex merriami leucogenys Osgood

Merriam's Shrew

1909. *Sorex leucogenys* Osgood, Proc. Biol. Soc. Washington, 22:52, April 17.

1939. *Sorex merriami leucogenys,* Benson and Bond, Jour. Mamm., 20:348, Aug. 14.

Type.—From mouth of the canyon of Beaver River, about 3 miles east of Beaver, Beaver Co., Utah.

Description.—Grayish or brownish-gray above; feet and underparts whitish; third unicuspid larger than fourth; crowded unicuspid row; unicuspids higher than long (anteroposteriorly) and lacking heavily pigmented internal ridges; according to Benson and Bond (1939:350), exceeding *S. m. merriami* in total length, condylobasal length, and cranial breadth.

Measurements.—See Table 5.

Distribution.—Expected in arid areas throughout state. See Fig. 5.

Remarks.—The specimen from Niobrara County is assigned to the subspecies *Sorex merriami leucogenys* on the basis of its great total length, long skull, and broad braincase. The five specimens consisting only of skull-fragments found in owl-pellets from Campbell County possess well-developed post-mandibular foramina (see Findley, 1955) characteristics of *S. merriami* and also are assigned to *S. m. leucogenys* (Long and Kerfoot, 1963). Measurements (Table 5) of the two specimens (KU 90998-9) from Albany County, obtained by A. B. Mickey, closely agree with those of *S. m. leucogenys* in total length, length of tail, length of head and body, and cranial breadth as listed by Benson and Bond (1939:348-351). The specimens from Albany County resemble *S. m. merriami* only in having a slightly shorter skull and hind foot than has *S. m. leucogenys.* Hoffmeister (1956:276) assigned a specimen of *S. merriami* from Owl Canyon, Colorado, approximately 50 miles to the southward, to *S. m. leucogenys.* Nevertheless, the first shrew reported from Wyoming, obtained in Albany County, was referred to *S. m. merriami* (Mickey and Steele, 1947:293); no reasons were given for the subspecific assignment, but probably it was based on geographic

TABLE 5.—EXTERNAL AND CRANIAL MEASUREMENTS OF MERRIAM'S SHREW FROM WYOMING.

Number	Sex and age	Locality	Total length	Length of tail	Length of hind foot	Condylobasal length	Palatal length	Cranial breadth	Least interorbital constriction	Maxillary breadth	Maxillary tooth-row
KU 90998	ad.	5 mi. N, 2 mi. E Laramie	....	40	11.5	16.1	6.8	8.1	3.7	5.2	6.0
KU 90999	ad.	3 mi. E Laramie	95	39	12.0	16.5	6.7	8.4	3.9	5.0	6.3
KU 19941	♂ ad.	Niobrara Co.	99	39	13.0	16.4	6.2	8.4	3.6	5.1	5.5

grounds. In my opinion, all specimens of S. *merriami* that I have seen are subspecifically allied with the specimens recorded from Colorado and are now referable to S. *m. leucogenys.* More specimens are needed of S. *merriami* in order to appraise its geographic variation if any occurs. It is understood that a specimen of S. *merriami* from Sweetwater County in western Wyoming is in the collection of the University of Wyoming.

Records of occurrence.—Specimens examined, 9, as follows: **Albany Co.:** 5 mi. N, 2 mi. E Laramie, 1; 3 mi. E Laramie, 7250 ft., 1; 3.6 mi. SW Laramie, 7190 ft., 1 Univ. Wyo. **Campbell Co.:** 42 mi. S, 13 mi. W Gillette, 5. **Niobrara Co.:** 10 mi. N Hat Creek P. O., 4300 ft., 1.

Family TALPIDAE—Moles

Largest of Insectivores in state; adapted for digging; eyes and ears not evident externally; zygomata present; known only from southeastern part of state.

Scalopus aquaticus caryi Jackson

Eastern Mole

1914. *Scalopus aquaticus caryi* Jackson, Proc. Biol. Soc. Washington, 27:20, Feb. 2.

Type.—From Neligh, Antelope Co., Nebraska.

Description.—Palest member of genus *Scalopus;* lacks ochraceous suffusion; pale grayish with shadows of purplish, yellowish, or silvery-gray often present; snout prominent and scantily haired.

Measurements.—Adult male (KU 14772), total length, 162; tail vertebrae, 25; hind foot, 22; condylobasal length, 34.1; cranial breadth, 17.4; least interorbital constriction, 7.6; length maxillary tooth-row, 13.2.

Distribution.—Southeastern part of state. See Fig. 5.

Records of occurrence.—Specimens examined, 2, as follows: **Goshen Co.:** Lingle, 1. **Laramie Co.:** Horse Creek, 3 mi. W Meriden, 5000 ft., 1.

Order CHIROPTERA

Bats

Modified for flight; digits elongated for supporting flight membrane; membrane extends from manus to hind limbs and from them to tail; pectoral girdle and musculature strongly developed; molars W-shaped; eyes minute; orientation by echolocation.

Family VESPERTILIONIDAE—Vespertilionid Bats

KEY TO BATS

1. Upper parts blackish having three conspicuous white spots; dentition i.$\frac{2}{3}$; c.$\frac{1}{1}$; p.$\frac{2}{2}$; m.$\frac{3}{3}$ . . . . **Euderma maculatum,** p. 535

1′. Upper parts lacking conspicuous white spots; dental formula not as in 1 2

2. Pinna more than 3 times as long as hind foot, **Plecotus townsendii pallescens**, p. 534

2′. Pinna less than 3 times as long as hind foot 3

3. Pelage dark and mottled with silver; dentition i. $\frac{2}{3}$; c. $\frac{1}{1}$; p. $\frac{2}{3}$; m. $\frac{3}{3}$, **Lasionycteris noctivagans**, p. 533

3′. Pelage lacking silver flecks or, if present, paler (more yellowish) and hoary; dentition not as in 3 4

4. Pelage having whitish or silver flecks; dentition i. $\frac{1}{3}$; c. $\frac{1}{1}$; p. $\frac{2}{2}$; m. $\frac{3}{3}$ ... 5

4′. Pelage lacking silvery or whitish flecks; dentition not as in 4 6

5. Upper parts reddish washed with silvery white; forearm less than 44, **Lasiurus borealis borealis**, p. 533

5′. Upper parts brownish washed with silvery white; forearm more than 44, **Lasiurus cinereus cinereus**, p. 534

6. Pelage pale yellowish or buffy (ears black); hind foot frequently less than 7 mm. .... **Myotis subulatus subulatus**, p. 530

6′. Pelage dark; hind foot longer than 7 7

7. Total length more than 106; dentition i. $\frac{2}{3}$; c. $\frac{1}{1}$; p. $\frac{1}{2}$; m. $\frac{3}{3}$, **Eptesicus fuscus pallidus**, p. 536

7′. Total length less than 106; dentition not as in 7 8

8. Pinna often more than 20 mm.; rostrum robust; pelage bright brown, **Myotis evotis evotis**, p. 532

8′. Pinna 19 or less; if more than 16, rostrum attentuate and pelage dull .. 9

9. Pinna long (up to 19 mm.); pelage dull; rostrum attenuate, **Myotis keenii septentrionalis**, p. 532

9′. Pinna short (less than 16); rostrum robust 10

10. Braincase rising abruptly from rostrum; pelage lacking brassy sheen, **Myotis volans interior**, p. 531

10′. Braincase rising gradually from rostrum; fresh pelage showing brassy sheen .. .. **Myotis lucifugus carissima**, p. 529

Myotis lucifugus carissima Thomas

Little Brown Myotis

1904. *Myotis carissima* Thomas, Ann. Mag. Nat. Hist., Ser. 7, 13:383, May.
1917. *Myotis lucifugus carissima*, Cary, N. Amer. Fauna, 42:43, October 3.

Type.—From Lake Hotel, Yellowstone National Park, Wyoming.

Description.—Upper parts Buckthorn Brown with more or less brassy sheen; dark bases of hairs on dorsum producing blackish effect in worn pelage; hairs of underparts dull Pale Pinkish Buff distally; small, except hind foot (see *Measurements*); interfemoral membrane naked; braincase rising gradually from robust rostrum.

Measurements.—Average and extreme external and cranial measurements of 18 adults from 2¼ miles northeast of Pinedale, 7500 ft., are as follows: Total length, 90.5 (83-103); tail vertebrae, 38.1 (31-44); hind foot, 11.0 (10-11); ear from tragus, 14.2 (13-15); greatest length of skull, 14.7 (14.2-15.1); breadth of braincase, 7.1 (6.9-7.5); maxillary tooth-row, 5.3 (5.1-5.5); interorbital breadth, 3.9 (3.7-4.0).

Distribution.—In suitable habitat throughout state. See Fig. 7.

Remarks.—A male (USNM 176760), incorrectly identified as *M. yumanensis* by Miller and Allen (1928:69), has an obscure brassy sheen and long skull (14.3 mm).

Records of occurrence.—Specimens examined, 86, as follows: **Albany Co.:** Laramie 1. **Big Horn Co.:** 7 mi. S Basin, 3900 ft., 4. **Campbell Co.:** Rockypoint, 3850 ft., 1. **Crook Co.:** 15 mi. ENE Sundance, 3825 ft., 1; *Sand Creek, 3750 ft., 2 USNM.* **Fremont Co.:** 17½ mi. S, 5 mi. E Lander, 6000 ft., 1. **Lincoln Co.:** 8 mi. N, 2 mi. E Sage, 6050 ft., 2; 14 mi. S, 1 mi. W Kemmerer, 6500 ft., 2. **Sublette Co.:** 7 mi. S Fremont Peak, 3 USNM; *2¾ mi. NE Pinedale, 7500 ft., 22.* **Teton Co.:** *Leeks Lodge, 1; 4 mi. N, 4 mi. E Moran, 4; Two Ocean Lake, 4;* Moran, 1 USNM, 3 Univ. Wyo.; *Jackson Hole Wildlife Park, 1 USNM; 1 mi. S Moran, 6; 1 mi. S, 3¾ mi. E Moran, 6200 ft., 2; 1 mi. S, 4 mi. E Moran, 6730 ft., 2;* Moose, 6225 ft., 1. **Yellowstone Nat'l Park:** Mammoth Hot Springs, 18 USNM; type locality, 1 USNM; *no specific locality, 3 USNM.*

Additional record (Miller and Allen, 1928:32).—**Yellowstone Nat'l Park:** Geyser Basin.

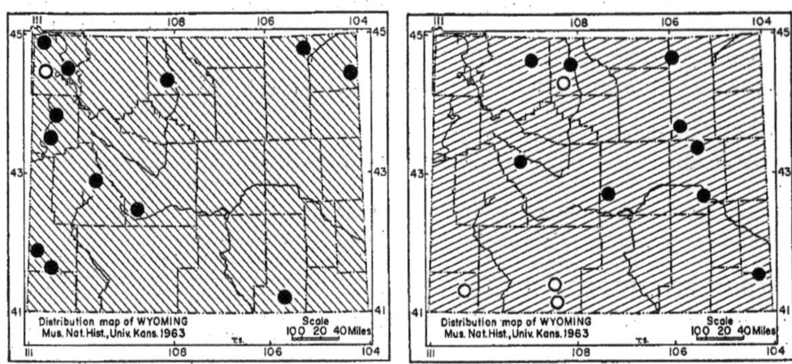

FIG. 7. *Myotis lucifugus carissima.* FIG. 8. *Myotis subulatus subulatus.*

Myotis subulatus subulatus (Say)

Small-footed Myotis

1823. *V[espertilio], subulatus* Say. *In* Long, Account of an exped. . . . to the Rocky Mountains. . . . 2:65.

1897. *Myotis subulatus*, Miller. N. Amer. Fauna, 13:75, October 16.

Type.—From Arkansas River near present town of LaJunta, Otero County, Colorado.

Description.—Pelage buffiest of *Myotis* in Wyoming; hairs long; ears small and black; small with relatively short foot (see *Measurements*); skull delicate; braincase sloping gradually to short rostrum.

Measurements.—Average and extreme external and cranial measurements of five adults from three localities in Campbell County (see below) are as follows: Total length, 77.4 (68-86); tail vertebrae, 31.2 (25-37); hind foot, 6.8 (5-8); ear from notch, 13 (11-14); condylobasal length, 13.1 (12.9-13.2); zygomatic breadth, 8.2 (8.0-8.4); breadth of braincase, 6.2 (6.1-6.3); depth of braincase including bullae, 5.5 (5.2-5.7); maxillary tooth-row, 5.2 (4.9-5.4); interorbital breadth, 3.1 (2.9-3.2).

Distribution.—Probably occurs throughout state. See Fig. 8.

Records of occurrence.—Specimens examined, 15, as follows: **Big Horn Co.:** Greybull, 2 USNM. **Campbell Co.:** 3 mi. N, 7½ mi. W Spotted Horse, 2; NW side Middle Butte, 38 mi. S, 19 mi. W Gillette, 5200 ft., 1; *E Side Middle Butte, 39 mi. S, 19 mi. W Gillette, 5200 ft. 3.* **Converse Co.:** 12 mi.

N, 7 mi. W Bill, 4700 ft., 1; 8 mi. S Douglas, 1 Univ. Wyo. **Fremont Co.:** Bull Lake, 1 USNM. **Laramie Co.:** Horse Creek, 6½ mi W Meriden, 5200 ft., 1. **Natrona Co.:** Rattlesnake Mtns., 1 USNM. **Park Co.:** 13 mi. N, 1 mi. E Cody, 5200 ft., 1; *5 mi N Cody, 1.*

Additional records (Miller and Allen, 1928:169).—**Big Horn Co.:** Otto. **Sweetwater Co.:** Bitter Creek; Kinney Ranch. **Uinta Co.:** Ft. Bridger.

Myotis volans interior Miller

Long-legged Myotis

1914. *Myotis longicrus interior* Miller, Proc. Biol. Soc. Washington, 27:211, October 31.

1928. *Myotis volans interior,* Miller and G. M. Allen, Bull. Nat'l Mus., 144: 142, May 25.

Type.—From five miles south of Twining, 11,300 ft., Taos County, New Mexico.

Description.—Upper parts and underparts slightly darker than in *Myotis lucifugus,* lacking brassy sheen; underside of wing furred markedly distal to elbow; ears short; size average for genus (see *Measurements*); rostrum short; occiput much elevated.

Measurements.—External and cranial measurements of three adults from within eight miles of Moran, Teton County, are as follows: Total length, 97, 95, 102; tail vertebrae, 48, 43, 47; hind foot, 8, 10.5, 10; ear from notch, 13, 13, 14; condylobasal length, 13.8, 13.6, —; zygomatic breadth, 8.6, 8.3, —; breadth of braincase, 7.4, 7.3, —; depth of braincase, 5.5, 5.3, —; interorbital breadth, 4.0, 3.9, 3.9.

Distribution.—Upper Sonoran and Transition life-zones probably throughout state. See Fig. 9.

Records of occurrence.—Specimens examined, 19, as follows: **Albany Co.:** Univ. Wyoming Sci. Camp, 1, 1 Univ. Wyo.; Laramie, 1, 1 USNM. **Big Horn Co.:** 1 mi. N, 11 mi. E Lovell, 1. **Hot Springs Co.:** 10 mi. S, 3 mi. E Thermopolis, 4600 ft., 1. **Lincoln Co.:** Afton, 1 USNM. **Natrona Co.:** Rattlesnake Mtns., 1 USNM. **Teton Co.:** Moran, 1; *1 mi. S, 4 mi. E Moran, 6730 ft., 1; 8 mi. S Moran, 7000 ft., 3 Univ. Wyo., 1; 2 mi. N Blacktail Butte, 3;* Moose, 6225 ft., 1. **Weston Co.:** ½ mi. E Buckhorn, 1.

Additional records (Miller and Allen, 1928:144, unless otherwise noted).—**Big Horn Co.:** Otto. **Fremont Co.:** Lake Fork. **Niobrara Co.:** S-Bar Creek (Quay, 1948:181). **Teton Co.:** *Jackson Hole (Findley, 1954:434).*

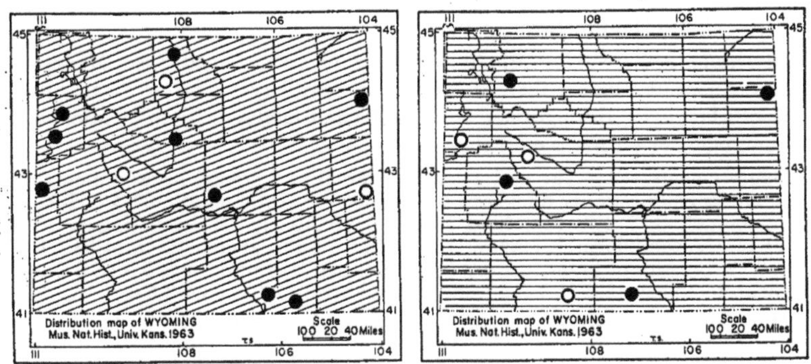

FIG. 9. *Myotis volans interior.* FIG. 10. *Myotis evotis evotis.*

Myotis evotis evotis (H. Allen)

Long-eared Myotis

1864. *Vespertilio evotis* H. Allen, Smithsonian Misc. Coll., 7 (165):48, June.

1896. *Vespertilio chrysonotus* J. A. Allen, Bull. Amer. Mus. Nat. Hist., 8:240, November 21, type from Kinney Ranch, Sweetwater County, Wyoming.

1897. *Myotis evotis*, Miller. N. Amer. Fauna, 13:77, October 16.

Type.—Type locality restricted (see Dalquest, 1943:2) to Monterey, California.

Description.—Upper parts Saccardo Brown to bright Buckthorn Brown; underparts creamy Pinkish Buff to Cinnamon-Buff; size small to average for genus (see *Measurements*); pinna of ear elongated, often more than 20 mm; pinnae not confluent across crown of head; rostrum less attenuate than in *Myotis keenii;* other characters as for genus.

Measurements.—Average and extreme external and cranial measurements of three adults (two females) examined from Park, Sublette, and Weston counties (see specimens examined) are as follows: Total length, 91.3 (87-96); length of tail, 41.3 (39-43); hind foot, 9.7 (9-11); ear from tragus, 20.3 (17-23); condylobasal length, 14.8 (14.8-14.8); zygomatic breadth, 9.0 (8.8-9.1); breadth of braincase, 7.0 (6.7-7.2); depth of braincase, 6.3 (6.1-6.5); maxillary tooth-row, 6.0 (6.0-6.0); interorbital breadth, 3.8 (3.7-3.9).

Distribution.—To be expected throughout state. See Fig. 10.

Records of occurrence.—Specimens examined, 5, as follows: **Carbon Co.:** Bottle Creek Picnic Ground, Sierra Madre Mtns., 8700 ft., 2. **Park Co.: 15 mi.** S, 21 mi. W Cody, 6200 ft., 1. **Sublette Co.:** 6 mi. N, 3 mi. E Pinedale, 7500 ft., 1. **Weston Co.:** 1½ mi. E Buckhorn, 6150 ft., 1.

Additional records (Miller and Allen, 1928:118, unless otherwise noted).—**Fremont Co.:** Bull Lake. **Sweetwater Co.:** Kinney Ranch. **Teton Co.:** Jackson Hole (Findley, 1954:434).

Myotis keenii septentrionalis (Trouessart)

Keen's Myotis

1897. *[Vespertilio gryphus]* var. *septentrionalis* Trouessart, Catalogus Mammalium . . . fasc. 1, p. 131.

1928. *Myotis keenii septentrionalis,* Miller and G. M. Allen, Bull. U. S. Nat'l Museum, 144:105, May 25.

Type.—From Halifax, Nova Scotia.

Description.—Resembles *Myotis lucifugus* in color; small (see *Measurements*); duller and darker than *Myotis evotis;* maxillary tooth-row longer than in *Myotis lucifugus;* distinguished from *evotis* and *lucifugus* by more attenuate rostrum; pinna of ear never more than 19.

Measurements.—Average and extreme external and cranial measurements of three adults from ½ mi. E Buckhorn are as follows: Total length, 87.3 (84-92); length of tail, 39.7 (38-41); hind foot, 9.0 (9.0-9.0); ear from tragus, 15.3 (15.0-16.0); condylobasal length, 14.3 (14.0-14.5); zygomatic breadth, 8.6 (8.6-8.7); breadth of braincase, 6.9 (6.7-7.0); depth of braincase, 5.9 (5.9-6.0); maxillary tooth-row, 5.8 (5.7-5.9); interorbital breadth, 3.5 (3.4-3.5).

Distribution.—Known only from northeastern Wyoming. See Fig. 11.

Records of occurrence.—Specimens examined, 3, as follows: **Weston Co.:** ½ mi. E Buckhorn, 6100 ft., 3.

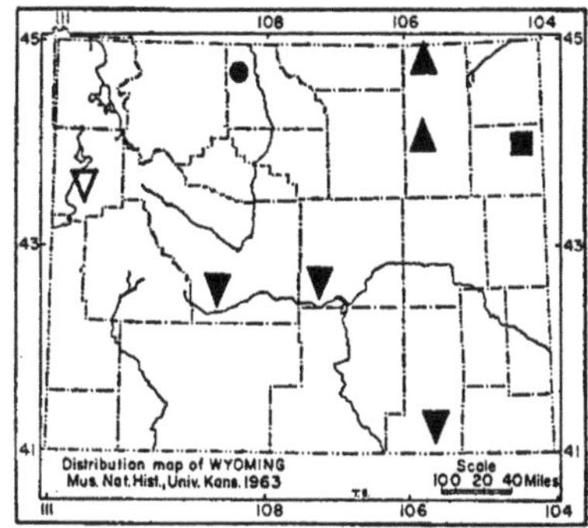

FIG. 11. Map showing the geographic distribution as now known of *Myotis keenii septentrionalis*, *Lasionycteris noctivagans*, *Lasiurus cinereus cinereus*, and *Euderma maculatum*.

GUIDE TO SPECIES

Square, *Myotis keenii septentrionalis*.
Triangles, *Lasiurus c. cinereus*.
Inverted Triangles, *Lasionycteris noctivagans*.
Circle, *Euderma maculatum*.

Lasionycteris noctivagans (Le Conte)

Silver-haired Bat

1831. *V[espertilio]. noctivagans* Le Conte, *In* McMurtrie, The animal kingdom . . . by Cuvier. 1:431.

1894. *Lasionycteris noctivagans*, H. Allen, Bull. U. S. Nat'l Mus., 43:105, March 14.

Type.—From eastern United States.

Description.—Upper parts and underparts black to Chestnut-Brown and mottled silver (distal tips of hairs silver), although mottling sometimes obscure; interfemoral membrane furred; zygomata delicate as in *Myotis;* p1 and p2 smaller than p3.

Measurements.—External and cranial measurements of an adult male from Split Rock are as follows: Total length, 92; tail vertebrae, 38; hind foot, 9; greatest length of skull, 15.7; zygomatic breadth, 9.3; breadth of braincase, 7.5; depth of braincase, 5.0.

Distribution.—Transition and Canadian life-zones, probably throughout state, especially in northwestern Wyoming. See Fig. 11.

Records of occurrence.—Specimens examined, 4, as follows: **Albany Co.:** Laramie, 1; *"near" Laramie, 1 USNM.* **Fremont Co.:** South Pass City, 1 USNM; Split Rock, 6200 ft., 1 USNM.

Additional record.—**Teton Co.:** Gros Ventre River (Negus and Findley, 1959:374).

Lasiurus borealis borealis (Müller)

Red Bat

1776. *Vespertilio borealis* Müller, Des Ritters Carl von Linné . . . vollständiges Natursystem . . ., Suppl., p. 20.

1897. *Lasiurus borealis*, Miller, N. Amer. Fauna, 13:105, October 16.

Type.—From New York.

Dsecription.—Smaller than *Lasiurus cinereus,* which it resembles in cranial characters; slightly larger than *Myotis lucifugus;* upper parts reddish and washed with silvery white and having buffy patch on shoulder; pelage brighter above and brighter in males than in females; length of forearm usually 37-43 mm; anterior upper premolar small and obscure as in *L. cinereus*.

Measurements.—Length of head and body, approximately 55; total length, 91-112; length of forearm, 37-43 (Hall and Kelson, 1959:188).

Distribution.—Southeastern Wyoming, perhaps in northeastern Wyoming. Not mapped.

Professor A. B. Mickey called to my attention the recorded occurrence of this species in Wyoming. Specimen No. 5264 was listed by H. Allen (1864:20, and 1894:153) from Laramie Peak. Now (1964) the specimen can not be found in the United States National Museum.

Lasiurus cinereus cinereus (Palisot de Beauvois)

Hoary Bat

1796. *Vespertilio cinereus* (Misspelled *linereus*) Palisot de Beauvois, Catalogue raisonné du muséum de Mr. C. W. Peale, Philadelphia, p. 18.

1864. *Lasiurus cinereus* H. Allen, Smithsonian Misc. Coll., 7:21, June.

Type.—From Pennsylvania, probably from Philadelphia.

Description.—Large (see *Measurements*); brownish and whitish or silvery distal tips of hairs on dorsum and sides producing hoary or frosted effect; pelage of dorsum usually mottled silvery-white, mahogany brown, and yellowish brown; flight-membrane, posterior rim of pinna of ear, and mouth dark brown; interfemoral membrane furred; skull robust with broad rostrum.

Measurements.—External and cranial measurements of two lactating females from Campbell County (see below) are as follows: Total length, 143 (KU 32326), 130 (KU 32325); length of tail, 54, 46; length of hind foot, 11, 11; ear from notch, 15, 16; forearm (dry), 55.2, 51.3; condylobasal length, 17.4, 16.4; zygomatic breadth, 13.0, 12.5; interorbital breadth, 5.3, 4.9; breadth of braincase, 9.4, 8.5; depth of braincase (including auditory bullae), 9.5, 9.1; maxillary tooth-row, 6.3, 6.1; postpalatal length, 8.0, 7.3; weight, 30, 29 gms.

Distribution.—Probably throughout state; known only from Campbell County. See Fig. 11.

Remarks.—Cary (1917:43) listed the hoary bat as occurring in the Canadian and Transition life-zones of the state but did not publish actual records. This tree-living bat has a wide range in North America.

Records of occurrence.—Specimens examined, 2, as follows: **Campbell Co.:** 3 mi. N, 7½ mi. W Spotted Horse, 1; NW side Middle Butte, 38 mi. S, 19 mi. W Gillette, 5200 ft., 1.

Plecotus townsendii pallescens (Miller)

Townsend's Big-eared Bat

1897. *Corynorhinus macrotis pallescens* Miller, N. Amer. Fauna, 13:52, October 16.

1959. *Plecotus townsendii pallescens*, Handley, Proc. U. S. Nat'l Mus. 110: 190, September 3.

Type.—From Keam Canyon, Navajo County, Arizona.

Description.—Upper parts between Sayal Brown and Cinnamon; underparts paler; ears greatly elongated (more than 300 per cent of hind foot) and scantily haired; interfemoral membrane naked; rostrum markedly concave dorsally; frontals noticeably convex in lateral view; dental formula: i.$\frac{2}{3}$; c.$\frac{1}{1}$; p.$\frac{2}{3}$; m $\frac{3}{3}$.

Measurements.—Average and extreme external and cranial measurements of eight females from Sand Creek are as follows: Total length, 105 (100-107); tail vertebrae, 49.3 (45-52); hind foot, 11.6 (11-12); greatest length of skull, 16.8 (16.1-17.1); zygomatic breadth, 8.7 (8.2-9.1); interorbital breadth, 3.5 (3.5-3.6); breadth of braincase, 7.8 (7.3-8.0); depth of braincase, 5.8 (5.5-6.0).

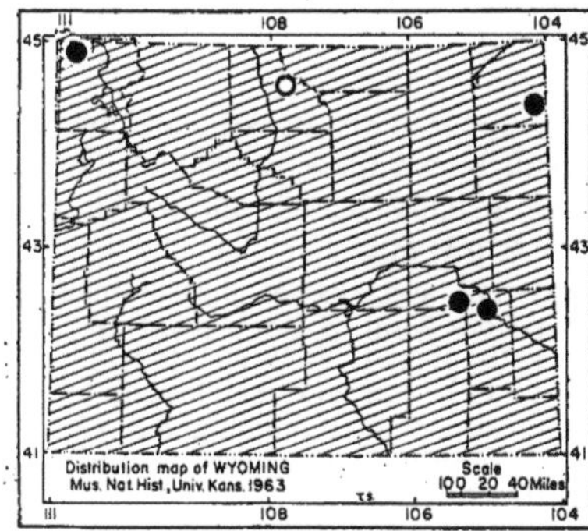

FIG. 12. Distribution of *Plecotus townsendii pallescens.*

Distribution.—Probably throughout state, especially in arid areas, but known only from some northern and eastern counties. See Fig. 12.

Records of occurrence.—Specimens examined, 19, as follows: **Converse Co.:** 8 mi. S Douglas, 1 Univ. Wyo. **Crook Co.:** Sand Creek, Black Hills, 3750 ft., 10 USNM. **Platte Co.:** 25 mi. NW Wheatland, 2. **Yellowstone Nat'l Park:** Mammoth Hot Spring, 6 USNM.

Additional record.—**Big Horn Co.:** 25 mi. NE Greybull (Handley, 1959).

Euderma maculatum (J. A. Allen)

Spotted Bat

1891. *Histiotus maculatus* J. A. Allen, Bull. Amer. Mus. Nat. Hist., 3:195, February 20.

1894. *Euderma maculata,* H. Allen, Bull. U. S. Nat'l Mus., 43:61, March 14.

Type.—From near Piru, Ventura County, California.

Description.—Large; ears exceedingly long; upper parts blackish with two white "saddle spots" and white spot at base of tail (all spots distinct and large); underparts whitish; skull low and rounded with large braincase; zygoma expanded at middle; pinnae of ears joined by membrane.

Measurements.—External measurements of a female from Salt Lake County, Utah, are as follows: Total length, 115; tail vertebrae, 47; hind foot, 12; ear, 47; length of tragus, 17; length of forearm, 51 (Durrant, 1952:59).

Distribution.—Known only from Bighorn Basin. See Fig. 11.

Remarks.—This bat usually occurs in arid habitats. The rare species is known at this writing from only 18 specimens, some of which are not well preserved. Mickey (1961:401) recorded the only specimen known from Wyoming.

Record of occurrence.—Specimen examined, 1, as follows: **Big Horn Co.:** Byron, 4020 ft., 1 Univ. Wyo.

Eptesicus fuscus pallidus Young

Big Brown Bat

1908. *Eptesicus pallidus* Young, Proc. Acad. Nat. Sci. Philadelphia, 60:408, October 14.

1912. *Eptesicus fuscus pallidus*, Miller, Bull. U. S. Nat. Mus., 79:62, December 31.

Type.—From Boulder, Colorado.

Description.—Large (see *Measurements*); upper parts Buckthorn Brown; underparts paler and often showing buff; dark brown mask present on face; pinna of ear black; interfemoral membrane scantily haired proximally; I1 markedly larger than I2; dental formula: i.$\frac{2}{3}$; c.$\frac{1}{1}$; p.$\frac{1}{2}$; m.$\frac{3}{3}$.

Measurements.—Average and extreme external and cranial measurements and weights of nine females of *E. f. pallidus* from ½ mi. S Rockypoint are as follows: Total length, 113.9 (109-118); length of tail, 44.9 (41-48); hind foot, 11.4 (10-13); ear from notch, 16.2 (14-19); condylobasal length 17.7 (17.2-18.4); zygomatic breadth, 12.4 (11.7-13.0); cranial breadth, 8.5 (8.1-8.9); least interorbital breadth, 4.3 (4.1-4.5); depth of braincase (including bullae), 7.2 (6.9-7.4); weight, 17.9 (16.0-20.0) grams.

Distribution.—Probably throughout state, but not known from southwestern parts. See Fig. 13.

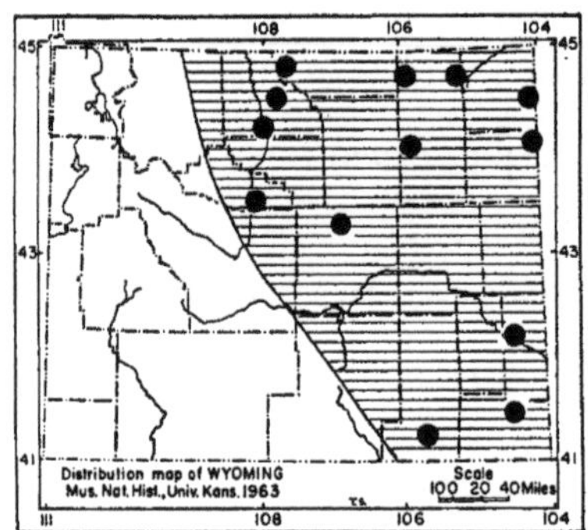

FIG. 13. Distribution of *Eptesicus fuscus pallidus.*

Records of occurrence.—Specimens examined, 34, as follows: **Albany Co.:** Laramie, 7200 ft., 1 Univ. Wyo. **Big Horn Co.:** Shell Creek, 1 mi. NW Shell, 2; *Greybull, 2 USNM;* 7 mi. S Basin, 3900 ft., 1. **Campbell Co.:** ½ mi. S Rockypoint, 3850 ft., 9; 3 mi. N, 7½ mi. W Spotted Horse, 2; NW side Middle Butte, 38 mi. S, 19 mi. W Gillette, 5200 ft., 1. **Crook Co.:** Sand Creek, 1 USNM. **Goshen Co.:** Ft. Laramie, 1 USNM. **Hot Springs Co.:** 9 mi. S, 2 mi. E Thermopolis, 4500 ft., 1. **Laramie Co.:** Horse Creek, 6½ mi. W Meriden, 5200 ft., 8. **Natrona Co.:** 27 mi. N, 1 mi. E Powder River, 6075 ft., 1. **Sheridan Co.:** 38 mi. E Lovell, 3. **Weston Co.:** 1½ mi. E Buckhorn, 6150 ft., 1.

Order LAGOMORPHA

Pikas, Rabbits, and Hares

Four upper incisors present; fibula ankylosed to tibia; tibia articulating with calcaneum; I1 grooved on anterior face; herbivorous.

KEY TO FAMILIES OF LAGOMORPHA

1. Hind legs slightly longer than forelegs; hind foot less than 40 mm. in length; supraorbital process absent; nasals widest anteriorly; five cheek-teeth above.............................. Family OCHOTONIDAE, p. 537

1′. Hind legs noticeably longer than forelegs; hind foot more than 40 mm. in length; supraorbital process present; nasals widest posteriorly; six cheek-teeth above................................ Family LEPORIDAE, p 541

Family OCHOTONIDAE—Pikas

Ochotona princeps (Richardson)

Pika

Owing to differences in time of molt, uncertainty as to the number of molts (pikas have at least two molts per year), and different colors of different pelages, pikas are difficult to use in assessing geographic variation. In my study, adults of the same season (or, when possible, of the same date) in the same condition of molt were so used. The most useful phase of molt had new hair anteriorly; both old pelage and new pelage could then be compared. No significant differences between sexes were noted; hence sexes were combined in the morphometric analyses. Specimens were identified as adults on the basis of wear of the cheek-teeth.

Pikas vary geographically in Wyoming. The differences among the numerous disjunct populations are slight and often clinal.

In south-central Wyoming *O. p. figginsi* and *O. p. saxatilis* have

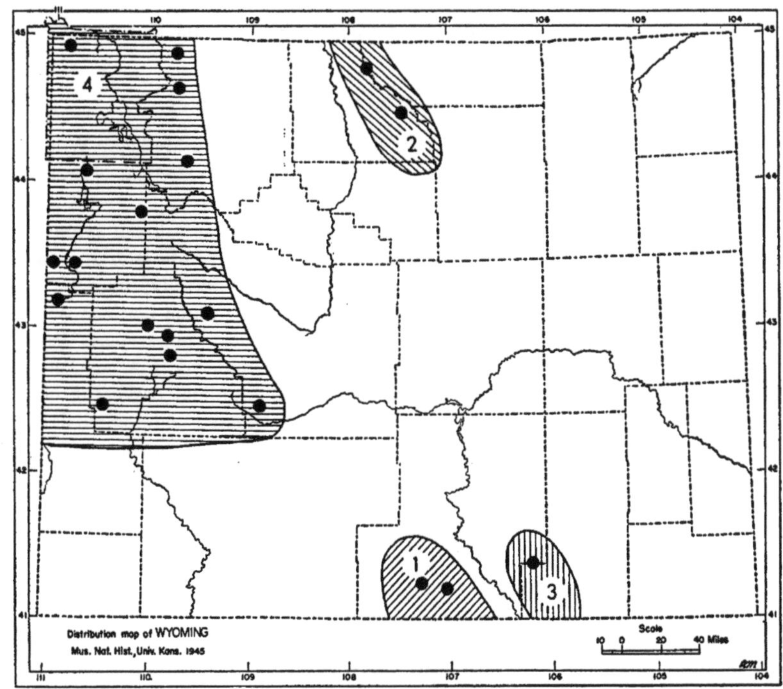

FIG. 14. Distribution of *Ochotona princeps*.
1. *O. p. figginsi* 3. *O. p. saxatilis*
2. *O. p. obscura* 4. *O. p. ventorum*

been recognized. Analysis of their cranial and external measurements yielded no differences considered to be significant when topotypes of both subspecies were also considered. This is indicative of intergradation, but the subspecies can be separated geographically throughout their range on the basis of color. The assignment herein of specimens from south-central Wyoming to these subspecies is in accordance with that of Hall (1951a).

Specimens from parts of Wyoming included previously within the geographic range of *O. p. ventorum* show clinal variation southwestward in several characters (length of external ear, length of nasals, width of palatal bridge, and zygomatic breadth) from the isolated populations of the Bighorn Mountains. The southwestern populations show some resemblance to *O. p. clamosa* Hall and Bowlus. Topotypes of *O. p. ventorum* and specimens from within three miles of the type locality are intermediate in this clinal variation. Inasmuch as the northernmost populations from the Bighorn Mountains seem to be as different from near topotypes of *O. p. ventorum* as is *O. p. clamosa* at the other end of this stepped cline, the northern population from the Bighorn Mountains is named as follows:

Ochotona princeps obscura, new subspecies

Type.—Male, subadult, skin and skull; No. 32918 KU; from Medicine Wheel Ranch, 28 miles east of Lovell, 9000 ft., Big Horn County, Wyoming; obtained by J. W. Twente, original No. 232.

Description.—Size medium for species; pinna of ear large. Fresh summer pelage: upper parts Cinnamon-Buff or Light Ochraceous-Buff laterally, intermixed with black and becoming Ochraceous-Tawny dorsally; underparts Pinkish Buff; foot Pinkish Buff faintly washed with yellow; pinna blackish with whitish margin. Old pelage on summer-taken specimens: upper parts Pale Ochraceous-Buff flecked with dark brown; foot and underparts as in fresh summer pelage but more nearly white; middorsal, longitudinal darkening obscure or absent. Sphenopterygoid canal small and situated anteromedial to ventral process of alisphenoid; bullae only slightly inflated; rostrum robust; nasals markedly long, expanded posterolaterally, and extending posterior to anterior margin of orbit; palatal foramen long.

Comparisons.—*O. p. obscura* is slightly paler than near topotypes of *O. p. ventorum.* An anteroposterior, middorsal dark line is obscure or absent in *O. p. obscura* but usually present in *O. p. ventorum.* The new subspecies differs further from *O. p. ventorum* in having longer ears, longer nasals, greater occipitonasal length, and longer posterior palatine foramina, smaller and less inflated auditory bullae, posterolaterally more expanded nasals, and a smaller sphenopterygoid canal situated more anteromedially to the ventral process of the alisphenoid. Differences from *O. p. saxatilis* and *O. p. figginsi* of south-central Wyoming are even greater in *O. p. obscura*. *O. p. saxatilis* is broader interorbitally and across the palatal bridge than *O. p. obscura;* the former also has shorter nasals, shorter palatal foramen, and is more yellowish. From *O. p. figginsi, O. p. obscura* differs as follows: upper parts paler and less richly cinnamon, interorbital region narrower and hind foot shorter.

Measurements.—External and cranial measurements of four adults from 19 miles eastward from Shell are as follows: Total length, 189.6 (185-197) three adults; hind foot, 33.8 (32-33); ear from notch, 24 (21-25); occipitonasal length, 47.0 (44.9-48.7); zygomatic breadth, 22.5 (22.3-23.1); interorbital breadth, 5.1 (5.0-5.1); length of nasals, 16.8 (16.1-17.5); width of palatal bridge, 2.1 (2.0-2.3); palatal foramen, width 8.2 (7.7-8.5), length 5.5 (5.0-5.8)

Distribution.—Bighorn Mountains. See Fig. 14.

Remarks.—Pikas eastward from Shell have larger auditory bullae than do topotypes of *O. p. obscura,* resembling *O. p. ventorum* in this character. Pikas from the southern Bighorn Mountains seem more closely related to *O. p. ventorum* than do those from the type locality of *O. p. obscura.*

Records of occurrence.—Specimens examined, 18, as follows: **Big Horn Co.:** type locality, 6; 4½ mi. S, 19 mi. E Shell, 9600 ft., 5; *Hd. Trappers Creek, 10,500 ft., 7 USNM.*

Ochotona princeps figginsi J. A. Allen

1912. *Ochotona figginsi* J. A. Allen, Bull. Amer. Mus. Nat. History, 31:103, May 28.

1924. *Ochotona princeps figginsi,* A. H. Howell, N. Amer. Fauna, 47:21, September 23.

Type.—From Pagoda Peak, Rio Blanco Co., Colorado.

Comparisons.—Darkest of pikas in Wyoming, Ochraceous-Tawny, grizzled blackish upper parts in fresh, summer pelage; Sayal Brown or Cinnamon-Brown upper parts in worn, summer pelage; underparts Cinnamon or Pinkish Cinnamon; pinna of ear blackish margined with white; foot Pinkish Cinnamon washed with yellowish; palatal bridge narrow; resembling *saxatilis,* the nearest subspecies, cranially and in external measurements; resembling *ventorum* in having short palatal foramen (topotypes of *figginsi* average 7.4 mm in length of this foramen).

Measurements.—Average and extreme external and cranial measurements of four adults from 7½ mi. N, 18½ mi. E Savery, are as follows: Total length, 187.3 (170-204); hind foot, 30.8 (30-33); ear, 20? (20-20? [in topotypes of *figginsi* the mean value is 23.3 and in nearby populations of pikas, this measurement is longer]); occipitonasal length, 45. 6 (44.8-46.7); zygomatic breadth, 21.9 (21.7-22.1); interorbital breadth, 5.2 (5.0-5.4); length of nasals, 15.8 (15.0-16.5); palatal bridge, 1.8 (1.6-2.0); palatal foramen, breadth 5.4 (5.1-5.9), length 8.0 (7.9-8.3).

Distribution.—Montane areas in Carbon County in south-central Wyoming. See Fig. 14.

Remarks.—Cranially, specimens of *O. p. figginsi* from Wyoming differ only slightly from topotypes of *figginsi* and from nearby specimens of *O. p. saxatilis.* Specimens of *figginsi* from Wyoming are intermediate in length of palatal foramen between topotypes of *figginsi* and southern populations of *O. p. ventorum.* Brownish pelage seems to be the most convenient character for identifying *figginsi.*

Records of occurrence.—Specimens examined, 10, as follows: **Carbon Co.:** Bridger Peak, 11,000 ft., 2 USNM; 7½ mi. N, 18½ mi. E Savery, 8400 ft,. 7; *W. slope Sierra Madre Mtns., 1.*

Ochotona princeps saxatilis Bangs

1899. *Ochotona saxatilis* Bangs, Proc. New England Zool. Club, 1:41, June 5.

1924. *Ochotona princeps saxatilis,* A. H. Howell, N. Amer. Fauna, 47:23, September 23.

Type.—From Montgomery, "near" Mt. Lincoln, Park County, Colorado.

Comparisons.—Differs from all other subspecies in Wyoming in being Clay Color or Honey Yellow above (often grizzled blackish) in fresh, late summer pelage; worn winter pelage more grayish; underparts having more yellowish Light Ochraceous-Buff; interorbital region broader than in any other Wyoming subspecies.

Measurements.—Average and extreme external and cranial measurements of nine specimens from within three miles of Browns Peak are as follows: Total length, 168.3 (162-183); hind foot, 28.5 (27.3-32.5); pinna, 23.4 (22-29); occipitonasal length, 45.1 (44.7-45.4); zygomatic breadth, 21.7 (21.6-22.0); interorbital breadth, 5.2 (4.8-5.6); nasals, 15.0 (14.2-15.8); width of palatal bridge, 2.3 (1.8-2.6); palatal foramen, breadth 4.9 (4.5-5.2), length 8.2 (7.7-8.8).

Distribution.—Montane areas of Albany and Carbon counties in south-central Wyoming. See Fig. 14.

Records of occurrence.—Specimens examined, 19, as follows: **Albany Co.:** 3 mi. E Browns Peak, 10,700 ft., 8; *Medicine Bow Peak, 11,500 ft., 1 USNM; ½ mi. E Medicine Bow Peak, 11,000 ft., 1; 3 mi. ESE Browns Peak, 10,000 ft., 2; Medicine Bow Mtns., 4 Univ. Wyo.* **Carbon Co.:** *2 mi. S, 1 mi. W Medicine Bow Peak, 10,600 ft., 1; Lake Marie, 10,440 ft., 1.* **County not specified:** *Medicine Bow Mtns., 1 USNM.*

Ochotona princeps ventorum A. H. Howell

1919. *Ochotona uinta ventorum* A. H. Howell, Proc. Biol. Soc. Washington, 32:106, May 20.

1924. *Ochotona princeps ventorum,* A. H. Howell, N. Amer. Fauna, 47:18, September 23.

Type.—From Fremont Peak, 11,500 feet, Wind River Mountains, Sublette County, Wyoming.

Comparisons.—Differs from *O. p. obscura* in being slightly darker especially along middorsal, longitudinal line; having shorter nasals; having sphenopterygoid canal larger and situated more posteriorly; having larger auditory bullae, shorter ear, and lesser occipitonasal length (see account of *obscura*).

Differs from *O. p. figginsi* in being paler in any pelage; having wider palatal bridge; and having shorter ear (see account of *figginsi*). Differs from *O. p. saxatilis* in lacking yellowish pelage (more grayish and less grizzled blackish) and being narrower interorbitally (see also account of *O. p. saxatilis*).

Measurements.—Average and extreme external and cranial measurements of 24 adult near topotypes (see specimens labeled with reference to Fremont Peak) are as follows: Total length, 187.0 (172-209); hind foot, 33.4 (30-35); ear, 23.9 (22-26); occipitionasal length, 44.0 (41.6-45.7); zygomatic breadth, 21.7 (21.1-22.5); interorbital breadth, 5.1 (4.7-5.5); length of nasals, 15.3 (14.1-16.5); width of palatal bridge, 2.1 (1.6-2.7); palatal foramen, breadth 5.3 (4.7-6.1), length 7.5 (6.0-8.9).

Distribution.—Montane areas in extreme western and northwestern Wyoming. See Fig. 14.

Remarks.—A pregnant female (KU 15872) containing 4 embryos 48 mm. in length was obtained on August 1, 1945, from the east end of Island Lake, 10,600 feet, by A. B. Leonard. Another (KU 15887) was obtained containing 4 embryos 47 mm. in length on August 2, 1945, from the west side of Seneca Lake, 10,300 feet, by H. H. Hall.

Records of occurrence.—Specimens examined, 121, as follows: **Fremont Co.:** Lake Fork, 9500-11,800 ft., 13 USNM; *14 mi. S, 8½ mi. W Lander, 1;* 17 mi. S, 6½ mi. W Lander, 8450 ft., 1.

Lincoln Co.: 2 mi. N, 8 mi. E Alpine, 9200 ft., 1.

Park Co.: 30 mi. N, 18 mi. W Cody, 10,500 ft., 4; Pahaska, 3 USNM; Needle Mtn., 11,000 ft., 1 USNM.

Sublette Co.: 2 mi. NE Kendall, 11,000 ft., 3 USNM; *type locality, 1 USNM;* Island Lake, 15 mi. N, 13 mi. E Pinedale, 10,600 ft., 1; *E End Island Lake, 3 mi. S Fremont Peak, 10,600 ft., 19; 4 mi. S Fremont Peak, 8 USNM; Barbara Lake, 11 mi. N, 10 mi. E Pinedale, 10,300 ft., 2; W Side Seneca Lake, 10,300 ft., 5 mi. S, 2 mi. W Fremont Peak, 6;* Middle Piney Lake, 9000 ft., 1 USNM; 19 mi. W, 2 mi. S Big Piney, 7700 ft., 31.

Teton Co.: 18 mi. N, 9 mi. W Moran, 1; ¼ mi. N, 1 mi. E Togwotee Pass, 9800 ft., 3; *Togwotee Pass, 10,800-11,000 ft., 2 USNM; Moose Creek, 10,000 ft., 12 USNM;* Jackson, 6500 ft., 3 USNM; Teton Pass, 7800 ft., 2 USNM.

Yellowstone Nat'l Park: Mammoth Hot Springs, 1 USNM; *Golden Gate, 1 USNM.*

Family LEPORIDAE—Rabbits, Hares, and Jack Rabbits

KEY TO GENERA OF LEPORIDAE

1. Interparietal fused with parietals; hind foot often more than 105 mm.; in juveniles, interptergoid space wider than 7.0 mm.... Genus **Lepus,** p. 546

1′. Interparietal not fused with parietals; hind foot often less than 105; in juveniles, interptergoid space narrower than 7.0... Genus **Sylvilagus,** p. 542

KEY TO THE SPECIES OF SYLVILAGUS

1. Greatest diameter of external auditory meatus less than 4.9 mm., **Sylvilagus floridanus,** p. 542

1′. Greatest diameter of external auditory meatus more than 4.9........ 2

2. Ear usually longer than 70 mm.; greatest diameter of external auditory meatus often more than 6.0 mm.............. **Sylvilagus audubonii,** p. 544

2′. Ear usually shorter than 70 mm.; greatest diameter of external auditory meatus usually less than 6.0 mm................ **Sylvilagus nuttallii,** p. 543

Sylvilagus floridanus similis Nelson

Eastern Cottontail

1907. *Sylvilagus floridanus similis* Nelson, Proc. Biol. Soc. Washington, 20:82, July 22.

Type.—From Valentine, Cherry County, Nebraska.

Description.—Size medium for genus (see *Measurements*); upper parts dark, Cinnamon intermixed with black in fresh pelage, bleached Pinkish Buff intermixed with black and white in worn pelage; underparts whitish except buffy throat; feet Light Pinkish Cinnamon to Pinkish Buff; tail white below, dorsally confluent in color with back; pinna of ear short and having buffy hairs; diameter of external opening of auditory meatus small (see *Measurements*). Darker and having shorter ears than other cottontails in state.

Measurements.—Average and extreme measurements of four adult females from Laramie County are as follows: Total length, 418.5 (407-445); tail vertebrae, 42.8 (33-53); hind foot, 97.5 (93-103); ear from notch, 50.0 (48-52); basilar length, 53.5 (52.3-55.8); zygomatic breadth, 35.3 (35.1-35.8); postorbital breadth, 12.1 (11.7-12.3); length of nasals, 29.9 (28.4-31.2); alveolar length of maxillary tooth-row, 14.4 (14.0-15.1); least width (anteroposteriorly) of palatal bridge, 6.2 (5.6-6.9); interpterygoid width between last molars, 6.2 (5.7-6.8); greatest diameter of opening of auditory meatus, 4.3 (4.1-4.5).

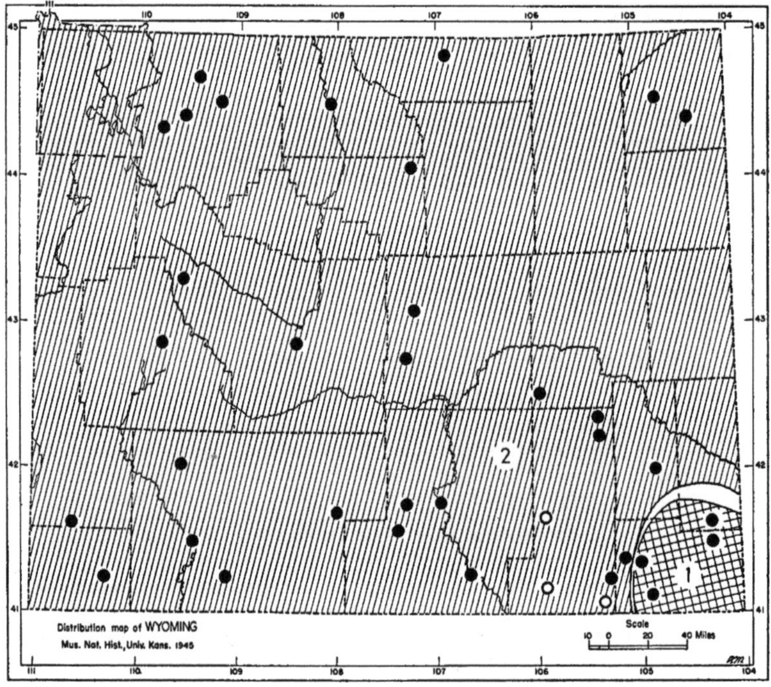

FIG. 15. Distribution of *Sylvilagus floridanus* and *Sylvilagus nuttallii*.
1. *S. f. similis* 2. *S. n. grangeri*

Distribution.—Extreme southeastern part of state. See Fig. 15.

Remarks.—According to Nelson (1909:173) this subspecies is confined to wooded borders "of streams and prairies on the arid plains." The largest number of embryos recorded in the state is eight in a pregnant female (KU 15940) shot by H. H. Hall on July 13, 1945, 6 miles north and 3 miles west of Cheyenne.

Records of occurrence.—Specimens examined, 8, as follows: **Goshen Co.:** 3½ mi. W Lagrange, 4600 ft., 1. **Laramie Co.:** Horse Creek, 3 mi. W Meriden, 5000 ft., 2; *3 mi. E Horse Creek* P. *O., 6400 ft., 3;* 6 mi. E Horse Creek P. O., 1; 6 mi. N, 3 mi. W Cheyenne, 6150 ft., 1.

Additional record from Nelson (1909:174).—**Laramie Co.:** *Meriden.*

Sylvilagus nuttallii grangeri (J. A. Allen)

Nuttall's Cottontail

1895. *Lepus sylvaticus grangeri* J. A. Allen, Bull. Amer. Mus. Nat. Hist., 7:264, Aug. 21.

1909. *Sylvilagus nuttallii grangeri,* Nelson, N. Amer. Fauna, 29:204, August 31.

Type.—From Hill City, Black Hills, Pennington County, South Dakota.

Description.—Size medium (see *Measurements*); color of upper parts variable ranging from Pale Ochraceous-Buff intermixed with blackish olive in juvenal pelage to Ochraceous-Tawny intermixed with blackish and scatterings of brown-tipped white hairs in adult pelage, but two specimens taken in October paler dorsally and laterally, with rump showing more white and less buff; underparts whitish except buffy throat; feet Pale Ochraceous-Buff, excepting white hind feet in October-taken specimens; pinna of ear long and margined by thin line of black distally; rostrum narrow; tympanic bullae notably inflated; diameter of external auditory meatus more than in *Sylvilagus floridanus similis.*

Measurements.—External and cranial measurements of four adults, two females and two males, from Park County are, respectively, as follows: Total length, 415, 413, 414, 395; tail vertebrae, 25, 52, 57, 52; hind foot, 95, 90, 94, 93; ear from notch, 60, 63, 65, 62; basilar length, 55.4, 52.9, 52.7, 52.7; zygomatic breadth, 35.0, 35.5, 36.6, 34.1; postorbital breadth, 12.1, 11.7, 10.3, 9.9; length of nasals, 30.9, 30.2, 31.2, 30.6; alveolar length of maxillary tooth-row, 13.5, 13.0, 13.7, 13.5; least width (anteroposteriorly) of palatal bridge, 6.3, 5.8, 6.4, 6.6; interpterygoid breadth between last molars, 6.3, 5.8, 5.0, 5.0; diameter of external auditory meatus, 5.7, 5.4, 5.5, 5.4.

Distribution.—Throughout state except extreme southeastern part. See Fig. 15.

Remarks.—*S. nuttallii* has never been recorded as being sympatric with *S. floridanus.* Some mammalogists have suspected that the two species intergrade along the eastern base of the Rocky Mountains. Hall and Kelson (1951:52-53) have concluded that no such intergradation occurs, with which conclusion I agree. Nelson (1909:206) indicated that *S. n. grangeri* intergrades with *S. n. pinetis* near the southern boundary of Wyoming. All specimens examined by me from southeastern Wyoming are referable to *S. n. grangeri* and reveal no indications of intergradation between *grangeri* and *pinetis*.

A pregnant female (KU 38008) containing eight embryos was obtained on June 15, 1950, from 13½ miles north and 12 miles west of Cody by R. H. Baker.

Records of occurrence.—Specimens examined, 44, as follows: **Albany Co.:** Springhill, 1 USNM; Laramie Peak, 1 USNM; 2 mi. SE Pole Mtn., 8250 ft., 1; *Laramie Mtns., 2 USNM.*

Big Horn Co.: Greybull, 1 USNM.

Carbon Co.: Rawlins, 1; Bridgers Pass, 1 USNM; Ft. Steele, 1; 6 mi. N, 6 mi. E Encampment, 6800 ft., 1.

Converse Co.: Deer Creek, 1 USNM.

Laramie Co.: 1 mi. N, 5 mi. W Horse Creek P. O., 7200 ft., 1; *2 mi. W Horse Creek* P. O., *6600 ft., 1.*

Crook Co.: Devils Tower, 3 USNM; Sundance, 1 USNM.

Fremont Co.: Jackeys Creek, 1 USNM; Wind River Basin, 1 USNM.

Lincoln Co.: 14 mi. S, 1 mi. W Kemmerer, 6550 ft., 1.

Natrona Co.: Rattlesnake Mtns., 3 USNM; Arminto, 1.

Park Co.: *13½ mi. N, 13 mi. W Cody, 2;* 13½ mi. N, 12 mi. W Cody, 1; 3½ mi. W Cody, 5100 ft., 1; 4 mi. S, 15 mi. W Cody, 1; 16 mi. S, 20 mi. W Cody, 1.

Platte Co.: Wheatland, 1 USNM.

Sublette Co.: 2¼ mi. NE Pinedale, 7500 ft., 1.

Sheridan Co.: 1 mi. E Acme, 1.

Sweetwater Co.: Eden, 1; *18 mi. NW Rock Springs, 1;* 2.5 mi. N Wamsutter, 1; Green River, 4 USNM; 21 mi. S, 18 mi. E Rock Springs, 1.

Uinta Co.: Ft. Bridger, 2 USNM.

Washakie Co.: ½ mi. S, 1 mi. W Tensleep, 4300 ft., 1.

Additional records (From Nelson, 1909:207).—**Albany Co.:** Rock Creek; Woods P. O.; Sherman.

Sylvilagus audubonii baileyi (Merriam)

Desert Cottontail

1897. *Lepus baileyi* Merriam, Proc. Biol. Soc. Washington, 11:148, June 9.
1908. *Sylvilagus audubonii baileyi,* Lantz, Trans. Kansas Acad. Sci., 22:336.

Type.—From Spring Creek, east side of Bighorn Basin, Washakie County, Wyoming.

Description.—Upper parts Light Ochraceous Buff dorsally fading to Light Buff laterally intermixed with black, producing olive-buff effect in juvenal pelage, becoming paler laterally owing to increase of white and darker dorsally owing to intermixture of black, Ochraceous-Buff, and Mummy Brown; underparts white except Ochraceous Tawny throat; tympanic bullae large; ears long; cheek-teeth large; lateral process of jugal often acuminate anteriorly and prominent.

Comparisons.—*S. audubonii* is paler and has longer ears and larger bullae than *S. floridanus. S. audubonii* is paler, has slightly larger bullae, and usually has a more acuminate, prominent lateral process of the jugal than *S. nuttallii.*

Measurements.—Average and extreme external and cranial measurements of four females from Washakie and Park counties and of one male from Park County are as follows: Total length, females 402.3 (390-415), male 400; tail vertebrae, 51.5 (48-59), 46; hind foot, 72.0 (62-78), 110; ear from notch of females, 72.0 (62-78), from crown of male 85; basilar length, 53.2 (51.0-55.4),

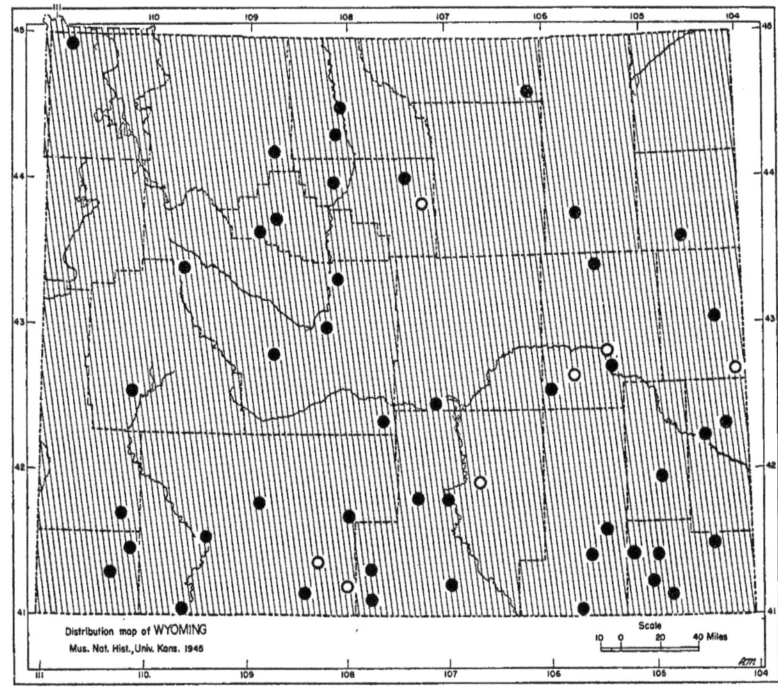

FIG. 16. Distribution of *Sylvilagus audubonii baileyi.*

52.1; zygomatic breadth, 35.5 (34.6-36.9), 35.1; postorbital least width, 11.0 (9.9-11.7), 11.0; length of nasals, 30.3 (28.7-31.7), 29.8; alveolar length of maxillary tooth-row, 13.6 (12.5-14.6), 13.3; palatal breadth, 5.8 (5.1-6.4), 4.9; width of interpterygoid space at level of last molars, 4.7 (4.4-5.0), 4.5; greatest diameter of auditory meatus, 6.2 (5.7-6.6), 5.5.

Distribution.—Throughout state except along Idaho-Wyoming boundary in extreme western part of state. See Fig. 16.

Remarks.—This rabbit frequents the open prairies more often than *S. floridanus* and *S. nuttallii*, with which two species it is sympatric.

Records of occurrence.—Specimens examined, 120, as follows: **Albany Co.:** 26 mi. N, 4½ mi. E Laramie, 6960 ft., 2; 5 mi. N Laramie, 4; *4½ mi. N Laramie, 1;* 2 mi. N Colorado line, Hwy. 281, 1.

Big Horn Co.: Greybull, 3 USNM; 7 mi. S Basin, 1.

Campbell Co.: South Butte, 40½ mi. S, 17½ mi. W Gillette, 6000 ft., 1.

Carbon Co.: *4 mi. W Rawlins, 1 USNM;* Rawlins, 1; Ft. Steele, 1 USNM; 12 mi. N Baggs, 1; *10 mi. N Baggs, 3; 7 mi. N Baggs, 1;* Riverside, 1 USNM; *6 mi. N Baggs, 1; 3 mi. N Baggs, 1;* 2 mi. N Baggs, 1.

Converse Co.: 12 mi. N, 6 mi. W Bill, 4800 ft., 3; Douglas, 6 USNM; Deer Creek, 3 USNM.

Fremont Co.: 7 mi. N Shoshoni, 4700 ft., 1; 2 mi. N, 6 mi. W Burris, 6450 ft., 1; Wind River Basin, 5 USNM; Lander, 7 USNM; Green Mtns., 1 USNM.

Goshen Co.: *Rawhide Buttes, 1 USNM;* Muskrat Canyon, 2; Ft. Laramie, 1 USNM.

Hot Springs Co.: 3 mi. N, 27 mi. W Thermopolis, 1; Owl Creek Mtns., 1 USNM.

Laramie Co.: *Horse Creek, 6½ mi. W Meriden, 5200 ft., 4;* Horse Creek, 3 mi. W Meriden, 1; 1½ mi. W Horse Creek P. O., 1; *½ mi. W Horse Creek P. O., 1; Horse Creek, 1 USNM;* 8 mi. E Horse Creek P. O., 1; *22 mi. NE Cheyenne, 1;* Islay, 1 USNM; *11 mi. N, 5½ mi. E Cheyenne, 5950 ft., 6; Camp Carling, 1 USNM;* Cheyenne, 2 USNM.

Lincoln Co.: Opal, 1 USNM.

Natrona Co.: 3 mi. E Independence Rock, 3.

Niobrara Co.: 10 mi. N Hat Creek P. O., 4300 ft., 5.

Park Co.: 6 mi. E Meeteeste, 5750 ft., 1.

Platte Co.: Wheatland, 1 USNM.

Sheridan Co.: Arvada, 3 USNM.

Sublette Co.: Big Piney, 2 USNM.

Sweetwater Co.: Superior, 1 USNM; Wamsutter, 5 USNM; Green River, 3 USNM; Kinney Ranch, 1 USNM; Henrys Fork, 1 USNM.

Uinta Co.: Church Buttes, 12 mi. NE Lyman, 1; *2.5 mi. N, 6 mi. E Lyman, 1;* Ft. Bridger, 1 USNM.

Washakie Co.: *Tensleep, 1 USNM;* ½ mi. S, 1 mi. W Tensleep, 4300 ft., 1; 4 mi. S, 7 mi. W Worland, 4100 ft., 1; *10 mi. S, 9 mi. W Worland, 4100 ft., 1.*

Weston Co.: 23 mi. SW Newcastle, 4500 ft., 8.

Yellowstone Nat'l Park: Mammoth Hot Springs, 1 USNM.

Additional records (Nelson, 1909:234).—**Carbon Co.:** Percy; *Aurora.* **Converse Co.:** Ft. Fetterman; Beaver. **Fremont Co.:** *Circle.* **Niobrara Co.:** Van Tassal Creek. **Sweetwater Co.:** Bitter Creek; 30 mi. S Wamsutter. **Washakie Co.:** Spring Creek, Bighorn Basin.

KEY TO SPECIES AND SUBSPECIES OF LEPUS

1. Tail black above; upper parts never white in winter; known only from southeastern part of state............. **Lepus californicus melanotis**, p. 552

1′. Tail not black above; upper parts of most individuals white in winter; known throughout most of state.................................. 2

2. Tail all white lacking narrow, dorsal, gray line; summer pelage pale and brassy colored...................... **Lepus townsendii campanius**, p. 550

2′. Tail not all white; summer pelage dark and brassy colored or pale and lacking brassy color.. 3

3. Tail all white excepting narrow, longitudinal, dorsal gray line; upper parts in summer pale and grayish or golden instead of brassy or buffy, **Lepus townsendii townsendii**, p. 551

3′. Tail bicolored, concolor with back above and white below; upper parts in summer dark and flecked with golden or brassy color............ 4

4. Upper parts grayish and reddish golden; nasals nearly straight-sided as seen in dorsal view..................... **Lepus americanus bairdii**, p. 547

4′. Upper parts grayish and flecked with yellowish-golden hairs; nasals markedly convex anteriorly in dorsal view, **Lepus americanus seclusus**, p. 548

Lepus americanus Erxleben

Snowshoe Rabbit

In *Lepus americanus*, females are significantly larger than males especially in total length, hind foot, ear, and interpterygoid space. Consequently, the sexes were not combined for morphometric analysis. Measurements were made as defined by Baker and Hankins (1950), excepting the least width of the palatal bridge (anteroposteriorly), which they did not record. Excepting on the tips of the ears the pelage changes from brownish agouti in summer to white in winter in most or all individuals.

Lepus americanus bairdii Hayden

1869. *Lepus americanus bairdii* Hayden, Amer. Nat., 3:115, May.

Type.—From Columbia Valley, Wind River Mountains, Fremont County, Wyoming.

Comparison.—From *L. a. seclusus, L. a. bairdii* differs as follows (in summer pelage): top and sides of head Cinnamon Buff instead of Cinnamon Brown; upper parts of body more reddish owing to intermixed hairs of reddish-gold instead of yellowish-gold; nasals nearly straight-sided as seen in dorsal view instead of markedly convex; interpterygoid space wider anteriorly.

Measurements.—Average and extreme external and cranial measurements of three adult males and six adult females from Uinta County are, respectively, as follows: Total length of males, 418.3 (385-470), of females, 450.0 (432-484); tail vertebrate, 32.7 (28-40), 31.7 (25-38); hind foot, 139.3 (135-146), 142.8 (134-155); ear 81.0 (80-83), 81.8 (78-90); basilar length, 57.0 (55.2-58.0), 57.8 (56.6-58.6); zygomatic breadth, 38.0 (36.8-38.7), 37.6 (36.0-38.9); postorbital breadth, 11.5 (11.2-11.8), 11.5 (10.4-13.2); length of

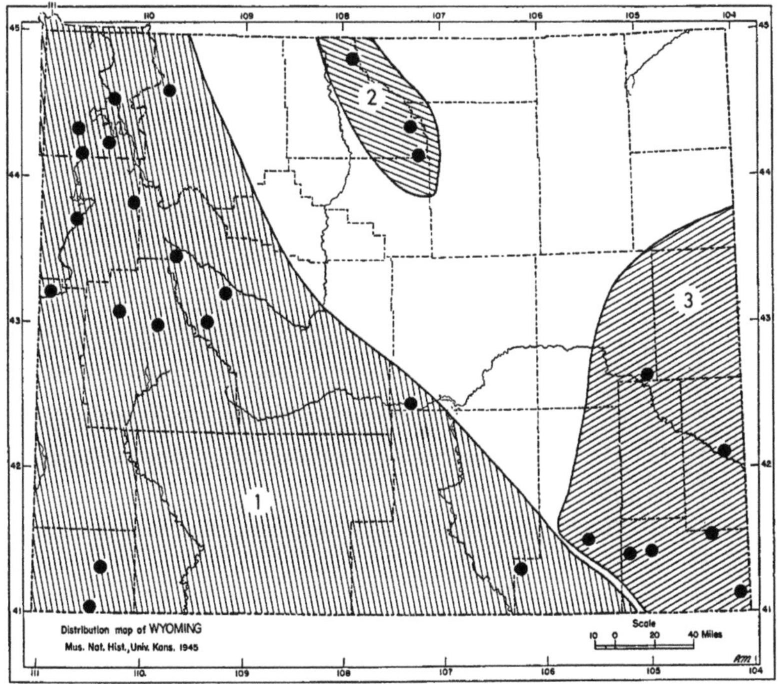

FIG. 17. Distribution of *Lepus americanus* and *Lepus californicus*.
Guide to kinds 2. *L. a. seclusus*
1. *L. a. bairdii* 3. *L. c. melanotis*

nasals, 29.6 (28.0-31.4), 31.2 (29.2-33.7); alveolar length of maxillary tooth-row, 13.7 (13.5-14.0), 13.6 (12.9-14.3); least width (anteroposteriorly) of palatal bridge, 5.8 (5.0-6.3), 5.7 (5.0-6.7); interpterygoid space between last molars, 6.8 (6.3-7.7), 7.0 (6.2-7.6).

Distribution.—Throughout state in boreal life-zones except in Bighorn Mountains and eastward thereof. See Fig. 17.

Records of occurrence.—Specimens examined, 41, as follows: **Albany Co.:** ¼ mi. S Univ. Wyo. Sci. Camp, 1 Univ. Wyo.; *3 mi. ESE Browns Peak, 10,000 ft., 2.* **Fremont Co.:** Jackeys Creek, 1 USNM; Bull Lake, 7700 ft., 1 USNM; Lake Fork, 9600 ft., 6 USNM. **Lincoln Co.:** 1 mi. N, 4 mi. E Alpine, 5650 ft., 1. **Natrona Co.:** 3 mi. E Independence Rock, 1. **Park Co.:** Whirlwind Peak, 1 USNM. **Sublette Co.:** 12 mi. N Kendall, 8000 ft., 2 USNM; 12 mi. NE Pinedale, 8000 ft., 2; *9 mi. N, 5 mi. E Pinedale, 1; LaBarge Creek, 2 USNM; Wind River Mtns., 3 USNM.* **Teton Co.:** Near Togwotee Pass, 1; Timbered Island, 4 mi. N Moose, 6750 ft., 1. **Uinta Co.:** Ft. Bridger, 1 USNM; 9½ mi. S, 2½ mi. E Robertson, 8700 ft., 7; *11½ mi. S, 2 mi. E Robertson, 3.* **Yellowstone Nat'l Park:** Yellowstone Lake, 1 USNM; Lewis Lake, 1 USNM; Hart Lake, 1 USNM; Snake River, 1 USNM.

Additional record (Nelson, 1909:112).—**Yellowstone Nat'l Park:** *Shoshone Lake.*

Lepus americanus seclusus Baker and Hankins

1950. *Lepus americanus seclusus* Baker and Hankins, Proc. Biol. Soc. Washington, 63:63, May 25.

1959. *Lepus americanus setzeri* Baker, Jour. Mamm., 40:145, February 20. [Proposed as a substitute name on the assumption that *Lepus a. seclusus* Baker and Hankins 1950 was preoccupied by *Lepus timidus seclusus* Degerbøl 1940. Actually, Degerbøl's name was expressly for a variety, and therefore an infrasubspecific name. Such a name does not preoccupy a species-group name (see Art. 45 of International Code Zool. Nomenclature, November 6, 1961).]

Type.—Male, adult, skin and skull, No. 20897, KU; from two miles north and 12 miles east of Shell, 7900 ft., Bighorn Mountains, Big Horn County, Wyoming.

Comparison.—See account of *L. a. bairdii.*

Measurements.—Average and extreme external and cranial measurements of six adults from Bighorn Mountains are as follows: Total length of four males, 408.3 (381-420), of two females, 455.0 (450-460); length of tail vertebrae, 30.0 (22-38), 23.5 (22-25); hind foot, 132.3 (129-138), 135.5 (131-140); ear from notch, 74.3 (70-77), 80.5 (80-81); basilar length, 57.5 (55.2-60.9), 59.1 (57.1-61.0); zygomatic breadth, 37.2 (34.4-39.5), 38.1 (36.7-39.4); postorbital constriction, 11.2 (10.7-12.0), 10.3 (10.1-10.5); length of nasals, 30.0 (29.4-30.5), 32.1 (30.3-32.9); alveolar length of maxillary tooth-row, 13.5 (13.2-14.1), 14.7 (14.6-14.7); least width (anteroposteriorly) of palatal bridge, 5.8 (5.6-6.0), 6.3 (5.9-6.7); interpterygoid space at level of last molars, 6.9 (6.4-7.4), 6.7 (6.7-6.7).

Distribution.—Known only from Bighorn Mountains, Big Horn County. See Fig. 17.

Remarks.—Baker and Hankins (1950) correctly recognized that *L. a. seclusus* differs from *L. a. bairdii* and *L. a. americanus* and more closely resembles *americanus* than *bairdii.* Those authors noted intermediacy in width of the interpterygoid space in their specimens; I also noted it in females of *seclusus,* but in four males, not all available to Baker and Hankins, that I measured, this space averaged wider than it did in any of the three subspecies listed by Baker and Hankins. The truncate posteriormost extensions of the nasals of *seclusus* and *bairdii* indicate relationship. Marginal white on the pinna of the ear of *seclusus* is less prominent than in either of the other subspecies considered.

A pregnant female (KU 20896) obtained on July 10, 1947, from the type locality by J. W. Schmaus, contained seven embryos.

Records of occurrence.—Specimens examined, 12, as follows: **Big Horn Co.:** Medicine Wheel Ranch, 28 mi. E Lovell, 9000 ft., 4; type locality, 2; *Hd. Trappers Creek, 8500 ft., 1 USNM;* 9 mi. N, 9 mi. E Tensleep, 8200 ft., 1; *Bighorn Mtns., 8400 ft., 4 USNM.*

Lepus townsendii Bachman

White-tailed Jack Rabbit

White-tailed jack rabbits are easy to distinguish from black-tailed jack rabbits by color of tail. The longer ears and larger size of the white-tailed jack rabbits distinguish them from the snowshoe rabbits; both species are white in winter in Wyoming. Females average slightly larger than males.

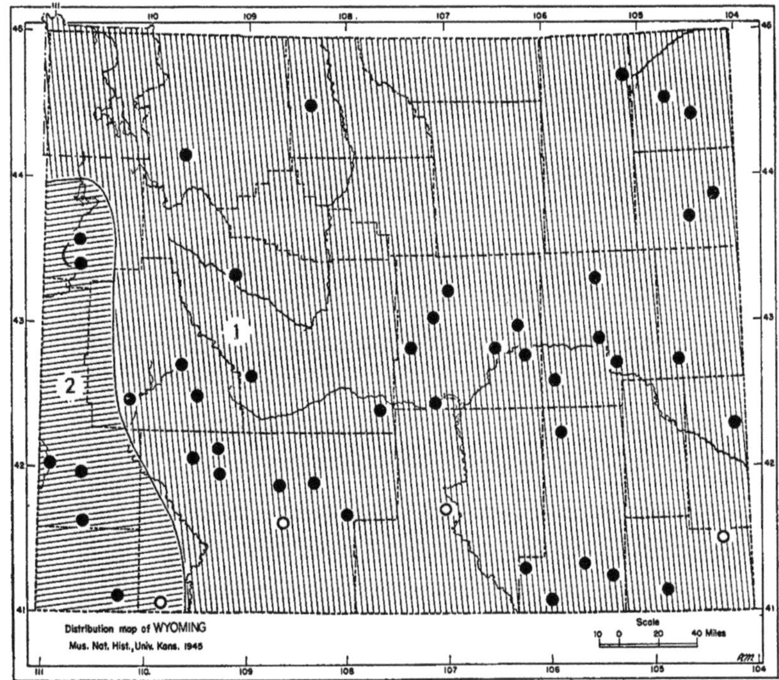

FIG. 18. Distribution of *Lepus townsendii*.
1. *L. t. campanius* 2. *L. t. townsendii*

Geographic variation in this species in Wyoming is slight. (Few males were available.) Clinal variation in females from north to south in decreasing width of the interpterygoid space is shown by the following measurements: *Park County*, width, 12.2 (1 specimen): *Washakie County*, 12.3 (1); *Weston County*, 12.4 (1); *Crook County*, 12.0 (1). One specimen from *Albany County* measured 11.4; the remainder of the southern specimens measured less, as follows: *Sweetwater County*, 11.0 (4); *Natrona County*, 11.1 (3); *Converse County*, 10.8 (1). Another cline was found in increasing width (anteroposteriorly) of palatal bridge from northeast to southwest, as follows: *Weston County*, 4.8 (1); *Crook County*, 5.2 (1); *Uinta County*, 7.0 (1). All other specimens were intermediate geographically and in average width of the palatal bridge. For example, four females from *Sweetwater County* averaged 5.7, and three from *Natrona County* averaged 5.9.

Lepus townsendii campanius Hollister

1837. *Lepus campestris* Bachman, Jour. Acad. National Sci. Philadelphia, 7:349.

1915. *Lepus townsendii campanius* Hollister, Proc. Biol. Soc. Washington, 28:70, March 12.

Type.—From plains of Saskatchewan.

Comparisons.—*L. t. campanius* differs from *L. t. townsendii*, in so far as I know, only in that the tail of *campanius* is ordinarily all white and that the upper parts in summer are paler and more brassy. Nelson (1909:79) mentioned obscure black on the pinna of the ear in *townsendii*.

Measurements.—Average and extreme measurements of three males and four females from Sweetwater County are as follows: Total length of males, 554.5 (554-555), of females, 562.3 (530-592); tail vertebrae, 72 (66-78), 64.7 (63-67); hind foot, 148 (146-150), 146 (138-150); ear from notch, 116.5 (114-119), 113 (111-114); basilar length, 69.9 (68.8-71.1), 69.9 (61.9-70.8); zygomatic breadth, 45.2 (44.4-45.6), 44.7 (43.8-45.6); postorbital breadth, 15.2 (13.5-16.9), 13.7 (12.9-14.8); length of nasals, 43.4 (40.6-48.2), 37.4 (33.4-41.2); alveolar length of tooth-row, 17.2 (16.7-17.6), 16.3 (15.2-16.8); least width (anteroposteriorly) of palatal bridge, 5.5 (5.3-5.6), 5.7 (5.5-5.8); interpterygoid width between last molars, 10.8 (10.5-11.4), 11.0 (10.4-12.0).

Distribution.—Throughout state except southwestern part. See Fig. 18.

Remarks.—The basis for the assignment of specimens to subspecies is discussed in *Remarks* under the account of *L. t. townsendii.*

Records of occurrence.—Specimens examined, 68, as follows: **Albany Co.:** Marshall, 1 USNM; 5 mi. N Laramie, 2; Libby Flats, 3 mi. N, 7 mi. W Centennial, 10,720 ft., 1; 1 mi. SSE Pole Mtn., 8350 ft., 1; Woods P. O., 1 USNM; *Medicine Bow Mtns., 1 USNM.*

Big Horn Co.: Germania, 2 USNM.

Campbell Co.: 7 mi. S, 5 mi. E Rockypoint, 1.

Converse Co.: 12 mi. N, 6 mi. W Bill, 4800 ft., 1; Ft. Fetterman, 3 USNM; Douglas, 1 USNM; *Deer Creek, 3 USNM;* 19 mi. S, 22 mi. W Douglas, 1.
Crook Co.: Devils Tower, 2 USNM; Sundance, 1 USNM.
Fremont Co.: Crowheart, 1 USNM; 2 mi. E Sweetwater Sta., 1; *1½ mi. E Oregon Trail Crossing of Sweetwater River, 1;* 1 mi. N, 8 mi. W Splitrock, 1.
Goshen Co.: Spoon Butte, 1 USNM.
Laramie Co.: Cheyenne, 1 USNM.
Natrona Co.: 19 mi. N, 8.5 mi. W Powder River, 1; *9 mi. N, 2½ mi. W Waltman, 6050 ft., 1; 7 mi. N, 2½ mi. W Waltman, 6100 ft., 1; 5 mi. N, 1½ mi. W Waltman, 6250 ft., 1; 7½ mi. N, 12 mi. W Powder River, 2;* ½ mi. N Waltman, 6100 ft., 1; 17 mi. N, 11 mi. W Casper, 1; *16 mi. N, 11 mi. W Casper, 1; 15 mi. N, 10 mi. W Casper, 1; 4 mi. S, 4 mi. W Waltman, 6075 ft., 1;* 2 mi. N, 12 mi. W Casper, 4800 ft., 1; Rattlesnake Mtns., 1 USNM; 1½ mi. S Casper, 1; 3 mi. E Independence Rock, 1.
Niobrara Co.: 3 mi. W Manville, 5250 ft., 1.
Park Co.: 24 mi. S, 27 mi. W Cody, 6400 ft., 1.
Sublette Co.: 3 mi. S Boulder, 1; 23 mi. S, 12 mi. E Pinedale, 7300 ft., 1; Big Piney, 1 USNM.
Sweetwater Co.: 5 mi. N, 12 mi. E Farson, 7600 ft., 1; Farson, 6580 ft., 1; 13 mi. S, 7 mi. E Farson, 6850 ft., 1; 25 mi. N, 38 mi. E Rock Springs, 6700 ft., 6; 2½ mi. N Wamsutter, 1; *Wamsutter, 2 USNM;* 20 mi. E Rock Springs, 2.
Washakie Co.: Winchester, 1.
Weston Co.: Newcastle, 2 USNM; 23 mi. SW Newcastle, 4500 ft., 2.
Yellowstone Nat'l Park: *Glenn Creek,* 1 USNM (not plotted because engraving for Fig. 18 was made before locality was found).

Additional records (Nelson, 1909:78).—**Carbon Co.:** *Percy;* Ft. Steele. **Laramie Co.:** Meriden. **Sweetwater Co.:** Bitter Creek.

Lepus townsendii townsendii Bachman

1839. *Lepus townsendii* Bachman, Jour. Acad. National Sci. Philadelphia, 8 (part 1): 90, pl. 2.

Type.—From "on the Walla-walla, one of the sources of the Columbia River," Washington (Bachman, 1839:90). Fort Walla Walla, type locality of *Lepus artemisia* [= *Sylvilagus nuttallii*] has often been incorrectly listed as the type locality of *L. t. townsendii.*

Comparisons.—Characters by means of which *L. t. townsendii* can be distinguished from *L. t. campanius* in Wyoming include a thin, dark, dorsal caudal line (absent in *campanius*) on the tail and dorsal hairs brassy-colored distally (pale gray proximally) instead of golden distally (dark-gray proximally).

Measurements.—One adult from Uinta County has the following external and cranial measurements: Total length, 650; tail vertebrae, 110; hind foot, 159; ear from notch, 115; basilar length, 73.1; zygomatic breadth, 46.9; postorbital breadth, 13.5; length of nasals, 41.6; alveolar length of maxillary toothrow, 16.5; least width (anteroposteriorly) of palatal bridge, 7.0; interpterygoid width between last molars, 11.2.

Distribution.—Southwestern Wyoming. See Fig. 18.

Remarks.—In speaking of *L. t. townsendii* and *L. t. campanius,* Nelson (1909:80) stated that, "many skulls of the two forms are practically indistinguishable." In my study more resemblance was noted between Wyoming skulls of *L. t. townsendii* and *L. t. campanius* than between *townsendii* from Nevada and from Wyoming.

White-tailed jack rabbits of both subspecies from Wyoming average smaller than specimens of *townsendii* and *campanius* examined from outside the state. As mentioned by Nelson, there is microgeographic variation in both subspecies. That variation may explain why color of pelage of *L. t. townsendii* from Wyoming differs from that of the same subspecies in Nevada.

Records of occurrence.—Specimens examined, 10, as follows: **Lincoln Co.:** Cokeville, 2 USNM; Hams Fork, 1 USNM; 12 mi. S, 2 mi. W Kemmerer, 6700 ft., 1. **Teton Co.:** "Near" Kelley P. O., 1; Jackson, 1 USNM. **Uinta Co.:** 8½ mi. S Robertson, 8000 ft., 1; *8 mi. S, 2½ mi. E Robertson, 8300 ft., 2; 16 mi. S, 2 mi. W Kemmerer, 1.*

Additional record (Nelson, 1909:82).—**Sweetwater Co.:** Henrys Fork.

Lepus californicus melanotis Mearns

Black-tailed Jack Rabbit

1890. *Lepus melanotis* Mearns, Bull. Amer. Mus. Nat. Hist., 2:297, February 21.

1909. *Lepus californicus melanotis,* Nelson, N. Amer. Fauna, 29:146, August 31.

Type.—From Independence, Kansas.

Description.—Large (see *Measurements*); upper parts Ochraceous-Tawny intermixed with whitish hairs in spring or blackish hairs in summer; underparts whitish excepting Ochraceous-Buff on neck; tail markedly blackish above; pinna of ear tipped distally with blackish posteriorly but margined posterolaterally with white, anterolaterally with buff; anteriormost extensions of frontals acuminate instead of broadly v-shaped as in *L. townsendii.*

Measurements.—External and cranial measurements of an adult female and three adult males all from Laramie County are as follows: Total length, female 598, males 557, 518, 555; length of tail, 81, 87, 77, 55; hind foot, 143, 136, 134, 130; ear from notch, 115, 120, 112, 110; basilar length of Hensel, 78.6, 69.2, 75.5, 76.1; zygomatic breadth, 45.8, 43.1, 42.8, 42.7; postorbital breadth, 12.3, 11.4, 12.2, 13.0; length of nasals, 44.1, 38.9, 41.1, 39.2; alveolar length of maxillary tooth-row, 17.3, 15.9, 18.1, 16.9; least width (anteroposteriorly) of palatal bridge, 6.1, 6.5, 6.1, 7.3; interpterygoid width between last molars, 10.1, 8.5, 10.5, 9.8.

Distribution.—Southeastern part of state. See Fig. 17.

Records of occurrence.—Specimens examined, 9, as follows: **Albany Co.:** 18 mi. N [Laramie?], 7100 ft., 1 Univ. Wyo. **Converse Co.: 1 mi. N, 3 mi. E** Orin, 4725 ft., 1. **Goshen Co.:** 5 mi. NW Torrington, 1. **Laramie Co.:** Horse Creek, 6½ mi. W Meriden, 5200 ft., 1; *Horse Creek, 3 mi. W Meriden, 1;* 1½ mi. W Horse Creek, 1; 8 mi. E Horse Creek, 1; 1 mi. W Pine Bluffs, 5000 ft., 2.

Order RODENTIA

Rodents

Usually small, clawed, pentadacyl, and having one pair of chisel-like, curved, continually-growing incisors in both upper and lower jaws; incisors separated from other teeth by wide diastema; canines lacking; usually herbivorous, at least in part; uterus bicornuate.

KEY TO FAMILIES OF RODENTS IN WYOMING

1. Body and tail provided with quills; infraorbital foramen larger than foramen magnum . . . . ERETHIZONTIDAE, p. 668
1′. Body and tail lacking quills; infraorbital foramen smaller than foramen magnum 2
2. Tail flattened or depressed dorsoventrally, scaly; teeth having 8-10 transverse ridges . . . . CASTORIDAE, p. 621
2′. Tail not flattened dorsoventrally; teeth lacking 8-10 transverse ridges . . 3
3. External fur-lined cheek pouches present 4
3′. External fur-lined cheek pouches lacking 5
4. Tail much shorter than head and body; front feet larger than hind; tympanic bullae not evident in dorsal view of skull . . . . . GEOMYIDAE, p. 600
4′. Tail usually longer than head and body; front feet smaller than hind; tympanic bullae evident in dorsal view of skull . . . . . HETEROMYIDAE, p. 613
5. Cheek teeth more than three 6
5′. Cheek teeth three in normal individuals 7
6. Postorbital processes absent . . . . ZAPODIDAE, p. 664
6′. Postorbital processes present . . . . SCIURIDAE, p. 553
7. Upper molariform teeth having three longitudinal rows of cusps, MURIDAE, p. 663
7′. Upper molars having only two longitudinal rows of cusps or showing numerous triangles and transverse folds in occlusal view, CRICETIDAE, p. 624

Family SCIURIDAE—Squirrels

Fossorial, terrestrial, or arboreal mammals often scansorial and in *Glaucomys* volant; infraorbital foramina small; third premolar small or lacking; some kinds hibernate.

KEY TO GENERA OF SCIURIDAE IN WYOMING

1. Membrane present between forelimbs and hind limbs modified for gliding . . . . **Glaucomys**, p. 599
1′. Membrane lacking between forelimbs and hind limbs 2
2. Antorbital canal lacking, the antorbital foramen piercing the zygomatic plate of the maxillary; head striped . . . . **Eutamias**, p. 553
2′. Antorbital canal present; head not striped 3
3. Zygomatic breadth more than 48; anterior lower premolar having paraconulid . . . . **Marmota**, p. 568
3′. Zygomatic breadth less than 48; anterior lower premolar lacking paraconulid 4
4. Tail less than one-fourth total length . . . . **Cynomys**, p. 590
4′. Tail more than one-fourth total length 5
5. Zygomata converging anteriorly; tail short usually; often striped or spotted . . . . **Spermophilus**, p. 572
5′. Zygomata nearly parallel; tail long and well-haired; lacking spots and/or stripes 6
6. Baculum well-developed; third premolar well developed . . . . . **Sciurus**, p. 598
6′. Baculum spiculelike; third premolar vestigial or absent . . **Tamiasciurus**, p. 593

Genus **Eutamias** Trouessart

Western Chipmunks

Small; stripes on head and dorsum.

KEY TO SPECIES OF CHIPMUNKS IN WYOMING (AFTER WHITE, 1953:589)

1. Dorsal stripes obscure; upper parts grayish....... **Eutamias dorsalis**, p. 564
1'. Dorsal stripes distinct; upper parts tawny (not grayish)............ 2
2. Venter yellowish or buff; tip of baculum more than 30 per cent of length of shaft; shaft not widened at base............. **Eutamias amoenus**, p. 562
2'. Venter white; tip of baculum less than 29 per cent of length of shaft or if as much as, or more than, 29 per cent, shaft widened at base........ 3
3. Size small to medium; greatest length of skull less than 34 mm.; shaft of baculum not widened at base; outermost dorsal dark stripe never obsolete..................................... **Eutamias minimus**, p. 554
3'. Size large; greatest length of skull rarely less than 34 mm.; shaft of baculum widened at base; outermost dorsal dark stripe often obsolete, never strongly evident...................... **Eutamias umbrinus**, p. 565

Eutamias minimus (Bachman)

Least Chipmunk

Smallest of Wyoming chipmunks; outermost dark stripes distinct; venter usually white; skull small to medium for genus; tip of baculum of adult males less than 28 per cent of length of shaft.

The least chipmunk in Wyoming has been studied by White (1953), and I agree with his findings as to number and distinctness of subspecies and that effects of glaciation were a major cause of subspeciation.

Chipmunks of the two western subspecies (*minimus* and *consobrinus*) are small, whereas chipmunks of the four subspecies to the eastward (*confinis, pallidus, operarius,* and *silvaticus*) are large (White, 1953:609-610).

Eutamias minimus confinis Howell

1925. *Eutamias minimus confinis* Howell, Jour. Mamm., 6:52, February 15.

Type.—From head of Trappers Creek, west slope of the Bighorn Mountains, Big Horn County, Wyoming.

Description.—Large (see *Measurements*); upper parts generally grayish brown; crown Clay Color mixed with Pale Smoke Gray; upper facial stripe Fuscous Black; other facial stripes Fuscous Black slightly mixed with Tawny; anterior margin of ear Yellow Ocher or Ochraceous-Orange; hairs inside posterior part of pinna Yellow Ocher or Ochraceous-Orange; posterior margin of ear Smoke Gray; postauricular patch buffy white or Smoke Gray; dorsal dark stripes black or Fuscous Black more or less mixed with Tawny or Tawny-Olive; dorsal light stripes creamy white, sometimes washed with Pale Smoke Gray; sides Raw Sienna or Cinnamon-Buff; rump and thighs Pale Smoke Gray mixed with Tawny-Olive; dorsal surface of tail black mixed with Clay Color; ventral surface of tail Clay Color, black along margin and Light Buff or Light Ochraceous Buff along outermost edge; antipalmar and antiplantar surfaces of feet Pinkish Buff; underparts creamy white sometimes with grayish underfur; skull large; baculum large (after White, 1953:596).

Comparisons.—From *E. m. silvaticus,* the subspecies of the Black Hills to the eastward, *E. m. confinis* differs in having paler, less reddish and more grayish upper parts and less tawny ventral surface of tail as mentioned by White (1952:261). In a later paper, White (1953:596) mistakenly listed the characteristics of *E. m. silvaticus* as those of *E. m. confinis.*

From *E. m. operarius,* the subspecies from south-central Wyoming, *E. m. confinis* differs in having darker rump, sides, and thighs, and more grayish upper parts.

From *E. m. minimus,* the subspecies in the Red Desert and other southwestern arid areas in Wyoming, *E. m. confinis* differs in being larger, having longer and broader skull; longer and deeper mandible; longer baculum; and being darker.

From *E. m. consobrinus,* the westernmost subspecies in Wyoming, *E. m. confinis* differs in having more grayish upper parts; darker underside of tail; broader and longer skull; longer baculum; and being larger.

From *E. m. pallidus,* the subspecies that occurs nearest to the westward and eastward, *E. m. confinis* differs in having darker upper parts and darker sides.

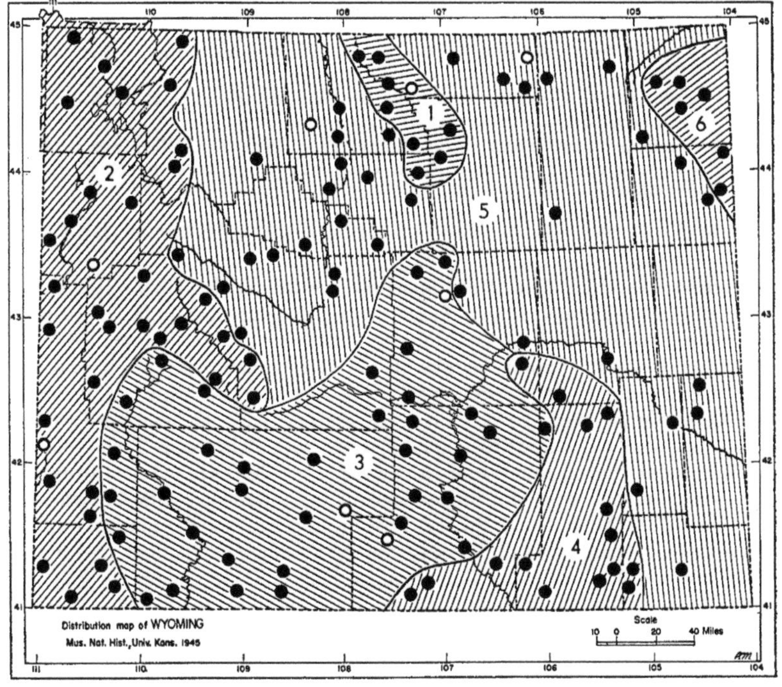

FIG. 19. Distribution of *Eutamias minimus.*

1. *E. m. confinis*
2. *E. m. consobrinus*
3. *E. m. minimus*
4. *E. m. operarius*
5. *E. m. pallidus*
6. *E. m. silvaticus*

Measurements.—Average and extreme external and cranial measurements of seven adult females and measurements of two adult males from Medicine Wheel Ranch, 28 miles east of Lovell, are, respectively, as follows: Total length, 204.4 (191-216), 180, 201; length of tail, 84.1 (70-91), 76, 89; hind foot, 31.4 (31-33), 30, 33; greatest length of skull, 32.6 (32.4-32.9), 31.5, 32.0; zygomatic breadth, 18.5 (18.4-18.6), 17.6, 17.7; length of nasals, 9.8 (9.6-10.4), 9.3, 9.6; cranial depth, 11.0 (10.7-11.2), 10.6, 11.0; maxillary toothrow, 5.3 (5.2-5.5), 5.1, 5.2.

Distribution.—Bighorn Mountains. See Fig. 19.

Records of occurrence.—Specimens examined, 113, as follows: **Big Horn Co.:** Medicine Wheel Ranch, 9000 ft., 28 mi. E Lovell, 11; Granite Creek Campground, Big Horn Mtns., 1; *2 mi. N, 13 mi. E Shell, 8500 ft., 2; 13 mi. E Shell, 8300 ft., 1;* 2 mi. S, 12 mi. E Shell, 7900 ft., 2; *3 mi. S, 17 mi. E Shell, 9000 ft., 8; 4½ mi. S, 17¼ mi. E Shell, 8500 ft., 11; 4½ mi. S, 19 mi. E Shell, 9600 ft., 1; Hd. Trappers Creek, 12 USNM; 8 mi. S Hd. Trappers Creek, 1 USNM;* 9 mi. N, 9 mi. E Tensleep, 8200 ft., 4. **Johnson Co.:** 1 mi. S, 7½ mi. W Buffalo, 6500 ft., 2; *1 mi. S, 5½ mi. W Buffalo, 6500 ft., 4; Hd. Canyon Creek, 6300-9300 ft., 11 USNM;* Hd. N Fork Powder River, 1 USNM. **Sheridan Co.:** 38 mi. E Lovell, Bighorn Nat'l Forest, 9600 ft., 10; *1½ mi. S, 5½ mi. W Jct. Hwys. U. S. 14 and Wyo. 14, 8480 ft., 2; "near" Sheridan, 2 USNM.* **Washakie Co.:** *4 mi. N, 9 mi. E Tensleep, 7000 ft., 26;* 3 mi. SE Tensleep, 4300 ft., 1.

Eutamias minimus consobrinus (J. A. Allen)

1890. *Tamias minimus consobrinus* J. A. Allen, Bull. Amer. Mus. Nat. Hist., 3:112, June.

1901. *Eutamias minimus consorbrinus,* Miller and Rehn, Proc. Boston Soc. Nat. Hist., 30:42, December 27.

1918. *Eutamias consobrinus clarus* Bailey, Proc. Biol. Soc. Washington, 31: 31, May 16. Type from Swan Lake Valley, Yellowstone Nat'l Park, Wyoming.

Type.—From near Barclay, Parley's Canyon, Wasatch Mountains, Salt Lake County, Utah.

Description.—Small (see *Measurements*); crown Smoke Gray intermixed with Ochraceous-Tawny; upper facial stripe Fuscous; other facial stripes Fuscous or Fucous-Black intermixed with Tawny; anterior margin of ear Ochraceous-Tawny; posterior margin of ear and postauricular patch grayish white; median dorsal dark stripe black with Ochraceous-Tawny along margins; other dorsal dark stripes black intermixed with Ochraceous-Tawny; median pair of dorsal pale stripes grayish white with Ochraceous-Tawny along margins; lateral pair of pale dorsal stripes whitish; sides Ochraceous-Tawny or Light Sayal Brown; rump and thighs Smoke Gray intermixed with Cinnamon-Buff; underside of tail Sayal Brown, Fuscous Black along margin, and Cinnamon-Buff or Ochraceous Buff along outermost edge; antipalmar and antiplantar surfaces Light Pinkish Cinnamon or Pinkish Buff; underparts grayish white intermixed with Buff or whitish; skull and baculum small (after White, 1953:593).

E. m. consobrinus is smaller than the other subspecies in Wyoming except *E. m. minimus,* from which *E. m. consobrinus* can be distinguished by darker upper parts and underside of tail.

Measurements.—Average and extreme external and cranial measurements of three females and three males from 13 miles south and two miles east of Robertson are, respectively, as follows: Total length, 192.3 (187-197), 184.0

(182-186); length of tail, 79.0 (74-83), 80.0 (77-84); hind foot, 28.3 (26-30), 26.3 (25-27); greatest length of skull, 30.1 (29.9-30.6), 30.5 (30.2-30.9); zygomatic breadth, 17.1 (16.6-17.3), 16.7 (16.6-16.8); length of nasals, 9.1 (8.9-9.5), 9.1 (8.7-9.3); cranial depth, 10.7 (10.4-11.0), 10.4 (10.2-10.6); maxillary tooth-row, 5.1 (4.9-5.4), 4.9 (4.4-5.4).

Distribution.—Western part of state. See Fig. 19.

Remarks.—As pointed out by White (1953:593), specimens from along the western boundary of Sweetwater County are integrades between *E. m. consobrinus* and *E. m. minimus*. The resemblance between these two subspecies is greater than that between either of them and any other subspecies in the state.

Specimens from the western part of Lincoln County were once assigned to *E. m. pictus* (Howell, 1929:41), and later to *E. m. scrutator* (Hall and Hatfield, 1934:Fig. 1) when *E. m. pictus* was restricted to Utah. These specimens are referred here to *E. m. consobrinus* on the basis of dark upper parts and on geographic grounds.

Records of occurrence.—Specimens examined, 263, as follows: **Fremont Co.:** Lake Fork, 9800-10,000 ft., 4 USNM; Moccasin Lake, 4 mi. N, 19 mi. W Lander, 10,100 ft., 1; *2½ mi. N, 17½ mi. W Lander, 1;* 4 mi. S, 8½ mi. W Lander, 9200 ft., 1; *16 mi. S, 5½ mi. W Lander, 8650 ft., 1; 23½ mi. S, 5 mi. W Lander, 8600 ft., 1;* ½ mi. N, 3 mi. E S. Pass City, 7900 ft., 7; *South Pass City, 8000 ft., 2 USNM.*

Lincoln Co.: 3 mi. N, 11 mi. E Alpine, 5650 ft., 2; 13 mi. N, 2 mi. W Afton, 6100 ft., 2; *Thane, 1 USNM; 10 mi. N, 2 mi. W Afton, 6100 ft., 2;* 15 mi. N Cokeville, 3 USNM; 6 mi. N, 2 mi. E Sage, 6050 ft., 1; Kemmerer, 1, 9 USNM; Cumberland, 14 mi. S, 1 mi. W Kemmerer, 6550 ft., 6.

Park Co.: Beartooth Lake, 8 USNM; Whirlwind Peak, 9000-11,000 ft., 1, 9 USNM; Valley, 1 USNM; Needle Mtn., 10,000-10,500 ft., 5 USNM.

Sublette Co.: 31 mi. N Pinedale, 8025 ft., 2; *12 mi. N Kendall, 4 USNM; 12 mi. NW Kendall, 4 USNM;* 4 mi. S Fremont Peak, 10,800-11,500 ft., 3 USNM; *5 mi. S Fremont Peak, 10,600 ft., 1 USNM;* 8 mi. NW Merna, 10,000 ft., 1 USNM; Merna, 8000 ft., 8 USNM; *12 mi. NE Pinedale, 8000 ft., 1 USNM; 9 mi. N, 5 mi. E Pinedale, 9100 ft., 13; Half Moon Lake, 7900 ft., 7;* 10 mi. NE Pinedale, 8000 ft., 1; *2¼ mi. NE Pinedale, 7500 ft., 3; 2 mi. NE Pinedale, 8000 ft., 2 USNM; 4 mi. W Pinedale, 7200 ft., 2; 5 mi. N, 3 mi. E Pinedale, 7500 ft., 3;* Pinedale, 4 USNM; 2 mi. S, 19 mi. W Big Piney, 7700 ft., 1; *3 mi. N Stanley, 4 USNM; 3 mi. W Stanley, 4 USNM;* Big Sandy, 6 USNM.

Teton Co.: *Moran, 4 USNM;* 1 mi. S, 3¾ mi. E Moran, 6210 ft., 2; *Jackson Hole Wildlife Park, 1 USNM; 5 mi. S Moran, 1; Elk, 1 USNM;* Togwotee Pass, 3, 2 USNM; 2½ mi. NE Moose, 2; *1 mi. E Moose, 1; 3¾ mi. E Moose, 6300 ft., 3; 3 mi. E Kelly P. O., 6800 ft., 3;* Teton Pass, 7 USNM.

Uinta Co.: *½ mi. S Cumberland, 1; 8½ mi. W Ft. Bridger, 6700 ft., 17;* 2 mi. W Ft. Bridger, 6070 ft., 1; *Ft. Bridger, 8 USNM; "near" Ft. Bridger, 2 USNM; Evanston, 9 USNM; Mountainview, 6 USNM; ½ mi. S Mountainview, 6900 ft., 2;* Sage Creek, 10 mi. N Lonetree, 7 USNM; *4½ mi. N, 4 mi. E Robertson, 8025 ft., 1; 6 mi. S, 2½ mi. E Robertson, 8200 ft., 3;* 8 mi. S, 2½ mi. E Robertson, 8300 ft., 1; *9 mi. S Robertson, 8000 ft., 5; 9½ mi. S, 1 mi. W Robertson, 8600 ft., 2; 10 mi. S, 1 mi. W Robertson, 8700 ft., 4; 11½ mi. S, 2 mi. E Robertson, 9200 ft., 1; 12 mi. S, 2 mi. E Robertson, 9000 ft., 1; 13 mi. S, 2 mi. E Robertson, 9200 ft., 7;* 4 mi. S Lonetree, 4 USNM.

Yellowstone Nat'l Park: Snow Pass, Mammoth Hot Springs, 1 USNM; Bunsen Peak, 1 USNM; Canyon, 3 USNM; Lake Sta., 1 USNM; Summit Lake, 2 USNM.

Additional records (White, 1953:594).—**Lincoln Co.:** Border; *Fossil.* **Teton Co.:** *Jenny Lake; Jackson Hole; Flat Creek Pass;* Flat Creek-Crystal Creek divide; *Flat Creek-Granite Creek divide; Sheep Creek.* **Yellowstone Nat'l Park:** *Swan Lake Valley.*

Eutamias minimus minimus (Bachman)

1839. *Tamias minimus* Bachman, Jour. Acad. Nat. Sci. Philadelphia, 8 (pt. 1):71.

1901. *Eutamias minimus,* Miller and Rehn, Proc. Boston Soc. Nat. Hist., 30:42, December 27.

Type.—From the Green River, near mouth of Big Sandy Creek, Sweetwater County, Wyoming.

Description.—Small (see *Measurements*); crown Pinkish Buff intermixed with grayish white; facial stripes Snuff-Brown intermixed with black; anterior margin of ear Drab suffused with Cinnamon; posterior margin of pinna of ear and postauricular patch grayish white; median and lateral dorsal dark stripes black or Sayal Brown more or less intermixed with Fuscous Black; pale dorsal stripes grayish white tinged with Buff; rump and thighs Smoke Gray; dorsal surface of tail Fuscous Black intermixed with Cinnamon-Buff; underside of tail Sayal Brown or Clay Color margined with Blackish Brown intermixed with Cinnamon-Buff; antiplanter and antipalmar surfaces Pale Pinkish Buff; underparts creamy white; skull and baculum small (after White, 1953:591). *E. minimus minimus* is smaller than any other subspecies in Wyoming except *E. m. consobrinus,* which is about the same size.

Measurements.—External and cranial measurements of seven males and four females from Kinney Ranch, 21 miles south of Bitter Creek, Sweetwater County, are, respectively, as follows: Total length, 187.9 (183-195), 190.5 (178-200); length of tail, 87.4 (81-91), 84.8 (82-87); hind foot, 30.4 (28-33), 30.0 (30-30); ear from notch, 13.4 (11-15), 13.8 (13-15); occipitonasal length, 30.5 (29.8-31.0), 30.7 (30.2-31.1); zygomatic breadth, 16.9 (16.5-17.4), 17.2 (16.7-17.6); length of nasals, 8.8 (8.4-9.6), 8.8 (8.4-9.3); cranial depth, 10.8 (10.5-11.1), 10.5 (10.2-11.0); maxillary tooth-row, 4.7 (4.3-4.9), 4.8 (4.5-4.9).

Distribution.—Central and southwest Wyoming. See Fig. 19.

Remarks.—*E. m. minimus* intergrades with *E. m. consobrinus* to the westward (see account of that subspecies), with *E. m. operarius* in the vicinities of Savery and Medicine Bow Peak (White, 1953:598), and with *E. m. pallidus* in a broad zone in northern Fremont County and the southern part of the Bighorn Basin. Intergrades (referred to *E. m. operarius*) between *E. m. minimus* and the larger *E. m. operarius* resemble the former in size and the latter in coloration, whereas intergrades (referred here to *E. m. pallidus*) between *E. m. minimus* and *E. m. pallidus* resemble *E. m. pallidus* in size and *E. m. minimus* in pallor (often they are slightly paler than either subspecies). *E. m. minimus* is adapted for living in sage-covered desert.

Records of occurrence.—Specimens examined, 207, as follows: **Albany Co.:** Spring Creek, 10 mi. W Marshall, 1 USNM.

Carbon Co.: 12 mi. S Alcova, 7000 ft., 2; Ferris Mtns., 7800 ft., 4 USNM; 8 mi. SE Lost Soldier, 1 USNM; 24 mi. N, 12 mi. E Sinclair, 6600 ft., 1; *18 mi. NNE Sinclair, 6500 ft., 2; Medicine Bow River, 1 Univ. Wyo.;* Rawlins, 1; Ft. Steele, 3 USNM; 30 mi. E Rawlins, 6750 ft., 2; Bridgers Pass, 18 mi. SW Rawlins, 7500 ft., 2, 6 USNM; Saratoga, 2 USNM.

Fremont Co.: Granite Mtns., 6; *Mt. Crooks, 8600 ft., 6;* 8 mi. E Rongis, 8000 ft., 1 USNM.

Lincoln Co.: Fontenelle, 6500 ft., 7 USNM; Opal, 3 USNM.

Natrona Co.: 27 mi. N, 1 mi. E Powder River, 6075 ft., 2; 15 mi. N, 1 mi. W Waltman, 1; *9 mi. S, 9 mi. W Waltman, 6950 ft., 1;* Casper, 5 USNM; *Rattlesnake Mtns., 7000-8000 ft., 9 USNM;* 16 mi. S, 11 mi. W Waltman, 6950 ft., 2; 1 mi. N, 9 mi. W Independence Rock, 1; *Sun Ranch, 5 mi. W Independence Rock, 6000 ft., 4; Independence Rock, 1 USNM; 1 mi. S, 5 mi. W Independence Rock, 2; Sun, 6400 ft., 3 USNM.*

Sublette Co.: Jct. Green River and New Fork, 7 USNM; 2 mi. SE Big Sandy, 1; Big Piney, 1 USNM.

Sweetwater Co.: Farson, 6580 ft., 11; *5 mi. E Farson, 1; Eden, 1 USNM;* 27 mi. N, 37 mi. E Rock Springs, 6700 ft., 1; *25 mi. N, 38 mi. E Rock Springs, 6700 ft., 3;* Steamboat Mtn., 5 USNM; Jct. Big Sandy and Green River, 7; Superior, 2 USNM; *17 mi. N, 6 mi. W Rock Springs, 7000 ft., 1;* Between Tipton and Tablerock, 6 Univ. Wyo.; Green River, 16 USNM; 13 mi. S, 14 mi. E Rock Springs, 6650 ft., 2; *Bitter Creek, 1 USNM; 18 mi. S Bitter Creek, 2; 22 mi. SSW Bitter Creek, 5;* 26 mi. S, 21 mi. W Rock Springs, 3; Kinney Ranch, 21 mi. S Bitter Creek, 6800 ft., 15, 3 USNM; *30 mi. S Bitter Creek, 2;* 32 mi. S, 22 mi. W Rock Springs, 1; 32 mi. S, 22 mi. E Rock Springs, 7025 ft., 12; 33 mi. S Bitter Creek, 6900 ft., 6; *30 mi. SE Ft. Bridger, 3 USNM;* Mouth Burnt Fort, Henry's Fork, 1 USNM; *3 mi. W Green River and 2 mi. N Utah boundary, 1; Green River, 4 mi. N Linwood, Utah, 5800 ft., 1 USNM; ½ mi. N Jct. Henrys Fork and Utah boundary, 2.*

Uinta Co.: 15 mi. WSW Granger, 1; *10 mi. SW Granger, 1 USNM.*

Additional records (White, 1953:592).—**Albany Co.:** *Sheep Creek.* **Carbon Co.:** Shirley; *Shirley Mtns.; Sulphur Springs.* **Natrona Co.:** Bitter Creek, near Powder River. **Sublette Co.:** Muddy Creek, near Big Sandy Creek. **Sweetwater Co.:** *27 mi. N Tablerock; Thayer; Tablerock;* Wamsutter. **County not certain:** *Little Sandy River.*

Eutamias minimus operarius Merriam

1905. *Eutamias amoenus operarius* Merriam, Proc. Biol. Soc. Washington, 18:164, June 29.

1922. *Eutamias minimus operarius,* Howell, Jour. Mamm., 3:183, August 4.

Type.—From Gold Hill, 7400 ft., Boulder County, Colorado.

Description.—Size large (see *Measurements*); crown Cinnamon-Buff intermixed with Pale Smoke Gray; facial stripes Fuscous Black intermixed with Sayal Brown; anterior margin of pinna Cinnamon-Buff; posterior margin of pinna of ear and postauricular patch Pale Smoke Gray; dorsal dark stripes black with Ochraceous-Tawny along margins; median pale stripes Pale Smoke Gray with Ochraceous-Tawny along margins; lateral dorsal pale stripes whitish; sides Tawny or Ochraceous-Tawny; rump and thighs Light Grayish Olive; dorsal surface of tail Sayal Brown or Ochraceous-Tawny with Fuscous Black along margin and Clay Color along outermost edge; antipalmar and antiplantar surfaces of feet Ochraceous-Buff; underparts grayish white, often washed with Buff; skull and baculum large (after White, 1953:598).

Comparisons.—For comparisons of *E. m. operarius* with *E. m. confinis, E. m. consobrinus,* and *E. m. minimus* see accounts of those subspecies. From *E. m. pallidus, E. m. operarius* differs in having darker upper parts, sides, and underside of tail. From *E. m. silvaticus, E. m. operarius* differs in having upper parts less grayish, sides darker, and underside of tail darker.

Measurements.—Average and extreme external and cranial measurements of four females and three males from three miles ESE Browns Peak are as follows: Total length, 199.8 (192-206), 197.2 (192-202); length of tail, 86.0 (83-92), 87.0 (84-90); hind foot, 30.8 (30-31), 29.3 (29-30); greatest length of skull, 31.4 (31.0-31.6), 31.3 (31.1-31.7); zygomatic breadth, 17.4 (16.8-17.3), 17.5 (17.2-17.8); length of nasals, 9.2 (8.5-10.0), 9.3 (8.6-9.7); cranial depth, 10.9 (10.5-11.1), 10.6 (10.5-10.9); maxillary tooth-row, 4.9 (4.8-5.1), 5.0 (4.6-5.2).

Distribution.—Mountainous south-central part of state. See Fig. 19.

Records of occurrence.—Specimens examined, 166, as follows: **Albany Co.:** Springhill, 9 USNM; 3 mi. SW Eagle Peak, 7500 ft., 6 USNM; 27 mi. N, 7½ mi. E Laramie, 6960 ft., 12; 9 mi. N, 13 mi. E Laramie, 7700 ft., 2; *8 mi. N, 4 mi. E Laramie, 1; 8 mi. NE Laramie, 1 Univ. Wyo.; 6 mi. N, 4 mi. E Laramie, 7500 ft., 1;* Medicine Bow Peak, 10,900 ft., 1 USNM; *Univ. Wyo. Science Camp, 9900 ft., 3; Nash Fork Creek, Medicine Bow Mtns., 4, 3 Univ. Wyo.; Snowy Range, 2 USNM; Holmes Camp Ground, Medicine Bow Mtns., 1; Hq. Park, Medicine Bow Mtns., 5 USNM; 3 mi. E Laramie, 7300 ft., 1, 2 Univ. Wyo.; 6 mi. E Laramie, 7500 ft., 1 Univ. Wyo.; 10 mi E Laramie, 9000 ft., 6 USNM;* 11 mi. E Laramie, 8500 ft., 1 Univ. Wyo.; *4 mi. S, 8 mi. E Laramie, 8600 ft., 1; 5½ mi. ESE Laramie, 8500 ft., 1; Pole Mtn., 1 Univ. Wyo.; 1 mi. SSE Pole Mtn., 3; 2 mi. SE Pole Mtn., 19; 3 mi. S Pole Mtn., 2; 10 mi. S, 11 mi. E Laramie, 8500 ft., 1 Univ. Wyo.;* 15 mi. SE Laramie, 2 USNM; *3 mi. E Browns Peak, 2; 2¼ mi. ESE Browns Peak, 1; 3 mi. ESE Browns Peak, 10,000 ft., 15; 2.2 mi. S Albany, 1 Univ. Wyo.;* 3½ mi. S Woods Landing, 7700 ft., 1.

Carbon Co.: 6 mi. S, 10 mi. E Saratoga, 8800 ft., 1; *Lake Marie, 10,440 ft., 1; 2 mi. S, ½ mi. W Medicine Bow Peak, 10,400 ft., 1; 2 mi. S, 2 mi. W Medicine Bow Peak, 10,700 ft., 1; 10 mi. N, 14 mi. E Encampment, 8000 ft., 2; 8 mi. N, 14 mi. E Encampment, 8400 ft., 2; 8 mi. N, 16 mi. E Encampment, 8400 ft., 3; 8 mi. N, 21½ mi. E Encampment, 9400 ft., 2; Riverside, 5 USNM;* S Base Bridger Peak, 3 USNM; *8 mi. N, 19½ mi. E Savery, 10,128 ft., 1; 6 mi. N, 14 mi. E Savery, 8400 ft., 1;* 5 mi. N, 5 mi. E Savery, 6900 ft., 2.

Converse Co.: 21½ mi. S, 24½ mi. W Douglas, 7600 ft., 10.

Laramie Co.: 5 mi. W Horse Creek P. O., 7200 ft., 2; *3½ mi. W Horse Creek P. O., 7000 ft., 3; 2 mi. W Horse Creek P. O., 6600 ft., 1;* 6 mi. W Islay, 4 USNM.

Natrona Co.: *6 mi. S, 2 mi. W Casper, 5900 ft., 1;* 7 mi. S, 2 mi. W Casper, 6370 ft., 2; *7 mi. S Casper, 7500 ft., 3 USNM; 10 mi. S Casper, 7750 ft., 3.*

Eutamias minimus pallidus (J. A. Allen)

1874. *T [amias]. quadrivitatus,* var. *pallidus* J. A. Allen, Proc. Boston Soc. Nat. Hist., 16:281.

1922. *Eutamias minimus pallidus,* Howell, Jour. Mamm., 3:183, August 4.

Lectotype.—From Camp Thorne, near Glendive, Dawson County, Montana.

Description.—Size large (see *Measurements*); crown Pale Smoke Gray intermixed with Clay Color; facial stripes Fuscous Black intermixed with Clay Color; anterior margin of pinna Pale Pinkish Buff; posterior margin of pinna of ear and postauricular patch grayish white; dorsal dark stripes black or Fuscous intermixed with Clay Color; median pair of dorsal pale stripes Pale Smoke Gray; lateral pair of dorsal pale stripes creamy white; sides Cinnamon-Buff; rump and thighs Smoke Gray intermixed with Pale Buff; dorsal surface of tail Fuscous Black slightly tinged with Warm Buff; ventral surface of tail Pinkish Cinnamon or Pinkish Buff, with Fuscous Black along margin and Warm

Buff along outermost edge; antipalmar and antiplantar surfaces Pinkish Buff, Warm Buff or Pale Yellow-Orange; underparts whitish with plumbeous underfur; skull and baculum large (after White, 1953:594-595).

From *E. m. silvaticus*, *E. m. pallidus* differs in paler upper parts, sides, and underside of tail. See accounts of other subspecies for other comparisons.

Measurements.—External and cranial measurements of an old adult female from Muskrat Canyon and another from five miles north and eight miles west of Spotted Horse are, respectively, as follows: Total length, 195, 214; length of tail, 85, 91; hind foot, 33, 33; greatest length of skull, 33.6, 32.3; zygomatic breadth, 18.4, 17.9; length of nasals, 10.5, 9.5; cranial depth, 11.7, 11.2; maxillary tooth-row, 5.1, 5.2.

Distribution.—Low areas of northern and eastern Wyoming, usually in riparian or wooded habitats. See Fig. 19.

This subspecies is widely distributed, and minor geographic variation is not uncommon in it. I have attempted to keep this variability in mind, and perhaps for this reason I have drawn the boundary between *E. m. pallidus* and *E. m. minimus* closer to typical *E. m. minimus* than did White (1953:590), but as did Howell (1929:44). The large size of old adults from within 14 miles of Shoshone warrants their reference to *E. m. pallidus.* Several specimens from the Big Horn Basin (for example, those from seven miles south of Basin) are slightly paler than any *E. m. pallidus* or *E. m. minimus*, but are here referred to *E. m. pallidus.*

White (1953:595) has referred intergrades between *E. m. pallidus* and *E. m. confinis* from within eight miles of Buffalo, Johnson County, to *E. m. confinis*, and I concur. *E. m. pallidus* probably interbreeds with *E. m. silvaticus* in the vicinity of Sundance (but specimens from the nearby Bear Lodge Mountains clearly are *silvaticus*). Intergrades (referred to *E. m. operarius*) between *E. m. operarius* and *E. m. pallidus* were obtained from the Laramie Mountains in southern Converse County (White, 1953:595).

Records of occurrence.—Specimens examined, 137, as follows: **Big Horn Co.:** *Shell Creek, 1 mi. NW Shell, 8; 6 mi. NW Greybull, 3800 ft., 6; Tributary Big Horn River, 1 mi. W Greybull, 2;* Greybull, 4 USNM; 4 mi. N Hyattville, 1 USNM; 7 mi. S Basin, 3900 ft., 5.

Campbell Co.: 4 mi. S, 6 mi. W Rockypoint, 4200 ft., 1; *4 mi. S, 3 mi. W Rockypoint, 5;* 5 mi. N, 8 mi. W Spotted Horse, 15; Middle Butte, 38 mi. S, 19 mi. W Gillette, 6010 ft., 3; *"near" Pumpkin Butte, Powder River Basin, 2 USNM; South Butte, 40½ mi. S, 17½ mi. W Gillette, 6000 ft., 2.*

Converse Co.: Douglas, 6 USNM.

Crook Co.: Moorcroft, 8 USNM.

Fremont Co.: Sheep Creek, 1 USNM; 40 mi. E Dubois, 1; 3 mi. S Dubois, 3 USNM; 12 mi. N, 3 mi. W Shoshone, 4650 ft., 2; *9 mi. N, 3 mi. E Shoshone, 4700 ft., 2; 7 mi. N, 3 mi. E Shoshone, 4700 ft., 3;* Meadow Creek, 1 USNM; 2½ mi. W Shoshone, 4800 ft., 1; Ft. Washakie, 6 USNM.

Goshen Co.: Rawhide Buttes, 12 mi. S, 1 mi. W Lusk, 1, 3 USNM; *Muskrat Canyon, 2;* Pine Ridge, 3 USNM.

Hot Springs Co.: Kirby Creek, 2 USNM; Hd. Bridger Creek, 2 USNM; 10 mi. SW Thermopolis, 1 USNM.

Laramie Co.: Locality not specified, 1.

Natrona Co.: Powder River, 3 USNM.

Park Co.: 2 mi. S, 2 mi. E Meteetse, 5750 ft., 3.

Platte Co.: Guernsey, 4700 ft., 1 USNM; 15 mi. SW Wheatland, 1 USNM.

Sheridan Co.: Sheridan, 3 USNM; 5 mi. NE Clearmont, 3900 ft., 1; Arvada, 7 USNM.

Washakie Co.: 10 mi. S Manderson, 1 USNM; 15 mi. W Tensleep, 1 USNM; 8 mi. S, 8 mi. W Worland, 1; Otter Creek, 2 USNM; *10 mi. S Tensleep, 2 USNM.*
Weston Co.: Thornton, 2 USNM; *Upton, 2 USNM;* Newcastle, 1 USNM.
County Not Certain: Owl Creek Mtns., 3 USNM.

Additional records (White, 1953:596).—**Big Horn Co.:** Otto. **Sheridan Co.:** Powder River, mouth of Clear Creek.

Eutamias minimus silvaticus White

1952. *Eutamias minimus silvaticus* White, Univ. Kansas Publ., Mus. Nat. Hist., 5(19):259-262, April 10.

Type.—From 3 mi. NW Sundance, 5900 ft., Crook County, Wyoming.

Description.—Size large (see *Measurements*); crown Sayal Brown suffused with Cinnamon-Buff; facial stripes Fuscous Black intermixed with Clay Color; anterior margin of ear Ochraceous-Orange; posterior margin of ear and postauricular patch grayish white; dorsal dark stripes Fuscous Black more or less intermixed with Ochraceous-Buff; medial dorsal pale stripes grayish white or white margined with Ochraceous-Buff; sides Ochraceous-Buff; rump and thighs Smoke Gray washed with Ochraceous-Buff; dorsal surface of tail black intermixed with Ochraceous-Buff; underside of tail Ochraceous-Orange with black along margin and Light Ochraceous-Buff along outermost edge; antiplantar and antipalmer surfaces Light Buff; underparts creamy white, often suffused more or less strongly with Ochraceous-Buff; skull and baculum large (after White, 1953:597). See accounts of other subspecies for comparisons.

Measurements.—Average and extreme measurements of three adult males and 11 adult females of *E. m. silvaticus* from the type locality are, respectively, as follows: Total length, 190 (189-190), 207 (202-220); length of tail, 85 (81-90), 97 (82-105); length of hind foot, 31 (30-33), 32 (31-34); length of ear, 14 (13-16), 15 (14-17); greatest length of skull, 32.0 (31.5-32.6), 32.3 (31.5-33.1); zygomatic breadth, 18.5 (18.5-18.5), 18.6 (18.2-19.0); least interorbital constriction, 6.9 (6.8-7.1), 7.0 (6.4-8.1); length of nasals, 9.4 (9.2-9.6), 9.6 (9.3-10.1) (White, 1952:262).

Distribution.—Bear Lodge Mountains and Black Hills in northeastern part of state. See Fig. 19.

Records of occurrence.—Specimens examined, 60, as follows: **Crook Co.:** Devils Tower, 4 USNM; 15 mi. N Sundance, Black Hills Nat'l Forest, 5500 ft., 6; *15 mi. ENE Sundance, 3825 ft., 1; Bear Lodge Mtns., 3 USNM;* type locality, 16; *1 mi. N Sundance, Black Hills Nat'l Forest, 1; Sundance, 8 USNM.* **Weston Co.:** 1½ mi. E Buckhorn, 6150 ft., 19; Newcastle, 2 USNM.

Eutamias amoenus luteiventris (J. A. Allen)

Yellow-pine Chipmunk

1890. *Tamias quadrivittatus luteiventris* J. A. Allen, Bull. Amer. Mus. Nat. Hist., 3:101, June.

1922. *Eutamias amoenus luteiventris,* A. H. Howell, Jour. Mamm., 3:179, August 4.

Type.—From "Chief Mountain Lake" [=Waterton Lake], 3½ mi. N U.S.-Canadian Boundary, Alberta, obtained on August 24, 1874.

Description.—Size medium (see *Measurements*); color (after White, 1953: 602-603) on crown "Cinnamon mixed with Smoke Gray; upper two facial stripes black; submalar stripe Fuscous or Fuscous Black mixed with Ochraceous-

Tawny; anterior margin of ear Ochraceous-Tawny; posterior margin of ear and postauricular patch Light Buff or buffy white; hairs inside posterior part of pinna of ear Ochraceous-Tawny; median dorsal dark stripe black; lateral pair of dorsal dark stripes black and mixed with Tawny, frequently brownish; median pair of dorsal light stripes white tinged with Pale Smoke Gray; lateral pair of dorsal light stripes creamy white; sides Tawny or Ochraceous-Tawny; rump and thighs Dark Smoke Gray strongly mixed with Cinnamon-Buff; dorsal surface of tail Fuscous Black mixed with Clay Color; ventral surface of tail Light Ochraceous Tawny, with Fuscous Black around margin and Clay Color around outermost edge; antipalmar and antiplantar surfaces of feet Cinnamon or Cinnamon-Buff; underparts Cinnamon-Buff or Light Ochraceous-Buff"; skull medium in size and moderately narrow across zygomata; according to White (1953:603) baculum "slender and not noticeably broadened at base; tip more than 30 per cent of length of shaft."

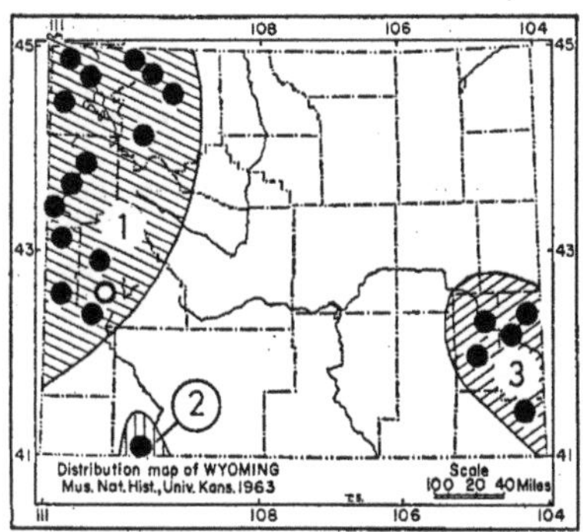

FIG. 20. Three species of sciurids.
1. *Eutamias amoenus luteiventris*
2. *Eutamias dorsalis utahensis*
3. *Spermophilus spilosoma obsoletus*

Measurements.—Average and extreme external and cranial measurements of eight males and five females, respectively, from within five miles of Moran are as follows: Total length, 213.6 (210-215), 215.4 (203-225); length of tail vertebrae, 96.0 (90-100), 88.4 (81-91); hind foot, 32.3 (32-33), 32.4 (32-33); ear from notch, 18.0 (17-19), 17.6 (17-19); occipitonasal length, 33.9 (32.9-35.5), 33.9 (33.6-34.7); zygomatic breadth, 18.4 (18.0-18.6), 18.5 (18.2-18.7); cranial depth, 11.2 (11.0-11.5), 11.2 (10.9-11.4); length of nasals, 10.6 (10.0-12.1), 11.0 (10.9-11.2); maxillary tooth-row, 5.2 (5.1-5.4), 5.2 (5.1-5.3).

Distribution.—Western third of state exclusive of extreme southwestern part and Wind River Mountains. Occurrence of *amoenus* in the Wind River Mountains has been suggested by White (1953:603), but no specimens have been obtained from there. See Fig. 20.

Remarks.—*Eutamias amoenus luteiventris*, *E. m. consobrinus*, and *E. umbrinus fremonti* may occur together (White, 1953:603); but whereas the last-mentioned chipmunk is found in forests and *E. m. consobrinus* in open country, *E. a. luteiventris* is most frequently taken and observed in intermediate habitat (forest-edge).

Records of occurrence.—Specimens examined, 193, as follows: **Lincoln Co.:** 3 mi. N, 11 mi. E Alpine, 5650 ft., 2; 10 mi. SE Afton, 7500 ft., 5.

Park Co.: *Clarks Fork, 11 USNM;* 31½ mi. N, 36 mi. W Cody, 6900 ft., 6; 29 mi. N, 31 mi. W Cody, 7200 ft., 1; *28 mi. N, 30 mi. W Cody, 7200 ft., 1;* 16¾ mi. N, 17 mi. W Cody, 5625 ft., 3; 25 mi. S, 28 mi. W Cody, 6350 ft., 2; *Valley, 6500-7500 ft., 3 USNM.*

Sublette Co.: Merna, 8000 ft., 6 USNM; *Hd. Dry Creek, Salt River Mtns., 1 USNM;* LaBarge Creek, 1 USNM.

Teton Co.: *Two Ocean Lake, 6850 ft., 3 Univ. Wyo., 2; Two Ocean Lake*

Road, 3; Lake Emma Matilda, 1; 3 mi. N Moran, 6800 ft., 2 Univ. Wyo., 1; 2½ mi. N, 3½ mi. E Moran, 7225 ft., 1; 2 mi. N Moran, 1 Univ. Wyo., 1 USNM; ½ mi. N, 1½ mi. E Moran, 6750 ft., 2 Univ. Wyo.; ½ mi. N, 2½ mi. E Moran, 6750 ft., 2 Univ. Wyo.; ¼ mi. N, 2½ mi. E Moran, 6230 ft., 7; Moran, 16 USNM, 1; 2 mi. E Moran, 1; Pacific Creek Road, 2½ mi. E Moran, 1; 2½ mi E Moran, 6220 ft., 1; 3 mi. E Moran, 6750 ft., 2 Univ. Wyo.; ¼ mi. S, 3 mi. E Moran, 6200 ft., 1; Jackson Hole Wildlife Park, 27 USNM; *1 mi. S, 3¾ mi. E Moran, 6200 ft., 8; 1 mi. S, 4 mi. E Moran, 6750 ft., 1 Univ. Wyo.; Teton Nat'l Park, 4; Snake River, 2 USNM;* Timbered Island, 4 mi. N Moose, 6750 ft., 5; *Bar BC Ranch, 2½ mi. NE Moose, 6500 ft., 9; Signal Mtn.* Pond, *1; Moose Creek, 6800-10,000 ft., 18 USNM;* Teton Pass, 16 USNM; *14 mi. S Wilson, 1 USNM;* Exact locality not specified, 2 USNM.

Yellowstone Nat'l Park: Mammoth Hot Springs, 2 USNM; *Bunsen Peak, 1 USNM; Roaring Mtn., 1 USNM;* Canyon, 3 USNM; Old Faithful, 2 USNM.

Additional records (From White, 1953:603).—**Sublette Co.:** Stanley. **Teton Co.:** *Jct. Two Ocean Lake Road and U. S. Hwy. 187; Leigh Lake; Signal Mtn. Road; Grand Teton, 9000 ft.* **Yellowstone Nat'l Park:** *Yancey; Apollinaris Spring; Yellowstone Lake; Upper Geyser Basin.*

Eutamias dorsalis utahensis Merriam

Cliff Chipmunk

1897. *Eutamias dorsalis utahensis* Merriam, Proc. Biol. Soc. Washington, 11:210, July 1.

Type.—From Ogden, Utah.

Description.—Size large (see *Measurements*); crown Pale Smoke Gray intermixed with Cinnamon; dark facial stripes Sayal Brown intermixed with Fuscous and enclosing areas of white; anterior margin of pinna of ear Ochraceous-Tawny; posterior margin of pinna and postauricular patch whitish; median dorsal dark stripe Fuscous or black; other longitudinal dorsal dark stripes obscure and gray intermixed with black or Fuscous; dorsal pair of pale longitudinal stripes Smoke Gray; lateral pair pale stripes whitish and often obscure; rump and flanks Pale Smoke Gray intermixed with Cinnamon; Pale Smoke Gray laterally except dorsal extension of bright Cinnamon extending almost to lateral pale stripe; dorsal surface of tail Fuscous Black intermixed with Tilleul Buff; underside of tail Cinnamon-Buff bordered first with Fuscous Black, bordered on outermost edge by Tilleul Buff; feet Cinnamon-Buff above; underparts whitish or creamy white; braincase well inflated; zygomata robust; infraorbital foramen narrowly oval or slitlike; and according to White (1953:605) baculum small with keel approximately one third of length of tip.

Measurements (Cranial measurements from White, 1953:600).—Average and extreme measurements of four adult males and two females are, respectively, as follows: Total length, 197 (191-203), females, 211 (210-212); tail vertebrae, 84.5 (81-88), 88.0 (86-90); hind foot, 31.0 (30-32), 32.5 (32-33); ear from notch, 18.0 (17-19), 19.5 (18-21); greatest length of skull, 34.7 (34.7-34.8), 36.0 (35.5-36.6); zygomatic breadth, 18.9 (18.7-19.2), 19.5 (19.4-19.7); cranial breadth, 16.4 (16.4-16.4), 16.3 (16.2-16.4); length of nasals, 10.8 (10.5-11.1), 11.3 (11.3-11.4); length of lower tooth-row, 5.08 (5.00-5.15), 5.25 (5.22-5.28); condylo-alveolar length of mandible, 17.91 (17.77-18.06), 18.87 (18.73-19.02).

Distribution.—Southern part of Sweetwater County. See Fig. 20.

Records of occurrence.—Specimens examined, 11, as follows: **Sweetwater Co.:** W side Green River, 1 mi. N Utah border, 6; *Green River, 4 mi. NE Linwood, Utah, 5800 ft., 5 USNM.*

Eutamias umbrinus (J. A. Allen)

Uinta Chipmunk

Size large; upper parts generally dark in comparison with those of other Wyoming species; outermost dorsal dark stripe lacking or poorly evident; skull rarely shorter than 34 mm. in adults; baculum widened at base; underparts whitish.

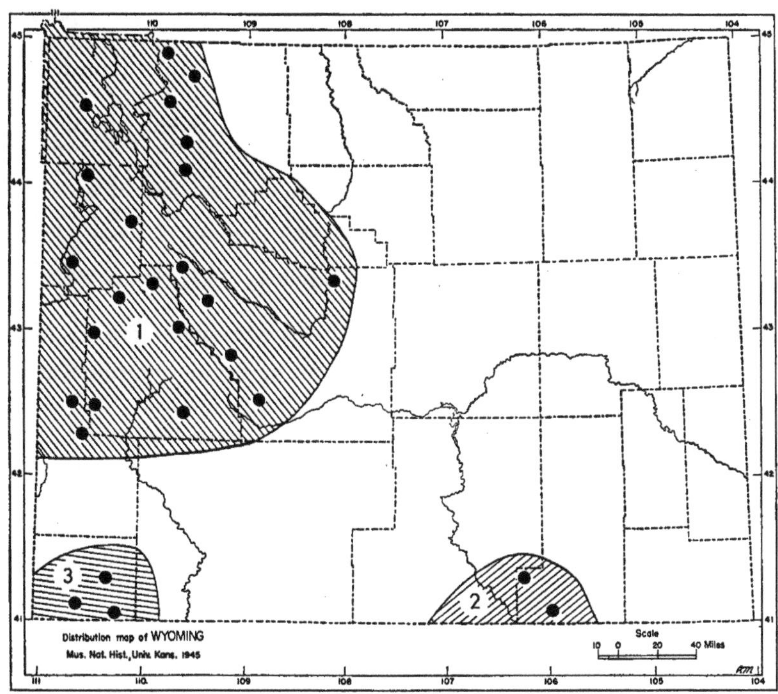

FIG. 21. Distribution of *Eutamias umbrinus*.
Guide to subspecies
1. *E. u. fremonti*
2. *E. u. montanus*
3. *E. u. umbrinus*

Eutamias umbrinus fremonti White

1953. *Eutamias umbrinus fremonti* White, Univ. Kansas Publ., Mus. Nat. Hist., 5:576, December 1.

Type.—From 31 mi. N Pinedale, 8025 ft., Sublette County, Wyoming.

Description.—Crown Cinnamon-Buff mixed with gray; upper facial stripe Sepia; ocular stripe Chaetura-Drab; submalar stripe Fuscous Black mixed with Sayal Brown; ears black; anterior margin of ear Mars Yellow; posterior margin of ear grayish white; hairs inside posterior part of pinna of ear Dresden Brown; postauricular patch Pale Smoke Gray; median dorsal dark stripe black; lateral dorsal dark stripes black mixed with Sayal Brown; outermost dorsal dark stripes Buckthorn Brown mixed with black or sometimes absent; median pair of dorsal light stripes grayish mixed with Buckthorn Brown; outer pair of dorsal light stripes creamy white; sides Buckthorn Brown; rump and thighs Pale Smoke Gray mixed with Saccardo's Umber; dorsal surface of tail black mixed with Buckthorn Brown; ventral surface of tail Sayal Brown, with Fuscous Black around margin and white or Light Buff around outermost edge; antipalmar and antiplantar surfaces of feet Warm Buff; underparts creamy white with dark underfur; skull large with robust zygomata; baculum widened at base tapering sharply to tip (afer White, 1953:607).

From *E. u. montanus*, *E. u. fremonti* differs in having darker upper parts; darker underside of tail; darker feet; and darker sides. From *E. u. umbrinus*, *E. u. fremonti* differs in having paler upper parts with darker, more reddish sides; darker feet; and slightly larger skull.

Measurements.—Average and extreme measurements listed by White (1953:601) for eight males and six females, respectively, from Togwotee Pass are as follows: Total length, 223 (216-243), 229 (223-239); length of tail, 99.0 (95-111), 101.0 (92-110); greatest length of skull, 35.6 (35.2-36.5), 35.3 (34.5-36.0); zygomatic breadth, 19.3 (18.9-19.7), 19.6 (19.3-20.0); cranial breadth, 15.9 (15.8-16.1), 15.9 (15.7-16.5); length of nasals, 11.4 (11.1-11.8), 11.3 (10.9-12.0); length of lower tooth-row, 5.34 (5.22-5.57), 5.40 (5.35-5.44); condylo-alveolar length of mandible, 19.17 (18.72-19.78), 19.02 (18.37-19.51).

Distribution.—Western part of state, especially timbered areas. See Fig. 21.

Remarks.—The distinguishing characters of *E. u. umbrinus* are less evident in specimens from Wyoming than in those from Utah, thereby indicating intergradation between *E. u. umbrinus* and *E. u. fremonti* in southwestern Wyoming; these characters are slightly darker sides and feet and slightly larger cranium in *E. u. fremonti* than in *E. u. umbrinus*. At present the geographic ranges of the two subspecies seem not to meet; the two are separated by low, arid areas.

Records of occurrence.—Specimens examined, 108, as follows: **Fremont Co.:** Jackeys Creek, 4 mi. SW Dubois, 8500 ft., 3; 12 mi. N, 3 mi. W Shoshone, 4650 ft., 1; Bull Lake, 3 USNM; *Lake Fork, 9600 ft., 9 USNM;* Mosquito Park Ranger Sta., 2½ mi. N, 17½ mi. W Lander, 9500 ft., 2; *½ mi. S, 17½ mi. W Lander, 1;* 17 mi. S, 6½ mi. W Lander, 8450 ft., 4.

Lincoln Co.: 8 mi. W Stanley, 8500 ft., 4 USNM; LaBarge Creek, 2 USNM.

Park Co.: Hd. Clarks Fork, 1 USNM; 16¼ mi. N, 17 mi. W Cody, 5625 ft., 2; Whirlwind Peak, 8500-11,000 ft., 10 USNM; *Crow Peak, Pahaska, 9000 ft., 3 USNM; Mouth Grinnell Creek, 2 USNM;* Valley, 3 USNM; Needle Mtn., 1 USNM.

Sublette Co.: 31 mi. N Pinedale, 8025 ft., 1; *12 mi. NW Kendall, 10,000 ft., 4 USNM;* 7 mi. S Fremont Peak, 10,400 ft., 6 USNM; *W side Barbara Lake, 8 mi. S, 3 mi. W Fremont Peak, 5;* Merna, 1 USNM; *Dry Creek, 8500-9000 ft., 4 USNM;* 2 mi. S, 19 mi. W Big Piney, 7700 ft., 5; Big Sandy, 3 USNM.

Teton Co.: Upper Arizona Creek, 3 USNM; *S. Moose Creek, 10,000 ft., 4 USNM; ¼ mi. N, 1 mi. E Togwotee Pass, 9800 ft., 2;* Togwotee Pass, 9500-11,000 ft., 7, 9 USNM; Jackson, 1 USNM.

Yellowstone Nat'l Park: 2.

Additional records (White, 1953:607).—**Park Co.:** *Beartooth Lake.* **Teton Co.:** *Amphitheater Lake, Teton Park; Flat Creek; Hd. Cache Creek; Flat Creek-Granite Creek divide; Flat Creek Pass; Flat Creek-Gravel Creek divide.*

Eutamias umbrinus montanus White

1953. *Eutamias umbrinus montanus* White, Univ. Kansas Publs., Mus. Nat. Hist., 5:576, December 1.

Type.—From 3 mi. S, ½ mi. E Ward, 9400 ft., Boulder County, Colorado.

Description.—Crown Raw Sienna mixed with gray; upper facial stripe and ocular stripe black mixed with Sepia; submalar stripe Snuff Brown mixed with black; ear black or Sepia; anterior margin of ear Ochraceous-Tawny; posterior margin of ear and postauricular patch grayish white; hairs inside posterior part of pinna of ear Cinnamon-Buff; median dorsal dark stripe black with Sayal Brown along margins; lateral pair of dorsal dark stripes black mixed with Sayal Brown; outermost pair of dorsal dark stripes Sayal Brown mixed with black or sometimes larcking; median pair of dorsal light stripes Pale Smoke Gray mixed with Clay Color; outer pair of dorsal light stripes creamy white; sides Clay Color; rump and thighs Neutral Gray; dorsal surface of tail black mixed with Cinnamon-Buff; ventral surface of tail Ochraceous-Tawny along outermost edge; antipalmar and antiplantar surfaces of feet Cinnamon-Buff; underparts creamy white with dark underfur; skull large with robust, arched zygomata; baculum widened at base tapering sharply to tip (after White, 1953:608).

For comparisons with *E. u. umbrinus* and *E. u. fremonti*, see accounts of those subspecies.

Measurements.—External and cranial measurements of three adult males and one female, respectively, from Corner Mountain are as follows: Total length, 226.3 (225-227), 231; length of tail 104.7 (102-108), 109; length of hind foot, 33.0 (32-34), 33; greatest length of skull, 35.8 (35.4-36.5), 35.0; zygomatic breadth, 19.1 (18.8-19.7), 19.0; length of nasals, 10.8 (10.5-11.0), 10.0; cranial depth, 11.5 (11.4-11.6), 11.4; maxillary tooth-row, 5.7 (5.6-5.8), 5.7.

Distribution.—South-central Wyoming. See Fig. 21.

Records of occurrence.—Specimens examined, 14, as follows: **Albany Co.:** *Corner Mtn., 8500 ft., 4;* Nash Fork Creek, 1; *NW Laramie, sec. 31 or 32, T. 17N, R. 74W, 1; 2 mi. N, 1 mi. W Centennial, 1; 4 mi. N Centennial, 8500 ft., 1; N. Fork Camp Ground, sec. 28, T. 16N, R. 78W, 8500 ft., 3; 3 mi. ESE Browns* Peak, *10,000 ft., 2;* 3½ mi. S Woods Landing, 1.

Eutamias umbrinus umbrinus (J. A. Allen)

1890. *Tamias umbrinus* J. A. Allen, Bull. Amer. Mus. Nat. Hist., 3:96, June.
1901. *Eutamias umbrinus,* Miller and Rehn, Proc. Boston Soc. Nat. Hist., 30(1):45, December 27.

Type.—From Blacks Fork, approximately 8000 ft., Uinta Mtns., Utah.

Description.—Crown Pale Smoke Gray; facial stripes Fuscous Black or Snuff Brown; ears Fuscous Black; posterior margin of ear and postauricular patch grayish white; median dorsal dark stripe black with Sayal Brown along margins; lateral pair of dorsal dark stripes Fuscous Black mixed with Sayal Brown, or entirely Sayal Brown; outermost pair of dorsal dark stripes Sayal Brown mixed with Fuscous Black or lacking; sides Sayal Brown mixed with Cinnamon; rump and thighs Sayal Brown mixed with Smoke Gray; antipalmar and antiplantar surfaces of feet Cinnamon-Buff, underside of tail Ochraceous-Tawny or Sayal Brown, with Fuscous Black around margin and Pinkish Buff around outermost edge; underparts creamy white with dark gray underfur;

braincase inflated with robust zygomata and slightly smaller than in other subspecies in Wyoming; baculum widened at base tapering to somewhat truncate tip (after White, 1953:606).

Comparisons.—From *E. u. fremonti, E. u. umbrinus* differs in having darker upper parts with paler, less reddish sides; paler feet; and slightly smaller skull. From *E. u. montanus, E. u. umbrinus* differs in having duller upper parts (more tawny); sides less tawny; skull slightly smaller. See *Remarks* under account of *E. u. fremonti.*

Measurements.—Average and extreme measurements listed by White (1963: 601) of 11 males and four females from 9-13 miles southward of Robertson, Uinta County, are, respectively, as follows: Total length, 218 (215-228), 224 (204-234); length of tail, 96.2 (81-112), 96.4 (90-100); greatest length of skull, 34.7 (34.3-35.2), 35.1 (34.9-35.4); zygomatic breadth, 18.9 (18.3-19.4), 19.4 (19.2-20.0); cranial depth, 15.7 (15.6-16.0), 15.9 (15.7-16.2); length of nasals, 10.9 (10.3-11.7), 11.0 (10.3-11.8); length of lower tooth-row, 5.13 (4.79-5.42), 5.17 (5.11-5.22); condylo-alveolar length of mandible, 18.04 (17.57-18.59), 18.46 (18.31-18.98).

Distribution.—Extreme southwestern part of state. See Fig. 21 and *Remarks* under account of *E. u. fremonti.*

Records of occurrence.—Specimens examined, 35, as follows: **Uinta Co.:** Ft. Bridger, 2 USNM; *9 mi. S Robertson, 8000 ft., 15;* 10 mi. S, 1 mi. W Robertson, 8700 ft., 5; *11½ mi. S, 2 mi. E Robertson, 9200 ft., 1; 12 mi. S, 2 mi. E Robertson, Ashley Nat'l Forest, 1; 13 mi. S, 2 mi. E Robertson, 9200 ft., 1; 5 mi. W Lonetree, Henrys Fork, 8000 ft., 4 USNM;* Beaver Creek, 4 mi. S Lonetree, 3 USNM; *Uinta Mtns., "near" Ft. Bridger, 3 USNM.*

Marmota flaviventris (Audubon and Bachman)

Yellow-bellied Marmot

The species *Marmota flaviventris* was revised by A. H. Howell (1915), who had named two of the three subspecies he reported from Wyoming. Whereas Howell assigned a geographic range extending southwestward to the Laramie Mountains and even to Bridger's Pass to *M. f. dacota,* Hall and Kelson (1959:324) restricted the distribution of *dacota* to the Black Hills. The difference in range depends on subspecific assignment of the marmots of south-central Wyoming. Specimens of marmots characteristic of any of the three subspecies can be found in south-central Wyoming. This indicates intergradation among the three subspecies. Assignment of specimens to subspecies from that area is difficult because relatively few specimens are available for comparison and because of the high degree of individual variation in marmots.

M. f. luteola has been characterized (A. H. Howell, 1915:50) on the basis of yellowish venter. Three types of yellowish venter have been observed in specimens known by the name *M. f. luteola.* Dark specimens from the Grand Mesa of Colorado show some golden-tipped hairs on their venters. A few specimens from south-central Wyoming that are more reddish also show this character. Other specimens (for example, KU 41642) from between the type locality of *M. f. luteola* and the Bighorn Mountains have yellowish and whitish spots on their venters. A specimen (KU 25300) from two miles south and one mile west of Medicine Bow Peak has a pale-yellowish, mid-ventral stripe. Many specimens and live marmots that I have observed from south-central Wyoming (for example, three topotypes of *M. f. luteola*) have dark, reddish venters, a characteristic *M. f. luteola* is supposed to lack. Cranial differences between *M. f. luteola* and *M. f. nosophora* were regarded by Howell as slight (1914a:15). I find no cranial differences that certainly are more than individual variation in specimens from Wyoming.

The subspecies *M. f. luteola,* the range of which is mostly in Colorado, occupies also south-central Wyoming. The more reddish upper parts of specimens in southern Wyoming closely resemble those of *M. f. dacota* of northeastern Wyoming. A better character than golden color on the venter by means of which *M. f. luteola* in Wyoming can be identified is dark, yellowish-brown upper parts

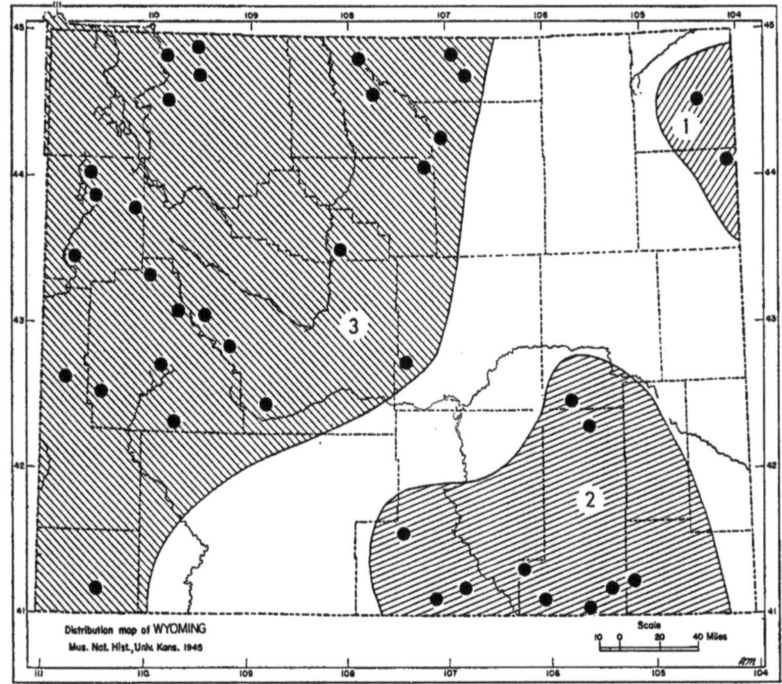

FIG. 22. Distribution of *Marmota flaviventris.*
Guide to subspecies
1. *M. f. dacota*
2. *M. f. luteola*
3. *M. f. nosophora*

that merge indistinctly with the dark areas around the mouth on specimens having golden-reddish or reddish venters.

M. f. nosophora and *M. f. dacota* are in western and northeastern Wyoming, respectively. *M. f. nosophora* tends toward *M. f. luteola* in dark upper parts in Uinta County. The one specimen (US 176898) of *nosophora* available from 12 miles north of Kendall is indistinguishable from the holotype of *M. f. dacota,* and is considered to be merely an individual variant.

Marmota flaviventris dacota (Merriam)

1889. *Arctomys dacota* Merriam, N. Amer. Fauna, 2:8, October 30.
1914. *M [armota]. f [laviventer]. dacota,* A. H. Howell, Proc. Biol. Soc. Washington, 27:15, February 2.
1915. *[M.] flaviventris dacota,* A. H. Howell, N. Amer. Fauna, 37:7, April 7.

Type.—From Custer, Custer County, Black Hills, South Dakota.

Description.—Large; brightly colored; reddish (chestnut) hairs intermixed with black and yellowish hairs on upper parts; black around mouth often extending back to level of ears; tail often reddish.

Comparisons.—Upper parts paler and brighter than yellowish-brown of *M. f. luteola;* contrast in color between upper parts and blackish muzzle more pronounced than in *luteola;* upper parts darker and brighter and tail more reddish than in most specimens of *M. f. nosophora.*

Measurements.—External and cranial measurements of an adult male from one mile south of Warren Peak Lookout and another from Weston County are, respectively, as follows: Total length, 660, 670; length of tail, 225, 173; hind foot, 82, 84; ear from notch, 29, 34; condylobasal length, 87.6, 93.6; length of nasals, 37.0, 32.4; zygomatic breadth, 56.5, 61.0; least interorbital constriction, 17.7, 15.9.

Distribution.—Northeastern Wyoming, especially the Black Hills. See Fig. 22.

Records of occurrence.—Specimens examined, 11, as follows: **Crook Co.**: 1 mi. S Warren Peak Lookout, 6000 ft., 8; *Bear Lodge Mtns., 1 USNM.* **Weston Co.**: ½ mi. E Buckhorn, 6100 ft., 1; *1½ mi. E Buckhorn, 6100 ft., 1.*

Marmota flaviventris luteola A. H. Howell

1914. *Marmota flaviventer luteola* A. H. Howell, Proc. Biol. Soc. Washington, 27:15, February 2.

Type.—From Woods Post Office in Medicine Bow Mountains, "about 7,500 ft.," Albany County, Wyoming.

Comparisons.—Smaller than *M. f. dacota,* but resembling it in color although more yellowish-brownish. See comparisons under *Marmota flaviventris dacota* and *M. f. nosophora.* J. A. Allen (1874:57) reported melanism in this subspecies.

Measurements.—External and cranial measurements of three adult female topotypes are as follows: Total length, 582, 595, 580; length of tail, 164, 162, 156; hind foot, 80, 76, 76; ear from notch, 20, 25, 24; condylobasal length, 87.1, 86.9, 82.8; length of nasals, 35.5, 35.2, 31.5; zygomatic breadth, 52.5, 52.3, 52.9; least interorbital breadth, 15.0, 15.1, 18.1.

Distribution.—South-central part of state. See Fig. 22.

Records of occurrence.—Specimens examined, 42, as follows: **Albany Co.:** 3 mi. W Eagle Peak, 7500 ft., 2 USNM; Univ. Wyo. Summer Camp, 1, 1 Univ. Wyo.; *14 mi. SE Laramie, 1 USNM;* Pole Mtn., 15 mi. SE Laramie, 8340-8700 ft., 3 USNM; *16 mi. SE Laramie, 3 USNM;* 1 mi. W Woods Landing (P. O.), 7700 ft., 5; *16 mi. S Laramie, 1 USNM;* Tie Siding, 2 USNM; *Laramie Mtns., 5 USNM.* **Carbon Co.:** Bridgers Pass, 3 USNM; *1½ mi. S, 2½ mi. W Medicine Bow Peak, 9450 ft., 1; 2 mi. S, 1 mi. W Medicine Bow Peak, 10,800 ft., 1; Riverside, 2 USNM; Lake Marie, 1 Univ. Wyo.;* 12 mi. N, 13 mi. E Encampment, 7300 ft., 2; 7½ mi. N, 18½ mi. E Savery, 8300 ft., 2. **Converse Co.:** 21½ mi. S, 24½ mi. W Douglas, 7600 ft., 4. **Laramie Co.:** 17 mi. E Laramie, 8300 ft., 2 USNM.

Additional record (A. H. Howell, 1915:52).—**Albany Co.:** *Sherman.*

Marmota flaviventris nosophora A. H. Howell

1914. *Marmota flaviventer nosophora* A. H. Howell, Proc. Biol. Soc. Washington, 27:15, February 2.

Type.—From Willow Creek, seven miles east of Corvallis, 4000 ft., Ravalli County, Montana.

Description.—Small (see *Measurements*); upper parts pale reddish; brownish washed with white around mouth; underparts reddish; skull resembling that of *M. f. luteola.*

Measurements.—External and cranial measurements of three adult females (20 mi. S and 3½ mi. W Lander; 17 mi. N and 18 mi. W Cody; 30 mi. N and 18 mi. W Cody) are, respectively, as follows: Total length, 618, 580, 610; tail vertebrae, 155, 165, 210; hind foot, 80, 75, 65; ear from notch, 25, 30, 25; condylobasal length, 90.5, 82.0, 85.6; length of nasals, 38.0, 32.9, 34.2; zygomatic breadth, 56.8, 51.3, 52.2; least interorbital breadth, 14.4, 16.9, 16.0.

Distribution.—Western Wyoming. See Fig. 22.

Remarks.—*M. f. nosophora* tends to be darker, especially around the mouth, in southwestern Wyoming, and may intergrade with *M. f. luteola* in south-central Wyoming (Carbon County) inasmuch as the northernmost examples of *luteola* are small and pale; they are whitish around the mouth as is *M. f. nosophora* of northwestern Wyoming.

Melanism has been reported in this subspecies only from the Teton Mountains, by Fryxell (1928:336-337), Murie (1934:323), and Armitage (1961:100-101).

Records of occurrence.—Specimens examined, 65, as follows: **Big Horn Co.:** Medicine Wheel Ranch, 28 mi. E Lovell, 9000 ft., 1; Granite Creek Camp Grounds, 1; *3 mi. N, 11 mi. E Shell, 7000 ft., 1; 2 mi. N, 12 mi. E Shell, 7500 ft., 1; 13 mi. E Shell, 8300 ft., 2; Hd. Trappers Creek, 8500 ft., 4 USNM.*

Fremont Co.: *Lost Cabin near Lonetree, 1 USNM;* Lake Fork, 10,600 ft., 3 USNM; ½ mi. N, 17 mi. W Lander, 1; 20 mi. S, 3½ mi. W Lander, 9000 ft., 1.

Hot Springs Co.: 5 mi. S, 1 mi. E Thermopolis, 4475 ft., 1.

Johnson Co.: 17 mi. W Buffalo, 1.

Lincoln Co.: 10 mi. SE Afton, 7000 ft., 1 USNM.

Natrona Co.: 16 mi. S, 11 mi. W Waltman, 6950 ft., 1.

Park Co.: 30 mi. N, 18 mi. W Cody, 10,500 ft., 1; 29 mi. N, 32 mi. W Cody, 6850 ft., 1; 17 mi. N, 18 mi. W Cody, 6250 ft., 1; 3 mi. S, 42 mi. W Cody, 6400 ft., 1.

Sheridan Co.: 1 mi. E Acme, 1; Sheridan, 1 USNM.

Sublette Co.: 12 mi. N Kendall, 7700 ft., 1 USNM; E end Island Lake, 3 mi. S Fremont Peak, 10,600 ft., 1; *W side Seneca, 5 mi. S, 2 mi. W Fremont Peak, 10,300 ft., 4; 5 mi. S Fremont Peak, 10,600 ft., 1 USNM; 7 mi. S Fremont Peak, 10,400 ft., 1 USNM; 9 mi. N, 5 mi. E Pinedale, 9100 ft., 1; N side Halfmoon Lake, 7900 ft., 1;* 2 mi. S, 19 mi. W Big Piney, 7700 ft., 5; New Fork Green River, 1 USNM; Little Sandy Creek, 3 USNM.

Teton Co.: Pacific Creek, 1 USNM; 3 mi. N, 2 mi. W Moran, 6800 ft., 1 Univ. Wyo.; 2 mi. W summit, Togwotee Pass, 9250 ft., 1 Univ. Wyo., 1 USNM; Jackson, 2 USNM; *Jackson Hole, 4 USNM.*

Uinta Co.: 9 mi. S Robertson, 8000 ft., 8; *13 mi. S, 2 mi. E Robertson, 9200 ft., 1.*

Washakie Co.: 5 mi. N, 9 mi. E Tensleep, 7400 ft., 1.

KEY TO SPECIES OF SPERMOPHILUS

1. Upper parts usually having 13 longitudinal stripes (some stripes break up into distinct spots) .... **Spermophilus tridecemlineatus,** p. 577
1′. Upper parts lacking 13 stripes, although either stripes or spots may be present 2
2. Lateral stripes present .... **Spermophilus lateralis,** p. 583
2′. Lateral stripes lacking 3
3. Metaloph on P4 continuous; upper parts uniform grayish or brownish, often dappled with fine, obscure spots 4
4. Underside of tail grayish; nasals deflected ventrad anteriorly, **Spermophilus armatus,** p. 575
4′. Underside of tail buffy or orange; nasals nearly straight-sided, **Spermophilus richardsonii,** p. 572
3′. Metaloph on P4 not continuous; upper parts reddish dappled, usually with distinct spots .... **Spermophilus spilosoma,** p. 577

Spermophilus richardsonii elegans Kennicott

Richardson's Ground Squirrel

1863. *Spermophilus elegans* Kennicott, Proc. Acad. Nat. Sci. Philadelphia, 15, 158.

1938. *Citellus richardsonii elegans,* A. H. Howell, N. Amer. Fauna, 56:76, May 18.

1959. *Spermophilus richardsonii elegans,* Hall and Kelson, The Mammals of North America, The Ronald Press, New York City, p. 339, March 31.

Type.—No holotype; eight cotypes and fragments of bone from Fort Bridger, Wyoming, in U. S. National Museum.

Description.—Resembles *Spermophilus armatus* cranially and in color, but smaller in western Wyoming (see *Measurements*), paler, and having underside of tail buffy instead of gray; tail relatively longer; front of face Cinnamon, cheeks and neck Pale Smoke Gray; legs Cinnamon Buff with antiplantar surfaces whitish, yellowish, or reddish; venter whitish or buffy; tail Fuscous Black above and below intermixed with Pale Buff; nasals nearly straight, lacking distinct notch anteriorly (see account of S. *armatus*).

Measurements.—Average and extreme external and cranial measurements of eight adult males (KU 25327, 25337, 25348, 25350, 25363, 25364, 25365, 25370) and five females (25335, 25349, 25351, 25355, 25362) from the Sierra Madre of Carbon County are, respectively, as follows: Total length, 277.0 (257-300), 288.0 (271-300); length of tail, 72.6 (59-83), 75.2 (73-79); length of hind foot, 41.4 (38-45), 42.4 (39-45); ear from notch, 16.1 (14-18), 15.6 (14-18); greatest length of skull, 44.2 (43.0-46.8), 44.4 (43.3-45.3);

zygomatic breadth, 28.7 (27.4-30.8), 29.2 (28.7-30.0); postorbital constriction, 10.7 (10.2-11.2), 10.7 (10.0-11.0); length of nasals, 15.7 (14.7-17.1), 15.3 (14.7-15.8); maxillary tooth-row, 10.6 (10.0-11.1), 10.4 (9.9-11.1).

Distribution.—Southern half of state, in Transition and Upper Sonoran life-zones. See Fig. 23.

Remarks.—One specimen (KU 91080) has markedly reddish upper parts; its venter is also reddish, resembling somewhat in color the venter of *Marmota flaviventris.* This color of venter is lacking in other individuals of *Spermophilus richardsonii,* although many specimens of this species have a paler, buffy or cinnamon suffusion over a whitish venter. The specimen was obtained by A. B. Mickey on May 19, 1947, one mile east of Laramie.

Spermophilus richardsonii elegans and S. *armatus* occur together even on "common ground" in some places in western Wyoming (for example at Pinedale and Cokeville, A. H. Howell, 1938:80). S. *armatus* usually occurs at higher elevations in Wyoming than S. *r. elegans,* which may account for the distribution of *armatus* in the mountains of northwestern Wyoming. S. *richardsonii* seems limited

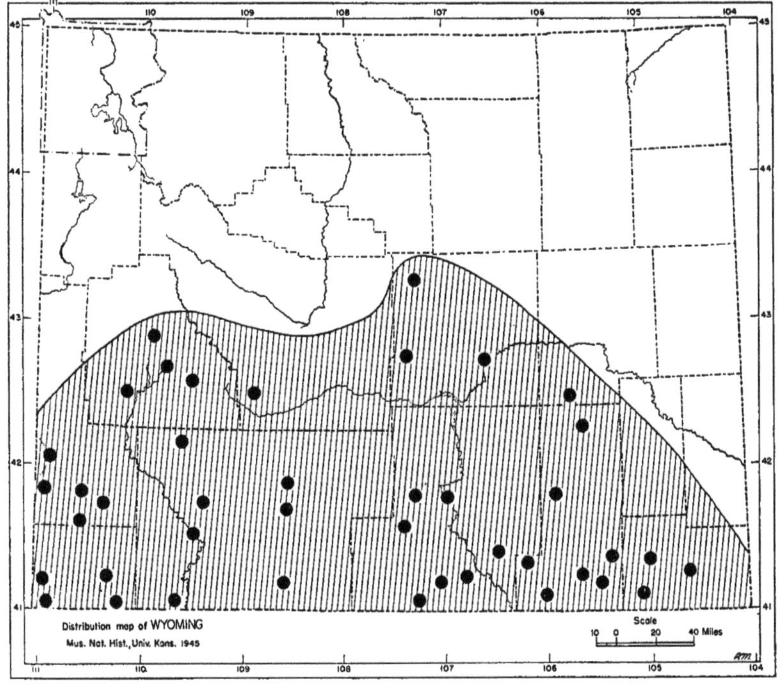

FIG. 23. Distribution of *Spermophilus richardsonii elegans.*

in northward distribution by these mountainous areas. *S. richardsonii* does occur at high elevations in the Medicine Bow Mountains, but usually in the Transition Life-zone.

The geographic range of *S. r. elegans* may limit the range of *S. armatus* and *vice versa.* These species closely resemble each other morphologically. Also the animals are not sympatric in a large geographic area, but only along an arid, mountainous area paralleling the western boundary of Wyoming. Furthermore there is some evidence of "character displacement" (see Brown and Wilson, 1956) in these species. Series of specimens of *S. r. elegans* from the Green River Watershed are slightly smaller and slightly paler than those from localities to the eastward. These differences seem to follow a gradual cline from east to west, and there are no statistical grounds for recognizing two subspecies of *S. richardsonii* along it. But the western specimens differ more from *armatus* than do specimens of *elegans* from eastern Wyoming. Hall (1943:377-378) described two specimens from Pinedale of *Spermophilus* considered by him to be hybrids or possibly intergrades between *richardsonii* and *S. armatus.* In 1945 Hall collected a large series of both species from the same locality and found no specimens of mixed ancestry. Hansen (1957) showed that the two specimens reported as hybrids were instead *Spermophilus richardsonii.*

Records of occurrence.—Specimens examined, 241, as follows: **Albany Co.:** 3 mi. SW Eagle Peak, 1 USNM; 12 mi. SE Rock River, 1; 8½ mi. W Horse Creek P. O., 2; *Laramie River, 1 USNM;* W side Sheep Mtn., 1 Univ. Wyo.; *5 mi. N Laramie, 1;* Laramie, 1 Univ. Wyo., 3 USNM; *1 mi. E Laramie, 7300 ft., 2, 1 Univ. Wyo.; 2 mi. E Laramie, 1 Univ. Wyo.; 2½ mi. SE Laramie, 7100 ft., 1 Univ. Wyo.; 3 mi. SE Laramie, 1 Univ. Wyo.;* 1 mi. SSE Pole Mtn., 8350 ft., 1; *Laramie Mtns., 3 USNM;* 16 mi. S, 8 mi. W Laramie, 7350 ft., 1; *Medicine Bow Mtns., 2 USNM; Woods P. O., 4 USNM; "near" Lake Hattie, 1;* not found: Lake Owen, 1.

Carbon Co.: Rawlins, 6744 ft., 3 USNM; *Little Medicine Bow River, 1 USNM;* Ft. Steele, 2 USNM; Bridgers Pass, 7 USNM; 6 mi. S, 14 mi. E Saratoga, 1; *8 mi. S, 4 mi. E Saratoga, 9; 8 mi. S, 6 mi. E Saratoga, 1; 12 mi. N, 13 mi. E Encampment, 10; 11 mi. N, 3 mi. E Encampment, 1; 10 mi. N, 3 mi. E Encampment, 6; 10 mi. N, 11 mi. E Encampment, 3; 9 mi. N Encampment, 1; 9 mi. N, 8 mi. E Encampment, 1; 9 mi. N, 9 mi. E Encampment, 3; 6 mi. N Encampment, 2;* 4 mi. N Encampment, 2; Bridgers Peak, 8800 ft., 1 USNM; *Riverside, 1 USNM;* 8 mi. E Savery, 5.

Converse Co.: 17 mi. S, 20 mi. W Douglas, 5.

Fremont Co.: South Pass City, 1 USNM.

Laramie Co.: *Horse Creek P. O., 6500 ft., 1; 6 mi. E Horse Creek P. O., 6500 ft., 1;* 8 mi. E Horse Creek P. O., 6500 ft., 1; *15 mi. N, 15 mi. W Cheyenne, 1;* 20 mi. NE Cheyenne, 1; *6 mi. W Islay, 1 USNM; Islay, 7000 ft., 1 USNM; 11 mi. N, 5½ mi. E Cheyenne, 5950 ft., 2;* 7 mi. W Cheyenne, 6500 ft., 1; *6 mi. W Cheyenne, 6500 ft., 2; W. edge Cheyenne, 6100 ft., 2, 7 USNM; Camp Carling, 1 USNM.*

Lincoln Co.: Cokeville, 6 USNM; Sage, 4 USNM; *Fossil, 8 USNM; Kemmerer, 1 USNM;* Opal, 1 USNM; 2 mi. S, 2 mi. W Kemmerer, 6700 ft., 1; Cumberland, 9 USNM.

Natrona Co.: 14 mi. N, 1 mi. W Waltman, 6225 ft., 1; *16 mi. S, 11 mi. W Waltman, 6950 ft., 1;* 19 mi. S, 9 mi. W Waltman, 7400 ft., 3; Poison Spider Creek, 1 USNM.

Sublette Co.: 1½ mi. NE Pinedale, 7400 ft., 3; *2 mi. W Pinedale, 1; 1½ mi. W Pinedale, 7200 ft., 6; Pinedale, 7300 ft:, 2 USNM; New Fork and Green River, 9 USNM;* 3 mi. S Boulder, 4; Big Sandy, 8000 ft., 2 USNM; Big Piney, 1 USNM.

Sweetwater Co.: Farson, 6580 ft., 2; 5 mi. SE Sand Dunes, Red Desert, 3; *27 mi. N, 37 mi. E Rock Springs, 6700 ft., 1;* 20 mi. SSE Sand Dunes, Red Desert, 1; *Superior, 4 USNM;* 13 mi. N, 6 mi. W Rock Springs, 6950 ft., 1; Green River, 3 USNM; Kinney Ranch, 21 mi. S Bitter Creek, 6800 ft., 3, 3 USNM; 32 mi. S, 22 mi. E Rock Springs, 7025 ft., 2; *3 mi. W Green River, 2 mi. N Utah boundary, 1.*

Uinta Co.: Evanston, 4 USNM; *Ft. Bridger, 18 USNM; Mountainview, 2 USNM;* ½ mi. S Mountainview, 6900 ft., 1; *10 mi. N Lonetree, 2 USNM; 2 mi. E Robertson, 7200 ft., 1;* 14½ mi. S, 1 mi. E Evanston, 6900 ft., 1; Lonetree, 7 USNM.

Additional records (A. H. Howell, 1938:77).—**Albany Co.:** *Pole Mtn., 15 mi. SE Laramie.*

Spermophilus armatus Kennicott

Uinta Ground Squirrel

1863. *Spermophilus armatus* Kennicott, Proc. Acad. Nat. Sci. Philadelphia, 15:158.

Type.—No holotype; eight cotypes from foothills of Uinta Mountains, "near" Fort Bridger, Wyoming.

Description.—Large for genus (see *Measurements*); resembles *S. richardsonii* but usually darker and larger in cranial and external dimensions; underside of tail gray and lacking Ochraceous-Buff; anteriormost lateral borders of nasals deflected ventrad.

Measurements.—Average and extreme external and cranial measurements of 11 males and eight females, respectively, from within three miles of Pinedale, Sublette County, are as follows: Total length, 281.8 (257-308), 269.7 (256-284); length of tail, 68.3 (62-74), 65.7 (61-72); length of hind foot, 42.3 (40-45), 41.6 (40-43); greatest length of skull, 45.0 (43.5-47.7), 43.6 (42.8-44.2); zygomatic breadth, 29.0 (27.7-30.7), 28.4 (27.7-30.1); postorbital breadth, 11.3 (10.6-12.5), 11.8 (11.2-13.0); length of nasals, 15.7 (14.0-16.9), 15.4 (14.8-15.9); maxillary tooth-row, 9.8 (9.2-10.3), 9.8 (9.6-10.3).

Distribution.—Western third of state, usually occurring in Canadian, Transition, and Upper Sonoran life-zones. At least one specimen (US 55429) from Wyoming Peak, Lincoln County, is from the Hudsonian Life-zone (10,900 ft.). See Fig. 24.

Remarks.—A melanistic, lactating female and a melanistic young were obtained on the same day (May 29, 1912) from the same place (Mountainview).

Records of occurrence.—Specimens examined, 178, as follows: **Fremont Co.:** 1 mi. S, 3 mi. E Dubois, 6900 ft., 1; *Jackeys Creek, 1 USNM.*

Lincoln Co.: 3 mi. N, 11 mi. E Alpine, 5650 ft., 1; 12 mi. N, 2 mi. W Afton, 6100 ft., 3; *9 mi. N, 2 mi. W Afton, 6100 ft., 1;* Afton, 10 USNM; Wyoming Peak, 10,900 ft., 1 USNM; Border, 5 USNM; *Cokeville, 6 USNM;* Hamsfork, 2 USNM; Kemmerer. 7 USNM; Opal, 4 USNM.

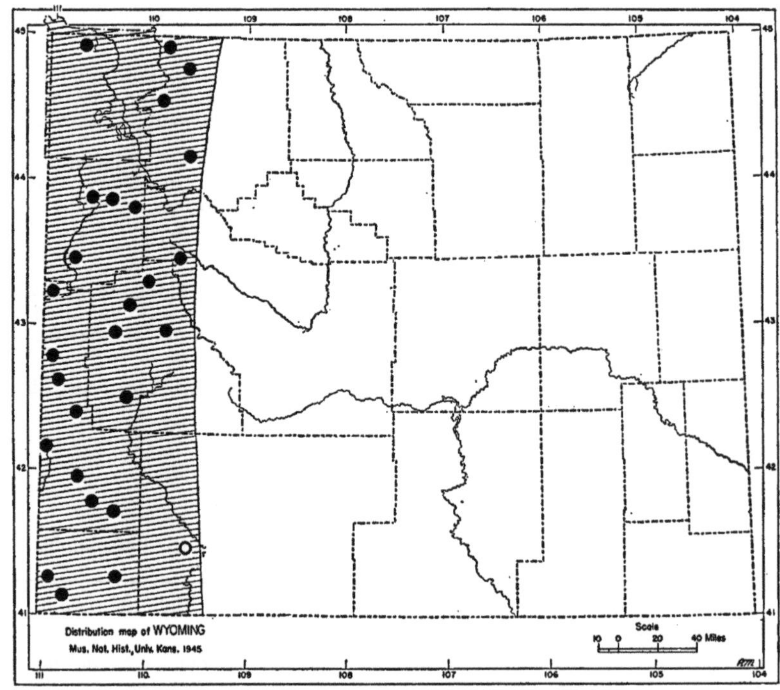

FIG. 24. Distribution of *Spermophilus armatus*.

Park Co.: 29 mi. N, 32 mi. W Cody, 7200 ft., 1; 16¾ mi. N, 17 mi. W Cody, 5625 ft., 7; *14¾ mi. N, 12 mi. W Cody, 5625 ft., 1; 12 mi. N, 8 mi. W Cody, 5625 ft., 1;* Mouth Grinnell Creek, 1 USNM; *25 mi. S, 28 mi. W Cody, 6350 ft., 2;* Valley, 6500 ft., 4 USNM.

Sublette Co.: 31 mi. N Pinedale, 8025 ft., 1; *12 mi. N Kendall, 7700 ft., 2 USNM;* 15 mi. N, 7 mi. W Pinedale, 8025 ft., 2; *8 mi. N, 5 mi. E Pinedale, 8900 ft., 2; 5 mi. N, 3 mi. E Pinedale, 7500 ft., 1;* 4 mi. N, 3 mi. E Pinedale, 7600 ft., 1; *W. end Half Moon Lake, 7900 ft., 3; 2¾ mi. NE Pinedale, 7500 ft., 18; 1½ mi. NE Pinedale, 7400 ft., 16; 1½ mi. W Pinedale, 7200 ft., 1;* Merna, 1 USNM; 3 mi. W Stanley, 1 USNM.

Teton Co.: *Pilgrim Creek, 3 mi. N Moran, 6800 ft., 1 Univ. Wyo.; ½ mi. N, 1 mi. E Moran, 6742 ft., 1;* Moran, 2 USNM; *Jackson Hole Wildlife Park, 9 USNM;* 9 mi. E Moran, 6794 ft., 1, 3 Univ. Wyo.; *1 mi. S, 4¾ mi. E Moran, 6800 ft., 2; 2 mi. S, 5 mi. E Moran, 6800 ft., 4 Univ. Wyo.;* 2 mi. S, 19 mi. E Moran, 1; 1 mi. E Elk, 7000 ft., 1, 1 Univ. Wyo.; Bar BC Ranch, 2½ mi. NE Moose, 6500 ft., 4; N Fork Gros Ventre River, 1 USNM; Jackson, 1 USNM; *Jackson Hole, 2 USNM;* Not found: *"near" Elk Ranch, 3; 1 mi. S Moosehead Ranch, 1.*

Uinta Co.: *Ft. Bridger, 7 USNM; Mountainview, 7 USNM;* 1/10 mi. S Mountainview, 1; *Springvalley, 2 USNM;* Evanston, 4 USNM; 7 mi. S, 6 mi. E Evanston, 7400 ft., 1; *2 mi. E Robertson, 7200 ft., 1; 14½ mi. S, 1 mi. E Evanston, 6900 ft., 1; 5 mi. S, 2½ mi. E Robertson, 8000 ft., 1; 8 mi. S, 2½ mi. E Robertson, 8300 ft., 2.*

Yellowstone Nat'l Park: Mammoth Hot Springs, 1 USNM; *12 mi. NW Tower Falls, 1;* Locality not specified, 3.

Additional records (A. H. Howell, 1938:80-81, unless otherwise noted).—

Park Co.: *Clarks Fork (opposite Crandall Creek).* **Sublette Co.:** *Daniel.* **Sweetwater Co.:** The Green River, 5 mi. W Green River (R. D. Svihla, 1931:260). **Teton Co.:** *Gros Ventre Mtns., Water Dog Lake.*

Spermophilus spilosoma obsoletus Kennicott

Spotted Ground Squirrel

1863. *Spermophilus obsoletus* Kennicott, Proc. Acad. Nat. Sci. Philadelphia, 15:157.

1938. *Citellus spilosoma obsoletus,* A. H. Howell, N. Amer. Fauna, No. 56, p. 130, May 18.

1955. *Spermophilus spilosoma obsoletus,* Hall, Univ. Kansas Mus. Nat. Hist., Misc. Publ., 7:94, December 13.

Type.—No holotype. Lyon and Osgood (1909:169) list seven specimens as those named by Kennicott. Howell (May 18, 1938:131) designated U. S. N. M. 3222/37998 as lectotype and restricted the type locality thereby to "50 miles west of Fort Kearny, Nebraska."

Description.—Size small for genus (see *Measurements*); Avellaneous or Light Pinkish Cinnamon upper parts having numerous small whitish spots; whitish underparts; antiplantar surfaces whitish or yellowish-white; tail and area immediately posterior to nostrils more reddish than dorsum; pelage over-all markedly pale but slightly more reddish when fresh or juvenal; skull smaller and relatively narrower than in *S. richardsonii* but relatively broader than in *S. tridecemlineatus;* supraoccipital often extending more posteriorly than exoccipital condyles; tympanic bullae large; posterior margin of postorbital process often rugose.

Measurements.—External and cranial measurements of three adult females and one adult male, respectively, from 3 miles W Guernsey, 4450 ft., are as follows: Total length, 228, 221, 233, 225; length of tail, 67, 64, 68, 71; hind foot, 30, 29, 31, 33; ear from notch, 8, 6, 7, 8; greatest length of skull (anteriormost part of nasals to posteriormost part of supraoccipital), 40.2, 40.0, —, 39.9; zygomatic breadth, 23.8, 24.2, —, 23.2; breadth of braincase, 17.7, 18.2, —, 18.3; postorbital breadth, 13.3, 13.7, —, 13.3; length of nasals, 14.1, 13.5, —, 14.0; maxillary tooth-row, 7.3, 7.0, —, 7.7.

Distribution.—Prairies in extreme eastern Wyoming. See Fig. 20.

Remarks.—One female (KU 91086) taken on June 6, 1950, three miles west of Guernsey by A. B. Mickey contained nine embryos measuring 9 mm. in length. Another from there was lactating on August 22, 1951.

Records of occurrence.—Specimens examined, 12, as follows: **Goshen Co.:** 6 mi. SW Spoon Butte, 4800 ft., 1 USNM; Fort Laramie, 2 USNM. **Laramie Co.:** Little Bear Creek [almost] 20 mi. SE Chugwater, 5500 ft., 1 USNM. **Platte Co.:** 3 mi. W Guernsey, 4450 ft., 7; 3 mi. E Wheatland, 1 USNM.

Spermophilus tridecemlineatus (Mitchill)

Thirteen-lined Ground Squirrel

The Thirteen-lined ground squirrel is the only rodent in the state having longitudinal stripes and spots on the upper parts. Pale stripes alternate with dark, brownish stripes on the dorsum; a row of nearly square, whitish spots occurs in each of the dark stripes. The venter is whitish or buffy. The supraoccipital bone is the posteriormost part of the skull. See accounts of subspecies.

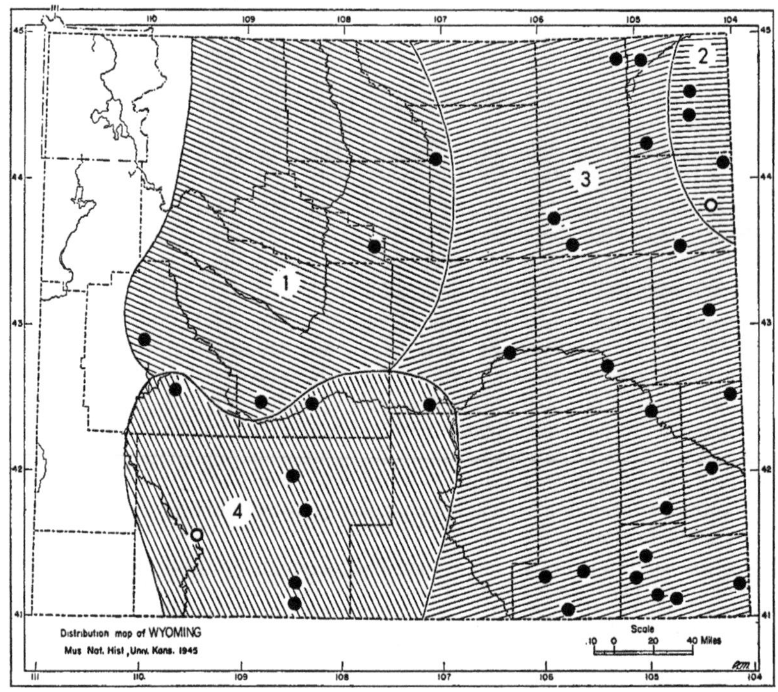

FIG. 25. Distribution of *Spermophilus tridecemlineatus*.
1. *S. t. alleni* 3. *S. t. pallidus*
2. *S. t. olivaceus* 4. *S. t. parvus*

Spermophilus tridecemlineatus alleni Merriam

1898. *Spermophilus tridecemlineatus alleni* Merriam, Proc. Biol. Soc. Washington, 12:71, March 24.

Type.—From near head of Canyon Creek, west slope of Bighorn Mountains, 8000 ft., Wyoming. The type (US 56050), male, is only subadult.

Comparisons and description.—Smaller in external and cranial dimensions than *S. t. pallidus;* skull relatively longer with smaller auditory bullae and markedly longer nasals than in *S. t. parvus;* darkest subspecies in Wyoming; dark dorsal stripes Mummy Brown; pale spots and stripes grayish white; front of face Cinnamon Buff; sides of face Pinkish Buff washed with Fuscous; antiplantar surfaces Pinkish Buff to Tilleul Buff; thighs Cinnamon Buff washed with Snuff Brown; pattern and color on upper side of tail at base confluent with that of dorsum, but becoming Fuscous Black margined with buff or whitish hairs distally; underside of tail Cinnamon or Pinkish Cinnamon, intermixed with whitish and Fuscous Black hairs; venter whitish washed with Pinkish Buff (after A. H. Howell, 1938:115).

Measurements.—Average and extreme external and cranial measurements of one male and two females ("adults" and subadult) from Bighorn Mountains and Bighorn Basin are as follows: Total length, 206.3 (203-211); tail vertebrae, 74 (73-75); hind foot, 31 (30-32); greatest length of skull, 36.4 (35.8-36.8); palatal length, 16.4 (16.2-16.5); zygomatic breadth, 19.9 (19.5-20.1); cranial breadth, 16 (16.0-16.1); interorbital breadth, 7.5 (7.1-7.7); postorbital constriction, 11.3 (10.8-11.6); length of nasals, 13.2 (12.0-14.5); maxillary tooth-row, 6.6 (6.4-6.8) (after A. H. Howell, 1938:115).

Distribution.—Only six specimens have been reported from four localities in western Wyoming, all from the Transition Life-zone. A. H. Howell (1938: 114) states: "The Bighorn ground squirrel [= S. *t. alleni*] is an inhabitant of mountains and foothills, and is decidedly darker than the races living on the plains. The limits of its range are not well known." See Fig. 25.

Remarks.—The status of S. *t. alleni* has not been clarified by field and laboratory studies since A. H. Howell's study (1938:114-115). In fact, intensive collecting of vertebrates within the geographic range of this subspecies by field parties from the Museum of Natural History of the University of Kansas has yielded no additional specimens. E. R. Hall (personal comm.) told me he thought that poisoning of mammals in Wyoming may have exterminated this subspecies.

Records of occurrence.—Specimens examined, 6, as follows: **Fremont Co.:** Miners Delight, near head of Twin Creek, 1 USNM. **Hot Springs Co.:** Head of Kirby Creek, 1 USNM. **Johnson Co.:** West slope of Bighorn Mtns., near head of Canyon Creek, 2 USNM. **Sublette Co.:** New Fork of Green River (Lander Road), 2 USNM.

Spermophilus tridecemlineatus olivaceus J. A. Allen

1895. *Spermophilus tridecemlineatus olivaceus* J. A. Allen, Bull. Amer. Mus. Nat. Hist., 7:337, November 8.

Type.—From Custer, Custer County, Black Hills, South Dakota.

Comparison and description.—Resembling *Spermophilus tridecemlineatus pallidus* in size, but pinna of ear slightly longer; darker than S. *t. pallidus*, showing more olivaceous on dorsum as follows: dorsal dark stripes brownish intermixed with reddish, golden-olive, and Fuscous Black; pale dorsal stripes pale golden-olive or yellowish-buff (one specimen, KU 41420, reddish olive); dorsal spots resembling pale dorsal stripes in color and less prominent or distinct than in S. *t. pallidus;* postauricular patches pale olive-buff to Cinnamon-Buff; ring around eye Cinnamon-Buff, olive-buff to Light Ochraceous-Buff; tail above brownish, strongly intermixed with Fuscous Black and margined with Light Ochraceous-Buff to Pinkish Cinnamon; venter Light Ochraceous-Buff to Pinkish Buff (but more reddish in KU 41420); underside of tail resembling dorsal side, except often rich ochraceous medially. Skull resembling that of S. *t. pallidus* except nasals of S. *t. olivaceus* averaging slightly shorter relatively and auditory bullae of S. *t. olivaceus* markedly larger and more nearly diskshaped (see *Measurements*) than in specimens examined of S. *t. pallidus* from extreme southeastern Wyoming.

Measurements.—External and cranial measurements of five adults (KU 41418, 41675-41678) from the Black Hills in Wyoming are as follows: Total length, 239.0 (234-242); length of tail, 72.0 (65-80); hind foot, 31.8 (30-34); ear from notch, 8.4 (7-9); greatest length of skull, 38.7 (37.5-40.5); zygomatic breadth, 22.7 (22.1-23.6); postorbital breadth, 10.9 (10.0-11.5); length of nasals, 13.0 (12.1-14.0); maxillary tooth-row, 7.2 (6.4-7.5); greatest diameter of auditory bulla, 8.16 (8.0-8.7).

Distribution.—Black Hills in northeastern Wyoming. See Fig. 25.

Remarks.—A. H. Howell (1938:113) placed *S. t. olivaceus* in the synonymy of *S. t. pallidus.* He stated: "The type series of *olivaceus* from Custer, S. Dak., has been compared with a large series of typical *pallidus* and is found to agree closely with it." It should be pointed out that Howell himself fixed the type locality of *S. t. pallidus,* designating the "mouth of the Yellowstone" as the type locality to be used instead of the "plains of the Lower Yellowstone River" designated by J. A. Allen. When Allen named *S. t. pallidus,* he commented on its geographic range as follows (1895:338): ". . . I would restrict *pallidus* to the arid region of the Plains, from the Upper Missouri southward to eastern Colorado, western Kansas, etc., and designate its type region the plains of the Lower Yellowstone River."

When Howell wrote "a large series of typical *pallidus,*" I think he was referring to specimens from throughout the geographic range of *S. t. pallidus* having the chief characteristics, especially paleness, of this subspecies. He refers to a series of topotypes as a "type series" instead of referring to the series as merely typical *pallidus.* Also, no large series of topotypes of *S. t. pallidus* existed. Howell mentioned only four specimens from the vicinity of the type locality as follows: a single specimen from the type locality (the lectotype) and three specimens from 26 miles "above" the type locality (A. H. Howell, 1938:114).

I have compared a large series (17 specimens) of spermophiles from the Black Hills (in Wyoming) and closely adjacent areas westward and southwestward with series of specimens of *S. t. pallidus* from Nebraska, southeastern Wyoming, extreme northwestern South Dakota, and one specimen from southeastern Montana. My comparison reveals that spermophiles from the Black Hills and nearby areas differ from *S. t. pallidus* as J. A. Allen (1895:337) said they did, namely in olivaceous, darker color (see *Comparisons*). Additionally, I find the spermophiles from the Black Hills to have larger auditory bullae and larger ears but relatively shorter nasals on the average than do specimens of *S. t. pallidus* from the surrounding plains. Slightly longer external ears may be related to larger auditory bullae in the spermophiles from the Black Hills. Therefore, in order to represent variation observed in all of these specimens, I choose to assign the dark, olivaceous specimens from Wyoming to *Spermophilus tridecemlineatus olivaceus* J. A. Allen, having its type locality at Custer, Black Hills, South Dakota.

The most noticeable characteristic of *S. t. olivaceus* is its dark color, which also characterizes *Eutamias minimus silvaticus* White, another sciurid endemic to the Black Hills. *S. t. olivaceus* occurs in the Transition Life-zone of the Black Hills. I have seen *S. t. olivaceus* and *E. m. silvaticus* so closely associated in Custer County, South Dakota, that the latter could be chased into the burrows of the former.

Spermophiles of this species examined from Moorcroft and southern Campbell County, Wyoming, seem to be intergrades between *S. t. olivaceus* and *S. t. pallidus* with regard to color. They are assigned to *S. t. pallidus*.

Records of occurrence.—Specimens examined, 20, as follows: **Crook Co.:** 15 mi. N Sundance, 5500 ft., 3; 1½ mi. NW Sundance, 5000 ft., 6; *1 mi. N Sundance, 1; Sundance, 6000 ft., 5 USNM; Bear Lodge Mtns., 1 USNM.* **Weston Co.:** 1½ mi. E Buckhorn, 6150 ft., 4.

Additional record (A. H. Howell, 1938:114).—**Weston Co.:** Newcastle.

Spermophilus tridecemlineatus pallidus J. A. Allen

1874. *[Spermophilus tridecemlineatus]* var. *pallidus* J. A. Allen, Proc. Boston Soc. Nat. Hist., 16:291, February 4. (A *nomen nudum*).

1877. *[Spermophilus tridecemlineatus]* var. *pallidus* J. A. Allen, *In* Coues and Allen, Monogr. N. Amer. Rodentia, p. 872, August.

Type.—None; type locality designated as "Plains of Lower Yellowstone River," Montana, by J. A. Allen (1877:872); specimen from mouth of Yellowstone River [in North Dakota] designated as lectotype by A. H. Howell (1938:112).

Comparisons.—Large for spermophiles of Wyoming (see *Measurements*); resembling *Spermophilus tridecemlineatus olivaceus* in size except pinna of ear, which is shorter, and auditory bulla, which is smaller but more inflated; larger than *S. t. alleni* or *S. t. parvus;* paler than *S. t. alleni;* lacking reddish of *S. t. parvus;* lacking olivaceous above and usually lacking buffy underparts (being whitish) of *S. t. olivaceus;* pale dorsal stripes and spots distinct (see accounts of other subspecies of *S. tridecemlineatus*).

Measurements.—Average and extreme external and cranial measurements of three adult males (KU 14921, 25305, 25308) and two adult females (KU 25312 and 91094) from Laramie and Goshen counties are as follows: Total length, 240.0 (237-243); length of tail, 77.2 (72-83); hind foot, 32.8 (32-34); ear from notch, 7.6 (6-9); greatest length of skull, 38.6 (37.3-39.8); zygomatic breadth, 22.8 (21.5-23.5); postorbital breadth, 10.7 (10.5-11.1); length of nasals, 13.8 (12.9-14.6); alveolar length of maxillary tooth-row, 6.8 (6.6-7.1); cranial depth, from parietals to base of cranium, 12.3 (11.7-12.7); greatest diameter auditory bulla, 7.4 (6.6-7.9).

Distribution.—Eastern Wyoming, especially the plains, except in the Black Hills area; intergrades with *S. t. parvus* in the area of confluence of the Sweetwater River with the North Platte, and tends toward that subspecies in size in the vicinity of the Laramie Plains; intergrades with *S. t. olivaceus* in southern Campbell County and at nearby Moorcroft. See Fig. 25.

Remarks.—One female (Univ. Wyo., no number), from 16 miles south and 8 miles west of Laramie, 7350 ft., contained 12 embryos.

Records of occurrence.—Specimens examined, 73, as follows: **Albany Co.:** 5 mi. N Laramie, 7200 ft., 2; *5 mi. N Laramie, 7400 ft., 2; 2 mi. N Laramie, 7250 ft., 1;* 22 mi. W Laramie, 7200 ft., 1; *Laramie, 1; Centennial, 1; 2½ mi. SE Laramie, 7100 ft., 2; Laramie Plains, 1;* 16 mi. S, 8 mi. W Laramie, 7350 ft., 1. **Campbell Co.:** 1½ mi. N, ½ mi. E Rockypoint, 3800 ft., 1; *Rockypoint, 1;* 42 mi. S, 16 mi. W Gillette, 5375 ft., 1; *44½ mi. S, 15½ mi. W Gillette, 5432 ft., 3; Belle Fourche River, 45 mi. S, 13 mi. W Gillette, 5350 ft., 1;* 23 mi. N, 6 mi. W Bill, 1. **Converse Co.:** Douglas, 2 USNM. **Crook Co.:** *4 mi. S, 7 mi. E Rockypoint,* 1; 5 mi. S, 6 mi. E Rockypoint, 3900 ft., 1; Moorcroft, 5 USNM. **Goshen Co.:** 6 mi. N Spoon Butte, 4800 ft., 1 USNM; 4 mi. S Lingle, 1. **Laramie Co.:** 8 mi. E Horse Creek P. O., 6400 ft., 1; Islay, 1 USNM; *Pinebluffs, 3 USNM;* 2 mi. S Pinebluffs, 5200 ft., 7; *7 mi. W Cheyenne, 6500 ft., 1;* Cheyenne, 6100 ft., 1, 14 USNM; 2 mi. S, 9½ mi. E Cheyenne, 5200 ft., 1. **Natrona Co.:** Casper, 4700 ft., 4 USNM. **Niobrara Co.:** 17 mi. N, 4 mi. E Lusk, 5200 ft., 1. **Platte Co.:** Cassa, 1 USNM; Chugwater, 5500 ft., 1 USNM. **Weston Co.:** 23 mi. SW Newcastle, 4500 ft., 3, 3 USNM.

Spermophilus tridecemlineatus parvus J. A. Allen

1895. *Spermophilus tridecemlineatus parvus* J. A. Allen, Bull. Amer. Mus. Nat. Hist., 7:337, November 8.

Type.—From Kennedys Hole, Uncompahgre Indian Reservation, 20 miles northeast of Ouray, Uintah County, Utah.

Comparisons.—Smaller than *Spermophilus tridecemlineatus pallidus* or *S. t. olivaceus,* but resembling *S. t. alleni* in size (see *Measurements*) except skull relatively shorter with larger auditory bullae and markedly shorter nasals; more reddish upper parts (especially dark dorsal lines) than on other spermophiles in state; paler than *S. t. alleni.*

Measurements.—External and cranial measurements of an adult male and adult female from 25 mi. N, 38 mi. E Rock Springs are, respectively, as follows: Total length, 202, 205; length of tail, 69, 70; hind foot, 29, 30; ear from notch, 9, 10; greatest length of skull, 34.4, 35.2; zygomatic breadth, 20.0, 19.4; postorbital breadth, 10.7, 11.0; length of nasals, 9.2, 9.6; maxillary tooth-row, 5.9, 6.3.

Distribution.—Southwestern part of state except extreme western portion (see Fig. 25); frequently occurs below Transition Life-zone.

Records of occurrence.—Specimens examined, 17, as follows: **Fremont Co.:** Myersville, 1 USNM. **Natrona Co.:** Independence Rock, 1 USNM; *Sun, 1 USNM.* **Sublette Co.:** Big Sandy, 7500 ft., 1 USNM; *Big Sandy Creek, 1 USNM.* **Sweetwater Co.:** 27 mi. N, 37 mi. E Rock Springs, 6700 ft., 1; *25 mi. N, 38 mi. E Rock Springs, 6700 ft., 5;* 20 mi. SSE Sand Dunes, Red Desert, 1; Kinney Ranch, 21 mi. S Bitter Creek, 6800 ft., 3; 33 mi. S Bitter Creek, 6900 ft., 2.

Additional record (A. H. Howell, 1938:119).—**Sweetwater Co.:** Green River.

Spermophilus lateralis (Say)

Golden-mantled Ground Squirrel

Medium sized ground squirrel with dorsolateral whitish stripe on each side, bordered by dark stripes (excepting *Spermophilus lateralis wortmani* and *S. l. lateralis,* which lack inner, dark dorsal stripes); dorsum brownish, grayish, or tawny; sides immediately ventral to lateral stripes grayish, buffy, tawny, or pale brownish; venter buffy or pale tawny; underside of tail ochraceous, tawny, or buffy; tail above brownish or intermixed with Fuscous Black (paler in *S. l. wortmani*); shoulders and posterior portion of head washed strongly with tawny, Cinnamon-Buff, to russet (on so-called mantle); supraoccipital forming posteriormost part of skull.

See accounts of subspecies for detailed descriptions. These squirrels resemble chipmunks. Many persons confuse the two genera. Golden-mantled ground squirrels can be distinguished from chipmunks by larger size and lack of stripes on the head.

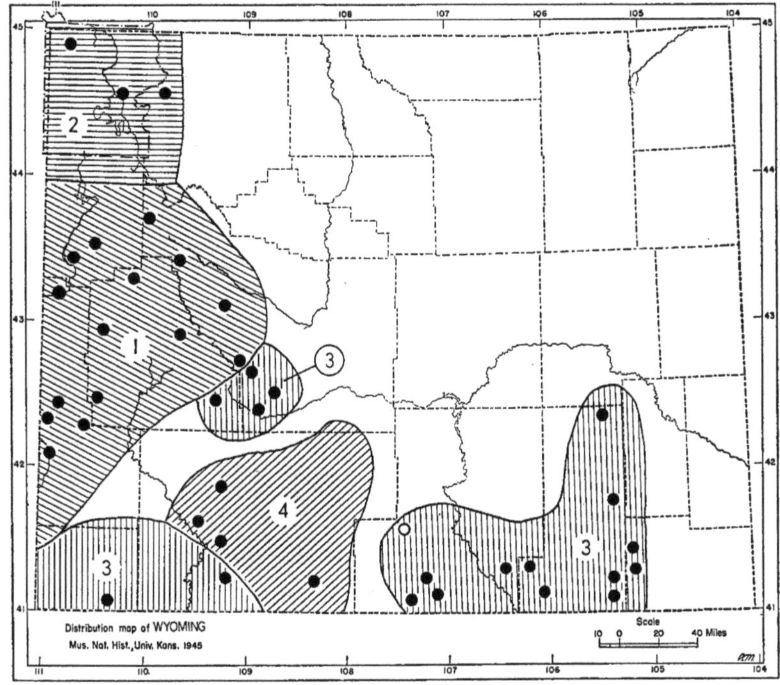

FIG. 26. Distribution of *Spermophilus lateralis.*

1. *S. l. castanurus* 3. *S. l. lateralis*
2. *S. l. cinerascens* 4. *S. l. wortmani*

Spermophilus lateralis castanurus (Merriam)

1890. *Tamias castanurus* Merriam, N. Amer. Fauna, 4:19, October 8.

1917. *Callospermophilus lateralis caryi* A. H. Howell, Proc. Biol. Soc. Washington, 30:105, May 23. Type from 7 mi. S Fremont Pk., Wind River Mtns., Wyoming.

1938. *Citellus lateralis castanurus*, A. H. Howell, N. Amer. Fauna, 56:201, May 18.

1959. *Spermophilus lateralis castanurus*, Hall and Kelson, Mammals of N. America, vol. 1, p. 361, March 31.

Type.—From Park City, Wasatch Mountains, Summit County, Utah.

Comparisons.—Differs from *S. l. cinerascens* in shorter hind foot and lesser size otherwise (see *Measurements*); pelage less grayish; underside of tail usually darker than in surrounding subspecies, but variable and paler in Wind River Mountains and closely surrounding areas; tail relatively shorter than in *S. l. cinerascens;* inner dark stripes distinct instead of reduced or absent as in *S. l. lateralis* and *S. l. wortmani;* skull smaller than in surrounding subspecies.

Measurements.—See Table 6.

Distribution.—Western part of state, extending eastward into Wind River Mountains (see Fig. 26).

TABLE 6.—EXTERNAL AND CRANIAL MEASUREMENTS OF *Spermophilus lateralis castanurus.*

Sex	Catalog number or number of individuals averaged	Total length	Tail vertebrae	Hind foot	Ear from notch	Greatest length of skull	Zygomatic breadth	Postorbital breadth	Length of nasals	Maxillary tooth-row
		Mosquito Park Ranger Sta., 2½ mi. N, 17½ mi. W Lander (old adults)								
♂	KU 32372...	269	120	40	19	43.2	26.2	12.3	14.5	9.4
♂	KU 32378...	270	95	42	17	42.9	25.2	12.8	14.3	8.2
♂	KU 32377...	248	110	36	15	....	....	....	14.6	...
		9 mi. N, 5 mi. E Pinedale, 9100 ft., Sublette Co.								
♀♀	Average of 3 old adults...	280.7	98.3	41.7	20	42.5	25.8	12.1	13.9	7.9
	Max........	288	104	42	20	42.9	26.5	13.0	14.2	8.0
	Min.........	275	92	41	20	41.9	24.8	11.5	13.5	7.9
		Gros Ventre Slide, 3 mi. E Kelly P.O., Teton Co.								
♂	KU 91074...	291	103	42	22	43.0	27.1	13.2	14.2	8.5
♂	KU 91073...	268	90	41	18	42.0	26.4	13.2	13.1	8.2
		Near topotypes from Utah								
2♂♂ 2♀♀	Average of 4 old adults...	277.5	93.0	39.3	19	42.8	26.3	12.5	13.9	8.2
	Max........	287	97	40	24	45.0	28.0	13.5	14.8	8.5
	Min.........	270	89	38	14	41.3	25.5	11.8	13.4	8.0

Remarks.—Because the geographic range of *S. l. castanurus* extends eastward into the Wind River Mountains, this subspecies is interposed between two other subspecies, *S. l. cinerascens* to the northward and *S. l. lateralis* to the southward. *S. l. lateralis* occurs in the Wind River Mountains as a relict population, isolated from a larger, wide-spread population of the same subspecies to the southward. Howell (1938:193-194) mentioned a series of four intergrades between this relict population of *S. l. lateralis* and *S. l. castanurus* of the Wind River Mountains (Howell regarded these latter squirrels as another subspecies, which he named *S. l. caryi*). My comparison of near topotypes of *S. l. castanurus* from Utah with several series of squirrels formerly known as *S. l. caryi* leads me to conclude that the squirrels known as *S. l. caryi* are referable to the subspecies *S. l. castanurus.* Howell (1917:105, 1938:197) mentioned only three characters as differentiating *S. l. caryi* from *S. l. castanurus:* dorsum more grayish; pale stripes often white instead of yellowish especially in winter pelage; and underside of tail paler. The skull may be slightly broader in topotypes of *S. l. castanurus* than in topotypes of *S. l. caryi,* and the maxillary tooth-row averages slightly longer (see *Measurements*).

With regard to the pale stripes, it is true that specimens of *S. l. castanurus* lack the white stripes of topotypes of *S. l. caryi.* These white stripes occur most frequently in winter pelage; in summer pelage of specimens previously referred to as *S. l. caryi* that I have examined, the white stripes may be half replaced by more yellowish stripes owing to molt, and in summer pelage yellowish stripes may occur (Howell, 1938:197). The change from new yellowish hairs to old white hairs probably results from exposure and wear.

The more grayish dorsum of *S. l. caryi* was well described by Howell (1938:197) when he stated that *S. l. caryi* was, "similar to *C. l. castanurus,* but paler and more grayish on the back. Compared with *C. l. lateralis*: Head and shoulders in summer pelage darker and more extensively tawny." *S. l. caryi* differs from *S. l. castanurus* only to the degree that some specimens of *caryi* resemble those of *S. l. lateralis,* and I consider this resemblance to result from intergradation between *S. l. castanurus* and *S. l. lateralis.*

The underside of the tail in *S. l. caryi* is not nearly so constantly pale as Howell thought. Furthermore, this pallor, too, may result from intergradation between *S. l. castanurus* and *S. l. lateralis,* and perhaps *S. l. cinerascens.* In both *S. l. lateralis* and *S. l. cinerascens* the underside of the tail is paler than in *S. l. castanurus.* Howell (1938:193-194) mentioned that in three of four intergrades (*S. l.*

lateralis X *S. l. caryi*) from Big Sandy, Wyoming, the underside of the tail was paler than in *S. l. caryi* and resembled that of *S. l. lateralis.* I observed several comparable degrees of darkness and pallor in near topotypes of *S. l. castanurus* and *S. l. caryi.* Scaled from one to five, on the basis of palest to darkest undersides of tail, the following average values for pallor or darkness increased toward darkness of tail in direct ratio to distance away from the area of intergradation mentioned by Howell: 4 specimens from ½ mi. S, 17½ mi. W Lander, mean 1.3, range in paleness, 1-2; 13 from 9 mi. N, 5 mi. E Pinedale, 2.5 (1-5); 8 from 2 mi. S, 19 mi. W Big Piney, Lincoln County, 2.8 (1-5); 2 from Gros Ventre Slide, Teton County, 3.0 (1-5); 6 from Brooks Mtn. and Togwotee Pass, 9800 ft., 3.5 (2-5); 14 near topotypes from Utah, 4.0 (1-5).

All specimens formerly known as *S. l. caryi* are here assigned to *S. l. castanurus.* The subspecific boundary between *S. l. castanurus* and *S. l. lateralis* can be conveniently drawn in Wyoming (and in other areas) between a locality from which specimens have dark, inner dorsal stripes and a locality from which specimens lack inner, dark stripes.

A female (KU 14949) taken on August 8, 1945, by E. R. Hall and H. H. Hall nine miles north and five miles east of Pinedale has moderately worn teeth and the usual number of teeth for this species, but each P4 is fused anteriorly to rooted dental material, which presents an added, flattened, sigmoid loop in occlusal view. The remaining teeth appear to be normal.

Records of occurrence.—Specimens examined, 74, as follows: **Fremont Co.:** *Brooks Mtn., 1;* ½ mi. N, 1 mi. E Togwotee Pass, 9800 ft., 1; *Jackeys Creek, 4 mi. S Dubois, 8500 ft., 1 USNM;* 5 mi. S Dubois, 8500-9500 ft., 2 USNM; Bull Lake, 9600 ft., 1 USNM; *Lake Fork, 10,600-10,800 ft., 3;* Mosquito Park Ranger Sta., 2½ mi. N, 17½ mi. W Lander, 9500 ft., 5; *½ mi. S, 17½ mi. W Lander, 3.*

Lincoln Co.: 3 mi. N, 9 mi. E Alpine, 1; 10 mi. SE Afton, 7000 ft., 2 USNM; *Hd. Dry Creek, Salt River Mtns., 9500 ft., 2 USNM;* Hd. Smiths Fork, 3 USNM; La Barge Creek, 9000 ft., 2 USNM; Cokeville, 1 USNM.

Sublette Co.: 12 mi. NW Kendall, 10,000 ft., 1 USNM; Merna, 1 USNM; *5 mi. S Fremont Peak, 10,600 ft., 1 USNM;* 7 mi. S Fremont Peak, 10,400 ft., 1 USNM; *8 mi. S, 3 mi. W Fremont Peak, 10,300 ft., 3; 9 mi. N, 5 mi. E Pinedale, 9100 ft., 13; Middle Piney Lake, 9000 ft., 1 USNM;* 2 mi. S, 19 mi. W Big Piney, 7700 ft., 4.

Teton Co.: *Bobcat Ridge, 2 USNM; S Moose Creek, 10,000 ft., 1 USNM; Togwotee Pass, 4, 3 USNM;* Gros Ventre Slide, 3 mi. E Kelly P. O., 6800 ft., 3, 3 Univ. Wyo.; *Sheep Mtn., 1;* Jackson, 3 USNM; *Flat Creek, 1 USNM.*

Spermophilus lateralis cinerascens (Merriam)

1890. *Tamias cinerascens* Merriam, N. Amer. Fauna, 4:20, October 8.

1938. *Citellus lateralis cinerascens,* A. H. Howell, N. Amer. Fauna, 56:198, May 18.

1959. *Spermophilus lateralis cinerascens,* Hall and Kelson, Mammals of N. America, vol. 1, p. 362, March 31.

Type.—From Helena, Montana.

Comparisons.—Resembling *Spermophilus lateralis castanurus*, but more grayish; sides of face and neck more ochraceous; underside of tail paler (less Tawny); hind foot longer; possessing inner dark stripes (absent or reduced in *S. l. lateralis* or *S. l. wortmani*); nasals shorter and skull larger than in *S. l. castanurus* (after A. H. Howell, 1938:198).

Measurements.—Average and extreme external measurements of one male and three females from "Yellowstone Park and vicinity" are as follows: Total length, 286 (270-297); tail vertebrae, 107 (95-118); hind foot, 43.6 (41-46); ear from notch (dry), 14.9 (14-16). Average and extreme cranial measurements of four adult females from Montana and Yellowstone Park are as follows: greatest length of skull, 43.9 (32.6-45); zygomatic breadth, 27.5 (26.3-29.3); cranial breadth, 20.1 (19.3-20.8); postorbital constriction, 13.3 (12.9-13.7); length of nasals, 14.9 (14.5-15.5); maxillary tooth-row, 8.9 (8.7-9.2) (after A. H. Howell, 1938:199).

Distribution.—Extreme northwestern part of state. See Fig. 26.

Records of occurrence.—Specimens examined, 8, as follows: Park Co.: Pahaska Tepee (Whirlwind Peak), 3 USNM. Yellowstone Nat'l Park: Golden Gate, 7000 ft., 4 USNM; Yellowstone Lake, 1 USNM.

Spermophilus lateralis lateralis (Say)

1823. *S[ciurus]. lateralis* Say, *in* Long, Account of an expedition . . . to the Rocky Mtns. . . ., 2:46.

1831. *Spermophilus lateralis*, Cuvier, Supplément à l'histoire naturelle génerale et particulière de Buffon, 1:335.

Type.—None designated by Say; description based on specimen obtained on Arkansas River, near Canyon City, Colorado (26 miles below Canyon City according to Merriam, Proc. Biol. Soc. Washington, 18:163, June 29, 1905).

Comparison.—Average in size (see *Measurements*) for the species; inner dark dorsal stripes reduced or absent; whitish stripes often tending toward Light Ochraceous-Buff or Light Buff; lateral dark stripe Mummy Brown or more commonly Fuscous Black; sides below lateral dark stripes grayish, tawny, or Light Ochraceous-Buff; mid-dorsal pelage of brownish, agouti hairs flecked with Light Buff, Ochraceous-Tawny, and Fuscous intermixed with Fuscous guard hairs (especially in fresh pelage) becoming somewhat Tawny or Ochraceous-Tawny on rump; venter Light Buff or whitish; tail above Fuscous Black intermixed with Light Buff or Ochraceous-Buff hairs; underside of tail Light Ochraceous-Buff or Ochraceous-Buff (not Tawny) margined by Fuscous Black; head and shoulders washed with Tawny or Ochraceous-Tawny, or occasionally yellowish Ochraceous-Buff; antiplantar surfaces Light Buff or whitish; juvenal pelage usually paler and more buffy; KU 14935 has much dark pigmentation lacking in pelage thereby being markedly reddish, and paler dorsally; nasals long, extending posterior to premaxillae. See accounts of other subspecies.

Measurements.—See Table 7.

Distribution.—Southeasternmost part of Wind River Mountains, extreme southwestern part of state, and mountainous areas in south-central Wyoming (Medicine Bow Mountains and Laramie Mountains). See Fig. 26 and *Remarks.*

Remarks.—*S. l. lateralis* and *S. l. wortmani* differ from the other subspecies of *S. lateralis* in small size, or absence, of the inner pair of dark dorsal stripes. *S. l. wortmani* differs from *S. l. lateralis* chiefly in being paler; cranially the two subspecies closely resemble each other.

S. l. wortmani is known only from the eastern part of the Green River watershed in Wyoming; in this low, arid area the pallor would seem to be an adaptation for survival. *S. l. lateralis* occurs in higher habitats (even in the Canadian Life-zone). Populations of *S. l. lateralis* almost surround *S. l. wortmani,* and the two subspecies seem to be more closely related to one another than to any of the other nearby subspecies.

One specimen (KU 14935) lacks much of the dark pigmentation from the dorsal pelage, having a more reddish, paler appearance than normal. It was obtained by H. W. Setzer on July 17, 1945, 2⅗ miles west of Horse Creek Post Office.

TABLE 7.—MEASUREMENTS OF THREE ISOLATED POPULATIONS OF *Spermophilus lateralis lateralis* IN WYOMING.

Locality, No. individuals, and sex	Total length	Tail vertabrae	Hind foot	Greatest length of skull	Zygomatic breadth	Postorbital breadth	Length of nasals	Maxillary tooth-row
Fremont County								
Average, adults, 3 females.....	274.3	89.0	41.7	44.8	27.5	12.1	16.2	8.6
Maximum......	279	100	43	45.6	28.0	12.2	16.4	8.8
Minimum......	267	82	41	44.3	27.0	11.8	15.9	8.3
Albany, Carbon and *Laramie counties*								
Average, adults, 1 male, and 4 females.....	276.8	88.3	41.3	43.4	26.8	12.3	15.1	8.4
Maximum......	286	99	43	43.8	27.8	12.7	15.5	8.7
Minimum......	260	78	40	42.7	25.9	11.8	14.6	8.1
Uinta County								
♂ subadult, KU 25374....	264	85	40	....	26.5	13.3	15.9	8.4
♀ subadult, KU 25373....	269	95	40	41.9	26.4	11.9	14.4	8.1

Records of occurrence.—Specimens examined, 67, as follows: **Albany Co.:** Springhill, 2 USNM; Laramie River, 1 USNM; *North Fork Camp Ground, sec. 28, T. 16N, R. 78W, 8500 ft., 1; 7 mi. W Centennial, 1 Univ. Wyo.;* Nash Fork Creek, 2 mi. W Centennial, 1; *Nash Fork Picnic Ground, sec. 13, T. 16N, R. 79W, 9850 ft., 2; ½ mi. E Barber Lake, 8700 ft., 1; 2 mi. S Browns Peak, 10,600 ft., 1; 3 mi. ESE Browns* Peak, *10,000 ft., 2;* Vedauwoo Glen, 10 mi. S,

11 mi. E Laramie, 8500 ft., 3 Univ. Wyo., 2; *Pole Mtn., 14-15 mi. SE Laramie, 8200-8300 ft., 3 USNM;* 3 mi. S Pole Mtn., 8100 ft., 1; *Woods P. O., 5 USNM;* 3½ mi. S Woods Landing, 8000 ft., 1; *Laramie Mtns., 1 USNM; Medicine Bow Mtns., 1 USNM.*

Carbon Co.: Lake Marie, 10,440 ft., 2; *1½ mi. S, 3 mi. W Medicine Bow Peak, 9400 ft., 1; 8 mi. N, 22 mi. E Encampment, 10,000 ft., 1;* Bridger Peak, 8800 ft., 3 USNM; *3½ mi. S, 10 mi. W Encampment, 9600 ft., 1;* 8 mi. N, 19½ mi. E Savery, 8800 ft., 7; 5 mi. N, 5 mi. E Savery, 6900 ft., 3.

Fremont Co.: 14 mi. S, 8.5 mi. W Lander, 1; *17 mi. S, 6½ mi. W Lander, 8450 ft., 2;* Miners Delight, 8400 ft., 1 USNM; *22 mi. S, 5½ mi. W Lander, 8800 ft. 1; 25 mi. S, 3 mi. W Lander, 9200 ft., 1; ½ mi. N, 3 mi. E South Pass City, 7900 ft., 2;* South Pass City, 1 USNM.

Laramie Co.: 2⅕ mi. W Horse Creek P. O., 6600 ft., 1; 6 mi. W Islay, 1 USNM.

Sublette Co.: Big Sandy, 6 USNM.

Sweetwater Co.: 5 mi. SW Maxon, 8000 ft., 1 USNM.

Uinta Co.: 9 mi. S Robertson, 8000 ft., 1; *10½ mi. S, 2 mi. E Robertson, 1.*

Additional records (A. H. Howell, 1938:194).—**Albany Co.:** *4 mi. N Jelm; Sherman.* **Carbon Co.:** Bridgers Pass.

Spermophilus lateralis wortmani (J. A. Allen)

1895. *Tamias wortmani* J. A. Allen, Bull. Amer. Mus. Nat. Hist., 7:335, November 8.

1911. *Callospermophilus lateralis wortmani*, Cary, N. Amer. Fauna, 33:84.

1959. *Spermophilus lateralis wortmani*, Hall and Kelson, The Mammals of North America, The Ronald Press, vol. 1, p. 363, March 31.

Type.—From Kinney Ranch, Bitter Creek, Sweetwater County, Wyoming.

Comparisons.—Much paler and more buffy than *Spermophilus lateralis lateralis;* differs from all other subspecies of *S. lateralis* (except *S. l. lateralis*) in lacking prominent, inner dark stripes; skull resembles that of *S. l. lateralis.*

Measurements.—Average and extreme external and cranial measurements of six adult topotypes (two males, four females) are as follows: Total length, 280 (271-289); tail vertebrae, 95 (87-101); hind foot, 43.2 (41-44); ear from notch (dry), 17.1 (16-18); greatest length of skull, 44.1 (43.4-46); palatilar length, 20.4 (20-21); zygomatic breadth, 27.9 (27.4-28.5); cranial breadth, 20.4 (20.2-20.7); postorbital constriction, 13 (12.5-13.8); length of nasals, 15.6 (15.2-16.2); maxillary tooth-row, 8.7 (8.3-9.1) (after A. H. Howell, 1938:195).

Distribution.—South-central part of state, especially Green River watershed. See Fig. 26.

Remarks.—In 1945 E. R. Hall (field notes) found no *S. lateralis* or habitat suitable for it at Kinney Ranch [headquarters] or nearer there than a point 12 miles to the northwest where he took two specimens in a stand of Douglas fir on a north facing slope of a butte.

Records of occurrence.—Specimens examined, 7, as follows: **Sweetwater Co.:** Superior, 1 USNM; Green River, 1 Univ. Wyo.; 10 mi. S, 6 mi. E Rock Springs, 2 Univ. Wyo.; *15 mi. S, 2 mi. E Rock Springs, 1 Univ. Wyo.;* 22 mi. SSW Bitter Creek, 2.

Additional record (A. H. Howell, 1938:195).—**Sweetwater Co.:** *Kinney Ranch (42 specimens).*

Cynomys ludovicianus ludovicianus (Ord)

Black-tailed Prairie Dog

1815. *Arctomys ludoviciana* Ord, *in* Guthrie, A new geogr. hist. com. grammar. . . . Philadelphia, Amer. ed. 2, 2:292 (description on p. 302).

1858. *Cynomys ludovicianus*, Baird, Mammals, *in* Repts. Expl. and Surv. . . . 8(1):xxxix, 331, July 14.

Type.—From "Upper Missouri River."

Description.—Size medium for *Cynomys*, excepting long tail; upper parts dark Pinkish Cinnamon in summer, yellowish (faded) in winter; pelage intermixed with buff and black hairs above; whitish about mouth and eye; color of tail above confluent with that of dorsum, although often more ochraceous; underparts whitish or pale buff except Pinkish Cinnamon underside of tail; skull robust, especially jugal; cheek-teeth large and expanded laterally; auditory bullae small.

The species can be distinguished from *Cynomys leucurus* on the basis of slightly larger size; darker cinnamon summer pelage; whitish about eye; longer tail tipped with blackish instead of whitish; skull more robust and slightly longer; jugal more robust, usually flaring less anteriorly in ventral view; auditory bullae relatively smaller and usually smaller in adults of *C. ludovicianus* than in subadults of *C. leucurus*. Furthermore, *C. ludovicianus* usually lives at lower elevations (for example, 3890-7400 ft.) than does *C. leucurus*, and is more often found in large colonies.

Measurements.—Average external and cranial measurements of four specimens from Wyoming measured by Hollister (1916:36) and certain measurements of two adults (4½ mi. E Farthing Station; 8½ mi. W Horse Creek P. O.) from Laramie County, respectively, are as follows: Total length, 393, 390, 365; tail vertebrae, 86, 73, 85; hind foot, 63, 62, 60; condylobasal length (to premaxillaries between incisors), 60.7, 62.8, 59.9; zygomatic breadth, 45.3, 47.8, —; mastoid breadth, 28.3, 29.0, 27.9; length of nasals, 24.0, 24.9, 23.8; maxillary tooth-row, 16.8, 16.7, 15.8; transverse width of third cheek-tooth from inner base to labial crown, —, 6.7, —.

Distribution.—Northeastern half of state. See Fig. 27.

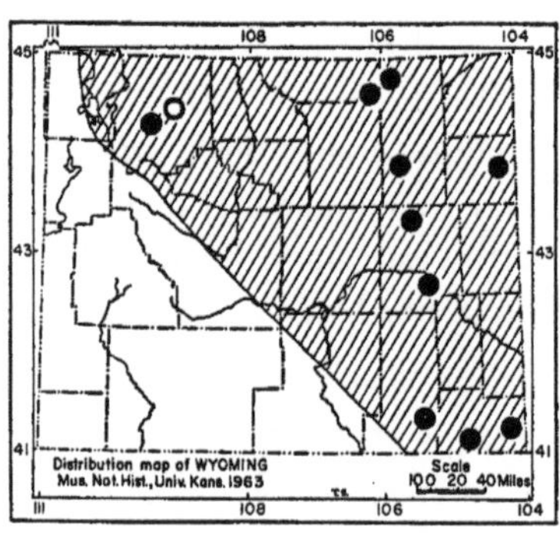

FIG. 27. *Cynomys ludovicianus.*

Remarks.—A specimen (KU 32383) taken in June has fresh pelage on the head and shoulders and in a small spot on the dorsum. The old pelage is faded, worn, and yellowish.

Hall and Kelson (1959:vol. 1, 364) mention two to 10 offspring in this species born from April to July, and they mention that members of this species may not hibernate.

Records of occurrence.—Specimens examined, 36, as follows: **Albany Co.:** 8½ mi. W Horse Creek P. O., 7400 ft., 3. **Campbell Co.:** 3 mi. N, 7½ mi. W Spotted Horse, 1; *2 mi. N, 6½ mi. W Spotted Horse, 1;* 44½ mi. S, 15½ mi. W

Gillette, 5432 ft., 3. **Converse Co.:** 10 mi. N, 7 mi W Bill, 4750 ft., 1; *Ft. Fetterman, 2 USNM;* Douglas, 3 USNM; *Deer Creek, 1 USNM.* **Laramie Co.:** *3½ mi. W Horse Creek* P. *O., 7000 ft., 1; E Farthing Sta., 1; Divide, 1 Univ. Wyo.; Pole [Lodgepole] Creek, 5 USNM;* Pinebluffs, 2 USNM; Cheyenne, 3 USNM. **Park Co.:** Ishawooa, 4 USNM. **Sheridan Co.:** Arvada, 2 USNM. **Weston Co.:** Newcastle, 2 USNM.

Additional records (from Baird, 1858:333-334, unless otherwise noted).—**Park Co.:** Sage Creek (Hollister, 1916:19). **Yellowstone Nat'l Park:** *Yellowstone River.* **County not certain:** *Medicine Bow Butte.*

Cynomys leucurus Merriam

White-tailed Prairie Dog

1890. *Cynomys leucurus* Merriam, N. Amer. Fauna, 3:59, September 11.

Type.—From Fort Bridger, Uinta County, Wyoming.

Comparisons.—Size slightly smaller than in *Cynomys ludovicianus;* paler and buffier except in winter pelage on dorsum; tail whitish intermixed with fine blackish hairs and shorter than that of *C. ludovicianus;* supraorbital brownish spot distinct; cheek-teeth large; skull less robust and usually smaller than in *C. ludovicianus* with jugal flaring anteriorly more often in ventral view; larger auditory bullae.

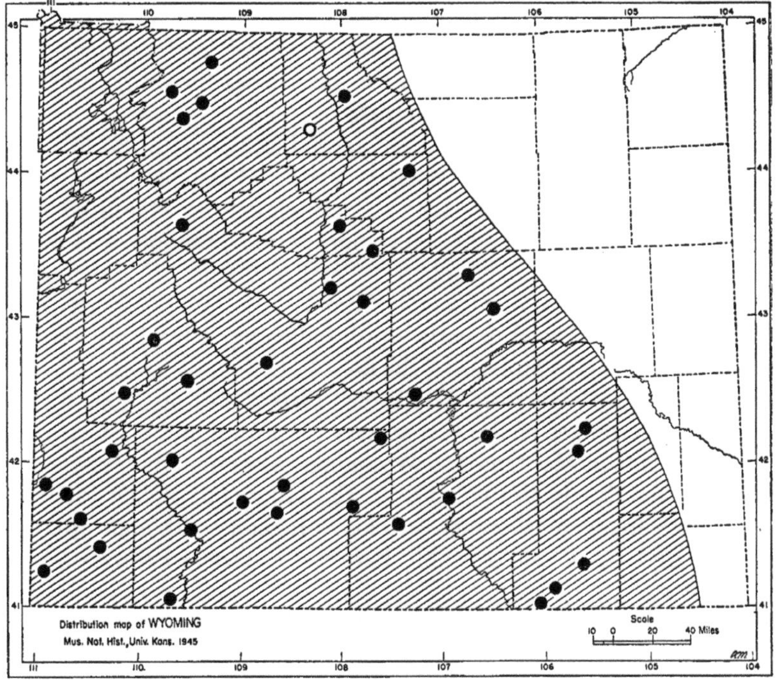

FIG. 28. Distribution of *Cynomys leucurus.*

Measurements.—External and cranial measurements of two adult males from one mile west of Savery are, respectively, as follows: Total length, 362, 340; tail vertebrae, 48, 50; hind foot, 60, 55; condylobasal length, 58.4, 58.1;

zygomatic breadth, 44.6, 44.1; mastoid breadth, 29.0, 29.0; length of nasals, 22.5, 22.5.

Distribution.—Western and south-central part of state, usually in Transition Life-zone (Hollister, 1916:25). See Fig. 28.

Remarks.—This prairie dog often occurs at higher elevations than does *C. ludovicianus*, is less colonial in some places, and is reported to hibernate (Hall and Kelson, 1959, 1:366). But Hollister (1916: 26) mentions that "In the Green River Basin, Wyoming, Vernon Bailey saw *Cynomys leucurus* eating sage-brush tips on snow a foot deep in zero (Fahrenheit) weather, and also after a night when the temperature had fallen to 22° below zero." Hollister (1916:25-26) described molting in *C. leucurus* from Wyoming as follows: "Specimens taken before May 10 are still in the old winter coat, with little evidence of molt. Skins collected from May 20 to June 1 have renewed [the pelage] over most of the underparts and somewhat on the head and shoulders. Numerous examples taken from June 1 to 10 are all in fresh coat except on the lower rump and tail. Skins collected July 15 to 30 are in full summer coat. By August 10 there is much evidence of wear over the forward half of body, and by early September the fall renewal has commenced. As in the case of *C. ludovicianus*, this progresses forward, and by September 25 to October 1 is complete."

Stockard (1929a) mentions that reproductive maturity in *C. leucurus* from Wyoming is attained in a year. Copulation was reported (Stockard, 1929b:209-212) as occurring from March 26 to April 5, and he stated that the gestation period probably was between 27 and 33 days. Numbers of embryos from 67 pregnant females averaged 5.48, and ranged from 1?-10. Males reportedly lost their ability to breed in early April, indicating a brief breeding period. According to Stockard (1929b-212) *C. ludovicianus* breeds earlier than does *C. leucurus*. Hollister (1916:26-27) mentions that the geographic ranges of the two species meet and, in fact, overlap at several points in Montana and Wyoming. Occasionally members of one species live in the colonies of the other; but Bailey found both species living in Montana in close proximity although apart and separated by no recognizable barrier (Hollister, 1916:26-27).

Records of occurrence.—Specimens examined, 79, as follows: **Albany Co.:** 3 mi. S Eagle Peak, 7500 ft., 1 USNM; Garrett, 2 USNM; *Laramie Plains, 1;* Laramie, 1 Univ. Wyo.; 16 mi. SW Laramie, 7500 ft. 1 Univ. Wyo.; Woods P. O., 5 USNM; *Laramie Mtns., 2 USNM.*

Big Horn Co.: 2 mi. N, 6 mi. E Greybull, 1.

Carbon Co.: Shirley Mtns., 7400 ft., 1 USNM; Ft. Steele, 1 USNM; Bridgers Pass, 18 mi. SW Rawlins, 7500 ft., 1, 1 USNM; *1 mi. W Savery, 6600 ft., 3.*

Fremont Co.: 10 mi. N Dubois, 1 USNM; 4 mi. N, 1 mi. W Shoshone, 4700 ft., 1; 1 mi. N, 4 mi. W Moneta, 1; *1 mi. S, 2 mi. W Hudson, 5200 ft., 1;* Lander, 3 USNM.

Hot Springs Co.: Kirby Creek, Bighorn Basin, 1 USNM; 15 mi. N Lost Cabin, 1 USNM.

Lincoln Co.: Fontenelle, 1 USNM; 6 mi. N, 2 mi. E Sage, 6050 ft., 2; Fossil, 1 USNM; Cumberland, 2 USNM.

Natrona Co.: 34½ mi. N, 19½ mi. W Casper, 1; *32 mi. N, 17 mi. W Casper, 5050 ft., 1; 26 mi. N, 12 mi. W Casper, 4625 ft., 1;* 17 mi. NNW Casper, 5050 ft., 2; 8 mi. W Independence Rock, 1; *1 mi. S, 9 mi. W Independence Rock, 5; 2 mi. S, 9 mi. W Independence Rock, 1.*

Park Co.: 9½ mi. N, 6 mi. W Cody, 1; 22 mi. W Cody, 6000 ft., 1; Ishawooa, 1 USNM; 18 mi. S, 20 mi. W Cody, 6600 ft., 1.

Sublette Co.: 1½ mi. W Pinedale, 7200 ft., 2; New Fork, Green River, 4 USNM; 2 mi. N Big Piney, 6900 ft., 2; *Big Piney, 1 USNM.*

Sweetwater Co.: 25 mi. SW Independence Rock, 1 USNM; 7 mi. N, 4 mi. E Big Sandy and Green rivers, 6500 ft., 1; 14 mi. N, 39 mi. E Rock Springs, 6600 ft., 1; 7 mi. ENE Wamsutter, 1; 14 mi. ENE Rock Springs, 1; 6 mi. N Bitter Creek, 1; Green River, 4 USNM; *32 mi. S, 22 mi. E Rock Springs, 7025 ft., 1; 35 mi. S, 20 mi. W Rock Springs, 1;* W side Green River, 1 mi. N Utah boundary, 1.

Uinta Co.: *16 mi. S, 2 mi. W Kemmerer, 6700 ft., 1;* 2 mi. S Carter, 6650 ft., 1; Evanston, 1 USNM.

Washakie Co.: *½ mi. W Tensleep, 4200 ft., 1;* Tensleep, 4200 ft., 1.

Additional records (Hollister, 1916:27).—**Albany Co.:** *Bear Creek.* **Big Horn Co.:** Otto. **Laramie Co.:** *Pole [Lodgepole] Creek.* **Sublette Co.:** *Big Sandy.* **Uinta Co.:** *Ft. Bridger.* **Washakie Co.:** *Bighorn Basin.* **County not certain:** *W Cheyenne.*

Tamiasciurus hudsonicus (Erxleben)

Red Squirrel or Chickaree

This small tree squirrel is common in Wyoming, and the geographic variation, especially in color, is astonishing. Detailed descriptions are listed under the accounts of subspecies. The dental formula is i.$\frac{1}{1}$; c.$\frac{0}{0}$; p.$\frac{1}{1}$; m.$\frac{3}{3}$.

Tamiasciurus hudsonicus baileyi (J. A. Allen)

1898. *Sciurus hudsonicus baileyi* J. A. Allen, Bull. Amer. Mus. Nat. Hist., 10:261, July 22.

1940. *Tamiasciurus hudsonicus baileyi*, Hayman and Holt, *In* Ellerman, The families and genera of living rodents, British Mus., 1:346, June 8.

Type.—From Bighorn Mountains, near head of Kirby Creek, 8400 ft., Washakie County, Wyoming.

Description.—Size medium for species (see *Measurements*); upper parts grayish yellow suffused with reddish olive, but to lesser extent than in *Tamiasciurus hudsonicus ventorum;* lateral dark stripe often present; underparts white or buffy; whitish ring around eye; antiplantar surfaces ochraceous or brownish; tail above reddish, almost Russet, margined with inner blackish and outer buffy or cinnamon borders; underside of tail yellowish gray with numerous blackish hairs, resembling underside of tail in *T. h. ventorum,* but paler; skull resembling that of typical *T. h. ventorum*. Comparisons are listed under accounts of other subspecies.

Measurements.—Average and extreme external and cranial measurements of four old adult males and of six old adult females from 12-18 miles eastward of Shell are, respectively, as follows: Total length, 339.3 (323-359), 347.6 (331-352); tail vertebrae, 134.5 (125-144), 137.2 (114-149); hind foot, 52.3 (49-55), 52.3 (51-54); ear from notch, 25.3 (24-27), 23.8 (18-28); occipito-nasal length, 49.0 (47.5-50.7), 49.3 (47.7-50.0); zygomatic breadth, 28.9 (27.8-29.8), 28.9 (28.5-29.1); length of nasals, 14.6 (13.8-15.6), 14.5 (14.0-15.5); maxillary tooth-row, 8.4 (8.2-8.5), 8.4 (7.8-8.9); postorbital breadth, 14.8 (14.5-15.2), 14.8 (14.3-15.2).

Distribution.—Forested areas from the Bighorn Mountains in north-central Wyoming southward to Green Mountains, and Laramie Mountains, thence to southern border of state. See Fig. 29.

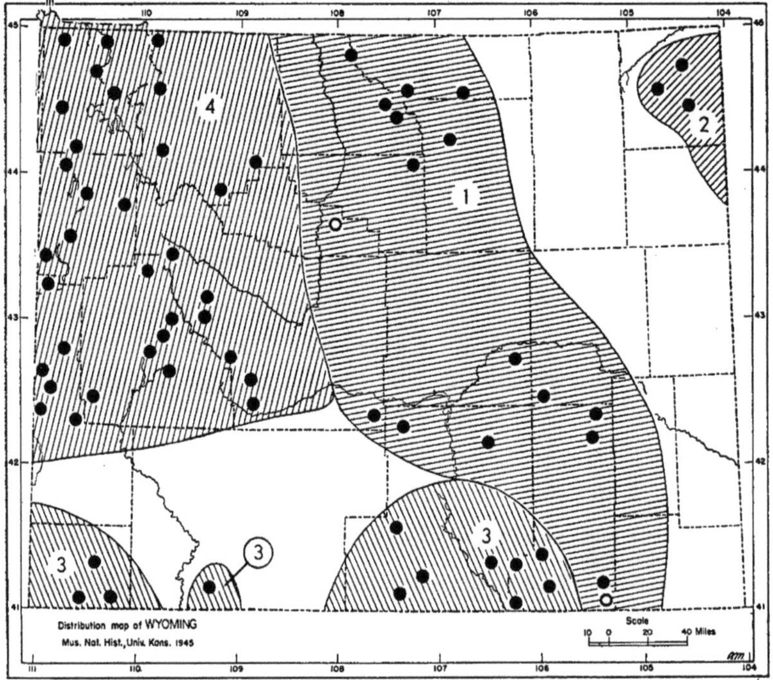

FIG. 29. Distribution of *Tamiasciurus hudsonicus*.
1. *T. h. baileyi* 3. *T. h. fremonti*
2. *T. h. dakotensis* 4. *T. h. ventorum*

Remarks.—This subspecies is closely allied to the darker *T. h. ventorum*, which occurs in more mountainous western Wyoming. J. A. Allen named *T. h. baileyi* only on the basis of color; examination of specimens subsequently accumulated has yielded little new information concerning this variable subspecies, except that smaller size is a better character than less reddish color for distinguishing

T. h. baileyi from *T. h. dakotensis.* Several specimens here assigned to *baileyi,* from the Casper Mountains, are as reddish as *dakotensis.* An adult male (USNM 66446) obtained "near Sheridan" on June 27, 1894, is molting and has dark brownish pelage and pale reddish pelage (fresh).

Records of occurrence.—Specimens examined, 167, as follows: **Albany Co.:** Springhill, 6300 ft., 6 USNM; *12 mi. N Laramie Peak, 6300 ft., 7 USNM;* Laramie Peak, 8800 ft., 3 USNM; 8 mi. E Laramie, 8000 ft., 1 Univ. Wyo.; *10 mi. E Laramie, 9000 ft., 1 USNM; 6½ mi. S, 8¾ mi. E Laramie, 8200 ft., 2; 15 mi. SE Laramie, 1 USNM; 2 mi. ESE Pole Mtn., 8300 ft., 1; 3 mi. S Pole Mtn., 8100 ft., 1.*

Big Horn Co.: Medicine Wheel Ranch, 28 mi. E Lovell, 9000 ft., 9; 2 mi. N, 12 mi. E Shell, 7500 ft., 13; *Hd. Trappers Creek, 8500 ft., 14 USNM; 12 mi. E Shell, 7500 ft., 4; 3 mi. S, 18 mi. E Shell, 9100 ft., 1;* 4½ mi. S, 17½ mi. E Shell, 8500 ft., 7; *9 mi. N, 9 mi. E Tensleep, 8200 ft., 1; Bighorn Mtns., 8400-9000 ft., 5 USNM.*

Carbon Co.: Ferris Mtns., 8500 ft., 6 USNM; Shirley Mtns., 7600 ft., 12 USNM.

Converse Co.: 21½-22½ mi. S, 24½ mi. W Douglas, 7700 ft., 6.

Fremont Co.: 8 mi. E Rongis, Green Mtns., 7 USNM.

Johnson Co.: 1 mi. S, 6 mi. W Buffalo, 5600 ft., 13; *1 mi. S, 4½ mi. W Buffalo, 5440 ft., 3; 2 mi. S, 6½ mi. W Buffalo, 5620 ft., 3.*

Natrona Co.: Casper Mtns., 7 mi. S Casper, 7500 ft., 11 USNM.

Sheridan Co.: *"Near" Wolf, 2 USNM; "Near" Sheridan, 3 USNM; Hd. Prospect Creek, W Sheridan, 3 USNM; 10 mi. N, 31 mi. E Greybull, 7580 ft., 1;* 6 mi. E Granite Pass, Bighorn Nat'l Forest, 1; Storey, 4800 ft., 1 Univ. Wyo.

Washakie Co.: *8 mi. N, 9 mi. E Tensleep, 7800 ft., 1;* 4 mi. N, 9 mi. E Tensleep, 7000 ft., 17.

Additional records (J. A. Allen, 1898:262).—**Albany Co.:** Sherman, southern end Laramie Mtns. **Hot Springs Co.:** Hd. Kirby Creek, Bighorn Mtns.

Tamiasciurus hudsonicus dakotensis (J. A. Allen)

1894. *Sciurus hudsonicus dakotensis* J. A. Allen, Bull. Amer. Mus. Nat. Hist., 6:325, November 7.

1940. *Tamiasciurus hudsonicus dakotensis,* Hayman and Holt, *In* Ellerman, The families and genera of living rodents, British Mus., 1:346, June 8.

Type.—From Squaw Creek, Black Hills, Custer County, South Dakota.

Comparisons.—Large for *Tamiasciurus hudsonicus* (see *Measurements*); upper parts Ochraceous-Tawny or Tawny Olive to Orange-Cinnamon intermixed with fine hairs of Fuscous Black; upper parts becoming more ochraceous laterally; tail Tawny to richly ochraceous above with submarginal Fuscous Black (not conspicuous) and marginal Ochraceous-Tawny or Light Buff hairs; underparts whitish or Light Buff; underside of tail Ochraceous Buff to Clay Color to Ochraceous-Tawny; skull large (see *Measurements*); auditory bullae large but not greatly inflated; teeth large.

Compared with *T. h. baileyi,* which occurs to the westward of *T. h. dakotensis,* the latter differs in being larger and in having more reddish, paler upper parts.

Measurements.—Average and extreme external and cranial measurements of four old adult females and measurements of two old adult males from three miles northwest Sundance are, respectively, as follows: Total length, 360.8 (340-390), 360, —; length of tail, 137.3 (122-165), 150, —; hind foot, 55.5 (52-60), 59, 53; ear from notch, 28.0 (27-30), 27, 26; exoccipital length, 50.2

(50.1-51.9), 51.3-50.5; zygomatic breadth, 29.8 (29.6-30.0), —, 29.8; length of nasals, 15.8 (15.3-16.3), 15.5, 15.8; maxillary tooth-row, 8.8 (8.7-8.9), 8.8, 8.6; postorbital breadth, 14.8 (14.4-15.0), —, 14.5.

Distribution.—Black Hills in northeastern part of state. See Fig. 29.

Records of occurrence.—Specimens examined, 30, as follows: **Crook Co.:** 15 mi. N Sundance, 5500 ft., 1; Devils Tower, 4 USNM; *15 mi. ENE Sundance, 3825 ft., 7; 3 mi. NW Sundance, 5900 ft., 16;* Sundance, 2 USNM.

Additional record (J. A. Allen, 1898:260).—**Crook Co.:** *Belle Fourche, western edge Black Hills.*

Tamiasciurus hudsonicus fremonti (Audubon and Bachman)

1853. *Sciurus fremonti* Audubon and Bachman, The viviparous quadrupeds of North America, 3(30):pl. 149, Fig. 2; text 3:237.

1950. *T[amiasciurus]. hudsonicus fremonti*, Hardy, Proc. Biol. Soc. Washington, 63:14, April 26.

Type.—Probably from the park region of central Colorado, in the "Rocky Mountains."

Comparisons.—Small for *Tamiasciurus* in Wyoming (see *Measurements*); upper parts Dresden Brown; white ring around eye; dark lateral stripe often present; underparts Pale Buff or white; underside of tail buffy-gray or buffy-black; tail above dark brown or black but showing numerous hairs tipped with whitish; bullae slightly less inflated than in *T. h. baileyi*.

Cranial differences, by means of which Durrant and Hansen (1954:90-91) separated *T. h. ventorum* from *T. h. fremonti* in Utah, are not reliable in Wyoming. In the latter state the two subspecies are more easily distinguished by means of color of tail and upper parts, *T. h. fremonti* being darker and having a dark tail showing whitish-tipped hairs. These two subspecies evidently intergrade in the southern Wind River Mountains; the intergrades are referable to *T. h. ventorum* on the basis of color characters and approach *T. h. fremonti* in small size. No intergrades are known between *T. h. fremonti* and *T. h. baileyi* in southeastern Wyoming; these subspecies can also be distinguished from each other on the basis of darker color and smaller size in *T. h. fremonti*.

Measurements.—Average and extreme external and cranial measurements of four old adult males and four old adult females from the southern third of Carbon County (Medicine Bow Mountains) are, respectively, as follows: Total length, 343.5 (325-368), 328.8 (320-337); tail vertebrae, 146.0 (127-185), 143.5 (131-178); hind foot, 48.3 (44-54), 48.3 (43-52); ear from notch, 25.5 (22-29), 25.3 (22-29); occipitonasal length, 49.1 (48.7-50.0), 48.0 (47.5-48.6); zygomatic breadth, 29.0 (28.5-29.9), 28.2 (27.6-29.0); length of nasals, 15.8 (15.3-16.2), 15.4 (15.0-16.1); maxillary tooth-row, 8.2 (7.7-8.4), 8.0 (7.7-8.2); postorbital breadth, 14.9 (14.7-15.1), 15.2 (14.5-16.3).

Distribution.—Southern part of Wyoming ranging into extreme southwestern part of state west of Green River and into forested mountains of south-central Wyoming (Medicine Bow Mountains). See Fig. 29.

Records of occurrence.—Specimens examined, 55, as follows: **Albany Co.:** Sec. 28, T. 16N, R. 78W, N Fork Little Laramie River, 8500 ft., 1; Libby Creek, Snowy Range, 9500 ft., 1 Univ. Wyo.; *Univ. Wyo. Science Camp, 2 Univ. Wyo.; 1 mi. W Ranger Sta., Medicine Bow Nat'l Forest, 1; Hq. Park, 10,200 ft., 1 USNM; Centennial, 1 Univ. Wyo.; Albany, 1 Univ. Wyo.; 3 mi. ESE Browns Peak, 10,000 ft., 1;* 3 mi. above Woods Landing, 8000 ft., 1 Univ. Wyo.; *Woods P. O., 5 USNM; 36 mi. SW Laramie, 7600 ft., 1 Univ. Wyo.;* 45 mi. SW Laramie, 7800 ft., 1 Univ. Wyo.
Carbon Co.: Bridgers Pass, 1 USNM; 6 mi. S, 14 mi. E Saratoga, 8800 ft., 4; *8 mi. N, 14 mi. E Encampment, 8400 ft., 1; 8 mi. N, 21½ mi. E Encamp-*

ment, 9400 ft., 1; 8 mi. N, 18½ mi. E Savery, 8600 ft., 1; *Bridgers Peak, 8800 ft., 1 USNM; 8 mi. N, 19½ mi. E Savery, 8800 ft., 3; 7 mi. N, 18 mi. E Savery, 8400 ft., 3; 6½ mi. N, 16 mi. E Savery, 8300 ft., 1;* 6 mi. N, 12½ mi. E Savery, 8400 ft., 1.

Sweetwater Co.: Maxon, 1 USNM.

Uinta Co.: Ft. Bridger, 6650 ft., 2, 3 USNM; 9 mi. S Robertson, 8000 ft., 12; *10 mi. S, 1 mi. W Robertson, 8700 ft., 1; Henrys Fork, 5 mi. W Lonetree, 8000 ft., 1 USNM;* 4 mi. S Lonetree, 1 USNM.

Tamiasciurus hudsonicus ventorum (J. A. Allen)

1898. *Sciurus hudsonicus ventorum* J. A. Allen, Bull. Amer. Mus. Nat. Hist., 10:263, July 22.

1939. *Tamiasciurus hudsonicus ventorum*, Davis, The Recent mammals of Idaho, Caxton Printers, Caldwell, Idaho, p. 229, April 5.

Type.—From South Pass City, Wind River Mountains, Fremont County, Wyoming.

Comparisons.—*T. h. ventorum* resembles *T. h. baileyi* but is darker, more reddish-brown, and more olivaceous on the dorsum. Also the tail is darker brown above and darker gray below. See accounts of other subspecies.

Measurements.—Average and extreme external and cranial measurements of four old adult females and measurements of two old adult males all from the type locality are, respectively, as follows: Total length, 308.3 (290-342), 305, 306; tail vertebrae, 135.3 (111-160), 123, 116; hind foot, 48.0 (44-50), 45, 49; ear from notch, 26.5 (25-30), 25, 25; occipitonasal length, 48.1 (47.9-48.2), 46.5, 49.4; zygomatic breadth, 27.9 (27.6-28.0), 27.1, 28.7; length of nasals, 15.3 (14.7-15.8), 13.9, 15.1; maxillary tooth-row, 8.1 (7.8-8.2), 8.5, 7.9; postorbital breadth, 15.1 (14.4-15.9), 14.4, 14.3.

Distribution.—Mountainous northwestern part of state. See Fig. 29.

Remarks.—*T. h. ventorum* and *T. h. fremonti* intergrade in southwestern Wyoming. In Uinta County specimens are referred to *T. h. fremonti* on the basis of color; in the southern Wind River Mountains, specimens are referred to *T. h. ventorum* on the basis of color of tail and dorsal pelage although they approach *T. h. fremonti* in size. Specimens more nearly typical of *T. h. ventorum* occur in the Teton Mountains, northern Wind River Mountains, and Yellowstone National Park.

Records of occurrence.—Specimens examined, 195, as follows: **Fremont Co.:** Jackeys Creek, 3-5 mi. S Dubois, 10,000 ft., 2 USNM; *Crowheart, 1 USNM;* Bull Lake, 7700 ft., 5 USNM; Lake Fork, 9600-10,000 ft., 10 USNM; *Moccasin Lake, 4 mi. N, 19 mi. W Lander, 10,200 ft., 1;* Mosquito Park Ranger Sta., 2½ mi. N, 17½ mi. W Lander, 9500 ft., 4; *Middle Lake, 2 mi. S, 20½ mi. W Lander, 1;* 12 mi. N, 4 mi. W South Pass City, 7800 ft., 2; *Fiddlers Lake, 1 Univ. Wyo.; 18½ mi. S, 5½ mi. W Lander, 9000 ft., 1; ½ mi. N, 3 mi. E South Pass City, 9;* South Pass City, 8000 ft., 11 USNM.

Lincoln Co.: 3 mi. N, 11 mi. E Alpine, 5650 ft., 1; Mouth Lynx Creek, Greys River, 6300 ft., 2; Afton, 1 USNM; 10 mi. SE Afton, 1 USNM; *30 mi. SW Daniel, 8000 ft., 2 USNM;* Hd. Smiths Fork, 2 USNM; LaBarge Creek, 9200 ft., 2 USNM.

Park Co.: 30½ mi. N, 36 mi. W Cody, 6900 ft., 2; Southwest slope Whirlwind Peak, 9000 ft., 2; *Mouth Grinnell Creek, 6300 ft., 11 USNM; 27 mi. W Cody, 6100 ft., 1; 25 mi. S, 28 mi. W Cody, 6400 ft., 1;* 30 mi. S, 30 mi. W Cody, 6450 ft., 1; *Needle Mtn., 10,000 ft., 6 USNM;* 15½ mi. S, 3 mi. W Meeteetse, 1; 38 mi. S, 6 mi. W Cody, 8000 ft., 1.

Sublette Co.: 31 mi. N Pinedale, 8025 ft., 1; *12 mi. N Kendall, 7700 ft., 2 USNM;* Fremont Peak, 10,500 ft., 1 USNM; *15 mi. NE Cora, 1 USNM;*

Sweeney Park, 9 mi. N, 8 mi. E Pinedale, 8500 ft., 6; 12 mi. NE Pinedale, 8000 ft., 1 USNM; *8 mi. N, 5 mi. E Pinedale, 8500 ft., 8; 7 mi. N, 5 mi. E Pinedale, 8200 ft., 1; 10 mi. N, 6 mi. E Pinedale, 9400 ft., 4; 10 mi. NE Pinedale, 8000 ft., 1; W side Barbara Lake, 8 mi. S, 3 mi. W Fremont Peak, 3; W end Half Moon Lake, 7900 ft., 1; 2¼ mi. NE Pinedale, 7500 ft., 2; 2 mi. NE Pinedale, 8000 ft., 1 USNM;* Pinedale, 7300 ft., 1 USNM; *Wind River Mountains, 2 USNM;* Big Sandy, 2 USNM; 2 mi. S, 19 mi. W Big Piney, 7700 ft., 2.

Teton Co.: 18 mi. N, 9 mi. W Moran, 1 Univ. Wyo.; *Upper Arizona Creek, 4 USNM; Whetstone Creek, 4 USNM; Pilgrim Creek, 3 mi. N Moran, 6800 ft., 2 Univ. Wyo.; Two Ocean Lake, 2½ mi. N, 3½ mi. E Moran, 6900-7225 ft., 2; ¼ mi. N, 2½ mi. E Moran, 6200 ft., 8;* Moran, 2, 3 USNM; *3 mi. E Moran, 6750 ft., 1; 4 mi. E Moran, 6750 ft., 2 Univ. Wyo.; 1 mi. S, 3¾ mi. E Moran, 7; 2 mi. S, 1 mi. W Moran, 6950 ft., 1 Univ. Wyo.; Moose Creek, 6800 ft., 3 USNM; S Moose Creek, 10,000 ft., 4 USNM;* Togwotee Pass, 9500 ft., 1 Univ. Wyo., 2 USNM; *Bar BC Ranch, 2½ mi. NE Moose, 6500 ft., 3;* Moose, 6225 ft., 1; *3 mi. S Moose, 6700 ft., 2 Univ. Wyo.;* Teton Pass, 7000-7200 ft., 4 USNM; locality not specified, 3 USNM.

Yellowstone Nat'l Park: Mammoth Hot Springs, 3 USNM; Tower Falls, 2; 1 mi. S Canyon, 7500 ft., 1; Yellowstone Lake, 3 USNM; Firehole River, 1 USNM; Snake River, 1 USNM; *locality not specified, 8000 ft., 1 USNM.*

Sciurus niger rufiventer É. Geoffroy St.-Hilaire

Fox Squirrel

1803. *Sciurus rufiventer* É. Geoffroy St.-Hilaire, Catalogue des mammifères du Museum National d'Histoire Naturelle, Paris, p. 176.

1907. *Sciurus niger rufiventer,* Osgood, Proc. Biol. Soc. Washington, 20:44, April 18.

Type.—Probably between southern Illinois and central Tennessee, Mississippi Valley (Osgood, 1907:44).

Description.—Larger than *Tamiasciurus,* smaller than *Cynomys* (see *Measurements*); tail long; upper parts brownish, grayish agouti with pinnae of ears markedly ochraceous; upper side of tail slightly darker and more reddish than dorsum; underparts whitish or buffy strongly washed with cinnamon, Pale Pinkish Buff, or Pinkish Cinnamon; underside of tail strongly ochraceous and often margined with fine Fuscous Black; skull compared with that of *Cynomys leucurus,* longer, relatively narrower, with zygomata hardly extended laterally; four instead of five cheek-teeth on each side of upper jaw; cheek-teeth smaller and narrower transversely; auditory bullae smaller.

Measurements.—External measurements of an adult male (Univ. Wyo. 243) from Laramie, perhaps introduced by man, are as follows: Total length, 492; tail vertebrae, 224; hind foot, 71; ear from notch, 31; weight, 647 gm.

Distribution.—Six fox squirrels observed by me in Cheyenne; expected to occur in wooded ravines and stream valleys at lower elevations in eastern Wyoming, especially in extreme southeastern part of state. Not mapped.

Remarks.—This species probably would not occur in Wyoming except for man's planting of trees in prairie areas. It seems to have become established in the southeastern part of the state. Occurrence there is not surprising in the light of the extensive immigration of this species into northeastern Colorado (Hoover and Yeager, 1953).

Records of occurrence.—**Specimen examined, 1 (probably introduced), from Albany Co.:** *Laramie, 7185 ft.*

Glaucomys sabrinus bangsi (Rhoads)

Northern Flying Squirrel

1897. *Sciuropterus alpinus bangsi* Rhoads, Proc. Acad. Nat. Sci. Philadelphia, 49:321, July 19.

1918. *Glaucomys sabrinus bangsi*, A. H. Howell, N. Amer. Fauna, 44:38, June 13.

Type.—From Idaho County, Idaho. In much of the most recent literature pertaining to *Glaucomys sabrinus bangsi* its type locality is listed as Raymond, Bear Lake County, Idaho. Earlier references list Raymond, Idaho County, Idaho, or they list only Idaho County, Idaho (for example, A. H. Howell, 1918:38). According to the original description the holotype was obtained from Idaho County, and the men who obtained it probably lived in the mining camp that was then known as Raymond, Idaho (see Davis, 1939:70).

Description.—Resembling *Tamiasciurus* in size (see *Measurements*); flight membrane extending from manus to pes, bordered by whitish or Pinkish Cinnamon hairs, and showing much Dark Neutral Gray submarginally; tail well-haired and flattened dorsoventrally apparently to aid in gliding; upper parts Tawny-Olive to Ochraceous-Tawny, often paler and often showing Dark Neutral Gray of basal parts of hairs; antiplantar surfaces grayish or brownish; tail above like dorsum but more grayish approaching Wood Brown and Olive-Brown (being darker distally); underparts whitish washed with Cinnamon-Buff or Pinkish Buff especially mid-ventrally; top of braincase markedly convex in lateral view; rostrum narrow; nasals flaring anteriorly; auditory bullae large; teeth small; dental formula, i.$\frac{1}{1}$, c.$\frac{0}{0}$, p.$\frac{2}{1}$, m.$\frac{3}{3}$; premaxillaries extending noticeably beyond incisors; supraoccipital extending posterior to exoccipital condyles.

Measurements.—External and cranial measurements of one old adult female (KU 37518) from Park County and one old adult female and two old adult males from Teton County are, respectively, as follows: Total length, 338, 330, 320, 335; length of tail, 149, 153, 148, 156; hind foot, 43, 41, 40, 42; ear from notch, 29, 27, 27, 26; greatest length of skull, 41.0, 41.0, 39.8, 40.5; zygomatic breadth, 25.5, 25.1, 25.3, 24.8; postorbital breadth, 8.9, 8.7, 9.1, 8.5; length of nasals, 12.4, 12.5, 12.9, 12.9; cranial depth, 13.6, 13.9, 13.9, 13.5; maxillary tooth-row, 8.0, 8.2, 8.0, 7.7.

Distribution.—Northern Wyoming, southward into Wind River Mountains. See Fig. 30.

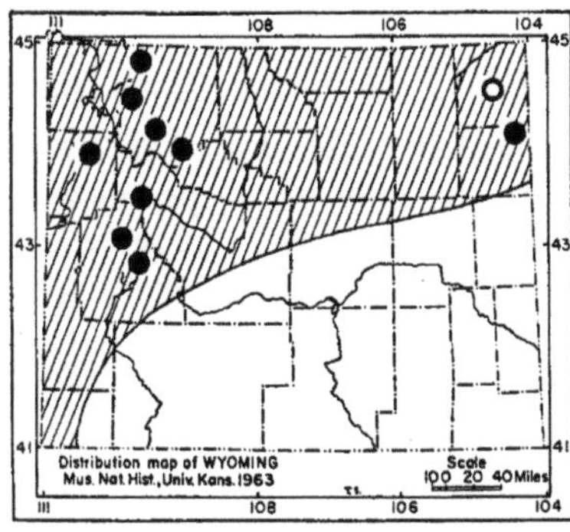

FIG. 30. *Glaucomys sabrinus bangsi.*

Remarks.—This squirrel is nocturnal and scansorial. It glides from one tree to the base of another by means of the membrane stretched from foreleg to hind leg. I agree with King (1951:469) that specimens from the Black Hills are referable to the subspecies *Glaucomys sabrinus bangsi.*

Records of occurrence.—Specimens examined, 30, as follows: **Fremont Co.:** Mtns. "near" Dubois, 2 USNM. **Park Co.:** 31½ mi. N, 36 mi. W Cody, 6900 ft., 1; Mouth Grinnell Creek, Pahaska, 2 USNM; 19 mi. S, 19 mi. W Cody, 2; 38 mi. S, 6 mi. W Cody, 8000 ft., 1. **Sublette Co.:** 12 mi. N Kendall, 7700 ft., 3 USNM; 10 mi. NE Pinedale, 8000 ft., 1. **Teton Co.:** Pacific Creek, 1 USNM; *S bank Snake River, ½ mi. E Moran, 6740 ft., 3; 1 mi. E Moran, 1, 1 Univ. Wyo.; 2 mi. E Moran, 1; 3 mi. E Moran, 6750 ft., 2 Univ. Wyo.; ¼ mi. S. 3½ mi. E Moran, 6200 ft. 1; 2 mi. S, 1 mi. W Moran, 6950 ft., 1, 1 Univ. Wyo.;* exact locality not specified, 1 USNM. **Weston Co.:** ½ mi. N, 3 mi. E Buckhorn, 6200 ft., 5.

Additional record (A. H. Howell, 1918:38, 40).—**Crook Co.:** Middle Fork Hay Creek, Bear Lodge Mtns.

Family GEOMYIDAE—Pocket Gophers

KEY TO SPECIES OF GEOMYIDS

1. Pale yellowish buff to dark brown; incisors smooth or grooved by indistinct line on anterior faces................. **Thomomys talpoides,** p. 600

1′. Pale yellowish tan; teeth distinctly grooved on anterior faces, **Geomys bursarius,** p. 612

Thomomys talpoides (Richardson)

Northern Pocket Gopher

The large number of subspecies of this and one or two other species of pocket gophers is looked upon with suspicion by several biologists who have not acquainted themselves with the facts. The great degree of geographic variation in *Thomomys* requires the naming of a large number of subspecies if the Linnaean system of classification is used. This complicates analysis of the variation, and many workers prefer to keep their problems uncomplicated. But it is paradoxical that much of the criticism directed toward the taxonomy of pocket gophers is from persons professing great interest in evolution and speciation. In gophers, as might be expected, many examples of exceptional and unusual interactions between populations occur. For example, a pair of sympatric subspecies occurs in Wyoming. Ecological separation of subspecies also is seen. It is a fact that the population concept of evolution, or the studying of large series, needs be applied to gophers in order best to arrive at the most meaningful classification although the constancy of characters over a given area that is uniform in soil and climate permits the ready identification of individual specimens. Conventional methods are used here to classify *Thomomys;* and accumulations of many specimens give a person little excuse for confusing differences and resemblances of taxonomic worth with those caused by ontogeny, sex, season, and individual variation. Even so, many of the conclusions drawn here concerning *Thomomys* in Wyoming are tentative or arbitrary compromises.

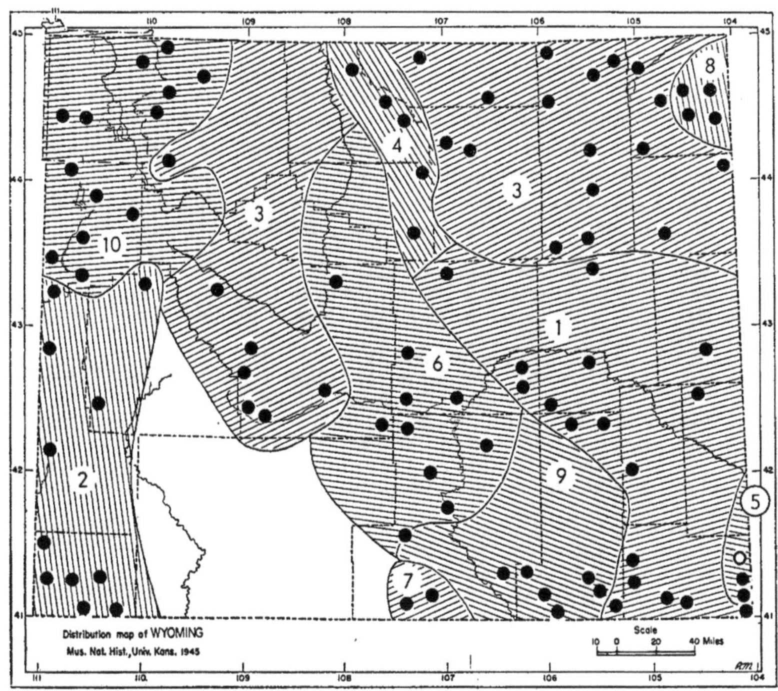

FIG. 31. Distribution of *Thomomys talpoides* excepting two subspecies.

1. *T. t. attenuatus*
2. *T. t. bridgeri*
3. *T. t. bullatus*
4. *T. t. caryi*
5. *T. t. cheyennensis*
6. *T. t. clusius*
7. *T. t. meritus*
8. *T. t. nebulosus*
9. *T. t. rostralis*
10. *T. t. tenellus*

Thomomys talpoides attenuatus Hall and Montague

1951. *Thomomys talpoides attenuatus* Hall and Montague, Univ. Kansas Publs., Mus. Nat. Hist., 5(3):29, February 28.

Type.—From 3½ miles west of Horse Creek Post Office, 7000 ft., Laramie County, Wyoming.

Comparisons.—Smaller (see *Measurements*) than *T. t. bullatus, T. t. rostralis,* and *T. t. cheyennensis;* paler than *T. t. rostralis;* reddish in northern part of its range becoming paler throughout much of the Laramie Mountains owing to arid, low habitats or intergradation with *T. t. cheyennensis,* and tending toward brownish in southern part of the Laramie Mountains where approaching *T. t. rostralis,* with which *attenuatus* intergrades.

Intergradation with *rostralis* is expressed in pale color of specimens (referred to *rostralis*) on the Laramie Plains and in small size of specimens (referred to *attenuatus*) in the southern Laramie Mountains. White-spotting is common in both subspecies. *T. t. attenuatus* has larger bullae and is darker than *T. t. pierreicolus* Swenk of Nebraska.

Measurements.—External and cranial measurements of two males and two females from the type locality are, respectively, as follows: Total length, 202, 189, 203, 192; length of tail, 61, 56, 59, 69; length of hind foot, 26, 24, 26, 26; basilar length, 30.1, 29.7, 30.0, 28.8; zygomatic breadth, 21.2, 20.1, —, 19.8; interorbital breadth, 6.6, 5.7, 6.1, 5.5; length of nasals, 13.6, 12.4, 14.1, 12.0; maxillary tooth-row, 7.0, 6.9, 6.8, 7.3 (after Hall and Montague, 1951:32).

Distribution.—Southeastern Wyoming, southeastward from northern Natrona County to southern Laramie County. See Fig. 31.

Records of occurrence.—Specimens examined, 71, as follows: **Albany Co.:** Springhill, 2 USNM; 3 mi. SW Eagle Peak, 7500 ft., 1 USNM; Sherman, 2 USNM.

Converse Co.: 12 mi. N, 6 mi. W Bill, 4800 ft., 11; *11 mi. N, 6 mi. W Bill, 4800 ft., 1;* Ft. Fetterman, 2 USNM; *Douglas, 2 USNM;* 21 mi. S, 24 mi. W Douglas, 7400 ft., 2; *22 mi. S, 25 mi. W Douglas, 7800-8000 ft., 2; 23 mi. S, 25 mi. W Douglas, 7800 ft., 2.*

Goshen Co.: Rawhide Buttes, 5400 ft., 1.

Laramie Co.: *20 mi. SE Chugwater, 1 USNM; 1 mi. N, 5 mi. W Horse Creek, 7200 ft., 1; 3½ mi. W Horse Creek* P. O., *7000-7200 ft., 5; 2-2½ mi. W Horse Creek* P. O., *3;* Horse Creek P. O., 6500 ft., 1; *3 mi. E Horse Creek, 6400 ft., 5; Beaver, 1 USNM;* 6 mi. W Islay, 2 USNM; *7 mi. W Cheyenne, 5200 ft., 1;* Cheyenne, 5 USNM; *Ft. D. A. Russell, 1 USNM;* 2 mi. S, ½ mi. E Pinebluffs, 1; *1 mi. S, 4½ mi. E Cheyenne, 5200 ft., 1;* 2 mi. S, 9½ mi. E Cheyenne, 5200 ft., 3.

Natrona Co.: 27 mi. N, 2 mi. E Powder River, 6025 ft., 2; 4 mi. E Casper, 5100 ft., 5; 10 mi. S Casper, 7750 ft., 3.

Niobrara Co.: 10 mi. N Hatcreek P. O., 4300 ft., 1.

Platte Co.: 15 mi. SW Wheatland, 5200 ft., 1 USNM.

Thomomys talpoides bridgeri Merriam

1901. *Thomomys bridgeri* Merriam, Proc. Biol. Soc. Washington, 14:113, July 19.

1939. *Thomomys talpoides bridgeri* Goldman, Jour. Mamm., 20:234, May 14.

Type.—From Mountainview (Hooper, 1943) or Harveys Ranch, on Smiths Fork, six miles southwest of Fort Bridger, Uinta County, Wyoming.

Description.—Large for *Thomomys* (see *Measurements*); upper parts dark brownish; underparts brownish or tawny; feet often whitish; skull robust excepting auditory and mastoid bullae, which are small and not inflated; braincase broad and zygomata parallel laterally, not inclined anteriorly, giving skull squarish appearance in dorsal view; marked, parallel temporal ridges; high lambdoidal crest; interpterygoid space V-shaped; openings of cheek-pouches usually black; baculum short, thick basally, and flaring distally.

Comparisons.—Much larger and darker than *T. t. pygmaeus.* Larger and darker than *T. t. ocius;* much more robust skull with more nearly parallel and more robust zygomata; smaller, more flattened bullae (auditory and mastoid).

Measurements.—Average and extreme external and cranial measurements of 13 adult females and five adult males from Fort Bridger are, respectively, as follows: Total length, 230.6 (214-246), 245.8 (237-252); length of tail, 69.2 (58-78), 72.0 (67-76); hind foot, 31.3 (29-33), 33.4 (32-34); occipitonasal length, 40.7 (37.5-42.3), 42.4 (40.3-42.9); zygomatic breadth, 24.5 (22.9-25.6), 25.4 (23.6-27.0); interorbital breadth, 6.6 (6.2-7.2), 6.6 (6.4-6.9); length of nasals, 15.2 (12.9-16.0), 15.6 (14.7-16.7); maxillary tooth-row, 8.2 (7.6-8.9), 8.4 (8.0-9.0).

Distribution.— Extreme southwestern Wyoming. See Fig. 31.

Remarks.—*T. t. bridgeri* is sympatric with *T. t. ocius* and especially so with *T. t. pygmaeus*. Three subspecies occurring in one area present one of the most interesting and unexpected problems in taxonomic mammalogy. The three subspecies interact as species in Wyoming. Both *T. t. bridgeri* and *T. t. pygmaeus* occur at the type locality (in southeastern Idaho) of *pygmaeus*. Members of each of the three subspecies intergrade with members of other subspecies of *T. talpoides* elsewhere (Davis, 1939:252; Whitlow and Hall, 1933:256; Durrant, 1952:165, 169). The pale *T. t. pygmaeus* and the dark *T. t. bridgeri* may be mechanically isolated because *bridgeri* is so much larger than *pygmaeus*. Furthermore, their bacula differ greatly. As would be expected, ecological separation or isolation is evident between *T. t. bridgeri* and *T. t. pygmaeus;* and at the type locality of the latter it occurs in sage-covered areas (*bridgeri* at that place, as it does elsewhere, occurs in more moist soil along the stream). To add to the complexity of the problem, *T. t. bridgeri* varies from a large, dark brownish lowland form to a smaller, more reddish upland form. Only the dark lowland form is known from Wyoming.

The type locality of both *T. t. ocius* and *T. t. bridgeri* is along Smiths Fork, six miles southwest of Fort Bridger. There, *T. t. bridgeri* occurs along the stream and *T. t. ocius* occurs in arid, sage-covered soils a few yards from the stream. These subspecies reproduce during the same time according to field catalogues belonging to the U. S. Fish and Wildlife Service. I doubt whether ecological separation accounts for the lack of gene flow and suspect behavioral or mechanical isolating mechanisms do so. No intergrades are known between *T. t. pygmaeus* and *T. t. ocius;* essentially, their geographic ranges are mutually exclusive.

A specimen (KU 15086) of *T. t. tenellus* from Jackson may be an intergrade with *T. t. bridgeri,* and a series of four immature gophers from three miles north and 11 miles east of Alpine, adjacent to the range of *tenellus,* is tentatively assigned to *T. t. bridgeri* on the basis of large size in a subadult male (KU 37524), which has a total length of 220 mm. and a squarish skull owing in part to nearly parallel zygomata.

Records of occurrence.—Specimens examined, 176, as follows: **Lincoln Co.:** 3 mi. N, 11 mi. E Alpine, 4; 13 mi. N, 2 mi. W Afton, 6100 ft., 6; *10 mi. N, 2 mi. W Afton, 6100 ft., 6; 9 mi. N, 2 mi. W Afton, 6100 ft., 4; 7 mi. N, 1 mi. W Afton, 6100 ft., 1; 6 mi. N, 1 mi. W Afton, 6100 ft., 1; Border, 1 USNM;* Cokeville, 8 USNM.

Sublette Co.: 34 mi. N, 4 mi. W Pinedale, 7925 ft., 1; *33 mi. N, 2 mi. W Pinedale, 7950 ft., 1; 31 mi. N Pinedale, 8025 ft., 11;* 2 mi. S, 19 mi. W Big Piney, 7700 ft., 5.

Uinta Co.: 14 mi. N Evanston, 6600-6800 ft., 5 USNM; *1 mi. W Ft. Bridger, 6650 ft., 3;* Ft. Bridger, 6650 ft., 22, 24 USNM; *Mountainview, 20 USNM;* 6½ mi. S, 6½ mi. E Evanston, 3; 4.1 mi. W Robertson, 6125 ft., 1; *4 mi. W Robertson, 6125 ft., 4; 8 mi. S, 2½ mi. E Robertson, 8300 ft., 1; 9 mi. S Robertson, 8000 ft., 16; 9 mi. S, 2½ mi. E Robertson, 1; 10 mi. S, 1 mi. W Robertson, 8700 ft., 1; 10 mi. S, 2 mi. E Robertson, 8900 ft., 5; 10 mi. S, 2½ mi. E Robertson, 8900 ft., 5; 11½ mi. S, 2 mi. E Robertson, 9200 ft., 1;* 13 mi. S, 2 mi. E Robertson, 9200 ft., 9; *5 mi. W Lonetree, 8000 ft., 1 USNM; Lonetree, 7400 ft., 3 USNM;* 4 mi. S Lonetree, 2 USNM.

Thomomys talpoides bullatus V. Bailey

1914. *Thomomys talpoides bullatus* V. Bailey, Proc. Biol. Soc. Washington, 27:115, July 10.

Type.—From Powderville, Custer County, Montana.

Comparisons.—Large (see *Measurements*), exceeding *Thomomys talpoides tenellus, T. t. caryi, T. t. clusius,* and *T. t. attenuatus;* compared with *T. t. bridgeri, T. t. tenellus, T. t. caryi,* and *T. t. nebulosus, bullatus* is much paler; mastoid bullae larger in *bullatus* than in *bridgeri* and auditory bullae also larger than in *bridgeri, tenellus,* and *clusius.*

T. t. bullatus resembles *T. t. nebulosus* of the Black Hills, which is slightly larger and much darker.

Measurements.—External and cranial measurements of the holotype (adult male) and an adult female from 2½ miles north and 17½ miles west of Lander are, respectively, as follows: Total length, 238, 222; length of tail, 72, 61; length of hind foot, 30, 27; occipitonasal length, 41.6, 37.8; zygomatic breadth, 24.4, 22.6; length of nasals, 15.9, 12.4; interorbital breadth, 6.8, 6.4; cranial depth; 12.5, 11.4; maxillary tooth-row, 8.0, 6.8.

Distribution.—Lowlands of northern Wyoming, southward from the Bighorn Basin into the Wind River Basin along the Wind River Mountains and from the Great Plains of Montana southward into the basins and valleys lying between the Black Hills and Bighorn Mountains; intergrading with *T. t. attenuatus* along the northern margin of geographic range of that subspecies. See Fig. 31.

Remarks.—*T. t. bullatus* presents several problems. First, in Wyoming it is slightly paler and slightly smaller than at the type locality. Gophers from the range of this subspecies are characterized mainly by large auditory and mastoid bullae, pale color, and large size. Second, specimens from the Wind River Mountains are difficult to identify. Hall and Kelson (1952:363) pointed up the problem by assigning an adult male (USNM 147347) from Meadow Creek and a young female from Sage Creek, eight miles northwest of Fort Washakie, Wyoming, to *T. t. clusius* on account of rosaceous color and narrow skull. On the basis of large size of the mentioned male, the large size (222 mm. total length) of a lactating female (KU 32526) from 2½ miles north and 17½ miles west of Lander, and the large size (225 mm. total length) of an adult male (USNM 56161) from Myersville, Wyoming, I am assigning these gophers to *T. t.*

bullatus. It is to be noted that *T. t. bridgeri* is another large gopher in this area. Specimens from approximately 30 miles north of Pinedale, referred to *bridgeri*, differ from topotypes of *bridgeri* in possessing a low occiput that lacks a lambdoidal crest. Gophers on the east side of the Wind River Mountains also show this characteristic. Furthermore, in the one specimen (KU 32525) having a baculum preserved, it is similar to that of *bridgeri*. Narrowness of the interorbital region in these specimens from the east side might be considered a character of *bridgeri*, but several differences from *bridgeri* are lesser size, paler and more reddish color, lower supraorbital ridges, less nearly parallel zygomata, larger auditory bullae and larger mastoid bullae. Additional collecting may reveal that an unnamed subspecies occurs on the east side of the Wind River Mountains. Third, in eastern Wyoming *bullatus* closely resembles *T. t. nebulosus*, which is slightly larger and darker. Color is the criterion used to delimit the ranges of these subspecies. Finally, *bullatus* is replaced to the southward by, and intergrades with, the smaller and darker *T. t. attenuatus*. See *Remarks* under the account of *T. t. attenuatus*.

Records of occurrence.—Specimens examined, 76, as follows: **Campbell Co.:** Powder River Crossing, 4100 ft., 2 USNM; 1⅜ mi. S, ¼ mi. W Rockypoint, 1; *2 mi. S, ½ mi. W Rockypoint, 1; 2¼ mi. S Rockypoint, 1;* 5 mi. S, 9 mi. W Rockypoint, 1; *5¾ mi. S, ⅞ mi. E Rockypoint, 1;* 3 mi. N, 7-7½ mi. W Spotted Horse, 2; *2½ mi. N, 7 mi. W Spotted Horse, 3; 1½ mi. N, 6½ mi. W Spotted Horse, 1; 5 mi. N Gillette, 1;* 2 mi. S Gillette, 1; 21 mi. S Gillette, 5000 ft., 1; 41 mi. S Gillette, 3; 43½ mi. S, 15½ mi. W Gillette, 1; *Belle Fourche River, 43½ mi. S, 13 mi. W Gillette, 5350 ft., 3.*

Crook Co.: *5 mi. S, 2 mi. E Rockypoint, 3800 ft., 2;* 4 mi. S, 6 mi. E Rockypoint, 1; *Little Missouri River, 1 USNM;* Devils Tower, 1 USNM; Moorcroft, 3 USNM.

Fremont Co.: Meadow Creek on Wind River, 5900 ft., 1 USNM; Sage Creek, 8 mi. NW Ft. Washakie, 1 USNM; 2½ mi. N, 17½ mi. W Lander, 9500 ft., 8; 17 mi. S, 6½ mi. W Lander, 8450 ft., 4; Miners Delight, 1 USNM; Myersville, 3 USNM.

Johnson Co.: 2 mi. S, 6½ mi. W Buffalo, 5620 ft., 1; 1 mi. S Pine Tree, 48 mi. W Gillette, 7.

Park Co.: Ishawooa Creek, 6300 ft., 4 USNM.

Sheridan Co.: Dayton, 4500 ft., 1 USNM; *Pass, 2 USNM; 5 mi. NE Clearmont, 3900 ft., 3;* Clearmont, 1 USNM.

Weston Co.: Buckhorn, 6150 ft., 2; *S Newcastle, 5 USNM;* 23 mi. SW Newcastle, 1.

Thomomys talpoides caryi V. Bailey

1914. *Thomomys talpoides caryi* V. Bailey, Proc. Biol. Soc. Washington, 27:115, July 10.

Type.—From head of Trappers Creek, 9500 ft., Bighorn Mountains, Big Horn County, Wyoming.

Comparisons.—Smaller (see *Measurements*) and darker than *T. t. bullatus*, which occurs to the eastward and westward; pelage paler and much more frequently white-spotted, nasals usually more nearly straight-sided, and zygomata less robust than in *T. t. tenellus* of northwestern Wyoming.

Measurements.—Measurements of the type and two females and average and extreme cranial and external measurements of four males, all except the type from 12 miles eastward of Shell, Big Horn County, are, respectively, as follows: Total length, 196, 230, 214, 212.8 (210-220); length of tail, 54, 67, 60, 58.3 (55-60); hind foot, 26, 31, 28, 27.5 (27-29); occipitonasal length, 35.6, 39.7, 39.2, 38.8 (37.0-40.2); zygomatic breadth, 20.6, 22.1, 23.8, 22.5 (21.1-23.3); length of nasals, 12.5, 14.9, 15.0, 13.8 (12.9-14.8); interorbital breadth, 6.4, 6.7, 6.7, 6.1 (5.8-6.5); cranial depth, 11.2, 11.9, 11.5, 11.4 (10.8-11.7); maxillary tooth-row, 6.8, 7.4, 7.2, 7.0 (6.4-7.3).

Distribution.—Bighorn Mountains and as far southward as Red Bank, Washakie County. See Fig. 31.

Remarks.—Of 39 specimens, 17 have fine white-spotting on the dorsum. One (KU 20238) has a white splotch from its nose to the level of its eyes.

Records of occurrence.—Specimens examined, 39, as follows: **Big Horn Co.:** Medicine Wheel Ranch, 28 mi. E Lovell, 9000 ft., 6; *38 mi. E Lovell, 3; Granite Creek Camp Ground, Bighorn Mtns., 3; 2 mi. N, 12 mi. E Shell, 7500 ft., 4; 1 mi. N, 12 mi. E Shell, 7600 ft., 6;* 12 mi. E Shell, 7500 ft., 1; 4½ mi. S, 17½ mi. E Shell, 8500 ft., 1; *Hd. Trappers Creek, 9500 ft., 2 USNM; Bighorn Mtns., 2 USNM.* **Washakie Co.:** 4 mi. N, 9 mi. E Tensleep, 7000 ft., 9; Red Bank, 4350 ft., 2 USNM.

Thomomys talpoides cheyennensis Swenk

1941. *Thomomys talpoides cheyennensis* Swenk, Missouri Valley Fauna, 4:5, March 1.

Type.—From two miles south of Dalton, Cheyenne County, Nebraska.

Comparison.—Resembles *Thomomys talpoides attenuatus* in color above (pale brownish) and below (usually white) but larger with more robust skull. Paler than *T. t. rostralis* or *T. t. bullatus.*

Measurements.—Average and extreme external and cranial measurements of six adult females and measurements of one adult male (KU 25657) all from one mile west of Pine Bluffs are, respectively, as follows: Total length, 225.3 (220-233), 224; length of tail, 59.5 (57-62), 60; length of hind foot, 27.7 (27-28), 29; occipitonasal length, 39.4 (37.8-40.5), 40.2; zygomatic breadth, 23.8 (23.2-24.0 five specimens), 23.2; interorbital breadth, 6.4 (6.0-6.6), 6.4; length of nasals, 14.2 (12.0-14.8), 12.1; cranial depth, 12.2 (11.8-12.5), 12.1; maxillary tooth-row, 7.2 (7.0-7.7), 6.8.

Distribution.—Southeastern part of state. See Fig. 31.

Remarks.—*T. t. cheyennensis* is the palest subspecies of a stepped cline tending eastward toward darker color, excepting *T. t. attenuatus* that is almost equally pale.

Records of occurrence.—Specimens examined, 19, as follows: **Laramie Co.:** 12 mi. N, ½ mi. W Pine Bluffs, 9; 1 mi. W Pine Bluffs, 5000 ft., 8; *Pine Bluffs, 2 USNM.*

Thomomys talpoides clusius Coues

1875. *Thomomys clusius* Coues, Proc. Acad. Nat. Sci. Philadelphia, 27:138, June 15.

1915. *Thomomys talpoides clusius,* V. Bailey, N. Amer. Fauna, 39:100, November 15.

Type.—From Bridger Pass, 18 miles southwest of Rawlins, Carbon County, Wyoming. The holotype was obtained on July 28, 1857, by Dr. W. A. Hammond, a member of a party led by Lt. F. T. Bryan, United States Army. Subsequent collecting of topotypes (by Bailey, in 1890 and 1888, and by parties from the Univ. Kansas Museum) reveals that they differ from the holotype in being darker and larger. The holotype, a lactating adult, is extremely pale, and its skin has been remade (Lyon and Osgood, 1909:63).

Three explanations to account for the pallor and small size of the holotype are as follows: First, it may have been obtained a short distance northward or westward of Bridger Pass; pale gophers occur in the central lowlands of Wyoming. Second, the pale gophers mentioned in the preceding sentence may be sympatric with dark gophers at Bridger Pass. Third, *T. t. clusius* may be extremely variable at Bridger Pass. Knowledge of the routes traveled by Bryan and Hammond in 1856 and 1858 (there is no account, to my knowledge, of their route in 1857) and study of other specimens obtained in June, July, and August, 1857, lead me to doubt that the holotype was taken a great distance from Bridger Pass. I favor the first or second explanation. Acceptance of either explanation enables the student easily to classify the gophers of the general area and is in harmony with what is known of gophers elsewhere. I assume that the holotype was a member of the population of pale gophers from the lowlands of central Wyoming, which is known as *T. t. clusius.*

Comparisons.—*Thomomys talpoides clusius* resembles *Thomomys talpoides ocius* in color and size (*ocius* slightly smaller) but usually lacks distinct black peripinnal areas, has less procumbent incisors, and pelage usually less yellowish dorsally. Compared with *T. t. rostralis, clusius* is much paler and smaller. Compared with *T. t. meritus, clusius* is much paler and has larger auditory bullae and a shorter rostrum. From *T. t. tenellus, clusius* differs in being much paler, and from *T. t. bullatus,* in paler upper parts, smaller auditory bullae, and smaller size.

Measurements.—External and cranial measurements of an adult male and an adult female from 18 miles north-northeast of Sinclair are, respectively, as follows: Total length, 202, 211; length of tail, 53, 53; length of hind foot, 24, 26; occipitonasal length, 37.0, 35.7; zygomatic breadth, 20.8, 21.2; interorbital breadth, 6.1, 6.2; length of nasals, 13.2, 13.1; maxillary tooth-row, 8.0, 7.5.

Distribution.—Central Wyoming in lowlands and some mountain ranges including Green Mountains, Rattlesnake Mountains, Ferris Mountains, and Shirley Mountains. See Fig. 31.

Records of occurrence.—Specimens examined, 37, as follows: **Carbon Co.:** Ferris Mtns., 5 USNM; Shirley Mtns., 8800 ft., 4 USNM; 18 mi. NNE Sinclair, 6500 ft., 2; Ft. Steele, 3 USNM; Bridger Pass, 1 USNM. **Fremont Co.:** 4 mi. N, 1 mi. W Shoshone, 4700 ft., 1; 8 mi. E Rongis, Green Mtns., 8000 ft., 2 USNM. **Natrona Co.:** Rattlesnake Mtns., 2 USNM; 40 mi. SW Casper, 4 USNM; *Sun, 3 USNM;* Sun Ranch, 5 mi. W Independence Rock, 5; *1 mi. S, 5 mi. W Independence Rock, 5.*

Thomomys talpoides meritus Hall

1951. *Thomomys talpoides meritus* Hall, Univ. Kansas Publs., Mus. Nat. Hist., 5(13):219-222, December 15.

Type.—From eight miles north and 19½ miles east of Savery, 8800 feet, Carbon County, Wyoming.

Comparisons.—Resembles *Thomomys talpoides rostralis,* but darker above and below, slightly smaller, smaller auditory bullae, and narrower skull. Much paler than *T. t. clusius.*

Measurements.—Average and extreme external and cranial measurements of seven adult males and five adult females from the type locality are, respectively, as follows: Total length, 204 (193-226), 207 (193-210); length of tail, 56 (46-68), 56 (50-63); length of hind foot, 27.6 (26-30), 27.4 (27-28); basilar length, 30.7 (29.0-33.0), 30.1 (29.5-30.7); zygomatic breadth, 20.4 (18.9-21.6), 19.5 (18.8-20.0); least interorbital breadth, 6.2 (5.8-6.6), 6.1 (5.9-6.3); mastoidal breadth, 17.9 (16.9-18.5), 17.2 (16.7-17.6); length of nasals, 13.7 (12.4-14.7), 13.2 (12.8-13.9); breadth of rostrum, 7.0 (6.5-7.5), 6.9 (6.7-7.3); length of rostrum, 16.3 (15.3-17.5), 15.8 (15.3-16.1); alveolar length of maxillary tooth-row, 7.1 (6.9-7.3), 7.1 (6.8-7.5) (Hall, 1951b:222).

Distribution.—Southern part of Sierra Madre Mountains in south-central Wyoming. See Fig. 31.

Records of occurrence.—Specimens examined, 23, as follows: **Carbon Co.:** 8 mi. N, 19½ mi. E Savery, 8800 ft., 12; *7 mi. N, 17 mi. E Savery, 8300 ft., 1; 6 mi. N, 12½ mi. E Savery, 8400 ft., 1; 6 mi. N. 13½ mi. E Savery, 8400 ft., 2; 6 mi. N, 14½ mi. E Savery, 8350 ft., 1; 5 mi. N, 3 mi. E Savery, 6800 ft., 1; 4 mi. N, 8 mi. E Savery, 7300 ft., 2;* 4 mi. N, 10 mi. E Savery, 7800 ft., 3.

Thomomys talpoides nebulosus V. Bailey

1914. *Thomomys talpoides nebulosus* V. Bailey, Proc. Biol. Soc. Washington, 27:116, July 10.

Type.—From Jack Boyden's Ranch, 3750 feet, Sand Creek Canyon, 15 miles northeast of Sundance, Crook County, Wyoming.

Description.—Large (see *Measurements*); dark upper parts; skull large with inflated auditory bullae. Resembles *T. t. bullatus* in large size.

Measurements.—External and cranial measurements of three females and three males from one half mile north and one mile east of Beulah are, respectively, as follows: Total length, 231, 243, 224, 237, 233, 235; length of tail, 71, 73, 60, 70, 53, 54; hind foot, 30, 32, 32, 31, 31, 32; occipitonasal length, 39.5, 40.9, 39.9, 40.6, 40.8, 43.1; zygomatic breadth, 23.8, 22.5, 22.9, 23.7, 24.0, 24.9; length of nasals, 14.5, 15.1, 14.1, 14.5, 14.5, 16.2; interorbital breadth, 6.3, 6.3, 7.0, 6.5, 6.7, 6.4; cranial depth, 11.7, 11.6, 12.4, 12.5, 11.5, 12.3; maxillary tooth-row, 7.2, 7.0, 6.7, 7.2, 7.9, 7.8.

Distribution.—Black Hills. See Fig. 31.

Remarks.—A large white spot on one shoulder was observed in two specimens (USNM 65883, KU 20226) from Crook County.

Records of occurrence.—Specimens examined, 33, as follows: **Crook Co.:** Bear Lodge Mtns., 6 USNM; ½ mi. N, 1 mi. E Beulah, 3550 ft., 14; *Beulah, 3500 ft., 1; type locality, 1 USNM; Sand Creek, 4 USNM; 1½ mi. NW Sundance, 5000 ft., 3;* Sundance, 2 USNM; Rattlesnake Creek, 2 USNM.

Thomomys talpoides ocius Merriam

1901. *Thomomys clusius ocius* Merriam, Proc. Biol. Soc. Washington, 14:114, July 19.

1946. *Thomomys talpoides ocius,* Durrant, Univ. Kansas Publs., Mus. Nat. Hist., 1:17, August 15.

Type.—From Mountainview, Smiths Fork, four miles southeast of Fort Bridger, Uinta County, Wyoming; specimen labtled as from Ft. Bridger, Wyoming.

Comparisons.—Small (see *Measurements*) but larger than *Thomomys talpoides pygmaeus;* upper parts paler than any nearby subspecies excepting, perhaps, *T. t. clusius,* which is less yellowish and usually lacks blackish peripinnal areas; underparts white as in *clusius;* skull weak and delicate resembling that of *clusius* or *pygmaeus* but having more procumbent incisors; baculum long (KU 17050, 21.0 mm.) and slender resembling that of *pygmaeus* and differing markedly from short bacula of *T. t. bridgeri* and *T. t. tenellus.* See accounts of other subspecies.

Measurements.—Average and extreme external and cranial measurements of six adult, lactating females from "Ft. Bridger" and measurements of the holotype (adult male) are, respectively, as follows: Total length, 197.2 (185-205), 204; length of tail, 57.7 (52-64), 60; length of hind foot, 24.8 (23-25.5), 26; occipitonasal length, 36.1 (35.8-36.6), 36.0; zygomatic breadth, 20.6 (19.8-21.5), 20.2; length of nasals, 13.3 (12.6-13.7), 12.1; interorbital breadth, 6.2 (6.0-6.7), 5.5; cranial depth, 10.8 (10.3-11.7), 10.5; maxillary tooth-row, 7.1 (6.4-7.5), 7.3.

Distribution.—Mainly from watershed of Green River in southwestern Wyoming. See Fig. 32.

Remarks.—See accounts of *T. t. bridgeri* and *T. t. pygmaeus* for remarks on relation of *ocius* to those subspecies.

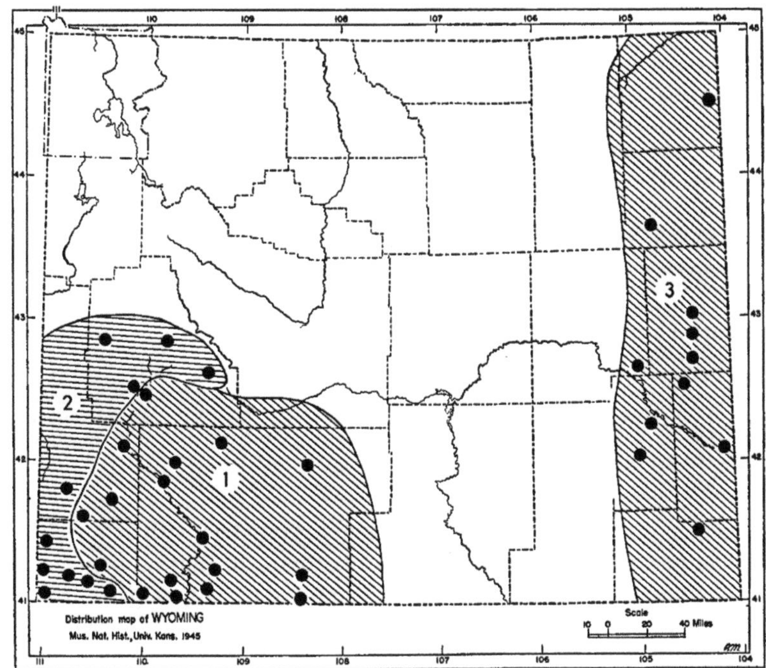

FIG. 32. Distribution of *Thomomys talpoides pygmaeus, Thomomys talpoides ocius,* and *Geomys bursarius.*

Guide to pocket gophers
1. *T. t. ocius*
2. *T. t. pygmaeus*
3. G. *b. lutescens*

Specimens from five miles southwest of Maxon are slightly darker than other specimens examined of *ocius* and may be tending toward *Thomomys talpoides ravus* to the southward or *T. t. meritus* to the eastward.

Records of occurrence.—Specimens examined, 104, as follows: **Lincoln Co.:** Fontenelle, 6500 ft., 1 USNM; Opal, 1 USNM; Cumberland, 6500 ft., 2.
Sublette Co.: Jct. Green River and New Fork, 3 USNM.
Sweetwater Co.: Farson, 6580 ft., 10; *Eden, 1 USNM;* 7 mi. N, 4 mi. E Jct. Big Sandy and Green Rivers, 6500 ft., 3; Jct. Big Sandy and Green Rivers, 6400 ft., 1; 27 mi. N, 37 mi. E Rock Springs, 5; *13 mi. S, 14 mi. E Rock Springs, 3;* 26 mi. S, 19 mi. W Rock Springs, 6850 ft., 1; 26 mi. S, 2 mi. E Rock Springs, 1; 32 mi. S, 22 mi. W Rock Springs, 1; 5 mi. SW Maxon, 7400 ft., 8 USNM; Kinney Ranch, 21 mi. S Bitter Creek, 9; 3 mi. W Green River, 2 mi. N Utah Boundary, 19; Henrys Fork, 12 mi. W Linwood, 6600 ft., 1 USNM; 33 mi. S Bitter Creek, 6900 ft., 4.
Uinta Co.: *2½ mi. W Ft. Bridger, 1;* 3 mi. WSW Ft. Bridger, 21; *"Ft. Bridger," 8.*

Thomomys talpoides pygmaeus Merriam

1901. *Thomomys pygmaeus* Merriam, Proc. Biol. Soc. Washington, 14:115, July 19.

1939. *Thomomys talpoides pygmaeus*, Davis, The Recent mammals of Idaho, The Caxton Printers, Caldwell, Idaho, p. 252, April 5.

Type.—From sage-covered areas along Montpelier Creek, 6700 ft., 6-10 miles northeast of Montpelier, Bear Lake County, Idaho.

Description.—Small (smallest gopher in genus *Thomomys*); dark hazel-brownish, lustrous upper parts; skull delicate and small; hind foot 23 mm. or less; zygomatic breadth less than 18.4 mm.; baculum long and slender as in *T. t. ocius.*

Comparisons.—Smaller than any nearby subspecies; paler brownish upper parts than in *T. t. bridgeri;* darker upper parts than in *T. t. ocius.*

Measurements.—Average and extreme external and cranial measurements of six subadult and adult males from the type locality (1), Henrys Fork (1), Merna (1), and Bear River (3) and measurements of two adult females from Merna and Fossil are, respectively, as follows: Total length, 176.5 (167-187), 175, 167; length of tail, 50.2 (46-55), 45, 52; hind foot, 22.5 (22-23), 23, 20; occipitonasal length, 31.3 (30.1-32.4, 5 specimens), 31.9, 29.3; zygomatic breadth, 16.7 (16.0-17.7), 18.3, 16.5; length of nasals, 10.6 (9.6-11.5, 5 specimens), 11.0, 10.4; interorbital breadth, 5.3 (5.0-5.5), 5.1, 5.0; cranial depth, 9.4 (9.0-9.9), 9.5, 9.2; maxillary tooth-row, 6.1 (6.0-6.5), 6.1, 5.8; mastoid breadth, 15.2 (14.5-16.3), 15.4, 14.6.

Distribution.—Southwestern Wyoming. See Fig. 32.

Remarks.—This subspecies is seemingly sympatric with *T. t. bridgeri*, and perhaps with *T. t. ocius.* See accounts of those subspecies. *T. t. pygmaeus* shows a cline of increasingly dark upper parts from the northern part of its geographic range in Wyoming toward the type locality; specimens from the eastern part of its known range also are dark.

Records of occurrence.—Specimens examined, 36, as follows: **Lincoln Co.:** *Bear River, 6600-7500 ft., 8 USNM;* Fossil, 1 USNM. **Sublette Co.:** Merna, 7800-8000 ft., 3 USNM; 12 mi. NE Pinedale, 1 USNM; Big Sandy, 1 USNM; Big Piney, 1 USNM. **Uinta Co.:** Bear River, 14 mi. N Evanston, 2 USNM;

1 mi. S, 3½ mi. W Evanston, 6900 ft., 2; *6½ mi. S, 6½ mi. E Evanston, 7400 ft., 1;* 7½ mi. S, 11 mi. E Evanston, 8600 ft., 3; 14½ mi. S, 1 mi. E Evanston, 8100 ft., 3; *1½ mi. N, 8½ mi. W Robertson, 7300 ft., 2; 1½ mi. N, 7 mi. W Robertson, 7500 ft., 1; ½ mi. S, 14½ mi. W Robertson, 7600 ft., 3;* 4½ mi. S, 2½ mi. E Robertson, 8025 ft., 1; *8 mi. S, 2½ mi. E Robertson, 8300 ft., 1; Henrys Fork, 8000 ft., 5 mi. N Lonetree, 1 USNM;* Lonetree, 7400 ft., 1 USNM.

Thomomys talpoides rostralis Hall and Montague

1951. *Thomomys talpoides rostralis* Hall and Montague, Univ. Kansas Publs., Mus. Nat. Hist., 5(3):27, February 28.

Type.—From one mile east of Laramie, 7164 feet, Albany County, Wyoming.

Comparisons.—Brownish upper parts and usually cinnamon-washed underparts darker than in *Thomomys talpoides clusius* and *T. t. cheyennensis;* upper parts often white-spotted in *rostralis* as well as in *T. t. meritus, T. t. attenuatus,* and *T. t. cheyennensis;* at high elevations *rostralis* large and dark but paler at lower elevations, east and west of Medicine Bow Mountains. See *Comparisons* in accounts of other subspecies.

Topotypes of *rostralis* are much paler than specimens from high peaks of the Medicine Bow Mountains. In color, but not in size or cranial characteristics, there is a cline of increasingly darker color from the east (*cheyennensis*) westward (through the ranges of *attenuatus, rostralis,* and *meritus*). See *Remarks* under accounts of other subspecies.

Measurements.—Average and extreme external and cranial measurements of nine adult females and measurements of three males all from the type locality are, respectively, as follows: Total length, 214 (198-230), 220, 228, 212; length of tail, 56 (45-72), 56, 68, 56; length of hind foot, 27.1 (25-28.5), 28, 30, 27; basilar length, 31.6 (30.0-33.5), 33.2, 33.3, 33.0; zygomatic breadth, 22.4 (20.7-23.3), 23.7, —, 22.8; interorbital breadth, 6.5 (6.2-7.0), 6.4, 6.5, 6.5; length of nasals, 14.4 (13.2-14.9), 15.5, 15.0, 14.2; maxillary tooth-row, 7.9 (7-1-8.4), 8.2, 7.3, 7.6 (After Hall and Montague, 1951:32).

Distribution.—Medicine Bow Mountains and Transition Life-zone surrounding them. See Fig. 31. See also the account of *T. t. clusius.*

Records of occurrence.—Specimens examined, 151, as follows: **Albany Co.:** 5 mi. N Laramie, 7200 ft., 1; ⅜ mi. S, 2 mi. E Medicine Bow Peak, 10,800 ft., 3; *Nash Fork Creek, 4; 2 mi. S Browns Peak, 10,600 ft., 7; 2¼ mi. ESE Browns Peak, 10,300 ft., 7;* 3 mi. ESE Browns Peak, 10,000 ft., 6; *1 mi. E Laramie, 7160-7164 ft., 18;* 5½ mi. ESE Laramie, 8500 ft., 4; *4 mi. S, 8 mi. E Laramie, 8600 ft., 5; 6 mi. S, 8 mi. E Laramie, 1;* Pole Mtn., 5; *1 mi. SSE Pole Mtn., 8250 ft., 10; 1 mi. S Pole Mtn., 2;* 2 mi. SW Pole Mtn., 8300-8900 ft., 6; *2½ mi. S Pole Mtn., 8340 ft., 1; 3 mi. S Pole Mtn., 1; 15 mi. SW Laramie, 7300 ft., 1;* Woods P. O., 3 USNM.
Carbon Co.: Bridger Pass, 18 mi. SW Rawlins, 7500 ft., 23, 16 USNM; 6 mi. S, 13 mi. E Saratoga, 8500 ft., 2; *6 mi. S, 14 mi. E Saratoga, 8800 ft., 1; 7 mi. S, 11 mi. E Saratoga, 8000 ft., 5; 8 mi. S, 6 mi. E Saratoga, 9; 10 mi. N, 14 mi. E Encampment, 8000 ft., 2; 10 mi. N, 16 mi. E Encampment, 1; 8 mi. N, 16 mi. E Encampment, 8400 ft., 7.*

Thomomys talpoides tenellus Goldman

1939. *Thomomys talpoides tenellus* Goldman, Jour. Mamm., 20:238, May 15.

Type.—From Whirlwind Peak, 10,500 feet, Absaroka Mountains, Park County, Wyoming.

Comparisons.—Smaller and darker (see *Measurements*) than *Thomomys talpoides bullatus, T. t. bridgeri,* and *T. t. clusius.* Pelage darker and baculum much shorter than in *T. t. pygmaeus* and *T. t. ocius.* Mastoid bullae smaller, lambdoidal crest lower, and zygomata inclined more gradually anteriorly (less parallel laterally and less nearly a right angle anteriorly) than in *T. t. bridgeri*. From *bullatus, tenellus* differs in smaller auditory bullae and less robust rostrum; darker than subspecies of eastern Wyoming (except *nebulosus*, which is much larger) and rarely (instead of frequently) spotted dorsally with white-spots.

Measurements.—External and cranial measurements of the holotype (female), two other adult females, and an adult male all from the type locality are, respectively, as follows: Total length, 200, 195, 193, 210; length of tail, 57, 63, 62, 61; length of hind foot, 25, 26, 25, 28; occipitonasal length, 33.6, 34.4, 32.8, 35.9; zygomatic breadth, 18.1, 19.2, 18.1, 19.6; interorbital breadth, 5.9, 6.2, 5.6, 6.3; length of nasals, 11.8, 11.7, 11.1, 12.7; maxillary tooth-row, 7.0, 6.6, 6.7, 7.5. Baculum of KU 91138, 14.4 mm. in length.

Distribution.—Mountainous northwestern part of state. See Fig. 31.

Remarks.—To the southward this subspecies seemingly intergrades with the larger *T. t. bridgeri*. Both are dark and have a short baculum. Specimens from the Wind River Mountains are referable to *bridgeri* and *bullatus* instead of to *tenellus*.

Records of occurrence.—Specimens examined, 108, as follows: **Park Co.:** 31½ mi. N, 36 mi. W Cody, 6900 ft., 1; Black Mtn., 4 USNM; Whirlwind Peak, 10,500 ft., 1 USNM; *S. W. slope Whirlwind Peak, 9000 ft., 19;* 6 mi. S, 3 mi. W Whirlwind Peak, 6700 ft., 3; *Mouth Grinnell Creek, 6300-7500 ft., 15 USNM; 2 mi. S, 42 mi. W Cody, 6400 ft., 1.*
Teton Co.: 2 mi. S, 1 mi. W Snake River, 2; 18 mi. N, 9 mi. W Moran, 4 Univ. Wyo.; *Upper Arizona Creek, 5 USNM; Moose Creek, 6800 ft., 2 USNM; S Moose Creek, 10,000 ft., 4 USNM; Pacific Creek, 6800 ft., 2 USNM; Two Ocean Lake, 2; Emma Matilda Lake, 2; Jackson Hole Wildlife Park, 3 USNM; 3 mi. N Moran, 6800 ft., 1; ¼ mi. N, 2½ mi. E Moran, 6230 ft., 5;* Moran, 2 USNM, 3 Univ. Wyo.; *3¾ mi. E Moran, 6300 ft., 2; ½ mi. E Moran, 6742 ft., 1 Univ. Wyo.; ¾ mi. S, 3¾ mi. E Moran, 6200 ft., 2; 1 mi. S, 3¾ mi. E Moran, 6200 ft., 1; 2 mi. S, 1 mi. W Moran, 6950 ft., 1; 7 mi. S Moran, 1;* Togwotee Pass, 1 USNM, 1 Univ. Wyo. 2½ mi. NE Moose, 6500 ft., 5; *Elk Ranch, 1;* Teton Pass, 7200 ft., 4 USNM; Jackson, 6300 ft., 2.
Yellowstone Nat'l Park: Lamar River, 2 USNM; W. end Yellowstone Lake, 2 USNM; Old Faithful, 1 USNM.

Geomys bursarius lutescens Merriam

Plains Pocket Gopher

1890. *Geomys bursarius lutescens* Merriam, N. Amer. Fauna, 4:51, October 8.

Type.—From sandhills on Birdwood Creek, Lincoln County, Nebraska.

Description.—Large (see *Measurements*); pale yellowish brown; claws long and robust; skull large with robust rostrum and grooved upper incisors (one deep groove medially and one shallow groove near inner margin).

Measurements.—Average and extreme external and cranial measurements of four adult males and measurements of an adult female all from 23 miles southwest of Newcastle are, respectively, as follows: Total length, 266.0 (254-285), 247; length of tail, 76.0 (69-88), 71; hind foot, 33.5 (33-34), 30; condylobasal length, 44.8 (42.8-46.2), 41.5; condylonasal length, 46.6 (44.1-48.3), 43.2; zygomatic breadth, 28.5 (25.7-30.5), 25.6; interorbital breadth, 6.7 (6.1-7.2), 6.5; length of nasals, 16.6 (14.8-17.3), 15.1; maxillary tooth-row, 8.7 (8.3-9.0), 8.4.

Distribution.—Eastern fifth of state. See Fig. 32.

Records of occurrence.—Specimens examined, 84, as follows: **Converse Co.:** 3 mi. N, 5 mi. E Orin, 4725 ft., 5. **Crook Co.:** ½ mi. N, 1 mi. E Beulah, 3550 ft., 1. **Goshen Co.:** Rawhide Buttes, 12 mi. S, 1 mi. W Lusk, 1; *5 mi. N Muskrat Canyon, 1;* 8 mi. SE Torrington, 4045 ft., 3. **Laramie Co.:** Horse Creek, 6½ mi. W Meriden, 5200 ft., 18; *Horse Creek, 6 mi. W Meriden, 5200 ft., 2; 4 mi. W Meriden, 3; 2½ mi. SW Meriden, 1; 20 mi. SE Chugwater, 2 USNM;* not found, 1 mi. W Galeo, 3. **Niobrara Co.:** 10 mi. N Hat Creek P. O., 4300 ft., 18; Hat Creek P. O., 4300 ft., 4; *Lusk, 3 USNM;* 2 mi. S, ½ mi. E Lusk, 5000 ft., 3. **Platte Co.:** 3 mi. W Guernsey, 4550 ft., 1; Uva, 2 USNM. **Weston Co.:** 23 mi. SW Newcastle, 4500 ft., 13.

Family HETEROMYIDAE

KEY TO HETEROMYID RODENTS

1. Plantar surfaces well-haired; interparietal less than ¼ greatest width of skull . . . . **Dipodomys ordii,** p. 618
1′. Plantar surfaces naked; interparietal more than ¼ greatest width of skull . . . . Genus **Perognathus,** p. 613

Genus Perognathus Wied-Neuwied

Pocket Mice

Mostly small having large auditory bullae; rooted molars; distinctly grooved upper incisors; and a dental formula of i.$\frac{1}{1}$, c.$\frac{0}{0}$, p.$\frac{1}{1}$, m.$\frac{3}{3}$.

KEY TO SPECIES OF PEROGNATHUS

1. Length of tail usually more than half of total length of animal (in adults); hind foot 21-23 mm.; occurring in southwestern Wyoming, *Perognathus parvus,* p. 617
1′. Length of tail usually less than half of total length of animal; hind foot shorter than 21 or longer than 23 mm. 2
2. Hind foot more than 23.5 mm. (24–28); upper parts brownish, *Perognathus hispidus,* p. 617
2′. Hind foot less than 20.5 (17–19) mm.; upper parts olivaceous or pale grayish 3
3. Interparietal bone more than 4 mm. wide (transverse to long axis of skull); upper parts olivaceous . . . . *Perognathus fasciatus,* p. 613
3′. Interparietal bone less than 4 mm. wide (transverse to long axis of skull); upper parts buffy . . . . *Perognathus flavus,* p. 616

Perognathus fasciatus Wied-Neuwied

Olive-backed Pocket Mouse

Ordinarily the olive-backed pocket mouse can be distinguished from the silky pocket mouse by smaller postauricular pale patches. When the ear is appressed to the side of the head, the exposed part of the pale postauricular patch is no longer than the ear, as measured from the notch, whereas in P. *flavus* the patch is longer than the ear in almost all specimens.

Perognathus fasciatus callistus Osgood

1900. *Perognathus callistus* Osgood, N. Amer. Fauna, 18:28, September 20.
1953. *Perognathus fasciatus callistus,* Jones, Univ. Kansas Publs., Mus. Nat. Hist., 5(29):517, 524, August 1.

Type.—From Kinney Ranch, Sweetwater County, Wyoming.

Comparisons.—Characterized chiefly by large, elongate auditory bullae and large size, *Perognathus fasciatus callistus* differs from the smaller *P. f. litus* in olivaceous instead of silvery upper parts, larger mastoid and otic bullae, longer hind foot, and usually narrower interparietal.

From *P. f. olivaceogriseus* of eastern Wyoming, *P. f. callistus* differs in usually larger size, larger auditory bullae, narrower interparietal, longer hind foot, and paler, more yellowish olivaceous upper parts.

Measurements.—External and cranial measurements of the holotype, three other adult males, and a single adult female, all from the type locality, are, respectively, as follows: Total length, 135, 131, 132, 137, 127; length of tail, 63, 68, 58, 64, 54; length of hind foot, 18, 18, 18, 19, 18; occipitonasal length, 22.8, 22.5, —, 22.7, 22.7; zygomatic breadth, 12.4, 11.8, 12.3, 12.6, 12.3; interorbital breadth, 4.7, 4.9, 5.1, 5.0, 5.1; length of nasals, 7.9, 8.2, —, 8.2, 8.0; maxillary tooth-row, 3.1, 3.2, 3.2, 3.3, 3.1.

Distribution.—Arid habitats in southwestern Wyoming. See Fig. 33.

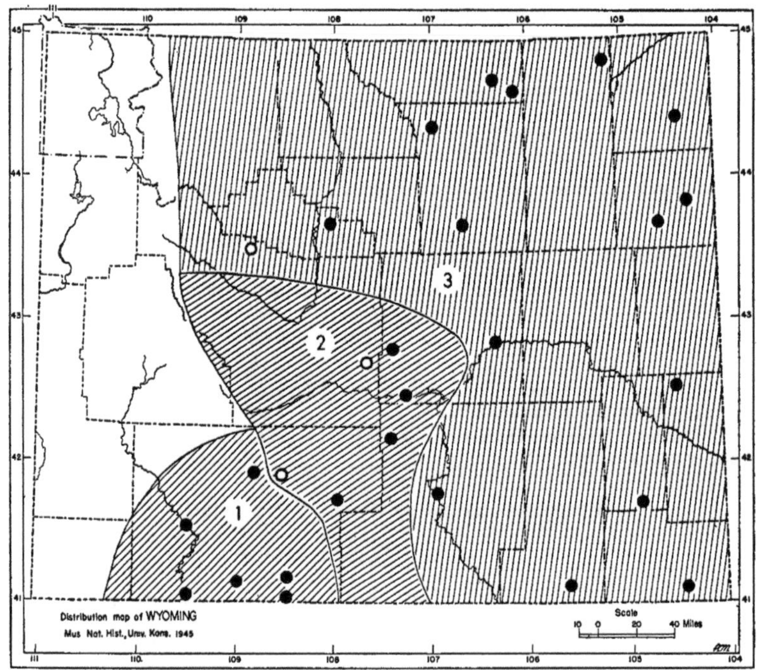

FIG. 33. Distribution of *Perognathus fasciatus.*
Guide to subspecies
1. *P. f. callistus*
2. *P. f. litus*
3. *P. f. olivaceogriseus*

Remarks.—I follow Jones (1953:524) in arranging P. *callistus* as a subspecies of P. *fasciatus* on account of observed intergradation between *callistus* and P. *f. litus* (in a series of specimens from 27 miles north and 37 miles east of Rock Springs).

Records of occurrence.—Specimens examined, 28, as follows: **Sweetwater Co.:** 27 mi. N, 37 mi. E Rock Springs, 6700 ft., 1; *25 mi. N, 38 mi. E Rock Springs, 6700 ft., 4;* Green River, 2 USNM; *18 mi. S Bitter Creek, 6800 ft., 3;* Kinney Ranch, 6800 ft., 4 USNM, 6; 32 mi. S, 22 mi. E Rock Springs, 7025 ft., 2; *30 mi. S Bitter Creek, 6900 ft., 2;* 33 mi. S Bitter Creek, 6900 ft., 2; The Green River, 4 mi. ENE Linwood, Utah, 2.

Perognathus fasciatus litus Cary

1911. *Perognathus fasciatus litus* Cary, Proc. Biol. Soc. Washington, 24:61, March 22.

Type.—From Sun, Sweetwater Valley, Natrona County, Wyoming.

Comparisons.—For comparison with *Perognathus fasciatus callistus* see account of that subspecies. Compared with P. *f. olivaceogriseus,* P. *f. litus* is smaller and paler; the upper parts appear silvery and lack the olivaceous color of *olivaceogriseus.*

Measurements.—External and cranial measurements of the holotype (subadult female) and another female (same age) from the type locality are, respectively, as follows: Total length, 128, 127; length of tail, 59, 57; length of hind foot, 18, 18; occipitonasal length, 22.0, 22.0; zygomatic breadth, —, 11.1; interorbital breadth, 4.6, 5.0; length of nasals, 7.3, 8.2; cranial depth, 7.0, 7.3; maxillary tooth-row, 3.0, 3.2.

Distribution.—Sweetwater Valley and nearby parts of Red Desert. See Fig. 33.

Records of occurrence.—Specimens examined, 9, as follows: **Carbon Co.:** 8 mi. SE Lost Soldier, 6700 ft., 1. **Natrona Co.:** 16 mi S, 11 mi. W Waltman, 6950 ft., 1; *5 mi. W Independence Rock, 6000 ft., 4;* Sun., 2 USNM. **Sweetwater Co.:** 2½ mi. N Wamsutter, 1.

Additional records (Jones, 1953:524).—**Fremont Co.:** Granite Mtns. **Sweetwater Co.:** 27 mi. N Table Rock.

Perognathus fasciatus olivaceogriseus Swenk

1940. *Perognathus flavescens olivaceogriseus* Swenk, Missouri Valley Fauna, 3:6, June 5.

1953. *Perognathus fasciatus olivaceogriseus,* Jones, Univ. Kansas Publs., Mus. Nat. Hist., 5:520, August 1.

Type.—From Little Bordeaux Creek, sec. 14, T. 33 N, R. 48 W, 3 mi. E. Chadron, Dawes County, Nebraska (Jones, 1953:520-522).

Comparisons.—Characterized chiefly by dark, Cream Buff upper parts finely lined with black, distinct ochraceous lateral line, and small auditory bullae. See *Comparisons* under accounts of other subspecies.

Measurements.—External and cranial measurements of an adult female and an adult male from 23 miles southwest of Newcastle and of two adult males from two miles south and 6½ miles west of Buffalo are, respectively, as follows: Total length, 140, 134, 126, 132; length of tail, 64, 64, 59, 64; length of hind foot, 17, 17, 17, 18; occipitonasal length, 23.7, 23.2, 22.2, 22.1; zygomatic breadth,

12.5, 12.6, 12.0, 11.5; interorbital breadth, 4.7, 4.9, 4.9, 5.0; length of nasals, 8.8, 8.3, 8.4, 7.9; mastoid breadth, 11.1, 11.2, 10.4, 10.4; maxillary tooth-row, 3.3, 3.0, 3.1, 3.3.

Distribution.—Mainly in rocky soils in northeastern half of state. See Fig. 33.

Remarks.—Two specimens (KU 87747-8) from 15 miles east-southeast of Cheyenne are markedly darker dorsally than specimens to the northward (for example, from Muskrat Canyon, Goshen County). The specimens from the vicinity of Cheyenne also differ from all specimens of P. *fasciatus* from Wyoming in having slight tinges of buff or cinnamon on their venters (some of which have disappeared since preservation). In both characters these specimens approach P. *f. infraluteus,* a subspecies occurring to the southward with type locality at Loveland, Colorado, and may be intergrades between that subspecies and *olivaceogriseus.* Topotypes of *infraluteus* vary in color of venter from white to cinnamon. However, the skulls of KU 87747-8 are indistinguishable from skulls of *olivaceogriseus,* including moderately large auditory bullae.

Records of occurrence.—Specimens examined, 26, as follows: **Albany Co.:** Tie Siding Picnic Ground, 8595 ft., 1. **Campbell Co.:** 1¼ mi. N, ½ mi. E Rockypoint, 3850 ft., 1. **Carbon Co.:** Ft. Steele, 1 USNM. **Crook Co.:** Sundance, 1 USNM. **Goshen Co.:** Muskrat Canyon, 1. **Hot Springs Co.:** Kirby Creek, Bighorn Basin, 5000 ft., 2 USNM. **Johnson Co.:** 2 mi. S, 6½ mi. W Buffalo, 5620 ft., 3; 1 mi. WSW Kaysee, 4700 ft., 1. **Laramie Co.:** 15 mi. ESE Cheyenne, 2. **Natrona Co.:** *1 mi. NE Casper, 5150 ft., 2;* Casper, 1 USNM. **Platte Co.:** 2½ mi. S Chugwater, 1. **Sheridan Co.:** 5 mi. NE Clearmont, 3900 ft., 1; Arvada, 3 USNM. **Weston Co.:** Newcastle, 1 USNM; 23 mi. SW Newcastle, 4500 ft., 4.

Additional record (Jones, 1953:522).—**Fremont Co.:** 40 mi. E Dubois.

Perognathus flavus piperi Goldman

Silky Pocket Mouse

1917. *Perognathus flavus piperi* Goldman, Proc. Biol. Soc. Washington, 30:148, July 27.

Type.—From 23 miles southwest of Newcastle, Weston County, Wyoming.

Comparison.—Compared with *Perognathus fasciatus olivaceogriseus,* which is comparable in size, P. *flavus piperi* differs in lacking olivaceous on its Light Buff to Light Ochraceous-Buff upper parts and in having conspicuous, larger auricular pale patches. Also, the lateral line is less distinct, although present. In both kinds of mice the venters are white and the upper parts are finely lined with black.

Measurements.—External and cranial measurements of the holotype are as follows: Total length, 113; length of tail, 51; length of hind foot, 17; occipitonasal length, 22.0; mastoidal breadth, 12.4; interorbital breadth, 4.6; length of nasals, 8.1; maxillary tooth-row, 3.4 (Goldman, 1917:148).

Distribution.—Known in Wyoming only from the type locality, but probably occurs elsewhere in Transition Life-zone of eastern fourth of state. Not mapped.

Record of occurrence.—Specimen examined, 1, as follows: **Weston Co.:** *23 mi. SW Newcastle, 1 USNM (not plotted).*

Perognathus parvus clarus Goldman

Great Basin Pocket Mouse

1917. *Perognathus parvus clarus* Goldman, Proc. Biol. Soc. Washington, 30:147, July 27.

Type.—From Cumberland, Lincoln County, Wyoming.

Description.—Large (see *Measurements*) with unusually long tail; brownish or yellowish brown above, whitish below; hind foot long, usually more than 21 mm; nasals long and straight-sided; upper incisors robust; zygomata slender; tympanic bullae well inflated.

Comparison.—Compared with *Perognathus fasciatus callistus*, which occurs also in southwestern Wyoming, *Perognathus parvus clarus* differs in having longer tail and hind foot, averaging larger in most measurements, and possessing more grizzled (blackish) upper parts.

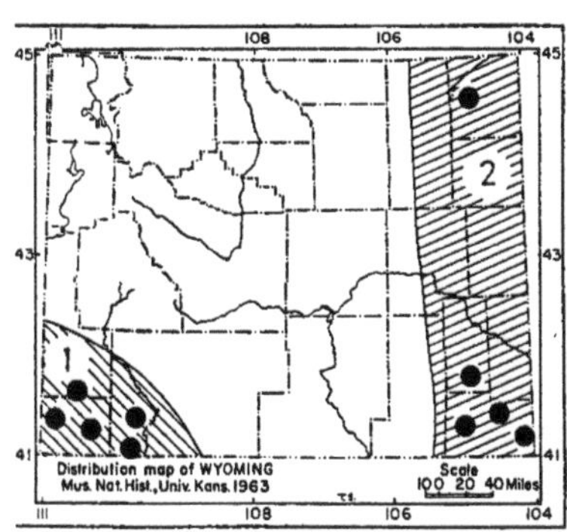

FIG. 34. Two species of *Perognathus*.
1. *P. parvus clarus*
2. *P. hispidus paradoxus*

Measurements.—External and cranial measurements of an old adult male and an old adult female from 26 miles south and 21 miles west of Rock Springs are, respectively, as follows: Total length, 182, 162; length of tail, 96, 82; length of hind foot, 24.5, 22; ear from notch, 8.2, 7.5; occipitonasal length, 24.5, 26.9; basilar length, 17.1, 18.2; zygomatic breadth, 12.0, 13.5; interorbital breadth, 5.7, 5.7; length of nasals, 9.6, 11.1; mastoidal breadth, 11.9, 13.2; maxillary tooth-row, 3.6, 3.8. Average and extreme measurements of the holotype and seven adult topotypes, all males, are as follows: Total length, 173 (160-186); length of tail, 90 (83-97); length of hind foot, 22 (21-23); occipitonasal length, 25.4 (25.1-26.0, six specimens); zygomatic breadth, 12.4 (12.2-12.5, four specimens); interorbital breadth, 5.7 (5.5-5.9); length of nasals, 9.8 (9.5-10.0); cranial depth, 7.7 (7.6-8.0, six specimens); maxillary tooth-row, 3.8 (3.6-3.9).

Distribution.—Desert areas in southwestern Wyoming. See Fig. 34.

Records of occurrence.—Specimens examined, 31, as follows: **Lincoln Co.:** Cumberland, 6550 ft., 2, 9 USNM. **Sweetwater Co.:** 26 mi. S, 21 mi. W Rock Springs, 6; *32 mi. S, 22 mi. W Rock Springs, 5;* ½ mi. N Jct. Henrys Fork and Utah Boundary, 2. **Uinta Co.:** Bear River, 14 mi. N Evanston, 6800 ft., 2 USNM; *Ft. Bridger, 1 USNM; 2½ mi. WSW Ft. Bridger, 6800 ft., 2;* Mountainview, 2 USNM.

Perognathus hispidus paradoxus Merriam

Hispid Pocket Mouse

1889. *Perognathus paradoxus* Merriam, N. Amer. Fauna, 1:24, October 25.

1900. *Perognathus hispidus paradoxus,* Osgood, N. Amer. Fauna, 18:44, September 20.

Type.—From Banner, Trego County, Kansas.

Comparisons.—Characterized chiefly by large size (hind foot, 25-28 mm.) and strongly grizzled (blackish) upper parts, *Perognathus hispidus paradoxus* is much larger and more brownish than *P. fasciatus* and *P. flavus*. Also, in *P. hispidus* mastoid bullae relatively smaller and auricular patches less conspicuous. Ochraceous lateral line distinctive in all three species.

Measurements.—External and cranial measurements of an old female (KU 18255) and an old male (KU 18257) from 2½ miles south of Chugwater and of an old male (KU 25672) from two miles south of Pine Bluffs are, respectively, as follows: Total length, 220, 221, 220; length of tail, 108, —, 105; length of hind foot, 28, 28, 27; ear from notch, 12, 12, 13; occipitonasal length, 30.5, 33.2, 34.0; zygomatic breadth, 16.4, 17.0, 17.2; interorbital breadth, 7.4, 8.0, 7.4; length of nasals, 11.4, 13.0, 12.9; mastoidal breadth, 14.6, 14.5, 14.9; maxillary tooth-row, 5.0, 4.7, 5.2.

Distribution.—Prairie habitat of eastern fifth of state. See Fig. 34.

Records of occurrence.—Specimens examined, 19, as follows: **Crook Co.:** 2 mi. N, 14 mi. W Hulett, 1. **Laramie Co.:** 5½ mi. S Glenys, 5200 ft., 1; Horse Creek, 6½ mi. W Meriden, 5200 ft., 1; 1 mi. S Pine Bluffs, 1; *2 mi. S Pine Bluffs, 5200 ft., 7.* **Platte Co.:** Chugwater, 1 USNM; *2½ mi. S Chugwater, 7.*

Dipodomys ordii Woodhouse

Ord's Kangaroo Rat

Adapted for saltatorial locomotion having greatly elongated hind limbs and long tail. Auditory bullae, as in many leapers, greatly enlarged. Cheek-teeth persistently growing; anterior root of zygomatic arch enlarged; color pale Ochraceous-Buff above and white below. External cheek-pouches large. Known only from open deserts or plains. Dental formula = i.$\frac{1}{1}$, c.$\frac{0}{0}$, p.$\frac{1}{1}$, m.$\frac{3}{3}$.

Dipodomys ordii luteolus (Goldman)

1917. *Perodipus ordii luteolus* Goldman, Proc. Biol. Soc. Washington, 30: 112, May 23.

1921. *Dipodomys ordii luteolus*, Grinnell, Jour. Mamm., 2:96, May 2.

Type.—From Casper, Natrona County, Wyoming.

Comparisons.—From specimens of *Dipodomys ordii priscus*, topotypes of *D. o. luteolus* differ little in color (upper parts between Light Ochraceous-Buff and Ochraceous-Buff, venter white), although specimens of *luteolus* from eastern Wyoming are slightly darker than topotypes of *luteolus* to the westward; *luteolus* is larger except that its hind foot is shorter and the auditory bullae are less inflated than in *priscus*.

From *D. o. terrosus*, *D. o. luteolus* differs in paler upper parts, smaller size excepting longer tail and pinna of ear, less inflated auditory bullae and smaller skull.

Measurements.—Average and extreme external and cranial measurements of 12 adult males and seven adult females from one mile northeast of Casper are, respectively, as follows: Total length, 265.6 (254-281), 260.7 (250-269); length of tail, 152.2 (145-163), 148.0 (139-153); length of hind foot, 42.2 (42-43), 41.0 (40-43); greatest length of skull, 38.9 (37.5-39.5), 38.6 (37.6-40.5); greatest breadth across bullae, 24.1 (23.8-25.0), 24.2 (23.0-25.7);

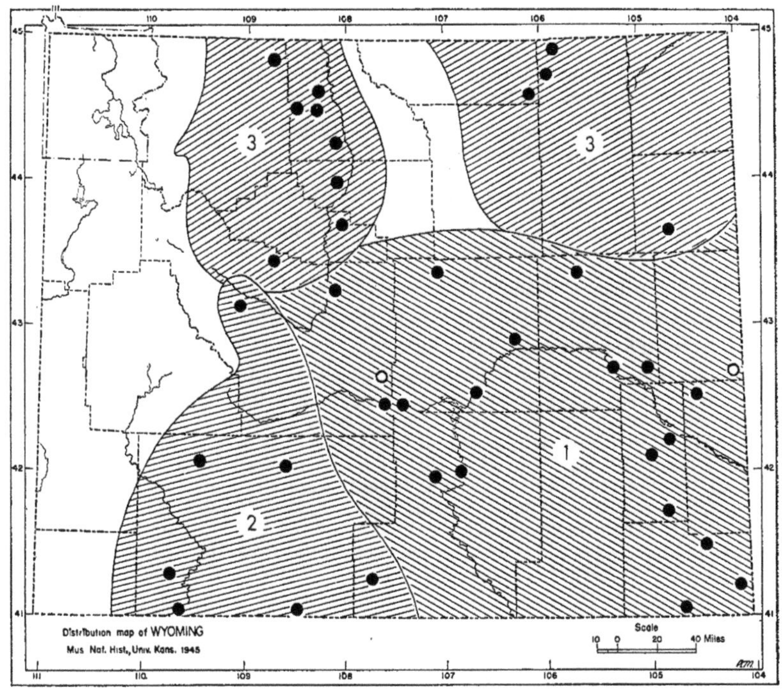

FIG. 35. Distribution of Dipodomys ordii
Guide to subspecies 2. D. o. priscus
1. D. o. luteolus 3. D. o. terrosus

breadth across maxillary arches, 20.8 (19.9-22.1), 20.9 (20.0-21.9); width of rostrum, 4.3 (4.2-4.4), 4.3 (4.2-4.4); length of nasals, 13.9 (13.0-14.5), 13.9 (13.3-14.9); least interorbital breadth, 13.0 (12.5-13.7), 12.9 (12.5-13.8); basilar length, 24.6 (24.0-25.7), 24.7 (24.0-25.5) (Setzer, 1949:567).

Distribution.—Throughout central lowlands of Wyoming ranging eastward and southward across state boundaries. See Fig. 35.

Remarks.—*D. o. luteolus* and *D. o. terrosus* intergrade in extreme northeastern Wyoming. In 11 pregnant females taken in the period July 19-22, 1945, six miles west of Meridan, the number of embryos averaged 3.1 (2-4).

Records of occurrence.—Specimens examined, 187, as follows: **Carbon Co.:** 24 mi. N, 12 mi. E Sinclair, 6600 ft., 9; 18 mi. NNE Sinclair, 6500 ft., 4; *Ft. Steele, 2 USNM.*

Converse Co.: 12 mi. N, 6 mi. W Bill, 4800 ft., 1; Douglas, 3 USNM; 3 mi. N, 5 mi. E Orin, 4725 ft., 1. **Fremont Co.:** *5 mi. N, 1 mi. W Shoshoni, 4700 ft., 4;* 2½ mi. W Shoshoni, 4800 ft., 42; 3 mi. W Splitrock, 7400 ft., 4. **Goshen Co.:** Rawhide Buttes, 1 USNM. **Laramie Co.:** Horse Creek, 6 mi. W Meriden, 5200 ft., 39; *Horse Creek, 4 mi. W Meriden, 5050 ft., 1; 5½ mi. S Glenys (Glenys = 3 mi. N Horse Creek), 5200 ft., 4;* Pine Bluffs, 1 USNM; *1 mi. S Pine Bluffs, 4; 2 mi. S Pine Bluffs, 5200 ft., 6;* 15 mi. ESE Cheyenne, 3.

Natrona Co.: 27 mi. N, 1 mi. E Powder River, 6075 ft., 1; 1 mi. NE Casper, 5150 ft., 35; *Casper, 3 USNM;* Alcova, 5180 ft., 2; 5 mi. W Independence Rock, 6500 ft., 2; *1 mi. S, 9 mi. W Independence Rock, 2; 1 mi. S, 5 mi. W Independence Rock, 1; Sun, 8 USNM.* **Platte Co.:** 3 mi. W Guernsey, 4550 ft., 1; Jetsam, 2 USNM; 2½ mi. S Chugwater, 5300 ft., 1.

Additional records (Setzer, 1949:534).—**Fremont Co.:** Granite Mtns. **Niobrara Co.:** Van Tassel Creek.

Dipodomys ordii priscus Hoffmeister

1942. *Dipodomys ordii priscus* Hoffmeister, Proc. Biol. Soc. Washington, 55:167, December 31.

Type.—From Kinney Ranch, Sweetwater County, Wyoming.

Comparisons.—Smaller and paler than D. *o. terrosus;* for comparison with D. *o. luteolus* see account of that subspecies.

Measurements.—Average and extreme external and cranial measurements (after Setzer, 1949:568) of seven adult males and four adult females from 33 miles south of Bitter Creek are, respectively, as follows: Total length, 259.0 (251-265), 257.0 (249-264); length of tail, 148.0 (144-152), 147.0 (138-152); length of hind foot, 44.0 (43-45), 43.0 (40-45); greatest length of skull, 39.1 (38.0-40.4), 39.4 (38.1-40.4); greatest breadth across bullae, 24.3 (23.7-25.1), 24.6 (23.5-25.2); breadth across maxillary arches, 20.7 (20.0-21.2), 20.8 (20.1-21.9); width across rostrum, 4.1 (4.0-4.3), 4.2 (4.1-4.3); length of nasals, 14.3 (13.8-15.2), 14.3 (14.0-14.9); least interorbital breadth, 13.1 (12.7-13.6), 13.1 (12.7-13.3); basilar length, 24.9 (23.7-25.5), 24.7 (24.0-25.2).

Distribution.—High arid plateaus and low desert areas of Green River Watershed and northward into Wind River Basin. See Fig. 35.

Records of occurrence.—Specimens examined, 82, as follows: **Carbon Co.:** 10½-11½ mi. N Baggs, 4. **Fremont Co.:** *Wind River, 1 USNM;* 7 mi. N Ft. Washakie, 1 USNM; *Ft. Washakie, 2 USNM.* **Sweetwater Co.:** Eden, 1 USNM; 27 mi. N, 37 mi. E Rock Springs, 6700 ft., 1; *25 mi. N, 38 mi. E Rock Springs, 6700 ft., 11;* 26 mi. S, 21 mi. W Rock Springs, 16; *32 mi. S, 22 mi. W Rock Springs, 11; 30 mi. S Bitter Creek, 3;* 33 mi. S Bitter Creek, 28; W. side Green River, 3.

Dipodomys ordii terrosus Hoffmeister

1942. *Dipodomys ordii terrosus* Hoffmeister, Proc. Biol. Soc. Washington, 55:165, December 31.

Type.—From Yellowstone River, five miles west Forsyth, 2750 ft., Rosebud County, Montana.

Comparisons.—For comparisons of this large, dark Cinnamon-Buff kangaroo rat with *Dipodomys ordii luteolus* and D. *o. priscus,* see accounts of those subspecies.

Measurements.—Average and extreme external measurements of 13 adults (10 females) from six miles northwest of Greybull are as follows: Total length, 275.6 (263-290); length of tail, 155.5 (140-166); length of hind foot, 43.5 (42-45). Measurements of three males and two females from Jordan, Montana, have been listed (Setzer, 1949:566) as follows: Total length, 280, 267, 279, 265, 273; length of tail, 155, 155, 162, 149, 154; length of hind foot, 44, 40, 41, 40.5, 41; greatest length of skull, 42.7, 40.5, 40.8, 41.4, 41.3; greatest breadth across bullae, 26.5, 24.8, 25.7, 25.4, 25.1; breadth across maxillary arches, 23.6, 21.1, 21.6, 22.2, 22.4; width of rostrum, 4.0, 4.0, 4.4, 4.3, 4.0; length of nasals, 15.2, 14.5, 14.6, 14.7, 14.9; least interorbital breadth, 14.5, 13.1, 13.4, 13.0, 13.8; basilar length, 27.0, 25.1, 25.9, 26.8, 26.0.

Distribution.—Lowlands of northern Wyoming. See Fig. 35.

Remarks.—*D. o. terrosus* intergrades with *D. o. luteolus* in Weston, Sheridan, and Campbell counties. The intergrades are referable to *D. o. terrosus.*

Records of occurrence.—Specimens examined, 57, as follows: **Big Horn Co.:** 6 mi. NW Greybull, 3800 ft., 13; 10 mi. W Germania, 2 USNM; 3 mi. E Germania, 1 USNM; *Greybull, 4 USNM; Bighorn River, 1 USNM;* 7 mi. S Basin, 3900 ft., 2. **Campbell Co.:** Powder River Crossing, 1 USNM; *Little Powder River, 3 USNM;* 2 mi. N, 7 mi. W Spotted Horse, 3890 ft., 2; *1½ mi. N, 6½ mi. W Spotted Horse, 1.* **Fremont Co.:** Sheep Creek, S. base Owl Creek Mtns., 6000 ft., 1 USNM. **Hot Springs Co.:** Kirby Creek, 1 USNM. **Park Co.:** 4 mi. N Garland, 3. **Sheridan Co.:** Arvada, 18 USNM. **Washakie Co.:** 4 mi. S, 7 mi. W Worland, 4100 ft., 1; *7 mi. S, 5 mi. W Worland, 4100 ft., 1.* **Weston Co.:** 23 mi. SW Newcastle, 4500 ft., 1, 1 USNM.

Family CASTORIDAE

Castor canadensis Kuhl

Beaver

Largest rodent in United States, adults weighing from 40 to almost 100 pounds; hind feet large, adapted for swimming; tail scaly and depressed, paddle shaped; upper parts brown; underparts paler brown; pelage dense consisting of fine underfur and long coarse guard hairs; dental formula: c. $\frac{1}{1}$, i. $\frac{0}{0}$, p. $\frac{1}{1}$, m. $\frac{3}{3}$.

According to Bagley (*in* Grasse and Putnam, 1955) the beaver was abundant in Wyoming prior to the appearance of white man. About 1820, the trappers and fur-traders entered Wyoming from several directions, trapping beavers even at the headwaters of several of the great river systems and seriously depleting the beaver populations in Wyoming, until the beaver was considered, for practical purposes, extinct. Every major rendezvous but two of the fur-trappers in the period 1823-1840 was held in Wyoming.

The beaver in Wyoming has had limited legal protection since sometime in 1899, and has responded to protection by becoming abundant again. The aim of beaver management in Wyoming (see Grasse and Putnam, 1955:38-39) was to plant brood stock (usually "nuisance" beaver) in every Wyoming stream that furnished adequate habitat and that would be "improved" by the introduction or re-introduction of beaver. Usually beaver are removed from streams in lowlands and planted in headwaters. This would cause more "gene flow" than would occur otherwise. Subspecific characters would be expected to be altered where crossbreeding between two subspecies occurs. However, the beaver in Wyoming, except the population in Carbon County, belong to one subspecies.

Fortunately, specimens of beaver taken before and around the turn of the century are available. They and specimens taken since 1940 are indistinguishable.

Beavers are valuable furbearers. Also, they conserve water and change arid habitats to lush riparian habitats. Much information on the life history and ecology of beaver in Yellowstone has been provided by Warren (1926). In shallow streams beavers build dams and lodges that represent extraordinary engineering feats. Beaver feed on the living bark of trees. According to Warren (1926:167) the aspen is preferred in northwestern Wyoming, but cottonwood, willow, and other trees are also used.

Castor canadensis missouriensis Bailey

1919. *Castor canadensis missouriensis* Bailey, Jour. Mamm., 1:32, November 28.

Type.—From Apple Creek, seven miles east of Bismarck, Burleigh County, North Dakota.

Comparison.—Compared with *Castor canadensis concisor, C. c. missouriensis* differs in upper parts slightly paler, postorbital process of jugal lower (orbit more nearly open), nasals narrower, and postorbital process of jugal thicker and truncated.

Measurements.—Cranial measurements of an adult male and of two adult females, all from the type locality, are, respectively, as follows: Occipitonasal length, 139, 130, 131; zygomatic breadth, 98, 91, 93; interorbital breadth, 25.5, 23.9, 27.7; length of nasals, 49.8, 48.5, 47.8; maxillary tooth-row, 31.8, 30.4, 30.9; greatest width of nasals, 24.9, 24.1, 24.5. Cranial measurements (average and extreme) of four specimens (two males, two of unknown sex) from western Park County and of one specimen (male) from the Gros Ventre River, Teton County, all taken since 1931, are as follows: Occipitonasal length, 137 (132-144); zygomatic breadth, 100 (97-102); interorbital breadth, 24.0 (22.6-25.8); length of nasals, 50.3 (48.4-52.4); maxillary tooth-row, 30.9 (29.0-32.2). Cranial measurements of an adult male from Bighorn Mountains, of an adult male from Warm Springs, and of an adult of unknown sex from Teton Canyon, all taken before the turn of the century, are, respectively, as follows: Occipitonasal length, —, 142, 130; zygomatic breadth, 95, 102, 91; interorbital breadth, 23.1, 26.4, 23.1; length of nasals, 46.4, 51.0, 45.0; maxillary tooth-row, 30.4, 30.6, 28.5; greatest breadth of nasals, 24.0, 26.2, 21.5. External measurements of two adult females and one adult male from Little Laramie River Valley are as follows: Total length, 918, 1087, 1027; tail 251, 266, 276; hind foot, 160, 185, 169; ear, 35, 36, 37.

Distribution.—Throughout most of Wyoming in suitable habitat excepting the mountainous south-central part occupied by *C. c. concisor.* See account of that subspecies. Not mapped.

Remarks.—Specimens taken before the year 1900 and those taken after 1946 from northwestern Wyoming do not differ significantly from topotypes of *missouriensis.* Pelages from Wyoming are not paler than in topotypes from North Dakota, although not comparable as to season. Two subadults from Fort Bridger seem referable to *missouriensis.* Recently taken (1946-1958) beaver from Chugwater Creek, Pole Creek, and Horse Creek, all in southeastern Wyoming, resemble topotypes of *missouriensis* except that the nasals are markedly wider (therein approaching *C. c. concisor*). Beaver taken in 1947 from the Little Laramie River Valley have narrow nasals, but some specimens approach *concisor* in shape of the jugals.

According to the label on beaver No. 5433 USNM from Swan Lake Flat, the animal had been eaten by a mountain lion.

Records of occurrence.—Specimens examined, 127, as follows: **Albany Co.:** Little Laramie River Valley, 7; 1½ mi. SE Pole Mtn., 8200 ft., 1.

Big Horn Co.: Bighorn Mtns., 2 USNM.

Fremont Co.: 7 mi. S South Pass City, 3.

Laramie Co.: Upper Horse Creek, 9 mi. W Horse Creek P. O., 5; Stone Creek, 3 mi. W Horse Creek P. O., 5; Horse Creek P. O., 6400 ft., 20; *3 mi. E Horse Creek* P. *O., 6400 ft., 22; Pole Creek, 5-6 mi. E Federal* P. *O., 21;* Locality not specified, 7.

Park Co.: Pat O'hara Creek, 15 mi. N, 8 mi. W Cody, 2; 26 mi. S, 25 mi. W Cody, 1; 30 mi. S, 43 mi. W Cody, 1.

Platte Co.: Warm Springs, Guernsey, 1 USNM; Chugwater Creek, 16.

Teton Co.: Emma Matilda Lake, 1; 1 mi. S, 3¾ mi. E Moran, 2; Teton Canyon, 1 USNM; Teton Basin, N. Fk. Teton River, 1 USNM; 3 mi. NW Kelly, 2; Gros Ventre Rd., E Kelly, 2.

Uinta Co.: Ft. Bridger, 8000 ft., 1; 9 mi. S Robertson, 8000 ft., 1.

Yellowstone Nat'l Park: Yanceys Ranch, 1 USNM; Swan Lake Flat, 1 USNM.

Castor canadensis concisor Warren and Hall

1939. *Castor canadensis concisor* Warren and Hall, Jour. Mamm., 20:358, August 14.

Type.—From Monument Creek, southwest of Monument, El Paso County, Colorado.

Comparison.—See account of *Castor canadensis missouriensis.*

Measurements.—Cranial measurements of the type and a topotype (sex unknown) are, respectively, as follows: Occipitonasal length, 141.5, 131; basilar length, 127.4, 119.4; zygomatic breadth, 102.2, 96.5; mastoidal breadth, 69.2, 61.1; least interorbital breadth, 24.5, 25.5; length of nasals, 54.4, 46.2; width of nasals, 26.4, 23.6; maxillary tooth-row, 31.0, 31.8 (Warren and Hall, 1939:361).

Distribution.—Known only from the Sierra Madre in south-central Wyoming. May occur also in the Medicine Bow Mountains. Not mapped.

Remarks.—See account of *Castor canadensis missouriensis.* The first specimen (KU 26633, a skull only) listed under "Specimens examined" is referable to *C. c. concisor* and differs from *C. c. missouriensis* in relatively broader zygomata, open orbit, and in the shape of the jugal. Its nasals are narrow as in *missouriensis.* The second specimen, listed below, a subadult, is referred to *concisor* on geographic grounds.

Records of occurrence.—Specimens examined, 2, as follows: **Carbon Co.:** 8 mi. N, 16 mi. E Encampment, 8400 ft., 1; 7½ mi. N, 18½ mi. E Savery, 8400 ft., 1.

Family CRICETIDAE

KEY TO CRICETID RODENTS

1. Cheek-teeth cusped; no occlusal lakes of dentine surrounded by enamel, 2
2. Tail less than 60 per cent of length of head and body; coronoid process more than 1½ times as high as wide....... **Onychomys leucogaster,** p. 635
2′. Tail more than 60 per cent of length of head and body; coronoid process less than 1½ times as high as wide.............................. 3
3. Anterior faces of incisors grooved; total length of animal less than 150.. 4
4. Total length less than 125; length of tail less than 55; pelage pale except middorsally; pinna of ear having two spots one above the other, **Reithrodontomys montanus,** p. 625
4′. Total length more than 125; length of tail more than 55; pelage dark agouti as in *Mus*; pinna of ear lacking spots, **Reithrodontomys megalotis,** p. 626
3′. Anterior faces of incisors smooth; total length of animal more than 150, 5
5. Tail as long as head and body (rarely less, usually more)........... 6
6. Tooth-row usually shorter than 4 mm.; pinna of ear shorter than hind foot.................................... **Peromyscus crinitus,** p. 627
6′. Tooth-row longer than 4 mm.; pinna of ear longer than hind foot, **Peromyscus truei,** p. 634
5′. Tail shorter than head and body.............................. 7
7. Tail not sharply bicolored; hind foot 22 or longer, **Peromyscus leucopus,** p. 634
7′. Tail sharply bicolored; hind foot usually shorter than 22, **Peromyscus maniculatus,** p. 628
1′. Cheek-teeth lack cusps; occlusal lakes of dentine surrounded by enamel.. 8
8. Tail bushy; ears nearly naked (scantily haired).... **Neotoma cinerea,** p. 638
8′. Tail not bushy; ears haired.................................. 9
9. Total length more than 480; tail compressed laterally and scaly, **Onadatra zibethicus,** p. 661
9′. Total length less than 480; tail round and haired............... 10
10. Molars rooted in adults...................................... 11
11. Upper parts reddish; in lower molars inner re-entrant angles little if any deeper than outer re-entrant angles..... **Clethrionomys gapperi,** p. 642
11′. Upper parts buffy gray; in lower molars inner re-entrant angles deeper than outer re-entrant angles............ **Phenacomys intermedius,** p. 646
10′. Molars rootless in adults.................................... 12
12. Tail shorter than 27; m3 having at least 4 prisms; auditory bullae cancellous................................ **Lagurus curtatus,** p. 660
12′. Tail longer than 27; m3 having fewer than 4 prisms; auditory bullae noncancellous.. 13
13. Tail less than 30 per cent of head and body.................... 14
14. Venter washed with ochraceous or cinnamon; upper parts reddish brown............................... **Microtus ochrogaster,** p. 659
14′. Venter usually whitish; upper parts brownish or grayish.......... 15
15. Venter occasionally washed with cinnamon; upper middle molar having fifth posterior loop.................... **Microtus pennsylvanicus,** p. 647
15′. Venter whitish, upper middle molar lacking fifth loop, **Microtus montanus,** p. 649
13′. Tail about 30 per cent of head and body or longer............... 16
16. Tail longer than 70; hind foot longer than 25; pelage dark, **Microtus richardsoni,** p. 656
16′. Tail usually shorter than 70; hind foot shorter than 25; pelage olivaceous-brown; sides pale gray.............. **Microtus longicaudus,** p. 653

Genus Reithrodontomys Giglioli

Harvest Mice

The two species of harvest mice so closely resemble mice of the genus *Peromyscus* that specimens of *Reithrodontomys* frequently are mistakenly identified as young *Peromyscus*. Occasionally they are misidentified as *Mus musculus*. A certain means for distinguishing the harvest mice is the longitudinal groove on the front of each upper incisor tooth. The incisor teeth of *Peromyscus* and *Mus* lack the groove. Pocket mice, genus *Perognathus*, have grooved upper incisors, but unlike harvest mice they have four instead of three cheek teeth, an ochraceous lateral line on each side of the body, and external, fur-lined cheek pouches.

Reithrodontomys montanus albescens Cary

Plains Harvest Mouse

1903. *Reithrodontomys albescens* Cary, Proc. Biol. Soc. Washington, 16:53, May 6.

1911. *Reithrodontomys montanus albescens* Cary, N. Amer. Fauna, 33:110, August 17.

Type.—From 18 miles northwest Kennedy, Cherry County, Nebraska.

Description.—Small (see *Measurements*); upper parts brownish gray usually becoming blackish mid-dorsally; pinna showing two spots, one above the other, at base; tail sharply bicolor, but dorsal blackish confined to narrow line; hind foot short (see *Measurements*); skull small but braincase well inflated; upper incisors distinctly grooved.

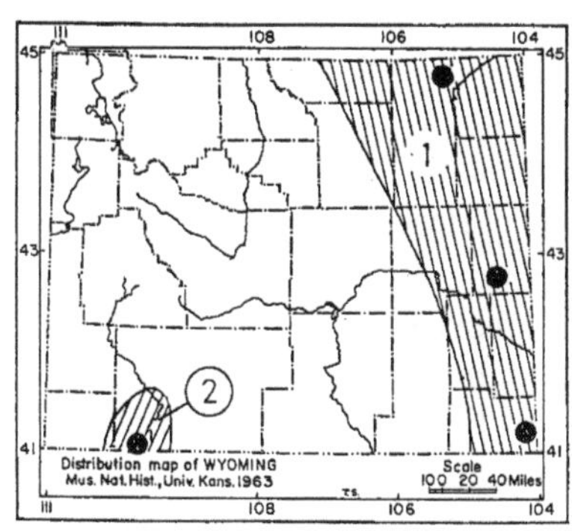

FIG. 36.
1. *Reithrodontomys montanus albescens*
2. *Peromyscus crinitus doutti*

Comparisons.—From *Reithrodontomys megalotis*, *R. montanus* differs in paler (more grayish and less brownish) upper parts; shorter tail, hind foot, and pinna of ear; spots (present instead of absent) at base of pinna of ear; narrow (instead of broad) line of blackish on top side of tail; smaller skull; slightly more inflated braincase; and less plumbeous underparts (the tips of the hairs of the venter are white in both species).

Measurements.—External and cranial measurements of the subadult male and lactating female from Campbell County and the subadult female from Niobrara County are, respectively, as follows: Total length, 111, 115, 118; length of tail, 49, 44, 50; length of hind food, 15, 20, 16; length of ear, 13, 12, 12; greatest length of skull, —, 18.8, 17.9; zygomatic breadth, 10.0, 10.2, 9.6; interorbital breadth, 3.0, 2.9, 2.8; length of nasals, —, 7.5, 6.2; cranial depth, 6.3, 6.4, 6.2; maxillary tooth-row, 3.2, 3.0, 3.1.

Distribution.—Known only from prairie areas in eastern fifth of state. See Fig. 36.

Records of occurrence.—Specimens examined, 4, as follows: **Campbell Co.:** *4/10 mi. N, 3/10 mi. E Rockypoint, 3800 ft., 1;* Rockypoint, 1. **Niobrara Co.:** 2 mi. S, ½ mi. E Lusk, 5000 ft., 1. **Laramie Co.:** 1 mi. S Pine Bluffs, 1.

Reithrodontomys megalotis dychei J. A. Allen

Western Harvest Mouse

1895. *Reithrodontomys dychei* J. A. Allen, Bull. Amer. Mus. Nat. Hist., 7:120, May 21.

1914. *Reithrodontomys megalotis dychei,* A. H. Howell, N. Amer. Fauna, 36:30, June 5.

Type.—From Lawrence, Douglas County, Kansas.

Description.—Large for *Reithrodontomys* (see *Measurements*); upper parts brownish washed with ochraceous; venter whitish but showing plumbeous (grayish) of basal parts of hairs; tail sharply bicolor being broadly blackish above; skull large; incisors grooved. For comparison with *R. montanus,* see account of that species.

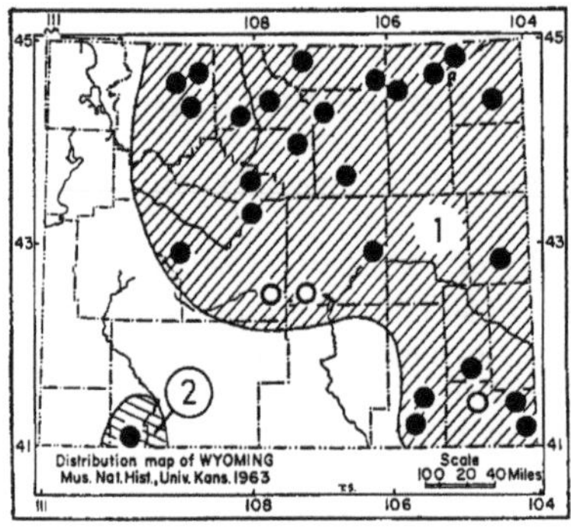

FIG. 37.
1. *Reithrodontomys megalotis dychei*
2. *Peromyscus truei truei*

Measurements.—External and cranial measurements of one adult male and four pregnant females from one mile west-southwest of Kaycee are, respectively, as follows: Total length, 134, 134, 128, 144, 143; length of tail, 65, 59, —, 68, 63; length of hind foot, 16, 17, 17, 17, 17; length of ear from notch, 13, 13, 12, 12, 15; greatest length of skull, —, 21.0, 19.7, 21.2, 21.0; zygomatic breadth, 10.2, 10.8, 10.4, 10.9, 10.9; interorbital breadth, 3.1, 3.2, 3.0, 2.9, 3.0; length of nasals, 7.5, 7.1, 7.0, 8.4, 7.6; cranial depth, 6.8, 6.9, 6.7, 6.7, 6.5; maxillary toothrow, 3.2, 3.1, 3.2, 3.2, 3.1.

Distribution.—Great Plains in northeastern Wyoming. See Fig. 37.

Remarks.—Eight pregnant females from Wyoming contained an average of 5.5 (3-7) embryos. Seven embryos were recorded from each of two females (KU 43938, 43945) from one mile west-southwest of Kaycee. This subspecies has been recently studied by Jones and Mursaloğlu (1961:9-27).

Records of occurrence.—Specimens examined, 84, as follows: **Albany Co.:** 27 mi. N, 8 mi. E Laramie, 6420 ft., 2; 1½ mi. N, 1 mi. W Laramie, 1; *2 mi. W Laramie, 1 Univ. Wyo.; 1 mi. S, 3 mi. W Laramie, 7200 ft., 2 Univ. Wyo.; 3.6 mi. SW Laramie, 1.*

Big Horn Co.: 7½ mi. E Greybull, 4050 ft., 1; 7 mi. S, ½ mi. E Basin, 1.

Campbell Co.: *1½ mi. N, ¾ mi. E Rockypoint, 1; Rockypoint, 5;* 5 mi. S, 4 mi. W Rockypoint, 1; 5 mi. N, 8 mi. W Spotted Horse, 2.

Crook Co.: 4 mi. N, 3 mi. E Rockypoint, 3800 ft., 3; 1½ mi. NW Sundance, 5000 ft., 3.

Fremont Co.: 12 mi. N, 3 mi. W Shoshoni, 4650 ft., 1; 3/10 mi. NW Milford, 5357 ft., 1; *Milford, 5357 ft., 1.*

Hot Springs Co.: 3 mi. N, 10 mi. W Thermopolis, 4900-4950 ft., 7.

Johnson Co.: *8/10 mi. S, 1 mi. W Buffalo, 4800 ft., 4;* 2 mi. S, 6½ mi. W Buffalo, 5620 ft., 4; 1 mi. WSW Kaycee, 4700 ft., 8.

Laramie Co.: Horse Creek, 3 mi. W Meriden, 5000 ft., 1; 1 mi. N, ½ mi. W Pine Bluffs, 5040 ft., 4; *1 mi. S Pine Bluffs, 5100 ft., 1; 2 mi. S Pine Bluffs, 5200 ft., 2.*

Natrona Co.: 1 mi. NE Casper, 5150 ft., 1; *2¼ mi. S, 1½ mi. W Casper, 1.*

Niobrara Co.: 2 mi. S, ½ mi. E Lusk, 5000 ft., 1.

Park Co.: 4 mi. N Garland, 2; 13 mi. N, 1 mi. E Cody, 5200 ft., 2; 6/10 mi. S, 32/10 mi. E Cody, 5020 ft., 1.

Platte Co.: 2½ mi. S Chugwater, 5300 ft., 4.

Sheridan Co.: 3 mi. WNW Monarch, 4; 5 mi. NE Clearmont, 3900 ft., 5.

Washakie Co.: 1 mi. N, 3 mi. E Tensleep, 4350 ft., 5.

Additional records (A. H. Howell, 1914b:32).—**Fremont Co.:** Splitrock. **Laramie Co.:** Meadows; *Pole Creek.* **Natrona Co.:** Sun. **Sheridan Co.:** ***Arvada.***

Peromyscus crinitus doutti Goin

Canyon Mouse

1944. *Peromyscus crinitus doutti* Goin, Jour. Mamm., 25:189, May 26.

Type.—From Antelope Canyon, 20 miles southeast of Duchesne, 7200 ft., Duchesne County, Utah.

Description.—Medium-sized for *Peromyscus* (see *Measurements*) except tail more than half total length; tail well-haired; upper parts buffy ochraceous, grizzled slightly with blackish; underparts whitish and often possessing pectoral spot of pale or dark ochraceous; maxillary tooth-row usually less than 4.0 mm. long.

Measurements.—External and cranial measurements of an adult male and adult female from four miles northeast of Linwood, Utah, in Wyoming, are, respectively, as follows: Total length, 165, 178; length of tail, 87, 90; hind foot, 22, 22; greatest length of skull, 24.8, 26.1; zygomatic breadth, 12.2, —; interorbital breadth, 4.4, 4.5; length of nasals, 8.8, 10.0; cranial depth, 8.0, 7.6; maxillary tooth-row, 3.8, 3.5.

Distribution.—Known from arid, rocky areas along the Green River in extreme southern, southwestern Wyoming. See Fig. 36.

Remarks.—Goin (1944:189-190) mentions that P. *c. doutti* is brighter than P. *c. auripectus* to the southward, and usually differs further from that subspecies in lacking a pectoral spot. She mentions a specimen from Sweetwater County, Wyoming, having a pale pectoral spot. Of four other specimens examined by me from essentially the same place, one (USNM 177481) has a bright pectoral spot and another (USNM 177484) has a pale spot. The other two specimens lack pectoral spots. All four are referable to P. *c. doutti* because they are brighter than typical specimens of P. *c. auripectus.*

Records of occurrence.—Specimens examined, 4, as follows: **Sweetwater Co.:** 4 mi. NE Linwood, Utah, in Wyoming, 3800 ft., 4 USNM.

Additional records (Goin, 1944:191): **Sweetwater Co.:** *1 mi. N Linwood, Utah.*

Peromyscus maniculatus (Wagner)

Deer Mouse

Peromyscus maniculatus is better represented in collections from Wyoming than is any other species of mammal. It occurs in all life-zones, and is usually abundant in all places except, perhaps, those in the two uppermost life-zones. The species is usually darker in high habitats and paler in low habitats, possibly because the latter are more arid. Size and tail length vary from place to place, but the greatest variation is in color of upper parts. They vary from dark brownish to pale yellowish buff and to bright ochraceous red. Even so, all populations are referable to a single subspecies, P. *m. nebrascensis,* except the brownish populations (P. *m. artemisiae*) from extreme northwestern Wyoming.

There are, then, in Wyoming two subspecies of the deer mouse. One occurs in western Wyoming. The eastern one ranges westward through the Wind River Mountains.

Peromyscus maniculatus artemisiae (Rhoads)

1894. *Sitomys americanus artemisiae* Rhoads, Proc. Acad. Nat. Sci. Philadelphia, 46:260, October.

1909. *Peromyscus maniculatus artemisiae,* Osgood, N. Amer. Fauna, 28:58, April 17.

Type.—From Ashcroft, British Columbia.

Description.—Medium in size for *Peromyscus* (see *Measurements*); upper parts dark brownish; pinna of ear blackish brown often margined with whitish; tail sharply bicolored, blackish brown above and whitish below; underparts whitish; auditory bullae small.

Comparisons.—Compared with *Peromyscus truei* and P. *crinitus,* P. *m. artemisiae* has a shorter tail and is darker, lacking ochraceous. From *Peromyscus leucopus,* P. *maniculatus artemisiae* differs in shorter tail, smaller size (especially of skull), and more sharply bicolored tail. From *Peromyscus maniculatus nebrascensis,* P. *m. artemisiae* differs in larger size (on the average), darker upper parts, and darker pinnae.

Measurements.—External and cranial measurements of two adult females and three adult males from Jackson Hole Wildlife Park are, respectively, as follows: Total length, 157, 170, 158, 173, 150; length of tail, 71, 70, 71, 81, 65; length of hind foot, 20, 20, 20, 20, 20; greatest length of skull, 24.8, —, 25.6, —, 24.6; zygomatic breadth, 12.9, —, —, —, 12.9; interorbital breadth, 4.0, 4.0, 4.0, 3.7, 3.9; length of nasals, 10.2, —, 9.9, 11.0, 9.9; cranial depth, 7.3, 7.4, 7.9, —, —; maxillary tooth-row, 3.8, 4.0, 3.5, 3.5, —.

Distribution.—Known from Yellowstone Plateau and mountains immediately eastward thereof southward through Jackson Hole. See Fig. 38.

Remarks.—Known breeding records are in June and July; nine pregnant females contained an average of 5.6 (4-7) embryos.

Records of occurrence.—Specimens examined, 222, as follows: **Lincoln Co.:** *3 mi. N, 11 mi. E Alpine, 13; 13 mi. N, 2 mi. W Afton, 5;* 10 mi. SE Afton, 7500 ft., 3 USNM; Border, 11 USNM.

Park Co.: 28-31½ mi. N, 30-36 mi. W Cody, 9; SW Slope Whirlwind Peak, 6700 ft., 2; *2 mi. S, 42 mi. W Cody, 6400 ft., 8; 6 mi. S, 3 mi. W Whirlwind Peak, 6700 ft., 1.*

Teton Co.: *18 mi. N, 9 mi. W Moran, 6743 ft., 3, 1 Univ. Wyo.; Whetstone Creek, 1; Pacific Creek, 1; Two Ocean Lake, 1, 1 Univ. Wyo.; 1 mi. N Moran, 1 USNM; ¼-½ mi. N, 2½ mi. E Moran, 6230-6750 ft., 1 Univ. Wyo., 5; Jackson Hole Wildlife Park, 25 USNM; Moran, 6244 ft., 3, 17 USNM; ½ mi. E Moran, 6740 ft., 1 Univ. Wyo.; 2½ mi. E Moran, 6750 ft., 1 Univ. Wyo.; 3-3¾ mi. E Moran, 6230-6300 ft., 8; ¼ mi. S, 3 mi. E Moran, 6200 ft., 1; 1 mi. S, 3¾-4 mi. E Moran, 6200-6750 ft., 8, 1 Univ. Wyo.; S Moran, 1; 5 mi. S Moran, 1 USNM;* Togwotee Pass, 1, 15 USNM, 1 Univ. Wyo.; *Timbered Island, 4 mi. N Moose, 6750 ft., 7; Bar BC Ranch, 2½ mi. NE Moose, 6500 ft., 39; N. end Blacktail Butte, 1 mi. E Moose, 6600 ft., 8; Jackson, 1.*

Yellowstone Nat'l Park: Mammoth Hot Springs, 2; *Bunsen Peak, 1; Locality not specified, 13.*

Additional record (Osgood, 1909:61).—**Lincoln Co.:** LaBarge Creek.

Peromyscus maniculatus nebrascensis (Coues)

1877. *Hesperomys sonoriensis* var. *nebrascensis* Coues, *In* Coues and Allen, Monograph N. Amer. Rodentia, p. 79, August.

1909. *Peromyscus maniculatus nebrascensis,* Osgood, N. Amer. Fauna, 28:75, April 17.

1911. *Peromyscus maniculatus osgoodi* Mearns, Proc. Biol. Soc. Washington, 24:102, May 15, type from Calf Creek, Custer County, Montana.

1958. *Peromyscus maniculatus nebrascensis,* Jones, Proc. Biol. Soc. Washington, 71:107-110, July 16.

Type.—From Deer Creek, Converse County, Wyoming (see Jones, 1958: 107).

Description.—Size medium for *Peromyscus* (see *Measurements*); upper parts varying from pale buff to reddish ochraceous and to brownish and olivaceous buff; pinna of ear brownish and margined with whitish; underparts white; feet whitish; tail sharply bicolored, brownish above (sometimes black) and whitish below, and shorter than half total length animal; skull medium in size (see *Measurements*); auditory bullae moderately inflated.

Comparisons.—For comparisons with other species and subspecies of *Peromyscus,* see accounts of those taxa.

Measurements.—Average and extreme external and cranial measurements of eight old adult males and measurements of three old adult females from along Horse Creek, Laramie County, are, respectively, as follows: Total length, 153.5 (144-164), 164, 155, 155; length of tail, 64.6 (55-69), 66, 58, 65; length of hind foot, 21.0 (20-22), 20, 19, 21; length of ear, 15.5 (14-17), 15, 15, 17; greatest length of skull, 25.3 (24.5-26.3), 26.0, 25.9, 24.4; zygomatic breadth, 13.3 (12.8-13.7), 13.1, 13.4, 13.0; interorbital breadth, 4.0 (3.8-4.1), 4.0, 3.9, 3:8; length of nasals, 10.2 (9.8-11.0), 10.8, 10.3, 9.6; cranial depth, 7.7 (7.5-8.0), 8.1, 7.9, 7.4; maxillary tooth-row, 3.6 (3.4-3.9), 3.5, 3.8, 3.7. Average and extreme external and cranial measurements of eight old adult males and measurements of one old adult female from 4½ miles south and four miles east of Robertson are, respectively, as follows: Total length, 153.0 (137-162), 170; length of tail, 63.8 (56-72), 65; length of hind foot, 20.0 (17-22), 17; length of ear, 17.8 (16-19), 17; greatest length of skull, 25.2 (24.1-25.6), 26.7; zygomatic breadth, 12.8 (12.6-13.0), 13.6; interorbital breadth, 3.8 (3.6-4.0), 3.9; length of nasals, 10.1 (9.2-10.7), 10.5; cranial depth, 7.6 (7.2-8.0), 7.6; maxillary tooth-row, 3.6 (3.5-3.7), 3.3.

Distribution.—Most habitats, from low desert areas to high alpine areas, throughout most of state excepting the area occupied by P. *m. artemisiae.* See Fig. 38.

Remarks.—There has been much confusion in the past 70 years concerning the name correctly to be applied to the deer mouse occupying most of Wyoming. Jones (1958:107-111) unraveled the truth of the matter by ascertaining first that the type specimen of P. *m. nebrascensis* came from Deer Creek in Converse County, Wyoming, instead of from the Deer Creek much farther east in Nebraska. Consequently, specimens from throughout most of Wyoming now correctly take the name P. *m. nebrascensis,* although before 1958 P. *m. osgoodi* was the name used for almost 50 years for these same mice. A syntype of P. *m. nebrascensis* has subsequently been found that was formerly thought to be no longer in existence (Jones and Mursaloğlu, 1961:101-103).

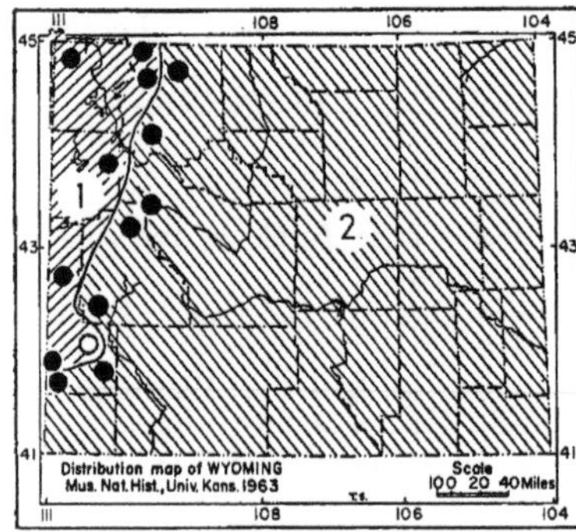

FIG. 38. Marginal records of *Peromyscus maniculatus.*
1. *P. m. artemisiae*
2. *P. m. nebrascensis*

Blair (1953) has discussed gene exchange in populations of P. *maniculatus* in the light of recent work on dispersal and home range and in the light of studies on variation in deer mice undertaken by Dice and his associates. Dark mice from northwestern Wyoming (referred in this paper to P. *m. artemisiae*) differ from other deer mice in Wyoming mainly in color. Two geographic areas exert influences on the latter mice mentioned by Blair. The sand hills of western Nebraska and extreme eastern Wyoming because of aridity and pale soils effect fairly high frequencies of pale mice not seen from surrounding areas. High frequencies of medium-dark mice in the same hills are attributed to the influx of hereditary materials from the surrounding populations. In the Black Hills, dark soils effect a fairly high frequency of dark mice not seen in surrounding areas except in the mountains of northwest

Wyoming. But even in the Black Hills many mice are only medium-dark. My observations of Wyoming deer mice confirm these findings in part. Many medium-dark mice occur in the Black Hills, some of which show blackish color mid-dorsally. On the whole the specimens are somewhat more olivaceous than mice from the lowlands of Wyoming. Specimens from the sandy soils of eastern Wyoming are on the average paler than those from the Black Hills and also from the mountains in south-central Wyoming. Jones and Mursoloğlu (1961) noted that some mice from these mountains were more reddish than specimens from "adjacent arid plains" of Wyoming, but one specimen (the syntype of P. *m. nebrascensis*) is also reddish.

In examing large series from many places in Wyoming I conclude that several colors of pelage often are found in each population. Pale mice occur among dark mice in, say, forested areas, and dark mice or reddish mice occur with pale mice in arid areas. In general, the color of the mice conforms to Gloger's Rule in being predominately paler in arid areas and darker in more humid (higher) areas. But all mice examined from these areas (except extreme western Wyoming) are referable to *nebrascensis* on the basis of morphological resemblances as, I suppose on geographic grounds, are the mice listed by Quay (1948:181) from Niobrara County.

The number of embryos in 29 pregnant females from Carbon County averages 5.6 (2-9).

Records of occurrence.—Specimens examined, 1,797, as follows: **Albany Co.:** *Springhill, 6300 ft., 7 USNM; 29 mi. N, 8¾ mi. E Laramie, 6420 ft., 1; 27 mi. N, 8 mi. E Laramie, 6420 ft., 2; 27 mi. N, 5 mi. E Laramie, 6900 ft., 2; 26½ mi. N, 12 mi. E Laramie, 6100 ft., 1; 9 mi. NE Laramie, 7500 ft., 3 Univ. Wyo.; 8 mi. N Laramie, 7300 ft., 1 Univ. Wyo.; 8 mi. NE Laramie, 7500 ft., 8 Univ. Wyo.; 6 mi. NW Laramie, 7250 ft., 1 Univ. Wyo.; 6 mi. NE Laramie, 7250 ft., 1 Univ. Wyo.; 5 mi. N Laramie, 7400 ft., 2; 4½ mi. N Laramie, 7500 ft., 2 Univ. Wyo.; Univ. Wyo. Sci. Camp, 1 Univ. Wyo.; Nash Fork Creek, 2 mi. W Centennial, 2; ½ mi. S, 2 mi. E Medicine Bow Peak, 10,800 ft., 1; 3 mi. NE Laramie, 7350 ft., 2 Univ. Wyo.; 2-2½ mi. NW Laramie, 7175-7250 ft., 4 Univ. Wyo., 1; 1½ mi. N Laramie, 7200 ft., 1 Univ. Wyo.; 4 mi. W Laramie, 7275 ft., 1 Univ. Wyo.; Laramie, 7100-7150 ft., 8 Univ. Wyo.; 1 mi. E Laramie, 7164-7300 ft., 6 Univ. Wyo., 3; 1¾-2 mi. E Laramie, 7100-7300 ft., 6 Univ. Wyo., 8; 2½ mi. E Laramie, 7250 ft., 2 Univ. Wyo.; 3 mi. E Laramie, 7250 ft., 4 Univ. Wyo., 2; 4 mi. E Laramie, 7200 ft., 2 Univ. Wyo.; 5 mi. E Laramie, 1 Univ. Wyo.; 9 mi. E Laramie, 7400 ft, 1 Univ. Wyo.; 10 mi. E Laramie, 1 USNM; 2 mi. S Browns Peak, 10,600 ft., 1; 2¼ mi. ESE Browns Peak, 10,300 ft., 3; 3 mi. ESE Browns Peak, 10,000 ft., 11; 4 mi. SW Laramie, 7175 ft., 2 Univ. Wyo., 3; 6½ mi. S, 8¾ mi. E Laramie, 8200 ft., 10; N. Fork Camp Grounds, 8500 ft., 2; Wallace Picnic Ground, 8350 ft., 1; 1 mi. SSE Pole Mtn., 8350 ft., 8; 2 mi. SW Pole Mtn., 8300 ft., 2; 2 mi. SE Pole Mtn., 8200 ft., 2; 3 mi. S Pole Mtn., 8100 ft., 3; Red Buttes, 12 mi. S Laramie, 1 Univ. Wyo.; 15 mi. SE Laramie, 8200 ft., 5 USNM; Tie City [Tie Siding], 8175 ft., 2; 2 mi. N Colorado Boundary, Hwy. 287, 2.*

Big Horn Co.: *Medicine Wheel Ranch, 28 mi. E Lovell, 9000 ft., 12; Granite Creek Camp Grounds, 21; 2 mi. N, 12 mi. E Shell, 7500 ft., 14; Shell Creek, 1 mi. NW Shell, 3; 12 mi. E Shell, 7500 ft., 3; 6 mi. NW Greybull, 3800 ft., 4; 10 mi. W Germania, 1 USNM; Germania, 1 USNM; 3 mi. E Germania, 1 USNM; Greybull, 12 USNM; Hd. Trappers Creek, 8500 ft., 9 USNM; Bighorn River, 1 USNM; Bighorn Mtns., 1 USNM; 7 mi. S Basin, 3900 ft., 11.*

Campbell Co.: Powder *River Crossing, 4100 ft., 2 USNM; Little* Powder *River, 3 USNM; 3 mi. N, 3 mi. W Rockypoint, 3800 ft., 4; Rockypoint, 1; 2 mi. S, 1½ mi. W Rockypoint, 1; 4 mi. S, 6 mi. W Rockypoint, 4200 ft., 2; 6 mi. S, 4 mi. W Rockypoint, 4200 ft., 2; 6½ mi. S, 5 mi. W Rockypoint, 4200 ft., 2; Ivy Creek, 5 mi. N, 8 mi. W Spotted Horse, 3; 2½ mi. N, 7 mi. W Spotted Horse, 3800 ft., 10; Middle Butte, 38 mi. S, 19 mi. W Gillette, 6010 ft., 8; S. Butte, 40½ mi. S, 17½ mi. W Gillette, 6000 ft., 4; 45½ mi. S, 15½ mi. W Gillette, 5342 ft., 3; 45½ mi. S, 13½ mi. W Gillette, 5340 ft., 3.*

Carbon Co.: *12 mi. S Alcova, 2 USNM; Ferris Mtns., 8500 ft., 6 USNM; Shirley Mtns., 7600 ft., 12 USNM; 33 mi. W Medicine Bow, 6400 ft., 4; 18 mi. NNE Sinclair, 6500 ft., 31; Ft. Steele, 12 USNM; 30 mi. E Rawlins, 6700 ft., 4; Bridger* Pass, *18 mi. SW Rawlins, 7500 ft., 16; Saratoga, 5 USNM; 6 mi. S, 13 mi. E Saratoga, 8500 ft., 2; 6 mi. S, 14 mi. E Saratoga, 8800 ft., 11; 10 mi. N, 14 mi. E Encampment, 8000 ft., 13; 10 mi. N, 16 mi. E Encampment, 8000 ft., 2; 9½-10 mi. N, 10½-12 mi. E Encampment, 7200-9200 ft., 23; 9 mi. N, 3 mi. E Encampment, 6500 ft., 3; 9 mi. N, 8 mi. E Encampment, 7000 ft., 5; 8 mi. N, 14½ mi. E Encampment, 8100 ft., 6; 8 mi. N, 16 mi. E Encampment, 8400 ft., 15; 8 mi. N, 18 mi. E Encampment, 8900 ft., 1; 8 mi. N, 21½ mi. E Encampment, 9400 ft., 2; Riverside, 5 USNM; 2 mi. S Bridger* Peak, *9300 ft., 11; 10 mi. N Baggs, 4 Univ. Wyo.; 8 mi. N Baggs, 1; 8 mi. N, 19½-20 mi. E Savery, 8800 ft., 17; 7½ mi. N, 18½ mi. E Savery, 8400 ft., 1; 7 mi. N, 17 mi. E Savery, 8300 ft., 1; 6 mi. N, 12½-13½ mi. E Savery, 8400 ft., 7; 6 mi. N, 14-14½ mi. E Savery, 8350-8400 ft., 6; 6 mi. N, 15 mi. E Savery, 8500 ft., 1; 5 mi. N, 10½ mi. E Savery, 8000 ft., 2; 4 mi. N, 10 mi. E Savery, 4; ½ mi. N Baggs, 6500 ft., 2; 1 mi. SW* Dixon, *6600 ft., 4.*

Converse Co.: *13 mi. N, 2 mi. W Bill, 3800 ft., 1; 12 mi. N, 6 mi. W Bill, 2; 12 mi. N, 2 mi. W Bill, 4800 ft., 9;* Deer Creek, *2½ mi. S, 1 mi. W Glenrock, 1; 4½ mi. S, 3 mi. W* Glenrock, *1; 3 mi. N, 5 mi. E Orin, 4725 ft., 6; 21½-22 mi. S, 24½ mi. W* Douglas, *7700 ft., 33; 23 mi. S, 25 mi. W Douglas, 7800 ft., 2.*

Crook Co.: *2 mi. S Colony, 3500 ft., 3; Little Missouri River, 7 USNM; 2 mi. N, 15 mi. W Hulett, 2; 2 mi. N, 13 mi. W Hulett, 1; Devils Tower, 3350 ft., 2 USNM; 15 mi. ENE Sundance, 3825 ft., 11; 3 mi. NW Sundance, 5900 ft., 12; 2 mi. NW Sundance, 11; 1½ mi. NW Sundance, 5000 ft., 1; Sundance, 10 USNM; Belle Fourche River, Moorcroft, 2; Moorcroft, 10 USNM.*

Fremont Co.: Jackeys Creek, 3 mi. S Dubois, 11 USNM; *12 mi. N, 3 mi. W Shoshoni, 4650 ft., 8; 7 mi. N Shoshoni, 4700 ft., 5; 5 mi. N, 1 mi. W Shoshoni, 4700 ft., 1; 2½ mi. W Shoshoni, 4800 ft., 2; Bull Lake Creek, Wind River, 1 USNM; Bull Lake, 3 USNM; Lake Fork, 9600-10,000 ft., 3 USNM; 2 mi. N, 6 mi. W Burris, 6450 ft., 5; Sage Creek, 8 mi. NW Ft. Washakie, 4 USNM;* Ft. Washakie, 3 USNM; *Moccasin Lake, 4 mi. N, 19 mi. W Lander, 10,100 ft., 9; Mosquito* Park *Ranger Sta., 2½ mi. N, 17½ mi. W Lander, 9000-9500 ft., 3; 3/10 mi. SE Milford, 5357 ft., 2; Middle Lake, 2 mi. S, 20½ mi. W Lander, 1; ½ mi. N, 3 mi. E South* Pass *City, 2; South* Pass *City, 8000 ft., 2 USNM; 17 mi. S, 6½ mi. W Lander, 7; 22-23 mi. S, 5-5½ mi. W Lander, 8800 ft., 2; Splitrock, 6200 ft., 2 USNM; Sweetwater, 2 Univ. Wyo.; 8 mi. E Rongis, Green Mtns., 8000 ft., 4 USNM; Mt. Crooks and Crooks Gap, 33.*

Goshen Co.: *Muskrat Canyon, 7; Rawhide Buttes, 4 USNM.*

Hot Springs Co.: *Kirby Creek, 1 USNM; 3 mi. N, 10 mi. W Thermopolis, 1; 2 mi. S Thermopolis, 4350 ft., 1; 10 mi. S, 3 mi. E Thermopolis, 4600 ft., 12; 24 mi. SE Thermopolis, 1.*

Johnson Co.: *14 mi. W Buffalo, 1; 8/10 mi. S, 1 mi. W Buffalo, 1; 1 mi. S, 7½ mi. W Buffalo, 6500 ft., 1; 1 mi. S, 5½ mi. W Buffalo, 4800-5600 ft., 2; 1 mi. S, 4½ mi. W Buffalo, 5420 ft., 2; 2 mi. S, 6½ mi. W Buffalo, 5620 ft., 7; 3 mi. W Klondike, 2; 1 mi. S, 4 mi. W Klondike, 6500 ft., 3; 1 mi. WSW Kaycee, 4700 ft., 43.*

Laramie Co.: *12 mi. N Horse Creek* P. *O., 1 Univ. Wyo.; Horse Creek, 6 mi. W Meriden, 5200 ft., 2; Horse Creek, 3 mi. W Meriden, 5000 ft., 15; 5 mi. W Horse Creek* P. *O., 7200 ft., 8; 3½ mi. W Horse Creek* P. *O., 7000 ft., 13; 2½ mi. W Horse Creek* P. *O., 6600 ft., 2; 2.2 mi. W Horse Creek* P. *O., 6600 ft., 3; Horse Creek* P. *O. 6500 ft., 17; 3 mi. E Horse Creek* P. *O., 6400 ft., 3; 5½ mi. S Glenys, 5200 ft., 2; 6 mi. W Islay, 15 USNM; 11 mi. N, 5½ mi. E Cheyenne, 5950 ft., 2; 1 mi. S, ½ mi. E Pine Bluffs, 5200 ft., 5; 2 mi. S Pine Bluffs, 5200 ft., 45; 7 mi. W Cheyenne, 6500 ft., 9; 15 mi. ESE Cheyenne, 6.*

Lincoln Co.: *6 mi. N, 2 mi. E Sage, 3;* Fontenelle, 6500 ft., 2 USNM; *Kemmerer, 6 USNM;* Cokeville, 9 USNM; *Cumberland, 18 USNM, 12.*

Natrona Co.: *27 mi. N, 1 mi. E Powder River, 6075-6100 ft., 3; Arminto, 2; 1 mi. NE Casper, 5150 ft., 3; 3 mi. W Casper, 5; Casper, 1 USNM; 4 mi. E Casper, 5100 ft., 1; 4½ mi. S, 1½ mi. W Casper, 5550 ft., 2; 6 mi. S, 2 mi. W Casper, 1; 16 mi. S, 11 mi. W Waltman, 6950 ft., 6; Rattlesnake Mtns., 7500 ft., 7 USNM; 7 mi. S Casper, 6000 ft., 4 USNM; 7 mi. S, 2 mi. W Casper, 6370 ft., 6; 10 mi. S Casper, 7750 ft., 4; Dry Creek, 12 mi. N Sun, 6400 ft., 7 USNM; 1 mi. N, 9 mi. W Independence Rock, 5; Sun Ranch, 5 mi. W Independence Rock, 6000 ft., 13.*

Niobrara Co.: *10 mi. N Hatcreek* P. *O., 4300 ft., 15; 2 mi. S, ½ mi. E Lusk, 5000 ft., 1.*

Park Co.: *Clarks Fork, 1 USNM;* 16¼ mi. N, 17 mi. W Cody, 5625 ft., 13; *4 mi. N Garland, 35; 13 mi. N, 1 mi. E Cody, 5200 ft., 4; 5 mi N Cody, 9 USNM; 4 mi. S, 12 mi. W Cody, 1; 4 mi. S, 5 mi. W Cody, 11; 25 mi. S, 28 mi. W Cody, 6350 ft., 9;* Valley, 6500-7500 ft., 23 USNM; *Needle Mtn., 10 USNM; 2 mi. S, 2 mi. E Meeteetse, 5750 ft., 11; 15½ mi. S, 13 mi. W Meeteetse, 7.*

Platte Co.: *2½ mi. S Chugwater, 5300 ft., 4.*

Sheridan Co.: *4 mi. NNE Banner, 2; Wolf, 4 USNM; 3 mi. WNW Monarch, 11; Sheridan, 5 USNM; 4-5 mi. NE Clearmont, 30; Arvada, 9 USNM.*

Sublette Co.: 31 mi. N Pinedale, 8025 ft., 11; *N. Side Halfmood Lake, 7900 ft., 5; 5 mi. N, 3 mi. E Pinedale, 7500 ft., 4; 2¼ mi. NE Pinedale, 7500 ft., 3; Big Sandy, 6 USNM;* 2 mi. S, 19 mi. W Big Piney, 7700 ft., 10; *Jct. Green River and New Fork, 11 USNM.*

Sweetwater Co.: *Farson, 6; 5 mi. SE Sand Dunes, 1; Jct. Big Sandy and Green rivers, 2; Superior, Steamboat Mtn., 7 USNM; Between Tipton and Tablerock, 1 Univ. Wyo.; 2½ mi. N Wamsutter, 9; Green River, 5; ¼ mi. S Green River, 6090 ft., 1 Univ. Wyo.; 26 mi S, 21 mi. W Rock Springs, 3; 32 mi. S, 22 mi. W Rock Springs, 23; 18 mi. S Bitter Creek, 6800 ft., 4; Kinney Ranch, 21 mi. S Bitter Creek, 6800 ft., 20; 22 mi. SW Bitter Creek, 3; 30 mi. S Bitter Creek, 2; 33 mi. S Bitter Creek, 4; 4 mi. NE Linwood 5800 ft., 10 USNM; W side Green River, 1 mi. N Utah Border, 6; ½ mi. N Henrys Fork and Utah Boundary, 1.*

Uinta Co.: *½ mi. S, 1½ mi. W Cumberland, 6500 ft., 3; Bear River, 14 mi. N Evanston, 6600-6800 ft., 7 USNM; 1 mi. N Ft. Bridger, 6650 ft., 10; 8½ mi. W Ft. Bridger, 7100 ft., 4; 2 mi. W Ft. Bridger, 6700 ft., 1; Ft. Bridger, 6650 ft., 13 USNM, 3; 2½ mi. WSW Ft. Bridger, 6800 ft., 5; Mountainview, 11 USNM; Evanston, 14 USNM; 1½ mi. S, 2 mi. E Robertson, 1; Sage Creek, 10 mi. N Lonetree, 2; 4½ mi. S, 2½ mi. E Robertson, 8075 ft., 6; 4½ mi. S, 4 mi. E Robertson, 8025 ft., 19; 6 mi. S, 2½ mi. E Robertson, 12; 8 mi. S, 2½ mi. E Robertson, 8300 ft., 13; 9 mi. S Robertson, 8000 ft., 76; 10 mi. S, 1 mi. W Robertson, 1; 10 mi. S, 2½ mi. E Robertson, 8900 ft., 2; 10½-11½ mi. S, 2 mi. E Robertson, 7200 ft., 10; 13 mi. S, 2 mi. E Robertson, 9200 ft., 1; Lonetree, 2 USNM; Beaver Creek, 4 mi. S Lonetree, 5 USNM.*

Washakie Co.: *5 mi. N, 9 mi. E Tensleep, 7400 ft., 1; 4 mi. N, 9 mi. E Tensleep, 7000 ft., 28; 1 mi. N, 3 mi. E Tensleep, 4350 ft., 1; ½ mi. S, 1 mi. W Tensleep, 4300 ft., 3; 4 mi. S, 7 mi. W Worland, 4100 ft., 7; 8 mi. S, 8 mi. W Worland, 4200 ft., 1; 10 mi. S, 9 mi. W Worland, 4100 ft., 9.*

Weston Co.: *1½ mi. E Buckhorn, 2; 9 mi. N, 1 mi. E Newcastle, 5; Newcastle, 12 USNM; 23 mi. SW Newcastle, 4500 ft., 18.*

Additional records (Osgood, 1909:77 unless otherwise noted).—**Albany Co.:** *Sheep Creek, 17 mi. W Toltec; Sherman.* **Big Horn Co.:** *Otto.* **Converse Co.:** *Ft. Fetterman.* **Niobrara Co.:** *Eastern part* (Quay, 1948:181).

Peromyscus leucopus aridulus Osgood

White-footed Mouse

1909. *Peromyscus leucopus aridulus* Osgood, N. Amer. Fauna, 28:122, April 17.

Type.—From Fort Custer, Big Horn County, Montana.

Comparison.—Compared with *Peromyscus maniculatus, P. leucopus* differs in averaging larger (see *Measurements*); being darker above; having more robust and larger skull; having less sharply bicolored tail (underside often brownish). Subadults can be confused with *P. maniculatus,* but few of them have the hind foot shorter than 21 mm.

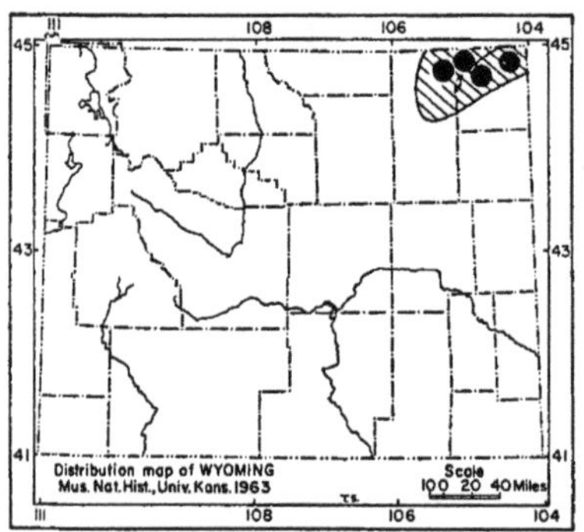

FIG. 39. Distribution of *Peromyscus leucopus aridulus.*

Measurements.—External and cranial measurements of an adult male and an old adult male from within 11 miles of Rockypoint (Crook County) and external measurements of a pregnant (3 embryos) female (skin only) from two miles north and 13 miles west of Hulett are, respectively, as follows: Total length, 174, 186, 188; length of tail, 71, 78, 86; length of hind foot, 22, 23, 22; length of ear, 15, 19, 16; greatest length of skull, 27.0, 27.8; zygomatic breadth, 13.4, 14.7; interorbital breadth, 4.1, 4.5; length of nasals, 10.0, 11.0; cranial depth, 8.2, 8.4; maxillary tooth-row, 3.8, 4.1.

Distribution.—Northeastern part of state; may be found elsewhere in eastern Wyoming in brushy or wooded habitat. See Fig. 39.

Records of occurrence.—Specimens examined, 6, as follows: **Campbell Co.:** 3 mi. N, 3 mi. W Rockypoint, 1. **Crook Co.:** 2 mi. S Colony, 3500 ft., 1; *3 mi. S, 5 mi. E Rockypoint, 1;* 7 mi. S, 8 mi. E Rockypoint, 3900 ft., 1; 2 mi. N, 13 mi. W Hulett, 2.

Peromyscus truei truei (Shufeldt)

Piñon Mouse

1885. *Hesperomys truei* Shufeldt, Proc. U. S. Nat. Mus., 8:407, September 14.

1894. P *[eromyscus]. Truei,* Thomas, Ann. Mag. Nat. Hist., ser. 6, 14:365, November.

Type.—From Wingate, McKinley County, New Mexico.

Description.—Large, especially pinnae of ears (see *Measurements*); tail well-haired and long; upper parts grayish ochraceous with ochraceous lateral line; underparts whitish; tooth-row long; auditory bullae large.

Comparisons.—Long, well-haired tail and ochraceous lateral line distinguish this mouse from other kinds of *Peromyscus* in Wyoming except *P. crinitus.* Compared with *P. crinitus, P. t. truei* has longer ears, longer tooth-rows (more than 8.0 mm.), larger auditory bullae, and narrower skull.

Measurements.—Measurements of one male and three females (all adults) from four miles northeast of Linwood, Utah (in Wyoming), are as follows: Total length, 171, 171, 173, 172; length of tail, 85, 87, 89, 86; hind foot, 23, 23, 23, 24; greatest length of skull, 26.2, 25.9, —, 25.8; zygomatic breadth, 12.9, 12.6, —, —; interorbital breadth, 4.3, 4.4, —, 4.3; length of nasals, 9.3, 9.3, 9.8, 9.1; cranial depth, 8.5, 8.3, —, —; maxillary tooth-row, 4.1, 4.2, 4.1, 3.9.

Distribution.—Arid area along Green River in southwestern part of state. See Fig. 37.

Records of occurrence.—Specimens examined, 4, as follows: **Sweetwater Co.**: 4 mi. NE Linwood, Utah, 5800 ft., 4 USNM.

Onychomys leucogaster (Wied-Neuwied)

Northern Grasshopper Mouse

Medium in size for Cricetidae except tail, which is short (usually less than twice the length of the hind foot); upper parts brownish gray to Pinkish Cinnamon; underparts and feet white; tail tipped with white; cheek-teeth (unworn) showing sharp, elongate cusps; coronoid process of mandible elongate. Food consists principally of insects.

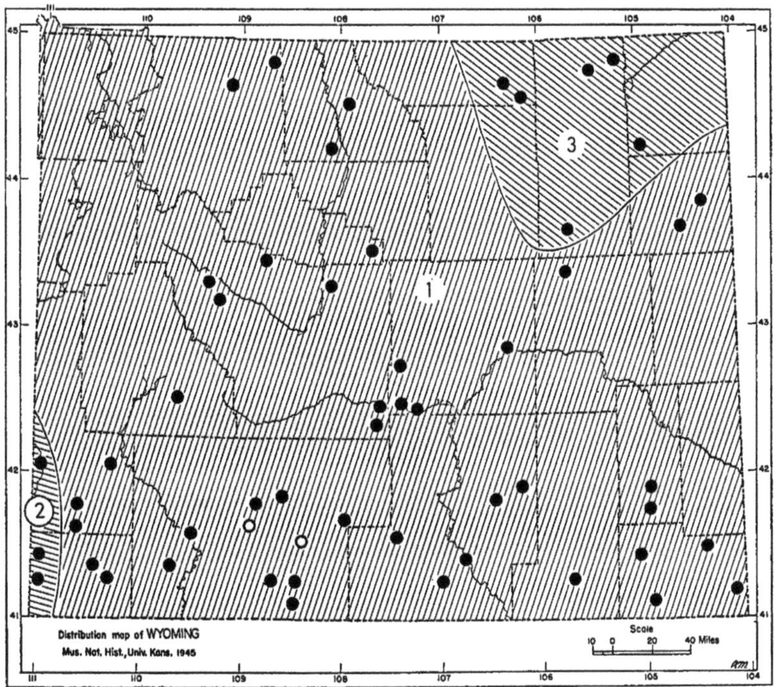

FIG. 40. Distribution of *Onychomys leucogaster*.
Guide to subspecies
1. *O. l. arcticeps*
2. *O. l. brevicaudus*
3. *O. l. missouriensis*

Onychomys leucogaster arcticeps Rhoads

1898. *Onychomys arcticeps* Rhoads, Proc. Acad. Nat. Science Philadelphia, 50:194, May 3.

1914. *Onychomys leucogaster arcticeps,* Hollister, Proc. U. S. Nat'l Mus., 47:439, October 29.

Type.—From Clapham, Union County, New Mexico.

Comparisons.—Compared with *Onychomys leucogaster brevicaudus, O. l. arcticeps* differs in larger size and paler upper parts. From *O. l. missouriensis, O. l. arcticeps* differs in paler, buffier upper parts.

Measurements.—External and cranial measurements of four adults, all males (7 mi. S Basin; 6 mi. NW Greybull; 13 mi. N and 1 mi. E Cody; 5 mi. N and 1 mi. W Shoshoni) are, in order by locality, as follows: Total length, 136, 144, 139, 132; length of tail, 41, 41, 43, 34; hind foot, 20, 21, 20, 17; ear from notch, 17, 18, 18, 16; greatest length of skull, 28.0, 27.7, 27.0, 27.2; zygomatic breadth, 15.5, 15.5, 14.2, —; interorbital breadth, 4.7, 4.7, 4.3, 4.8; length of nasals, 11.7, 10.0, 10.1, 10.2; cranial depth, 8.6, 8.6, 8.0, 8.5; maxillary tooth-row, 4.3, —, 4.1, 4.2. External and cranial measurements of an adult male and an adult female from two miles south of Pine Bluffs, Laramie County, are as follows: Total length, 139, 140; length of tail, 35, 38; hind foot, 19, 19; ear from notch, 22, 14; greatest length of skull, 28.8, 28.3; zygomatic breadth, —, 15.3; interorbital breadth, 4.8, 4.6; length of nasals, 11.0, 10.8; cranial depth, 8.2, 8.4; maxillary tooth-row, 4.6, 4.1.

Distribution.—Throughout most of state in proper habitat, excepting areas occupied by *O. l. missouriensis* and *O. l. brevicaudus.* Occurs in arid plains and deserts. See Fig. 40.

Remarks.—In an adult male (KU 18283) the last (third) upper molars are absent. Five females from Laramie County, taken from 16 June to 22 July, contained an average of 4.2 (3-6) embryos.

Records of occurrence.—Specimens examined, 181, as follows: **Albany Co.:** 1½ mi N Laramie, 7200 ft., 2 Univ. Wyo.; *Laramie, 7100 ft., 2 Univ. Wyo.*

Big Horn Co.: 6 mi. NW Greybull, 3800 ft., 4; *Greybull, 1 USNM;* 7 mi. S Basin, 3900 ft., 2.

Carbon Co.: 3.3 mi. W Medicine Bow, 1; 30 mi. E Rawlins, 6750 ft., 4; Bridgers Pass, 2 USNM; Saratoga, 4 USNM; 1 mi. N Encampment, 7100 ft., 1.

Converse Co.: 12 mi. N, 7 mi. W Bill, 4700 ft., 1.

Fremont Co.: Sheep Creek, S. Base Owl Creek Mtns., 1 USNM; 2 mi. N, 6 mi. W Burris, 6450 ft., 2; 5 mi. N, 1 mi. W Shoshoni, 3; Bull Lake, 1 USNM; 3 mi. W Splitrock, 7400 ft., 1 Univ. Wyo.; Crooks Gap, 7000 ft., 1; *Mt. Crooks, 8600 ft., 2.*

Hot Springs Co.: Hd. Bridgers Creek, 6500 ft., 1 USNM.

Laramie Co.: Horse Creek, 6-6½ mi. W Meriden, 5200 ft., 8; *Horse Creek, 4 mi. W Meriden, 5200 ft., 2;* 5½ mi. S Glenys, 5200 ft., 2; 2 mi. S Pine Bluffs, 5200 ft., 6; Cheyenne, 1 USNM.

Lincoln Co.: Fontenelle, 3 USNM; Kemmerer, 3 USNM; Cumberland, 1 USNM.

Natrona Co.: 16 mi. S, 11 mi. W Waltman, 6950 ft., 2; 1 mi. NE Casper, 5150 ft., 14; *Casper, 5 USNM;* 1 mi. N, 9 mi. W Independence Rock, 1; *Dry Creek, 2 mi. N Sun, 6400 ft., 1 USNM;* Sun, 4 USNM.

Park Co.: 13 mi. N, 1 mi. E Cody, 5200 ft., 1; 4 mi. N Garland, 2.

Platte Co.: Bordeaux, 1 USNM; Chugwater, 1 USNM.

Sublette Co.: Big Sandy, 7 USNM.

Sweetwater Co.: 27 mi. N, 37 mi. E Rock Springs, 6700 ft., 1; *25 mi. N, 38 mi. E Rock Springs, 6700 ft., 1;* Superior, Steamboat Mtn., 1 USNM; 2½ mi. N Wamsutter, 1; Green River, 1 USNM; 26 mi. S, 21 mi. W Rock Springs, 4; 32 mi. S, 22 mi. W Rock Springs, 2; *32 mi. S, 22 mi. E Rock Springs, 7025* *ft., 1;* 18 mi. S Bitter Creek, 8; *21 mi. SW Bitter Creek, 1; Kinney Ranch, 21 mi. S Bitter Creek, 6800 ft., 18, 1 USNM;* 30 mi. S Bitter Creek, 11; *33 mi. S Bitter Creek, 6900 ft., 22.*

Uinta Co.: Mountainview, 4 USNM; Ft. Bridger, 1 USNM.

Weston Co.: Newcastle, 1 USNM; 23 mi. SW Newcastle, 4500 ft., 4.

Additional records (Hollister, 1914:441).—**Sweetwater Co.:** Superior; Bitter Creek. **Carbon Co.:** *Aurora.*

Onychomys leucogaster brevicaudus Merriam

1891. *Onychomys leucogaster brevicaudus* Merriam, N. Amer. Fauna, 5:52, July 30.

Type.—From Blackfoot, Bingham County, Idaho.

Comparisons.—For comparison with *O. l. arcticeps,* see account of that subspecies. Compared with *O. l. missouriensis, O. l. brevicaudus* is smaller and slightly darker.

Measurements.—External and cranial measurements of the type (adult male), two adult females from Evanston, and an adult female and two males (subadults) from Bear River, 14 miles north of Evanston, are, respectively, as follows: Total length, 139, 135, 146, 140, 142, —; length of tail, 38, 41, 41, 35, 42, —; length of hind foot, 19.5, 20, 20, 20, 21, —; total length of skull, —, —, —, 26.9, 27.3, 26.7; zygomatic breadth, 14.5, —, 15.1, 14.1, 14.4, 14.7; interorbital breadth, 4.8, 4.4, 4.4, 4.6, 4.5, 4.5; length of nasals, 10.0, —, —, 10.7, 10.2, 10.4; cranial depth, —, 8.3, 8.9, 8.1, 8.5, 8.3; maxillary tooth-row, 4.0, 4.4, 4.3, 4.3, 4.4, 4.2.

Distribution.—Along Bear River in Lincoln County. See Fig. 40.

Remarks.—This subspecies differs more from *O. l. arcticeps* than the latter subspecies does from *O. l. missouriensis.*

Records of occurrence.—Specimens examined, 10, as follows: **Lincoln Co.:** Cokeville, 1 USNM. **Uinta Co.:** Bear River, 14 mi. N Evanston, 6600 ft., 3 USNM; Evanston, 6 USNM.

Onychomys leucogaster missouriensis (Audubon and Bachman)

1851. *Mus missouriensis* Audubon and Bachman, The viviparous quadrupeds of North America, 2:327.

1914. *Onychomys leucogaster missouriensis,* Hollister, Proc. U. S. Nat'l Mus., 47:438, October 29.

Type.—From Fort Union, near present town of Buford, Williams County, North Dakota.

Comparisons.—See accounts of *O. l. arcticeps* and *O. l. brevicaudus.*

Measurements.—External and cranial measurements of an old female from the Little Powder River and of three females (subadults) and one subadult male from Arvada are, respectively, as follows: Total length, 160, 138, 140, 145, 137; length of tail, 43, 39, 38, 37, 41; length of hind foot, 20, 20, 20, 20, 20; total length of skull, 27.8, 26.9, 27.6, 27.4, 26.8; zygomatic breadth, —, 14.3, 14.9, 14.2, —; interorbital breadth, 4.7, 4.5, 4.5, 4.3, 4.5; length of nasals, 10.7, 10.0, 11.2, 10.4, 10.2; cranial depth, 8.8, 8.3, 8.6, 8.5, 8.2; maxillary tooth-row, 4.7, 4.4, 4.3, 4.5, 4.5.

Distribution.—That part of northeastern Wyoming drained by northward flowing rivers. See Fig. 40.

Remarks.—Specimens referred to *O. l. missouriensis* are intergrades between that subspecies and *O. l. arcticeps*. On Figure 40 the boundary between the two subspecies is nearly the same as that drawn by Hollister (1914:435). Bone fragments of *O. leucogaster* from Middle and Pumpkin buttes, Campbell County, reported by Long and Kerfoot (1963), are here referred on geographic grounds to *O. l. missouriensis* because the localities in Campbell County are in the same river valley (Belle Fourche) as is Moorcroft, from which place specimens are referable to *O. l. missouriensis*.

Records of occurrence.—Specimens examined, 15, as follows: **Campbell Co.:** Little Powder River, 1 USNM; 2 mi. S, 1½ mi. W Rockypoint, 2; 42 mi. S, 13 mi. W Gillette, 2 (from owl pellet, discarded). **Crook Co.:** Moorecroft, 5 USNM. **Sheridan Co.:** 5 mi. NE Clearmont, 3900 ft., 1; Arvada, 4 USNM.

Neotoma cinerea (Ord)

Bushy-tailed Wood Rat

A large rodent having a bushy, squirrellike tail; upper parts varying from dark blackish ochraceous to pale Cinnamon-Buff; venter ordinarily white; tail grayish or ochraceous above, usually

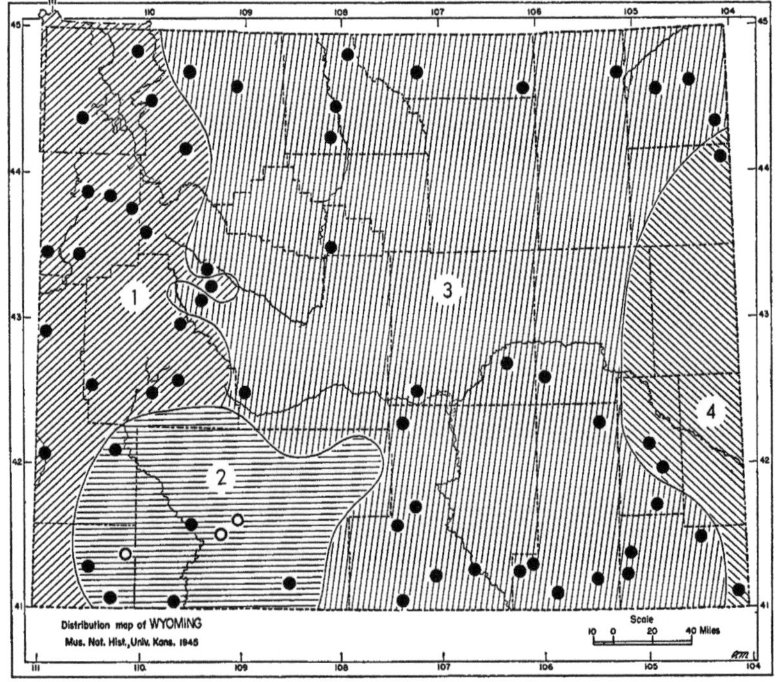

FIG. 41. Distribution of *Neotoma cinerea*.

1. *Neotoma cinerea cinerea*
2. *Neotoma cinerea cinnamomea*
3. *Neotoma cinerea orolestes*
4. *Neotoma cinerea rupicola*

whitish below; rostrum long and slender with nasals flaring or straight-sided; supraorbital ridges often conspicuous; palatal slits elongate; M1 with deep anterointernal re-entrant angle; M3 with anterior closed triangle and two confluent posterior loops; dental formula as in *Peromyscus*.

Neotoma cinerea cinerea (Ord)

1815. *Mus cinereus* Ord, *In* Guthrie, A new geogr., hist., comm. grammar . . ., Philadelphia, 2nd Amer. ed., 2:292.

1858. *Neotoma cinerea*, Baird, Mammals, *in* Repts., Expl. Surv. . . ., 8(1):499, July 14.

Type.—From near Great Falls, Cascade County, Montana.

Comparisons.—Compared with other subspecies of *N. cinerea* in Wyoming, *N. c. cinerea* is darkest (grayish or blackish instead of buffy); is as large as *orolestes;* has small auditory bullae as has that subspecies; and is unique in having some specimens lacking sphenopalatal vacuities; nasals flare anteriorly more than in other subspecies.

Measurements.—External and cranial measurements of adult males from Moran, Teton Pass, Salt River, and Lamar River are, respectively, as follows: Total length, 396, 380, 406, 433; length of tail, 180, 162, 178, 194; hind foot, 49, 48, 49, 47; condylonasal length, 52.6, 51.9, 54.3, 54.5; zygomatic breadth, 27.2, 26.4, 28.3, 27.6; interorbital breadth, 5.6, 5.3, 5.6, 5.6; length of nasals, 20.0, 19.9, 20.5, 20.4; cranial depth, 14.7, 13.7, 13.9, 14.0; maxillary toothrow, 9.6, 9.9, 10.0, 10.6; palatal slits, 12.4, 11.6, 12.8, 12.4.

Distribution.—Western Wyoming mainly in the mountains of northwestern part of state. See Fig. 41.

Remarks.—Specimens from Cokeville are intergrades between *N. c. cinerea* and paler wood rats to the eastward, but are referable to *N. c. cinerea* on account of grizzled blackish upper parts in most of the specimens.

Records of occurrence.—Specimens examined, 42, as follows: **Fremont Co.:** 5 mi. W Union, 1 USNM; Bull Lake, 7700 ft., 5 USNM; Not found, Little Sheep Mtn., 8600 ft., 1 Univ. Wyo. **Lincoln Co.:** Salt River, 10 mi. N Afton, 6200 ft., 3 USNM; Cokeville, 5 USNM. **Park Co.:** Mouth Grinnell Creek, Pahaska, 1 USNM; *6 mi. S, 3 mi. W Whirlwind Peak, 6700 ft., 2;* 25 mi. S, 28 mi. W Cody, 6350 ft., 1. **Sublette Co.:** Halfmoon Lake, 7900 ft., 3; 3 mi. W Stanley, 8000 ft., 1 USNM; Big Sandy, 2 USNM; Jct. Green River and New Fk., 1 USNM. **Teton Co.:** Jackson Hole Wildlife Park, 3 USNM; *Moran, 6742 ft., 1, 1 USNM; 1 mi. E Moran, 6600 ft., 1, 1 Univ. Wyo.;* 8 mi. E Moran, 1 USNM; Togwotee Pass, 11,000 ft., 1 USNM; Gros Ventre, 1 USNM; Teton Pass, 7200 ft., 2 USNM. **Yellowstone Nat'l Park:** *Camp Thorne, 1 USNM;* Lamar River, 2 USNM; Old Faithful, 1 USNM.

Neotoma cinerea cinnamomea J. A. Allen

1895. *Neotoma cinnamomea* J. A. Allen, Bull. Amer. Mus. Nat. Hist., 7:331, November 8.

1910. *Neotoma cinerea orolestes,* Goldman, N. Amer. Fauna, 31:104, October 19.

1944. *Neotoma cinerea cinnamomea,* Hooper, Jour. Mamm., 25; 415, December 12.

Type.—From Kinney Ranch, Sweetwater County, Wyoming.

Comparisons.—Compared with *Neotoma cinerea orolestes,* most specimens of *N. c. cinnamomea* are paler (less ochraceous) above; tail above ordinarily paler grayish; auditory bullae often larger. Compared with *N. c. cinerea,* specimens of *N. c. cinnamomea* are much paler (less grayish); have less flaring (anteriorly) nasals; ordinarily have larger bullae; and have small to large sphenopalatal vacuities (absent to small in *cinerea*). As mentioned by Goldman (1910:104), *N. c. cinnamomea* approaches *N. c. rupicola* in size and pallor.

Measurements.—Measurements of a male topotype (USNM 88297) and of two males and one female from four miles south of Lonetree are, respectively, as follows: Total length, 399, 353, 397, 373; length of tail, 173, 144, 176, 162; hind foot, 43, 44, 44, 42; condylonasal length, 51.2, 48.5, 51.5, —; zygomatic breadth, 26.3, 23.9, 26.5, 25.4; interorbital breadth, 5.9, 5.7, 6.0, 5.7; length of nasals, 20.2, 18.5, 19.6, 18.5; cranial depth, 13.3, 12.7, 12.8, —; maxillary tooth-row, 9.3, 10.0, 10.0, 9.5; length of palatal slits, 11.0, 10.9, 10.5, 10.8.

Distribution.—Deserts of southwestern Wyoming in suitable habitat. See Fig. 41.

Remarks.—There is good reason to question the validity of this subspecies. Its characters overlap those of *N. c. orolestes. N. c. cinnamomea* has been regarded as a valid subspecies by Hooper (1944:415) and as a synonym of *N. c. orolestes* by Goldman (1910: 104). The latter mentioned the resemblance (probably convergent responses to aridity) of *cinnamomea* to *rupicola.* Given sufficient time *cinnamomea* probably will become a more strongly marked subspecies; it is well on the way, having high frequencies for pale upper parts and large auditory bullae.

Records of occurrence.—Specimens examined, 15, as follows: **Lincoln Co.:** Fontenelle, 1 USNM. **Sweetwater Co.:** Green River, 4 USNM; Type locality, 1 USNM; W side Green River, 1 mi. N Utah boundary, 4; *4 mi. NE Linwood, Utah, 5800 ft., 1 USNM.* **Uinta Co.:** Ft. Bridger, 1 USNM; Beaver Creek, 4 mi. S Lonetree, 3 USNM.

Additional records (Hooper, 1944:416). **Sweetwater Co.:** *Thayer Jct.;* 6 mi. S Point of Rocks; 5 mi. E Rock Springs. **Uinta Co.:** 10 mi. SW Granger.

Neotoma cinerea orolestes Merriam

1894. *Neotoma orolestes* Merriam, Proc. Biol. Soc. Washington, 9:128, July 2.

1910. *Neotoma cinerea orolestes,* Goldman, N. Amer. Fauna, 31:104, October 19.

Type.—From Saguache Valley, 20 miles west of Saguache, Saguache County, Colorado.

Comparisons.—See accounts of other subspecies of *Neotoma cinerea.*

Measurements.—External and cranial measurements of an adult female from Horse Creek, of an adult male from Devils Tower, and of two females and one male (all adults) from six miles west of Islay are, respectively, as follows: Total length, 405, 371, 380, 362, 386; length of tail, 173, 168, 161, 154, 164; hind foot, 42, 44, 44, 42, 44; condylobasal length, 51.9, 49.2, 49.0, 48.4, 53.3; zygomatic breadth, 27.8, 26.7, 26.1, 26.2, 27.1; interorbital breadth, 6.1, 6.2, 5.6, 5.6, 6.0; length of nasals, 21.0, 18.5, 19.4, 18.8, 21.4; cranial depth, 14.5, 13.8, 12.7, 12.6, 13.6; maxillary tooth-row, 9.6, 9.2, 9.6, 10.2, 9.8; length of palatal slits, 11.1, 11.6, 10.7, 10.4, 12.2.

Distribution.—Throughout most of Wyoming in suitable habitat, ranging through the central lowlands of the state south into the Medicine Bow and Sierra Madre Mountains and northward into the Bighorn Mountains and Basin and most of the Black Hills. See Fig. 41.

Remarks.—An adult (KU 15739) from Horse Creek has the nasals, rostrum, and palatal slits all curved to the left. The distance from the lacrimal to the anterior tip of the nasal is 20.9 mm. on the left, 22.5 mm. on the right.

An adult male (USNM 202720) from Arvada has markedly inflated bullae and wide interpterygoid fossa.

Four of the five specimens (USNM 159741-159745) from six miles west of Islay have pinkish or Cinnamon Buff venters. The pinkish color is not the least ochraceous and is unknown to me on other mammals. Many of the pinkish hairs lack plumbeous bases, suggesting that the color has not resulted from staining by juices or soil. Wood rats taken more recently from nearby Horse Creek lack the pinkish color.

Specimens from 3 to 6½ miles west of Meriden vary greatly in size (a pregnant female, KU 15729, is only 300 mm. in total length, whereas a male is 428 mm.). They are somewhat buffy and are referred to *N. c. orolestes* instead of *N. c. rupicola.*

Records of occurrence.—Specimens examined, 88, as follows: **Albany Co.:** Laramie Pk., 8800 ft., 1 USNM; *Corner Mtn., sec. 28, T. 16N, R. 78W, 6, 2 Univ. Wyo.;* 4 mi. N Centennial, 8500 ft., 1; *Lake Owen, 3 Univ. Wyo.; 2 mi. N, 1 mi. W Centennial, 8500 ft., 1 Univ. Wyo.; 2 mi. N Centennial, 1 Univ. Wyo.; Silver Rung Camp, Snowy Range, 10,000 ft., 1 Univ. Wyo.; Libby Creek Camp, Snowy Range, 8700 ft., 1;* 32 mi. W Laramie, 1; Pole Mtn., 15-16 mi. SE Laramie, 8200-8300 ft., 4 USNM; Woods P. O., 1 USNM.

Big Horn Co.: Medicine Wheel Ranch, 28 mi. E Lovell, 9000 ft., 1; Greybull, 5 USNM; 7 mi. S Basin, 3900 ft., 1.

Campbell Co.: 6 mi. S, 4 mi. W Rockypoint, 4200 ft., 1.

Carbon Co.: Ferris Mtns., 7800 ft., 2 USNM; Sinclair, 1 Univ. Wyo.; *Lake Marie, 10,440 ft., 4;* Bridger Pass, 1 USNM; Bridger Pk., 1 USNM; Riverside, 2 USNM; 4 mi. N, 8 mi. E Savery, 7300 ft., 3.

Converse Co.: Deer Creek, 2 USNM; *no specific locality, 3.*

Crook Co.: Devils Tower, 2 USNM; 15 mi. N Sundance, 1; Sand Creek, 1 USNM.

Fremont Co.: Crowheart, 1 USNM; Lake Fork, 9600-11,400 ft., 3; ½ mi. N, 3 mi. E South Pass City, 1.

Hot Springs Co.: 10 mi. S, 3 mi. E Thermopolis, 1.

Laramie Co.: 6-6½ mi. W Meriden, 5200 ft., 2; *3 mi. W Meriden, 5000 ft., 2;* 2½ mi. W Horse Creek, 3; *Horse Creek, 6500 ft., 7;* 6 mi. W Islay, 5 USNM; *THD Ranch, sec. 3, T. 17N, R. 65W, 1.*

Natrona Co.: 7 mi. S Casper, 1 USNM; *Platte River, 18 mi. SW Casper, 1 USNM;* Sun Ranch, 5 mi. W Independence Rock, 6000 ft., 1, 1 USNM.

Park Co.: Black Mtn., Hd. Pat O'hara Creek, 1 USNM; 13 mi. N, 1 mi. E Cody, 1.

Platte Co.: 2½ mi. S Chugwater, 5300 ft., 1.

Sheridan Co.: Wolf, Eaton's Ranch, 3 USNM; Arvada, 1 USNM.

Neotoma cinerea rupicola J. A. Allen

1894. *Neotoma rupicola* J. A. Allen, Bull. Amer. Mus. Nat. Hist., 6:323, November 7.

1910. *Neotoma cinerea rupicola*, Goldman, N. Amer. Fauna, 31:107, October 19.

Type.—From Corral Draw, 3700 feet elevation, Pine River Indian Reservation, Black Hills, South Dakota.

Comparisons.—See accounts of *Neotoma cinerea cinerea, N. c. orolestes,* and *N. c. cinnamomea. N. c. rupicola* is characterized by small size and pallid coloration.

Measurements.—External and cranial measurements of an adult female from Uva, Platte County, are as follows: Total length, 368; length of tail, 152; length of hind foot, 42; condylonasal length, 48.0; zygomatic breadth, 24.7; interorbital breadth, 5.3; length of nasals, 21.5; cranial depth, 12.3; maxillary tooth-row, 9.4; palatal slits (length), 9.9.

Distribution.—Arid southeastern lowlands and the southern part of the Black Hills. See Fig. 41.

Records of occurrence.—Specimens examined, 6, as follows: **Laramie Co.:** 2 mi. S Pine Bluffs, 5200 ft., 2. **Platte Co.:** Uva, 1 USNM; Bordeaux, 1 USNM. **Weston Co.:** ½ mi. E Buckhorn, 6100 ft., 2.

Clethrionomys gapperi (Vigors)

Gapper's Red-backed Vole

Approximately the size of deer mouse with short tail; upper parts reddish brown, reddish chestnut, or yellowish ochraceous; sides yellowish or grayish or reddish brown; underparts plumbeous gray washed with conspicuous white or, occasionally, with cinnamon; ear blackish but often margined with rich brownish or chestnut; face grayish or pale brownish; skull small with rounded zygomata; teeth small, distinctly grooved on sides of molars; exoccipital condyles posterior to supraoccipital in adults.

Cockrum and Fitch (1952:283-287) described geographic variation in this species in Wyoming. Briefly, a subspecies (*Clethrionomys gapperi brevicaudus*) occurs in the Black Hills. *Clethrionomys gapperi galei* ranges along the Rocky Mountain chain from northwestern Wyoming to south-central Wyoming; populations along these mountains vary in color of pelage but not in cranial characters. Specimens from the central lowlands of Wyoming and from the Bighorn Mountains are paler than dark mice from the Medicine Bow Mountains and from Jackson Hole. Mice in the Uinta Mountains were regarded as closely related to *C. g. galei*, more closely related than *galei* is to other adjacent subspecies. In fact, only grayer head and cheeks, paler upper parts, and more nearly white belly were mentioned as distinctive and were made the basis for the subspecific name *Clethrionomys gapperi uintaensis.*

No cranial differences were noted between the mice of the Uinta Mountains and *C. g. galei* by Cockrum and Fitch. Four topotypes of *C. g. uintaensis* (Univ. Utah 5917, 6134, 6136, 6150), examined by me, taken in August are paler than specimens of *C. g. galei* from the Medicine Bow Mountains, Wyoming, but are no paler than specimens of that subspecies from Weston County, Big Horn County, and Mount Crooks, Fremont County. The specimens, which possess also grayish heads and white venters, from Mount Crooks were not available to Cockrum and Fitch. Neither were specimens from San Pete County, Utah, which are as dark as any specimen of *C. galei* that I have examined from Wyoming. Considering the extent of variation in *C. g. galei* and the absence of cranial characters of diagnostic value in specimens from the Uinta Mountains, it seems best to refer specimens from the Uinta Mountains to *C. g. galei,* and to arrange the name *C. g. uintaensis* (type locality, Paradise Park, 45 miles northwest of Vernal, Uintah County, Utah) as a synonym of *C. g. galei.*

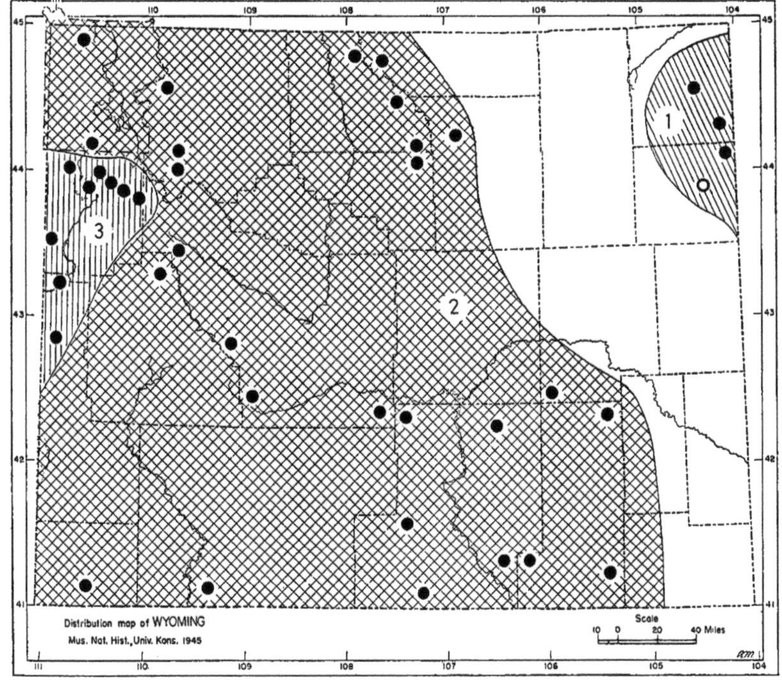

FIG. 42. Distribution of *Clethrionomys gapperi.*
Guide to subspecies
1. *C. g. brevicaudus*
2. *C. g. galei*
3. *C. g. idahoensis*

Specimens from the Teton Mountains and Jackson Hole are provisionally referred to the subspecies *C. g. idahoensis,* which is closely related to *galei* but darker and said to have narrower nasals (Merriam, 1891). No cranial differences were observed between mice from Jackson Hole and populations of *C. g. galei;* but inasmuch as the Jackson Hole area would probably be an area in which intergradation between the two subspecies would occur, it seems best, in my opinion, to refer specimens from that place to *idahoensis* on the basis of their darker color.

Clethrionomys gapperi brevicaudus (Merriam)

1891. *Evotomys gapperi brevicaudus* Merriam, N. Amer. Fauna, 5:119, July 30.

1942. *Clethrionomys gapperi brevicaudus,* Bole and Moulthrop, Sci. Publ. Cleveland Mus. Nat. Hist., 5:153, September 11.

Type.—From three miles north of Custer, 6000 feet, South Dakota.

Comparisons.—Tail shorter, hind foot averaging longer, and nasals longer than in other subspecies in Wyoming; interorbital region broader than in *Clethrionomys gapperi galei.*

Measurements.—External and cranial measurements of the type (probably subadult, male) and an adult male from Rattlesnake Creek are, respectively, as follows: Total length, 125, 127; length of tail, 31, 30; hind foot, 19, 19; occipitonasal length, 23.5, 24.3; zygomatic breadth, 12.5, 12.8; interorbital breadth, 3.8, 3.8; length of nasals, 6.8, 7.1; cranial depth, 7.1, 7.5; maxillary tooth-row, 5.3, 5.2. Measurements of 20 adults (presumably males and females) from Pennington, South Dakota, of which 11 are preserved as skins but lack skulls, are listed by Cockrum and Fitch (1952:287) as follows: Total length, 142 (123-155); length of tail, 35 (30-39); hind foot, 19.5 (18.6-21.0); basal length, 23.3 (21.7-24.5); condylobasilar length, 23.3 (21.9-24.5); zygomatic breadth, 13.7 (12.9-14.7); lambdoidal breadth, 11.7 (11.3-12.9); alveolar length upper cheek-teeth, 5.5 (5.2-5.8); interorbital breadth, 3.9 (3.6-4.1); length of nasals, 7.7 (7.1-8.5); breadth of rostrum, 3.2 (2.9-3.6); length of incisive foramina, 5.0 (4.6-5.3).

Distribution.—The Black Hills. See Fig. 42.

Records of occurrence.—Specimens examined, 27, as follows: **Crook Co.:** 3 mi. NW Sundance, 5900 ft., 3; Rattlesnake Creek, 6100 ft., 3. **Weston Co.:** 1½ mi. E Buckhorn, 6150 ft., 21.

Additional record.—**Weston Co.:** 12 mi. SE Newcastle (Cockrum and Fitch, 1952:288).

Clethrionomys gapperi galei (Merriam)

1890. *Evotomys galei* Merriam, N. Amer. Fauna, 4:23, October 8.

1897. *Evotomys gapperi galei,* Bailey, Proc. Biol. Soc. Washington, 11:126, May 13.

1931. *Clethrionomys gapperi galei,* Hall, Univ. California Publ. Zool., 37:6, April 10.

1941. *Clethrionomys gapperi uintaensis* Doutt, Proc. Biol. Soc. Washington, 54:161, December 8. Type from Paradise Park, 45 miles NW Vernal, Uintah County, Utah.

Type.—From Ward, 9500 feet, Boulder County, Colorado.

Comparisons.—For comparisons with *Clethrionomys gapperi brevicaudus* and *C. g. idahoensis,* see accounts of those subspecies.

Measurements.—Average and extreme external and cranial measurements of 24 adults (10 males) from three miles east-southeast of Browns Peak are as follows: Total length, 143.9 (130-157); length of tail, 39.0 (34-45); hind foot, 18.2 (17-19); ear from notch, 13.5 (13-15); greatest length of skull, 24.1 (22.8-25.5); zygomatic breadth, 12.9 (12.2-13.7); interorbital breadth, 3.6 (3.3-3.9); length of nasals, 7.3 (6.5-8.3); lambdoidal breadth, 11.2 (10.7-12.0); maxillary tooth-row, 5.0 (4.6-5.4).

Distribution.—Most montane and riparian habitats from Yellowstone Nat'l Park and from the Bighorn Mountains southward (excepting Jackson Hole) into the Uinta Mountains and the Medicine Bow and Laramie mountains. See Fig. 42.

Remarks.—Although specimens from the Medicine Bow Mountains are darker than specimens of this subspecies from northern counties, the type (from Boulder County, Colorado) is pale. See discussion under account of *Clethrionomys gapperi.*

Records of occurrence.—Specimens examined, 250, as follows: **Albany Co.:** Laramie Peak, 8000-8800 ft., 5 USNM; *sec. 19, T. 16N, R. 78W, 9500 ft., 1;* Upper Libby Creek Campground, 9700 ft., 1; *Holmes Campground, 9700 ft., 1; Nash Fork Picnic Ground, 9850 ft., 5; Nash Fork Creek, 2 mi. W Centennial, 7; N Fork Campground, sec. 28, T. 16N, R. 78W, 8500 ft., 4; 2¼ mi. ESE Browns Peak, 10,300 ft., 2; 3 mi. ESE Browns Peak, 10,000 ft., 56;* 10 mi. E Laramie, 9000 ft., 1 USNM.

Big Horn Co.: Medicine Wheel Ranch, 28 mi. E Lovell, 22; 4½ mi. S, 17½ mi. E Shell, 8500 ft., 1; *Head Trappers Creek, 8400-9500 ft., 23 USNM;* 9 mi. N, 9 mi. E Tensleep, 8000 ft., 1.

Carbon Co.: Ferris Mtns., 8000-8500 ft., 15 USNM; Shirley Mtns., 7600 ft., 5 USNM; Bridgers Pass, 18 mi. SW Rawlins, 7500 ft., 2; 8 mi. N, 19½ mi. E Encampment, 9150 ft., 4; *Ridge above Battle, 10,000 ft., 1 USNM; 8 mi. N, 19½ mi. E Savery, 8800 ft., 1;* 6 mi. N, 14 mi. E Savery, 8400 ft., 2.

Converse Co.: 21½-22 mi. S, 24½ mi. W Douglas, 7600 ft., 3.

Fremont Co.: Jackeys Creek, 3 mi. S Dubois, 2 USNM; Moccasin Lake, 4 mi. N, 19 mi. W Lander, 10,000 ft., 3; *3 mi. N, 18 mi. W Lander, 1; Mosquito Park Ranger Sta., 2½ mi. N, 17½ mi. W Lander, 9500 ft., 10; Middle Lake, 2 mi. S, 20½ mi. W Lander, 1; 17 mi. S, 6½ mi. W Lander, 8450 ft., 4;* 22 mi. S, 5½ mi. W Lander, 8800 ft., 3; 8 mi. E Rongis, Green Mtns., 8000 ft., 11 USNM; *Mt. Crooks, 8600 ft., 6.*

Johnson Co.: 2 mi. S, 6½ mi. W Buffalo, 5600 ft., 1; *1 mi. S, 4 mi. W Klondike, 6500 ft., 1.*

Park Co.: Grinnell Creek, Pahaska, 7000 ft., 4 USNM; *Crows Peak, 9500 ft., 2 USNM;* Valley, 7500 ft., 3 USNM; Needle Mtn., 10,000-10,500 ft., 12 USNM.

Sheridan Co.: 38 mi. E Lovell, 1.

Sublette Co.: 31 mi. N Pinedale, 8025 ft., 1.

Sweetwater Co.: 5 mi. SW Maxon, 9000 ft., 2 USNM.

Uinta Co.: 9 mi. S Robertson, 8000-8400 ft., 5; *9 mi. S, 2 mi. E Robertson, 1; 11½ mi. S, 2 mi. E Robertson, 9200 ft., 1; 14 mi. S, 2 mi. E Robertson, 9000 ft., 1.*

Washakie Co.: 4 mi. N, 9 mi. E Tensleep, 7000 ft., 1.

Yellowstone Nat'l Park: Snow Pass, Mammoth Hot Springs, 6 USNM; *Glen Creek, 7000 ft., 2 USNM;* Harebell Creek, 1 USNM; No specific locality, 1 USNM.

Clethrionomys gapperi idahoensis (Merriam)

1891. *Evotomys idahoensis* Merriam, N. Amer. Fauna, 5:66, July 30.

1933. *Clethrionomys gapperi idahoensis*, Whitlow and Hall, Univ. California Publ. Zool., 40:265, September 30.

Type.—From Sawtooth (=Alturas) Lake, east Base Sawtooth Mountains, 7200 feet, Blaine County, Idaho.

Comparisons.—Resembles *Clethrionomys gapperi galei*, but upper parts darker. For comparison with *C. g. brevicaudus*, see account of that subspecies.

Measurements.—External and cranial measurements of three adults (all females) from the northern half of Teton County, are as follows: Total length, 155, 143, 157; length of tail, 59, 42, 49; hind foot, 18, 19, 17; ear from notch, 17, 15, 15; greatest length of skull, 24.2, 22.7, 23.9; zygomatic breadth, 13.2, 12.2, 12.5; interorbital breadth, 3.6, 3.4, 3.5; length of nasals, 6.8, 6.2, 7.2; lambdoidal breadth, 11.2, 10.4, 11.1; maxillary tooth-row, 5.1, 5.1, 5.4.

Distribution.—Teton County. See Fig. 42, and account of the species.

Records of occurrence.—Specimens examined, 84, as follows: **Lincoln Co.:** 3 mi. N, 11 mi. E Alpine, 5650 ft., 1; 10 mi. SE Afton, 4 USNM. **Teton Co.:** NW Corner Teton Nat'l Forest, 18 mi. N, 9 mi. W Moran, 1; *Whetstone Creek, 5 USNM;* Upper Arizona Creek, 2 USNM; *Trib. Pacific Creek, 2 USNM;* Pacific Creek, 12 USNM; *Big Game Ridge, 1 USNM;* Jackson Hole Wildlife Park, 17 USNM, 1; *1 mi. N Moran, 2 USNM; Moran, 7, 3 USNM; 0-1 mi. S, 3¾ mi. E Moran, 6200 ft., 11; Throughfare Creek, 1 USNM;* Black Rock Meadows, 3 mi. S, 17 mi. E Moran, 8600 ft., 1; *Black Rock Creek, 2 mi. W Pass, 2 USNM;* Togwotee Pass, 10,000 ft., 4 USNM, 4; *Moose Creek, 6800 ft., 10 USNM; S Moose Creek, 9500 ft., 4 USNM;* Teton Pass, 7200 ft., 4 USNM; No specific locality, 2 USNM.

Phenacomys intermedius intermedius Merriam

Heather Vole

1889. *Phenacomys intermedius* Merriam, N. Amer. Fauna, 2:32, October 30.

1894. *Phenacomys truei* J. A. Allen, Bull. Amer. Mus. Nat. Hist., 6:331, November 7. Type from "Black Hills" (=Laramie Mountains), Wyoming.

Type.—From 20 miles north-northwest Kamloops, 5500 feet, British Columbia.

Description.—Medium in size for a microtine; upper parts dusky tan; venter whitish; tail brownish or grayish above, whitish below, and short (see *Measurements*); pinnae of ear bearing Ochraceous hairs; occlusal pattern of molars distinct in having deep re-entrant angles (zig-zag appearance); cheek-teeth rooted in adults.

Measurements.—External and cranial measurements of an adult male and an adult female from Beartooth Lake are as follows: Total length, 145, 156; length of tail, 31, 33; hind foot, 18, 18.5; condylonasal length skull, 25.8, 26.1; zygomatic breadth, 15.3, 15.4; interorbital breadth, 3.6, 3.7; length of nasals, 8.0, 7.8; cranial depth, 7.4, 7.1; maxillary tooth-row, 5.8, 6.5.

Distribution.—Montane areas of western and south-central Wyoming. Not known from Bighorn Mountains or Black Hills. See Fig. 43.

Remarks.—This vole is widely distributed altitudinally; its apparent absence in some montane areas is not yet understood.

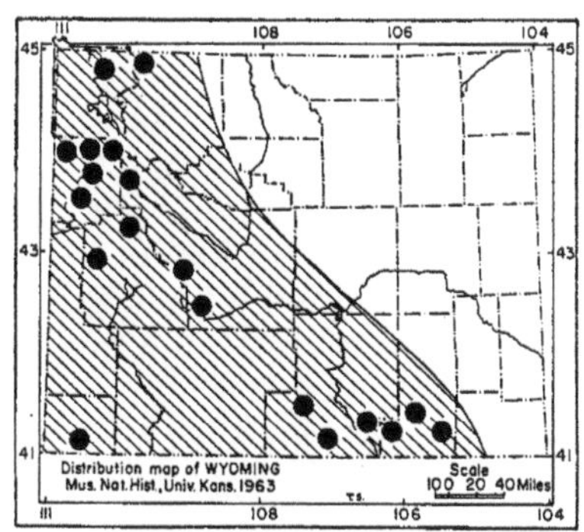

FIG. 43. Distribution of *Phenacomys intermedius intermedius.*

Records of occurrence.—Specimens examined, 63, as follows: **Albany Co.:** Little Laramie River, sec. 28, T. 16N, R. 78 W, 8500 ft., 1; Libby Creek, sec. 24, T. 16N, R. 79W, 9500 ft., 1; *Nash Fork Picnic Ground, sec. 13, T. 16N, R. 79W, 9850 ft., 1; sec. 24, T. 16N, R. 79W, 9800 ft., 1; Nelson Park, Medicine Bow Mtns., 1; 2¼ mi. ESE Browns Peak, 10,300 ft., 1; 3 mi. ESE Browns Peak, 10,000 ft., 12;* 10 mi. E Laramie, 8500 ft., 2 USNM; *"Black Hills" (Laramie Mtns.), 1 USNM.* **Carbon Co.:** Bridgers Pass, 18 mi. SW Rawlins, 7500 ft., 1; *1 mi. NW Silver Lake, 8280 ft., 1;* 8 mi. N, 18 mi. E Encampment, 8900 ft., 1; 2 mi. S Bridgers Peak, 9300 ft., 1. **Fremont Co.:** 2 mi. E Togwotee Pass, 1; 4 mi. N, 19 mi. W Lander, 10,100 ft., 1; *3 mi. N, 18 mi. W Lander, 1; 2 mi. S, 20½ mi. W Lander, 1;* 17 mi. S 6½ mi. W Lander, 1. **Park Co.:** Beartooth Lake, 9 USNM. **Sublette Co.:** 31 mi. N Pinedale, 8025 ft., 1; Merna, 8000 ft., 1 USNM. **Teton Co.:** Whetstone Creek, 1 USNM; Two Ocean Lake, 1 USNM; 18 mi. N, 9 mi. W Moran, 1; *Trib. Pacific Creek, 1 USNM; ¼ mi. N, 2½ mi. E Moran, 6230 ft., 3; Jackson Hole Wildlife Park, 2 USNM;* Moran, 2 USNM; *3¾ mi. E Moran, 6300 ft., 3; ¾-1 mi. S, 3¾ mi. E Moran, 6200 ft., 5;* 2½ mi. NE Moose, 6500 ft., 1. **Uinta Co.:** 11½ mi. S, 2 mi. E Robertson, 9200 ft., 1. **Yellowstone Nat'l Park:** Tower Falls, 1 USNM.

Microtus pennsylvanicus (Ord)

Meadow Vole

Best characterized by fifth posterior loop (separated from fourth) on middle upper molar; feet usually dark; upper parts usually darker than those of *M. longicaudus;* venter occasionally washed with cinnamon as is that of *Microtus ochrogaster.*

Microtus pennsylvanicus insperatus (J. A. Allen)

1894. *Arvicola insperatus* J. A. Allen, Bull. Amer. Mus. Nat. Hist., 6:347, December 7.

1943. *Microtus pennsylvanicus insperatus,* R. Anderson, Canadian Field-Nat., 57:92, October 17.

Type.—From Custer, Custer County, South Dakota.

Comparison.—From *Microtus pennsylvanicus pullatus, M. p. insperatus* differs in paler upper parts, relatively and actually shorter tail, and longer upper molar tooth-row.

Measurements.—Average and extreme external and cranial measurements of 11 adults (five females) from within two miles of Buffalo are as follows: Total length, 172.3 (160-184); length of tail, 50.1 (45-54); hind foot, 21.2 (20.5-22); ear from notch, 13.1 (11-14); total length of skull, 27.5 (26.5-28.5); zygomatic breadth, 15.7 (14.6-17.0); interorbital breadth, 3.5 (3.3-3.7); length of nasals, 7.7 (7.4-8.2); cranial depth, 8.3 (7.9-8.9); maxillary tooth-row, 6.5 (6.3-6.7).

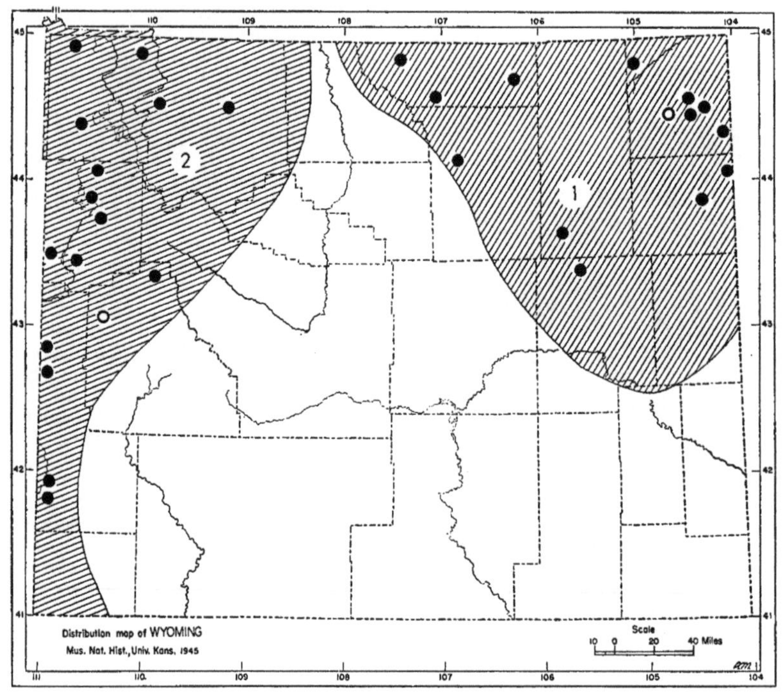

FIG. 44. Distribution of *Microtus pennsylvanicus*.
1. *M. p. insperatus* 2. *M. p. pullatus*

Distribution.—Known from northeastern Wyoming. See Fig. 44.

Remarks.—This subspecies is sympatric with *M. longicaudus* in the Black Hills, from which hills *M. montanus* is absent.

Records of occurrence.—Specimens examined, 158, as follows: **Campbell Co.:** *5 mi. S, 6 mi. W Rockypoint, 1;* 42 mi. S, 13 mi. W Gillette, 4 (from owl pellets, discarded); *Belle Fourche River, 45 mi. S, 13 mi. W Gillette, 5350 ft., 2.*

Converse Co.: 10 mi. N, 6 mi. W Bill, 8 (from owl pellets, discarded).

Crook Co.: *4 mi. N, 3 mi. E Rockypoint, 1;* 3 mi. S, 2 mi. E Rockypoint, 3800 ft., 3; *3 mi. S, 5 mi. E Rockypoint, 2;* 4 mi. S, 7 mi. E Rockypoint, 2; 15 mi. N Sundance, 5500 ft., 3; 15 mi. ENE Sundance, 3825 ft., 6; *Bear Lodge Mtns., 2 USNM;* 3 mi. NW Sundance, 5900 ft., 1; *1½ mi. NW Sundance, 5000 ft., 4; Sundance, 16 USNM;* Sand Creek, 3750 ft., 2 USNM; *Rattlesnake Creek, 1 USNM.*

Johnson Co.: 4/5 mi. S, 1 mi. W Buffalo, 4800 ft., 36; *1 mi. S, 5½ mi. W Buffalo, 4800 ft., 1; 1½ mi. S, 5½ mi. W Buffalo, 1; ¼ mi. E Klondike, 5160 ft., 1.*

Sheridan Co.: 3 mi. WNW Monarch, 3800 ft., 4; 5 mi. NE Clearmont, 3900 ft., 3; 4 mi. NNE Banner, 4100 ft., 26.

Weston Co.: 1½ mi. E Buckhorn, 6150 ft., 26; Newcastle, 2 USNM.

Additional record (S. Anderson, 1956:102).—**Crook Co.:** Bear Lodge Mtns., 6½ mi. SSE Alva.

Microtus pennsylvanicus pullatus S. Anderson

1956. *Microtus pennsylvanicus pullatus* S. Anderson, Univ. Kansas Publ., Mus. Nat. Hist., 9:97, May 10.

Type.—From twelve miles north and two miles east of Sage, 6100 feet, Lincoln County, Wyoming.

Comparison.—See account of *Microtus pennsylvanicus insperatus.*

Measurements.—External and cranial measurements of two adult males and four adult females (three pregnant) from within four miles of Cody, Park County, are, respectively, as follows: Total length, 170, 174, 171, 180, 185, 173; length of tail, 40, 48, 45, 49, 46, 48; hind foot, 19.5, 20, 20.5, 21, 20.5, 21; ear from notch, 12, 13, 13, 14, 13, 13; total length of skull, 28.7, —, —, —, —, 27.3; zygomatic breadth, 16.5, 16.9, 16.5, 17.3, 17.1, 16.0; interorbital breadth, 3.4, 3.3, 3.5, 3.2, 3.5, 3.6; length of nasals, 8.3, 7.7, 7.4, —, 8.0, 8.0; cranial depth, 9.0, —, —, 8.8, —, 9.1; maxillary tooth-row, 6.5, 6.5, 6.1, 6.7, 6.6, 6.4.

Distribution.—Montane areas of western Wyoming. See Fig. 44.

Remarks.—A cline of increasing pallor extends eastward or northeastward in this subspecies toward the pale *M. p. insperatus.* The subspecific boundary was drawn by Anderson (1956:97-98) at the "most distinct break in this cline," in the Bighorn Basin of Wyoming.

Records of occurrence.—Specimens examined, 140, as follows: **Lincoln Co.:** 9-9½ mi. N, 2 mi. W Afton, 6100 ft., 4; *7 mi. N, 1 mi. W Afton, 11;* Salt River, 10 mi. W Afton, 6200 ft., 1 USNM; *15 mi. N, 3 mi. E Sage, 6100 ft., 1; 12 mi. N, 2 mi. E Sage, 4;* Cokeville, 2 USNM; 6 mi. N, 2 mi. E Sage, 2. **Park Co.:** ⅗ mi. S, 3⅕ mi. E Cody, 5020 ft., 15; Pahaska, 1 USNM. **Sublette Co.:** 34 mi. N, 4 mi. W Pinedale, 7950 ft., 2. **Teton Co.:** Whetstone Creek, 4 USNM; Jackson Hole Wildlife Park, 3; *Within 4 mi. Moran, 6200 ft., 54; 9 USNM;* Elk, 2 USNM; Teton Pass, 7200 ft., 2 USNM; Jackson, 1 USNM. **Yellowstone Nat'l Park:** Mammoth Hot Springs, 15 USNM; Lamar River, 7000 ft., 2 USNM; Old Faithful, 5 USNM.

Additional records (S. Anderson, 1956:99).—**Sublette Co.:** Kendall. **Teton Co.:** *5 mi. N Moran; Trappers Lake; Jenny Lake; String Lake; Sheep Creek.*

Microtus montanus (Peale)

Montane Vole

Upper parts brownish with more or less reddish or yellowish; underparts whitish; middle upper molar lacking fifth (posterior) loop. Known from boreal habitat throughout Wyoming except in the Black Hills.

Microtus montanus codiensis S. Anderson

1954. *Microtus montanus codiensis* S. Anderson, Univ. Kansas Publs., Mus. Nat. Hist., 7(7):497, July 23.

Type.—From 3⅕ miles east and ⅗ mile south of Cody, 5020 feet elevation, Park County, Wyoming.

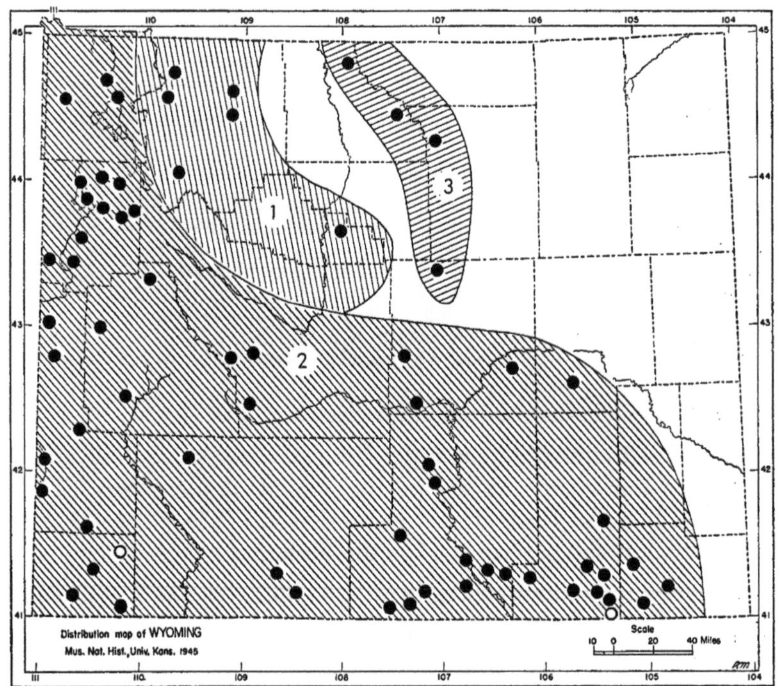

FIG. 45. Distribution of *Microtus montanus*.
Guide to subspecies 2. *M. m. nanus*
1. *M. m. codiensis* 3. *M. m. zygomaticus*

Comparisons.—Compared with *Microtus montanus nanus* from Idaho, specimens of *M. m. codiensis* differ in larger size, relatively long alveolobasilar length, relatively long alveolar length of upper molar tooth-row, relatively widespreading zygomatic arches, relatively long tail, and paler color.

Compared with *M. m. zygomaticus* of the Bighorn Mountains, *M. m. codiensis* has a relatively longer tail that is also actually longer, longer hind foot, longer upper molar tooth-row, slightly paler and less grizzled pelage. The auditory bullae are larger and less flattened, the angle formed at the suture between the basioccipital and basisphenoid bones is less acute, the suture is less elevated between the bullae when viewed from below (ventral view), and the incisive foramina are less constricted posteriorly.

Measurements.—Average and extreme external and cranial measurements of 34 males and females, 27 from the type locality and seven from other localities in the range assigned to this subspecies, are as follows: Total length, 165 (146-186); length of tail, 44.2 (35-55); hind foot, 19.6 (17-21); condylobasilar length, 25.5 (24.0-27.5); zygomatic breadth, 15.6 (14.7-16.6); alveolar length of upper tooth-row, 6.6 (6.2-7.0); prelambdoidal breadth, 8.8 (8.1-9.5); lambdoidal breadth, 12.0 (11.2-12.8) (Anderson, 1954:500).

Distribution.—Montane areas west of the Bighorn Basin and east of the geographic range of *M. m. nanus*, in northwestern Wyoming. See Fig. 45.

Remarks.—The distribution of *M. m. codiensis,* in fact its existence, is difficult to explain. Anderson (1954:500) pointed out that two other species of *Microtus* occur at the type locality of *codiensis,* and he posed the question, "Could interspecific hybridization between 'good species' of *Microtus* take place in nature and possibly alter the characteristics of a local population?" He did not develop the idea further. It can be pointed out that *M. montanus nanus* occurs with three other species of *Microtus* in Jackson Hole (Findley, 1951:119) and that there is no indication there of the infusion of any blood of the other species of *M. montanus.* Another explanation for the differentiation of *codiensis* is given on page 732 beyond.

Records of occurrence.—Specimens examined, 47, as follows: **Hot Springs Co.:** 3 mi. N, 10 mi. W Thermopolis, 4950 ft., 3. **Park Co.:** Black Mtn., hd. Pat O'hara Creek, 3 USNM; 13 mi. N, 1 mi. E Cody, 5200 ft., 1; SW slope of Whirlwind Peak, 9000 ft., 1; *5 mi. N Cody, 1 USNM;* type locality, 31; *Ishawooa Creek, 2 USNM; Valley, 1 USNM;* Needle Mtn., 4 USNM.

Microtus montanus nanus (Merriam)

1891. *Arvicola (Mynomes) nanus* Merriam, N. Amer. Fauna, 5:63, July 30.
1938. *Microtus montanus nanus,* Hall, Proc. Biol. Soc. Washington, 51:133, August 23.
1917. *Microtus montanus caryi* V. Bailey, Proc. Biol. Soc. Washington, 30: 29, February 21. Type from Milford, Fremont County, Wyoming; regarded as inseparable from *M. m. nanus* by S. Anderson. Univ. Kansas Publ., Mus. Nat. Hist., 7:494, July 23, 1954.

Type.—From Pahsimeroi Mountains, head of Pahsimeroi River, 9350 feet, Custer County, Idaho.

Comparisons.—Compared with *Microtus montanus zygomaticus, M. m. nanus* differs in longer tail, longer upper molar tooth-row, relatively narrower zygomatic breadth, and larger, more inflated auditory bullae. For comparison with *M. m. codiensis,* see account of that subspecies.

Measurements.—Average and extreme measurements of six adult males from near Pocatello, Bannock County, Idaho, and nine adult males from within 15 miles of Afton, Lincoln County, Wyoming, are, respectively, as follows: Total length, 143 (135-150), 163 (143-179); length of tail, 35.1 (33-38), 42.8 (36-49); hind foot, 18.9 (18-20), 18.8 (17-20); condylobasilar length, 24.4 (24.0-26.0), 25.6 (24.5-26.2); alveolobasilar length, 14.1 (13.7-14.5), 14.6 (13.8-15.0); palatilar length, 13.2 (12.9-13.6), 13.8 (13.2-14.5); alveolar length of upper molar tooth-row, 6.3 (6.1-6.5), 6.3 (6.0-6.6); depth of braincase, 7.7 (7.5-7.9), 8.0 (7.7-8.3); zygomatic breadth, 14.3 (13.8-14.7), 15.3 (14.4-16.3); interorbital breadth, 3.6 (3.5-3.7), 3.5 (3.3-3.7) (after S. Anderson, 1954:494).

Distribution.—Montane habitats in southern half of state ranging northward into extreme northwestern part. See Fig. 45.

Records of occurrence.—Specimens examined, 691, as follows: **Albany Co.:** 30 mi. N, 10 mi. E Laramie, 6760 ft., 6; *29¾ mi. N, 9½ mi E Laramie, 6350 ft., 1; 26¾ mi. N, 6½ mi. E Laramie, 6700 ft., 3; 26 mi. N, 4½ mi. E Laramie, 6960 ft., 8;* 8 mi. N, 13 mi. E Laramie, 7500 ft., 1; 5 mi. N Laramie, 7400 ft., 15; *2½ mi. NW Laramie, 2;* Hq. Park, 10,200 ft., 4 USNM; *½ mi. S, 2 mi. E*

Medicine Bow Peak, 10,800 ft., 1; Centennial, 8120 ft., 1; 2¼ mi. ESE Browns Peak, 10,300 ft., 3; 3 mi. ESE Browns Peak, 10,000 ft., 12; 2 mi. S Browns Peak, 10,600 ft., 1; 1 mi. E Laramie, 7160 ft., 4; 4 mi. SW Laramie, 2; 6 mi. SW Laramie, 1; 6½ mi. S, 8¾ mi. E Laramie, 8200 ft., 1; Pole Mtn., 14-15 mi. SE Laramie, 8300 ft., 4, 3 USNM; *1 mi. SSE Pole Mtn., 8350 ft., 4; 2 mi. SW Pole Mtn., 8300 ft., 13; 3 mi. S Pole Mtn., 8100 ft., 1; Therkildsons Ranch, SW Laramie, 1; Blair Camp Ground, Pole Mtn., 1; Pole Mtn., 2; Wallis Picnic Ground, 8350 ft., 1; Douglas Creek, 8250 ft., sec. 29, T. 13N, R. 79W, 2.*

Carbon Co.: 27 mi. N Sinclair, 6200 ft., 1 Univ. Wyo.; 18 mi. NNE Sinclair, 6500 ft., 10; Bridgers Pass, 18 mi. SW Rawlins, 7500 ft., 7; Saratoga, 2 USNM; 6 mi. S, 13 mi. E Saratoga, 5; *6 mi. S, 14 mi. E Saratoga, 8800 ft., 1;* Lake Marie, 10,440 ft., 2; *1 mi. S Lake Marie, 2; 10 mi. N, 14 mi. E Encampment, 8400 ft., 14; 9½-10 mi. N, 11½-12 mi. E Encampment, 7200 ft., 5; 9 mi. N, 3-4 mi. E Encampment, 6500 ft., 23; 9 mi. N, 8 mi. E Encampment, 7000 ft., 3; 8 mi. N, 14-14½ mi. E Encampment, 8400 ft., 11; 8 mi. N, 16 mi. E Encampment, 8400 ft., 11; 8 mi. N, 18 mi. E Encampment, 8900 ft., 1; 8 mi. N, 21½ mi. E Encampment, 9400 ft., 5;* ¼ mi. N Riverside, 7380 ft., 2; S. base Bridger Peak, 8800 ft., 1 USNM; *2 mi. S Bridger Peak, 9300 ft., 2;* 8 mi. N, 19½ mi. E Savery, 8800 ft., 27; *7½ mi. N, 18½ mi. E Savery, 8400 ft., 1; 7 mi. N, 17 mi. E Savery, 8300 ft., 1;* 6 mi. N, 12½-13½ mi. E Savery, 8400 ft., 10; *6 mi. N, 14-15 mi. E Savery, 8500 ft., 18; 5 mi. N, 10½ mi. E Savery, 8000 ft., 12; 4 mi. N, 8 mi. E Savery, 7300 ft., 3.*

Converse Co.: Beaver, 1 USNM.

Fremont Co.: *Brooks Peak, 5; within one mi. Milford, 27, 4 USNM, 3 Univ. Wyo.;* 2½ mi. N, 17½ mi. W Lander, 9500 ft., 3; *2 mi. N, 17 mi. W Lander, 9300 ft., 4;* 15½ mi. S, 7½ mi. W Lander, 9200 ft., 1; South Pass City, 8000 ft., 19 USNM; *23½ mi. S, 5 mi. W Lander, 8600 ft., 7.*

Laramie Co.: 5 mi. N, 1 mi. W Horse Creek P. O., 7200 ft., 1; 11 mi. N, 5½ mi. E Cheyenne, 5450 ft., 7; *6 mi. W Cheyenne, 6500 ft., 1;* 7 mi. W Cheyenne, 6500 ft., 8; *Cheyenne, 4 USNM.*

Lincoln Co.: 13 mi. N, 2 mi. W Afton, 1; *10 mi. N, 2 mi. W Afton, 4; 9½ mi. N, 2 mi. W Afton, 3; 9 mi. N, 2 mi. W Afton, 9; 7 mi. N, 1 mi. W Afton, 12;* Afton, 1 USNM; LaBarge Creek, 1 USNM; Border, 6 USNM; Cokeville, 2 USNM; 6 mi. N, 2 mi. E Sage, 1; Cumberland, 5 USNM; *½ mi. S, 1½ mi. W Cumberland, 6.*

Natrona Co.: 16 mi. S, 11 mi. W Waltman, 6950 ft., 44; 6 mi. S, 2 mi. W Casper, 5900 ft., 4; *6½ mi. S, 2 mi. W Casper, 6100 ft., 1; 7 mi. S, 2 mi. W Casper, 6370 ft., 3; 10 mi. S Casper, 7750 ft., 33;* Sun, 2 USNM; *5 mi. W Independence Rock, 6000 ft., 4; 1 mi. S, 5 mi. W Independence Rock, 2.*

Sublette Co.: 34 mi. N, 4 mi. W Pinedale, 1; *33 mi. N, 2 mi. W Pinedale, 6; 32 mi. N, 1 mi. W Pinedale, 1; 31 mi. N Pinedale, 4;* Merna, 6 USNM; Big Piney, 1 USNM.

Sweetwater Co.: Farson, 3; Kinney Ranch, 21 mi. S Bitter Creek, 6800 ft., 9, 2 USNM; 32 mi. S, 22 mi. E Rock Springs, 7025 ft., 15.

Teton Co.: Upper Arizona Creek, 1 USNM; Pacific Creek, 1 USNM; *Big Game Ridge, 2 USNM; Two Ocean Lake, 3 USNM;* 18 mi. N, 9 mi. W Moran, 1; Jackson Hole Wildlife Park, 2; *¼ mi. N, 2½ mi. E Moran, 6230 ft., 3; Moran, 16; 1 mi. E Moran, 1 Univ. Wyo.;* ¾-1 mi. S, 3¾ mi. E Moran, 6200 ft., 6; Togwotee Pass, 2; Black Rock Creek, 2 USNM; Bar BC Ranch, 2½ mi. NE Moose, 6500 ft., 1; Teton Pass, 7200 ft., 1 USNM; Jackson, 1.

Uinta Co.: *16 mi. S, 2 mi. W Kemmerer, 6700 ft., 3;* Ft. Bridger, 6650 ft., 25, 9 USNM; 9 mi. S Robertson, 8000 ft., 9; *9½ mi. S, ½ mi. W Robertson, 8600 ft., 1; 10 mi. S, 1 mi. W Robertson, 8700 ft., 25; 14 mi. S, 2 mi. E Robertson, 9000 ft., 5;* 4 mi. S Lonetree, 1 USNM.

Yellowstone Nat'l Park: Canyon Camp, 1 USNM; Lower Geyser Basin, 1 USNM; N. end Lake, 3 USNM.

Additional records (S. Anderson, 1954:496, unless otherwise noted).—**Albany Co.:** *7 mi. N, 2 mi. E Laramie; Laramie; 7-7/10 mi. SSW Laramie;* Sherman. **Carbon Co.:** *10 mi. S, 25 mi. E Saratoga, 9800 ft.* (S. Anderson, 1959:462). **Lincoln Co.:** *Sheep Creek.* **Sublette Co.:** *Dell Creek.* **Sweet-**

water Co.: *Bitter Creek.* **Teton Co.:** *Whetstone Creek; S. Fork Buffalo River; Jenny Lake.* **Uinta Co.:** 10 mi. SW Granger. **Yellowstone Nat'l Park:** *Upper Yellowstone River.*

Microtus montanus zygomaticus S. Anderson

1954. *Microtus montanus zygomaticus* S. Anderson, Univ. Kansas Publs., Mus. Nat. Hist., 7:500, July 23.

Type.—From Medicine Wheel Ranch, 28 miles east of Lovell, 9000 feet elevation, Big Horn County, Wyoming.

Comparisons.—See accounts of *Microtus montanus codiensis* and *M. m. nanus.*

Measurements.—Average and extreme external and cranial measurements of 12 adult male topotypes are as follows: Total length, 159 (144-174); length of tail, 36.4 (30-41); hind foot, 18.2 (16-20); condylobasilar length, 25.8 (24.7-26.7); alveolobasilar length, 14.8 (13.8-15.3); palatilar length, 13.8 (12.7-14.2); alveolar length of upper molar tooth-row, 6.4 (5.9-6.6); zygomatic breadth, 15.9 (15.0-16.7); interorbital breadth, 3.6 (3.4-3.7); lambdoidal breadth, 12.1 (11.5-12.5); prelambdoidal breadth, 8.6 (8.3-8.9); depth of braincase, 8.0 (7.6-8.3).

Distribution.—Bighorn Mountains. See Fig. 45.

Remarks.—According to S. Anderson (1954:502), *M. m. zygomaticus* deviates from the norm for the species the most of any subspecies, followed next by *M. m. codiensis* and then by several other populations (including two subspecies). *M. m. nanus* is wide-ranging and approaches nearest to the norm.

Records of occurrence.—Specimens examined, 48, as follows: **Big Horn Co.:** type locality, 24; head of Trappers Creek, 9500 ft., 2 USNM. **Johnson Co.:** 1 mi. S, 7½ mi. W Buffalo, 6500 ft., 3; *Bighorn Mtns., 3 USNM.* **Natrona Co.:** Buffalo Creek, 27 mi. N, 1 mi. E Powder River, 6075 ft., 16.

Microtus longicaudus longicaudus (Merriam)

Long-tailed Vole

1888. *Arvicola (Mynomes) longicaudus* Merriam, Amer. Nat., 22:934, October.

1895. *Microtus (Mynomes) longicaudus,* J. A. Allen, Bull. Amer. Mus. Nat. Hist., 7:266, August 21.

1891. *Arvicola (Mynomes) mordax* Merriam, N. Amer. Fauna, 5:61, July 30 (=*Microtus longicaudus mordax,* Goldman, 1938. Jour. Mamm., 19:491, November 14, type from Sawtooth or Alturus Lake, 7200 feet, Blaine County, Idaho).

Type.—From Custer, Black Hills, 5500 feet, Custer County, South Dakota.

Description.—Medium in size for *Microtus* (see *Measurements*) except tail which is long (usually 3 to 3½ times length of hind foot); upper parts Grayish Bister to Dark Sepia Brown often with reddish tinge or olivaceous wash; sides frequently worn and often paler than dorsum; middle upper molar having four closed triangles; skull relatively smooth compared to that of *M. richardsoni;* rostrum, nasals, and maxillary tooth-row short (see *Measurements*); auditory bullae large and well inflated.

Measurements.—External and cranial measurements of an adult female from Merna (Horse Creek), of the type of *M. longicaudus* (lactating female), and of three adult males from the Mouth of Grinnell Creek are, respectively, as follows: Total length, 205, 185, 190, 180, 190; length of tail, 69, 65, 66, 64, 70; length of hind foot, 20, 21, 22, 22, 22; condylonasal length, 27.8, 26.8, —, 27.8, 28.9; zygomatic breadth, 15.3, —, —, 15.4, 15.8; interorbital breadth, 3.8, 3.6, —, 3.7, 3.6; length of nasals, 8.2, 7.8, 7.9, 7.9, 8.0; depth of skull, 8.6, 7.9, —, 8.1, 8.0; maxillary tooth-row, 6.6, 6.5, 6.3, 6.6, 6.8. Average and extreme external and cranial measurements of 17 adults (11 males) from nine to ten miles southward of Robertson, Uinta County, are as follows: Total length, 183.4 (173-195); length of tail, 63.8 (56-72); hind foot, 20.8 (20-23); ear from notch, 15.4 (14-20); condylonasal length, 27.6 (26.6-28.5); zygomatic breadth, 15.2 (14.3-16.2); interorbital breadth, 3.7 (3.5-3.9); length of nasals, 8.2 (7.9-9.0); cranial depth, 8.0 (7.8-8.6); maxillary tooth-row, 6.4 (6.2-6.7).

Distribution.—Montane areas generally, but in many riparian habitats in lower, arid areas. Occurs in suitable habitat throughout state. See Fig. 46.

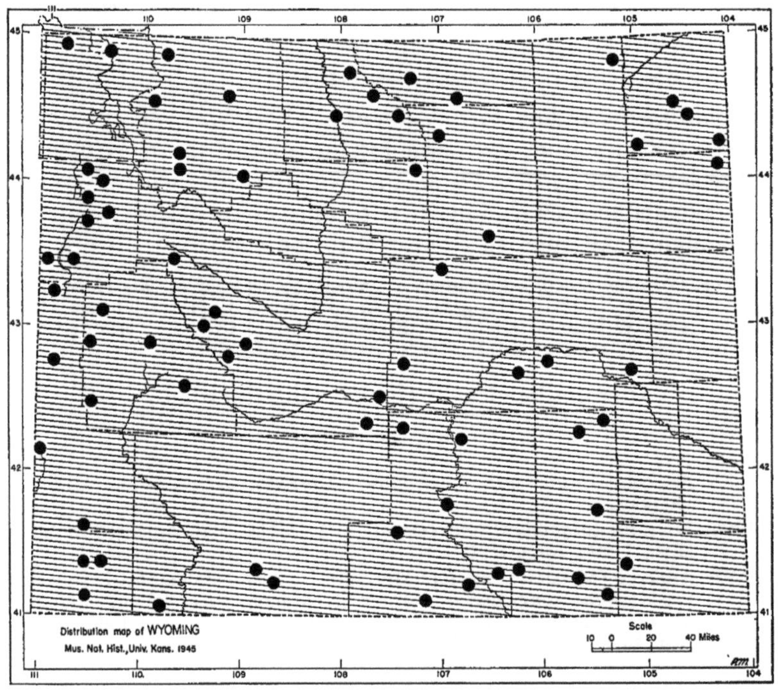

FIG. 46. Distribution of *Microtus longicaudus longicaudus.*

Remarks.—The population of long-tailed voles in the Black Hills is isolated to an unknown degree from other Wyoming populations of this vole to the westward and southward. To Bailey (1900:48) the long-tailed vole was unknown between the Black Hills and the Bighorn Mountains. Now specimens are available from the Belle Fourche River at Moorcroft only 40 some miles west of the Black

Hills, and a much longer distance by way of the riparian habitat of the Belle Fourche River. At the time I visited the locality the river was dry except for a pool of water behind a beaver dam. Specimens examined from Campbell, Johnson, and Sheridan counties show that the voles in the Black Hills (*M. l. longicaudus* of Merriam) are not so isolated as Bailey thought.

C. H. Merriam (1888 and 1891) named both *M. l. longicaudus* and *M. l. mordax*. He chose as the holotype of *M. l. longicaudus* an extremely dark vole having blackish feet and long black ears. Most of the voles of this species in the Black Hills are not so dark and have short brownish ears and somewhat paler feet. Bailey (1900) examined large series of topotypes and near topotypes of *mordax* and *longicaudus;* he recognized that Merriam's description of the population in the Black Hills and his comparison of it with the Idaho voles were inaccurate. Bailey mentioned that *mordax* and *longicaudus* are "very similar" in size and cranial dimensions. Differences noted by him were "slightly longer, slenderer rostrum and nasals; slenderer zygomata, and longer condylar ramus of mandible" in *M. mordax*. A few voles in the Black Hills have slightly larger tympanic bullae than those of *mordax,* but the character does not warrant subspecific recognition. Examination of specimens from Idaho and from throughout Wyoming has revealed that *M. longicaudus* does not vary much, although many specimens from Jackson Hole have a reddish tinge on their upper parts, and occasional specimens from the Black Hills are darker than most Wyoming specimens.

Records of occurrence.—Specimens examined, 557, as follows: **Albany Co.:** Springhill, 6300 ft., 5 USNM; 3 mi. SW Eagle Peak, 7500 ft., 1; 30 mi. N, 10 mi. E Laramie, 6760 ft., 1; *29 mi. N, 8¾ mi. E Laramie, 6420 ft., 3; 26¾-27¾ mi. N, 6½-6¾ mi. E Laramie, 2;* Nash Fork Creek, 1; *Snowy Range, 2 USNM; N. Fk. Little Laramie River, sec. 28, T. 16N, R. 78W, 8500 ft., 1;* W. *Base Corner Mtn., sec. 28, T. 16N, R. 78W, 8500 ft., 2; N. Fk. Campground, sec. 28, T. 16N, R. 78W, 8500 ft., 2; Class Lake, sec. 17, T. 16N, R. 79W, 10,750 ft., 1; 3 mi. ESE Browns Peak, 10,000 ft., 7;* Laramie, 1 USNM; *1 mi. E Laramie, 7164 ft., 1; 7 mi. E Laramie, 7900 ft., 1 Univ. Wyo.; Vedauwoo Picnic Ground, 1; Tie City Picnic Ground, 8575 ft., 1; Wallis Picnic Ground, 8350 ft., 1; Laramie Mtns., 8500 ft., 5 USNM; 1 mi. SSE Pole Mtn., 8350 ft., 8;* 2 mi. SW Pole Mtn., 8300 ft., 13; *3 mi. S Pole Mtn., 8100 ft., 4.*

Big Horn Co.: Medicine Wheel Ranch, 28 mi. E Lovell, 9000 ft., 9; Granite Creek Campground, 3; *2 mi. N, 12 mi. E Shell, 7500 ft., 2; Hd. Trappers Creek, 8400-8500 ft., 10 USNM;* 4½ mi. S, 17½ mi. E Shell, 4; Greybull, 2 USNM.

Campbell Co.: 1¼ mi. N, ½ mi. E Rockypoint, 1.

Carbon Co.: Ferris Mtns., 8500 ft., 3 USNM; Shirley Mtns., 4 USNM; Ft. Steele, 1 USNM; Bridgers Pass, 18 mi. SW Rawlins, 7500 ft., 35, 13 USNM; *1 mi NNW Silver Lake, 1;* 10 mi. N, 14 mi. E Encampment, 8; *8 mi. N, 14-14½ mi. E Encampment, 8000-8100 ft., 6; 8 mi. N, 16 mi. E Encampment, 8400 ft., 6; Lake Marie, 10,440 ft., 1;* Riverside, 1 USNM; *Sierra Madre Mtns., above Battle, 2 USNM;* 8 mi. N, 19-20 mi. E Savery, 8800 ft., 8; *7 mi. N,*

17 mi. E Savery, 8300 ft., 3; 6 mi. N, 14 mi. E Savery, 2; 5 mi. N, 10½ mi. E Savery, 8400 ft., 1; 4 mi. N, 8 mi. E Savery, 7300 ft., 4.

Converse Co.: 4½ mi. S, 3 mi. W Glenrock, 1; 3 mi. N, 5 mi. E Orin, 4725 ft., 1.

Crook Co.: Warren Peak, Bear Lodge Mtns., 6000 ft., 1 USNM; 3 mi. NW Sundance, 5900 ft., 7; *1½ mi. NW Sundance, 5000 ft., 1; Sundance, 12 USNM;* Belle Fourche Riv., Moorcroft, 3; Rattlesnake Creek, 6000 ft., 2 USNM.

Fremont Co.: Jackeys Creek, 1 USNM; Bull Lake, 3 USNM; Lake Fork, 11,400 ft., 1 USNM; Milford, 5400 ft., 2; *Moccasin Lake, 4 mi. N, 19 mi. W Lander, 1;* Mosquito Park Ranger Sta., 2½ mi. N, 17½ mi. W Lander, 3; Splitrock, 6200 ft., 1 USNM; Mt. Crooks, Green Mtns., 8600 ft., 24; *Green Mtns., 2 USNM.*

Johnson Co.: 1½ mi. S, 5½ mi. W Buffalo, 4600-5520 ft., 4; *2 mi. S, 6½ mi. W Buffalo, 5620 ft., 1; 2 mi. W Klondike, 5980 ft., 1;* 1 mi. WSW Kaycee, 4700 ft., 1.

Laramie Co.: 1 mi. N, 5 mi. W Horse Creek P. O., 6600 ft., 1; *5 mi. W Horse Creek P. O., 7200 ft., 3; 3½ mi. W Horse Creek P. O., 3.*

Lincoln Co.: 3 mi. N, 11 mi. E Alpine, 2; 10 mi. SE Afton, 2 USNM; Border, 4 USNM; Cumberland, 6500 ft., 1.

Natrona Co.: 27 mi. N, 1 mi. E Powder River, 6075 ft., 2; 16 mi. S, 11 mi. W Waltman, 6950 ft., 27; *Rattlesnake Mtns., 7000-7500 ft., 7 USNM;* 10 mi. S Casper, 7750 ft., 2; *7 mi. S, 2 mi. W Casper, 6300 ft., 1; 7 mi. S Casper, 6000 ft., 5 USNM.*

Park Co.: Clarks Fork, 3 USNM; 5 mi. N Cody, 1 USNM; Mouth Grinnell Creek, 6300-7300 ft., 10 USNM; Needle Mtn., 2 USNM; Valley, 6500-7500 ft., 13; 15½ mi. S, 13 mi. W Meeteetse, 2.

Sheridan Co.: Wolf, 1 USNM; 4 mi. NNE Banner, 4100 ft., 5.

Sublette Co.: Merna (Horse Creek), 7800 ft., 1 USNM; 12 mi. N Kendall, 7700 ft., 1 USNM; *Halfmoon Lake, 7900 ft., 2;* 10 mi. NE Pinedale, 8000 ft., 1; *5 mi. N, 3 mi. E Pinedale, 7500 ft., 5; 2½ mi. NE Pinedale, 7500 ft., 3;* Big Sandy, 1 USNM; 2 mi. S, 19 mi. W Big Piney, 5.

Sweetwater Co.: 32 mi. S, 22 mi. E Rock Springs, 7025 ft., 6; 22 mi. SSW Bitter Creek, 7; ½ mi. N Jct. Henrys Fork and Utah Boundary, 1.

Teton Co.: Whetstone Creek, 1 USNM; Pacific Creek, 4 USNM; 1 mi. N Moran, 7 USNM; *Two Ocean Lake, 1 Univ. Wyo.; Moran, 1; 3¾ mi. E Moran, 6300 ft., 4; ¾-1 mi. S, 3¾ mi. E Moran, 6200 ft., 19;* Gros Ventre Slide, 3 mi. E Kelly P. O., 6800 ft., 1; Bar BC Ranch, 2½ mi. NE Moose, 6500 ft., 1; Teton Pass, 7200 ft., 14 USNM; Jackson, 1 USNM.

Uinta Co.: 8½ mi. W Ft. Bridger, 7100 ft., 1; Ft. Bridger, 6650 ft., 11, 1 USNM; *Mountainview, 1 USNM; 8 mi. S, 2½ mi. E Robertson, 8300 ft., 1;* 9 mi. S Robertson, 8400 ft., 41; *9 mi. S, 2½ mi. E Robertson, 8600 ft., 2; 10 mi. S, 1 mi. W Robertson, 8700 ft., 11; 10 mi. S, 2½ mi. E Robertson, 8900 ft., 5; 13 mi. S, 2 mi. E Robertson, 9200 ft., 1; 14 mi. S, 2 mi. E Robertson, 9000 ft., 1.*

Washakie Co.: 4 mi. N, 9 mi. E Tensleep, 7000 ft., 8; *1 mi. N, 3 mi. E Tensleep, 4350 ft., 3.*

Weston Co.: 1½ mi. E Buckhorn, 6150 ft., 14.

Yellowstone Nat'l Park: Snow Pass, Mammoth Hot Springs, 3 USNM; Tower Falls, 1 USNM; no specific locality, 1 USNM.

Microtus richardsoni macropus (Merriam)

Richardson's Vole

1891. *Arvicola (Mynomes) macropus* Merriam, N. Amer. Fauna, 5:60, July 30.

1900. *Microtus richardsoni macropus,* V. Bailey, N. Amer. Fauna, 17:61, June 6.

Type.—From Pahsimeroi Mountains, 9300 feet, Custer County, Idaho.

Description.—Large (most adults longer than 200 mm., some more than 250 mm. in total length); dark brownish upper parts, whitish or grayish underparts; skull robust with robust rostrum but small, flattened tympanic bullae.

Comparisons.—This species is easily identified on account of its dark color and the large size of adults. Subadults might be confused with *Microtus pennsylvanicus* (which differs in having a 5th closed loop on the second upper molar tooth), with *M. montanus* (which has relatively and actually larger auditory bullae), and with *M. longicaudus* (which also has relatively and actually larger bullae). The upper molar row in most individuals of *M. richardsoni* is longer than in the other species.

Measurements.—External and cranial measurements of the holotype (adult? male), four lactating females, and one adult male from 10 miles southeast of Afton are, respectively, as follows: Total length, 220, 233, 225, 220, 230, 240; length of tail, 71, 76, 73, 70, 76, 72; hind foot, 26, 27, 26, 26, 26, 29; condylonasal length, 31.5, —, 33.2, —, —, 34.2; zygomatic breadth, 19.6, 19.9, 19.8 —, —, —; interorbital breadth, 4.9, 4.7, 4.7, 4.3, —, 5.0; length of nasals, 9.0, 9.5, 9.7, 9.0, —, 10.0; length of bulla from pterygoid process to auditory meatus, 7.3, 7.7, 7.6, 7.4, 8.0, 8.0.

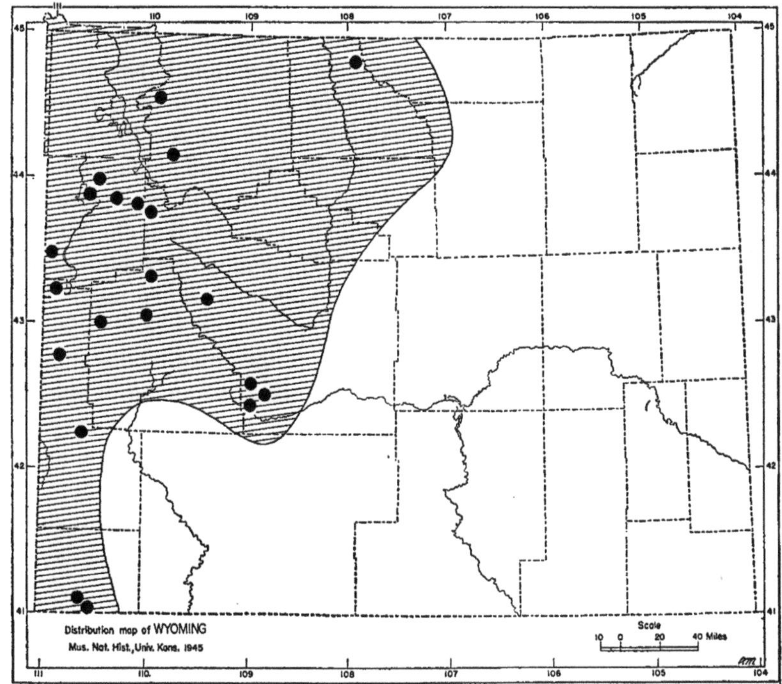

FIG. 47. Distribution of *Microtus richardsoni macropus.*

Distribution.—Montane and riparian habitat in western Wyoming. See Fig. 47.

Remarks.—Voles are difficult to use in ascertaining geographic variation (see Anderson, 1959:433) because each large sample is likely to be made up of individuals all of the same age. Microtines are especially difficult to age.

Recently, Rasmussen and Chamberlain (1959) named a new subspecies, *M. richardsoni myllodontus*, from Utah and assigned an indefinite portion of Wyoming to the geographic range of the new subspecies. The holotype of *M. r. macropus*, not examined by Rasmussen and Chamberlain, possesses two characters (irregular shape of infraorbital foramen and offset tooth-row), which were listed as absolutely diagnostic for the subspecies *M. r. myllodontus*. The right mandibular molariform teeth of USNM 31451 are not aligned as figured by Rasmussen and Chamberlain (1959:Fig. 1), but differ in that the third molar is set in (offset) much as in Utah specimens. Irregularly shaped infraorbital foramina and offset tooth-rows are found in most specimens examined of *M. richardsoni*. Some characteristics attributed to *M. r. myllodontus* (for example, wide nasals) may reflect stage of growth. Lack of adults from the type locality of *macropus* may account for the "larger" size ascribed to *myllodontus*. Differences in coloration between the two subspecies are slight. *M. r. myllodontus* has larger auditory bullae and more procumbent incisors but in my opinion these two differences do not warrant subspecific recognition of *myllodontus*.

Examination of topotypes and near topotypes of *myllodontus* from the collection of the University of Utah reveals that the mandibular tooth-rows differ from those of Wyoming specimens and of near topotypes of *macropus*. The difference is not so much in the alignment of the teeth but in the shape of the labial surface of the mandible, which is markedly inset at the level of the third molar instead of weakly inset. The infraorbital foramina of *myllodontus* is irregularly shaped less frequently in Wyoming specimens and near topotypes of *macropus*. The auditory bullae are larger in *myllodontus*. It now seems reasonable to regard *M. r. myllodontus* as a weakly differentiated subspecies, but to restrict its range to the following localities: *Idaho*, Hd. Crow Creek, Caribou County (1 USNM); *Utah*, all localities listed by Rasmussen and Chamberlain, 1959. Subadult specimens from Uinta County, Wyoming, lack procumbent incisors, have dark grayish venters, and are assigned tentatively to *M. r. macropus*.

Records of occurrence.—Specimens examined, 132, as follows: **Big Horn Co.:** Medicine Wheel Ranch, 28 mi. E Lovell, 9000 ft., 1.

Fremont Co.: Brooks Pk., 4; Lake Fork, 9600-11,400 ft., 10 USNM; 15½ mi. S, 7½ mi. W Lander, 9200 ft., 5; 23½ mi. S, 5 mi. W Lander, 7; South Pass City, 8000 ft., 1 USNM.

Lincoln Co.: 3 mi. N, 11 mi. E Alpine, 5650 ft., 1; 10 mi. SE Afton, 7500-9000 ft., 15 USNM; LaBarge Creek, 9000 ft., 5 USNM.

Park Co.: Pahaska Tepee, 6500 ft., 1 USNM; Valley, 7000-7500 ft., 5 USNM.

Sublette Co.: 31 mi. N Pinedale, 8025 ft., 7; 2 mi. N. Kendall, 7700 ft., 2 USNM; 12 mi. N Pinedale, 8000 ft., 1 USNM; *E. end Island Lake, 10,600 ft.*, 2.

Teton Co.: *Whetstone Creek, 8 USNM; Arizona Creek, 7 USNM;* Pacific Creek, 8 USNM; *Big Game Ridge, 2 USNM; Gravel Creek, 3 USNM;* Jackson Hole Wildlife Park, 1; *Blackrock Meadows, 8600 ft., 3;* Blackrock Creek, 3 USNM; Togwotee Pass, 5; Teton Pass, 7200 ft., 14 USNM; *Moose Creek, 6800 ft., 4 USNM.*

Uinta Co.: 9 mi. S Robertson, 8400 ft., 4; *10 mi. S, 1 mi. W Robertson, 8700 ft., 2;* 14 mi. S, 2 mi. E Robertson, 9000 ft., 1.

Microtus ochrogaster haydenii (Baird)

Prairie Vole

1958. *Arvicola (Pedomys) haydenii* Baird, Mammals, *in* Repts. Expl. Surv. . . . 8(1):543, July 14.

1907. *Microtus ochrogaster haydeni*, Osgood, Proc. Biol. Soc. Washington, 20:48, April 18.

Type.—From Fort Pierre, Stanley County, South Dakota.

Description.—Medium in size for *Microtus;* upper parts brownish agouti with more or less reddish wash; underparts grayish strongly washed with Ochraceous-Buff or cinnamon especially on perianal region; skull strongly arched in lateral view; tooth-row long with wide re-entrant angles in grooves of molars; auditory bullae showing ventral ridges (not greatly inflated); posterior loop of middle upper molar not cut off (see account of *Microtus pennsylvanicus*).

Measurements.—Average and extreme external and cranial measurements of 28 adults (14 males) from 28-32 miles north and 7-12½ miles east of Laramie are as follows: Total length, 158.6 (144-168); length of tail, 40.3 (33-44); hind foot, 20.5 (19-22); ear from notch, 12.2 (11-13); condylobasal length, 28.0 (26.0-29.6); zygomatic breadth, 16.1 (14.8-17.2); lambdoidal breadth, 12.3 (11.7-12.7); maxillary tooth-row, 6.5 (5.9-7.3); cranial depth, 11.3 (10.7-12.2); breadth of rostrum, 5.8 (5.1-6.3). Measurements of a pregnant female from four miles north of Garland, Park County, are as follows: Total length, 153; length of tail, 21; hind foot, 19; ear from notch, 12; condylobasal length, 27.9; zygomatic breadth, 15.9; lambdoidal breadth, 12.3; maxillary tooth-row, 6.2; cranial depth, 10.6; rostral breadth, 6.0.

Distribution.—Prairie areas of the Great Plains. See Fig. 48.

Remarks.—Voles from the arid areas of extreme eastern Wyoming adjacent to the Sand Hills of Nebraska are notably paler than other individuals of this species in Wyoming. Even those from Crook County are paler than those from western counties. Ten adults from Niobrara County are notably large (total length, 171.7, 160-180) as well as pale. Even so, no significant cranial differences (even in size) were noted between specimens from extreme eastern Wyoming and those from more western and northwestern counties. Perhaps the one adult known from Park County makes it necessary to qualify that statement; the specimen has a shorter tail and shallower cranium (see *Measurements*) than other specimens from Wyoming.

Records of occurrence.—Specimens examined, 274, as follows: **Albany Co.:** 32 mi. N, 12½ mi. E Laramie, 6080 ft., 4; *31¼-31½ mi. N, 12 mi. E Laramie, 6160 ft., 19; 29-29¾ mi. N, 8¼-19½ mi. E Laramie, 6350-6420 ft., 31; 28¾ mi. N, 7¼ mi. E Laramie, 6500 ft., 7; 28 mi. N, 8¾ mi. E Laramie, 6420 ft., 11; 26 mi. N, 4½ mi. E Laramie, 6960 ft., 1.*

Big Horn Co.: Greybull, 9 USNM; *8/10-1 mi. S Greybull, 3788-3795 ft., 3.*
Campbell Co.: Ivy Creek, 5 mi. N, 8 mi. W Spotted Horse, 4.
Converse Co.: 6 mi. N, 4 mi. W Bill, 4800 ft., 1; 3 mi. N, 5 mi. E Orin, 4725 ft., 9; Beaver, 1 USNM.
Crook Co.: 15 mi. ENE Sundance, 3825 ft., 1; 1½ mi. NW Sundance, 5000 ft., 1; *Sundance, 2 USNM;* Sand Creek, 1 USNM.
Fremont Co.: 12 mi. N, 3 mi. W Shoshoni, 10.
Goshen Co.: Rawhide Butte, 5800 ft., 4 USNM; 1 mi. SW Torrington, 3.
Johnson Co.: 8/10 mi. S, 1 mi. W Buffalo, 4800 ft., 5; *1 mi. S, 5½ mi. W Buffalo, 4800 ft., 1; 2 mi. S, 6½ mi. W Buffalo, 5620 ft., 3;* ¼ mi. E Klondike, 5160 ft., 2; 1 mi. WSW Kaycee, 4700 ft., 5.
Laramie Co.: 6½ mi. W Meriden, 5200 ft., 2; *Horse Creek, 3 mi. W Meriden, 5000 ft., 1; JHD Ranch, sec. 3, T. 17N, R. 65W, 2; 4 mi. S Hillsdale, 2; 2 mi. N Pine Bluffs, 5050 ft., 6; 1 mi. N, ½ mi. W Pine Bluffs, 5040 ft., 9; 1 mi. S, ½ mi. E Pine Bluffs, 5200 ft., 7;* 2 mi. S Pine Bluffs, 5200 ft., 23.
Natrona Co.: 4 mi. E Casper, 5100 ft., 1; *2¼ mi. S, 1½ mi. W Casper, 5250 ft., 3; 6 mi. S, 2 mi. W Casper, 5900 ft., 1; 7 mi. S, 2 mi. W Casper, 6370 ft., 1;* Alcova, 5180 ft., 1; Sun, 1 USNM.
Niobrara Co.: 10 mi. N Hatcreek P. O., 9; 1 mi. W Lusk, 5000 ft., 17; *2 mi. S, ½ mi. E Lusk, 5000 ft., 2.*
Park Co.: 4 mi. N Garland, 2.
Platte Co.: *35¼ mi. N, 18 mi. E Laramie, 5500 ft., 6;* 2½ mi. S Chugwater, 5300 ft., 12.
Sheridan Co.: Dayton, 1 USNM; *3 mi. WNW Monarch, 3800 ft., 3;* 4 mi. NNE Banner, 4100 ft., 4; 5 mi. NE Clearmont, 3900 ft., 12.
Washakie Co.: 1 mi. N, 3 mi. E Tensleep, 4350 ft., 1.
Weston Co.: Newcastle, 1 USNM; 23 mi. SW Newcastle, 4500 ft., 2.
County not certain: Pass, 4 USNM.

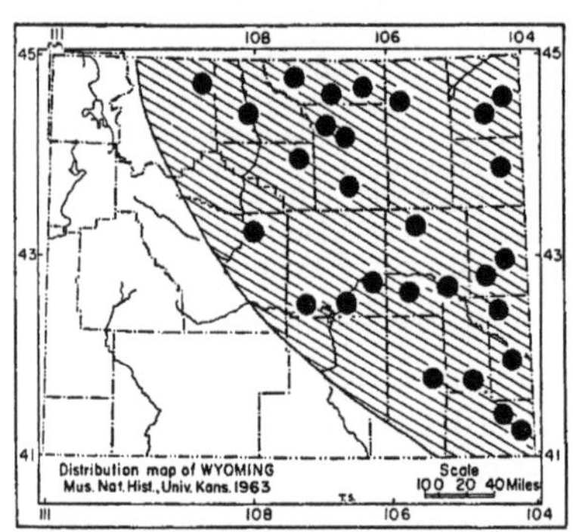

FIG. 48. *Microtus ochrogaster haydenii.*

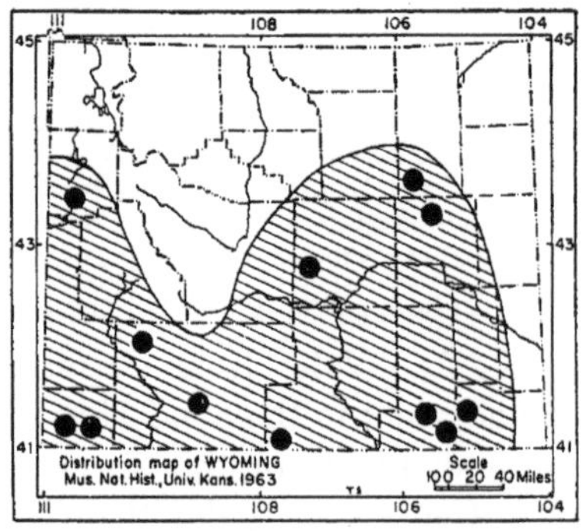

FIG. 49. *Lagurus curtatus levidensis.*

Lagurus curtatus levidensis (Goldman)

Sagebrush Vole

1941. *Lemmiscus curtatus levidensis* Goldman, Proc. Biol. Soc. Washington, 54:70, July 31.

1951. *Lagurus curtatus levidensis,* Kelson, Jour. Mamm., 32:114, February 15.

Type.—From five miles east of Canadian River, west base Medicine Bow Range, east of Walden, North Park, 8000 ft., Jackson County, Colorado.

Description.—Small for Microtinae, with short tail; upper parts dusky and brownish tan or grayish buff; venter whitish or creamy white often showing plumbeous gray; tail whitish or bicolored brownish or buffy above and whitish below; pinna of ear brownish; skull flattened and frontal region dished; cheek-teeth rootless; m3 with at least four closed triangles. Winter pelage of two specimens from within a mile of Pole Mountain is more lax and more yellowish than summer pelage.

Measurements.—Averages and extremes for external measurements of seven adults and cranial measurements of six adults (two females) from within 17 miles of Laramie are as follows: Total length, 123.3 (116-127); length of tail, 21.1 (16-23); hind foot, 15.1 (11-16); ear from notch, 10.8 (10-11); occipitonasal length, 24.1 (23.1-25.0); zygomatic breadth, 14.2 (13.3-14.5); interorbital breadth, 3.3 (3.1-3.5); length nasals, 6.6 (6.4-7.0); maxillary toothrow, 6.0 (5.6-6.1).

Distribution.—Central, western, and southern Wyoming mainly from Transition Life-zone. See Fig. 49.

Records of occurrence.—Specimens examined, 60, as follows: **Albany Co.:** *N. Fork Little Laramie River, sec. 28, T. 16N, R. 78W, 8600 ft., 2;* 4.5 mi. N Laramie, 2; *1 mi. E Laramie, 1; 3 mi. E Laramie, 7300 ft., 2; 4.4 mi. E Laramie, 1 USNM;* 1 mi. SSE Pole Mtn., 8350 ft., 6. **Campbell Co.:** *40½ mi. S, 17½ mi. W Gillette, 1* (owl pellet); 42 mi. S, 13 mi. W Gillette, 1 (owl pellets). **Carbon Co.:** 1 mi. SW Dixon, 6600 ft., 1. **Converse Co.:** 10 mi. N, 6 mi. W Bill, 9 (from owl pellets, discarded). **Laramie Co.:** 5 mi. W Horse Creek P. O., 1. **Natrona Co.:** 16 mi. S, 11 mi. W Waltman, 6950 ft., 1. **Sweetwater Co.:** Farson, 6580 ft., 1; 32 mi. S, 22 mi. E Rock Springs, 7025 ft., 9. **Teton Co.:** Jackson, 1 USNM. **Uinta Co.:** 8 mi. S, 2½ mi. E Robertson, 8300 ft., 3; *9 mi. S Robertson, 12;* Smiths Fork, 20 mi. S Ft. Bridger, 6 USNM.

Ondatra zibethicus (Linnaeus)

Muskrat

Large for Cricetidae (see *Measurements*); upper parts brownish or reddish brown; underparts fawn or silvery gray washed with reddish; tail scaly and almost naked; tail compressed laterally; skull as in *Microtus* but larger and more angular; nasals long, broad, and flaring anteriorly; interorbital region narrow; semi-aquatic.

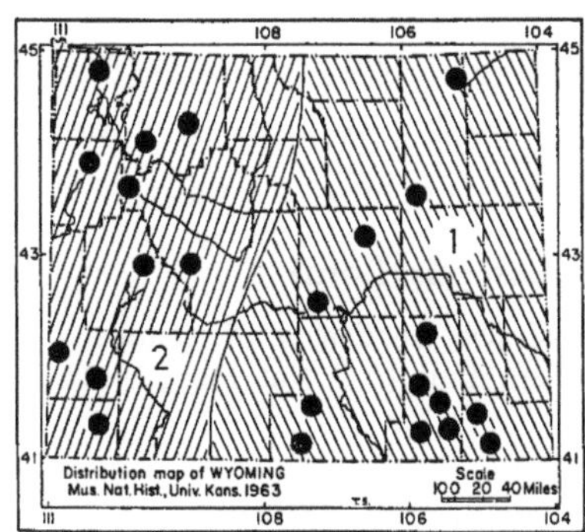

FIG. 50. *Ondatra zibethicus.*
1. *O. z. cinnamominus*
2. *O. z. osoyoosensis*

The muskrat is probably the most valuable furbearer in Wyoming. It is the largest kind of Microtine and usually feeds on vegetable matter. Two subspecies occur in the state.

Ondatra zibethicus cinnamominus (Hollister)

1910. *Fiber zibethicus cinnamominus* Hollister, Proc. Biol. Soc. Washington, 23:125, September 2.

1912. *Ondatra zibethicus cinnamomina,* Miller, Bull. U. S. Nat. Mus., 79: 232, December 31.

Type.—From Wakeeney, Trego County, Kansas.

Comparison.—Compared with *Ondatra zibethicus osoyoosensis, O. z. cinnamominus* averages paler, is much smaller cranially, and has a notably shorter maxillary tooth-row. The pelage of *cinnamomina* is more reddish or buffy instead of brownish.

Measurements.—No significant differences in size were noted between males and females. External measurements of two adult males from Sun, and an adult female from a half mile south of Rockypoint are, respectively, as follows: Total length, 490, 485, 554; length of tail, 215, 210, 214; hind foot, 81, 78, 76. Cranial measurements of 20 adults taken in winter and early spring from 6-6.1 miles southwest of Laramie, are as follows: Condylonasal length, 58.4 (55.5-63.0); zygomatic breadth, 36.3 (33.3-38.6); interorbital breadth, 6.0 (5.5-6.5); length of nasals, 18.2 (17.2-19.9); maxillary tooth-row, 15.0 (14.1-15.8).

Distribution.—In general the Great Plains. See Fig. 50.

Remarks.—Intergradation between this subspecies and *Ondatra zibethicus osoyoosensis* occurs along the eastern base of the Rocky Mountains, but specimens are needed from the Bighorn, Medicine Bow, and Sierra Madre ranges in order the better to judge which subspecies occurs in each of those mountain ranges.

Records of occurrence.—Specimens examined, 72, as follows: **Albany Co.:** 3 mi. SW Eagle Peak, 7500 ft., 1 USNM; Rock Creek, 2 USNM; 18 mi. N Laramie, 1 Univ. Wyo.; *Vic. Bosler, 7000 ft., 1; Olson Ranch, Little Laramie River Valley, 2; Little Laramie River, 1;* 14 mi. W Laramie, 1 Univ. Wyo.; Laramie, 7150 ft., 1; *Near Osterman Lake, 7225 ft., 1, 1 Univ. Wyo.; 2 mi E Laramie, 1 Univ. Wyo.; Big Laramie River, 7100-7150 ft., 1, 1 Univ. Wyo.; 5 mi. S, 12 mi. W Laramie, 7225 ft., 1, 1 Univ. Wyo.; Caldwell Lake, 3; 7 mi. W Fox Park, sec. 11, T. 13N, R. 79W, 8500 ft., 6; 6-6.1 mi. SW Laramie, 7180 ft., 21; 16 mi. SW Laramie, 7200 ft., 1 Univ. Wyo.*

Campbell Co.: ½ mi. S Rockypoint, 3800 ft., 1; 38 mi. S, 19 mi. W Gillette, 1.

Carbon Co.: Bridgers Pass, 18 mi. SW Rawlins, 7500 ft., 1; 7½ mi. N, 8½ mi. E Savery, 8400 ft., 1.

Laramie Co.: Horse Creek, 1 mi. E Horse Creek P. O., 1; *3.1 mi. S Horse Creek, 1;* 7 mi. W Cheyenne, 1.

Natrona Co.: 28 mi. N, 12 mi. W Casper, 4700 ft., 3; Sun, 3 USNM.

Ondatra zibethicus osoyoosensis (Lord)

1863. *Fiber osoyoosensis* Lord. Proc. Zool. Soc. London, p. 97, October.

1912. *Ondatra zibethica osoyoosensis,* Miller. Bull. U. S. Nat. Mus., 79:231, December 31.

Type.—From Lake Osoyoos, British Columbia.

Comparison.—See account of *Ondatra zibethicus cinnamominus.*

Measurements.—External measurements of an adult male and adult female from Fort Bridger, of an adult male from 25 miles south and 28 miles west of Cody, and of an adult male and adult female from within a mile of Moran are, respectively, as follows: Total length, 578, 570, 563, 561, 553; length of tail, 263, 259, 260, 260, 245; hind foot, 78, 74, 82, 80, 80; ear from notch, 24, 27, 27, 23, 25. Cranial measurements of the male from 25 mi. S and 28 mi. W Cody, of the male and female from within a mile of Moran, and of an adult female from within three miles of Pinedale are, respectively, as follows: Condylonasal length, 67.0, 64.2, 64.0, 63.9; zygomatic breadth, 42.2, 38.8,

39.6, 38.5; interorbital breadth, 6.6, 6.2, 6.9, 7.0; length of nasals, 23.2, 21.5, 21.4, 20.8; maxillary tooth-row, 15.3, 16.1, 15.3, 14.8.

Distribution.—Western Wyoming. See Fig. 50.

Records of occurrence.—Specimens examined, 21, as follows: **Fremont Co.:** Brooks Lake, 1 Univ. Wyo.; 4 mi NW Milford, 5357 ft., 1. **Lincoln Co.:** Cokeville, 1 USNM; *Fossil, 1 USNM;* Opal, 2 USNM. **Park Co.:** 12 mi. S, 15 mi. W Cody, 1; *25 mi. S, 28 mi. W Cody, 1; 30 mi. S, 43 mi. W Cody, 1;* Valley, 1 USNM. **Sublette Co.:** 7 mi. S Fremont Peak, 10,400 ft., 2 USNM; *2¼ mi. NE Pinedale, 3.* **Teton Co.:** ¼ mi. E Moran, 6740 ft., 2. **Uinta Co.:** 1 mi. N Ft. Bridger, 2. **Yellowstone Nat'l Park:** Yellowstone River, 1 USNM. **Not found:** Pass Creek, 1 USNM (Park Co.?).

Family MURIDAE Gray

KEY TO MURID RODENTS INTRODUCED INTO WYOMING

1. Hind foot longer than 25; tail longer than 115. . . . **Rattus norvegicus,** p. 663
1′. Hind foot shorter than 25; tail shorter than 115. . . . . **Mus musculus,** p. 663

Rattus norvegicus norvegicus (Berkenhout)

Norway Rat

1769. *Mus norvegicus* Berkenhout, Outlines of the natural history of Great Britain and Ireland, 1:5.
1916. *Rattus norvegicus,* Hollister, Proc. Biol. Soc. Washington, 29:126, June 6.

Type.—From England.

Description.—Large (see *Measurements*); upper parts grayish brown, but melanistic and albinistic skins frequently seen in collections; tail sparsely haired; skull weakly developed; supraorbital ridges distinct; molars provided with small tubercles arranged in three longitudinal rows; dentition, i.$\frac{1}{1}$, c.$\frac{0}{0}$, p.$\frac{0}{0}$, m.$\frac{3}{3}$.

Measurements.—Total length, 316-460; length of tail, 122-215; hind foot, 30-45; occipitonasal length, 41.0-51.5; zygomatic breadth, 21.0-25.2. (After Hall and Kelson, 1959:769.)

Distribution.—Not native to Wyoming; introduced by man; occurs in and around habitations of man. Not mapped.

Record of occurrence.—Specimen examined, 1, as follows: **Albany Co.:** Woods Landing, 1 Univ. Wyo.

Mus musculus domesticus Rutty

House Mouse

1772. *Mus domesticus* Rutty, Essay Nat. Hist. County Dublin, 1:281.
1943. *Mus musculus domesticus,* Schwarz and Schwarz, Jour. Mamm. 24:65, February 20.

Type.—From Dublin, Ireland.

Description.—Small (see *Measurements*) resembling *Reithrodontomys megalotis* in size and color (but anterior faces of incisors not grooved as in *R. megalotis*); upper parts grayish brown-agouti; venter grayish; tail scantily haired and dark grayish; eyes small; occlusal surface of incisors showing distinct posterior notch in lateral view; three longitudinal rows of tubercles on each molar tooth-row.

Measurements.—Total length, 130-198; length of tail, 63-102; hind foot, 14-21; ear from notch, 11-18; occipitonasal length, 20.1-22.9; zygomatic breadth, 10.6-12.1. (After Hall and Kelson, 1959:770.)

Distribution.—In and about man's habitations, but occasionally feral populations are found in summer. Not mapped.

Remarks.—This non-native species, in somewhat the same fashion as the Norway rat, is frequently injurious to man and to the native fauna. Application here of the subspecific name *M. m. domesticus* to specimens from Wyoming is in accordance with the suggestion of Schwarz and Schwarz (1943) and is not the result of critical study by myself of subspecific variation in *Mus musculus.*

Records of occurrence.—Specimens examined, 60, as follows: **Albany Co.:** 10 mi. NE Laramie, 7300 ft., 1 Univ. Wyo.; 4 mi. W Laramie, 7275 ft., 1 Univ. Wyo.; Laramie, 7300 ft., 3 Univ. Wyo. **Big Horn Co.:** 7 mi. S Basin, 1. **Campbell Co.:** 2½ mi. N, 7 mi. W Spotted Horse, 2; 45½ mi. S, 15 mi. W Gillette, 1; 53½ mi. S, 15½ mi. W Gillette, 5340 ft., 1. **Crook Co.:** 1½ mi. NW Sundance, 5000 ft., 1. **Laramie Co.:** 6½ mi. W Meriden, 5200 ft., 1; 11 mi. N, 5½ mi. E Cheyenne, 5950 ft., 2; 2 mi. N Pine Bluffs, 5050 ft., 2. **Park Co.:** 4 mi. N Garland, 6. **Sheridan Co.:** 3 mi. WNW Monarch, 3800 ft., 15; 4 mi. NNE Banner, 4100 ft., 13; 5 mi. NE Clearmont, 3900 ft., 6. **Sweetwater Co.:** ¼ mi. S Green River, 6090 ft,. 3. **Uinta Co.:** Ft. Bridger, 6650 ft., 1.

Family ZAPODIDAE

KEY TO ZAPODID RODENTS

1. Palatal breadth at M3 less than 4.2; total length less than 225, **Zapus hudsonius**, p. 664

1′. Palatal breadth at M3 more than 4.4; total length usually more than 225, **Zapus princeps**, p. 665

Zapus hudsonius (Zimmermann)

Meadow Jumping Mouse

Smaller than *Zapus princeps;* upper parts lacking distinct dorsal blackish band, although sides paler than ochraceous or brownish back; lateral line ochraceous; tail long, brownish above and slightly paler below; hind foot long; baculum shorter than in Z. *princeps;* incisors grooved anteriorly; P4 small; nasals long; dentition, i.$\frac{1}{1}$, c.$\frac{0}{0}$, p.$\frac{1}{0}$, m.$\frac{3}{3}$; saltatorial.

Occurs in moist areas in Canadian and Transition life-zones.

Zapus hudsonius campestris Preble

1899. *Zapus hudsonius campestris* Preble, N. Amer. Fauna, 15:20, August 8.

Type.—From Bear Lodge Mountains, Crook County, Wyoming.

Comparison.—From *Zapus hudsonius preblei, Z. h. campestris* differs in brighter ochraceous (and more blackish) upper parts; dorsal dark band less obscure; skull averaging larger; frontal region less inflated.

Measurements.—External measurements of 18 adults (8 males) from three miles northwest of Sundance are averaged as follows: Total length, 210.7 (191-

224); length of tail, 126.2 (118-136); hind foot, 29.9 (29-31); ear from notch, 12.4 (11-16). Krutzsch (1954:462) lists measurements of 19 adults from the same locality as follows: Condylobasal length, 19.9 (19.2-20.8); breadth braincase, 9.7 (9.4-10.0); interorbital breadth, 4.3 (3.8-4.5); mastoidal breadth, 10.4 (10.1-10.9); length maxillary tooth-row, 3.6 (3.4-3.8); occipitonasal length, 23.2 (22.4-24.2); palatal length, 10.0 (9.5-10.5); zygomatic breadth, 11.1 (10.7-11.8).

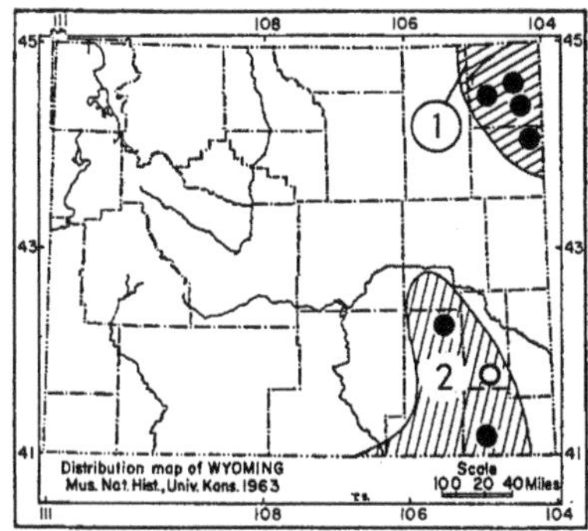

FIG. 51. *Zapus hudsonius.*
1. *Z. h. campestris*
2. *Z. h. preblei*

Distribution.—Black Hills and closely surrounding areas. May occur in the Bighorn Mountains; recorded from their northern base in Montana. See Fig. 51.

Records of occurrence.—Specimens examined, 40, as follows: **Crook Co.:** *Bear Lodge Mtns., 6 USNM; Warren Peak, 6000 ft., 2 USNM;* 15 mi. N Sundance, 5500 ft., 2; Devils Tower, Floodplain Belle Fourche River, 1 USNM; 3 mi NW Sundance, 19; *Sundance, 3 USNM.* **Weston Co.:** 1½ mi. E Buckhorn, 6150 ft., 7.

Zapus hudsonius preblei Krutzsch

1954. *Zapus hudsonius preblei* Krutzsch, Univ. Kansas Publs., Mus. Nat. Hist., 7:452, April 21.

Type.—From Loveland, Larimer County, Colorado.

Comparison.—For comparison with *Zapus hudsonius campestris,* see account of that subspecies.

Measurements.—External and cranial measurements of three adults from Springhill are as follows: Total length, 200, 210, 200; length of tail, 117, 128, 126; hind foot, 29, 31, 31; occipitonasal length, —, 23.4, —; zygomatic breadth, —, 11.5, —; interrorbital breadth, —, 4.3, —; length of nasals, —, 8.9, —; cranial depth, —, 7.5, —; maxillary tooth-row, —, 3.7, —.

Distribution.—Rare, but known from southeastern Wyoming and mountains of south-central Wyoming. See Fig. 51.

Records of occurrence.—Specimens examined, 6, as follows: **Albany Co.:** Springhill, 6300 ft., 5 USNM. **Laramie Co.:** Cheyenne, 1 USNM.

Additional record.—**Platte Co.:** Chugwater (Krutzsch, 1954:453).

Zapus princeps J. A. Allen

Western Jumping Mouse

Large for *Zapus;* upper parts yellowish gray or ochraceous laterally intermixed with brownish becoming dark brownish middorsally (usually with dark brownish middorsal stripe); ochraceous, thin lateral line usually present between brownish gray of sides and whitish of venter; underparts white; tail longer than body, naked and dark grayish; hind foot long; nasals long; incisors grooved; upper premolars larger than in *Zapus hudsonius;* baculum elongate in Z. *princeps.*

This mammal is structurally adapted for saltation and occurs in the Transition and Canadian life-zones usually near water.

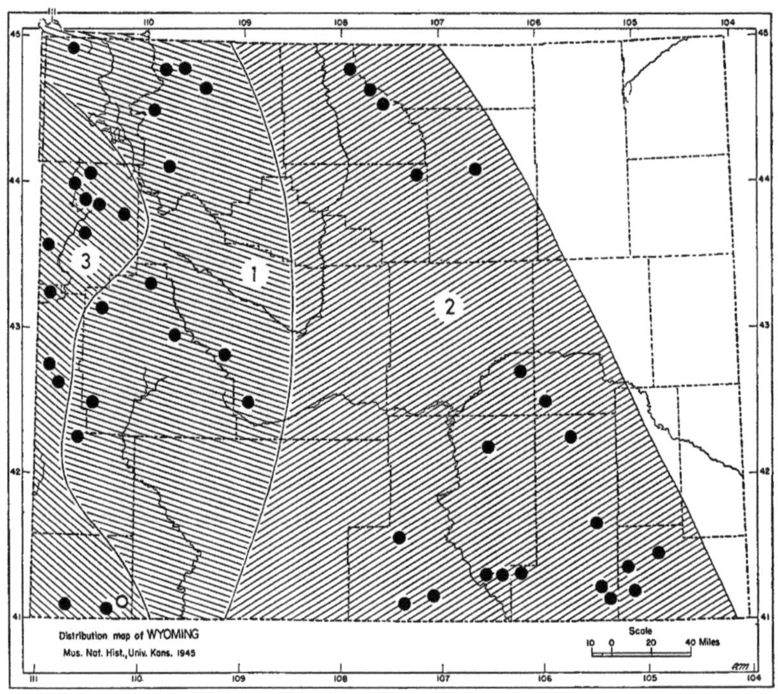

FIG. 52. Distribution of *Zapus princeps.*
Guide to subspecies
1. Z. *p. idahoensis*
2. Z. *p. princeps*
3. Z. *p. utahensis*

Zapus princeps idahoensis Davis

1934. *Zapus princeps idahoensis* Davis, Jour. Mamm., 15:221, August 10.

Type.—From five miles east of Warm Lake, 7000 feet, Valley County, Idaho.

Comparisons.—From *Zapus princeps princeps,* Z. *p. idahoensis* differs in lesser size; darker upper parts; lesser zygomatic breadth; and more nearly parallel upper tooth-rows. From *Zapus princeps princeps,* Z. *p. idahoensis* differs in lesser size; paler, less ochraceous upper parts; more obscure lateral line; slightly buffy instead of white underparts; and smaller skull.

Measurements.—External measurements of 20 adults (nine males) from 31 miles north of Pinedale average as follows: Total length, 227.0 (223-250); length of tail, 139.1 (126-149); hind foot, 31.3 (30-33); ear from notch, 17.1 (15-19). Cranial measurements of 24 adult males and females from western Park County westward of Cody are listed by Krutzsch (1954:456) as follows: Condylobasal length, 21.4 (20.6-22.6); breadth of braincase, 10.5 (9.9-11.2); interorbital breadth, 4.6 (4.3-5.2); mastoidal breadth, 11.0 (10.5-11.4); length of maxillary tooth-row, 4.0 (3.9-4.2); occipitonasal length, 24.7 (24.1-25.6); palatal length, 10.9 (10.2-11.5); zygomatic breadth, 12.6 (12.0-13.3).

Distribution.—Yellowstone Park, eastern Park County, and the Wind River Mountains. See Fig. 52.

Records of occurrence.—Specimens examined, 138, as follows: **Fremont Co.:** 4 mi. N, 19 mi. W Lander, 10,000 ft., 1; 23½ mi. S, 5 mi. W Lander, 8600 ft., 4.

Lincoln Co.: LaBarge Creek, 9 USNM.

Park Co.: Black Mtn., 3 USNM; 31.5 mi. N, 36 mi. W Cody, 6900 ft., 7; *28 mi. N, 30 mi. W Cody, 7200 ft., 1;* 16¾ mi. N, 17 mi. W Cody, 5625 ft., 14; 2 mi. S, 42 mi. W Cody, 6400 ft., 4; *Pahaska, 4 USNM; Mouth Grinnell Creek, 7000 ft., 17 USNM;* 25 mi. S, 28 mi. W Cody, 6350 ft., 5; *Valley, 7500 ft., 5 USNM.*

Sublette Co.: 31 mi. N Pinedale, 8025 ft., 27; 12 mi. N Kendall, 7700 ft., 7 USNM; *Merna, 7800 ft., 3 USNM; Halfmoon Lake, 7900 ft., 5; 12 mi. NE Pinedale, 2 USNM;* 10 mi. NE Pinedale, 8000 ft., 1; *8 mi. N, 5 mi. E Pinedale, 8900 ft., 1; 5 mi. N, 3 mi. E Pinedale, 7500 ft., 4;* 2 mi. S, 19 mi. W Big Piney, 7700 ft., 3; *3 mi. W Stanley, 8000 ft., 3.*

Yellowstone Nat'l Park: Glen Creek, 2 USNM; *no certain locality, 6 USNM.*

Additional record (Krutzsch, 1954:404).—**Park Co.:** *12 mi. W Wapiti.*

Zapus princeps princeps J. A. Allen

1893. *Zapus princeps* J. A. Allen, Bull. Amer. Mus. Nat. Hist., 5:71, April 28.

Type.—From Florida, La Plata County, Colorado.

Comparisons.—From *Zapus princeps utahensis, Z. p. princeps* differs in upper parts more ochraceous; lateral line broader; skull smaller; and upper tooth-rows more nearly parallel. For comparison with *Z. p. idahoensis,* see account of that subspecies.

Measurements.—Average and extreme external measurements of 22 adults (six males) from 10 miles north and 14 miles east of Encampment are as follows: Total length, 233.0 (222-250); length of tail, 136.1 (127-145); hind foot, 31.3 (28-33); ear from notch, 14.4 (13-16). Cranial measurements of 11 adult males and females from eight miles north and 19½ miles east of Savery are as follows: Condylobasal length, 21.2 (20.8-21.8); breadth of braincase, 10.2 (10.0-10.5); interorbital breadth, 4.5 (4.2-4.7); mastoidal breadth, 10.9 (10.6-11.1); length maxillary tooth-row, 3.9 (3.7-4.1); occipitonasal length, 24.5 (23.7-25.0); palatal length, 10.8 (10.5-11.1); zygomatic breadth, 12.2 (12.0-12.5).

Distribution.—Suitable habitats in Bighorn Mountains, mountain ranges in Central Lowlands, and mountains of south-central Wyoming. See Fig. 52.

Records of occurrence.—Specimens examined, 253, as follows: **Albany Co.:** 3 mi. SW Eagle Peak, 7500 ft., 3 USNM; 30 mi. N, 10 mi. E Laramie, 6760 ft., 1; *29 mi. N, 8¾ mi. E Laramie, 6420 ft., 7; 26¾ mi. N, 6½ mi. E Laramie, 6700 ft., 1;* Nash Fork Creek, 2 mi. W Centennial, 1; *Centennial, 8140 ft., 1; 2 mi. S Browns Peak, 10,600 ft., 2; 3 mi. ESE Browns Peak, 10,000 ft., 8; 4 mi. S, 8 mi. E Laramie, 8600 ft., 2;* 6 mi. S, 8 mi. E Laramie, 8500 ft., 1; 1 mi. SSE Pole Mtn., 8350 ft., 3; *1½ mi. SE Pole Mtn., 8200 ft., 1; 2 mi. SW Pole Mtn., 8300 ft., 3.*

Big Horn Co.: Medicine Wheel Ranch, 28 mi. E Lovell, 9000 ft., 37; Granite Creek Campground, Bighorn Mtns., 3; *2 mi. N, 12 mi. E Shell, 7500 ft., 12; 12 mi. E Shell, 7500 ft., 1;* 4½ mi. S, 17½ mi. E Shell, 6.

Carbon Co.: Shirley Mtns., 7600 ft., 4 USNM; Bridgers Pass, 7500 ft., 6; Lake Marie, 10,440 ft., 1; 6 mi. S, 14 mi. E Saratoga, 8800 ft., 5; *10 mi. N,*

12 mi. E Encampment, 7200 ft., 2; 10 mi. N, 14 mi. E Encampment, 8000 ft., 28; 10 mi. N, 16 mi. E Encampment, 8000 ft., 1; *9 mi. N, 3 mi. E Encampment, 2; 8 mi. N, 14½ E Encampment, 8100-8400 ft., 20; 8 mi. N, 16 mi. E Encampment, 8400 ft., 6; 8 mi. N, 18 mi. E Encampment, 8900 ft., 1; 8 mi. N, 19½ mi. E Savery, 7; 7-7½ mi. N, 18-18½ mi. E Savery, 8400 ft., 3; 6 mi. N, 13½-14½ mi. E Savery, 11;* 4 mi. N, 8 mi. E Savery, 7300 ft., 2.

Converse Co.: 21 mi. S, 24 mi. W Douglas, 7700 ft., 5; *21.5 mi. S, 24.5 mi. W Douglas, 18.*

Johnson Co.: .8 mi. S, 1 mi. W Buffalo, 4800 ft., 1; *1 mi. S, 5½ mi. W Buffalo, 1; 1½ mi. S, 5½ mi. W Buffalo, 2; 2 mi. S, 6½ mi. W Buffalo, 5620 ft., 4.*

Laramie Co.: Meadow, 2 USNM; 1 mi. N, 5 mi. W Horse Creek P. O., 7200 ft., 2; *5 mi. W Horse Creek* P. O., *1;* Islay, 1 USNM.

Natrona Co.: 7 mi. S, 2 mi. W Casper, 6370 ft., 2; *7 mi. S Casper, 6000 ft., 1 USNM; 23 mi. S, 25 mi. W Casper, 7800 ft., 8.*

Washakie Co.: 5 mi. N, 9 mi. E Tensleep, 7400 ft., 2; *4 mi. N, 9 mi. E Tensleep, 7000 ft., 5.*

Zapus princeps utahensis Hall

1934. *Zapus princeps utahensis* Hall, Occas. Papers Mus. Zool., Univ. Michigan, 296:3, November 2.

Type.—From Beaver Creek, 19 miles south of Manila, Daggett County, Utah.

Comparisons.—See accounts of *Zapus princeps idahoensis* and *Z. p. princeps.*

Measurements.—Average and extreme external measurements of 16 adults (7 males) from nine miles south of Robertson are as follows: Total length, 246.1 (228-268); length of tail, 145.6 (131-165); hind foot, 33.5 (31-35); ear from notch, 15.3 (13-17). Cranial measurements of 15 adult males and females from southward of Robertson are listed by Krutzsch (1954:459) as follows: Condylobasal length, 22.0 (21.0-22.6); breadth braincase, 10.7 (10.3-11.1); interorbital breadth, 5.0 (4.7-5.1); mastoidal breadth, 11.2 (10.8-11.6); length maxillary tooth-row, 4.1 (3.9-4.2); occipitonasal length, 25.4 (24.6-26.4); palatal length, 11.1 (10.8-11.7); zygomatic breadth, 13.2 (12.4-14.0).

Distribution.—Southwestern Wyoming. See Fig. 52.

Records of occurrence.—Specimens examined, 171, as follows: **Lincoln Co.:** 3 mi. N, 11 mi. E Alpine, 5650 ft., 39; *Greys River, 6000 ft., 2 Univ. Wyo.;* Afton, 2 USNM; 10 mi. SE Afton, 7500 ft., 8 USNM.

Teton Co.: Upper Arizona Creek, 1 USNM; *Whetstone Creek, 36 USNM; Gravel Creek, 1 USNM; Two Ocean Lake, 1;* 18 mi. N, 9 mi. W Moran, 1, 6 Univ. Wyo.; Jackson Hole Wildlife Park, 2; *Moran, 1, 2 USNM;* 1¼ mi. E Moran, 4, 6 Univ. Wyo.; 3 mi. S, 17 mi. E Moran, 8600 ft., 2; *Black Rock Meadows, 2 Univ. Wyo.;* 2½ mi. NE Moose, 6225 ft., 1; *Moose, 6225 ft., 1;* Moose Creek, 6800 ft., 2 USNM.

Uinta Co.: 9 mi. S Robertson, 8000-8400 ft., 22; *9½ mi. S, 1 mi. W Robertson, 8600 ft., 3; 10 mi. S, 1 mi. W Robertson, 8700 ft., 19; 13 mi. S, 1 mi. E Robertson, 9000 ft., 4;* 4 mi. S Lonetree, 3 USNM.

Additional record (Krutzsch, 1954:420).—**Uinta Co.:** 5 mi. E Lonetree.

Family ERETHIZONTIDAE Porcupines

Erethizon dorsatum (Linnaeus)

Porcupine

Second only to the beaver in size among American rodents north of Panama; upper parts thickly set with black and yellow quills; underparts scantily haired; four toes on front feet and five on hind feet; diastema longer than maxillary tooth-row; infraorbital foramen larger than foramen magnum; four cheek-teeth on each side of each jaw (except in juveniles in which the last molar has not yet erupted).

This large slow-moving rodent is an able climber and usually feeds on the bark of trees or on other vegetal matter. Two subspecies occur in Wyoming.

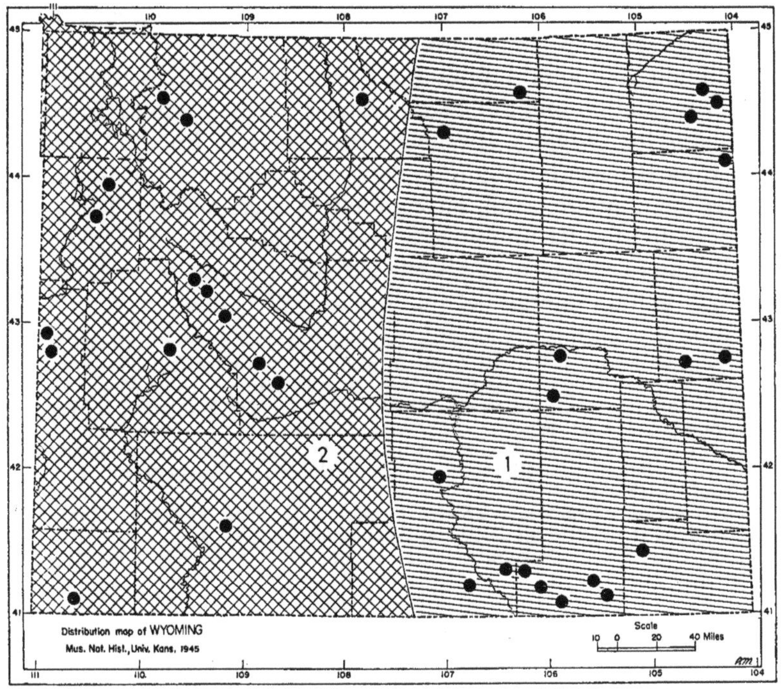

FIG. 53. Distribution of *Erethizon dorsatum*.
1. *E. d. bruneri* 2. *E. d. epixanthum*

Erethizon dorsatum bruneri Swenk

1916. *Erethizon epixanthum bruneri* Swenk, Univ. Nebraska Studies, 16:117, November 21.

1947. *Erethizon dorsatum bruneri*, Anderson, Bull. Nat. Mus. Canada, 102:173, January 24.

Type.—From three miles east of Mitchell, Scotts Bluff County, Nebraska.

Comparison.—Compared with *Erethizon dorsatum epixanthum*, *E. d. bruneri* is smaller (see *Measurements*), has frontals notably arched in lateral view (not dished), and has auditory bullae less inflated.

Measurements.—Cranial measurements of three adult males from Arvada and one adult female from Manville are, respectively, as follows: Occipitonasal length, 97, 95, 98, 93; condylobasal length, 110, 105, 104, 107; zygomatic breadth, 75, 71, 66, 66; length of nasals, 37, 38, 38, 34; maxillary toothrow, 25.7, 24.0, 26.3, 27.3; cranial depth, 30, 30, 29, 29; width of auditory bulla to lateral projection of meatus, 22.0, 21.5, 18.4, 21.0.

Distribution.—Great Plains and montane areas closely adjacent. See Fig. 53.

Remarks.—Skulls from Arvada have small bullae, raised or arched frontals, and are small. A subadult from Johnson County is referred to *E. d. bruneri* because of the geographic nearness of the locality of the specimen to that of specimens from Arvada and because all of the specimens are in the same watershed.

No adults are known from the Sierra Madre in south-central Wyoming, and a single subadult having slightly raised frontals gives basis for assigning specimens from there to *bruneri*. Specimens from the nearby Medicine Bow Mountains are undoubtedly *bruneri* for they are small, have raised frontals, and small bullae. A specimen from 24 miles north and 12 miles east of Sinclair is subadult and is referred to *bruneri* only on the basis of geographic proximity to the mountains of south-central Wyoming.

Records of occurrence.—Specimens examined, 44, as follows: **Albany Co.:** Medicine Bow Mtns., 1 USNM; *Laramie Mtns., 1 USNM;* 25 mi. W Laramie, 7500 ft., 1 Univ. Wyo.; Laramie, 1 USNM; Pole Mtn., 1 Univ. Wyo.; *Near Wallace Camp Ground, SE Laramie, 8500 ft., 1; 2 mi. SW Pole Mtn., 1;* Woods Creek, 8000 ft., 1 Univ. Wyo.

Carbon Co.: 24 mi. N, 12 mi. E Sinclair, 6600 ft., 1 Univ. Wyo.; 8 mi. N, 14 mi. E Encampment, 8400 ft., 1; *8 mi. N, 16 mi. E Encampment, 1;* Encampment, 2 USNM.

Converse Co.: Glenrock, 1 USNM; *Boxelder, 1 USNM;* 21 mi. S, 24 mi. W Douglas, 7600 ft., 1; *22.5 mi. S, 25 mi. W Douglas, 7800 ft., 1.*

Crook Co.: 15 mi. N Sundance, 1; 15 mi. ENE Sundance, 3825 ft., 1; Sundance, 1 USNM.

Johnson Co.: 1 mi. S, 4½ mi. W Buffalo, 5440 ft., 1.

Laramie Co.: Horse Creek, 6500 ft., 1.

Natrona Co.: *10 mi. S Casper, 7750 ft., 1* (marginal but not plotted on map because engraving for Fig. 53 was made before specimen was seen).

Niobrara Co.: Kirtley, 1 USNM; Manville, 8 USNM.

Sheridan Co.: Arvada, 12 USNM.

Weston Co.: 1½ mi. E Buckhorn, 6150 ft., 1.

Erethizon dorsatum epixanthum Brandt

1835. *Erethizon epixanthus* Brandt, Mem. Acad. Imp. Sci., St. Petersbourg, ser. 6, Sci. Math. Phys. et Nat., 3:390.

1884. *Erethrizon dorsatus epixanthus,* True, Proc. U. S. Nat. Mus., 7(App., Circ. 29):600, November 29.

Type.—From California.

Comparison.—See account of *Erethizon dorsatum bruneri.*

Measurements.—External measurements of a subadult female from two and a fourth miles northeast of Pinedale are: Total length, 795; length of tail, 220; hind foot, 90; ear, 35. Average and extreme cranial measurements of seven adult males from Elk (2) and along the Snake River (5) and measurements of two adult females from along the Snake River are, respectively, as follows: Occipitonasal length, 101.6 (95-105), 102, 92; condylobasal length, 113.0 (110-115), 112, 98; zygomatic breadth, 72.0 (70-76), 74, 68; length of nasals, 41.0 (36-47), 41, 39; cranial depth, 29.6 (28-32), 28.3, 28.0; maxillary tooth-row, 27.1 (24.9-29.3), 26.8, 25.7.

Distribution.—Mountains and basins of western Wyoming. See Fig. 53.

Remarks.—In Wyoming, characters of this subspecies are best expressed in the mountainous northwestern part of the state. An adult male from Shell has the frontals raised as in *bruneri,* but its skull is exceptionally long (condylobasal length, 115 mm.). Owing to its large size, this specimen is referred to *E. d. epixanthum.* In subadults the frontals are often dished and the bullae are large. Specimens from the Wind River Basin and Green River Basin have dished frontals and for this reason are referred to *epixanthum.*

Records of occurrence.—Specimens examined, 66, as follows: **Big Horn Co.:** Shell, 19 USNM. **Fremont Co.:** 2 mi. N, 6 mi. W Burris, 6450 ft., 1; Crowheart, 2 USNM; *Meadow Creek, 4 USNM;* Sage Creek, 3 USNM; Lander, 1 USNM; Hailey, 1 USNM. **Lincoln Co.:** 9 mi. N, 2½ mi. W Afton, 2; Afton, 1 USNM; *Greys River, 1.* **Park Co.:** 2 mi. S, 42 mi. W Cody, 6400 ft., 1; 18 mi. S, 23 mi. W Cody, 6400 ft., 1; *Ishawooa Creek, 19 mi. S, 19 mi. W Cody, 1.* **Sublette Co.:** *N. side Halfmoon Lake, 1;* 2¼ mi. NE Pinedale, 7500 ft., 1. **Sweetwater Co.:** Rock Springs, 1 USNM. **Teton Co.:** *N. Fork Snake River, 7 USNM; Middle Fork Snake River, 7 USNM; Snake River, 1 USNM;* Pacific Creek, 1 mi. N, 5 mi. E Moran, 6800 ft., 1 Univ. Wyo.; *½ mi. N Moran, 6742 ft., 1 Univ. Wyo.;* Elk, 6 USNM. **Uinta Co.:** 9 mi. S Robertson, 8000 ft., 1; *10½ mi. S, 2 mi. E Robertson, 1.*

Order CARNIVORA

Carnivores

KEY TO FAMILIES OF CARNIVORES

1. Hind foot five-toed.. 2
2. Three lower molars; length of head and body more than 41 inches, URSIDAE, p. 682
2′. Two lower molars; length of head and body less than 41 inches..... 3
3. Teeth, 40; tail annulated.........................PROCYONIDAE, p. 687
3′. Teeth fewer than 40; tail not annulated.............MUSTELIDAE, p. 688
1′. Hind foot four-toed.. 4
4. Four digits on forefoot; claws non-retractile; teeth, 42.....CANIDAE, p. 671
4′. Five digits on forefoot; claws retractile; teeth 28-30........FELIDAE, p. 704

Family CANIDAE

KEY TO CANIDS

1. Basilar length usually less than 147; frontals concave dorsally (dished).. 2
2. Back of pinna of ear blackish; tail lacking dorsal black stripe; inferior margin of mandible lacking prominent step...................... 3
3. Tip of tail white; ears black on outer surface of pinna; upper parts reddish..**Vulpes vulpes,** p. 678
3′. Tip of tail blackish; ears grayish on outer surface of pinna; upper parts yellowish gray intermixed dorsally with golden buff....**Vulpes velox,** p. 680

2′. Back of pinna of ear reddish or rufous; tail having continuous dorsal black stripe; inferior margin of mandible having prominent step, **Urocyon cinereoargenteus**, p. 682

1′. Basilar length usually more than 147; frontals not concave dorsally .. 4

4. Anteroposterior length of upper canine less than 11; upper carnassial shorter than 23.4.................................**Canis latrans**, p. 672

4′. Anteroposterior length of upper canine more than 11; upper carnassial usually longer than 23.4.............................**Canis lupus**, p. 675

Canis latrans Say

Coyote

Larger than a fox and smaller than a wolf; upper parts Pinkish Buff to Cinnamon Buff intermixed with gray and black; tail concolor with back but paler below; forelegs, muzzle, and ears Pinkish Buff, Cinnamon Buff, or even darker; underparts buffy; skull resembling that of *Canis lupus* but less robust with slenderer rostrum and weaker, less robust, dentition; canines elongate; tail held down when animal runs; dental formula as in *C. lupus.*

Two subspecies of the coyote occur in Wyoming and intergrade in a broad zone at the eastern base of the Rockies. Young (Young and Jackson, 1951:Fig. 10) shows dispersal distances of tagged, released coyote pups plotted on a map of Wyoming. One coyote moved 100 miles, another 90 miles, but most moved shorter distances. It is interesting that no coyotes were known to have moved across the subspecies-boundary either westward or eastward (boundary shown in Fig. 54). Long dispersal distances would seemingly broaden a zone of intergradation.

Some persons favor preserving the coyote but many sheep owners would exterminate it even in the National Parks in order to prevent the animals from straying from the protected areas onto lands where sheep are grazed. At one time in Wyoming, when parts of the National Forests were allocated to individual fur-trappers for their use, less poison was placed there by predatory-mammal control agents than was the case later. Probably in reference to the period in which individual fur-trappers had registered trap-lines in the National Forests, Young (*in* Young and Jackson, 1951:73) has written that a "barrage of effective control poison stations" around the "border line of the Wyoming National Forest" lessened predation by coyotes on sheep.

Sperry (1941) examined the stomachs of 8,339 coyotes from 17 western states taken in all months of the year over a five-year period. He found that the coyote fed mainly on rabbits (33 per cent of food items, and remains were found in 43 per cent of the stomachs examined). Carrion was the second most important food. The third most important food was rodents. Remains of sheep and goats were found in about one-fifth of the stomachs. The proportion of sheep and goats that was carrion could not be ascertained. The manner in which stomachs were obtained, and the practices of the men charged with collecting stomachs had been questioned (see A. B. Howell, 1930:381-382). In my opinion, the remainder of food items lacks importance as evidence for or against the coyote. It is true that the coyote preys on antelope and deer, but predation may be of benefit to a population of deer or antelope.

Therefore, only in sheep-raising areas are coyotes in important competition with man. The control measures used today and in the past are questionable, and in many cases terrible (see Symposium on Predatory Animal Control, Various authors, 1930:325-389; see also Young, *in* Young and Jackson, 1951:Plate 16, Plate 18, pp. 171-224). In Yellowstone National Park, coyote control was practiced for many years. In two years (1914-1915) almost two

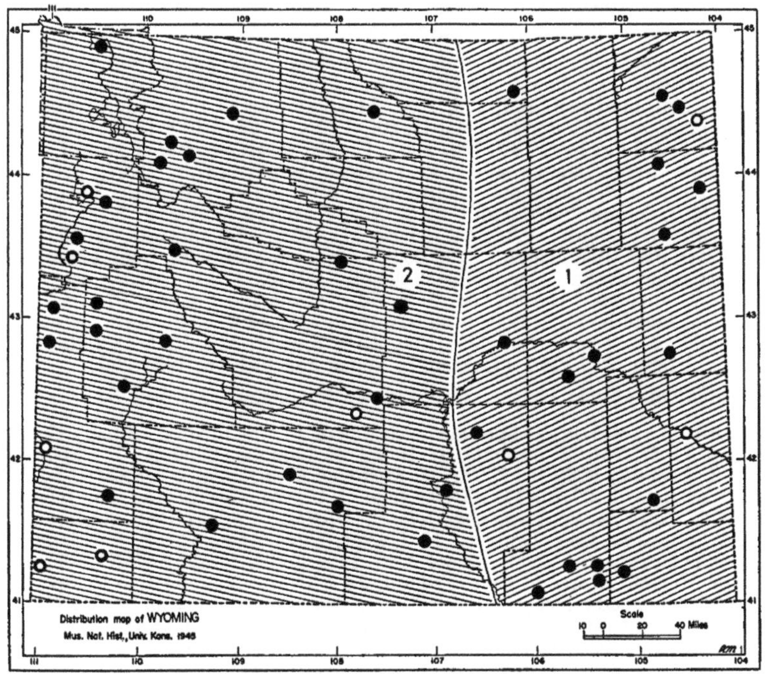

FIG. 54. Distribution of *Canis latrans*.
1. *Canis latrans latrans* 2. *Canis latrans lestes*

hundred coyotes were killed in the Park, and park rangers and trappers killed 223 coyotes in 1922 (Bailey, 1930:141). In 1923, no less than 226 were killed; and control practices continued for years thereafter, but the coyote remained abundant.

Now coyotes are protected in the Park. I observed one in the northern part begging for food from the tourists! Everywhere in its range the coyote has persisted in spite of concentrated efforts to exterminate it. Adolph Murie (1940) wrote an excellent account of the coyote in Yellowstone National Park.

Canis latrans latrans Say

1823. *Canis latrans* Say, *in* Long, Account of an expedition . . . to the Rocky Mountains. . . , 1:168.

Type.—From Engineer Cantonment, about 12 miles southeast of the present town of Blair, Washington County, Nebraska.

Comparison.—Compared with *Canis latrans lestes, C. l. latrans* averages slightly smaller; has much paler upper parts; muzzle, ears, and legs near Pinkish Buff to Pinkish Cinnamon instead of near Cinnamon to Sayal Brown (almost Snuff Brown); skull smaller and dentition weaker. This subspecies intergrades in a wide zone with *Canis latrans lestes.*

Measurements.—External measurements of three adult males from Laramie are, respectively, as follows: Total length, 1,210, 1,205, 1,255; tail vertebrae, 330, 370, 380; hind foot, 195, 195, 195. Measurements of three adult females from Laramie are as follows: Total length, 1,125, 1,160, 1,150; tail vertebrae, 317, 330, 330; hind foot, 180, 190, 195. Cranial measurements of two of the adult males from Laramie are: Condylobasal length, 187.8, 185.7; palatal length, 96.4, 96.8; squamosal constriction, 60.8, 61.0; zygomatic breadth, 102.5, 103.3; interorbital breadth, 32.3, 32.3; maxillary tooth-row, 88.4, 86.9; upper carnassial length, 20.8, 21.6; length M1, 12.3, 13.2; breadth M1, 17.1, 16.9; lower carnassial length, 22.2, 23.0. Two adult females from Laramie measure, respectively: Condylobasal length, 176.2, 178.1; palatal length, 93.5, 92.1; squamosal constriction, 56.6, 56.1; zygomatic breadth, 97.1, 95.8; interorbital breadth, 31.5, 32.0; maxillary tooth-row, 83.2, 83.3; upper carnassial length, 20.2, 19.5; length M1, 12.6, 12.9; breadth M1, 15.9, 16.4; lower carnassial length, 22.1, 22.1 (Jackson, *in* Young and Jackson, 1951:259).

Distribution.—Great Plains east of Bighorn Mountains. See Fig. 54.

Records of occurrence.—Specimens examined, 131, as follows: **Albany Co.:** Laramie, 11 USNM; *E Laramie, 1;* 16 mi. E Laramie, 8000 ft., 7 USNM; *17 mi. E Laramie, 8000 ft., 7 USNM; Medicine Bow Range, 3 USNM;* 14-15 mi. SE Laramie, 8300 ft., 11 USNM; Jelm, 1 USNM; *Red Mtns., near Jelm, 1 USNM; Horse Ranch Pass, near Jelm, 1 USNM.* **Carbon Co.:** Shirley, 1 USNM. **Converse Co.:** Douglas, 10 USNM; Moss Agate Creek, 1 USNM; No specific locality, 3 USNM. **Crook Co.:** Bear Lodge Mtns., 2 USNM; Sundance, 2 USNM. **Laramie Co.:** Federal, 14 USNM. **Natrona Co.:** Casper, 1 USNM; No specific locality, 2 USNM. **Niobrara Co.:** Manville, 16 USNM. **Platte Co.:** Chugwater, 1 USNM. **Sheridan Co.:** Arvada, 2 USNM. **Weston Co.:** Upton, 2 USNM; Hampshire, 12 USNM; Newcastle, 2 USNM; *Howard, 14 USNM (not found); Howard, Wildlife Creek, 3 USNM (not found).*

Additional records (Jackson, *in* Young and Jackson, 1951:262).—**Carbon Co.:** Aurora. **Crook Co.:** *Grand Canyon, near Sundance;* Rattlesnake Canyon, near Sundance. **Goshen Co.:** Ft. Laramie.

Canis latrans lestes Merriam

1897. *Canis lestes* Merriam, Proc. Biol. Soc. Washington, 11:25, March 15.
1913. *Canis latrans lestes,* Grinnell, Proc. California Acad. Sci., ser. 4, 3:285, August 28.

Type.—From Toyabe Mountains, near Cloverdale, Nye County, Nevada.

Comparison.—See account of *Canis latrans latrans.*

Measurements.—Average and extreme cranial measurements of 111 young adult and adult males from throughout the range of the subspecies (11 from Wyoming) are as follows: Condylobasal length, 187.1 (176.2-199.0); palatal length, 97.2 (89.8-105.0); squamosal breadth, 60.2 (55.0-65.0); zygomatic breadth, 99.5 (93.1-106.4); interorbital breadth, 33.4 (29.4-37.3); maxillary tooth-row, 88.1 (82.1-94.0); upper carnassial length, 19.6 (17.4-22.2); length M1, 12.7 (11.3-14.1); breadth M1, 16.7 (15.0-18.3); length lower carnassial, 22.5 (20.3-25.8). Measurements of 82 young adult and adult females from throughout the geographic range of the subspecies (11 from Wyoming) are as follows: Condylobasal length, 176.1 (164.4-189.9); palatal length, 91.5 (84.0-99.6); squamosal breadth, 57.2 (54.4-60.3); zygomatic breadth, 94.2 (88.4-99.8); interorbital breadth, 31.5 (27.6-34.9); maxillary tooth-row, 83.1 (75.8-89.3); upper carnassial length, 18.4 (16.6-20.0); length M1, 12.3 (11.1-14.1); breadth M1, 15.9 (14.1-17.4); lower carnassial length, 21.3 (19.6-23.4) (Jackson, *in* Young and Jackson, 1951:281-282).

Distribution.—Western half of Wyoming. See Fig. 54.

Records of occurrence.—Specimens examined, 183, as follows: **Big Horn Co.:** Shell, 14 USNM; *Shell Creek Basin, 5 USNM.*

Carbon Co.: 16 mi. N, 2 mi. W Saratoga, 5; 20 mi. SW Saratoga, 2; *5½ mi. N Big Creek* P. *O., 1 USNM; Big Creek* P. *O., 20 USNM; 3-3½ mi. E. Big Creek* P. *O., 2 USNM.*

Fremont Co.: Dubois, 1; *Vic. Dubois, 1;* 20 mi. SE Mouth Kirby Creek, 1 USNM; *Pinyon Ridge, 1 USNM; Sweetwater River, 17 USNM;* Splitrock, 3 USNM.

Lincoln Co.: 18 mi. N. 9 mi. E Afton, 1; Afton, 4 USNM; *Afton, Grays River, 2 USNM;* Opal, 24 USNM.

Natrona Co.: 8 mi. N, 2½ mi. W Waltman, 6050 ft., 1.

Park Co.: Cody, 7 USNM; 14 mi. S, 26½ mi. W Cody, 1; 19 mi. S, 19 mi. W Cody, Ishawooa Creek, 3; Valley, 1 USNM.

Sublette Co.: *Wagon Creek, 1 USNM;* 12 mi. N Kendall, 1 USNM; Kendall, 11 USNM; Pinedale, 4 USNM; Big Piney, 2 USNM.

Sweetwater Co.: 5 mi. SE Sand Dunes, 1; 2½ mi. N Wamsutter, 1; Rock Springs, 2 USNM.

Teton Co.: 3 mi. SW Black Rock Creek, 1 USNM; Kelly, 4 USNM.

Yellowstone Nat'l Park: Yanceys, 1 USNM; *Fort Yellowstone, 3 USNM; Near Mt. Gardiner, 1 USNM; Rose Creek, 7 USNM; No specific locality, 26.*

Additional records (Jackson, *in* Young and Jackson, 1951:287).—**Fremont Co.:** *Longs Creek;* Rongis. **Lincoln Co.:** Cokeville; *Sheep Range.* **Sublette Co.:** *Pole Creek.* **Teton Co.:** *Crystal;* Moran; *Elk;* Jackson. **Uinta Co.:** Ft. Bridger; Evanston.

Canis lupus Linnaeus

Gray Wolf

Larger than coyote, some individuals exceeding 100 pounds; pelage of upper parts of thick, dense fur varying from white, through various agouti or grizzled grayish brown phases, to pure black; underparts of white wolves white, or black wolves black, but of others pale to medium buff or gray; tail and ears concolor generally with dorsum; skull large and massive; teeth robust; upper carnassial actually and relatively broader than in *C. latrans;* tail held high in running instead of low as in *C. latrans.*

Gray wolves formerly preyed principally on the bison, culling aged and ill-adapted individuals from the vast herds. With the coming of the white man wolves were reduced in numbers because the bison were all but exterminated and because professional wolf-killers were active. Co-operation of the Federal government with cattlemen in World War I extirpated the wolf. In Yellowstone Park, where many animals made successful last stands, the wolf was shown no mercy; and it was extirpated.

Reviewing the status of the wolf it can be seen that it was destructive to stock and game animals. The following is taken from Seton (1929:260):

"R. M. Allen, the manager of the Ames Cattle Company (Nebraska), writing to *Recreation,* Sept., 1897, p. 207, says his range is the northeast part of Crook County, Wyo. (5,435 square miles). Since the spring of 1895, they have killed on it about 500 Gray-wolves, and they seem as numerous as ever." From this report Seton estimated the number of wolves in Wyoming at that time as 10,000.

Bounties were paid on 4,281 wolves in the years of 1897-1898, 2,140 per year (Seton, 1929:261). Seton estimated a minimum of 8,000 wolves in the state at that time.

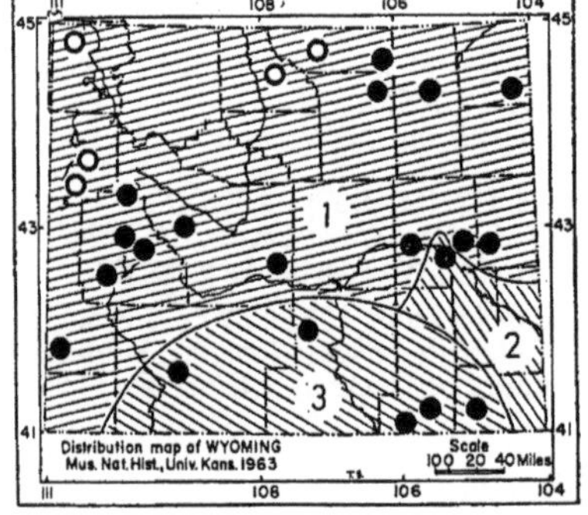

FIG. 55. Distribution of *Canis lupus.*
1. *C. l. irremotus*
2. *C. l. nubilus*
3. *C. l. youngi*

In the eleven years prior to 1908, Seton (1929:261) states that bounties were paid on 20,819 wolves. Probably twice as many were killed as were reported because half of those killed never are found.

Bailey (1930:135-137) described wolves as occasionally numerous in Yellowstone Park, and mentioned that they preyed mainly on elk; one wolf, he thought, killed one game animal every 24 hours, preferring fresh-killed meat to carrion. According to him, "It is therefore evident that wolves and game cannot be successfully maintained on the same range." Bailey also mentions the trapping of several wolves in 1915-1916 in the Park.

Cahalane (1948:251) knew of three reliable reports of the wolf in the Park in 1941, 1944, and 1946, all referring to a single wolf. He states, "Presumably the animal (or animals) left the protected area for nothing has been heard of it during the past year."

Canis lupus irremotus Goldman

1937. *Canis lupus irremotus* Goldman, Jour. Mamm., 18:41, February 14.

Type.—From Red Lodge, Carbon County, Montana.

Comparisons.—Compared with *Canis lupus youngi, C. l. irremotus* is of about the same size, but is paler on the average (upper parts less overlain with black), and narrower in the frontal region of the skull. Compared with *Canis lupus nubilus, C. l. irremotus* is larger, paler, and narrower in the frontal region especially relative to the width of the rostrum. *C. l. irremotus* shows more resemblance to *C. l. youngi* of the southern Rockies than to *C. l. nubilus* of the Great Plains.

Measurements.—External measurements of the type (male) and an unusually large male from Gallatin County, Montana, are as follows: Total length, 1870, 1834; length of tail, 410, —; hind foot, 240, —; weight, —, 106 pounds. External measurements of two adult females from Soda Springs, Idaho, are as follows: 1929, 2046; length of tail, 480, 440; hind foot, 236, 254. Cranial measurements of the type and an adult male topotype are, respectively, as follows: Greatest length, 259.2, 262; condylobasal length, 237, 241; zygomatic breadth, 144.9, 142.7; squamosal constriction, 81.1, 81; width of rostrum, 47.7, 49.5; interorbital breadth, 44.6, 43.1; postorbital constriction, 34.7, 35.5; length of mandible, 186, 193.5; height of rostrum, 74.4, 82.6; maxillary tooth-row, 105.7, 106.2; upper carnassial, outer crown length, 25.7, 26.1; crown width, 13.9, 15.3. Cranial measurements of two adult females from Dillon, Montana, are, respectively: Greatest length, 254.5, 244.5; zygomatic breadth, 127, 123.3; maxillary tooth-row, 109.2, 102.3; outer crown length upper carnassial, 26, 25.2 (after Goldman, *in* Young and Goldman, 1944:446-447).

Distribution.—Western part of state and northern two-thirds of the remainder. See Fig. 55.

Records of occurrence.—Specimens examined, 34, as follows: **Campbell Co.:** Gillette, 1 USNM. **Converse Co.:** Glenrock, 1 USNM; 10 mi. N Lost Springs, 3 USNM; No specific locality, 4 USNM. **Crook Co.:** Near Sand Creek Canyon, Black Hills, 1 USNM. **Fremont Co.:** Ft. Washakie, 1 USNM; *Lenore, 2 USNM;* 8 mi. NW Splitrock, 2 USNM. **Johnson Co.:** Barber, 2 USNM. **Lincoln Co.:** Cokeville, 1 USNM. **Niobrara Co.:** 15 mi. NW Manville, 1 USNM. **Sheridan Co.:** Arvada, 2 USNM. **Sublette Co.:** Wagon Creek, southwestern Wyoming, 2 USNM; Cora, 5 USNM; Pinedale, 1 USNM; Big Piney, 4 USNM. **Not found:** *Howard, 1 USNM* (Weston Co.?).

Additional records (Young and Goldman, 1944:448-449).—**Big Horn Co.:** Shell. **Sheridan Co.:** Otto. **Teton Co.:** Elk; Kelly. **Yellowstone Nat'l Park:** Hell Roaring Creek; *no specific locality, 3.*

Canis lupus nubilus Say

1823. *Canis nubilus* Say, *In* Long, Account of an expedition . . . to the Rocky Mountains . . . , 1:169.

1829. *Canis lupus* var. *nubilus,* Richardson, Fauna Boreali-Americana, p. 69.

Type.—From Engineer Cantonment, near present town of Blair, Washington County, Nebraska.

Comparisons.—For comparison with *Canis lupus irremotus*, see account of that subspecies. Compared with *Canis lupus youngi*, fewer specimens are as buffy. *C. l. nubilus* is larger, has a longer rostrum and palate, and a broader supraoccipital shield that rises more steeply. The shield in most skulls lacks the descending terminal hook so frequently present in *nubilus* and does not project so far posteriorly over the foramen magnum.

Measurements.—Cranial measurements of two males (old adults) from Douglas and one subadult-adult male from Natrona County are, respectively, as follows: Length of palate, 120, 124, 119; length of nasals, 101, —, 94; cranial depth, 57, 62, 57; cranial depth plus height of sagittal crest, 75, 74, 68; zygomatic breadth, 138, 138, 128; outer crown length of upper carnassial, 24.5, 23.0, 25.0; maxillary tooth-row, 105, 105, 102.

Distribution.—Southeastern Wyoming. See Fig. 55.

Records of occurrence.—Specimens examined, 15, as follows: **Converse Co.:** Douglas, 12 USNM. **Natrona Co.:** *No specific locality, 2 USNM.* **No specific locality:** *"Wyoming," 1 USNM.*

Canis lupus youngi Goldman

1937. *Canis lupus youngi* Goldman, Jour. Mamm., 18:40, February 11.

Type.—From Harts Draw, north slope of Blue Mountains, 20 miles northwest of Monticello, San Juan County, Utah.

Comparisons.—See accounts of *Canis lupus irremotus* and *C. l. nubilus.*

Measurements.—Cranial measurements of an adult male and an adult female from Laramie are as follows: Length of palate, 124, 118; length of nasals, 104, 94; cranial depth, 62, 60; cranial depth plus height of sagittal crest, 77, 70; zygomatic breadth, 142, 132; outer crown length of upper carnassial, 24.3, 19.0; maxillary tooth-row, 108.4, 101.2.

Distribution.—Mountains along southern boundary of Wyoming and from Green River watershed. See Fig. 55.

Records of occurrence.—Specimens examined, 16, as follows: **Albany Co.:** Laramie, 2 USNM; Jelm, 3 USNM. **Laramie Co.:** Federal, 7 USNM. **Sweetwater Co.:** Rock Springs, 4 USNM.

Additional record (Young and Goldman, 1944:463).—**Carbon Co.:** Dry Lake, 15 mi. N Rawlins.

Vulpes vulpes (Linnaeus)

Red Fox

Smaller than coyote and larger than swift fox; long hairs in pelage give animal appearance of being larger than it actually is; upper parts yellowish red; sides paler yellowish red; venter white or plumbeous; chin plumbeous or whitish; muzzle blackish; outer surface of pinna of ear black and inner surface white; tail long, grayish, reddish, or yellowish-red tipped with white; feet blackish; legs reddish proximally; skull having temporal ridges V-shaped and narrowly separated, not lyrate or U-shaped as in gray fox.

The red fox feeds mainly on mice, pocket gophers, ground squirrels, rabbits, and such birds as it is able to catch (see Bailey, 1930: 142). Some foxes feed on chickens or other fowl of careless farmers. The animal has a valuable fur, but fur-farming, fortunately for wild foxes, has reduced the value of fox pelts, and in response to decreased pressure from fur-trappers the red fox seems to be increasing in numbers.

When white man first visited Wyoming the red fox was probably abundant there, but because of its valuable fur the species was trapped in great numbers. The red fox population was reduced also because of the poisoning campaigns carried out in Wyoming; in fact, instead of the target species, the coyote, being exterminated, the three species of Wyoming foxes (all furbearers) became rare although the coyote remained abundant (and poisoning still is carried on in Wyoming!).

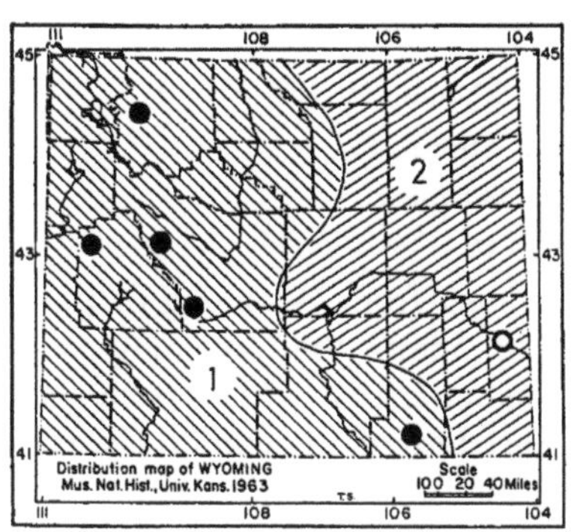

FIG. 56. Distribution of *Vulpes vulpes*.
1. *V. v. macroura*
2. *V. v. regalis*

The red fox is polychromatic having black, cross, and silver phases as well as red; but in Wyoming I have examined only the red phase. Almost certainly the red fox of the Great Plains, *Vulpes vulpes regalis*, ranged into eastern Wyoming, but all specimens that I have examined are from farther westward and are referable to *V. v. macroura*, which is a smaller, more yellowish fox. One specimen from Laramie (KU 91363) is large for a female. I follow Churcher (1959) in using the specific name *Vulpes vulpes* (Linnaeus) for American red foxes.

Vulpes vulpes macroura Baird

1852. *Vulpes macrourus* Baird, in Stansbury, Exploration and survey of the Valley of the Great Salt Lake of Utah. . . . (Spec. Sess., U. S. Senate, Exec. No. 3), App. C, p. 309, June.

1936. *Vulpes fulva macroura*, V. Bailey, Nature Mag., 28(5):317, November.

Type.—From Wasatch Mountains, bordering the Great Salt Lake, Utah.

Comparison.—Smaller than *Vulpes vulpes regalis* and more yellowish instead of golden reddish.

Measurements.—External measurements of an adult male from Lake Fork are as follows: Total length, 1015; length of tail, 461; hind foot, 72. This specimen was taken in 1893. Cranial measurements of the male from Park County, and of five specimens (sexes unknown, two are subadults) from 17 miles south and 6½ miles west of Lander, are, respectively, as follows: Condylonasal length, —, 130, 128, —, 124, 124; zygomatic breadth, 71.5, 77, —, —, 71, —; postorbital processes of frontals, —, 35.4, 30.2, 35.1, 33.8, —; greatest length of nasals, 55.3, 53.5, 54.5, 53.5, 53.0, 50.8; maxillary tooth-row, 61.7, 64.4, 62.8, 63.9, 60.9, 61.2; outer crown length upper carnassial, 12.7, 14.0, 14.4, 14.0, 12.3, 13.3.

Distribution.—Western and most of southern part of state.

Remarks.—The specimen (KU 91363) from Laramie is large for a female, perhaps because it is an intergrade between *macroura* and *V. v. regalis*. The label mentions that this specimen may have escaped from a fur farm.

Records of occurrence.—Specimens examined, 10, as follows: **Albany Co.:** Laramie, 1. **Fremont Co.:** Lake Fork, Wind River Mtns., 10,000 ft., 1 USNM; 17 mi. S, 16½ mi. W Lander, 8450 ft., 6. **Park Co.:** 3½ mi. S, 27 mi. W Cody, 1. **Sublette Co.:** Cliff Creek, Hoback Mtns., 1 Univ. Wyo.

Vulpes vulpes regalis Merriam

1900. *Vulpes regalis* Merriam, Proc. Washington Acad. Sci., 2:672, December 28.

1929. *Vulpes fulva regalis*, B. Bailey, Jour. Mamm., 10:157, May 9.

Type.—From Elk River, Sherburne County, Minnesota.

Camparison.—See account of *Vulpes vulpes macroura*.

Distribution.—Great Plains; possibly extirpated. See Fig. 56.

Remarks.—I have examined no specimen from Wyoming of this Great Plains subspecies. It occurs just across the Wyoming-Nebraskan boundary (J. K. Jones, Jr., manuscript). The specimen recorded by Baird (1858:133) from Fort Laramie is tentatively assigned to *V. v. regalis* on geographic grounds.

Record of occurrence.—**Goshen Co.:** Fort Laramie, 1 (Baird, 1858:133).

Vulpes velox velox (Say)

Swift Fox

1823. *Canis velox* Say, *In* Long, Account of an exped. . . . to the Rocky Mountains. . . ., 1.487.

1851. *Vulpes velox*, Audubon and Bachman, The viviparous quadrupeds of North America, 2:13.

Type.—From South Platte River, probably in Logan County, Colorado.

Description.—Small for Canidae; upper parts grayish buff or yellowish gray; tail tipped with black; venter paler than dorsum; pelage flecked with golden or yellow streaks; skull small with temporal ridges poorly developed; feet concolor with dorsum (not black as in *Vulpes vulpes*).

Measurements.—No measurements are available for Wyoming specimens, except the length (112 mm.) of hind foot measured dry on a specimen from Bridgers Pass. The dried feet in two specimens from Pueblo, Colorado, are 109 and 113 mm. long.

Distribution.—Probably once occurred throughout Wyoming in valleys and on prairies. Known only from three localities. See Fig. 57.

Remarks.—Only five specimens, to my knowledge, of the swift fox have been preserved from Wyoming, of which three (from Aurora, June 1898) are young. One adult (skin only) was taken by the early exploring party of Lt. F. T. Bryan and W. S. Wood on August 14, 1856, at Bridgers Pass. My notes state, "The pelage is an unusual color, extremely buffy except middorsally where the hairs are predominately brownish. . . . The muzzle of the Wyo. specimen is not as dark as [in] *hebes* nor as pale as [in] Pueblo specimens." The length of the hind foot is short (dry) as in the specimens from Pueblo differing from the type of *hebes* (120 mm.). In size, in brownish instead of grayish pelage, and on geographic grounds the swift fox in Wyoming of territorial days is referable to *V. v. velox*.

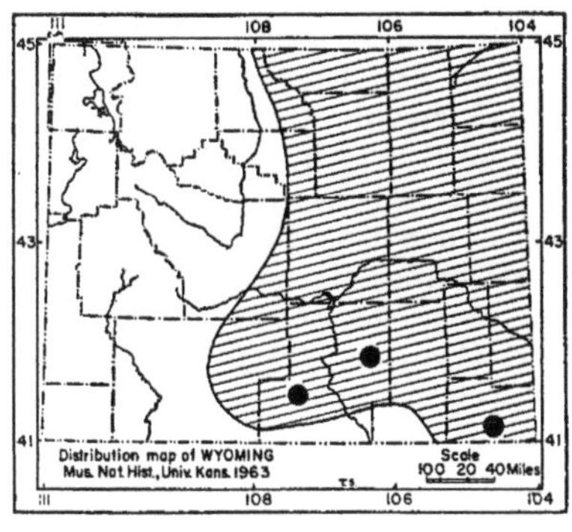

FIG. 57. Distribution of *Vulpes velox velox*.

Likewise, a swift fox taken in November of 1958 at a point five miles east and two miles south of Archer, Laramie County, is pale and does not differ significantly in color from the specimen taken by Bryan and Wood in 1856. The recently taken specimen is referred to *V. v. velox*. The three young swift foxes from Aurora are referred to *V. v. velox* on geographic grounds. The swift fox is one of the furbearers (G. M. Allen, 1942:195-196) that is almost or entirely beneficial to man. Poisoned baits placed for the coyote and wolf extirpated the swift fox on most parts of the Great Plains. Swift foxes more readily take poisoned baits than do some other carnivores (Bailey, 1905:179). Of course, trapping of furbearers, modification of the original environment, and capture by dogs (G. M. Allen, *op. cit.*) helped to extirpate the species. Poisoning carried on today probably will prevent the swift fox from regaining its former abundance in Wyoming although it is reappearing in some of the prairie states. The swift fox deserves full protection.

Records of occurrence.—Specimens examined, 5, as follows: **Carbon Co.:** Aurora, 3 Amer. Mus. Nat. Hist.; Bridgers Pass, 1 USNM. **Laramie Co.:** 2 mi. S, 5 mi. E Archer, 1.

Urocyon cinereoargenteus ocythous Bangs

Gray Fox

1899. *Urocyon cinereoargenteus ocythous* Bangs, Proc. New England Zool. Club, 1:43, June 5.

Type.—From Platteville, Grant County, Wisconsin.

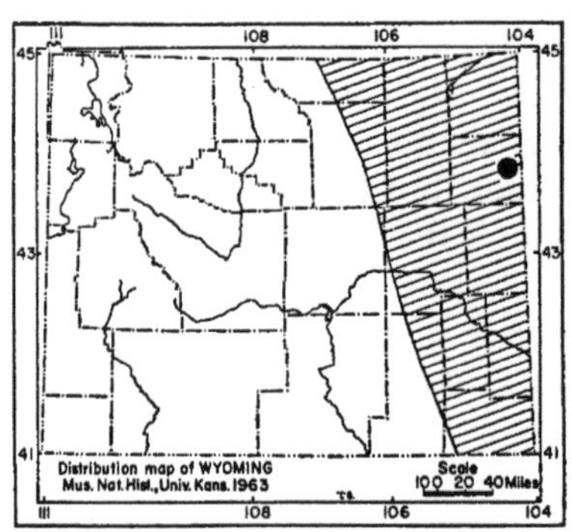

FIG. 58. Occurrence of *Urocyon cineroargenteus ocythous.*

Description.—Smaller than coyote, about the size of red fox; upper parts grizzled grayish; throat white; face gray; sides of neck, lower flanks, and underside of tail rusty; tail with median dorsal streak of black hair; ventral border of mandible with distinct step; longitudinal (temporal) ridges (two) on parietals and frontals U-shaped and more widely separated than in *Vulpes.*

Measurements. — Measurements listed by Hall and Kelson (1959:860) are as follows: Total length, 800-1125; length of tail, 275-443; hind foot, 100-150; condylobasal length of skull, 110-130. External measurements of the only specimen known from Wyoming are as follows: Total length, 1016; length of tail, 407.

Distribution.—Probably occurs only sparingly on the Great Plains. See Fig. 58.

Remarks.—A large adult female (USNM 167892) from Owen [= Owens] in Weston County provides the only record of the gray fox in the state and the westernmost record of occurrence of the subspecies.

Family URSIDAE—Bears

KEY TO THE BEARS

1\. Upper parts grizzled with golden buff; mane or "roach" of long hair on shoulders; upper M2 longer than 30.................. **Ursus arctos,** p. 685

1′. Upper parts not grizzled with golden buff; mane lacking; upper M2 shorter than 30............................... **Ursus americanus,** p. 682

Ursus americanus cinnamomum Audubon and Bachman

Black Bear

1854. *Ursus americanus* var. *cinnamomum* Audubon and Bachman, The viviparous quadrupeds of North America, 3:125.

Type.—From Lower Clearwater River, Camp Chopunnish, near mouth of Jim Ford Creek, Clearwater County, western Idaho (see Bailey, 1936:319).

Description.—Large for Carnivora but smaller than grizzly bear (see *Measurements*); colors of pelage variable as follows: USNM 227661, from Elk Creek on Grays River, Lincoln County, is pale yellowish brownish, darker on crown. USNM 227925, from same locality, is pale brownish but not yellowish and has small white spot on throat. USNM 227926, from Shell Basin, is glossy black everywhere except on nose, which is brownish. USNM 283630, from Yellowstone Park, also is glossy black, but has a small white spot on throat. USNM 13297, from Yellowstone Park, is deep reddish brown except for a small white spot on throat. USNM 197059, probably from Yellowstone Park, is tinged with reddish everywhere and shows a golden sheen especially on feet. Skull of black bear much as in grizzly bear except smaller; M2 much smaller. Claws of forefeet smaller than in grizzly bear.

Measurements.—Cranial measurements of an adult male from the head of Jose Creek, of an adult labeled male from 16 miles south of Dubois, of a subadult male from 5 miles south and 16 miles east of Saratoga, of an adult-subadult female from 24 miles south and 36 miles west of Cody, of an adult female from the head of Big Sandy Creek, and of an adult female from Cottonwood are, respectively, as follows: Greatest length of skull, 291, 248, 250, 242, 277, 257; zygomatic breadth, 180, 151, 135, 144, 165, 160; breadth postorbital processes frontals, 96.2, 82.7, 76.3, 81.9, 91.5, 89.3; length of nasals, 87.6, 66.2, 63.4, 60.8, 71.9, 68.4; maxillary tooth-row, 102.0, 88.0, 85.9, 90.2, 97.6, 87.4; breadth M2, 15.3, —, 14.0, 14.1, 13.6, 13.6; length M2, 25.5, —, 22.8, 25.8, 24.1, 21.1.

Distribution.—Along Rocky Mountain chain. See Fig. 59.

Remarks.—Bears in Wyoming are identified as *Ursus americanus cinnamomum* on the basis of small molars and arched frontals. The measurements of Wyoming black bears agree closely with those listed by Durrant 1952:410) for *cinnamomum*. Two other subspecies might be expected to occur in Wyoming, namely *U. a. americanus* in eastern Wyoming and *U. a. amblyceps* in southern Wyoming. But, in spite of great variation among specimens, Wyoming bears seem referable to a single subspecies; none of the bears is so broad across the zygomata or has the skull dished or so much flattened as does *amblyceps* (see Durrant, *loc. cit.*). I have seen no specimens from the Great Plains of Wyoming, where *U. a. americanus* might be expected to occur.

Bears were and are most abundant in mountainous areas in Wyoming. Early reports and historical accounts frequently mention grizzly bears in Wyoming, but I know of only a few early references to the black bear. A specimen sent to the National Zoological Park from Yellowstone Park is preserved and labeled 1892, and another specimen in the National Museum is from Warm Springs Creek, Teton County, and was taken in 1893. To what degree the extensive poisoning campaigns in Wyoming adversely affected bears I do not know. Whether the black bear increased in numbers in areas where the grizzlies were killed out, I do not know.

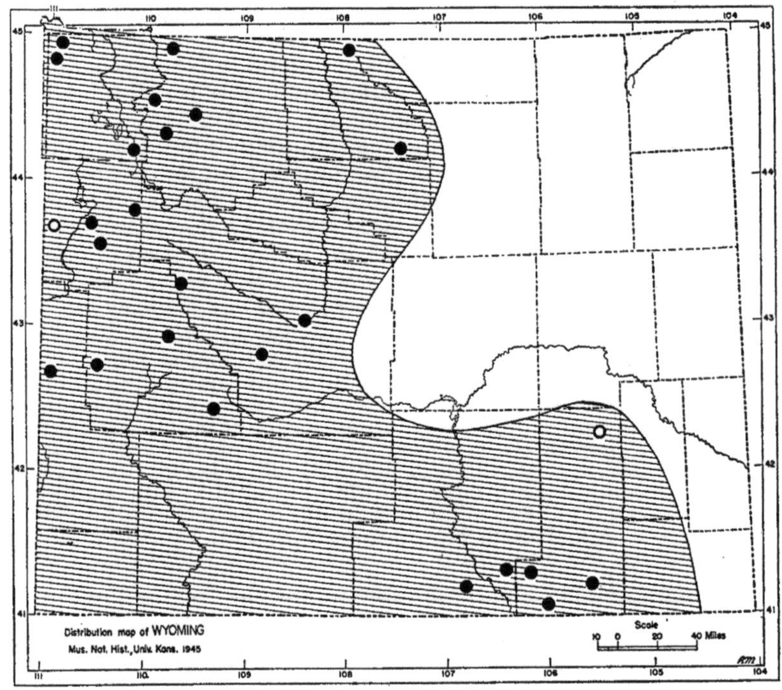

FIG. 59. Distribution of *Ursus americanus cinnamomum.*

Jackson (1944:11) reported 3,512 black bears in Wyoming in the year 1941, showing that the black bear was seemingly plentiful. Wyoming ranked twelfth among the states in numbers of black bears; but among 14 states having at least 2,000 bears, Wyoming was the only state having no black bears listed on state and private lands; its bears all were listed as on federal lands. Nearly all of the bears that tourists see in Yellowstone National Park are black bears.

Records of occurrence.—Specimens examined, 48, as follows: **Albany Co.:** Centennial, 2 Univ. Wyo.; Laramie, 1 Univ. Wyo.; S Jelm, 1 Univ. Wyo.

Big Horn Co.: Northern Bighorn Mtns., 2 USNM; *Shell Basin, 1 USNM;* 16 mi. N Tensleep, 1.

Carbon Co.: 5 mi. S, 16 mi. E Saratoga, 2; vic. Encampment, 1 Univ. Wyo.

Fremont Co.: 16 mi. S Dubois, 1; Riverton, 1 USNM; Lander, 1 USNM; *6 mi. S, ½ mi. E Lander, 1.*

Lincoln Co.: Afton, 1 USNM; *Grays River, 3 USNM; E. Fork Hams Fork, 1.*

Park Co.: 31 mi. N, 36 mi. W Cody, 1; 4 mi. N, 4 mi. W Pahaska, 1; 15 mi. S, 23 mi. W Cody, 1; 24 mi. S, 36 mi. W Cody, 1.

Sublette Co.: *Hd. Jose Creek, trib. Green River, N Pinedale, 1;* Pinedale, 3 USNM; 27 mi. E Afton, 1 USNM; 18 mi. S, 30 mi. W Lander, 1.

Teton Co.: *Squirrel Meadows, 2 USNM;* Warm Springs Creek, 1 USNM; *Pacific Creek, "Elk," 1 USNM; Black Rock, 2 USNM;* Jackson Hole, 2 USNM; Gros Ventre River, 1 USNM.

Yellowstone Nat'l Park: Mammoth Hot Springs, 1 USNM; Upper Yellowstone River, 1 USNM; Gallatin River, 1 USNM; No specific locality, 5 USNM.
Co. not certain: *Northwest Wyoming, 1 USNM.*

Additional records.—**Albany Co.:** Laramie Peak (Honess and Frost, 1942: 30), in 1876. **Teton Co.:** Teton Creek, Teton Canyon (Fryxell, 1943:81), in 1879.

Ursus arctos imperator Merriam

Grizzly Bear

1914. *Ursus imperator* Merriam, Proc. Biol. Soc. Washington, 27:180, August 13.
1916. *Ursus washake* Merriam, Proc. Biol. Soc. Washington, 29:152, September 6. Type from Shoshone River, Wyoming.
1918. *Ursus mirus* Merriam, N. Amer. Fauna, 41:40, February 9. Type from Slough Creek, Yellowstone National Park, Wyoming.
1918. *Ursus rogersi rogersi* Merriam, N. Amer. Fauna, 41:66, February 9. Type from Greybull River, Absaroka Mountains, Wyoming.
1918. *Ursus rogersi bisonophagus* Merriam, N. Amer. Fauna, 41:66, February 9. Type from Bear Lodge, Sundance National Forest, Black Hills, Crook County, Wyoming.

Type.—From Yellowstone National Park, Wyoming.

Description.—Large for Carnivora (some males exceed 600 pounds; Seton [1929, 2:8] mentioned a male killed in Yellowstone that weighed 916 pounds); pelage on back long, especially on shoulders forming there a "roach"; upper parts varying from blackish to yellowish or brownish, always with more on less (depending to some extent on wear) of a golden or brassy sheen effected by yellowish or whitish tips of long hairs; underparts resembling upper parts in color; feet usually darker than upper parts; claws of forefeet twice length of claws of hind feet; skull large (see *Measurements*); upper last molar (M2) large, robust (larger than in black bear); number of premolars vary but when all teeth present dental formula is i.$\frac{3}{3}$; c.$\frac{1}{1}$; p.$\frac{4}{4}$; m.$\frac{2}{3}$.

Measurements.—The following measurements are from manuscript by Prof. E. Raymond Hall. Cranial measurements of the holotype of *imperator* (male), of an adult male topotype, and of two adult female topotypes are, respectively, as follows: Basal length, 336, 331, 293, 294; condylobasal length, 355, 357, 312, 311; basilar length, 332, 325, 287, 289; occipitonasal length, 318, 319, 274, 285; palatal length, 198, 187, 163, 171; zygomatic breadth, 233, 216, 192, 193; depth of skull, 129, 125, 105, 113; interorbital breadth, 89.3, 79.2, 76.6, 79.0; length P4-M2, 80.8, 79.7, 69.1, 74.5; M2 length, 41.1, 38.5, 33.3, 37.5; M2 breadth, 21.5, 19.7, 18.0, 19.7.

Distribution.—Known formerly throughout most of state, but now confined to Yellowstone National Park and closely adjacent areas. See Fig. 60.

Remarks.—I follow Erdbrink (1953) in regarding the North American grizzly bear as conspecific with the Eurasian bear, *Ursus arctos.* In my opinion, a single subspecies, *Ursus arctos imperator* Merriam 1914, of the grizzly bear occurs in the Rocky Mountains of Wyoming. Possibly another subspecies occurs on the Great Plains, described as a buffalo-killing grizzly and known as the "White Bear" of Lewis and Clark (Merriam, 1918:18). All of the specimens examined by me from Wyoming are tentatively referred to *Ursus arctos imperator;* holotypes from Wyoming for names that I regard as junior synonyms of *Ursus imperator* have been seen. The junior synonyms are listed in the synonymy of *imperator.*

Records (see beyond) show that in Wyoming grizzlies were abundant in pioneer times, perhaps more abundant than the black bear. Considered, with some justification, a threat to livestock and to man, the grizzly was slain in great numbers. An excerpt from the journal of E. Willard Smith (Barry, 1943:287-297) states that on October 10, 1839, 100 buffalo and six grizzlies were killed "quite near camp" at the mouth of Muddy Creek on the Snake River, Carbon County.

Seton (1929:21) reported that "between 60 and 100 Grizzly-bears" existed in Yellowstone Park, and added: "Each year the number of hunters increases; each year more deadly traps, subtler poisons and more irresistible guns are out to get the Grizzly. He has no chance at all of escape. There is no closed season, no new invulnerability to meet the new perils. He is absolutely at the mercy of those who know no mercy; and before five years more, I expect to learn that there are no Grizzlies left in the United States, except in Yellowstone Park. . . ."

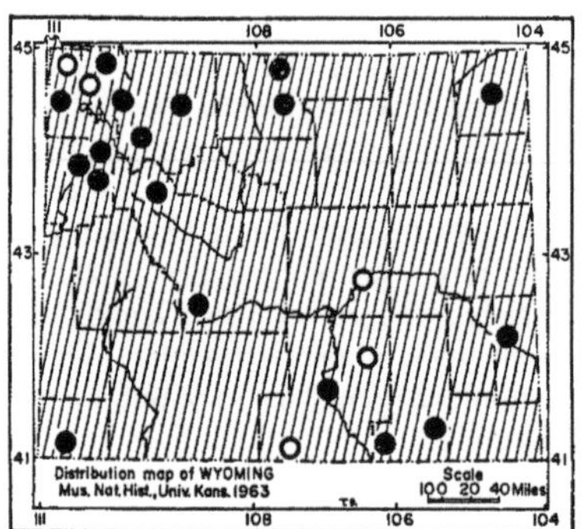

FIG. 60. *Ursus arctos imperator.*

Jackson (1944:11) reported 533 grizzlies in Wyoming in the years 1937-1942. Cahalene (1948:248) reported that almost one-third of the surviving grizzlies in the United States lived in Yellowstone National Park (Wyoming) and Glacier National Park (Montana). The grizzly is in danger of being extirpated in Wyoming.

Records of occurrence.—Specimens examined, 67, as follows: **Albany Co.:** Laramie Mtns., 3 USNM; Medicine Bow Mtns., 1 USNM.

Big Horn Co.: Bighorn Mtns., 1 USNM.

Carbon Co.: Ft. Steele, 1 USNM.

Crook Co.: Bear Lodge Mtns., 1 USNM; *Sundance Nat'l Forest, 3 USNM.*

Fremont Co.: 16 mi. NE Dubois, 1 USNM; South Pass City, 7900 ft., 1.

Goshen Co.: Ft. Laramie, 1 USNM.

Lincoln Co.: *Grays River, 1 USNM; Afton, Deadman Creek, 1 USNM.*

Park Co.: *Canfield Creek, E Park, 3 USNM;* 2 mi. S, 6 mi. W Cody, 1; 4 mi. N, 4 mi. W Pahaska, 2; *Absaroka Mtns., 1 USNM; N. Fork Shoshoni River, Absaroka Mtns., 4 USNM;* Valley, 1 USNM; *Upper Greybull River, 1 USNM.*

Sheridan Co.: Hd. Little Bighorn River, 1 USNM.

Sublette Co.: *N. Fork Teton River, 1 USNM.*

Teton Co.: Bridger Lake, 1 USNM; *Squirrel Meadows, 1 USNM; Pacific Creek, 2 mi. N Road Slide, 1 USNM; Specimen Ridge, 1 USNM; Arizona Creek, 1; Hd. Box Creek Gorge, 1;* Two Ocean Lake, 1 Univ. Wyo.; *Jackson Hole, 1 USNM;* Black Rock Creek, 1 USNM; *Upper Gros Ventre Valley, N. Fish Creek, 1 USNM; Fish Creek, 1 USNM.*

Uinta Co.: Northern foothills Uintah Mtns., 1 USNM.

Yellowstone Nat'l Park: *Yellowstone River, 3 USNM;* Slough Creek, 2 USNM; *Lake Jct., 3 Univ. Wyo.;* Old Faithful, 1; *Lake Hotel, 2 USNM; no specific locality, 8 USNM.*

Co. not certain: *Mtns. near Fort Laramie, 1 USNM; Medicine Bow Mtns., near Fort Laramie, 1 USNM; Northwestern Wyoming, 1 USNM; Wyoming Territory, 1 USNM; Bighorn Creek, 1 USNM.*

Not found: *Del Norte Creek, 1 USNM.*

Additional records.—**Carbon Co.:** Medicine Bow Butte (Baird, 1858:220); Mouth Muddy Creek, Snake River (Barry, 1943:291), in 1839. **Natrona Co.:** Poison Spider Creek, opposite Casper Mtn. (Coutant, 1899:112), in 1812. **Yellowstone Nat'l Park** (Bailey, 1930:166-175): Mammoth Hot Springs; Canyon Hotel; *Yellowstone River, Yellowstone Lake; Near Buffalo Ranch; Between Lonestar Geyser and Shoshoni Lake; Quadrant Mtn.*

Family PROCYONIDAE—Raccoons and Allies

KEY TO THE PROCYONIDAE

1. Tail longer than body; hard palate extends posteriorly about as far as last upper molar; hind foot less than 80 mm.....**Bassariscus astutus, p. 687**
1′. Tail shorter than body; hard palate extends posteriorly behind last upper molar for a distance of more than combined lengths of M1 and 2; hind foot more than 90................................**Procyon lotor, p. 688**

Bassariscus astutus arizonensis Goldman

Ringtail

1932. *Bassariscus astutus arizonensis* Goldman, Proc. Biol. Soc. Washington, 45:87, June 21.

Type.—From Cosper Ranch, about 12 miles south of Blue, 5000 ft., Greenlee Co., Arizona.

Description.—Size medium (see *Measurements*); tail ringed with alternating dark and pale bands and longer than body; brownish gray upper parts; face brownish; lacking gray spot immediately anterior to pinna and paler upper parts of *B. a. nevadensis;* smaller, darker of face, and lacking dark-brownish color dorsally and buff laterally of *B. a. flavus* (see *Remarks*).

Measurements.—Available external and cranial measurements of the specimen from Wyoming are as follows: Total length, 734; length of tail, 365; length of hind foot, 23 [?]; ear (from notch), 37; alveolar length of maxillary tooth-row, 28.1; alveolar length of mandibular tooth-row, 32.2.

Distribution.—Known only from one locality in the southwestern part of state, but may occur elsewhere. See Fig. 61.

Remarks.—Long and House (1961:274-275) assigned the only specimen known from Wyoming to the subspecies *B. a. arizonensis.* It was obtained on October 18, 1957, by K. Demick.

Record of occurrence.—Specimen examined as follows: **Lincoln Co.:** 3 mi. N, 24 mi. W Kemmerer, 6300 ft., 1 Univ. Wyo.

Procyon lotor hirtus Nelson and Goldman

Raccoon

1930. *Procyon lotor hirtus* Nelson and Goldman, Jour. Mamm., 11:455, November 11.

Type.—From Elk River, Sherburne County, Minnesota.

Description.—Size medium for carnivores (see *Measurements*); mask of black on face; tail annulated, black and brownish gray; upper parts grayish brown; venter grayish; baculum bilobed anteriorly; stance plantigrade or nearly so; postorbital processes distinct, acuminate; teeth large; molariform teeth with robust conical cusps; canines robust and moderately long; dental formula, i.$\frac{3}{3}$, c.$\frac{1}{1}$, p.$\frac{4}{4}$, m.$\frac{2}{2}$.

Measurements.—External and cranial measurements of an adult male from two miles north of Wheatland, an adult, lactating female from Muskrat Canyon, and cranial measurements of an adult female from 16 miles west of Hulett are, respectively, as follows: Total length, 940, 765; length of tail, 315, 248; hind foot, 130, 120; ear from notch, 65, 66; condylobasal length, 122, 111, 113; zygomatic breadth, 81, 76, 75; breadth across postorbital processes of frontals, 25.2, 30.3, 31.5; length of nasals, 36.4, 38.8, 35.2; maxillary tooth-row, 45.1, 43.2, 42.4; cranial depth, 34.1, 35.9, 39.0.

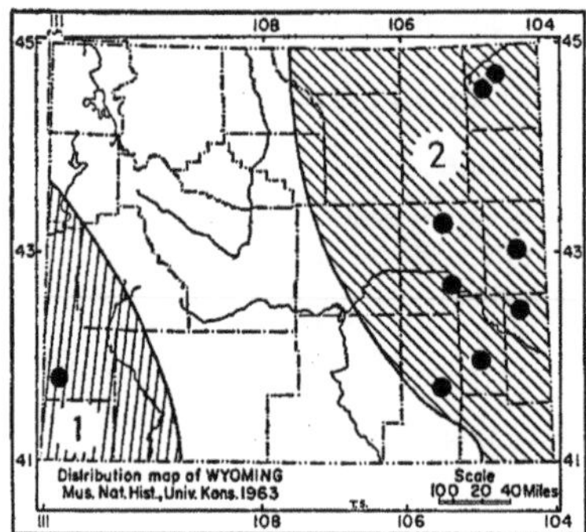

FIG. 61. Two procyonids.
1. *Bassariscus astutus arizonensis*
2. *Procyon lotor hirtus*

Distribution.—Known from eastern fifth of state. See Fig. 61.

Records of occurrence.—Specimens examined, 11, as follows: **Albany Co.:** Sybelle Creek, 30 mi. N, 15 mi. E Laramie, 1 Univ. Wyo.; *Sybelle Canyon, 1 Univ. Wyo.* **Converse Co.:** 3 mi. E Bill, 4700 ft., 1; Douglas, 3 Univ. Wyo. **Crook Co.:** New Haven, 1 USNM; Little Missouri River, 2 mi. N, 16 mi. W Hulett, 1. **Goshen Co.:** Muskrat Canyon, 1. **Niobrara Co.:** 6½ mi. SW Hat Creek, 1 Univ. Wyo. **Platte Co.:** 2 mi. N Wheatland, 1.

Family MUSTELIDAE

KEY TO MUSTELID CARNIVORES

1. Premolars $\frac{4}{4}$ 2
2. Obscure brownish stripes laterally . . . . **Gulo gulo, p.** 697
2′. Stripes lacking 3
3. Tail more than 290; m1 more than 11 . . . . **Martes pennanti, p.** 691
3′. Tail less than 290; m1 less than 11 . . . . **Martes americana, p.** 689
1′. Premolars fewer than $\frac{4}{4}$ 4
4. Premolars $\frac{4}{3}$ . . . . **Lutra canadensis, p.** 703
4′. Premolars $\frac{3}{3}$ 5
5. Talonid of m1 trenchant . . . . **Mustela, p.** 692
5′. Talonid of m1 basined or dished 6

6. Single whitish longitudinal stripe on top of head and neck; basilar length more than 80 mm.......................... **Taxidea taxus**, p. 699
6′. Single white neck stripe lacking; basilar length less than 80 mm...... 7
7. Upper parts black with broken white stripes or spots, **Spilogale putorius**, p. 700
7′. Upper parts black with continuous white stripes (two) or entirely black except white on top of head.................... **Mephitis mephitis**, p. 702

Martes americana (Turton)

Marten

Much smaller than *Martes pennanti* (tail less than 290 in *americana*); upper parts pale brownish; underparts brownish with spots (especially on chin and throat) of orange or occasionally yellow; feet and tail brownish or blackish brown.

The marten has been studied by Hagmeier (1961), who hesitates to use the subspecies concept. He presents data obtained from many specimens, and I have attempted to draw conclusions (or make judgments) from the assembled data with my own data from specimens that I have examined. From southern Wyoming, only one adult male is available to me, and it differs in several respects as shown below from specimens from northwestern Wyoming. Hagmeier's data contain several interesting items.

Hagmeier (1961:130) found that the martens from northwestern Wyoming resembled those from Colorado in condylobasal length, but that martens from Utah were longer in this dimension than either specimens from northwestern Wyoming (known as *M. a. vulpina*) or specimens from Colorado (known as *origenes*). I found that specimens from northwestern Wyoming were longer in this dimension than the one adult from southern Wyoming, and from the measurements listed by Hagmeier for either *origenes* or *vulpina*. From these data I would judge Utah specimens to be more closely related to *M. a. vulpina* than to *M. a. origenes*. Durrant (1952:424) had only four skulls from Utah when he assigned Utah specimens to *origenes*. He found them to be intermediate between the two subspecies but referable to *origenes*. He mentions that the size of the inner cusp of the upper carnassial in Utah specimens resembles this character in *origenes*. Smaller teeth, especially the carnassial, is the most striking difference between the specimen from southern Wyoming and those from northwestern Wyoming. Davis (1939:130) suggested that *vulpina* (= *caurina* of his nomenclature) might range into Utah from the north. In Wyoming it seems reasonable to recognize two subspecies each characterized as mentioned beyond.

Martes americana origenes (Rhoads)

1902. *Mustela caurina origenes* Rhoads, Proc. Acad. Nat. Sci. Philadelphia, 54:458, September 30.

1953. *Martes americana origenes,* Wright, Jour. Mamm., 34:84, February 9.

Type.—From Marvine Mountain, Garfield County, Colorado.

Comparison.—Compared with *Martes americana vulpina, M. a. origenes* has smaller teeth (see especially labial length of upper carnassial in Hagmeier, 1961:136); shorter skull on the average; smaller auditory bullae on the average; and slightly paler pelage with less white on margin of pinna of ear (in skin of KU 16787).

Measurements.—External and cranial measurements of an adult male from three miles east-southeast of Browns Peak are as follows: Total length, 652; length of tail, 218; hind foot, 94; ear from notch, 42; condylobasal length, 79.5; zygomatic breadth, 48.6; breadth across postorbital processes of frontals, 24.2; maxillary tooth-row, 29.7; labial crown length of upper carnassial, 8.3; breadth M1, 8.5.

Distribution.—Mountains of south-central Wyoming. See Fig. 62.

Record of occurrence.—Specimens examined, 2, as follows: **Albany Co.:** *Little Brooklyn Lake, Medicine Bow Mtns., 10,350 ft., 1 Univ. Wyo.;* 3 mi. ESE Browns Peak, 10,000 ft., 1.

Martes americana vulpina (Rafinesque)

1819. *Mustela vulpina* Rafinesque, Amer. Jour. Sci., p. 82.

1959. *Martes americana vulpina,* Hall and Kelson, The Mammals of North America, The Ronald Press, 2:901, March 31.

Type.—From regions watered by the Missouri, presumably Montana.

Comparison.—See account of *Martes americana origenes.*

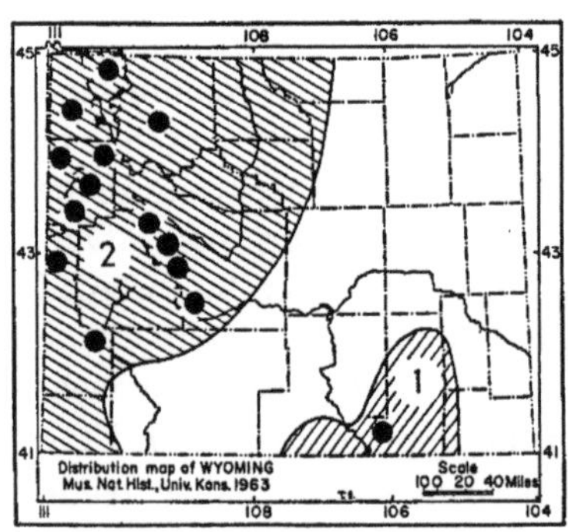

FIG. 62. *Martes americana.*
1. *M. a. origenes*
2. *M. a. vulpina*

Measurements.—External measurements of an adult male (KU 32329) from Middle Lake, Fremont County, are as follows: Total length, 620; length of tail, 200; hind foot, 91; ear from notch, 50. Average and extreme cranial measurements of five males and four females (all subadults) from Thorofare Creek are as follows: Condylobasal length, 82.4 (81.6-84.1), 72.2 (70.0-73.4); zygomatic breadth, 46.7 (44.1-51.1), 41.6 (40.4-43.1); breadth across postorbital processes of frontals, 23.4 (22.4-24.3), 20.9 (20.2-22.0); maxillary tooth-row, 29.6 (29.1-30.4), 25.7 (24.8-26.5); labial crown length of upper carnassial, 8.7 (8.6-8.9), 7.4 (6.8-7.9); breadth M1, 8.8 (8.3-9.2), 7.4 (6.8-7.9).

Distribution—Northwestern Wyoming. See Fig. 62.

Records of occurrence.—Specimens examined, 102, as follows: **Fremont Co.:** Crowheart, 2 USNM; Lake Fork, 1 USNM; Middle Lake, 2 mi. S, 20½ mi. W Lander, 1; 17 mi. S, 6½ mi. W Lander, 8450 ft., 23. **Lincoln Co.:** Greys River, 1; LaBarge Creek, 9000 ft., 1 USNM. **Park Co.:** *Dry forks, Ishawooa Creek, 19 mi. S, 21 mi. W Cody, 3;* Ishawooa Creek, 19 mi. S, 19 mi. W Cody, 2; *24 mi. S, 36 mi. W Cody, 10; Yellow Creek Mesa, 25 mi. S, 40 mi. W Cody, 2; Pass Creek, Yellowstone Watershed, 25 mi. S, 40 mi. W Cody, 5; Thorofare Creek, 27 mi. S, 42 mi. W Cody, 12; no specific locality, 1.* **Teton Co.:** Bridger Lake, 1 USNM; 10 mi. N, 16 mi. E Moran, 22; *Lava Creek, 2;* Spread Creek, 1 USNM; *Bear Creek, Gros Ventre River, 4 USNM;* Jackson, 3 USNM; *Jackson Hole, 3 USNM.* **Yellowstone Nat'l Park:** Lamar River, 1 USNM; W. shore Yellowstone Lake, 1 Univ. Wyo.

Martes pennanti columbiana Goldman

Fisher

1935. *Martes pennanti columbiana* Goldman, Proc. Biol. Soc. Washington, 48:176, November 15.

Type.—From Stuart Lake, near headwaters of Fraser River, British Columbia.

Description.—Much larger than *Martes americana* (tail more than 290 in *pennanti*); dark brown, grayish on foreparts, rump blackish; tail entirely black; M1 more nearly square than in *M. americana;* skull much larger than in *M. americana.*

Measurements.—Hall and Kelson (1959:901) list measurements as follows: Males total length, 990-1033; length of tail, 381-422; weight up to 15 pounds, possibly 18 pounds. Females 830-900; length of tail, 340-380; hind foot, 89-115; weight up to 6 pounds; basilar length, 87-108.

Distribution.—Recorded once from Yellowstone National Park, possibly from the part of the park that is in Wyoming. Not mapped.

Remarks.—Seton (1929:457) remarks: "When I lived in the Yellowstone Park in the summer of 1897, I made all inquiries, and was told by the hunters that the Fisher was unknown; Capt. George S. Anderson told me that about 1893, he had confiscated a skin taken by a poacher in the Park. All hunters assure me that the Fisher is never found in Jackson's Hole."

This record in Seton (*loc. cit.*) was assigned to the nominate subspecies because the subspecies to the westward (*M. p. columbiana*) had not yet been recognized. With no evidence available for or against this assignment, it has persisted. Owing to the fact that *M. p. columbiana* occurs in Idaho and the northern Rockies, the skin recorded from Yellowstone Park is here listed under *columbiana* on geographic grounds.

Mr. Earl M. Thomas (*Wyoming Wildlife,* Game and Fish Dept., Wyoming, July 1952) states that "in the early 1920's" Lars Scorr trapped two fishers on the Beartooth Plateau.

Record of occurrence.—**Yellowstone Nat'l Park:** No specific locality, 1 skin seen by Capt. George S. Anderson, about 1893 (Seton, 1929:457).

KEY TO SPECIES OF MUSTELA

1. Length of upper tooth-rows less than 20 in males and 17.8 in females... 2
2. Postglenoidal length of skull more than 47 per cent of condylobasal length..**Mustela erminea,** p. 692
2′. Postglenoidal length of skull less than 47 per cent of condylobasal length..**Mustela frenata,** p. 693
1′. Length of upper tooth-rows more than 20 in males and 17.8 in females.. 3
3. Pelage creamy or dusky grayish white; black facial mask present; m1 lacking metaconid............................**Mustela nigripes,** p. 697
3′. Pelage dark brown (white spot on chin or throat); mask on face lacking; m1 having incipient metaconid....................**Mustela vison,** p. 695

Mustela erminea muricus (Bangs)

Ermine

1899. *Putorius (Arctogale) muricus* Bangs, Proc. New England Zool. Club, 1:71, July 31.

1945. *Mustela erminea murica,* Hall, Jour. Mamm., 26:84, February 27.

Type.—From Echo, 7500 feet, El Dorado County, California.

Description.—Small (smallest carnivore in state); upper parts brown; underparts yellowish white (not the least cinnamon); winter pelage white except that tip of tail is black in all seasons; skull small; auditory bullae small; teeth exceedingly small. Males larger than females.

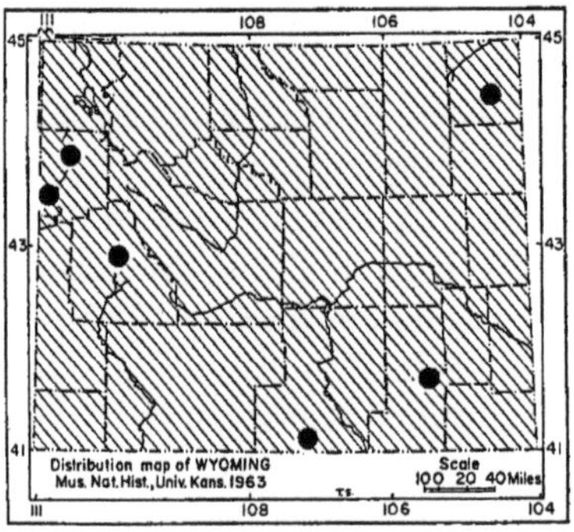

FIG. 63. *Mustela eminea muricus.*

Measurements.—External and cranial measurements of a female from ¼ mile east of Moran and of four adult males (one each from 8 mi. N and 19½ mi. E Savery; 2¼ mi. NE Pinedale; 3 mi. NW Sundance; and 30 mi. N and 10 mi. E Laramie) are, respectively, as follows: Total length, 214, 228, 220, 232, 221; length of tail, 55, 61, 55, 61, 62; hind foot, 26, 27.5, 28, 30, 27; ear from notch, 16, 13, 15, 16, 15; condylobasal length, 32.0, 34.5, 32.7, 34.9, —; zygomatic breadth, 16.1, 18.0, 17.4, 18.0, 17.0; breadth postorbital processes of frontals, 8.8, 9.9?, 9.8, 9.8, 9.5; cranial breadth, 16.5, 16.1, 17.0, 16.6, —; cranial depth, 11.0, 11.3, 11.2, 10.4, —; maxillary tooth-row, 8.0, 9.0, 8.8, 9.0, 8.4; outer crown length of upper carnassial, 3.4, 3.5, 3.5, 3.5, 3.4.

Distribution.—State-wide in suitable habitats. See Fig. 63.

Records of occurrence.—Specimens examined, 13, as follows: **Albany, Co.:** 30 mi. N, 10 mi. E Laramie, 6560 ft., 1; *26 mi. N, 4½ mi. E Laramie, 6960 ft., 1.* **Carbon Co.:** 8 mi. N, 19½ mi. E Savery, 8800 ft., 2. **Crook Co.:** 3 mi. NW Sundance, 5900 ft., 1. **Sublette Co.:** 2¼ mi. NE Pinedale, 7500 ft., 1. **Teton Co.:** *¼ mi. E Moran, 6700 ft., 1;* Jackson Hole Wildlife Park, 6740 ft., 3, 1 USNM; Teton Pass, 1 USNM; *No specific locality, 1 USNM.*

Mustela frenata Lichtenstein

Long-tailed Weasel

Larger than *Mustela erminea;* upper parts brown; underparts whitish but usually washed strongly with cinnamon or reddish brown; winter pelage white except tip of tail which is black in all seasons; skull larger than in *M. erminea* and much smaller than in *M. vison.* See accounts of *Mustela erminea, M. vison,* and *M. nigripes.* Four subspecies occur in Wyoming.

Mustela frenata alleni (Merriam)

1896. *Putorius alleni* Merriam, N. Amer. Fauna, 11:24, June 30.
1936. *Mustela frenata alleni,* Hall, Carnegie Inst. Washington Publ. 473:106, November 20.

Type.—From Custer, Custer County, South Dakota.

Comparisons.—From *Mustela frenata longicauda, M. f. alleni* differs in lesser size (basilar length, less than 43.5 in males, 40 in females; total length, less than 400 in males, 375 in females). *M. f. alleni* averages slightly smaller than Wyoming specimens of *M. f. nevadensis* and is near Clay Color instead of near Brussels Brown on upper parts in summer (*alleni* is paler brown in summer).

Measurements.—External and cranial measurements of an adult female from Sundance are as follows: Total length, 352; length of tail, 123; hind foot, 41; zygomatic breadth, 23.4; interorbital breadth, 7.8; maxillary tooth-row, 13.0.

Distribution.—Occurs in the Black Hills. See Fig. 64.

Record of occurrence.—Specimen examined, 1, as follows: Crook Co.: Sundance, 1 USNM.

Mustela frenata longicauda Bonaparte

1838. *Mustela longicauda* Bonaparte, Charlesworth's Mag. Nat. Hist., 2:38, January.
1936. *Mustela frenata longicauda,* Hall, Carnegie Inst. Washington Publ. 473:105, November 20.
1877. *Putorius culbertsoni* Coues, Dept. Int., U. S. Geol. Surv. Terr., Misc. Publ. No. 8, p. 136. Type from Ft. Laramie, Wyoming.

Type.—Possibly from Carlton House, Saskatchewan.

Comparisons.—Paler (near Clay Color instead of Brussels Brown) than *Mustela frenata nevadensis;* smaller and paler than *M. f. oribasus;* larger than *M. f. alleni.*

Measurements.—Average and extreme cranial measurements listed by Hall (1951c:420-421) of five adult males from Alberta and of five adult females from Alberta (3), North Dakota (1), and Saskatchewan (1) are, respectively, as follows: Basilar length, 46.0 (44.7-46.5), 42.3 (40.0-43.7); zygomatic breadth, 30.3 (29.4-31.0), 25.9 (24.5-26.7); cranial depth, 15.4 (15.0-16.0), 13.7 (12.8-14.3); length of tooth-rows, 17.9 (17.2-18.6), 16.3 (15.1-16.8); outer crown length of upper carnassial, 6.0 (5.4-6.3), 5.4 (5.0-5.9); breadth M1, 4.6 (4.3-5.0), 4.3 (4.0-4.8).

Distribution.—Expected to occur only in eastern Wyoming south of the Black Hills. See Fig. 64.

Record of occurrence (Hall, 1951c:269).—**Goshen Co.:** Fort Laramie.

Mustela frenata nevadensis Hall

1936. *Mustela frenata nevadensis* Hall, Carnegie Inst., Washington Publ. 473:91, November 20.

Type.—From three miles east of Baker, White Pine County, Nevada.

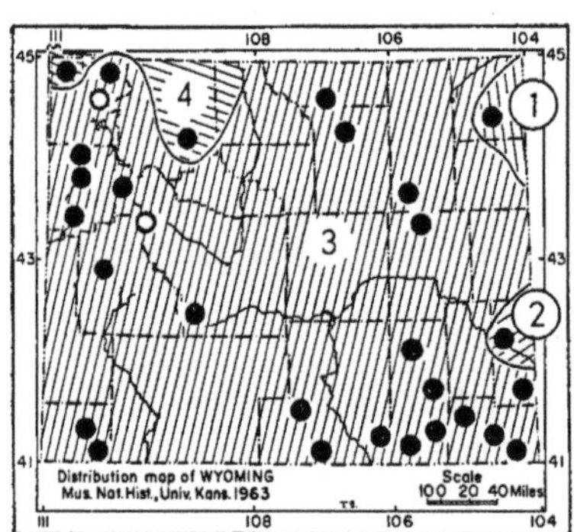

FIG. 64. *Mustela frenata.*
1. *M. frenata alleni*
2. *M. frenata longicauda*
3. *M. frenata nevadensis*
4. *M. frenata oribasus*

Comparisons.—See accounts of other subspecies.

Measurements.—External measurements of an adult male from three miles east of Dubois and of another from Woods Landing are as follows: Total length, 469, 423; length of tail, 171, 160; hind foot, 46, 48; length of ear, 24, 22. Cranial measurements of three adult males from 10 miles south and 9 miles west of Worland, from 12 miles north and two miles west of Afton, and from 27 miles north and 7½ miles east of Laramie and of an adult female from three miles west-southwest of Fort Bridger are, respectively, as follows: Condylobasal length, 48.5, 48.4, 47.8, 40.8; zygomatic breadth, 29.6, 28.6, 27.0, 23.5; postorbital processes of frontals, 14.2, 13.0, 12.5, 10.8; cranial breadth (braincase), 21.5, 20.8, 22.3, 20.5; cranial depth, 14.8, 14.4, 15.8, 14.2; maxillary tooth-row, 14.9, 14.7, 14.1, 12.6; outer crown length upper carnassial, 5.5, 5.6, 5.2, 4.8.

Distribution.—Throughout most of state in suitable habitats. See Fig. 64.

Records of occurrence.—Specimens examined, 36, as follows: **Albany Co.:** Garrett, 1 USNM; 27 mi. N, 7½ mi. E Laramie, 6960 ft., 1; Jack Rabbit Canyon, E Laramie, 1; 2 mi. W Univ. Wyo. Sci. Camp, 1 Univ. Wyo.; *2¼ mi. ESE Browns Peak, 10,300 ft., 1;* 8 mi. SW Laramie, 1; *Woods Landing, 1; Laramie River, 1 USNM.*

Campbell Co.: 42 mi. S, 13 mi. W Gillette, 1 (from owl pellet, discarded).

Carbon Co.: Bridgers Pass, 1 USNM; 7 mi. N, 17 mi. E Savery, 8300 ft., 1.

Converse Co.: 10 mi. N, 6 mi. W Bill, 1.

Fremont Co.: 1 mi. S, 3 mi. E Dubois, 6900 ft., 1; 17 mi. S, 6½ mi. W Lander, 1.

Goshen Co.: Hawk Springs Reservoir, 1.

Laramie Co.: Horse Creek, 3 mi. W Horse Creek P. O., 1; *3$\frac{6}{10}$ mi. S Horse Creek, 6600 ft., 1;* 11 mi. N, 5½ mi. E Cheyenne, 5950 ft., 1; *30 mi. W Pine Bluffs, 1;* 15 mi. E Cheyenne, 5630 ft., 1.

Lincoln Co.: 12 mi. N, 2 mi. W Afton, 6100 ft., 1; *9 mi. N, 2 mi. W Afton, 6100 ft., 1.*

Sheridan Co.: Story, 1.

Teton Co.: Arizona Creek, 2; Jackson Hole Wildlife Park, 3 USNM; *Crystal Creek, Gros Ventre River, 2 USNM;* Jackson, 1 USNM.

Uinta Co.: 3 mi. WSW Ft. Bridger, 6650 ft., 1; Lonetree, 1 USNM.

Yellowstone Nat'l Park: Lamar River, 1 USNM.

Washakie Co.: 10 mi. S, 9 mi. W Worland, 4100 ft., 1.

Co. not certain: *"Wyoming Territory," 1 USNM.*

Additional records (Hall, 1951C:290-291).—**Albany Co.:** *3 mi. SW Laramie; 12 mi. S Laramie.* **Carbon Co.:** *Medicine Bow Mtns.; 15 mi. SE Parco.* **Fremont Co.:** 20 mi. NW Dubois. **Johnson Co.:** Buffalo. **Park Co.:** *Greybull River.* **Sublette Co.:** Bronx. **Teton Co.:** *Whetstone Creek.* **Yellowstone Nat'l Park:** Yellowstone Lake.

Mustela frenata oribasus (Bangs)

1899. *Putorius (Arctogale) longicauda oribasus* Bangs, Proc. New England Zool. Club, 1:81, December 27.

1936. *Mustela frenata oribasa,* Hall, Carnegie Inst. Washington Publ., 473:105, November 20.

Type.—From source of Kettle River, 7500 feet, on the summit between middle fork of Kettle River and Cherry Creek at Pinnacles, British Columbia.

Comparison.—Compared with *Mustela frenata nevadensis,* with which *M. f. oribasus* intergrades in northwestern Wyoming, *M. f. oribasus* differs in larger size.

Measurements.—Cranial measurements listed by Hall (1951c:422-423) of five adult and subadult males and two adult females, all from British Columbia, are, respectively, as follows: Basilar length, 48.8, 48.8, 46.6, 47.5, 45.0, 41.7, 42.0; zygomatic breadth, 32.2, 31.7, 30.7, 30.1, 31.0, 26.7, 27.0; interorbital breadth, 13.0, 12.0, 10.1, 10.3, 11.8, 10.5, 10.2; depth of skull, 15.5, 15.5, 15.0, 15.0, 15.0, 14.0, 14.0; length of tooth-rows, 19.1, 19.5, 17.7, 18.5, 17.5, 16.4, 16.4; lateral (= outer crown length) upper carnassial, 6.2, 6.3, 5.8, 5.9, 5.6, 5.5, 5.5; M1, breadth 5.3, 5.3, 4.6, 4.6, 4.7, 4.5, 4.3.

Distribution.—Northwestern Wyoming. See Fig. 64.

Records of occurrence (Hall, 1951c:274).—**Park Co.:** Four Bears. **Yellowstone Nat'l Park:** Glen Creek, Mammoth Hot Springs.

Mustela vison Schreber

Mink

Larger than long-tailed weasel, smaller than striped skunk; upper parts rich, blackish brown; underparts only slightly paler than upper parts, feet and tail concolor with back; white markings usually present on chin; skull flattened as in *Spilogale;* auditory bullae well inflated; m1 having incipient metaconid; skull larger with more robust zygomata than in *Mustela frenata.* Females about 10 per cent smaller in linear measurements than males and only half as heavy. The mink is at home on land (where rabbits and small rodents are frequently preyed upon) or in water (in which the mink catches fish and frogs). The fur has a high value, and mink farming at this writing is big business.

Mustela vison energumenos (Bangs)

1896. *Putorius vison energumenos* Bangs, Proc. Boston Soc. Nat. Hist., 27:5, March.

1912. *Mustela vison energumenos,* Miller, Bull. U. S. Nat. Mus., 79:101, December 31.

Type.—From Sumas, British Columbia.

Comparison.—Compared with *Mustela vison letifera, M. v. energumenos* is smaller; darker; skull smaller, less robust; teeth smaller. (See measurements.)

Measurements.—External and cranial measurements of two adult males and one adult female from Woods Post Office are, respectively, as follows: Total length, 555, 570, 520; length of tail, 186, 191, 176; hind foot, 70, 68, 57; condylobasal length, 64.4, 63.5, 58.5; zygomatic breadth, 36.2, 36.3, 32.7; interorbital breadth, 13.2, 12.3, 10.8; length of nasals, 12.5, 11.9, —; maxillary tooth-row, 20.5, 19.8, 17.7. Measurements of an adult male from Teton County, of another male from 12 miles south and 15 miles west of Cody, of an adult female from 15½ miles south and 13 miles west of Meeteetse, and of a lactating, adult female from 6450 ft., on the Wind River, are, respectively, as follows: Total length, 562, —, 527, 472; length of tail, 197, —, 174, 145; hind foot, 69, —, 64, 57; ear from notch, 24, —, 26, 23; condylobasal length, 63.9, 64.8, 60.3, 58.4; zygomatic breadth, 37.3, 37.3, 35.2, 34.5; interorbital breadth, 10.1, 11.0, 11.7, 10.9; cranial depth, 19.7, 20.5, 18.5, 18.8; maxillary tooth-row, 19.6, 19.3, 18.6, 17.9; outer crown length upper carnassial, 7.2, 7.4, 7.2, 6.1; breadth M1, 6.0, 6.1, 5.9, 5.4.

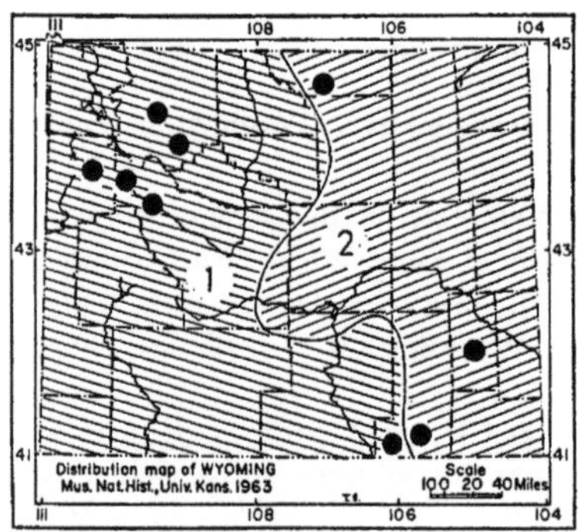

FIG. 65. Distribution of *Mustela vison.*
1. *M. v. energumenos*
2. *M. v. letifera*

Distribution.—Rocky Mountains and southwestern Wyoming. See Fig. 65.

Records of occurrence.—Specimens examined, 8, as follows: **Albany Co.:** Woods P. O., 3 USNM. **Fremont Co.:** Brooks Lake, 1 USNM; Wind River, 3 mi. N, 6 mi. W Burris, 6450 ft., 1. **Park Co.:** 12 mi. S, 15 mi. W Cody, 1; 15½ mi. S, 13 mi. W Meeteetse, 1. **Teton Co.:** 1 mi. S, 3¾ mi. E Moran, 6200 ft., 1.

Mustela vison letifera Hollister

1913. *Mustela vison letifera* Hollister, Proc. U. S. Nat. Mus., 44:475, April 18.

Type.—From Elk River, Sherburne County, Minnesota.

Comparison.—For comparison with *Mustela vison energumenos,* see account of that subspecies.

Measurements.—External measurements of an adult male from 2 miles north of Wheatland are as follows: Total length, 615; length of tail, 205; hind foot, 65; ear from notch, 26. Cranial measurements of an adult male (skull only) from Story are as follows: Condylobasal length, 75.2; zygomatic breadth, 44.7; interorbital breadth, 12.0; cranial depth, 20.9; maxillary tooth-row, 23.9; outer crown length of upper carnassial, 8.0; breadth M1, 7.3.

Distribution.—Great Plains in eastern Wyoming. See Fig. 65.

Records of occurrence.—Specimens examined, 6, as follows: **Albany Co.:** *Laramie, 1;* 6½ mi. SW Laramie, 2; *Olson River, 1.* **Platte Co.:** 2 mi. N Wheatland, 1 Univ. Wyo. **Sheridan Co.:** Story, 1.

Mustela nigripes (Audubon and Bachman)

Black-footed Ferret

1851. *Putorius nigripes* Audubon and Bachman. The viviparous quadrupeds of North America, 2:297.

1912. *Mustela nigripes*, Miller, Bull. U. S. Nat. Mus., 79:102, December 31.

Type.—From Fort Laramie, Goshen County, Wyoming.

Description.—About size of mink (see *Measurements*); upper parts yellowish buff, occasionally whitish, especially on face and venter; black mask on face; black feet; tail tipped with black; typical mustelid skull, but mastoid processes notably angular.

Measurements.—External measurements of an adult male from Laramie and of an adult (probably male) from "Wyoming" are as follows: Total length, 422, 496; length of tail, 96, 117; hind foot, 56, 56. Cranial measurements of an adult male from Manville and an adult (probably male) from within 12 miles of Cheyenne are, respectively, as follows: Condylobasal length, 67.6, 66.5; zygomatic breadth, 43.5, —; interorbital breadth, 12.0, 16.0; alveolar length maxillary tooth-row, 20.2, 19.6; labial alveolar length upper carnassial, 7.7, 6.9.

Distribution.—Great Plains and closely adjacent areas. See Fig. 66.

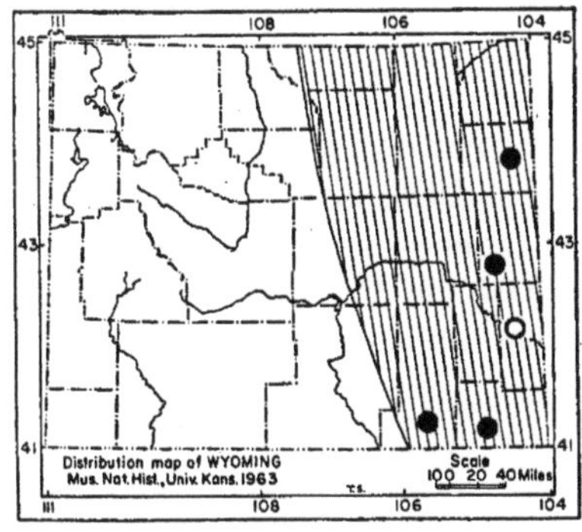

FIG. 66. *Musela nigripes.*

Remarks.—The black-footed ferret lived in prairie dog towns, and the geographic ranges of *Cynomys* and this species in general coincided. Poisoning of ground squirrels and prairie dogs greatly decreased the numbers of these sciurids and thereby indirectly, and possibly directly through action of the poison, may have extirpated the black-footed ferret in the state.

Records of occurrence.—Specimens examined, 8, as follows: **Albany Co.:** Laramie, 1 Univ. Wyo.; *5 mi. W Laramie, 1 USNM.* **Laramie Co.:** Cheyenne, 1 USNM; *within 12 mi. Cheyenne, 1 USNM.* **Niobrara Co.:** Manville, 1 USNM. **Weston Co.:** Newcastle, 2 USNM. **County not specified:** *"Wyoming" 1 Univ. Wyo.*

Additional record.—**Goshen Co.:** Type locality.

Gulo gulo luscus (Linnaeus)

Wolverine

1758. *[Ursus] luscus* Linnaeus, Syst. Nat., ed. 10, 1:47.

1823. *Gulo luscus*, Sabine. *In* Franklin, Narrative of a journey to the shores of the Polar Sea in . . . 1819-22, p. 650.

1959. *Gulo gulo luscus*, Kurtén and Rausch, Acta Arctica, Fasc. 11, p. 19. Kurtén and Rausch arrange Eurasian and American wolverines as two subspecies of a single species, but the evidence does not support the implication of Kurtén and Rausch that American wolverines comprise a single subspecies.

Type.—From Hudson Bay.

Description.—Large for Mustelidae, reaching 40 pounds; upper parts blackish brown; broad pale brown stripe extends from neck to base of tail on each side of body; tympanic bullae moderately inflated and separate from paroccipital processes; M1 expanded lingually in occlusal view; trigonid longer than talonid in m1, metaconid absent; dental formula, i. $\frac{3}{3}$, c. $\frac{1}{1}$, p. $\frac{4}{4}$, m. $\frac{1}{2}$.

Measurements.—Seton (1929:406) lists external measurements of four males and seven females as follows: "Average length," 1051, 940.7; length of tail, 212.5, 183.5; hind foot, 198.5, 177.5. Weights are recorded by Seton of almost 30 pounds and mention is made of some wolverines weighing more than 30 pounds.

Distribution.—Owing to extensive wandering, may occur in any high mountains in Wyoming; seems to persist in mountains of northwestern Wyoming. Not mapped.

Remarks.—Cahalane (1948:249) mentioned two sight records "in the past ten years" from the Grand Teton, and a single sight record from the south entrance of Yellowstone Park in 1946.

The December, 1955, number of *Wyoming Wildlife*, Wyoming Game and Fish Commission, carries a report of a sight record from "east of Warren Bridge in the winter of 1953, according to Gene Peterson" said to be in the same area from which the species was reported 13 years earlier. Don Newhart (in the same magazine, for May, 1962) reported seeing a mother wolverine and two young along Eagle Creek, Shoshone Mountains.

The only preserved specimen of a wolverine that probably was obtained in Wyoming was sent to the National Zoological Park from Yellowstone Park by Captain George S. Anderson, U. S. A., Acting Superintendent of the Park. The animal was kept for an unknown length of time in the National Zoological Park in Washington, D. C. When the wolverine died its skull came to the United States National Museum (on July 16, 1895). Presumably the individual was captured in the Wyoming part of the Park instead of in Idaho or Montana.

Records of occurrence.—Specimen examined, 1, as follows: **Yellowstone Nat'l Park:** *No specific locality, 1 USNM*.

Additional record.—**Park Co.:** About 50 mi. W Cody and S Shoshone River (N. A. Wood, 1921:234; see also *National Geographic*, 19:353, May 1908). **Yellowstone Nat'l Park(?):** According to Coues (1877:49), "Mr. C. H. Merriam. . . . procured a specimen [of *Gulo*] on the Yellowstone River, Wyoming, in August, 1872."

Taxidea taxus taxus (Schreber)

Badger

1778. *Ursus taxus* Schreber, Die Säugthiere . . ., 3:520.
1894. *Taxidea taxus*, Rhoads, Amer. Nat., 28:524, June.

Type.—From Labrador and Hudson Bay.

Description.—Large for Mustelidae (see *Measurements*); upper parts an intermixture of white, cream, blackish, and gray, with white stripe extending a variable distance from nose middorsally onto back; face marked with black; feet brownish black; venter grayish with mid-ventral stripe or blotch of white; claws long and powerful, especially on forelimbs; legs short; skull large; zygomata robust; molars robust; P4 with accessory cusp behind deuterocone; adapted for digging; feeds mainly on burrowing rodents.

Measurements.—External measurements of an adult male from the Platte River, Carbon County, and another from six miles south of Laramie are as follows: Total length, 742, 755; length of tail, 117, —; hind foot, 95, 115; ear from notch, 50, 60. The former weighed 15 pounds. Cranial measurements of these two males, of another from 12 miles north of Laramie, and another from Laramie, and of an adult female from 16 miles west of Laramie are, respectively, as follows: Condylobasal length, 126.9, 130.6, 131.5, 130.0, 127.4; zygomatic breadth, 82.0, 91.3, 84.6, 88.8, 80.3; interorbital breadth, 27.1, 30.7, 27.8, 26.7, 26.9; breadth across postorbital processes of frontals, 35.4, 41.1, 40.0, 37.7, 37.6; maxillary tooth-row, 42.0, 40.0, 42.7, 42.6, 41.4; cranial depth, 43.0, 44.4, 44.5, 44.2, 45.5; outer crown length of M1, 13.1, 12.8, 14.2, 13.8, 12.9; greatest breadth of M1, 9.6, 10.7, 9.6, 10.5, 9.9.

Distribution.—Mainly in lower life-zones, throughout most of state. See Fig. 67.

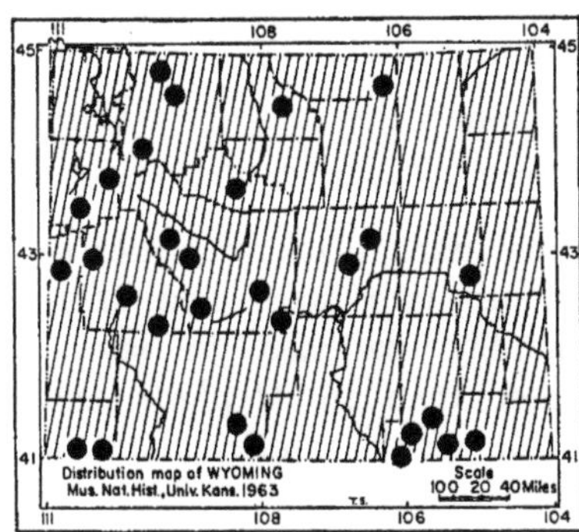

FIG. 67. Distribution of *Taxidea taxus taxus*.

Remarks.—All of the specimens that I have examined are referable to a single subspecies; they show great individual variation in color, color pattern, and size. I tentatively follow Hall (1936) and Hall and Kelson (1959:925-927) in assigning the aforementioned specimens to *Taxidea taxus taxus*. However, Schantz (1946:81, 1950:90) named two subspecies from South Dakota and Montana with type localities much nearer Wyoming than the type locality of the nominate subspecies. I have examined specimens from areas within the geographic ranges of each of the new subspecies, and in my opinion they and the Wyoming specimens are inseparable on account of greatly overlapping variation. The description for *T. t. montana* Schantz seems more appropriate for Wyoming specimens than does that of *T. t. dacotensis* Schantz, but the latter name was proposed first. The name *Taxidea taxus jeffersonii* Harlan (Fauna Americana, pp. 309-310, 1825), in my opinion, has priority over *T. t. montana* and *T. t. dacotensis* unless the type locality of *jeffersonii* is found to be within the range of *T. t. neglecta* of the Pacific Coast, in which case *neglecta* would become a synonym. The subspecies of *Taxidea taxus* are in need of revision.

Records of occurrence.—Specimens examined, 58, as follows: **Albany Co.:** *Lookout, 1 USNM;* 12 mi. N Laramie, 1; 16 mi. W Laramie, 7650 ft., 1; *Laramie, 1, 1 USNM; Laramie, 15 mi. SE Green Mtn., 8200 ft., 1 USNM; Laramie, 20 mi. SE Green Mtn., 8200 ft., 1 USNM; 4 mi. E Laramie, 7500 ft., 1 Univ. Wyo.; 16-17 mi. E Laramie, 8000 ft., 2 USNM; 6 mi. S Laramie, 1;* Pole Mtn., 16 mi. SE Laramie, 1 USNM; Jelm, 1 USNM; *Tie Siding, 1.* **Big Horn Co.:** Shell, 1 USNM; *Bighorn Basin, 1 USNM.* **Carbon Co.:** *Big Creek, 4 USNM; E. bank Platte River, 7700 ft., 1.* **Converse Co.:** 1 mi. S, 4 mi. W Shawnee, 4900 ft., 1. **Fremont Co.:** Lake Fork, 10,000 ft., 1 USNM; Sage Creek, Lenore, 1 USNM; 17 mi. S, 6½ mi. W Lander, 8450 ft., 1; 7 *mi. E Sweetwater Sta., 1; Sweetwater River, 2 USNM;* Sweetwater River, 25 mi. NW Splitrock, 1 USNM; *South Pass City, 1 USNM;* Crooks Gap., 7000 ft., 1; *Pacific, S S. Pass City, 2 USNM.* **Hot Springs Co.:** 2 mi. N, 6 mi. W Thermopolis, 1. **Laramie Co.:** Federal, 3 USNM. **Lincoln Co.:** 12 mi. N, 2 mi. W Afton, 6100 ft., 1; *Grover, near Afton, 1 Univ. Wyo.* **Natrona Co.:** 15 mi. SW Midwest, 1 Univ. Wyo.; 18 mi. W Casper, 1 USNM. **Park Co.:** 16¼ mi. N, 17 mi. W Cody, 2; 3 mi. S, 14 mi. W Cody, 1; Valley, 1 USNM. **Sheridan Co.:** Arvada, 1 USNM. **Sublette Co.:** Kendall, 1 USNM; New Fork, Green River, 1 USNM; *Big Sandy Creek, 1 USNM;* 41 mi. S, 16 mi. E Pinedale, 1. **Sweetwater Co.:** 2½ mi. S. Bitter Creek, 1; 45 mi. S Wamsutter, 2 USNM. **Teton Co.:** Warm Spring Creek, 1 USNM; *W. slope Uhl, 1;* Jackson, 1 USNM. **Uinta Co.:** 9 mi. S Robertson, 8000 ft., 1; Lonetree Creek, 1 USNM.

Spilogale putorius (Linnaeus)

Spotted Skunk

Smaller than *Mephitis mephitis;* upper parts black with four to six white stripes, frequently broken up into spots and short stripes; underparts black; skull notably flattened in lateral view; auditory bullae relatively larger than in *Mephitis* and upper molar less nearly square; skull much smaller than in *Mephitis;* dentition, i.$\frac{3}{3}$, c.$\frac{1}{1}$, p.$\frac{3}{3}$, m.$\frac{1}{2}$.

This small skunk is more agile than the striped skunk and frequently climbs among rock outcroppings and sometimes in trees. The spotted skunk preys mainly on insects and rodents. Two subspecies occur in Wyoming, but there is no evidence of direct intergradation between them. Indeed, the two kinds seemingly meet without intergrading at Iron Mountain where Mr. Elmer Smith, of Horse Creek, in the course of trapping furbearers in the winter of

1943, took an individual of S. *p. gracilis* and on another day an individual of S. *p. interrupta* in the same set. In the winter of 1944, he caught two additional individuals at Iron Mountain, both of which are S. *p. gracilis* (photographs of the fur-catches of 1943 and 1944 by Elmer Smith, on file in field notes of E. R. Hall in K. U. Mus. Nat. Hist.).

Spilogale putorius gracilis Merriam

1890. *Spilogale gracilis* Merriam, N. Amer. Fauna, 3:83, September 11.
1959. *Spilogale putorius gracilis,* Van Gelder, Bull. Amer. Mus. Nat. Hist., 117(5):279, June 15.
1890. *Spilogale saxatilis* Merriam, N. Amer. Fauna, 4:13, October 8. Type from Provo, Utah County, Utah.

Type.—From Grand Canyon of the Colorado River, north of San Francisco Mountain, 3500 feet, Arizona.

Comparison.—From *Spilogale putorius interrupta,* S. *p. gracilis* differs as follows: skull relatively broader and shorter (basilar length of adult male less than 50.1 mm.); lateral white stripe larger; tip of tail white instead of black.

Measurements.—External measurements of an adult male and adult female are listed by Merriam (1890:13) as follows: Total length, 450, 400; tail vertebrae, 176, 163; hind foot, 49, 41. Measurements of a subadult female from three miles south of Hudson are as follows: Total length, 351; tail, 138; hind foot, 40; ear from notch, 23; basilar length, 45.4; zygomatic breadth, 31.7; breadth upper carnassial, 18.0. Basilar length of an old male from Splitrock is 49.0 mm.

Distribution.—Lower life-zones in western two-thirds of state. See Fig. 68.

Records of occurrence.—Specimens examined, 4, as follows: **Carbon Co.:** Fort Steele, 1 USNM. **Fremont Co.:** 3 mi. S Hudson near Lander, 1 Univ. Wyo.; Splitrock, 2 USNM.

Additional record.—**Laramie Co.:** Iron Mtn. (Hall, field notes, see comments under account of *S. putorius*).

Spilogale putorius interrupta (Rafinesque)

1820. *Mephitis interrupta* Rafinesque, Ann. Nat., No. 1, p. 3.
1959. *Spilogale putorius interrupta,* Van Gelder, Bull. Amer. Mus. Nat. Hist., 117(5):273, June 15.
1902. *Spilogale tenuis* A. H. Howell, Proc. Biol. Soc. Washington, 15:241, December 16. Type from Arkins, Larimer County, Colorado.

Type.—From Upper Missouri River (see Lichtenstein, 1838:281).

Comparison.—For comparison with *Spilogale putorius gracilis,* see account of that subspecies.

Measurements.—External and cranial measurements of an old adult male from Arkins, Colorado, are as follows: Total length, 450; length of tail, 165; hind foot, 51; basilar length, 52; occipitonasal length, 52.3; zygomatic breadth, 34.7; least interorbital breadth, 14.3; cranial depth, 15.5 (after A. H. Howell, 1906:22, 36).

Distribution.—Lowlands of eastern Wyoming ranging northward into the Black Hills. See Fig. 68.

Remarks.—Only one specimen of this subspecies is known from the state. Cary listed additional occurrences (additional records, below) in his unpublished manuscript, "Mammals of Wyoming." In my possession is a copy of a letter from Mr. Stanley P. Young, U. S. Fish and Wildlife Service, to Prof. A. B. Mickey, dated January 27, 1950, to which are attached three copied sheets of the account on the spotted skunk written by Cary. In 1962, when I inquired about examining and possibly using Cary's manuscript, thought to contain many distributional data and early records concerning other species, it was found to be missing from its file with no record of its disposition.

Records of occurrence.—Specimen examined, 1 USNM, from: **Platte Co.**: Chugwater.

Additional records (Cary, unpublished, see *Remarks*, unless otherwise noted).—**Albany Co.**: base of Laramie Peak; *Horseshoe Creek, near Springhill, more than 6000 ft.*; North Laramie River, north of Garrett. **Johnson Co.**: On the Powder River, between Kaycee and Sussex (assumed to be *tenuis* on the word of J. W. Stevenson, Buffalo taxidermist). **Laramie Co.**: Iron Mtn. (Hall, field notes). **Weston Co.**: Near Newcastle (rare).

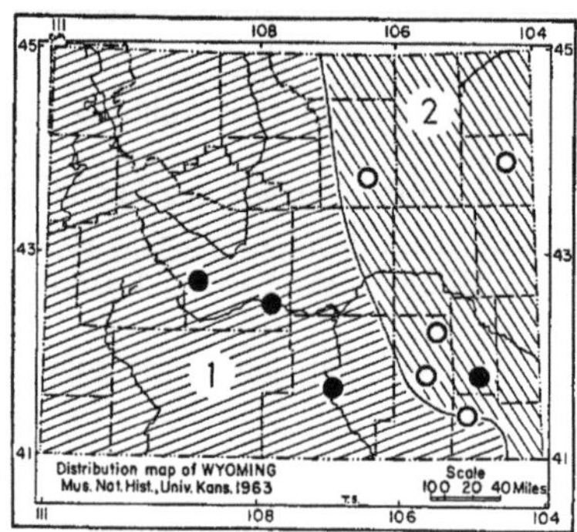

FIG. 68. *Spilogale putorius*.
1. *S. p. gracilis*
2. *S. p. interrupta*

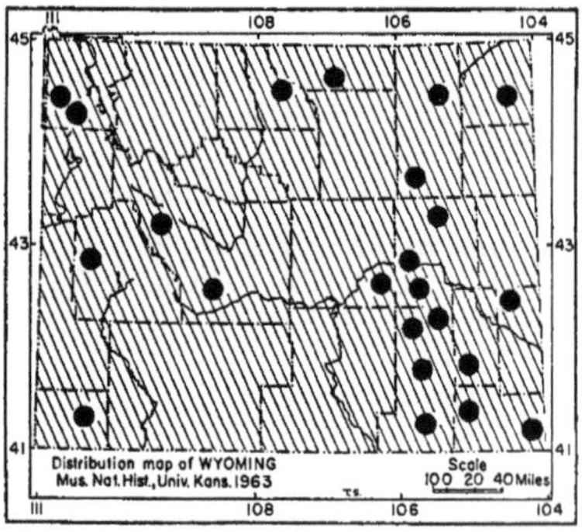

FIG. 69. Distribution of *Mephitis mephitis hudsonica*.

Mephitis mephitis hudsonica Richardson

Striped Skunk

1829. *Mephitis americana* var. *hudsonica* Richardson, Fauna Boreali-Americana, 1:55.

1934. *Mephitis mephitis hudsonica*, Hall, Univ. California Publs. Zool., 40:368, November 5.

Type.—From plains of the Saskatchewan, Canada.

Description.—Larger than *Spilogale* (see *Measurements*); upper parts black with narrow white line extending from nose mid-dorsally to level of pinnae of ears and with two dorsolateral white stripes extending from base of tail to neck where the stripes coalesce, expand laterad, and are truncated abruptly behind the level of the ears; tail black but strongly intermixed with blotches of white above and below; underparts black; skull much as in *Spilogale*, but auditory bullae smaller (relatively), dimensions longer, teeth more robust, upper molar nearly square in occlusal view with cusp arrangement almost petaloid; upper carnassial having sharp, elongate inner cusp; rostrum short dipping abruptly from arched or elevated frontals; lambdoidal and sagittal crests conspicuous in adult males; dental formula same as in *Spilogale*.

Measurements.—External measurements of a male and female from Laramie, of two males from Fort Bridger, and of a male and female from six miles north and four miles west of Bill (all adults) are, respectively, as follows: Total length, 640, 640, 670, 786, 704, 540; length of tail, 255, 260, 260, 248, 295, 215; hind foot, 75, 74, 81, 77, 80, 62; ear from notch, —, 29, 20, 34, 33, 24. Cranial measurements of five males and two females from within five miles of Laramie are, respectively, listed as follows: Condylobasal length, 76.0, 77.8, 77.3, 80.3, 73.5, 72.4, 69.7; zygomatic breadth, —, 47.3, 49.0, 54.3, 46.3, 46.6, 47.3; interorbital breadth, 19.6, 18.9, 18.9, 20.0, 17.8, 18.2, 19.0; maxillary tooth-row, 22.4, 22.3, 23.8, 24.3, 22.3, 22.4, 21.5; outer crown length of upper carnassial, —, 7.2, 7.2, 8.3, 7.8, 7.2, 7.5; breadth of upper molar, —, 8.9, 8.9, 9.4, 8.9, 8.6, 8.9.

Distribution.—Lower life-zones probably throughout state. See Fig. 69.

Remarks.—The striped skunk feeds mainly on insects and vegetable material in summer, and in winter mainly on mice and carrion.

Records of occurrence.—Specimens examined, 41, as follows: **Albany Co.:** Springhill, 6300 ft., 2 USNM; Eagle Peak, 1 USNM; Garrett, 1 USNM; 4.5 mi. N Laramie, 7275 ft., 3; *Laramie River, Laramie, 7150 ft., 1; 2 mi. W Laramie, 7200 ft., 1 Univ. Wyo.; 1 mi. W Laramie, 7100 ft., 1 Univ. Wyo.; Laramie, 7150 ft., 1, 1 USNM, 1 Univ. Wyo.; 4 mi. SW Laramie, 7175 ft., 3; JHD Ranch, sec. 3, T. 17N, R. 65W, 1 Univ. Wyo.*

Big Horn Co.: Bighorn Mtns., 1 USNM.

Campbell Co.: 13½ mi. N, 2½ mi. E Gillette, 1; 40½ mi. S, 17½ mi. W Gillette, 1.

Converse Co.: 6 mi. N, 4 mi. W Bill, 4800 ft., 2; Glenrock, 1 USNM; 17 mi. SW Douglas, 2.

Crook Co.: 3½ mi. NW Sundance, 5000 ft., 1.

Fremont Co.: Bull Lake Creek, 1 USNM; 10 mi. SE Lander, 1.

Goshen Co.: Rawhide Buttes, 5200 ft., 1 USNM.

Laramie Co.: 2 mi. W Horse Creek P. O., 6550 ft., 1; 2 mi. S Pine Bluffs, 5200 ft., 1.

Natrona Co.: Casper Mtns., 7 mi. S Casper, 6000 ft., 1 USNM.

Platte Co.: 15 mi. SW Wheatland, 5200 ft., 1 USNM.

Sheridan Co.: *Rona* (not found), 1 USNM; 4 mi. NNE Banner, 4100 ft., 1.

Sublette Co.: Kendall, 1 USNM.

Uinta Co.: Ft. Bridger, 3.

Yellowstone Nat'l Park: Lower Geyser Basin, 1 USNM; Shoshoni Lake, 1 USNM.

Lutra canadensis nexa Goldman

River Otter

1935. *Lutra canadensis nexa* Goldman, Proc. Biol. Soc. Washington, 48:182, November 15.

Type.—From near Deeth, Humboldt River, Elko County, Nevada.

Description.—Large for Mustelidae (see *Measurements*); upper parts brown; venter paler brown; neck thick; skull flattended; legs short and toes webbed; tail long, thick at base, and oval in cross-section; ears small, capable of being closed; opening of auditory meatus large; M1 rhomboid, robust; P4 having hollowed or dished deuterocone; tympanic bullae flattened; dentition, i.$\frac{3}{3}$, c.$\frac{1}{1}$, p.$\frac{4}{3}$, m.$\frac{1}{2}$; aquatic.

Measurements.—Hall (1946:196) lists external measurements for a subadult female of this subspecies from Bull Head Ranch, Nevada, as follows: Total length, 1,006; length of tail, 393; length of hind foot, 121. Females are slightly smaller than males.

Distribution.—The only specimen known to me from Wyoming is from Yellowstone National Park. Not mapped.

Remarks.—Most of Wyoming has been thought to be occupied by *Lutra canadensis canadensis* (see Hall and Kelson, 1959:Map 477) because Bailey (1930:142) referred a record to the nominate subspecies before *L. c. nexa* was named and before the river otter was nearly extirpated in the state. Only one specimen is available for examination, and it is a skin lacking the skull. Inasmuch as *L. c. nexa* occurs in Idaho in montane situations (even in nearby Teton County, Idaho, according to Davis, 1939:140), it probably occurred also in the mountainous western part of Wyoming. Thus, for geographic reasons the river otter of Wyoming is referred to *L. c. nexa.*

Jones (1962:94) exhaustively reviewed scientific and non-scientific literature of early observers in Nebraska and found that the otter was widely distributed in that state before the species was extirpated there. Swenk (1920:2) described from Nebraska a subspecies, *L. c. interior,* which if distinct or valid probably ranged across the Great Plains into eastern Wyoming. Swenk described *L. c. interior* as pale, large, and having a short hind foot and widely spaced teeth.

Durrant (1952:436) mentioned reports of otters in the Uinta Mountains, the foothills of which are in southwestern Wyoming; he saw, in 1949, an Indian from Uintah County wearing otter fur braided into his hair (the otter was said to have been taken along the Green River). Durrant suggested that *L. c. nexa* probably intergraded with *L. c. sonora* in the Uinta Mountains.

Records of occurrence.—Specimen examined, 1, as follows: **Yellowstone Nat'l Park:** *Upper Yellowstone River, 1 USNM.*

Additional records.—**Yellowstone Nat'l Park** (Skinner, 1927:199): *Gardiner River; Slough Creek; Willow Park; Obsidian Creek; Gibbon Canyon; Riverside Geyser; Lake Outlet; Wolf Point; Yellowstone Lake.* **Teton Co.:** *Vic. S. Fork Snake River, Teton Mountains* (Fryxell, 1943:84).

Family FELIDAE

KEY TO FELIDS

1. Tail more than 30 per cent of total length; 3 upper premolars present on each side . . . . **Felis concolor,** p. 705

1′. Tail less than 30 per cent of total length; 2 upper premolars present on each side 2
2. Tail tipped with black; tail less than ½ length of hind foot, **Lynx canadensis**, p. 707
2′. Tail tipped with black above but white below; tail more than ½ length of hind foot . . . . **Lynx rufus**, p. 708

Felis concolor Linnaeus

Mountain Lion

Large for genus *Felis* (larger than lynx); tail more than twice as long as hind foot; upper parts brownish or grayish (sometimes reddish); venter whitish often overlain with buff; ears blackish; chin and throat white; tail tipped with brownish or black; young spotted with black on buff ground color.

The mountain lion is nearly extirpated in Wyoming. The animal has a large home range and lacks the ability to discern the boundaries of areas in which it is protected. Unless protective measures are strictly maintained and increased the mountain lion will surely be extirpated, as it was in Yellowstone National Park.

The mountain lion feeds on deer and occasionally livestock. Many persons are pleased that the lion is now rare, and absent in many areas. Persons who realize the value of predators on big game and who favor the preservation of all species of wildlife have been unable so far to have this species protected or to have the areas enlarged in which it is protected. The mountain lion has nearly disappeared in Wyoming.

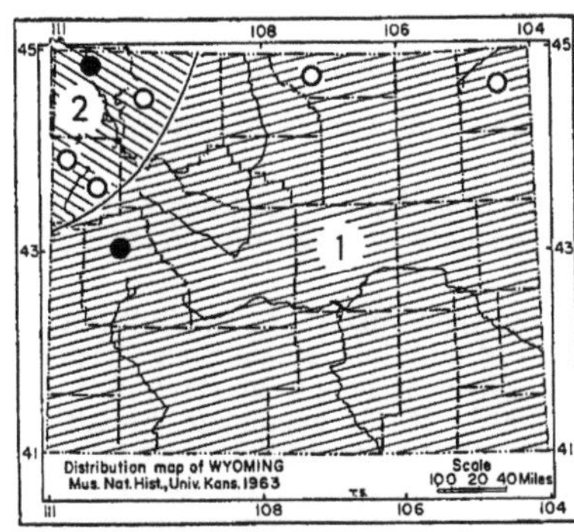

FIG. 70. Distribution of *Felis concolor*.
1. *F. c. hippolestes*
2. *F. c. missoulensis*

Young (in Young and Goldman, 1946:40) wrote that: "Nineteen of these predators have been removed from the stock and game ranges of the State during 25 years of cooperative state-federal predator control campaigns. Yellowstone National Park, where for years past it has been given full protection, probably contains the largest concentrations of puma at the present time, but up to the year 1930 Bailey (129-151) reported the animal 'few and far between' even there." In 1950 Prof. A. B. Mickey obtained the skull and leg bones of an adult male (KU 96904) labeled as taken on

February 3 of that year on "Squaw Creek [not found on any map] . . . E. of Saratoga, Carbon Co."

Felis concolor hippolestes Merriam

1897. *Felis hippolestes* Merriam, Proc. Biol. Soc. Washington, 11:219, July 15.

1929. *Felis concolor hippolestes*, Nelson and Goldman, Jour. Mamm., 10:347, November 11.

Type.—From near Cora, head of Wind River, Sublette County, Wyoming.

Comparison.—From *Felis concolor missoulensis*, *F. c. hippolestes* differs in longer skull, rostrum, palate, and diastema behind lower canine, and lesser zygomatic breadth. Probably largest of all subspecies of *F. concolor.*

Measurements.—Three adult males and three adult females from Meeker, Colorado, have, respectively, the following lengths: 2,438, 2,336, 2,286, 2,134, 2,058, 2,006. Of these six adults cranial measurements are given for two males and two females, which are listed, respectively, with those of the type: Greatest length of skull, 237, 213.2, 203, 199.2, type 227.4; condylobasal length, 210, 197.6, 184.4, 178.9, 204.4; zygomatic breadth, 162.5, 147.9, 133.7, 131.4, 160; cranial depth, 85, 79.5, 70.9, 68.5, 81.8; interorbital breadth, 46.2, 45.5, 38, 37.7, 48.5; postorbital processes of frontals, 83.2, 81, 69.3, 72.4, 83.5; "maxillary" tooth-row (to P4), 66.9, 62.8, 57.6, 58.5, 66.5; outer crown length of upper carnassial, 22.9, 23.9, 20.2, 21.6, 23 (Young and Goldman, 1946:210).

Distribution.—Occurred throughout most of state, except in extreme northwestern Wyoming. See Fig. 70.

Records of occurrence.—Specimens examined, 3, as follows: **Carbon Co.:** *"Squaw Creek"* [not found on any map] . . . *E. of Saratoga, 1.* **Sublette Co.:** Cora, 2 USNM.

Additional records (Young and Goldman, 1946:211).—**Crook Co.:** Hd. Bear Creek, Bear Lodge Mtns. **Sheridan Co.:** Wolf. **Sublette Co.:** *Type locality.* **Co. not certain:** *Green Valley.*

Felis concolor missoulensis Goldman

1943. *Felis concolor missoulensis* Goldman, Jour. Mamm., 24:229, June 7.

Type.—From Sleeman Creek, about 10 miles southwest of Missoula, Montana County, Montana.

Comparison.—See account of *Felis concolor hippolestes.* Specimens of this subspecies in Wyoming are intergrades between *Felis concolor hippolestes* and *missoulensis.*

Measurements.—Cranial measurements of the type (male), of an adult male from 12 miles east of Hamilton, Montana, and of an adult female from Glacier National Park are, respectively, as follows: Greatest length of skull, 221.8, 222.5, 189.4; condylobasal length, 203.5, 200.3, 169.6; zygomatic breadth, 164.3, 162.3, 135.3; cranial depth, 81.5, 79.3, 67.3; interorbital constriction, 45.4, 47.8, 36; postorbital processes of frontals, 78.7, 85.2, 67.4; "maxillary" tooth-row (to P4), 65.7, 65.3, 57.2; outer crown length of upper carnassial, 22.7, 23, 21.6 (Young and Goldman, 1946:207).

Distribution.—Northwestern part of state. See Fig. 70.

Records of occurrence.—Specimens examined, 4 USNM, from: **Yellowstone Nat'l Park:** Blacktail Creek.

Additional records (Young and Goldman, 1946:208).—**Park Co.:** Wapiti River; no certain locality. **Teton Co.:** Jackson Lake; Buffalo River, 20 mi. SE Jackson Lake. **Yellowstone Nat'l Park:** No certain locality.

Lynx canadensis canadensis Kerr

Lynx

1792. *Lynx canadensis* Kerr, The animal kingdom . . . , 1:157.

Type.—From Eastern Canada, probably Quebec.

Description.—Resembles *Lynx rufus* in size, but hind foot longer, ears more strongly tufted; upper parts Ochraceous-Tawny intermixed with brown, black, white, and sometimes reddish; eyelids white; pinna of ear buffy at base, having central white spot, and black tufts; skull larger than in *L. rufus*, except auditory bullae, which are smaller; presphenoid broad posteriorly (instead of narrow as in *L. rufus*); anterior condyloid foramen separated from foramen lacerum posterior; and tail tipped with black.

Measurements.—External measurements in inches listed by Halloran and Blanchard (1959:450-451) for an adult male (USNM 287768) from Hoback Canyon, Sublette County, are as follows: Total length, 38⅛; length of tail, 5; hind foot, 10¼; and ear, 3⅜. The weight of this specimen was 21¾ pounds. Cranial measurements of a skull found in 1949 at Cottonwood Lake, and now in the collection of the Museum of Natural History, are as follows: Occipito-nasal length, 128, zygomatic breadth, 94; interorbital breadth, 29.1; post-orbital breadth, 39.0; breadth across orbital processes of frontals, 57.5; length of nasals, 40.0; maxillary tooth-row, 42.2; greatest breadth presphenoid, 9.0; length of canine, crown-root 39.6, crown-alveolar along anterior surface, 21.4.

Distribution.—Formerly occurred at high elevations along Rocky Mountain chain, but now confined to high, inaccessible (to man) ranges of northwestern Wyoming, if not extirpated at the time of this writing. See Fig. 71.

Remarks.—Aside from the fact that the lynx is a rare, beautiful member of the timberline fauna and an inspiration to any naturalist lucky enough to see one, this cat should be protected because it rarely interferes with activities of man. The lynx seldom descends to elevations as low as 8000 feet, rarely leaves the forest, and feeds mainly on snowshoe rabbits. Occasionally the lynx preys on grouse, other birds, and small mammals, thereby rendering each or any of these animals a service by destroying sickly or poorly adapted stock. But the lynx is classed as a predator and destroyed whenever possible in Wyoming. Two paragraphs written by Halloran and Blanchard (1959:450-451) are as follows:

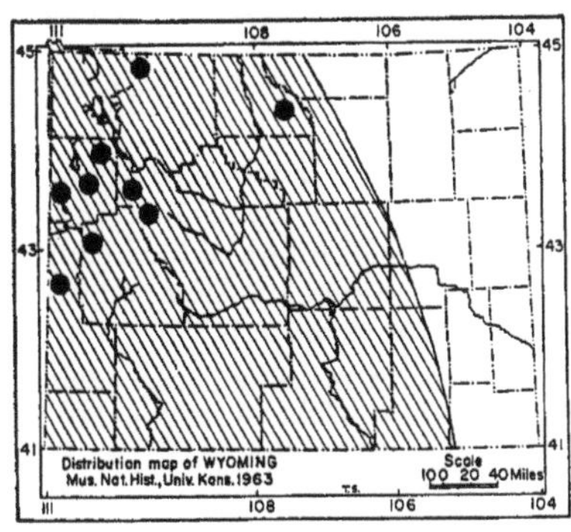

FIG. 71. *Lynx canadensis canadensis.*

"Thomas (Bull. 7, Wyo. Game & Fish Dept. 1954) stated, 'In Wyoming, the lynx is classified as a predator and may be trapped at any time with no report to the Wyoming Game and Fish Department . . . The animal is not abundant in any area of the state. . . .'

"Howard J. Martley, District Agent, writes (per. comm., 9/30/57): "There are not many true Canada Lynx in Wyoming and those that inhabit the state are found almost all together in the extreme western portions, in the high mountain country. From 1952 until 1955 we know of five positive Canada Lynx catches made by our predatory animal hunters. . . .'"

About all that can be said is that the lynx, as any other species of mammal, should be preserved; and that since the animal rarely does harm from man's point of view, it should be protected.

Records of occurrence.—Specimens examined, 14, as follows: **Big Horn Co.:** Shell, 1 USNM. **Fremont Co.:** Dubois, 1 USNM, Hd. Wind River, 1 USNM. **Lincoln Co.:** Cottonwood Lake, 6 mi. S, 6 mi. W Afton, 1. **Park Co.:** Painter, 1 USNM. **Sublette Co.:** *Hoback Canyon, Hoback River, 6800 ft., 1 USNM;* Kendall, 1 USNM. **Teton Co.:** Bridger Lake, 1 USNM; Elk, 4 USNM; Moose Creek, 15⅜ mi. NW Jackson, 1 USNM. **Yellowstone Nat'l Park:** No specific locality, 1 USNM.

Additional record.—**Albany Co.:** 8 mi. SE Laramie, Laramie Mtns., 1 specimen taken in December of 1963 (letter of March 11, 1964, from Larry N. Brown to Charles A. Long), not plotted on map.

Lynx rufus pallescens Merriam

Bobcat

1899. *Lynx fasciatus pallescens* Merriam, N. Amer. Fauna, 16:104, October 28.

1901. [*Lynx rufa*] *pallescens*, Elliot, Field Columb. Mus., Publ. 45, Zool. Series, 2:297, March 6.

1902. *Lynx uinta* Merriam. Proc. Biol. Soc. Washington, 15:71, March 22. Type from Bridger Pass, Carbon County, Wyoming. Regarded as inseparable from *pallescens* by Grinnell and Dixon, Univ. California Publs. Zool., 21:350, January 24, 1924.

Type.—From south side of Mount Adams, near Trout Lake, Skamania County, Washington.

Comparison.—Differs from *Lynx canadensis* as follows: smaller; tail tipped with black only above instead of also below; presphenoid bone usually narrower; auditory bullae larger; tail more (instead of less) than half length of hind foot; anterior condyloid foramen confluent with foramen lacerum posterior instead of separated. Occupies lower life-zones than does *Lynx canadensis.*

Measurements.—Cranial measurements of three adult males and one adult female from Manville are, respectively, as follows: Condylobasal length, 120, 120, 117, 118; zygomatic breadth, 94, 88, 91, 89; postorbital processes of frontals, 66.6, 59.0, 64.9, 60.5; length of nasals, 36.2, 33.5, 35.0, 31.5; alveolar length of maxillary tooth-row, 40.4, 40.2, 39.6, 38.8; labial alveolar length of carnassial, 14.2, 14.5, 14.4, 12.9. Measurements of three adult males from Big Creek, Albany County, are as follows: Condylobasal length, 118, 117, 117; zygomatic breadth, 91, 90, 90; postorbital processes, 60.0, 66.0, 62.4; length of nasals, 34.7, 36.7, 33.8; maxillary tooth-row, 38.5, 37.2, 38.9; length of carnassial, 13.6, 13.8, 13.6. Measurements of three adult males and three adult females from along the Sweetwater River, Fremont County, are, respectively, as follows: Condylobasal length, 125, 116, 120, 110, 109, 112; zygomatic breadth, 93, 89, 93, 84, 85, 86; breadth across postorbital processes, 64.0?, 63.5, 61.2, 57.8, 60.0, 61.2; length of nasals, 36.7, 33.0, 34.6, 32.1, 31.4, 30.6; maxillary tooth-row, 41.5, 38.1, 39.6, 37.1, 36.8, 36.3; length carnassial,

15.5, 14.0, 13.2, 13.5, 13.8, 13.4. Measurements of three adult males from Federal, Laramie County, are as follows: Condylobasal length, 112, 118, 116; zygomatic breadth, 84, 86, 91; postorbital processes, 57.5, 55.6, 61.6; length of nasals, 30.7, 31.5, 33.1; maxillary tooth-row, 37.9, 39.8, 37.7; length carnassial, 13.4, 14.2, 13.6.

Distribution.—Most of state, especially in valleys and basins. Said by Cahalane (1948:252) to have been extirpated in Yellowstone and Teton national parks (by 1947). I have examined no specimens from Yellowstone National Park. See Fig. 72.

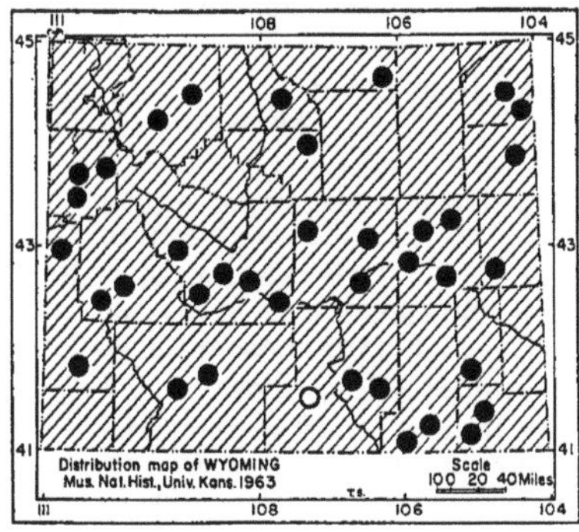

FIG. 72. *Lynx rufus pallescens.*

Remarks.—Many specimens from south-central and southeastern Wyoming are slightly more reddish than specimens from other parts of Wyoming. Nevertheless, all Wyoming bobcats clearly belong to the single subspecies *Lynx rufus pallescens* Merriam.

The bobcat remains abundant in Wyoming in spite of predator control measures.

Records of occurrence.—Specimens examined, 160, as follows: **Albany Co.:** *Rodgers Canyon, 1 Univ. Wyo.;* Laramie, 4 USNM; Jelm, 1 USNM.

Big Horn Co.: Shell, 2 USNM; *Shell Creek Basin, 1 USNM.*

Carbon Co.: 16 mi. N, 2 mi. W Saratoga, 1; 15 mi. S Medicine Bow, 1 Univ. Wyo.

Converse Co.: 3 mi. E Bill, 4700 ft., 3; 28 mi. NE Casper, 1; *Big Creek, near Douglas, 3 USNM;* Glenrock, 3 USNM; Douglas, 5 USNM.

Crook Co.: Sundance, 2 USNM; Black Hills, Sand Creek, 1 USNM.

Fremont Co.: *Lenore, 2 USNM;* Sage Creek, 1 USNM; Hailey, 1 USNM; Sweetwater River, 25 mi. NW Slitrock, 1 USNM; *Vic. Lander, 1;* Atlantic City, 7655 ft., 2; Sweetwater River, near Rongis, 2 USNM; *Sweetwater River, Splitrock, 2 USNM; Sweetwater River, 3 USNM; "Fremont County," 1.*

Laramie Co.: *Horse Creek, 3 mi. W Horse Creek* P. O., *16; Horse Creek, 6400 ft., 8; 3 mi. E Horse Creek* P. O., *25;* 5 mi. E Horse Creek P. O., 1; *9 mi. NW Federal, 1 USNM;* Federal, 7 USNM; *"Laramie County,"* 9.

Lincoln Co.: 18 mi. N, 9 mi. E Afton, 1; Kemmerer, 6927 ft., 1.

Natrona Co.: 8 mi. N, 2½ mi. W Waltman, 6050 ft., 1; 18 mi. N, 6 mi. W Casper, 1; 18 mi. SW Casper, 1.

Niobrara Co.: Manville, 8 USNM.

Park Co.: *Cody, 3 USNM; 3 mi. S, 14 mi. W Cody, 1;* 4 mi. S, 25 mi. W Cody, 3; Ishawooa Creek, 19 mi. S, 19 mi. W Cody, 3.

Platte Co.: Chugwater Creek, 20 mi. N Horse Creek, 13.

Sheridan Co.: Arvada, 1 USNM.

Sublette Co.: Green River, Jct. with New Fork, 1 USNM; Big Piney, 2 USNM.

Sweetwater Co.: Point of rocks, 6509 ft., 1; Rock Springs, 2 USNM.

Teton Co.: Elk, 1 USNM; 2 mi. N, 3 mi. W Summit, Togwotee Pass, 9000 ft., 1 Univ. Wyo.; Nat'l Elk Refuge, 1 USNM.

Washakie Co.: ½ mi. S, 1 mi. W Tensleep, 4300 ft., 1.

Weston Co.: Newcastle, 1 USNM.

Additional record.—**Carbon Co.:** Bridgers Pass (Merriam, 1902:71).

Order ARTIODACTYLA

KEY TO SPECIES OF ARTIODACTYLA

1. Horn sheaths have bony cores; horns present in both sexes (occasionally absent in females of **Antilocapra**) 2
2. Horns not branched or forked; lateral hooves present ...... BOVIDAE, p. 717
3. Tail longer than 150 4
4. Length of skull less than 350; length of maxillary tooth-row less than 120 .... **Oreamnos americanus**, p. 720

2′. Horns in males branched or forked; lateral hooves lacking, ANTILOCAPRIDAE, **Antilocapra americana**, p. 716

1′. Bony antlers present; antlers in males only .... CERVIDAE.

5. Antlers palmate; pendulous "bell" suspended from throat, **Alces alces**, p. 715

5′. Antlers not palmate; bell lacking 6

4′. Length of skull more than 350; length of maxillary tooth-row more than 120 .... **Bison bison**, p. 717

3′. Tail shorter than 150 .... **Ovis canadensis**, p. 720

6. Posterior narial cavity not completely divided by vomer; canines present above .... **Cervus canadensis**, p. 710

6′. Posterior narial cavity divided by vomer; canines lacking above 7

7. Antlers having one main beam on each side from which tines rise vertically; metatarsal gland less than 25 mm. long; tail brown above, fringed with white .... **Odocoileus virginianus**, p. 713

7′. Antlers branch dichotomously; metatarsal gland more than 25 mm. long; tail tipped with black above .... **Odocoileus hemionus**, p. 712

Family CERVIDAE

Cervus canadensis nelsoni V. Bailey

Wapiti or American Elk

1935. *Cervus canadensis nelsoni* V. Bailey, Proc. Biol. Soc. Washington, 38:188, November 15.

Type.—From Yellowstone National Park, Wyoming.

Description.—Size large for Artiodactyla (see *Measurements*); upper parts in summer Light Buffy Fawn to Creamy Buff; upper parts in winter near Tawny; rump patch Creamy Buff or whitish; head, neck, legs, and belly brownish; antlers (occur in males only) large with long points usually numbering up to nine; antlers deciduous; four hooves on each foot; upper canines short and tusklike; lower canines incisiform; cheek-teeth selenodont; posterior nasal cavity not completely divided by vomer; skull long, but shorter than in *Alces.*

Measurements.—Cranial measurements of the type (male) and an adult female topotype are listed by Bailey (1935:188) as follows: Basal length, 430, 410; length of nasals, 170, 170; upper molar series, 140, 140; mastoid breadth, 165, 150; zygomatic breadth, 200, 180; exorbital breadth, 210, 195; rostrum at canines, 90, 80; antlers (in type only) over beam, right 1260, left, 1250; spread of beams, 1000; of tips, 920. Cranial measurements of an adult female from 18 miles south and 25 miles west of Cody and of another from Jackson Hole, south of Moran, are, respectively, as follows: Greatest length of skull, 423, 436; zygomatic breadth, —, 181; length of nasals, 147, 163; length molariform tooth-row, 125, 134.

Distribution.—Formerly statewide; extirpated in many areas. See Fig. 73.

Remarks.—Many elk formerly migrated from the mountains of Wyoming into the Red Desert and other lowlands to winter, and a few still do. The elk is abundant in Jackson Hole and in Yellowstone Park. From this center of abundance many animals have been reintroduced into new areas in and out of Wyoming.

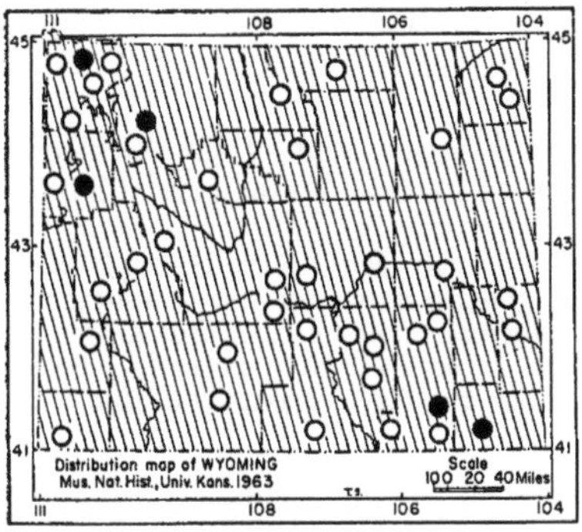

In pioneer times the subspecies of the east, *Cervus canadensis canadensis*, may have ranged into Wyoming from the east. But this subspecies was extirpated before it could be studied. If it occurred in Wyoming, it probably was never abundant.

One specimen (USNM 231542, foot only) has approximately 75 porcupine quills imbedded in the soft part of a foot about two inches above the dew-hooves.

FIG. 73. Distribution of *Cervus canadensis nelsoni.*

Records of occurrence.—Specimens examined, 11, as follows: **Albany Co.:** Cheyenne Pass, 1 USNM.

Laramie Co.: Ft. D. A. Russell, 1 USNM.

Park Co.: 18 mi. S. 25 mi. W Cody, 1.

Teton Co.: Hd. Gros Ventre River, 1 USNM; *Jackson Hole, S Moran, 1.*

Weston Co.: *1.5 mi. E Buckhorn, 6150 ft., 1* (perhaps not natural occurrence, skull sawed in half, found 1951).

Yellowstone Nat'l Park: Deer Creek, 1 USNM; *Meadows Creek, 1 USNM;* no certain locality, 2 USNM.

Co. uncertain: *"Northwestern Wyoming," 1 USNM.* An undetermined number of antlers are in the U. S. National Museum: from Mammoth Hot Springs, some from "Yellowstone Park," and some labeled merely "Wyoming."

Additional records (Murie, 1951:42-46, unless otherwise noted).—**Albany Co.:** Laramie Peak, Toltec; *Between Douglas and Rock Creek, in 1888;* Medicine Bow Mtns.; Pole Mtn.

Big Horn Co.: Bighorn Mtns.

Campbell Co.: Belle Fourche River, in 1894.

Carbon Co.: Ferris Mtns.; Seminole Mtns., in 1877; Shirley Mtns.; *Elk Mtns., in 1864 and 1888;* Medicine Bow River (Baird, 1858:643), in 1856; Sierra Madre Mtns.

Converse Co.: Douglas.

Crook Co.: *Black Hills, before 1890, reintroduced 1912-13;* Bear Lodge Mtns.; Sundance Mtn. in 1885; *Stockard Creek, in 1885.*

Fremont Co.: Wind River Mtns.; Granite Mtns.; Green Mtns.

Goshen Co.: Rawhide Butte; Ft. Laramie, in 1846.

Hot Springs Co.: Owl Creek Mtns.

Lincoln Co.: *Hoback Mtns. and Basin; Wyoming Range;* Near Fontenelle.

Natrona Co.: Casper, in 1886; Rattlesnake Mtns.

Park Co.: Absaroka Range.

Platte Co.: *North Platte River, in 1832.*

Sheridan Co.: Sheridan, in 1834.

Sublette Co.: Vic. Pinedale; Big Piney.

Sweetwater Co.: Red Desert; Bitter Creek, 1868.
Teton Co.: *Webb Canyon; Moose Creek;* Teton Mtns. (Coutant, 1899:105), 1812.
Uinta Co.: Foothills Uinta Mtns., 1886.
Washakie Co.: Tensleep, 1896.
Yellowstone Nat'l Park: Valleys of the following streams: *Lamar,* Yellowstone, *Gardiner,* Gallatin, Slough, and Snake (Bailey, 1930:43-49).

Odocoileus hemionus hemionus (Rafinesque)

Mule Deer

1817. *Cervus hemionus* Rafinesque, Amer. Monthly Mag., 1:436, October.
1898. *Odocoileus hemionus,* Merriam, Proc. Biol. Soc. Washington, 12:100, April 30.
1902. *Dama hemionus,* J. A. Allen, Bull. Amer. Mus. Nat. Hist., 16:20, February 1. *Dama* is the correct generic name for the white-tailed deer according to the rules of zoological nomenclature, but Opinion 581, International Commission on Zoological Nomenclature [Bull. Zool. Nomen., 17:267-275, September 6, 1960] recommends the use of *Odocoileus.*)

Type.—From mouth of the Big Sioux River, South Dakota (see Bailey, 1927:41).

Description.—Averaging larger than *Odocoileus virginianus;* tail white, tipped with black; upper parts reddish or yellowish tawny, in winter more grayish or brownish; throat, buttocks, and abdomen whitish; antlers forking dichotomously; hooves blackish; diastema present in lower jaw; lacking upper canines; lacrimal pits (in lacrimal bones) deep; nasals averaging narrower, and lower incisors averaging smaller than in *O. virginianus.*

Measurements.—Seton (1929:325) listed external measurements of a buck as follows: Total length, 1,673; length of tail, 177; hind foot, 495; ear from notch, 254. Cranial measurements of an adult male and an adult female from Weston County are as follows: Greatest length of skull, 295, 280; zygomatic breadth, 135, 122; length nasals, 91, 83; maxillary tooth-row, 70, 74; length diastema, 71.5, 72.

Distribution.—State-wide in suitable habitat. See Fig. 74.

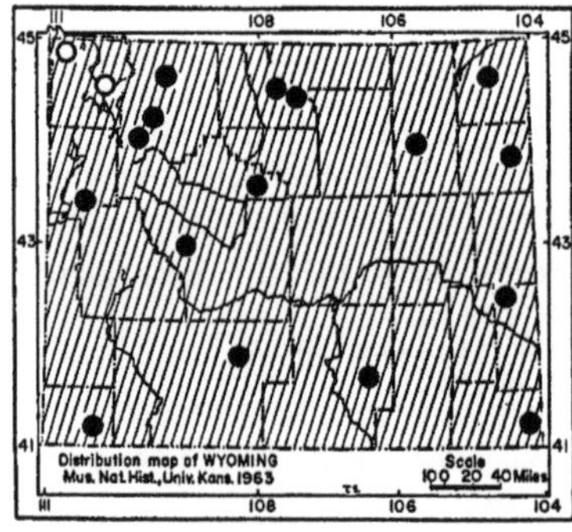

FIG. 74. Distribution of *Odocoileus hemionus hemionus.*

Remarks.—Usually found in forest areas, but often occurs in meadows or on the prairies of Transition Life-zone, and occasionally ranges into all life-zones.

Baird (1858:658) mentions a great abundance of mule deer along the Yellowstone River. Seton (1929:250) writes that in the fall of 1920 the number of deer killed in Wyoming was 16,166, and of these numbers Seton says they were chiefly mule deer. Jackson (1944:8) listed 61,022 mule deer as occurring in Wyoming, exceeded in number by nine states. This species still is abundant in the state.

Records of occurrence.—Specimens examined, 30, as follows: **Albany Co.:** *no specific locality, 1.* **Big Horn Co.:** 12 mi. N Shell, 7500 ft., 1; Hd. Trappers Creek, 8500 ft., 1 USNM. **Campbell Co.:** 36 mi. S, 20 mi. W Gillette, 1. **Carbon Co.:** Percy, 2 USNM. **Crook Co.:** Bear Lodge Mtns., 9524 ft., 3. **Fremont Co.:** Ft. Washakie, 2 USNM. **Goshen Co.:** Muskrat Canyon, 1 (antler, discarded). **Hot Springs Co.:** 9 mi. S, 3 mi. E Thermopolis, 4600 ft., 1. **Laramie Co.:** 1 mi. W Pine Bluffs, 1; *2 mi. S Pine Bluffs, 5200 ft., 1.* **Natrona Co.:** *½ mi. N, 5½ mi. W Powder River, 1.* **Park Co.:** 16¾ mi. N, 17 mi. W Cody, 5625 ft., 1; 25 mi. S, 28 mi. W Cody, 6350 ft., 1; Valley, 2 USNM; *Shoshoni Mtns., 2 USNM.* **Sweetwater Co.:** 5 mi. ESE Sand Dunes, 1 (antler, discarded). **Teton Co.:** Jackson, Hoback Canyon, 2 USNM. **Uinta Co.:** 4 mi. N, 2 mi. W Lonetree, 1. **Weston Co.:** 3 mi. S, 1 mi. E Newcastle, 1; *no specific locality, 3.*

Additional record.—**Yellowstone Nat'l Park** (Bailey, 1930:60-65): Mammoth Hot Springs, *Gardiner;* Yellowstone Lake.

Odocoileus virginianus (Zimmermann)

White-tailed Deer

Smallest cervid in Wyoming; upper parts grayish or chestnut (reddish tawny); underparts white; tail brownish above but whitish below; chin, throat, ring around eye, and band around muzzle whitish; antlers with long sub-basal snag; beam curving forward, with unbranched prongs or tines rising from beams; antlers not present in females; ears relatively shorter than in *Odocoileus hemionus;* metatarsal gland less than 25 mm. long; lacrimal pits shallow; incisors small.

Never so abundant as mule deer in Wyoming, white-tailed deer were nevertheless abundant in various places there in pioneer times. Since then the numbers have dwindled. Skinner (1929:107) attributes the "extermination" of this species in Yellowstone National Park to hunters along the boundaries and also to wolves, "accidents, cougars, coyotes, and emigration." According to Bailey (1930), more than 100 white-tailed deer were in the Park but were extirpated there not by competition with elk, not by predation by coyotes or wolves, not by poaching, but mainly by the hand of hunters hunting along the boundaries of the Park, to which places the deer "migrated." White-tailed deer still occur in fair numbers in Wyoming; and the number listed by Jackson (1944:7) is 3,000. Probably only one (*O. v. dacotensis*) of the two subspecies that formerly occurred in Wyoming is there today.

Odocoileus virginianus dacotensis Goldman and Kellogg

1940. *Odocoileus virginianus dacotensis* Goldman and Kellogg, Proc. Biol. Soc. Washington, 53:82, June 28.

Type.—From White Earth River, Mountrail County, North Dakota.

Comparison.—*O. v. dacotensis* differs from *Odocoileus virginianus ochroura* in larger size and heavier dentition.

Measurements.—Cranial measurements of a large adult male listed by Kellogg (*in* Taylor, 1956:40) are as follows: Total length, 1981; tail, 305; hind foot, 508; heigth at shoulder, 1067; condylobasal length of skull, 322.

Distribution.—Occurred on Great Plains (including Black Hills) and in southern Wyoming in suitable habitat. Usually found in Canadian or Transition life-zones. See Fig. 75.

Remarks.—Formerly abundant along the Upper Missouri and Upper Platte rivers, but never so abundant as mule deer. Extirpated in many places in Wyoming, but still occurs in the Black Hills.

Records of occurrence.—Specimens examined, 16, as follows: **Crook Co.:** Bear Lodge Mtns., 13 USNM; *Lost Canyon, Black Hills, 5000 ft., 1 Univ. Wyo.* **Weston Co.:** *no specific locality, 2.*

Additional records (Baird, 1858:652-653).—**Carbon Co.:** Medicine Bow Creek. **Counties not certain:** *Upper Missouri and Upper Platte Rivers.*

Odocoileus virginianus ochrourus V. Bailey

1932. *Odocoileus virginianus ochrouris* V. Bailey, Proc. Biol. Soc. Washington, 45:43, April 2.

Type.—From Coolin, South end of Priest Lake, Bonner County, Idaho.

Comparison.—For comparison with *Odocoileus virginianus dacotensis,* see account of that subspecies.

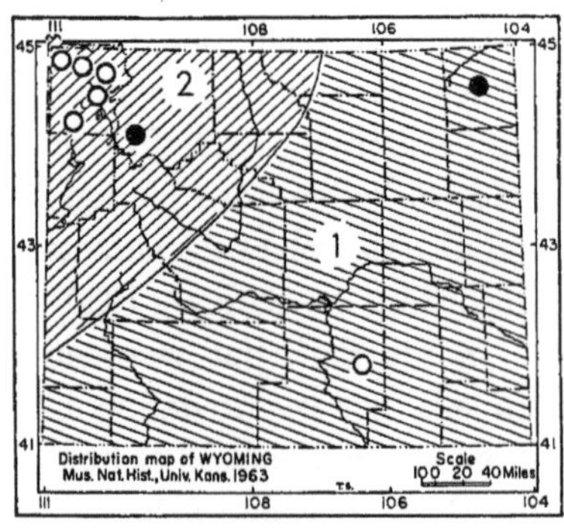

FIG. 75. *Odocoileus virginianus.*
1. *O. v. dacotensis*
2. *O. v. ochrourus*

Measurements. — Cranial measurements of an adult male from Valley are as follows: Length of nasals, 95; width between bases of antlers, 75; length first prong, 105; upper molariform tooth-row, 81; zygomatic breadth, 127.

Distribution. — Northwestern Wyoming. See Fig. 75.

Remarks. — This subspecies is thought to have been extirpated in Wyoming. Exactly how abundant this deer was when white man first visited the area is unknown.

Records of occurrence.—Specimens examined, 3, as follows: **Park Co.:** Valley, 3 USNM.

Additional records (Bailey, 1930:66).—**Yellowstone Nat'l Park:** Mammoth Hot Springs; Yanceys; Tower Falls; *Gardiner;* Yellowstone Lake; Soldier Sta.

Alces alces shirasi Nelson

Moose

1914. *Alces americanus shirasi* Nelson, Proc. Biol. Soc. Washington, 27:72, April 25.

1952. *Alces alces shirasi,* Peterson, Contrib. Royal Ontario Mus. Zool. and Palaeo., 34:23, October 15.

Type.—From Snake River, four miles south of Yellowstone Park, Teton County, Wyoming.

Description.—Large (weight more than 900 pounds); pelage dark brown except lower half of legs, muzzle, and pinnae of ears, which are paler; ears long; muzzle long; shoulders taller than rump; beard or "bell" suspended from chin or throat; antlers in males only; antlers palmate with short points radiating from antlers proper; skull and mandible long and slender; dentition, i.$\frac{0}{3}$, c.$\frac{0}{1}$, p.$\frac{3}{3}$, m.$\frac{3}{3}$.

Measurements.—External measurements of type listed by Nelson (1914: 73), are as follows: Total length, 2540; length of hind foot, 762; greatest length of front hoof, 130.

Distribution.—Occurred in Rocky Mountains in abundance but was almost extirpated; the animal was abundant in Yellowstone Park (its presence not suspected there) and upon its discovery (1908-1910) and under protection the moose has increased and now occupies much of its former range. Introduced into the Bighorn Mountains (Simon, 1951:3). See Fig. 76.

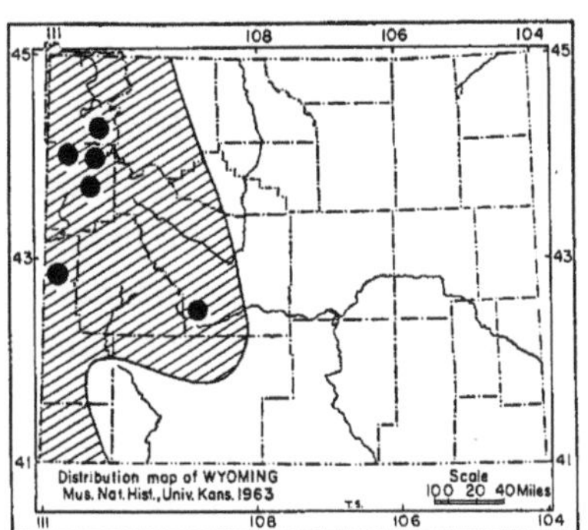

FIG. 76. Distribution of *Alces alces shirasi.*

Remarks.—Cahalane (1948: 256) stated that the moose was common in Teton National Forest but that "hunting in the surrounding region during 1945 and 1946 caused a noticeable reduction in the number of park moose" at that time.

Antlers are dropped by males in winter (November to March) and the antlers are of full size in the period August to September (Peterson, 1955:93).

Records of occurrence.—Specimens examined, 8, as follows: **Fremont Co.:** South Pass, 7550 ft., 1. **Lincoln Co.:** Moose Flat, Greys River, 6800 ft., 1. **Park Co.:** *Ft. Shonshone, near Cody, 1 USNM.* **Teton Co.:** type locality, 1 USNM; Pacific Creek (winter droppings); Elk, 2 USNM. **Yellowstone Nat'l Park:** Yellowstone River, S. E. Yellowstone Park, 1 USNM; *no specific locality, 1 USNM.*

Additional records (Peterson, 1952:24).—**Teton Co.:** *Teton Canyon; near Jackson.* **Yellowstone Nat'l Park:** *Bridge L. (= Bridger Lake?); Near Hawks Nest.*

Antilocapra americana americana (Ord)

Pronghorn

1815. *Antilope americana* Ord, *In* Guthrie, A new geogr., hist. and comm. grammar. . . ., Philadelphia, ed. 2, 2:292 (described on p. 308).

1818. *Antilocapra americana* Ord, Jour. Phys. Chim. Hist. Nat. et Arts, 87:149.

Type.—From plains and highlands of the Missouri River.

Description.—Small for Artiodactyla (see *Measurements*); upper parts tan or yellowish brown; mane on neck black; rump, underparts, and two bands across throat white; horns in both sexes, deciduous, composed of fused hairs; horns in males forked; bone-cores not forked; lateral hoofs absent; frontal sinuses opening to outside by two large fossae in dorsal surface of frontal bones; cheek-teeth hypsodont, selenodont, and rootless; p4 with closed anterior fossette; mandible slender and recurved dorsad (not so nearly straight as in deer); nasals long and slender with lateral margins nearly parallel; dentition, i. $\frac{0}{3}$, c. $\frac{0}{1}$, p. $\frac{3}{3}$, m. $\frac{3}{3}$.

Measurements.—Weights of many males exceed 100 pounds; females are approximately 10 per cent smaller. Cranial measurements of two adult males from "Weston County" and one adult female from 4½ miles east of Moneta are, respectively, as follows: Length of nasals, 101.1, 106.5, 105.1; breadth from tip to tip of horn-cores. —, 240, —; maxillary tooth-row, 69.6, 67.5, 75.6; length supraoccipital to posteriormost part of nasals, 115.1, 117.6, 116.0; length of lower diastema, 67.0, 76.2, 67.1.

Distribution.—Valleys, prairies, and arid lowlands of state. See Fig. 77.

Remarks.—The pronghorn in pioneer times was almost as abundant on the prairies as the buffalo (some persons claimed the pronghorn was more abundant). Activities of man (Nelson mentions occupancy of the pronghorn's territory, 1925:3) reduced the numbers from thirty to forty millions in the 19th century to approximately 30,000 in 1922-1924; of these 30,000 pronghorns in existence no less than 6,977 were counted in Wyoming, first in number of pronghorns among the states. However, Nelson (*op. cit.*:57) states that in 1885 about 30,000 pronghorns ranged along the Big Sandy River and vast numbers grazed on Wyoming's prairies. Today in Wyoming the pronghorn is more abundant than in 1922-1924.

Bryant (1885:132) writes that during July, 1846, the pronghorns along the Sweetwater River were compelled to shelter themselves from wolves by mingling with bison.

Records of occurrence.—Specimens examined, 43, as follows: **Albany Co.:** Sybylee, 1 USNM.
Campbell Co.: Belle Fourche Valley, 1 USNM; Pumpkin Buttes, 1 USNM.
Carbon Co.: Bridgers Pass, 1 USNM.
Crook Co.: Sundance, 5 USNM.
Fremont Co.: 4½ mi. E Moneta, 1; Continental Peak, 1 USNM.
Goshen Co.: Ft. Laramie, 1 USNM.
Lincoln Co.: 10 mi. N, 15 mi. E Kemmerer, 6600 ft., 1.
Natrona Co.: 19 mi. S, 9 mi. W Waltman, 1; 1 mi. NE Casper, 3; 1 mi.

N, 9 mi. W Independence Rock, 1; *Sweetwater Valley, 1 USNM; Sweetwater River, 2 USNM; 5 mi. W Independence Rock, 6000 ft., 1.*

Platte Co.: Warm Springs (Guernsey), 1 USNM.

Sweetwater Co.: *Red Desert, 2;* 35 mi. E Farson, 2 USNM.

Teton Co.: Jackson Hole, 1 USNM.

Weston Co.: 3 mi. S, 1 mi. E Newcastle, 1; *no specific locality, 2.*

Yellowstone Nat'l Park: *No specific locality, 5 USNM.*

Co. not certain: "Northwest Wyoming," 2 USNM; "Wyoming," 4 USNM.

Not found: *E Orange Butte, 1 USNM.*

Additional records.—*All counties except Washakie and Teton* (Nelson, 1925: Fig. 18) (not plotted).

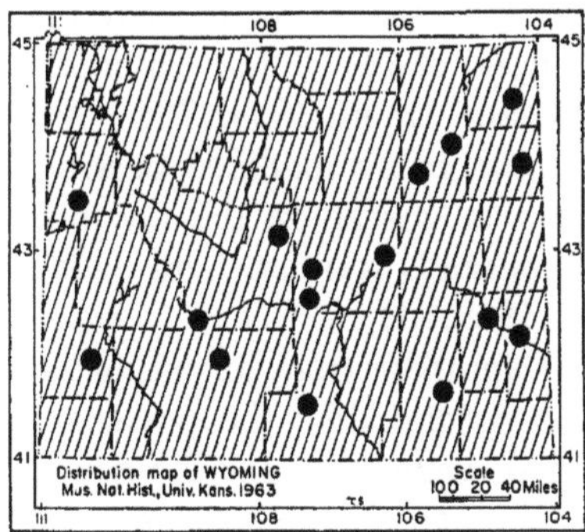

FIG. 77. Distribution of *Antilocapra americana americana.*

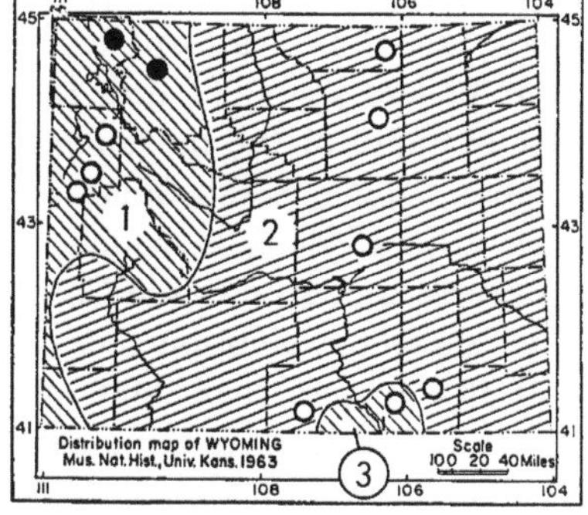

FIG. 78. Distribution of *Bison bison.*
1. *Bison bison athabascae.*
2. *Bison bison bison.*

Family BOVIDAE

Bison bison (Linnaeus)

Bison or Buffalo

Large, total length more than 2000 mm. and greatest length of skull more than 350; horns present on both sexes, smooth and conical with bony core; pelage shaggy on head, shoulders, and chin; pelage dark brown; forehead short and broad; nasals acuminate, not reaching premaxillae; molars with style between anterior and posterior lobes; dentition, i. $\frac{0}{3}$, c. $\frac{0}{1}$, p. $\frac{3}{3}$, m. $\frac{3}{3}$.

The status of the buffalo in Wyoming is complicated. Before white man entered Wyoming, the bison was amazingly abundant in the lowlands and present also in the high mountain ranges. The bison of the lowlands "absolutely blackened" the prairies a journey of a day or two from the north fork of the Platte River in 1832, according to Captain Bonneville (Garretson, 1938:59). Robert Stuart wrote of killing 32 bison on November 2, 1812, along Poison Spider Creek across from Casper Mountain (Coutant, 1899:112) and mentioned numerous herds of bison along the North Platte River. E. Willard Smith mentioned the killing of 100 buffalo on the Snake River at the mouth of the Muddy on October 10, 1839 (Barry, 1943:291).

While the lowland bison was being slaughtered in great numbers, herds of it seemingly retreated into the high mountains, formerly occupied by a large, dark subspecies (*Bison bison athabascae*). Probably the mountain subspecies was also greatly decimated by man. In any case, there is mention that the bison crossed the Rocky Mountains and entered the upper plains of the Columbia River "of late years" (Baird, 1858:681, and Davis, 1939:376). Another example of the movements of bison being modified possibly to avoid danger is the failure of bison to winter on the Laramie Plains after 1844-45, in which year thousands starved there owing to a severe snowstorm (see Garretson, 1938:69).

The slaughter of bison reduced the numbers from an estimated 60,000,000 to less than 1,000 and the species, like the passenger pigeon, would have become extinct had it not been for the action of William T. Hornaday and others who caused the species to be preserved. According to Bailey (1930:18), the bison in Wyoming was extirpated except for a small herd in Yellowstone Park. That herd, belonging to the mountain subspecies, often called the Woods Bison, was small when white man first entered the area, numbered 600 in 1880, reached a low of 25 in 1901, and gradually increased in number until 1927. According to Garretson (1938:207) the total number of buffalo in Yellowstone Park about 1903 was 29 including 21 of the subspecies of the lowlands (*Bison bison bison*) introduced from the Flathead Indian Reservation, Montana. Bailey (*op cit.*, p. 21) states that the introduction was in 1902 and included three bulls from northern Texas. In his discussion of the woods bison, thought to have occurred in the high mountain ranges of western United States, G. M. Allen (1942) does not mention its presence in Yellowstone Park. Roe (1951) has written a valuable account of the bison.

Bison bison athabascae Rhoads

1898. *Bison bison athabascae* Rhoads, Proc. Acad. Nat. Sci. Philadelphia, 49:498, January 18.

Type.—Fifty or fewer miles southwest of Fort Resolution, Mackenzie.

Comparison.—Compared with *Bison bison bison* of the plains, *B. b. athabascae* is darker, longer haired, larger and has longer, less curved horns. *B. b. athabascae* is broader across the frontals and the rim of its orbit is elongated anterolaterally.

Measurements.—Cranial measurements of an adult male (cranium only) from 22 miles west of Cody are as follows: supraoccipital-premaxillary length, 536; shortest length of horn core, 80; greatest length of horn core, 94; greatest diameter of horn core, 99.5; distance from base of core to mid-line of frontals, 147; greatest diameter of orbital foramen, 66.1; outer rim of orbit to posterior margin of nasals, 168; alveolar length of maxillary tooth-row, 148.

Distribution.—Formerly high mountain ranges in state, now extirpated. See Fig. 78.

Remarks.—Fryxell (1928:135-137) found skeletons of bison at nearly 12,000 feet elevation on the Snowy Range, Albany County; they presumably were *B. b. athabascae,* which occurred also in the nearby mountains of northern Colorado at high elevations. Fryxell also mentions skeletons from Jackson Hole and Yellowstone Park. The skeletons in the park were in some cases overlain by travertine, and were of animals that according to Fryxell undoubtedly occurred in the park before 1881.

Bison with record lengths of horns (elongate horns are characteristic of *athabascae*) are reported from Yellowstone Park (*Wyoming Wildlife,* December 1939); dates of collection were not recorded.

Records of occurrence.—Specimens examined, 2, as follows: **Park Co.:** 22 mi. W Cody, 6000 ft., 1. **Yellowstone Nat'l Park:** Lamar River, Lamar Valley, 1.

Additional records (Fryxell, 1928:129-139).—**Albany Co.:** Snowy Range, 9000-12,000 ft. **Teton Co.:** Buffalo River; Hoback River; Gros Ventre Range, E Jackson; *Jackson Hole.*

Bison bison bison (Linnaeus)

1758. *Bos bison* Linnaeus, Syst. nat., ed. 10, 1:72.
1888. *Bison bison,* Jordan, Manual of the vertebrate animals . . . , ed. 5, p. 337.

Type.—From "Quivira," central Kansas (see Hershkovitz, Proc. Biol. Soc. Washington, 70:32, June 28, 1957).

Comparison.—For comparison with *Bison bison athabascae,* see account of that subspecies.

Measurements.—External measurements (extremes) of adult males and females listed by Seton (1929:641-642) are, respectively, as follows: Height at shoulder, males 1673-1825, female(s) 1521; total length, 3042-3803, 2123; length of tail, 507-913, 381-508; hind foot, 584-661, 508. Seton lists weights for bulls as 1800-2200 pounds.

Distribution.—Known in pioneer times from the Great Plains, central lowlands, and Green River watershed in great numbers. Nearly extirpated before 1900. Introduced into Yellowstone Park about 1902-1903, where the species now occurs. See Fig. 78.

Records of occurrence.—Specimens examined, 13, as follows: **Yellowstone Nat'l Park:** Turbid Lake, 1 USNM; no specific locality, 12 USNM (dates of collection, 1909-1916).

Additional records.—**Albany Co.:** Laramie Plains, in 1844-1845 (Garretson, 1938:69). **Carbon Co.:** Snake River, mouth Muddy Creek (present-day town of Baggs), in 1839 (Barry, 1943:291). **Fremont Co.:** *W N. Fork Platte*

River, 1838 (Garretson, 1938:59). **Johnson Co.:** Powder River, 1838 (Haffen and Young, 1938:53). **Natrona Co.:** Poison Spider Creek, opposite Mt. Casper, 1812 (Coutant, 1899:112). **Sheridan Co.:** Powder River, 1838 (Haffen and Young, 1938:53). **Yellowstone Nat'l Park:** Lamar River, 1908, introduced (Bailey, 1930:21; Cahalane, 1948:256).

Oreamnos americanus missoulae J. A. Allen

Mountain Goat

1904. *Oreamnos montanus missoulae* J. A. Allen, Bull. Amer. Mus. Nat. Hist., 20:20, February 10.

1912. *Oreamnos americanus missoulae*, Hollister, Proc. Biol. Soc. Washington, 25:186, December 24.

Type.—From Missoula, Missoula County, Montana.

Description.—Medium in size for Artiodactyla (weight, 150-300 pounds); pelage whitish; beard present; hump on withers; tail short; horns present in both sexes, unbranched, conical, ridged at bases, diverging laterally, inclining posteriorly, spikelike and slenderer than in *Ovis;* skull slender; lacrimal pits absent; interdigital cleft deep; dentition, i. $\frac{0}{3}$, c. $\frac{0}{1}$, p. $\frac{3}{3}$, m. $\frac{3}{3}$.

Measurements.—Hall and Kelson (1959:1026-1027) list measurements as follows: Total length, 1521-1787; length of tail, 152-203; hind foot, 330-368 (all measurements are of males). Females are 10-20 per cent smaller than males in some dimensions.

Distribution.—Not native to Wyoming. Cahalane (1948:258) states that, "Goats may be expected to invade Yellowstone Park from a planting made about 1942 by the Montana Game Commission on the Beartooth Plateau outside of the northeast corner of the park." In a letter dated 19 April 1963, Mr. D. L. Pattie wrote that the mountain goat is now present in Wyoming and that he collected a male and female on the Montana-Wyoming state line, approximately 30 miles east of Yellowstone Park. Not mapped.

Remarks.—Introductions of many species of non-native animals have been made many times in the history of man. It is unusual when introductions are successful and species are established; whenever an introduction is successful, it is almost always at the expense of some native species.

Ovis canadensis Shaw

Mountain Sheep or Bighorn

Large for Artiodactyla, approaching 350 pounds in some rams; females smaller than males; upper parts varying from dark brownish to creamy white; underparts whitish, creamy; rump patch whitish; horns grayish brown or yellowish brown, notably large (circumference, 19 inches around base of one in a ram), and coiled in old adults; hooves on forefeet larger than on hind feet.

Two subspecies occurred in Wyoming, one, *O. c. canadensis,* in the Rocky Mountains and the other, *O. c. auduboni,* in the Black

Hills and on the Great Plains east of the Rocky Mountains. Honess and Frost (1942:3-4) cite opinions of several persons that the subspecies of eastern Wyoming is invalid because the animals were indistinguishable from those of the earlier named Rocky Mountain subspecies. I follow Cowan (1940:542-543) in recognizing the eastern subspecies, *Ovis canadensis auduboni,* because he revised the sheep of North America and critically examined specimens from the range of the subspecies in question. *O. c. auduboni* is, however, weakly differentiated according to Cowan.

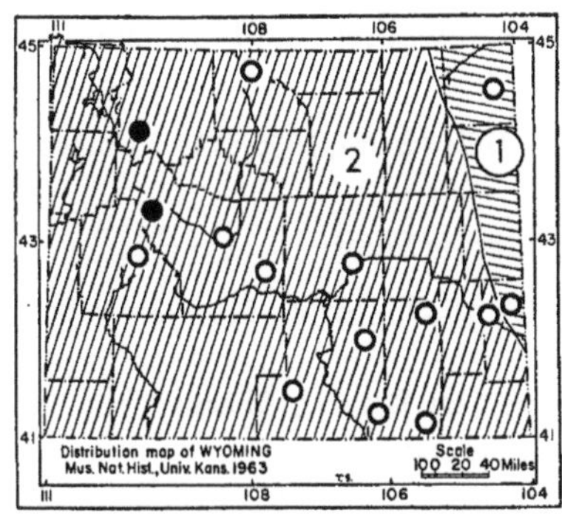

FIG. 79.
1. *Ovis canadensis auduboni.*
2. *Ovis canadensis canadensis.*

Ovis canadensis auduboni Merriam

1901. *Ovis canadensis auduboni* Merriam, Proc. Biol. Soc. Washington, 14: 31, April 5.

Type.—From Upper Missouri, probably badlands between Cheyenne and White rivers, South Dakota.

Comparison.—From *Ovis canadensis canadensis, O. c. auduboni* differed as follows: wider across nasals and maxillae; possibly wider across mastoids in females; basioccipital narrower and upper tooth-row possibly longer in males. Cowan (1940:543) did not find the depth of the lower mandible, from dental rim to lower margin, in *auduboni* to be deeper as Merriam suggested (1901:31).

Measurements.—Cranial measurements listed by Cowan (1940:544) of two adult males and one adult female and a young adult female, all from North Dakota, are respectively, as follows: Basilar length, 285, —, 261, 234; nasal length, 109, —, 101, —; nasal width, 49, —, 44, 48; orbital width, 116, 120, 115, 106; zygomatic breadth, 129, 134, 121, 120; maxillary breadth, 92, 98, 94, 91; mastoid breadth, 92, 89, 87, 80; palatal breadth M3, 59, 58, 56, 52; palatal breadth, P2, 34, 32, 35, 32; upper molar series, 94, 92, 90, 77; breadth of basioccipital, 28, 28, 26, 25.

Distribution.—Black Hills and Great Plains of eastern Wyoming. See Fig. 79.

Remarks.—Cowan (1940:543) states: "*O. c. auduboni* based as it is on slight cranial characters presented by a small number of specimens is to be regarded as a weak race. The indications are however that the characteristics exhibited by these individuals are not the result of chance and that sheep so characterized inhabited a definite range in the badlands to the east of the Rocky Mountains."

The subspecies is now extinct (Cowan, 1940:542). Introduction (Jackson, 1944:28) of the nearby and morphologically similar subspecies, *O. c. canadensis,* accounts for recent appearances of mountain sheep in the Black Hills, where I saw them in 1961, and in Nebraska (Jones, 1962:98).

Records of occurrence.—**Crook Co.:** Black Hills (Baird, 1858;679). **Goshen Co.:** Spoon Butte (Cook, 1931:171).

Ovis canadensis canadensis Shaw

1804. *Ovis canadensis* Shaw, Naturalists' Miscl., 51:pl. 610, December 1803?

Type.—From mountains along Bow River, near Exshaw, Alberta.

Comparison.—See account of *Ovis canadensis auduboni.*

Measurements.—External measurements listed by Cowan (1940:533) of an adult male from Wiggins Fork, Colorado [= Wyoming?], of another adult ram from Alberta, and of a five-year-old ram and an adult female from Glacier County, Montana, are, respectively, as follows: Total length, 1953, 1600, 1726, 1490; length of tail, 127, 100, 95, 80; hind foot, 394, 440, 482, 406. Cranial measurements of 12 males and nine females from northern Montana and Alberta are listed by Cowan (1940:516-517) as follows: Basilar length, 283 (270-294), 252 (242-262); nasal length, 117 (108-125) (11 specimens), 98 (89-110); nasal width, 52 (42-57) (11 specimens), 38 (35-42); zygomatic breadth, 134 (123-140), 120 (116-125); mastoid breadth, 96 (91-101), 79 (76-83); upper molar series, 86 (78-91), 85 (78-87); width of basioccipital, 32 (29-35), 26 (24-28); length of horn, 826 (642-1030), 273 (222-310).

Distribution.—Formerly known throughout state except in extreme eastern Wyoming and in arid lowlands (Upper Sonoran Life-zone). Now known from Yellowstone Park, Jackson Hole, and high mountain ranges of northwestern Wyoming. See Fig. 79.

Remarks.—The mountain sheep is adapted to live in high mountainous terrain, but in pioneer times, especially in winter, descended to lower life-zones, even into the Transition Life-zone. Early observers stated that the sheep frequently mingled with antelope and deer on the prairies.

Records of occurrence.—Specimens examined, 17, as follows: **Fremont Co.:** Crowheart, 6 USNM; *Wind River Mtns., 1 USNM.* **Park Co.:** 31 mi. S, 31 mi. W Cody, 1; *Carter Mtns., 1 USNM; Needle Mtn., 3 USNM; Shonshoni Mtns., 2 USNM.* **Sublette Co.:** *Hd. Roaring Fork, Green River watershed, 1 Univ. Wyo.* **Yellowstone Nat'l Park:** *No specific locality, 1 USNM.* **Co. not certain:** *"Wyoming," 1 USNM.*

Additional records (Honess and Frost, 1942:28-30, 62, unless otherwise noted).—**Albany Co.:** Laramie Peak; Medicine Bow Peak (Seton, 1929:539); Sherman. **Big Horn Co.:** Crystal Creek. **Carbon Co.:** *N U. P. Rail Road, S N. Platte, W Deer Creek* (Grinnell, 1928:1-9); *Percy* (Cowan, 1940:541); Medicine Bow Creek (Baird, 1858:679); Bridgers Pass (Baird, 1858:679); *Sheep Mtn., near Little Snake River* (Anonymous, 1940:64-69). **Fremont Co.:** *Wiggins Fork* (Seton, 1929:520); Near Riverton (Cowan, 1940:541); *Dinwoody Mtns.* (Cowan, 1940:541); *Eleanor Creek, Wind River Mtns.* (Seton, 1929:520); Granite Mtns. **Goshen Co.:** Ft. Laramie (Baird, 1858:679). **Natrona Co.:** 1 mi. below mouth Poison Spider Creek. **Platte Co.:** *Between*

Wendover and Guernsey. **Sublette Co.:** Fremont Lake. **Yellowstone Nat'l Park:** *Gardiner* (Cowan, 1940:541); *Eastern part of park* (Grinnell, 1928:1-9). **Co. not certain:** *North Platte; Goshen hole above Guernsey.*

DISCUSSION

Wyoming, characterized by varied topographic features and great differences in altitude from place to place, is an area where the life-zone concept of Merriam (1899) has merit. This concept is based on temperature; cold climate is at high elevations and polar latitudes and cold-adapted faunas and floras are found in both of these places. Warmth-adapted faunas and floras are found in lowlands and in lower (equatorial) latitudes. Kendeigh (1954:152-171) reviewed criticisms of the life-zone concept. Although some kinds of mammals are confined to high (cold) life-zones and others are confined to low (warm) life-zones, many kinds range through more than one life-zone. In some species, subspecies of one species occupy different life-zones. More than 60 kinds of mammals are known to occur in the Upper Sonoran Life-zone in Wyoming.

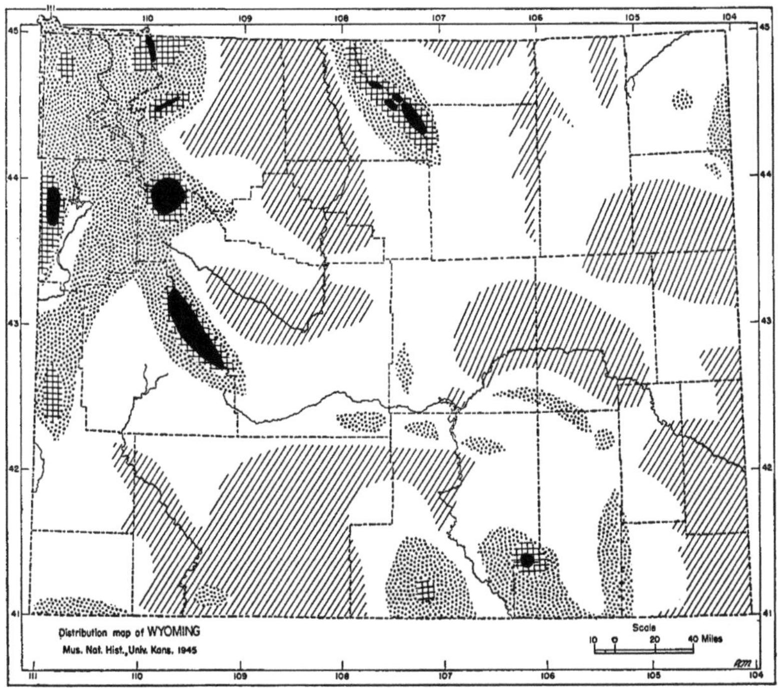

FIG. 80. Life-zones of Wyoming. Upper Sonoran— diagonal lines. Transition— No markings. Canadian— Stippled. Hudsonian— cross-hatched. Arctic-Alpine—solid black. Explanation in text; after Cary (1917).

Many of the 60 kinds occur also in at least the next two higher life-zones (Transition and Canadian). Furthermore, the Transition Life-zone, extensive in Wyoming, is made up of several diverse habitats. Desert covered by sagebrush, prairie covered by grasses, hills covered by yellow-pine and riparian situations supporting deciduous trees all are classified as Transition Life-zone. Thus, the concept of life-zones is complex. For determining the distribution (past and present) and the speciation of Wyoming mammals, the concept is useful only in a general way. Faunal areas based on patterns of distribution are more useful in this respect. The concept of life-zones is useful, nevertheless, in analyzing such patterns of distribution because one life-zone is frequently a barrier to dispersal of mammals occurring in another. For this reason, the life-zones as shown by Merrit Cary (1917) are reviewed in the light of accumulated additional evidence.

Varied climate in Wyoming, according to Cary (1917:12), results "from a difference in altitude within its borders of nearly 10,700 feet; and in a lesser degree to a difference in latitude of 4 degrees, and a wide range of local physiographic conditions." Five life-zones, lowermost to uppermost, are recognizable by certain plant indicators as follows (Cary, 1917:12-13): *Upper Sonoran*, broad-leaved cottonwood, juniper, saltbush, and yucca, occupying most valleys and low plains; *Transition*, yellow pine, narrow-leaved cottonwood, grasses, sagebrush, found on high plains, basal slopes of mountains, and most foothills; *Canadian*, spruce, fir, lodgepole pine, and aspen, in Boreal forests on mountain slopes; *Hudsonian*, white-barked pine, dwarfed spruce and fir in the timberline region; and *Arctic-Alpine*, small willows, grasses, sedges, on the mountain peaks above timberline.

As mentioned above, more than 60 kinds of mammals are known from the Upper Sonoran Life-zone. Four kinds of mammals confined to this zone are *Peromyscus crinitus* and *P. truei* from along the Green River in southwestern Wyoming and *Scalopus aquaticus* and *Spermophilus spilosoma* of southeastern Wyoming.

The Transition Life-zone covering about half of the area of the state (Cary, 1917:31) in most places is treeless and made up of the vast prairies and sage-covered plains. Few, if any, mammals are confined strictly to this zone; some kinds usually found in the zone include *Thomomys talpoides bridgeri*, *T. t. pygmaeus*, and *Spermophilus armatus* of southwestern Wyoming and *Lagurus curtatus* that occurs throughout much of Wyoming.

Of the many kinds of mammals in the Canadian Life-zone, many range into higher or lower zones. Some examples usually found in this zone are *Sorex vagrans, Lepus americanus, Tamiasciurus hudsonicus fremonti, Glaucomys, sabrinus, Clethrionomys gapperi, Microtus richardsoni,* and *M. montanus.*

Kinds frequently found in the Hudsonian Life-zone or at timberline are *Ochotona princeps* and *Ovis canadensis.* Numerous other mammals range into this and the higher Arctic-Alpine Life-zone from lower zones. No mammal is confined to the Arctic-Alpine Life-zone. Life-zones are shown in Fig. 80.

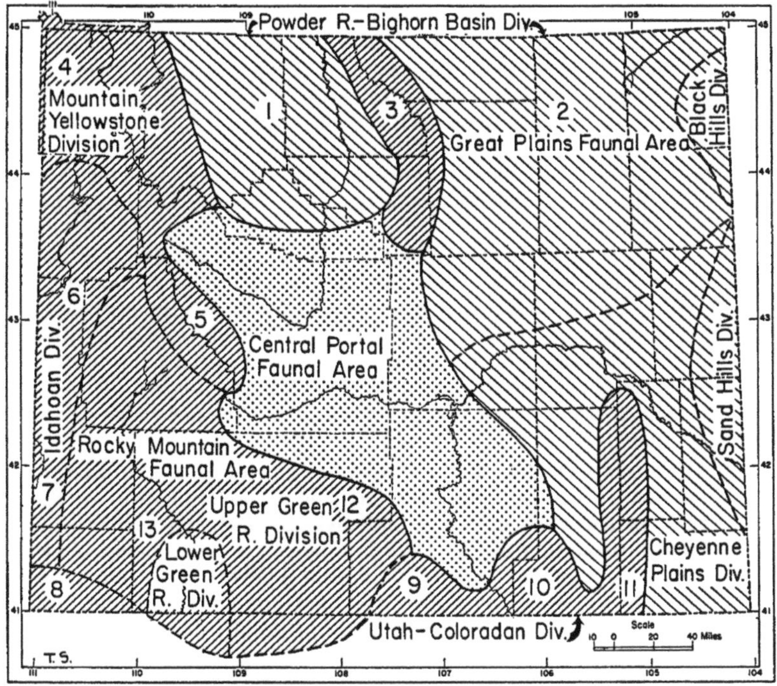

FIG. 81. Faunal Areas (names on map), Faunal Divisions (names on map), and Faunal Subdivisions (numerals on map refer to numerals below, in this legend). Description in text.

1. Bighorn Basin Faunal Subdivision
2. Powder River Faunal Subdivision
3. Bighorn Mountain Faunal Subdivision (part of Mountain-Yellowstone Plateau Faunal Division)
4. Yellowstone Plateau Faunal Subdivision
5. Wind River Mountains Faunal Subdivision
6. Snake River Faunal Subdivision
7. Bear River Faunal Subdivision
8. Uinta Mountains Faunal Subdivision
9. Sierra Madre Mountains Faunal Subdivision
10. Medicine Bow Mountains Faunal Subdivision
11. Laramie Mountains Faunal Subdivision
12. Red Desert Faunal Subdivision
13. Upper Green River Faunal Subdivision

Faunal Areas

The following arrangement of faunal areas is based on patterns of geographic distribution. See Fig. 81.

I. *Great Plains Faunal Area.*—*Microtus ochrogaster* and *Reithrodontomys megalotis* occupy this area. See beyond.

A. POWDER RIVER VALLEY-BIGHORN BASIN FAUNAL DIVISION.—Extensions of the Great Plains southward between mountainous areas of northern Wyoming comprise this zone. Mammals almost confined to this zone are *Thomomys talpoides bullatus* and *Dipodomys ordii terrosus*. They break through the low mountains rimming the Bighorn Basin on the south and occur on southward into southeastern Wyoming.

1. *Bighorn Basin Faunal Subdivision.*—This arid area merits mention because of the pallor its inhabitants often possess (populations of *Eutamias minimus pallidus*) and because of the endemic subspecies, *Microtus montanus codiensis*. Furthermore, in conjunction with the Wind River Basin, this arid basin isolates montane mammals of the Bighorn Mountains from those of the main Rocky Mountain chain. This faunal subdivision is in the Upper Sonoran Life-zone (Cary, 1917:23).

2. *Powder River Faunal Subdivision.*—This faunal subdivision belongs mainly to the Transition Life-zone of Cary (1917:31) and is inhabited by the mammals that live in both of the aforementioned subdivisions and by many prairie and even boreal mammals. *Microtus longicaudus*, ordinarily boreal, ranges along streams into this subdivision, as does *Sylvilagus nuttallii*, which is not boreal. This arid area isolates the forested Black Hills from other mountains of Wyoming.

B. CHEYENNE PLAINS FAUNAL DIVISION.—This division is in a sense a southern extension of the immediately preceding faunal division. Several mammals are common to both divisions (*Eutamias minimus pallidus*, *Spermophilus tridecemlineatus pallidus*, and *Lepus townsendii* are examples). In the Cheyenne Plains Faunal Division *Lepus californicus* and *Sylvilagus floridanus* enter the state, and are barred from ranging northward and westward probably not by physiographic features alone, but in part by *L. townsendii* and *L. nuttallii* respectively. Two subspecies of *Thomomys talpoides* differ from the one subspecies in the more northern Powder River-Bighorn Basin Faunal Division and vary clinally from the Nebraskan boundary westward into the Medicine Bow Mountains, then into the Sierra Madre Mountains farther to the westward, in increasingly dark color.

C. SANDHILLS FAUNAL DIVISION.—This division is based on the presence of mammals not found elsewhere in the state that live in the sandy soils extending westward from the Sand Hills of Nebraska. Examples are *Spermophilus spilosoma*, *Neotoma cinerea rupicola*, and *Mustela frenata longicauda*. Several mammals become slightly paler in this division (for example, individuals of *Spermophilus tridecemlineatus* and *Microtus ochrogaster*).

D. BLACK HILLS FAUNAL DIVISION.—This is delimited by the margin of the forested Black Hills. Several subspecies are endemic to this division (for example, *Eutamias minimus silvaticus*, *Spermophilus tridecemlineatus olivaceus*, *Mustela frenta alleni*, and perhaps, in Wyoming, *Sorex cinereus haydeni*). None of these kinds of mammals differs greatly from close relatives to the westward. Some montane mammals of the Black Hills belong to subspecies occurring westward (for example, *Glaucomys sabrinus* and *Microtus longicaudus*). *M. l. longicaudus* of the Black Hills was previously thought of as not belonging to a western subspecies and was thought to be characterized by dark upper parts. One specimen is dark and possesses long ears. A dark individual of *Spermophilus tridecemlineatus* is also known. Probably evolution toward dark subspecies such as *Eutamias minimus silvaticus* is in progress.

II. ***Rocky Mountain Faunal Area.***—This is made up of the Rocky Mountains, the Bighorn Mountains, and the basins and plateaus west of the Continental Divide. Four subspecies of *Spermophilus lateralis* occur in this area, but none is in the Bighorn Mountains. *Sorex cinereus cinereus* and *Sorex palustris* occur throughout most of the area but are absent from basins and other lowlands that are highly arid.

E. MOUNTAIN-YELLOWSTONE PLATEAU FAUNAL DIVISION.—Many montane mammals occur throughout this mountainous division. Some examples are *Lepus americanus*, *Marmota flaviventris nosophora*, and *Zapus princeps*.

3. *Bighorn Mountain Faunal Subdivision.*—Several montane mammals have differentiated only slightly from those found in the mountains of the main Rocky Mountain chain; some others have differentiated more and warrant subspecific recognition. Examples of the latter are *Ochotona princeps obscura*, *Lepus americanus seclusus*, *Eutamias minimus confinis*, *Thomomys talpoides caryi*, and *Microtus montanus zygomaticus*. All of these are endemic to the Bighorn Mountains.

4. *Yellowstone Plateau Faunal Subdivision.*—Some subspecies and species seem limited in southward distribution at the south margin of the Yellowstone Plateau, although some of these range farther southward on the Teton Range. Examples are *Spermophilus lateralis cinerascens*, *Martes pennanti*, and *Gulo gulo*. *Eutamias amoenus* and *Peromyscus maniculatus artemisiae* also occupy this subdivision but are absent from the Wind River Mountains.

5. *Wind River Mountains Faunal Subdivision.*—Most montane mammals in this faunal subdivision closely resemble those in mountains farther northward. A population of *Spermophilus lateralis lateralis* seems to be a relict population of the southern part of the Wind River Mountains, separated from larger populations of the same subspecies to the southward by the range of a pale subspecies, *Spermophilus lateralis wortmani*. *S. l. castanurus* also occurs in these mountains intergrading with *S. l. lateralis*. Some mammals found in mountain ranges northward and westward are absent from this subdivision, namely *Eutamias amoenus*, *Gulo gulo*, and *Martes pennanti*.

F. IDAHOAN FAUNAL DIVISION.—Many mammals occupying this faunal division belong to taxa occurring mainly in Idaho. One example is *Lepus townsendii townsendii.*

6. *Snake River Faunal Subdivision.*—Several subspecies that occur mainly in this subdivision are *Sorex vagrans vagrans*, *Peromyscus maniculatus artemisiae*, and *Neotoma cinerea cinerea.*
7. *Bear River Faunal Subdivision.*—This subdivision is an arid extension of the Snake River Faunal Subdivision; aridity limits the extension southward of some northern montane mammals. Some montane mammals of the Uinta Mountains to the southward occur no farther northward than the southern boundary of this low arid subdivision. *Lepus townsendii townsendii*, *Thomomys talpoides pygmaeus*, and *Onychomys leucogaster brevicaudus* occur mainly in this subdivision, and *Bassariscus astutus* is known only from there.

G. COLORADAN FAUNAL DIVISION.—This includes the mountains along the southern boundary of Wyoming. *Spermophilus lateralis lateralis* occurs throughout the division. *Tamiasciurus hudsonicus fremonti* occurs throughout most of the division except the easternmost mountain range, the Laramie Mountains.

8. *Uinta Mountains Faunal Subdivision.*—Several mammals are endemic to the Uinta Mountains; not all of them occur in Wyoming; one kind that occurs in Wyoming is *Eutamias umbrinus umbrinus.*
9. *Sierra Madre Mountains Faunal Subdivision.*—Only two kinds of mammals are endemic to this subdivision; these are the weakly differentiated subspecies, *Thomomys talpoides meritus* and *Ochotona princeps figginsi.* The fauna of these mountains greatly resembles that of the Medicine Bow Mountains to the eastward.
10. *Medicine Bow Mountains Faunal Subdivision.*—Mammals of this faunal subdivision show affinities with those of the Coloradan Rockies to the southward. Examples of subspecies that occur in these mountains and farther south are *Marmota flaviventris luteolus*, *Tamiasciurus hudsonicus fremonti*, and *Spermophilus lateralis lateralis.* *Thomomys talpoides rostralis* and *Ochotona princeps saxatilis* do not occur in the nearby Sierra Madre Mountains to the west, and *T. h. fremonti* does not occur in the nearby Laramie Mountains to the east.
11. *Laramie Mountains Faunal Subdivision.*—The Laramie Mountains extend northward farther than do the Medicine Bow Mountains and are lower in elevation. Pikas are unknown from the Laramie Mountains possibly because of their lower elevation. These mountains are connected with the Medicine Bow Mountains by high ground along the southern border of the state and chiefly for this reason possess a mammalian fauna resembling that of the Medicine Bow Mountains. The absence of pikas is one striking difference, and the presence of a different subspecies of red squirrel, *Tamiasciurus hudsonicus baileyi*, which ranges northward through small isolated ranges (for example, the Casper Mountains), then through the Bighorn Mountains, is another. Aridity may account for the unexpected occurrence of this subspecies of red squirrel in

these mountains. Being lower and more arid than the Medicine Bow Mountains occupied by the dark *T. h. fremonti*, the Laramie Mountains may favor the existence there of a paler subspecies.

H. LOWER GREEN RIVER FAUNAL DIVISION.—Where the Green River leaves the state the land is arid and barren. Entering the state at this low point are several mammals not known elsewhere in Wyoming. These are *Eutamias dorsalis*, *Peromyscus crinitus*, and *Peromyscus truei*.

I. UPPER GREEN RIVER FAUNAL DIVISION.—Arid basins and plateaus are the chief physiographic features of this faunal division. Kinds of mammals endemic to it are usually paler than their closest relatives but otherwise little different. Examples of such mammals are *Thomomys talpoides ocius*, *Perognathus fasciatus callistus*, and *Neotoma cinerea cinnamomea*. *Spermophilus richardsonii* becomes paler here than it is to the eastward and also slightly smaller, but the differences are slight and clinal.

12. *Red Desert Faunal Subdivision.*—One distinctly colored (pale) subspecies is endemic to this subdivision, namely *Spermophilus lateralis wortmani*.

13. *Upper Green River Faunal Subdivision.*—This faunal subdivision is little different from the Red Desert subdivision. *Eutamias minimus minimus*, adapted for life in sage-covered areas, and *Spermophilus tridecemlineatus parvus*, which is a small reddish squirrel, range throughout this subdivision; they range through the Red Desert Faunal Subdivision into the Wind River Basin of the Central Portal Area (see below). *Dipodomys ordii priscus* has a similar distribution.

III. ***Central Portal Faunal Area.***—This area is the lowlands mainly of the Wind River Basin, to a lesser extent of the northern margin of the Red Desert along the continental divide, and of prairies of the watersheds of central Wyoming, where extensive intergradation occurs between or among subspecies of prairie mammals. For example, *Thomomys talpoides bullatus* intergrades with *T. t. clusius* in this area. The geographic range of *T. t. clusius*, in the main, lies in this area. The low relief has permitted many lowland mammals to cross the barrier of the Rocky Mountain chain, except in glacial periods. These lowlands are, however, a barrier of great effectiveness isolating the boreal mammals of northwestern Wyoming from those of south-central Wyoming.

Speciation, Subspeciation, and Zoogeography

During the height of Wisconsin glaciation, the highlands of Wyoming were covered by glaciers and permanent snow fields. It is reasonable to assume that some of the cold-adapted species occurring today in the Arctic regions of the Far North occurred in periglacial areas of Wyoming in the Pleistocene. Some species that occur in the Arctic and in Wyoming today are as follows: *Sorex cinereus; S. vagrans; Erethizon dorsatum; Canis latrans; C. lupus;*

Vulpes vulpes; Ursus americanus; U. arctos; Mustela erminea; M. vison; Gulo gulo; Lutra canadensis; Lynx canadensis; and *Alces alces.* Even these species would have been affected by glaciation; they could not, of course, have existed on the extensive glaciers and snow fields. Most of the kinds of mammals now in Wyoming immigrated into the state after the height of glaciation.

North versus South

The mountains of northwestern Wyoming lack continuity with the mountains of south-central Wyoming. The discontinuity is caused by lowlands referred to in this paper as the Central Portal Faunal Area and the Upper and Lower Green River Faunal divisions. Considering species, these mountainous areas (northern and southern) show marked affinities in their respective faunas; each of several species has one subspecies in the northern mountains and another subspecies in the southern mountains. These differentiated pairs are montane kinds, and indicate that a boreal bridge or no arid barrier was between these mountainous areas. The distribution of *Spermophilus lateralis lateralis* supports this theory (there is a relict population of this subspecies in the Wind River Mountains). Three examples of differentiated north-south pairs are two subspecies of *Martes americana,* two of *Marmota flaviventris,* and two of *Tamiasciurus hudsonicus.* The presence of this arid barrier and its effects were mentioned by Findley and Anderson (1956:80-82). They noticed also affinities between montane mammals of south-central Wyoming and mammals of the Uinta Mountains (see distributions of *Tamiasciurus hudsonicus fremonti* and *Spermophilus lateralis lateralis*). The effects of the Green River barrier between those two mountainous areas are discussed beyond, but I concur with Findley and Anderson that the effectiveness of that river as a barrier is slightly less than is the barrier termed by me the Central Portal Area, which is effective against only boreal mammals.

A pollen profile (Hansen, 1951) from this low area south of the Wind River Mountains (Eden Valley) provides an indication of "postglacial" climate that is consistent with the hypothesis outlined above. Grasses replaced conifers only to be succeeded by composites and other plants of the Transition Life-zone. Probably when conifers were abundant, boreal mammals had continuous distributions from the Wind River Mountains into the mountains of southern Wyoming. Since then this low area has become an arid barrier isolating boreal mammals in northwestern Wyoming from those in southern Wyoming.

East versus West

There are, in my opinion, two reasons for the high frequency (at least 18 pairs) of one taxon in western Wyoming and another closely related taxon in eastern Wyoming. First, eastern Wyoming is mainly prairie; the mountain ranges there are separated from the Rocky Mountain chain by grassland barriers of lower elevation. Western Wyoming is mountainous, and its lowlands are not so effective in barring dispersion of mammals from one mountain range to another. The pattern may, therefore, be caused by a kind of mammal adapted for life on the Great Plains and a closely related montane kind. Possibly, the occurrence of cold, moist periods (evidenced by glaciations) drove warmth-adapted kinds away from the mountain chain, but some cold-adapted forms managed to survive nearer it and being isolated by climate from warmth-adapted kinds differentiated further. During warm periods the mountains were a refuge for boreal forms, as is true today. Several examples of pairs of mammals showing mountain *versus* prairie modes of life are as follows: *Sorex cinereus haydeni, S. c. cinereus; Canis latrans lestes, C. l. latrans;* and *Bison bison athabascae, B. b. bison.* Often the kind found on the prairie is pale (for example, *Microtus pennsylvanicus insperatus*), whereas the montane form is dark (*M. p. pullatus*). Of *Eutamias minimus,* a large, pale prairie subspecies now lives on the Great Plains, while the heavily forested Black Hills support a large, dark subspecies. Another large, dark subspecies lives on the forested slopes of the Bighorn Mountains. Presumably these three subspecies evolved *in situ.*

Second, assuming that Wyoming had few, if any, present-day kinds during the height of Pleistocene glaciation, certain mammals adapted for a warmer climate could approach the Rocky Mountain chain during interglacial warm periods. These kinds could occupy Wyoming by invading it from the east, or west, or from both directions. But when cold climate (evidenced by recurring glaciation) again occurred, the kinds probably were again driven away from the Rocky Mountain Chain. Glaciation would separate what had been a continuously distributed population into east and west parts. Intermittent warm periods and successively warmer glacial periods would permit more and more recontact between the populations. Now some kinds react to each other as good species and others interbreed freely. Two closely related ground squirrels that act as species are *Spermophilus armatus* and *S. richadsonii.* The geographic ranges of these kinds overlap in western Wyoming. The fossil record of S. *richardsonii* shows that this species occurred farther eastward and southward in the Pleistocene, as far south as Kansas (Hibbard, 1937:233-237).

Two closely related species the geographic ranges of which are not known to overlap seem to be ecologically isolated. These species are *Sylvilagus nuttallii* and S. *floridanus*. They have been taken within five miles of one another in eastern Wyoming.

The specific distinctness of the aforementioned examples and of other such pairs strongly suggests that isolation, not semiisolation, of these species occurred in the past.

The extremely variable subspecies *Peromyscus maniculatus nebrascensis* occurs over much of Wyoming but in the west is replaced by P. *m. artemisiae*, a darker subspecies occurring in the Teton Mountains and in Jackson Hole. The variability of *P. m. nebrascensis* and the fact that it occupies many habitats throughout most of Wyoming suggest that the glaciated Rocky Mountains isolated *nebrascensis* from the other subspecies, which invaded Wyoming from the west after glacial ice receded in western Wyoming.

Cameron (1958:83) mentioned that the Appalachian Mountains acted as a barrier between microtines in the northeastern United States permitting subspecific differentiation. The much higher Rocky Mountains would seem to be an effective barrier between lowland populations of mammals, especially during glacial advances. Therefore, I attribute much of the speciation and subspeciation that has occurred in Wyoming to the isolating properties of the glaciated Rocky Mountain chain.

Other Effects of Glaciation

During the dying phases of glaciation it would seem that even in the lowland of western Wyoming some refuges would have existed that maintained warmth-adapted kinds. Some examples of subspeciation on a small scale would be expected. *Microtus montanus codiensis* on the western side of the Bighorn Basin may be one; it is in contact with *M. m. nanus* in the same mountains. The population now differentiated as *codiensis* probably was isolated in the Bighorn Basin long enough to be differentiated.

Lepus americanus seclusus probably followed rising life-zones into the Bighorn Mountains and the warming lowlands severed the connection of *seclusus* with its parent stock. (See p. 736 for explanation.)

Furthermore, there are migration routes known to have been followed by bison and elk in historic times from the mountains of western Wyoming into the lowlands south of the Wind River Mountains in the Red Desert Area, again suggesting that not too long ago this low area served as a refuge for kinds of mammals. The former existence of forests in the treeless lowlands of Wyoming that now act as barriers to dispersion of montane mammals is suggested by the fact that faunas of well isolated mountain ranges show strong affinities with faunas of the Rocky Mountain chain. Some kinds of montane mammals in the Black Hills are indistinguishable from kinds occurring in the Rocky Mountains. One example is the flying squirrel, *Glaucomys sabrinus.* Some kinds in the Black Hills, and more in the Bighorn Mountains, strongly resemble the Rocky Mountain kinds, occasionally being referable to the same subspecies. One example is *Microtus longicaudus longicaudus.* On the other hand, many kinds of mammals have differentiated into more or less distinct subspecies in the Black Hills.

Gloger's Rule

Gloger's Rule refers to a condition of great importance and high frequency in the mammals of Wyoming. The rule states that, in mammals and birds (and we now know also in insects), races that inhabit warm, humid regions have more melanin pigmentation than races of the same species in cooler, drier regions; arid regions are characterized by faunas showing yellow and reddish-brown "phaeomelanin" (see Dobzhansky, 1951:152) pigmentation. In Wyoming many examples demonstrate this rule. These examples can be discussed from several points of view.

Northwestern Wyoming is mountainous, and owing to much precipitation is heavily forested. Several montane mammals reach southern distributional limits at the margins of this mountainous area. *Microtus richardsoni* is one; it is notably dark among voles. *Microtus pennsylvanicus pullatus* is another example, also notably dark.

Considering the closely related east-west pairs of mammals mentioned above, most are characterized by dark, montane, western kinds and pale, prairie-dwelling, eastern kinds. One example is the mink, *Mustela vison,* which has a dark western subspecies and a pale eastern one.

Among rabbits, the species that is the most dependant on riparian habitat is the darkest. Compare *Sylvilagus floridanus* (dark), S. *nuttallii* (intermediate), and S. *audubonii* (pale).

The Black Hills, well isolated by surrounding grasslands, have a fauna that is a mixture of prairie and montane kinds. Many of these show more or less differentiation as kinds; in most cases the subspecies confined to the Black Hills are dark subspecies. Two striking examples are *Tamiasciurus hudsonicus dakotensis* and *Eutamias minimus silvaticus*. On the plains surrounding the Black Hills, Stebler (1939) found a positive correlation between color of pelage of some mammals and color of exposed (by lack of vegetation) soils.

In fact, the species *T. hudsonicus* and *E. minimus* provide excellent measures of Gloger's Rule. Dark *T. hudsonicus ventorum* and *T. h. fremonti* occur in the humid mountains of nothwestern Wyoming and in the Medicine Bow Mountains of southern Wyoming, respectively. The Bighorn Mountains (somewhat more arid than the northwestern mountains) and the Laramie Mountains (a less humid spur of the Medicine Bow Mountains) and the semi-arid mountain ranges of central Wyoming are all occupied by a pale subspecies, *T. h. baileyi*. The humid Black Hills have a large, dark reddish, endemic subspecies. Of course, no red squirrels occur in the arid grasslands.

Eutamias minimus pallidus (of the Great Plains) is pale; *E. m. minimus* (Green River Watershed and Central lowlands, inhabiting sage-covered desert) is pale reddish or pale buff; four montane subspecies of *minimus* are dark, and all four are confined to boreal habitats.

Dice (1941, 1942) and Blair (1953) described variation in color in *Peromyscus maniculatus*, and showed that montane mice usually average darker than mice on arid areas. The latter mice are variable but frequently buffy-ochraceous. This variation is in accord with Gloger's Rule.

Some specimens of *Microtus ochrogaster* and *Spermophilus tridecemlineatus* are exceptionally pale in the arid, extreme eastern part of Wyoming adjacent to the Sand Hills of Nebraska. Two of the warmest and driest faunal divisions (Upper Green River Faunal Zone and Sand Hills Zone) have pale, endemic subspecies, exemplified respectively by *Spermophilus lateralis wortmani* and *Neotoma cinerea rupicola*.

There are numerous other kinds of mammals in Wyoming that demonstrate Gloger's Rule. The Wyoming mammalian fauna seemingly does not demonstrate significant support for Bergman's Rule or Allen's Rule, probably owing to a maximum difference in latitude of only four degrees.

Rivers

The Snake, Yellowstone, Missouri, Bighorn, Powder, North Platte, and Green-Colorado rivers are essentially headwater streams in Wyoming. This means, of course, that the rivers are youthful, characterized by downcutting and by V-shaped, narrow valleys. Such streams have little effect as barriers to dispersal of montane or lowland mammals.

Thus owing to a high base-level and to youthful streams, the latter have had no appreciable influence on the speciation of mammals in Wyoming. This is in contrast to the marked influence on speciation and subspeciation exerted by the Green-Colorado River in Utah (Durrant, 1952:514-515). In Utah, even so, the Green-Colorado River is less effective as a barrier upstream than downstream (in southern Utah).

In the lowlands, away from the mountain-sources of streams, some of the rivers have cut fairly broad valleys. The low valleys, which are now arid, act as barriers to dispersal of some (especially boreal) mammals. Thus the boreal mammals of the Bighorn Mountains are isolated by surrounding basins drained by rivers that carried away sediments of the basins. The Black Hills are isolated in part by the Belle Fourche and other rivers. The valley of the Green River helps to isolate montane mammals in the northwestern mountains from those in the mountains of south-central Wyoming. Arid river-valleys, it can be seen, act as barriers to dispersal of boreal mammals between isolated mountain masses.

Although not important as barriers that hasten speciation, rivers in Wyoming are important as routes of immigration of numerous mammals (of the Transition and Upper Sonoran life-zones) in the fauna of Wyoming. Some mammals that probably have followed the riparian vegetation of streams and rivers into Wyoming are *Scalopus aquaticus, Sylvilagus floridanus, Thomomys talpoides bridgeri, Peromyscus crinitus, P. truei* and *P. leucopus.*

The Process of Evolution

Sumner (1930, 1932) demonstrated genetic differences between subspecies of mammals. Such differences, of course, are necessary to the hypotheses outlined beyond. Evolution of subspecies and perhaps of a few species since the ice ages results from mutations of genes, perhaps of chromosomes. Aside from Clark's (1938) work showing that the bright buff-ochraceous color of the upper parts that occurs in deer mice is a gene-character dominant over the recessive dark, dull color frequently found in montane mice, little is known of the genetics of mammals of Wyoming. In these deer mice, the expression of these two genes (the phenotypes) certainly substantiates the idea (see Huxley, 1955) that populations respond to different environments with varying gene-frequencies. Many natural populations do not now show polymorphism distinctly (Dobzhansky, 1951:74). In them, variability, present and stored among the individuals of the population as numerous recessives, is held in check by the factors that control the evolution of the population, namely agents of natural selection.

If one accepts the idea that the fixation of a particular kind of character, or at least a high frequency (approaching unity) of this character, will occur only in an isolated population, the importance of the barriers discussed earlier cannot be overemphasized. Natural selection or perhaps genetic drift caused the differentiation of these isolated populations. Genetic drift is thought of as having been effective only in small populations that were driven into refugia in the basins or valleys in periods of cold climate or in small populations driven up the mountain slopes when the climate became warm. Dobzhanksy (1951:76-82 and 156-157) has discussed selection and genetic drift.

Inasmuch as there is little reliable evidence in Wyoming that can be used for dating glacial or post-glacial events, determination of rates of evolution or differentiation is difficult. Nevertheless, it is possible to hypothesize relative rates of evolution in some kinds of mammals, and in others the approximate rates can be deduced.

The east-west pairs discussed previously, which were isolated by the glaciated Rocky Mountain system, show a high degree of differentiation. In fact, several distinct species evolved, namely *Sylvilagus nuttallii* and *S. floridanus*, *Cynomys leucurus* and *C. ludovicianus*, and *Spermophilus armatus* and *S. richardsonii*. Also, two subspecies of *Spilogale putorius* seemingly do not intergrade in Wyoming. Marked cranial differences are seen between species or subspecies belonging to many east-west pairs. Judging from the degree of differentiation of these pairs and assuming that glaciation isolated the members of each pair from one another, the kinds must be at least as old as late Wisconsin, probably older than most subspecies discussed beyond.

A relict population of the snowshoe rabbit (*Lepus americanus seclusus*) occurring in the Bighorn Mountains of north-central Wyoming resembles *Lepus americanus americanus* that occurs far to the northward (no nearer than North Dakota) more than *Lepus a. bairdii* of the Rocky Mountains.

Probably the ancestral stock of *L. a. americanus* retreated northward as the ice of the Valders readvance (continental glaciation occurring approximately 11,000 years ago) receded, and a relict population (now *seclusus*) remained in the vicinity of the Bighorn Mountains isolated from *L. a. americanus* by warming temperatures. The degree of differentiation (described in the account of *L. a. seclusus*) between *L. a. americanus* and *L. a. seclusus* probably was attained in 11,000 years or less. Work on radioactive travertine (discussed previously) indicates that glacial ice covered areas in Yellowstone Park approximately at the time of the Valders re-advance.

Probably the pale *Spermophilus lateralis wortmani,* endemic to the Red Desert, descended from the darker *S. l. lateralis,* the geographic range of which nearly surrounds that of *S. l. wortmani.* A pollen profile (Hansen, 1951) shows that in post-glacial time the area occupied by *wortmani* became dryer and warmer. Increasing aridity would seemingly favor the evolution of a pale subspecies (in accordance with Gloger's Rule), and the age of *wortmani* can, therefore, be deduced as no older than post-glacial time and probably younger. Findley and Anderson (1956) suggested that the north-south pairs of mammals, discussed previously, probably also differentiated since the end of the Pleistocene.

The Black Hills were never glaciated. The isolation of the kinds of mammals confined to the Black Hills by the Powder River Subdivision (or Transition Life-zone) might be correlated with the appearance of grasses in the pollen profile in the Eden Valley of southwestern Wyoming (Hansen, 1951), but the absence of significant differentiation in several boreal mammals (for example, *Sorex vagrans, Glaucomys sabrinus* and *Microtus longicaudus*) indicates that the fauna of the Black Hills was more recently isolated. *Clethrionomys gapperi brevicaudus,* which is endemic to the Black Hills, differs from *C. g. galei* of the Rocky Mountains in no less than seven characters as follows: shorter tail; longer hind foot; notably longer nasals; greater zygomatic and lambdoidal breadths; greater basal and condylobasal lengths. In fact, there is some doubt that *C. g. brevicaudus* should be regarded as a subspecies of *C. gapperi;* formerly *brevicaudus* was considered to be a distinct species. It seems that the two subspecies (*brevicaudus* and *galei*) are members of an east-west pair that were isolated no later than in late Wisconsin time (late Pleistocene).

Habitat in the Black Hills is seemingly suitable for *Sorex palustris, Microtus montanus,* and *Lepus americanus,* but these species have not been found there. Their absence indicates an enduring and effective barrier between the Black Hills and the Rocky Mountains. Lack of subspecific differentiation in the Black Hills in *Microtus longicaudus* and *Glaucomys sabrinus* could indicate that those two species reached the Black Hills relatively recently. Marked differentiation of some other subspecies in these hills indicates that these subspecies were long isolated there. But none of the kinds in the Black Hills, in line with the thesis outlined above, would be older than late Pleistocene.

Differentiation

Spermophilus armatus and *S. richardsonii* are closely related species as also are *Sylvilagus floridanus* and *S. nuttallii*. Each of these pairs of species seems to have evolved only slightly beyond the subspecific level. Some subspecies differ much in size, color of pelage, and cranial characters. For example, *Tamiasciurus hudsonicus fremonti* differs markedly from *T. h. baileyi*. In one area the two are separated by the narrow Laramie Plains, which are uninhabited by *Tamiasciurus*. Two other well-differentiated subspecies are *Eutamias minimus pallidus* and *E. m. silvaticus*. They differ mainly in color, but the difference is remarkably constant.

In appraising the mammals in the Black Hills it is quickly ascertained that the well-differentiated subspecies there are dark, probably owing to high humidity in this environment. Probably most of the other mammals there are actively evolving in color and are becoming darker. Dichromatism in the deer mice there was described by Dice (1942) and Blair (1953). Therefore, I looked for dichromatism, dark and pale phenotypes, in other species there. The holotype of *Microtus longicaudus longicaudus* of the Black Hills is extremely dark, but most individuals are indistinguishable from those in the Rocky Mountains. A dark individual (not melanistic) of *Spermophilus tridecemlineatus* was seen, and there are numerous individuals of both pale and moderately dark squirrels of this species in these hills.

In the arid Upper Green River Faunal Division of southwestern Wyoming where an endemic, pale golden-mantled squirrel occurs, a dichromatic woodrat (*Neotoma cinerea cinnamomea*) also occurs; it is probably evolving into a pale population. The dichromatism in this woodrat is not extreme, but shows both pale and moderately dark ochraceous phenotypes. In arid, eastern Wyoming the deer mouse is dichromatic and has a high frequency of pale phenotypes. Thus, there is evidence, with regard to pale and dark pelage, supporting the idea that differentiation occurs owing to an increased frequency of adaptive phenotypes under pressures of natural selection.

SOME EFFECTS OF MAN IN HISTORIC TIME

The effects of man in historic time are and were drastic on the native fauna of Wyoming. When white man first entered the state, big-game mammals, fur-bearing mammals, and smaller mammals were abundant. Fur being in demand, the fur-bearers became the object of exploitation. Rich harvests of furs were reaped with no regard to the future until the fur-bearers became rare. After several explorations into Wyoming resulted in the discovery of a pass through the Rocky Mountains, a stream of people passed through the state, and some settled there. The abundant big game was utilized as food by them. The coming of the railroad hastened the extirpation of the bison, thousands of which were wastefully slaughtered. The wolves and later the coyotes were poisoned by persons attempting to improve conditions for cattle and sheep on the vast grazing lands taken from countless elk, bison, antelope, deer, and mountain sheep. These poisoners have wasted fur resources of Wyoming because many fur-bearing mammals other than wolves were (and are) killed by poisoning. Farming increased, and rodents were poisoned. Fortunately, several wise men led a movement that established national parks and national forests in Wyoming. These have served to *preserve* wildlife, but the wolf was extirpated and the mountain lion is nearly so. The work of the predatory animal killers should have then come to an end, but as yet has not. In spite of a great effort against the coyote, it persists and in many places is abundant and continues to prey mainly on rabbits and rodents (and occasionally on sheep). The lynx and the swift fox, two species that are beneficial to man, are nearly extirpated (but not protected); and the status of some other fur-bearing mammals is not good. Poisoning was carried on even in the national parks and still is in national forests and on much of the other public land. Scientists have called for "sane" predator control and some have called for a complete stop to it. The beaver, protected by the state, has increased in numbers and is now again a fur resource. A fraction of the big game that occurred in Wyoming in the past is now permitted to exist there. The magnificent grizzly bear, however, will eventually disappear from the state, probably even from Yellowstone Park, unless this bear is afforded protection outside the park. Grizzlies are protected in other states that still have them. Many of the other remaining kinds of mammals in Wyoming will probably also disappear owing to the activities of man unless given special protection. In addition to extirpating some mammals in Wyoming, man has added several kinds by direct introduction, by altering habitats, or by both methods. Kinds that occur in Wyoming owing to man's activities are: *Didelphis marsupialis, Sciuris niger, Rattus norvegicus, Mus musculus,* and *Oreamnos americanus.* Remarks concerning the results of some of these introductions are in the accounts of these species.

SUMMARY

One hundred and one species of mammals, including 172 subspecies and monotypic species, occur, or formerly occurred in Wyoming. A few kinds are thought to have been extirpated, but they may reinvade the state. The 101 species belong to 58 genera, which make up 22 families comprising seven orders. Five species (*Didelphis marsupialis, Sciurus niger, Rattus norvegicus, Mus musculus* and *Oreamnos americanus*) have been introduced directly or indirectly by man. There are 167 kinds, 96 species, 53 genera, 20 families, and six orders of native mammals in Wyoming.

Approximately 45 species occur in the Upper Sonoran Life-zone; at least 60 occur in the Transition; at least 50 occur in the Canadian; approximately 28 occur in the Hudsonian; and approximately 12 occur in the high Arctic-Alpine Life-zone. No species is confined to the two uppermost life-zones. Many of the enumerated species range through two or more life-zones. *Eutamias minimus, Thomomys talpoides, Canis latrans,* and *Ovis canadensis* occur in all life-zones, and seemingly *Bison bison* did so. Each of many species, including the four mentioned immediately above, is made up of more than one subspecies, which show adaptation to particular life-zones. The most notable adaptation is dark pelage in humid (high) life-zones and pale pelage in arid (low) life-zones—in accordance with Gloger's Rule. *Peromyscus maniculatus* is the most abundant mammal in the state, is wide-ranging, and seemingly responds locally to selective pressures by developing varying frequencies of pelage-colors, also in accordance with Gloger's Rule. Three faunal areas in Wyoming are 1) the Great Plains, 2) the Rocky Mountains, and 3) the Central Portal.

Wyoming was extensively glaciated along the Rocky Mountains during several of at least four cold glacial epochs in the Pleistocene (a fifth advance is now considered Recent). Perhaps as many as 14 species of Wyoming's Recent mammals occurred in peri-glacial areas in the cold glacial epochs, but most species are thought to have invaded the state since the height of glaciation. Because each Quaternary glaciation seems to have been less extensive than the preceding one, an increasing number of contacts were made between invaders from eastward and westward. At least 18 well-differentiated east-west pairs of closely related mammals occur in the state today. Sporadic, weak glacial advances, probably at the end of the Wisconsin, probably drove small populations of boreal mammals only short distances into lowlands nearby, instead of far away from the mountains, and perhaps account for the differentiation, possibly by genetic drift, of *Microtus montanus codiensis. Lepus americanus bairdii* of the Rocky Mountains differs from *L. a. seclusus* of the Bighorn Mountains. *L. a. seclusus* probably is a relict population that has undergone slight differentiation on account of isolation from *Lepus a. americanus* that now occurs only farther to the northward. The present geographic separation of the two subspecies probably resulted from the recession of the Valder's Advance and the warming temperature in the area between the subspecies.

In Post-glacial Time, slight differentiation has occurred in numerous species in Wyoming, mainly as a result of adaptation of the mammals to the "new" habitats—forested mountains and warm, arid lowlands. For example, *Eutamias minimus confinis* has differentiated in a "newly" forested area, and *Spermophilus lateralis wortmani* has differentiated in an area that has only recently (about 10,000 years ago) become notably arid.

SPECIES AND SUBSPECIES OF UNVERIFIED OCCURRENCE

1. *Microsorex hoyi washingtoni* Jackson.—This shrew has been reported from Montana to the northward (Hall and Kelson, 1959:51) and from Colorado (as *M. hoyi*) to the southward (Pettus and Lechleitner, 1963: 119).
2. *Myotis yumanensis sociabilis* Grinnell.—Miller and Allen (1928:69) recorded this bat from Wyoming, but I have examined the specimen, which is *Myotis lucifugus*. *M. yumanensis*, characterized by rich coppery brown pelage, is known from Idaho and Montana (Hall and Kelson, 1959:163), and may occur in western Wyoming.
3. *Myotis yumanensis yumanensis* H. Allen.—This subspecies is known from northern Utah (Krutzsch and Heppenstall, 1955:126) and may occur in southwestern Wyoming.
4. *Myotis thysanodes thysanodes* Miller.—This bat occurs west, south, and east of Wyoming (Hall and Kelson, 1959:170) and probably occurs in the state.
5. *Myotis californicus californicus* Audubon and Bachman.—This bat is known from western Montana (Bell, Moore, Raymond, and Tibbs, 1962) and from Colorado and Idaho (Hall and Kelson, 1959:173); the bat may occur in Wyoming and is recognizable by its high occiput.
6. *Pipistrellus hesperus hesperus* H. Allen.—This bat is known from Utah (Durrant, 1952:53) and may occur in southwestern Wyoming.

7. *Tadarida brasiliensis mexicana* (Saussure).—This bat occurs south of Wyoming (Hall and Kelson, 1959:206) and may occur in southwestern Wyoming (in low, warm life-zones).
8. *Antrozous pallidus pallidus* (LeConte).—This bat occurs in northern Utah (Durrant, 1952:62) and may occur in southwestern Wyoming.
9. *Sylvilagus idahoensis* (Merriam).—This small rabbit may range into western Wyoming.
10. *Lepus californicus deserticola* Mearns.—This rabbit may range into western Wyoming, from Idaho where it has been recorded (Hall and Kelson, 1959:284).
11. *Eutamias quadrivittatus quadrivittatus* (Say).—This chipmunk has been recorded (Hall and Kelson, 1959:312) from less than 30 miles south of Wyoming and may occur in southeastern Wyoming.
12. *Spermophilus variegatus grammurus* (Say).—A specimen of the rock squirrel in the United States National Museum is labeled from southeastern Wyoming, but the label on the specimen has been retied and a skull by the same number belongs to *Spermophilus richardsonii.* The rock squirrel does occur in northern Colorado and may occur in southeastern Wyoming, although intensive collecting there has yielded no specimen.
13. *Sciurus aberti ferreus* True.—Abert's squirrel is known from the mountains of northern Colorado (Hall and Kelson, 1959:385) and probably occurs in south-central Wyoming.
14. *Thomomys talpoides pierreicolus* Swenk.—This gopher occurs in northwestern Nebraska (Hall and Kelson, 1959:443) and may occur in Wyoming immediately south of the Black Hills.
15. *Perognathus flavescens flavescens* Merriam.—This pocket mouse may range into eastern Wyoming from Nebraska where it is known in the Sand Hills (Jones, 1953).
16. *Reithrodontomys megalotis megalotis* (Baird).—This harvest mouse may range across a boreal barrier along the western boundary of Wyoming into western Wyoming from Utah and Idaho where it has been recorded within 44 miles of the Wyoming boundary (Hall and Kelson, 1959:587).
17. *Peromyscus boylii utahensis* Durrant.—Durrant (1952:319) records this mouse from northern Utah from where it may range into southwestern Wyoming (in warm, dry habitats).
18. *Microtus pennsylvanicus uligocola* S. Anderson.—Although described from northern Colorado this vole has not been taken in southeastern Wyoming, where it may occur.
19. *Urocyon cinereoargenteus scottii* Mearns.—This subspecies may occur in southeastern Wyoming; it has been reported from northern Colorado (Hall and Kelson, 1959:863).
20. *Bassariscus astutus flavus* Rhoads.—This subspecies of ringtail is known from northern Colorado (Hall and Kelson, 1959:880) and may range or wander into southern Wyoming.
21. *Procyon lotor excelsus* Nelson and Goldman.—This raccoon is known from Idaho (Hall and Kelson, 1959:886-887) and may occur along streams in western Wyoming.

TYPE LOCALITIES IN WYOMING

ALBANY COUNTY

1. One mile east of Laramie, 7164 feet.
   Thomomys talpoides rostralis Hall and Montague 1951.
2. "Black Hills" [=Laramie Mountains].
   Phenacomys truei J. A. Allen 1894 [*Phenacomys intermedius intermedius*].
3. Woods Post Office, Medicine Bow Mountains, about 7500 feet.
   Marmota flaviventer luteola A. H. Howell 1914.

BIG HORN COUNTY

4. Medicine Wheel Ranch, 28 miles east of Lovell, 9000 feet.
   Microtus montanus zygomaticus S. Anderson 1954.
   Ochotona princeps obscura Long, present publication.
5. Two miles north and 12 miles east of Shell, 7900 feet, Bighorn Mountains.
   Lepus americanus seclusus Baker and Hankins 1950.
   Lepus americanus setzeri Baker [*Lepus americanus seclusus*].
6. Head of Trappers Creek, 9500 feet, Bighorn Mountains.
   Thomomys talpoides caryi V. Bailey 1914.
   Eutamias minimus confinis A. H. Howell 1925.

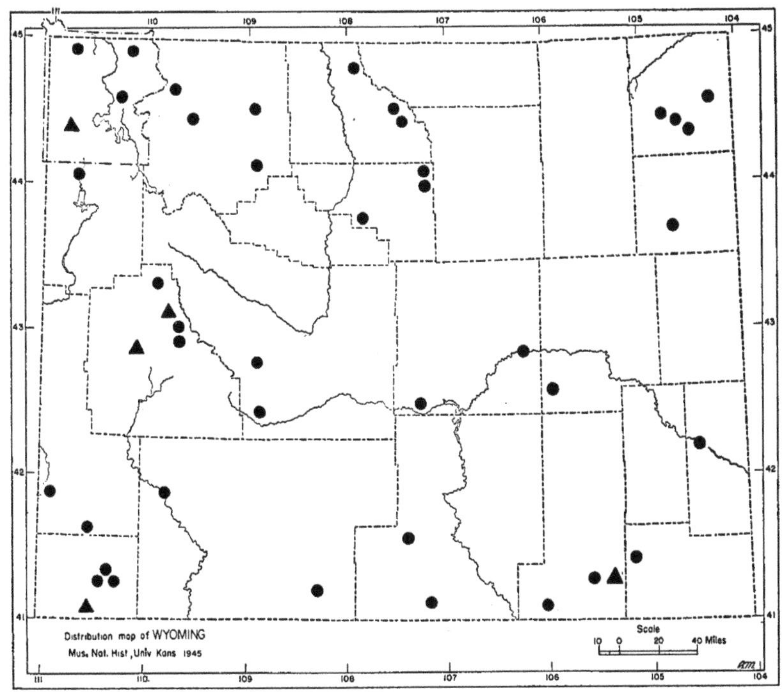

FIG. 82. Type localities. Solid (black) circles— precise type localities. Solid triangles— no certain locality.

CARBON COUNTY

7. Bridgers Pass, 18 miles southwest of Rawlins.
   Thomomys clusius Coues 1875 [*Thomomys talpoides clusius*].
   Lynx uinta Merriam 1902 [*Lynx rufus pallescens*].
8. Eight miles north and 19½ miles east of Savery, 8800 feet.
   Thomomys talpoides meritus Hall 1951.

CONVERSE COUNTY

9. Deer Creek.
   Hesperomys sonoriensis var. *nebrascensis* Coues 1877 [*Peromyscus maniculatus nebrascensis*].

CROOK COUNTY

10. Jack Boyden's Ranch, Sand Creek Canyon, 3750 feet, 15 miles northeast Sundance.
    Thomomys talpoides nebulosus V. Bailey 1914.
11. Bear Lodge, Sundance National Forest, Black Hills.
    Ursus rogersi bisonophagus Merriam 1918 [*Ursus arctos imperator*].
12. Bear Lodge Mountains.
    Zapus hudsonius campestris Preble 1899.
13. Three miles northwest Sundance, 5900 feet.
    Eutamias minimus silvaticus White 1952.

FREMONT COUNTY

14. Milford.
    Microtus montanus caryi V. Bailey 1917 [*Microtus montanus nanus*].
15. South Pass City, Wind River Mountains.
    Sciurus hudsonicus ventorum J. A. Allen 1898 [*Tamiasciurus hudsonicus ventorum*].

GOSHEN COUNTY

16. Fort Laramie.
    Putorius nigripes Audubon and Bachman 1851 [*Mustela nigripes*].
    Putorius culbertsoni Goues 1877 [*Mustela frenata longicauda*].

LARAMIE COUNTY

17. Three and one-half miles west of Horse Creek Post Office, 7000 feet.
    Thomomys talpoides attenuatus Hall and Montague 1951.

LINCOLN COUNTY

18. Twelve miles north and 2 miles east of Sage, 6100 feet.
    Microtus pennsylvanicus pullatus S. Anderson 1956.
19. Cumberland.
    Perognathus parvus clarus Goldman 1917.

NATRONA COUNTY

20. Casper.
    Perodipus ordii luteolus Goldman 1917 [*Dipodomys ordii luteolus*].
21. Sun.
    Perognathus fasciatus litus Cary 1911.

PARK COUNTY

22. Whirlwind Peak, 10,500 feet.
    Thomomys talpoides tenellus Goldman 1939.
23. Three-fifths of a mile south and 3⅕ miles east of Cody, 5020 feet.
    Microtus montanus codiensis S. Anderson 1954.

24. Shoshone River.
Ursus washakie Merriam 1916 [*Ursus arctos imperator*].
25. Greybull River, Absaroka Mountains.
Ursus rogersi rogersi Merriam 1918 [*Ursus arctos imperator*].

SUBLETTE COUNTY

26. Thirty-one miles north of Pinedale.
Eutamias umbrinus fremonti White 1953.
27. Columbia Valley, Wind River Mountains.
Lepus americanus bairdii Hayden 1869.
28. Fremont Peak, 11,500 feet, Wind River Mountains.
Ochotona uinta ventorum A. H. Howell 1919 [*Ochotona princeps ventorum*].
29. Seven miles south of Fremont Peak, Wind River Mountains.
Callospermophilus lateralis caryi A. H. Howell 1917 [*Spermophilus lateralis castanurus*].
30. Near Cora, head of Wind River.
Felis hippolestes Merriam 1897 [*Felis concolor hippolestes*].

SWEETWATER COUNTY

31. Near mouth Big Sandy Creek on Green River.
Tamias minimus Bachman 1839 [*Eutamias minimus minimus*].
32. Kinney Ranch, south of Bitter Creek.
Tamias wortmani J. A. Allen 1895 [*Spermophilus lateralis wortmani*].
Neotoma cinnamomea J. A. Allen 1895 [*Neotoma cinerea cinnamomea*].
Vespertilio chrysonotus J. A. Allen 1896 [*Myotis evotis evotis*].
Perognathus callistus Osgood 1900 [*Perognathus fasciatus callistus*].
Dipodomys ordii priscus Hoffmeister 1942.

TETON COUNTY

33. Snake River, 4 miles south of Yellowstone Park.
Alces americanus shirasi Nelson 1914 [*Alces alces shirasi*].

UINTA COUNTY

34. Fort Bridger.
Spermophilus elegans Kennicott 1863 [*Spermophilus richardsonii elegans*].
Cynomys leucurus Merriam 1890.
35. Mountainview, Smiths Fork, 4 miles southeast of Fort Bridger.
Thomomys clusius ocius Merriam 1901 [*Thomomys talpoides ocius*].
36. Mountainview, Harvey's Ranch, on Smiths Fork, 6 miles southwest of Fort Bridger.
Thomomys bridgeri Merriam 1901 [*Thomomys talpoides bridgeri*].
37. Near Fort Bridger, foothills of Uinta Mountains.
Spermophilus armatus Kennicott 1863.

WASHAKIE COUNTY

38. Near head Canyon Creek, west slope of the Bighorn Mountains, 8000 feet.
Spermophilus tridecemlineatus alleni Merriam 1898.
39. Spring Creek, east side of the Bighorn Basin.
Lepus baileyi Merriam 1897 [*Sylvilagus audubonii baileyi*].
40. Near head of Kirby Creek, 8400 feet.
Sciurus hudsonicus baileyi J. A. Allen 1898 [*Tamiasciurus hudsonicus baileyi*].

WESTON COUNTY

41. Twenty-three miles southwest of Newcastle.
    Perognathus flavus piperi Goldman 1917.

YELLOWSTONE NATIONAL PARK

42. Swan Lake Valley.
    Eutamias consobrinus clarus Bailey 1918 [*Eutamias minimus consobrinus*].
43. Slough Creek.
    Ursus mirus Merriam 1918 [*Ursus arctos imperator*].
44. Lake Hotel.
    Myotis carissima Thomas 1904 [*Myotis lucifugus carissima*].
45. No certain locality.
    Ursus imperator Merriam 1914 [*Ursus arctos imperator*].
    Cervus canadensis nelsoni V. Bailey 1935.

LITERATURE CITED

ALDEN, W. C.
1926. Glaciation and Physiography of Wind River Mountains, Wyoming. Jour. Washington Acad. Sci., 16:73.
1928. Yellowstone National Park and its environs in the Great Ice Age. Ranger Naturalists Manual. Not seen.

ALLEN, G. M.
1916. Bats of the genus Corynorhinus. Bull. Mus. Comp. Zool., 60(9): 331-356, 1 pl., April.

ALLEN, H.
1864. Monograph of the bats of North America. Smithsonian Misc. Coll. (165), 7:xxiii + 85, 68 figs. June.
1894. A monograph of the bats of North America. Bull. U. S. Nat. Mus., 43:ix + 198, 38 pls. March 14.
1942. Extinct and vanishing mammals of the Western Hemisphere. . . . . , Special Publ. No. 11, Amer. Comm. Int. Wildlife Protection, xv + 620 pp., illus.

ALLEN, J. A.
1874. Notes on the mammals of portions of Kansas, Colorado, Wyoming, and Utah. Bull. Essex Inst., 6(4):43-66, April.
1875. Synopsis of the American Leporidae. Proc. Boston Soc. Nat. Hist., 17:431, February 17.
1895. Descriptions of new American mammals. Bull. Amer. Mus. Nat. Hist., 7:327-340, November 8.
1898. Revision of the chickarees of North American red squirrels (subgenus *Tamiasciurus*). Bull. Amer. Mus. Nat. Hist., 10-249-298, July 22.
1912. A new pika from Colorado. Bull. Amer. Mus. Nat. Hist., 31:103, 104, May 28.

ANDERSON, S.
1954. Subspeciation in the meadow mouse, Microtus montanus, in Wyoming and Colorado. Univ. Kansas Publs., Mus. Nat. Hist., 7:489-506, 2 figs., July 23.
1956. Subspeciation in the meadow mouse, Microtus pennsylvanicus, in Wyoming, Colorado, and adjacent areas. Univ. Kansas Publs., Mus. Nat. Hist., 9:85-104, 2 figs., May 10.
1959. Distribution, variation, and relationships of the montane vole, Microtus montanus. Univ. Kansas Publs., Mus. Nat. Hist., 9:415-511, 12 figs., August 1.

ANONYMOUS.
1940. Hunting experiences of early days. W. A. Richards diary. Annals of Wyoming, 12:64-69, January.

ARMITAGE, K. B.
1961. Frequency of melanism in the golden-mantled marmot. Jour. Mamm., 42:100-101, February 20.

ATWOOD, W. W., JR.
1937. Records of Pleistocene glaciers in the Medicine Bow and Park ranges. Jour. Geol., 45:113-140.

ATWOOD, W. W., SR.
1909. Glaciation of the Uinta and Wasatch mountains. U. S. Geol. Surv. Prof. Paper 61, 96 pp., illus.

BACHMAN, J.
1839. Additional remarks on the genus Lepus. . . ., Jour. Acad. Nat. Sci. Philadelphia, 8(1):95-101.

BAILEY, V.
1900. Revision of American voles of the genus Microtus. N. Amer. Fauna, 17:1-88, 5 pls. 17 figs., June 6.

1905. Biological Survey of Texas. N. Amer. Fauna 25:1-222, 16 pls., 24 figs., October 24.

1927. A biological survey of North Dakota. N. Amer. Fauna, 49:vi + 226, frontis., 21 pls., 8 figs., January 8.

1930. Animal life of Yellowstone National Park. Charles C. Thomas, Springfield, Illinois, 241 pp., 69 figs.

1935. A new name for the Rocky Mountain elk. Proc. Biol. Soc. Washington, 48:187-189, November 15.

1936. The mammals and life zones of Oregon. N. Amer. Fauna, 55:1-416, 51 pls., 102 figs., 1 map, August 29.

BAIRD, S. F.

1858. Explorations and surveys for a railroad route. . . . Part I, xvii + 757 pp., 60 pls., July 14.

BAKER, R. H. and HANKINS, R. M.

1950. A new subspecies of snowshoe rabbit from Wyoming. Proc. Biol. Soc. Washington, 63:63-64, May 25.

BARRY, J. N.

1943. An excerpt from the journal of E. Willard Smith, 1839-1840. Annals Wyoming, 15:287-297.

BELL, J. F., MOORE, J., RAYMOND, G. H., and TIBBS, C. E.

1962. Characteristics of rabies in bats in Montana. Amer. Jour. Public Health, 52(8):1293-1301, August.

BENSON, S. B. and BOND, R. M.

1939. Notes on Sorex merriami Dobson. Jour. Mamm., 20:348-351. 2 figs., August 14.

BLACKWELDER, E.

1915. Post-Cretaceous history of the mountains of central western Wyoming. Jour. Geol., 23:97-117, 193-267, 307-340.

BLAIR, W. F.

1953. Factors affecting gene exchange between populations in the *Peromyscus maniculatus* group. Texas Jour. Sci., 5:17-33, 1 fig., March.

BRADLEY, F. H.

1872. Geological report of the Snake River Division. *In* sixth Annual report, U. S. Geological Survey of the Territories . . . Washington, pp. 191-271.

BRADLEY, W. H.

1936. Geomorphology of the north flank of the Uinta Mountains. U. S. Geol. Surv. Prof. Papers, 185-1, pp. 163-199, illus.

BRANSON, E. B. and BRANSON, C. C.

1945. Glaciation in southern part of the Wind River Mountains and their foothills. Bull. Geol. Soc. Amer., 56:1148-1149.

BROWN, W. L., JR., and WILSON, E. O.

1956. Character displacement. Soc. Syst. Zool., 5:49-64, 6 figs., June.

BRYAN, F. T.

1858. Report of Francis T. Bryan to Col. J. J. Abert, Chief Corps Top. Engs., U. S. A. *In* Exec. Documents, U. S. Senate, First Session, 35th Congress and Special Session 1858.

BRYANT, E.

1885. Rocky Mountain adventures. New York, pp. 1-452. Not seen.

BURT, W. H.

1959. The history and affinities of the Recent land mammals of western North America. Zoogeography, Amer. Assoc. Advancement Sci., pp. 131-154, 4 figs.

CAHALANE, V. H.

1948. The status of mammals in the U. S. National Park System, 1947. Jour. Mamm., 29:247-259. August.

CAMERON, A. W.
1958. Mammals of the islands in the Gulf of St. Lawrence. Nat. Mus. Canada, Bull. No. 154, iii + 165 pp., frontis., 8 figs., 29 maps.

CARY, M.
1917. Life-zone investigations in Wyoming. N. Amer. Fauna, 42:1-95, 17 figs., 1 map, October 3.

CHURCHER, C. S.
1959. The specific status of the new world red fox. Jour. Mamm., 40: 513-520, 1 fig., November 20.

CLARK, F. H.
1938. Inheritance and linkage relations of mutant characters in the deer-mouse, *Peromyscus maniculatus*. Contr. Lab. Vert. Genetics, Univ. Michigan, 7:1-11.

COCKRUM, E. L. and FITCH, K. L.
1952. Geographic variation in red-backed mice (genus Clethrionomys) of the Rocky Mountain Region. Univ. Kansas Publs., Mus. Nat. Hist. 5(22):281-292, 1 fig.

CONAWAY, C. H.
1952. Life history of the water shrew (*Sorex palustris navigator*). Amer. Midl. Nat., 48:219-248.

COOK, H. J.
1931. A mountain sheep record for Nebraska. Jour. Mamm., 12:170-171, May 14.

COUES, E.
1877. Fur-bearing animals: a monograph of North American Mustelidae . . . U. S. Geol. Surv. Territories, Misc. Publ. No. 8, xiv + 348, 20 pls.

COUTANT, C. G.
1899. The history of Wyoming. . . . Laramie, Wyoming, Chaplin Spafford and Mathison Co., xxiv + 712 pp., illus.

COWAN, I. McT.
1940. Distribution and variation in the native sheep of North America. Amer. Midl. Nat., 24:505-580, 4 pls., 1 map, November.

DAKE, C. L.
1919. Glacial features on the south side of Beartooth Plateau, Wyoming. Jour. Geol., 27:128-131.

DALQUEST, W. W.
1943. The systematic status of the races of the little big-eared bat *Myotis evotis* H. Allen. Proc. Biol. Soc. Washington, 56:1-2, February 25.

DARTON, N. H.
1906. Geology of the Owl Creek Mountains. . . . U. S. Geol. Surv., 48 pp., illus.

DARWIN, C.
1859. The origin of species by means of natural selection . . . 6th ed., New York, Random House, 386 pp., illus., reprinted, 1936.

DAVIS, W. B.
1939. The Recent mammals of Idaho. The Caxton Printers, Caldwell, Idaho, 400 pp., frontis., 33 figs., April 5.

DEEVEY, E. S., JR.
1949. Biogeography of the Pleistocene. Bull. Geol. Soc. Amer., 60:1315-1416, illus.

DICE, L. R.
1941. Variation of the deer-mouse (*Peromyscus maniculatus*) on the sand hills of Nebraska and adjacent areas. Contr. Lab. Vert. Genetics, Univ. Michigan, 15:1-19.

1942. Variation of the deer-mouse *(Peromyscus maniculatus)* of the Bad Lands and Black Hills of South Dakota and Wyoming. Contr. Lab. Vert. Genetics, Univ. Michigan, 19:1-10.

DOBZHANSKY, T.
1951. Genetics and the origin of species, 3rd ed. New York, Columbia Univ. Press, x + 364 pp., illus.

DORF, E.
1959. Climatic changes of the past and present. Mus. Paleontol., Univ. Michigan, 13(8):181-210, illus.

DURRANT, S. D.
1952. Mammals of Utah. . . . Univ. Kansas Publs., Mus. Nat. Hist., 6:1-549, 91 figs., August 10.

DURRANT, S. D. and HANSEN, R. M.
1954. Taxonomy of the chickarees (Tamiasciurus) of Utah. Jour. Mamm., 35:87-95, February 10.

DURRANT, S. D. and LEE, M. R.
1955. Rare shrews from Utah and Wyoming. Jour. Mamm., 36:560-561, December 14.

ENGELS, W. L.
1936. Distribution of races of the brown bat (Eptesicus) in Western North America. Amer. Midl. Nat., 17:653-660, May.

ERDBRINK, D. P.
1953. A review of fossil and Recent bears of the old world. Deventer-Drukkerij Jan de Lange, 2 vols., 597 pp., illus.

FENNEMAN, N. M.
1931. Physiography of western United States. McGraw-Hill, New York, xiii + 534 pp., illus.

FINDLEY, J. S.
1951. Habitat preferences of four species of *Microtus* in Jackson Hole, Wyoming. Jour. Mamm., 32:118-120, February 15.
1954. Reproduction in two species of *Myotis* in Jackson Hole, Wyoming. Jour. Mamm., 35:434, August 20.
1955. Speciation of the wandering shrew. Univ. Kansas Publs., Mus. Nat. Hist., 9:1-68, 18 figs., December 10.

FINDLEY, J. S. and ANDERSON, S.
1956. Zoogeography of the montane mammals of Colorado. Jour. Mamm., 37:80-82, 1 fig., February 28.

FLINT, R. F.
1947. Glacial geology and the Pleistocene epoch. John Wiley, New York, xviii + 589 pp., illus.

FRYXELL, F. M.
1928. The former range of the bison in the Rocky Mountains. Jour. Mamm., 9:129-139, May 9.
1928. Melanism among the marmots of the Teton Range, Wyoming. Jour. Mamm., 9:336-337, November 13.
1930. Glacial features of Jackson Hole, Wyoming. Augustana Library Publs., No. 13, Rock Id. Illinois, xii + 128 pp., illus.
1943. Thomas Moran's journey to the Tetons. Annals Wyoming, 15: 71-84.

FRYXELL, F. M., HORBERG, L., and EDMUND, R.
1941. Geomorphology of the Teton Range and adjacent basins, Wyoming-Idaho. Bull. Geol. Soc. Amer., 52:1903.

GARRETSON, M. S.
1938. The American bison. New York Zoological Soc., New York, xii + 254 pp., frontis., illus.

GOIN, O. B.
1944. A new race of the canyon mouse. Jour. Mamm., 25:189-191, May 26.

GOLDMAN, E. A.
1910. Revision of the wood rats of the genus Neotoma. N. Amer. Fauna, 31:1-124, 8 pls., October 19.
1917. Two new pocket mice from Wyoming. Proc. Biol. Soc. Washington, 30:147-148, July 27.

GRASSE, J. E. and PUTNAM, E. F.
1955. Beaver management and ecology in Wyoming, 2nd ed. Wyoming Game and Fish Comm., Bull., No. 6, 75 pp., illus.

GRINNELL, G. B.
1928. Mountain sheep. Jour. Mamm., 9:1-9, February 9.

GUTHE, O. E.
1935. The Black Hills of South Dakota and Wyoming. Papers Michigan Acad. Sci., Arts, and Letters, 20:343-376, illus.

HAFFEN, L. R. and YOUNG, F. M.
1938. Fort Laramie and the pageant of the West, 1834-1890. Arthur H. Clarke Co., Glendale, California, 429 pp., illus.

HAGMEIER, E. M.
1961. Variation and relationships in North American marten. Canadian Field-Nat., 75:122-138, 6 figs., July-September.

HALL, E. R.
1936. Mustelid mammals from the Pleistocene of North America. . . ., Carnegie Inst., Washington Publ. No. 473, pp. 41-119, 6 figs., November 20.
1943. Intergradation versus hybridization in ground squirrels of the western United States. Amer. Midl. Nat., 29:375-378, 1 fig., March.
1946. Mammals of Nevada. Univ. California Press, Berkeley and Los Angeles, xi + 710 pp., frontis., 485 figs., July 1.
1951a. A synopsis of the North American Lagomorpha. Univ. Kansas Publs., Mus. Nat. Hist., 5:119-202, 68 figs., December 15.
1951b. A new pocket gopher (genus Thomomys) from Wyoming and Colorado. Univ. Kansas Publs., Mus. Nat. Hist., 5(13):219-222, December 15.
1951c. American weasels. Univ. Kansas Publs., Mus. Nat. Hist., 4:1-446, 41 pls., 31 figs., December 27.

HALL, E. R. *et al.*
1957. Vernacular names for North American mammals north of Mexico. Univ. Kansas, Mus. Nat. Hist., Misc. Publ. No. 14, pp. 1-16, June 19.

HALL, E R. and HATFIELD, D. M.
1934. A new race of chipmunk from the Great Basin of western United States. Univ. California Publs. Zool., 40(6):321-326, 1 fig., February 12.

HALL, E. R. and KELSON, K.
1951. Comments on the taxonomy and geographic distribution of some North American rabbits. Univ. Kansas Publs., Mus. Nat. Hist., 5(5):49-58, October 1.
1952. Comments on the taxonomy and geographic distribution of some North American rodents. Univ. Kansas Publs., Mus. Nat. Hist., 5(26):343-371, December 15.
1959. The mammals of North America. New York, The Ronald Press, 2 vols., 1083 + xxx + viii + 158 pp., 553 figs., March 31.

HALL, E. R. and MONTAGUE, H. G.
1951. Two new pocket gophers from Wyoming and Colorado. Univ. Kansas Publs., Mus. Nat. Hist., 5(3):25-32, February 28.

HALLORAN, A. F. and BLANCHARD, W. E.
1959. Lynx from western Wyoming. Jour. Mamm., 40:450-451, August 20.

HAMILTON, W. J., JR.
1940. The biology of the smoky shrew (*Sorex fumeus fumeus* Miller). Zoologica, 25:473-492.

HANDLEY, C. O., JR.
1959. A revision of American bats of the genera Euderma and Plecotus. Proc. U. S. National Mus., 110:95-246, illus., September 3.

HANSEN, H. P.
1951. Pp. 111-118, *in* Moss, J. H. *et al.*, Early man in the Eden Valley. Univ. Pennsylvania Mus. Monographs, 124 pp., illus.

HANSEN, R. M.
1957. Remarks on reported hybrid ground squirrels, *Citellus.* Jour. Mamm., 37:550-552, January 9. For 1956.

HARES, C. J.
1948. Striated boulders on the Medicine Bow Mountains, Wyoming. Bull. Geol. Soc. Amer., 59:1329.

HARLAN, R.
1825. Fauna Americana being a description of the mammiferous animals inhabiting North America. Anthony Finley Publ., Philadelphia, x + 319 pp.

HIBBARD, C. W.
1937. Notes on some vertebrates from the Pleistocene of Kansas. Trans. Kansas Acad. Sci., 40:233-237, 1 pl.

HOFFMAN, R. S. and TABOR, R. D.
1960. Notes on *Sorex* in the northern Rocky Mountain Alpine zone. Jour. Mamm., 41:230-234, May 20.

HOFFMEISTER, D. F.
1956. A record of *Sorex merriami* from northeastern Colorado. Jour. Mamm., 37:276, June 9.

HOLLISTER, N.
1914. A systematic account of the grasshopper mice. Proc. U. S. Nat. Mus., 47:427-489, 1 pl., 2 figs., October 29.
1916. A systematic account of the prairie-dogs. N. Amer. Fauna, 40: 1-37, 7 pls., June 20.

HOLMES, G. W. and MOSS, J. H.
1955. Pleistocene geology of the southwestern Wind River Mountains, Wyoming. Bull. Geol. Soc. Amer., 66:629-654, illus.

HOLMES, W. H.
1881. Glacial phenomena in Yellowstone National Park. Amer. Nat., 15:203-208.

HONESS, R. F. and FROST, N. M.
1942. A Wyoming bighorn sheep study. Wyoming Game and Fish Dept., Bull. No. 1, vi + 126 pp., 45 figs., 1 map.

HOOPER, E. T.
1943. The type locality of Thomomys bridgeri and Thomomys ocius. Jour. Mamm., 24:503, November 20.
1944. The name Neotoma cinerea cinnamomea Allen applied to woodrats from southwestern Wyoming. Jour. Mamm., 25:415, December 12.

HOOVER, R. L. and YEAGER, L. E.
1953. Status of the fox squirrel in northeastern Colorado. Jour. Mamm., 34:359, August 14.

HOWARD, E. B. and HACK, J. T.
1943. The Finley Site. Amer. Antiquity, 8:224-241.

HOWELL, A. B.
1930. At the cross-roads. Jour. Mamm., 11:377-389, August 9.

HOWELL, A. H.
1906. Revision of the skunks of the genus Spilogale. N. Amer. Fauna, 26:1-55, 10 pls., November 24.

1914a. Ten new marmots from North America. Proc. Biol. Soc., Washington, 27:13-18, February 2.

1914b. Revision of the American harvest mice (Genus Reithrodontomys). N. Amer. Fauna, 36:1-97, 7 pls., June 5.

1915. Revision of the American marmots. N. Amer. Fauna, 37:1-80, 15 pls., April 7.

1917. Description of a new race of Say's ground squirrel from Wyoming. Proc. Biol. Soc. Washington, 30:105-106, May 23.

1918. Revision of the American flying squirrels. N. Amer. Fauna, 44: 1-64, illus., June 13.

1929. Revision of the American chipmunks (genera *Tamias* and *Eutamias*). N. Amer. Fauna, 52:1-157, 10 pls., 9 maps, November 30.

1938. Revision of the North American ground squirrels, with a classification of the North Amercian Sciuridae. N. Amer. Fauna, 56:1-256, illus., April.

HUXLEY, J.

1955. Morphism and evolution. Heredity, 9:1-52, April.

JACKSON, H. H. T.

1925. The Sorex arcticus and Sorex articus cinereus of Kerr. Jour. Mamm., 6:55-56, February.

1928. A taxonomic review of the American long-tailed shrews (genera *Sorex* and *Microsorex*). N. Amer. Fauna, 51:vi + 218, 13 pls., 24 figs., July 24.

1944. Big-game resources of the United States 1937-1942. Research Rept. 8, U. S. Dept. Interior, ii + 56 pp., 31 figs.

JONES, J. K., JR.

1953. Geographic distribution of the pocket mouse, Perognathus fasciatus. Univ. Kansas Publs., 5(29):515-526, 7 figs., August 1.

1954. Distribution of some Nebraskan mammals. Univ. Kansas Publs., Mus. Nat. Hist., 7(6):479-487, April 21.

1958. The type locality and nomenclatorial status of Peromyscus maniculatus nebrascensis (Coues). Proc. Biol. Soc. Washington, 71:107-111, July 16.

1962. Early records of some mammals from Nebraska. Bull. Univ. Nebraska State Mus., 4(6):89-100, November.

JONES, J. K., JR., and MURSALOGLU, B.

1961. A syntype of *Peromyscus maniculatus nebrascensis* (Coues). Proc. Biol. Soc. Washington, 74:101-104, May 19.

1961. Geographic variation in the harvest mouse, Reithrodontomys megalotis, on the Central Great Plains and in adjacent regions. Univ. Kansas Publs., Mus. Nat. Hist., 14(2):9-27, 1 fig., July 24.

KEEFER, W. R.

1957. Geology of the Du Noir area, Fremont County, Wyoming. U. S. Geol. Surv. Prof. Paper 294E, pp. 155-221, illus.

KENDEIGH, S. C.

1954. History and evaluation of various concepts of plant and animal communities in North America. Ecology, 35:152-171, 8 figs., April.

KING, J. A.

1951. The subspecific identity of the Black Hills flying squirrels (*Glaucomys sabrinus*). Jour. Mamm., 32:469-470, November 19.

KRUTZSCH, P. H.

1954. North American jumping mice (genus Zapus). Univ. Kansas Publs., Mus. Nat. Hist., 7(4):349-472, 47 figs., April 21.

KRUTZSCH, P. H. and HEPPENSTALL, C. A.

1955. Additional distributional records of bats in Utah. Jour. Mamm., 36:126-127, February 28.

LICHTENSTEIN, H.
1838. Über die Gattung *Mephitis*. Abhandl. K. Akad. Wiss., Berlin, pp. 249-313.

LONG, C. A.
1961. A dental abnormality in *Sorex palustris*. Jour. Mamm., 42:527-528, 1 fig., November 20.

LONG, C. A. and HOUSE, H. B.
1961. *Bassariscus astutus* in Wyoming. Jour. Mamm., 42:274-275, May 20.

LONG, C. A. and KERFOOT, W. C.
1963. Mammalian remains from owl-pellets in eastern Wyoming. Jour. Mamm., 44:129-131, February 20.

LYON, M. W., JR., and OSGOOD, W. H.
1909. Catalogue of the type specimens of mammals in the United States National Museum, including the Biological Survey collection. U. S. Nat. Mus. Bull., 62:325 pp., January 28.

MATTHES, F. E.
1900. Glacial sculpture of the Bighorn Mountains Wyoming. U. S. Geol. Surv., 21st Annual Rept., pp. 167-190.

MEARS, B.
1953. Quaternary features of the Medicine Bow Mountains, Wyoming. *In* Wyoming Geol. Assoc. Guidebook, pp. 81-84, illus.

MERRIAM, C. H.
1888. Description of a new species of meadow mouse from the Black Hills of Dakota. Amer. Nat., 22:934-935, 1 fig., October.
1890. Contribution toward a revision of the little striped skunks of the genus Spilogale. N. Amer. Fauna, 4:1-15, 1 inserted folder, October.
1891. Results of a biological reconnaissance of Idaho, south of Latitude 45° and east of the thirty-eighth meridian, made during the summer of 1890, with annotated lists of the mammals and birds, and descriptions of new species. N. Amer. Fauna, 5:1-132, 4 pls., July 30.
1899. Life zones and crop zones of the United States. U. S. Dept. Agric. Bull. No. 10, pp. 7-79.
1901. Two new bighorns and a new antelope from Mexico and the United States. Proc. Biol. Soc. Washington, 14:29-32, April 5.
1902. A new bobcat (Lynx uinta) from the Rocky Mountains. Proc. Biol. Soc. Washington, 15:71-72, March 22.
1918. Revision of the grizzly and big brown bears of North America (genus Ursus) with description of a new genus, Vetularctos. N. Amer. Fauna, 41:1-36, 16 pls., February 9.

MICKEY, A. B.
1948. A record of the shrew *Sorex nanus* for Wyoming. Jour. Mamm., 29:294-295, August 31.
1961. Record of the spotted bat from Wyoming. Jour. Mamm., 42:401-402, August 21.

MICKEY, A. B. and STEELE, C. N.
1947. A record of *Sorex merriami merriami* for southeastern Wyoming. Jour. Mamm., 28:293, August 20.

MILLER, G. S., JR.
1897. Revision of the bats of the family Vespertilionidae. N. Amer. Fauna, 13:1-140, 3 pls., 40 figs., October 16.

MILLER, G. S., JR., and ALLEN, G. M.
1928. The American bats of the genera *Myotis* and *Pizonyx*. Bull. U. S. Nat. Mus. No. 144, pp. viii + 218, 1 pl., 1 fig., May 25.

MINER, N. A. and APFEL, E. T.
1946. Three stages of glaciation identified in Wind River Valley, Wyoming. Bull. Geol. Soc. Amer., 57:1218-1219.

MINER, N. A. and DELO, D. M.
1943. Glaciation of the Du Noir Valley, Fremont County, Wyoming. Jour. Geol., 51:131-137.

MONTAGNE, J. de la and LOVE, J. D.
1957. Giant glacial grooves and their significance in the Jackson Hole area, Wyoming. Bull. Geol. Soc. Amer., 68:1861.

MOSS, J. H.
1949. Glaciation in the southern Wind River Mountains, Wyoming. Bull. Geol. Soc. Amer., 60:1911.
1951. Late glacial advances in the southern Wind River Mountains, Wyoming. Amer. Jour. Sci., 249:865-883.

MURIE, A.
1940. Ecology of the coyote in the Yellowstone. Fauna Nat'l Parks U. S. Bull. No. 4:x + 206 pp., illus.

MURIE, O. J.
1934. Melanism in an Alaskan vole. Jour. Mamm., 15:323, November 15.
1951. The elk of North America. Stackpole Co., Harrisburg, Pennsylvania, 376 pp., 32 figs., frontis.

NEGUS, N. C. and FINDLEY, J. S.
1959. Mammals of Jackson Hole, Wyoming. Jour. Mamm., 40:371-381. 1 fig., August 20.

NELSON, E. W.
1909. The rabbits of North America. N. Amer. Fauna, 29:1-314, 13 pls., 19 figs., August 31.
1914. Description of a new subspecies of moose from Wyoming. Proc. Biol. Soc. Washington, 27:71-74, April 25.
1925. Status of the pronghorned antelope, 1922-1924. U. S. Dept. Agric Bull. No. 1346, 64 pp., frontis., 21 figs., August.

OSGOOD, W. H.
1907. Some unrecognized and misapplied names of American mammals. Proc. Biol. Soc. Washington, 20:43-52, April 18.
1909. Revision of the mice of the American genus Peromyscus. N. Amer. Fauna, 28:1-285, April 17.

PARSONS, W. H.
1939. Glacial geology of the Sunlight area, Park County, Wyoming. Jour. Geol., 47:737-747.

PEARSON, O. P.
1945. Longevity of the short-tailed shrew. Amer. Midl. Nat., 34:531-546.

PETERSON, R. L.
1952. A review of the living representatives of the genus *Alces*. Contrs. Royal Ontario Mus. Zool. and Paleontol., 34:1-30, 8 figs., October 15.
1955. North American Moose. Univ. Toronto Press, Toronto. xi + 280 pp., 66 figs.

PETTUS, D. and LECHLEITNER, R. R.
1963. *Microsorex* in Colorado. Jour. Mamm., 44:119, February 20.

PRUITT, W. O., JR.
1954. Aging in the masked shrew, *Sorex cinereus cinereus* Kerr. Jour. Mamm., 35:35-39, 1 fig., February 10.

QUAY, W. B.
1948. Notes on some bats from Nebraska and Wyoming. Jour. Mamm., 29:181-182, May 14.

RASSMUSSEN, D. I. and CHAMBERLAIN, N. V.
1959. A new Richardson's meadow mouse from Utah. Jour. Mamm., 40:53-56, February 20.

RAY, L. L.
1940. Glacial chronology of the southern Rocky Mountains. Bull. Geol. Soc. Amer., 51:1851-1918.

RICHMOND, G. M.
1941. Multiple glaciation of the Wind River Mountains, Wyoming. Bull. Geol. Soc. Amer., 52:1929-1930.
1948. Modification of Blackwelder's sequence of Pleistocene glaciation in the Wind River Mountains, Wyoming. Bull. Geol. Soc. Amer., 59:1400-1401.
1953. Pleistocene field conference in Rocky Mountain National Park. Science, 117(3034):177-178.

RIDGWAY, R.
1912. Color standards and color nomenclature. Washington, D. C., iv + 43 pp., 53 pls.

ROE, F. G.
1951. The North American buffalo . . . , Univ. Toronto Press, viii + 957 pp., illus.

SALISBURY, R. D.
1906. Glacial geology of the Bighorn Mountains. U. S. Geol. Surv. Prof. Paper 51, pp. 71-90, illus.

SALISBURY, R. D. and BLACKWELDER, E.
1903. Glaciation in the Bighorn Mountains. Jour. Geol., 11:216-223.

SCHANTZ, V. S.
1946. A new badger from South Dakota. Proc. Biol. Soc. Washington, 59:81-82, June 19.
1950. A new badger from Montana. Jour. Mamm., 31:90-92, February 21.

SCHLUNDT, H. and MOORE, R. B.
1909. Radioactivity of the thermal waters of Yellowstone National Park. Bull. U. S. Geol. Surv. No. 395, 35 pp.

SCHWARZ, E. and SCHWARZ, H. K.
1943. The wild and commensal stocks of the house mouse, *Mus musculus* Linnaeus. Jour. Mamm., 24:59-72, February 20.

SETON, E. T.
1929. Lives of game animals. Garden City, New York, Doubleday Doran and Co., 4 vols. (2 parts each vol.), illus.

SETZER, H. W.
1949. Subspeciation in the Kangaroo Rat, Dipodomys ordii. Univ. Kansas Publs., Mus. Nat. Hist., 1(23):473-573, 27 figs., December 27.

SIMON, J. R.
1951. The Wyoming or Yellowstone moose. Wildlife series No. 2, 7 pp., 3 figs., August.

SIMPSON, G. G.
1945. The principles of classification of mammals and a classification of mammals. Bull. Amer. Mus. Nat. Hist., 85:xvi + 350 pp., October 5.

SINCLAIR, W. J.
1912. Some glacial deposits east of Cody . . . . Science, 35:314-315.

SKINNER, M. P.
1927. The predatory and fur-bearing animals of the Yellowstone National Park. Roosevelt Wild Life Bull., 4(2):159-381, 51 figs., 1 fold-out map, June.
1929. White-tailed deer formerly in the Yellowstone Park. Jour. Mamm., 10:101-115, 2 pls., May 9.

SPERRY, C. C.
1941. Food habits of the coyote. Wildlife Research Bull., 4, U. S. Dept. Interior, pp. iv + 70, frontis., 3 pls.

ST. JOHN, O.
1877. Geological report of the Teton division. *In* Eleventh Annual Rept., U. S. Geol. Surv. of the Territories. . . . Washington, pp. 321-508.

STEBLER, A. M.
1939. An ecological study of the mammals of the badlands and the Black Hills of South Dakota and Wyoming. Ecology, 20:382-393, 3 figs.

STEVENSON, J.
1871. Preliminary report of the U. S. Geological Survey of Wyoming and portion of contiguous territories. Gov't Printing Office, Washington, D. C., ix + 1091 pp., January 1.

STOCKARD, A. H.
1929a. Papers Michigan Acad. Sci., Arts. and Letters, 11:471-479.

1929b. Observations on reproduction in the white-tailed prairie-dog (Cynomys leucurus). Jour. Mamm., 10:209-212, August 10.

SUMNER, F. B.
1930. Genetics and distributional studies of three subspecies of *Peromyscus*. Jour. Genetics, 23:275-276.

1932. Genetics, distribution, and evolutionary studies of the subspecies of deer mice *(Peromyscus)*. Bibligr. Genetica, 9:1-116.

SVIHLA, R. D.
1931. Mammals of the Uinta Mountains region. Jour. Mamm., 12:256-266, August 24.

SWENK, M. H.
1920. The birds and mammals of Nebraska. Nebraska Blue Book. . . . pp. 392-411, December.

SWENSON, F. A.
1949. Geology of the northwest flank of the Gros Ventre Mountains, Wyoming. Augustana Library Publs., No. 21, 75 pp., illus.

TAYLOR, W. P., EDITOR.
1956. The deer of North America. Telegraph Press, Harrisburg, Pennsylvania, xix + 668 pp., illus.

VARIOUS AUTHORS.
1930. Symposium on predatory animal control. . . . Jour. Mamm., 11:325-389, August 9.

WARREN, E. R.
1926. A study of the beaver in the Yancey region of Yellowstone National Park. Roosevelt Wild Life Annals, 1:1-191, illus., October.

1942. The mammals of Colorado. . . . Univ. Oklahoma Press, Norman, xviii + 330 pp., frontis., 50 pls.

WARREN, E. R. and HALL, E. R.
1939. A new subspecies of beaver from Colorado. Jour. Mamm., 20: 358-362, 1 map, August 14.

WEED, W. H.
1893. The glaciation of the Yellowstone Valley north of the Park. U. S. Geol. Surv. Bull., 104:1-41.

WENTWORTH, C. K. and DELO, D. M.
1931. Dinwoody glaciers, Wind River Mountains, Wyoming. . . . Bull. Geol. Soc. Amer., 42:605-620.

WHITE, J. A.
1952. A new chipmunk (genus Eutamias) from the Black Hills. Univ. Kansas Publs., Mus. Nat. Hist., 5(19):259-262, April 10.

1953. Taxonomy of the chipmunks, Eutamias quadrivittatus and Eutamias umbrinus. Univ. Kansas Publs., Mus. Nat. Hist., 5(33):563-582, 6 figs., December 1.

1953. Geographic distribution and taxonomy of the chipmunks of Wyoming. Univ. Kansas Publs., Mus. Nat. Hist., 5(34):583-610, December 1.

WHITLOW, W. and HALL, E. R.
1933. Mammals of the Pocatello region of southeastern Idaho. Univ. California Publs. Zool., 40(3):235-276, 3 figs., September 30.

WOOD, N. A.
1921. A wolverine in a tree. Jour. Mamm., 2:234, November 29.

YOUNG, S. P. and GOLDMAN, E. A.
1944. The wolves of North America. Amer. Wildlife Inst., Washington, D. C., 2 parts, 636 pp., frontis., 131 pls., 15 figs.

1946. The puma mysterious American cat. The Amer. Wildlife Inst., Washington, D. C., two parts, 358 pp., 93 pls.

YOUNG, S. P. and JACKSON, H. H. T.
1951. The clever coyote. Stackpole Co., Harrisburg, Pa., two parts, 411 pp., frontis., 81 pls., 11 figs.

Transmitted February 24, 1964 (with some additions as of March 19, 1964).

□

30-2329

(Continued from inside of front cover)

l. 13. 1. Five natural hybrid combinations in minnows (Cyprinidae). By Frank B. Cross and W. L. Minckley. Pp. 1-18. June 1, 1960.

2. A distributional study of the amphibians of the Isthmus of Tehuantepec, México. By William E. Duellman. Pp. 19-72, pls. 1-8, 3 figures in text. August 16, 1960. 50 cents.

3. A new subspecies of the slider turtle (Pseudemys scripta) from Coahuila, México. By John M. Legler. Pp. 73-84, pls. 9-12, 3 figures in text. August 16, 1960.

4. Autecology of the copperhead. By Henry S. Fitch. Pp. 85-288, pls. 13-20, 26 figures in text. November 30, 1960.

5. Occurrence of the garter snake, Thamnophis sirtalis, in the Great Plains and Rocky Mountains. By Henry S. Fitch and T. Paul Maslin. Pp. 289-308, 4 figures in text. February 10, 1961.

6. Fishes of the Wakarusa river in Kansas. By James E. Deacon and Artie L. Metcalf. Pp. 309-322, 1 figure in text. February 10, 1961.

7. Geographic variation in the North American cyprinid fish, Hybopsis gracilis. By Leonard J. Olund and Frank B. Cross. Pp. 323-348, pls. 21-24, 2 figures in text. February 10, 1961.

8. Descriptions of two species of frogs, genus Ptychohyla; studies of American hylid frogs, V. By William E. Duellman. Pp. 349-357, pl. 25, 2 figures in text. April 27, 1961.

9. Fish populations, following a drought, in the Neosho and Marais des Cygnes rivers of Kansas. By James Everett Deacon. Pp. 359-427, pls. 26-30, 3 figures. August 11, 1961. 75 cents.

10. Recent soft-shelled turtles of North American (family Trionychidae). By Robert G. Webb. Pp. 429-611, pls. 31-54, 24 figures in text. February 16, 1962. $2.00.

Index. Pp. 613-624.

l. 14. 1. Neotropical bats from western México. By Sydney Anderson. Pp. 1-8. October 24, 1960.

2. Geographic variation in the harvest mouse, Reithrodontomys megalotis, on the central Great Plains and in adjacent regions. By J. Knox Jones, Jr., and B. Mursaloglu. Pp. 9-27, 1 figure in text. July 24, 1961.

3. Mammals of Mesa Verde National Park, Colorado. By Sydney Anderson. Pp. 29-67, pls. 1 and 2, 3 figures in text. July 24, 1961.

4. A new subspecies of the black myotis (bat) from eastern Mexico. By E. Raymond Hall and Ticul Alvarez. Pp. 69-72, 1 figure in text. December 29, 1961.

5. North American yellow bats, "Dasypterus," and a list of the named kinds of the genus Lasiurus Gray. By E. Raymond Hall and J. Knox Jones, Jr. Pp. 73-98, 4 figures in text. December 29, 1961.

6. Natural history of the brush mouse (Peromyscus boylii) in Kansas with description of a new subspecies. By Charles A. Long. Pp. 99-111, 1 figure in text. December 29, 1961.

7. Taxonomic status of some mice of the Peromyscus boylii group in eastern Mexico, with description of a new subspecies. By Ticul Alvarez. Pp. 113-120, 1 figure in text. December 29, 1961.

8. A new subspecies of ground squirrel (Spermophilus spilosoma) from Tamaulipas, Mexico. By Ticul Alvarez. Pp. 121-124. March 7, 1962.

9. Taxonomic status of the free-tailed bat, Tadarida yucatanica Miller. By J. Knox Jones, Jr., and Ticul Alvarez. Pp. 125-133, 1 figure in text. March 7, 1962.

10. A new doglike carnivore, genus Cynarctus, from the Clarendonian Pliocene, of Texas. By E. Raymond Hall and Walter W. Dalquest. Pp. 135-138, 2 figures in text. April 30, 1962.

11. A new subspecies of wood rat (Neotoma) from northeastern Mexico. By Ticul Alvarez. Pp. 139-143. April 30, 1962.

12. Noteworthy mammals from Sinaloa, Mexico. By J. Knox Jones, Jr., Ticul Alvarez, and M. Raymond Lee. Pp. 145-159, 1 figure in text. May 18, 1962.

13. A new bat (Myotis) from Mexico. By E. Raymond Hall. Pp. 161-164, 1 figure in text. May 21, 1962.

14. The mammals of Veracruz. By E. Raymond Hall and Walter W. Dalquest. Pp. 165-362, 2 figures in text. May 20, 1963. $2.00.

15. The Recent mammals of Tamaulipas, México. By Ticul Alvarez. Pp. 363-473, 5 figures in text. May 20, 1963. $1.00.

16. A new subspecies of the fruit-eating bat, Sturnira ludovici, from western Mexico. By J. Knox Jones, Jr., and Gary L. Phillips. Pp. 475-481, 1 figure in text. March 2, 1964.

17. Records of the fossil mammal Sinclairella, Family Apatemyidae, from the Chadronian and Orellan. By William C. Clemens, Jr. Pp. 483-491, March 2, 1964.

18. The mammals of Wyoming. By Charles A. Long. Pp. 493-758, 82 figures in text. July 6, 1965. $3.00.

Index to come.

(Continued on outside of back cover)

(Continued from inside of back cover)

Vol. 15. 1. The amphibians and reptiles of Michoacán, México. By William E. Duellman. Pp. 1-148, pls. 1-6, 11 figures in text. December 20, 1961. $1.50.
2. Some reptiles and amphibians from Korea. By Robert G. Webb, J. Knox Jones Jr., and George W. Byers. Pp. 149-173. January 31, 1962.
3. A new species of frog (Genus Tomodactylus) from western México. By Robert G. Webb. Pp. 175-181, 1 figure in text. March 7, 1962.
4. Type specimens of amphibians and reptiles in the Museum of Natural History, the University of Kansas. By William E. Duellman and Barbara Berg. Pp. 183-204. October 26, 1962.
5. Amphibians and reptiles of the rainforests of southern El Petén, Guatemala. By William E. Duellman. Pp. 205-249, pls. 7-10, 6 figures in text. October 4, 1963.
6. A revision on snakes of the genus Conophis (Family Colubridae, from Middle America). By John Wellman. Pp. 251-295, 9 figures in text. October 4, 1963.
7. A review of the Middle American tree frogs of the genus Ptychohyla. By William E. Duellman. Pp. 297-349, pls. 11-18, 7 figures in text. October 18, 1963. 50 cents.
8. Natural history of the racer Coluber constrictor. By Henry S. Fitch. Pp. 351-468, pls. 19-22, 20 figures in text. December 30, 1963. $1.00.
9. A review of the frogs of the Hyla bistincta group. By William E. Duellman. Pp. 469-491, 4 figures in text. March 2, 1964.
10. An ecological study of the garter snake, Thamnophis sirtalis. By Henry S. Fitch. Pp. 493-564, pls. 23-25, 14 figures in text. May 17, 1965.
11. Breeding cycle in the ground skink, Lygosoma laterale. By Henry S. Fitch and Harry W. Greene. Pp. 565-575, 3 figures in text. May 17, 1965.
12. Amphibians and reptiles from the Yucatan Peninsula, México. By William E. Duellman. Pp. 577-614, 1 figure in text. June 22, 1965.
More numbers will appear in volume 15.

Vol. 16. 1. Distribution and taxonomy of mammals of Nebraska. By J. Knox Jones, Jr. Pp. 1-356, plates 1-4, 82 figures in text. October 1, 1964. $3.50.
2. Synopsis of the lagomorphs and rodents of Korea. By J. Knox Jones, Jr., and David H. Johnson. Pp. 357-407. February 12, 1965.
3. Mammals from Isla Cozumel, Mexico, with description of a new species of harvest mouse. By J. Knox Jones, Jr., and Timothy E. Lawlor. Pp. 409-410, 1 figure in text. April 13, 1965.
More numbers will appear in volume 16.

Vol. 17. 1. Location of fossil vertebrates in the Niobrara Formation (Cretaceous) of Kansas. By David Bardack. Pp. 1-14. January 22, 1965.
2. Chorda tympani branch of the facial nerve in the middle ear of tetrapods. By Richard C. Fox. Pp. 15-21. June 22, 1965.
More numbers will appear in volume 17.

INDEX TO VOLUME 14

New systematic names are in **boldface type**

aberti, Sciurus, 40, 742
Abert's squirrel, 40
acridens, Cynarctus, 138
affinis, Peromyscus, 304
agilis, Vespertilio, 247
Agouti
nelsoni, 324
paca, 324
agouti, Mexican, 326
albescens,
Felis, 463
Reithrodontomys, 625
albigula, Neotoma, 141, 157, 450
albigularis, Thomomys, 275
Alces
alces, 715
americanus, 715
shirasi, 715
alces, Alces, 715
alfaroi, Oryzomys, 290, 437
Alfaro's rice rat, 290
allamandi, Galictis, 340
alleni,
Liomys, 286, 433
Mustela, 693
Putorius, 693
Sciurus, 424
Spermophilus, 578
Allen's
big-eared bat, 415
short-tailed bat, 233
squirrel, 424
Aloutta
mexicana, 258
villosa, 258
alpinus, Sciuropterus, 599
alstoni, Neotomodon, 315
altamirae, Lepus, 420
Alvarez, Ticul
A new subspecies of ground squirrel (Spermophilus spilosoma) from Tamaulipas, Mexico, 121
A new subspecies of wood rat (Neotoma) from northeastern Mexico, 139
Taxonomic status of some mice of the Peromyscus boylii group in eastern Mexico, with description of a new subspecies, 111
The Recent mammals of Tamaulipas, Mexico, 363
Alvarez, Ticul with Hall, E. Raymond. A new subspecies of black myotis (bat) from eastern Mexico . . ., 69
Alvarez, Ticul with Jones, J. Knox, Jr. Taxonomic status of the free-tailed bat,
Tadarida yucatanica Miller, 125
Alvarez, Ticul with Jones, J. Knox, Jr. and Lee, M. Raymond. Noteworthy mammals from Sinaloa, Mexico, 145
amata fig, 337
ambiguus, Peromyscus, **118**, 443
amblyceps, Ursus, 59, 683
American elk, 710
americana,
Antilope, 716
Antilocapra, 467, 716
Lepus, 65
Martes, 65, 689
Mazama, 355, 466
Mephitis, 702
americanus,
Alces, 715
Homo, 262
Lepus, 546
Oreamnos, 720
Ursus, 59, 456, 682
amoenus, Eutamias, 559, 562
Ammospermophilus leucogaster, 66
amotus, Myotis, 247
amplus, Peromyscus, 307
anahuacaus, Crotalus, 316
analogus, Baiomys, 311
Anderson, Sydney
Mammals of Mesa Verde National Park, Colorado, 29
Neotropical bats from western México, 1
angulatus,
Dicotyles, 465
Tayassu, 465
angustapalata, Neotoma, 451
angustirostris, Peromyscus, 309
annectens,
Lutra, 343
Spermophilus, 123
Anoura, 8
geoffroyi, 229
lasiopyga, 229
anteater, two-toed, 264
anteburro, 348
Anthony's bat, 234
Antilocapra,
americana, 467, 716
mexicana, 467
Antilope americana, 716

Antrozous pallidus, 415, 742
apache, Felis, 464
Apatemyidae from Chadronian, and Orellan, 483
aquaticus,
  Oryzomys, 435
  Scalopus, 528
arcticeps, Onychomys, 636
arcticus, Sorex, 517
Arctogale, 692
Arctomys
  dacota, 570
  ludovicianus, 590
arctos, Ursus, 685
ardilla
  barcina, 424
  chica, 270, 424
  colorado, 423
  montañero, 270
  negra, 272
  pinta, 272, 423
argentatus, Myotis, 247
argentinus, Lasiurus, 94
aridulus, Peromyscus, 634
arizonae, Neotoma, 54
arizonensis, Bassariscus, 687
armadillo, nine-banded, 264, 418
armatus, Spermophilus, 575
artemisiae, Peromyscus, 628
Artibeus
  aztecus, 3, 153, 403
  cinereus, 3, 153, 238
  hirsutus, 3, 5, 7
  jamaicensis, 3, 4, 234, 402, 478, 480
  lituratus, 3, 4, 153, 237, 402, 480
  nanus, 3, 480
  palmarum, 3, 153, 237, 402, 480
  phaeotis, 3, 238
  toltecus, 153, 154, 238, 403, 480
  turpis, 3, 238, 480
Arvicola
  haydenii, 659
  insperatus, 647
  longicaudus, 653
  macropus, 656
  nanus, 651
astutus, Bassariscus, 60, 330, 456, 687
Atalapha, 78
  brasiliensis, 95
  cinerea, 95
  egregia, 75
  mexicana, 75
  pallescens, 95
  teliotis, 412
Ateles
  geoffroyi, 260, 417
  vellerosus, 260, 417
ater, Molossus, 158, 255, 417
athabascae, Bison, 718
atronasus, Dipodomys, 432
atrovarius, Thomomys, 156
attenuatus, Thomomys, 601
attwateri, Peromyscus, 101, 445
auduboni, Ovis, 721
audubonii, Sylvilagus, 40, 156, 268, 418, 544
aureogaster, Sciurus, 270, 423
aureus, Thomomys, 47
auriculus, Myotis, 247, 408
auripectus, Peromyscus, 49
aurispinosa,
  Dysopes, 415
  Tadarida, 131, 415
auritus, Chrotopterus, 225
Aztec
  fruit-eating bat, 403
  mouse, 305
azteca,
  Carollia, 230
  Felis, 463
aztecus,
  Artibeus, 3, 153, 403
  Caluromys, 201
  Myotis, 247
  Peromyscus, 113, 305
  Potos, 335, 458
  Reithrodontomys, 12, 19, 48

badger, 61, 459, 699
baileyi,
  Lepus, 544
  Lynx, 61
  Sciurus, 593
  Sylvilagus, 544
  Tamisciurus, 593
Baiomys
  analogus, 311
  brunneus, 311
  musculus, 311
  taylori, 311, 447
bairdii,
  Lepus, 547
  Tapirus, 348
Baird's tapir, 348
Balantiopteryx
  io, 215
  pallida, 151
  plicata, 151, 214
bangsi,
  Glaucomys, 599
  Sciuropterus, 599
barbara, Eria, 339, 459
Bassaricyon, 138
Bassariscus,
  arizonensis, 687
  astutus, 60, 330, 456, 687
  flavus, 60, 456, 687
  nevadensis, 687
  sumichrasti, 331
bat,
  Allen's big-eared, 415
  Allen's short-tailed, 233
  Anthony's, 234

bat—*Continued*
Aztec fruit-eating, 403
big brown, 39, 250, 410, 536
big fruit-eating, 237, 402
big-eared, 39, 252, 415
Brazilian brown, 250
Brazilian free-tailed, 39, 254, 415
Brazilian long-nosed, 208
Brazilian small-eared, 221, 400
brown, 39
brown small-eared, 222
Cozumel spear-nosed, 223
Davy's naked-backed, 218, 398
Dobson's mustached, 216
dog-faced, 254
doglike, 212, 213
dwarf fruit-eating, 238
evening, 252, 413
false vampire, 225, 226
free-tailed, 39, 125, 254, 415, 416
fringe-lipped, 224
fruit-eating, 234, 238, 247, 402, 403, 404, 435, 477
funnel-eared, 242
Geoffroy's free-tailed, 416
Geoffroy's tailless, 229
Gervais' fruit-eating, 238
greater doglike, 213
greater white-lined, 211
hairy-legged vampire, 241, 406
hoary, 251, 412, 534
Isthmian, 234
Jamaican fruit-eating, 234, 402
leaf-chinned, 219, 399
lesser doglike, 212
Linnaeus' false vampire, 226
little fruit-eating, 404
little yellow, 252, 414
long-nosed, 208, 229, 401
long-tongued, 229, 399, 400
mastiff, 255, 416
Mexican big-eared, 252
Mexican dog-faced, 254
Mexican funnel-eared, 242, 407
Mexican long-tongued, 399
mustached, 216, 217, 398
naked-backed, 218, 398
northern yellow, 251, 412
pale spear-nosed, 224
Pallas' long-tongued, 226, 400
pallid, 415
Parnell's mustached, 217
Peal's free-tailed, 415
Peter's, 214
Peter's false vampire, 225
Peter's leaf-chinned, 219, 399
red, 251, 411, 533
red mastiff, 255, 417
Seba's short-tailed, 230

bat—*Concluded*
Seminole, 251
short-tailed, 230, 233
silver-haired, 533
small-eared, 222, 400
southern yellow, 252, 413
spear-nosed, 223
spotted, 535
Thomas', 214
Thomas' sac-winged, 215
Toltec fruit-eating, 238, 403
Townsend's, 534
Townsend's big-eared, 39
Underwood's long-tongued, 229
vampire, 225, 226, 239, 241, 405, 406
Wagner's, 255
white-lined, 211
wrinkle-faced, 239, 404
yellow, 251, 252, 412, 413, 414
yellow-shouldered, 234, 401
bats,
Neotropical from western México, 1
yellow, 73
beatae, Peromyscus, 113, 146, 305
bear,
black, 59, 456, 682
grizzly, 685
beaver, 434, 621
berlandieri,
Cryptotis, 396
Sigmodon, 449
Taxidea, 61, 460
big
brown bat, 39, 250, 410, 536
fruit-eating bat, 237, 402
pocket gopher, 278
big-eared bat, 39, 252, 415
bilineata, Saccopteryx, 211
Bison
athabascae, 718
bison, 717
bison, 717
Bison, 717
Bos, 719
bisonophagus, Ursus, 685
black
bear, 59, 456, 682
myotis, 69, 248, 409
rat, 321
black-eared
mouse, 302, 440
rice rat, 289
black-footed ferret, 697
blackish deer mouse, 308
blackjack oak, 106
black-tailed
jack rabbit, 39, 420, 552
prairie dog, 590
blandus, Peromyscus, 440

Blarina, 516
blossevillii, Lasiurus, 94
boa constrictor, 311
bobcat, 61, 347, 464, 708
bonariensis, Vespertilio, 94
borealis,
  Lasiurus, 78, 92, 93, 155, 251, 411, 480, 533
  Vespertilio, 411, 533
Bos bison, 719
bottae, Thomomys, 47
Botta's pocket gopher, 47
boylii, Peromyscus, 49, 101, 113, 443, 742
brachyotis, Lasiurus, 94
brasiliensis,
  Atalapha, 95
  Eptesicus, 250
  Sylvilagus, 266, 418
  Tadarida, 39, 254, 415, 742
Brazilian
  brown bat, 250
  free-tailed bat, 39, 254, 415
  long-nosed bat, 208
  small-eared bat, 221, 400
brazo fuerte, 263
brevicaudus,
  Clethrionomys, 644
  Evotomys, 644
  Onychomys, 637
brevirostris, Mus, 322
bridgeri, Thomomys, 602
brocket, red, 355, 466
brown
  bat, 39, 536
  big brown bat, 39, 536
  small-eared bat, 222
bruneri, Erethizon, 669
brunneus, Baiomys, 311
brush mouse, 49, 101, 304, 443
bullaris, Tylomys, 292
bullatus,
  Peromyscus, 306
  Thomomys, 604
bursarius, Geomys, 612
bushy-tailed wood rat, 54

cabeza de viejo, 339
cabreri, Spermophilus, 124
cacomitli, Felis, 347, 464
cacomixtle, Tropical, 331
cagottis, Canis, 328
California myotis, 247, 408
californica, Didelphis, 195, 393
californicus,
  Lepus, 39, 552, 742
  Myotis, 38, 163, 247, 741
callistus, Perognathus, 614
Callospermophilus
  caryi, 584
Callospermophilus—*Concluded*
  lateralis, 584
  wortmani, 589
Caluromys
  aztecus, 201
  derbianus, 201
campanius, Lepus, 550
campestris,
  Lepus, 550
  Zapus, 664
canadensis,
  Castor, 48, 621
  Cervus, 63, 710
  Lutra, 703
  Lynx, 65, 707
  Ovis, 64, 720
canaster, Galicitis, 340
cansensis, Peromyscus, 101
Canidae, 138
Canis
  cagottis, 328
  irremotus, 677
  latrans, 59, 328, 454, 672
  lestes, 675
  lupus, 455, 675
  mearnsi, 59
  microdon, 454
  monstrabilis, 455
  nebrascensis, 455
  nubilus, 677
  texensis, 455
  velox, 680
  youngi, 678
canyon mouse, 49, 627
capuchins, 323
Cariacus
  mexicanus, 466
  virginianus, 466
carissima, Myotis, 529
Carollia
  azteca, 230
  castanea, 233
  perspicillata, 230
  subrufa, 233
carrorum, Oryzomys, 436
caryi,
  Callospermophilus, 584
  Microtus, 651
  Reithrodontomys, 19
  Scalopus, 528
  Thomomys, 605
castanea, Carollia, 233
castaneus, Lasiurus, 94
castanops, Cratogeomys, 428
castanurus,
  Citellus, 584
  Spermophilus, 584
  Tamias, 584

Castor
canadensis, 48, 621
concisor, 48, 623
mexicanus, 434
missouriensis, 622
caudatus, Lasiurus, 94
caurina, Mustela, 690
cave myotis, 244, 407
centralis,
Centronycteris, 214
Diphylla, 241
Centronycteris
centralis, 214
maximiliani, 214
Centurio senex, 239, 404, 480
cerreti, 326
Cervus
canadensis, 63, 710
hemionus, 712
nelsoni, 63, 710
temana, 466
Chadronian,
Cynarctus from, 135
Sinclairella from, 483
chango, 260
chapmani,
Lepus, 419
Oryzomys, 290
Sylvilagus, 419
cheyennensis, Thomomys, 606
Chilonycteris, 217
davyi, 398
fulvus, 398
mexicana, 398
parnellii, 7
chipmunk,
cliff, 564
Colorado, 46
least, 46, 554
Uinta, 565
yellow pine, 562
Chiroderma, 3, 7, 8
isthmicum, 234
jesupi, 234
salvini, 480
villosum, 234
Choeronycteris mexicana, 8, 153, 399
Chrotopterus auritus, 225
chrysonotus, Vespertilio, 532
chrysopsis, Reithrodontomys, 297
cinerascens,
Citellus, 586
Spermophilus, 586
Tamias, 586
cinerea,
Atalapha, 95
Neotoma, 54, 639
cinereoargenteus, Urocyon, 59, 138, 329, 455, 682, 742
cinereus,
Artibeus, 3, 153, 238
Lasiurus, 92, 95, 251, 412, 534
Mus, 639
Sorex, 65, 517
Vespertilio, 411, 534
cinnamomea,
Neotoma, 639
Ondatra, 661
cinnamominus, Fiber, 661
cinnamomum, Ursus, 682
cinnamon myotis, 246
cirrhosus, Trachops, 224
Citellus
castanurus, 584
cinerascens, 586
elegans, 572
lateralis, 584
obsoletus, 577
richardsonii, 572
spilosoma, 577
Clarendonian, Pliocene, a new doglike carnivore from, 135
clarus,
Eutamias, 556
Perognathus, 617
Clemens, William A., Jr.
Records of the fossil mammal Sinclariella, family Apatemyidae, from the Chadronian and Orellan, 483
Clethrionomys
brevicaudus, 644
galei, 644
gapperi, 65, 642
idahoensis, 646
uintaensis, 642
cliff chipmunk, 564
climbing rat, naked-tailed, 292
clusius, Thomomys, 606
coati, 333, 458
coatimundi, 333
codiensis, Microtus, 649
Coendou mexicanus, 322
coffini, Trachops, 224
collared peccary, 350, 465
colliaei, Sciurus, 156
collinus, Peromyscus, 444
Colorado chipmunk, 46
colorado, zorro, 202
columbiana, Martes, 691
comadreja, 196
commissarisi, Glossophaga, 480
compactus, Dipodomys, 431
concavus, Heterogeomys, 427
conchuelas, 350
concisor, Castor, 48, 623
concolor, Felis, 61, 345, 462, 705
conejo, 267

conepatl, Conepatus, 342
Conepatus
  conepatl, 342
  leuconotus, 341, 462
  mearnsi, 462
  mesoleucus, 462
  semistriatus, 342
  texensis, 314, 462
confinis, Eutamias, 554
connectens,
  Lepus, 419,
  Sylvilagus, 268, 419
consobrinus,
  Eutamias, 556
  Peromyscus, 445
  Tamias, 556
cookei,
  Ixodes, 105
Corynorhinus
  macrotis, 534
  mexicanus, 253
  pallescens, 534
  phyllotis, 415
Cozumel spear-nosed bat, 223
cozumelae, Mimon, 223
cotton rat, hispid, 312, 448
cottontail,
  desert, 40, 268, 418, 544
  eastern, 267, 419, 542
  Mexican, 268
  Nuttall's, 40, 543
coyote, 58, 328, 454, 672
coyotl, 328, 454
couchii, Spermophilus, 422
couesi,
  Erethizon, 55
  Oryzomys, 287, 288, 435
crassus, Tayassu, 350
Cratogeomys
  castanops, 428
  estor, 280
  fulvescens, 280
  perotensis, 278, 280
  planifrons, 428
  subluteus, 280
  tamaulipensis, 428
crawfordi, Notiosorex, 147
Crawford's shrew, 397
crinitus, Peromyscus, 49, 627
crooki,
  Dorcelaphus, 465
  Odocoileus, 465
Crotalus
  anahuacaus, 316
  triseriatus, 316
crucidens, Cynarctus, 138
Cryptotis
  berlandieri, 396
  madrea, 396
Cryptotis—*Concluded*
  mexicana, 205, 396
  micrura, 207
  nelsoni, 206
  obscura, 206
  parva, 396
  pergracilis, 396
  pueblensis, 396
culbertsoni, Putorius, 693
cunicularius, Sylvilagus, 268
curtatus,
  Lagurus, 660
  Lemmiscus, 660
curti, Lepus, 420
Cyclopes, 295,
  didactylus, 264
  mexicanus, 264
Cynarctus
  acridens, 138
  crucidens, 138
  fortidens, 137
Cynomops malagai, 251
Cynomys
  gunnisoni, 42
  leucurus, 591
  ludovicianus, 590
  zuniensis, 42

dacota,
  Marmota, 570
  Arctomys, 570
dacotensis,
  Odocoileus, 714
  Taxidea, 700
dakotensis,
  Sciurus, 270, 595
  Sinclairella, 485
  Tamiasciurus, 595
Dalquest, Walter W. with Hall, E. Raymond
  A new doglike carnivore, genus Cynarctos, from the Clarendonian, Pliocene, of Texas, 135
  The mammals of Veracruz, 165
dalquesti, Myotis, 71, 248, 409
Dama, 353,
  hemionus, 712
  texensis, 466
  virginiana, 466
Dasyprocta
  mexicana, 326
  punctata, 326
Dasypterus, 75
  ega, 75, 413
  floridanus, 75
  fuscatus, 90
  intermedius, 80
  panamensis, 75, 91
  punensis, 94
  xanthinus, 75, 413

Dasypus
mexicanus, 155, 264, 418
novemcinctus, 155, 264, 418
texianus, 265
davyi,
Chilonycteris, 398
Pteronotus, 218, 398
Davy's naked-backed bat, 218, 398
deer
mouse, 49, 300, 441, 628
mule, 62, 712
white-tailed, 713
degelidus, Lasiurus, 94
Dendragapus obscurus, 34
deppei, Sciurus, 270
Deppe's squirrel, 270, 424
derbianus, Caluromys, 201
Dermacentor variabilis, 105
desert cottontail, 40, 268, 418, 544
deserticola, Lepus, 742
Desmodus
murinus, 7, 239, 405
rotundus, 7, 239, 405
Dicotyles angulatus, 465
didactylus, Cyclopes, 264
Didelphis
Californica, 195, 393
marsupialis, 195, 393
tabascensis, 195
texensis, 394
virginiana, 515
difficilis,
Peromyscus, 53, 307, 446
Reithrodontomys, 298
Diphylla
centralis, 241
ecaudata, 241, 406
Dipodomys
atronasus, 432
compactus, 431
durranti, 431
fuscus, 431
longipes, 48
luteolus, 618
merriami, 432
ordii, 48, 431, 618
parvabullatus, 431
perotensis, 282
phillipsi, 282
priscus, 620
terrosus, 620
discolor, Phyllostomus, 224
distincta, Neotoma, 317
Dobson's mustached bat, 216
dog-faced bat, 254
doglike bat, 212, 213
domesticus, Mus, 663
Dorcelaphus
crooki, 465
Dorcelaphus—*Concluded*
virginianus, 465
texanus, 466
dorsalis, Eutamias, 564
dorsatum, Erethizon, 55, 668
Douglas fir, 33
doutti, Peromyscus, 627
durranti, Dipodomys, 431
dusky
grouse, 34
shrew, 206
dwarf
fruit-eating bat, 238
shrew, 523
dychei, Reithrodontomys, 21, 626
Dysopes aurispinosa, 415

eastern
cottontail, 268, 419, 542
hog-nosed skunk, 462
mole, 528
pipistrelle, 249, 409
spotted skunk, 461
ecaudata, Diphylla, 241, 406
edulis,
Mephitis, 461
Pinus, 33
ega,
Dasypterus, 75, 91, 94, 413
Lasiurus, 90, 94, 252, 413
Nycticejus, 75
egragia, Atalapha, 75
egregius, Lasiurus, 75, 94
Eira
barbara, 339, 459
senex, 339, 459
El Carrizo deer mouse, 446
elegans,
Citellus, 572
Myotis, **163**, 248
Spermophilus, 572
elk, American, 710
Enchistenes hartii, 404
energumenos,
Mustela, 60, 695
Putorius, 695
enslenii, Lasiurus, 94
epixanthum, Erethizon, 669
epixanthus, Erethizon, 670
Eptesicus
brasiliensis, 250
fuscus, 39, 250, 410, 480, 536
mirandorensis, 250, 410, 480
pallidus, 39, 536
propinquus, 250
eremicoides, Peromyscus, 445
eremicus,
Peromyscus, 157
Ursus, 456

Erethizon
bruneri, 669
couesi, 55
dorsatum, 55, 668
epixanthum, 669
epixanthus, 670
ermine, 692
erminea, Mustela, 65, 692
escuinapa, Lynx, 347
estor,
Cratogeomys, 280
Mephitis, 60
Euderma maculatum, 535
Eumops glaucinus, 255
europs, Tadarida, 131
Eutamias
amoenus, 559, 562
clarus, 556
confinis, 554
consobrinus, 556
dorsalis, 564
fremonti, 565
hopiensis, 46
luteiventris, 562
minimus, 46, 554, 558
montanus, 567
operarius, 559
pallidus, 560
quadrivittatus, 46, 742
silvaticus, 562
umbrinus, 565, 567
utahensis, 564
evening bat, 252, 413
evides, Peromyscus, 114
evotis,
Myotis, 38, 532
Notiosorex, 148
Vespertilio, 532
Evotomys
brevicaudus, 644
galei, 644
gapperi, 644
idahoensis, 646
excelsus, Procyon, 742
eximius, Mephitis, 314
extremus, Myotis, 71

false vampire, 225, 226
fasciatus,
Lynx, 708
Perognathus, 613
Felis
albescens, 463
apache, 464
azteca, 463
cacomitli, 347, 464
concolor, 61, 345, 462, 705
fossata, 347
glaucula, 464
hippolestes, 61, 705

Felis—*Concluded*
mayensis, 345
missoulensis, 706
oaxacensis, 346, 464
onca, 344, 463
pardalis, 346, 463
stanleyana, 462
veraecrucis, 344, 463
wiedii, 346, 464
yagouaroundi, 346, 464
femorosacca, Tadarida, 127
ferret, black-footed, 697
ferreus, Sciurus, 742
ferruginea, Tadarida, 127, 132, 416
Fiber
cinnamominus, 661
osoyoosensis, 662
zibethicus, 661
fig,
amata, 337
strangler, 197
figginsi, Ochotona, 539
fir, Douglas, 33
fisher, 691
flavescens, Perognathus, 615, 742
flaviventris, Marmota, 41, 568
flavus,
Bassariscus, 60, 456, 687
Perognathus, 66, 281, 616
Potos, 335, 458
flordiana, Neotoma, 105
floridanus,
Dasypterus, 75
Lasiurus, 94
Lepus, 419
Sylvilagus, 267, 419, 542
flying squirrel, 275, 425,
northern, 599
forest rabbit, 267, 418
fortidens,
Cynarctus, 138
Myotis, 154, 246
fossata, Felis, 347
fossilis, Lasiurus, 95
four-eyed opossum, 196, 394
fox,
gray, 59, 329, 455, 682
red, 58, 678
squirrel, 598
swift, 680
frantzi, Lasiurus, 94
free-tailed bat, 39, 125, 254, 415, 416
fremonti,
Eutamias, 565
Sciurus, 596
Tamiasciurus, 41, 596
frenata, Mustela, 60, 196, 338, 458, 693
fringed myotis, 39, 247

fringe-lipped bat, 224
fruit-eating bat, 234, 238, 247, 402, 403, 404, 435, 477
frumentor, Sciurus, 275
fuerte, brazo, 263
fulva, Vulpes, 679
fulvescens,
  Cratogeomys, 280
  Hesperomys, 438
  Oryzomys, 291, 437, 438
  Reithrodontomys, 298, 438
fulvous
  harvest mouse, 298, 438
  pocket gopher, 280
fulvus,
  Chilonycteris, 398
  Peromyscus, 300
  Pteronotus, 218, 398
fumeus, sorex, 516
funebris, Lasiurus, 93
funnel-eared bat, 242
furvus, Peromyscus, 308
fuscatus,
  Dasypterus, 90
  Lasiurus, 94
fuscipes, Procyon, 457
fuscogriseus, Metachirus, 394
fuscus,
  Eptesicus, 39, 250, 410, 480, 536
  Dipodomys, 431
fusus, Microtus, 55

galei,
  Clethrionomys, 644
  Evotomys, 644
Galictis
  allamandi, 340
  canaster, 340
gapperi,
  Clethrionomys, 65, 642
  Evotomys, 646
Gapper's red-backed vole, 642
gato
  montez, 347
  rabón, 465
geoffroyi,
  Anoura, 229
  Ateles, 260, 417
Geoffroy's
  free-tailed bat, 416
  spider monkey, 260
  tailless bat, 229
Geomys
  bursarius, 612
  personatus, 425, 426
  lutescens, 612
  tropicalis, 426
Gervais' fruit-eating bat, 238
glasgowi, Haemolaelops, 105
glaucinus, Eumops, 255
Glaucomys
  bangsi, 599
  herreranus, 275, 425
  sabrinus, 599
  volans, 275, 425
glaucula, Felis, 464
Glossophaga
  commissarisi, 480
  leachii, 7, 158, 226, 400, 478, 480
  soricina, 7, 158, 226, 400, 478
goat, mountain, 720
golden-mantled ground squirrel, 583
goldmani,
  Peromyscus, 156
  Sylvilagus, 156
golliheri, Lasiurus, 94
graceful myotis, 248
gracilis,
  Nyctinomus, 127
  Rhogeëssa, 480
  Spilogale, 60, 701
  Tadarida, 131
grammurus, Spermophilus, 46, 742
grangeri,
  Lepus, 543
  Sylvilagus, 543
grasshopper mouse, 447, 448, 635
gray
  fox, 59, 329, 455, 682
  wolf, 455, 675
grayi, Lasiurus, 95
Great Basin pocket mouse, 617
greater
  doglike bat, 213
  white-lined bat, 211
griseoflavus, Reithrodontomys, 438
grisón, 340
grizzly bear, 685
ground squirrel, 269, 421
  a new subspecies of, from Mexico, 121
  golden-mantled, 45, 583
  Mexican, 421
  Perote, 269
  Richardson's, 572
  spotted, 422, 577
  thirteen-lined, 577
  Uinta, 575
grouse, dusky, 34
gryphus, Vespertilio, 532
Guatemalan small-eared shrew, 206
Gulo
  gulo, 697
  luscus, 697
gulo, Gulo, 697
gunnisoni, Cynomys, 42
Gunnison's prairie dog, 42
gymnurus, Tylomys, 292

Haemolaelops glasgowi, 105
hairy-legged vampire, 241, 406
Hall, E. Raymond.
  A new bat (Myotis) from Mexico, 161
Hall, E. Raymond and Dalquest, Walter W.
  A new doglike carnivore, genus Cynarctus, from the Clarendonian, Pliocene, of Texas, 135
  The mammals of Veracruz, 165
Hall, E. Raymond and Alvarez, Ticul.
  A new subspecies of black myotis (bat) from eastern Mexico . . ., 69
Hall, E. Raymond and Jones, J. Knox, Jr.
  North American yellow bats . . ., 73
hartii, Enchistenes, 404
harvest mouse, 48, 296, 438, 626
  geographic variation in, on Great Plains, 9
  western, 48
haydeni, Sorex, 519
haydenii,
  Arvicola, 659
  Microtus, 659
heather vole, 646
hemionus,
  Dama, 712
  Cervus, 712
  Odocoileus, 62, 465, 712
hernandezii, Procyon, 333, 457
herreranus, Glaucomys, 275, 425
Hesperomys
  fulvescens, 438
  nebrascensis, 629
  sonoriensis, 629
  taylori, 447
  texana, 441
  toltecus, 450
  truei, 634
hesperus, Pipistrellus, 410, 741
Heterogeomys
  concavus, 427
  hispidus, 275, 427
  isthmicus, 278
  lanius, 278
  latirostris, 278
  negatus, 427
  torridus, 278
Heteromys
  lepturus, 286
  temporalis, 287
hippolestes, Felis, 61, 705
hirsutus, Artibeus, 3, 5, 7
hirtus, Procyon, 688
hispid
  cotton rat, 312, 448
  pocket gopher, 275, 427
  pocket mouse, 429, 617
hispidus,
  Heterogeomys, 275, 427
  Perognathus, 429, 617
  Sigmodon, 104, 312, 448
  Vespertilio, 78
Histiotus maculatus, 535
hoary bat, 251, 412, 534
hog-nosed
  eastern hog-nosed skunk, 341, 462
  skunk, 341, 462
  striped hog-nosed skunk, 342
Homo
  americanus, 262
  sapiens, 262
hondurensis, Sturnira, 481
hooded skunk, 340, 461
hopiensis, Eutamias, 46
hormiguero, oso, 263
house mouse, 57, 322, 663
howler monkey, 258
hoyi, Microsorex, 741
huastecae, Oryzomys, 437
hudsonica, Mephitis, 702
hudsonicus,
  Sciurus, 593
  Tamiasciurus, 41, 593
hudsonius, Zapus, 664
humeralis,
  Nycticeius, 252, 413
  Vespertilio, 413
humilis, Reithrodontomys, 14
Hylonycteris underwoodi, 229

idahoensis,
  Clethrionomys, 646
  Evotomys, 646
  Sylvilagus, 742
  Zapus, 666
Idionycteris mexicanus, 415
imperator, Ursus, 685
incautus, Myotis, 407
incensus, Peromyscus, 304
inflatus, Scalopus, 397
inopinata, Neotoma, 54
insperatus,
  Arvicola, 647
  Microtus, 647
insularis, Lasiurus, **85**, 94
interior,
  Lutra, 704
  Myotis, 39, 531
intermedius,
  Lasiurus, 75, 78, 84, 94, 251, 412
  Phenacomys, 65, 646
  Reithrodontomys, 439
interrupta,
  Mephitis, 461, 701
  Spilogale, 461, 701

io, Balantiopteryx, 215
irremotus, Canis, 677
irroratus, Liomys, 284, 432
Ischnoglossa nivalis, 401
Isthmian bat, 234
isthmicum,
  Chiroderma, 234
  Sturnira, 7
isthmicus, Heterogeomys, 278
Ixodes cookei, 105

jack rabbit,
  black-tailed, 39, 420, 552
  white-tailed, 549
jaguar, 344, 356, 463
jaguarundi, 346
jalapae, Mus, 322
Jamaican fruit-eating bat, 234, 402
jamaicensis, Artibeus, 3, 4, 234, 402, 478, 480
javalin, 351
javalina, 351
jeffersonii, Taxidea, 700
jesupi, Chiroderma, 234
Jico mouse, 308
Jones, J. Knox, Jr., and Alvarez, Ticul. Taxonomic status of the free-tailed bat, Tadarida yucatanica Miller, 125
Jones, J. Knox, Jr., Alvarez, Ticul, and Lee, M. Raymond. Noteworthy mammals from Sinaloa, Mexico, 145
Jones, J. Knox, Jr., with Hall, E. Raymond.
  North American yellow bats . . ., 73
Jones, J. Knox, Jr., Mursaloğlu, B. Geographic variation in the harvest mouse . . ., 9
Jones, J. Knox, Jr., and Phillips, Gary L. A new subspecies of the fruit-eating bat, Sturnira ludovici, from western Mexico, 475
juguarundi, 346
jumping mouse,
  meadow, 664
  western, 665
juniper, Utah, 33
Juniperus osteosperma, 33

kangaroo rat, 282
  Merriam's, 432
  Ord's, 48, 431, 618
  Phillips', 282
kappleri, Peropteryx, 213
keeni, Myotis, 247, 408, 532
Keen's myotis, 247, 408, 532
king snake, 311
kinkajou, 355, 458

Lagomorpha, 536
Lagurus
  curtatus, 660
  levidensis, 660
Lampropeltis polyzona, 311
lanius, Heterogeomys, 278
large-toothed shrew, 204
Lasionycteris noctivagans, 533
lasiopyga, Anoura, 229
Lasiurus
  argentinus, 94
  blossevillii, 94
  borealis, 93, 155, 251, 411, 480, 533
  brachyotis, 94
  castaneus, 94
  caudatus, 94
  cinereus, 92, 95, 251, 412, 534
  degelidus, 94
  ega, 90, 91, 94, 252, 413
  egregius, 75, 94
  enslenii, 94
  floridanus, 82, 84, 94
  fossilis, 95
  frantzi, 94
  funebris, 93
  fuscatus, 94
  golliheri, 94
  grayi, 95
  insularis, 85, 94
  intermedius, 75, 78, 84, 94, 251, 412
  minor, 94
  ornatus, 94, 241
  panamensis, 91, 94, 252
  pfeifferi, 94
  salinae, 94
  seminolus, 94, 241
  semotus, 95
  teliotis, 94, 155, 241, 412, 480
  varius, 94
  villosissimus, 95
  xanthinus, 94, 252, 413
lasiurus, Vespertilio, 93
lateralis,
  Callospermophilus, 584
  Citellus, 584
  Sciurus, 587
  Spermophilus, 45, 583, 587
laticaudata, Tadarida, 127, 132, 416
latirostris,
  Heterogeomys, 278
  Manatus, 465
  Trichechus, 348, 465
latrans, Canis, 58, 328, 454, 672
leachii,
  Glossophaga, 7, 158, 226, 440, 478, 480
  Monophyllus, 400
leaf-chinned bat, 219, 399
least
  chipmunk, 46, 554
  shrew, 396
Lee, M. Raymond with Jones, J. Knox, Jr., and Alvarez, Ticul. Noteworthy mammals from Sinaloa, Mexico, 145

Lemmiscus
curtatus, 660
levidensis, 660
Leptonycteris, 8
nivalis, 229, 401
Lepus
altamirae, 420
americana, 65
americanus, 546
baileyi, 544
bairdii, 547
californicus, 39, 552, 742
campanius, 550
campestris, 550
chapmani, 419
connectens, 419
curti, 420
deserticola, 742
floridanus, 419
grangeri, 543
melanotis, 552
merriami, 421
parvulus, 418
seclusus, 548
setzeri, 548
sylvaticus, 543
texianus, 39
townsendii, 549, 551
truei, 418
lepturus, Heteromys, 286
lesser doglike bat, 212
lestes, Canis, 675
letifera, Mustela, 696
leucodon, Neotoma, 141
leucogaster, Onychomys, 66, 447, 635
leucogenys, Sorex, 37, 526
leuconotus, Conepatus, 341, 462
leucoparia, Spilogale, 461
leucopus, Peromyscus, 104, 393, 441, 634
leucurus,
Ammospermophilus, 66
Cynomys, 591
levidensis,
Lagurus, 660
Lemmiscus, 660
levipes, Peromyscus, 113, 115, 305, 443
lilium, Sturnirum, 3, 7, 8, 153, 234, 401, 478
Linneaus' false vampire, 226
Liomys, 8,
alleni, 286, 433
irroratus, 284, 432
obscurus, 284
pictus, 283
pretiosus, 286
texensis, 433
torridus, 286
veraecrucis, 284
lion, mountain, 61, 345, 705
little
brown myotis, 529
fruit-eating bat, 404
yellow bat, 252, 414
littoralis,
Neotoma, 453
Taxidea, 460,
lituratus, Artibeus, 3, 4, 153, 237, 402, 480
litus, Perognathus, 615
Long, Charles A.
Natural history of the brush mouse (Peromyscus boylii) in Kansas with description of a new subspecies . . ., 99
The mammals of Wyoming, 493
long-eared myotis, 38, 532
longicauda,
Mustela, 693
Putorius, 695
longicaudus,
Arvicola, 653
Microtus, 54, 653
Phenacomys, 296
longicrus, Myotis, 531
longipes,
Dipodomys, 48
Onychomys, 447
long-legged myotis, 39, 247, 531
long-nosed bat, 208, 229, 401
long-tailed
vole, 54, 653
weasel, 60, 693
long-tongued bat, 229, 399, 400
lotor, Procyon, 59, 333, 457, 688, 742
lucifugus, Myotis, 529
ludovici, Sturnira, 234, 477
ludovicianus,
Arctomys, 590
Cynomys, 590
lupus, Canis, 455, 675
luscus,
Gulo, 697
Ursus, 697
luteiventris,
Eutamias, 562
Tamias, 562
luteola, Marmota, 41, 570
luteolus,
Dipodomys, 618
Perodipus, 618
lutescens, Geomys, 612
Lutra
annectens, 343
canadensis, 703
interior, 704
nexa, 703
sonora, 704
Lynx
baileyi, 61
canadensis, 65, 707
escuinapa, 347

Lynx—*Concluded*
fasciatus, 708
pallescens, 708
rufus, 61, 347, 464, 708
texensis, 464
uinta, 708
lynx, 707

macrodon, Sorex, 204
macrophonius, Mustela, 338
macropus,
Arvicola, 656
Microtus, 656
macrotis,
Corynorhinus, 534
Peropteryx, 213
macroura,
Mephitis, 60, 461
Vulpes, 58, 679
maculatum, Euderma, 535
maculatus, Histiotus, 535
madrea, Cryptotis, 396
malagai, Cynomops, 251
man, 262
manatee, 348, 465
Manatus, latirostris, 465
manatus, Trichechus, 348, 465
mango, 337
maniculatus, Peromyscus, 16, 22, 49, 300, 440, 628
margay, 346, 464
marilandica, Quercus, 106
marina, 352
Marmosa
mexicana, 199, 395
murina, 395
marmot,
yellow-bellied, 41, 568
Marmota
dacota, 570
flaviventris, 41, 568
luteola, 41, 570
nosophara, 571
marsupialis, Didelphis, 195, 393
marsh rice rat, 287, 434
marta, 335
marten, 689
Martes
americana, 65, 689
columbiana, 691
origenes, 690
pennanti, 691
vulpina, 690
martucha, 335
masked shrew, 517
mastiff bat, 255, 416
maximiliani, Centronycteris, 214
mayensis, Felis, 345
Mazama
americana, 355, 466
sartorii, 355
temana, 355, 466
meadow
jumping mouse, 664
vole, 647
mearnsi,
Canis, 58
Conepatus, 462
megalophylla, Mormoops, 219, 399
megalotis,
Micronycteris, 221, 400
Reithrodontomys, 9, 48, 298, 626, 742
melanorhinus, Myotis, 38
melanotis
Lepus, 552
Oryzomys, 289, 436, 437
Peromyscus, 302
melanophrys, Peromyscus, 445
melanura, Neotoma, 157
menziesii, Pseudotsuga, 33
Mephitis
americana, 702
edulis, 461
estor, 60
eximius, 341
hudsonica, 702
interrupta, 461, 701
macroura, 340, 461
mephitis, 60, 461, 702
varians, 461
meritus, Thomomys, 607
merriami,
Dipodomys, 432
Lepus, 421
Perognathus, 429
Peromyscus, 156
Sorex, 37, 526
Merriam's
pocket mouse, 429
shrew, 37, 526
mesoleucus, Conepatus, 462
Mesa Verde National Park, Colorado, mammals of, 29
mesomelas, Peromyscus, 304
mesquite, 445
Metachirus
fuscogriseus, 394
pallidus, 394
Mexican
agouti, 326
big-eared bat, 252
cottontail, 268
deer mouse, 309
dog-faced bat, 254
funnel-eared bat, 242, 407
ground squirrel, 421
harvest mouse, 299, 440
long-tongued bat, 399
mouse opossum, 199, 395
opossum, 199, 395
porcupine, 322
shrew, 205, 396
small-eared shrew, 205, 396

Mexican—*Concluded*
spiny pocket mouse, 284, 432
vole, 55, 318, 454
wood rat, 54, 317
mexicana,
Aloutta, 258
Antilocapra, 467
Atalapha, 95
Chilonycteris, 7, 398
Choeronycteris, 153, 399
Cryptotis, 205, 396
Dasyprocta, 326
Marmosa, 199, 395
Micronycteris, 221, 400
Neotoma, 54, 317
Pteronotus, 217, 398
Tadarida, 39, 254, 415, 742
Tamandua, 262
mexicanus,
Cariacus, 466
Castor, 434
Coendou, 322
Corynorhinus, 253
Cyclopes, 264
Dasypus, 155, 264, 418
Idionycteris, 415
Microtus, 55, 318
Molossus, 415
Myotis, 163, 247, 407
Nycticeius, 252, 413
Oryzomys, 435
Perognathus, 281
Peromyscus, 309
Plecotus, 252
Reithrodontomys, 299, 439
Spermophilus, 421
Vespertilio, 247, 408
mico de noche, 264
microdon, Canis, 454
Micronycteris,
megalotis, 221, 400
mexicana, 221, 400
sylvestris, 222
micropus, Neotoma, 453
Microsorex
hoyi, 741
washingtoni, 741
Microtus
caryi, 651
codiensis, 649
fusus, 55
haydenii, 659
insperatus, 647
longicaudus, 54, 653
macropus, 656
mexicanus, 55, 318
mogollonensis, 55
montanus, 55, 649
mordax, 54, 653
myllodontus, 658
nanus, 651
ochrogaster, 105, 659
Microtus—*Concluded*
pennsylvanicus, 647, 742
pullatus, 649
quasiater, 320
richardsoni, 656
subsimus, 454
uligocola, 742
zygomaticus, 653
micrura, Cryptotis, 207
Mimon cozumelae, 223
mimus, Sciurus, 40
minimus,
Eutamias, 554, 558
Tamias, 558
mink, 60, 695
minor, Lasiurus, 94
miquihuanensis, Odocoileus, 466
miradorensis,
Eptesicus, 250, 410, 480
Scotophilus, 410
mirus, Ursus, 685
missoulae, Oreamnos, 720
missoulensis, Felis, 706
missouriensis,
Castor, 622
Mus, 637
Onychomys, 637
mogollonensis, Microtus, 55
molaris, Nasua, 335, 458
mole,
eastern, 528
Tamaulipan, 397
molossa, Tadarida, 131
Molossus
ater, 155, 255, 417
mexicanus, 415
nigricans, 155, 255, 417
rufus, 255
monachus, Vespertilio, 93
monkey,
Geoffroy's spider, 260
howler, 258
spider, 260, 417
mono, 260
Monophyllus leachii, 400
monstrabilis, Canis, 455
montana, Taxidea, 700
montane vole, 55, 649
montanus,
Eutamias, 567
Microtus, 55, 649
Oreamnos, 720
Reithrodontomys, 625
monticola, Sorex, 37
moose, 715
mordax, Microtus, 54, 653
Mormoops
megalophylla, 219, 399
senicula, 219, 400
moto, 269
mountain
goat, 720

mountain—*Concluded*
lion, 61, 345, 705
sheep, 720
mouse,
Aztec, 305
black-eared, 302, 440
blackish deer, 308
brush, 49, 99, 101, 304, 443
canyon, 49, 627
deer, 49, 300, 440, 628
El Carrizo deer, 446
fulvous harvest, 298, 438
Gapper's red-backed, 642
grasshopper, 447, 635
Great Basin pocket, 617
harvest, 48, 296, 625
hispid, 429
hispid pocket, 617
house, 57, 322, 663
Jico, 308
jumping, 664
meadow jumping, 664
Merriam's pocket, 429
Mexican deer, 309
Mexican harvest, 299, 440
Mexican spiny pocket, 284, 432
Motzorongo spiny pocket, 287
narrow-nosed, 309
Nelson's deer, 311
Nelson's pocket, 430
northern grasshopper, 447, 635
northern pygmy, 311, 447
olive-backed pocket, 613
opossum, 199
painted spiny pocket, 283
Perote, 306
piñon, 634
pinyon, 53
plains, 625
plateau, 445
pocket, 281, 287
pygmy, 311, 447
red-backed, 642
rock, 53
Santo Domingo spiny, 286
silky pocket, 281
southern grasshopper, 448
southern pygmy, 311
spiny, 283, 429, 432
Sumichrast's harvest, 297
volcano, 315
volcano harvest, 297
western harvest, 48, 296, 438, 626
western jumping, 665
white-ankled, 244
white-footed, 303, 414, 634
Zacatecan deer, 307, 446
mule deer, 62, 712
muricus,
Mustela, 692
Putorius, 692
murina, Marmosa, 395
murinus, Desmodus, 7, 239, 405
Mursaloğlu, B. with Jones, J. Knox, Jr.
Geographic variation in the harvest mouse . . ., 9
Mus
brevirostris, 322
cinereus, 639
domesticus, 663
jalapae, 322
missouriensis, 637
musculus, 57, 106, 148, 322, 663
norvegicus, 663
musculus,
Baiomys, 311
Mus, 57, 106, 148, 322, 663
muskrat, 54, 661
mustached bat, 216, 217, 398
Mustela
alleni, 693
caurina, 690
energumenos, 60, 695
erminea, 65, 692
frenata, 60, 196, 338, 458, 693
letifera, 696
longicauda, 693
macrophonius, 338
muricus, 692
nevadensis, 60, 694
nigripes, 697
oribasus, 695
origenes, 690
perda, 338
perote, 338
tropicalis, 338, 459,
vison, 60, 695
vulpina, 690
myllodontus, Microtus, 658
Mynomes, 651
Myotis
amotus, 247
argentatus, 249
auriculus, 247, 408
aztecus, 247
californicus, 38, 163, 247, 741
carissima, 529
dalquesti, 71, 248, 409
elegans, 163, 248
evotis, 38, 532
extremus, 71
fortidens, 154, 246
incautus, 407
interior, 39, 531
keeni, 247, 408, 532
longicrus, 531
lucifugus, 529
melanorhinus, 38
mexicanus, 163, 247, 407
nigracans, 71, 163, 248, 409
occultus, 154
quebecensis, 93
septentrionalis, 532
sociabilis, 741

Myotis—*Concluded*
stephensi, 38
subulatus, 38, 163, 530
thysanodes, 39, 247, 741
velifer, 153, 244, 407
volans, 247, 531
yumanensis, 741
myotis,
black, 69, 248, 409
California, 247, 408
cave, 244, 407
cinnamon, 246
fringed, 39, 247
graceful, 248
Keen's, 247, 408, 532
little brown, 529
long-eared, 38, 532
long-legged, 39, 247, 531
silvery-haired, 249
small-footed, 38, 530

Nahuatl, 328
naked-backed bat, 218, 398
nanus,
Artibeus, 3, 480
Arvicola, 651
Microtus, 651
Sorex, 523
narica, Nasua, 335, 458
narrow-nosed mouse, 309
naso, Rhynchonycteris, 208
Nasua, 138,
molaris, 335, 458
narica, 333, 458
nasutus, Peromyscus, 53
navigator,
Neosorex, 524
Sorex, 524
nebrascensis,
Canis, 455
Hesperomys, 629
Peromyscus, 629
Reithrodontomys, 12, 21, 22
nebulosus, Thomomys, 608
negatus, Heterogeomys, 427
neglecta, Taxidea, 700
negligens, Sciurus, 272, 424
nelsoni,
Agouti, 324
Cervus, 63, 710
Cryptotis, 206
Neotoma, 317
Perognathus, 430
Peromyscus, 311
Vampyrum, 226
Nelson's
pocket mouse, 430
small-eared shrew, 206
wood rat, 317
Neosorex navigator, 524
Neotoma
albigula, 141, 157, 450
Neotoma—*Concluded*
angustapalata, 451
arizonae, 54
cinerea, 54, 639
cinnamomea, 639
distincta, 317
floridana, 105
inopinata, 54
leucodon, 141
littoralis, 453
melanura, 157
mexicana, 54, 317
micropus, 453
nelsoni, 317
orolestes, 639, 640
rupricola, 642
subsolana, 141, 450
torquata, 317
Neotomodon
alstoni, 315
perotensis, 315
Neotropical bats from western México, 1
nevadensis,
Bassariscus, 687
Mustela, 60
nexa, Lutra, 703
niger, Sciurus, 598
nigricans,
Molossus, 158, 255, 417
Myotis, 71, 163, 248, 409
nigripes,
Mustela, 697
Putorius, 697
nine-banded armadillo, 264, 418
nivalis,
Ischnoglossa, 401
Leptonycteris, 229, 401
noctivagans,
Lasionycteris, 533
Vespertilio, 533
nopal, 444
northern
flying squirrel, 599
grasshopper mouse, 635
pocket gopher, 600
pygmy mouse, 311, 447
yellow bat, 251, 412
norvegicus,
Mus, 663
Rattus, 321, 663
Norway rat, 321, 663
nosophora, Marmota, 571
Notiosorex
crawfordi, 147
evotis, 148
noveboracensis, Vespertilio, 93
novemcinctus, Dasypus, 155, 264, 418
nubilis, Canis, 677
nuttallii, Sylvilagus, 40, 543
Nuttall's cottontail, 40, 543

Nycteris, 76, 78,
hispida, 78
Nycticeius
humeralis, 252, 413
mexicanus, 252, 413
Nycticejus ega, 75
Nycticeus poepingii, 94
Nyctinomops, yucatanicus, 132
Nyctinomus gracilis, 127
Nyctomys sumichrasti, 295

oak, blackjack, 106
oaxacensis, Felis, 346, 464
obscura,
Cryptotis, 206
Ochotona, **538**
obscurus,
Dendragapus, 34
Liomys, 284
Sorex, 37, 521
obsoletus,
Citellus, 577
Spermophilus, 577
ocelot, 346, 463
occidentalis, Sturnira, **477**
occultus, Myotis, 154
ochraventer, Peromyscus, 446
ocius, Thomomys, 608
oculatus, Sciurus, 275
Ochotona
figginsi, 539
obscura, 538
princeps, 65, 537
saxatilis, 540
uinta, 540
ventorum, 540
ochrogaster, Microtus, 105, 659
ochrouris, Odocoileus, 714
ochrourus, Odocoileus, 714
ocythous, Urocyon, 682
Odocoileus
crooki, 465
dacotensis, 714
hemionus, 62, 465, 712
miquihuanensis, 466
ochrouris, 714
ochrourus, 714
texanus, 466
texensis, 466
thomasi, 354
toltecus, 354
veraecrucis, 355, 466
virginianus, 353, 465, 713
olivaceogriseus, Perognathus, 615
olivaceus, Spermophilus, 579
olive-backed pocket mouse, 613
onca, 346,
Felis, 344, 463
Ondatra
cinnamominus, 661
osoyoosensis, 54, 662
zibethicus, 54, 661
Onychomys
arcticeps, 636
brevicaudus, 637
leucogaster, 66, 447, 635
longipes, 447
missouriensis, 637
subrufus, 448
torridus, 157, 448
yakiensis, 157
operarius, Eutamias, 559
opossum, 192, 199, 393, 395, 515
four-eyed, 196, 394
Mexican, 199, 395
mouse, 199, 395
Philander, 195, 394
wooly, 201
ordii,
Dipodomys, 48, 431, 618
Perodipus, 618
Ord's kangaroo rat, 48, 431, 618
Oreamnos
americanus, 720
missoulae, 720
montanus, 720
Orellan, Sinclairella from, 483
oribasus,
Mustela, 695
Putorius, 695
oricolus, Spermophilus, **123,** 422
origenes,
Martes, 690
Mustela, 690
orinomus, Urocyon, 330
orizabae,
Sorex, 204
Sylvilagus, 268
ornatus, Lasiurus, 94, 241
orolestes, Neotoma, 639, 640
Oryzomys
alfaroi, 290, 437
aquaticus, 435
carrorum, 436
chapmani, 290
couesi, 287, 288, 435
fulvescens, 291, 437, 438
huastecae, 437
melanotis, 289, 436, 437
mexicanus, 435
palatinus, 291, 435
palustris, 287, 288, 434
peragrus, 288, 435
rostratus, 289, 436, 437
osgoodi, Peromyscus, 629
oso hormiguero, 263
osoyoosensis,
Fiber, 662
Ondatra, 54, 662
osteosperma, Juniperus, 33
otter,
river, 343, 703
southern river, 343

Ovis
  auduboni, 720
  canadensis, 64, 721

paca,
  Agouti, 324
  spotted, 324
pack rat, 107
painted spiny pocket mouse, 283
palatinus, Oryzomys, 291, 435
pale spear-nosed bat, 224
Pallas' long-tongued bat, 226, 400
pallescens,
  Atalapha, 95
  Corynorhinus, 534
  Lynx, 708
  Plecotus, 39, 534
  Spermophilus, 123
pallid bat, 415
pallida, Balantiopteryx, 151
pallidus,
  Antrozous, 415, 742
  Eptesicus, 39, 536
  Eutamias, 560
  Metachirus, 394
  Philander, 195, 394
  Procyon, 59
  Spermophilus, 581
  Tamias, 560
  Vespertilio, 415
palmarum, Artibeus, 3, 153, 237, 402, 480
palustris,
  Oryzomys, 287, 288, 434
  Sorex, 65, 524
panamensis,
  Dasypterus, 75, 91
  Lasiurus, 91, 94, 252
paradoxus, Perognathus, 617
pardalis, Felis, 346, 463
parnellii,
  Chilonycteris, 7
  Pteronotus, 152, 217
Parnell's mustached bat, 217
parva, Cryptotis, 396
parvabullatus, Dipodomys, 431
parvidens,
  Spermophilus, 421
  Sturnira, 7, 153, 234, 401, 478
parvulus,
  Lepus, 418
  Sylvilagus, 268, 418
parvus,
  Perognathus, 617
  Spermophilus, 582
Peal's free-tailed bat, 415
peccary,
  collared, 350, 465
  white-lipped, 352
pectoralis,
  Peromyscus, 444
  Reithrodontomys, 11
Pedomys, 659
pennanti, Martes, 691
pennsylvanicus, Microtus, 647, 742
peragrus, Oryzomys, 288, 435
perda, Mustela, 338
pergracilis, Cryptotis, 396
pernix, Perognathus, 148
Perodipus
  luteolus, 618
  ordii, 618
Perognathus
  callistus, 614
  clarus, 617
  fasciatus, 613
  flavescens, 615, 742
  flavus, 66, 281, 616
  hispidus, 429, 617
  litus, 615
  merriami, 429
  mexicanus, 281
  nelsoni, 430
  olivaceogriseus, 615
  paradoxus, 617
  parvus, 617
  pernix, 148
  piperi, 616
Peromyscus
  affinis, 304
  ambiguus, 118, 443
  amplus, 307
  angustirostris, 309
  aridulus, 634
  artemisiae, 628
  attwateri, 101, 445
  auripectus, 49
  aztecus, 113, 305,
  beatae, 113, 146, 305,
  blandus, 440
  boylii, 49, 101, 113, 443, 742
  bullatus, 306,
  cansensis, 101
  collinus, 444
  consobrinus, 445
  crinitus, 49, 627
  difficilis, 53, 307, 446
  doutti, 627
  eremicoides, 445
  eremicus, 157
  evides, 114
  fulvus, 300
  furvus, 308
  goldmani, 156
  incensus, 304,
  leucopus, 104, 393, 441, 634
  levipes, 113, 115, 305, 443,
  maniculatus, 16, 22, 49, 300, 440, 628
  melanophrys, 445
  melanotis, 302
  merriami, 156
  mesomelas, 304
  mexicanus, 309

Peromyscus—*Concluded*
nasutus, 53
nebrascensis, 629
nelsoni, 311
ochraventer, 446
osgoodi, 629
pectoralis, 444
petricola, 446
rowleyi, 49, 101
rufinus, 49
saxicola, 308
simulatus, 308
sonoriensis, 440
teapensis, 311
texanus, 441
totontepecus, 311
truei, 50, 53, 634
utahensis, 742
Peropteryx
kappleri, 213
macrotis, 212
Perote
ground squirrel, 269
mouse, 306
pocket gopher, 278
perotae, Mustela, 338
perotensis,
Cratogeomys, 278, 280
Dipodomys, 282
Neotomodon, 315
Reithrodontomys, 297
Spermophilis, 269
personatus, Geomys, 425, 426
perspicillata, Carollia, 230
Peters'
bat, 214
false vampire, 225
leaf-chinned bat, 219, 399
squirrel, 275
petricola, Peromyscus, 446
pfeifferi, Lasiurus, 94
phaeotis, Artibeus, 3, 238
Phenacomys
intermedius, 65, 646
longicaudus, 296
truei, 646
Philander
opossum, 195, 394
pallidus, 195, 394
Phillips, Gary L. with Jones, J. Knox, Jr.
A new subspecies of the fruit-eating bat, Sturnira ludovici, from western Mexico, 475
Phillips' kangaroo rat, 282
phillipsi, Dipodomys, 282
Phyllostomus
discolor, 224
verrucosus, 224
phyllotis,
Corynorhinus, 415
Plecotus, 415
pictus, Liomys, 283
pierreicolus, Thomomys, 742
pika, 537
pine,
pinyon, 33
ponderosa, 34
vole, 320
pine vole, Jalapan, 320
pinetus, Sylvilagus, 40
piñon mouse, 634
Pinus
edulis, 33
ponderosa, 34
pinyon
mouse, 53
pine, 33
piperi, Perognathus, 616
pipistrelle
eastern, 249, 409
western, 410
Pipistrellus
hesperus, 410, 741
potosinus, 410
subflavus, 249, 409
veraecrucis, 249
plains
harvest mouse, 625
pocket gopher, 612
planifrons, Cratogeomys, 428
plateau mouse, 445
Plecotus
mexicanus, 252
pallescens, 39, 534
phyllotis, 415
townsendii, 39, 534
plicata, Balantiopteryx, 151, 214
pocket
gopher, 275, 280
mouse, 281, 287
pocket gopher,
big, 278
Botta's, 47
fulvous, 280
hispid, 275, 427
northern, 600
Perote, 278
plains, 612
southern, 275
Texas, 425
tropical, 426
yellow-faced, 428
pocket mouse,
Great Basin, 617
hispid, 617
olive-backed, 613
silky, 616

poepingii, Nycticeus, 94
polyzona, Lampropeltis, 311
porcupine, 55, 668
  Mexican, 322
Potos
  aztecus, 335, 458
  flavus, 335, 458
potosinus, Pipistrellus, 410
prairie vole, 659
prairie dog,
  black-tailed, 590
  Gunnison's, 42
  white-tailed, 591
preblei, Zapus, 665
pretiosus, Liomys, 286
princeps,
  Ochotona, 65, 537
  Zapus, 65, 665
priscus, Dipodomys, 620
Procyon
  excelsus, 742
  fuscipes, 457
  hernandezii, 333, 457
  hirtus, 688
  lotor, 59, 333, 457, 688, 742
  pallidus, 59
  shufeldti, 333
Procyonidae, 138
pronghorn, 467, 716
propinquus, Eptesicus, 250
pruniosus, Vespertilio, 95
Pseudotsuga menziesii, 33
psilotis, Pteronotus, 216
Pteronotus
  davyi, 218, 398
  fulvus, 218, 398
  mexicana, 217, 398
  parnellii, 152, 217
  psilotis, 216
  rubiginosus, 398
pueblensis, Cryptotis, 396
pullatus, Microtus, 649
puma, 462
punctata, Dasyprocta, 326
punensis, Dasypterus, 94
Putorius
  alleni, 693
  culbertsoni, 693
  energumenos, 695
  longicauda, 695
  muricus, 692
  nigripes, 697
  oribasus, 695
  vison, 695
putorius, Spilogale, 60, 461, 700
pygmaea, Spilogale, 157
pgymaeus, Thomomys, 610
pygmy
  mouse, 311, 447
  rice rat, 291, 437
quadrivitatus, Tamias, 560
quadrivittatus,
  Eutamias, 46, 742
  Tamias, 562
quasiater, Microtus, 320
quebecensis, Myotis, 93
Quercus marilandica, 106

rabbit,
  black-tailed jack, 420, 552
  forest, 267, 418
  jack, 420
  snowshoe, 546
  white-tailed jack, 549
rabón, gato, 465
raccoon, 59, 332, 457, 688
rat,
  Alfaro's rice, 290
  black, 321
  black-eared rice, 289, 436
  bushy-tailed wood, 54, 638
  climbing, 292
  cotton, 312
  hispid cotton, 312, 448
  kangaroo, 48, 282
  marsh rice, 287, 434
  Merriam's kangaroo, 432
  Mexican wood, 317
  naked-tailed climbing, 292
  Nelson's wood, 317
  Norway, 321, 663
  Ord's kangaroo, 48, 431, 618
  pack, 107
  Phillips' kangaroo, 282
  pgymy, 291, 437
  rice, 287, 289
  southern plains wood, 453
  Sumichrast's vesper, 295
  Tamaulipan wood, 451
  vesper, 295
  white-throated wood, 450
  wood, 54, 638
ratón
  de cola grande, 282
  tlacuache, 196, 199
rattlesnake, 316
Rattus
  norvegicus, 321, 663
  rattus, 321
rattus, Rattus, 321
red
  bat, 251, 411, 533
  brocket, 355
  fox, 58, 678
  mastiff bat, 255, 417
  squirrel, 41, 593
red-backed vole, Gapper's, 642
red-bellied squirrel, 272, 423
regalis, Vulpes, 680

Reithrodontomys, 625
albescens, 625
aztecus, 12, 19, 48
caryi, 19
chrysopsis, 297
difficilis, 298
dychei, 21, 626
fulvescens, 298, 438
griseoflavus, 438
humilis, 14
intermedius, 439
megalotis, 9, 48, 298, 626, 742
mexicanus, 299, 439
montanus, 625
nebrascensis, 12, 21, 22
pectoralis, 11
perotensis, 297
saturatus, 296
sumichrasti, 12, 297,
tropicalis, 298, 439, 440
Rhogeëssa
gracilis, 480
tumida, 252, 414
Rhynchonycteris naso, 208
rice rat,
Alfaro's, 290
black-eared, 289, 436
marsh, 287
pgymy, 290, 437
richardsoni, Microtus, 656
richardsonii,
Citellus, 572
Spermophilus, 572
Richardson's
ground squirrel, 572
vole, 656
ringtail, 60, 330, 456, 687
river otter, 343, 703
rock
mouse, 53
squirrel, 46, 269, 422
rogersi, Ursus, 685
rostralis, Thomomys, 611
rostratus, Oryzomys, 289, 436, 437
rotundus, Desmodus, 7, 239, 405
rowleyi, Peromyscus, 49, 101
rubellus, Vespertilio, 93
rubiginosus, Pteronotus, 216
rubra, Vespertilio, 93
rufinus, Peromyscus, 49
rufiventer, Sciurus, 598
rufus,
Lynx, 61, 347, 464, 708
Molossus, 255
Vespertilio, 93
rupricola, Neotoma, 642
russatus, Sylvilagus, 268

sabrinus, Glaucomys, 599
Saccopteryx bilineata, 211
sagebrush vole, 660
salinae, Lasiurus, 94
salvini, Chiroderma, 480
Santo Domingo spiny pocket mouse, 286
sapiens, Homo, 262
sartorii, Mazama, 355
saturatus,
Reithrodontomys, 296
Sigmodon, 314, 448
saussureri, Sorex, 205, 396
Saussure's shrew, 205, 396
saxatilis,
Ochotona, 540
Spilogale, 701
saxicola, Peromyscus, 308
Scalopus
aquaticus, 528
caryi, 528
inflatus, 397
Sciuropterus,
alpinus, 599
bangsi, 599
Sciurus
aberti, 40, 742
alleni, 424
aureogaster, 270, 423
baileyi, 593
colliaei, 156
dakotensis, 595
deppei, 270
ferreus, 742
fremonti, 596
fumentor, 275
hudsonicus, 593
lateralis, 587
mimus, 40
negligens, 272, 424
niger, 598
oculatus, 275
rufiventer, 598
sinaloensis, 156
truei, 156
ventorum, 597
Scotophilus miradorensis, 410
scottii, Urocyon, 59, 330, 455, 742
Seba's short-tailed bat, 230
seclusus, Lepus, 548
Seminole bat, 251
seminolus, Lasiurus, 94, 241
semistriatus, Conepatus, 342
semotus, Lasiurus, 95
senex,
Centurio, 239, 404, 480
Eira, 339, 459
senicula, Mormoops, 219, 400
septentrionalis, 532
Myotis, 532
Vespertilio, 532

setzeri, Lepus, 548
sheep, mountain, 720
shirasi, Alces, 715
short-tailed bat, 230, 233
shrew,
Crawford's, 397
dusky, 206
dwarf, 523
Guatemalan small-eared, 206
large-toothed, 204
least, 396
masked, 517
Merriam's, 37, 526
Mexican small-eared, 205, 396
Nelson's small-eared, 206
Saussure's, 205, 396
slender small-eared, 396
small-eared, 205, 396
vagrant, 204, 520
water, 524
shufeldti, Procyon, 333
Sigmodon
berlandieri, 449
hispidus, 104, 312, 448
saturatus, 314, 448
solus, 450
toltecus, 314, 450
silky pocket mouse, 281, 616
silvaticus, Eutamias, 562
silver-haired bat, 533
silvery-haired myotis, 249
similis,
Sorex, 521
Sylvilagus, 542
simulatus, Peromyscus, 308
Sinaloa, Mexico, noteworthy mammals from, 145
sinaloensis, Sciurus, 156
Sinclairella dakotensis, 485
skunk,
eastern hog-nosed, 462
hog-nosed, 341
hooded, 340, 461
spotted, 60, 461, 700
striped, 60, 461, 702
slender small-eared shrew, 396
small-eared
bat, 222, 400
shrew, 205, 396
small-footed myotis, 38, 530
snowshoe rabbit, 546
sociabilis, Myotis, 741
solus, Sigmodon, 450
sonora, Lutra, 704
sonoriensis,
Hesperomys, 629
Peromyscus, 440,
Sorex
arcticus, 517
cinereus, 65, 517
Sorex—*Concluded*
fumeus, 516
haydeni, 519
leucogenys, 37, 526
macrodon, 204
merriami, 37, 526
monticola, 37
nanus, 523
navigator, 524
obscurus, 37, 521
orizabae, 204
palustris, 65, 524
saussurei, 205, 396
similis, 521
tenellus, 523
vagrans, 37, 204, 520
veraecrucis, 205
soricina, Glossophaga, 7, 158, 226, 400, 478
southern
flying squirrel, 275, 425
grasshopper mouse, 448
plains wood rat, 453
pocket gopher, 275
pygmy mouse, 311
river otter, 343
yellow bat, 252, 413
spear-nosed bat, 223
spectrum, Vampyrum, 226
Spermophilus
alleni, 578
annectens, 123
armatus, 575
cabreri, 124
castanurus, 584
cinerascens, 586
couchii, 422
elegans, 572
grammurus, 46, 742
lateralis, 45, 583, 587
mexicanus, 421
obsoletus, 577
olivaceus, 579
oricolus, 123, 422
pallescens, 123
pallidus, 581
parvidens, 421
parvus, 582
perotensis, 269
richardsonii, 572
spilosoma, 121, 422, 577
tridecemlineatus, 577
variegatus, 46, 269, 422, 742
wortmani, 589
spider monkey, 260, 417
Spilogale
gracilis, 60, 701
interrupta, 461, 701
leucoparia, 461
putorius, 60, 461, 700

Spilogale—*Concluded*
pygmaea, 157
saxatilis, 701
tenuis, 701
spilosoma,
Citellus, 577
Spermophilus, 121, 422, 577
spotted
bat, 535
paca, 324
ground squirrel, 422, 577
skunk, 60, 461, 700
squirrel,
Abert's, 40
Allen's, 424
Deppe's, 270, 424
flying, 275, 425
fox, 598
golden-mantled ground, 583
ground, 269, 421, 572
Peters', 275
red, 41, 593
red-bellied, 272, 423
Richardson's ground, 572
rock, 46, 269, 422
southern flying, 275, 425
spotted ground, 422, 577
thirteen-lined, 577
Uinta ground, 575
stanleyana, Felis, 462
stephensi, Myotis, 38
strangler fig, 197
striped
hog-nosed skunk, 342
skunk, 60, 461, 702
Sturnira
hondurensis, 481
isthmicum, 7
lilum, 3, 7, 8, 153, 234, 401, 478
ludovici, 234, 477
occidentalis, 477
parvidens, 7, 153, 234, 401, 478
subflavus,
Pipistrellus, 249, 409
Vespertilio, 409
subluteus, Cratogeomys, 280
subrufa, Carollia, 233
subrufus, Onychomys, 448
subsimus, Microtus, 454
subsolana, Neotoma, **141**, 450
subulatus,
Myotis, 38, 163, 530
Vespertilio, 530
sumichrasti,
Bassariscus, 331
Nyctomys, 295
Reithrodontomys, 298, 439, 440
Sumichrast's
harvest mouse, 297
vesper rat, 295
swift fox, 680
sylvaticus, Lepus, 543
sylvestris, Micronycteris, 222
Sylvilagus
audubonii, 40, 156, 268, 418, 544
baileyi, 544
brasiliensis, 266, 418
chapmani, 419
connectens, 268, 419
cunicularius, 268
floridanus, 267, 419, 542
goldmani, 156
grangeri, 543
idahoensis, 742
nuttallii, 40, 543
orizabae, 268
parvulus, 268, 418
pinetus, 40
russatus, 268
similis, 542
truei, 260, 418
warreni, 40

tabascensis, Didelphis, 195
Tadarida
aurispinosa, 131, 415
brasiliensis, 39, 254, 415, 742
europs, 131
femorosacca, 127
ferruginea, 127, 132, 416
gracilis, 131
laticaudata, 127, 132, 416
mexicana, 39, 254, 415, 741
molossa, 131
yucatanica, 127, 132
tajacu, Tayassu, 350, 465
talpoides, Thomomys, 600, 742
Tamandua, 263
mexicana, 262
tetradactyla, 262
Tamaulipan
mole, 397
wood rat, 451
Tamaulipas, the Recent mammals of, 363
tamaulipensis, Cratogeomys, 428
Tamias
castanurus, 584
cinerascens, 586
consobrinus, 556
luteiventris, 562
minimus, 558
pallidus, 560
quadrivitatus, 560
quadrivittatus, 562
umbrinus, 567
wortmani, 589
Tamiasciurus
baileyi, 593
dakotensis, 595

Tamiasciurus—*Concluded*
fremonti, 41, 596
hudsonius, 41, 593
ventorum, 597
tapir, Baird's, 348
Tapirus bairdii, 348
Tayassu
angulatus, 465
crassus, 350
tajacu, 350, 465
taylori,
Baiomys, 311, 447
Hesperomys, 447
tayra, 339, 459
Taxidea
berlandieri, 61, 460
dacotensis, 700
jeffersonii, 700
littoralis, 460
montana, 700
neglecta, 700
taxus, 61, 460, 699
taxus,
Taxidea, 61, 460, 699
Ursus, 699
teapensis, Peromyscus, 311
tejon, 333
teliotis,
Atalapha, 412
Lasiurus, 94, 155, 241, 412, 480
temana,
Cervus, 466
Mazama, 466
temazate, 355
temporalis, Heteromys, 287
tenellus,
Sorex, 523
Thomomys, 611
tenuis, Spilogale, 701
tepezcuintle, 324
terrosus, Dipodomys, 620
tesselatus, Vespertilio, 93
tetradactyla, Tamandua, 262
texana, Hesperomys, 441
Texas pocket gopher, 425
texanus,
Dorcelaphus, 466
Odocoileus, 466
Peromyscus, 441
texensis,
Canis, 455
Conepatus, 341, 462
Dama, 466
Didelphis, 394
Liomys, 433
Lynx, 464
Odocoileus, 466
texianus,
Dasypus, 265
Lepus, 39
thirteen lined ground squirrel, 577
Thomas'
bat, 214
sac-winged bat, 215
thomasi, Odocoileus, 354
Thomomys
albigularis, 275
atrovarius, 156
attenuatus, 601
aureus, 47
bottae, 47
bridgeri, 602
bullatus, 604
caryi, 605
cheyennensis, 606
clusius, 606
meritus, 607
nebulosus, 608
ocius, 608
pierreicolus, 742
pygmaeus, 610
rostralis, 611
talpoides, 600, 742
tenellus, 611
umbrinus, 156, 275,
thysanodes, Myotis, 39, 247, 741
ticks, 350
tigre, real, 344
tigrillo, 346
tlacuache, 192
Toltec fruit-eating bat, 238, 403
toltecus,
Artibeus, 153, 154, 238, 403, 480
Hesperomys, 450
Odocoileus, 354
Sigmodon, 314, 450
torquata, Neotoma, 317
torridus,
Heterogeomys, 278
Liomys, 286
Onychomys, 157, 448
totontepecus, Peromyscus, 311
townsendii,
Lepus, 549, 551
Plecotus, 39, 534
Townsend's
bat, 534
big-eared bat, 39
Trachops
cirrhosus, 224
coffini, 224
Trichechus
latirostris, 348, 465
manatus, 348, 465
tridecemlineatus, Spermophilus, 577
triseriatus, Crotalus, 316
tropical
cacomixtle, 331
pocket gopher, 426

tropicalis,
  Geomys, 426
  Mustela, 338, 459
  Reithrodontomys, 298, 439, 440
truei,
  Hesperomys, 634
  Lepus, 418
  Peromyscus, 50, 53, 634
  Phenacomys, 646
  Sciurus, 156
  Sylvilagus, 260, 418
tumbalensis, Tylomys, 292
tumida, Rhogeëssa, 252, 414
turpis, Artibeus, 3, 238, 480
tuza, 275,
  real, 324
two-toed anteater, 264
Tylomys
  bullaris, 292
  gymnurus, 292
  tumbalensis, 292

Uinta
  chipmunk, 565
  ground squirrel, 575
uinta,
  Lynx, 708
  Ochotona, 540
uintaensis, Clethrionomys, 642
uligocola, Microtus, 742
umbrinus,
  Eutamias, 565, 567
  Tamias, 567
  Thomomys, 156, 275
underwoodi, Hylonycteris, 229
Underwood's long-tongued bat, 229
Urocyon
  cinereoargenteus, 59, 138, 329, 455, 682, 742
  ocythous, 682
  orinomus, 330
  scottii, 59, 330, 455, 742
  virginianus, 455
Ursus
  amblyceps, 59, 683
  americanus, 59, 456, 682
  arctos, 685
  bisonophagus, 685
  cinnamomum, 682
  eremicus, 456
  imperator, 685
  luscus, 697
  mirus, 685
  rogersi, 685
  taxus, 699
  washake, 685
Utah juniper, 33
utahensis,
  Eutamias, 564
utahensis—*Concluded*
  Peromyscus, 742
  Zapus, 668

vagrans, Sorex, 37, 204, 520
vagrant shrew, 204
vampire, 225, 226, 239, 241, 405, 406
vampiro, 239
Vampyrum
  nelsoni, 226
  spectrum, 226
variabilis, Dermacentor, 105
varians, Mephitis, 461
variegatus, Spermophilus, 46, 269, 422, 742
varius, Lasiurus, 94
velifer, Myotis, 153, 244, 407
vellerosus, Ateles, 260, 417
velox,
  Canis, 680
  Vulpes, 680
venado, 353
ventorum,
  Ochotona, 540
  Sciurus, 597
  Tamiasciurus, 597
veraecrucis,
  Felis, 344, 463
  Liomys, 284
  Odocoileus, 355, 466
  Pipistrellus, 249
  Sorex, 205
Veracruz, the mammals of, 165
verrucosus, Phyllostomus, 224
vesper rat, 295
Vespertilio, 78
  agilis, 247
  bonariensis, 94
  borealis, 78, 411, 533
  cinereus, 411, 534
  chrysonotus, 532
  evotis, 532
  gryphus, 532
  hispidus, 78
  humeralis, 413
  lasiurus, 93
  mexicanus, 247, 408
  monachus, 93
  noctivagans, 533
  noveboracensis, 93
  pallidus, 415
  pruniosus, 95
  rubellus, 93
  rubra, 93
  rufus, 93
  septentrionalis, 532
  subflavus, 409, 530
  subulatus, 530
  tesselatus, 93
villosa, Aloutta, 258

villosissimus, Lasiurus, 95
villosum, Chiroderma, 234
virginiana,
  Dama, 465
  Didelphis, 515
virginianus,
  Cariacus, 466
  Dorcelaphus, 465
  Odocoileus, 353, 465, 713
  Urocyon, 455
vison,
  Mustela, 695
  Putorius, 695
volans,
  Glaucomys, 275, 425
  Myotis, 247, 531
volcano
  harvest mouse, 297
  mouse, 315
vole,
  Gapper's red-backed, 642
  heather, 646
  Jalapan pine, 320
  long-tailed, 54, 653,
  meadow, 647
  Mexican, 55, 318, 454
  montane, 55, 649
  pine, 320
  prairie, 659
  red-backed, 642
  Richardson's, 656
  sagebrush, 660
Vulpes
  fulva, 679
  macroura, 58, 679
  regalis, 680
  velox, 680
  vulpes, 58, 678
vulpes, Vulpes, 58, 678
vulpina,
  Martes, 690
  Mustela, 690

Wagner's bat, 255
wapiti, 63, 710
warreni, Sylvilagus, 40
washake, Ursus, 685
washingtoni, Microsorex, 741
water shrew, 524
weasel, long-tailed, 60, 338, 458, 693
western
  harvest mouse, 48, 296, 438, 626
  jumping mouse, 665
  pipistrelle, 410
white-ankled mouse, 444
white-footed mouse, 303, 441, 634
white-lined bat, 211
white-lipped peccary, 352
white-tailed
  deer, 353, 465, 713
white-tailed—*Concluded*
  jack rabbit, 549
  prairie dog, 591
white-throated wood rat, 450
wiedii, Felis, 346, 464
wolf, gray, 455, 675
wolverine, 697
wood rat,
  a new subspecies of, from northeastern Mexico, 139
  bushy-tailed, 54, 638
  Mexican, 54, 317,
  Nelson's, 317
  plains, 453
  southern plains, 453
  Tamaulipan, 451
  white-throated, 450
wooly opossum, 201
wortmani,
  Callospermophilus, 589
  Spermophilus, 589
  Tamias, 589
wrinkle-faced bat, 239, 404
Wyoming, the mammals of, 493

xanthinus,
  Dasypterus, 75, 413
  Lasiurus, 91, 94, 252, 413

yagouaroundi, Felis, 346,
yaguaroundi, Felis, 464
yakiensis, Onychomys, 157
yellow bat, 251, 252, 412, 413, 414
yellow-bellied marmot, 41, 568
yellow-faced pocket gopher, 428
yellow-pine chipmunk, 562
yellow-shouldered bat, 234, 401
youngi, Canis, 678
yucatanica, Tadarida, 127, 132
yucatanicus, Nyctinomops, 132
yumanensis, Myotis, 741

Zacatecan deer mouse, 307, 446
zacaton, 316
Zapus
  campestris, 664
  hudsonius, 664
  idahoensis, 666
  preblei, 665
  princeps, 65, 665
  utahensis, 668
zibethicus,
  Fiber, 661
  Ondatra, 54, 661
zorillo, 341
zorro colorado, 202
zuniensis, Cynomys, 42
zygomaticus, Microtus, 653

Lightning Source UK Ltd.
Milton Keynes UK
UKHW020101140219
337249UK00009B/465/P